PHYSICALSCIENCE

EIGHTH EDITION

PHYSICAL SCIENCE

BILL W. TILLERY

ARIZONA STATE UNIVERSITY

Boston   Burr Ridge, IL   Dubuque, IA   New York   San Francisco   St. Louis
Bangkok   Bogotá   Caracas   Kuala Lumpur   Lisbon   London   Madrid   Mexico City
Milan   Montreal   New Delhi   Santiago   Seoul   Singapore   Sydney   Taipei   Toronto

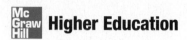
Higher Education

PHYSICAL SCIENCE, EIGHTH EDITION

Printed in China

2 3 4 5 6 7 8 9 0 CTP/CTP 0 9

ISBN 978–0–07–340452–3
MHID 0–07–340452–7

Publisher: *Thomas Timp*
Sponsoring Editor: *Debra B. Hash*
Director of Development: *Kristine Tibbetts*
Senior Developmental Editor: *Mary E. Hurley*
Senior Marketing Manager: *Lisa Nicks*
Senior Project Manager: *Vicki Krug*
Senior Production Supervisor: *Sherry L. Kane*
Lead Media Project Manager: *Judi David*
Manager, Creative Services: *Michelle D. Whitaker*
Cover/Interior Designer: *Elise Lansdon*
(USE) Cover Image: *© Paul Souders/Corbis*
Senior Photo Research Coordinator: *John C. Leland*
Photo Research: *David Tietz/Editorial Image, LLC*
Supplement Producer: *Mary Jane Lampe*
Compositor: *Aptara, Inc.*
Typeface: *10/12 Minion*
Printer: *CTPS*

The credits section for this book begins on page 666 and is considered an extension of the copyright page.

Library of Congress Cataloging-in-Publication Data

Tillery, Bill W.
    Physical science / Bill W. Tillery. — 8th ed.
        p.   cm.
Includes index.
ISBN 978–0–07–340452–3 — ISBN 0–07–340452–7 (hard copy : alk. paper)
1. Physical sciences. I. Title.

Q158.5.T55 2009
500.2—dc22

                                                                    2008022993

www.mhhe.com

BRIEF CONTENTS

CONTENTS

CHEMISTRY

8 Atoms and Periodic Properties 203

9 Chemical Bonds 229

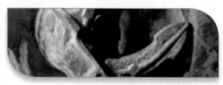

10 Chemical Reactions 251

11 Water and Solutions 275

12 Organic Chemistry 299

13 Nuclear Reactions 327

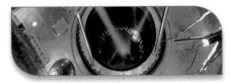

ASTRONOMY

14 The Universe 355

15 The Solar System 379

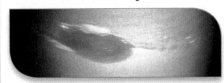

16 Earth in Space 405

EARTH SCIENCE

17 Rocks and Minerals 431

18 Plate Tectonics 451

19 Building Earth's Surface 471

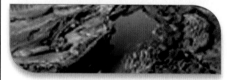

20 Shaping Earth's Surface 493

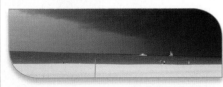

PREFACE

Physical Science is a straightforward, easy-to-read but substantial introduction to the fundamental behavior of matter and energy. It is intended to serve the needs of nonscience majors who are required to complete one or more physical science courses. It introduces basic concepts and key ideas while providing opportunities for students to learn reasoning skills and a new way of thinking about their environment. No prior work in science is assumed. The language, as well as the mathematics, is as simple as can be practical for a college-level science course.

ORGANIZATION

The *Physical Science* sequence of chapters is flexible, and the instructor can determine topic sequence and depth of coverage as needed. The materials are also designed to support a conceptual approach or a combined conceptual and problem-solving approach. With laboratory studies, the text contains enough material for the instructor to select a sequence for a two-semester course. It can also serve as a text in a one-semester astronomy and earth science course or in other combinations.

> "The text is excellent. I do not think I could have taught the course using any other textbook. I think one reason I really enjoy teaching this course is because of the text. I could say for sure that this is one of the best textbooks I have seen in my career. . . . I love this textbook for the following reasons: (1) it is comprehensive, (2) it is very well written, (3) it is easily readable and comprehendible, (4) it has good graphics."
> —Ezat Heydari, Jackson State University

MEETING STUDENT NEEDS

Physical Science is based on two fundamental assumptions arrived at as the result of years of experience and observation from teaching the course: (a) that students taking the course often have very limited background and/or aptitude in the natural sciences; and (b) that this type of student will better grasp the ideas and principles of physical science if they are discussed with minimal use of technical terminology and detail. In addition, it is critical for the student to see relevant applications of the material

to everyday life. Most of these everyday-life applications, such as environmental concerns, are not isolated in an arbitrary chapter; they are discussed where they occur naturally throughout the text.

Each chapter presents historical background where appropriate, uses everyday examples in developing concepts, and follows a logical flow of presentation. The historical chronology, of special interest to the humanistically inclined nonscience major, serves to humanize the science being presented. The use of everyday examples appeals to the nonscience major, typically accustomed to reading narration, not scientific technical writing, and also tends to bring relevancy to the material being presented. The logical flow of presentation is helpful to students not accustomed to thinking about relationships between what is being read and previous knowledge learned, a useful skill in understanding the physical sciences. Worked examples help students to integrate concepts and understand the use of relationships called equations. They also serve as a model for problem solving; consequently, special attention is given to *complete* unit work and to the clear, fully expressed use of mathematics. Where appropriate, chapters contain one or more activities, called Concepts Applied, that use everyday materials rather than specialized laboratory equipment. These activities are intended to bring the science concepts closer to the world of the student. The activities are supplemental and can be done as optional student activities or as demonstrations.

"It is more readable than any text I've encountered. This has been my first experience teaching university physical science; I picked up the book and found it very user-friendly. The level of detail is one of this text's greatest strengths. It is well suited for a university course."
—Richard M. Woolheater, Southeastern Oklahoma State University

"The author's goals and practical approach to the subject matter are exactly what we are looking for in a textbook. . . . The practical approach to problem solving is very appropriate for this level of student."
—Martha K. Newchurch, Nicholls State University

". . . the book engages minimal use of technical language and scientific detail in presenting ideas. It also uses everyday examples to illustrate a point. This approach bonds with the mindset of the nonscience major who is used to reading prose in relation to daily living."
—Ignatius Okafor, Jarvis Christian College

"I was pleasantly surprised to see that the author has written a textbook that seems well suited to introductory physical science at this level. . . . *Physical Science* seems to strike a nice balance between the two—avoiding unnecessary complications while still maintaining a rigorous viewpoint. I prefer a textbook that goes beyond what I am able to cover in class, but not too much. Tillery seems to have done a good job here."
—T. G. Heil, University of Georgia

NEW TO THIS EDITION

Numerous revisions have been made to the text to add new topic areas, update the content on current events and the most recent research topics, and make the text even more user-friendly and relevant for students:

Two new elements have been added to this edition to further enhance the text's focus on developing concepts:

- Core and Supporting Concepts have been added to the chapter openers to help integrate the chapter concepts and the chapter outline. The Core and Supporting Concepts outline and emphasize the concepts at a chapter level. The concepts list is designed to help students focus their studies by identifying the most important topics in the outline.
- A new Appendix D: Solutions for Follow-Up Example Exercises has been added to provide solutions to the unanswered, second example problems within chapters to further assist students in grasping the concepts as well as the quantitative work conveyed in the in-chapter examples.

The end-of-chapter Applying the Concepts multiple-choice self-tests have been revised to better focus on the concepts covered within each chapter.

The list below provides chapter-specific updates:

Chapter 1: Text information on pseudoscience has been updated and expanded.
Chapter 2: Five new beginning "one-step problems" have been added to the Parallel Exercises Set A and B sections. Also, information on GPS has been added.
Chapter 3: There is a new section on Conserving Energy.
Chapter 5: A Closer Look on hearing problems has been added.
Chapter 7: A new section on special relativity and general relativity (with cross references to astronomy) has been added.
Chapter 11: A Closer Look on decompression sickness is now included.
Chapter 14: A Myth, Mistakes, and Misunderstanding box on seeing stars during day from bottom of a well has been added.
Chapter 15: The information on Pluto and the definition of planets have been updated, along with the most recent data on space exploration, space probes, and new probes. Sections with more detail on Kepler's Laws, the geocentric model, and the heliocentric model have been included. A Myth, Mistakes, and Misunderstanding box on blue moons has also been added.
Chapter 16: The material on global warming has been updated and revised.
Chapter 17: A new Science and Society box on using mineral resources now appears.
Chapter 19: A new photo of a cinder cone volcano has been added.
Chapter 21: Photos of actual fossils have been added. Also, the sections on Geologic Periods and Typical Fossils, Mass Extinctions, and Interpreting Geologic History—A Summary have been expanded.
Chapter 23: Sections on Climate Change, Causes of Global Climate Change, and Global Warming along with a discussion of hurricane Katrina have been added.
Chapter 24: A Closer Look on rogue waves is now included.

THE LEARNING SYSTEM

Physical Science has an effective combination of innovative learning aids intended to make the student's study of science more effective and enjoyable. This variety of aids is included to help students clearly understand the concepts and principles that serve as the foundation of the physical sciences.

OVERVIEW

Chapter 1 provides an *overview* or orientation to what the study of physical science in general and this text in particular are all about. It discusses the fundamental methods and techniques used by scientists to study and understand the world around us. It also explains the problem-solving approach used throughout the text so that students can more effectively apply what they have learned.

CHAPTER OPENING TOOLS
Core Concept and Supporting Concepts

New! Core and Supporting Concepts integrate the chapter concepts and the chapter outline. The Core and Supporting Concepts outline and emphasize the concepts at a chapter level. The concepts list is designed to help students focus their studies by identifying the most important topics in the chapter outline.

Chapter Outline

The chapter outline includes all the major topic headings and subheadings within the body of the chapter. It gives you a quick glimpse of the chapter's contents and helps you locate sections dealing with particular topics.

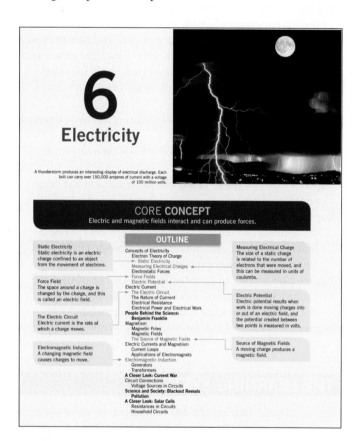

Chapter Overview

Each chapter begins with an introductory overview. The overview previews the chapter's contents and what you can expect to learn from reading the chapter. It adds to the general outline of the chapter by introducing you to the concepts to be covered, facilitating in the integration of topics, and helping you to stay focused and organized while reading the chapter for the first time. After reading the introduction, browse through the

chapter, paying particular attention to the topic headings and illustrations so that you get a feel for the kinds of ideas included within the chapter.

> **"Tillery does a much better job explaining concepts and reinforcing them. I believe his style of presentation is better and more comfortable for the student. His use of the overviews and examples is excellent!"**
> —George T. Davis, Jr., Mississippi Delta Community College

EXAMPLES

Each topic discussed within the chapter contains one or more concrete, worked *Examples* of a problem and its solution as it applies to the topic at hand. Through careful study of these examples, students can better appreciate the many uses of problem solving in the physical sciences.

> **"I feel this book is written well for our average student. The images correlate well with the text, and the math problems make excellent use of the dimensional analysis method. While it was a toss-up between this book and another one, now that we've taught from the book for the last year, we are extremely happy with it."**
> —Alan Earhart, Three Rivers Community College

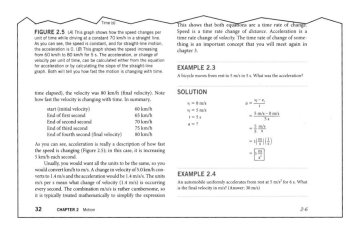

FIGURE 2.5 (A) This graph shows how the speed changes per unit of time while driving at a constant 70 km/h in a straight line. As you can see, the speed is constant, and for straight-line motion, the acceleration is 0. (B) This graph shows the speed increasing from 60 km/h to 80 km/h for 5 s. The acceleration, or change of velocity per unit of time, can be calculated either from the equation for acceleration or by calculating the slope of the straight-line graph. Both will tell you how fast the motion is changing with time.

time elapsed), the velocity was 80 km/h (final velocity). Note how fast the velocity is changing with time. In summary,

start (initial velocity)	60 km/h
End of first second	65 km/h
End of second second	70 km/h
End of third second	75 km/h
End of fourth second (final velocity)	80 km/h

As you can see, acceleration is really a description of how fast the speed is changing (Figure 2.5); in this case, it is increasing 5 km/h each second.

Usually, you would want all the units to be the same, so you would convert km/h to m/s. A change in velocity of 5.0 km/h converts to 1.4 m/s and the acceleration would be 1.4 m/s/s. The units m/s per s mean what change of velocity (1.4 m/s) is occurring every second. The combination m/s/s is rather cumbersome, so it is typically treated mathematically to simplify the expression

This shows that both equations are a time rate of change. Speed is a time rate of change of *distance*. Acceleration is a time rate change of *velocity*. The time rate of change of something is an important concept that you will meet again in chapter 3.

EXAMPLE 2.3

A bicycle moves from rest to 5 m/s in 5 s. What was the acceleration?

SOLUTION

$v_i = 0$ m/s
$v_f = 5$ m/s
$t = 5$ s
$a = ?$

$$a = \frac{v_f - v_i}{t}$$

$$= \frac{5 \text{ m/s} - 0 \text{ m/s}}{5 \text{ s}}$$

$$= \frac{5}{5}\,\frac{\text{m/s}}{\text{s}}$$

$$= 1\left(\frac{m}{s}\right)\left(\frac{1}{s}\right)$$

$$= \boxed{1\,\frac{m}{s^2}}$$

EXAMPLE 2.4

An automobile uniformly accelerates from rest at 5 m/s² for 6 s. What is the final velocity in m/s? (Answer: 30 m/s)

A Closer Look

A Bicycle Racer's Edge

Galileo was one of the first to recognize the role of friction in opposing motion. As shown in Figure 2.9, friction with the surface and air friction combine to produce a net force that works against anything that is moving on the surface. This article is about air friction and some techniques that bike riders use to reduce that opposing force—perhaps giving them an edge in a close race.

The bike riders in Box Figure 2.1 are forming a single-file line, called a *pace-line*, because the slipstream reduces the air resistance for a closely trailing rider. Cyclists say that riding in the slipstream of another cyclist will save much of their energy. They can move up to 5 mi/h faster than they would expending the same energy riding alone.

In a sense, riding in a slipstream means that you do not have to push as much air out of your way. It has been estimated that at 20 mi/h, a cyclist must move a little less than half a ton of air out of the way every minute. Along with the problem of moving air out of the way, there are two basic factors related to air resistance. These are (1) a

BOX FIGURE 2.1 The object of the race is to be in the front, to finish first. If this is true, why are these racers forming a single-file line?

turbulent versus a smooth flow of air and (2) the problem of frictional drag. A turbulent flow of air contributes to air resistance because it causes the air to separate slightly on the back side, which increases the pressure on the front of the moving object. This is why racing cars, airplanes, boats, and other racing vehicles are streamlined to a teardroplike shape. This shape is not as likely to have the lower-pressure-producing air turbulence behind (and resulting greater pressure in front) because it smoothes, or streamlines, the air flow.

The frictional drag of air is similar to the frictional drag that occurs when you push a book across a rough tabletop. You know that smoothing the rough tabletop will reduce the frictional drag on the book. Likewise, the smoothing of a surface exposed to moving air will reduce air friction. Cyclists accomplish this "smoothing" by wearing smooth Lycra clothing and by shaving hair from arm and leg surfaces that are exposed to moving air. Each hair contributes to the overall frictional drag, and removal of the arm and leg hair can thus result in seconds saved. This might provide enough of an edge to win a close race. Shaving legs and arms, together with the wearing of Lycra or some other tight, smooth-fitting garments, are just a few of the things a cyclist can do to gain an edge. Perhaps you will be able to think of more ways to reduce the forces that oppose motion.

APPLYING SCIENCE TO THE REAL WORLD

Concepts Applied

Each chapter also includes one or more *Concepts Applied* boxes. These activities are simple investigative exercises that students can perform at home or in the classroom to demonstrate important concepts and reinforce understanding of them. This feature also describes the application of those concepts to everyday life.

Science and Society

These readings relate the chapter's content to current societal issues. Many of these boxes also include Questions to Discuss that provide an opportunity to discuss issues with your peers.

Myths, Mistakes, and Misunderstandings

These brief boxes provide short, scientific explanations to dispel a societal myth or a home experiment or project that enables you to dispel the myth on your own.

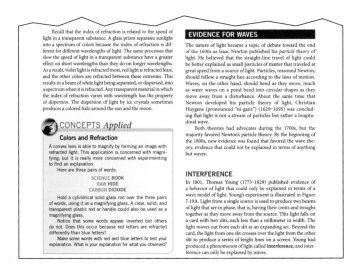

Recall that the index of refraction is related to the speed of light in a transparent substance. A glass prism separates sunlight into a spectrum of colors because the index of refraction is different for different wavelengths of light. The same processes that slow the speed of light in a transparent substance have a greater effect on short wavelengths than they do on longer wavelengths. As a result, violet light is refracted most, red light is refracted least, and the other colors are refracted between these extremes. This results in a beam of white light being separated, or dispersed, into a spectrum when it is refracted. Any transparent material in which the index of refraction varies with wavelength has the property of *dispersion*. The dispersion of light by ice crystals sometimes produces a colored halo around the sun and the moon.

CONCEPTS *Applied*

Colors and Refraction

A convex lens is able to magnify by forming an image with refracted light. This application is concerned with magnifying, but it is really more concerned with experimenting to find an explanation.
Here are three pairs of words:

SCIENCE BOOK
RAW HIDE
CARBON DIOXIDE

Hold a cylindrical solid glass rod over the three pairs of words, using it as a magnifying glass. A clear, solid, and transparent plastic rod or handle could also be used as a magnifying glass.

Notice that some words appear inverted but others do not. Does this occur because red letters are refracted differently than blue letters?

Make some words with red and blue letters to test your explanation. What is your explanation for what you observed?

EVIDENCE FOR WAVES

The nature of light became a topic of debate toward the end of the 1600s as Isaac Newton published his *particle theory* of light. He believed that the straight-line travel of light could be better explained as small particles of matter that traveled at great speed from a source of light. Particles, reasoned Newton, should follow a straight line according to the laws of motion. Waves, on the other hand, should bend as they move, much as water waves on a pond bend into circular shapes as they move away from a disturbance. About the same time that Newton developed his particle theory of light, Christian Huygens (pronounced "ni-ganz") (1629–1695) was concluding that light is not a stream of particles but rather a longitudinal wave.

Both theories had advocates during the 1700s, but the majority favored Newton's particle theory. By the beginning of the 1800s, new evidence was found that favored the wave theory, evidence that could not be explained in terms of anything but waves.

INTERFERENCE

In 1801, Thomas Young (1773–1829) published evidence of a behavior of light that could only be explained in terms of a wave model of light. Young's experiment is illustrated in Figure 7.19A. Light from a single source is used to produce two beams of light that are in phase, that is, having their crests and troughs together as they move away from the source. This light falls on a card with two slits, each less than a millimeter in width. The light moves out from each slit as an expanding arc. Beyond the card, the light from one slit crosses over the light from the other slit to produce a series of bright lines on a screen. Young had produced a phenomenon of light called **interference**, and interference can only be explained by waves.

Closer Look

One or more boxed *Closer Look* features can be found in each chapter of *Physical Science*. These readings present topics of special human or environmental concern (the use of seat belts, acid rain, and air pollution, for example). In addition to environmental concerns, topics are presented on interesting technological applications (passive solar homes, solar cells, catalytic converters, etc.) or on the cutting edge of scientific research (for example, El Niño and dark energy). All boxed features are informative materials that are supplementary in nature. The *Closer Look* readings serve to underscore the relevance of physical science in confronting the many issues we face daily.

Science and Society

Geothermal Energy

Geothermal energy means earth (*geo*) heat (*thermal*) energy, or energy in the form of heat from Earth. Beneath the surface of Earth is a very large energy resource in the form of hot water and steam that can be used directly for heat or converted to electricity.

Most of the U.S. geothermal resources are located in the western part of the nation where the Juan de Fuca Plate is subducted beneath the continental lithosphere. The Juan de Fuca partially melts, forming magma that is buoyed toward Earth's surface, erupting as volcanoes (see Figure 19.25). This subduction is the source of heating for the hot water and steam resources in ten western states. There are also geothermal resources in Hawaii from a different geothermal source.

One use of geothermal energy is to generate electricity. Geothermal power plants are located in California (ten sites), Hawaii, Nevada (ten sites), Oregon, and Utah (two sites). These sites have a total generating capacity of 2,700 megawatts (MW), which is enough electricity to supply the needs of 3.5 million people. The world's largest geothermal power plant is located at the Geysers Power Plant in northern California. Dry steam provides the energy for twenty-three units at this site, which generate more than 1,700 MW of electrical power. All of the other sites use hot water rather than steam to generate electricity, with a total generating capacity of 1,000 MW.

In addition to producing electricity, geothermal hot water is used directly for space heating. Space heating in individual houses is accomplished by piping hot water from one geothermal well. District systems, on the other hand, pipe hot water from one or more geothermal wells to several buildings, houses, or blocks of houses. Currently, geothermal hot water is used in individual and district space-heating systems at more than 120 locations. There are more than 1,200 potential geothermal sites that could be developed to provide hot water to more than 370 cities in eight states. The creation of such geothermal districts could result in a savings of up to 50 percent over the cost of natural gas heating.

Geothermal hot water is also used directly in greenhouses and aquaculture facilities. There are more than thirty-five large geothermal-energized greenhouses raising vegetables and flowers and more than twenty-five geothermal-energized aquaculture facilities raising fish in Arizona, California, Colorado, Idaho, Montana, Nevada, New York, Oregon, South Dakota, Utah, and Wyoming (see http://geoheat.oit.edu/drays.htm. A food dehydration facility in Nevada, for example, uses geothermal energy to process more than 15 million pounds of dried onions and garlic per year. Other uses of geothermal energy include laundries, swimming pools, spas, and resorts. Over two hundred resorts are using geothermal hot water in the United States.

Geothermal energy is considered to be one of the renewable energy resources since the energy supply is maintained by plate tectonics. Currently, geothermal energy production is ranked third behind hydroelectricity and biomass but ahead of solar and wind. It has been estimated that known geothermal resources could supply thousands of megawatts more power beyond current production, and development of the potential direct-use applications could displace the use—and greenhouse gas emissions—of 18 million barrels of oil per year.

QUESTIONS TO DISCUSS

Discuss with your group the following questions concerning the use of geothermal energy:

1. Why is the development of geothermal energy not proceeding more rapidly?

2. Should the government provide incentives for developing geothermal resources? Give reasons for your answer.

3. What are the advantages and disadvantages of a government-controlled geothermal energy industry?

4. As other energy supplies become depleted, who should be responsible for developing new energy supplies, investor-owned industry or government agencies?

Myths, Mistakes, & Misunderstandings

Bye Bye California?

It is a myth that California will eventually fall off the continent into the ocean. The San Andreas fault is the boundary between the Pacific and North American Plates. The Pacific Plate is moving northwest along the North American Plate at 45 mm per year (about the rate your fingernails grow). The plates are moving horizontally by each other, so there is no reason to believe California will fall into the ocean. However, some 15 million years and millions of earthquakes from now, Los Angeles might be across the bay from San Francisco. See "Earthquakes, Mega Quakes, and the Movies" at http://earthquake.usgs.gov/learning/topics.php?topicID=36+topic=Myths

being gathered and evaluated, and the exact number of plates and their boundaries are yet to be determined with certainty. The major question that remains to be answered is what drives the plates, moving them apart, together, and by each other? One explanation is that slowly turning *convective cells* in the plastic asthenosphere drive the plates (Figure 18.18). According to this hypothesis, hot mantle materials rise at the diverging boundaries. Some of the material escapes to form new crust, but most of it spreads out beneath the lithosphere. As it moves beneath the lithosphere, it drags the overlying plate with it. Eventually, it cools and sinks back inward under a subduction zone.

There is uncertainty about the existence of convective cells in the asthenosphere and their possible role because of a lack of clear evidence. Seismic data is not refined enough to show convective cell movement beneath the lithosphere. In addition,

People Behind the Science

Many chapters also have fascinating biographies that spotlight well-known scientists, past or present. From these *People Behind the Science* biographies, students learn about the human side of the science: physical science is indeed relevant, and real people do the research and make the discoveries. These readings present physical science in real-life terms that students can identify with and understand.

"The People Behind the Science features help relate the history of science and the contributions of the various individuals."
—Richard M. Woolheater, Southeastern Oklahoma State University

Source: Modified from the Hutchinson Dictionary of Scientific Biography © Research Machines plc 2006. All Rights Reserved. Helicon Publishing is a division of Research Machines.

END-OF-CHAPTER FEATURES

At the end of each chapter, students will find the following materials:

- *Summary:* highlights the key elements of the chapter.
- *Summary of Equations* (chapters 1–13): reinforces retention of the equations presented.
- *Key Terms:* gives page references for finding the terms defined within the context of the chapter reading.
- *Applying the Concepts:* tests comprehension of the material covered with a multiple-choice quiz.
- *Questions for Thought:* challenges students to demonstrate their understanding of the topics.
- *Parallel Exercises* (chapters 1–13): reinforces problem-solving skills. There are two groups of parallel exercises, Group A and Group B. The Group A parallel exercises have complete solutions worked out, along with useful comments, in appendix E. The Group B parallel exercises are similar to those in Group A but do not contain answers in the text. By working through the Group A

parallel exercises and checking the solutions in appendix E, students will gain confidence in tackling the parallel exercises in Group B and thus reinforce their problem-solving skills.

- *For Further Analysis:* exercises include analysis or discussion questions, independent investigations, and activities intended to emphasize critical thinking skills and societal issues, and develop a deeper understanding of the chapter content.
- *Invitation to Inquiry:* exercises that consist of short, open-ended activities that allow you to apply investigative skills to the material in the chapter.

"The most outstanding feature of Tillery's *Physical Science* is the use of the Group A Parallel Exercises. Prior to this text, I cannot count the number of times I have heard students state that they understood the material when presented in class, but when they tried the homework on their own, they were unable to remember what to do. The Group A problems with the complete solution were the perfect reminder for most of the students. I also believe that Tillery's presentation of the material addresses the topics with a rigor necessary for a college-level course but is easily understandable for my students without being too simplistic. The material is challenging but not too overwhelming."
—J. Dennis Hawk, Navarro College

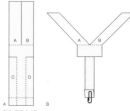

FIGURE 1.17 Pattern for a paper helicopter.

END-OF-TEXT MATERIALS

Appendices providing math review, additional background detail, solubility and humidity charts, solutions for the in-chapter follow-up examples, and solutions for the Group A Parallel Exercises can be found at the back of the text. There is also a glossary of all key terms, an index, and special tables printed on the inside covers for reference use.

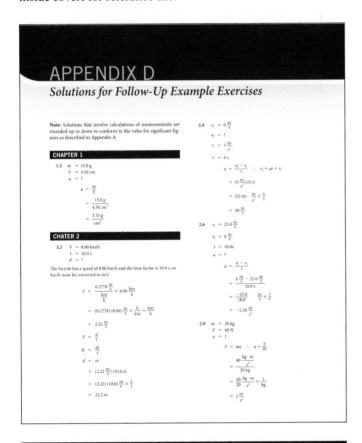

SUPPLEMENTS

Physical Science is accompanied by a variety of multimedia supplementary materials, including an interactive ARIS site with testing software containing multiple-choice test items and other teacher resources. The supplement package also includes a laboratory manual, both student and instructor's editions, by the author of the text.

MULTIMEDIA SUPPLEMENTARY MATERIALS
McGraw-Hill's ARIS—Assessment, Review, and Instruction System

McGraw-Hill's ARIS for *Physical Science* is a complete, online electronic homework and course management system designed for greater ease of use than any other system available. Available with the *Physical Science* eighth edition text, instructors can create and share course materials and assignments with colleagues with a few clicks of the mouse. All PowerPoint lectures, assignments, quizzes, an instructor's lab manual, text images, an instructor's manual, test bank questions, clicker questions, animations, and more are directly tied to text-specific materials in

Physical Science, but instructors can also edit questions, import their own content, and create announcements and due dates for assignments. ARIS has automatic grading and reporting of easy-to-assign homework, quizzing, and testing. All student activity within McGraw-Hill's ARIS is automatically recorded and available to the instructor through a fully integrated grade book that can be downloaded to Excel.

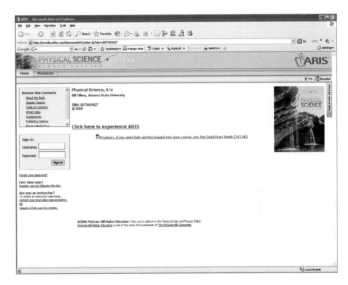

Personal Response Systems

Personal Response Systems ("clickers') can bring interactivity into the classroom or lecture hall. Wireless response systems give the instructor and students immediate feedback from the entire class. The wireless response pads are essentially remotes that are easy to use and engage students. Clickers allow instructors to motivate student preparation, interactivity, and active learning. Instructors receive immediate feedback to gauge which concepts students understand. Questions covering the content of the *Physical Science* text and formatted in PowerPoint are available on ARIS for *Physical Science.*

Computerized Test Bank Online

A comprehensive bank of test questions is provided within a computerized test bank powered by McGraw-Hill's flexible electronic testing program EZ Test Online (www.eztestonline.com). EZ Test Online allows instructors to create paper and online tests or quizzes in this easy-to-use program!

Imagine being able to create and access your test or quiz anywhere, at any time without installing the testing software. Now, with EZ Test Online, instructors can select questions from multiple McGraw-Hill test banks or author their own and then either print the test for paper distribution or give it online.

Test Creation
- Author/edit questions online using the fourteen different question type templates.
- Create printed tests or deliver online to get instant scoring and feedback.

- Create question pools to offer multiple versions online—great for practice.
- Export tests for use in WebCT, Blackboard, PageOut, and Apple's iQuiz.
- Compatible with EZ Test Desktop tests already created.
- Sharing tests with colleagues, adjuncts, TAs is easy.

Online Test Management
- Set availability dates and time limits for the quiz or test.
- Control how the test will be presented.
- Assign points by question or question type with drop-down menu.
- Provide immediate feedback to students or delay until all finish the test.
- Create practice tests online to enable students mastery.
- Your roster can be uploaded to enable student self-registration.

Online Scoring and Reporting
- Automated scoring for most of EZ Test's numerous question types
- Allows manual scoring for essay and other open response questions
- Manual rescoring and feedback is also available.
- EZ Test's grade book is designed to easily export to your grade book.
- View basic statistical reports

Support and help
- User's Guide and built-in, page-specific help
- Flash tutorials for getting started on the support site
- Support website—www.mhhe.com/eztest
- Product specialist available at 1-800-331-5094
- Online training: http://auth.mhhe.com/mpss/workshops/

PRESENTATION CENTER

Complete set of electronic book images and assets for instructors.

Build instructional materials wherever, whenever, and however you want!

Accessed from your textbook's ARIS website, **Presentation Center** is an online digital library containing photos, artwork, animations, and other media types that can be used to create customized lectures, visually enhanced tests and quizzes, compelling course websites, or attractive printed support materials. All assets are copyrighted by McGraw-Hill Higher Education but can be used by instructors for classroom purposes. The visual resources in this collection include:

- **Art and Photo Library:** Full-color digital files of all of the illustrations and many of the photos in the text can be readily incorporated into lecture presentations, exams, or custommade classroom materials.
- **Worked Example Library, Table Library, and Numbered Equations Library:** Access the worked examples, tables, and equations from the text in electronic format for inclusion in your classroom resources.

- **Animations Library:** Files of animations and videos covering the many topics in *Physical Science* are included so that you can easily make use these animations in a lecture or classroom setting.

Also residing on your textbook's ARIS website are:
- **PowerPoint Slides:** For instructors who prefer to create their lectures from scratch, all illustrations, photos, and tables are preinserted by chapter into blank PowerPoint slides.
- **Lecture Outlines:** Lecture notes, incorporating illustrations and animated images, have been written to the eighth edition text. They are provided in PowerPoint format so that you may use these lectures as written or customize them to fit your lecture.

"I find Physical Science to be superior to either of the texts that I have used to date. . . . The animations and illustrations are better than those of other textbooks that I have seen, more realistic and less trivial."
—T. G. Heil, University of Georgia

ELECTRONIC BOOKS

If you or your students are ready for an alternative version of the traditional textbook, McGraw-Hill brings you innovative and inexpensive electronic textbooks. By purchasing e-books from McGraw-Hill, students can save as much as 50 percent on selected titles delivered on the most advanced e-book platforms available.

E-books from McGraw-Hill are smart, interactive, searchable, and portable, with such powerful tools as detailed searching, highlighting, note taking, and student-to-student or instructor-to-student note sharing. E-books from McGraw-Hill will help students to study smarter and quickly find the information they need. Students will also save money. Contact your McGraw-Hill sales representative to discuss e-book packaging options.

PRINTED SUPPLEMENTARY MATERIAL
Laboratory Manual

The *laboratory manual*, written and classroom tested by the author, presents a selection of laboratory exercises specifically written for the interests and abilities of nonscience majors. There are laboratory exercises that require measurement, data analysis, and thinking in a more structured learning environment. Alternative exercises that are open-ended "Invitations to Inquiry" are provided for instructors who would like a less structured approach. When the laboratory manual is used with *Physical Science*, students will have an opportunity to master basic scientific principles and concepts, learn new problem-solving and thinking skills, and understand the nature of scientific inquiry from the perspective of hands-on experiences. The *instructor's edition of the laboratory manual* can be found on the *Physical Science* ARIS site.

ACKNOWLEDGMENTS

We are indebted to the reviewers for the eighth edition for their constructive suggestions, new ideas, and invaluable advice. Special thanks and appreciation goes out to the eighth edition reviewers:

S. Ray Bullock, *Grambling State University*
Brian Carter, *California State at San Marcos*
Pamela M. Clevenger, *Hinds Community College*
Tom Colbert, *Augusta State University*
George T. Davis, Jr., *Mississippi Delta Community College*
J. Dennis Hawk, *Navarro College*
Asaad Istephan, *Madonna University*
Andrew Kiruluta, *Harvard University*
Cynthia M. Lamberty, *Nicholls State University*
Eric C. Martell, *Millikin University*
Pamela Ray, *Chattahoochee Valley Community College*

This revision of *Physical Science* has also been made possible by the many users and reviewers of its previous editions. The author and publisher are grateful to the following reviewers of previous editions for their critical reviews, comments, and suggestions:

Lawrence H. Adams, *Polk Community College*
Miah M. Adel, *University of Arkansas at Pine Bluff*
Adedoyin M. Adeyiga, *Cheyney University of Pennsylvania*
John Akutagawa, *Hawaii Pacific University*
Arthur L. Alt, *University of Great Falls*
Brian Augustine, *James Madison University*
Richard Bady, *Marshall University*
David Benin, *Arizona State University*
Michael Berheide, *Berea College*
Rao Bidthanapally, *Oakland University*
Ignacio Birriel, *Morehead State University*
Charles L. Bissell, *Northwestern State University of Louisiana*
Charles Blatchley, *Pittsburg State University*
W. H. Breazeale, Jr., *Francis Marion College*
William Brown, *Montgomery College*
Peter E. Busher, *Boston University*
Steven Carey, *Mobile College*
Darry S. Carlston, *University of Central Oklahoma*
Stan Celestian, *Glendale Community College*
Pamela M. Clevenger, *Hinds Community College*
Randel Cox, *Arkansas State University*
Paul J. Croft, *Jackson State University*
Keith B. Daniels, *University of Wisconsin–Eau Claire*
Valentina David, *Bethune-Cookman College*
Carl G. Davis, *Danville Area Community College*
Joe D. DeLay, *Freed-Hardeman University*
Renee D. Diehl, *Pennsylvania State University*
Karim Diff, *Santa Fe Community College*
Bill Dinklage, *Utah Valley State College*
Paul J. Dolan, Jr., *Northeastern Illinois University*
Thomas A. Dooling, *University of North Carolina at Pembroke*
Laurencin Dunbar, *Livingstone College*
Alan D. Earhart, *Three Rivers Community College*
Dennis Englin, *The Master's College*

Carl Frederickson, *University of Central Arkansas*
Steven S. Funck, *Harrisburg Area Community College*
Lucille B. Garmon, *State University of West Georgia*
Peter K. Glanz, *Rhode Island College*
Nova Goosby, *Philander Smith College*
D. W. Gosbin, *Cumberland County College*
Omar Franco Guerrero, *University of Delaware*
Floretta Haggard, *Rogers State College*
Robert G. Hamerly, *University of Northern Colorado*
Eric Harms, *Brevard Community College*
Louis Hart, *West Liberty State College*
J. Dennis Hawk, *Navarro College*
T. G. Heil, *University of Georgia*
L. D. Hendrick, *Francis Marion College*
Ezat Heydari, *Jackson State University*
C. A. Hughes, *University of Central Oklahoma*
Christopher Hunt, *Prince George's Community College*
Judith Iriarte-Gross, *Middle Tennessee State University*
Booker Juma, *Fayetteville State University*
Alice K. Kolalowska, *Mississippi State University*
Linda C. Kondrick, *Arkansas Tech University*
Abe Korn, *New York City Tech College*
Eric T. Lane, *University of Tennessee at Chattanooga*
Lauree G. Lane, *Tennessee State University*
Robert Larson, *St. Louis Community College*
William Luebke, *Modesto Junior College*
Douglas L. Magnus, *St. Cloud State University*
Stephen Majoros, *Lorain County Community College*
L. Whit Marks, *Central State University*
Richard S. Mitchell, *Arkansas State University*
Jesse C. Moore, *Kansas Newman College*
Michael D. Murphy, *Northwest Alabama Community College*
Martha K. Newchurch, *Nicholls State University*
Gabriel Niculescu, *James Madison University*
Christopher Kivuti Njue, *Shaw University*
Ignatius Okafor, *Jarvis Christian College*
Kale Oyedeji, *Morehouse College*
Oladayo Oyelola, *Lane College*
Harold Pray, *University of Central Arkansas*
Virginia Rawlins, *University of North Texas*
Antonie H. Rice, *University of Arkansas at Pine Bluff*
Karen Savage, *California State University at Northridge*
R. Allen Shotwell, *Ivy Tech State College*
Michael L. Sitko, *University of Cincinnati*
K. W. Trantham, *Arkansas Tech University*
R. Steven Turley, *Brigham Young University*
David L. Vosburg, *Arkansas State University*
Ling Jun Wang, *University of Tennessee at Chattanooga*
Martha Reherd Weller, *Middle Tennessee State University*
J. S. Whitaker, *Boston University*
Donald A. Whitney, *Hampton University*
Linda Wilson, *Middle Tennessee State University*
David Wingert, *Georgia State University*
Richard M. Woolheater, *Southeastern Oklahoma State University*
Heather Woolverton, *University of Central Arkansas*
Michael Young, *Mississippi Delta Community College*

We would also like to thank the following contributors to the eighth edition:

J. Dennis Hawk, Navarro College, for his knowledge of student conceptual understandings, used in developing and revising the personal response system questions to accompany *Physical Science,* eighth edition.

Pamela M. Clevenger, Hinds Community College, for her creativity in revising the multimedia PowerPoint lecture outlines to accompany *Physical Science,* Eighth Edition.

Lee E. Evinger, Missouri Western State University, for his detailed and informative review of the Applying the Concepts end-of-chapter multiple-choice self-tests, which helped to greatly improve this set of questions for the eighth edition.

Zdeslay Hrepic, Fort Hays State University and **Melinda Huff,** Northeastern Oklahoma A & M College, for their detailed reviews of the *Physical Science* Laboratory Manual.

Last, I wish to acknowledge the very special contributions of my wife, Patricia Northrop Tillery, whose assistance and support throughout the revision were invaluable.

MEET THE AUTHOR

BILL W. TILLERY

Bill W. Tillery is professor emeritus of Physics at Arizona State University, where he was a member of the faculty from 1973 to 2006. He earned a bachelor's degree at Northeastern State University (1960), and master's and doctorate degrees from the University of Northern Colorado (1967). Before moving to Arizona State University, he served as director of the Science and Mathematics Teaching Center at the University of Wyoming (1969–73) and as an assistant professor at Florida State University (1967–69). Bill served on numerous councils, boards, and committees, and was honored as the "Outstanding University Educator" at the University of Wyoming in 1972. He was elected the "Outstanding Teacher" in the Department of Physics and Astronomy at Arizona State University in 1995.

During his time at Arizona State, Bill taught a variety of courses, including general education courses in science and society, physical science, and introduction to physics. He received more than forty grants from the National Science Foundation, the U.S. Office of Education, from private industry (Arizona Public Service), and private foundations (The Flinn Foundation) for science curriculum development and science teacher inservice training. In addition to teaching and grant work, Bill authored or coauthored more than sixty textbooks and many monographs, and served as editor of three separate newsletters and journals between 1977 and 1996.

Bill has attempted to present an interesting, helpful program that will be useful to both students and instructors. Comments and suggestions about how to do a better job of reaching this goal are welcome. Any comments about the text or other parts of the program should be addressed to:

Bill W. Tillery
e-mail: bill.tillery@asu.edu

1
What Is Science?

Physical science is concerned with your physical surroundings and your concepts and understanding of these surroundings.

CORE **CONCEPT**
Science is a way of thinking about and understanding your environment.

OUTLINE

Objects and Properties
Properties are qualities or attributes that can be used to describe an object or event.

Data
Data is measurement information that can be used to describe objects, conditions, events, or changes.

Scientific Method
Science investigations include collecting observations, developing explanations, and testing explanations.

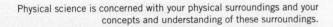

Objects and Properties
Quantifying Properties
Measurement Systems
Standard Units for the Metric System
  Length
  Mass
  Time
Metric Prefixes
Understandings from Measurements
  Data
  Ratios and Generalizations
  The Density Ratio
  Symbols and Equations
  How to Solve Problems
The Nature of Science
  The Scientific Method
  Explanations and Investigations
Science and Society: Basic and Applied Research
  Laws and Principles
  Models and Theories
People Behind the Science: Florence Bascom

Quantifying Properties
Measurement is used to accurately describe properties of objects or events.

Symbols and Equations
An equation is a statement of a relationship between variables.

Laws and Principles
Scientific laws describe relationships between events that happen time after time, describing *what* happens in nature.

Models and Theories
A scientific theory is a broad working hypothesis based on extensive experimental evidence, describing *why* something happens in nature.

1

OVERVIEW

Have you ever thought about your thinking and what you know? On a very simplified level, you could say that everything you know came to you through your senses. You see, hear, and touch things of your choosing, and you can also smell and taste things in your surroundings. Information is gathered and sent to your brain by your sense organs. Somehow, your brain processes all this information in an attempt to find order and make sense of it all. Finding order helps you understand the world and what may be happening at a particular place and time. Finding order also helps you predict what may happen next, which can be very important in a lot of situations.

This is a book on thinking about and understanding your physical surroundings. These surroundings range from the obvious, such as the landscape and the day-to-day weather, to the not so obvious, such as how atoms are put together. Your physical surroundings include natural things as well as things that people have made and used (Figure 1.1). You will learn how to think about your surroundings, whatever your previous experience with thought-demanding situations. This first chapter is about "tools and rules" that you will use in the thinking process.

OBJECTS AND PROPERTIES

Physical science is concerned with making sense out of the physical environment. The early stages of this "search for sense" usually involve *objects* in the environment, things that can be seen or touched. These could be objects you see every day, such as a glass of water, a moving automobile, or a blowing flag. They could be quite large, such as the Sun, the Moon, or even the solar system, or invisible to the unaided human eye. Objects can be any size, but people are usually concerned with objects that are larger than a pinhead and smaller than a house. Outside these limits, the actual size of an object is difficult for most people to comprehend.

As you were growing up, you learned to form a generalized mental image of objects called a *concept*. Your concept of an object is an idea of what it is, in general, or what it should be according to your idea (Figure 1.2). You usually have a word stored away in your mind that represents a concept. The word *chair*, for example, probably evokes an idea of "something to sit on." Your generalized mental image for the concept that goes with the word *chair* probably includes a four-legged object with a backrest. Upon close inspection, most of your (and everyone else's) concepts are found to be somewhat vague. For example, if the word *chair* brings forth a mental image of something with four legs and a backrest (the concept), what is the difference between a "high chair" and a "bar stool"? When is a chair a chair and not a stool? These kinds of questions can be troublesome for many people.

Not all of your concepts are about material objects. You also have concepts about intangibles such as time, motion, and relationships between events. As was the case with concepts of material objects, words represent the existence of intangible concepts. For example, the words *second, hour, day,* and *month* represent concepts of time. A concept of the pushes and pulls that come with changes of motion during an airplane flight might be represented with such words as *accelerate* and *falling*. Intangible concepts might seem to be more abstract since they do not represent material objects.

By the time you reach adulthood, you have literally thousands of words to represent thousands of concepts. But most,

FIGURE 1.1 Your physical surroundings include naturally occurring and manufactured objects such as sidewalks and buildings.

you would find on inspection, are somewhat ambiguous and not at all clear-cut. That is why you find it necessary to talk about certain concepts for a minute or two to see if the other person has the same "concept" for words as you do. That is why

FIGURE 1.2 What is your concept of a chair? Are all of these pieces of furniture chairs? Most people have concepts, or ideas of what things in general should be, that are loosely defined. The concept of a chair is one example of a loosely defined concept.

FIGURE 1.3 Could you describe this rock to another person over the telephone so that the other person would know *exactly* what you see? This is not likely with everyday language, which is full of implied comparisons, assumptions, and inaccurate descriptions.

when one person says, "Boy, was it hot!" the other person may respond, "How hot was it?" The meaning of *hot* can be quite different for two people, especially if one is from Arizona and the other from Alaska!

The problem with words, concepts, and mental images can be illustrated by imagining a situation involving you and another person. Suppose that you have found a rock that you believe would make a great bookend. Suppose further that you are talking to the other person on the telephone, and you want to discuss the suitability of the rock as a bookend, but you do not know the name of the rock. If you knew the name, you would simply state that you found a "_____." Then you would probably discuss the rock for a minute or so to see if the other person really understood what you were talking about. But not knowing the name of the rock and wanting to communicate about the suitability of the object as a bookend, what would you do? You would probably describe the characteristics, or **properties,** of the rock. Properties are the qualities or attributes that, taken together, are usually peculiar to an object. Since you commonly determine properties with your senses (smell, sight, hearing, touch, and taste), you could say that the properties of an object are the effect the object has on your senses. For example, you might say that the rock is a "big, yellow, smooth rock with shiny gold cubes on one side." But consider the mental image that the other person on the telephone forms when you describe these properties. It is entirely possible that the other person is thinking of something very different from what you are describing (Figure 1.3)!

As you can see, the example of describing a proposed bookend by listing its properties in everyday language leaves much to be desired. The description does not really help the other person form an accurate mental image of the rock. One problem with the attempted communication is that the description of any property implies some kind of *referent.* The word **referent** means that you *refer to,* or think of, a given property in terms of another, more familiar object. Colors, for example, are sometimes stated with a referent. Examples are "sky blue," "grass green," or "lemon yellow." The referents for the colors blue, green, and yellow are, respectively, the sky, living grass, and a ripe lemon.

Referents for properties are not always as explicit as they are with colors, but a comparison is always implied. Since the comparison is implied, it often goes unspoken and leads to assumptions in communications. For example, when you stated that the rock was "big," you assumed that the other person knew that you did not mean as big as a house or even as big as a bicycle. You assumed that the other person knew that you meant that the rock was about as large as a book, perhaps a bit larger.

Another problem with the listed properties of the rock is the use of the word *smooth.* The other person would not know if you meant that the rock *looked* smooth or *felt* smooth. After all, some objects can look smooth and feel rough. Other objects can look rough and feel smooth. Thus, here is another assumption, and probably all of the properties lead to implied comparisons, assumptions, and a not-very-accurate communication. This is the nature of your everyday language and the nature of most attempts at communication.

QUANTIFYING PROPERTIES

Typical day-to-day communications are often vague and leave much to be assumed. A communication between two people, for example, could involve one person describing some person, object, or event to a second person. The description is made by using referents and comparisons that the second person may or may not have in mind. Thus, such attributes as "long" fingernails or "short" hair may have entirely different meanings to different people involved in a conversation. Assumptions and vagueness can be avoided by using **measurement** in a description. Measurement is a process of comparing a property to a well-defined and agreed-upon referent. The well-defined and agreed-upon referent is used as a standard called a **unit.** The measurement process involves three steps: (1) *comparing* the referent unit to the property being described, (2) following a *procedure,* or operation, that specifies how the comparison is made, and (3) *counting* how many standard units describe the property being considered.

The measurement process uses a defined referent unit, which is compared to a property being measured. The *value* of the property is determined by counting the number of referent units. The name of the unit implies the procedure that results in the number. A measurement statement always contains a *number* and *name* for the referent unit. The number answers the question "How much?" and the name answers the question "Of what?" Thus, a measurement always tells you "how much of what." You will find that using measurements will sharpen your communications. You will also find that using measurements is one of the first steps in understanding your physical environment.

MEASUREMENT SYSTEMS

Measurement is a process that brings precision to a description by specifying the "how much" and "of what" of a property in a particular situation. A number expresses the value of the property, and the name of a unit tells you what the referent is as well as implies the procedure for obtaining the number. Referent units must be defined and established, however, if others are to understand and reproduce a measurement. When standards are established, the referent unit is called a **standard unit** (Figure 1.4). The use of standard units makes it possible to communicate and duplicate measurements. Standard units are usually defined and established by governments and their agencies that are created for that purpose. In the United States, the agency concerned with measurement standards is the National Institute of Standards and Technology. In Canada, the Standards Council of Canada oversees the National Standard System.

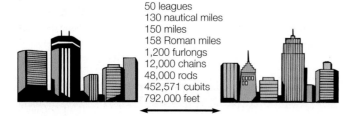

50 leagues
130 nautical miles
150 miles
158 Roman miles
1,200 furlongs
12,000 chains
48,000 rods
452,571 cubits
792,000 feet

FIGURE 1.4 Any of these units and values could have been used at some time or another to describe the same distance between these hypothetical towns. Any unit could be used for this purpose, but when one particular unit is officially adopted, it becomes known as the *standard unit.*

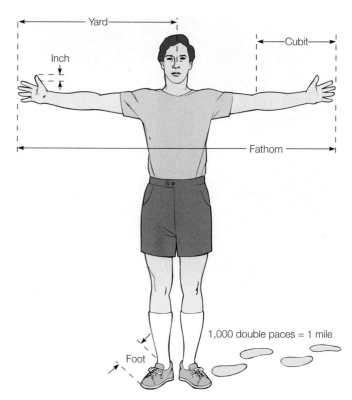

FIGURE 1.5 Many early units for measurement were originally based on the human body. Some of the units were later standardized by governments to become the basis of the English system of measurement.

There are two major *systems* of standard units in use today, the *English system* and the *metric system.* The metric system is used throughout the world except in the United States, where both systems are in use. The continued use of the English system in the United States presents problems in international trade, so there is pressure for a complete conversion to the metric system. More and more metric units are being used in everyday measurements, but a complete conversion will involve an enormous cost. Appendix A contains a method for converting from one system to the other easily. Consult this section if you need to convert from one metric unit to another metric unit or to convert from English to metric units or vice versa. Conversion factors are listed inside the front cover.

People have used referents to communicate about properties of things throughout human history. The ancient Greek civilization, for example, used units of *stadia* to communicate about distances and elevations. The *stadium* was a unit of length of the racetrack at the local stadium (*stadia* is the plural of *stadium*), based on a length of 125 paces. Later civilizations, such as the ancient Romans, adopted the stadia and other referent units from the ancient Greeks. Some of these same referent units were later adopted by the early English civilization, which eventually led to the **English system** of measurement. Some adopted units of the English system were originally based on parts of the human body, presumably because you always had these referents with you (Figure 1.5). The inch, for example, used the end joint of the thumb for a referent. A foot,

TABLE 1.1		
The SI base units		
Property	**Unit**	**Symbol**
Length	meter	m
Mass	kilogram	kg
Time	second	s
Electric current	ampere	A
Temperature	kelvin	K
Amount of substance	mole	mol
Luminous intensity	candela	cd

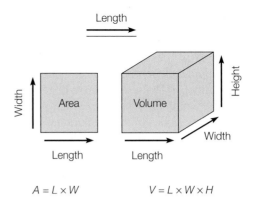

$$A = L \times W \qquad V = L \times W \times H$$

FIGURE 1.6 Area, or the extent of a surface, can be described by two length measurements. Volume, or the space that an object occupies, can be described by three length measurements. Length, however, can be described only in terms of how it is measured, so it is called a *fundamental property.*

naturally, was the length of a foot, and a yard was the distance from the tip of the nose to the end of the fingers on an arm held straight out. A cubit was the distance from the end of an elbow to the fingertip, and a fathom was the distance between the fingertips of two arms held straight out. As you can imagine, there were problems with these early units because everyone had different-sized body parts. Beginning in the 1300s, the sizes of the various units were gradually standardized by English kings.

The **metric system** was established by the French Academy of Sciences in 1791. The academy created a measurement system that was based on invariable referents in nature, not human body parts. These referents have been redefined over time to make the standard units more reproducible. The *International System of Units,* abbreviated *SI,* is a modernized version of the metric system. Today, the SI system has seven base units that define standards for the properties of length, mass, time, electric current, temperature, amount of substance, and light intensity (Table 1.1). All units other than the seven basic ones are *derived* units. Area, volume, and speed, for example, are all expressed with derived units. Units for the properties of length, mass, and time are introduced in this chapter. The remaining units will be introduced in later chapters as the properties they measure are discussed.

STANDARD UNITS FOR THE METRIC SYSTEM

If you consider all the properties of all the objects and events in your surroundings, the number seems overwhelming. Yet, close inspection of how properties are measured reveals that some properties are combinations of other properties (Figure 1.6). Volume, for example, is described by the three length measurements of length, width, and height. Area, on the other hand, is described by just the two length measurements of length and width. Length, however, cannot be defined in simpler terms of any other property. There are four properties that cannot be described in simpler terms, and all other properties are combinations of these four. For this reason, they are called the **fundamental properties.** A fundamental property cannot be defined in simpler terms other than to describe how it is

measured. These four fundamental properties are (1) *length,* (2) *mass,* (3) *time,* and (4) *charge.* Used individually or in combinations, these four properties will describe or measure what you observe in nature. Metric units for measuring the fundamental properties of length, mass, and time will be described next. The fourth fundamental property, charge, is associated with electricity, and a unit for this property will be discussed in chapter 6.

LENGTH

The standard unit for length in the metric system is the **meter** (the symbol or abbreviation is m). The meter is defined as the distance that light travels in a vacuum during a certain time period, 1/299,792,458 second. The important thing to remember, however, is that the meter is the metric *standard unit* for length. A meter is slightly longer than a yard, 39.3 inches. It is approximately the distance from your left shoulder to the tip of your right hand when your arm is held straight out. Many doorknobs are about 1 meter above the floor. Think about these distances when you are trying to visualize a meter length.

MASS

The standard unit for mass in the metric system is the **kilogram** (kg). The kilogram is defined as the mass of a certain metal cylinder kept by the International Bureau of Weights and Measures in France. This is the only standard unit that is still defined in terms of an object. The property of mass is sometimes confused with the property of weight since they are directly proportional to each other at a given location on the surface of Earth. They are, however, two completely different properties and are measured with different units. All objects tend to maintain their state of rest or straight-line motion, and this property is called "inertia." The *mass* of an object is a measure of the inertia of an object. The *weight* of the object is a measure of the force of gravity on it. This distinction between weight and mass will be discussed in detail in chapter 2. For now, remember that weight and mass are not the same property.

TIME

The standard unit for time is the **second** (s). The second was originally defined as 1/86,400 of a solar day (1/60 × 1/60 × 1/24). Earth's spin was found not to be as constant as thought, so this old definition of one second had to be revised. Adopted in 1967, the new definition is based on a high-precision device known as an *atomic clock*. An atomic clock has a referent for a second that is provided by the characteristic vibrations of the cesium-133 atom. The atomic clock that was built at the National Institute of Standards and Technology in Boulder, Colorado, will neither gain nor lose a second in 20 million years!

METRIC PREFIXES

The metric system uses prefixes to represent larger or smaller amounts by factors of 10. Some of the more commonly used prefixes, their abbreviations, and their meanings are listed in Table 1.2. Suppose you wish to measure something smaller than the standard unit of length, the meter. The meter is subdivided into ten equal-sized subunits called *decimeters*. The prefix *deci-* has a meaning of "one-tenth of," and it takes 10 decimeters to equal the length of 1 meter. For even smaller measurements, each decimeter is divided into ten equal-sized subunits called *centimeters*. It takes 10 centimeters to equal 1 decimeter and 100 to equal 1 meter. In a similar fashion, each prefix up or down the metric ladder represents a simple increase or decrease by a factor of 10 (Figure 1.7).

When the metric system was established in 1791, the standard unit of mass was defined in terms of the mass of a certain volume of water. A cubic decimeter (dm^3) of pure water at 4°C was *defined* to have a mass of 1 kilogram (kg). This definition

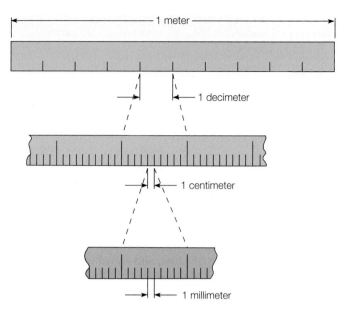

FIGURE 1.7 Compare the units shown here. How many millimeters fit into the space occupied by 1 centimeter? How many millimeters fit into the space of 1 decimeter? How many millimeters fit into the space of 1 meter? Can you express all of this as multiples of 10?

was convenient because it created a relationship between length, mass, and volume. As illustrated in Figure 1.8, a cubic decimeter is 10 cm on each side. The volume of this cube is therefore 10 cm × 10 cm × 10 cm, or 1,000 cubic centimeters (abbreviated as cc or cm^3). Thus, a volume of 1,000 cm^3 of water has a mass of 1 kg. Since 1 kg is 1,000 g, 1 cm^3 of water has a mass of 1 g.

The volume of 1,000 cm^3 also defines a metric unit that is commonly used to measure liquid volume, the **liter** (L). For smaller amounts of liquid volume, the milliliter (mL) is used. The relationship between liquid volume, volume, and mass of water is therefore

$$1.0 \text{ L} \Rightarrow 1.0 \text{ } dm^3 \text{ and has a mass of } 1.0 \text{ kg}$$

or, for smaller amounts,

$$1.0 \text{ mL} \Rightarrow 1.0 \text{ } cm^3 \text{ and has a mass of } 1.0 \text{ g}$$

TABLE 1.2

Some metric prefixes			
Prefix	**Symbol**	**Meaning**	**Unit Multiplier**
exa-	E	quintillion	10^{18}
peta-	P	quadrillion	10^{15}
tera-	T	trillion	10^{12}
giga-	G	billion	10^{9}
mega-	M	million	10^{6}
kilo-	k	thousand	10^{3}
hecto-	h	hundred	10^{2}
deka-	da	ten	10^{1}
unit			
deci-	d	one-tenth	10^{-1}
centi-	c	one-hundredth	10^{-2}
milli-	m	one-thousandth	10^{-3}
micro-	μ	one-millionth	10^{-6}
nano-	n	one-billionth	10^{-9}
pico-	p	one-trillionth	10^{-12}
femto-	f	one-quadrillionth	10^{-15}
atto-	a	one-quintillionth	10^{-18}

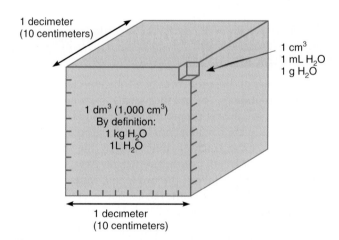

FIGURE 1.8 A cubic decimeter of water (1,000 cm^3) has a liquid volume of 1 L (1,000 mL) and a mass of 1 kg (1,000 g). Therefore, 1 cm^3 of water has a liquid volume of 1 mL and a mass of 1 g.

UNDERSTANDINGS FROM MEASUREMENTS

One of the more basic uses of measurement is to *describe* something in an exact way that everyone can understand. For example, if a friend in another city tells you that the weather has been "warm," you might not understand what temperature is being described. A statement that the air temperature is 70°F carries more exact information than a statement about "warm weather." The statement that the air temperature is 70°F contains two important concepts: (1) the numerical value of 70 and (2) the referent unit of degrees Fahrenheit. Note that both a numerical value and a unit are necessary to communicate a measurement correctly. Thus, weather reports describe weather conditions with numerically specified units; for example, 70° Fahrenheit for air temperature, 5 miles per hour for wind speed, and 0.5 inches for rainfall (Figure 1.9). When such numerically specified units are used in a description, or a weather report, everyone understands *exactly* the condition being described.

DATA

Measurement information used to describe something is called **data.** Data can be used to describe objects, conditions, events, or changes that might be occurring. You really do not know if the weather is changing much from year to year until you compare the yearly weather data. The data will tell you, for example, if the weather is becoming hotter or dryer or is staying about the same from year to year.

Let's see how data can be used to describe something and how the data can be analyzed for further understanding. The cubes illustrated in Figure 1.10 will serve as an example. Each cube can be described by measuring the properties of size and surface area.

First, consider the size of each cube. Size can be described by **volume,** which means *how much space something occupies.* The volume of a cube can be obtained by measuring and multiplying the length, width, and height. The data is

volume of cube *a*	1 cm³
volume of cube *b*	8 cm³
volume of cube *c*	27 cm³

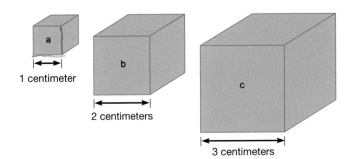

FIGURE 1.10 Cube *a* is 1 centimeter on each side, cube *b* is 2 centimeters on each side, and cube *c* is 3 centimeters on each side. These three cubes can be described and compared with data, or measurement information, but some form of analysis is needed to find patterns or meaning in the data.

Now consider the surface area of each cube. **Area** means *the extent of a surface,* and each cube has six surfaces, or faces (top, bottom, and four sides). The area of any face can be obtained by measuring and multiplying length and width. The data for the three cubes describes them as follows:

	Volume	Surface Area
cube *a*	1 cm³	6 cm²
cube *b*	8 cm³	24 cm²
cube *c*	27 cm³	54 cm²

RATIOS AND GENERALIZATIONS

Data on the volume and surface area of the three cubes in Figure 1.10 describes the cubes, but whether it says anything about a relationship between the volume and surface area of a cube is difficult to tell. Nature seems to have a tendency to camouflage relationships, making it difficult to extract meaning from raw data. Seeing through the camouflage requires the use of mathematical techniques to expose patterns. Let's see how such techniques can be applied to the data on the three cubes and what the pattern means.

One mathematical technique for reducing data to a more manageable form is to expose patterns through a **ratio.** A ratio is a relationship between two numbers that is obtained when one number is divided by another number. Suppose, for example, that an instructor has 50 sheets of graph paper for a laboratory group of 25 students. The relationship, or ratio, between the number of sheets and the number of students is 50 papers to 25 students, and this can be written as 50 papers/25 students. This ratio is *simplified* by dividing 25 into 50, and the ratio becomes 2 papers/1 student. The 1 is usually understood (not stated), and the ratio is written as simply 2 papers/student. It is read as 2 papers "for each" student, or 2 papers "per" student. The concept of simplifying with a ratio is an important one, and you will see it time and time again throughout science. It is important that you understand the meaning of "per" and "for each" when used with numbers and units.

Weather Report

Friday (24 hours ended at 5 P.M.)
Highs—airport 73°F, downtown 76°F
Lows—airport 68°F, downtown 70°F
Rainfall .... 0.26 in
Average wind speed .... 5.2 mph
Relative humidity .... High 85%
Low 75%
Rainfall ± normal to date.....+0.94 in

FIGURE 1.9 A weather report gives exact information, data that describes the weather by reporting numerically specified units for each condition being described.

Applying the ratio concept to the three cubes in Figure 1.10, the ratio of surface area to volume for the smallest cube, cube *a*, is 6 cm² to 1 cm³, or

$$\frac{6 \text{ cm}^2}{1 \text{ cm}^3} = 6 \frac{\text{cm}^2}{\text{cm}^3}$$

meaning there are 6 square centimeters of area *for each* cubic centimeter of volume.

The middle-sized cube, cube *b,* had a surface area of 24 cm² and a volume of 8 cm³. The ratio of surface area to volume for this cube is therefore

$$\frac{24 \text{ cm}^2}{8 \text{ cm}^3} = 3 \frac{\text{cm}^2}{\text{cm}^3}$$

meaning there are 3 square centimeters of area *for each* cubic centimeter of volume.

The largest cube, cube *c,* had a surface area of 54 cm² and a volume of 27 cm³. The ratio is

$$\frac{54 \text{ cm}^2}{27 \text{ cm}^3} = 2 \frac{\text{cm}^2}{\text{cm}^3}$$

or 2 square centimeters of area *for each* cubic centimeter of volume. Summarizing the ratio of surface area to volume for all three cubes, you have

small cube	*a*−6:1
middle cube	*b*−3:1
large cube	*c*−2:1

Now that you have simplified the data through ratios, you are ready to generalize about what the information means. You can generalize that the surface-area-to-volume ratio of a cube *decreases* as the volume of a cube becomes larger. Reasoning from this generalization will provide an explanation for a number of related observations. For example, why does crushed ice melt faster than a single large block of ice with the same volume? The explanation is that the crushed ice has a larger surface-area-to-volume ratio than the large block, so more surface is exposed to warm air. If the generalization is found to be true for shapes other than cubes, you could explain why a log chopped into small chunks burns faster than the whole log. Further generalizing might enable you to predict if large potatoes would require more or less peeling than the same weight of small potatoes. When generalized explanations result in predictions that can be verified by experience, you gain confidence in the explanation. Finding patterns of relationships is a satisfying intellectual adventure that leads to understanding and generalizations that are frequently practical.

THE DENSITY RATIO

The power of using a ratio to simplify things, making explanations more accessible, is evident when you compare the simplified ratio 6 to 3 to 2 with the hodgepodge of numbers that you would have to consider without using ratios. The power of using the ratio technique is also evident when considering other properties of matter. Volume is a property that is sometimes confused with mass. Larger objects do not necessarily contain more matter than smaller objects. A large balloon, for example,

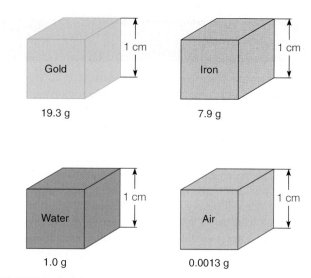

FIGURE 1.11 Equal volumes of different substances do not have the same mass, as these cube units show. Calculate the densities in g/cm³. Do equal volumes of different substances have the same density? Explain.

is much larger than this book, but the book is much more massive than the balloon. The simplified way of comparing the mass of a particular volume is to find the ratio of mass to volume. This ratio is called **density,** which is defined as *mass per unit volume.* The *per* means "for each" as previously discussed, and *unit* means one, or each. Thus, "mass per unit volume" literally means the "mass of one volume" (Figure 1.11). The relationship can be written as

$$\text{density} = \frac{\text{mass}}{\text{volume}}$$

or

$$\rho = \frac{m}{V}$$

(ρ is the symbol for the Greek letter rho.)

equation 1.1

As with other ratios, density is obtained by dividing one number and unit by another number and unit. Thus, the density of an object with a volume of 5 cm³ and a mass of 10 g is

$$\text{density} = \frac{10 \text{ g}}{5 \text{ cm}^3} = 2 \frac{\text{g}}{\text{cm}^3}$$

The density in this example is the ratio of 10 g to 5 cm³, or 10 g/5 cm³, or 2 g to 1 cm³. Thus, the density of the example object is the mass of *one* volume (a unit volume), or 2 g *for each* cm³.

Any unit of mass and any unit of volume may be used to express density. The densities of solids, liquids, and gases are usually expressed in grams per cubic centimeter (g/cm³), but the densities of liquids are sometimes expressed in grams per milliliter (g/mL). Using SI standard units, densities are expressed as kg/m³. Densities of some common substances are shown in Table 1.3.

TABLE 1.3

Densities (ρ) of some common substances	
	g/cm^3
Aluminum	2.70
Copper	8.96
Iron	7.87
Lead	11.4
Water	1.00
Seawater	1.03
Mercury	13.6
Gasoline	0.680

If matter is distributed the same throughout a volume, the *ratio* of mass to volume will remain the same no matter what mass and volume are being measured. Thus, a teaspoonful, a cup, and a lake full of freshwater at the same temperature will all have a density of about 1 g/cm^3 or 1 kg/L. A given material will have its own unique density; example 1.1 shows how density can be used to identify an unknown substance. For help with significant figures, see appendix A (p. 605).

CONCEPTS *Applied*

Density Matters—Sharks and Cola Cans

What do a shark and a can of cola have in common? Sharks are marine animals that have an internal skeleton made entirely of cartilage. These animals have no swim bladder to adjust their body density in order to maintain their position in the water; therefore, they must constantly swim or they will sink. The bony fish, on the other hand, have a skeleton composed of bone and most also have a swim bladder. These fish can regulate the amount of gas in the bladder to control their density. Thus, the fish can remain at a given level in the water without expending large amounts of energy.

Have you ever noticed the different floating characteristics of cans of the normal version of a carbonated cola beverage and a diet version? The surprising result is that the normal version usually sinks and the diet version usually floats. This has nothing to do with the amount of carbon dioxide in the two drinks. It is a result of the increase in density from the sugar added to the normal version, while the diet version has much less of an artificial sweetener that is much sweeter than sugar. So, the answer is that sharks and regular cans of cola both sink in water.

EXAMPLE 1.1

Two blocks are on a table. Block A has a volume of 30.0 cm^3 and a mass of 81.0 g. Block B has a volume of 50.0 cm^3 and a mass of 135 g. Which block has the greater density? If the two blocks have the same density, what material are they? (See Table 1.3.)

SOLUTION

Density is defined as the ratio of the mass of a substance per unit volume. Assuming the mass is distributed equally throughout the volume, you could assume that the ratio of mass to volume is the same no matter what quantity of mass and volume are measured. If you can accept this assumption, you can use equation 1.1 to determine the density.

Block A

mass (m) = 81.0 g
volume (V) = 30.0 cm$_3$
density = ?

$$\rho = \frac{m}{V}$$

$$= \frac{81.0 \text{ g}}{30.0 \text{ cm}^3}$$

$$= 2.70 \frac{\text{g}}{\text{cm}^3}$$

Block B

mass (m) = 135 g
volume (V) = 50.0 cm$_3$
density = ?

$$\rho = \frac{m}{V}$$

$$= \frac{135 \text{ g}}{50.0 \text{ cm}^3}$$

$$= 2.70 \frac{\text{g}}{\text{cm}^3}$$

As you can see, both blocks have the same density. Inspecting Table 1.3, you can see that aluminum has a density of 2.70 g/cm3, so both blocks must be aluminum.

EXAMPLE 1.2

A rock with a volume of 4.50 cm^3 has a mass of 15.0 g. What is the density of the rock? (Answer: 3.33 g/cm^3)

CONCEPTS *Applied*

A Dense Textbook?

What is the density of this book? Measure the length, width, and height of this book in cm, then multiply to find the volume in cm^3. Use a scale to find the mass of this book in grams. Compute the density of the book by dividing the mass by the volume. Compare the density in g/cm^3 with other substances listed in Table 1.3.

Myths, Mistakes, & Misunderstandings

Tap a Can?

Some people believe that tapping on the side of a can of carbonated beverage will prevent it from foaming over when the can is opened. Is this true or a myth? Set up a controlled experiment (see p. 15) to compare opening cold cans of carbonated beverage that have been tapped with cans that have not been tapped. Are you sure you have controlled all the other variables?

SYMBOLS AND EQUATIONS

In the previous section, the relationship of density, mass, and volume was written with symbols. Density was represented by ρ, the lowercase letter rho in the Greek alphabet, mass was represented by m, and volume by V. The use of such symbols is established and accepted by convention, and these symbols are like the vocabulary of a foreign language. You learn what the symbols mean by use and practice, with the understanding that *each symbol stands for a very specific property or concept.* The symbols actually represent **quantities,** or *measured properties.* The symbol m thus represents a quantity of mass that is specified by a number and a unit, for example, 16 g. The symbol V represents a quantity of volume that is specified by a number and a unit, such as 17 cm^3.

Symbols

Symbols usually provide a clue about which quantity they represent, such as m for mass and V for volume. However, in some cases, two quantities start with the same letter, such as volume and velocity, so the uppercase letter is used for one (V for volume) and the lowercase letter is used for the other (v for velocity). There are more quantities than upper- and lowercase letters, however, so letters from the Greek alphabet are also used, for example, ρ for mass density. Sometimes a subscript is used to identify a quantity in a particular situation, such as v_i for initial, or beginning, velocity and v_f for final velocity. Some symbols are also used to carry messages; for example, the Greek letter delta (Δ) is a message that means "the change in" a value. Other message symbols are the symbol $\therefore$, which means "therefore," and the symbol $\propto$, which means "is proportional to."

Equations

Symbols are used in an **equation,** a statement that describes a relationship where *the quantities on one side of the equal sign are identical to the quantities on the other side.* Identical refers to both the numbers and the units. Thus, in the equation describing the property of density, $\rho = m/V$, the numbers on both sides of the equal sign are identical (e.g., 5 = 10/2). The units on both sides of the equal sign are also identical (e.g., $g/cm^3 = g/cm^3$).

Equations are used to (1) *describe a property,* (2) *define a concept,* or (3) *describe how quantities change relative to each other.* Understanding how equations are used in these three classes is basic to successful problem solving and comprehension of physical science. Each class of uses is considered separately in the following discussion.

Describing a property. You have already learned that the compactness of matter is described by the property called density. Density is a ratio of mass to a unit volume, or $\rho = m/V$. The key to understanding this property is to understand the meaning of a ratio and what "per" or "for each" means. Other examples of properties that can be defined by ratios are how fast something is moving (speed) and how rapidly a speed is changing (acceleration).

Defining a concept. A physical science concept is sometimes defined by specifying a measurement procedure. This is called an *operational definition* because a procedure is established that defines a concept as well as tells you how to measure it. Concepts of what is meant by force, mechanical work, and mechanical power and concepts involved in electrical and magnetic interactions can be defined by measurement procedures.

Describing how quantities change relative to each other. The term **variable** refers to a specific quantity of an object or event that can have different values. Your weight, for example, is a variable because it can have a different value on different days. The rate of your heartbeat, the number of times you breathe each minute, and your blood pressure are also variables. Any quantity describing an object or event can be considered a variable, including the conditions that result in such things as your current weight, pulse, breathing rate, or blood pressure.

As an example of relationships between variables, consider that your weight changes in size in response to changes in other variables, such as the amount of food you eat. With all other factors being equal, a change in the amount of food you eat results in a change in your weight, so the variables of amount of food eaten and weight change together in the same ratio. A *graph* is used to help you picture relationships between variables (see "Simple Line Graph" on p. 611).

When two variables increase (or decrease) together in the same ratio, they are said to be in **direct proportion.** When two variables are in direct proportion, *an increase or decrease in one variable results in the same relative increase or decrease in a second variable.* Recall that the symbol $\propto$ means "is proportional to," so the relationship is

$$\text{amount of food consumed} \propto \text{weight gain}$$

Variables do not always increase or decrease together in direct proportion. Sometimes one variable *increases* while a second variable *decreases* in the same ratio. This is an **inverse proportion** relationship. Other common relationships include one variable increasing in proportion to the *square* or to the *inverse square* of a second variable. Here are the forms of these four different types of proportional relationships:

Direct	$a \propto b$
Inverse	$a \propto 1/b$
Square	$a \propto b^2$
Inverse square	$a \propto 1/b^2$

Proportionality Statements

Proportionality statements describe in general how two variables change relative to each other, but a proportionality statement is *not* an equation. For example, consider the last time you filled your fuel tank at a service station (Figure 1.12). You could say that the volume of gasoline in an empty tank you are filling is directly proportional to the amount of time that the fuel pump was running, or

$$\text{volume} \propto \text{time}$$

This is not an equation because the numbers and units are not identical on both sides. Considering the units, for example,

Inverse Square Relationship

An inverse square relationship between energy and distance is found in light, sound, gravitational force, electric fields, nuclear radiation, and any other phenomena that spread equally in all directions from a source.

Box Figure 1.1 could represent any of the phenomena that have an inverse square relationship, but let us assume it is showing a light source and how the light spreads at a certain distance (d), at twice that distance ($2d$), and at three times that distance ($3d$). As you can see, light twice as far from the source is spread over four times the area and will therefore have one-fourth the intensity. This is the same as $\frac{1}{2^2}$, or $\frac{1}{4}$.

Light three times as far from the source is spread over nine times the area and will therefore have one-ninth the intensity. This is the same as $\frac{1}{3^2}$, or $\frac{1}{9}$, again showing an inverse square relationship.

You can measure the inverse square relationship by moving an overhead projector so its light is shining on a wall (see distance d in Box Figure 1.1). Use a light meter or some other way of measuring the intensity of light. Now move the projector to double the distance from the wall. Measure the increased area of the projected light on the wall, and again measure the intensity of the light. What relationship did you find between the light intensity and distance?

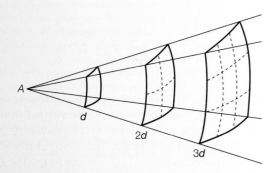

BOX FIGURE 1.1 How much would light moving from point *A* spread out at twice the distance ($2d$) and three times the distance ($3d$)? What would this do to the brightness of the light?

FIGURE 1.12 The volume of fuel you have added to the fuel tank is directly proportional to the amount of time that the fuel pump has been running. This relationship can be described with an equation by using a proportionality constant.

In the example, the constant is the flow of gasoline from the pump in L/min (a ratio). Assume the rate of flow is 40 L/min. In units, you can see why the statement is now an equality.

$$L = (\text{min})\left(\frac{L}{\text{min}}\right)$$

$$L = \frac{\text{min} \times L}{\text{min}}$$

$$L = L$$

A proportionality constant in an equation might be a **numerical constant,** a constant that is without units. Such numerical constants are said to be dimensionless, such as 2 or 3. Some of the more important numerical constants have their own symbols; for example, the ratio of the circumference of a circle to its diameter is known as π (pi). The numerical constant of π does not have units because the units cancel when the ratio is simplified by division (Figure 1.13). The value of π is usually rounded to 3.14, and an example of using this numerical constant in an equation is that the area of a circle equals π times the radius squared ($A = \pi r^2$).

The flow of gasoline from a pump is an example of a constant that has dimensions (40 L/min). Of course the value of this constant will vary with other conditions, such as the particular fuel pump used and how far the handle on the pump hose is depressed, but it can be considered to be a constant under the same conditions for any experiment.

HOW TO SOLVE PROBLEMS

The activity of problem solving is made easier by using certain techniques that help organize your thinking. One such technique is to follow a format, such as the following procedure:

Step 1: Read through the problem and *make a list* of the variables with their symbols on the left side of the page, including the unknown with a question mark.

it should be clear that minutes do not equal liters; they are two different quantities. To make a statement of proportionality into an equation, you need to apply a **proportionality constant,** which is sometimes given the symbol k. For the fuel pump example, the equation is

$$\text{volume} = (\text{time})(\text{constant})$$

or

$$V = tk$$

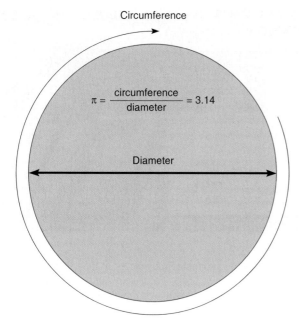

Circumference

$$\pi = \frac{\text{circumference}}{\text{diameter}} = 3.14$$

Diameter

FIGURE 1.13 The ratio of the circumference of *any* circle to the diameter of that circle is always π, a numerical constant that is usually rounded to 3.14. Pi does not have units because they cancel in the ratio.

Step 2: Inspect the list of variables and the unknown, and identify the equation that expresses a relationship between these variables. A list of equations discussed in each chapter is found at the end of that chapter. *Write the equation* on the right side of your paper, opposite the list of symbols and quantities.

Step 3: If necessary, *solve the equation* for the variable in question. This step must be done before substituting any numbers or units in the equation. This simplifies things and keeps down confusion that might otherwise result. If you need help solving an equation, see the section on this topic in appendix A.

Step 4: If necessary, *convert unlike units* so they are all the same. For example, if a time is given in seconds and a speed is given in kilometers per hour, you should convert the km/h to m/s. Again, this step should be done at this point in the procedure to avoid confusion or incorrect operations in a later step. If you need help converting units, see the section on this topic in appendix A.

Step 5: Now you are ready to *substitute the number value and unit* for each symbol in the equation (except the unknown). Note that it might sometimes be necessary to perform a "subroutine" to find a missing value and unit for a needed variable.

Step 6: Do the indicated *mathematical operations* on the numbers and on the units. This is easier to follow if you first separate the numbers and units, as shown in the example that follows and in the examples throughout this text. Then perform the indicated operations on the numbers and units as separate steps, showing

all work. If you are not sure how to read the indicated operations, see the section on "Symbols and Operations" in appendix A.

Step 7: Now ask yourself if the number seems reasonable for the question that was asked, and ask yourself if the unit is correct. For example, 250 m/s is way too fast for a running student, and the unit for speed is not liters.

Step 8: *Draw a box* around your answer (numbers and units) to communicate that you have found what you were looking for. The box is a signal that you have finished your work on this problem.

For an example problem, use the equation from the previous section describing the variables of a fuel pump, $V = tk$, to predict how long it will take to fill an empty 80-liter tank. Assume $k = 40$ L/min.

Step 1

$V = 80$ L	$V = tk$	**Step 2**
$k = 80$ L/min	$\dfrac{V}{k} = \dfrac{tk}{k}$	**Step 3**
$t = ?$		
	$t = \dfrac{V}{k}$	

(no conversion needed for this problem) **Step 4**

$$t = \frac{80 \text{ L}}{40 \, \frac{\text{L}}{\text{min}}}$$ **Step 5**

$$= \frac{80}{40} \, \frac{\text{L}}{1} \times \frac{\text{min}}{\text{L}}$$ **Step 6**

$$= \boxed{2 \text{ min}}$$ **Step 7**

Note that procedure step 4 was not required in this solution.

This formatting procedure will be demonstrated throughout this text in example problems and in the solutions to problems found in appendix E. Note that each of the chapters with problems has parallel exercises. The exercises in groups A and B cover the same concepts. If you cannot work a problem in group B, look for the parallel problem in group A. You will find a solution to this problem, in the previously described format, in appendix E. Use this parallel problem solution as a model to help you solve the problem in group B. If you follow the suggested formatting procedures and seek help from the appendix as needed, you will find that problem solving is a simple, fun activity that helps you to learn to think in a new way. Here are some more considerations that will prove helpful.

1. Read the problem carefully, perhaps several times, to understand the problem situation. Make a sketch to help you visualize and understand the problem in terms of the real world.
2. Be alert for information that is not stated directly. For example, if a moving object "comes to a stop," you know that the final velocity is zero, even though this was not stated outright. Likewise, questions about

"how far?" are usually asking a question about distance, and questions about "how long?" are usually asking a question about time. Such information can be very important in procedure step 1, the listing of quantities and their symbols. Overlooked or missing quantities and symbols can make it difficult to identify the appropriate equation.

3. Understand the meaning and concepts that an equation represents. An equation represents a *relationship* that exists between variables. Understanding the relationship helps you to identify the appropriate equation or equations by inspection of the list of known and unknown quantities (procedure step 2). You will find a list of the equations being considered at the end of each chapter. Information about the meaning and the concepts that an equation represents is found within each chapter.

4. Solve the equation before substituting numbers and units for symbols (procedure step 3). A helpful discussion of the mathematical procedures required, with examples, is in appendix A.

5. Note whether the quantities are in the same units. A mathematical operation requires the units to be the same; for example, you cannot add nickels, dimes, and quarters until you first convert them all to the same unit of money. Likewise, you cannot correctly solve a problem if one time quantity is in seconds and another time quantity is in hours. The quantities must be converted to the same units before anything else is done (procedure step 4). There is a helpful section on how to use conversion ratios in appendix A.

6. Perform the required mathematical operations on the numbers and the units as if they were two separate problems (procedure step 6). You will find that following this step will facilitate problem-solving activities because the units you obtain will tell you if you have worked the problem correctly. If you just write the units that you think should appear in the answer, you have missed this valuable self-check.

7. Be aware that not all learning takes place in a given time frame and that solutions to problems are not necessarily arrived at "by the clock." If you have spent a half an hour or so unsuccessfully trying to solve a particular problem, move on to another problem or do something entirely different for a while. Problem solving often requires time for something to happen in your brain. If you move on to some other activity, you might find that the answer to a problem that you have been stuck on will come to you "out of the blue" when you are not even thinking about the problem. This unexpected revelation of solutions is common to many real-world professions and activities that involve thinking.

Example Problem

Mercury is a liquid metal with a mass density of 13.6 g/cm^3. What is the mass of 10.0 cm^3 of mercury?

Solution

The problem gives two known quantities, the mass density (ρ) of mercury and a known volume (V), and identifies an unknown quantity, the mass (m) of that volume. Make a list of these quantities:

$$\rho = 13.6 \text{ g/cm}^3$$
$$V = 10.0 \text{ cm}^3$$
$$m = ?$$

The appropriate equation for this problem is the relationship between density (ρ), mass (m), and volume (V):

$$\rho = \frac{m}{V}$$

The unknown in this case is the mass, m. Solving the equation for m, by multiplying both sides by V, gives:

$$V\rho = \frac{mV}{V}$$
$$V\rho = m, \text{ or}$$
$$m = V\rho$$

Now you are ready to substitute the known quantities in the equation:

$$m = \left(13.6 \, \frac{\text{g}}{\text{cm}^3}\right)(10.0 \text{ cm}^3)$$

And perform the mathematical operations on the numbers and on the units:

$$m = (13.6)(10.0) \left(\frac{\text{g}}{\text{cm}^3}\right)(\text{cm}^3)$$
$$= 136 \frac{\text{g·cm}^3}{\text{cm}^3}$$
$$= \boxed{136 \text{ g}}$$

THE NATURE OF SCIENCE

Most humans are curious, at least when they are young, and are motivated to understand their surroundings. These traits have existed since antiquity and have proven to be a powerful motivation. In recent times, the need to find out has motivated the launching of space probes to learn what is "out there," and humans have visited the moon to satisfy their curiosity. Curiosity and the motivation to understand nature were no less powerful in the past than today. Over two thousand years ago, the Greeks lacked the tools and technology of today and could only make conjectures about the workings of nature. These early seekers of understanding are known as *natural philosophers,* and they observed, thought about, and wrote about the workings of all of nature. They are called philosophers because their understandings came from reasoning only, without experimental evidence. Nonetheless, some of their ideas were essentially correct and are still in use today. For example, the idea of matter being composed of *atoms* was first reasoned by certain Greeks in the fifth century B.C. The idea of *elements,* basic components that make up matter, was developed much earlier but refined by the ancient Greeks in the fourth century B.C. The concept of what the elements are and the concept of the nature of atoms have changed over time, but the ideas first came from ancient natural philosophers.

THE SCIENTIFIC METHOD

Some historians identify the time of Galileo and Newton, approximately three hundred years ago, as the beginning of modern science. Like the ancient Greeks, Galileo and Newton were interested in studying all of nature. Since the time of Galileo and Newton, the content of physical science has increased in scope and specialization, but the basic means of acquiring understanding, the scientific investigation, has changed little. A *scientific investigation* provides understanding through *experimental evidence* as opposed to the conjectures based on the "thinking only" approach of the ancient natural philosophers. In chapter 2, for example, you will learn how certain ancient Greeks described how objects fall toward Earth with a thought-out, or reasoned, explanation. Galileo, on the other hand, changed how people thought of falling objects by developing explanations from both creative thinking and precise measurement of physical quantities, providing experimental evidence for his explanations. Experimental evidence provides explanations today, much as it did for Galileo, as relationships are found from precise measurements of physical quantities. Thus, scientific knowledge about nature has grown as measurements and investigations have led to understandings that lead to further measurements and investigations.

What is a scientific investigation, and what methods are used to conduct one? Attempts have been made to describe scientific methods in a series of steps (define problem, gather data, make hypothesis, test, make conclusion), but no single description has ever been satisfactory to all concerned. Scientists do similar things in investigations, but there are different approaches and different ways to evaluate what is found. Overall, the similar things might look like this:

1. Observe some aspect of nature.
2. Propose an explanation for something observed.
3. Use the explanation to make predictions.
4. Test predictions by doing an experiment or by making more observations.
5. Modify explanation as needed.
6. Return to step 3.

The exact approach used depends on the individual doing the investigation and on the field of science being studied.

Another way to describe what goes on during a scientific investigation is to consider what can be generalized. There are at least three separate activities that seem to be common to scientists in different fields as they conduct scientific investigations, and these generalizations look like this:

- Collecting observations
- Developing explanations
- Testing explanations

No particular order or routine can be generalized about these common elements. In fact, individual scientists might not even be involved in all three activities. Some, for example, might spend all of their time out in nature, "in the field" collecting data and generalizing about their findings. This is an acceptable means of investigation in some fields of science. Other scientists might spend all of their time indoors at computer terminals developing theoretical equations to explain the generalizations made by others. Again, the work at a computer terminal is an acceptable means of scientific investigation. Thus, many of today's specialized scientists never engage in a five-step process. This is one reason why many philosophers of science argue that there is no such thing as *the* scientific method. There are common activities of observing, explaining, and testing in scientific investigations in different fields, and these activities will be discussed next.

EXPLANATIONS AND INVESTIGATIONS

Explanations in the natural sciences are concerned with things or events observed, and there can be several different ways to develop or create explanations. In general, explanations can come from the results of experiments, from an educated guess, or just from imaginative thinking. In fact, there are even several examples in the history of science of valid explanations being developed from dreams.

Explanations go by various names, each depending on intended use or stage of development. For example, an explanation in an early stage of development is sometimes called a *hypothesis*. A **hypothesis** is a tentative thought- or experiment-derived explanation. It must be compatible with observations and provide understanding of some aspect of nature, but the key word here is *tentative*. A hypothesis is tested by experiment and is rejected, or modified, if a single observation or test does not fit.

The successful testing of a hypothesis may lead to the design of experiments, or it could lead to the development of another hypothesis, which could, in turn, lead to the design of yet more experiments, which could lead to. . . . As you can see, this is a branching, ongoing process that is very difficult to describe in specific terms. In addition, it can be difficult to identify an endpoint in the process that you could call a conclusion. The search for new concepts to explain experimental evidence may lead from hypothesis to new ideas, which results in more new hypotheses. This is why one of the best ways to understand scientific methods is to study the history of science. Or do the activity of science yourself by planning, then conducting experiments.

Testing a Hypothesis

In some cases, a hypothesis may be tested by simply making some simple observations. For example, suppose you hypothesized that the height of a bounced ball depends only on the height from which the ball is dropped. You could test this by observing different balls being dropped from several different heights and recording how high each bounced.

Another common method for testing a hypothesis involves devising an experiment. An **experiment** is a recreation of an event or occurrence in a way that enables a scientist to support or disprove a hypothesis. This can be difficult, since an event can be influenced by a great many different things. For example, suppose someone tells you that soup heats to the boiling point faster

Science is the process of understanding your environment. It begins with making observations, creating explanations, and conducting research experiments. New information and conclusions are based on the results of the research.

There are two types of scientific research: basic and applied. *Basic research* is driven by a search for understanding and may or may not have practical applications. Examples of basic research include seeking understandings about how the solar system was created, finding new information about matter by creating a new element in a research lab, or mapping temperature variations on the bottom of the Chesapeake Bay. Such basic research expands our knowledge but will not lead to practical results.

Applied research has a goal of solving some practical problem rather than just looking for answers. Examples of applied research include the creation and testing of a new highly efficient fuel cell to run cars on hydrogen fuel, improving the energy efficiency of the refrigerator, or creating a faster computer chip from new materials.

Whether research is basic or applied depends somewhat on the time frame. If a practical use cannot be envisioned in the future, then it is definitely basic research. If a practical use is immediate, then the work is definitely applied research. If a practical use is developed some time in the future, then the research is partly basic and partly practical. For example, when the laser was invented, there was no practical use for it. It was called "an answer waiting for a question." Today, the laser has many, many practical applications.

Knowledge gained by basic research has sometimes resulted in the development of technological breakthroughs. On the other hand, other basic research—such as learning how the solar system formed—has no practical value other than satisfying our curiosity.

QUESTIONS TO DISCUSS

1. Should funding priorities go to basic research, applied research, or both?
2. Should universities concentrate on basic research and industries concentrate on applied research, or should both do both types of research?
3. Should research–funding organizations specify which types of research should be funded?

---

than water. Is this true? How can you find the answer to this question? The time required to boil a can of soup might depend on a number of things: the composition of the soup, how much soup is in the pan, what kind of pan is used, the nature of the stove, the size of the burner, how high the temperature is set, environmental factors such as the humidity and temperature, and more factors. It might seem that answering a simple question about the time involved in boiling soup is an impossible task. To help unscramble such situations, scientists use what is known as a *controlled experiment*. A **controlled experiment** compares two situations in which all the influencing factors are identical except one. The situation used as the basis of comparison is called the *control group* and the other is called the *experimental group*. The single influencing factor that is allowed to be different in the experimental group is called the *experimental variable.*

The situation involving the time required to boil soup and water would have to be broken down into a number of simple questions. Each question would provide the basis on which experimentation would occur. Each experiment would provide information about a small part of the total process of heating liquids. For example, in order to test the hypothesis that soup will begin to boil before water, an experiment could be performed in which soup is brought to a boil (the experimental group), while water is brought to a boil in the control group. Every factor in the control group is *identical* to the factors in the experimental group except the experimental variable—the soup factor. After the experiment, the new data (facts) are gathered and analyzed. If there were no differences between the two groups, you could conclude that the soup variable evidently did not have a cause-and-effect relationship with the time needed to come to a boil (i.e., soup was not responsible for the time to boil). However, if there were a difference, it would be likely that

this variable was responsible for the difference between the control and experimental groups. In the case of the time to come to a boil, you would find that soup indeed does boil faster than water alone. If you doubt this, why not do the experiment yourself?

Accept Results?

Scientists are not likely to accept the results of a single experiment, since it is possible that a random event that had nothing to do with the experiment could have affected the results and caused people to think there was a cause-and-effect relationship when none existed. For example, the density of soup is greater than the density of water, and this might be the important factor. A way to overcome this difficulty would be to test a number of different kinds of soup with different densities. When there is only one variable, many replicates (copies) of the same experiment are conducted, and the consistency of the results determines how convincing the experiment is.

Furthermore, scientists often apply statistical tests to the results to help decide in an impartial manner if the results obtained are *valid* (meaningful; fit with other knowledge), *reliable* (give the same results repeatedly), and show cause-and-effect or if they are just the result of random events.

Other Considerations

As you can see from the discussion of the nature of science, a scientific approach to the world requires a certain way of thinking. There is an insistence on ample supporting evidence by numerous studies rather than easy acceptance of strongly stated opinions. Scientists must separate opinions from statements of fact. A scientist is a healthy skeptic.

Careful attention to detail is also important. Since scientists publish their findings and their colleagues examine their work, there is a strong desire to produce careful work that can be easily defended. This does not mean that scientists do not speculate and state opinions. When they do, however, they take great care to clearly distinguish fact from opinion.

There is also a strong ethic of honesty. Scientists are not saints, but the fact that science is conducted out in the open in front of one's peers tends to reduce the incidence of dishonesty. In addition, the scientific community strongly condemns and severely penalizes those who steal the ideas of others, perform shoddy science, or falsify data. Any of these infractions could lead to the loss of one's job and reputation.

Science is also limited by the ability of people to pry understanding from the natural world. People are fallible and do not always come to the right conclusions, because information is lacking or misinterpreted, but science is self-correcting. As new information is gathered, old, incorrect ways of thinking must be changed or discarded. For example, at one time people were sure that the Sun went around Earth. They observed that the Sun rose in the east and traveled across the sky to set in the west. Since they could not feel Earth moving, it seemed perfectly logical that the Sun traveled around Earth. Once they understood that Earth rotated on its axis, people began to understand that the rising and setting of the Sun could be explained in other ways. A completely new concept of the relationship between the Sun and Earth developed.

Although this kind of study seems rather primitive to us today, this change in thinking about the Sun and Earth was a very important step in understanding the universe and how the various parts are related to one another. This background information was built upon by many generations of astronomers and space scientists, and it finally led to space exploration.

People also need to understand that science cannot answer all the problems of our time. Although science is a powerful tool, there are many questions it cannot answer and many problems it cannot solve. The behavior and desires of people generate most of the problems societies face. Famine, drug abuse, and pollution are human-caused and must be resolved by humans. Science may provide some tools for social planners, politicians, and ethical thinkers, but science does not have, nor does it attempt to provide, answers for the problems of the human race. Science is merely one of the tools at our disposal.

Pseudoscience

Pseudoscience (*pseudo*– means false) is a deceptive practice that uses the appearance or language of science to convince, confuse, or mislead people into thinking that something has scientific validity when it does not. When pseudoscientific claims are closely examined, they are not found to be supported by unbiased tests. For example, although nutrition is a respected scientific field, many individuals and organizations make claims about their nutritional products and diets that cannot be supported. Because of nutritional research, we all know that we must obtain certain nutrients such as vitamins and minerals from the food that we eat or we may become ill. Many scientific experiments reliably demonstrate

the validity of this information. However, in most cases, it has not been proven that the nutritional supplements so vigorously promoted are as useful or desirable as advertised. Rather, selected bits of scientific information (vitamins and minerals are essential to good health) have been used to create the feeling that additional amounts of these nutritional supplements are necessary or that they can improve your health. In reality, the average person eating a varied diet will obtain all of these nutrients in adequate amounts and will not require nutritional supplements.

Another related example involves the labeling of products as organic or natural. Marketers imply that organic or natural products have greater nutritive value because they are organically grown (grown without pesticides or synthetic fertilizers) or because they come from nature. Although there are questions about the health effects of trace amounts of pesticides in foods, no scientific study has shown that a diet of natural or organic products has any benefit over other diets. The poisons curare, strychnine, and nicotine are all organic molecules that are produced in nature by plants that could be grown organically, but we would not want to include them in our diet.

Absurd claims that are clearly pseudoscience sometimes appear to gain public acceptance because of promotion in the media. Thus, some people continue to believe stories that psychics can really help solve puzzling crimes, that perpetual energy machines exist, or that sources of water can be found by a person with a forked stick. Such claims could be subjected to scientific testing and disposed of if they fail the test, but this process is generally ignored. In addition to experimentally testing such a claim that appears to be pseudoscience, here are some questions that you should consider when you suspect something is pseudoscience:

1. What is the background and scientific experience of the person promoting the claim?
2. How many articles have been published by the person in peer-reviewed scientific journals?
3. Has the person given invited scientific talks at universities and national professional organization meetings?
4. Has the claim been researched and published by the person in a peer-reviewed scientific journal, *and* have other scientists independently validated the claim?
5. Does the person have something to gain by making the claim?

 CONCEPTS *Applied*

Seekers of Pseudoscience

See what you can find out about some recent claims that might not stand up to direct scientific testing. Look into the scientific testing—or lack of testing—behind claims made in relation to cold fusion, cloning human beings, a dowser carrying a forked stick to find water, psychics hired by police departments, Bigfoot, the Bermuda Triangle, and others you might wish to investigate. One source to consider is www.randi.org/jr/archive.html

LAWS AND PRINCIPLES

Sometimes you can observe a series of relationships that seem to happen over and over again. There is a popular saying, for example, that "if anything can go wrong, it will." This is called Murphy's law. It is called a *law* because it describes a relationship between events that seems to happen time after time. If you drop a slice of buttered bread, for example, it can land two ways, butter side up or butter side down. According to Murphy's law, it will land butter side down. With this example, you know at least one way of testing the validity of Murphy's law.

Another "popular saying" type of relationship seems to exist between the cost of a houseplant and how long it lives. You could call it the "law of houseplant longevity" that the life span of a houseplant is inversely proportional to its purchase price. This "law" predicts that a ten-dollar houseplant will wilt and die within a month, but a fifty-cent houseplant will live for years. The inverse relationship is between the variables of (1) cost and (2) life span, meaning the more you pay for a plant, the shorter the time it will live. This would also mean that inexpensive plants will live for a long time. Since the relationship seems to occur time after time, it is called a "law."

A **scientific law** describes an important relationship that is observed in nature to occur consistently time after time. Basically, scientific laws describe *what* happens in nature. The law is often identified with the name of a person associated with the formulation of the law. For example, with all other factors being equal, an increase in the temperature of the air in a balloon results in an increase in its volume. Likewise, a decrease in the temperature results in a decrease in the total volume of the balloon. The volume of the balloon varies directly with the temperature of the air in the balloon, and this can be observed to occur consistently time after time. This relationship was first discovered in the latter part of the eighteenth century by two French scientists, A.C. Charles and Joseph Gay-Lussac. Today, the relationship is sometimes called *Charles' law* (Figure 1.14). When you read about a scientific *law*, you should remember that a law is a statement that means something about a relationship that you can observe time after time in nature.

Have you ever heard someone state that something behaved a certain way *because* of a scientific principle or law? For example, a big truck accelerated slowly *because* of Newton's laws of motion. Perhaps this person misunderstands the nature of scientific principles and laws. Scientific principles and laws do not dictate the behavior of objects; they simply describe it. They do not say how things ought to act but rather how things *do* act. A scientific principle or law is *descriptive;* it describes how things act.

A **scientific principle** describes a more specific set of relationships than is usually identified in a law. The difference between a scientific principle and a scientific law is usually one of the extent of the phenomena covered by the explanation, but there is not always a clear distinction between the two. As an example of a scientific principle, consider Archimedes' principle. This principle is concerned with the relationship between an object, a fluid, and buoyancy, which is a specific phenomenon.

Verbal: The volume of a gas is directly proportional to the (absolute) temperature for a given amount if the pressure is constant.

Equation: $\Delta V = \Delta T k$

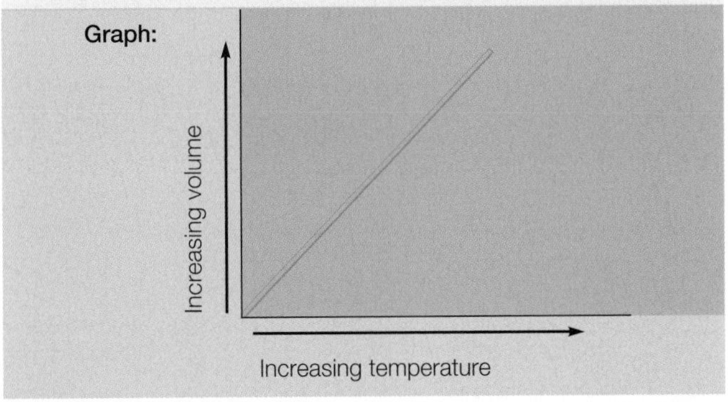

FIGURE 1.14 A relationship between variables can be described in at least three different ways: (1) verbally, (2) with an equation, and (3) with a graph. This figure illustrates the three ways of describing the relationship known as Charles' law.

MODELS AND THEORIES

Often the part of nature being considered is too small or too large to be visible to the human eye, and the use of a *model* is needed. A **model** (Figure 1.15) is a description of a theory or idea that accounts for all known properties. The description can come in many different forms, such as a physical model, a computer model, a sketch, an analogy, or an equation. No one has ever seen the whole solar system, for example, and all you can see in the real world is the movement of the Sun, moon, and planets against a background of stars. A physical model or sketch of the solar system, however, will give you a pretty good idea of what the solar system might look like. The physical model and the sketch are both models, since they both give you a mental picture of the solar system.

At the other end of the size scale, models of atoms and molecules are often used to help us understand what is happening in this otherwise invisible world. A container of small, bouncing rubber balls can be used as a model to explain the relationships of Charles' law. This model helps you see what happens to invisible particles of air as the temperature, volume, or pressure of the gas changes. Some models are better than others are, and models constantly change as our understanding evolves. Early twentieth-century models of atoms, for example, were based on a "planetary model," in which electrons moved around the nucleus like planets around the Sun. Today, the model has changed as our under-

A

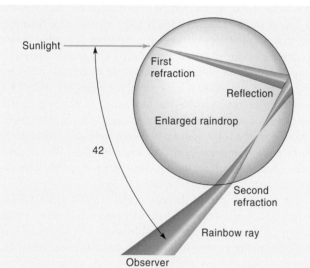

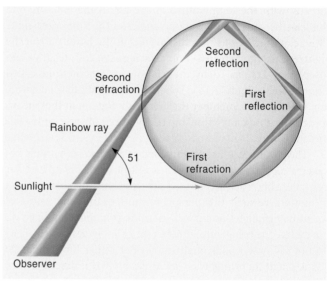

B

FIGURE 1.15 A model helps you visualize something that cannot be observed. You cannot observe what is making a double rainbow, for example, but models of light entering the upper and lower surfaces of a raindrop help you visualize what is happening. The drawings in *B* serve as a model that explains how a double rainbow is produced (also see "The Rainbow" in chapter 7).

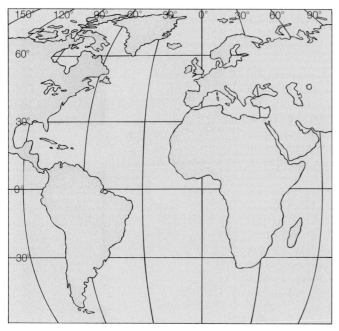

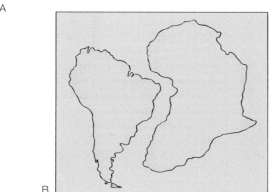

FIGURE 1.16 (*A*) Normal position of the continents on a world map. (*B*) A sketch of South America and Africa, suggesting that they once might have been joined together and subsequently separated by continental drift.

standing of the nature of atoms has changed. Electrons are now pictured as vibrating with certain wavelengths, which can make standing waves only at certain distances from the nucleus. Thus, the model of the atom changed from one that views electrons as solid particles to one that views them like vibrations on a string.

The most recently developed scientific theory was refined and expanded during the 1970s. This theory concerns the surface of Earth, and it has changed our model of what Earth is like. At first, the basic idea of today's accepted theory was pure and simple conjecture. The term *conjecture* usually means an explanation or idea based on speculation, or one based on trivial grounds without any real evidence. Scientists would look at a map of Africa and South America, for example, and mull over how the two continents look like pieces of a picture puzzle that had moved apart (Figure 1.16). Any talk of moving continents was considered conjecture, because it was not based on anything acceptable as real evidence.

Many years after the early musings about moving continents, evidence was collected from deep-sea drilling rigs that the ocean floor becomes progressively older toward the African and South American continents. This was good enough evidence to establish the "seafloor spreading hypothesis" that described the two continents moving apart.

If a hypothesis survives much experimental testing and leads, in turn, to the design of new experiments with the generation of new hypotheses that can be tested, you now have a working *theory*. A **theory** is defined as a broad working hypothesis that is based on extensive experimental evidence. A scientific theory tells you *why* something happens. For example, the plate tectonic theory describes how the continents have moved apart, just like pieces of a picture puzzle. Is this the same idea that was once considered conjecture? Sort of, but this time it is supported by experimental evidence.

The term *scientific theory* is reserved for historic schemes of thought that have survived the test of detailed examination for long periods of time. The *atomic theory,* for example, was developed in the late 1800s and has been the subject of extensive investigation and experimentation over the last century. The atomic theory and other scientific theories form the framework of scientific thought and experimentation today. Scientific theories point to new ideas about the behavior of nature, and these ideas result in more experiments, more data to collect, and more explanations to develop. All of this may lead to a slight modification of an existing theory, a major modification, or perhaps the creation of an entirely new theory. These activities are all part of the continuing attempt to satisfy our curiosity about nature.

SUMMARY

Physical science is a search for order in our physical surroundings. People have *concepts,* or mental images, about material *objects* and intangible *events* in their surroundings. Concepts are used for thinking and communicating. Concepts are based on *properties,* or attributes that describe a thing or event. Every property implies a *referent* that describes the property. Referents are not always explicit, and most communications require assumptions. Measurement brings precision to descriptions by using numbers and standard units for referents to communicate "exactly how much of exactly what."

Measurement is a process that uses a well-defined and agreed-upon *referent* to describe a *standard unit.* The unit is compared to the property being defined by an *operation* that determines the *value* of the unit by *counting.* Measurements are always reported with a *number,* or value, and a *name* for the unit.

The two major *systems* of standard units are the *English system* and the *metric system.* The English system uses standard units that were originally based on human body parts, and the metric system uses standard units based on referents found in nature. The metric system also uses a system of

People Behind the Science

Florence Bascom (1862–1945)

Florence Bascom, a U.S. geologist, was an expert in the study of rocks and minerals and founded the geology department at Bryn Mawr College, Pennsylvania. This department was responsible for training the foremost women geologists of the early twentieth century.

Born in Williamstown, Massachusetts, in 1862, Bascom was the youngest of the six children of suffragist and schoolteacher Emma Curtiss Bascom and William Bascom, professor of philosophy at Williams College. Her father, a supporter of suffrage and the education of women, later became president of the University of Wisconsin, to which women were admitted in 1875. Florence Bascom enrolled there in 1877 and with other women was allowed limited access to the facilities but was denied access to classrooms filled with men. In spite of this, she earned a B.A. in 1882, a B.Sc. in 1884, and an M.S. in 1887. When Johns Hopkins University graduate school opened to women in 1889, Bascom was allowed to enroll to study geology on the condition that she sat behind a screen to avoid distracting the male students. With the support of her advisor, George Huntington Williams, and her father, she managed in 1893 to become the second woman to gain a Ph.D. in geology (the first being Mary Holmes at the University of Michigan in 1888).

Bascom's interest in geology had been sparked by a driving tour she took with her father and his friend Edward Orton, a geology professor at Ohio State. It was an exciting time for geologists with new areas opening up all the time. Bascom was also inspired by her teachers at Wisconsin and Johns Hopkins, who were experts in the new fields of metamorphism and crystallography. Bascom's Ph.D. thesis was a study of rocks that had previously been thought to be sediments but that she proved to be metamorphosed lava flows.

While studying for her doctorate, Bascom became a popular teacher, passing on her enthusiasm and rigor to her students. She taught at the Hampton Institute for Negroes and American Indians and at Rockford College before becoming an instructor and associate professor at Ohio State University in geology from 1892 to 1895. Moving to Bryn Mawr College, where geology was considered subordinate to the other sciences, she spent two years teaching in a storeroom while building a considerable collection of fossils, rocks, and minerals. While at Bryn Mawr, she took great pride in passing on her knowledge and training to a generation of women who would become successful. At Bryn Mawr, she rose rapidly, becoming reader (1898), associate professor (1903), professor (1906), and finally professor emeritus from 1928 till her death in 1945 in Northampton, Massachusetts.

Bascom became, in 1896, the first woman to work as a geologist on the U.S. Geological Survey, spending her summers mapping formations in Pennsylvania, Maryland, and New Jersey, and her winters analyzing slides. Her results were published in Geographical Society of America bulletins. In 1924, she became the first woman to be elected a fellow of the Geographical Society and went on, in 1930, to become the first woman vice president. She was associate editor of the *American Geologist* (1896–1905) and achieved a four-star place in the first edition of *American Men and Women of Science* (1906), a sign of how highly regarded she was in her field.

Bascom was the author of over forty research papers. She was an expert on the crystalline rocks of the Appalachian Piedmont, and she published her research on Piedmont geomorphology. Geologists in the Piedmont area still value her contributions, and she is still a powerful model for women seeking status in the field of geology today.

---

prefixes to express larger or smaller amounts of units. The metric standard units for length, mass, and time are the *meter, kilogram,* and *second.*

Measurement information used to describe something is called *data.* One way to extract meanings and generalizations from data is to use a *ratio,* a simplified relationship between two numbers. Density is a ratio of mass to volume, or $\rho = m/V.$

Symbols are used to represent *quantities,* or measured properties. Symbols are used in *equations,* which are shorthand statements that describe a relationship where the quantities (both number values and units) are identical on both sides of the equal sign. Equations are used to (1) *describe* a property, (2) *define* a concept, or (3) *describe* how *quantities change* together.

Quantities that can have different values at different times are called *variables.* Variables that increase or decrease together in the same ratio are said to be in *direct proportion.* If one variable increases while the other decreases in the same ratio, the variables are in *inverse proportion.* Proportionality statements are not necessarily equations. A *proportionality constant* can be used to make such a statement into an equation. Proportionality constants might have numerical value only, without units, or they might have both value and units.

Modern science began about three hundred years ago during the time of Galileo and Newton. Since that time, *scientific investigation* has been used to provide *experimental evidence* about nature. *Methods* used to conduct scientific investigations can be generalized as *collecting observations, developing explanations,* and *testing explanations.*

A *hypothesis* is a tentative explanation that is accepted or rejected based on experimental data. Experimental data can come from *observations* or from a *controlled experiment.* The controlled experiment compares two situations that have all the influencing factors identical except one. The single influencing variable being tested is called the *experimental variable,* and the group of variables that form the basis of comparison is called the *control group.*

An accepted hypothesis may result in a *principle,* an explanation concerned with a specific range of phenomena, or a *scientific law,* an explanation concerned with important, wider-ranging phenomena. Laws are sometimes identified with the name of a scientist and can be expressed verbally, with an equation, or with a graph.

A *model* is used to help understand something that cannot be observed directly, explaining the unknown in terms of things already understood. Physical models, mental models, and equations are all examples of models that explain how nature behaves. A *theory* is a broad, detailed explanation that guides development and interpretations of experiments in a field of study.

SUMMARY OF EQUATIONS

1.1

$$\text{density} = \frac{\text{mass}}{\text{volume}}$$

$$\rho = \frac{m}{V}$$

KEY TERMS

area (p. 7)
controlled experiment (p. 15)
data (p. 7)
density (p. 8)
direct proportion (p. 10)
English system (p. 4)
equation (p. 10)
experiment (p. 14)
fundamental properties (p. 5)
hypothesis (p. 14)
inverse proportion (p. 10)
kilogram (p. 5)
liter (p. 6)
measurement (p. 4)
meter (p. 5)
metric system (p. 5)
model (p. 17)
numerical constant (p. 11)
properties (p. 3)
proportionality constant (p. 11)
pseudoscience (p. 16)
quantities (p. 10)
ratio (p. 7)
referent (p. 3)
scientific law (p. 17)
scientific principle (p. 17)
second (p. 6)
standard unit (p. 4)
theory (p. 20)
unit (p. 4)
variable (p. 10)
volume (p. 7)

APPLYING THE CONCEPTS

1. A generalized mental image of an object is a (an)
   a. definition.
   b. impression.
   c. concept.
   d. mental picture.

2. Which of the following is the best example of the use of a referent?
   a. A red bicycle
   b. Big as a dump truck
   c. The planet Mars
   d. Your textbook

3. A well-defined and agreed-upon referent used as a standard in all systems of measurement is called a (an)
   a. yardstick.
   b. unit.
   c. quantity.
   d. fundamental.

4. The system of measurement based on referents in nature, but not with respect to human body parts, is the
   a. natural system.
   b. English system.
   c. metric system.
   d. American system.

5. A process of comparing a property to a well-defined and agreed-upon referent is called a
   a. measurement.
   b. referral.
   c. magnitude.
   d. comparison.

6. One of the following is **not** considered to be a fundamental property:
   a. weight.
   b. length.
   c. time.
   d. charge.

7. How much space something occupies is described by its
   a. mass.
   b. volume.
   c. density.
   d. weight.

8. The relationship between two numbers that is usually obtained by dividing one number by the other is called a (an)
   a. ratio.
   b. divided size.
   c. number tree.
   d. equation.

9. The ratio of mass per volume of a substance is called its
   a. weight.
   b. weight-volume.
   c. mass-volume.
   d. density.

10. After identifying the appropriate equation, the next step in correctly solving a problem is to
    a. substitute known quantities for symbols.
    b. solve the equation for the variable in question.
    c. separate number and units.
    d. convert all quantities to metric units.

CHAPTER 1 What Is Science? **21**

11. Suppose a problem situation describes a speed in km/h and a length in m. What conversion should you do before substituting quantities for symbols? Convert
   a. km/h to km/s.
   b. m to km.
   c. km/h to m/s.
   d. In this situation, no conversions should be made.

12. An equation describes a relationship where
   a. the numbers and units on both sides are proportional but not equal.
   b. the numbers on both sides are equal but not the units.
   c. the units on both sides are equal but not the numbers.
   d. the numbers and units on both sides are equal.

13. The equation $\rho = \frac{m}{V}$ is a statement that
   a. describes a property.
   b. defines how variables can change.
   c. describes how properties change.
   d. identifies proportionality constant.

14. Measurement information that is used to describe something is called
   a. referents.
   b. properties.
   c. data.
   d. a scientific investigation.

15. If you consider a very small portion of a material that is the same throughout, the density of the small sample will be
   a. much less.
   b. slightly less.
   c. the same.
   d. greater.

16. The symbol Δ has a meaning of
   a. "is proportional to."
   b. "the change in."
   c. "therefore."
   d. "however."

17. A model is
   a. a physical copy of an object or system made at a smaller scale.
   b. a sketch of something complex used to solve problems.
   c. an interpretation of a theory by use of an equation.
   d. All of the above are models.

18. The use of a referent in describing a property always implies
   a. a measurement.
   b. naturally occurring concepts.
   c. a comparison with a similar property of another object.
   d. that people have the same understandings of concepts.

19. A 5-km span is the same as how many meters?
   a. 0.005 m
   b. 0.05 m
   c. 500 m
   d. 5,000 m

20. One-half liter of water is the same volume as
   a. 5,000 mL.
   b. 0.5 cc.
   c. 500 cm^3.
   d. 5 dm^3.

21. Which of the following is not a measurement?
   a. 24°C
   b. 65 mph
   c. 120
   d. 0.50 ppm

22. What happens to the surface-area-to-volume ratio as the volume of a cube becomes larger?
   a. It remains the same.
   b. It increases.
   c. It decreases.
   d. The answer varies.

23. If one variable increases in value while a second, related variable decreases in value, the relationship is said to be
   a. direct.
   b. inverse.
   c. square.
   d. inverse square.

24. What is needed to change a proportionality statement into an equation?
   a. Include a proportionality constant.
   b. Divide by unknown to move symbol to left side of equal symbol.
   c. Add units to one side to make units equal.
   d. Add numbers to one side to make both sides equal.

25. A proportionality constant
   a. always has a unit.
   b. never has a unit.
   c. might or might not have a unit.

26. A scientific investigation provides understanding through
   a. explanations based on logical thinking processes alone.
   b. experimental evidence.
   c. reasoned explanations based on observations.
   d. diligent obeying of scientific laws.

27. Statements describing how nature is observed to behave consistently time after time are called scientific
   a. theories.
   b. laws.
   c. models.
   d. hypotheses.

28. A controlled experiment comparing two situations has all identical influencing factors except the
   a. experimental variable.
   b. control variable.
   c. inverse variable.
   d. direct variable.

29. In general, scientific investigations have which activities in common?
   a. state problem, gather data, make hypothesis, test, make conclusion
   b. collect observations, develop explanations, test explanations
   c. observe nature, reason an explanation for what is observed
   d. observe nature, collect data, modify data to fit scientific model

30. Quantities, or measured properties, that are capable of changing values are called
   a. data.
   b. variables.
   c. proportionality constants.
   d. dimensionless constants.

31. A proportional relationship that is represented by the symbols $a \propto 1/b$ represents which of the following relationships?
   a. direct proportion
   b. inverse proportion
   c. direct square proportion
   d. inverse square proportion

32. A hypothesis concerned with a specific phenomenon is found to be acceptable through many experiments over a long period of time. This hypothesis usually becomes known as a
   a. scientific law.
   b. scientific principle.
   c. theory.
   d. model.

33. A scientific law can be expressed as
   a. a written concept.
   b. an equation.
   c. a graph.
   d. all of the above.

34. The symbol ∝ has a meaning of
   a. "almost infinity."
   b. "the change in."
   c. "is proportional to."
   d. "therefore."

35. Which of the following symbols represents a measured property of the compactness of matter?
   a. m
   b. ρ
   c. V
   d. Δ

36. A candle with a certain weight melts in an oven and the resulting weight of the wax is
   a. less.
   b. the same.
   c. greater.
   d. The answer varies.

37. An ice cube with a certain volume melts and the resulting volume of water is
   a. less.
   b. the same.
   c. greater.
   d. The answer varies.

38. Compare the density of ice to the density of water. The density of the ice is
   a. less.
   b. the same.
   c. greater.
   d. The answer varies.

39. A beverage glass is filled to the brim with ice-cold water (0°C) and ice cubes. Some of the ice cubes are floating above the water level. When the ice melts, the water in the glass will
   a. spill over the brim.
   b. stay at the same level.
   c. be less full than before the ice melted.

40. What is the proportional relationship between the volume of juice in a cup and the time the juice dispenser has been running?
   a. direct
   b. inverse
   c. square
   d. inverse square

41. What is the proportional relationship between the number of cookies in the cookie jar and the time you have been eating the cookies?
   a. direct
   b. inverse
   c. square
   d. inverse square

42. A movie projector makes a 1 m by 1 m image when projecting 1 m from a screen, a 2 m by 2 m image when projecting 2 m from the screen, and a 3 m by 3 m image when projecting 3 m from the screen. What is the proportional relationship between the distance from the screen and the area of the image?
   a. direct
   b. inverse
   c. square
   d. inverse square

43. A movie projector makes a 1 m by 1 m image when projecting 1 m from a screen, a 2 m by 2 m image when projecting 2 m from the screen, and a 3 m by 3 m image when projecting 3 m from the screen. What is the proportional relationship between the distance from the screen and the intensity of the light falling on the screen?
   a. direct
   b. inverse
   c. square
   d. inverse square

44. According to the scientific method, what needs to be done to move beyond conjecture or simple hypotheses in a person's understanding of his or her physical surroundings?
   a. Make an educated guess.
   b. Conduct a controlled experiment.
   c. Find an understood model with answers.
   d. Search for answers on the Internet.

Answers

1. c **2.** b **3.** b **4.** c **5.** a **6.** a **7.** b **8.** a **9.** d **10.** b **11.** c **12.** d **13.** a **14.** c **15.** c **16.** b **17.** d **18.** c **19.** d **20.** c **21.** c **22.** c **23.** b **24.** a **25.** c **26.** b **27.** b **28.** a **29.** b **30.** b **31.** b **32.** a **33.** d **34.** c **35.** b **36.** b **37.** a **38.** a **39.** b **40.** a **41.** b **42.** c **43.** d **44.** b

QUESTIONS FOR THOUGHT

1. What is a concept?
2. What are two components of a measurement statement? What does each component tell you?
3. Other than familiarity, what are the advantages of the English system of measurement?
4. Define the metric standard units for length, mass, and time.
5. Does the density of a liquid change with the shape of a container? Explain.
6. Does a flattened pancake of clay have the same density as the same clay rolled into a ball? Explain.
7. What is an equation? How are equations used in the physical sciences?
8. Compare and contrast a scientific principle and a scientific law.
9. What is a model? How are models used?
10. Are all theories always completely accepted or completely rejected? Explain.

FOR FURTHER ANALYSIS

1. Select a statement that you feel might represent pseudoscience. Write an essay supporting *and* refuting your selection, noting facts that support one position or the other.
2. Evaluate the statement that science cannot solve human-produced problems such as pollution. What does it mean to

say pollution is caused by humans and can only be solved by humans? Provide evidence that supports your position.

3. Make an experimental evaluation of what happens to the density of a substance at larger and larger volumes.

4. If your wage were dependent on your work-time squared, how would it affect your pay if you double your hours?

5. Merriam-Webster's 11th *Collegiate Dictionary* defines science, in part, as "knowledge or a system of knowledge covering general truths or the operation of general laws especially as obtained and tested through scientific method." How would you define science?

6. Are there any ways in which scientific methods differ from commonsense methods of reasoning?

7. The United States is the only country in the world that does not use the metric system of measurement. With this understanding, make a list of advantages and disadvantages for adopting the metric system in the United States.

INVITATION TO INQUIRY

Paper Helicopters

Construct paper helicopters and study the effects that various variables have on their flight. After considering the size you wish to test, copy the patterns shown in Figure 1.17 on a sheet of notebook paper. Note that solid lines are to be cut and dashed lines are to be folded. Make three scissor cuts on the solid lines. Fold A toward you and B away from you to form the wings. Then fold C and D inward to overlap, forming the body. Finally, fold up the bottom on the dashed line and hold it together with a paper clip. Your finished product should look like the

helicopter in Figure 1.17. Try a preliminary flight test by standing on a chair or stairs and dropping it.

Decide what variables you would like to study to find out how they influence the total flight time. Consider how you will hold everything else constant while changing one variable at a time. You can change the wing area by making new helicopters with more or less area in the A and B flaps. You can change the weight by adding more paper clips. Study these and other variables to find out who can design a helicopter that will remain in the air the longest. Who can design a helicopter that is most accurate in hitting a target?

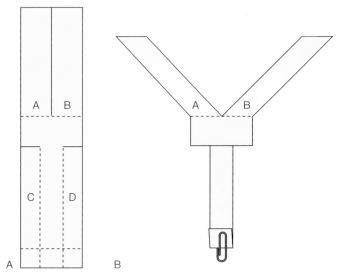

FIGURE 1.17 Pattern for a paper helicopter.

PARALLEL EXERCISES

The exercises in groups A and B cover the same concepts. Solutions to group A exercises are located in appendix E.

Note: *You will need to refer to Table 1.3 to complete some of the following exercises.*

Group A

1. What is your height in meters? In centimeters?
2. What is the density of mercury if 20.0 cm³ has a mass of 272 g?
3. What is the mass of a 10.0 cm³ cube of lead?
4. What is the volume of a rock with a density of 3.00 g/cm³ and a mass of 600 g?
5. If you have 34.0 g of a 50.0 cm³ volume of one of the substances listed in Table 1.3, which one is it?
6. What is the mass of water in a 40 L aquarium?
7. A 2.1 kg pile of aluminum cans is melted, then cooled into a solid cube. What is the volume of the cube?
8. A cubic box contains 1,000 g of water. What is the length of one side of the box in meters? Explain your reasoning.
9. A loaf of bread (volume 3,000 cm³) with a density of 0.2 g/cm³ is crushed in the bottom of a grocery bag into a volume of 1,500 cm³. What is the density of the mashed bread?
10. According to Table 1.3, what volume of copper would be needed to balance a 1.00 cm³ sample of lead on a two-pan laboratory balance?

Group B

1. What is your mass in kilograms? In grams?
2. What is the density of iron if 5.0 cm³ has a mass of 39.5 g?
3. What is the mass of a 10.0 cm³ cube of copper?
4. If ice has a density of 0.92 g/cm³, what is the volume of 5,000 g of ice?
5. If you have 51.5 g of a 50.0 cm³ volume of one of the substances listed in Table 1.3, which one is it?
6. What is the mass of gasoline ($\rho = 0.680$ g/cm³) in a 94.6 L gasoline tank?
7. What is the volume of a 2.00 kg pile of iron cans that are melted, then cooled into a solid cube?
8. A cubic tank holds 1,000.0 kg of water. What are the dimensions of the tank in meters? Explain your reasoning.
9. A hot dog bun (volume 240 cm³) with a density of 0.15 g/cm³ is crushed in a picnic cooler into a volume of 195 cm³. What is the new density of the bun?
10. According to Table 1.3, what volume of iron would be needed to balance a 1.00 cm³ sample of lead on a two-pan laboratory balance?

2
Motion

Information about the mass of a hot air balloon and forces on the balloon will enable you to predict if it is going to move up, down, or drift across the river. This chapter is about such relationships among force, mass, and changes in motion.

CORE **CONCEPT**
A net force is required for any change in a state of motion.

OUTLINE

Forces
Inertia is the tendency of an object to remain in unchanging motion when the net force is zero.

Newton's First Law of Motion
Every object retains its state of rest or straight-line motion unless acted upon by an unbalanced force.

Newton's Third Law of Motion
A single force does not exist by itself; there is always a matched and opposite force that occurs at the same time.

Describing Motion
Measuring Motion
   Speed
   Velocity
   Acceleration
Science and Society: Transportation and the Environment
   Forces
Horizontal Motion on Land
Falling Objects
A Closer Look: A Bicycle Racer's Edge
Compound Motion
   Vertical Projectiles
   Horizontal Projectiles
A Closer Look: Free Fall
Three Laws of Motion
   Newton's First Law of Motion
   Newton's Second Law of Motion
   Weight and Mass
   Newton's Third Law of Motion
Momentum
   Conservation of Momentum
   Impulse
Forces and Circular Motion
Newton's Law of Gravitation
   Earth Satellites
A Closer Look: Gravity Problems
   Weightlessness
People Behind the Science: Isaac Newton

Falling Objects
The force of gravity uniformly accelerates falling objects.

Newton's Second Law of Motion
The acceleration of an object depends on the net force applied and the mass of the object.

Newton's Law of Gravitation
All objects in the universe are attracted to all other objects in the universe.

25

In chapter 1, you learned some "tools and rules" and some techniques for finding order in your physical surroundings. Order is often found in the form of patterns, or relationships between quantities that are expressed as equations. Recall that equations can be used to (1) describe properties, (2) define concepts, and (3) describe how quantities change relative to each other. In all three uses, patterns are quantified, conceptualized, and used to gain a general understanding about what is happening in nature.

In the study of physical science, certain parts of nature are often considered and studied together for convenience. One of the more obvious groupings involves *movement*. Most objects around you spend a great deal of time sitting quietly without motion. Buildings, rocks, utility poles, and trees rarely, if ever, move from one place to another. Even things that do move from time to time sit still for a great deal of time. This includes you, automobiles, and bicycles (Figure 2.1). On the other hand, the Sun, the Moon, and starry heavens seem to always move, never standing still. Why do things stand still? Why do things move?

Questions about motion have captured the attention of people for thousands of years. But the ancient people answered questions about motion with stories of mysticism and spirits that lived in objects. It was during the classic Greek culture, between 600 B.C. and 300 B.C., that people began to look beyond magic and spirits. One particular Greek philosopher, Aristotle, wrote a theory about the universe that offered not only explanations about things such as motion but also offered a sense of beauty, order, and perfection. The theory seemed to fit with other ideas that people had and was held to be correct for nearly two thousand years after it was written. It was not until the work of Galileo and Newton during the 1600s that a new, correct understanding about motion was developed. The development of ideas about motion is an amazing and absorbing story. You will learn in this chapter how to describe and use some properties of motion. This will provide some basic understandings about motion and will be very helpful in understanding some important aspects of astronomy and the earth sciences, as well as the movement of living things.

DESCRIBING MOTION

Motion is one of the more common events in your surroundings. You can see motion in natural events such as clouds moving, rain and snow falling, and streams of water moving, all in a never-ending cycle. Motion can also be seen in the activities of people who walk, jog, or drive various machines from place to place. Motion is so common that you would think everyone would intuitively understand the concepts of motion, but history indicates that it was only during the past three hundred years or so that people began to understand motion correctly. Perhaps the correct concepts are subtle and contrary to common sense, requiring a search for simple, clear concepts in an otherwise complex situation. The process of finding such order in a multitude of sensory impressions by taking measurable data and then inventing a concept to describe what is happening is the activity called *science*. We will now apply this process to motion.

What is motion? Consider a ball that you notice one morning in the middle of a lawn. Later in the afternoon, you notice that the ball is at the edge of the lawn, against a fence, and you wonder if the wind or some person moved the ball. You do not know if the wind blew it at a steady rate, if many gusts of wind moved it, or even if some children kicked it all over the yard. All you know for sure is that the ball has been moved because it is

in a different position after some time passed. These are the two important aspects of motion: (1) a change of position and (2) the passage of time.

If you did happen to see the ball rolling across the lawn in the wind, you would see more than the ball at just two locations. You would see the ball moving continuously. You could consider, however, the ball in continuous motion to be a series of individual locations with very small time intervals. Moving involves a change of position during some time period. Motion is the act or process of something changing position.

The motion of an object is usually described with respect to something else that is considered to be not moving. (Such a stationary object is said to be "at rest.") Imagine that you are traveling in an automobile with another person. You know that you are moving across the land outside the car since your location on the highway changes from one moment to another. Observing your fellow passenger, however, reveals no change of position. You are in motion relative to the highway outside the car. You are not in motion relative to your fellow passenger. Your motion, and the motion of any other object or body, is the process of a change in position *relative* to some reference object or location. Thus, *motion* can be defined as the act or process of changing position relative to some reference during a period of time.

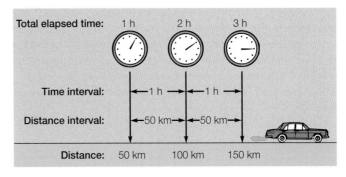

FIGURE 2.2 If you know the value of any two of the three variables of distance, time, and speed, you can find the third. What is the average speed of this car? Two ways of finding the answer are in Figure 2.3.

FIGURE 2.1 The motion of this windsurfer, and of other moving objects, can be described in terms of the distance covered during a certain time period.

MEASURING MOTION

You have learned that objects can be described by measuring certain fundamental properties such as mass and length. Since motion involves (1) a change of *position* and (2) the passage of *time,* the motion of objects can be described by using combinations of the fundamental properties of length and time. These combinations of measurement describe three properties of motion: *speed, velocity,* and *acceleration.*

SPEED

Suppose you are in a car that is moving over a straight road. How could you describe your motion? You need at least two measurements: (1) the distance you have traveled and (2) the time that has elapsed while you covered this distance. Such a distance and time can be expressed as a ratio that describes your motion. This ratio is a property of motion called **speed,** which is a measure of how fast you are moving. Speed is defined as distance per unit of time, or

$$\text{speed} = \frac{\text{distance}}{\text{time}}$$

The units used to describe speed are usually miles/hour (mi/h), kilometers/hour (km/h), or meters/second (m/s).

Let's go back to your car that is moving over a straight highway and imagine you are driving to cover equal distances in equal periods of time. If you use a stopwatch to measure the time required to cover the distance between highway mile markers (those little signs with numbers along major highways), the time intervals will all be equal. You might find, for example, that one minute lapses between each mile marker. Such a uniform straight-line motion that covers equal distances in equal periods of time is the simplest kind of motion.

If your car were moving over equal distances in equal periods of time, it would have a *constant speed* (Figure 2.2). This means that the car is neither speeding up nor slowing down. It is usually difficult to maintain a constant speed. Other cars and distractions such as interesting scenery cause you to reduce your speed. At other times you increase your speed. If you calculate your speed over an entire trip, you are considering a large distance between two places and the total time that elapsed. The increases and decreases in speed would be averaged. Therefore, most speed calculations are for an *average speed.* The speed at any specific instant is called the *instantaneous speed.* To calculate the instantaneous speed, you would need to consider a very short time interval—one that approaches zero. An easier way would be to use the speedometer, which shows the speed at any instant.

Constant, instantaneous, or average speeds can be measured with any distance and time units. Common units in the English system are miles/hour and feet/second. Metric units for speed are commonly kilometers/hour and meters/second. The ratio of any distance to time is usually read as distance per time, such as miles per hour. The *per* means "for each."

It is easier to study the relationships between quantities if you use symbols instead of writing out the whole word. The letter v can be used to stand for speed, the letter d can be used to stand for distance, and the letter t, to stand for time. A bar over the v ($\bar{v}$) is a symbol that means average (it is read "v-bar" or "v-average"). The relationship between average speed, distance, and time is therefore

$$\bar{v} = \frac{d}{t}$$

equation 2.1

This is one of the three types of equations that were discussed on page 10, and in this case, the equation defines a motion property. You can use this relationship to find average speed. For example, suppose a car travels 150 km in 3 h. What was the average speed? Since $d = 150$ km, and $t = 3$ h, then

$$\bar{v} = \frac{150 \text{ km}}{3 \text{ h}}$$

$$= 50 \frac{\text{km}}{\text{h}}$$

As with other equations, you can mathematically solve the equation for any term as long as two variables are known (Figure 2.3). For example, suppose you know the speed and the time but want to find the distance traveled. You can solve this by first writing the relationship

$$\bar{v} = \frac{d}{t}$$

and then multiplying both sides of the equation by t (to get d on one side by itself),

$$(\bar{v})(t) = \frac{(d)(t)}{t}$$

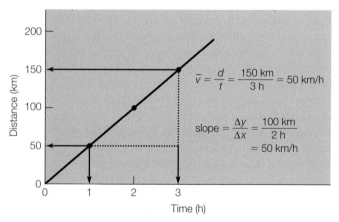

FIGURE 2.3 Speed is distance per unit of time, which can be calculated from the equation or by finding the slope of a distance-versus-time graph. This shows both ways of finding the speed of the car shown in Figure 2.2.

and the t's on the right cancel, leaving

$$\bar{v}t = d \qquad \text{or} \qquad d = \bar{v}t$$

If the $\bar{v}$ is 50 km/h and the time traveled is 2 h, then

$$d = \left(50 \frac{\text{km}}{\text{h}}\right)(2\text{h})$$

$$= (50)(2)\left(\frac{\text{km}}{\text{h}}\right)(\text{h})$$

$$= 100 \frac{(\text{km})(\text{h})}{\text{h}}$$

$$= 100 \text{ km}$$

Notice how both the numerical values and the units were treated mathematically. See "How to Solve Problems" in chapter 1 for more information.

EXAMPLE 2.1

The driver of a car moving at 72.0 km/h drops a road map on the floor. It takes him 3.00 seconds to locate and pick up the map. How far did he travel during this time?

SOLUTION

The car has a speed of 72.0 km/h and the time factor is 3.00 s, so km/h must be converted to m/s. From inside the front cover of this book, the conversion factor is 1 km/h = 0.2778 m/s, so

$$\bar{v} = \frac{0.2778 \frac{\text{m}}{\text{s}}}{\frac{\text{km}}{\text{h}}} \times 72.0 \frac{\text{km}}{\text{h}}$$

$$= (0.2778)(72.0) \frac{\text{m}}{\text{s}} \times \frac{\text{h}}{\text{km}} \times \frac{\text{km}}{\text{h}}$$

$$= 20.0 \frac{\text{m}}{\text{s}}$$

The relationship between the three variables, $\bar{v}, t$, and d, is found in equation 2.1: $\bar{v} = d/t$.

$$\bar{v} = 20.0 \frac{\text{m}}{\text{s}} \qquad\qquad \bar{v} = \frac{d}{t}$$

$$t = 3.00 \text{ s} \qquad\qquad \bar{v}t = \frac{dt}{t}$$

$$d = ? \qquad\qquad d = \bar{v}t$$

$$= \left(20.0 \frac{\text{m}}{\text{s}}\right)(3.00 \text{ s})$$

$$= (20.0)(3.00) \frac{\text{m}}{\text{s}} \times \frac{\text{s}}{1}$$

$$= \boxed{60.0 \text{ m}}$$

EXAMPLE 2.2

A bicycle has an average speed of 8.00 km/h. How far will it travel in 10.0 seconds? (Answer: 22.2 m)

CONCEPTS *Applied*

Style Speeds

Observe how many different styles of walking you can identify in students walking across the campus. Identify each style with a descriptive word or phrase.

Is there any relationship between any particular style of walking and the speed of walking? You could find the speed of walking by measuring a distance, such as the distance between two trees, then measuring the time required for a student to walk the distance. Find the average speed for each identified style of walking by averaging the walking speeds of ten people.

Report any relationships you find between styles of walking and the average speed of people with each style. Include any problems you found in measuring, collecting data, and reaching conclusions.

CONCEPTS *Applied*

How Fast Is a Stream?

A stream is a moving body of water. How could you measure the speed of a stream? Would timing how long it takes a floating leaf to move a measured distance help?

What kind of relationship, if any, would you predict for the speed of a stream and a recent rainfall? Would you predict a direct relationship? Make some measurements of stream speeds and compare your findings to recent rainfall amounts.

VELOCITY

The word *velocity* is sometimes used interchangeably with the word *speed,* but there is a difference. **Velocity** describes the *speed and direction* of a moving object. For example, a speed might be described as 60 km/h. A velocity might be described as 60 km/h to the west. To produce a change in velocity, either the speed or the direction is changed (or both are changed). A satellite moving with a constant speed in a circular orbit around Earth does not have a constant velocity since its direction of movement is constantly changing. Velocity can be represented graphically with arrows. The lengths of the arrows are proportional to the magnitude, and the arrowheads indicate the direction (Figure 2.4).

ACCELERATION

Motion can be changed in three different ways: (1) by changing the speed, (2) by changing the direction of travel, or (3) combining both of these by changing both the speed and direction of travel at the same time. Since velocity describes both the speed and the direction of travel, any of these three

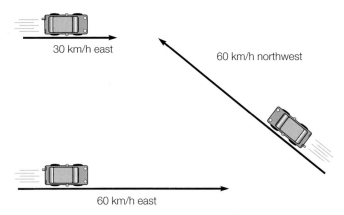

FIGURE 2.4 Here are three different velocities represented by three different arrows. The length of each arrow is proportional to the speed, and the arrowhead shows the direction of travel.

changes will result in a change of velocity. You need at least one additional measurement to describe a change of motion, which is how much time elapsed while the change was taking place. The change of velocity and time can be combined to define the *rate* at which the motion was changed. This rate is called **acceleration.** Acceleration is defined as a change of velocity per unit time, or

$$\text{acceleration} = \frac{\text{change of velocity}}{\text{time elapsed}}$$

Another way of saying "change in velocity" is the final velocity minus the initial velocity, so the relationship can also be written as

$$\text{acceleration} = \frac{\text{final velocity} - \text{initial velocity}}{\text{time elapsed}}$$

Acceleration due to a change in speed only can be calculated as follows. Consider a car that is moving with a constant, straight-line velocity of 60 km/h when the driver accelerates to 80 km/h. Suppose it takes 4 s to increase the velocity of 60 km/h to 80 km/h. The change in velocity is therefore 80 km/h minus 60 km/h, or 20 km/h. The acceleration was

$$\text{acceleration} = \frac{80\frac{\text{km}}{\text{h}} - 60\frac{\text{km}}{\text{h}}}{4\text{ s}}$$

$$= \frac{20\frac{\text{km}}{\text{h}}}{4\text{ s}}$$

$$= 5\frac{\text{km/h}}{\text{s}}\text{ or}$$

$$= 5\,\text{km/h/s}$$

The average acceleration of the car was 5 km/h for each ("per") second. This is another way of saying that the velocity increases an average of 5 km/h in each second. The velocity of the car was 60 km/h when the acceleration began (initial velocity). At the end of 1 s, the velocity was 65 km/h. At the end of 2 s, it was 70 km/h; at the end of 3 s, 75 km/h; and at the end of 4 s (total

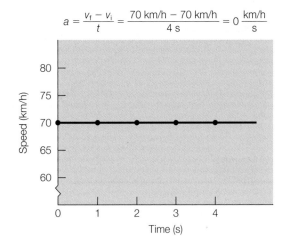

$$a = \frac{v_f - v_i}{t} = \frac{70 \text{ km/h} - 70 \text{ km/h}}{4 \text{ s}} = 0 \frac{\text{km/h}}{\text{s}}$$

A

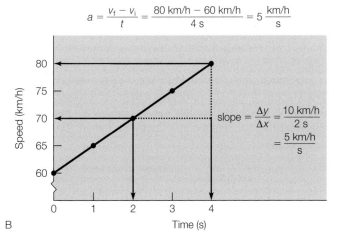

$$a = \frac{v_f - v_i}{t} = \frac{80 \text{ km/h} - 60 \text{ km/h}}{4 \text{ s}} = 5 \frac{\text{km/h}}{\text{s}}$$

$$\text{slope} = \frac{\Delta y}{\Delta x} = \frac{10 \text{ km/h}}{2 \text{ s}}$$
$$= \frac{5 \text{ km/h}}{\text{s}}$$

B

FIGURE 2.5 (*A*) This graph shows how the speed changes per unit of time while driving at a constant 70 km/h in a straight line. As you can see, the speed is constant, and for straight-line motion, the acceleration is 0. (*B*) This graph shows the speed increasing from 60 km/h to 80 km/h for 5 s. The acceleration, or change of velocity per unit of time, can be calculated either from the equation for acceleration or by calculating the slope of the straight-line graph. Both will tell you how fast the motion is changing with time.

time elapsed), the velocity was 80 km/h (final velocity). Note how fast the velocity is changing with time. In summary,

start (initial velocity)	60 km/h
End of first second	65 km/h
End of second second	70 km/h
End of third second	75 km/h
End of fourth second (final velocity)	80 km/h

As you can see, acceleration is really a description of how fast the speed is changing (Figure 2.5); in this case, it is increasing 5 km/h each second.

Usually, you would want all the units to be the same, so you would convert km/h to m/s. A change in velocity of 5.0 km/h converts to 1.4 m/s and the acceleration would be 1.4 m/s/s. The units m/s per s mean what change of velocity (1.4 m/s) is occurring every second. The combination m/s/s is rather cumbersome, so it is typically treated mathematically to simplify the expression

(to simplify a fraction, invert the divisor and multiply, or m/s × 1/s = m/s^2). Remember that the expression 1.4 m/s^2 means the same as 1.4 m/s per s, a change of velocity in a given time period.

The relationship among the quantities involved in acceleration can be represented with the symbols a for average acceleration, v_f for final velocity, v_i for initial velocity, and t for time. The relationship is

$$a = \frac{v_f - v_i}{t}$$

equation 2.2

As in other equations, any one of these quantities can be found if the others are known. For example, solving the equation for the final velocity, v_f, yields:

$$v_f = at + v_i$$

In problems where the initial velocity is equal to zero (starting from rest), the equation simplifies to

$$v_f = at$$

Recall from chapter 1 that the symbol Δ means "the change in" a value. Therefore, equation 2.1 for speed could be written

$$\bar{v} = \frac{\Delta d}{t}$$

and equation 2.2 for acceleration could be written

$$a = \frac{\Delta v}{t}$$

This shows that both equations are a time rate of change. Speed is a time rate change of *distance*. Acceleration is a time rate change of *velocity*. The time rate of change of something is an important concept that you will meet again in chapter 3.

EXAMPLE 2.3

A bicycle moves from rest to 5 m/s in 5 s. What was the acceleration?

SOLUTION

$$v_i = 0 \text{ m/s}$$
$$v_f = 5 \text{ m/s}$$
$$t = 5 \text{ s}$$
$$a = ?$$

$$a = \frac{v_f - v_i}{t}$$
$$= \frac{5 \text{ m/s} - 0 \text{ m/s}}{5 \text{ s}}$$
$$= \frac{5}{5} \frac{\text{m/s}}{\text{s}}$$
$$= 1 \left(\frac{\text{m}}{\text{s}}\right)\left(\frac{1}{\text{s}}\right)$$
$$= \boxed{1 \frac{\text{m}}{\text{s}^2}}$$

EXAMPLE 2.4

An automobile uniformly accelerates from rest at 5 m/s^2 for 6 s. What is the final velocity in m/s? (Answer: 30 m/s)

Environmental science is an interdisciplinary study of the earth's environment. The concern of this study is the overall problem of human degradation of the environment and remedies for that damage. As an example of an environmental topic of study, consider the damage that results from current human activities involving the use of transportation. Researchers estimate that overall transportation activities are responsible for about one-third of the total U.S. carbon emissions that are added to the air every day. Carbon emissions are a problem because they are directly harmful in the form of carbon monoxide. They are also indirectly harmful because of the contribution of carbon dioxide to possible global warming and the consequences of climate change.

Here is a list of things that people might do to reduce the amount of environmental damage from transportation:

A. Use a bike, carpool, walk, or take public transportation whenever possible.

B. Combine trips to the store, mall, and work, leaving the car parked whenever possible.

C. Purchase hybrid electric or fuel cell-powered cars or vehicles whenever possible.

D. Move to a planned community that makes the use of cars less necessary and less desirable.

QUESTIONS TO DISCUSS

Discuss with your group the following questions concerning connections between thought and feeling:

1. What are your positive or negative feelings associated with each item in the list?

2. Would your feelings be different if you had a better understanding of the global problem?

3. Do your feelings mean that you have reached a conclusion?

4. What new items could be added to the list?

So far, you have learned only about straight-line, uniform acceleration that results in an increased velocity. There are also other changes in the motion of an object that are associated with acceleration. One of the more obvious is a change that results in a decreased velocity. Your car's brakes, for example, can slow your car or bring it to a complete stop. This is *negative acceleration*, which is sometimes called *deceleration*. Another change in the motion of an object is a change of direction. Velocity encompasses both the rate of motion and direction, so a change of direction is an acceleration. The satellite moving with a constant speed in a circular orbit around Earth is constantly changing its direction of movement. It is therefore constantly accelerating because of this constant change in its motion. Your automobile has three devices that could change the state of its motion. Your automobile therefore has three accelerators—the gas pedal (which can increase magnitude of velocity), the brakes (which can decrease magnitude of velocity), and the steering wheel (which can change direction of velocity). (See Figure 2.6.) The important thing to remember is that acceleration results from any *change* in the motion of an object.

The final velocity (v_f) and the initial velocity (v_i) are different variables than the average velocity ($\bar{v}$). You cannot use an initial or final velocity for an average velocity. You may, however, calculate an average velocity ($\bar{v}$) from the other two variables as long as the acceleration taking place between the initial and final velocities is uniform. An example of such a uniform change would be an automobile during a constant,

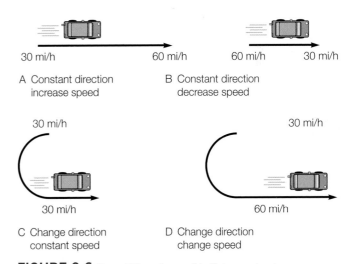

FIGURE 2.6 Four different ways (*A–D*) to accelerate a car.

straight-line acceleration. To find an average velocity *during* a uniform acceleration, you add the initial velocity and the final velocity and divide by 2. This averaging can be done for a uniform acceleration that is increasing the velocity or for one that is decreasing the velocity. In symbols,

$$\bar{v} = \frac{v_f + v_i}{2}$$

equation 2.3

EXAMPLE 2.5

An automobile moving at 25.0 m/s comes to a stop in 10.0 s when the driver slams on the brakes. How far did the car travel while stopping?

SOLUTION

The car has an initial velocity of 25.0 m/s (v_i) and the final velocity of 0 m/s (v_f) is implied. The time of 10.0 s (t) is given. The problem asked for the distance (d). The relationship given between $\bar{v}$, t, and d is given in equation 2.1, $\bar{v} = d/t$, which can be solved for d. The average velocity ($\bar{v}$), however, is not given but can be found from equation 2.3,

$$\bar{v} = \frac{v_f + v_i}{2}$$

$v_i = 25.0 \text{ m/s}$

$v_f = 0 \text{ m/s}$

$t = 10.0 \text{ s}$

$\bar{v} = ?$

$d = ?$

$$\bar{v} = \frac{d}{t} \quad \therefore \quad d = \bar{v} \cdot t$$

Since $\bar{v} = \dfrac{v_f + v_i}{2}$,

you can substitute $\left(\dfrac{v_f + v_i}{2}\right)$ for $\bar{v}$, and

$$d = \left(\frac{v_f + v_i}{2}\right)(t)$$

$$= \left(\frac{0\frac{\text{m}}{\text{s}} + 25.0\frac{\text{m}}{\text{s}}}{2}\right)(10.0 \text{ s})$$

$$= 12.5 \times 10.0 \frac{\text{m}}{\text{s}} \times \text{s}$$

$$= 125 \frac{\text{m} \cdot \cancel{\text{s}}}{\cancel{\text{s}}}$$

$$= \boxed{125 \text{ m}}$$

EXAMPLE 2.6

What was the deceleration of the automobile in example 2.5? (Answer: -2.50 m/s^2)

 CONCEPTS *Applied*

Acceleration Patterns

Suppose the radiator in your car has a leak and drops of fluid fall constantly, one every second. What pattern would the drops make on the pavement when you accelerate the car from a stoplight? What pattern would they make when you drive at a constant speed? What pattern would you observe as the car comes to a stop? Use a marker to make dots on a sheet of paper that illustrate (1) acceleration, (2) constant speed, and (3) negative acceleration. Use words to describe the acceleration in each situation.

FORCES

The Greek philosopher Aristotle considered some of the first ideas about the causes of motion back in the fourth century B.C. However, he had it all wrong when he reportedly stated that a dropped object falls at a constant speed that is determined by its weight. He also incorrectly thought that an object moving across Earth's surface requires a continuously applied force in order to continue moving. These ideas were based on observing and thinking, not measurement, and no one checked to see if they were correct. It would take about two thousand years before people began to correctly understand motion.

Aristotle did recognize an association between force and motion, and this much was acceptable. It is partly correct because a force is closely associated with *any* change of motion, as you will see. This section introduces the concept of a force, which will be developed more fully when the relationship between forces and motion is considered.

A **force** is a push or a pull that is acting on an object. Consider, for example, the movement of a ship from the pushing of two tugboats (Figure 2.7). Tugboats can vary the strength of the force exerted on a ship, but they can also push in different directions. What effect does direction have on two forces acting on an object? If the tugboats were side by side, pushing in the same direction, the overall force is the sum of the two forces. If they act in exactly opposite directions, one pushing on each side of the ship, the overall force is the difference between the strength of the two forces. If they have the same strength, the overall effect is to cancel each other without producing any motion. The **net force** is the sum of all the forces acting on an object. Net force means "final," after the forces are added (Figure 2.8).

When two parallel forces act in the same direction, they can be simply added. In this case, there is a net force that is equivalent to the sum of the two forces. When two parallel forces act in opposite directions, the net force is the difference in the direction of the larger force. When two forces act neither in a way that is exactly together nor exactly opposite each other, the result will be like a new, different net force having a new direction and strength.

Forces have a strength and direction that can be represented by force arrows. The tail of the arrow is placed on the object that feels the force, and the arrowhead points in the direction in which the force is exerted. The length of the arrow is proportional to the strength of the force. The use of force arrows helps you visualize and understand all the forces and how they contribute to the net force.

There are four **fundamental forces** that *cannot* be explained in terms of any other force. They are the gravitational, electromagnetic, weak, and strong nuclear forces. Gravitational forces act between all objects in the universe—between you and Earth, between Earth and the Sun, between the planets in the solar systems, and, in fact, hold stars in large groups called galaxies. Switching scales from the very large galaxy to inside an atom, we find electromagnetic forces acting between electrically charged parts of atoms, such as electrons and protons. Electromagnetic forces are responsible for the structure of atoms, chemical change, and electricity and magnetism. Weak and strong forces act inside the nucleus of an atom, so they are not as easily observed at work as are gravitational and electromagnetic forces. The weak force is involved in certain nuclear reactions. The strong nuclear force is involved in close-range holding of the nucleus together. In

FIGURE 2.7 The rate of movement and the direction of movement of this ship are determined by a combination of direction and size of force from each of the tugboats. Which direction are the two tugboats pushing? What evidence would indicate that one tugboat is pushing with a greater force? If the tugboat by the numbers is pushing with a greater force and the back tugboat is keeping the back of the ship from moving, what will happen?

Forces Applied   **Net Force**

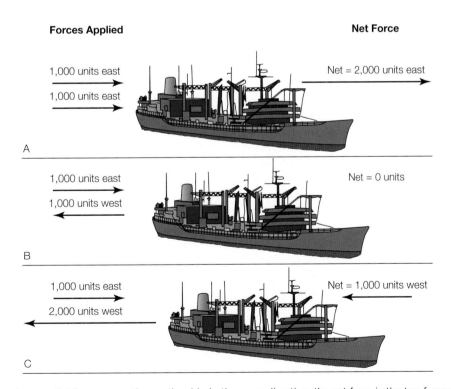

FIGURE 2.8 (*A*) When two parallel forces are acting on the ship in the same direction, the net force is the two forces added together. (*B*) When two forces are opposite and of equal size, the net force is zero. (*C*) When two parallel forces in opposite directions are not of equal size, the net force is the difference in the direction of the larger force.

general, the strong nuclear force between particles inside a nucleus is about 10^2 times stronger than the electromagnetic force and about 10^{39} times stronger than the gravitation force. The fundamental forces are responsible for everything that happens in the universe, and we will learn more about them in chapters on electricity, light, nuclear energy, chemistry, geology, and astronomy.

HORIZONTAL MOTION ON LAND

Everyday experience seems to indicate that Aristotle's idea about horizontal motion on Earth's surface is correct. After all, moving objects that are not pushed or pulled do come to rest in a short period of time. It would seem that an object keeps moving only if a force continues to push it. A moving automobile will slow and come to rest if you turn off the ignition. Likewise, a ball that you roll along the floor will slow until it comes to rest. Is the natural state of an object to be at rest, and is a force necessary to keep an object in motion? This is exactly what people thought until Galileo published his book *Two New Sciences* in 1638, which described his findings about motion. The book had three parts that dealt with uniform motion, accelerated motion, and projectile motion. Galileo described details of simple experiments, measurements, calculations, and thought experiments as he developed definitions and concepts of motion. In one of his thought experiments, Galileo presented an argument against Aristotle's view that a force is needed to keep an object in motion. Galileo imagined an object (such as a ball) moving over a horizontal surface without the force of friction. He concluded that the object would move forever with a constant velocity as long as there was no unbalanced force acting to change the motion.

Why does a rolling ball slow to a stop? You know that a ball will roll farther across a smooth, waxed floor such as a bowling lane than it will across a floor covered with carpet. The rough carpet offers more resistance to the rolling ball. The resistance of the floor friction is shown by a force arrow, F_{floor}, in Figure 2.9. This force, along with the force arrow for air resistance, F_{air}, opposes the forward movement of the ball. Notice the dashed line arrow in part A of Figure 2.9. There is no other force applied to the ball, so the rolling speed decreases until the ball finally comes to a complete stop. Now imagine what force you would need to exert by pushing with your hand, moving along with the ball to keep it rolling at a uniform rate. An examination of the forces in part B of Figure 2.9 can help you determine the amount of force. The force you apply, F_{applied}, must counteract the resistance forces. It opposes the forces that are slowing down the ball as illustrated by the direction of the arrows. To determine how much force you should apply, look at the arrow equation. F_{applied} has the same length as the sum of the two resistance forces, but it is in the opposite direction of the resistance forces. Therefore, the overall force, F_{net}, is zero. The ball continues to roll at a uniform rate when you *balance* the force opposing its motion. It is reasonable, then, that if there were no opposing forces, you would not need to apply a force

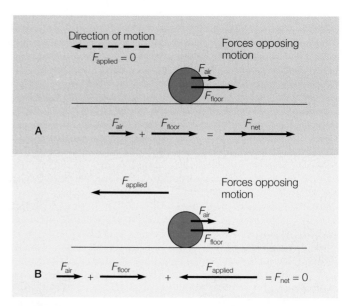

FIGURE 2.9 The following focus in on horizontal forces only: (*A*) This ball is rolling to your left with no forces in the direction of motion. The sum of the force of floor friction (F_{floor}) and the force of air friction (F_{air}) results in a net force opposing the motion, so the ball slows to a stop. (*B*) A force is applied to the moving ball, perhaps by a hand that moves along with the ball. The force applied (F_{applied}) equals the sum of the forces opposing the motion, so the ball continues to move with a constant velocity.

to keep it rolling. This was the kind of reasoning that Galileo did when he discredited the Aristotelian view that a force was necessary to keep an object moving. Galileo concluded that a moving object would continue moving with a constant velocity if no unbalanced forces were applied, that is, if the net force were zero.

It could be argued that the difference in Aristotle's and Galileo's views of forced motion is really a degree of analysis. After all, moving objects on Earth do come to rest unless continuously pushed or pulled. But Galileo's conclusion describes *why* they must be pushed or pulled and reveals the true nature of the motion of objects. Aristotle argued that the natural state of objects is to be at rest, and he tried to explain why objects move. Galileo, on the other hand, argued that it is just as natural for objects to be moving, and he tried to explain why they come to rest. Galileo called the behavior of matter that causes it to persist in its state of motion **inertia**. Inertia is the *tendency of an object to remain in unchanging motion whether actually at rest or moving in the absence of an unbalanced force* (friction, gravity, or whatever). The development of this concept changed the way people viewed the natural state of an object and opened the way for further understandings about motion. Today, it is understood that a spacecraft moving through free space will continue to do so with no unbalanced forces acting on it (Figure 2.10A). An unbalanced force is needed to slow the spacecraft (Figure 2.10B), increase its speed (Figure 2.10C), or change its direction of travel (Figure 2.10D).

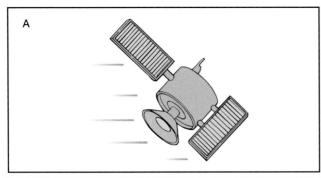

A

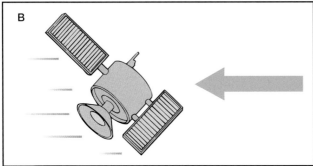

B

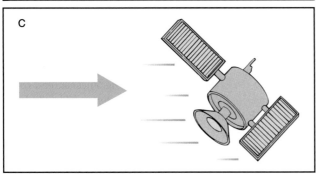

C

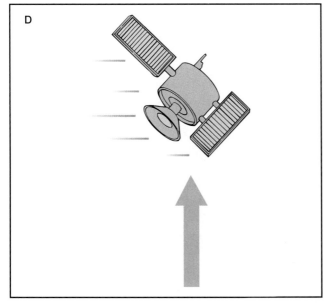

D

FIGURE 2.10 Examine the four illustrations and explain how together they illustrate inertia.

Walk or Run in Rain?

Is it a mistake to run in rain if you want to stay drier? One idea is that you should run because you spend less time in the rain, so you will stay drier. On the other hand, this is true only if the rain lands on the top of your head and shoulders. If you run, you will end up running into more raindrops on the larger surface area of your face, chest, and front of your legs.

Two North Carolina researchers looked into this question with one walking and the other running over a measured distance while wearing cotton sweatsuits. They then weighed their clothing and found that the walking person's sweatsuit weighed more. This means you should run to stay drier.

FALLING OBJECTS

Did you ever wonder what happens to a falling rock during its fall? Aristotle reportedly thought that a rock falls at a uniform speed that is proportional to its weight. Thus, a heavy rock would fall at a faster uniform speed than a lighter rock. As stated in a popular story, Galileo discredited Aristotle's conclusion by dropping a solid iron ball and a solid wooden ball simultaneously from the top of the Leaning Tower of Pisa (Figure 2.11). Both balls, according to the story, hit the ground nearly at the same time. To do this, they would have to fall with the same velocity. In other words, the velocity of a falling object does not depend on its weight. Any difference in freely falling bodies is explainable by air resistance. Soon after the time of Galileo, the air pump was invented. The air pump could be used to remove the air from a glass tube. The effect of air resistance on falling objects could then be demonstrated by comparing how objects fall in the air with how they fall in an evacuated glass tube. You know that a coin falls faster than a feather when they are dropped together in the air. A feather and heavy coin will fall together in the near vacuum of an evacuated glass tube because the effect

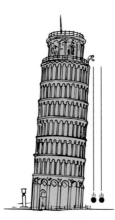

FIGURE 2.11 According to a widespread story, Galileo dropped two objects with different weights from the Leaning Tower of Pisa. They reportedly hit the ground at about the same time, discrediting Aristotle's view that the speed during the fall is proportional to weight.

of air resistance on the feather has been removed. When objects fall toward Earth without considering air resistance, they are said to be in **free fall**. Free fall considers only gravity and neglects air resistance.

 CONCEPTS *Applied*

Falling Bodies

Galileo concluded that all objects fall together, with the same acceleration, when the upward force of air resistance is removed. It would be most difficult to remove air from the room, but it is possible to do some experiments that provide some evidence of how air influences falling objects.

1. Take a sheet of paper and your textbook and drop them side by side from the same height. Note the result.
2. Place the sheet of paper on top of the book and drop them at the same time. Do they fall together?
3. Crumple the sheet of paper into a loose ball and drop the ball and book side by side from the same height.
4. Crumple a sheet of paper into a very tight ball and again drop the ball and book side by side from the same height.

   Explain any evidence you found concerning how objects fall.

Galileo concluded that light and heavy objects fall together in free fall, but he also wanted to know the details of what was going on while they fell. He now knew that the velocity of an object in free fall was *not* proportional to the weight of the object. He observed that the velocity of an object in free fall *increased* as the object fell and reasoned from this that the velocity of the falling object would have to be (1) somehow proportional to the *time* of fall and (2) somehow proportional to the *distance* the object fell. If the time and distance were both related to the velocity of a falling object at a given time and distance, how were they related to one another? To answer this question, Galileo made calculations involving distance, velocity, and time and, in fact, introduced the concept of acceleration. The relationships between these variables are found in the same three equations that you have already learned. Let's see how the equations can be rearranged to incorporate acceleration, distance, and time for an object in free fall.

Step 1: Equation 2.1 gives a relationship between average velocity ($\bar{v}$), distance (d), and time (t). Solving this equation for distance gives

$$d = \bar{v}t$$

Step 2: An object in free fall should have uniformly accelerated motion, so the average velocity could be calculated from equation 2.3,

$$\bar{v} = \frac{v_f + v_i}{2}$$

Substituting this equation in the rearranged equation 2.1, the distance relationship becomes

$$d = \left(\frac{v_f + v_i}{2}\right)(t)$$

Step 3: The initial velocity of a falling object is always zero just as it is dropped, so the v_i can be eliminated,

$$d = \left(\frac{v_f}{2}\right)(t)$$

Step 4: Now you want to get acceleration into the equation in place of velocity. This can be done by solving equation 2.2 for the final velocity (v_f), then substituting. The initial velocity (v_i) is again eliminated because it equals zero.

$$a = \frac{v_f - v_i}{t}$$

$$v_f = at$$

$$d = \left(\frac{at}{2}\right)(t)$$

Step 5: Simplifying, the equation becomes

$$d = \frac{1}{2}at^2$$

equation 2.4

Thus, Galileo reasoned that a freely falling object should cover a distance *proportional to the square of the time of the fall* ($d \propto t^2$). In other words the object should fall 4 times as far in 2 s as in 1 s ($2^2 = 4$), 9 times as far in 3 s ($3^2 = 9$), and so on. Compare this prediction with Figure 2.12.

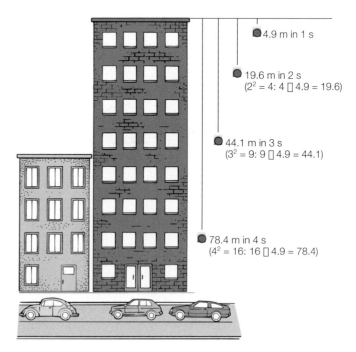

4.9 m in 1 s

19.6 m in 2 s
($2^2 = 4$: 4 ☐ 4.9 = 19.6)

44.1 m in 3 s
($3^2 = 9$: 9 ☐ 4.9 = 44.1)

78.4 m in 4 s
($4^2 = 16$: 16 ☐ 4.9 = 78.4)

FIGURE 2.12 An object dropped from a tall building covers increasing distances with every successive second of falling. The distance covered is proportional to the square of the time of falling ($d \propto t^2$).

Galileo was one of the first to recognize the role of friction in opposing motion. As shown in Figure 2.9, friction with the surface and air friction combine to produce a net force that works against anything that is moving on the surface. This article is about air friction and some techniques that bike riders use to reduce that opposing force—perhaps giving them an edge in a close race.

The bike riders in Box Figure 2.1 are forming a single-file line, called a *pace-line,* because the slipstream reduces the air resistance for a closely trailing rider. Cyclists say that riding in the slipstream of another cyclist will save much of their energy. They can move up to 5 mi/h faster than they would expending the same energy riding alone.

In a sense, riding in a slipstream means that you do not have to push as much air out of your way. It has been estimated that at 20 mi/h, a cyclist must move a little less than half a ton of air out of the way every minute. Along with the problem of moving air out of the way, there are two basic factors related to air resistance. These are (1) a

BOX FIGURE 2.1 The object of the race is to be in the front, to finish first. If this is true, why are these racers forming a single-file line?

turbulent versus a smooth flow of air and (2) the problem of frictional drag. A turbulent flow of air contributes to air resistance because it causes the air to separate slightly on the back side, which increases the pressure on the front of the moving object. This is why racing cars, airplanes, boats, and other racing vehicles are streamlined to a teardroplike shape. This shape is not as

likely to have the lower-pressure-producing air turbulence behind (and resulting greater pressure in front) because it smoothes, or streamlines, the air flow.

The frictional drag of air is similar to the frictional drag that occurs when you push a book across a rough tabletop. You know that smoothing the rough tabletop will reduce the frictional drag on the book. Likewise, the smoothing of a surface exposed to moving air will reduce air friction. Cyclists accomplish this "smoothing" by wearing smooth Lycra clothing and by shaving hair from arm and leg surfaces that are exposed to moving air. Each hair contributes to the overall frictional drag, and removal of the arm and leg hair can thus result in seconds saved. This might provide enough of an edge to win a close race. Shaving legs and arms, together with the wearing of Lycra or some other tight, smooth-fitting garments, are just a few of the things a cyclist can do to gain an edge. Perhaps you will be able to think of more ways to reduce the forces that oppose motion.

Galileo checked this calculation by rolling balls on an inclined board with a smooth groove in it. He used the inclined board to slow the motion of descent in order to measure the distance and time relationships, a necessary requirement since he lacked the accurate timing devices that exist today. He found, as predicted, that the falling balls moved through a distance proportional to the square of the time of falling. This also means that the *velocity of the falling object increased at a constant rate,* as shown in Figure 2.13. Recall that a change of velocity during some time period is called *acceleration.* In other words, a falling object *accelerates* toward the surface of Earth.

Since the velocity of a falling object increases at a constant rate, this must mean that falling objects are *uniformly accelerated* by the force of gravity. *All objects in free fall experience a constant acceleration.* During each second of fall, the object on Earth gains 9.8 m/s (32 ft/s) in velocity. This gain is the acceleration of the falling object, 9.8 m/s^2 (32 ft/s^2).

The acceleration of objects falling toward Earth varies slightly from place to place on the surface because of Earth's shape and spin. The acceleration of falling objects decreases from the poles to the equator and also varies from place to place because Earth's mass is not distributed equally. The value of 9.8 m/s^2 (32 ft/s^2) is an approximation that is fairly close to, but not

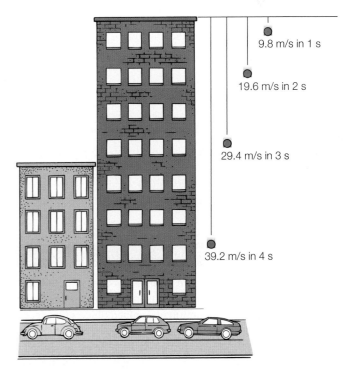

FIGURE 2.13 The velocity of a falling object increases at a constant rate, 9.8 m/s^2.

exactly, the acceleration due to gravity in any particular location. The acceleration due to gravity is important in a number of situations, so the acceleration from this force is given a special symbol, **g.**

EXAMPLE 2.7

A rock that is dropped into a well hits the water in 3.0 s. Ignoring air resistance, how far is it to the water?

SOLUTION 1

The problem concerns a rock in free fall. The time of fall (t) is given, and the problem asks for a distance (d). Since the rock is in free fall, the acceleration due to the force of gravity (g) is implied. The metric value and unit for g is 9.8 m/s^2, and the English value and unit is 32 ft/s^2. You would use the metric g to obtain an answer in meters and the English unit to obtain an answer in feet. Equation 2.4, $d = 1/2\ at^2$, gives a relationship between distance (d), time (t), and average acceleration (a). The acceleration in this case is the acceleration due to gravity (g), so

$$t = 3.0\ \text{s} \qquad d = \frac{1}{2}gt^2 (a = g = 9.8\ \text{m/s}^2)$$
$$g = 9.8\ \text{m/s}^2$$
$$d = ? \qquad d = \frac{1}{2}(9.8\ \text{m/s}^2)(3.0\ \text{s})^2$$
$$= (4.9\ \text{m/s}^2)(9.0\ \text{s}^2)$$
$$= 44\ \frac{\text{m}\cdot\cancel{\text{s}^2}}{\cancel{\text{s}^2}}$$
$$= \boxed{44\ \text{m}}$$

SOLUTION 2

You could do each step separately. Check this solution by a three-step procedure:

1. Find the final velocity, v_f, of the rock from $\bar{v}_f = at$;

2. Calculate the average velocity (v) from the final velocity

$$\bar{v} = \frac{v_f + v_i}{2}$$

then;

3. Use the average velocity ($\bar{v}$) and the time (t) to find distance (d), $d = \bar{v}t$.

Note that the one-step procedure is preferred over the three-step procedure because fewer steps mean fewer possibilities for mistakes.

COMPOUND MOTION

So far we have considered two types of motion: (1) the horizontal, straight-line motion of objects moving on the surface of Earth and (2) the vertical motion of dropped objects that accelerate toward the surface of Earth. A third type of motion occurs when an object is thrown, or projected, into the air. Essentially, such a projectile (rock, football, bullet, golf ball, or whatever) could be directed straight upward as a vertical projection, directed straight out as a horizontal projection, or directed at

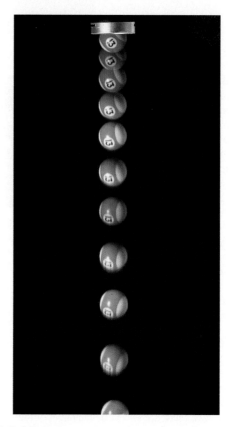

FIGURE 2.14 High-speed, multiflash photograph of a freely falling billiard ball.

some angle between the vertical and the horizontal. Basic to understanding such compound motion is the observation that (1) gravity acts on objects *at all times,* no matter where they are, and (2) the acceleration due to gravity (g) is *independent of any motion* that an object may have.

VERTICAL PROJECTILES

Consider first a ball that you throw straight upward, a vertical projection. The ball has an initial velocity but then reaches a maximum height, stops for an instant, then accelerates back toward Earth. Gravity is acting on the ball throughout its climb, stop, and fall. As it is climbing, the force of gravity is continually reducing its velocity. The overall effect during the climb is deceleration, which continues to slow the ball until the instantaneous stop. The ball then accelerates back to the surface just like a ball that has been dropped (Figure 2.14). If it were not for air resistance, the ball would return with the same speed in the opposite direction that it had initially. The velocity arrows for a ball thrown straight up are shown in Figure 2.15.

HORIZONTAL PROJECTILES

Horizontal projectiles are easier to understand if you split the complete motion into vertical and horizontal parts. Consider, for example, an arrow shot horizontally from a bow. The force of gravity accelerates the arrow downward, giving it an increasing downward velocity as it moves through the air. This increasing

There are two different meanings for the term *free fall*. In physics, *free fall* means the unconstrained motion of a body in a gravitational field, without considering air resistance. Without air resistance, all objects are assumed to accelerate toward the surface at 9.8 m/s².

In the sport of skydiving, *free fall* means falling within the atmosphere without a drag-producing device such as a parachute. Air provides a resisting force that opposes the motion of a falling object, and the net force is the difference between the downward force (weight) and the upward force of air resistance. The weight of the falling object depends on the mass and acceleration from gravity, and this is the force down-ward. The resisting force is determined by at least two variables: (1) the area of the object exposed to the airstream and (2) the speed of the falling object. Other variables such as streamlining, air temperature, and turbulence play a role, but the greatest effect seems to be from exposed area and the increased resistance as speed increases.

A skydiver's weight is constant, so the downward force is constant. Modern skydivers typically free-fall from about 3,650 m (about 12,000 ft) above the ground until about 750 m (about 2,500 ft), where they open their parachutes. After jumping from the plane, the diver at first accelerates toward the surface, reaching speeds up to about 185 to 210 km/h (about 115 to 130 mi/h). The air resistance increases with increased speed and the net force becomes less and less. Eventually, the downward weight force will be balanced by the upward air resistance force, and the net force becomes zero. The person now falls at a constant speed, and we say the terminal velocity has been reached. It is possible to change your body position to vary your rate of fall up or down to 32 km/h (about 20 mi/h). However, by diving or "standing up" in free fall, experienced skydivers can reach speeds of up to 290 km/h (about 180 mi/h). The record free fall speed, done without any special equipment, is 517 km/h (about 321 mi/h). Once the parachute opens, a descent rate of about 16 km/h (about 10 mi/h) is typical.

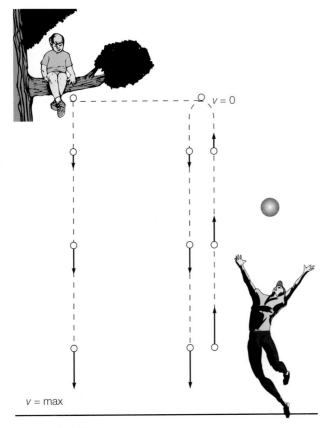

FIGURE 2.15 On its way up, a vertical projectile is slowed by the force of gravity until an instantaneous stop; then it accelerates back to the surface, just as another ball does when dropped from the same height. The straight up and down moving ball has been moved to the side in the sketch so we can see more clearly what is happening. Note that the falling ball has the same speed in the opposite direction that it had on the way up.

downward velocity is shown in Figure 2.16 as increasingly longer velocity arrows (v_v). There are no forces in the horizontal direction if you can ignore air resistance, so the horizontal velocity of the arrow remains the same, as shown by the v_h velocity arrows. The combination of the increasing vertical (v_v) motion and the unchanging horizontal (v_h) motion causes the arrow to follow a curved path until it hits the ground.

An interesting prediction that can be made from the shot arrow analysis is that an arrow shot horizontally from a bow will hit the ground at the same time as a second arrow that is simply dropped from the same height (Figure 2.16). Would this be true of a bullet dropped at the same time as one fired horizontally from a rifle? The answer is yes; both bullets would hit the ground at the same time. Indeed, without air resistance, all the bullets and arrows should hit the ground at the same time if dropped or shot from the same height.

Golf balls, footballs, and baseballs are usually projected upward at some angle to the horizon. The horizontal motion of these projectiles is constant as before because there are no horizontal forces involved. The vertical motion is the same as that of a ball projected directly upward. The combination of these two motions causes the projectile to follow a curved path called a *parabola*, as shown in Figure 2.17. The next time you have the opportunity, observe the path of a ball that has been projected at some angle. Note that the second half of the path is almost a reverse copy of the first half. If it were not for air resistance, the two values of the path would be exactly the same. Also note the distance that the ball travels as compared to the angle of projection. An angle of projection of 45° results in the maximum distance of travel if air resistance is ignored and if the launch point and the landing are at the same elevation.

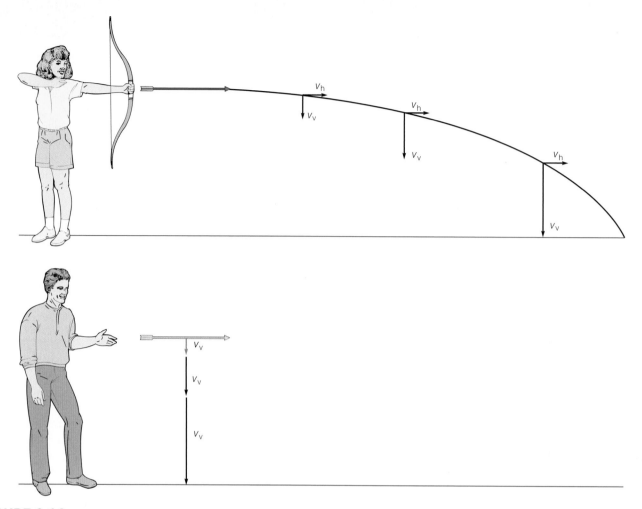

FIGURE 2.16 A horizontal projectile has the same horizontal velocity throughout the fall as it accelerates toward the surface, with the combined effect resulting in a curved path. Neglecting air resistance, an arrow shot horizontally will strike the ground at the same time as one dropped from the same height above the ground, as shown here by the increasing vertical velocity arrows.

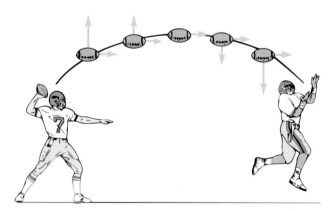

FIGURE 2.17 A football is thrown at some angle to the horizon when it is passed downfield. Neglecting air resistance, the horizontal velocity is a constant, and the vertical velocity decreases, then increases, just as in the case of a vertical projectile. The combined motion produces a parabolic path. Contrary to statements by sportscasters about the abilities of certain professional quarterbacks, it is impossible to throw a football with a "flat trajectory" because it begins to accelerate toward the surface as soon as it leaves the quarterback's hand.

THREE LAWS OF MOTION

In the previous sections, you learned how to describe motion in terms of distance, time, velocity, and acceleration. In addition, you learned about different kinds of motion, such as straight-line motion, the motion of falling objects, and the compound motion of objects projected up from the surface of Earth. You were also introduced, in general, to two concepts closely associated with motion: (1) that objects have inertia, a tendency to resist a change in motion and (2) that forces are involved in a change of motion.

The relationship between forces and a change of motion is obvious in many everyday situations (Figure 2.18). When a car, bus, or plane starts moving, you feel a force on your back. Likewise, you feel a force on the bottoms of your feet when an elevator starts moving upward. On the other hand, you seem to be forced toward the dashboard if a car stops quickly, and it feels as if the floor pulls away from your feet when an elevator drops rapidly. These examples all involve patterns between forces and motion, patterns that can be quantified, conceptualized,

FIGURE 2.18 In a moving airplane, you feel forces in many directions when the plane changes its motion. You cannot help but notice the forces involved when there is a change of motion.

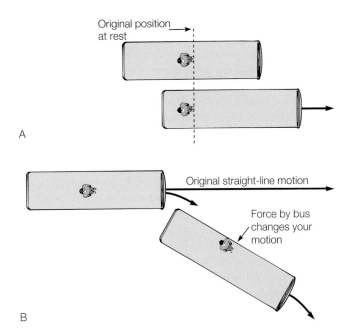

FIGURE 2.19 Top view of a person standing in the aisle of a bus. (*A*) The bus is at rest and then starts to move forward. Inertia causes the person to remain in the original position, appearing to fall backward. (*B*) The bus turns to the right, but inertia causes the person to retain the original straight-line motion until forced in a new direction by the side of the bus.

and used to answer questions about why things move or stand still. These patterns are the subject of Newton's three laws of motion.

NEWTON'S FIRST LAW OF MOTION

Newton's first law of motion is also known as the *law of inertia* and is very similar to one of Galileo's findings about motion. Recall that Galileo used the term *inertia* to describe the tendency of an object to resist changes in motion. Newton's first law describes this tendency more directly. In modern terms (not Newton's words), the **first law of motion** is as follows:

> **Every object retains its state of rest or its state of uniform straight-line motion unless acted upon by an unbalanced force.**

This means that an object at rest will remain at rest unless it is put into motion by an unbalanced force; that is, the net force must be greater than zero. Likewise, an object moving with uniform straight-line motion will retain that motion unless a net force causes it to speed up, slow down, or change its direction of travel. Thus, Newton's first law describes the tendency of an object to resist *any* change in its state of motion.

Think of Newton's first law of motion when you ride standing in the aisle of a bus. The bus begins to move, and you, being an independent mass, tend to remain at rest. You take a few steps back as you tend to maintain your position relative to the ground outside. You reach for a seat back or some part of the bus. Once you have a hold on some part of the bus, it supplies the forces needed to give you the same motion as the bus and you no longer find it necessary to step backward. You now have the same motion as the bus, and no forces are involved, at least until the bus goes around a curve. You now feel a tendency to move to the side of the bus. The bus has changed its straight-line motion, but you, again being an independent mass, tend to move straight ahead. The side of the seat forces you into following the curved motion of the bus. The forces you feel

when the bus starts moving or turning are a result of your tendency to remain at rest or follow a straight path until forces correct your motion so that it is the same as that of the bus (Figure 2.19).

 CONCEPTS *Applied*

First Law Experiment

Place a small ball on a flat part of the floor in a car, SUV, or pickup truck. First, predict what will happen to the ball in each of the following situations: (1) The vehicle moves forward from a stopped position. (2) The vehicle is moving at a constant speed. (3) The vehicle is moving at a constant speed, then turns to the right. (4) The vehicle is moving at a constant speed, then comes to a stop. Now, test your predictions, and then explain each finding in terms of Newton's first law of motion.

NEWTON'S SECOND LAW OF MOTION

Newton had successfully used Galileo's ideas to describe the nature of motion. Newton's first law of motion explains that any object, once started in motion, will continue with a constant velocity in a straight line unless a force acts on the moving object. This law not only describes motion but establishes the role of a force as well. A change of motion is therefore *evidence* of the action of net force. The association of forces and a change of motion is common in your everyday experience. You have felt forces on your back in an accelerating automobile, and you have felt other forces as the automobile turns or stops. You have also learned about gravitational forces that accelerate objects toward the surface of Earth. Unbalanced forces and acceleration are involved in any change of motion. Newton's second law of motion is a relationship between *net force, acceleration,* and *mass* that describes the cause of a change of motion.

Consider the motion of you and a bicycle you are riding. Suppose you are riding your bicycle over level ground in a straight line at 10 miles per hour. Newton's first law tells you that you will continue with a constant velocity in a straight line as long as no external, unbalanced force acts on you and the bicycle. The force that you *are* exerting on the pedals seems to equal some external force that moves you and the bicycle along (more on this later). The force exerted as you move along is needed to *balance* the resisting forces of tire friction and air resistance. If these resisting forces were removed, you would not need to exert any force at all to continue moving at a constant velocity. The net force is thus the force you are applying minus the forces from tire friction and air resistance. The *net force* is therefore zero when you move at a constant speed in a straight line (Figure 2.20).

If you now apply a greater force on the pedals, the *extra* force you apply is unbalanced by friction and air resistance. Hence, there will be a net force greater than zero, and you will accelerate. You will accelerate during, and *only* during, the time that the net force is greater than zero. Likewise, you will slow down if you apply a force to the brakes, another kind of resisting friction. A third way to change your velocity is to apply a force on the handlebars, changing the direction of your velocity. Thus, *unbalanced forces* on you and your bicycle produce an *acceleration.*

Starting a bicycle from rest suggests a relationship between force and acceleration. You observe that the harder you push on the pedals, the greater your acceleration. Recall that when quantities increase or decrease together in the same ratio, they

FIGURE 2.21 More mass results in less acceleration when the same force is applied. With the same force applied, the riders and bike with twice the mass will have half the acceleration, with all other factors constant. Note that the second rider is not pedaling.

are said to be *directly proportional.* The acceleration is therefore directly proportional to the net force applied.

Suppose that your bicycle has two seats, and you have a friend who will ride with you but not pedal. Suppose also that the addition of your friend on the bicycle will double the mass of the bike and riders. If you use the same net force as before, the bicycle will undergo a much smaller acceleration. In fact, with all other factors equal, doubling the mass and applying the same extra force will produce an acceleration of only half as much (Figure 2.21). An even more massive friend would reduce the acceleration even more. Recall that when a relationship between two quantities shows that one quantity increases as another decreases, in the same ratio, the quantities are said to be *inversely proportional.* The acceleration of an object is therefore inversely proportional to its mass.

If we express force in appropriate units, we can combine these relationships as an equation,

$$a = \frac{F}{m}$$

By solving for *F,* we rearrange the equation into the form in which it is most often expressed,

$$F = ma$$

equation 2.5

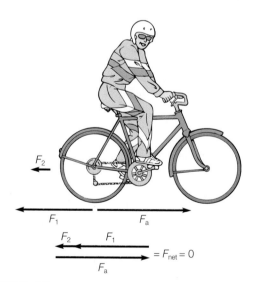

FIGURE 2.20 At a constant velocity, the force of tire friction (F_1) and the force of air resistance (F_2) have a sum that equals the force applied (F_a). The net force is therefore 0.

In the metric system, you can see that the units for force will be the units for mass (m) times acceleration (a). The unit for mass is kg and the unit for acceleration is m/s^2. The combination of these units, (kg)(m/s^2), is a unit of force called the **newton** (N) in honor of Isaac Newton. So,

$$1 \text{ newton} = 1\text{N} = 1\frac{\text{kg·m}}{\text{s}^2}$$

Newton's second law of motion is the essential idea of his work on motion. According to this law, there is always a relationship between the acceleration, a net force, and the mass of an object. Implicit in this statement are three understandings: (1) that we are talking about the net force, meaning total external force acting on an object, (2) that the motion statement is concerned with acceleration, not velocity, and (3) that the mass does not change unless specified.

The acceleration of an object depends on *both* the *net force applied* and the *mass* of the object. The **second law of motion** is as follows:

> **The acceleration of an object is directly proportional to the net force acting on it and inversely proportional to the mass of the object.**

Until now, equations were used to *describe properties* of matter such as density, velocity, and acceleration. This is your first example of an equation that is used to *define a concept,* specifically the concept of what is meant by a force. Since the concept is defined by specifying a measurement procedure, it is also an example of an *operational definition.* You are told not only what a newton of force is but also how to go about measuring it. Notice that the newton is defined in terms of mass measured in kg and acceleration measured in m/s^2. Any other units must be converted to kg and m/s^2 before a problem can be solved for newtons of force.

EXAMPLE 2.8

A 60 kg bicycle and rider accelerate at 0.5 m/s^2. How much extra force was applied?

SOLUTION

The mass (m) of 60 kg and the acceleration (a) of 0.5 m/s^2 are given. The problem asked for the extra force (F) needed to give the mass the acquired acceleration. The relationship is found in equation 2.5, $F = ma$.

$$m = 60 \text{ kg}$$
$$a = 0.5 \frac{\text{m}}{\text{s}^2}$$
$$F = ?$$

$$F = ma$$
$$= (60\text{kg})\left(0.5 \frac{\text{m}}{\text{s}^2}\right)$$
$$= (60)(0.5)(\text{kg})\left(\frac{\text{m}}{\text{s}^2}\right)$$
$$= 30 \frac{\text{kg.m}}{\text{s}^2}$$
$$= \boxed{30 \text{ N}}$$

An *extra* force of 30 N beyond that required to maintain constant speed must be applied to the pedals for the bike and rider to maintain an acceleration of 0.5 m/s^2. (Note that the units kg·m/s^2 form the definition of a newton of force, so the symbol N is used.)

EXAMPLE 2.9

What is the acceleration of a 20 kg cart if the net force on it is 40 N? (Answer: 2 m/s^2)

 CONCEPTS *Applied*

Second Law Experiment

Tie one end of a string to a book and the other end to a large rubber band. With your index finger curled in the loop of the rubber band, pull the book across a smooth tabletop. How much the rubber band stretches will provide a rough estimate of the force you are applying. (1) Pull the book with a constant velocity across the tabletop. Compare the force required for different constant velocities. (2) Accelerate the book at different rates. Compare the force required to maintain the different accelerations. (3) Use a different book with a greater mass and again accelerate the book at different rates. How does more mass change the results?

Based on your observations, can you infer a relationship between force, acceleration, and mass?

WEIGHT AND MASS

What is the meaning of weight—is it the same concept as mass? Weight is a familiar concept to most people, and in everyday language, the word is often used as having the same meaning as mass. In physics, however, there is a basic difference between weight and mass, and this difference is very important in Newton's explanation of motion and the causes of motion.

Mass is defined as the property that determines how much an object resists a change in its motion. The greater the mass, the greater the *inertia,* or resistance to change in motion. Consider, for example, that it is easier to push a small car into motion than to push a large truck into motion. The truck has more mass and therefore more inertia. Newton originally defined mass as the "quantity of matter" in an object, and this definition is intuitively appealing. However, Newton needed to measure inertia because of its obvious role in motion, and he redefined mass as a measure of inertia.

You could use Newton's second law to measure a mass by exerting a force on the mass and measuring the resulting acceleration. This is not very convenient, so masses are usually measured on a balance by comparing the force of gravity acting on a standard mass compared to the force of gravity acting on the unknown mass.

The force of gravity acting on a mass is the *weight* of an object. Weight is a force and has different units (N) than mass (kg). Since weight is a measure of the force of gravity acting on an object, the force can be calculated from Newton's second law of motion,

$$F = ma$$

or

downward force = (mass)(acceleration due to gravity)

or

$$\text{weight} = (\text{mass})\,(g)$$
$$\text{or} \quad w = mg$$

equation 2.6

You learned in the section on falling objects that g is the symbol used to represent acceleration due to gravity. Near Earth's surface, g has an approximate value of 9.8 m/s². To understand how g is applied to an object that is not moving, consider a ball you are holding in your hand. By supporting the weight of the ball, you hold it stationary, so the upward force of your hand and the downward force of the ball (its weight) must add to a net force of zero. When you let go of the ball, the gravitational force is the only force acting on the ball. The ball's weight is then the net force that accelerates it at g, the acceleration due to gravity. Thus, $F_{net} = w = ma = mg$. The weight of the ball never changes in a given location, so its weight is always equal to $w = mg$, even if the ball is not accelerating.

In the metric system, *mass* is measured in kilograms. The acceleration due to gravity, g, is 9.8 m/s². According to equation 2.6, weight is mass times acceleration. A kilogram multiplied by an acceleration measured in m/s² results in kg·m/s², a unit you now recognize as a force called a newton. The *unit of weight* in the metric system is therefore the *newton* (N).

In the English system, the pound is the unit of *force*. The acceleration due to gravity, g, is 32 ft/s². The force unit of a pound is defined as the force required to accelerate a unit of mass called the *slug*. Specifically, a force of 1.0 lb will give a 1.0 slug mass an acceleration of 1.0 ft/s².

The important thing to remember is that *pounds* and *newtons* are units of *force* (Table 2.1). A *kilogram,* on the other hand, is a measure of *mass*. Thus, the English unit of 1.0 lb is comparable to the metric unit of 4.5 N (or 0.22 lb is equivalent to 1.0 N). Conversion tables sometimes show how to convert from pounds (a unit of weight) to kilograms (a unit of mass). This is possible because weight and mass

are proportional in a given location on the surface of Earth. Using conversion factors from inside the front cover of this book, see if you can express your weight in pounds and newtons and your mass in kg.

EXAMPLE 2.10

What is the weight of a 60.0 kg person on the surface of Earth?

SOLUTION

A mass (m) of 60.0 kg is given, and the acceleration due to gravity (g) 9.8 m/s² is implied. The problem asked for the weight (w). The relationship is found in equation 2.6, $w = mg$, which is a form of $F = ma$.

$$m = 60.0 \text{ kg} \qquad w = mg$$
$$g = 9.8 \frac{m}{s^2} \qquad = (60.0 \text{ kg}) \left(9.8 \frac{m}{s^2}\right)$$
$$w = ? \qquad = (60.0)(9.8)\,(\text{kg})\left(\frac{m}{s^2}\right)$$
$$= 588 \frac{\text{kg·m}}{s^2}$$
$$= \boxed{590 \text{ N}}$$

EXAMPLE 2.11

A 60.0 kg person weighs 100.0 N on the Moon. What is the acceleration of gravity on the Moon? (Answer: 1.67 m/s²)

NEWTON'S THIRD LAW OF MOTION

Newton's first law of motion states that an object retains its state of motion when the net force is zero. The second law states what happens when the net force is *not* zero, describing how an object with a known mass moves when a given force is applied. The two laws give one aspect of the concept of a force; that is, if you observe that an object starts moving, speeds up, slows down, or changes its direction of travel, you can conclude that an unbalanced force is acting on the object. Thus, any change in the state of motion of an object is *evidence* that an unbalanced force has been applied.

Newton's third law of motion is also concerned with forces. First, consider where a force comes from. A force is always produced by the interaction of two objects. Sometimes we do not know what is producing forces, but we do know that they always come in pairs. Anytime a force is exerted, there is always a matched and opposite force that occurs at the same time. For example, if you push on the wall, the wall pushes back with an equal and opposite force. The two forces are opposite and balanced, and you know this because $F = ma$ and neither you nor the wall accelerated. If the acceleration is zero, then you know from $F = ma$ that the net force is zero (zero equals zero). Note also that the two forces were between two different objects, you and the wall. Newton's third law always describes what happens between two different objects. To simplify the many interactions that occur on Earth, consider a spacecraft in space. According to Newton's second law ($F = ma$), a force must be applied to change the state of motion of the spacecraft. What is a possible source of such a force? Perhaps an

TABLE 2.1

Units of mass and weight in the metric and English systems of measurement

	Mass	×	Acceleration	=	Force
Metric system	kg	×	$\dfrac{m}{s^2}$	=	$\dfrac{\text{kg·m}}{s^2}$ (newton)
English system	$\left(\dfrac{\text{lb}}{\text{ft/s}^2}\right)$	×	$\dfrac{\text{ft}}{s^2}$	=	lb (pound)

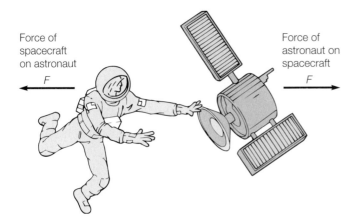

Force of
spacecraft
on astronaut
F ⟵

Force of
astronaut on
spacecraft
⟶ F

FIGURE 2.22 Forces occur in matched pairs that are equal in magnitude and opposite in direction.

astronaut pushes on the spacecraft for 1 second. The spacecraft would accelerate *during* the application of the force, then move away from the original position at some constant velocity. The astronaut would also move away from the original position but in the opposite direction (Figure 2.22). A *single* force *does not exist* by itself. There is always a matched and opposite force that occurs at the same time. Thus, the astronaut exerted a momentary force on the spacecraft, but the spacecraft evidently exerted a momentary force back on the astronaut as well, for the astronaut moved away from the original position in the opposite direction. Newton did not have astronauts and spacecraft to think about, but this is the kind of reasoning he did when he concluded that forces always occur in matched pairs that are equal and opposite. Thus, the **third law of motion** is as follows:

> **Whenever two objects interact, the force exerted on one object is equal in size and opposite in direction to the force exerted on the other object.**

The third law states that forces always occur in matched pairs that act in opposite directions and on two *different* bodies. You could express this law with symbols as

$$F_{\text{A due to B}} = F_{\text{B due to A}}$$

equation 2.7

where the force on the astronaut, for example, would be "A due to B," and the force on the satellite would be "B due to A."

Sometimes the third law of motion is expressed as follows: "For every action, there is an equal and opposite reaction," but this can be misleading. Neither force is the cause of the other. The forces are at every instant the cause of each other, and they appear and disappear at the same time. If you are going to describe the force exerted on a satellite by an astronaut, then you must realize that there is a simultaneous force exerted on the astronaut by the satellite. The forces (astronaut on satellite and satellite on astronaut) are equal in magnitude but opposite in direction.

Perhaps it would be more common to move a satellite with a small rocket. A satellite is maneuvered in space by firing a rocket in the direction opposite to the direction someone wants to move the satellite. Exhaust gases (or compressed gases) are

FIGURE 2.23 The football player's foot is pushing against the ground, but it is the ground pushing against the foot that accelerates the player forward to catch a pass.

accelerated in one direction and exert an equal but opposite force on the satellite that accelerates it in the opposite direction. This is another example of the third law.

Consider how the pairs of forces work on Earth's surface. You walk by pushing your feet against the ground (Figure 2.23). Of course you could not do this if it were not for friction. You would slide as on slippery ice without friction. But since friction does exist, you exert a backward horizontal force on the ground, and, as the third law explains, the ground exerts an equal and opposite force on you. You accelerate forward from the net force as explained by the second law. If Earth had the same mass as you, however, it would accelerate backward at the same rate that you were accelerated forward. Earth is much more massive than you, however, so any acceleration of Earth is a vanishingly small amount. The overall effect is that you are accelerated forward by the force the ground exerts on you.

Return now to the example of riding a bicycle that was discussed previously. What is the source of the *external* force that accelerates you and the bike? Pushing against the pedals is not external to you and the bike, so that force will *not* accelerate you and the bicycle forward. This force is transmitted through the bike mechanism to the rear tire, which pushes against the ground. It is the ground exerting an equal and opposite force against the system of you and the bike that accelerates you forward. You must consider the forces that act on the system of the bike and you before you can apply $F = ma$. The only forces that will affect the forward motion of the bike system are the force of the ground pushing it forward and the frictional forces that oppose the forward motion. This is another example of the third law.

EXAMPLE 2.12

A 60.0 kg astronaut is freely floating in space and pushes on a freely floating 120.0 kg spacecraft with a force of 30.0 N for 1.50 s. (a) Compare the forces exerted on the astronaut and the spacecraft, and (b) compare the acceleration of the astronaut to the acceleration of the spacecraft.

SOLUTION

(a) According to Newton's third law of motion (equation 2.7),

$$F_{A \text{ due to } B} = F_{B \text{ due to } A}$$
$$30.0 \text{ N} = 30.0 \text{ N}$$

Both feel a 30.0 N force for 1.50 s but in opposite directions.

(b) Newton's second law describes a relationship between force, mass, and acceleration, $F = ma$.
For the astronaut:

$m = 60.0 \text{ kg}$ $\quad F = ma \quad \therefore \quad a = \dfrac{F}{m}$

$F = 30.0 \text{ N}$

$a = ?$

$$a = \frac{30.0 \frac{\text{kg·m}}{\text{s}^2}}{60.0 \text{ kg}}$$

$$= \frac{30.0}{60.0}\left(\frac{\text{kg·m}}{\text{s}^2}\right)\left(\frac{1}{\text{kg}}\right)$$

$$= 0.500 \frac{\cancel{\text{kg}}\text{·m}}{\cancel{\text{kg}}\text{·s}^2} = \boxed{0.500 \frac{\text{m}}{\text{s}^2}}$$

For the spacecraft:

$m = 120.0 \text{ kg}$ $\quad F = ma \quad \therefore \quad a = \dfrac{F}{m}$

$F = 30.0 \text{ N}$

$a = ?$

$$a = \frac{30.0 \frac{\text{kg·m}}{\text{s}^2}}{120.0 \text{ kg}}$$

$$= \frac{30.0}{120.0}\left(\frac{\text{kg·m}}{\text{s}^2}\right)\left(\frac{1}{\text{kg}}\right)$$

$$= 0.250 \frac{\cancel{\text{kg}}\text{·m}}{\cancel{\text{kg}}\text{·s}^2} = \boxed{0.250 \frac{\text{m}}{\text{s}^2}}$$

EXAMPLE 2.13

After the interaction and acceleration between the astronaut and spacecraft described in example 2.12, they both move away from their original positions. What is the new speed for each? (Answer: astronaut $v_f = 0.750$ m/s; spacecraft $v_f = 0.375$ m/s) (Hint: $v_f = at + v_i$)

MOMENTUM

Sportscasters often refer to the *momentum* of a team, and newscasters sometimes refer to an election where one of the candidates has *momentum*. Both situations describe a competition where one side is moving toward victory and it is difficult to stop. It seems appropriate to borrow this term from the physical sciences because momentum is a property of movement. It takes a longer time to stop something from moving when it has a lot of momentum. The physical science concept of momentum is closely related to Newton's laws of motion. **Momentum** (p) is defined as the product of the mass (m) of an object and its velocity (v),

$$\text{momentum} = \text{mass} \times \text{velocity}$$

or

$$p = mv$$

equation 2.8

The astronaut in example 2.12 had a mass of 60.0 kg and a velocity of 0.750 m/s as a result of the interaction with the spacecraft. The resulting momentum was therefore (60.0 kg) (0.750 m/s), or 45.0 kg·m/s. As you can see, the momentum would be greater if the astronaut had acquired a greater velocity or if the astronaut had a greater mass and acquired the same velocity. Momentum involves both the inertia and the velocity of a moving object.

CONSERVATION OF MOMENTUM

Notice that the momentum acquired by the spacecraft in example 2.12 is *also* 45.0 kg·m/s. The astronaut gained a certain momentum in one direction, and the spacecraft gained the *very same momentum in the opposite direction*. Newton originally defined the second law in terms of a rate of change of momentum being proportional to the net force acting on an object. Since the third law explains that the forces exerted on both the astronaut and the spacecraft were equal and opposite, you would expect both objects to acquire equal momentum in the opposite direction. This result is observed any time objects in a system interact and the only forces involved are those between the interacting objects (Figure 2.24). This statement leads to a particular kind of relationship called a *law of conservation*. In this case, the law applies to momentum and is called the **law of conservation of momentum:**

> **The total momentum of a group of interacting objects remains the same in the absence of external forces.**

Conservation of momentum, energy, and charge are among examples of conservation laws that apply to everyday situations. These situations always illustrate two understandings: that (1) each conservation law is an expression that describes a physical principle that can be observed and (2) each law holds regardless of the details of an interaction or how it took place. Since the conservation laws express something that always occurs, they tell us what might be expected to happen and what might be expected not to happen in a given situation. The conservation laws also allow unknown quantities to be found by analysis. The law of conservation of momentum, for example, is useful in analyzing motion in simple systems of collisions such as those of billiard balls, automobiles, or railroad cars. It is

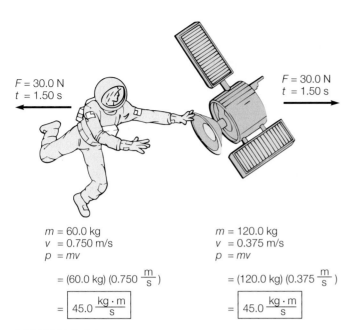

$F = 30.0$ N
$t = 1.50$ s

$F = 30.0$ N
$t = 1.50$ s

$m = 60.0$ kg
$v = 0.750$ m/s
$p = mv$

$= (60.0$ kg$) (0.750 \frac{\text{m}}{\text{s}})$

$= \boxed{45.0 \frac{\text{kg} \cdot \text{m}}{\text{s}}}$

$m = 120.0$ kg
$v = 0.375$ m/s
$p = mv$

$= (120.0$ kg$) (0.375 \frac{\text{m}}{\text{s}})$

$= \boxed{45.0 \frac{\text{kg} \cdot \text{m}}{\text{s}}}$

FIGURE 2.24 Both the astronaut and the spacecraft received a force of 30.0 N for 1.50 s when they pushed on each other. Both then have a momentum of 45.0 kg·m/s in the opposite direction. This is an example of the law of conservation of momentum.

also useful in measuring action and reaction interactions, as in rocket propulsion, where the backward momentum of the exhaust gases equals the momentum given to the rocket in the opposite direction. When this is done, momentum is always found to be conserved.

The firing of a bullet from a rifle and the concurrent "kick" or recoil of the rifle is often used as an example of conservation of momentum where the interaction between objects results in momentum in opposite directions (Figure 2.25). When the rifle is fired, the exploding gunpowder propels the bullet with forward momentum. At the same time, the force from the exploding gunpowder pushes the rifle backward with a momentum opposite that of the bullet. The bullet moves forward with a momentum of $(mv)_b$ and the rifle moves in an opposite direction to the bullet, so its momentum is $-(mv)_r$. According

$-(mv)_r$ $=$ $(mv)_b$

FIGURE 2.25 A rifle and bullet provide an example of conservation of momentum. Before being fired, a rifle and bullet have a total momentum ($p = mv$) of zero since there is no motion. When fired, the bullet is then propelled in one direction with a forward momentum $(mv)_b$. At the same time, the rifle is pushed backward with a momentum opposite to that of the bullet, so its momentum is shown with a minus sign, or $-(mv)_r$. Since $(mv)_b$ plus $-(mv)_r$ equals zero, the total momentum of the rifle and bullet is zero after as well as before the rifle is fired.

to the law of conservation of momentum, the momentum of the bullet $(mv)_b$ must equal the momentum of the rifle $-(mv)_r$ in the opposite direction. If the bullet and rifle had the same mass, they would each move with equal velocities when the rifle was fired. The rifle is much more massive than a bullet, however, so the bullet has a much greater velocity than the rifle. The momentum of the rifle is nonetheless equal to the momentum of the bullet, and the recoil can be significant if the rifle is not held firmly against the shoulder. When held firmly against the shoulder, the rifle and the person's body are one object. The increased mass results in a proportionally smaller recoil velocity.

EXAMPLE 2.14

A 20,000 kg railroad car is coasting at 3 m/s when it collides and couples with a second, identical car at rest. What is the resulting speed of the combined cars?

SOLUTION

Moving car $\rightarrow m_1 = 20,000$ kg, $v_1 = 3$ m/s

Second car $\rightarrow m_2 = 20,000$ kg, $v_2 = 0$

Combined cars $\rightarrow v_{1\&2} = ?$ m/s

Since momentum is conserved, the total momentum of the cars should be the same before and after the collision. Thus,

$$\text{momentum before} = \text{momentum after}$$
$$\text{car 1} + \text{car 2} = \text{coupled cars}$$
$$m_1v_1 + m_2v_2 = (m_1 + m_2)v_{1\&2}$$
$$v_{1\&2} = \frac{m_1v_1}{(m_1 + m_2)}$$
$$v_{1\&2} = \frac{(20,000 \text{ kg}) \left(3 \frac{\text{m}}{\text{s}}\right)}{(20,000 \text{ kg}) + (20,000 \text{ kg})}$$
$$= \frac{20,000 \text{ kg}. \, 3 \frac{\text{m}}{\text{s}}}{40,000 \text{ kg}}$$
$$= 0.5 \times 3 \frac{\text{kg} \cdot \text{m}}{\text{s}} \times \frac{1}{\text{kg}}$$
$$= 1.5 \frac{\text{m}}{\text{s}}$$
$$= \boxed{2 \frac{\text{m}}{\text{s}}}$$

(Answer rounded to one significant figure.)
Car 2 had no momentum with a velocity of zero, so m_2v_2 on the left side of the equation equals zero. When the cars couple, the mass is doubled $(m + m)$, and the velocity of the coupled cars will be 2 m/s.

EXAMPLE 2.15

A student and her rowboat have a combined mass of 100.0 kg. Standing in the motionless boat in calm water, she tosses a 5.0 kg rock out the back of the boat with a velocity of 5.0 m/s. What will be the resulting speed of the boat? (Answer: 0.25 m/s)

IMPULSE

Have you ever heard that you should "follow through" when hitting a ball? When you follow through, the bat is in contact with the ball for a longer period of time. The force of the hit is important, of course, but both the force and how long the force is applied determine the result. The product of the force and the time of application is called **impulse.** This quantity can be expressed as

$$\text{impulse} = Ft$$

where F is the force applied during the time of contact t. The impulse you give the ball determines how fast the ball will move and thus how far it will travel.

Impulse is related to the change of motion of a ball of a given mass, so the change of momentum (mv) is brought about by the impulse. This can be expressed as

$$\text{change of momentum} = (\text{applied force})(\text{time of contact})$$
$$\Delta p = Ft$$

equation 2.9

where Δp is a change of momentum. You "follow through" while hitting a ball in order to increase the contact time. If the same force is used, a longer contact time will result in a greater impulse. A greater impulse means a greater change of momentum, and since the mass of the ball does not change, the overall result is a moving ball with a greater velocity. This means following through will result in more distance from hitting the ball with the same force. That's why it is important to follow through when you hit the ball.

Now consider bringing a moving object to a stop by catching it. In this case, the mass and the velocity of the object are fixed at the time you catch it, and there is nothing you can do about these quantities. The change of momentum is equal to the impulse, and the force and time of force application *can* be manipulated. For example, consider how you would catch a raw egg that is tossed to you. You would probably move your hands with the egg as you caught it, increasing the contact time. Increasing the contact time has the effect of reducing the force since $\Delta p = Ft$. You change the force applied by increasing the contact time, and, hopefully, you reduce the force sufficiently so the egg does not break.

Contact time is also important in safety. Automobile airbags, the padding in elbow and knee pads, and the plastic barrels off the highway in front of overpass supports are examples of designs intended to increase the contact time. Again, increasing the contact time reduces the force since $\Delta p = Ft$. The impact force is reduced and so are the injuries. Think about this the next time you see a car that was crumpled and bent by a collision. The driver and passengers were probably saved from more serious injuries since more time was involved in stopping the car that crumpled. A car that crumples is a safer car in a collision.

FORCES AND CIRCULAR MOTION

Consider a communications satellite that is moving at a uniform speed around Earth in a circular orbit. According to the first law of motion, there *must be* forces acting on the satellite, since it does *not* move off in a straight line. The second law of motion also indicates forces, since an unbalanced force is required to change the motion of an object.

Recall that acceleration is defined as a rate of change in velocity and that velocity has both magnitude and direction. The velocity is changed by a change in speed, direction, or both speed and direction. The satellite in a circular orbit is continuously being accelerated. This means that there is a continuously acting unbalanced force on the satellite that pulls it out of a straight-line path.

The force that pulls an object out of its straight-line path and into a circular path is a **centripetal** (center-seeking) **force.** Perhaps you have swung a ball on the end of a string in a horizontal circle over your head. Once you have the ball moving, the only unbalanced force (other than gravity) acting on the ball is the centripetal force your hand exerts on the ball through the string. This centripetal force pulls the ball from its natural straight-line path into a circular path. There are no outward forces acting on the ball. The force that you feel on the string is a consequence of the third law; the ball exerts an equal and opposite force on your hand. If you were to release the string, the ball would move away from the circular path in a *straight line* that has a right angle to the radius at the point of release (Figure 2.26). When you release the string, the centripetal force ceases, and the ball then follows its natural straight-

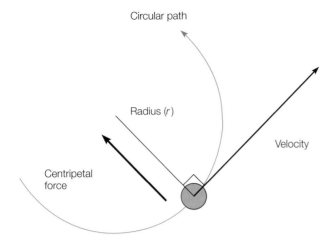

FIGURE 2.26 Centripetal force on the ball causes it to change direction continuously, or accelerate into a circular path. Without the unbalanced force acting on it, the ball would continue in a straight line.

line motion. If other forces were involved, it would follow some other path. Nonetheless, the apparent outward force has been given a name just as if it were a real force. The outward tug is called a **centrifugal force.**

The magnitude of the centripetal force required to keep an object in a circular path depends on the inertia, or mass, of the object and the acceleration of the object, just as you learned in the second law of motion. The acceleration of an object moving in a circle can be shown by geometry or calculus to be directly proportional to the square of the speed around the circle (v^2) and inversely proportional to the radius of the circle (r). (A smaller radius requires a greater acceleration.) Therefore, the acceleration of an object moving in uniform circular motion (a_c) is

$$a_c = \frac{v^2}{r}$$

equation 2.10

The magnitude of the centripetal force of an object with a mass (m) that is moving with a velocity (v) in a circular orbit of a radius (r) can be found by substituting equation 2.5 in $F = ma$, or

$$F = \frac{mv^2}{r}$$

equation 2.11

EXAMPLE 2.16

A 0.25 kg ball is attached to the end of a 0.5 m string and moved in a horizontal circle at 2.0 m/s. What net force is needed to keep the ball in its circular path?

SOLUTION

$$m = 0.25 \text{ kg}$$
$$r = 0.5 \text{ m}$$
$$v = 2.0 \text{ m/s}$$
$$F = ?$$

$$F = \frac{mv^2}{r}$$

$$= \frac{(0.25 \text{ kg}) (2.0 \text{ m/s}^2)}{0.5 \text{ m}}$$

$$= \frac{(0.25 \text{ kg})(4.0 \text{ m}^2/\text{s}^2)}{0.5 \text{ m}}$$

$$= \frac{(0.25)(4.0)}{0.5} \frac{\text{kg} \cdot \text{m}^2}{\text{s}^2} \times \frac{1}{\text{m}}$$

$$= 2 \frac{\text{kg} \cdot \text{m}^2}{\text{m} \cdot \text{s}^2}$$

$$= 2 \frac{\text{kg} \cdot \text{m}}{\text{s}^2}$$

$$= \boxed{2 \text{ N}}$$

EXAMPLE 2.17

Suppose you make the string in example 2.16 half as long, 0.25 m. What force is now needed? (Answer: 4.0 N)

NEWTON'S LAW OF GRAVITATION

You know that if you drop an object, it always falls to the floor. You define *down* as the direction of the object's movement and *up* as the opposite direction. Objects fall because of the force of gravity, which accelerates objects at $g = 9.8$ m/s^2 (32 ft/s^2) and gives them weight, $w = mg$.

Gravity is an attractive force, a pull that exists between all objects in the universe. It is a mutual force that, just like all other forces, comes in matched pairs. Since Earth attracts you with a certain force, you must attract Earth with an exact opposite force. The magnitude of this force of mutual attraction depends on several variables. These variables were first described by Newton in *Principia*, his famous book on motion that was printed in 1687. Newton had, however, worked out his ideas much earlier, by the age of 24, along with ideas about his laws of motion and the formula for centripetal acceleration. In a biography written by a friend in 1752, Newton stated that the notion of gravitation came to mind during a time of thinking that "was occasioned by the fall of an apple." He was thinking about why the Moon stays in orbit around Earth rather than moving off in a straight line as would be predicted by the first law of motion. Perhaps the same force that attracts the Moon toward Earth, he thought, attracts the apple to Earth. Newton developed a theoretical equation for gravitational force that explained not only the motion of the Moon but the motion of the whole solar system. Today, this relationship is known as the **universal law of gravitation:**

> **Every object in the universe is attracted to every other object with a force that is directly proportional to the product of their masses and inversely proportional to the square of the distances between them.**

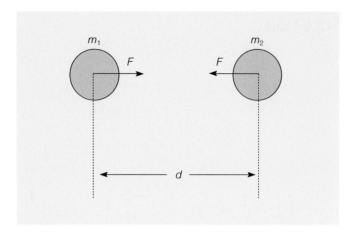

FIGURE 2.27 The variables involved in gravitational attraction. The force of attraction (F) is proportional to the product of the masses (m_1, m_2) and inversely proportional to the square of the distance (d) between the centers of the two masses.

In symbols, m_1 and m_2 can be used to represent the masses of two objects, d the distance between their centers, and G a constant of proportionality. The equation for the law of universal gravitation is therefore

$$F = G\frac{m_1 m_2}{d^2}$$

equation 2.12

This equation gives the magnitude of the attractive force that each object exerts on the other. The two forces are oppositely directed. The constant G is a universal constant, since the law applies to all objects in the universe. It was first measured experimentally by Henry Cavendish in 1798. The accepted value today is $G = 6.67 \times 10^{-11}$ N·m²/kg². Do not confuse G, the universal constant, with g, the acceleration due to gravity on the surface of Earth.

Thus, the magnitude of the force of gravitational attraction is determined by the mass of the two objects and the distance between them (Figure 2.27). The law also states that *every* object is attracted to every other object. You are attracted to all the objects around you—chairs, tables, other people, and so forth. Why don't you notice the forces between you and other objects? The answer is in example 2.18.

EXAMPLE 2.18

What is the force of gravitational attraction between two 60.0 kg (132 lb) students who are standing 1.00 m apart?

SOLUTION

$$G = 6.67 \times 10^{-11}\text{N·m}^2/\text{kg}^2$$
$$m_1 = 60.0\,\text{kg}$$
$$m_2 = 60.0\,\text{kg}$$
$$d = 1.00\,\text{m}$$
$$F = ?$$

$$F = G\frac{m_1 m_2}{d^2}$$

$$= \frac{(6.67 \times 10^{-11}\text{N·m}^2/\text{kg}^2)(60.0\,\text{kg})(60.0\,\text{kg})}{(1.00\,\text{m})^2}$$

$$= (6.67 \times 10^{-11})(3.60 \times 10^3)\,\frac{\text{N·m}^2\cdot\frac{\text{kg}^2}{\text{kg}^2}}{\text{m}^2}$$

$$= 2.40 \times 10^{-7}\,(\text{N·m}^2)\left(\frac{1}{\text{m}^2}\right)$$

$$= 2.40 \times 10^{-7}\,\frac{\text{N·m}^2}{\text{m}^2}$$

$$= \boxed{2.40 \times 10^{-7}\text{N}}$$

(Note: A force of 2.40×10^{-7} (0.00000024) N is equivalent to a force of 5.40×10^{-8} lb (0.00000005 lb), a force that you would not notice. In fact, it would be difficult to measure such a small force.)

As you can see in example 2.18, one or both of the interacting objects must be quite massive before a noticeable force results from the interaction. That is why you do not notice the force of gravitational attraction between you and objects that are not very massive compared to Earth. The attraction between you and Earth overwhelmingly predominates, and that is all you notice.

Newton was able to show that the distance used in the equation is the distance from the center of one object to the center of the second object. This does not mean that the force originates at the center but that the overall effect is the same as if you considered all the mass to be concentrated at a center point. The weight of an object, for example, can be calculated by using a form of Newton's second law, $F = ma$. This general law shows a relationship between *any* force acting *on* a body, the mass of a body, and the resulting acceleration. When the acceleration is due to gravity, the equation becomes $F = mg$. The law of gravitation deals *specifically with the force of gravity* and how it varies with distance and mass. Since weight is a force, then $F = mg$. You can write the two equations together,

$$mg = G\frac{mm_e}{d^2}$$

where m is the mass of some object on Earth, m_e is the mass of Earth, g is the acceleration due to gravity, and d is the distance between the centers of the masses. Canceling the m's in the equation leaves

$$g = G\frac{m_e}{d^2}$$

which tells you that on the surface of Earth, the acceleration due to gravity, 9.8 m/s², is a constant because the other two variables (mass of Earth and the distance to center of Earth) are constant. Since the m's canceled, you also know that the mass of an object does not affect the rate of free fall; all objects fall at the same rate, with the same acceleration, no matter what their masses are.

Example 2.19 shows that the acceleration due to gravity, g, is about 9.8 m/s² and is practically a constant for relatively short distances above the surface. Notice, however, that Newton's law of

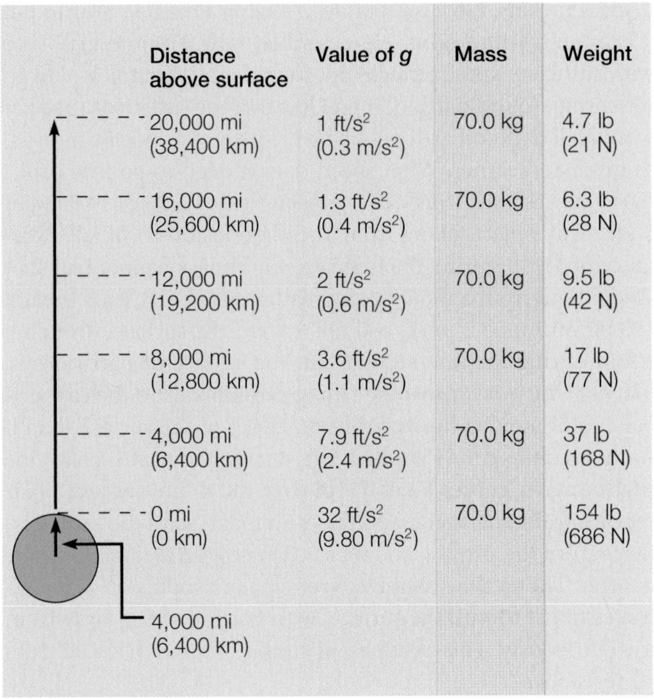

Distance above surface	Value of g	Mass	Weight
20,000 mi (38,400 km)	1 ft/s² (0.3 m/s²)	70.0 kg	4.7 lb (21 N)
16,000 mi (25,600 km)	1.3 ft/s² (0.4 m/s²)	70.0 kg	6.3 lb (28 N)
12,000 mi (19,200 km)	2 ft/s² (0.6 m/s²)	70.0 kg	9.5 lb (42 N)
8,000 mi (12,800 km)	3.6 ft/s² (1.1 m/s²)	70.0 kg	17 lb (77 N)
4,000 mi (6,400 km)	7.9 ft/s² (2.4 m/s²)	70.0 kg	37 lb (168 N)
0 mi (0 km)	32 ft/s² (9.80 m/s²)	70.0 kg	154 lb (686 N)

4,000 mi (6,400 km)

FIGURE 2.28 The force of gravitational attraction decreases inversely with the square of the distance from Earth's center. Note the weight of a 70.0 kg person at various distances above Earth's surface.

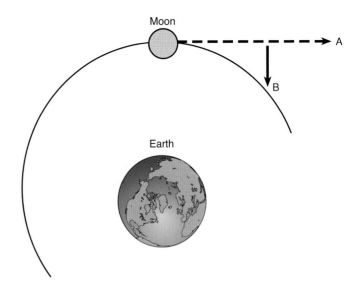

FIGURE 2.29 Gravitational attraction acts as a centripetal force that keeps the Moon from following the straight-line path shown by the dashed line to position A. It was pulled to position B by gravity (0.0027 m/s²) and thus "fell" toward Earth the distance from the dashed line to B, resulting in a somewhat circular path.

gravitation is an inverse square law. This means if you double the distance, the force is $1/(2)^2$ or 1/4 as great. If you triple the distance, the force is $1/(3)^2$ or 1/9 as great. In other words, the force of gravitational attraction and g decrease inversely with the square of the distance from Earth's center. The weight of an object and the value of g are shown for several distances in Figure 2.28. If you have the time, a good calculator, and the inclination, you could check the values given in Figure 2.28 for a 70.0 kg person by doing problems similar to example 2.19. In fact, you could even calculate the mass of Earth, since you already have the value of g.

Using reasoning similar to that found in example 2.19, Newton was able to calculate the acceleration of the Moon toward Earth, about 0.0027 m/s². The Moon "falls" toward Earth because it is accelerated by the force of gravitational attraction. This attraction acts as a *centripetal force* that keeps the Moon from following a straight-line path as would be predicted from the first law. Thus, the acceleration of the Moon keeps it in a somewhat circular orbit around Earth. Figure 2.29 shows that the Moon would be in position A if it followed a straight-line path instead of "falling" to position B as it does. The Moon thus "falls" around Earth. Newton was able to analyze the motion of the Moon quantitatively as evidence that it is gravitational force that keeps the Moon in its orbit. The law of gravitation was extended to the Sun, other planets, and eventually the universe. The quantitative predictions of observed relationships among the planets were strong evidence that all objects obey the same law of gravitation. In addition, the law provided a means to calculate the mass of Earth, the Moon, the planets, and the Sun. Newton's law of gravitation, laws of motion, and work with mathematics formed the basis of

most physics and technology for the next two centuries, as well as accurately describing the world of everyday experience.

EXAMPLE 2.19

The surface of Earth is approximately 6,400 km from its center. If the mass of Earth is 6.0×10^{24} kg, what is the acceleration due to gravity, g, near the surface?

$$G = 6.67 \times 10^{-11} \text{ N·m}^2/\text{kg}^2$$
$$m_e = 6.0 \times 10^{24} \text{ kg}$$
$$d = 6,400 \text{ km } (6.4 \times 10^6 \text{ m})$$
$$g = ?$$

$$g = \frac{Gm_e}{d^2}$$

$$= \frac{(6.67 \times 10^{-11} \text{N·m}^2/\text{kg}^2)(6.0 \times 10^{24} \text{kg})}{(6.4 \times 10^6 \text{m})^2}$$

$$= \frac{(6.67 \times 10^{-11})(6.0 \times 10^{24})}{4.1 \times 10^{13}} \frac{\frac{\text{N·m}^2·\text{kg}}{\text{kg}^2}}{\text{m}^2}$$

$$= \frac{4.0 \times 10^{14}}{4.1 \times 10^{13}} \frac{\frac{\text{kg·m}}{\text{s}^2}}{\text{kg}}$$

$$= \boxed{9.8 \text{ m/s}^2}$$

(Note: In the unit calculation, remember that a newton is a kg·m/s².)

EXAMPLE 2.20

What would be the value of g if Earth were less dense, with the same mass and double the radius? (Answer: $g = 2.4$ m/s²)

EARTH SATELLITES

As you can see in Figure 2.30, Earth is round-shaped and nearly spherical. The curvature is obvious in photographs taken from space but not so obvious back on the surface because Earth is so large. However, you can see evidence of the curvature in places on the surface where you can see with unobstructed vision for long distances. For example, a tall ship appears to "sink" on the horizon as it sails away, following Earth's curvature below your line of sight. The surface of Earth curves away from your line of sight or any other line tangent to the surface, dropping at a rate of about 4.9 m for every 8 km (16 ft in 5 mi). This means that a ship 8 km away will appear to drop about 5 m below the horizon and anything less than about 5 m tall at this distance will be out of sight, below the horizon.

Recall that a falling object accelerates toward Earth's surface at g, which has an average value of 9.8 m/s². Ignoring air resistance, a falling object will have a speed of 9.8 m/s at the end of 1 second and will fall a distance of 4.9 m. If you wonder why the object did not fall 9.8 m in 1 second, recall that the object starts with an initial speed of zero and has a speed of 9.8 m/s only during the last instant. The average speed was an average of the initial and final speeds, which is 4.9 m/s. An average speed of 4.9 m/s over a time interval of 1 second will result in a distance covered of 4.9 m.

Did you know that Newton was the first to describe how to put an artificial satellite into orbit around Earth? He did not discuss rockets, however, but described in *Principia* how to put a cannonball into orbit. He described how a cannonball shot with sufficient speed straight out from a mountaintop would go into orbit around Earth. If it had less that the sufficient speed, it would fall back to Earth following the path of projectile motion, as discussed earlier. What speed does it need to go into orbit? Earth curves away from a line tangent to the surface at 4.9 m per 8 km. Any object falling from a resting position will fall a distance of 4.9 m during the first second. Thus, a cannonball shot straight out from a mountaintop with a speed of 8 km/s (nearly 18,000 mi/hr, or 5 mi/s) will fall toward the surface, dropping 4.9 m during the first second. But the surface of Earth drops, too, curving away below the falling cannonball. So the cannonball is still moving horizontally, no closer to the surface than it was a second ago. As it falls 4.9 m during the next second, the surface again curves away 4.9 m over the 8 km distance. This repeats again and again, and the cannonball stays the same distance from the surface, and we say it is now an *artificial satellite* in orbit. The satellite requires no engine or propulsion as it continues to fall toward the surface, with Earth curving away from it continuously. This assumes, of course, no loss of speed from air resistance.

Today, an artificial satellite is lofted by rocket or rockets to an altitude of more than 320 km (about 200 mi), above the air friction of the atmosphere, before being aimed horizontally. The satellite is then "injected" into orbit by giving it the correct tangential speed. This means it has attained an orbital speed of at least 8 km/s (5 mi/s) but less than 11 km/s (7 mi/s). At a speed less than 8 km/s, the satellite would fall back to the surface in a parabolic path. At a speed more than 11 km/s, it will move faster than the surface curves away and will escape from Earth into space. But with the correct tangential speed, and above the atmosphere and air friction, the satellite follows a circular orbit for long periods of time without the need for any more propulsion. An orbit injection speed of more than 8 km/s (5 mi/s) would result in an elliptical rather than a circular orbit.

A satellite could be injected into orbit near the outside of the atmosphere, closer to Earth but outside the air friction that might reduce its speed. The satellite could also be injected far away from Earth, where it takes a longer time to complete one orbit. Near the outer limits of the atmosphere—that is, closer to the surface—a satellite might orbit Earth every 90 minutes or so. A satellite as far away as the Moon, on the other hand, orbits Earth in a little less than 28 days. A satellite at an altitude of 36,000 km (a little more than 22,000 mi) has a period of 1 day. In the right spot over the equator, such a satellite is called a **geosynchronous satellite,** since it turns with Earth and does not appear to move across the sky (Figure 2.31). The photographs of the cloud cover you see in weather reports were taken from one or more geosynchronous weather satellites. Communications networks are also built around geosynchronous satellites. One way to locate one of these geosynchronous satellites is to note the aiming direction of backyard satellite dishes that pick up television signals.

FIGURE 2.30 From space, this photograph of Earth shows that it is nearly spherical.

Gravity does act on astronauts in space-craft that are in orbit around Earth. Since gravity is acting on the astronaut and spacecraft, the term *zero gravity* is not an accurate description of what is happening. The astronaut, spacecraft, and everything in it are experiencing *apparent weightlessness* because they are continuously falling toward the surface. Everything seems to float because everything is falling together. But, strictly speaking, everything still has weight, because weight is defined as a gravitational force acting on an object ($w = mg$).

Whether weightlessness is apparent or real, however, the effects on people are the same. Long-term orbital flights have provided evidence that the human body changes from the effect of weightlessness. Bones lose calcium and other minerals, the heart shrinks to a much smaller size, and leg muscles shrink so much on prolonged flights that astronauts cannot walk when they return to the surface. These changes occur because on Earth, humans are constantly subjected to the force of gravity. The nature of the skeleton and the strength of the muscles are determined by how the body reacts to this force. Metabolic pathways and physiological processes that maintain strong bones and muscles evolved having to cope with a specific gravitational force. When we are suddenly subjected to a place where gravity is significantly different, these processes result in weakened systems.

If we lived on a planet with a different gravitational force, we would have muscles and bones that were adapted to the gravity on that planet. Many kinds of organisms have been used in experiments in space to try to develop a better understanding of how their systems work without gravity.

The problems related to prolonged weightlessness must be worked out before long-term weightless flights can take place. One solution to these problems might be a large, uniformly spinning spacecraft. The astronauts would tend to move in a straight line, and the side of the turning spacecraft (now the "floor") would exert a force on them to make them go in a curved path. This force would act as an artificial gravity.

WEIGHTLESSNESS

News photos sometimes show astronauts "floating" in the Space Shuttle or next to a satellite (Figure 2.32). These astronauts appear to be weightless but technically are no more weightless than a skydiver in free fall or a person in a falling elevator. Recall that weight is a gravitational force, a measure of the gravitational attraction between Earth and an object (mg). The weight of a cup of coffee, for example, can be measured by placing the cup on a scale. The force the cup of coffee exerts against the scale is its weight. You also know that the scale pushes back on the cup of coffee since it is not accelerating, which means the net force is zero.

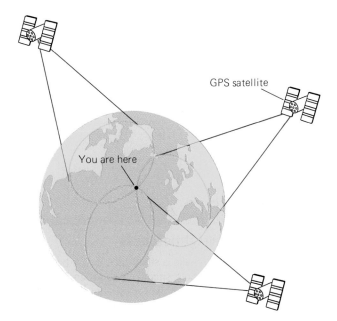

FIGURE 2.31 In the Global Positioning System (GPS), each of a fleet of orbiting satellites sends out coded radio signals that enable a receiver on the earth to determine both the exact position of the satellite in space and its exact distance from the receiver. Given this information, a computer in the receiver then calculates the circle on Earth's surface on which the receiver must lie. Data from three satellites give three circles, and the receiver must be located at the one point where all three intersect.

FIGURE 2.32 Astronauts in an orbiting space station may appear to be weightless. Technically, however, they are no more weightless than a skydiver in free fall or a person near or on the surface of Earth in a falling elevator.

People Behind the Science

Isaac Newton (1642-1727)

Isaac Newton was a British physicist who is regarded as one of the greatest scientists ever to have lived. He discovered the three laws of motion that bear his name and was the first to explain gravitation, clearly defining the nature of mass, weight, force, inertia, and acceleration. In his honor, the SI unit of force is called the newton. Newton also made fundamental discoveries in light, finding that white light is composed of a spectrum of colors and inventing the reflecting telescope.

Newton was born on January 4, 1643 (by the modern calendar). He was a premature, sickly baby born after his father's death, and his survival was not expected. When he was 3, his mother remarried, and the young Newton was left in his grandmother's care. He soon began to take refuge in things mechanical, making water clocks, kites bearing fiery lanterns aloft, and a model mill powered by a mouse, as well as innumerable drawings and diagrams. When Newton was 12, his mother withdrew him from school with the intention of making him into a farmer. Fortunately, his uncle recognized Newton's ability and managed to get him back into school to prepare for college.

Newton was admitted to Trinity College, Cambridge, and graduated in 1665, the same year that the university was closed because of the plague. Newton returned to his boyhood farm to wait out the plague, making only an occasional visit back to Cambridge. During this period, he performed his first prism experiments and thought about motion and gravitation.

Newton returned to study at Cambridge after the plague had run its course, receiving a master's degree in 1668 and becoming a professor at the age of only 26. Newton remained at Cambridge almost thirty years, studying alone for the most part, though in frequent contact with other leading scientists by letter and through the Royal Society in London. These were Newton's most fertile years. He labored day and night, thinking and testing ideas with calculations.

In Cambridge, he completed what may be described as his greatest single work, the *Philosophae Naturalis Principia Mathematica* (*Mathematical Principles of Natural Philosophy*). This was presented to the Royal Society in 1686, which subsequently withdrew from publishing it because of a shortage of funds. The astronomer Edmund Halley (1656–1742), a wealthy man and friend of Newton, paid for the publication of the *Principia* in 1687. In it, Newton revealed his laws of motion and the law of universal gravitation.

Newton's greatest achievement was to demonstrate that scientific principles are of universal application. In the *Principia Mathematica,* he built the evidence of experiment and observation to develop a model of the universe that is still of general validity. "If I have seen further than other men," he once said, "it is because I have stood on the shoulders of giants"; and Newton was certainly able to bring together the knowledge of his forebears in a brilliant synthesis.

No knowledge can ever be total, but Newton's example brought about an explosion of investigation and discovery that has never really abated. He perhaps foresaw this when he remarked, "To myself, I seem to have been only like a boy playing on the seashore, and diverting myself in now and then finding a smoother pebble or a prettier shell than ordinary, whilst the great ocean of truth lay all undiscovered before me."

With his extraordinary insight into the workings of nature and rare tenacity in wresting its secrets and revealing them in as fundamental and concise a way as possible, Newton stands as a colossus of science. In physics, only Archimedes (287–212 B.C.) and Albert Einstein (1879–1955), who also possessed these qualities, may be compared to him.

Source: Modified from the Hutchinson *Dictionary of Scientific Biography*. © Research Machines plc 2003. All Rights Reserved. Helicon Publishing is a division of Research Machines.

Now consider what happens if a skydiver tries to pour a cup of coffee while in free fall. Even if you ignore air resistance, you can see that the skydiver is going to have a difficult time, at best. The coffee, the cup, and the skydiver will all be falling together. Gravity is acting to pull the coffee downward, but gravity is also acting to pull the cup from under it at the same rate. The coffee, the cup, and the skydiver all fall together, and the skydiver will see the coffee appear to "float" in blobs. If the diver lets go of the cup, it too will appear to float as everything continues to fall together. However, this is only *apparent* weightlessness, since gravity is still acting on everything; the coffee, the cup, and the skydiver only *appear* to be weightless because they are all accelerating at *g*.

The astronauts in orbit are in free fall, falling toward Earth just as the skydiver, so they too are undergoing apparent weightlessness. To experience true weightlessness, the astronauts would have to travel far from Earth and its gravitational field, and far from the gravitational fields of other planets.

CONCEPTS *Applied*

Apparent Weightlessness

Use a sharp pencil to make a small hole in the bottom of a Styrofoam cup. The hole should be large enough for a thin stream of water to flow from the cup but small enough for the flow to continue for three or four seconds. Test the water flow over a sink.

Hold a finger over the hole in the cup as you fill it with water. Stand on a ladder or outside stairwell as you hold the cup out at arm's length. Move your finger, allowing a stream of water to flow from the cup, and at the same time drop the cup. Observe what happens to the stream of water as the cup is falling. Explain your observations. Also predict what you would see if you were falling with the cup.

Motion can be measured by speed, velocity, and acceleration. *Speed* is a measure of how fast something is moving. It is a ratio of the distance covered between two locations to the time that elapsed while moving between the two locations. The *average speed* considers the distance covered during some period of time, while the *instantaneous speed* is the speed at some specific instant. *Velocity* is a measure of the speed and direction of a moving object. *Acceleration* is the change of velocity during some period of time.

A *force* is a push or a pull that can change the motion of an object. The *net force* is the sum of all the forces acting on an object.

Galileo determined that a continuously applied force is not necessary for motion and defined the concept of *inertia*: an object remains in unchanging motion in the absence of a net force. Galileo also determined that falling objects accelerate toward Earth's surface independent of the weight of the object. He found the acceleration due to gravity, *g,* to be 9.8 m/s^2 (32 ft/s^2), and the distance an object falls is proportional to the square of the time of free fall ($d \propto t^2$).

Compound motion occurs when an object is projected into the air. Compound motion can be described by splitting the motion into vertical and horizontal parts. The acceleration due to gravity, *g,* is a constant that is acting at all times and acts independently of any motion that an object has. The path of an object that is projected at some angle to the horizon is therefore a parabola.

Newton's *first law of motion* is concerned with the motion of an object and the lack of a net force. Also known as the *law of inertia,* the first law states that an object will retain its state of straight-line motion (or state of rest) unless a net force acts on it.

The *second law of motion* describes a relationship between net force, mass, and acceleration. A *newton* of force is the force needed to give a 1.0 kg mass an acceleration of 1.0 m/s^2.

Weight is the downward force that results from Earth's gravity acting on the mass of an object. Weight is measured in *newtons* in the metric system and *pounds* in the English system.

Newton's *third law of motion* states that forces are produced by the interaction of *two different* objects. These forces always occur in matched pairs that are equal in size and opposite in direction.

Momentum is the product of the mass of an object and its velocity. In the absence of external forces, the momentum of a group of interacting objects always remains the same. This relationship is the *law of conservation of momentum. Impulse* is a change of momentum equal to a force times the time of application.

An object moving in a circular path must have a force acting on it, since it does not move in a straight line. The force that pulls an object out of its straight-line path is called a *centripetal force.* The centripetal force needed to keep an object in a circular path depends on the mass of the object, its velocity, and the radius of the circle.

The *universal law of gravitation* is a relationship between the masses of two objects, the distance between the objects, and a proportionality constant. Newton was able to use this relationship to show that gravitational attraction provides the centripetal force that keeps the Moon in its orbit.

SUMMARY OF EQUATIONS

2.1

$$\text{average speed} = \frac{\text{distance}}{\text{time}}$$

$$\bar{v} = \frac{d}{t}$$

2.2

$$\text{acceleration} = \frac{\text{change of velocity}}{\text{time}}$$

$$\text{acceleration} = \frac{\text{final velocity} - \text{initial velocity}}{\text{time}}$$

$$a = \frac{v_f - v_i}{t}$$

2.3

$$\text{average velocity} = \frac{\text{final velocity} + \text{initial velocity}}{2}$$

$$\bar{v} = \frac{v_f + v_i}{2}$$

2.4

$$\text{distance} = \frac{1}{2}(\text{acceleration})(\text{time})^2$$

$$d = \frac{1}{2}at^2$$

2.5

$$\text{force} = \text{mass} \times \text{acceleration}$$

$$F = ma$$

2.6

$$\text{weight} = \text{mass} \times \text{acceleration due to gravity}$$

$$w = mg$$

2.7

$$\text{force on object A} = \text{force on object B}$$

$$F_{A \text{ due to B}} = F_{B \text{ due to A}}$$

2.8

$$\text{momentum} = \text{mass} \times \text{velocity}$$

$$p = mv$$

2.9

$$\text{change of momentum} = \text{force} \times \text{time}$$

$$\Delta p = Ft$$

2.10

$$\text{centripetal acceleration} = \frac{\text{velocity squared}}{\text{radius of circle}}$$

$$a_c = \frac{v^2}{r}$$

2.11

$$\text{centripetal force} = \frac{\text{mass} \times \text{velocity squared}}{\text{radius of circle}}$$

$$F = \frac{mv^2}{r}$$

2.12

$$\text{gravitational force} = \text{constant} \times \frac{\text{one mass} \times \text{another mass}}{\text{distance squared}}$$

$$F = G\frac{m_1 m_2}{d^2}$$

KEY TERMS

acceleration (p. **29**)
centrifugal force (p. **49**)
centripetal force (p. **48**)
first law of motion (p. **41**)
force (p. **32**)
free fall (p. **36**)
fundamental forces (p. **32**)
g (p. **38**)
geosynchronous satellite (p. **52**)
impulse (p. **48**)
inertia (p. **34**)
law of conservation of momentum (p. **46**)
mass (p. **43**)
momentum (p. **46**)
net force (p. **32**)
newton (p. **43**)
second law of motion (p. **43**)
speed (p. **27**)
third law of motion (p. **45**)
universal law of gravitation (p. **49**)
velocity (p. **29**)

APPLYING THE CONCEPTS

1. A straight-line distance covered during a certain amount of time describes an object's
   a. speed.
   b. velocity.
   c. acceleration.
   d. Any of the above are correct.

2. How fast an object is moving in a particular direction is described by
   a. speed.
   b. velocity.
   c. acceleration.
   d. none of the above.

3. Acceleration occurs when an object undergoes
   a. a speed increase.
   b. a speed decrease.
   c. a change in the direction of travel.
   d. any of the above.

4. A car moving at 60 km/h comes to a stop in 10 s when the driver slams on the brakes. In this situation, what does 60 km/h represent?
   a. average speed
   b. final speed
   c. initial speed
   d. constant speed

5. A car moving at 60 km/h comes to a stop in 10 s when the driver slams on the brakes. In this situation, what is the final speed?
   a. 60 km/h
   b. 0 km/h
   c. 0.017 km/s
   d. 0.17 km/s

6. According to Galileo, an object moving without opposing friction or other opposing forces will
   a. still need a constant force to keep it moving at a constant speed.
   b. need an increasing force, or it will naturally slow and then come to a complete stop.
   c. continue moving at a constant speed.
   d. undergo a gradual acceleration.

7. In free fall, an object is seen to have a (an)
   a. constant velocity.
   b. constant acceleration.
   c. increasing acceleration.
   d. decreasing acceleration.

8. A tennis ball is hit, moving upward from the racket at some angle to the horizon before it curves back to the surface in the path of a parabola. While moving along this path,
   a. the horizontal speed remains the same.
   b. the vertical speed remains the same.
   c. both the horizontal and vertical speeds remain the same.
   d. both the horizontal and vertical speeds change.

9. A quantity of 5 m/s^2 is a measure of
   a. metric area.
   b. acceleration.
   c. speed.
   d. velocity.

10. An automobile has how many different devices that can cause it to undergo acceleration?
    a. none
    b. one
    c. two
    d. three or more

11. Ignoring air resistance, an object falling toward the surface of Earth has a velocity that is
    a. constant.
    b. increasing.
    c. decreasing.
    d. acquired instantaneously but dependent on the weight of the object.

12. Ignoring air resistance, an object falling near the surface of Earth has an acceleration that is
    a. constant.
    b. increasing.
    c. decreasing.
    d. dependent on the weight of the object.

13. Two objects are released from the same height at the same time, and one has twice the weight of the other. Ignoring air resistance,
    a. the heavier object hits the ground first.
    b. the lighter object hits the ground first.
    c. they both hit at the same time.
    d. whichever hits first depends on the distance dropped.

14. A ball rolling across the floor slows to a stop because
    a. there is a net force acting on it.
    b. the force that started it moving wears out.
    c. the forces are balanced.
    d. the net force equals zero.

15. The basic difference between instantaneous and average speed is that
    a. instantaneous speed is always faster than the average speed.
    b. average speed is for a total distance over a total time of trip.
    c. average speed is the sum of two instantaneous speeds divided by 2.
    d. the final instantaneous speed is always the fastest speed.

16. Does *any change* in the motion of an object result in an acceleration?
    a. Yes.
    b. No.
    c. It depends on the type of change.

17. A measure of how fast your speed is changing as you travel to campus is a measure of
    a. velocity.
    b. average speed.
    c. acceleration.
    d. the difference between initial and final speed.

18. Considering the forces on the system of you and a bicycle as you pedal the bike at a constant velocity in a horizontal straight line,
    a. the force you are exerting on the pedal is greater than the resisting forces.
    b. all forces are in balance, with the net force equal to zero.
    c. the resisting forces of air and tire friction are less than the force you are exerting.
    d. the resisting forces are greater than the force you are exerting.

19. Newton's first law of motion describes
    a. the tendency of a moving or stationary object to resist any change in its state of motion.
    b. a relationship between an applied force, the mass, and the resulting change of motion that occurs from the force.
    c. how forces always occur in matched pairs.
    d. none of the above.

20. You are standing freely on a motionless shuttle bus. When the shuttle bus quickly begins to move forward, you
    a. are moved to the back of the shuttle bus as you move forward over the surface of Earth.
    b. stay in one place over the surface of Earth as the shuttle bus moves from under you.
    c. move along with the shuttle bus.
    d. feel a force toward the side of the shuttle bus.

21. Mass is measured in kilograms, which is a measure of
    a. weight.
    b. force.
    c. inertia.
    d. quantity of matter.

22. Which metric unit is used to express a measure of weight?
    a. kg
    b. J
    c. N
    d. m/s^2

23. Newton's third law of motion states that forces occur in matched pairs that act in opposite directions between two different bodies. This happens
    a. rarely.
    b. sometimes.
    c. often but not always.
    d. every time two bodies interact.

24. If you double the unbalanced force on an object of a given mass, the acceleration will be
    a. doubled.
    b. increased fourfold.
    c. increased by one-half.
    d. increased by one-fourth.

25. If you double the mass of a cart while it is undergoing a constant unbalanced force, the acceleration will be
    a. doubled.
    b. increased fourfold.
    c. half as much.
    d. one-fourth as much.

26. Doubling the distance between the center of an orbiting satellite and the center of Earth will result in what change in the gravitational attraction of Earth for the satellite?
    a. one-half as much
    b. one-fourth as much
    c. twice as much
    d. four times as much

27. If a ball swinging in a circle on a string is moved twice as fast, the force on the string will be
    a. twice as great.
    b. four times as great.
    c. one-half as much.
    d. one-fourth as much.

28. A ball is swinging in a circle on a string when the string length is doubled. At the same velocity, the force on the string will be
    a. twice as great.
    b. four times as great.
    c. one-half as much.
    d. one-fourth as much.

29. Suppose the mass of a moving scooter is doubled and its velocity is also doubled. The resulting momentum is
    a. halved.
    b. doubled.
    c. quadrupled.
    d. the same.

30. Two identical moons are moving in identical circular paths, but one moon is moving twice as fast as the other is. Compared to the slower moon, the centripetal force required to keep the faster moon on the path is
    a. twice as much.
    b. half as much.
    c. four times as much.
    d. one-fourth as much.

31. Which undergoes a greater change of momentum, a golf ball or the head of a golf club when the ball is hit from a golf tee?
    a. The ball undergoes a greater change.
    b. The head of the club undergoes a greater change.
    c. Both undergo the same change but in opposite directions.
    d. The answer depends on how fast the club is moved.

32. Newton's law of gravitation tells us that
    a. planets are attracted to the Sun's magnetic field.
    b. objects and bodies have weight only on the surface of Earth.
    c. every object in the universe is attracted to every other object in the universe.
    d. only objects in the solar system are attracted to Earth.

33. An astronaut living on a space station that is orbiting around Earth will
    a. experience zero gravity.
    b. weigh more than she did on Earth.
    c. be in free fall, experiencing apparent weightlessness.
    d. weigh the same as she would on the Moon.

34. A measure of the force of gravity acting on an object is called
    a. gravitational force.
    b. weight.
    c. mass.
    d. acceleration.

35. You are at rest with a grocery cart at the supermarket when you see a checkout line open. You apply a certain force to the cart for a short time and acquire a certain speed. Neglecting friction, how long would you have to push with *half* the force to acquire the same final speed?
    a. one-fourth as long
    b. one-half as long
    c. twice as long
    d. four times as long

36. Once again you are at rest with a grocery cart at the supermarket when you apply a certain force to the cart for a short time and acquire a certain speed. Suppose you had bought more groceries, enough to double the mass of the groceries and cart. Neglecting friction, doubling the mass would have what effect on the resulting final speed if you used the same force for the same length of time? The new final speed would be
    a. one-fourth.
    b. one-half.
    c. doubled.
    d. quadrupled.

37. You are moving a grocery cart at a constant speed in a straight line down the aisle of a store. For this situation, the forces on the cart are
    a. unbalanced, in the direction of the movement.
    b. balanced, with a net force of zero.
    c. equal to the force of gravity acting on the cart.
    d. greater than the frictional forces opposing the motion of the cart.

38. You are outside a store, moving a loaded grocery cart down the street on a very steep hill. It is difficult, but you are able to pull back on the handle and keep the cart moving down the street in a straight line and at a constant speed. For this situation, the forces on the cart are
    a. unbalanced, in the direction of the movement.
    b. balanced, with a net force of zero.
    c. equal to the force of gravity acting on the cart.
    d. greater than the frictional forces opposing the motion of the cart.

39. Neglecting air resistance, a ball in free fall near Earth's surface will have
    a. constant speed and constant acceleration.
    b. increasing speed and increasing acceleration.
    c. increasing speed and decreasing acceleration.
    d. increasing speed and constant acceleration.

40. From a bridge, a ball is thrown straight up at the same time a ball is thrown straight down with the same initial speed. Neglecting air resistance, which ball would have a greater speed when it hits the ground?
    a. the one thrown straight up
    b. the one thrown straight down
    c. Both balls would have the same speed.

41. After being released, a ball thrown straight up from a bridge would have an acceleration of
    a. 9.8 m/s^2.
    b. zero.
    c. less than 9.8 m/s^2.
    d. more than 9.8 m/s^2.

42. A gun is aimed horizontally at the center of an apple hanging from a tree. The instant the gun is fired, the apple falls and the bullet
    a. hits the apple.
    b. arrives late, missing the apple.
    c. arrives early, missing the apple.
    d. may or may not hit the apple, depending on how fast it is moving.

43. According to the third law of motion, which of the following must be true about a car pulling a trailer?
    a. The car pulls on the trailer and the trailer pulls on the car with an equal and opposite force. Therefore, the net force is zero and the trailer cannot move.
    b. Since they move forward, this means the car is pulling harder on the trailer than the trailer is pulling on the car.
    c. The action force from the car is quicker than the reaction force from the trailer, so they move forward.
    d. The action-reaction forces between the car and trailer are equal, but the force between the ground and car pushes them forward.

44. A small sports car and a large SUV collide head on and stick together without sliding. Which vehicle had the larger momentum change?
    a. the small sports car
    b. the large SUV
    c. It would be equal for both.

45. Again consider the small sports car and large SUV that collided head on and stuck together without sliding. Which vehicle must have experienced the larger deceleration during the collision?
    a. the small sports car
    b. the large SUV
    c. It would be equal for both.

46. An orbiting satellite is moved from 10,000 to 30,000 km from Earth. This will result in what change in the gravitational attraction between Earth and the satellite?
    a. None—the attraction is the same.
    b. one-half as much
    c. one-fourth as much
    d. one-ninth as much

47. Newton's law of gravitation considers the *product* of two masses because
    a. the larger mass pulls harder on the smaller mass.
    b. both masses contribute equally to the force of attraction.
    c. the large mass is considered before the smaller mass.
    d. the distance relationship is one of an inverse square.

Answers

1. a **2.** b **3.** d **4.** c **5.** b **6.** c **7.** b **8.** a **9.** b **10.** d **11.** b **12.** a **13.** c **14.** a **15.** b **16.** a **17.** c **18.** b **19.** a **20.** b **21.** c **22.** c **23.** d **24.** a **25.** c **26.** b **27.** b **28.** c **29.** c **30.** c **31.** c **32.** c **33.** c **34.** b **35.**c **36.** b **37.** b **38.** b **39.** d **40.** c **41.** a **42.** a **43.** d **44.** c **45.** a **46.** d **47.** b

QUESTIONS FOR THOUGHT

1. An insect inside a bus flies from the back toward the front at 2 m/s. The bus is moving in a straight line at 20 m/s. What is the speed of the insect?

2. Disregarding air friction, describe all the forces acting on a bullet shot from a rifle into the air.

3. Can gravity act in a vacuum? Explain.

4. Is it possible for a small car to have the same momentum as a large truck? Explain.

5. Without friction, what net force is needed to maintain a 1,000 kg car in uniform motion for 30 minutes?

6. How can there ever be an unbalanced force on an object if every action has an equal and opposite reaction?

7. Why should you bend your knees as you hit the ground after jumping from a roof?

8. Is it possible for your weight to change as your mass remains constant? Explain.

9. What maintains the speed of Earth as it moves in its orbit around the Sun?

10. Suppose you are standing on the ice of a frozen lake and there is no friction whatsoever. How can you get off the ice? (Hint: Friction is necessary to crawl or walk, so that will not get you off the ice.)

11. A rocket blasts off from a platform on a space station. An identical rocket blasts off from free space. Considering everything else to be equal, will the two rockets have the same acceleration? Explain.

12. An astronaut leaves a spaceship that is moving through free space to adjust an antenna. Will the spaceship move off and leave the astronaut behind? Explain.

FOR FURTHER ANALYSIS

1. What are the significant similarities and differences between speed and velocity?

2. What are the significant similarities and differences between velocity and acceleration?

3. Compare your beliefs and your own reasoning about motion before and after learning Newton's three laws of motion.

4. Newton's law of gravitation explains that every object in the universe is attracted to every other object in the universe. Describe a conversation between yourself and another person who does not believe this law, as you persuade them that the law is indeed correct.

5. Why is it that your weight can change by moving from one place to another, but your mass stays the same?

6. Assess the reasoning that Newton's first law of motion tells us that centrifugal force does not exist.

INVITATION TO INQUIRY

The Domino Effect

The *domino effect* is a cumulative effect produced when one event initiates a succession of similar events. In the actual case of dominoes, a row is made by standing dominoes on their end so they stand face to face in a line. When the domino on the end is tipped over, it will fall into its neighbor, which falls into the next one, and so on until the whole row has fallen.

How should the dominoes be spaced so the row falls with maximum speed? Should one domino strike the next one as high as possible, in the center, or as low as possible? If you accept this invitation, you will need to determine how you plan to space the dominoes as well as how you will measure the speed.

PARALLEL EXERCISES

The exercises in groups A and B cover the same concepts. Solutions to group A exercises are located in appendix E.

Note: *Neglect all frictional forces in all exercises.*

Group A

1. What is the average speed in km/h of a car that travels 160 km for 2 h?

2. What is the average speed in km/h for a car that travels 50.0 km in 40.0 min?

3. What is the weight of a 5.2 kg object?

4. What net force is needed to give a 40.0 kg grocery cart an acceleration of 2.4 m/s^2?

5. What is the resulting acceleration when an unbalanced force of 100.0 N is applied to a 5.00 kg object?

6. What is the average speed, in km/h, for a car that travels 22 km in exactly 15 min?

7. Suppose a radio signal travels from Earth and through space at a speed of 3.0×10^8 m/s. How far into space did the signal travel during the first 20.0 minutes?

8. How far away was a lightning strike if thunder is heard 5.00 seconds after seeing the flash? Assume that sound traveled at 350.0 m/s during the storm.

Group B

1. What was the average speed in km/h of a car that travels 400.0 km in 4.5 h?

2. What was the average speed in km/h of a boat that moves 15.0 km across a lake in 45 min?

3. How much would a 80.0 kg person weight (a) on Mars, where the acceleration of gravity is 3.93 m/s^2, and (b) on Earth's Moon, where the acceleration of gravity is 1.63 m/s^2?

4. What force is needed to give a 6,000 kg truck an acceleration of 2.2 m/s^2 over a level road?

5. What is the resulting acceleration when a 300 N force acts on an object with a mass of 3,000 kg?

6. A boat moves 15.0 km across a lake in 30.0 min. What was the average speed of the boat in kilometers per hour?

7. If the Sun is a distance of 1.5×10^8 km from Earth, how long does it take sunlight to reach Earth if it moves at 3.0×10^8 m/s?

8. How many meters away is a cliff if an echo is heard 0.500 s after the original sound? Assume that sound traveled at 343 m/s on that day.

9. A car is driven at an average speed of 100.0 km/h for two hours, then at an average speed of 50.0 km/h for the next hour. What was the average speed for the three-hour trip?

10. What is the acceleration of a car that moves from rest to 15.0 m/s in 10.0 s?

11. How long will be required for a car to go from a speed of 20.0 m/s to a speed of 25.0 m/s if the acceleration is 3.0 m/s^2?

12. A bullet leaves a rifle with a speed of 720 m/s. How much time elapses before it strikes a target 1,609 m away?

13. A pitcher throws a ball at 40.0 m/s, and the ball is electronically timed to arrive at home plate 0.4625 s later. What is the distance from the pitcher to the home plate?

14. The Sun is 1.50×10^8 km from Earth, and the speed of light is 3.00×10^8 m/s. How many minutes elapse as light travels from the Sun to Earth?

15. An archer shoots an arrow straight up with an initial velocity magnitude of 100.0 m/s. After 5.00 s, the velocity is 51.0 m/s. At what rate is the arrow decelerated?

16. A ball thrown straight up climbs for 3.0 s before falling. Neglecting air resistance, with what velocity was the ball thrown?

17. A ball dropped from a building falls for 4 s before it hits the ground. (a) What was its final velocity just as it hit the ground? (b) What was the average velocity during the fall? (c) How high was the building?

18. You drop a rock from a cliff, and 5.00 s later you see it hit the ground. How high is the cliff?

19. What is the resulting acceleration when an unbalanced force of 100 N is applied to a 5 kg object?

20. What is the momentum of a 100 kg football player who is moving at 6 m/s?

21. A car weighing 13,720 N is speeding down a highway with a velocity of 91 km/h. What is the momentum of this car?

22. A 15 g bullet is fired with a velocity of 200 m/s from a 6 kg rifle. What is the recoil velocity of the rifle?

23. An astronaut and equipment weigh 2,156 N on Earth. Weightless in space, the astronaut throws away a 5.0 kg wrench with a velocity of 5.0 m/s. What is the resulting velocity of the astronaut in the opposite direction?

24. (a) What is the weight of a 1.25 kg book? (b) What is the acceleration when a net force of 10.0 N is applied to the book?

25. What net force is needed to accelerate a 1.25 kg book 5.00 m/s^2?

26. What net force does the road exert on a 70.0 kg bicycle and rider to give them an acceleration of 2.0 m/s^2?

27. A 1,500 kg car accelerates uniformly from 44.0 km/h to 80.0 km/h in 10.0 s. What was the net force exerted on the car?

28. A net force of 5,000.0 N accelerates a car from rest to 90.0 km/h in 5.0 s. (a) What is the mass of the car? (b) What is the weight of the car?

29. What is the weight of a 70.0 kg person?

30. How much centripetal force is needed to keep a 0.20 kg ball on a 1.50 m string moving in a circular path with a speed of 3.0 m/s?

31. On Earth, an astronaut and equipment weigh 1,960.0 N. While weightless in space, the astronaut fires a 100 N rocket backpack for 2.0 s. What is the resulting velocity of the astronaut and equipment?

9. A car has an average speed of 80.0 km/h for one hour, then an average speed of 90.0 km/h for two hours during a three-hour trip. What was the average speed for the three-hour trip?

10. What is the acceleration of a car that moves from a speed of 5.0 m/s to a speed of 15 m/s during a time of 6.0 s?

11. How much time is needed for a car to accelerate from 8.0 m/s to a speed of 22 m/s if the acceleration is 3.0 m/s^2?

12. A rocket moves through outer space at 11,000 m/s. At this rate, how much time would be required to travel the distance from Earth to the Moon, which is 380,000 km?

13. Sound travels at 348 m/s in the warm air surrounding a thunderstorm. How far away was the place of discharge if thunder is heard 4.63 s after a lightning flash?

14. How many hours are required for a radio signal from a space probe near the dwarf planet Pluto, 6.00×10^9 km away, to reach Earth? Assume that the radio signal travels at the speed of light, 3.00×10^8 m/s.

15. A rifle is fired straight up, and the bullet leaves the rifle with an initial velocity magnitude of 724 m/s. After 5.00 s, the velocity is 675 m/s. At what rate is the bullet decelerated?

16. A rock thrown straight up climbs for 2.50 s, then falls to the ground. Neglecting air resistance, with what velocity did the rock strike the ground?

17. An object is observed to fall from a bridge, striking the water below 2.50 s later. (a) With what velocity did it strike the water? (b) What was its average velocity during the fall? (c) How high is the bridge?

18. A ball dropped from a window strikes the ground 2.00 s later. How high is the window above the ground?

19. Find the resulting acceleration from a 300 N force that acts on an object with a mass of 3,000 kg.

20. What is the momentum of a 30.0 kg shell fired from a cannon with a velocity of 500 m/s?

21. What is the momentum of a 39.2 N bowling ball with a velocity of 7.00 m/s?

22. A 30.0 kg shell is fired from a 2,000 kg cannon with a velocity of 500 m/s. What is the resulting velocity of the cannon?

23. An 80.0 kg man is standing on a frictionless ice surface when he throws a 2.00 kg book at 10.0 m/s. With what velocity does the man move across the ice?

24. (a) What is the weight of a 5.00 kg backpack? (b) What is the acceleration of the backpack if a net force of 10.0 N is applied?

25. What net force is required to accelerate a 20.0 kg object to 10.0 m/s^2?

26. What forward force must the ground apply to the foot of a 60.0 kg person to result in an acceleration of 1.00 m/s^2?

27. A 1,000.0 kg car accelerates uniformly to double its speed from 36.0 km/h in 5.00 s. What net force acted on this car?

28. A net force of 3,000.0 N accelerates a car from rest to 36.0 km/h in 5.00 s. (a) What is the mass of the car? (b) What is the weight of the car?

29. How much does a 60.0 kg person weigh?

30. What tension must a 50.0 cm length of string support in order to whirl an attached 1,000.0 g stone in a circular path at 5.00 m/s?

31. A 200.0 kg astronaut and equipment move with a velocity of 2.00 m/s toward an orbiting spacecraft. How long will the astronaut need to fire a 100.0 N rocket backpack to stop the motion relative to the spacecraft?

3
Energy

The wind can be used as a source of energy. All you need is a way to capture the energy—such as these wind turbines in California—and to live somewhere where the wind blows enough to make it worthwhile.

CORE **CONCEPT**

Energy is transformed through working or heating, and the total amount remains constant.

OUTLINE

Work
When work is done on an object, it gains energy.

Energy Forms
Energy comes in various forms: mechanical, chemical, radiant, electrical, and nuclear.

Energy Sources Today
The main sources of energy today are petroleum, coal, nuclear, and moving water.

Work
   Units of Work
A Closer Look: Simple Machines
   Power
Motion, Position, and Energy
   Potential Energy
   Kinetic Energy
Energy Flow
   Work and Energy
   Energy Forms
   Energy Conversion
   Energy Conservation
   Energy Transfer
Energy Sources Today
Science and Society: Grow Your Own Fuel?
   Petroleum
   Coal
People Behind the Science: James Joule
   Moving Water
   Nuclear
   Conserving Energy
Energy Sources Tomorrow
   Solar Technologies
   Geothermal Energy
   Hydrogen

Motion, Position, and Energy
Potential energy is the energy an object has due to position. Kinetic energy is the energy an object has due to motion.

Energy Conservation
Energy is transformed from one form to another, and the total amount remains constant.

Energy Sources Tomorrow
Alternate sources of energy are solar, geothermal, and hydrogen.

61

The term *energy* is closely associated with the concepts of force and motion. Naturally moving matter, such as the wind or moving water, exerts forces. You have felt these forces if you have ever tried to walk against a strong wind or stand in one place in a stream of rapidly moving water. The motion and forces of moving air and moving water are used as *energy sources* (Figure 3.1). The wind is an energy source as it moves the blades of a windmill, performing useful work. Moving water is an energy source as it forces the blades of a water turbine to spin, turning an electric generator. Thus, moving matter exerts a force on objects in its path, and objects moved by the force can also be used as an energy source.

Matter does not have to be moving to supply energy; matter *contains* energy. Food supplied the energy for the muscular exertion of the humans and animals that accomplished most of the work before the twentieth century. Today, machines do the work that was formerly accomplished by muscular exertion. Machines also use the energy contained in matter. They use gasoline, for example, as they supply the forces and motion to accomplish work.

Moving matter and matter that contains energy can be used as energy sources to perform work. The concepts of work and energy and the relationship to matter are the topics of this chapter. You will learn how energy flows in and out of your surroundings as well as a broad, conceptual view of energy that will be developed more fully throughout the course.

WORK

You learned earlier that the term *force* has a special meaning in science that is different from your everyday concept of force. In everyday use, you use the term in a variety of associations such as police force, economic force, or the force of an argument. Earlier, force was discussed in a general way as a push or pull. Then a more precise scientific definition of force was developed from Newton's laws of motion—a force is a result of an interaction that is capable of changing the state of motion of an object.

The word *work* represents another one of those concepts that has a special meaning in science that is different from your everyday concept. In everyday use, work is associated with a task to be accomplished or the time spent in performing the task. You might work at understanding physical science, for example, or you might tell someone that physical science is a lot of work. You also probably associate physical work, such as lifting or moving boxes, with how tired you become from the effort. The definition of mechanical work is not concerned with tasks, time, or how tired you become from doing a task. It is concerned with the application of a force to an object and the distance the object moves as a result of the force. The **work** done on the object is defined as *the product of the applied force and the parallel distance through which the force acts:*

$$\text{work} = \text{force} \times \text{distance}$$

$$W = Fd$$

equation 3.1

Mechanical work is the product of a force and the distance an object moves as a result of the force. There are two important considerations to remember about this definition: (1) something *must move* whenever work is done, and (2) the movement must be in the *same direction* as the direction of the force. When you move a book to a higher shelf in a bookcase, you are doing work on the book. You apply a vertically upward force equal to the weight of the book as you move it in the same direction as the direction of the applied force. The work done on the book can therefore be calculated by multiplying the weight of the book by the distance it was moved (Figure 3.2).

UNITS OF WORK

The units of work can be obtained from the definition of work, $W = Fd$. In the metric system, a force is measured in newtons (N), and distance is measured in meters (m), so the unit of work is

$$W = Fd$$
$$W = (\text{newton})(\text{meter})$$
$$W = (\text{N})(\text{m})$$

The newton-meter is therefore the unit of work. This derived unit has a name. The newton-meter is called a **joule** (J) (pronounced "jool").

$$1 \text{ joule} = 1 \text{ newton meter}$$

The units for a newton are kg·m/s^2, and the unit for a meter is m. It therefore follows that the units for a joule are $\text{kg·m}^2/\text{s}^2$.

FIGURE 3.1 This is Glen Canyon Dam on the Colorado River between Utah and Arizona. The 216 m (about 710 ft) wall of concrete holds Lake Powell, which is 170 m (about 560 ft) deep at the dam when full. Water falls through 5 m (about 15 ft) diameter penstocks to generators at the bottom of the dam.

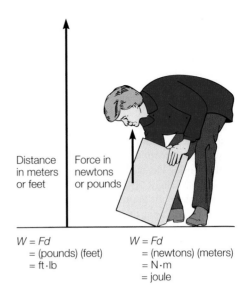

Distance in meters or feet | Force in newtons or pounds

$W = Fd$
$\quad$ = (pounds) (feet)
$\quad$ = ft·lb

$W = Fd$
$\quad$ = (newtons) (meters)
$\quad$ = N·m
$\quad$ = joule

FIGURE 3.3 Work is done against gravity when lifting an object. Work is measured in joules or in foot-pounds.

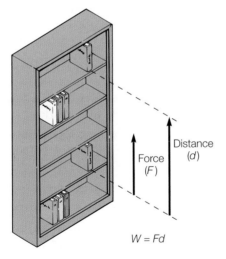

FIGURE 3.2 The force on the book moves it through a vertical distance from the second shelf to the fifth shelf, and work is done, $W = Fd$.

In the English system, the force is measured in pounds (lb), and the distance is measured in feet (ft). The unit of work in the English system is therefore the ft·lb. The ft·lb does not have a name of its own as the N·m does (Figure 3.3).

EXAMPLE 3.1

How much work is needed to lift a 5.0 kg backpack to a shelf 1.0 m above the floor?

SOLUTION

The backpack has a mass (m) of 5.0 kg, and the distance (d) is 1.0 m. To lift the backpack requires a vertically upward force equal to the weight of the backpack. Weight can be calculated from $w = mg$:

$m = 5.0$ kg
$g = 9.8$ m/s^2
$w = ?$

$w = mg$
$\quad = (5.0 \text{ kg})\left(9.8 \frac{\text{m}}{\text{s}^2}\right)$
$\quad = (5.0 \times 9.8) \text{ kg} \times \frac{\text{m}}{\text{s}^2}$
$\quad = 49 \frac{\text{kg·m}}{\text{s}^2}$
$\quad = \boxed{49\text{N}}$

The definition of work is found in equation 3.1,

$F = 49$ N
$d = 1.0$ m
$W = ?$

$W = Fd$
$\quad = (49 \text{ N})(1.0 \text{ m})$
$\quad = (49 \times 1.0)(\text{N·m})$
$\quad = 49\text{N·m}$
$\quad = \boxed{49 \text{ J}}$

EXAMPLE 3.2

How much work is required to lift a 50 lb box vertically a distance of 2 ft? (Answer: 100 ft·lb)

Simple Machines

Simple machines are tools that people use to help them do work. Recall that work is a force times a distance and you can see that the simple machine helps you do work by changing a force or a distance that something is moved. The force or distance advantage you gain by using the machine is called the *mechanical advantage*. The larger the mechanical advantage, the greater the effort that you would save by using the machine.

A lever is a simple machine, and Box Figure 3.1 shows what is involved when a lever reduces a force needed to do work. First, note there are two forces involved. The force that you provide by using the machine is called the *effort force*. You and the machine are working against the second force, called the *resistance force*. In the illustration, a 60 N effort force is used to move a resistance force of 300 N.

There are also two distances involved in using the lever. The distance over which your effort force acts is called the *effort distance*, and the distance the resistance moves is called the *resistance distance*. You would push down with an effort force of 60 N through an effort distance of 1 m. The 300 N rock, on the other hand, was raised a resistance distance of 0.2 m.

You did 60 N × 1 m, or 60 J of work on the lever. The work done on the rock by the lever was 300 N × 0.2 m, or 60 J of work. The work done by you on the lever is the same as the work done by the lever on the rock, so

$$\text{work input} = \text{work output}$$

Since work is force times distance, we can write this concept as

$$\text{effort force} \times \text{effort distance} =$$
$$\text{resistance force} \times \text{resistance distance}$$

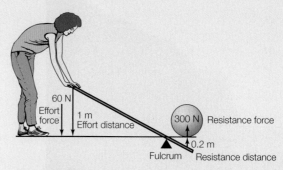

BOX FIGURE 3.1 The lever is one of six simple machines.

Ignoring friction, the work you get out of any simple machine is the same as the work you put into it. The lever enabled you to trade force for distance, and the mechanical advantage (MA) can be found from a ratio of the resistance force (F_R) divided by the effort force (F_E):

$$MA = \frac{F_R}{F_E}$$

Therefore, the example lever in Box Figure 3.1 had a mechanical advantage of

$$MA = \frac{F_R}{F_E}$$
$$= \frac{300 \text{ N}}{60 \text{ N}}$$
$$= 5$$

You can also find the mechanical advantage by dividing the effort distance (d_E) by the resistance distance (d_R):

$$MA = \frac{d_E}{d_R}$$

For the example lever, we find

$$MA = \frac{d_E}{d_R}$$
$$= \frac{1 \text{ m}}{0.2 \text{ m}}$$
$$= 5$$

So, we can use either the forces or distances involved in simple machines to calculate the mechanical advantage. In summary, a simple machine works for you by making it possible to apply a small force over a large distance to get a large force working over a small distance.

There are six kinds of simple machines: inclined plane, wedge, screw, lever, wheel and axle, and pulley. As you will see, the screw and wedge can be considered types of inclined planes; the wheel and axle and the pulley can be considered types of levers.

1. The *inclined plane* is a stationary ramp that is used to trade distance for force. You are using an inclined plane when

POWER

You are doing work when you walk up a stairway, since you are lifting yourself through a distance. You are lifting your weight (force exerted) the *vertical* height of the stairs (distance through which the force is exerted). Consider a person who weighs 120 lb and climbs a stairway with a vertical distance of 10 ft. This person will do (120 lb)(10 ft) or 1,200 ft·lb of work. Will the amount of work change if the person were to run up the stairs? The answer is no; the same amount of work is accomplished. Running up the stairs, however, is more tiring than walking up the stairs. You use the same amount of energy but at a greater *rate* when running. The rate at which energy is

CONCEPTS *Applied*

Book Work

Place a tied loop of string between the center pages of a small book. Pick up the loop so the string lifts the book, supporting it with open pages down. Use a spring scale to find the weight of the book in newtons. Measure the done work in lifting the book 1 m. Use the spring scale to measure the work done in pulling the book along a tabletop for 1 m. Is the amount of work done lifting the book the same as the amount of work done pulling the book along the tabletop? Why or why not?

you climb a stairway, drive a road that switches back and forth when going up a mountainside, or use a board to slide a heavy box up to a loading dock. Each use gives a large mechanical advantage by trading distance for force. For example, sliding a heavy box up a 10 m ramp to a 2 m-high loading dock raises the box with less force through a greater distance. The mechanical advantage of this inclined plane would be

$$MA = \frac{d_E}{d_R}$$

$$= \frac{10 \text{ m}}{2 \text{ m}}$$

$$= 5$$

Ignoring friction, a mechanical advantage of 5 means that a force of only 20 newtons would be needed to push a box weighing 100 newtons up the ramp.

2. The *wedge* is an inclined plane that moves. An ax is two back-to-back inclined planes that move through the wood it is used to split. Wedges are found in knives, axes, hatchets, and nails.

3. The *screw* is an inclined plane that has been wrapped around a cylinder, with the threads playing the role of the incline. A finely threaded screw has a higher mechanical advantage and requires less force to turn, but it also requires a greater effort distance.

4. The *lever* is a bar or board that is free to pivot about a fixed point called a fulcrum. There are three classes of levers based on the location of the fulcrum, effort force, and resistance force

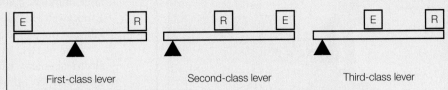

First-class lever Second-class lever Third-class lever

BOX FIGURE 3.2 The three classes of levers are defined by the relative locations of the fulcrum, effort, and resistance.

(Box Figure 3.2). A first-class lever has the fulcrum between the effort force and the resistance force. Examples are a seesaw, pliers, scissors, crowbars, and shovels. A second-class lever has the effort resistance between the fulcrum and the effort force. Examples are nutcrackers and twist-type jar openers. A third-class lever has the effort force between the resistance force and the fulcrum. Examples are fishing rods and tweezers.

A claw hammer can be used as a first-class lever to remove nails from a board. If the hammer handle is 30 cm and the distance from the nail slot to the fulcrum is 5 cm, the mechanical advantage will be

$$MA = \frac{d_E}{d_R}$$

$$= \frac{30 \text{ cm}}{5 \text{ cm}}$$

$$= 6$$

5. A *wheel and axle* has two circles, with the smaller circle called the axle and the larger circle called the wheel. The wheel and axle can be considered to be a lever that can move in a circle. Examples are a screwdriver, doorknob, steering wheel, and any application of a turning crank. The mechanical advantage is found from the radius of the wheel, where the effort is applied, to the radius of the axle, which is the distance over which the resistance moves. For example, a large screwdriver has a radius of 15 cm in the handle (the wheel) and 0.5 cm in the bit (the axle). The mechanical advantage of this screwdriver is

$$MA = \frac{d_E}{d_R}$$

$$= \frac{3 \text{ cm}}{0.5 \text{ cm}}$$

$$= 6$$

6. A *pulley* is a movable lever that rotates around a fulcrum. A single fixed pulley can only change the direction of a force. To gain a mechanical advantage, you need a fixed pulley and a movable pulley such as those found in a block and tackle. The mechanical advantage of a block and tackle can be found by comparing the length of rope or chain pulled to the distance the resistance has moved.

---

transformed or the rate at which work is done is called **power** (Figure 3.4). Power is measured as work per unit of time,

$$\text{power} = \frac{\text{work}}{\text{time}}$$

$$P = \frac{W}{t}$$

equation 3.2

Considering just the work and time factors, the 120 lb person who ran up the 10 ft height of stairs in 4 seconds would have a power rating of

$$P = \frac{W}{t} = \frac{(120 \text{ lb})(10 \text{ ft})}{4 \text{ s}} = 300 \frac{\text{ft·lb}}{\text{s}}$$

If the person had a time of 3 s on the same stairs, the power rating would be greater, 400 ft·lb/s. This is a greater *rate* of energy use, or greater power.

When the steam engine was first invented, there was a need to describe the rate at which the engine could do work. Since people at this time were familiar with using horses to do their work, the steam engines were compared to horses. James Watt, who designed a workable steam engine, defined **horsepower** as a power rating of 550 ft·lb/s (Figure 3.5A). To convert a power rating in the English units of ft·lb/s to horsepower, divide the power rating by 550 ft·lb/s/hp. For example, the 120 lb person who had a power rating of 400 ft·lb/s had a horsepower of 400 ft·lb/s ÷ 550 ft·lb/s/hp, or 0.7 hp.

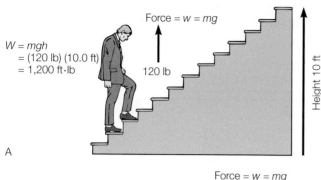

$W = mgh$
$= (120 \text{ lb}) (10.0 \text{ ft})$
$= 1,200 \text{ ft·lb}$

Force = $w = mg$

120 lb

Height 10 ft

A

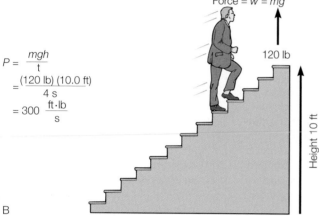

Force = $w = mg$

120 lb

$P = \dfrac{mgh}{t}$
$= \dfrac{(120 \text{ lb}) (10.0 \text{ ft})}{4 \text{ s}}$
$= 300 \dfrac{\text{ft·lb}}{\text{s}}$

Height 10 ft

B

FIGURE 3.4 (*A*) The work accomplished in climbing a stairway is the person's weight times the vertical distance. (*B*) The power level is the work accomplished per unit of time.

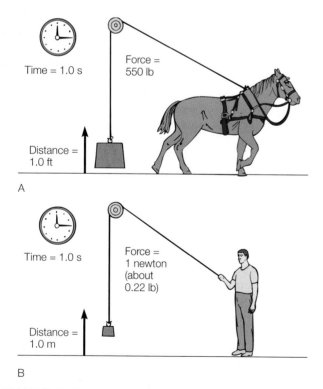

Time = 1.0 s

Force = 550 lb

Distance = 1.0 ft

A

Time = 1.0 s

Force = 1 newton (about 0.22 lb)

Distance = 1.0 m

B

FIGURE 3.5 (*A*) A horsepower is defined as a power rating of 550 ft·lb/s. (*B*) A watt is defined as a newton-meter per second, or joule per second.

In the metric system, power is measured in joules per second. The unit J/s, however, has a name. A J/s is called a **watt** (W).* The watt (Figure 3.5B) is used with metric prefixes for large numbers: 1,000 W = 1 kilowatt (kW) and 1,000,000 W = 1 megawatt (MW). It takes 746 W to equal 1 horsepower. One kilowatt is equal to about 1 1/3 horsepower.

The electric utility company charges you for how much electrical energy you have used. Electrical energy is measured by power (kW) times the time of use (h). The kWh is a unit of *work*, not power. Since power is

$$P = \frac{W}{t}$$

then it follows that

$$W = Pt$$

So power times time equals a unit of work, kWh. We will return to kilowatts and kilowatt-hours later when we discuss electricity.

EXAMPLE 3.3

An electric lift can raise a 500.0 kg mass a distance of 10.0 m in 5.0 s. What is the power of the lift?

SOLUTION

Power is work per unit time ($P = W/t$), and work is force times distance ($W = Fd$). The vertical force required is the weight lifted, and $w = mg$. Therefore, the work accomplished would be $W = mgh$, and the power would be $P = mgh/t$. Note that h is for height, a vertical distance (d).

$m = 500.0 \text{ kg}$

$g = 9.8 \text{ m/s}^2$

$h = 10.0 \text{ m}$

$t = 5.0 \text{ s}$

$P = ?$

$$P = \frac{mgh}{t}$$

$$= \frac{(500.0 \text{ kg})(9.8 \text{ m/s}^2)(10.0 \text{ m})}{5.0 \text{ s}}$$

$$= \frac{(500.0)(9.8)(10.0)}{5.0} \frac{\text{kg·}\frac{\text{m}}{\text{s}^2}\text{·m}}{\text{s}}$$

$$= 9,800 \frac{\text{N·m}}{\text{s}}$$

$$= 9,800 \frac{\text{J}}{\text{s}}$$

$$= 9,800 \text{ W}$$

$$= 9.8 \text{ kW}$$

The power in horsepower (hp) units would be

$$9,800 \text{ W} \times \frac{\text{hp}}{746 \text{ W}} = \boxed{13 \text{ hp}}$$

EXAMPLE 3.4

A 150 lb person runs up a 15 ft stairway in 10.0 s. What is the horsepower rating of the person? (Answer: 0.41 hp)

*Note that symbols for units, such as the watt, are not in italics. Symbols for quantities, such as work, are always in italics.

MOTION, POSITION, AND ENERGY

Closely related to the concept of work is the concept of **energy.** Energy can be defined as the *ability to do work.* This definition of energy seems consistent with everyday ideas about energy and physical work. After all, it takes a lot of energy to do a lot of work. In fact, one way of measuring the energy of something is to see how much work it can do. Likewise, when work is done *on* something, a change occurs in its energy level. The following examples will help clarify this close relationship between work and energy.

POTENTIAL ENERGY

Consider a book on the floor next to a bookcase. You can do work on the book by vertically raising it to a shelf. You can measure this work by multiplying the vertical upward force applied times the distance that the book is moved. You might find, for example, that you did an amount of work equal to 10 J on the book (see example 3.1).

Suppose that the book has a string attached to it, as shown in Figure 3.6. The string is threaded over a frictionless pulley and attached to an object on the floor. If the book is caused to fall from the shelf, the object on the floor will be vertically lifted through some distance by the string. The falling book exerts a force on the object through the string, and the object is moved through a distance. In other words, the *book* did work on the object through the string, $W = Fd$.

The book can do more work on the object if it falls from a higher shelf, since it will move the object a greater distance.

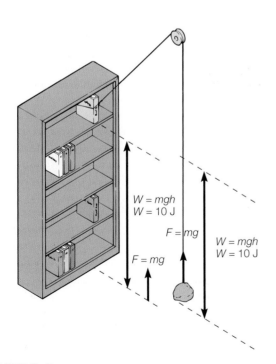

FIGURE 3.6 If moving a book from the floor to a high shelf requires 10 J of work, then the book will do 10 J of work on an object of the same mass when the book falls from the shelf.

The higher the shelf, the greater the *potential* for the book to do work. The ability to do work is defined as energy. The energy that an object has because of its position is called **potential energy** (*PE*). Potential energy is defined as *energy due to position.* This is called *gravitational potential energy,* since it is a result of gravitational attraction. There are other types of potential energy, such as that in a compressed or stretched spring.

Note the relationship between work and energy in the example. You did 10 J of work to raise the book to a higher shelf. In so doing, you increased the potential energy of the book by 10 J. The book now has the *potential* of doing 10 J of additional work on something else; therefore,

Work done on an object to change position	=	Increase in potential energy	=	Increase in work the object can do
work on book	=	potential energy of book	=	work by book
(10 J)		(10 J)		(10 J)

As you can see, a joule is a measure of work accomplished on an object. A joule is also a measure of potential energy. And, a joule is a measure of how much work an object can do. Both work and energy are measured in joules (or ft·lbs).

The gravitational potential energy of an object can be calculated, as described previously, from the work done *on* the object to change its position. You exert a force equal to its weight as you lift it some height above the floor, and the work you do is the product of the weight and height. Likewise, the amount of work the object *could* do because of its position is the product of its weight and height. For the metric unit of mass, weight is the product of the mass of an object times *g*, the acceleration due to gravity, so

$$\text{gravitational potential energy} = \text{weight} \times \text{height}$$
$$PE = mgh$$

equation 3.3

For English units, the pound *is* the gravitational unit of force, or weight, so equation 3.3 becomes $PE = (w)(h)$.

Under what conditions does an object have zero potential energy? Considering the book in the bookcase, you could say that the book has zero potential energy when it is flat on the floor. It can do no work when it is on the floor. But what if that floor happens to be the third floor of a building? You could, after all, drop the book out of a window. The answer is that it makes no difference. The same results would be obtained in either case since it is the *change of position* that is important in potential energy. The zero reference position for potential energy is therefore arbitrary. A zero reference point is chosen as a matter of convenience. Note that if the third floor of a building is chosen as the zero reference position, a book on ground level would have negative potential energy. This means that you would have to do work on the book to bring it back to the zero potential energy position (Figure 3.7). You will learn more about negative energy levels later in the chapters on chemistry.

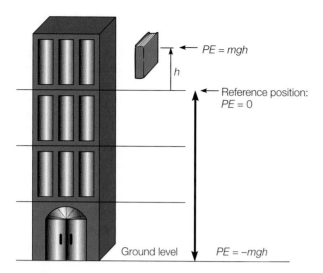

FIGURE 3.7 The zero reference level for potential energy is chosen for convenience. Here the reference position chosen is the third floor, so the book will have a negative potential energy at ground level.

EXAMPLE 3.5

What is the potential energy of a 2.14 kg book that is on a bookshelf 1.0 m above the floor?

SOLUTION

Equation 3.3, $PE = mgh$, shows the relationship between potential energy (PE), weight (mg), and height (h).

$$m = 2.14 \text{ kg} \qquad PE = mgh$$
$$h = 1.0 \text{ m}$$
$$PE = ? \qquad = (2.14 \text{ kg})\left(9.8 \frac{\text{m}}{\text{s}^2}\right)(1.0 \text{ m})$$
$$= (2.14)(9.8)(1.0) \frac{\text{kg} \cdot \text{m}}{\text{s}^2} \times \text{m}$$
$$= 21 \text{ Nm}$$
$$= \boxed{21 \text{ J}}$$

EXAMPLE 3.6

How much work can a 5.00 kg mass do if it is 5.00 m above the ground? (Answer: 250 J)

 CONCEPTS *Applied*

Work and Power

Power is the rate of expending energy or of doing work. You can find your power output by taking a few measurements.

First, let's find how much work you do in walking up a flight of stairs. Your work output will be approximately equal to the change in your potential energy (*mgh*), so you will need (1) to measure the vertical height of a flight

of stairs in metric units and (2) to calculate or measure your mass (conversion factors are located inside the front cover). Record your findings in your report.

Second, find your power output by climbing the stairs as fast as you can while someone measures the time with a stopwatch. Find your power output in watts. Convert this to horsepower by consulting the conversion factors inside the front cover. Did you develop at least 1 hp? Does a faster person always have more horsepower?

$$\text{walking power} = \frac{(\quad \text{kg})(9.8 \text{m/s}^2)(\quad \text{m})}{(\quad \text{s})}$$

$$\text{running power} = \frac{(\quad \text{kg})(9.8 \text{ m/s}^2)(\quad \text{m})}{(\quad \text{s})}$$

KINETIC ENERGY

Moving objects have the ability to do work on other objects because of their motion. A rolling bowling ball exerts a force on the bowling pins and moves them through a distance, but the ball loses speed as a result of the interaction (Figure 3.8). A moving car has the ability to exert a force on a small tree and knock it down, again with a corresponding loss of speed. Objects in motion have the ability to do work, so they have energy. The energy of motion is known as **kinetic energy.** Kinetic energy can be measured (1) in terms of the work done to put the object in motion or (2) in terms of the work the moving object will do in coming to rest. Consider objects that you put into motion by throwing. You exert a force on a football as you accelerate it through a distance before it leaves your hand. The kinetic energy that the ball now has is equal to the work (force times distance) that you did on the ball. You exert a force on a baseball through a distance as the ball increases its speed before it leaves your hand. The kinetic energy that the ball now has is equal to the work that you did on the ball. The ball exerts a force on the

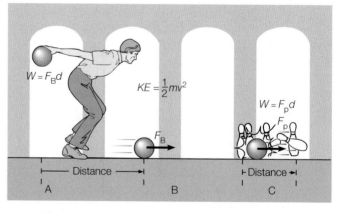

FIGURE 3.8 (A) Work is done on the bowling ball as a force (F_B) moves it through a distance. (B) This gives the ball a kinetic energy equal in amount to the work done on it. (C) The ball does work on the pins and has enough remaining energy to crash into the wall behind the pins.

hand of the person catching the ball and moves it through a distance. The net work done on the hand is equal to the kinetic energy that the ball had. Therefore,

$$\begin{array}{c}\text{Work done to}\\\text{put an object}\\\text{in motion}\end{array} = \begin{array}{c}\text{Increase in}\\\text{kinetic energy}\end{array} = \begin{array}{c}\text{Increase in}\\\text{work the object}\\\text{can do}\end{array}$$

A baseball and a bowling ball moving with the same velocity do not have the same kinetic energy. You cannot knock down many bowling pins with a slowly rolling baseball. Obviously, the more massive bowling ball can do much more work than a less massive baseball with the same velocity. Is it possible for the bowling ball and the baseball to have the same kinetic energy? The answer is yes, if you can give the baseball sufficient velocity. This might require shooting the baseball from a cannon, however. Kinetic energy is proportional to the mass of a moving object, but velocity has a greater influence. Consider two balls of the same mass, but one is moving twice as fast as the other. The ball with twice the velocity will do *four* times as much work as the slower ball. A ball with three times the velocity will do *nine* times as much work as the slower ball. Kinetic energy is proportional to the square of the velocity ($2^2 = 4$; $3^2 = 9$). The kinetic energy (*KE*) of an object is

$$\text{kinetic energy} = \frac{1}{2}(\text{mass})(\text{velocity})^2$$

$$KE = \frac{1}{2}mv^2$$

equation 3.4

The unit of mass is the kg, and the unit of velocity is m/s. Therefore, the unit of kinetic energy is

$$KE = (\text{kg})\left(\frac{\text{m}}{\text{s}}\right)^2$$

$$= (\text{kg})\left(\frac{\text{m}^2}{\text{s}^2}\right)$$

$$= \frac{\text{kg}\cdot\text{m}^2}{\text{s}^2}$$

which is the same thing as

$$\left(\frac{\text{kg}\cdot\text{m}}{\text{s}^2}\right)(\text{m})$$

or

N·m

or

joule (J)

Kinetic energy is measured in joules.

EXAMPLE 3.7

A 7.00 kg bowling ball is moving in a bowling lane with a velocity of 5.00 m/s. What is the kinetic energy of the ball?

SOLUTION

The relationship between kinetic energy (*KE*), mass (*m*), and velocity (*v*) is found in equation 3.4, $KE = 1/2\ mv^2$:

$$m = 7.00 \text{ kg} \qquad KE = \frac{1}{2}mv^2$$

$$v = 5.00 \text{ m/s}$$

$$KE = ? \qquad = \frac{1}{2}(7.00 \text{ kg})\left(5.00\ \frac{\text{m}}{\text{s}}\right)^2$$

$$= \frac{1}{2}(7.00 \times 25.0) \text{ kg} \times \frac{\text{m}^2}{\text{s}^2}$$

$$= \frac{1}{2}\ 175\ \frac{\text{kg}\cdot\text{m}^2}{\text{s}^2}$$

$$= 87.5\ \frac{\text{kg}\cdot\text{m}}{\text{s}^2}\cdot\text{m}$$

$$= 87.5 \text{ N}\cdot\text{m}$$

$$= \boxed{87.5 \text{ J}}$$

EXAMPLE 3.8

A 100.0 kg football player moving with a velocity of 6.0 m/s tackles a stationary quarterback. How much work was done on the quarterback? (Answer: 1,800 J)

ENERGY FLOW

The key to understanding the individual concepts of work and energy is to understand the close relationship between the two. When you do work on something, you give it energy of position (potential energy) or you give it energy of motion (kinetic energy). In turn, objects that have kinetic or potential energy can now do work on something else as the transfer of energy continues. Where does all this energy come from and where does it go? The answer to these questions is the subject of this section on energy flow.

WORK AND ENERGY

Energy is used to do work on an object, exerting a force through a distance. This force is usually *against* something (Figure 3.9), and here are five examples of resistance:

1. *Work against inertia.* A net force that changes the state of motion of an object is working against inertia. According to the laws of motion, a net force acting through a distance is needed to change the velocity of an object.
2. *Work against gravity.* Consider the force from gravitational attraction. A net force that changes the position of an object is a downward force from the acceleration due to gravity acting on a mass, $w = mg$. To change the position of an object, a force opposite to *mg* is needed to act through the distance of the position change. Thus, lifting an object requires doing work against the force of gravity.
3. *Work against friction.* The force that is needed to maintain the motion of an object is working against friction. Friction is always present when two surfaces in contact move over each other. Friction resists motion.

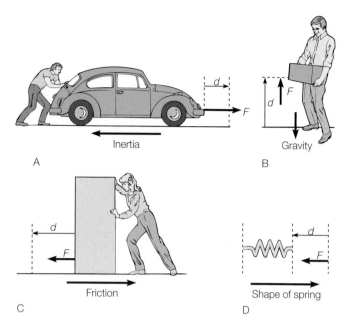

FIGURE 3.9 Examples of working against (*A*) inertia, (*B*) gravity, (*C*) friction, and (*D*) shape.

4. *Work against shape.* The force that is needed to stretch or compress a spring is working against the shape of the spring. Other examples of work against shape include compressing or stretching elastic materials. If the elastic limit is reached, then the work goes into deforming or breaking the material.

5. *Work against any combination of inertia, fundamental forces, friction, and/or shape.* It is a rare occurrence on Earth that work is against only one type of resistance. Pushing on the back of a stalled automobile to start it moving up a slope would involve many resistances. This is complicated, however, so a single resistance is usually singled out for discussion.

Work is done against a resistance, but what is the result? The result is that some kind of *energy change* has taken place. Among the possible energy changes are the following:

1. *Increased kinetic energy.* Work against inertia results in an increase of kinetic energy, the energy of motion.
2. *Increased potential energy.* Work against gravity and work against shape result in an increase of potential energy, the energy of position.
3. *Increased temperature.* Work against friction results in an increase in the temperature. Temperature is a manifestation of the kinetic energy of the particles making up an object, as you will learn in chapter 4.
4. *Increased combinations of kinetic energy, potential energy, and/or temperature.* Again, isolated occurrences are more the exception than the rule. In all cases, however, the sum of the total energy changes will be equal to the work done.

Work was done *against* various resistances, and energy was *increased* as a result. The object with increased energy can now do work on some other object or objects. A moving object

has kinetic energy, so it has the ability to do work. An object with potential energy has energy of position, and it, too, has the ability to do work. You could say that energy *flowed* into and out of an object during the entire process. The following energy scheme is intended to give an overall conceptual picture of energy flow. Use it to develop a broad view of energy. You will learn the details later throughout the course.

ENERGY FORMS

Energy comes in various forms, and different terms are used to distinguish one form from another. Although energy comes in various *forms*, this does not mean that there are different *kinds* of energy. The forms are the result of the more common fundamental forces—gravitational, electromagnetic, and nuclear—and objects that are interacting. Energy can be categorized into five forms: (1) *mechanical*, (2) *chemical*, (3) *radiant*, (4) *electrical*, and (5) *nuclear*. The following is a brief discussion of each of the five forms of energy.

Mechanical energy is the form of energy of familiar objects and machines (Figure 3.10). A car moving on a highway has kinetic mechanical energy. Water behind a dam has potential mechanical energy. The spinning blades of a steam turbine have kinetic mechanical energy. The form of mechanical energy is usually associated with the kinetic energy of everyday-sized objects and the potential energy that results from gravity. There are other possibilities (e.g., sound), but this description will serve the need for now.

Chemical energy is the form of energy involved in chemical reactions (Figure 3.11). Chemical energy is released in the chemical reaction known as *oxidation*. The fire of burning wood is an example of rapid oxidation. A slower oxidation releases energy from food units in your body. As you will learn in the chemistry unit, chemical energy involves electromagnetic forces between the parts of atoms. Until then, consider the following

FIGURE 3.10 Mechanical energy is the energy of motion, or the energy of position, of many familiar objects. This boat has energy of motion.

A

B

FIGURE 3.11 Chemical energy is a form of potential energy that is released during a chemical reaction. Both (*A*) wood and (*B*) coal have chemical energy that has been stored through the process of photosynthesis. The pile of wood might provide fuel for a small fireplace for several days. The pile of coal might provide fuel for a power plant for a hundred days.

comparison. Photosynthesis is carried on in green plants. The plants use the energy of sunlight to rearrange carbon dioxide and water into plant materials and oxygen. Leaving out many steps and generalizing, this reaction could be represented by the following word equation:

$$\text{energy} + \text{carbon dioxide} + \text{water} = \text{wood} + \text{oxygen}$$

The plant took energy and two substances and made two different substances. This is similar to raising a book to a higher shelf in a bookcase. That is, the new substances have more energy than the original ones did. Consider a word equation for the burning of wood:

$$\text{wood} + \text{oxygen} = \text{carbon dioxide} + \text{water} + \text{energy}$$

Notice that this equation is exactly the reverse of photosynthesis. In other words, the energy used in photosynthesis was released

FIGURE 3.12 This demonstration solar cell array converts radiant energy from the Sun to electrical energy, producing an average of 200,000 watts of electric power (after conversion).

during oxidation. Chemical energy is a kind of potential energy that is stored and later released during a chemical reaction.

Radiant energy is energy that travels through space (Figure 3.12). Most people think of light or sunlight when considering this form of energy. Visible light, however, occupies only a small part of the complete electromagnetic spectrum, as shown in Figure 3.13. Radiant energy includes light and all other parts of the spectrum (see chapter 7). Infrared radiation is sometimes called "heat radiation" because of the association with heating when this type of radiation is absorbed. For example, you feel the interaction of infrared radiation when you hold your hand near a warm range element. However, infrared radiation is another type of radiant energy. In fact, some snakes, such as rattlesnakes, copperheads, and water mocassins, have pits between their eyes that can detect infrared radiation emitted from warm animals where you see total darkness. Microwaves are another type of radiant energy that are used in cooking. As with other forms of energy, light, infrared, and microwaves will be considered in more detail later. For now, consider all types of radiant energy to be forms of energy that travel through space.

Electrical energy is another form of energy from electromagnetic interactions that will be considered in detail later. You are familiar with electrical energy that travels through wires to your home from a power plant (Figure 3.14), electrical energy that is generated by chemical cells in a flashlight, and electrical energy that can be "stored" in a car battery.

Nuclear energy is a form of energy often discussed because of its use as an energy source in power plants. Nuclear energy is another form of energy from the atom, but this time, the energy involves the nucleus, the innermost part of an atom, and nuclear interactions. This will be considered in detail in chapter 13.

ENERGY CONVERSION

Potential energy can be converted to kinetic energy and vice versa. The simple pendulum offers a good example of this conversion. A simple pendulum is an object, called a bob, suspended

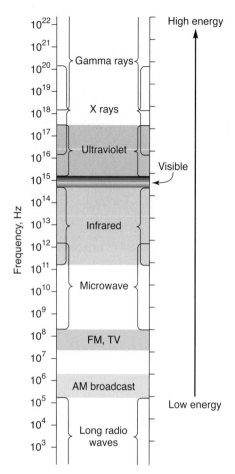

FIGURE 3.13 The frequency spectrum of electromagnetic waves. The amount of radiant energy carried by these waves increases with frequency. Note that visible light occupies only a small part of the complete spectrum.

by a string or wire from a support. If the bob is moved to one side and then released, it will swing back and forth in an arc. At the moment that the bob reaches the top of its swing, it stops for an instant, then begins another swing. At the instant of stopping, the bob has 100 percent potential energy and no kinetic energy. As the bob starts back down through the swing, it is gaining kinetic energy and losing potential energy. At the instant the bob is at the bottom of the swing, it has 100 percent kinetic energy and no potential energy. As the bob now climbs through the other half of the arc, it is gaining potential energy and losing kinetic energy until it again reaches an instantaneous stop at the top, and the process starts over. The kinetic energy of the bob at the bottom of the arc is equal to the potential energy it had at the top of the arc (Figure 3.15). Disregarding friction, the sum of the potential energy and the kinetic energy remains constant throughout the swing.

The potential energy lost during a fall equals the kinetic energy gained (Figure 3.16). In other words,

$$PE_{lost} = KE_{gained}$$

Substituting the values from equations 3.3 and 3.4,

$$mgh = \frac{1}{2} mv^2$$

Canceling the m and solving for v_f,

$$v_f = \sqrt{2gh}$$

equation 3.5

Equation 3.5 tells you the final speed of a falling object after its potential energy is converted to kinetic energy. This assumes, however, that the object is in free fall, since the effect of air resistance is ignored. Note that the m's cancel, showing again that the mass of an object has no effect on its final speed.

FIGURE 3.14 The blades of a steam turbine. In a power plant, chemical or nuclear energy is used to heat water to steam, which is directed against the turbine blades. The mechanical energy of the turbine turns an electric generator. Thus, a power plant converts chemical or nuclear energy to mechanical energy, which is then converted to electrical energy.

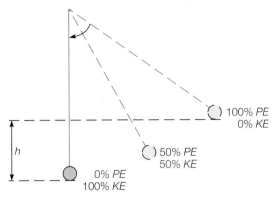

FIGURE 3.15 This pendulum bob loses potential energy (PE) and gains an equal amount of kinetic energy (KE) as it falls through a distance h. The process reverses as the bob moves up the other side of its swing.

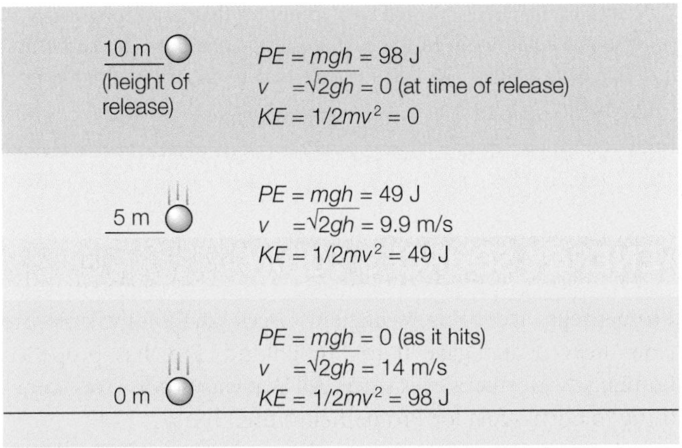

10 m ⃝ (height of release)	$PE = mgh = 98$ J $v = \sqrt{2gh} = 0$ (at time of release) $KE = 1/2mv^2 = 0$
5 m ⃝	$PE = mgh = 49$ J $v = \sqrt{2gh} = 9.9$ m/s $KE = 1/2mv^2 = 49$ J
0 m ⃝	$PE = mgh = 0$ (as it hits) $v = \sqrt{2gh} = 14$ m/s $KE = 1/2mv^2 = 98$ J

FIGURE 3.16 The ball trades potential energy for kinetic energy as it falls. Notice that the ball had 98 J of potential energy when dropped and has a kinetic energy of 98 J just as it hits the ground.

EXAMPLE 3.9

A 1.0 kg book falls from a height of 1.0 m. What is its velocity just as it hits the floor?

SOLUTION

The relationships involved in the velocity of a falling object are given in equation 3.5.

$$h = 1.0 \text{ m} \qquad v_f = \sqrt{2gh}$$
$$g = 9.8 \text{ m/s}^2 \qquad = \sqrt{(2)(9.8 \text{ m/s}^2)(1.0 \text{ m})}$$
$$v_f = ? \qquad = \sqrt{2 \times 9.8 \times 1.0 \frac{\text{m}}{\text{s}^2}\cdot\text{m}}$$
$$= \sqrt{19.6 \frac{\text{m}^2}{\text{s}^2}}$$
$$= \boxed{4.4 \text{ m/s}}$$

EXAMPLE 3.10

What is the kinetic energy of a 1.0 kg book just before it hits the floor after a 1.0 m fall? (Answer: 9.7 J)

Any *form* of energy can be converted to another form. In fact, most technological devices that you use are nothing more than *energy-form converters* (Figure 3.17). A lightbulb, for example, converts electrical energy to radiant energy. A car converts chemical energy to mechanical energy. A solar cell converts radiant energy to electrical energy, and an electric motor converts electrical energy to mechanical energy. Each technological device converts some form of energy (usually chemical or electrical) to another form that you desire (usually mechanical or radiant).

It is interesting to trace the *flow of energy* that takes place in your surroundings. Suppose, for example, that you are riding a bicycle. The bicycle has kinetic mechanical energy as it moves along. Where did the bicycle get this energy? From you, as you use the chemical energy of food units to contract your muscles and move the bicycle along. But where did your chemical energy come from? It came from your food, which consists of plants, animals who eat plants, or both plants and animals. In any case, plants are at the bottom of your food chain. Plants convert radiant energy from the Sun into chemical energy. Radiant energy comes to the plants from the Sun because of the nuclear reactions that took place in the core of the Sun. Your bicycle is therefore powered by nuclear energy that has undergone a number of form conversions!

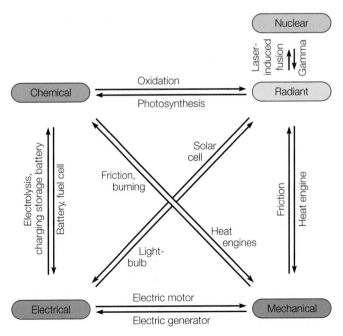

FIGURE 3.17 The energy forms and some conversion pathways.

ENERGY CONSERVATION

Energy can be transferred from one object to another, and it can be converted from one form to another form. If you make a detailed accounting of all forms of energy before and after a transfer or conversion, the total energy will be *constant.* Consider your bicycle coasting along over level ground when you apply the brakes. What happened to the kinetic mechanical energy of the bicycle? It went into heating the rim and brakes of your bicycle, then eventually radiated to space as infrared radiation. All radiant energy that reaches Earth is eventually radiated back to space (Figure 3.18). Thus, throughout all the form conversions and energy transfers that take place, the total sum of energy remains constant.

The total energy is constant in every situation that has been measured. This consistency leads to another one of the conservation laws of science, the **law of conservation of energy:**

> **Energy is never created or destroyed. Energy can be converted from one form to another, but the total energy remains constant.**

You may be wondering about the source of nuclear energy. Does a nuclear reaction create energy? Albert Einstein answered this question back in the early 1900s, when he formulated his now-famous relationship between mass and energy, $E = mc^2$. This relationship will be discussed in detail in chapter 13. Basically, the relationship states that mass *is* a form of energy, and this has been experimentally verified many times.

ENERGY TRANSFER

Earlier it was stated that when you do work on something, you give it energy. The result of work could be increased kinetic mechanical energy, increased gravitational potential energy, or an increase in the temperature of an object. You could summarize this by stating that either *working* or *heating* is always involved any time energy is transformed. This is not unlike your financial situation. In order to increase or decrease your financial status, you need some mode of transfer, such as cash or checks, as a means of conveying assets. Just as with cash

flow from one individual to another, energy flow from one object to another requires a mode of transfer. In energy matters, the mode of transfer is working or heating. Any time you see working or heating occurring, you know that an energy transfer is taking place. The next time you see heating, think about what energy form is being converted to what new energy form. (The final form is usually radiant energy.) Heating is the topic of chapter 4, where you will consider the role of heat in energy matters.

Myths, Mistakes, & Misunderstandings

Leave the Computer on?

It is a myth that leaving your computer on all the time uses less energy and makes it last longer. There is a very small surge when the computer is first turned on, but this is insignificant compared to the energy wasted by a computer that is running when it is not being used. In the past, cycling a computer on and off may have reduced its lifetime, but this is not true of modern computers.

ENERGY SOURCES TODAY

Prometheus, according to ancient Greek mythology, stole fire from heaven and gave it to humankind. Fire has propelled human advancement ever since. All that was needed was something to burn—fuel for Prometheus's fire.

Any substance that burns can be used to fuel a fire, and various fuels have been used over the centuries as humans advanced. First, wood was used as a primary source for heating. Then coal fueled the Industrial Revolution. Eventually, humankind roared into the twentieth century burning petroleum. According to a 2005 report on primary energy consumed in the United States, petroleum was the most widely used source of energy (Figure 3.19). It provided about 40 percent of the total energy used, and natural gas contributed about 23 percent of the total. The use of coal provided about 23 percent of the total. Biomass, which is any material formed by photosynthesis, contributed about 3 percent of the total. Note that petroleum, coal, biomass, and natural gas are all chemical sources of energy, sources that are mostly burned for their energy. These chemical sources supplied about 89 percent of the total energy consumed. About a third of this was burned for heating, and the rest was burned to drive engines or generators.

Nuclear energy and hydropower are the nonchemical sources of energy. These sources are used to generate electrical energy. The alternative sources of energy, such as solar and geothermal, provided about 1 percent of the total energy consumed.

The energy-source mix has changed from past years, and it will change in the future. Wood supplied 90 percent of the energy until the 1850s, when the use of coal increased. Then, by 1910, coal was supplying about 75 percent of the total energy

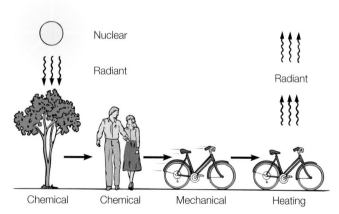

FIGURE 3.18 Energy arrives from the Sun, goes through a number of conversions, then radiates back into space. The total sum leaving eventually equals the original amount that arrived.

Have you heard of biodiesel? Biodiesel is a vegetable-based oil that can be used for fuel in diesel engines. It can be made from soy oils, canola oil, or even recycled deep-fryer oil from a fast-food restaurant. Biodiesel can be blended with regular diesel oil in any amount. Or it can be used 100 percent pure in diesel cars, trucks, and buses, or as home heating oil.

Why would we want to use vegetable oil to run diesel engines? First, it is a sustainable (or renewable) resource. It also reduces dependency on foreign oil, as well as cuts the trade deficit. It runs smoother, produces less exhaust smoke, and reduces the health risks associated with petroleum diesel. The only negative aspect seems to occur when recycled oil from fast-food restaurants

is used. People behind such a biodiesel-powered school bus complained that it smelled like fried potatoes, making them hungry.

There is a website maintained by some biodiesel users where you can learn how to produce your own biodiesel from algae. See www.biodieselnow.com and search for the term *algae*.

---

needs. Then petroleum began making increased contributions to the energy supply. Now increased economic and environmental constraints and a decreasing supply of petroleum are producing another supply shift. The present petroleum-based energy era is about to shift to a new energy era.

About 97 percent of the total energy consumed today is provided by four sources: (1) petroleum (including natural gas), (2) coal, (3) hydropower, and (4) nuclear. The following is a brief introduction to these four sources.

PETROLEUM

The word *petroleum* is derived from the Greek word *petra*, meaning rock, and the Latin word *oleum*, meaning oil. Petroleum is oil that comes from oil-bearing rock. Natural gas is universally associated with petroleum and has similar origins. Both petroleum and natural gas form from organic sediments, materials that have settled out of bodies of water. Sometimes a local condition

permits the accumulation of sediments that are exceptionally rich in organic material. This could occur under special conditions in a freshwater lake, or it could occur on shallow ocean basins. In either case, most of the organic material is from plankton—tiny free-floating animals and plants such as algae. It is from such accumulations of buried organic material that petroleum and natural gas are formed. The exact process by which these materials become petroleum and gas is not understood. It is believed that bacteria, pressure, appropriate temperatures, and time are all important. Natural gas is formed at higher temperatures than is petroleum. Varying temperatures over time may produce a mixture of petroleum and gas or natural gas alone.

Petroleum forms a thin film around the grains of the rock where it formed. Pressure from the overlying rock and water move the petroleum and gas through the rock until it reaches a rock type or structure that stops it. If natural gas is present, it occupies space above the accumulating petroleum. Such accumulations of petroleum and natural gas are the sources of supply for these energy sources.

Discussions about the petroleum supply and the cost of petroleum usually refer to a "barrel of oil." The *barrel* is an accounting device of 42 U.S. gallons. Such a 42-gallon barrel does not exist. When or if oil is shipped in barrels, each drum holds 55 U.S. gallons. The various uses of petroleum products are discussed in chapter 12.

The supply of petroleum and natural gas is limited. Most of the continental drilling prospects appear to be exhausted, and the search for new petroleum supplies is now offshore. In general, over 25 percent of our nation's petroleum is estimated to come from offshore wells. Imported petroleum accounts for more than half of the oil consumed, with most imported oil coming from Mexico, Canada, Venezuela, Nigeria, and Saudi Arabia.

Petroleum is used for gasoline (about 45 percent), diesel (about 40 percent), and heating oil (about 15 percent). Petroleum is also used in making medicine, clothing fabrics, plastics, and ink.

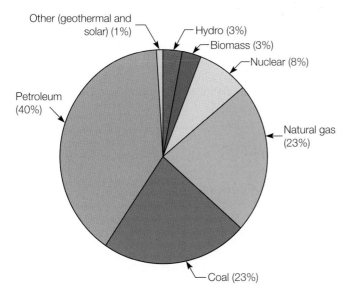

FIGURE 3.19 Primary energy consumed in the United States by source, 2005. *Source:* Energy Information Administration (www.epa.gov/owmitmet/basics.html)

COAL

Petroleum and natural gas formed from the remains of tiny organisms that lived millions of years ago. Coal, on the other hand, formed from an accumulation of plant materials that

People Behind the Science

James Prescott Joule (1818–1889)

James Prescott Joule

James Joule was a British physicist who helped develop the principle of conservation of energy by experimentally measuring the mechanical equivalent of heat. In recognition of Joule's pioneering work on energy, the SI unit of energy is named the joule.

Joule was born on December 24, 1818, into a wealthy brewing family. He and his brother were educated at home between 1833 and 1837 in elementary math, natural philosophy, and chemistry, partly by the English chemist John Dalton (1766–1844) (see p. 204). Joule was a delicate child and very shy, and apart from his early education, he was entirely self-taught in science. He does not seem to have played any part in the family brewing business, although some of his first experiments were done in the laboratory at the brewery.

Joule had great dexterity as an experimenter, and he could measure temperatures very precisely. At first, other scientists could not believe such accuracy and were skeptical about the theories that Joule developed to explain his results. The encouragement of Lord Kelvin from 1847 changed these attitudes, however, and Kelvin subsequently used Joule's practical ability to great advantage. By 1850, Joule was highly regarded by other scientists and was elected a fellow of the Royal Society. Joule's own wealth was able to fund his scientific career, and he never took an academic post. His funds eventually ran out, however. He was awarded a pension in 1878 by Queen Victoria, but by that time, his mental powers were going. He suffered a long illness and died on October 11, 1889.

Joule realized the importance of accurate measurement very early on, and exact data became his hallmark. His most active research period was between 1837 and 1847. In a long series of experiments, he studied the relationship between electrical, mechanical, and chemical effects and heat, and in 1843, he was able to announce his determination of the amount of work required to produce a unit of heat. This is called the mechanical equivalent of heat (4.184 joules per calorie).

One great value of Joule's work was the variety and completeness of his experimental evidence. He showed that the same relationship could be examined experimentally and that the ratio of equivalence of the different forms of energy did not depend on how one form was converted into another or on the materials involved. The principle that Joule had established is that energy cannot be created or destroyed but only transformed.

Joule lives on in the use of his name to measure energy, supplanting earlier units such as the erg and calorie. It is an appropriate reflection of his great experimental ability and his tenacity in establishing a basic law of science.

Source: Modified from the Hutchinson *Dictionary of Scientific Biography*. © Research Machines plc 2003. All Rights Reserved. Helicon Publishing is a division of Research Machines.

collected under special conditions millions of years ago. Thus, petroleum, natural gas, and coal are called **fossil fuels.** Fossil fuels contain the stored radiant energy of organisms that lived millions of years ago.

The first thing to happen in the formation of coal was that plants in swamps died and sank. Stagnant swamp water protected the plants and plant materials from consumption by animals and decomposition by microorganisms. Over time, chemically altered plant materials collected at the bottom of pools of water in the swamp. This carbon-rich material is *peat* (not to be confused with peat moss). Peat is used as a fuel in many places in the world. The flavor of Scotch (whisky) is the result of the peat fires used to brew the liquor. Peat is still being produced naturally in swampy areas today. Under pressure and at high temperatures peat will eventually be converted to coal. There are several stages, or *ranks,* in the formation of coal. The lowest rank is lignite (brown coal), and then subbituminous, then bituminous (soft coal), and the highest rank is anthracite (hard coal).

Each rank of coal has different burning properties and a different energy content. Coal also contains impurities of clay, silt, iron oxide, and sulfur. The mineral impurities leave an ash when the coal is burned, and the sulfur produces sulfur dioxide, a pollutant.

Most of the coal mined today is burned by utilities to generate electricity (about 80 percent). The coal is ground to a face-powder consistency and blown into furnaces. This greatly increases efficiency but produces *fly ash,* ash that "flies" up the chimney. Industries and utilities are required by the U.S. Clean Air Act to remove sulfur dioxide and fly ash from plant emissions. About 20 percent of the cost of a new coal-fired power plant goes into air pollution control equipment. Coal is an abundant but dirty energy source.

MOVING WATER

Moving water has been used as a source of energy for thousands of years. It is considered a renewable energy source, inexhaustible as long as the rain falls. Today, hydroelectric plants generate about 3 percent of the nation's *total* energy consumption at about 2,400 power-generating dams across the nation. Hydropower furnished about 40 percent of the United States' electric power in 1940. Today, dams furnish 9 percent of the electric power. It is projected that this will drop even lower, perhaps to 7 percent in the near future. Energy consumption has increased, but hydropower production has not kept pace because geography limits the number of sites that can be built.

Water from a reservoir is conducted through large pipes called penstocks to a powerhouse, where it is directed against turbine blades that turn a shaft on an electric generator. A rough approximation of the power that can be extracted from the falling water can be made by multiplying the depth of the water (in feet) by the amount of water flowing (in cubic feet per second), then dividing by 10. The result is roughly equal to the horsepower.

NUCLEAR

Nuclear power plants use nuclear energy to produce electricity. Energy is released as the nuclei of uranium and plutonium atoms split, or undergo a nuclear reaction called fission and form new

 CONCEPTS *Applied*

City Power

Compare amounts of energy sources needed to produce electric power. Generally, 1 MW (1,000,000 W) will supply the electrical needs of 1,000 people.

1. Use the population of your city to find how many megawatts of electricity are required for your city.
2. Use the following equivalencies to find out how much coal, oil, gas, or uranium would be consumed in one day to supply the electrical needs.

$$1 \text{ kWh of electricity} = \begin{cases} 1 \text{ lb of coal} \\ 0.08 \text{ gal of oil} \\ 9 \text{ cubic ft of gas} \\ 0.00013 \text{ g of uranium} \end{cases}$$

Example

Assume your city has 36,000 people. Then 36 MW of electricity will be needed. How much oil is needed to produce this electricity?

$$36 \text{ MW} \times \frac{1,000 \text{ kW}}{\text{MW}} \times \frac{24 \text{ h}}{\text{day}} \times \frac{0.08 \text{ gal}}{\text{kWh}} = \begin{array}{l} 69,120 \text{ or about} \\ 70,000 \text{ gal/day} \end{array}$$

Since there are 42 gallons in a barrel,

$$\frac{70,000 \text{ gal·day}}{42 \text{ gal·barrel}} = \frac{7,000}{42} \times \frac{\text{gal}}{\text{day}} \times \frac{\text{barrel}}{\text{gal}} = \begin{array}{l} 1,666, \text{ or about} \\ 2,000 \text{ barrel/day} \end{array}$$

elements (for the details, see chapter 13). The fissioning takes place in a large steel vessel called a *reactor*. Water is pumped through the reactor to produce steam, which is used to produce electrical energy, just as in the fossil fuel power plants. The nuclear processes are described in detail in chapter 13, and the process of producing electrical energy is described in detail in chapter 6. Nuclear power plants use nuclear energy to produce electricity, but some people oppose the use of this process. The electric utility companies view nuclear energy as *one* energy source used to produce electricity. They state that they have no allegiance to any one energy source but are seeking to utilize the most reliable and dependable of several energy sources. Petroleum, coal, and hydropower are also used as energy sources for electric power production. The electric utility companies are concerned that petroleum and natural gas are becoming increasingly expensive, and there are questions about long-term supplies. Hydropower has limited potential for growth, and solar energy is prohibitively expensive today. Utility companies see two major energy sources that are available for growth: coal and nuclear. There are problems and advantages to each, but the utility companies feel they must use coal and nuclear power until the new technologies, such as solar power, are economically feasible.

CONSERVING ENERGY

Conservation is not a way of generating energy, but it is a way of reducing the need for additional energy consumption and saves money for the consumer. Some conservation technologies are sophisticated, while others are quite simple. For example, if a small, inexpensive wood-burning stove were developed and used to replace open fires in the less-developed world, energy consumption in these regions could be reduced by 50 percent.

Many observers have pointed out that demanding more energy while failing to conserve is like demanding more water to fill a bathtub while leaving the drain open. To be sure, conservation and efficiency strategies by themselves will not eliminate demands for energy, but they can make the demands much easier to meet, regardless of what options are chosen to provide the primary energy. Energy efficiency improvements have significantly reduced the need for additional energy sources. Consider these facts, which are based primarily on data published by the U.S. Energy Information Administration:

- Total primary energy use per capita in the United States in 2003 was almost identical to that in 1973. Over the same thirty-year period, economic output (gross domestic product, or GDP) per capita increased 74 percent.
- National energy intensity (energy use per unit of GDP) fell 43 percent between 1973 and 2002. About 60 percent of this decline is attributable to energy efficiency improvements.
- If the United States had not dramatically reduced its energy intensity over the past thirty years, consumers and businesses would have spent at least $430 billion more on energy purchases in 2003.

Even though the United States is much more energy-efficient today than it was twenty-five years ago, the potential is still enormous for additional cost-effective energy savings. Some newer energy efficiency measures have barely begun to be adopted. Other efficiency measures could be developed and commercialized in coming years.

Much of the energy we consume is wasted. This statement is not meant as a reminder to simply turn off lights and lower furnace thermostats; it is a technological challenge. Our use of energy is so inefficient that most potential energy in fuel is lost as waste heat, becoming a form of environmental pollution.

The amount of energy wasted through poorly insulated windows and doors alone is about as much energy as the United States receives from the Alaskan pipeline each year. It is estimated that by using inexpensive, energy-efficient measures, the average energy bills of a single home could be reduced by 10 percent to 50 percent and the emissions of carbon dioxide into the atmosphere could be cut.

Many conservation techniques are relatively simple and highly cost-effective. More efficient and less energy-intensive industry and domestic practices could save large amounts of energy. Improved automobile efficiency, better mass transit, and increased railroad use for passenger and freight traffic are simple and readily available means of conserving transportation energy. In response to the 1970s' oil price shocks, automobile mileage averages in the United States more than doubled, from 5.55 km/L (13 mpg) in 1975 to 12.3 km/L (28.8 mpg) in 1988. Unfortunately, the oil glut and falling fuel prices of the late 1980s discouraged further conservation. Between 1990 and 1997, the average slipped to only 11.8 km/L (27.6 mpg). It remains to be seen if the sharp increase of gasoline prices in the early years of the twenty-first century will translate into increased miles per gallon in new car design.

Several technologies that reduce energy consumption are now available. Highly efficient fluorescent lightbulbs that can be used in regular incandescent fixtures give the same amount of light for 25 percent of the energy, and they produce less heat. Since lighting and air conditioning (which removes the heat from inefficient incandescent lighting) account for 25 percent of U.S. electricity consumption, widespread use of these lights could significantly reduce energy consumption. Low-emissive glass for windows can reduce the amount of heat entering a building while allowing light to enter. The use of this type of glass in new construction and replacement windows could have a major impact on the energy picture. Many other technologies, such as automatic dimming devices or automatic light-shutoff devices, are being used in new construction.

The shift to more efficient use of energy needs encouragement. Often, poorly designed, energy-inefficient buildings and machines can be produced inexpensively. The short-term cost is low, but the long-term cost is high. The public needs to be educated to look at the long-term economic and energy costs of purchasing poorly designed buildings and appliances.

Electric utilities have recently become part of the energy conservation picture. In some states, they have been allowed to make money on conservation efforts; previously, they could make money only by building more power plants. This encourages them to become involved in energy conservation education, because teaching their customers how to use energy more efficiently allows them to serve more people without building new power plants.

ENERGY SOURCES TOMORROW

An *alternative source of energy* is one that is different from the typical sources used today. The sources used today are the fossil fuels (coal, petroleum, and natural gas), nuclear, and falling water. Alternative sources could be solar, geothermal, hydrogen gas, fusion, or any other energy source that a new technology could utilize.

SOLAR TECHNOLOGIES

The term *solar energy* is used to describe a number of technologies that directly or indirectly utilize sunlight as an alternative energy source (Figure 3.20). There are eight main categories of these solar technologies:

1. **Solar cells.** A solar cell is a thin crystal of silicon, gallium, or some polycrystalline compound that generates electricity when exposed to light. Also called photovoltaic devices, solar cells have no moving parts and produce electricity directly, without the need for hot fluids or intermediate conversion states. Solar cells have been used extensively in space vehicles and satellites. Here on Earth, however, use has been limited to demonstration projects, remote site applications, and consumer specialty items such as solar-powered watches and calculators. The problem with solar cells today is that the manufacturing cost is too high (they are essentially handmade). Research is continuing on the development of highly efficient, affordable solar cells that could someday produce electricity for the home. See page 167 to find out how a solar cell is able to create a current.

2. **Power tower.** This is another solar technology designed to generate electricity. One type of planned power tower will have a 171 m (560 ft) tower surrounded by some 9,000 special mirrors called heliostats. The heliostats will focus sunlight on a boiler at the top of the tower where salt (a mixture of sodium nitrate and potassium nitrate) will be heated to about 566°C (about 1,050°F). This molten salt will be pumped to a steam generator, and the steam will be used to drive a generator, just like other power plants. Water could be heated directly in the power tower boiler. Molten salt is used because it can be stored in an insulated storage tank for use when the Sun is not shining, perhaps for up to twenty hours.

3. **Passive application.** In passive applications, energy flows by natural means, without mechanical devices such as motors, pumps, and so forth. A passive solar house would include such considerations as the orientation of a house to the Sun, the size and positioning of windows, and a roof overhang that lets sunlight in during the winter but keeps it out during the summer. There are different design plans to capture, store, and distribute solar energy throughout a house, and some of these designs are described on page 101.

4. **Active application.** An active solar application requires a solar collector in which sunlight heats air, water, or some liquid. The liquid or air is pumped through pipes in a house to generate electricity, or it is used directly for hot water. Solar water heating makes more economic sense today than the other applications.

5. **Wind energy.** The wind has been used for centuries to move ships, grind grain into flour, and pump water. The wind blows, however, because radiant energy from the Sun heats some parts of Earth's surface more than other parts. This differential heating results in pressure differences and the horizontal movement of air, which is called wind. Thus, wind is another form of solar energy. Wind turbines

FIGURE 3.20 Wind is another form of solar energy. This wind turbine generates electrical energy for this sailboat, charging batteries for backup power when the wind is not blowing. In case you are wondering, the turbine cannot be used to make a wind to move the boat. In accord with Newton's laws of motion, this would not produce a net force on the boat.

are used to generate electrical energy or mechanical energy. The biggest problem with wind energy is the inconsistency of the wind. Sometimes the wind speed is too great, and other times it is not great enough. Several methods of solving this problem are being researched.

6. **Biomass.** Biomass is any material formed by photosynthesis, including small plants, trees, and crops, and any garbage, crop residue, or animal waste. Biomass can be burned directly as a fuel, converted into a gas fuel (methane), or converted into liquid fuels such as alcohol. The problems with using biomass include the energy expended in gathering the biomass, as well as the energy used to convert it to a gaseous or liquid fuel.

7. **Agriculture and industrial heating.** This is a technology that simply uses sunlight to dry grains, cure paint, or do anything that can be done with sunlight rather than using traditional energy sources.

8. **Ocean thermal energy conversion (OTEC).** This is an electric generating plant that would take advantage of the approximately 22°C (about 40°F) temperature difference between the surface and the depths of tropical, subtropical, and equatorial ocean waters. Basically, warm water is drawn into the system to vaporize a fluid, which expands through a turbine generator. Cold water from the depths condenses the vapor back to a liquid form, which is then cycled back to the warm-water side. The concept has been tested and found to be technically successful. The greatest interest in using it seems to be among islands that have warm surface waters (and cold depths) such as Hawaii, Puerto Rico, Guam, and the Virgin Islands.

GEOTHERMAL ENERGY

Geothermal energy is energy from beneath Earth's surface. The familiar geysers, hot springs, and venting steam of Yellowstone National Park are clues that this form of energy exists. There is substantially more geothermal energy than is revealed in Yellowstone, however, and geothermal resources are more widespread than once thought. Earth has a high internal temperature, and recoverable geothermal resources may underlie most states. These resources occur in four broad categories of geothermal energy: (1) dry steam, (2) hot water, (3) hot, dry rock, and (4) geopressurized resources. Together, the energy contained in these geothermal resources represents about 15,000 times more energy than is consumed in the United States in a given year. The only problem is getting to the geothermal energy, then using it in a way that is economically attractive.

Most geothermal energy occurs as *hot, dry rock,* which accounts for about 85 percent of the total geothermal resource. Hot, dry rock is usually in or near an area of former volcanic activity. The problem of utilizing this widespread resource is how to get the energy to the surface. Research has been conducted by drilling wells, then injecting water into one well and extracting energy from the heated water pumped from the second well. There is more interest in the less widespread but better understood geothermal systems of hot water and steam.

Geopressurized resources are trapped underground reservoirs of hot water that contain dissolved natural gas. The water temperature is higher than the boiling point, so heat could be used as a source of energy as well as the dissolved natural gas. Such geopressurized reservoirs make up about 14 percent of the total accessible geothermal energy found on Earth. They are still being studied in some areas since there is concern over whether the reservoirs are large enough to be economically feasible as an energy source. More is known about recovering energy from other types of hot water and steam resources, so these seem more economically attractive.

Hot water and steam comprise the smallest geothermal resource category, together making up only about 1 percent of the total known resource. However, more is known about the utilization and recovery of these energy sources, which are estimated to contain an amount of energy equivalent to about half of the present known reserve of petroleum in the United States. *Steam* is very rare, occurring in only three places in the United States. Two of these places are national parks (Lassen and Yellowstone), so this geothermal steam cannot be used as an energy source. The third place is at the Geysers, an area of fumaroles near San Francisco, California. Steam from the Geysers is used to generate a significant amount of electricity.

Hot water systems make up most of the *recoverable* geothermal resources. Heat from deep volcanic or former volcanic sources create vast, slow-moving convective patterns in groundwater. If the water circulating back near the surface is hot enough, it can be used for generating electricity, heating buildings, or many other possible applications. Worldwide, geothermal energy is used to operate pulp and paper mills, cool hotels, raise fish, heat greenhouses, dry crops, desalt water, and dozens of other things. Thousands of apartments, homes, and businesses are today heated geothermally in Oregon and Idaho in the United States, as well as in Hungary, France, Iceland, and New Zealand. Today, understand that each British thermal unit supplied by geothermal energy does not have to be supplied by fossil fuels. Tomorrow, you will find geothermal resources becoming more and more attractive as the price and the consequences of using fossil fuels continue to increase.

HYDROGEN

Hydrogen is the lightest and simplest of all the elements, occurring as a diatomic gas that can be used for energy directly in a fuel cell or burned to release heat. Hydrogen could be used to replace natural gas with a few modifications of present natural gas burners. A big plus in favor of hydrogen as a fuel is that it produces no pollutants. In addition to the heat produced, the only emission from burning hydrogen is water, as shown in the following equation:

Hydrogen + oxygen → water + 68,300 calories

The primary problem with using hydrogen as an energy source is that *it does not exist* on or under Earth's surface in any but trace amounts! Hydrogen must therefore be obtained by a chemical reaction from such compounds as water. Water is a plentiful substance on Earth, and an electric current will cause decomposition of water into hydrogen and oxygen gas. Measurement of the electric current and voltage will show that:

Water + 68,300 calories → hydrogen + oxygen

Thus, assuming 100 percent efficiency, the energy needed to obtain hydrogen gas from water is exactly equal to the energy released by hydrogen combustion. So hydrogen cannot be used to produce energy, since hydrogen gas is not available, but it can be used as a means of storing energy for later use. Indeed, hydrogen may be destined to become an effective solution to the problems of storing and transporting energy derived from solar energy sources. In addition, hydrogen might serve as the transportable source of energy, such as that needed for cars and trucks, replacing the fossil fuels. In summary, hydrogen has the potential to provide clean, alternative energy for a number of uses, including lighting, heating, cooling, and transportation.

SUMMARY

Work is defined as the product of an applied force and the distance through which the force acts. Work is measured in newton-meters, a metric unit called a *joule*. *Power* is work per unit of time. Power is measured in *watts*. One watt is 1 joule per second. Power is also measured in *horsepower*. One horsepower is 550 ft· lb/s.

Energy is defined as the ability to do work. An object that is elevated against gravity has a potential to do work. The object is said to have *potential energy*, or *energy of position*. Moving objects have the ability to do work on other objects because of their motion. The *energy of motion* is called *kinetic energy*.

Work is usually done *against inertia, gravity, friction, shape*, or *combinations of these*. As a result, there is a gain of *kinetic energy, potential energy, an increased temperature*, or *any combination of these*. Energy comes in the *forms of mechanical, chemical, radiant, electrical*, or *nuclear*. Potential energy can be converted to kinetic, and kinetic can be *converted* to potential. Any form of energy can be *converted* to any other form. Most technological devices are *energy-form converters* that do work for you. Energy flows into and out of the surroundings, but the amount of energy is always constant. The *law of conservation of energy* states that *energy is never created or destroyed*. Energy conversion always takes place through *heating* or *working*.

The basic energy sources today are the chemical *fossil fuels* (petroleum, natural gas, and coal), *nuclear energy*, and *hydropower*. *Petroleum* and *natural gas* were formed from organic material of plankton, tiny free-floating plants and animals. A barrel of petroleum is 42 U.S. gallons, but such a container does not actually exist. *Coal* formed from plants that were protected from consumption by falling into a swamp. The decayed plant material, *peat*, was changed into the various *ranks* of coal by pressure and heating over some period of time. Coal is a dirty fuel that contains impurities and sulfur. Controlling air pollution from burning coal is costly. Water power and nuclear energy are used for the generation of electricity. An alternative source of energy is one that is different from the typical sources used today. Alternative sources could be *solar, geothermal*, or *hydrogen*.

SUMMARY OF EQUATIONS

3.1

$$\text{work} = \text{force} \times \text{distance}$$
$$W = Fd$$

3.2

$$\text{power} = \frac{\text{work}}{\text{time}}$$
$$P = \frac{W}{t}$$

3.3

$$\text{gravitational potential energy} = \text{weight} \times \text{height}$$
$$PE = mgh$$

3.4

$$\text{kinetic energy} = \frac{1}{2}(\text{mass})(\text{velocity})^2$$
$$KE = \frac{1}{2}mv^2$$

3.5

$$\text{final velocity} = \text{square root of } (2 \times \text{acceleration due to gravity} \times \text{height of fall})$$
$$v_f = \sqrt{2gh}$$

KEY TERMS

chemical energy (p. **70**)

electrical energy (p. **71**)

energy (p. **67**)

fossil fuels (p. **76**)

geothermal energy (p. **79**)

horsepower (p. **65**)

joule (p. **62**)

kinetic energy (p. **68**)

law of conservation of energy (p. **74**)

mechanical energy (p. **70**)

nuclear energy (p. **71**)

potential energy (p. **67**)

power (p. **65**)

radiant energy (p. **71**)

watt (p. **66**)

work (p. **62**)

APPLYING THE CONCEPTS

1. According to the definition of mechanical work, pushing on a rock accomplishes no work unless there is
   a. movement.
   b. a net force.
   c. an opposing force.
   d. movement in the same direction as the direction of the force.

2. The metric unit of a joule (J) is a unit of
   a. potential energy.
   b. work.
   c. kinetic energy.
   d. any of the above.

3. A Nm/s is a unit of
   a. work.
   b. power.
   c. energy.
   d. none of the above.

4. A kilowatt-hour is a unit of
   a. power.
   b. work.
   c. time.
   d. electrical charge.

5. A power rating of 550 ft-lb per s is known as a
   a. watt.
   b. newton.
   c. joule.
   d. horsepower.

6. A power rating of 1 joule per s is known as a
   a. watt.
   b. newton.
   c. joule.
   d. horsepower.

7. According to $PE = mgh$, gravitational potential energy is the same thing as
   a. exerting a force through a distance in any direction.
   b. the kinetic energy an object had before coming to a rest.
   c. work against a vertical change of position.
   d. the momentum of a falling object.

8. Two cars have the same mass, but one is moving three times as fast as the other is. How much more work will be needed to stop the faster car?
   a. the same amount.
   b. twice as much.
   c. three times as much.
   d. nine times as much.

9. Kinetic energy can be measured in terms of
   a. work done on an object to put it into in motion.
   b. work done on a moving object to bring it to rest.
   c. both a and b.
   d. neither a nor b.

10. Potential and kinetic energy are created when work is done to change a position (*PE*) or a state of motion (*KE*). Ignoring friction, how does the amount of work done to make the change compare to the amount of *PE* or *KE* created?
    a. Less energy is created.
    b. Both are the same.
    c. More energy is created.
    d. This cannot be generalized.

11. Many forms of energy in use today can be traced back to
    a. the Sun.
    b. coal.
    c. Texas.
    d. petroleum.

12. In all of our energy uses, we find that
    a. the energy used is consumed.
    b. some forms of energy are consumed but not others.
    c. more energy is created than is consumed.
    d. the total amount of energy is constant in all situations.

13. Any form of energy can be converted to another, but energy used on Earth usually ends up in what form?
    a. electrical
    b. mechanical
    c. nuclear
    d. radiant

14. Radiant energy can be converted to electrical energy using
    a. lightbulbs.
    b. engines.
    c. solar cells.
    d. electricity.

15. The "barrel of oil" mentioned in discussions about petroleum is
    a. 55 U.S. gallons.
    b. 42 U.S. gallons.
    c. 12 U.S. gallons.
    d. a variable quantity.

16. The amount of energy generated by hydroelectric plants in the United States as part of the total electrical energy is
    a. fairly constant over the years.
    b. decreasing because new dams are not being constructed.
    c. increasing as more and more energy is needed.
    d. decreasing as dams are destroyed because of environmental concerns.

17. Fossil fuels provide what percent of the total energy consumed in the United States today?
    a. 25 percent
    b. 50 percent
    c. 86 percent
    d. 99 percent

18. Alternative sources of energy include
    a. solar cells.
    b. wind.
    c. hydrogen.
    d. all of the above.

19. A renewable energy source is
    a. coal.
    b. biomass.
    c. natural gas.
    d. petroleum.

20. The potential energy of a box on a shelf, relative to the floor, is a measure of
    a. the work that was required to put the box on the shelf from the floor.
    b. the weight of the box times the distance above the floor.
    c. the energy the box has because of its position above the floor.
    d. all of the above.

21. A rock on the ground is considered to have zero potential energy. In the bottom of a well, the rock would be considered to have
    a. zero potential energy, as before.
    b. negative potential energy.
    c. positive potential energy.
    d. zero potential energy but will require work to bring it back to ground level.

22. Which quantity has the greatest influence on the amount of kinetic energy that a large truck has while moving down the highway?
    a. mass
    b. weight
    c. velocity
    d. size

23. Electrical energy can be converted to
    a. chemical energy.
    b. mechanical energy.
    c. radiant energy.
    d. any of the above.

24. Most all energy comes to and leaves Earth in the form of
    a. nuclear energy.
    b. chemical energy.
    c. radiant energy.
    d. kinetic energy.

25. A spring-loaded paper clamp exerts a force of 2 N on ten sheets of paper it is holding tightly together. Is the clamp doing work as it holds the papers together?
    a. Yes.
    b. No.

26. The force exerted when doing work by lifting a book bag against gravity is measured in units of
    a. kg.
    b. N.
    c. W.
    d. J.

27. The work accomplished by lifting an object against gravity is measured in units of
    a. kg.
    b. N.
    c. W.
    d. J.

28. An iron cannonball and a bowling ball are dropped at the same time from the top of a building. At the instant before the balls hit the sidewalk, the heavier cannonball has a greater
    a. velocity.
    b. acceleration.
    c. kinetic energy.
    d. All of these are the same for the two balls.

29. Two students are poised to dive off equal-height diving towers into a swimming pool below. Student B is twice as massive as student A. Which of the following is true?
    a. Student B will reach the water sooner than student A.
    b. Both students have the same gravitational *PE*.
    c. Both students will have the same *KE* just before hitting the water.
    d. Student B did twice as much work climbing the tower.

30. A car is moving straight down a highway. What factor has the greatest influence on how much work must be done on the car to bring it to a complete stop?
    a. how fast it is moving
    b. the weight of the car
    c. the mass of the car
    d. the latitude of the location

31. Two identical cars are moving straight down a highway under identical conditions, except car B is moving three times as fast as car A. How much more work is needed to stop car B?
    a. twice as much
    b. three times as much
    c. six times as much
    d. nine times as much

32. When you do work on something, you give it energy.
    a. often
    b. sometimes
    c. every time
    d. never

33. Which of the following is **not** the use of a solar energy technology?
    a. wind
    b. burning of wood
    c. photovoltaics
    d. water from a geothermal spring

34. Today, the basic problem with using solar cells as a major source of electricity is
    a. efficiency.
    b. manufacturing cost.
    c. reliability.
    d. that the Sun does not shine at night.

35. The solar technology that makes more economic sense today than the other applications is
    a. solar cells.
    b. power tower.
    c. water heating.
    d. ocean thermal energy conversion.

36. Petroleum is believed to have formed over time from buried
    a. pine trees.
    b. plants in a swamp.
    c. organic sediments.
    d. dinosaurs.

Answers

1. d 2. d 3. b 4. b 5. d 6. a 7. c 8. d 9. c 10. b 11. a 12. d 13. d 14. c
15. b 16. b 17. c 18. d 19. b 20. d 21. d 22. c 23. d 24. c 25. b 26. b
27. d 28. c 29. d 30. a 31. d 32. c 33. d 34. b 35. c 36. c

QUESTIONS FOR THOUGHT

1. How is work related to energy?

2. What is the relationship between the work done while moving a book to a higher bookshelf and the potential energy that the book has on the higher shelf?

3. Does a person standing motionless in the aisle of a moving bus have kinetic energy? Explain.

4. A lamp bulb is rated at 100 W. Why is a time factor not included in the rating?

5. Is a kWh a unit of work, energy, power, or more than one of these? Explain.

6. If energy cannot be destroyed, why do some people worry about the energy supplies?

7. A spring clamp exerts a force on a stack of papers it is holding together. Is the spring clamp doing work on the papers? Explain.

8. Why are petroleum, natural gas, and coal called *fossil fuels?*

9. From time to time, people claim to have invented a machine that will run forever without energy input and develops more energy than it uses (perpetual motion). Why would you have reason to question such a machine?

10. Define a joule. What is the difference between a joule of work and a joule of energy?

11. Compare the energy needed to raise a mass 10 m on Earth to the energy needed to raise the same mass 10 m on the Moon. Explain the difference, if any.

12. What happens to the kinetic energy of a falling book when the book hits the floor?

FOR FURTHER ANALYSIS

1. Evaluate the requirement that something must move whenever work is done. Why is this a requirement?

2. What are the significant similarities and differences between work and power?

3. Whenever you do work on something, you give it energy. Analyze how you would know for sure that this is a true statement.

4. Simple machines are useful because they are able to trade force for distance moved. Describe a conversation between yourself and another person who believes that you do less work when you use a simple machine.

5. Use the equation for kinetic energy to prove that speed is more important than mass in bringing a speeding car to a stop.

6. Describe at least several examples of negative potential energy and how each shows a clear understanding of the concept.

7. The forms of energy are the result of fundamental forces—gravitational, electromagnetic, and nuclear—and objects that are interacting. Analyze which force is probably involved with each form of energy.

8. Most technological devices convert one of the five forms of energy into another. Try to think of a technological device that does not convert an energy form to another. Discuss the significance of your finding.

9. Are there any contradictions to the law of conservation of energy in any area of science?

INVITATION TO INQUIRY

New Energy Source?

Is waste paper a good energy source? There are 103 waste-to-energy plants in the United States that burn solid garbage, so we know that waste paper would be a good source, too. The plants burn solid garbage to make steam that is used to heat buildings and generate electricity. Schools might be able to produce a pure waste paper source because waste paper accumulates near computer print stations and in offices. Collecting waste paper from such sources would yield 6,800 Btu/lb, which is about half the heat value of coal.

If you choose to do this invitation, start by determining how much waste paper is created per month in your school. Would this amount produce enough energy to heat buildings or generate electricity?

PARALLEL EXERCISES

The exercises in groups A and B cover the same concepts. Solutions to group A exercises are located in appendix E.

Note: *Neglect all frictional forces in all exercises.*

Group A

1. A force of 200 N is needed to push a table across a level classroom floor for a distance of 3 m. How much work was done on the table?

2. A 880 N box is pushed across a level floor for a distance of 5.0 m with a force of 440 N. How much work was done on the box?

3. How much work is done in raising a 10.0 kg backpack from the floor to a shelf 1.5 m above the floor

4. If 5,000 J of work is used to raise a 102 kg crate to a shelf in a warehouse, how high was the crate raised?

5. A 60.0 kg student runs up a 5.00 m-high stairway in a time of 3.92 seconds. How many watts of power did she develop?

Group B

1. How much work is done when a force of 800.0 N is exerted while pushing a crate across a level floor for a distance of 1.5 m?

2. A force of 400.0 N is exerted on a 1,250 N car while moving it a distance of 3.0 m. How much work was done on the car?

3. A 5.0 kg textbook is raised a distance of 30.0 cm as a student prepares to leave for school. How much work did the student do on the book?

4. An electric hoist does 196,000 J of work in raising a 250.0 kg load. How high was the load lifted?

5. What is the horsepower of a 1,500.0 kg car that can go to the top of a 360.0 m-high hill in exactly 1.00 minute?

Group A—Continued

6. (a) How many horsepower is a 1,400 W blow dryer? (b) How many watts is a 3.5 hp lawnmower?

7. What is the kinetic energy of a 2,000 kg car moving at 72 km/h?

8. How much work is needed to stop a 1,000.0 kg car that is moving straight down the highway at 54.0 km/h?

9. A horizontal force of 10.0 lb is needed to push a bookcase 5 ft across the floor. (a) How much work was done on the bookcase? (b) How much did the gravitational potential energy change as a result?

10. (a) How much work is done in moving a 2.0 kg book to a shelf 2.00 m high? (b) What is the change of potential energy of the book as a result? (c) How much kinetic energy will the book have as it hits the ground when it falls?

11. A 150 g baseball has a velocity of 30.0 m/s. What is its kinetic energy in J?

12. (a) What is the kinetic energy of a 1,000.0 kg car that is traveling at 90.0 km/h? (b) How much work was done to give the car this kinetic energy? (c) How much work must be done to stop the car?

13. A 60.0 kg jogger moving at 2.0 m/s decides to double the jogging speed. How did this change in speed change the kinetic energy?

14. A bicycle and rider have a combined mass of 70.0 kg and are moving at 6.00 m/s. A 70.0 kg person is now given a ride on the bicycle. (Total mass is 140.0 kg.) How did the addition of the new rider change the kinetic energy at the same speed?

15. A 170.0 lb student runs up a stairway to a classroom 25.0 ft above ground level in 10.0 s. (a) How much work did the student do? (b) What was the average power output in hp?

16. (a) How many seconds will it take a 20.0 hp motor to lift a 2,000.0 lb elevator a distance of 20.0 ft? (b) What was the average velocity of the elevator?

17. A ball is dropped from 9.8 ft above the ground. Using energy considerations only, find the velocity of the ball just as it hits the ground.

18. What is the velocity of a 1,000.0 kg car if its kinetic energy is 200 kJ?

19. A Foucault pendulum swings to 3.0 in above the ground at the highest points and is practically touching the ground at the lowest point. What is the maximum velocity of the pendulum?

20. An electric hoist is used to lift a 250.0 kg load to a height of 80.0 m in 39.2 s. (a) What is the power of the hoist motor in kW? (b) In hp?

Group B—Continued

6. (a) How many horsepower is a 250 W lightbulb? (b) How many watts is a 230 hp car?

7. What is the kinetic energy of a 30-gram bullet that is traveling at 200 m/s?

8. How much work will be done by a 30-gram bullet traveling at 200 m/s?

9. A force of 50.0 lb is used to push a box 10.0 ft across a level floor. (a) How much work was done on the box? (b) What is the change of potential energy as a result of this move?

10. (a) How much work is done in raising a 50.0 kg crate a distance of 1.5 m above a storeroom floor? (b) What is the change of potential energy as a result of this move? (c) How much kinetic energy will the crate have as it falls and hits the floor?

11. What is the kinetic energy in J of a 60.0 g tennis ball approaching a tennis racket at 20.0 m/s?

12. (a) What is the kinetic energy of a 1,500.0 kg car with a velocity of 72.0 km/h? (b) How much work must be done on this car to bring it to a complete stop?

13. The driver of an 800.0 kg car decides to double the speed from 20.0 m/s to 40.0 m/s. What effect would this have on the amount of work required to stop the car, that is, on the kinetic energy of the car?

14. Compare the kinetic energy of an 800.0 kg car moving at 20.0 m/s to the kinetic energy of a 1,600.0 kg car moving at an identical speed.

15. A 175.0 lb hiker is able to ascend a 1,980.0 ft high slope in 1 hour and 45 minutes. (a) How much work did the hiker do? (b) What was the average power output in hp?

16. (a) How many seconds will it take a 10.0 hp motor to lift a 2,000.0 lb elevator a distance of 20.0 feet? (b) What was the average velocity of the elevator?

17. A ball is dropped from 20.0 ft above the ground. (a) At what height is half of its energy kinetic and half potential? (b) Using energy considerations only, what is the velocity of the ball just as it hits the ground?

18. What is the velocity of a 60.0 kg jogger with a kinetic energy of 1,080.0 J?

19. A small sports car and a pickup truck start coasting down a 10.0 m hill together, side by side. Assuming no friction, what is the velocity of each vehicle at the bottom of the hill?

20. A 70.0 kg student runs up the stairs of a football stadium to a height of 10.0 m above the ground in 10.0 s. (a) What is the power of the student in kW? (b) In hp?

4

Heat and Temperature

Sparks fly from a plate of steel as it is cut by an infrared laser. Today, lasers are commonly used to cut as well as weld metals, so the cutting and welding are done by light, not by a flame or electric current.

CORE **CONCEPT**

A relationship exists between heat, temperature, and the motion and position of molecules.

OUTLINE

Kinetic Molecular Theory
All matter is made of molecules that move and interact.

Heat
Heat is a measure of internal energy that has been transferred or absorbed.

Heat Flow
Heat flow resulting from a temperature difference takes place as conduction, convection, and/or radiation.

The Kinetic Molecular Theory
   Molecules
   Molecules Interact
   Phases of Matter
   Molecules Move
Temperature
   Thermometers
   Temperature Scales
A Closer Look: Goose Bumps and Shivering
Heat
   Heat as Energy Transfer
      Heat Defined
      Two Heating Methods
   Measures of Heat
   Specific Heat
   Heat Flow
      Conduction
Science and Society: Require Insulation?
      Convection
      Radiation
Energy, Heat, and Molecular Theory
   Phase Change
A Closer Look: Passive Solar Design
   Evaporation and Condensation
Thermodynamics
   The First Law of Thermodynamics
   The Second Law of Thermodynamics
   The Second Law and Natural Processes
People Behind the Science: Count Rumford (Benjamin Thompson)

Temperature
Temperature is a measure of the average kinetic energy of molecules.

Measures of Heat
Heat may be increased by an energy form conversion, and the relationship between the energy form and the resulting heating is always the same.

Thermodynamics
The laws of thermodynamics describe a relationship between changes of internal energy, work, and heat.

OVERVIEW

Heat has been closely associated with the comfort and support of people throughout history. You can imagine the appreciation when your earliest ancestors first discovered fire and learned to keep themselves warm and cook their food. You can also imagine the wonder and excitement about 3000 B.C., when people put certain earthlike substances on the hot, glowing coals of a fire and later found metallic copper, lead, or iron. The use of these metals for simple tools followed soon afterward. Today, metals are used to produce complicated engines that use heat for transportation and that do the work of moving soil and rock, construction, and agriculture. Devices made of heat-extracted metals are also used to control the temperature of structures, heating or cooling the air as necessary. Thus, the production and control of heat gradually built the basis of civilization today (Figure 4.1).

The sources of heat are the energy forms that you learned about in chapter 3. The fossil fuels are *chemical* sources of heat. Heat is released when oxygen is combined with these fuels. Heat also results when *mechanical* energy does work against friction, such as in the brakes of a car coming to a stop. Heat also appears when *radiant* energy is absorbed. This is apparent when solar energy heats water in a solar collector or when sunlight melts snow. The transformation of *electrical* energy to heat is apparent in toasters, heaters, and ranges. *Nuclear* energy provides the heat to make steam in a nuclear power plant. Thus, all energy forms can be converted to heat.

The relationship between energy forms and heat appears to give an order to nature, revealing patterns that you will want to understand. All that you need is some kind of explanation for the relationships—a model or theory that helps make sense of it all. This chapter is concerned with heat and temperature and their relationship to energy. It begins with a simple theory about the structure of matter and then uses the theory to explain the concepts of heat, energy, and temperature changes.

THE KINETIC MOLECULAR THEORY

The idea that substances are composed of very small particles can be traced back to certain early Greek philosophers. The earliest record of this idea was written by Democritus during the fifth century B.C. He wrote that matter was empty space filled with tremendous numbers of tiny, indivisible particles called *atoms*. This idea, however, was not acceptable to most of the ancient Greeks, because matter seemed continuous, and empty space was simply not believable. The idea of atoms was rejected by Aristotle as he formalized his belief in continuous matter composed of Earth, air, fire, and water elements. Aristotle's belief about matter, like his beliefs about motion, predominated through the 1600s. Some people, such as Galileo and Newton, believed the ideas about matter being composed of tiny particles, or atoms, since this theory seemed to explain the behavior of matter. Widespread acceptance of the particle model did not occur, however, until strong evidence was developed through chemistry in the late 1700s and early 1800s. The experiments finally led to a collection of assumptions about the small particles of matter and the space around them. Collectively, the assumptions could be called the **kinetic molecular theory.** The following is a general description of some of these assumptions.

MOLECULES

The basic assumption of the kinetic molecular theory is that all matter is made up of tiny, basic units of structure called *atoms*.

Atoms are neither divided, created, nor destroyed during any type of chemical or physical change. There are similar groups of atoms that make up the pure substances known as chemical *elements*. Each element has its own kind of atom, which is different from the atoms of other elements. For example, hydrogen, oxygen, carbon, iron, and gold are chemical elements, and each has its own kind of atom.

In addition to the chemical elements, there are pure substances called *compounds* that have more complex units of structure (Figure 4.2). Pure substances, such as water, sugar, and alcohol, are composed of atoms of two or more elements that join together in definite proportions. Water, for example, has structural units that are made up of two atoms of hydrogen tightly bound to one atom of oxygen (H_2O). These units are not easily broken apart and stay together as small physical particles of which water is composed. Each is the smallest particle of water that can exist, a molecule of water. A *molecule* is generally defined as a tightly bound group of atoms in which the atoms maintain their identity. How atoms become bound together to form molecules is discussed in chapters 8–10.

Some elements exist as gases at ordinary temperatures, and all elements are gases at sufficiently high temperatures. At ordinary temperatures, the atoms of oxygen, nitrogen, and other gases are paired in groups of two to form *diatomic molecules*. Other gases, such as helium, exist as single, unpaired atoms at ordinary temperatures. At sufficiently high temperatures, iron, gold, and other metals vaporize to form gaseous, single, unpaired atoms. In the kinetic molecular theory, the

FIGURE 4.1 Heat and modern technology are inseparable. These glowing steel slabs, at over 1,100°C (about 2,000°F), are cut by an automatic flame torch. The slab caster converts 300 tons of molten steel into slabs in about 45 minutes. The slabs are converted to sheet steel for use in the automotive, appliance, and building industries.

term *molecule* has the additional meaning of the smallest, ultimate particle of matter that can exist. Thus, the ultimate particle of a gas, whether it is made up of two or more atoms bound together or of a single atom, is conceived of as a molecule. A single atom of helium, for example, is known as a *monatomic molecule*. For now, a **molecule** is defined as the

smallest particle of a compound or a gaseous element that can exist and still retain the characteristic properties of that substance.

MOLECULES INTERACT

Some molecules of solids and liquids interact, strongly attracting and clinging to each other. When this attractive force is between the same kind of molecules, it is called *cohesion*. It is a stronger cohesion that makes solids and liquids different from gases, and without cohesion, all matter would be in the form of gases. Sometimes one kind of molecule attracts and clings to a different kind of molecule. The attractive force between unlike molecules is called *adhesion*. Water wets your skin because the adhesion of water molecules and skin is stronger than the cohesion of water molecules. Some substances, such as glue, have a strong force of adhesion when they harden from a liquid state, and they are called adhesives.

PHASES OF MATTER

Three phases of matter are common on earth under conditions of ordinary temperature and pressure. These phases—or forms of existence—are solid, liquid, and gas. Each of these has a different molecular arrangement (Figure 4.3). The different characteristics of each phase can be attributed to the molecular arrangements and the strength of attraction between the molecules (Table 4.1).

Solids have definite shapes and volumes because they have molecules that are nearly fixed distances apart and bound by relatively strong cohesive forces. Each molecule is a nearly fixed distance from the next, but it does vibrate and move around an

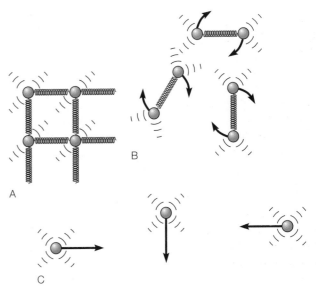

FIGURE 4.3 (*A*) In a solid, molecules vibrate around a fixed equilibrium position and are held in place by strong molecular forces. (*B*) In a liquid, molecules can rotate and roll over each other because the molecular forces are not as strong. (*C*) In a gas, molecules move rapidly in random, free paths.

FIGURE 4.2 Metal atoms appear in the micrograph of a crystal of titanium niobium oxide, magnified 7,800,000 times with the help of an electron microscope.

TABLE 4.1

The shape and volume characteristics of solids, liquids, and gases are reflections of their molecular arrangements*

	Solids	Liquids	Gases
Shape	Fixed	Variable	Variable
Volume	Fixed	Fixed	Variable

*These characteristics are what would be expected under ordinary temperature and pressure conditions on the surface of Earth.

equilibrium position. The masses of these molecules and the spacing between them determine the density of the solid. The hardness of a solid is the resistance of a solid to forces that tend to push its molecules farther apart.

Liquids have molecules that are not confined to an equilibrium position as in a solid. The molecules of a liquid are close together and bound by cohesive forces that are not as strong as in a solid. This permits the molecules to move from place to place within the liquid. The molecular forces are strong enough to give the liquid a definite volume but not strong enough to give it a definite shape. Thus, a liter of water is always a liter of water (unless it is under tremendous pressure) and takes the shape of the container holding it. Because the forces between the molecules of a liquid are weaker than the forces between the molecules of a solid, a liquid cannot support the stress of a rock placed on it as a solid does. The liquid molecules *flow,* rolling over each other as the rock pushes its way between the molecules. Yet, the molecular forces are strong enough to hold the liquid together, so it keeps the same volume.

Gases are composed of molecules with weak cohesive forces acting between them. The gas molecules are relatively far apart and move freely in a constant, random motion that is changed often by collisions with other molecules. Gases therefore have neither fixed shapes nor fixed volumes.

Gases that are made up of positive ions and negative electrons are called *plasmas*. Plasmas have the same properties as gases but also conduct electricity and interact strongly with magnetic fields. Plasmas are found in fluorescent and neon lights on Earth, the Sun, and other stars. Nuclear fusion occurs in plasmas of stars (see chapter 14), producing starlight as well as sunlight. Plasma physics is studied by scientists in their attempt to produce controlled nuclear fusion.

There are other distinctions between the phases of matter. The term *vapor* is sometimes used to describe a gas that is usually in the liquid phase. Water vapor, for example, is the gaseous form of liquid water. Liquids and gases are collectively called *fluids* because of their ability to flow, a property that is lacking in most solids.

MOLECULES MOVE

Suppose you are in an evenly heated room with no air currents. If you open a bottle of ammonia, the odor of ammonia is soon noticeable everywhere in the room. According to the kinetic

molecular theory, molecules of ammonia leave the bottle and bounce around among the other molecules making up the air until they are everywhere in the room, slowly becoming more evenly distributed. The ammonia molecules *diffuse,* or spread, throughout the room. The ammonia odor diffuses throughout the room faster if the air temperature is higher and slower if the air temperature is lower. This would imply a relationship between the temperature and the speed at which molecules move about.

The relationship between the temperature of a gas and the motion of molecules was formulated in 1857 by Rudolf Clausius. He showed that the temperature of a gas is proportional to the average kinetic energy of the gas molecules. This means that ammonia molecules have a greater average velocity at a higher temperature and a slower average velocity at a lower temperature. This explains why gases diffuse at a greater rate at higher temperatures. Recall, however, that kinetic energy involves the mass of the molecules as well as their velocity ($KE = 1/2\ mv^2$). It is the *average kinetic energy* that is proportional to the temperature, which involves the molecular mass as well as the molecular velocity. Whether the kinetic energy is jiggling, vibrating, rotating, or moving from place to place, the **temperature** of a substance is *a measure of the average kinetic energy of the molecules making up the substance* (Figure 4.4).

The kinetic molecular theory explains why matter generally expands with increased temperatures and contracts with decreased temperatures. At higher temperatures, the molecules of a substance move faster, with increased agitation; therefore, they move a little farther apart, thus expanding the substance. As the substance cools, the motion slows, and the molecular forces are able to pull the molecules closer together, thus contracting the substance.

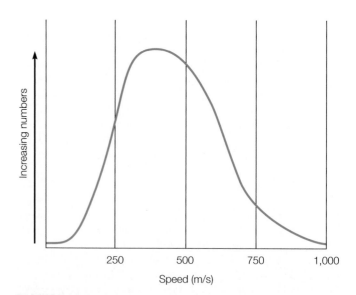

FIGURE 4.4 The number of oxygen molecules with certain speeds that you might find in a sample of air at room temperature. Notice that a few are barely moving and some have speeds over 1,000 m/s at a given time, but the *average* speed is somewhere around 400 m/s.

CONCEPTS *Applied*

Moving Molecules

Blow up a small balloon and make a knot in the neck so it will not leak air. Note the size of the balloon. Place the balloon in the freezer part of a refrigerator for an hour, then again note the size of the balloon. Immediately place the balloon in direct sunlight for an hour and again note the size of the balloon. Explain your observations by using the kinetic molecular theory.

TEMPERATURE

If you ask people about the temperature, they usually respond with a referent ("hotter than the summer of '89") or a number ("68°F or 20°C"). Your response, or feeling, about the referent or number depends on a number of factors, including a *relative* comparison. A temperature of 20°C (68°F), for example, might seem cold during the month of July but warm during the month of January. The 20°C temperature is compared to what is expected at the time, even though 20°C is 20°C, no matter what month it is.

When people ask about the temperature, they are really asking *how hot or how cold something is.* Without a thermometer, however, most people can do no better than *hot* or *cold,* or perhaps *warm* or *cool,* in describing a relative temperature. Even then, there are other factors that confuse people about temperature. Your body judges temperature on the basis of the net *direction* of energy flow. You sense situations in which heat is flowing into your body as *warm* and situations in which heat is flowing from your body as *cool.* Perhaps you have experienced having your hands in snow for some time, then washing your hands in cold water. The cold water feels warm. Your hands are colder than the water, energy flows into your hands, and they communicate "warm."

THERMOMETERS

The human body is a poor sensor of temperature, so a device called a *thermometer* is used to measure the hotness or coldness of something. Most thermometers are based on the relationship between some property of matter and changes in temperature. Almost all materials expand with increasing temperatures. A strip of metal is slightly longer when hotter and slightly shorter when cooler, but the change of length is too small to be useful in a thermometer. A more useful, larger change is obtained when two metals that have different expansion rates are bonded together in a strip. The bimetallic (*bi* = two; *metallic* = metal) strip will bend toward the metal with less expansion when the strip is heated (Figure 4.5). Such a bimetallic strip is formed into a coil and used in thermostats and dial thermometers (Figure 4.6).

The common glass thermometer is a glass tube with a bulb containing a liquid, usually mercury or colored alcohol,

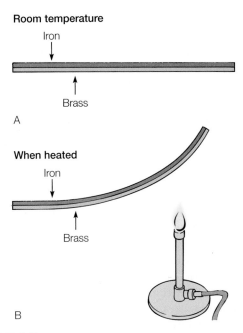

FIGURE 4.5 (*A*) A bimetallic strip is two different metals, such as iron and brass, bonded together as a single unit, shown here at room temperature. (*B*) Since one metal expands more than the other, the strip will bend when it is heated. In this example, the brass expands more than the iron, so the bimetallic strip bends away from the brass.

that expands up the tube with increases in temperature and contracts back toward the bulb with decreases in temperature. The height of this liquid column is used with a referent scale to measure temperature. Some thermometers, such as a fever thermometer, have a small constriction in the bore

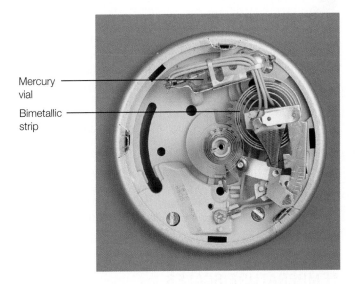

FIGURE 4.6 This thermostat has a coiled bimetallic strip that expands and contracts with changes in the room temperature. The attached vial of mercury is tilted one way or the other, and the mercury completes or breaks an electric circuit that turns the heating or cooling system on or off.

so the liquid cannot normally return to the bulb. Thus, the thermometer shows the highest reading, even if the temperature it measures has fluctuated up and down during the reading. The liquid must be forced back into the bulb by a small swinging motion, bulb-end down, then sharply stopping the swing with a snap of the wrist. The inertia of the mercury in the bore forces it past the constriction and into the bulb. The fever thermometer is then ready to use again.

Today, scientists have developed a different type of thermometer and a way around the problems of using a glass mercury fever thermometer. This new approach measures the internal core temperature by quickly reading infrared radiation from the eardrum. All bodies with a temperature above absolute zero emit radiation, including your body (see radiation on p. 98). The intensity of the radiation is a sensitive function of body temperature, so reading the radiation emitted will tell you about the temperature of that body.

The human eardrum is close to the hypothalamus, the body's thermostat, so a temperature reading taken here must be close to the temperature of the internal core. You cannot use a mercury thermometer in the ear because of the very real danger of puncturing the eardrum, along with obtaining doubtful readings from a mercury bulb. You can use a device to measure the infrared radiation coming from the entrance to the ear canal, however, to quickly obtain a temperature reading. A microprocessor chip is programmed with the relationship between the body temperature and the infrared radiation emitted. Using this information, it calculates the temperature by measuring the infrared radiation. The microprocessor then sends the temperature reading to an LCD display on the outside of the device, where it can be read almost immediately.

 CONCEPTS *Applied*

Human Thermometer?

Here is a way to find out how well the human body senses temperature. Obtain three containers that are large enough to submerge your hand in water. In one container, place enough ice water, including ice cubes, to cover a hand. In a second container, place enough water as hot as you can tolerate (without burning yourself) to cover a hand. Fill the third container with enough moderately warm water to cover a hand.

Submerge your right hand in the hot water and your left hand in the ice water for one minute. Dry your hands quickly, then submerge both in the warm water. How does the water feel to your right hand? How does it feel to your left hand? How well do your hands sense temperature?

TEMPERATURE SCALES

There are several referent scales used to define numerical values for measuring temperatures (Figure 4.7). The **Fahrenheit scale** was developed by the German physicist Gabriel D.

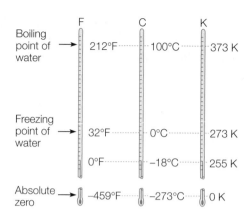

FIGURE 4.7 The Fahrenheit, Celsius, and Kelvin temperature scales.

Fahrenheit (1686–1736) in about 1715. Fahrenheit invented a mercury-in-glass thermometer with a scale based on two arbitrarily chosen reference points. The original Fahrenheit scale was based on the temperature of an ice and salt mixture for the lower reference point (0°) and the temperature of the human body as the upper reference point (about 100°). Thus, the original Fahrenheit scale was a centigrade scale with 100 divisions between the high and the low reference points. The distance between the two reference points was then divided into equal intervals called *degrees*. There were problems with identifying a "normal" human body temperature as a reference point, since body temperature naturally changes during a given day and from day to day. Some people "normally" have a higher body temperature than others. Some may have a normal body temperature of 99.1°F, while others have a temperature of 97°F. The average for a large population is 98.6°F. The only consistent thing about the human body temperature is constant change. The standards for the Fahrenheit scale were eventually changed to something more consistent, the freezing point and the boiling point of water at normal atmospheric pressure. The original scale was retained with the new reference points, however, so the "odd" numbers of 32°F (freezing point of water) and 212°F (boiling point of water under normal pressure) came to be the reference points. There are 180 equal intervals, or degrees, between the freezing and boiling points on the Fahrenheit scale.

The **Celsius scale** was invented by Anders C. Celsius (1701–1744), a Swedish astronomer, in about 1735. The Celsius scale uses the freezing point and the boiling point of water at normal atmospheric pressure, but it has different arbitrarily assigned values. The Celsius scale identifies the freezing point of water as 0°C and the boiling point as 100°C. There are 100 equal intervals, or degrees, between these two reference points, so the Celsius scale is sometimes called the *centigrade* scale.

There is nothing special about either the Celsius scale or the Fahrenheit scale. Both have arbitrarily assigned numbers, and neither is more accurate than the other. The Celsius scale is more convenient because it is a decimal scale and because it has a direct relationship with a third scale to be described shortly, the Kelvin scale. Both scales have arbitrarily assigned reference points and an arbitrary number line that indicates *relative* temperature changes. Zero is simply one of the points on each number line and does *not* mean that there is no temperature. Likewise, since the numbers are relative measures of temperature change, 2° is not twice as hot as a temperature of 1° and 10° is not twice as hot as a temperature of 5°. The numbers simply mean some measure of temperature *relative to* the freezing and boiling points of water under normal conditions.

You can convert from one temperature to the other by considering two differences in the scales: (1) the difference in the degree size between the freezing and boiling points on the two scales and (2) the difference in the values of the lower reference points.

The Fahrenheit scale has 180° between the boiling and freezing points (212°F − 32°F), and the Celsius scale has 100° between the same two points. Therefore, each Celsius degree is 180/100 or 9/5 as large as a Fahrenheit degree. Each Fahrenheit degree is 100/180 or 5/9 of a Celsius degree. You know that this is correct because there are more Fahrenheit degrees than Celsius degrees between freezing and boiling. The relationship between the degree sizes is 1°C = 9/5°F and 1°F = 5/9°C. In addition, considering the difference in the values of the lower reference points (0°C and 32°F) gives the equations for temperature conversion. (For a review of the sequence of mathematical operations used with equations, refer to the "Working with Equations" section in the Mathematical Review of appendix A.)

$$T_F = \frac{9}{5} T_C + 32°$$

equation 4.1

$$T_C = \frac{5}{9} (T_F - 32°)$$

equation 4.2

EXAMPLE 4.1

The average human body temperature is 98.6°F. What is the equivalent temperature on the Celsius scale?

SOLUTION

$$T_C = \frac{5}{9} (T_F - 32°)$$

$$= \frac{5}{9} (98.6° - 32°)$$

$$= \frac{5}{9} (66.6°)$$

$$= \frac{333°}{9}$$

$$= \boxed{37°C}$$

EXAMPLE 4.2

A bank temperature display indicates 20°C (room temperature). What is the equivalent temperature on the Fahrenheit scale? (Answer: 68°F)

There is a temperature scale that does not have arbitrarily assigned reference points, and zero *does* mean nothing. This is not a relative scale but an absolute temperature scale called the **Kelvin scale.** The Kelvin scale was proposed in 1848 by William Thompson (1824–1907), who became Lord Kelvin in 1892. The zero point on the Kelvin scale is thought to be the lowest limit of temperature. *Absolute zero* is the *lowest temperature possible,* occurring when all random motion of molecules was historically projected to cease. Absolute zero is written as 0 K. A degree symbol is not used, and the K stands for the SI standard scale unit, Kelvin. The Kelvin scale uses the same degree size as the Celsius scale, and −273°C = 0 K. Note in Figure 4.7 that 273 K is the freezing point of water, and 373 K is the boiling point. You could think of the Kelvin scale as a Celsius scale with the zero point shifted by 273°. Thus, the relationship between the Kelvin and Celsius scales is

$$T_K = T_C + 273$$

equation 4.3

A temperature of absolute zero has never been reached, but scientists have cooled a sample of sodium to 700 nanokelvins, or 700 billionths of a kelvin above absolute zero.

EXAMPLE 4.3

A science article refers to a temperature of 300.0 K. (a) What is the equivalent Celsius temperature? (b) The equivalent Fahrenheit temperature?

SOLUTION

(a) The relationship between the Kelvin scale and Celsius scale is found in equation 4.3, $T_K = T_C + 273$. Solving this equation for Celsius yields $T_C = T_K - 273$.

$$T_C = T_K - 273$$

$$= 300.0 - 273$$

$$= \boxed{27°C}$$

(b)

$$T_F = \frac{9}{5} T_C + 32°$$

$$= \frac{9}{5} 27.0° + 32°$$

$$= \frac{243°}{5} + 32°$$

$$= 48.6° + 32°$$

$$= \boxed{81°F}$$

A Closer Look

Goose Bumps and Shivering

For an average age and minimal level of activity, many people feel comfortable when the environmental temperature is about 25°C (77°F). Comfort at this temperature probably comes from the fact that the body does not have to make an effort to conserve or get rid of heat.

Changes that conserve heat in the body occur when the temperature of the air and clothing directly next to a person becomes less than 20°C or if the body senses rapid heat loss. First, blood vessels in the skin are constricted. This slows the flow of blood near the surface, which reduces heat loss by conduction. Constriction of skin blood vessels reduces body heat loss but may also cause the skin and limbs to become significantly cooler than the body core temperature (producing cold feet, for example).

Sudden heat loss, or a chill, often initiates another heat-saving action by the body. Skin hair is pulled upright, erected to slow heat loss to cold air moving across the skin.

Contraction of a tiny muscle attached to the base of the hair shaft makes a tiny knot, or bump, on the skin. These are sometimes called "goose bumps" or "chill bumps." Although goose bumps do not significantly increase insulation in humans, the equivalent response in birds and many mammals elevates feathers or hairs and greatly enhances insulation.

Further cooling after the blood vessels in the skin have been constricted results in the body taking yet another action. The body now begins to produce *more* heat, making up for heat loss through involuntary muscle contractions called "shivering." The greater the need for more body heat, the greater the activity of shivering.

If the environmental temperatures rise above about 25°C (77°F), the body triggers responses that cause it to *lose* heat. One response is to make blood vessels in the skin larger, which increases blood flow in the skin. This brings more heat from the core to be conducted through the skin, then

radiated away. It also causes some people to have a red blush from the increased blood flow in the skin. This action increases conduction through the skin, but radiation alone provides insufficient cooling at environmental temperatures above about 29°C (84°F). At about this temperature, sweating begins and perspiration pours onto the skin to provide cooling through evaporation. The warmer the environmental temperature, the greater the rate of sweating and cooling through evaporation.

The actual responses to a cool, cold, warm, or hot environment will be influenced by a person's level of activity, age, and gender, and environmental factors such as the relative humidity, air movement, and combinations of these factors. Temperature is the single most important comfort factor. However, when the temperature is high enough to require perspiration for cooling, humidity also becomes an important factor in human comfort.

HEAT

Suppose you have a bowl of hot soup or a cup of hot coffee that is too hot. What can you do to cool it? You can blow across the surface, which speeds evaporation and therefore results in cooling, but this is a slow process. If you were in a hurry, you would probably add something cooler, such as ice. Adding a cooler substance will cool the hot liquid.

You know what happens when you mix fluids or objects with a higher temperature with fluids or objects with a lower temperature. The warmer-temperature object becomes cooler and the cooler-temperature object becomes warmer. Eventually, both will have a temperature somewhere between the warmer and the cooler. This might suggest that something is moving between the warmer and cooler objects, changing the temperature. What is doing the moving?

The term **heat** is used to describe the "something" that moves between objects when two objects of different temperature are brought together. As you will learn in the section on heat as energy transfer, heat flow represents a form of energy transfer that takes place between objects. For now, we will continue to think of heat as "something"—an energy transfer—that moves between objects of different temperatures, such as your bowl of hot soup and a cold ice cube.

The relationship that exists between energy and temperature will help explain the concept of heat, so we will consider it first. If you rub your hands together a few times, they will feel a little warmer. If you rub them together vigorously for a

while, they will feel a lot warmer, maybe hot. A temperature increase takes place anytime mechanical energy causes one surface to rub against another (Figure 4.8). The two surfaces could be solids, such as the two blocks, but they can also be the surface of a solid and a fluid, such as air. A solid object moving through the air encounters air compression, which

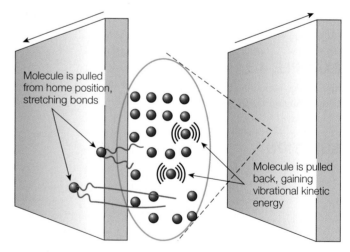

FIGURE 4.8 Here is how friction results in increased temperatures: Molecules on one moving surface will catch on another surface, stretching the molecular forces that are holding it. They are pulled back to their home position with a snap, resulting in a gain of vibrational kinetic energy.

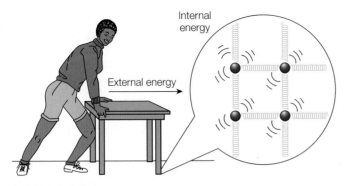

FIGURE 4.9 *External energy* is the kinetic and potential energy that you can see. *Internal energy* is the total kinetic and potential energy of molecules. When you push a table across the floor, you do work against friction. Some of the external mechanical energy goes into internal kinetic and potential energy, and the bottom surfaces of the legs become warmer.

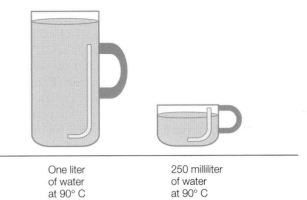

One liter of water at 90° C   250 milliliter of water at 90° C

FIGURE 4.10 Heat and temperature are different concepts, as shown by a liter of water (1,000 mL) and a 250 mL cup of water, both at the same temperature. You know the liter of water contains more internal energy because it will require more ice cubes to cool it to, say, 25°C than will be required for the cup of water. In fact, you will have to remove 48,750 *additional* calories to cool the liter of water.

results in a higher temperature of the surface. A high velocity meteor enters Earth's atmosphere and is heated so much from the compression that it begins to glow, resulting in the fireball and smoke trail of a "falling star."

To distinguish between the energy of the object and the energy of its molecules, we use the terms *external* and *internal* energy. **External energy** is the total potential and kinetic energy of an everyday-sized object. All the kinetic and potential energy considerations discussed in chapters 2 and 3 were about the external energy of an object.

Internal energy is the total kinetic and potential energy of the *molecules* of an object. The kinetic energy of a molecule can be much more complicated than straight-line velocity might suggest, however, because a molecule can have many different types of motion at the same time (pulsing, twisting, turning, etc.). Overall, internal energy is characterized by properties such as temperature, density, heat, volume, pressure of a gas, and so forth.

When you push a table across the floor, the observable *external* kinetic energy of the table is transferred to the *internal* kinetic energy of the molecules between the table legs and the floor, resulting in a temperature increase (Figure 4.9). The relationship between external and internal kinetic energy explains why the heating is proportional to the amount of mechanical energy used.

HEAT AS ENERGY TRANSFER

Temperature is a measure of the degree of hotness or coldness of a body, a measure that is based on the average molecular kinetic energy. Heat, on the other hand, is based on the *total internal energy* of the molecules of a body. You can see one difference in heat and temperature by considering a cup of water and a large tub of water. If both the small and the large amount of water have the same temperature, both must have the same average molecular kinetic energy. Now, suppose you wish to cool both by, say, 20°. The large tub of water would take much longer to cool, so it must be that the large amount of water has more internal energy (Figure 4.10). Heat is a measure based on the *total* internal energy of the molecules of a body, and there is more total energy in a large tub of water than in a cup of water at the same temperature.

Heat Defined

How can we measure heat? Since it is difficult to see molecules, internal energy is difficult to measure directly. Thus, heat is nearly always measured during the process of a body gaining or losing energy. This measurement procedure will also give us a working definition of heat:

> **Heat is a measure of the internal energy that has been absorbed or transferred from one body to another.**

The *process* of increasing the internal energy is called "heating", and the *process* of decreasing internal energy is called "cooling." The word *process* is italicized to emphasize that heat is energy in transit, not a material thing you can add or take away. Heat is understood to be a measure of internal energy that can be measured as energy flows into or out of an object.

Two Heating Methods

There are two general ways that heating can occur. These are (1) from a temperature difference, with energy moving from the region of higher temperature, and (2) from an object gaining energy by way of an energy-form conversion.

When a *temperature difference* occurs, energy is transferred from a region of higher temperature to a region of lower temperature. Energy flows from a hot range element, for example, to a pot of cold water on a range. It is a natural process for energy to flow from a region of higher temperature to a region of a lower temperature just as it is natural for a ball to roll downhill. The temperature of an object and the temperature of the surroundings determine if heat will be transferred to or from an object. The terms *heating* and *cooling* describe the direction of energy flow, naturally moving from a region of higher energy to one of lower energy.

The internal energy of an object can be increased during an *energy-form conversion* (mechanical, radiant, electrical, etc.), so we say that heating is taking place. The classic experiments by Joule showed an equivalence between mechanical energy and heating, electrical energy and heating, and other conversions.

On a molecular level, the energy forms are doing work on the molecules, which can result in an increase of internal energy. Thus, heating by energy-form conversion is actually a transfer of energy by *working*. This brings us back to the definition that "energy is the ability to do work." We can mentally note that this includes the ability to do work at the molecular level.

Heating that takes place because of a temperature difference will be considered in more detail after we consider how heat is measured.

MEASURES OF HEAT

Since heating is a method of energy transfer, a quantity of heat can be measured just like any quantity of energy. The metric unit for measuring work, energy, or heat is the *joule*. However, the separate historical development of the concepts of heat and the concepts of motion resulted in separate units, some based on temperature differences.

The metric unit of heat is called the **calorie** (cal). A calorie is defined as the *amount of energy (or heat) needed to increase the temperature of 1 gram of water 1 degree Celsius*. A more precise definition specifies the degree interval from 14.5°C to 15.5°C because the energy required varies slightly at different temperatures. This precise definition is not needed for a general discussion. A **kilocalorie** (kcal) is the *amount of energy (or heat) needed to increase the temperature of 1 kilogram of water 1 degree Celsius*. The measure of the energy released by the oxidation of food is the kilocalorie, but it is called the Calorie (with a capital C) by nutritionists (Figure 4.11). Confusion can be avoided by making sure that the scientific calorie is never capitalized (cal) and the dieter's Calorie is always capitalized. The best solution would be to call the Calorie what it is, a kilocalorie (kcal).

The English system's measure of heating is called the **British thermal unit** (Btu). A Btu is *the amount of energy (or heat) needed to increase the temperature of 1 pound of water 1 degree Fahrenheit*. The Btu is commonly used to measure the heating or cooling rates of furnaces, air conditioners, water heaters, and

so forth. The rate is usually expressed or understood to be in Btu per hour. A much larger unit is sometimes mentioned in news reports and articles about the national energy consumption. This unit is the *quad*, which is 1 quadrillion Btu (a million billion or 10^{15} Btu).

Heat is increased by an energy-form conversion, and the equivalence between energy and heating was first measured by James Joule. He found that the relationship between the energy form (mechanical, electrical radiant, etc.) and the resulting heating was always the same. For example, the relationship between mechanical work done and the resulting heating is always

$$4.184 \text{ J} = 1 \text{ cal}$$

or

$$4,184 \text{ J} = 1 \text{ kcal}$$

The establishment of this precise proportionality means that, fundamentally, mechanical work and heat are different forms of the same thing.

EXAMPLE 4.4

A 1,000.0 kg car is moving at 90.0 km/h (25.0 m/s). How many kilocalories are generated when the car brakes to a stop?

SOLUTION

The kinetic energy of the car is

$$KE = \frac{1}{2}mv^2$$

$$= \frac{1}{2}(1,000.0 \text{ kg})(25.0 \text{ m/s})^2$$

$$= (500.0)(625)\frac{\text{kg·m}^2}{\text{s}^2}$$

$$= 312,500 \text{ J}$$

You can convert this to kcal by using the relationship between mechanical energy and heat:

$$(312,500 \text{ J})\left(\frac{1 \text{ kcal}}{4,184 \text{ J}}\right)$$

$$\frac{312,500}{4,184}\frac{\text{J·kcal}}{\text{J}}$$

$$\boxed{74.7 \text{ kcal}}$$

(Note: The temperature increase from this amount of heating could be calculated from equation 4.4.)

SPECIFIC HEAT

You can observe a relationship between heat and different substances by doing an experiment in "kitchen physics." Imagine that you have a large pot of liquid to boil in preparing a meal. Three variables influence how much heat you need:

1. the initial temperature of the liquid;
2. how much liquid is in the pot; and,
3. the nature of the liquid (water or soup?).

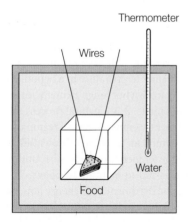

FIGURE 4.11 The Calorie value of food is determined by measuring the heat released from burning the food. If there is 10.0 kg of water and the temperature increased from 10° to 20°C, the food contained 100 Calories (100,000 calories). The food illustrated here would release much more energy than this.

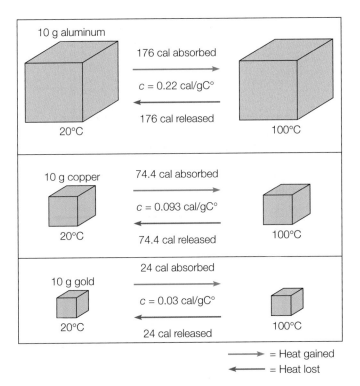

FIGURE 4.12 Of these three metals, aluminum needs the most heat per gram per degree when warmed and releases the most heat when cooled. Why are the cubes different sizes?

What this means specifically is

1. *Temperature change.* The amount of heat needed is proportional to the temperature change. It takes more heat to raise the temperature of cool water, so this relationship could be written as $Q \propto \Delta T$.
2. *Mass.* The amount of heat needed is also proportional to the amount of the substance being heated. A larger mass requires more heat to go through the same temperature change than a smaller mass. In symbols, $Q \propto m$.
3. *Substance.* Different materials require different amounts of heat to go through the same temperature range when their masses are equal (Figure 4.12). This property is called the **specific heat** of a material, which is defined as the amount of heat needed to increase the temperature of 1 gram of a substance 1 degree Celsius.

Considering all the variables involved in our kitchen physics cooking experience, we find the heat (Q) needed is described by the relationship

$$Q = mc\Delta T$$

equation 4.4

where c is the symbol for specific heat. Specific heat is related to the internal structure of a substance; some of the energy goes into the internal potential energy of the molecules, and some goes into the internal kinetic energy of the molecules. The difference in values for the specific heat of different substances is related to the number of molecules in a 1-gram sample of each and to the way they form a molecular structure.

Specific heat is responsible for the fact that air temperatures vary more over land than over a large body of water. Table 4.2 gives the specific heat of soil as 0.200 cal/gC° and the specific heat of water as 1.00 cal/gC°. Since specific heat is defined as the amount of heat needed to increase the temperature of 1 gram of a substance 1 degree, this means 1 gram of water exposed to 1 calorie of sunlight will warm 1°C. One gram of soil exposed to 1 calorie of sunlight, on the other hand, will be warmed by 5°C because it only takes 0.2 calories to warm the soil 1°C. Thus, the temperature is more even near large bodies of water because it is harder to change the temperature of the water.

EXAMPLE 4.5

How much heat must be supplied to a 500.0 g pan to raise its temperature from 20.0°C to 100.0°C if the pan is made of (a) iron and (b) aluminum?

SOLUTION

The relationship between the heat supplied (Q), the mass (m), and the temperature change (ΔT) is found in equation 4.4. The specific heats (c) of iron and aluminum can be found in Table 4.2.

(a) Iron:

$m = 500.0$ g

$c = 0.11$ cal/gC°

$T_f = 100.0$°C

$Q = ?$

$Q = mc\Delta T$

$= (500.0 \text{ g}) \left(0.11 \dfrac{\text{cal}}{\text{gC}°}\right)(80.0°\text{C})$

$= (500.0)(0.11)(80.0) \text{ g} \times \dfrac{\text{cal}}{\text{gC}°} \times °\text{C}$

$= 4{,}400 \ \dfrac{\text{g} \cdot \text{cal} \cdot \cancel{\text{C}°}}{\cancel{\text{g}}\cancel{\text{C}°}}$

$= 4{,}400$ cal

$= \boxed{4.4 \text{ kcal}}$

(b) Aluminum:

$m = 500.0$ g

$c = 0.22$ cal/gC°

$T_f = 100.0$°C

$T_i = 20.0$°C

$Q = ?$

$Q = mc\Delta T$

$= (500.0 \text{ g}) \left(0.22 \dfrac{\text{cal}}{\text{gC}°}\right)(80.0°\text{C})$

$= (500.0)(0.22)(80.0) \text{ g} \times \dfrac{\text{cal}}{\text{gC}°} \times °\text{C}$

$= 8{,}800 \ \dfrac{\text{g} \cdot \text{cal} \cdot \cancel{\text{C}°}}{\cancel{\text{g}}\cancel{\text{C}°}}$

$= 8{,}800$ cal

$= \boxed{8.8 \text{ kcal}}$

It takes twice as much heat energy to warm the aluminum pan through the same temperature range as an iron pan. Thus, with equal rates of energy input, the iron pan will warm twice as fast as an aluminum pan.

EXAMPLE 4.6

What is the specific heat of a 2 kg metal sample if 1.2 kcal are needed to increase the temperature from 20.0°C to 40.0°C? (Answer: 0.03 kcal/kgC°)

TABLE 4.2

The specific heat of selected substances

Substance	Specific Heat (cal/gC° or kcal/kgC°)
Air	0.17
Aluminum	0.22
Concrete	0.16
Copper	0.093
Glass (average)	0.160
Gold	0.03
Ice	0.500
Iron	0.11
Lead	0.0305
Mercury	0.033
Seawater	0.93
Silver	0.056
Soil (average)	0.200
Steam	0.480
Water	1.00

Note: To convert to specific heat in J/kgC°, multiply each value by 4,184. Also note that 1 cal/gC° = 1 kcal/kgC°.

 CONCEPTS *Applied*

More Kitchen Physics

Consider the following information as it relates to the metals of cooking pots and pans.

1. It is easier to change the temperature of metals with low specific heats.
2. It is harder to change the temperature of metals with high specific heats.

   Look at the list of metals and specific heats in Table 4.2 and answer the following questions:

1. Considering specific heat alone, which metal could be used for making practical pots and pans that are the most energy efficient to use?
2. Again considering specific heat alone, would certain combinations of metals provide any advantages for rapid temperature changes?

HEAT FLOW

In the "Heat as Energy Transfer" section, you learned the process of heating is a transfer of energy involving (1) a temperature difference or (2) energy-form conversions. Heat transfer that takes place because of a temperature difference takes place in three different ways: by conduction, convection, or radiation.

Conduction

Anytime there is a temperature difference, there is a natural transfer of heat from the region of higher temperature to the

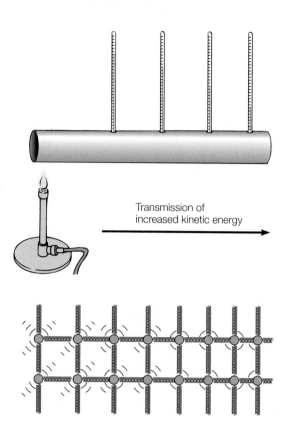

FIGURE 4.13 Thermometers placed in holes drilled in a metal rod will show that heat is conducted from a region of higher temperature to a region of lower temperature. The increased molecular activity is passed from molecule to molecule in the process of conduction.

region of lower temperature. In solids, this transfer takes place as heat is *conducted* from a warmer place to a cooler one. Recall that the molecules in a solid vibrate in a fixed equilibrium position and that molecules in a higher temperature region have more kinetic energy, on the average, than those in a lower temperature region. When a solid, such as a metal rod, is held in a flame, the molecules in the warmed end vibrate violently. Through molecular interaction, this increased energy of vibration is passed on to the adjacent, slower-moving molecules, which also begin to vibrate more violently. They, in turn, pass on more vibrational energy to the molecules next to them. The increase in activity thus moves from molecule to molecule, causing the region of increased activity to extend along the rod. This is called **conduction,** the transfer of energy from molecule to molecule (Figure 4.13).

Most insulating materials are good insulators because they contain many small air spaces (Figure 4.14). The small air spaces are poor conductors because the molecules of air are far apart, compared to a solid, making it more difficult to pass the increased vibrating motion from molecule to molecule. Styrofoam, glass wool, and wool cloth are good insulators because they have many small air spaces, not because of the material they are made of. The best insulator is a vacuum, since there are no molecules to pass on the vibrating motion (Table 4.3).

FIGURE 4.14 Fiberglass insulation is rated in terms of R-value, a ratio of the conductivity of the material to its thickness.

TABLE 4.3

Rate of conduction of materials*

Silver	0.97
Copper	0.92
Aluminum	0.50
Iron	0.11
Lead	0.08
Concrete	4.0×10^{-3}
Glass	2.5×10^{-3}
Tile	1.6×10^{-3}
Brick	1.5×10^{-3}
Water	1.3×10^{-3}
Wood	3.0×10^{-4}
Cotton	1.8×10^{-4}
Styrofoam	1.0×10^{-4}
Glass wool	9.0×10^{-5}
Air	6.0×10^{-5}
Vacuum	0

(Better Conductor ↑ / Better Insulator ↓)

*Based on temperature difference of 1°C per cm. Values are cal/s through a square centimeter of the material.

Wooden and metal parts of your desk have the same temperature, but the metal parts will feel cooler if you touch them. Metal is a better conductor of heat than wood and feels cooler because it conducts heat from your finger faster. This is the same reason that a wood or tile floor feels cold to your bare feet. You use an insulating rug to slow the conduction of heat from your feet.

 CONCEPTS *Applied*

Touch Temperature

Objects that have been in a room with a constant temperature for some time should all have the same temperature. Touch metal, plastic, and wooden parts of a desk or chair to sense their temperature. Explain your findings.

Convection

Convection is the transfer of heat by a large-scale displacement of groups of molecules with relatively higher kinetic energy. In conduction, increased kinetic energy is passed from molecule to molecule. In convection, molecules with higher kinetic energy are moved from one place to another place. Conduction happens primarily in solids, but convection happens only in liquids and gases, where fluid motion can carry molecules with higher kinetic energy over a distance. When molecules gain energy, they move more rapidly and push more vigorously against their surroundings. The result is an expansion as the region of heated molecules pushes outward and increases the volume. Since the same amount of matter now occupies a larger volume, the overall density has been decreased (Figure 4.15).

In fluids, expansion sets the stage for convection. Warm, less dense fluid is pushed upward by the cooler, more dense fluid around it. In general, cooler air is more dense; it sinks and flows downhill. Cold air, being more dense, flows out near the bottom of an open refrigerator. You can feel the cold, dense air pouring from the bottom of a refrigerator to your toes on the floor. On the other hand, you hold your hands *over* a heater because the warm, less dense air is pushed upward. In a room, warm air is pushed upward from a heater.

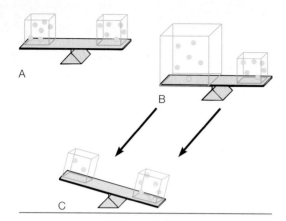

FIGURE 4.15 (*A*) Two identical volumes of air are balanced, since they have the same number of molecules and the same mass. (*B*) Increased temperature causes one volume to expand from the increased kinetic energy of the gas molecules. (*C*) The same volume of the expanded air now contains fewer gas molecules and is less dense, and it is buoyed up by the cooler, more dense air.

The warm air spreads outward along the ceiling and is slowly displaced as newly warmed air is pushed upward to the ceiling. As the air cools, it sinks over another part of the room, setting up a circulation pattern known as a *convection current* (Figure 4.16). Convection currents can also be observed in a large pot of liquid that is heating on a range. You can see the warmer liquid being forced upward over the warmer parts of the range element, then sink over the cooler parts. Overall, convection currents give the liquid in a pot the appearance of turning over as it warms.

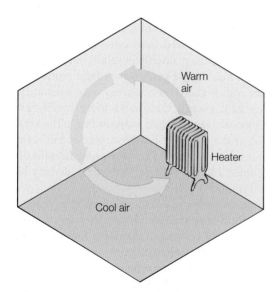

FIGURE 4.16 Convection currents move warm air throughout a room as the air over the heater becomes warmed, expands, and is moved upward by cooler air.

CONCEPTS *Applied*

How Convection Works

Convection takes place in fluids where a temperature difference exists. To see why this occurs, obtain a balloon filled with very cold water and a second balloon filled with the same volume of very hot water. Carefully put the balloon with cold water in a large container of hot water. Place the balloon filled with hot water into a large container of cold water. What happens in each container? What does this tell you about the relationship between the temperature and density of a fluid and how convection works?

Radiation

The third way that heat transfer takes place because of a temperature difference is called **radiation.** Radiation involves the form of energy called *radiant energy,* energy that moves through space. As you learned in chapter 3, radiant energy includes visible light and many other forms as well. All objects with a temperature above absolute zero give off radiant energy. The absolute temperature of the object determines the rate, intensity, and kinds of radiant energy emitted. You know that visible light is emitted if an object is heated to a certain temperature. A heating element on an electric range, for example, will glow with a reddish-orange light when at the highest setting, but it produces no visible light at lower temperatures, although you feel warmth in your hand when you hold it near the element. Your hand absorbs the nonvisible radiant energy being emitted from the element. The radiant energy does work on the molecules of your hand, giving them more kinetic energy. You sense this as an increase in temperature, that is, warmth.

All objects above absolute zero (0 K) emit radiant energy, but all objects also absorb radiant energy. A hot object, however, emits more radiant energy than a cold object. The hot object will emit more energy than it absorbs from the colder object, and the colder object will absorb more energy from the hot object than it emits. There is, therefore, a net energy transfer that will take place by radiation as long as there is a temperature difference between the two objects.

ENERGY, HEAT, AND MOLECULAR THEORY

The kinetic molecular theory of matter is based on evidence from different fields of physical science, not just one subject area. Chemists and physicists developed some convincing conclusions about the structure of matter over the past 150 years, using carefully designed experiments and mathematical calculations that explained observable facts about matter. Step by step, the detailed structure of this submicroscopic, invisible world of particles became firmly established. Today, an understanding

CONCEPTS *Applied*

Candle Wax, Bubbles, and a Lightbulb

Here are three experiments you can do on heat flow:

Conduction: Use melted candle wax to stick thumbtack heads to a long metal rod. Heat one end of the rod with a flame. Record any evidence you observe that heat moves across the rod by conduction.

Convection: Choose a calm, warm day with plenty of strong sunlight. Make soap bubbles to study convection currents between a grass field and an adjacent asphalt parking lot. Find other adjacent areas where you think you might find convection currents and study them with soap bubbles, too. Record your experiments, findings, and explanations for what you observed.

Radiation: Hold your hand under an unlighted electric lightbulb and then turn on the bulb. Describe evidence that what you feel traveled to you by radiation, not conduction or convection. Describe any experiments you can think of to prove you felt radiant energy.

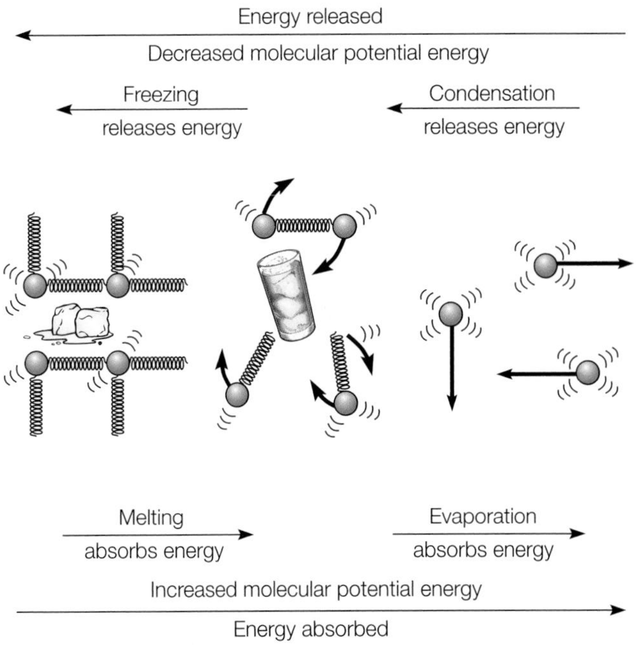

FIGURE 4.17 Each phase change absorbs or releases a quantity of latent heat,which goes into or is released from molecular potential energy.

of this particle structure is basic to physics, chemistry, biology, geology, and practically every other science subject. This understanding has also resulted in present-day technology.

PHASE CHANGE

Solids, liquids, and gases are the three common phases of matter, and each phase is characterized by different molecular arrangements. The motion of the molecules in any of the three common phases can be increased by (1) adding heat through a temperature difference or (2) the absorption of one of the five forms of energy, which results in heating. In either case, the temperature of the solid, liquid, or gas increases according to the specific heat of the substance, and more heating generally means higher temperatures.

More heating, however, does not always result in increased temperatures. When a solid, liquid, or gas changes from one phase to another, the transition is called a **phase change.** A phase change always absorbs or releases *a quantity of heat that is not associated with a temperature change.* Since the quantity of heat associated with a phase change is not associated with a temperature change, it is called *latent heat.* Latent heat refers to the "hidden" energy of phase changes, which is energy (heat) that goes into or comes out of *internal potential energy* (Figure 4.17).

There are three kinds of major phase changes that can occur: (1) *solid-liquid,* (2) *liquid-gas,* and (3) *solid-gas.* In each case, the phase change can go in either direction. For example, the solid-liquid phase change occurs when a solid melts to a liquid or when a liquid freezes to a solid. Ice melting to water and water freezing to ice are common examples of this phase change and its two directions. Both occur at a temperature called the *freezing point* or the *melting point,* depending on the direction of the phase change. In either case, however, the freezing and melting points are the same temperature.

The liquid-gas phase change also occurs in two different directions. The temperature at which a liquid boils and changes to a gas (or vapor) is called the *boiling point.* The temperature at which a gas or vapor changes back to a liquid is called the *condensation point.* The boiling and condensation points are the same temperature. There are conditions other than boiling under which liquids may undergo liquid-gas phase changes, and these conditions are discussed in the next section, "Evaporation and Condensation."

You probably are not as familiar with solid-gas phase changes, but they are common. A phase change that takes a solid directly to a gas or vapor is called *sublimation.* Mothballs and dry ice (solid CO_2) are common examples of materials that undergo sublimation, but frozen water, meaning common ice, also sublimates under certain conditions. Perhaps you have noticed ice cubes in a freezer become smaller with time as a result of sublimation. The frost that forms in a freezer, on the other hand, is an example of a solid-gas phase change that takes place in the other direction. In this case, water vapor forms the frost without going through the liquid state, a solid-gas phase change that takes place in an opposite direction to sublimation.

For a specific example, consider the changes that occur when ice is subjected to a constant source of heat (Figure 4.18). Starting at the left side of the graph, you can see that the temperature of the ice increases from the constant input of heat. The ice warms according to $Q = mc\Delta T$, where c is the specific

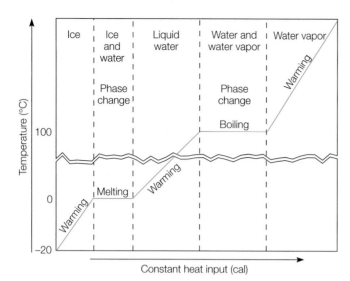

FIGURE 4.18 This graph shows three warming sequences and two phase changes with a constant input of heat. The ice warms to the melting point, then absorbs heat during the phase change as the temperature remains constant. When all the ice has melted, the now-liquid water warms to the boiling point, where the temperature again remains constant as heat is absorbed during this second phase change from liquid to gas. After all the liquid has changed to gas, continued warming increases the temperature of the water vapor.

heat of ice. When the temperature reaches the melting point (0°C), it stops increasing as the ice begins to melt. More and more liquid water appears as the ice melts, but the temperature *remains* at 0°C even though heat is still being added at a constant rate. It takes a certain amount of heat to melt all of the ice. Finally, when all the ice is completely melted, the temperature again increases at a constant rate between the melting and boiling points. Then, at constant temperature the addition of heat produces another phase change, from liquid to gas. The quantity of heat involved in this phase change is used in doing the work of breaking the molecule-to-molecule bonds in the solid, making a liquid with molecules that are now free to move about and roll over one another. Since the quantity of heat (Q) is absorbed without a temperature change, it is called the **latent heat of fusion** (*L*f). The latent heat of fusion is *the heat involved in a solid-liquid phase change in melting or freezing.* You learned in chapter 3 that when you do work on something, you give it energy. In this case, the work done in breaking the molecular bonds in the solid gave the molecules more *potential* energy (Figure 4.19). This energy is "hidden," or latent, since heat was absorbed but a temperature increase did not take place. This same potential energy is given up when the molecules of the liquid return to the solid state. A melting solid absorbs energy and a freezing liquid releases this *same amount* of energy, warming the surroundings. Thus, you put ice in a cooler because the melting ice absorbs the latent heat of fusion from the beverage cans, cooling them. Citrus orchards are flooded with water when freezing temperatures are expected because freezing water releases the latent heat of fusion, which warms the air around the trees. For water, the latent heat of fusion is 80.0 cal/g (144.0 Btu/lb). This means

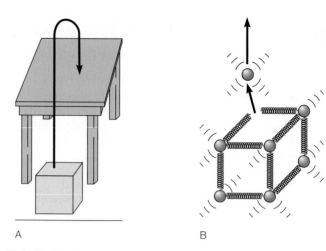

FIGURE 4.19 (*A*) Work is done against gravity to lift an object, giving the object more gravitational potential energy. (*B*) Work is done against intermolecular forces in separating a molecule from a solid, giving the molecule more potential energy.

that every gram of ice that melts in your cooler *absorbs* 80.0 cal of heat. Every gram of water that freezes *releases* 80.0 cal. The total heat involved in a solid-liquid phase change depends on the mass of the substance involved, so

$$Q = mL_f$$

equation 4.5

where L_f is the latent heat of fusion for the substance involved.

Refer again to Figure 4.18. After the solid-liquid phase change is complete, the constant supply of heat increases the temperature of the water according to $Q = mc\Delta T$, where *c* is now the specific heat of liquid water. When the water reaches the boiling point, the temperature again remains constant even though heat is still being supplied at a constant rate. The quantity of heat involved in the liquid-gas phase change again goes into doing the work of overcoming the attractive molecular forces. This time the molecules escape from the liquid state to become single, independent molecules of gas. The quantity of heat (Q) absorbed or released during this phase change is called the **latent heat of vaporization** (*L*v). The latent heat of vaporization is *the heat involved in a liquid-gas phase change where there is evaporation or condensation.* The latent heat of vaporization is the energy gained by the gas molecules as work is done in overcoming molecular forces. Thus, the escaping molecules absorb energy from the surroundings, and a condensing gas (or vapor) releases this *exact same amount of energy.* For water, the latent heat of vaporization is 540.0 cal/g (970.0 Btu/lb). This means that every gram of water vapor that condenses on your bathroom mirror releases 540.0 cal, which warms the bathroom. The total heating depends on how much water vapor condensed, so

$$Q = mL_v$$

equation 4.6

where L_v is the latent heat of vaporization for the substance involved. The relationships between the quantity of heat

Passive solar application is an economically justifiable use of solar energy today. Passive solar design uses a structure's construction to heat a living space with solar energy. There are few electric fans, motors, or other energy sources used. The passive solar design takes advantage of free solar energy; it stores and then distributes this energy through natural conduction, convection, and radiation.

In general, a passive solar home makes use of the materials from which it is constructed to capture, store, and distribute solar energy to its occupants. Sunlight enters the house through large windows facing south and warms a thick layer of concrete, brick, or stone. This energy "storage mass" then releases energy during the day and, more important, during the night. This release of energy can be by direct radiation to occupants, by conduction to adjacent air, or by convection of air across the surface of the storage mass. The living space is thus heated without special plumbing or forced air circulation. As you can imagine, the key to a successful passive solar home is to consider every detail of natural energy flow, including the materials of which floors and walls are constructed, convective air circulation patterns, and the size and placement of windows. In addition, a passive solar home requires a different lifestyle and living patterns. Carpets, for example, would defeat the purpose of a storage-mass floor, since it would insulate the storage mass from sunlight. Glass is not a good insulator, so windows must

have curtains or movable insulation panels to slow energy loss at night. This requires the daily activity of closing curtains or moving insulation panels at night and then opening curtains and moving panels in the morning. Passive solar homes, therefore, require a high level of personal involvement by the occupants.

There are three basic categories of passive solar design: (1) direct solar gain, (2) indirect solar gain, and (3) isolated solar gain.

A *direct solar gain* home is one in which solar energy is collected in the actual living space of the home (Box Figure 4.1). The advantage of this design is the large, open window space with a calculated overhang, which admits maximum solar energy in the winter but prevents solar gain in the summer. The disadvantage is that the occupants are living in the collection and storage components of the design and can place nothing (such as carpets and furniture) that would interfere with warming the storage mass in the floors and walls.

An *indirect solar gain* home uses a massive wall inside a window that serves as a storage mass. Such a wall, called a *Trombe wall*, is shown in Box Figure 4.2. The Trombe wall collects and stores solar energy, then warms the living space with radiant energy and convection currents. The disadvantage to the indirect solar gain design is that large windows are blocked by the Trombe wall. The advantage is that the occupants are not in direct contact with the solar collection and storage area, so they can place carpets

and furniture as they wish. Controls to prevent energy loss at night are still necessary with this design.

An *isolated solar gain* home uses a structure that is separated from the living space to collect and store solar energy. Examples of an isolated gain design are an attached greenhouse or sun porch (Box Figure 4.3). Energy flow between the attached structure and the living space can be by conduction, convection, and radiation, which can be controlled by opening or closing off the attached structure. This design provides the best controls, since it can be completely isolated, opened to the living space as needed, or directly used as living space when the conditions are right. Additional insulation is needed for the glass at night, however, and for sunless winter days.

It has been estimated that building a passive solar home would cost about 10 percent more than building a traditional home of the same size. Considering the possible energy savings, you might believe that most homes would now have a passive solar design. They do not, however, as most new buildings require technology and large amounts of energy to maximize comfort. Yet, it would not require too much effort to consider where to place windows in relation to the directional and seasonal intensity of the sun and where to plant trees. Perhaps in the future you will have an opportunity to consider using the environment to your benefit through the natural processes of conduction, convection, and radiation.

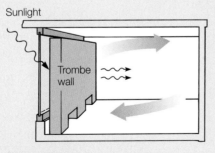

BOX FIGURE 4.1 The direct solar gain design collects and stores solar energy in the living space.

BOX FIGURE 4.2 The indirect solar gain design uses a Trombe wall to collect, store, and distribute solar energy.

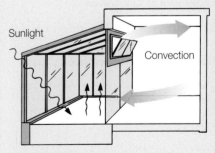

BOX FIGURE 4.3 The isolated solar gain design uses a separate structure to collect and store solar energy.

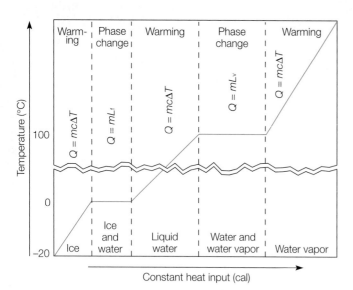

FIGURE 4.20 Compare this graph to the one in Figure 4.18. This graph shows the relationships between the quantity of heat absorbed during warming and phase changes as water is warmed from ice at −20°C to water vapor at some temperature above 100°C. Note that the specific heat for ice, liquid water, and water vapor (steam) has different values.

TABLE 4.4		
Some physical constants for water and heat		
Specific Heat (c)		
Water	$c = 1.00$ cal/gC°	
Ice	$c = 0.500$ cal/gC°	
Steam	$c = 0.480$ cal/gC°	
Latent Heat of Fusion		
L_f (water)	$L_f = 80.0$ cal/g	
Latent Heat of Vaporization		
L_v (water)	$L_v = 540.0$ cal/g	
Mechanical Equivalent of Heat		
1 kcal	4,184 J	

absorbed during warming and phase changes are shown in Figure 4.20. Some physical constants for water and heat are summarized in Table 4.4.

EXAMPLE 4.7

How much energy does a refrigerator remove from 100.0 g of water at 20.0°C to make ice at −10.0°C?

SOLUTION

This type of problem is best solved by subdividing it into smaller steps that consider (1) the heat added or removed and the resulting temperature changes *for each phase* of the substance and (2) the heat flow resulting from any *phase change* that occurs within the ranges of changes as identified by the problem (see Figure 4.20). The heat involved in each phase change and the heat involved in the heating or cooling of each phase are identified as Q_1, Q_2, and so forth. Temperature readings are calculated with *absolute values*, so you ignore any positive or negative signs.

1. Water in the liquid state cools from 20.0°C to 0°C (the freezing point) according to the relationship $Q = mc\Delta T$, where c is the specific heat of water, and

$$Q_1 = mc\Delta T$$
$$= (100.0 \text{ g})\left(1.00 \frac{\text{cal}}{\text{g°C}}\right)(0° - 20.0°C)$$
$$= (100.0)(1.00)(20.0°)\frac{\text{g·cal·°C}}{\text{g°C}}$$
$$= 2,000 \text{ cal}$$
$$Q_1 = 2.00 \times 10^3 \text{ cal}$$

2. The latent heat of fusion must now be removed as water at 0°C becomes ice at 0°C through a phase change, and

$$Q_2 = mL_f$$
$$= (100.0 \text{ g})\left(80.0 \frac{\text{cal}}{\text{g}}\right)$$
$$= (100.0)(80.0)\frac{\text{g·cal}}{\text{g}}$$
$$= 8,000 \text{ cal}$$
$$Q_2 = 8.00 \times 10^3 \text{ cal}$$

3. The ice is now at 0°C and is cooled to −10°C as specified in the problem. The ice cools according to $Q = mc\Delta T$, where c is the specific heat of ice. The specific heat of ice is 0.500 cal/gC°, and

$$Q_3 = mc\Delta T$$
$$= (100.0 \text{ g})\left(0.500 \frac{\text{cal}}{\text{gC°}}\right)(10.0° - 0°C)$$
$$= (100.0)(0.500)(10.0°)\frac{\text{g·cal·°C}}{\text{gC°}}$$
$$= 500 \text{ cal}$$
$$Q_3 = 5.00 \times 10^2 \text{ cal}$$

The total energy removed is then

$$Q_t = Q_1 + Q_2 + Q_3$$
$$= (2.00 \times 10^3 \text{ cal}) + (8.00 \times 10^3 \text{ cal}) + (5.00 \times 10^2 \text{ cal})$$
$$= 10.50 \text{ kcal}$$
$$Q_t = \boxed{10.50 \times 10^3 \text{ kcal}}$$

EVAPORATION AND CONDENSATION

Liquids do not have to be at the boiling point to change to a gas and, in fact, tend to undergo a phase change at any temperature when left in the open. The phase change occurs at any temperature but does occur more rapidly at higher

Average = $\frac{115}{10}$ = 11.5   Average = $\frac{65}{8}$ = 8.1

FIGURE 4.21 Temperature is associated with the average energy of the molecules of a substance. These numbered circles represent arbitrary levels of molecular kinetic energy that, in turn, represent temperature. The two molecules with the higher kinetic energy values [25 in (*A*)] escape, which lowers the average values from 11.5 to 8.1 (*B*). Thus, evaporation of water molecules with more kinetic energy contributes to the cooling effect of evaporation in addition to the absorption of latent heat.

temperatures. The temperature of the water is associated with the *average* kinetic energy of the water molecules. The word *average* implies that some of the molecules have a greater energy and some have less (refer to Figure 4.4). If a molecule of water that has an exceptionally high energy is near the surface and is headed in the right direction, it may overcome the attractive forces of the other water molecules and escape the liquid to become a gas. This is the process of *evaporation*. Evaporation reduces a volume of liquid water as water molecules leave the liquid state to become water vapor in the atmosphere (Figure 4.21).

Water molecules that evaporate move about in all directions, and some will return, striking the liquid surface. The same forces that they escaped from earlier capture the molecules, returning them to the liquid state. This is called the process of condensation. Condensation is the opposite of evaporation. In *evaporation,* more molecules are leaving the liquid state than are returning. In *condensation,* more molecules are returning to the liquid state than are leaving. This is a dynamic, ongoing process with molecules leaving and returning continuously. The net number leaving or returning determines whether evaporation or condensation is taking place (Figure 4.22).

When the condensation rate *equals* the evaporation rate, the air above the liquid is said to be *saturated.* The air immediately next to a surface may be saturated, but the condensation of water molecules is easily moved away with air movement. There is no net energy flow when the air is saturated, since the heat carried away by evaporation is returned by condensation. This is why you fan your face when you are hot. The moving air from the fanning action pushes away water molecules from the air near your skin, preventing the adjacent air from becoming saturated, thus increasing the rate of evaporation. Think about this process the next time you see someone fanning his or her face.

FIGURE 4.22 The inside of this closed bottle is isolated from the environment, so the space above the liquid becomes saturated. While it is saturated, the evaporation rate equals the condensation rate. When the bottle is cooled, condensation exceeds evaporation and droplets of liquid form on the inside surfaces.

There are four ways to increase the rate of evaporation. (1) An increase in the temperature of the liquid will increase the average kinetic energy of the molecules and thus increase the number of high-energy molecules able to escape from the liquid state. (2) Increasing the surface area of the liquid will also increase the likelihood of molecular escape to the air. This is why you spread out wet clothing to dry or spread out a puddle you want to evaporate. (3) Removal of water vapor from near the surface of the liquid will prevent the return of the vapor molecules to the liquid state and thus increase the net rate of evaporation. This is why things dry more rapidly on a windy day. (4) Reducing the atmospheric pressure will increase the rate of evaporation. The atmospheric pressure and the intermolecular forces tend to hold water molecules in the liquid state. Thus, reducing the atmospheric pressure will reduce one of the forces holding molecules in a liquid state. Perhaps you have noticed that wet items dry more quickly at higher elevations, where the atmospheric pressure is less.

Why Is It Called a "Pop" Can?

Obtain two empty, clean pop cans, a container of ice water with ice cubes, and a container of boiling water. You might want to "dry run" this experiment to make sure of the procedure before actually doing it.

   Place about 2 cm of water in a pop can and heat it on a stove until the water boils and you see evidence of steam coming from the opening. Using tongs, quickly invert the can halfway into a container of ice water. Note how much water runs from the can as you remove it from the ice water.

   Repeat this procedure, this time inverting the can halfway into a container of boiling water. Note how much water runs from the can as you remove it from the boiling water.

   Explain your observations in terms of the kinetic molecular theory, evaporation, and condensation. It is also important to explain any differences observed between what happened to the two pop cans.

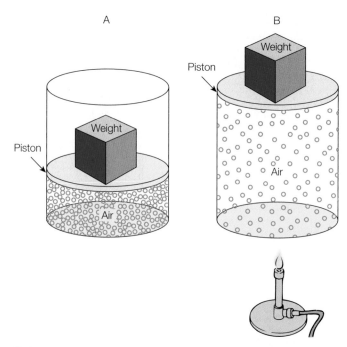

FIGURE 4.23 A very simple heat engine. The air in (*B*) has been heated, increasing the molecular motion and thus the pressure. Some of the heat is transferred to the increased gravitational potential energy of the weight as it is converted to mechanical energy.

THERMODYNAMICS

The branch of physical science called *thermodynamics* is concerned with the study of heat and its relationship to mechanical energy, including the science of heat pumps, heat engines, and the transformation of energy in all its forms. The *laws of thermodynamics* describe the relationships concerning what happens as energy is transformed to work and the reverse, also serving as useful intellectual tools in meteorology, chemistry, and biology.

   Mechanical energy is easily converted to heat through friction, but a special device is needed to convert heat to mechanical energy. A *heat engine* is a device that converts heat into mechanical energy. The operation of a heat engine can be explained by the kinetic molecular theory, as shown in Figure 4.23. This illustration shows a cylinder, much like a big can, with a closely fitting piston that traps a sample of air. The piston is like a slightly smaller cylinder and has a weight resting on it, supported by the trapped air. If the air in the large cylinder is now heated, the gas molecules will acquire more kinetic energy. This results in more gas molecule impacts with the enclosing surfaces, which results in an increased pressure. Increased pressure results in a net force, and the piston and weight move upward as shown in Figure 4.23B. Thus, some of the heat has now been transformed to the increased gravitational potential energy of the weight.

   Thermodynamics is concerned with the *internal energy* (U), the total internal potential and kinetic energies of molecules making up a substance, such as the gases in the simple heat engine. The variables of temperature, gas pressure, volume, heat, and so forth characterize the total internal energy, which is called the *state* of the system. Once the system is

identified, everything else is called the *surroundings*. A system can exist in a number of states since the variables that characterize a state can have any number of values and combinations of values. Any two systems that have the same values of variables that characterize internal energy are said to be in the same state.

THE FIRST LAW OF THERMODYNAMICS

Any thermodynamic system has a unique set of properties that will identify the internal energy of the system. This state can be changed two ways, by (1) heat flowing into (Q_{in}) or out (Q_{out}) of the system, or (2) by the system doing work (W_{out}) or by work being done on the system (W_{in}). Thus, work (W) and heat (Q) can change the internal energy of a thermodynamic system according to

$$JQ - W = U_2 - U_1$$

equation 4.7

where J is the mechanical equivalence of heat ($J = 4.184$ joule/calorie), Q is the quantity of heat, W is work, and ($U_2 - U_1$) is the internal energy difference between two states. This equation represents the **first law of thermodynamics,** which states that the energy supplied to a thermodynamic system in the form of heat, minus the work done by the system, is equal to the change in internal energy. The first law of thermodynamics is an application of the law of conservation of energy, which applies to all energy matters. The first law of thermodynamics is concerned specifically with a

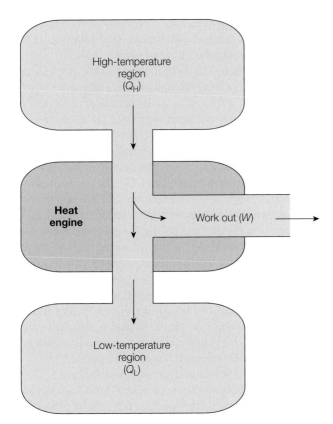

FIGURE 4.24 The heat supplied (Q_H) to a heat engine goes into the mechanical work (W), and the remainder is expelled in the exhaust (Q_L). The work accomplished is therefore the difference in the heat input and output ($Q_H - Q_L$), so the work accomplished represents the heat used, $W = J(Q_H - Q_L)$.

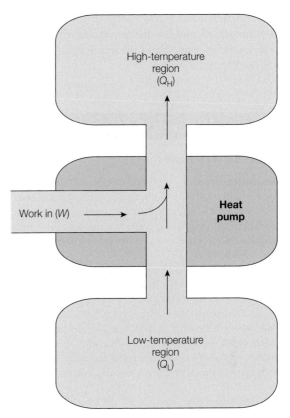

FIGURE 4.25 A heat pump uses work (W) to move heat from a low temperature region (Q_L) to a high temperature region (Q_H). The heat moved (Q_L) requires work (W), so $JQ_L = W$. A heat pump can be used to chill things at the Q_L end or warm things at the Q_H end.

thermodynamic system. As an example, consider energy conservation that is observed in the thermodynamic system of a heat engine (see Figure 4.24). As the engine cycles to the original state of internal energy ($U_2 - U_1 = 0$), all the external work accomplished must be equal to all the heat absorbed in the cycle. The heat supplied to the engine from a high temperature source (Q_H) is partly converted to work (W), and the rest is rejected in the lower-temperature exhaust (Q_L). The work accomplished is therefore the difference in the heat input and the heat output ($Q_H - Q_L$), so the work accomplished represents the heat used,

$$W = J(Q_H - Q_L)$$

equation 4.8

where J is the mechanical equivalence of heat ($J = 4.184$ joules/calorie). A schematic diagram of this relationship is shown in Figure 4.24. You can increase the internal energy (produce heat) as long as you supply mechanical energy (or do work). The first law of thermodynamics states that the conversion of work to heat is reversible, meaning that heat can be changed to work. There are several ways of converting heat to work, for example, the use of a steam turbine or gasoline automobile engine.

THE SECOND LAW OF THERMODYNAMICS

A heat pump is the *opposite* of a heat engine, as shown schematically in Figure 4.25. The heat pump does work (W) in compressing vapors and moving heat from a region of lower temperature (Q_L) to a region of higher temperature (Q_H). That work is required to move heat this way is in accord with the observation that heat naturally flows from a region of higher temperature to a region of lower temperature. Energy is required for the opposite, moving heat from a cooler region to a warmer region. The natural direction of this process is called the **second law of thermodynamics,** which is that heat flows from objects with a higher temperature to objects with a cooler temperature. In other words, if you want heat to flow from a colder region to a warmer one, you must *cause* it to do so by using energy. And if you do, such as with the use of a heat pump, you necessarily cause changes elsewhere, particularly in the energy sources used in the generation of electricity. Another statement of the second law is that it is impossible to convert heat completely into mechanical energy. This does not say that you cannot convert mechanical energy completely into heat, for example, in the brakes of a car when the brakes bring it to a stop. The law says that the reverse process is not possible, that you cannot convert 100 percent of a heat source into mechanical energy. Both of the preceding statements of the second law are concerned with a *direction* of thermodynamic processes, and the implications of this direction will be discussed next.

Myths, Mistakes, & Misunderstandings

It Makes Its Own Fuel?

Have you ever heard of a perpetual motion machine? A perpetual motion machine is a hypothetical device that would produce useful energy out of nothing. This is generally accepted as being impossible, according to laws of physics. In particular, perpetual motion machines would violate either the first or the second law of thermodynamics.

A perpetual motion machine that violates the first law of thermodynamics is called a "machine of the first kind." In general, the first law says that you can never get something for nothing. This means that without energy input, there can be no change in internal energy, and without a change in internal energy, there can be no work output. Machines of the first kind typically use no fuel or make their own fuel faster than they use it. If this type of machine appears to work, look for some hidden source of energy.

A "machine of the second kind" does not attempt to make energy out of nothing. Instead, it tries to extract either random molecular motion into useful work or useful energy from some degraded source, such as outgoing radiant energy. The second law of thermodynamics says this cannot happen any more than rocks can roll uphill on their own. This just does not happen.

The American Physical Society states, "The American Physical Society deplores attempts to mislead and defraud the public based on claims of perpetual motion machines or sources of unlimited useful free energy, unsubstantiated by experimentally tested established physical principles."

Visit www.phact.org/e/dennis4.html to see a historical list of perpetual motion and free energy machines.

THE SECOND LAW AND NATURAL PROCESSES

Energy can be viewed from two considerations of scale: (1) the observable *external energy* of an object and (2) the *internal energy* of the molecules, or particles that make up an object. A ball, for example, has kinetic energy after it is thrown through the air, and the entire system of particles making up the ball acts like a single massive particle as the ball moves. The motion and energy of the single system can be calculated from the laws of motion and from the equations representing the concepts of work and energy. All of the particles are moving together, in *coherent motion,* when the external kinetic energy is considered.

But the particles making up the ball have another kind of kinetic energy, with the movements and vibrations of internal kinetic energy. In this case, the particles are not moving uniformly together but are vibrating with motions in many different directions. Since there is a lack of net motion and a lack of correlation, the particles have a jumbled *incoherent motion,* which is often described as chaotic. This random, chaotic motion is sometimes called *thermal motion.*

Thus, there are two kinds of motion that the particles of an object can have: (1) a coherent motion where they move together, in step, and (2) an incoherent, chaotic motion of individual particles. These two types of motion are related to the two modes of energy transfer, working and heating. The relationship is that *work* on an object is associated with its *coherent motion,* while *heating* an object is associated with its internal *incoherent motion.*

The second law of thermodynamics implies a direction to the relationship between work (coherent motion) and heat (incoherent motion), and this direction becomes apparent as you analyze what happens to the motions during energy conversions. Some forms of energy, such as electrical and mechanical, have a greater amount of order since they involve particles moving together in a coherent motion. The term *quality of energy* is used to identify the amount of coherent motion. Energy with high order and coherence is called a *high-quality energy.* Energy with less order and less coherence, on the other hand, is called *low-quality energy.* In general, high-quality energy can be easily converted to work, but low-quality energy is less able to do work.

High-quality electrical and mechanical energy can be used to do work but then become dispersed as heat through energy-form conversions and friction. The resulting heat can be converted to do more work only if there is a sufficient temperature difference. The temperature differences do not last long, however, as conduction, convection, and radiation quickly disperse the energy even more. Thus, the transformation of high-quality energy into lower-quality energy is a natural process. Energy tends to disperse, both from the conversion of an energy form to heat and from the heat flow processes of conduction, convection, and radiation. Both processes flow in one direction only and cannot be reversed. This is called the *degradation of energy,* which is the transformation of high-quality energy to lower-quality energy. In every known example, it is a natural process of energy to degrade, becoming less and less available to do work. The process is *irreversible* even though it is possible to temporarily transform heat to mechanical energy through a heat engine or to upgrade the temperature through the use of a heat pump. Eventually, the upgraded mechanical energy will degrade to heat and the increased heat will disperse through the processes of heat flow.

The apparent upgrading of energy by a heat pump or heat engine is always accompanied by a greater degrading of energy someplace else. The electrical energy used to run the heat pump, for example, was produced by the downgrading of chemical or nuclear energy at an electrical power plant. The overall result is that the *total* energy was degraded toward a more disorderly state.

A *thermodynamic measure of disorder* is called **entropy.** Order means patterns and coherent arrangements. Disorder means dispersion, no patterns, and a randomized, or spread-out, arrangement. This leads to another statement about the second law of thermodynamics and the direction of natural change, that

the total entropy of the universe continually increases.

Note the use of the words *total* and *universe* in this statement of the second law. The entropy of a system can decrease (more

Count Rumford (Benjamin Thompson) (1753–1814)

Count Rumford was a U.S.-born physicist who first demonstrated conclusively that heat is not a fluid but a form of motion. He was born Benjamin Thompson in Woburn, Massachusetts, on March 26, 1753. At the age of 19, he became a schoolmaster as a result of much self-instruction and some help from local clergy. He moved to Rumford (now named Concord), New Hampshire, and almost immediately married a wealthy widow many years his senior.

Thompson's first activities seem to have been political. When the War of Independence broke out, he remained loyal to the British crown and acted as some sort of secret agent. Because of these activities, he had to flee to London in 1776 (having separated from his wife the year before). He was rewarded with government work and an appointment as a lieutenant colonel in a British regiment.

In 1791, Thompson was made a count of the Holy Roman Empire in recognition of civil administration work he did in Bavaria. He took his title from Rumford in his homeland, and it is by this name that we know him today—Count Rumford.

Rumford's early work in Bavaria combined social experiments with his lifelong interests concerning heat in all its aspects. When he employed beggars from the streets to manufacture military uniforms, he faced the problem of feeding them. A study of nutrition led him to recognize the importance of water and vegetables, and Rumford

decided that soups would fit his requirements. He devised many recipes and developed cheap food emphasizing the potato. Soldiers were employed in gardening to produce the vegetables. Rumford's enterprise of manufacturing military uniforms led to a study of insulation and to the conclusion that heat was lost mainly through convection. Therefore, he designed clothing to inhibit convection—sort of the first thermal clothing.

No application of heat technology was too humble for Rumford's experiments. He devised the domestic range—the "fire in a box"—and special utensils to go with it. In the interest of fuel efficiency, he devised a calorimeter to compare the heats of combustion of various fuels. Smoky fireplaces also drew his attention, and after a study of the various air movements, he produced designs incorporating all the features now considered essential

in open fires and chimneys, such as the smoke shelf and damper. His search for an alternative to alcoholic drinks led to the promotion of coffee and the design of the first percolator.

The work for which Rumford is best remembered took place in 1798. As military commander for the elector of Bavaria, he was concerned with the manufacture of cannons. These were bored from blocks of iron with drills, and it was believed that the cannons became hot because as the drills cut into the iron, heat was escaping in the form of a fluid called caloric. However, Rumford noticed that heat production increased as the drills became blunter and cut less into the metal. If a very blunt drill was used, no metal was removed, yet the heat output appeared to be limitless. Clearly, heat could not be a fluid in the metal but must be related to the work done in turning the drill. Rumford also studied the expansion of liquids of different densities and different specific heats, and showed by careful weighing that the expansion was not due to caloric taking up the extra space.

Rumford's contribution to science in demolishing the caloric theory of heat was very important, because it paved the way to the realization that heat is related to energy and work, and that all forms of energy can be converted to heat. However, it took several decades to establish the understanding that caloric does not exist and there was no basis for the caloric theory of heat.

Source: Modified from the Hutchinson *Dictionary of Scientific Biography.* © Research Machines plc 2003. All Rights Reserved. Helicon Publishing is a division of Research Machines.

order), for example when a heat pump cools and condenses the random, chaotically moving water vapor molecules into the more ordered state of liquid water. When the energy source for the production, transmission, and use of electrical energy is considered, however, the *total* entropy will be seen as increasing. Likewise, the total entropy increases during the growth of a plant or animal. When all the food, waste products, and products of metabolism are considered, there is again an increase in *total* entropy.

Thus, the *natural process* is for a state of order to degrade into a state of disorder with a corresponding increase in entropy. This means that all the available energy of the universe is gradually diminishing, and over time, the universe

should therefore approach a limit of maximum disorder called the *heat death* of the universe. The *heat death* of the universe is the theoretical limit of disorder, with all molecules spread far, far apart, vibrating slowly with a uniform low temperature.

The heat death of the universe seems to be a logical consequence of the second law of thermodynamics, but scientists are not certain if the second law should apply to the whole universe. What do you think? Will the universe with all its complexities of organization end with the simplicity of spread-out and slowly vibrating molecules? As has been said, nature is full of symmetry—so why should the universe begin with a bang and end with a whisper?

EXAMPLE 4.8

A heat engine operates with 65.0 kcal of heat supplied and exhausts 40.0 kcal of heat. How much work did the engine do?

SOLUTION

Listing the known and unknown quantities:

heat input	$Q_H = 65.0$ kcal
heat rejected	$Q_L = 40.0$ kcal
mechanical equivalent of heat	1 kcal = 4,184 J

The relationship between these quantities is found in equation 4.8 $W = J(Q_H - Q_L)$. This equation states a relationship between the heat supplied to the engine from a high-temperature source (Q_H), which is partly converted to work (W), with the rest rejected in a lower-temperature exhaust (Q_L). The work accomplished is therefore the difference in the heat input and the heat output ($Q_H - Q_L$), so the work accomplished represents the heat used, where J is the mechanical equivalence of heat (1 kcal = 4,184 J). Therefore,

$$W = J(Q_H - Q_L)$$

$$= 4,184 \frac{J}{kcal}(65.0 \text{ kcal} - 40.0 \text{ kcal})$$

$$= 4,184 \frac{J}{kcal}(25.0 \text{ kcal})$$

$$= 4,184 \times 25.0 \frac{J \cdot kcal}{kcal}$$

$$= 104,600 \text{ J}$$

$$= 105 \text{ kJ}$$

CONCEPTS *Applied*

Thermodynamics in Action

The laws of thermodynamics are concerned with changes in energy and heat. This application explores some of these relationships.

Obtain an electric blender and a thermometer. Fill the blender halfway with water, then let it remain undisturbed until the temperature is constant as shown by two consecutive temperature readings.

Remove the thermometer from the blender, then run the blender at the highest setting for a short time. Stop and record the temperature of the water. Repeat this procedure several times.

Explain your observations in terms of thermodynamics. See if you can think of other experiments that show relationships between changes in energy and heat.

SUMMARY

The kinetic theory of matter assumes that all matter is made up of *molecules*. Molecules can have *vibrational, rotational,* or *translational* kinetic energy. The *temperature* of an object is related to the *average kinetic energy* of the molecules making up the object. A measure of temperature tells how hot or cold an object is on a *temperature scale*. Zero on the *Kelvin scale* is the temperature at which all random molecular motion ceases to exist.

The *external energy* of an object is the observable mechanical energy of that object as a whole. The *internal energy* of the object is the mechanical energy of the molecules that make up the object. *Heat* refers to the total internal energy and is a transfer of energy that takes place because of (1) a *temperature difference* between two objects or (2) an *energy-form conversion*. An energy-form conversion is actually an energy conversion involving work at the molecular level, so all energy transfers involve *heating* and *working*.

A quantity of heat can be measured in *joules* (a unit of work or energy) or *calories* (a unit of heat). A *kilocalorie* is 1,000 calories, another unit of heat. A *Btu,* or *British thermal unit,* is the English system unit of heat. The *mechanical equivalent of heat* is 4,184 J or 1 kcal.

The *specific heat* of a substance is the amount of energy (or heat) needed to increase the temperature of 1 gram of a substance 1 degree Celsius. The specific heat of various substances is not the same because the molecular structure of each substance is different.

Heat transfer takes place through conduction, convection, or radiation. *Conduction* is the transfer of increased kinetic energy from

molecule to molecule. Substances vary in their ability to conduct heat, and those that are poor conductors are called *insulators*. Gases, such as air, are good insulators. The best insulator is a vacuum. *Convection* is the transfer of heat by the displacement of large groups of molecules with higher kinetic energy. Convection takes place in fluids, and the fluid movement that takes place because of density differences is called a *convection current*. *Radiation* is radiant energy that moves through space. All objects with an absolute temperature above zero give off radiant energy, but all objects absorb it as well.

The transition from one phase of matter to another that happens at a constant temperature is called a *phase change*. A phase change always absorbs or releases a quantity of *latent heat* not associated with a temperature change. Latent heat is energy that goes into or comes out of *internal potential energy*. The *latent heat of fusion* is absorbed or released at a solid-liquid phase change.

Molecules of liquids sometimes have a high enough velocity to escape the surface through the process called *evaporation*. Evaporation is a cooling process. Vapor molecules return to the liquid state through the process called *condensation*. Condensation is the opposite of evaporation and is a warming process. When the condensation rate equals the evaporation rate, the air is said to be *saturated*.

Thermodynamics is the study of heat and its relationship to mechanical energy, and the *laws of thermodynamics* describe these relationships. The *first law of thermodynamics* states that the energy supplied to a thermodynamic system in the form of heat, minus the work done by the system, is equal to the change in internal energy. The *second law of thermodynamics* states that heat flows from objects with a higher temperature to objects with a lower temperature. *Entropy* is a thermodynamic measure of disorder; it is seen as continually increasing in the universe and may result in the maximum disorder called the *heat death* of the universe.

SUMMARY OF EQUATIONS

4.1
$$T_F = \frac{9}{5}T_C + 32°$$

4.2
$$T_C = \frac{5}{9}(T_F - 32°)$$

4.3
$$T_K = T_C + 273$$

4.4 Quantity of heat = (mass)(specific heat)(temperature change)
$$Q = mc\Delta T$$

4.5 Heat absorbed or released = (mass)(latent heat of fusion)
$$Q = mL_f$$

4.6 Heat absorbed or released = (mass)(latent heat of vaporization)
$$Q = mL_v$$

4.7 (mechanical equivalence of heat)(quantity of heat) − (work) = internal energy difference between two states
$$JQ - W = U_2 - U_1$$

4.8 work = (mechanical equivalence of heat)(difference in heat input and heat output)
$$W = J(Q_H - Q_L)$$

KEY TERMS

British thermal unit (p. **94**)
calorie (p. **94**)
Celsius scale (p. **90**)
conduction (p. **96**)
convection (p. **97**)
entropy (p. **106**)
external energy (p. **93**)
Fahrenheit scale (p. **90**)
first law of thermodynamics (p. **104**)
heat (p. **92**)
internal energy (p. **93**)
Kelvin scale (p. **91**)
kilocalorie (p. **94**)
kinetic molecular theory (p. **86**)
latent heat of fusion (p. **100**)
latent heat of vaporization (p. **100**)
molecule (p. **87**)
phase change (p. **99**)

radiation (p. **98**)
second law of thermodynamics (p. **105**)
specific heat (p. **95**)
temperature (p. **88**)

APPLYING THE CONCEPTS

1. The Fahrenheit thermometer scale is
   a. more accurate than the Celsius scale.
   b. less accurate than the Celsius scale.
   c. sometimes more or less accurate, depending on the air temperature.
   d. no more accurate than the Celsius scale.

2. On the Celsius temperature scale
   a. zero means there is no temperature.
   b. 80° is twice as hot as 40°.
   c. the numbers relate to the boiling and freezing of water.
   d. there are more degrees than on the Fahrenheit scale.

3. Internal energy refers to the
   a. translational kinetic energy of gas molecules.
   b. total potential and kinetic energy of the molecules.
   c. total vibrational, rotational, and translational kinetic energy of molecules.
   d. average of all types of kinetic energy of the gas molecules.

4. External energy refers to the
   a. energy that changed the speed of an object.
   b. energy of all the molecules making up an object.
   c. total potential energy and kinetic energy of an object that you can measure directly.
   d. energy from an extraterrestrial source.

5. Heat is the
   a. total internal energy of an object.
   b. average kinetic energy of molecules.
   c. measure of potential energy of molecules.
   d. same thing as a very high temperature.

6. The specific heat of copper is 0.093 cal/g°C and the specific heat of aluminum is 0.22 cal/g°C. The same amount of energy applied to equal masses, say 50.0 g of copper and aluminum, will result in
   a. a higher temperature for copper.
   b. a higher temperature for aluminum.
   c. the same temperature for each metal.
   d. unknown results.

7. The specific heat of water is 1.00 cal/g°C and the specific heat of ice is 0.500 cal/g°C. The same amount of energy applied to equal masses, say 50.0 g of water and ice, will result in (assume the ice does not melt)
   a. a greater temperature increase for the water.
   b. a greater temperature increase for the ice.
   c. the same temperature increase for each.
   d. unknown results.

8. The transfer of heat that takes place by the movement of groups of molecules with higher kinetic energy is
   a. conduction.
   b. convection.
   c. radiation.
   d. sublimation.

9. The transfer of heat that takes place by energy moving through space is
   a. conduction.
   b. convection.
   c. radiation.
   d. sublimation.

10. The transfer of heat that takes place directly from molecule to molecule is
    a. conduction.
    b. convection.
    c. radiation.
    d. sublimation.

11. The evaporation of water cools the surroundings, and the condensation of this vapor
    a. does nothing.
    b. warms the surroundings.
    c. increases the value of the latent heat of vaporization.
    d. decreases the value of the latent heat of vaporization.

12. The heat involved in the change of phase from solid ice to liquid water is called
    a. latent heat of vaporization.
    b. latent heat of fusion.
    c. latent heat of condensation.
    d. none of the above.

13. The energy supplied to a system in the form of heat, minus the work done by the system, is equal to the change in internal energy. This statement describes the
    a. first law of thermodynamics.
    b. second law of thermodynamics.
    c. third law of thermodynamics.

14. If you want to move heat from a region of cooler temperature to a region of warmer temperature, you must supply energy. This is described by the
    a. first law of thermodynamics.
    b. second law of thermodynamics.
    c. third law of thermodynamics.

15. More molecules are returning to the liquid state than are leaving the liquid state. This process is called
    a. boiling.
    b. freezing.
    c. condensation.
    d. melting.

16. The temperature of a gas is proportional to the
    a. average velocity of the gas molecules.
    b. internal potential energy of the gas.
    c. number of gas molecules in a sample.
    d. average kinetic energy of the gas molecules.

17. The temperature known as room temperature is nearest to
    a. 0°C.
    b. 20°C.
    c. 60°C.
    d. 100°C.

18. Using the Kelvin temperature scale, the freezing point of water is correctly written as
    a. 0 K.
    b. 0°K.
    c. 273 K.
    d. 273°K.

19. The specific heat of soil is 0.20 kcal/kgC° and the specific heat of water is 1.00 kcal/kgC°. This means that if 1 kg of soil and 1 kg of water each receives 1 kcal of energy, ideally,
    a. the water will be warmer than the soil by 0.8°C.
    b. the soil will be 4°C warmer than the water.
    c. the soil will be 5°C warmer than the water.
    d. the water will warm by 1°C, and the soil will warm by 0.2°C.

20. Styrofoam is a good insulating material because
    a. it is a plastic material that conducts heat poorly.
    b. it contains many tiny pockets of air.
    c. of the structure of the molecules that make it up.
    d. it is not very dense.

21. The transfer of heat that takes place because of density difference in fluids is
    a. conduction.
    b. convection.
    c. radiation.
    d. none of the above.

22. Latent heat is "hidden" because it
    a. goes into or comes out of internal potential energy.
    b. is a fluid (caloric) that cannot be sensed.
    c. does not actually exist.
    d. is a form of internal kinetic energy.

23. As a solid undergoes a phase change to a liquid, it
    a. releases heat while remaining a constant temperature.
    b. absorbs heat while remaining a constant temperature.
    c. releases heat as the temperature decreases.
    d. absorbs heat as the temperature increases.

24. A heat engine is designed to
    a. move heat from a cool source to a warmer location.
    b. move heat from a warm source to a cooler location.
    c. convert mechanical energy into heat.
    d. convert heat into mechanical energy.

25. The work that a heat engine is able to accomplish is ideally equivalent to the
    a. difference between the heat supplied and the heat rejected.
    b. heat that was produced in the cycle.
    c. heat that appears in the exhaust gases.
    d. sum total of the heat input and the heat output.

26. Suppose ammonia is spilled in the back of a large room. If there were no air currents, how would the room temperature influence how fast you would smell ammonia at the opposite side of the room?
    a. Warmer is faster.
    b. Cooler is faster.
    c. There would be no influence.

27. Which of the following contains the most heat?
    a. A bucket of water at 0°C.
    b. A barrel of water at 0°C.
    c. Neither contains any heat since the temperature is zero.
    d. Both have the same amount of heat.

28. Anytime a temperature difference occurs, you can expect
    a. cold to move to where it is warmer, such as cold moving into a warm house during the winter.
    b. heat movement from any higher temperature region.
    c. no energy movement unless it is hot enough, such as the red-hot heating element on a stove.

29. The cheese on a hot pizza takes a long time to cool because it
   a. is stretchable and elastic.
   b. has a low specific heat.
   c. has a high specific heat.
   d. has a white color.

30. The specific heat of copper is roughly three times as great as the specific heat of gold. Which of the following is true for equal masses of copper and gold?
   a. If the same amount of heat is applied, the copper will become hotter.
   b. Copper heats up three times as fast as gold.
   c. A piece of copper stores three times as much heat at the same temperature.
   d. The melting temperature of copper is roughly three times that of gold.

31. Cooking pans made from which of the following metals would need less heat to achieve a certain cooking temperature?
   a. aluminum (specific heat 0.22 kcal/kg°C)
   b. copper (specific heat 0.093 kcal/kg°C)
   c. iron (specific heat 0.11 kcal/kg°C)

32. Conduction best takes place in a
   a. solid.
   b. fluid.
   c. gas.
   d. vacuum.

33. Convection best takes place in a
   a. solid.
   b. fluid.
   c. alloy.
   d. vacuum.

34. Radiation is the only method of heat transfer that can take place in a
   a. solid.
   b. liquid.
   c. gas.
   d. vacuum.

35. What form of heat transfer will warm your body without warming the air in a room?
   a. conduction
   b. convection
   c. radiation
   d. None of the above is correct.

36. When you add heat to a substance, its temperature
   a. always increases.
   b. sometimes decreases.
   c. might stay the same.
   d. might go up or down, depending on the temperature.

37. The great cooling effect produced by water evaporating comes from its high
   a. conductivity.
   b. specific heat.
   c. latent heat.
   d. transparency.

38. At temperatures above freezing, the evaporation rate can equal the condensation rate only at
   a. very high air temperatures.
   b. mild temperatures.
   c. low temperatures.
   d. any temperature.

39. The phase change from ice to liquid water takes place at
   a. constant pressure.
   b. constant temperature.
   c. constant volume.
   d. all of the above.

40. Which of the following has the greatest value for liquid water?
   a. latent heat of fusion
   b. latent heat of vaporization
   c. Both are equivalent.
   d. None of the above is correct.

41. Which of the following supports the second law of thermodynamics?
   a. Heat naturally flows from a low temperature region to a higher temperature region.
   b. All of a heat source can be converted into mechanical energy.
   c. Energy tends to degrade, becoming a lower and lower quality.
   d. A heat pump converts heat into mechanical work.

42. The second law of thermodynamics tells us that the amount of disorder, called entropy, is always increasing. Does the growth of a plant or animal violate the second law?
   a. Yes, a plant or animal is more highly ordered.
   b. No, the total entropy of the universe increased.
   c. The answer is unknown.

43. The heat death of the universe in the future is when the universe is supposed to
   a. have a high temperature that will kill all living things.
   b. have a high temperature that will vaporize all matter in it.
   c. freeze at a uniform low temperature.
   d. use up the universal supply of entropy.

Answers

1. d   2. c   3. b   4. c   5. a   6. a   7. b   8. b   9. c   10. a   11. b   12. b   13. a   14. b   15. c   16. d   17. b   18. c   19. b   20. b   21. b   22. a   23. b   24. d   25. a   26. a   27. b   28. b   29. c   30. c   31. b   32. a   33. b   34. d   35. c   36. c   37. c   38. d   39. b   40. b   41. c   42. b   43. c

QUESTIONS FOR THOUGHT

1. What is temperature? What is heat?

2. Explain why most materials become less dense as their temperature is increased.

3. Would the tight packing of more insulation, such as glass wool, in an enclosed space increase or decrease the insulation value? Explain.

4. A true vacuum bottle has a double-walled, silvered bottle with the air removed from the space between the walls. Describe how this design keeps food hot or cold by dealing with conduction, convection, and radiation.

5. Why is cooler air found in low valleys on calm nights?

6. Why is air a good insulator?

7. Explain the meaning of the mechanical equivalent of heat.

8. What do people really mean when they say that a certain food "has a lot of Calories"?

9. A piece of metal feels cooler than a piece of wood at the same temperature. Explain why.

10. Explain how the latent heat of fusion and the latent heat of vaporization are "hidden."

11. What is condensation? Explain, on a molecular level, how the condensation of water vapor on a bathroom mirror warms the bathroom.

12. Which provides more cooling for a Styrofoam cooler: 10 lb of ice at 0°C or 10 lb of ice water at 0°C? Explain your reasoning.

13. Explain why a glass filled with a cold beverage seems to "sweat." Would you expect more sweating inside a house during the summer or during the winter? Explain.

14. Explain why a burn from 100°C steam is more severe than a burn from water at 100°C.

15. Briefly describe, using sketches as needed, how a heat pump is able to move heat from a cooler region to a warmer region.

16. Which has more entropy: ice, liquid water, or water vapor? Explain your reasoning.

17. Suppose you use a heat engine to do the work to drive a heat pump. Could the heat pump be used to provide the temperature difference to run the heat engine? Explain.

FOR FURTHER ANALYSIS

1. Considering the criteria for determining if something is a solid, liquid, or gas, what is table salt, which can be poured?

2. What are the significant similarities and differences between heat and temperature?

3. Gas and plasma are phases of matter, yet gas runs a car and plasma is part of your blood. Compare and contrast these terms and offer an explanation for the use of similar names.

4. Analyze the table of specific heats (Table 4.2) and determine which metal would make an energy-efficient and practical pan, providing more cooking for less energy.

5. This chapter contains information about three types of passive solar home design. Develop criteria or standards of evaluation that would help someone decide which design is right for their local climate.

6. Could a heat pump move heat without the latent heat of vaporization? Explain.

7. Explore the assumptions on which "heat death of the universe" idea is based. Propose and evaluate an alternative idea for the future of the universe.

INVITATION TO INQUIRY

Who Can Last Longest?

How can we be more energy efficient? Much of our household energy consumption goes into heating and cooling, and much energy is lost through walls and ceilings. This invitation is about the insulating properties of various materials and their arrangement.

The challenge of this invitation is to create an insulated container that can keep an ice cube from melting. Decide on a maximum size for the container, and then decide what materials to use. Consider how you will use the materials. For example, if you are using aluminum foil, should it be shiny side in or shiny side out? If you are using newspapers, should they be folded flat or crumpled loosely?

One ice cube should be left outside the container to use as a control. Find out how much longer your insulated ice cube will outlast the control.

PARALLEL EXERCISES

The exercises in groups A and B cover the same concepts. Solutions to group A exercises are located in appendix E.

Note: *Neglect all frictional forces in all exercises.*

Group A

1. The average human body temperature is 98.6°F. What is the equivalent temperature on the Celsius scale?

2. An electric current heats a 221 g copper wire from 20.0°C to 38.0°C. How much heat was generated by the current? (c_{copper} = 0.093 kcal/kgC°)

3. A bicycle and rider have a combined mass of 100.0 kg. How many calories of heat are generated in the brakes when the bicycle comes to a stop from a speed of 36.0 km/h?

4. A 15.53 kg loose bag of soil falls 5.50 m at a construction site. If all the energy is retained by the soil in the bag, how much will its temperature increase? (c_{soil} = 0.200 kcal/kgC°)

5. A 75.0 kg person consumes a small order of french fries (250.0 Cal) and wishes to "work off" the energy by climbing a 10.0 m stairway. How many vertical climbs are needed to use all the energy?

6. A 0.5 kg glass bowl (c_{glass} = 0.2 kcal/kgC°) and a 0.5 kg iron pan (c_{iron} = 0.11 kcal/kgC°) have a temperature of 68°F when placed in a freezer. How much heat will the freezer have to remove from each to cool them to 32°F?

Group B

1. The Fahrenheit temperature reading is 98° on a hot summer day. What is this reading on the Kelvin scale?

2. A 0.25 kg length of aluminum wire is warmed 10.0°C by an electric current. How much heat was generated by the current? ($c_{aluminum}$ = 0.22 kcal/kgC°)

3. A 1,000.0 kg car with a speed of 90.0 km/h brakes to a stop. How many cal of heat are generated by the brakes as a result?

4. A 1.0 kg metal head of a geology hammer strikes a solid rock with a velocity of 5.0 m/s. Assuming all the energy is retained by the hammer head, how much will its temperature increase? (c_{head} = 0.11 kcal/kgC°)

5. A 60.0 kg person will need to climb a 10.0 m stairway how many times to "work off" each excess Cal (kcal) consumed?

6. A 50.0 g silver spoon at 20.0°C is placed in a cup of coffee at 90.0°C. How much heat does the spoon absorb from the coffee to reach a temperature of 89.0°C?

7. A sample of silver at 20.0°C is warmed to 100.0°C when 896 cal is added. What is the mass of the silver? (c_{silver} = 0.056 kcal/kgC°)

8. A 300.0 W immersion heater is used to heat 250.0 g of water from 10.0°C to 70.0°C. About how many minutes did this take?

9. A 100.0 g sample of metal is warmed 20.0°C when 60.0 cal is added. What is the specific heat of this metal?

10. How much heat is needed to change 250.0 g of ice at 0°C to water at 0°C?

11. How much heat is needed to change 250.0 g of water at 80.0°C to steam at 100.0°C?

12. A 100.0 g sample of water at 20.0°C is heated to steam at 125.0°C. How much heat was absorbed?

13. In an electric freezer, 400.0 g of water at 18.0°C is cooled, frozen, and the ice is chilled to −5.00°C. (a) How much total heat was removed from the water? (b) If the latent heat of vaporization of the Freon refrigerant is 40.0 cal/g, how many grams of Freon must be evaporated to absorb this heat?

14. A heat engine is supplied with 300.0 cal and rejects 200.0 cal in the exhaust. How many joules of mechanical work was done?

15. A refrigerator removes 40.0 kcal of heat from the freezer and releases 55.0 kcal through the condenser on the back. How much work was done by the compressor?

7. If the silver spoon placed in the coffee in problem 6 causes it to cool 0.75°C, what is the mass of the coffee? (Assume c_{coffee} = 1.0 cal/gC°)

8. How many minutes would be required for a 300.0 W immersion heater to heat 250.0 g of water from 20.0°C to 100.0°C?

9. A 200.0 g china serving bowl is warmed 65.0°C when it absorbs 2.6 kcal of heat from a serving of hot food. What is the specific heat of the china dish?

10. A 1.00 kg block of ice at 0°C is added to a picnic cooler. How much heat will the ice remove as it melts to water at 0°C?

11. A 500.0 g pot of water at room temperature (20.0°C) is placed on a stove. How much heat is required to change this water to steam at 100.0°C?

12. Spent steam from an electric generating plant leaves the turbines at 120.0°C and is cooled to 90.0°C liquid water by water from a cooling tower in a heat exchanger. How much heat is removed by the cooling tower water for each kg of spent steam?

13. Lead is a soft, dense metal with a specific heat of 0.028 kcal/kgC°, a melting point of 328.0°C, and a heat of fusion of 5.5 kcal/kg. How much heat must be provided to melt a 250.0 kg sample of lead with a temperature of 20.0°C?

14. A heat engine converts 100.0 cal from a supply of 400.0 cal into work. How much mechanical work was done?

15. A heat pump releases 60.0 kcal as it removes 40.0 kcal at the evaporator coils. How much work does this heat pump ideally accomplish?

5

Wave Motions and Sound

Compared to the sounds you hear on a calm day in the woods, the sounds from a waterfall can carry up to a million times more energy.

CORE **CONCEPT**

Sound is transmitted as increased and decreased pressure waves that carry energy.

Sometimes you can feel the floor of a building shake for a moment when something heavy is dropped. You can also feel prolonged vibrations in the ground when a nearby train moves by. The floor of a building and the ground are solids that transmit vibrations from a disturbance. Vibrations are common in most solids because the solids are elastic, having a tendency to rebound, or snap back, after a force or an impact deforms them. Usually you cannot see the vibrations in a floor or the ground, but you sense they are there because you can feel them.

There are many examples of vibrations that you can see. You can see the rapid blur of a vibrating guitar string (Figure 5.1). You can see the vibrating up-and-down movement of a bounced-upon diving board. Both the vibrating guitar string and the diving board set up a vibrating motion of air that you identify as a sound. You cannot see the vibrating motion of the air, but you sense it is there because you hear sounds.

There are many kinds of vibrations that you cannot see but can sense. Heat, as you have learned, is associated with molecular vibrations that are too rapid and too tiny for your senses to detect other than as an increase in temperature. Other invisible vibrations include electrons that vibrate, generating spreading electromagnetic radio waves or visible light. Thus, vibrations are not only observable motions of objects but are also characteristics of sound, heat, electricity, and light. The vibrations involved in all these phenomena are alike in many ways, and all involve energy. Therefore, many topics of physical science are concerned with vibrational motion. In this chapter, you will learn about the nature of vibrations and how they produce waves in general. These concepts will be applied to sound in this chapter and to electricity, light, and radio waves in later chapters.

FORCES AND ELASTIC MATERIALS

If you drop a rubber ball, it bounces because it is capable of recovering its shape when it hits the floor. A ball of clay, on the other hand, does not recover its shape and remains a flattened blob on the floor. An *elastic* material is one that is capable of recovering its shape after a force deforms it. A rubber ball is elastic and a ball of clay is not elastic. You know a metal spring is elastic because you can stretch it or compress it and it recovers its shape.

There is a direct relationship between the extent of stretching or compression of a spring and the amount of force applied to it. A large force stretches a spring a lot; a small force stretches it a little. As long as the applied force does not exceed the elastic limit of the spring, it will always return to its original shape when you remove the applied force. There are three important considerations about the applied force and the response of the spring:

1. The greater the applied force, the greater the compression or stretch of the spring from its original shape.
2. The spring appears to have an *internal restoring force,* which returns it to its original shape.
3. The farther the spring is pushed or pulled, the *stronger* the restoring force that returns the spring to its original shape.

FORCES AND VIBRATIONS

A **vibration** is a back-and-forth motion that repeats itself. A motion that repeats itself is called *periodic motion.* Such a motion is not restricted to any particular direction, and it can be in many different directions at the same time. Almost any solid can be made to vibrate if it is elastic. To see how forces are involved in vibrations, consider the spring and mass in Figure 5.2. The spring and mass are arranged so that the mass can freely move back and forth on a frictionless surface. When the mass has not been disturbed, it is at rest at an *equilibrium position* (Figure 5.2A). At the equilibrium position, the spring is not compressed or stretched, so it applies no force on the mass. If, however, the mass is pulled to the right (Figure 5.2B), the spring is stretched and applies a restoring force on the mass toward the left. The farther the mass is displaced, the greater the stretch of the spring and thus the greater the restoring force. The restoring force is proportional to the displacement and is in the opposite direction of the applied force.

If the mass is now released, the restoring force is the only force acting (horizontally) on the mass, so it accelerates back toward the equilibrium position. This force will continuously decrease until the moving mass arrives back at the equilibrium position, where the force is zero (Figure 5.2C). The mass will have a maximum velocity when it arrives, however, so it overshoots the equilibrium position and continues moving to the left (Figure 5.2D). As it moves to the left of the equilibrium position, it compresses the spring, which exerts an increasing force on the mass. The moving mass comes to a temporary halt, but now the restoring force again starts it moving back toward the equilibrium position. The whole process repeats itself again and again as the mass moves back and forth over the same path.

The vibrating mass and spring system will continue to vibrate for a while, slowly decreasing with time until the vibrations stop completely. The slowing and stopping is due to air resistance and internal friction. If these could be eliminated or compensated for with additional energy, the mass would continue to vibrate in periodic motion indefinitely.

FIGURE 5.1 Vibrations are common in many elastic materials, and you can see and hear the results of many in your surroundings. Other vibrations in your surroundings, such as those involved in heat, electricity, and light, are invisible to the senses.

The periodic vibration, or oscillation, of the mass is similar to many vibrational motions found in nature called *simple harmonic motion*. Simple harmonic motion is defined as the vibratory motion that occurs when there is a restoring net force opposite to and proportional to a displacement.

DESCRIBING VIBRATIONS

A motion of a vibrating mass is described by measuring three basic quantities called the *amplitude of vibration* (**amplitude** for short), the *period*, and the *frequency* of vibration (see Figure 5.3). The amplitude is the largest displacement from the equilibrium position (rest position) that the mass can have in this motion. All other displacements that you may see and measure, when observing a vibrating mass, are smaller than the amplitude.

A complete vibration is called a **cycle**. A cycle is the movement from some point, say the far left, all the way to the far right and back to the same point again, the far left in this example. The **period** (*T*) is the number of seconds per cycle. For example, suppose 0.1 s is required for an object to move through one complete cycle, to complete the motion from one point, then back to that point. The period of this vibration is 0.1 s. In other words, the period *T* is the time of one full cycle or one full vibration.

Sometimes it is useful to know how frequently a vibration completes a cycle every second. The number of cycles per second is called the **frequency** (*f*). For example, a vibrating object moves through 10 cycles in 1 s. The frequency of this vibration is 10 cycles per second. Frequency is measured in a unit called a **hertz** (Hz). The unit for a hertz is 1/s since a cycle does not have dimensions. Thus, a frequency of 10 cycles per second is referred to as 10 hertz or 10 1/s. In other words, frequency *f* tells you how many full vibrations (or full cycles) are performed in 1 second.

The period and frequency are two ways of describing the time involved in a vibration. Since the period (*T*) is the number of seconds per cycle and the frequency (*f*) is the number of

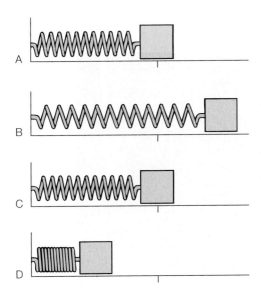

FIGURE 5.2 A mass on a frictionless surface is at rest at an equilibrium position (*A*) when undisturbed. When the spring is stretched (*B*) or compressed (*D*), then released (*C*), the mass vibrates back and forth because restoring forces pull opposite to and proportional to the displacement.

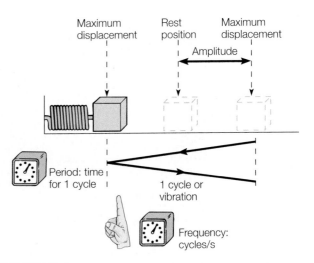

FIGURE 5.3 A vibrating mass attached to a spring is displaced from the rest or equilibrium position and then released. The maximum displacement is called the *amplitude* of the vibration. A cycle is one complete vibration. The period is the number of seconds per cycle. The frequency is a count of how many cycles are completed in 1 s.

cycles per second, the relationship is

$$T = \frac{1}{f}$$

equation 5.1

or

$$f = \frac{1}{T}$$

equation 5.2

EXAMPLE 5.1

A vibrating system has a period of 0.5 s. What is the frequency in Hz?

SOLUTION

$$T = 0.5 \text{ s} \qquad f = \frac{1}{T}$$
$$f = ?$$
$$= \frac{1}{0.5 \text{s}}$$
$$= \frac{1}{0.5} \frac{1}{\text{s}}$$
$$= 2 \frac{1}{\text{s}}$$
$$= \boxed{2 \text{ Hz}}$$

Simple harmonic motion of a vibrating object (such as the motion of a mass on a spring) can be represented by a graph. This graph illustrates how the displacement of the object is changing with time. From the graph you can usually read the amplitude, the period, and the frequency of the waves. If a pen is fixed to a vibrating mass and a paper is moved beneath it at a steady

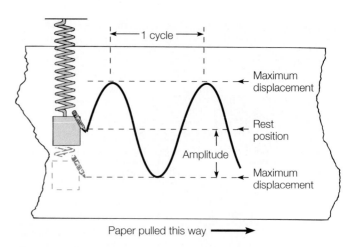

FIGURE 5.4 A graph of simple harmonic motion is described by a sinusoidal curve.

rate, it will draw a curve, as shown in Figure 5.4. The greater the amplitude of the vibrating mass, the greater the height of this curve. The greater the frequency, the closer together the peaks and valleys. Note the shape of this curve. This shape is characteristic of simple harmonic motion and is called a *sinusoidal,* or sine, graph. It is so named because it is the same shape as a graph of the sine function in trigonometry.

WAVES

Most people, when they hear the word *waves,* imagine water waves coming to a shore in an ongoing motion or perhaps the water ripples on a lake where a rock hits the still water surface (Figure 5.5). The water in such a lake is a medium in which the waves travel, but the medium itself does not travel. The rock created an isolated disturbance of the medium. Because of the disturbance, the molecules of the medium were set in periodic

FIGURE 5.5 A water wave moves across the surface. How do you know for sure that it is energy, not water, that is moving across the surface?

motion (vibrations), and this periodic motion moved out in the medium away from the point where the rock fell (the *center* or *source of disturbance*). The water did not move away from the center of disturbance, but the vibrations moved away. These traveling vibrations of the medium are called **waves.** Suppose you use your finger or some other object to disturb the still water surface at exactly one point not just once but many times, repeatedly, one time after another in regular time intervals; for example, you tap the surface every one-third of a sound. In this example, vibrations of your finger are the source of *periodic disturbances* that travel in the medium. In short, you generated waves. The world around us provides us with many examples of waves or *wave motion*.

KINDS OF MECHANICAL WAVES

If you could see the motion of an individual water molecule near the surface as a water wave passed, you would see it trace out a circular path as it moves up and over, down and back. This circular motion is characteristic of the motion of a particle reacting to a water wave disturbance. There are other kinds of mechanical waves, and each involves particles in a characteristic motion.

A **longitudinal wave** is a disturbance that causes particles to move closer together or farther apart in the same direction that the wave is moving. If you attach one end of a coiled spring to a wall and pull it tight, you will make longitudinal waves in the spring if you grasp the spring and then move your hand back and forth parallel to the spring. Each time you move your hand toward the length of the spring, a pulse of closer-together coils will move across the spring (Figure 5.6A). Each time you pull your hand back, a pulse of farther-apart coils will move across the spring. The coils move back and forth in the same direction that the wave is moving, which is the characteristic for a longitudinal wave.

You will make a different kind of mechanical wave in the stretched spring if you now move your hand up and down. This creates a **transverse wave.** A transverse wave is a disturbance that causes motion perpendicular to the direction that the wave is moving. Particles responding to a transverse wave do not move closer together or farther apart in response to the disturbance; rather, they vibrate up and then down in a direction perpendicular to the direction of the wave motion (see Figure 5.6B).

CONCEPTS *Applied*

Making Waves

Obtain a Slinky or a long, coiled spring and stretch it out on the floor. Have another person hold the opposite end stationary while you make waves move along the spring. Make longitudinal and transverse waves, observing how the disturbance moves in each case. If the spring is long enough, measure the distance, then time the movement of each type of wave. How fast were your waves?

Whether you make mechanical longitudinal or transverse waves depends not only on the nature of the disturbance creating the waves but also on the nature of the medium. Mechanical transverse waves can move through a material only if there is some interaction, or attachment, between the molecules making up the medium. In a gas, for example, the molecules move about freely without attachments to one another. A pulse can cause these molecules to move closer together or farther apart, so a gas can carry a longitudinal wave. But if a gas molecule is caused to move up and then down, there is no reason for other molecules to do the same, since they are not attached. Thus, a gas will carry mechanical longitudinal waves but not mechanical transverse waves. Likewise, a liquid will carry mechanical longitudinal waves but not mechanical transverse waves because the liquid molecules simply slide past one another. The surface of a liquid, however, is another story because of surface tension. A surface water wave is, in fact, a combination of longitudinal and transverse wave patterns that produce the circular motion of a disturbed particle. Solids can and do carry both longitudinal and transverse waves because of the strong attachments between the molecules.

WAVES IN AIR

Because air is fluid, mechanical waves in air can only be longitudinal; therefore, sound waves in air must be longitudinal waves. A familiar situation will be used to describe the nature of a mechanical longitudinal wave moving through air before we consider sound specifically. The situation involves a small room with no open windows and two doors that open into the room. When you open one door into the room, the other door closes. Why does this happen? According to the kinetic molecular theory, the room contains many tiny, randomly moving gas molecules that make up the air. As you opened the door, it pushed on these gas molecules, creating a jammed-together zone of molecules immediately adjacent to the door. This jammed-together zone of air now has a greater density and pressure, which immediately spreads outward from the door as a pulse. The disturbance is rapidly passed from molecule to

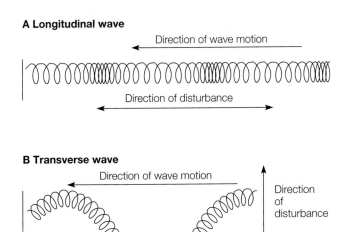

FIGURE 5.6 (*A*) Longitudinal waves are created in a spring when the free end is moved back and forth parallel to the spring. (*B*) Transverse waves are created in a spring when the free end is moved up and down.

molecule, and the pulse of compression spreads through the room. In the example of the closing door, the pulse of greater density and increased pressure of air reached the door at the other side of the room, and the composite effect of the molecules hitting the door, that is, the increased pressure, caused it to close.

If the door at the other side of the room does not latch, you can probably cause it to open again by pulling on the first door quickly. By so doing, you send a pulse of thinned-out molecules of lowered density and pressure. The door you pulled quickly pushed some of the molecules out of the room. Other molecules quickly move into the region of less pressure, then back to their normal positions. The overall effect is the movement of a thinned-out pulse that travels through the room. When the pulse of slightly reduced pressure reaches the other door, molecules exerting their normal pressure on the other side of the door cause it to move. After a pulse has passed a particular place, the molecules are very soon homogeneously distributed again due to their rapid, random movement.

If you were to swing a door back and forth, it would be a vibrating object. As it vibrates back and forth, it would have a certain frequency in terms of the number of vibrations per second. As the vibrating door moves toward the room, it creates a pulse of jammed-together molecules called a *condensation* (or compression) that quickly moves throughout the room. As the vibrating door moves away from the room, a pulse of thinned-out molecules called a *rarefaction* quickly moves throughout the room. The vibrating door sends repeating pulses of condensation (increased density and pressure) and rarefaction (decreased density and pressure) through the room as it moves back and forth (Figure 5.7). You know that the pulses transmit energy because they produce movement, or do work on, the other door. Individual molecules execute a harmonic motion about their equilibrium position and can do work on a movable object. Energy is thus transferred by this example of longitudinal waves.

 CONCEPTS *Applied*

A Splash of Air?

In a very still room with no air movement whatsoever, place a smoking incense, punk, or appropriate smoke source in an ashtray on a table. It should make a thin stream of smoke that moves straight up. Hold one hand flat, fingers together, and parallel to the stream of smoke. Quickly move it toward the smoke for a very short distance as if pushing air toward the smoke. Then pull it quickly away from the stream of smoke. You should be able to see the smoke stream move away from, then toward your hand. What is the maximum distance from the smoke that you can still make the smoke stream move? There are at least two explanations for the movement of the smoke stream: (1) pulses of condensation and rarefaction or (2) movement of a mass of air, such as occurs when you splash water. How can you prove one explanation or the other to be correct without a doubt?

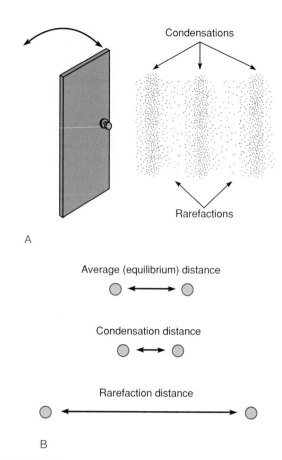

FIGURE 5.7 (*A*) Swinging the door inward produces pulses of increased density and pressure called *condensations*. Pulling the door outward produces pulses of decreased density and pressure called *rarefactions*. (*B*) In a condensation, the average distance between gas molecules is momentarily decreased as the pulse passes. In a rarefaction, the average distance is momentarily increased.

DESCRIBING WAVES

A tuning fork vibrates with a certain frequency and amplitude, producing a longitudinal wave of alternating pulses of increased-pressure condensations and reduced-pressure rarefactions. The concept of the frequency and amplitude of the vibrations is shown in Figure 5.8A, and a representation of the condensations and rarefactions is shown in Figure 5.8B. The wave pattern can also be represented by a graph of the changing air pressure of the traveling sound wave, as shown in Figure 5.8C. This graph can be used to define some interesting concepts associated with sound waves. Note the correspondence between (1) the amplitude, or displacement, of the vibrating prong, (2) the pulses of condensations and rarefactions, and (3) the changing air pressure. Note also the correspondence between the frequency of the vibrating prong and the frequency of the wave cycles.

Figure 5.9 shows the terms commonly associated with waves from a continuously vibrating source. The wave *crest* is the maximum disturbance from the undisturbed (rest) position. For a sound wave, this would represent the maximum

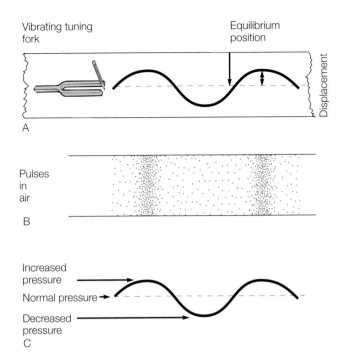

Vibrating tuning fork

Equilibrium position

Displacement

A

Pulses in air

B

Increased pressure

Normal pressure

Decreased pressure

C

FIGURE 5.8 Compare the (A) back-and-forth vibrations of a tuning fork with (B) the resulting condensations and rarefactions that move through the air and (C) the resulting increases and decreases of air pressure on a surface that intercepts the condensations and rarefactions.

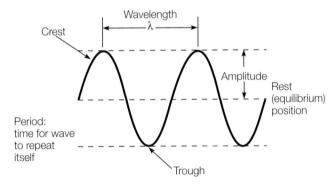

Wavelength λ

Crest

Amplitude

Rest (equilibrium) position

Period: time for wave to repeat itself

Trough

FIGURE 5.9

Here are some terms associated with periodic waves. The *wavelength* is the distance from a part of one wave to the same part in the next wave, such as from one crest to the next. The *amplitude* is the displacement from the rest position. The *period* is the time required for a wave to repeat itself, that is, the time for one complete wavelength to move past a given location.

increase of air pressure. The wave *trough* is the maximum disturbance in the opposite direction from the rest position. For a sound wave, this would represent the maximum decrease of air pressure. The *amplitude* of a wave is the maximum displacement from rest to the crest *or* from rest to the trough. A quantity called the **wavelength** of a wave can be measured as the distance between either two adjacent crests or two adjacent troughs, or any two identical points of adjacent

waves. The wavelength is denoted by the Greek letter λ (pronounced "lambda") and is measured in meters. The *period* (*T*) of the wave is the same as the period of one full vibration of one element of the medium. For example, this period can be determined by measuring the time that elapses between two moments when two adjacent crests are passing by you. It takes a time equal to one period *T* for a wave to move a distance equal to one wavelength λ. The frequency *f* of a wave is the same as the frequency of vibrations of the medium. You can determine the *frequency* of a wave by counting how many crests pass by you in a unit time.

There is a relationship between the wavelength, period, and speed of a wave. Recall that speed is

$$v = \frac{\text{distance}}{\text{time}}$$

Since it takes one period (*T*) for a wave to move one wavelength (λ), then the speed of a wave can be measured from

$$v = \frac{\text{one wavelength}}{\text{one period}} = \frac{\lambda}{T}$$

The *frequency,* however, is more convenient than the period for dealing with waves that repeat themselves rapidly. Recall the relationship between frequency (*f*) and the period (*T*) is

$$f = \frac{1}{T}$$

Substituting *f* for 1/*T* yields

$$v = \lambda f$$

equation 5.3

This equation tells you that the velocity of a wave can be obtained from the product of the wavelength and the frequency. Note that it also tells you that the wavelength and frequency are inversely proportional at a given velocity.

EXAMPLE 5.2

A sound wave with a frequency of 260 Hz has a wavelength of 1.27 m. With what speed would you expect this sound wave to move?

SOLUTION

$f = 260$ Hz
$\lambda = 1.27$ m
$v = ?$

$v = \lambda f$

$= (1.27 \text{ m})\left(260 \frac{1}{s}\right)$

$= 1.27 \times 260 \text{ m} \times \frac{1}{s}$

$= \boxed{330 \frac{\text{m}}{\text{s}}}$

EXAMPLE 5.3

In general, the human ear is most sensitive to sounds at 2,500 Hz. Assuming that sound moves at 330 m/s, what is the wavelength of sounds to which people are most sensitive? (Answer: 13 cm)

SOUND WAVES

SOUND WAVES IN AIR AND HEARING

You cannot hear a vibrating door because the human ear normally hears sounds originating from vibrating objects with a frequency between 20 and 20,000 Hz. Longitudinal waves with frequencies less than 20 Hz are called **infrasonic.** You usually *feel* sounds below 20 Hz rather than hearing them, particularly if you are listening to a good sound system. Longitudinal waves above 20,000 Hz are called **ultrasonic.** Although 20,000 Hz is usually considered the upper limit of hearing, the actual limit varies from person to person and becomes lower and lower with increasing age. Humans do not hear infrasonic or ultrasonic sounds, but various animals have different limits. Dogs, cats, rats, and bats can hear higher frequencies than humans. Dogs can hear an ultrasonic whistle when a human hears nothing, for example. Some bats make and hear sounds of frequencies up to 100,000 Hz as they navigate and search for flying insects in total darkness. Scientists discovered recently that elephants communicate with extremely low-frequency sounds over distances of several kilometers. Humans cannot detect such low-frequency sounds. This raises the possibility of infrasonic waves that other animals can detect that we cannot.

A tuning fork that vibrates at 260 Hz makes longitudinal waves much like the swinging door, but these longitudinal waves are called *audible sound waves* because they are within the frequency range of human hearing. The prongs of a struck tuning fork vibrate, moving back and forth. This is more readily observed if the prongs of the fork are struck, then held against a sheet of paper or plunged into a beaker of water. In air, the vibrating prongs first move toward you, pushing the air molecules into a condensation of increased density and pressure. As the prongs then move back, a rarefaction of decreased density and pressure is produced. The alternation of increased and decreased pressure pulses moves from the vibrating tuning fork and spreads outward equally in all directions, much like the surface of a rapidly expanding balloon (Figure 5.10). When the pulses reach your eardrum, it is forced in and out by the pulses. It now vibrates with the same frequency as the tuning fork. The vibrations of the eardrum are transferred by three tiny bones to a

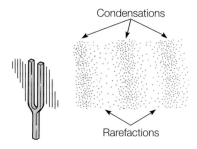

FIGURE 5.10 A vibrating tuning fork produces a series of condensations and rarefactions that move away from the tuning fork. The pulses of increased and decreased pressure reach your ear, vibrating the eardrum. The ear sends nerve signals to the brain about the vibrations, and the brain interprets the signals as sounds.

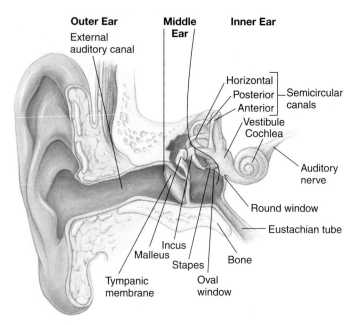

FIGURE 5.11 Anatomy of the ear. Sound enters the outer ear and, upon reaching the middle ear, impinges upon the tympanic membrane, which vibrates three bones (malleus, incus, and stapes). The vibrating stapes hits the oval window, and hair cells in the cochlea convert the vibrations into action potentials, which follow the auditory nerve to the brain. Hair cells in the semicircular canals and in the vestibule sense balance. The eustachian tube connects the middle ear to the throat, equalizing air pressure.

fluid in a coiled chamber (Figure 5.11). Here, tiny hairs respond to the frequency and size of the disturbance, activating nerves that transmit the information to the brain. The brain interprets a frequency as a sound with a certain **pitch.** High-frequency sounds are interpreted as high-pitched musical notes, for example, and low-frequency sounds are interpreted as low-pitched musical notes. The brain then selects certain sounds from all you hear, and you "tune" to certain ones, enabling you to listen to whatever sounds you want while ignoring the background noise, which is made up of all the other sounds.

MEDIUM REQUIRED

The transmission of a sound wave requires a medium, that is, a solid, liquid, or gas to carry the disturbance. Therefore, sound does not travel through the vacuum of outer space, since there is nothing to carry the vibrations from a source. The nature of the molecules making up a solid, liquid, or gas determines how well or how rapidly the substance will carry sound waves. The two variables are (1) the inertia of the molecules and (2) the strength of the interaction. Thus, hydrogen gas, with the least massive molecules, will carry a sound wave at 1,284 m/s (4,213 ft/s) when the temperature is 0°C. More massive helium gas molecules have more inertia and carry a sound wave at only 965 m/s (3,166 ft/s) at the same temperature. A solid, however, has molecules that are strongly attached, so vibrations are passed rapidly from molecule to molecule. Steel, for example, is highly elastic, and sound will move through a steel rail at 5,940 m/s

Three general areas of hearing problems are related to your age and the intensity and duration of sounds falling on your ears. These are (1) middle ear infections of young children; (2) loss of ability to hear higher frequencies because of aging; and (3) ringing and other types of noise heard in the head or ear.

Middle ear infections are one of the most common illnesses of young children. The middle ear is a small chamber behind the eardrum that has three tiny bones that transfer sound vibrations from the eardrum to the inner ear (see Figure 5.11). The middle ear is connected to the throat by a small tube named the eustachian tube. This tube allows air pressure to be balanced behind the eardrum. A middle ear infection usually begins with a cold. Small children have short eustachian tubes that can become swollen from a cold, and this traps fluid in the middle ear. Fluid buildup causes pain and discomfort, as well as reduced hearing ability. This condition often clears on its own in several weeks or more. For severe and recurring cases, small tubes are sometimes inserted through the eardrum to allow fluid drainage. These tubes eventually fall out, and the eardrum heals. Middle ear infections are less of a problem by the time a child reaches school age.

A normal loss of hearing occurs because of aging. This loss is more pronounced for higher frequencies and is also greater in males than females. The normal loss of hearing begins in the early twenties and then increases through the sixties and seventies. However, this process is accelerated by spending a lot of time in places with very loud sounds. Loud concerts and riding in a "boom car" are two examples of situations that will speed the loss of ability to hear higher frequencies. A "boom car" is one with loud music you can hear "booming" a half a block or more away.

Tinnitus is a sensation of sound, as a ringing or swishing, that seems to originate in the ears or head and can only be heard by the person affected. There are a number of different causes, but the most common is nerve damage in the inner ear. Exposure to loud noises, explosions, firearms, and loud bands are common causes of tinnitus. Advancing age and certain medications, such as aspirin, can also cause tinnitus. Stopping the medications can end the tinnitus, but there is no treatment for nerve damage to the inner ear.

TABLE 5.1

Speed of sound in various materials

Medium	m/s	ft/s
Carbon dioxide (0°C)	259	850
Dry air (0°C)	331	1,087
Helium (0°C)	965	3,166
Hydrogen (0°C)	1,284	4,213
Water (25°C)	1,497	4,911
Seawater (25°C)	1,530	5,023
Lead	1,960	6,430
Glass	5,100	16,732
Steel	5,940	19,488

(19,488 ft/s). Thus, there is a reason for the old saying, "Keep your ear to the ground," because sounds move through solids more rapidly than through a gas (Table 5.1).

VELOCITY OF SOUND IN AIR

Most people have observed that sound takes some period of time to move through the air. If you watch a person hammering on a roof a block away, the sounds of the hammering are not in sync with what you see. Light travels so rapidly that you can consider what you see to be simultaneous with what is actually happening for all practical purposes. Sound, however, travels much more slowly, and the sounds arrive late in comparison to what you are seeing. This is dramatically illustrated by seeing a flash of lightning, then hearing thunder seconds later. Perhaps you know of a way to estimate the distance to a lightning flash by timing the interval between the flash and boom. If not, you will learn a precise way to measure this distance shortly.

The air temperature influences how rapidly sound moves through the air. The gas molecules in warmer air have a greater kinetic energy than those of cooler air. The molecules of warmer air therefore transmit an impulse from molecule to molecule more rapidly. More precisely, the speed of a sound wave increases 0.600 m/s (2.00 ft/s) for *each* Celsius degree increase in temperature above 0°C. So, it will be easier for your car or airplane to break the sound barrier on a cold day than on a warm day. How much easier? In *dry* air at sea-level density (normal pressure) at 0°C, the speed of sound is about 331 m/s (1,087 ft/s). If the air temperature is 30°C, sound will travel at 0.600 m/s faster for each degree above 0°C, or (0.600 m/s per 0°C)(30°C) = 18 m/s. Adding this to the speed of sound at 0°C, you have 331 m/s + 18 m/s = 349 m/s. You would need to move at 349 m/s to travel at the speed of sound when the air temperature is 30°C, but you could also travel at the speed of sound at 331 m/s when the air temperature is 0°C.

The simple relationship of the speed of sound at 0°C plus the fractional increase per degree above 0°C can be combined as in the following equations:

$$v_{T_p}(\text{m/s}) = v_0 + \left(\frac{0.600 \text{ m/s}}{°\text{C}}\right)(T_p)$$

equation 5.4

where v_{T_p} is the velocity of sound at the present temperature, v_0 is the velocity of sound at 0°C, and T_p is the present temperature. This equation tells you that the velocity of a sound wave increases 0.6 m/s for each degree C above 0°C. For units of ft/s,

$$v_{T_p}(\text{ft/s}) = v_0 + \left(\frac{2.00 \text{ ft/s}}{°\text{C}}\right)(T_p)$$

equation 5.5

Equation 5.5 tells you that the velocity of a sound wave increases 2.0 ft/s for each degree Celsius above 0°C.

EXAMPLE 5.4

What is the velocity of sound in m/s at room temperature (20.0°C)?

SOLUTION

$$v_0 = 331 \text{ m/s}$$
$$T_p = 20.0°C$$
$$v_{T_p} = ?$$

$$v_{T_p} = v_0 + \left(\frac{0.600 \text{ m/s}}{°C}\right)(T_p)$$

$$= 331 \text{ m/s} + \left(\frac{0.600 \text{ m/s}}{°C}\right)(20.0 °C)$$

$$= 331 \text{ m/s} + (0.600 \times 20.0)\frac{\text{m/s}}{°C} \times °C$$

$$= 331 \text{ m/s} + 12.0 \text{ m/s}$$

$$= \boxed{343 \text{ m/s}}$$

EXAMPLE 5.5

The air temperature is 86.0°F. What is the velocity of sound in ft/s? (Note that °F must be converted to °C for equation 5.5.) (Answer: 1,147 ft/s)

REFRACTION AND REFLECTION

When you drop a rock into a still pool of water, circular patterns of waves move out from the disturbance. These water waves are on a flat, two-dimensional surface. Sound waves, however, move in three-dimensional space like a rapidly expanding balloon. Sound waves are *spherical waves* that move outward from the source. Spherical waves of sound move as condensations and rarefactions from a continuously vibrating source at the center. If you identify the same part of each wave in the spherical waves, you have identified a *wave front*. For example, the crest of each condensation could be considered a wave front. From one wave front to the next, therefore, identifies one complete wave or wavelength. At some distance from the source, a small part of a spherical wave front can be considered a *linear wave front* (Figure 5.12).

Waves move within a homogeneous medium such as a gas or a solid at a fairly constant rate but gradually lose energy to friction. When a wave encounters a different condition (temperature, humidity, or nature of material), however, drastic changes may occur rapidly. The division between two physical conditions is called a *boundary*. Boundaries are usually encountered (1) between different materials or (2) between the same materials with different conditions. An example of a wave moving between different materials is a sound made in the next room that moves through the air to the wall and through the wall to the air in the room where you are. The boundaries are air-wall and wall-air. If you have ever been in a room with "thin walls," it is obvious that sound moved through the wall and air boundaries.

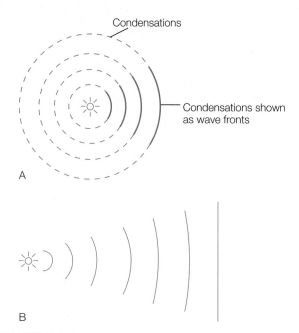

FIGURE 5.12 (*A*) Spherical waves move outward from a sounding source much like a rapidly expanding balloon. This two-dimensional sketch shows the repeating condensations as spherical wave fronts. (*B*) Some distance from the source, a spherical wave front is considered a linear, or plane, wave front.

Refraction

An example of sound waves moving through the same material with different conditions is found when a wave front moves through air of different temperatures. Since sound travels faster in warm air than in cold air, the wave front becomes bent. The bending of a wave front at boundaries is called **refraction.** Refraction changes the direction of travel of a wave front. Consider, for example, that on calm, clear nights, the air near Earth's surface is cooler than air farther above the surface. Air at rooftop height above the surface might be four or five degrees warmer under such ideal conditions. Sound will travel faster in the higher, warmer air than it will in the lower, cooler air close to the surface. A wave front will therefore become bent, or refracted, toward the ground on a cool night and you will be able to hear sounds from farther away than on warm nights (Figure 5.13A). The opposite process occurs during the day as Earth's surface becomes warmer from sunlight (Figure 5.13B). Wave fronts are refracted upward because part of the wave front travels faster in the warmer air near the surface. Thus, sound does not seem to carry as far in the summer as it does in the winter. What is actually happening is that during the summer, the wave fronts are refracted away from the ground before they travel very far.

Reflection

When a wave front strikes a boundary that is parallel to the front, the wave may be absorbed, transmitted, or undergo **reflection,**

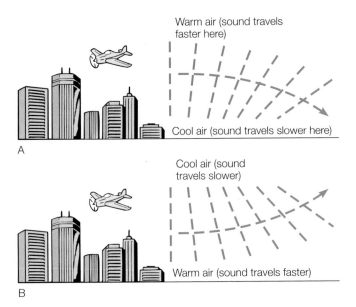

FIGURE 5.13 (*A*) Since sound travels faster in warmer air, a wave front becomes bent, or refracted, toward Earth's surface when the air is cooler near the surface. (*B*) When the air is warmer near the surface, a wave front is refracted upward, away from the surface.

FIGURE 5.14 This closed-circuit TV control room is acoustically treated by covering the walls with sound-absorbing baffles. Reverberation and echoes cannot occur in this treated room because absorbed sounds are not reflected.

depending on the nature of the boundary medium, or the wave may be partly absorbed, partly transmitted, partly reflected, or any combination thereof. Some materials, such as hard, smooth surfaces, reflect sound waves more than they absorb them. Other materials, such as soft, ruffly curtains, absorb sound waves more than they reflect them. If you have ever been in a room with smooth, hard walls and no curtains, carpets, or furniture, you know that sound waves may be reflected several times before they are finally absorbed.

Do you sing in the shower? Many people do because the tone is more pleasing than singing elsewhere. The walls of a shower are usually hard and smooth, reflecting sounds back and forth several times before they are absorbed. The continuation of many reflections causes a tone to gain in volume. Such mixing of reflected sounds with the original is called **reverberation.** Reverberation adds to the volume of a tone, and it is one of the factors that determine the acoustical qualities of a room, lecture hall, or auditorium. An open-air concert sounds flat without the reverberation of an auditorium and is usually enhanced electronically to make up for the lack of reflected sounds. Too much reverberation in a room or classroom is not good because the spoken word is not as sharp. Sound-absorbing materials are therefore used on the walls and floors where clear, distinct speech is important (Figure 5.14). The carpet and drapes you see in a movie theater are not decorator items but are there to absorb sounds.

If a reflected sound arrives after 0.10 s, the human ear can distinguish the reflected sound from the original sound. A reflected sound that can be distinguished from the original is called an **echo.** Thus, a reflected sound that arrives before 0.10 s is perceived as an increase in volume and is called a reverberation, but a sound that arrives after 0.10 s is perceived as an echo.

EXAMPLE 5.6

The human ear can distinguish a reflected sound pulse from the original sound pulse if 0.10 s or more elapses between the two sounds. What is the minimum distance to a reflecting surface from which we can hear an echo (see Figure 5.15A) if the speed of sound is 343 m/s?

SOLUTION

$$v = \frac{d}{t} \quad \therefore \quad d = vt$$

$t = 0.10\ \text{s}$
(minimum)

$v = 343\ \text{m/s}$

$d = ?$

$$= \left(343\ \frac{\text{m}}{\text{s}}\right)(0.10\ \text{s})$$

$$= 343 \times 0.10\ \frac{\text{m}}{\text{s}} \times \text{s}$$

$$= 34.3\ \frac{\text{m} \cdot \text{s}}{\text{s}}$$

$$= 34\ \text{m}$$

Since the sound pulse must travel from the source to the reflecting surface, then back to the source,

$$34\ \text{m} \times 1/2 = \boxed{17\ \text{m}}$$

The minimum distance to a reflecting surface from which we hear an echo when the air is at room temperature is therefore 17 m (about 56 ft).

EXAMPLE 5.7

An echo is heard exactly 1.00 s after a sound when the speed of sound is 1,147 ft/s. How many feet away is the reflecting surface? (Answer: 574 ft)

Sound wave echoes are measured to determine the depth of water or to locate underwater objects by a *sonar* device. The word *sonar* is taken from *so*und *na*vigation *r*anging. The device

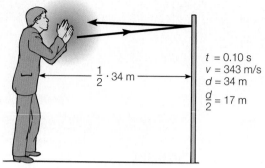

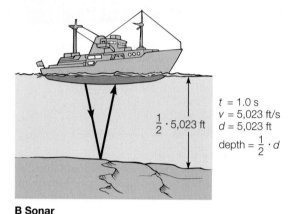

A Echo

B Sonar

FIGURE 5.15 (*A*) At room temperature, sound travels at 343 m/s. In 0.10 s, sound would travel 34 m. Since the sound must travel to a surface and back in order for you to hear an echo, the distance to the surface is one-half the total distance. (*B*) Sonar measures a depth by measuring the elapsed time between an ultrasonic sound pulse and the echo. The depth is one-half the round trip.

generates an underwater ultrasonic sound pulse, then measures the elapsed time for the returning echo. Sound waves travel at about 1,531 m/s (5,023 ft/s) in seawater at 25°C (77°F). A 1 s lapse between the ping of the generated sound and the echo return would mean that the sound traveled 5,023 ft for the round trip. The bottom would be half this distance below the surface (Figure 5.15B).

Myths, Mistakes, and Misunderstandings

A Duck's Quack Doesn't Echo?

You may have heard the popular myth that "a duck's quack doesn't echo, and no one knows why." An acoustic research experiment was carried out at the University of Salford in Greater Manchester, U.K., to test this myth. Acoustic experts first recorded a quacking duck in a special chamber that was constructed to produce no sound reflections. Simulations were then done in a reverberation chamber to match the effect of the duck quacking when flying past a cliff face. The tests found that a duck's quack indeed does echo, just like any other sound.

The quack researchers speculated that the myth might have resulted because:

1. The quack does echo, but it is usually too quiet to hear because a duck quacks too quietly.
2. Ducks quack near water, not near reflecting surfaces such as a mountain or building that would make an echo.
3. It is hard to hear the echo of a sound that fades in and fades out, as a duck quack does.

INTERFERENCE

Waves interact with a boundary much as a particle would, reflecting or refracting because of the boundary. A moving ball, for example, will bounce from a surface at the same angle it strikes the surface, just as a wave does. A particle or a ball, however, can be in only one place at a time, but waves can be spread over a distance at the same time. You know this since many different people in different places can hear the same sound at the same time.

Constructive and Destructive

When two traveling waves meet, they can interfere with each other, producing a new disturbance. This new disturbance has a different amplitude, which is the algebraic sum of the amplitudes of the two separate wave patterns. If the wave crests or wave troughs arrive at the same place at the same time, the two waves are said to be *in phase*. The result of two waves arriving in phase is a new disturbance with a crest and trough that has greater displacement than either of the two separate waves. This is called *constructive interference* (Figure 5.16A). If the trough of one wave arrives at the same place and time as the crest of another wave, the waves are completely *out of phase*. When two waves are completely out of phase, the crest of one wave (positive displacement) will cancel the trough of the other wave (negative displacement), and the result is zero total disturbance, or no wave. This is called *destructive interference* (Figure 5.16B). If the two sets of wave patterns do not have the same amplitudes or wavelengths, they will be neither completely in phase nor completely out of phase. The result will be partly constructive or destructive interference, depending on the exact nature of the two wave patterns.

Beats

Suppose that two vibrating sources produce sounds that are in phase, equal in amplitude, and equal in frequency. The resulting sound will be increased in volume because of constructive interference. But suppose the two sources are slightly different in frequency, for example, 350 and 352 Hz. You will hear a regularly spaced increase and decrease of sound known as **beats.** Beats occur because the two sound waves experience alternating constructive and destructive interferences (Figure 5.17). The phase relationship changes because of the difference in frequency, as you can see in Figure 5.17. These alternating constructive and destructive interference zones are moving from the source to the receiver, and the receiver hears the results as a rapidly rising and falling sound level. The beat

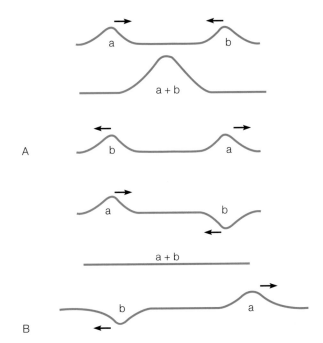

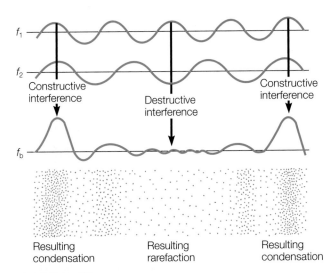

FIGURE 5.16 (*A*) Constructive interference occurs when two equal, in-phase waves meet. (*B*) Destructive interference occurs when two equal, out-of-phase waves meet. In both cases, the wave displacements are superimposed when they meet, but they then pass through one another and return to their original amplitudes.

frequency is the difference between the frequencies of the two sources. A 352 Hz source and 350 Hz source sounded together would result in a beat frequency of 2 Hz. Thus, the frequencies are closer and closer together, and fewer beats will be heard per second. You may be familiar with the phenomenon of beats if you have ever flown in an airplane with two engines. If one engine is running slightly faster than the other, you hear a slow beat. The beat frequency (f_b) is equal to

FIGURE 5.17 Two waves of equal amplitude but slightly different frequencies interfere destructively and constructively. The result is an alternation of loudness called a *beat*.

the absolute difference in frequency of two interfering waves with slightly different frequencies, or

$$f_b = f_2 - f_1$$

equation 5.6

ENERGY OF WAVES

All waves transport energy, including sound waves. The vibrating mass and spring in Figure 5.2 vibrate with an amplitude that depends on how much work you did on the mass in moving it from its equilibrium position. More work on the mass results in a greater displacement and a greater amplitude of vibration. A vibrating object that is producing sound waves will produce more intense condensations and rarefactions if it has a greater amplitude. The intensity of a sound wave is a measure of the energy the sound wave is carrying (Figure 5.18). *Intensity* is defined as the power (in watts) transmitted by a wave to a unit area (in square meters) that is perpendicular to the waves. Intensity is therefore measured in watts per square meter (W/m^2) or

$$\text{Intensity} = \frac{\text{Power}}{\text{Area}}$$

$$I = \frac{P}{A}$$

equation 5.7

HOW LOUD IS THAT SOUND?

The *loudness* of a sound is a subjective interpretation that varies from person to person. Loudness is also related to (1) the energy of a vibrating object, (2) the condition of the air the sound wave travels through, and (3) the distance between you and the vibrating source. Furthermore, doubling the amplitude of the vibrating source will quadruple the *intensity* of the resulting sound wave, but the sound will not be perceived as four times as loud. The relationship between perceived loudness and the intensity of a sound wave

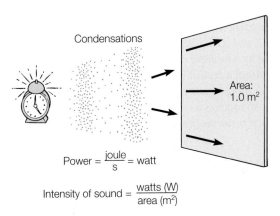

FIGURE 5.18 The intensity of a sound wave is the rate of energy transferred to an area perpendicular to the waves. Intensity is measured in watts per square meter, W/m^2.

TABLE 5.2

Comparison of noise levels in decibels with intensity

Example	Response	Decibels	Intensity W/m²
Least needed for hearing	Barely perceived	0	1×10^{-12}
Calm day in woods	Very, very quiet	10	1×10^{-11}
Whisper (15 ft)	Very quiet	20	1×10^{-10}
Library	Quiet	40	1×10^{-8}
Talking	Easy to hear	65	3×10^{-6}
Heavy street traffic	Conversation difficult	70	1×10^{-5}
Pneumatic drill (50 ft)	Very loud	95	3×10^{-3}
Jet plane (200 ft)	Discomfort	120	1

is not a linear relationship. In fact, a sound that is perceived as twice as loud requires ten times the intensity, and quadrupling the loudness requires a one-hundred-fold increase in intensity.

The human ear is very sensitive. It is capable of hearing sounds with intensities as low as 10^{-12} W/m² and is not made uncomfortable by sound until the intensity reaches about 1 W/m². The second intensity is a million million (10^{12}) times greater than the first. Within this range, the subjective interpretation of intensity seems to vary by powers of ten. This observation led to the development of the **decibel scale** to measure the intensity level. The scale is a ratio of the intensity level of a given sound to the threshold of hearing, which is defined as 10^{-12} W/m² at 1,000 Hz. In keeping with the power-of-ten subjective interpretations of intensity, a logarithmic scale is used rather than a linear scale. Originally, the scale was the logarithm of the ratio of the intensity level of a sound to the threshold of hearing. This definition set the zero point at the threshold of human hearing. The unit was named the *bel* in honor of Alexander Graham Bell (1847–1922). This unit was too large to be practical, so it was reduced by one-tenth and called a *decibel*. The intensity level of a sound is therefore measured in decibels (Table 5.2). Compare the decibel noise level of familiar sounds listed in Table 5.2, and note that each increase of ten on the decibel scale is matched by a *multiple* of ten on the intensity level. For example, moving from a decibel level of 10 to a decibel level of 20 requires *ten times* more intensity. Likewise, moving from a decibel level of 20 to 40 requires a one-hundred-fold increase in the intensity level. As you can see, the decibel scale is not a simple linear scale.

RESONANCE

You know that sound waves transmit energy when you hear a thunderclap rattle the windows. In fact, the sharp sounds from an explosion have been known not only to rattle but also break windows. The source of the energy is obvious when thunderclaps or explosions are involved. But sometimes energy transfer occurs through sound waves when it is not clear what is happening. A truck drives down the street, for example, and one window rattles but the others do not. A singer shatters a crystal water glass by singing a single note, but other objects remain undisturbed. A closer look at the nature of vibrating objects and the transfer of energy will explain these phenomena.

Almost any elastic object can be made to vibrate and will vibrate freely at a constant frequency after being sufficiently disturbed. Entertainers sometimes discover this fact and appear on late-night talk shows playing saws, wrenches, and other odd objects as musical instruments. All material objects have a *natural frequency* of vibration determined by the materials and shape of the objects. The natural frequencies of different wrenches enable an entertainer to use the suspended tools as if they were the bars of a xylophone.

If you have ever pumped a swing, you know that small forces can be applied at any frequency. If the frequency of the applied forces matches the natural frequency of the moving swing, there is a dramatic increase in amplitude. When the two frequencies match, energy is transferred very efficiently. This condition, when the frequency of an external force matches the natural frequency, is called **resonance.** The natural frequency of an object is thus referred to as the *resonant frequency,* that is, the frequency at which resonance occurs.

A silent tuning fork will resonate if a second tuning fork with the same frequency is struck and vibrates nearby (Figure 5.19). You will hear the previously silent tuning fork sounding if you stop the vibrations of the struck fork by touching it. The waves of condensations and rarefactions produced by the struck tuning fork produce a regular series of impulses that match the natural frequency of the silent tuning fork. This illustrates that at resonance, relatively little energy is required to start vibrations.

A truck causing vibrations as it is driven past a building may cause one window to rattle while others do not. Vibrations caused by the truck have matched the natural frequency of this window but not the others. The window is undergoing resonance from the sound wave impulses that matched its natural frequency. It is also resonance that enables a singer to break a water glass. If the tone is at the resonant frequency of the glass, the resulting vibrations may be large enough to shatter it.

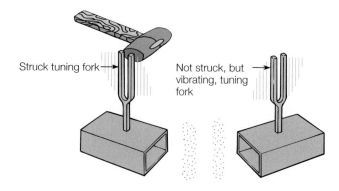

Struck tuning fork →

← Not struck, but vibrating, tuning fork

FIGURE 5.19 When the frequency of an applied force, including the force of a sound wave, matches the natural frequency of an object, energy is transferred very efficiently. The condition is called *resonance.*

CONCEPTS *Applied*

A Singing Glass

Did you ever hear a glass "sing" when the rim was rubbed? The trick to make the glass sing is to remove as much oil from your finger as possible. Then you lightly rub your wet finger around and on the top of the glass rim at the correct speed. Without oil, your wet finger will imperceptively catch on the glass as you rub the rim. With the appropriate pressure and speed, your catching finger might match the natural frequency of the glass. The resonant vibration will cause the glass to "sing" with a high-pitched note.

SOURCES OF SOUNDS

All sounds have a vibrating object as their source. The vibrations of the object send pulses or waves of condensations and rarefactions through the air. These sound waves have physical properties that can be measured, such as frequency and intensity. Subjectively, your response to frequency is to identify a certain pitch. A high-frequency sound is interpreted as a high-pitched sound, and a low-frequency sound is interpreted as a low-pitched sound. Likewise, a greater intensity is interpreted as increased loudness, but there is not a direct relationship between intensity and loudness as there is between frequency and pitch.

There are other subjective interpretations about sounds. Some sounds are bothersome and irritating to some people but go unnoticed by others. In general, sounds made by brief, irregular vibrations such as those made by a slamming door, dropped book, or sliding chair are called *noise*. Noise is characterized by sound waves with mixed frequencies and jumbled intensities (Figure 5.20). On the other hand, there are sounds made by very regular, repeating vibrations such as those made by a tuning fork. A tuning fork produces a *pure tone* with a sinusoidal curved pressure variation and regular frequency. Yet a tuning fork produces a tone that most people interpret as bland. You would not call a tuning fork sound a musical note! Musical sounds from instruments have a certain frequency and loudness, as do noise

and pure tones, but you can readily identify the source of the very same musical note made by two different instruments. You recognize it as a musical note, not noise and not a pure tone. You also recognize if the note was produced by a violin or a guitar. The difference is in the wave form of the sounds made by the two instruments, and the difference is called the *sound quality*. How does a musical instrument produce a sound of a characteristic quality? The answer may be found by looking at the instruments that make use of vibrating strings.

VIBRATING STRINGS

A stringed musical instrument, such as a guitar, has strings that are stretched between two fixed ends. When a string is plucked, waves of many different frequencies travel back and forth on the string, reflecting from the fixed ends. Many of these waves quickly fade away, but certain frequencies resonate, setting up patterns of waves. Before considering these resonant patterns in detail, keep in mind that (1) two or more waves can be in the same place at the same time, traveling through one another from opposite directions; (2) a confined wave will be reflected at a boundary, and the reflected wave will be inverted (a crest becomes a trough); and (3) reflected waves interfere with incoming waves of the same frequency to produce **standing waves.** Figure 5.21 is a graphic "snapshot" of what happens when reflected wave patterns meet incoming wave patterns. The incoming wave is shown as a solid line, and the reflected wave is shown as a dotted line. The result is (1) places of destructive interference, called *nodes,* which show no disturbance, and (2) loops of constructive interference, called *antinodes,* which take place where the crests and troughs of the two wave patterns produce a disturbance that rapidly alternates upward and downward. This pattern of alternating nodes and antinodes does not move along the string and is thus called a *standing wave.* Note that the standing wave for *one wavelength* will have a node at both ends and in the center and also two antinodes.

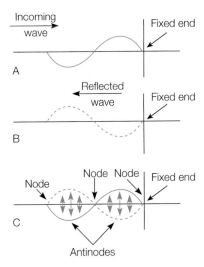

FIGURE 5.21 An incoming wave on a cord with a fixed end (*A*) meets a reflected wave (*B*) with the same amplitude and frequency, producing a standing wave (*C*). Note that a standing wave of one wavelength has three nodes and two antinodes.

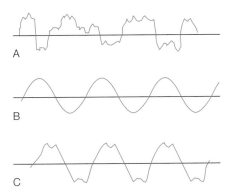

FIGURE 5.20 Different sounds that you hear include (*A*) noise, (*B*) pure tones, and (*C*) musical notes.

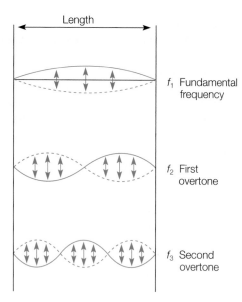

FIGURE 5.22 A stretched string of a given length has a number of possible resonant frequencies. The lowest frequency is the fundamental, f_1; the next higher frequencies, or overtones, shown are f_2 and f_3.

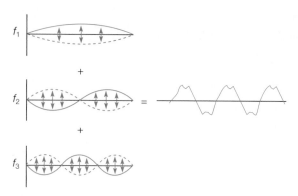

FIGURE 5.23 A combination of the fundamental and overtone frequencies produces a composite waveform with a characteristic sound quality.

Standing waves occur at the natural, or resonant, frequencies of the string, which are a consequence of the nature of the string, the string length, and the tension in the string. Since the standing waves are resonant vibrations, they continue as all other waves quickly fade away.

Since the two ends of the string are not free to move, the ends of the string will have nodes. The *longest* wave that can make a standing wave on such a string has a wavelength (λ) that is twice the length (L) of the string. Since frequency (f) is inversely proportional to wavelength ($f = v/\lambda$ from equation 5.3), this longest wavelength has the lowest frequency possible, called the **fundamental frequency.** The fundamental frequency has one antinode, which means that the length of the string has one-half a wavelength. The fundamental frequency (f_1) determines the pitch of the *basic* musical note being sounded and is called the first harmonic. Other resonant frequencies occur at the same time, however, since other standing waves can also fit onto the string. A higher frequency of vibration (f_2) could fit two half-wavelengths between the two fixed nodes. An even higher frequency (f_3) could fit three half-wavelengths between the two fixed nodes (Figure 5.22). Any whole number of halves of the wavelength will permit a standing wave to form. The frequencies (f_2, f_3, etc.) of these wavelengths are called the *overtones,* or harmonics. It is the presence and strength of various overtones that give a musical note from a certain instrument its characteristic quality. The fundamental and the overtones add together to produce the characteristic *sound quality,* (Figure 5.23) which is different for the same-pitched note produced by a violin and by a guitar.

Since nodes must be located at the ends, only half wavelengths ($1/2\ \lambda$) can fit on a string of a given length (L), so the fundamental frequency of a string is $1/2\ \lambda = L$, or $\lambda = 2L$. Substituting this value in the wave equation (solved for frequency, f) will give the relationship for finding the fundamental frequency and the overtones when the string length and velocity of waves on the string are known. The relationship is

$$f_n = \frac{nv}{2L}$$

equation 5.8

where $n = 1, 2, 3, 4\ldots$, and $n = 1$ is the fundamental frequency and $n = 2$, $n = 3$, and so forth are the overtones.

EXAMPLE 5.8

What is the fundamental frequency of a 0.5 m string if wave speed on the string is 400 m/s?

SOLUTION

The length (L) and the velocity (v) are given. The relationship between these quantities and the fundamental frequency ($n = 1$) is given in equation 5.8, and

$$L = 0.5 \text{ m}$$
$$v = 400 \text{ m/s}$$
$$f_1 = ?$$

$$f_n = \frac{nv}{2L} \quad \text{where } n = 1 \text{ for the fundamental frequency}$$

$$f_1 = \frac{1 \times 400 \text{ m/s}}{2 \times 0.5 \text{ m}}$$

$$= \frac{400}{1} \frac{\text{m}}{\text{s}} \times \frac{1}{\text{m}}$$

$$= 400 \frac{\text{m}}{\text{s}\cdot\text{m}}$$

$$= 400 \frac{1}{\text{s}}$$

$$= \boxed{400 \text{ Hz}}$$

EXAMPLE 5.9

What is the frequency of the first overtone in a 0.5 m string when the wave speed is 400 m/s? (Answer: 800 Hz)

The vibrating string produces a waveform with overtones, so instruments that have vibrating strings are called *harmonic instruments.* Instruments that use an air column as a sound

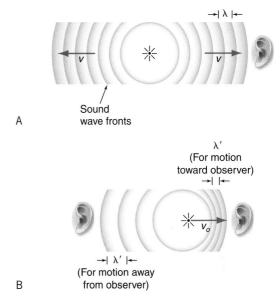

Science and Society

Laser Bug

Hold a fully inflated balloon lightly between your fingertips and talk. You will be able to feel the slight vibrations from your voice. Likewise, the sound waves from your voice will cause a nearby window to vibrate slightly. If a laser beam is bounced off the window, the reflection will be changed by the vibrations. The incoming laser beam is coherent; all the light has the same frequency and amplitude (see p. 191).

The reflected beam, however, will have different frequencies and amplitudes from the windowpane vibrating in and out. The changes can be detected by a receiver and converted into sound in a headphone.

You cannot see an infrared laser beam because infrared is outside the frequencies that humans can see. Any sound-sensitive target can be used by the laser bug, including a windowpane, inflated balloon,

hanging picture, or the glass front of a china cabinet.

QUESTIONS TO DISCUSS

1. Is it legal for someone to listen in on your private conversations?

2. Should the sale of technology such as the laser bug be permitted? What are the issues?

---

maker are also harmonic instruments. These include all the wind instruments such as the clarinet, flute, trombone, trumpet, pipe organ, and many others. The various wind instruments have different ways of making a column of air vibrate. In the flute, air vibrates as it moves over a sharp edge, while in the clarinet, saxophone, and other reed instruments, it vibrates through fluttering thin reeds. The air column in brass instruments, on the other hand, is vibrated by the tightly fluttering lips of the player.

The length of the air column determines the frequency, and woodwind instruments have holes in the side of a tube that are opened or closed to change the length of the air column. The resulting tone depends on the length of the air column and the resonant overtones.

SOUNDS FROM MOVING SOURCES

When the source of a sound is stationary, equally spaced sound waves expand from a source in all directions. But if the sounding source starts moving, then successive sound waves become displaced in the direction of movement, and this changes the pitch. For example, the siren of an approaching ambulance seems to change pitch when the ambulance passes you. The sound wave is "squashed" as the ambulance approaches you, and you hear a higher-frequency siren than the people inside the ambulance. When the ambulance passes you, the sound waves are "stretched" and you hear a lower-frequency siren (Figure 5.24). The overall effect of a higher pitch as a source approaches and then a lower pitch as it moves away is called the **Doppler effect.** The Doppler effect is evident if you stand by a street and an approaching car sounds its horn as it drives by you. You will hear a higher-pitched horn as the car approaches, which shifts to a lower-pitched horn as the waves go by you. The driver of the car, however, will hear the continual, true pitch of the horn because the driver is moving with the source.

A Doppler shift is also noted if the observer is moving and the source of sound is stationary. When the observer moves toward the source, the wave fronts are encountered more frequently than if the observer were standing still. As the observer moves away from the source, the wave fronts are encountered less frequently than if the observer were not moving. An observer on a moving

FIGURE 5.24 (A) Sound waves emitted by a stationary source and observed by a stationary observer. (B) Sound waves emitted by a source in motion toward the right. An observer on the right receives wavelengths that are shortened; an observer on the left receives wavelengths that are lengthened.

train approaching a crossing with a sounding bell thus hears a high-pitched bell that shifts to a lower-pitched bell as the train whizzes by the crossing. The Doppler effect occurs for all waves, including electromagnetic waves (see Figure 7.3 on p. 179).

When an object moves through the air at the speed of sound, it keeps up with its own sound waves. All the successive wave fronts pile up on one another, creating a large wave disturbance called a **shock wave** (Figure 5.25). The shock wave from a supersonic airplane is a cone-shaped shock wave of intense condensations trailing backward at an angle dependent on the speed of the aircraft. Wherever this cone of superimposed crests passes, a **sonic boom** occurs. The many crests have been added together, each contributing to the pressure increase. The human ear cannot differentiate between such a pressure wave created by a supersonic aircraft and a pressure wave created by an explosion.

People Behind the Science

Johann Christian Doppler (1803–1853)

Johann Christian Doppler

Johann Doppler was an Austrian physicist who discovered the Doppler effect, which relates the observed frequency of a wave to the relative motion of the source and the observer. The Doppler effect is readily observed in moving sound sources, producing a fall in pitch as the source passes the observer, but it is of most use in astronomy, where it is used to estimate the velocities and distances of distant bodies.

Doppler was born in Salzburg, Austria, on November 29, 1803, the son of a stonemason. He showed early promise in mathematics and attended the Polytechnic Institute in Vienna from 1822 to 1825. Despairing of ever obtaining an academic post, he decided to emigrate to the United States. Then, on the point of departure, he was offered a professorship of mathematics at the State Secondary School in Prague and changed his mind. He subsequently obtained professorships in mathematics at the State Technical Academy in Prague in 1841 and at the Mining Academy in Schemnitz in 1847. Doppler returned to Vienna the following year and, in 1850, became director of the new Physical Institute and Professor of Experimental Physics at the Royal Imperial University of Vienna. He died from a lung disease in Venice on March 17, 1853.

Doppler explained the effect that bears his name by pointing out that sound waves from a source moving toward an observer will reach the observer at a greater frequency than if the source is stationary, thus increasing the observed frequency and raising the pitch of the sound. Similarly, sound waves from a source moving away from the observer reach the observer more slowly, resulting in a decreased frequency and a lowering of pitch. In 1842, Doppler put forward this explanation and derived the observed frequency mathematically in Doppler's principle.

The first experimental test of Doppler's principle was made in 1845 at Utrecht in Holland. A locomotive was used to carry a group of trumpeters in an open carriage to and fro past some musicians able to sense the pitch of the notes being played. The variation of pitch produced by the motion of the trumpeters verified Doppler's equations.

Doppler correctly suggested that his principle would apply to any wave motion and cited light as an example as well as sound. He believed that all stars emit white light and that differences in color are observed on earth because the motion of stars affects the observed frequency of the light and hence its color. This idea was not universally true, as stars vary in their basic color. However, Armand Fizeau (1819–1896) pointed out in

1848 that shifts in the spectral lines of stars could be observed and ascribed to the Doppler effect and hence enable their motion to be determined. This idea was first applied in 1868 by William Huggins (1824–1910), who found that Sirius is moving away from the solar system by detecting a small redshift in its spectrum. With the linking of the velocity of a galaxy to its distance by Edwin Hubble (1889–1953) in 1929, it became possible to use the redshift to determine the distances of galaxies. Thus, the principle that Doppler discovered to explain an everyday and inconsequential effect in sound turned out to be of truly cosmological importance.

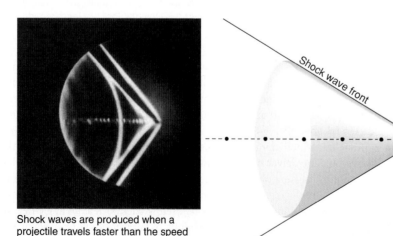

Shock waves are produced when a projectile travels faster than the speed of sound in air.

FIGURE 5.25 A sound source moves with velocity *greater* than the speed of sound in the medium. The envelope of spherical wavefront forms the conical shock wave.

The Doppler effect was named after the Austrian scientist Johann Doppler (1803–1853), who first demonstrated the effect using sound waves in 1842. The same principle applies to electromagnetic radiation as well as sound, but now the shifts are in the frequency of the radiation. A lower frequency is observed when a source of light is moving away, and this is called a "redshift." Also, a "blueshift" toward a higher frequency occurs when a source of light is moving toward an observer. Radio waves will also experience such shifts of frequency, and weather radar that measures frequency changes as a result of motion is called *Doppler radar*.

Weather radar broadcasts short radio waves from an antenna. When directed at a storm, the waves are reflected back to the antenna by rain, snow, and hail. Reflected radar waves are electronically converted and displayed on a monitor, showing the location and intensity of precipitation. A Doppler radar also measures frequency shifts in the reflected radio waves. Waves from objects moving toward the antenna show a higher frequency, and waves from objects moving away from the antenna show a lower frequency. These shifts of frequency are measured, then displayed as the speed and direction of winds that move raindrops and other objects in the storm.

Weather forecasters can direct a Doppler radar machine to measure different elevations of a storm system. This shows highly accurate information that can be used to identify, for example, where and when a tornado might form, the intensity of storm winds in a given area, and even how much precipitation fell from the storm.

Does a sonic boom occur just when an airplane breaks the sound barrier? The answer is no; an airplane traveling at or faster than the speed of sound produces the shock wave continuously, and a sonic boom will be heard everywhere the plane drags its cone-shaped shock wave. In addition, high-speed airplanes often produce two or more shock waves, which are associated with the nose, tail, and other projections on the aircraft. Can you find evidence of shock waves associated with projections on the airplane pictured in Figure 5.26?

The Austrian physicist Ernst Mach (1838–1916) published a paper in 1877 laying out the principles of supersonics. He also came up with the idea of using a ratio of the velocity of an object to the velocity of sound. Today, this ratio is called the *Mach number*. A plane traveling at the speed of sound has a Mach number of 1, a plane traveling at twice the speed of sound has a Mach number of 2, and so on. Ernst Mach was also the first to describe what is happening to produce a sonic boom, and he observed the existence of a conical shock wave formed by a projectile as it approached the speed of sound.

FIGURE 5.26 A cloud sometimes forms just as a plane accelerates to break the sound barrier. Moist air is believed to form the cloud water droplets as air pressure drops behind the shock wave.

SUMMARY

Elastic objects *vibrate,* or move back and forth, in a repeating motion when disturbed by some external force. The maximum value of the displacement of a vibration is called the *amplitude.* The *period* is the time required to complete one full vibration or cycle of the motion. The *frequency* is the number of vibrations that the object performs in a second. The unit of frequency is called *hertz.*

Traveling vibrations or disturbances of a medium are called *waves.* In a *transverse wave,* the elements of a medium vibrate in a direction perpendicular to the direction the wave is traveling. In a *longitudinal wave,* the elements of a medium vibrate in a direction parallel to the direction the wave is traveling. *Sound* is a longitudinal wave, where disturbances are periodic *condensations* (crests) and *rarefactions* (troughs). Sound waves do not travel in a vacuum because a vacuum does not have any medium. Sound waves of frequencies between 20 and 20,000 Hz are *audible sounds* that can be heard by humans. An audible sound of high frequency is perceived as a high-pitched sound, and low frequencies are perceived as low-pitched sounds.

The *amplitude of a wave* is the largest displacement from the equilibrium position that an element of the medium can experience when the wave is passing by. The *period* and *frequency* of a wave are the same as the period and frequency of vibrations on one element of the medium when the wave is passing by. *Wavelength* is the distance the wave travels during one period. All waves carry energy.

When a traveling wave encounters a boundary between two different media, it can be either *reflected* or *refracted* (or both) at the boundary. When two traveling waves meet, they interfere with each other. At any point they meet, this interference may be either *constructive* ("enhancement") or *destructive* ("disappearance").

Beats are interference patterns of two sound waves of slightly different frequencies.

Every elastic object or medium in nature has a characteristic *natural frequency* (or frequencies) of vibration. *Resonance* is a phenomenon where the energy transfer from one object (or medium) to another happens at natural frequencies of the object (or medium) that receives the energy. At resonance, the energy transfer is most efficient and the amplitude of vibrations grow very fast.

Sounds are compared by pitch, loudness, and quality. The *quality* is determined by the instrument sounding the note. Each instrument has its own characteristic quality because of the resonant frequencies that it produces. The basic or *fundamental frequency* is the longest standing wave that it can make. The fundamental frequency determines the basic note being sounded, and other resonant frequencies, or standing waves called *overtones* or *harmonics,* combine with the fundamental to give the instrument its characteristic quality.

A moving source of sound or a moving observer experiences an apparent shift of frequency called the *Doppler effect.* If the source is moving as fast as or faster than the speed of sound, the sound waves pile up into a *shock wave* called a *sonic boom.* A sonic boom sounds very much like the pressure wave from an explosion.

SUMMARY OF EQUATIONS

5.1

$$\text{period} = \frac{1}{\text{frequency}}$$

$$T = \frac{1}{f}$$

5.2

$$\text{frequency} = \frac{1}{\text{period}}$$

$$f = \frac{1}{T}$$

5.3

$$\text{velocity} = (\text{wavelength})\,(\text{frequency})$$

$$v = \lambda f$$

5.4

$$\begin{array}{lcl} \text{velocity of} & & \text{velocity} & & 0.600\ \text{m/s} & & \text{present} \\ \text{sound (m/s)} & = & \text{of sound} & + & \text{increase per} & \times & \text{temperature} \\ \text{at present} & & \text{at } 0°C & & \text{degree Celsius} & & \text{in } °C \\ \text{temperature} \end{array}$$

$$v_{T_p}(\text{m/s}) = v_0 + \left(\frac{0.600\ \text{m/s}}{°C}\right)(T_p)$$

5.5

$$\begin{array}{lcl} \text{velocity of} & & \text{velocity} & & 2.00\ \text{ft/s} & & \text{present} \\ \text{sound (ft/s)} & = & \text{of sound} & + & \text{increase per} & \times & \text{temperature} \\ \text{at present} & & \text{at } 0°C & & \text{degree Celsius} & & \text{in } °C \\ \text{temperature} \end{array}$$

$$v_{T_p}(\text{ft/s}) = v_0 + \left(\frac{2.00\ \text{ft/s}}{°C}\right)(T_p)$$

5.6

$$\text{beat frequency} = \text{one frequency} - \text{other frequency}$$

$$f_b = f_2 - f_1$$

5.7

$$\text{Intensity} = \frac{\text{Power}}{\text{Area}}$$

$$I = \frac{P}{A}$$

5.8

$$\text{resonant frequency} = \frac{\text{number} \times \text{velocity on string}}{2 \times \text{length of string}}$$

where number 1 = fundamental frequency, and numbers 2, 3, 4, and so on = overtones.

$$f_n = \frac{nv}{2L}$$

KEY TERMS

amplitude (p. **117**)
beat (p. **126**)
cycle (p. **117**)
decibel scale (p. **128**)
Doppler effect (p. **131**)
echo (p. **125**)
frequency (p. **117**)
fundamental frequency (p. **130**)
hertz (p. **117**)
infrasonic (p. **122**)
longitudinal wave (p. **119**)
period (p. **117**)
pitch (p. **122**)
reflection (p. **124**)
refraction (p. **124**)
resonance (p. **128**)
reverberation (p. **125**)
shock wave (p. **131**)
sonic boom (p. **131**)
standing waves (p. **129**)
transverse wave (p. **119**)
ultrasonic (p. **122**)
vibration (p. **116**)
waves (p. **119**)
wavelength (p. **121**)

APPLYING THE CONCEPTS

1. A back and forth motion that repeats itself is a
   a. spring.
   b. vibration.
   c. wave.
   d. pulse.
2. The number of vibrations that occur in 1s is called
   a. a period.
   b. frequency.
   c. amplitude.
   d. sinusoidal.

3. Frequency is measured in units of
   a. time.
   b. cycles.
   c. hertz.
   d. avis.

4. The maximum displacement from rest to the crest or from rest to the trough of a wave is called
   a. wavelength.
   b. period.
   c. equilibrium position.
   d. amplitude.

5. A wave with motion perpendicular to the direction that the wave is moving is classified as a
   a. longitudinal wave.
   b. transverse wave.
   c. water wave.
   d. compression wave.

6. Your brain interprets a frequency as a sound with a certain
   a. speed.
   b. loudness.
   c. pitch.
   d. harmonic.

7. Sound waves with frequencies greater than 20,000 Hz are
   a. infrasonic waves.
   b. supersonic waves.
   c. ultrasonic waves.
   d. impossible.

8. Generally, sounds travel faster in
   a. solids.
   b. liquids.
   c. gases.
   d. vacuums.

9. Sounds travel faster in
   a. warmer air.
   b. cooler air.
   c. Temperature does not influence the speed of sound.
   d. a vacuum.

10. The bending of a wave front between boundaries is
    a. reflection.
    b. reverberation.
    c. refraction.
    d. dispersion.

11. A reflected sound that reaches the ear within 0.1s after the original sound results in
    a. an echo.
    b. reverberation.
    c. refraction.
    d. confusion.

12. The wave front of a refracted sound bends toward
    a. warmer air.
    b. cooler air.
    c. the sky, no matter what the air temperature.
    d. the surface of Earth, no matter what the air temperature.

13. Two in-phase sound waves with the same amplitude and frequency arrive at the same place at the same time, resulting in
    a. higher frequency.
    b. refraction.
    c. a new sound wave with greater amplitude.
    d. reflection.

14. Two out-of-phase sound waves with the same amplitude and frequency arrive at the same place at the same time, resulting in
    a. a beat.
    b. cancellation of the two sound waves.
    c. a lower frequency.
    d. the bouncing of one wave.

15. Two sound waves of equal amplitude with slightly different frequencies will result in
    a. an echo.
    b. the Doppler effect.
    c. alternation of loudness of sound known as beats.
    d. two separate sounds.

16. Two sound waves of unequal amplitudes with different frequencies will result in
    a. an echo.
    b. the Doppler effect.
    c. alternation of loudness known as beats.
    d. two separate sounds.

17. The energy of a sound wave is proportional to the rate of energy transferred to an area perpendicular to the waves, which is called the sound
    a. intensity.
    b. loudness.
    c. amplitude.
    d. decibel.

18. A decibel noise level of 40 would be most likely found
    a. during a calm day in the forest.
    b. on a typical day in the library.
    c. in heavy street traffic.
    d. standing next to a pneumatic drill.

19. A resonant condition occurs when
    a. an external force matches a natural frequency.
    b. a beat is heard.
    c. two out-of-phase waves have the same frequency.
    d. a pure tone is created.

20. The fundamental frequency of a string is the
    a. shortest wavelength harmonic possible on the string.
    b. longest standing wave that can fit on the string.
    c. highest frequency possible on the string.
    d. shortest wavelength that can fit on the string.

21. The fundamental frequency on a vibrating string is what part of a wavelength?
    a. 1/4
    b. 1/2
    c. 1
    d. 2

22. Higher resonant frequencies that occur at the same time as the fundamental frequency are called
    a. standing waves.
    b. confined waves.
    c. oscillations.
    d. overtones.

23. A moving source of sound or a moving observer experiences the apparent shift in frequency called
    a. fundamental frequency.
    b. Doppler effect.
    c. wave front effect.
    d. shock waves.

24. Does the Doppler effect occur when the observer is moving and the source of sound is stationary?
    a. Yes, the effect is the same.
    b. No, the source must be moving.
    c. Yes, but the change of pitch effects is reversed in this case.

25. A rocket traveling at three times the speed of sound is traveling at
    a. sonic speed.
    b. Mach speed.
    c. Mach 3.
    d. supersonic speed.

26. A longitudinal mechanical wave causes particles of a material to move
    a. back and forth in the same direction the wave is moving.
    b. perpendicular to the direction the wave is moving.
    c. in a circular motion in the direction the wave is moving.
    d. in a circular motion opposite the direction the wave is moving.

27. A transverse mechanical wave causes particles of a material to move
    a. back and forth in the same direction the wave is moving.
    b. perpendicular to the direction the wave is moving.
    c. in a circular motion in the direction the wave is moving.
    d. in a circular motion opposite the direction the wave is moving.

28. Transverse mechanical waves will move only through
    a. solids.
    b. liquids.
    c. gases.
    d. All of the above are correct.

29. Longitudinal mechanical waves will move only through
    a. solids.
    b. liquids.
    c. gases.
    d. All of the above are correct.

30. A pulse of jammed-together molecules that quickly moves away from a vibrating object
    a. is called a condensation.
    b. causes an increased air pressure when it reaches an object.
    c. has a greater density than the surrounding air.
    d. All of the above are correct.

31. The characteristic of a wave that is responsible for what you interpret as pitch is the wave
    a. amplitude.
    b. shape.
    c. frequency.
    d. height.

32. Sound waves travel faster in
    a. solids as compared to liquids.
    b. liquids as compared to gases.
    c. warm air as compared to cooler air.
    d. All of the above are correct.

33. The difference between an echo and a reverberation is
    a. an echo is a reflected sound; reverberation is not.
    b. the time interval between the original sound and the reflected sound.
    c. the amplitude of an echo is much greater.
    d. reverberation comes from acoustical speakers; echoes come from cliffs and walls.

34. Sound interference is necessary to produce the phenomenon known as
    a. resonance.
    b. decibels.
    c. beats.
    d. reverberation.

35. The fundamental frequency of a standing wave on a string has
    a. one node and one antinode.
    b. one node and two antinodes.
    c. two nodes and one antinode.
    d. two nodes and two antinodes.

36. An observer on the ground will hear a sonic boom from an airplane traveling faster than the speed of sound
    a. only when the plane breaks the sound barrier.
    b. as the plane is approaching.
    c. when the plane is directly overhead.
    d. after the plane has passed by.

37. What comment is true about the statement that "the human ear hears sounds originating from vibrating objects with a frequency between 20 and 20,000 Hz"?
    a. This is true only at room temperature.
    b. About 95 percent hear in this range, while some hear outside the average limits.
    c. This varies, with females hearing frequencies above 20,000 Hz.
    d. Very few people hear this whole range, which decreases with age.

38. A sound wave that moves through the air is
    a. actually a tiny sound that the ear magnifies.
    b. pulses of increased and decreased air pressure.
    c. a transverse wave that carries information about a sound.
    d. a combination of longitudinal and transverse wave patterns.

39. During a track and field meet, the time difference between seeing the smoke from a starter's gun and hearing the bang would be less
    a. on a warmer day.
    b. on a cooler day.
    c. if a more powerful shell is used.
    d. if a less powerful shell is used.

40. What is changed by destructive interference of a sound wave?
    a. frequency
    b. phase
    c. amplitude
    d. wavelength

41. An airplane pilot hears a slow beat from the two engines of his plane. He increases the speed of the right engine and now hears a slower beat. What should the pilot now do to eliminate the beat?
    a. Increase the speed of the left engine.
    b. Decrease the speed of the right engine.
    c. Increase the speed of both engines.
    d. Increase the speed of the right engine

42. Resonance occurs when an external force matches the
    a. interference frequency.
    b. decibel frequency.
    c. beat frequency.
    d. natural frequency.

43. The sound quality is different for the same-pitched note produced by two different musical instruments, but you are able to recognize the basic note because of the same
    a. harmonics.
    b. fundamental frequency.
    c. node positions.
    d. standing waves.

44. What happens if the source of a sound is moving toward you at a high rate of speed?
   a. The sound will be traveling faster than from a stationary source.
   b. The sound will be moving faster only in the direction of travel.
   c. You will hear a higher frequency, but people in the source will not.
   d. All observers in all directions will hear a higher frequency.
45. What happens if you are moving at a high rate of speed toward some people standing next to a stationary source of a sound? You will hear
   a. a higher frequency than the people you are approaching will hear.
   b. the same frequency as the people you are approaching will hear.
   c. the same frequency as when you and the source are not moving.
   d. a higher frequency, as will all observers in all directions.

Answers

1. b 2. b 3. c 4. d 5. b 6. c 7. c 8. a 9. a 10. c 11. b 12. b 13. c 14. b 15. c 16. d 17. a 18. b 19. a 20. b 21. b 22. d 23. b 24. a 25. c 26. a 27. b 28. a 29. d 30. d 31. c 32. d 33. b 34. c 35. c 36. d 37. d 38. b 39. a 40. c 41. d 42. d 43. b 44. c 45. a

QUESTIONS FOR THOUGHT

1. What is a wave?
2. Is it possible for a transverse wave to move through air? Explain.
3. A piano tuner hears three beats per second when a tuning fork and a note are sounded together and six beats per second after the string is tightened. What should the tuner do next, tighten or loosen the string? Explain.
4. Why do astronauts on the Moon have to communicate by radio even when close to one another?
5. What is resonance?
6. Explain why sounds travel faster in warm air than in cool air.
7. Do all frequencies of sound travel with the same velocity? Explain your answer by using one or more equations.
8. What eventually happens to a sound wave traveling through the air?
9. What gives a musical note its characteristic quality?
10. Does a supersonic aircraft make a sonic boom only when it cracks the sound barrier? Explain.

11. What is an echo?
12. Why are fundamental frequencies and overtones also called resonant frequencies?

FOR FURTHER ANALYSIS

1. How would distant music sound if the speed of sound decreased with frequency?
2. What are the significant similarities and differences between longitudinal and transverse waves? Give examples of each.
3. Sometimes it is easer to hear someone speaking in a full room than in an empty room. Explain how this could happen.
4. Describe how you can use beats to tune a musical instrument.
5. Is sound actually destroyed in destructive interference?
6. Are vibrations the source of all sounds? Discuss whether this is supported by observations or is an inference.
7. How can sound waves be waves of pressure changes if you can hear several people talking at the same time?
8. Why is it not a good idea for a large band to march in unison across a bridge?

INVITATION TO INQUIRY

Does a Noisy Noise Annoy You?

There is an old question-answer game that children played that went like this, "What annoys an oyster?"

The answer was, "A noisy noise annoys an oyster."

You could do an experiment to find out how much noise it takes to annoy an oyster, but you might have trouble maintaining live oysters, as well as measuring how annoyed they might become. So, consider using different subjects, including humans. You could modify the question to, "What noise level effects how well we concentrate?"

If you choose to do this invitation, start by determining how you are going to make the noise, how you can control different noise levels, and how you can measure the concentration level of people. A related question could be, "Does listening to music while studying help or hinder students?"

PARALLEL EXERCISES

The exercises in groups A and B cover the same concepts. Solutions to group A exercises are located in appendix E.

Group A

1. A grasshopper floating in water generates waves at a rate of three per second with a wavelength of 2 cm. (a) What is the period of these waves? (b) What is the wave velocity?
2. The upper limit for human hearing is usually considered to be 20,000 Hz. What is the corresponding wavelength if the air temperature is 20.0°C?
3. A tone with a frequency of 440 Hz is sounded at the same time as a 446 Hz tone. What is the beat frequency?
4. Medical applications of ultrasound use frequencies up to 2.00×10^7 Hz. What is the wavelength of this frequency in air?

Group B

1. A water wave has a frequency of 6 Hz and a wavelength of 3 m. (a) What is the period of these waves? (b) What is the wave velocity?
2. The lower frequency limit for human hearing is usually considered to be 20.0 Hz. What is the corresponding wavelength for this frequency if the air temperature is 20.0°C?
3. A 520 Hz tone is sounded at the same time as a 516 Hz tone. What is the beat frequency?
4. The low range of frequencies used for medical applications is about 1,000,000 Hz. What is the wavelength of this frequency in air?

CHAPTER 5 Wave Motions and Sound

5. A baseball fan is 150.0 m from the home plate. How much time elapses between the instant the fan sees the batter hit the ball and the moment the fan hears the sound?

6. An echo is heard from a building 0.500 s after you shout "hello." How many feet away is the building if the air temperature is 20.0°C?

7. A sonar signal is sent from an oceangoing ship and the signal returns from the bottom 1.75 s later. How deep is the ocean beneath the ship if the speed of sound in seawater is 1,530 m/s?

8. A sound wave in a steel rail of a railroad track has a frequency of 660 Hz and a wavelength of 9.0 m. What is the speed of sound in this rail?

9. According to the condensed steam released, a factory whistle blows 2.5 s before you hear the sound. If the air temperature is 20.0°C, how many meters are you from the whistle?

10. Compare the distance traveled in 8.00 s as a given sound moves through (a) air at 0°C and (b) a steel rail.

11. A vibrating object produces periodic waves with a wavelength of 50 cm and a frequency of 10 Hz. How fast do these waves move away from the object?

12. The distance between the center of a condensation and the center of an adjacent rarefaction is 1.50 m. If the frequency is 112.0 Hz, what is the speed of the wave front?

13. Water waves are observed to pass under a bridge at a rate of one complete wave every 4.0 s. (a) What is the period of these waves? (b) What is the frequency?

14. A sound wave with a frequency of 260 Hz moves with a velocity of 330 m/s. What is the distance from one condensation to the next?

15. The following sound waves have what velocity?
    a. Middle C, or 256 Hz and 1.34 m λ
    b. Note A, or 440.0 Hz and 78.0 cm λ
    c. A siren at 750.0 Hz and λ of 45.7 cm
    d. Note from a stereo at 2,500.0 Hz and λ of 13.7 cm

16. What is the speed of sound, in ft/s, if the air temperature is:
    a. 0.0°C?
    b. 20.0°C?
    c. 40.0°C?
    d. 80.0°C?

17. An echo is heard from a cliff 4.80 s after a rifle is fired. How many feet away is the cliff if the air temperature is 43.7°F?

18. The air temperature is 80.00°F during a thunderstorm, and thunder was timed 4.63 s after lightning was seen. How many feet away was the lightning strike?

19. If the velocity of a 440 Hz sound is 1,125 ft/s in the air and 5,020 ft/s in seawater, find the wavelength of this sound in (a) air and (b) seawater.

5. How much time will elapse between seeing and hearing an event that happens 400.0 meters from you?

6. An echo bounces from a building exactly 1.00 s after you honk your horn. How many feet away is the building if the air temperature is 20.0°C?

7. A submarine sends a sonar signal, which returns from another ship 2.250 s later. How far away is the other ship if the speed of sound in seawater is 1,530.0 m/s?

8. A student under water clicks two rocks together and makes a sound with a frequency of 600.0 Hz and a wavelength of 2.5 m. What is the speed of this underwater sound?

9. You see condensed steam expelled from a ship's whistle 2.50 s before you hear the sound. If the air temperature is 20.0°C, how many meters are you from the ship?

10. Compare the distance traveled in 6.00 s as a given sound moves through (a) water at 25.0°C and (b) seawater at 25.0°C.

11. A tuning fork vibrates 440.0 times a second, producing sound waves with a wavelength of 78.0 cm. What is the velocity of these waves?

12. The distance between the center of a condensation and the center of an adjacent rarefaction is 65.23 cm. If the frequency is 256.0 Hz, how fast are these waves moving?

13. A warning buoy is observed to rise every 5.0 s as crests of waves pass by it. (a) What is the period of these waves? (b) What is the frequency?

14. Sound from the siren of an emergency vehicle has a frequency of 750.0 Hz and moves with a velocity of 343.0 m/s. What is the distance from one condensation to the next?

15. The following sound waves have what velocity?
    a. 20.0 Hz, λ of 17.2 m
    b. 200.0 Hz, λ of 1.72 m
    c. 2,000.0 Hz, λ of 17.2 cm
    d. 20,000.0 Hz, λ of 1.72 cm

16. How much time is required for a sound to travel 1 mile (5,280.0 ft) if the air temperature is:
    a. 0.0°C?
    b. 20.0°C?
    c. 40.0°C?
    d. 80.0°C?

17. A ship at sea sounds a whistle blast, and an echo returns from the coastal land 10.0 s later. How many km is it to the coastal land if the air temperature is 10.0°C?

18. How many seconds will elapse between seeing lightning and hearing the thunder if the lightning strikes 1 mi (5,280 ft) away and the air temperature is 90.0°F?

19. A 600.0 Hz sound has a velocity of 1,087.0 ft/s in the air and a velocity of 4,920.0 ft/s in water. Find the wavelength of this sound in (a) the air and (b) the water.

6
Electricity

A thunderstorm produces an interesting display of electrical discharge. Each bolt can carry over 150,000 amperes of current with a voltage of 100 million volts.

CORE **CONCEPT**
Electric and magnetic fields interact and can produce forces.

Chapters 2–5 have been concerned with *mechanical* concepts, explanations of the motion of objects that exert forces on one another. These concepts were used to explain straight-line motion, the motion of free fall, and the circular motion of objects on Earth as well as the circular motion of planets and satellites. The mechanical concepts were based on Newton's laws of motion and are sometimes referred to as Newtonian physics. The mechanical explanations were then extended into the submicroscopic world of matter through the kinetic molecular theory. The objects of motion were now particles, molecules that exert force on one another, and concepts associated with heat were interpreted as the motion of these particles. In a further extension of Newtonian concepts, mechanical explanations were given for concepts associated with sound, a mechanical disturbance that follows the laws of motion as it moves through the molecules of matter.

You might wonder, as did the scientists of the 1800s, if mechanical interpretations would also explain other natural phenomena such as electricity, chemical reactions, and light. A mechanical model would be very attractive because it already explained so many other facts of nature, and scientists have always looked for basic, unifying theories. Mechanical interpretations were tried, as electricity was considered a moving fluid, and light was considered a mechanical wave moving through a material fluid. There were many unsolved puzzles with such a model, and gradually it was recognized that electricity, light, and chemical reactions could not be explained by mechanical interpretations. Gradually, the point of view changed from a study of particles to a study of the properties of the space around the particles. In this chapter, you will learn about electric charge in terms of the space around particles. This model of electric charge, called the *field model*, will be used to develop concepts about electric current, the electric circuit, and electrical work and power. A relationship between electricity and the fascinating topic of magnetism is discussed next, including what magnetism is and how it is produced. The relationship is then used to explain the mechanical production of electricity (Figure 6.1), how electricity is measured, and how electricity is used in everyday technological applications.

CONCEPTS OF ELECTRICITY

You are familiar with the use of electricity in many electrical devices such as lights, toasters, radios, and calculators. You are also aware that electricity is used for transportation and for heating and cooling places where you work and live. Many people accept electrical devices as part of their surroundings, with only a hazy notion of how they work. To many people, electricity seems to be magical. Electricity is not magical, and it can be understood, just as we understand any other natural phenomenon. There are theories that explain observations, quantities that can be measured, and relationships between these quantities, or laws, that lead to understanding. All of the observations, measurements, and laws begin with an understanding of *electric charge.*

ELECTRON THEORY OF CHARGE

It was a big mystery for thousands of years. No one could figure out why a rubbed piece of amber, which is fossilized tree resin, would attract small pieces of paper (papyrus), thread, and hair. This unexplained attraction was called the "amber effect." Then about one hundred years ago, J. J. Thomson (1856–1940) found the answer while experimenting with electric currents. From these experiments, Thomson was able to conclude that negatively charged particles were present in all matter and in fact might be the stuff of which matter is made. The amber effect was traced to the movement of these particles, so they were called *electrons* after the Greek word for

amber. The word *electricity* is also based on the Greek word for amber.

Today, we understand that the basic unit of matter is the *atom*, which is made up of electrons and other particles such as *protons* and *neutrons*. The atom is considered to have a dense center part called a *nucleus* that contains the closely situated protons and neutrons. The electrons move around the nucleus at some relatively greater distance (Figure 6.2). Details on the nature of protons, neutrons, electrons, and models of how the atom is constructed will be considered in chapter 8. For understanding electricity, you need only consider the protons in the nucleus, the electrons that move around the nucleus, and the fact that electrons can be moved from an atom and caused to move to or from one object to another. Basically, the electrical, light, and chemical phenomena involve the *electrons* and not the more massive nucleus. The massive nuclei remain in a relatively fixed position in a solid, but some of the electrons can move about from atom to atom.

Electric Charge

Electrons and protons have a property called electric charge. Electrons have a *negative electric charge* and protons have a *positive electric charge*. The negative or positive description simply means that these two properties are opposite; it does not mean that one is better than the other. Charge is as fundamental to these subatomic particles as gravity is to masses. This means that you cannot separate gravity from a mass, and you cannot separate charge from an electron or a proton.

FIGURE 6.1 The importance of electrical power seems obvious in a modern industrial society. What is not so obvious is the role of electricity in magnetism, light, chemical change, and as the very basis for the structure of matter. All matter, in fact, is electrical in nature, as you will see.

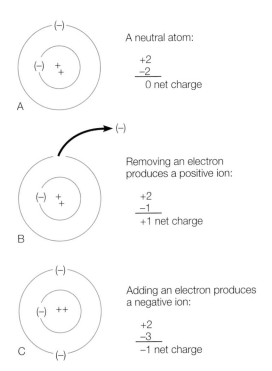

A neutral atom:

$$\begin{array}{r} +2 \\ -2 \\ \hline 0 \text{ net charge} \end{array}$$

A

Removing an electron produces a positive ion:

$$\begin{array}{r} +2 \\ -1 \\ \hline +1 \text{ net charge} \end{array}$$

B

Adding an electron produces a negative ion:

$$\begin{array}{r} +2 \\ -3 \\ \hline -1 \text{ net charge} \end{array}$$

C

FIGURE 6.3 (*A*) A neutral atom has no net charge because the numbers of electrons and protons are balanced. (*B*) Removing an electron produces a net positive charge; the charged atom is called a positive ion. (*C*) The addition of an electron produces a net negative charge and a negative ion.

Electric charges interact to produce what is called the *electrical force*. Like charges produce a repulsive electrical force as positive repels positive and negative repels negative. Unlike charges produce an attractive electrical force as positive and negative charges attract each other. You can remember how this happens with the simple rule of "*like charges repel and unlike charges attract.*"

Ordinary atoms are usually neutral because there is a balance between the number of positively charged protons and the number of negatively charged electrons. A number of different physical and chemical interactions can result in an atom gaining or losing electrons. In either case, the atom is said to be *ionized,* and *ions* are produced as a result. An atom that is ionized by losing electrons results in a *positive ion* because it has a net positive charge. An atom that is ionized by gaining electrons results in a *negative ion* because it has a net negative charge (Figure 6.3).

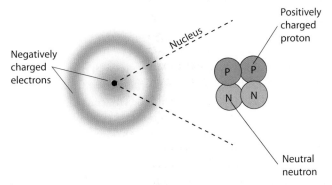

FIGURE 6.2 A very highly simplified model of an atom has most of the mass in a small, dense center called the *nucleus*. The nucleus has positively charged protons and neutral neutrons. Negatively charged electrons move around the nucleus at a much greater distance than is suggested by this simplified model. Ordinary atoms are neutral because there is balance between the number of positively charged protons and negatively charged electrons.

Static Electricity

Electrons can be moved from atom to atom to create ions. They can also be moved from one object to another by friction and by other means that will be discussed soon. Since electrons are negatively charged, an object that acquires an excess of electrons becomes a negatively charged body. The loss of electrons by another body results in a deficiency of electrons, which results in a positively charged object. Thus, *electric charges on objects result from the gain or loss of electrons.* Because the electric charge is confined to an object and is not moving, it is called an **electrostatic charge.** You probably call this charge *static electricity.* Static electricity is an accumulated electric charge at rest, that is, one that is not moving. When you comb your hair with a hard rubber comb, the comb becomes negatively charged because electrons are transferred *from* your hair to the comb. Your hair becomes positively charged with a charge equal in magnitude to the charge gained by the comb (Figure 6.4). Both the negative charge on the comb from an excess of electrons and the positive charge on your hair from a deficiency of electrons are charges that are momentarily at rest, so they are electrostatic charges.

Once charged by friction, objects such as the rubber comb soon return to a neutral, or balanced, state by the movement of electrons. This happens more quickly on a humid day because water vapor assists with the movement of electrons to or from charged objects. Thus, static electricity is more noticeable on dry days than on humid ones.

An object can become electrostatically charged (1) by *friction,* which transfers electrons from one object to another, (2) by

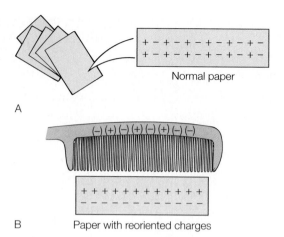

FIGURE 6.5 Charging by induction. The comb has become charged by friction, acquiring an excess of electrons. The paper (*A*) normally has a random distribution of (+) and (−) charges. (*B*) When the charged comb is held close to the paper, there is a reorientation of charges because of the repulsion of like charges. This leaves a net positive charge on the side close to the comb, and since unlike charges attract, the paper is attracted to the comb.

FIGURE 6.4 Arbitrary numbers of protons (+) and electrons (−) on a comb and in hair (*A*) before and (*B*) after combing. Combing transfers electrons from the hair to the comb by friction, resulting in a negative charge on the comb and a positive charge on the hair.

contact with another charged body, which results in the transfer of electrons, or (3) by *induction*. Induction produces a charge by a redistribution of charges in a material. When you comb your hair, for example, the comb removes electrons from your hair and acquires a negative charge. When the negatively charged comb is held near small pieces of paper, it repels some electrons in the paper to the opposite side of the paper. This leaves the side of the paper closest to the comb with a positive charge, and there is an attraction between the pieces of paper and the comb, since unlike charges attract. Note that no transfer of electrons takes place in induction; the attraction results from a reorientation of the charges in the paper (Figure 6.5). Note also that charge is transferred in all three examples; it is not created or destroyed.

 CONCEPTS *Applied*

Static Charge

1. This application works best when the humidity is low. Obtain a plastic drinking straw and several sheets of light tissue paper. Tear one of the sheets into small pieces and place them on your desk. Wrap a sheet of the tissue around the straw and push and pull the straw back and forth about ten times while holding it in the tissue. Touch the end of the straw to one of the pieces of tissue on your desk to see if it is attracted.

If it is, touch the attracted piece to another to see if it, too, is attracted. Depending on the humidity, you might be able to attract a long chain of pieces of tissue.

2. Suspend the straw from the edge of your desk with a length of cellophane tape. Try rubbing a plastic ballpoint pen, a black plastic comb, and other objects with a cotton cloth, flannel, fur, and other materials. Hold each rubbed object close to the straw and observe what happens.

3. Make a list of materials that seem to acquire a static charge and those that do not. See how many generalizations you can make about static electricity and materials. Describe any evidence you observed that two kinds of electric charge exist.

Electrical Conductors and Insulators

When you slide across a car seat or scuff your shoes across a carpet, you are rubbing some electrons from the materials and acquiring an excess of negative charges. Because the electric charge is confined to you and is not moving, it is an electrostatic charge. The electrostatic charge is produced by friction between two surfaces and will remain until the electrons can move away because of their mutual repulsion. This usually happens when you reach for a metal doorknob, and you know when it happens because the electron movement makes a spark. Materials like the metal of a doorknob are good **electrical conductors** because they have electrons that are free to move throughout the metal. If you touch plastic or wood, however, you will not feel a shock. Materials like plastic and wood do not have electrons that are free to move throughout the material, and they are called *electrical nonconductors*. Nonconductors are also called *electrical insulators* (Table 6.1). Electrons do not move easily through an insulator, but electrons can be added or removed, and the charge

TABLE 6.1

Electrical conductors and insulators

Conductors	Insulators
Silver	Rubber
Copper	Glass
Gold	Carbon (diamond)
Aluminum	Plastics
Carbon (graphite)	Wood
Tungsten	
Iron	
Lead	
Nichrome	

tends to remain. In fact, your body is a poor conductor, which is why you become charged by friction in the first place.

Materials vary in their ability to conduct charges, and this ability is determined by how tightly or loosely the electrons are held to the nucleus. Metals have millions of free electrons that can take part in the conduction of an electric charge. Materials such as rubber, glass, and plastics hold tightly to their electrons and are good insulators. Thus, metal wires are used to conduct an electric current from one place to another, and rubber, glass, and plastics are used as insulators to keep the current from going elsewhere.

There is a third class of materials, such as silicon and germanium, that sometimes conduct and sometimes insulate, depending on the conditions and how pure they are. These materials are called *semiconductors,* and their special properties make possible a number of technological devices such as the electrostatic copying machine, solar cells, and so forth.

MEASURING ELECTRICAL CHARGES

As you might have experienced, sometimes you receive a slight shock after walking across a carpet, and sometimes you are really zapped. You receive a greater shock when you have accumulated a greater electric charge. Since there is less electric charge at one time and more at another, it should be evident that charge occurs in different amounts, and these amounts can be measured. The size of an electric charge is identified with the number of electrons that have been transferred onto or away from an object. The quantity of such a charge (q) is measured in a unit called a **coulomb** (C). A coulomb unit is equivalent to the charge resulting from the transfer of 6.24×10^{18} of the charge carried by particles such as the electron. The coulomb is a metric unit of measure like the meter or second.

The coulomb is a *unit* of electric charge that is used with other metric units such as meters for distance and newtons for force. Thus, a quantity of charge (q) is described in units of coulomb (C). This is just like the process of a quantity of mass (m) being described in units of kilogram (kg). The concepts of charge and coulomb may seem less understandable than the concepts of mass and kilogram, since you cannot see charge or how it is measured. But charge does exist and it can be measured, so you can

understand both the concept and the unit by working with them. Consider, for example, that an object has a net electric charge (q) because it has an unbalanced number (n) of electrons (e^-) and protons (p^+). The net charge on you after walking across a carpet depends on how many electrons you rubbed from the carpet. The net charge in this case would be the excess of electrons, or

$$\text{quantity of charge} = (\text{number of electrons})(\text{electron charge})$$

or

$$q = ne$$

equation 6.1

Since 1.00 coulomb is equivalent to the transfer of 6.24×10^{18} particles such as the electron, the charge on one electron must be

$$e = \frac{q}{n}$$

where q is 1.00 C, and n is 6.24×10^{18} electrons,

$$e = \frac{1.00 \text{ coulomb}}{6.24 \times 10^{18} \text{ electron}}$$

$$= 1.60 \times 10^{-19} \frac{\text{coulomb}}{\text{electron}}$$

This charge, 1.60×10^{-19} coulomb, is the *smallest* common charge known (more exactly $1.6021892 \times 10^{-19}$ C). It is the **fundamental charge** of the electron. Every electron has a charge of -1.60×10^{-19} C, and every proton has a charge of $+1.60 \times 10^{-19}$ C. To accumulate a negative charge of 1 C, you would need to accumulate more than 6 billion billion electrons. All charged objects have multiples of the fundamental charge, so charge is said to be quantized. An object might have a charge on the order of about 10^{-8} to 10^{-6} C.

EXAMPLE 6.1

Combing your hair on a day with low humidity results in a comb with a negative charge on the order of 1.00×10^{-8} coulomb. How many electrons were transferred from your hair to the comb?

SOLUTION

The relationship between the quantity of charge on an object (q), the number of electrons (n), and the fundamental charge on an electron (e^-) is found in equation 6.1, $q = ne$.

$$q = 1.00 \times 10^{-8} \text{ C} \qquad q = ne \quad \therefore \quad n = \frac{q}{e}$$

$$e = 1.60 \times 10^{-19} \frac{\text{C}}{\text{e}} \qquad n = \frac{1.00 \times 10^{-8} \text{ C}}{1.60 \times 10^{-19} \frac{\text{C}}{\text{e}}}$$

$$n = ?$$

$$= \frac{1.00 \times 10^{-8}}{1.60 \times 10^{-19}} \cancel{\text{C}} \times \frac{\text{e}}{\cancel{\text{C}}}$$

$$= \boxed{6.25 \times 10^{-10} \text{ e}}$$

Thus, the comb acquired an excess of approximately 62.5 billion electrons. (Note that the convention in scientific notation is to express an answer with one digit to the left of the decimal. See appendix A for further information on scientific notation.)

ELECTROSTATIC FORCES

Recall that two objects with like charges, (−) and (−) or (+) and (+), produce a repulsive force, and two objects with unlike charges, (−) and (+), produce an attractive force. The size of either force depends on the amount of charge of each object and on the distance between the objects. The relationship is known as **Coulomb's law,** which is,

$$F = k \frac{q_1 q_2}{d^2}$$

equation 6.2

where k has the value of 9.00×10^9 newton-meters^2/coulomb^2 (9.00×10^9 N·m^2/C^2).

The force between the two charged objects is repulsive if q_1 and q_2 are the same charge and attractive if they are different (like charges repel, unlike charges attract). Whether the force is attractive or repulsive, you know that both objects feel equal forces, as described by Newton's third law of motion. In addition, the strength of this force decreases if the distance between the objects is increased (a doubling of the distance reduces the force to ¼ the original value).

EXAMPLE 6.2

Electrons carry a negative electric charge and revolve about the nucleus of the atom, which carries a positive electric charge from the proton. The electron is held in orbit by the force of electrical attraction at a typical distance of 1.00×10^{-10} m. What is the force of electrical attraction between an electron and proton?

SOLUTION

The fundamental charge of an electron (e^-) is 1.60×10^{-19} C, and the fundamental charge of the proton (p^+) is 1.60×10^{-19} C. The distance is given, and the force of electrical attraction can be found from equation 6.2:

$$q_1 = 1.60 \times 10^{-19} \text{ C}$$
$$q_2 = 1.60 \times 10^{-19} \text{ C}$$
$$d = 1.00 \times 10^{-10} \text{ m}$$
$$k = 9.00 \times 10^9 \text{ N·m}^2/\text{C}^2$$
$$F = ?$$

$$F = k \frac{q_1 q_2}{d^2}$$

$$= \frac{\left(9.00 \times 10^9 \frac{\text{N·m}^2}{\text{C}^2}\right)(1.60 \times 10^{-19} \text{ C})(1.60 \times 10^{-19} \text{ C})}{(1.00 \times 10^{-10} \text{ m})^2}$$

$$= \frac{(9.00 \times 10^9)(1.60 \times 10^{-19})(1.60 \times 10^{-19})}{1.00 \times 10^{-20}} \frac{\left(\frac{\text{N·m}^2}{\text{C}^2}\right)(\text{C}^2)}{\text{m}^2}$$

$$= \frac{2.30 \times 10^{-28}}{1.00 \times 10^{-20}} \frac{\text{N·m}^2}{\text{C}^2} \times \frac{\text{C}^2}{1} \times \frac{1}{\text{m}^2}$$

$$= \boxed{2.30 \times 10^{-8} \text{N}}$$

The electrical force of attraction between the electron and proton is 2.30×10^{-8} newton.

FORCE FIELDS

Does it seem odd to you that gravitational forces and electrical forces can act on objects that are not touching? How can gravitational forces act through the vast empty space between Earth and the Sun? How can electrical forces act through a distance to pull pieces of paper to your charged comb? Such questions have bothered people since the early discovery of small, light objects being attracted to rubbed amber. There was no mental model of how such a force could act through a distance without touching. The idea of "invisible fluids" was an early attempt to develop a mental model that would help people visualize how a force could act over a distance without physical contact. Then Newton developed the law of universal gravitation, which correctly predicted the magnitude of gravitational forces acting through space. Coulomb's law of electrical forces had similar success in describing and predicting electrostatic forces acting through space. "Invisible fluids" were no longer needed to explain what was happening, because the two laws seemed to explain the results of such actions. But it was still difficult to visualize what was happening physically when forces acted through a distance, and there were a few problems with the concept of action at a distance. Not all observations were explained by the model.

The work of Michael Faraday (1791–1867) and James Maxwell (1831–1879) in the early 1800s finally provided a new mental model for interaction at a distance. This new model did *not* consider the force that one object exerts on another one through a distance. Instead, it considered *the condition of space* around an object. The condition of space around an electric charge is considered to be changed by the presence of the charge. The charge produces a **force field** in the space around it. Since this force field is produced by an electrical charge, it is called an **electric field.** Imagine a second electric charge, called a *test charge,* that is far enough away from the electric charge that forces are negligible. As you move the test charge closer and closer, it will experience an increasing force as it enters the electric field. The test charge is assumed not to change the field that it is entering and can be used to identify the electric field that spreads out and around the space of an electric charge.

All electric charges are considered to be surrounded by an electric field. All *masses* are considered to be surrounded by a *gravitational field.* Earth, for example, is considered to change the condition of space around it because of its mass. A spaceship far, far from Earth does not experience a measurable force. But as it approaches Earth, it moves farther into Earth's gravitational field and eventually experiences a measurable force. Likewise, a magnet creates a *magnetic field* in the space around it. You can visualize a magnetic field by moving a magnetic compass needle around a bar magnet. Far from the bar magnet the compass needle does not respond. Moving it closer to the bar magnet, you can see where the magnetic field begins. Another way to visualize a magnetic field is to place a sheet of paper over a bar magnet, then sprinkle iron filings on the paper. The filings will clearly identify the presence of the magnetic field.

Another way to visualize a field is to make a map of the field. Consider a small positive test charge that is brought into an electric field. A *positive* test charge is always used by convention. As shown in Figure 6.6, a positive test charge is brought

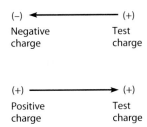

FIGURE 6.6 *A positive test charge* is used by convention to identify the properties of an electric field. The arrow points in the direction of the force that the test charge would experience.

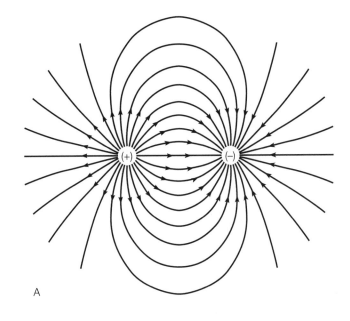

A

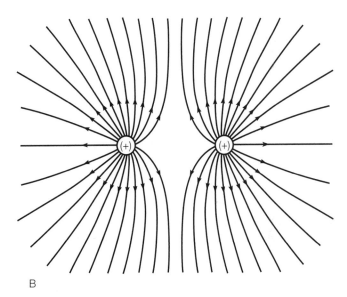

B

FIGURE 6.7 Lines of force diagrams for (*A*) a negative charge and (*B*) a positive charge when the charges are the same size as the test charge.

near a negative charge and a positive charge. The arrow points in the direction of the force that the *test charge experiences.* Thus, when brought near a negative charge, the test charge is attracted toward the unlike charge, and the arrow points that way. When brought near a positive charge, the test charge is repelled, so the arrow points away from the positive charge.

An electric field is represented by drawing *lines of force* or *electric field lines* that show the direction of the field. The arrows in Figure 6.7 show field lines that could extend outward forever from isolated charges, since there is always some force on a distant test charge. The field lines between pairs of charges in Figure 6.7 show curved field lines that originate on positive charges and end on negative charges. By convention, the field lines are closer together where the field is stronger and farther apart where the field is weaker.

The field concept explains some observations that were not explained with the Newtonian concept of action at a distance. Suppose, for example, that a charge produces an electric field. This field is not instantaneously created all around the charge, but it is seen to build up and spread into space. If the charge is suddenly neutralized, the field that it created continues to spread outward and then appears to collapse back at some speed, even though the source of the field no longer exists. Consider an example with the gravitational field of the Sun. If the mass of the Sun were to instantaneously disappear, would Earth notice this instantaneously? Or would the gravitational field of the Sun appear to collapse at some speed, say the speed of light, to be noticed by Earth some eight minutes later? The Newtonian concept of action at a distance did not consider any properties of space, so according to this concept, the gravitational force from the Sun would disappear instantly. The field concept, however, explains that the disappearance would be noticed after some period of time, about eight minutes. This time delay agrees with similar observations of objects interacting with fields, so the field concept is more useful than a mysterious action-at-a-distance concept, as you will see.

Actually there are three models for explaining how gravitational, electrical, and magnetic forces operate at a distance. (1) The *action-at-a-distance model* recognizes that masses are attracted gravitationally and that electric charges and magnetic poles attract and repel each other through space, but it gives no further explanation; (2) the *field model* considers a field to be a condition of space around a mass, electric charge, or magnet, and the properties of fields are described by field lines; and (3) the *field-particle model* is a complex and highly mathematical explanation of attractive and repulsive forces as

the rapid emission and absorption of subatomic particles. This model explains electrical and magnetic forces as the exchange of *virtual photons,* gravitational forces as the exchange of *gravitons,* and strong nuclear forces as the exchange of *gluons.*

ELECTRIC POTENTIAL

Recall from chapter 3 that work is accomplished as you move an object to a higher location on Earth, say by moving a book from the first shelf of a bookcase to a higher shelf. By virtue of its position, the book now has gravitational potential energy that can be measured by *mgh* (the force of the book's weight × distance), joules of gravitational potential energy. Using the field model, you could say that this work was accomplished against the gravitational

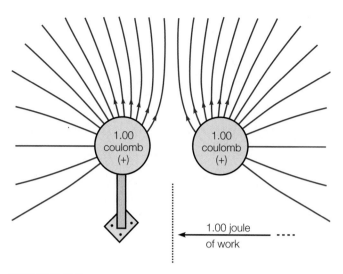

FIGURE 6.8 Electric potential results from moving a positive coulomb of charge into the electric field of a second positive coulomb of charge. When 1.00 joule of work is done in moving 1.00 coulomb of charge, 1.00 volt of potential results. A volt is a joule/coulomb.

field of Earth. Likewise, an electric charge has an electric field surrounding it, and work must be done to move a second charge into or out of this field. Bringing a like charged particle *into* the field of another charged particle will require work, since like charges repel, and separating two unlike charges will also require work, since unlike charges attract. In either case, the *electric potential energy* is changed, just as the gravitational potential energy is changed by moving a mass in Earth's gravitational field.

One useful way to measure electric potential energy is to consider the *potential difference* that occurs when a certain amount of work is used to move a certain quantity of charge. For example, suppose there is a firmly anchored and insulated metal sphere that has a positive charge (Figure 6.8). The sphere will have a positive electric field in the space around it. Suppose also that you have a second sphere that has exactly 1.00 coulomb of positive charge. You begin moving the coulomb of positive charge toward the anchored sphere. As you enter the electric field, you will have to push harder and harder to overcome the increasing repulsion. If you stop moving when you have done exactly 1.00 joule of work, the repulsion will *do* one joule of work if you now release the sphere. The sphere has potential energy in the same way that a compressed spring has potential energy. In electrical matters, the *potential difference that is created by doing 1.00 joule of work in moving 1.00 coulomb of charge is defined to be 1.00 volt.* The **volt** (*V*) is a measure of potential difference between two points, or

$$\text{electric potential difference} = \frac{\text{work to create potential}}{\text{charge moved}}$$

$$V = \frac{W}{q}$$

equation 6.3

In units,

$$1.00 \text{ volt (V)} = \frac{1.00 \text{ joule (J)}}{1.00 \text{ coulomb (C)}}$$

The potential difference can be measured by the *work that is done to move the charge* or by the *work that the charge can do* because of its position in the field. This is perfectly analogous to the work that must be done to give an object gravitational potential energy or to the work that the object can potentially do because of its new position. Thus, when a 12 volt battery is charging, 12.0 joules of work are done to transfer 1.00 coulomb of charge from an outside source against the electric field of the battery terminal. When the 12 volt battery is used, it does 12.0 joules of work for each coulomb of charge transferred from one terminal of the battery through the electrical system and back to the other terminal.

ELECTRIC CURRENT

So far, we have considered electric charges that have been instantaneously moved by friction but then generally stayed in one place. Experiments with static electricity played a major role in the development of the understanding of electricity by identifying charge, the attractive and repulsive forces between charges, and the field concept. Now, consider the flowing or moving of charge, an *electric current* (*I*). **Electric current** means a flow of charge in the same way that "water current" means a flow of water. Since the word "current" *means* flow, you are being redundant if you speak of "flow of current." It is the *charge* that flows, and the current is defined as the flow of charge.

THE ELECTRIC CIRCUIT

When you slide across a car seat, you are acquiring electrons on your body by friction. Through friction, you did *work* on the electrons as you removed them from the seat covering. You now have a net negative charge from the imbalance of electrons, which tend to remain on you because you are a poor conductor. But the electrons are now closer than they want to be, within a repulsive electric field, and there is an electrical potential difference between you and some uncharged object, say a metal door handle. When you touch the handle, the electrons will flow, creating a momentary current in the form of a spark, which lasts only until the charge on you is neutralized.

In order to keep an electric current going, you must maintain the separation of charges and therefore maintain the electric field (or potential difference), which can push the charges through a conductor. This might be possible if you could somehow continuously slide across the car seat, but this would be a hit-and-miss way of maintaining a separation of charges and would probably result in a series of sparks rather than a continuous current. This is how electrostatic machines work.

A useful analogy for understanding the requirements for a sustained electric current is the decorative waterwheel device (Figure 6.9). Water in the upper reservoir has a greater gravitational potential energy than water in the lower reservoir. As water flows from the upper reservoir, it can do work in turning the waterwheel, but it can continue to do this only as long as the pump does the work to maintain the potential difference between the two reservoirs. This "water circuit" will do work in turning the waterwheel as long as the pump returns the water to a higher potential continuously as the water flows back to the lower potential.

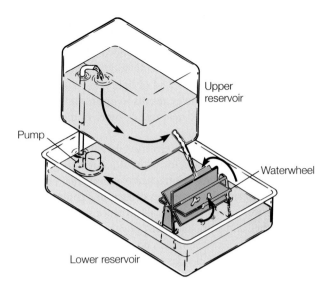

FIGURE 6.9 The falling water can do work in turning the waterwheel only as long as the pump maintains the potential difference between the upper and lower reservoirs.

So, by a water circuit analogy, a steady electric current is maintained by pumping charges to a higher potential, and the charges do work as they move back to a lower potential. The higher electric potential energy is analogous to the gravitational potential energy in the waterwheel example (Figure 6.9). An **electric circuit** contains some device, such as a battery or electric generator, that acts as a source of energy as it gives charges a higher potential against an electric field. The charges do work in another part of the circuit as they light bulbs, run motors, or provide heat. The charges flow through connecting wires to make a continuous path. An electric switch is a means of interrupting or completing this continuous path.

The electrical potential difference between the two connecting wires shown in Figure 6.10 is one factor in the work done *by* the device that creates a higher electrical potential (battery, for example) and the work done *in* some device (lamp, for example). Disregarding any losses due to the very small work done in moving electrons through a wire, the work done in both places would be the same. Recall that work done per unit of charge is

joules/coulomb, or volts (equation 6.3). The source of the electrical potential difference is therefore referred to as a *voltage source*. The device where the charges do their work causes a *voltage drop*. Electrical potential difference is measured in volts, so the term *voltage* is often used for it. Household circuits usually have a difference of potential of 120 or 240 volts. A voltage of 120 volts means that each coulomb of charge that moves through the circuit can do 120 joules of work in some electrical device.

Voltage describes the potential difference, in joules/coulomb, between two places in an electric circuit. By way of analogy to pressure on water in a circuit of water pipes, this potential difference is sometimes called an "electrical force" or "electromotive force" (emf). Note that in electrical matters, however, the potential difference is the *source* of a force rather than being a force such as water under pressure. Nonetheless, just as you can have a small water pipe and a large water pipe under the same pressure, the two pipes would have a different rate of water flow in gallons per minute. Electric current (I) is the rate at which charge (q) flows through a cross section of a conductor in a unit of time (t), or

$$\text{electric current} = \frac{\text{quantity of charge}}{\text{time}}$$

$$I = \frac{q}{t}$$

equation 6.4

The units of current are thus coulombs/second. A coulomb/second is called an **ampere** (A), or **amp** for short. In units, current is therefore

$$1.00 \text{ amp (A)} = \frac{1.00 \text{ coulomb (C)}}{1.00 \text{ second (s)}}$$

A 1.00 amp current is 1.00 coulomb of charge moving through a conductor each second, a 2.00 amp current is 2.00 coulombs per second, and so forth (Figure 6.11). Note in Table 1.2 (p. 6) that the ampere is a SI base unit.

Using the water circuit analogy, you would expect a greater rate of water flow (gallons/minute) when the water pressure is

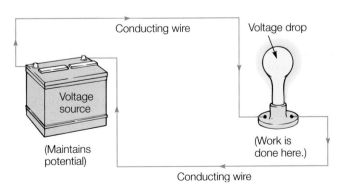

FIGURE 6.10 A simple electric circuit has a voltage source (such as a generator or battery) that maintains the electrical potential, some device (such as a lamp or motor) where work is done by the potential, and continuous pathways for the current to follow.

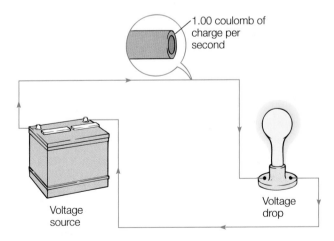

FIGURE 6.11 A simple electric circuit carrying a current of 1.00 coulomb per second through a cross section of a conductor has a current of 1.00 amp.

CHAPTER 6 Electricity **147**

produced by a greater gravitational potential difference. The rate of water flow is thus directly proportional to the difference in gravitational potential energy. In an electric circuit, the rate of current (coulombs/second, or amps) is directly proportional to the difference of electrical potential (joules/coulombs, or volts) between two parts of the circuit, $I \propto V$.

THE NATURE OF CURRENT

There are two ways to describe the current that flows outside the power source in a circuit: (1) a historically based description called *conventional current* and (2) a description based on a flow of charges called *electron current*. The *conventional current* describes current as positive charges moving from the positive to the negative terminal of a battery. This description has been used by convention ever since Ben Franklin first misnamed the charge of an object based on an accumulation, or a positive amount, of "electrical fluid." Conventional current is still used in circuit diagrams. The *electron current* description is in an opposite direction to the conventional current. The electron current describes current as the drift of negative charges that flow from the negative to the positive terminal of a battery. Today, scientists understand the role of electrons in a current, something that was unknown to Franklin. But conventional current is still used by tradition. It actually does not make any difference which description is used, since positive charges moving from the positive terminal are mathematically equivalent to negative charges moving from the negative terminal (Figure 6.12).

The description of an electron current also retains historical traces of the earlier fluid theories of electricity. Today, people understand that electricity is not a fluid but still speak of current, rate of flow, and resistance to flow (Figure 6.13). Fluid analogies can be helpful because they describe the overall electrical effects. But they can also lead to bad concepts such as

the following corrections: (1) in an electric current, electrons do not move through a wire just as water flows through a pipe; (2) electrons are not pushed out one end of the wire as more electrons are pushed in the other end; and (3) electrons do not move through a wire at the speed of light since a power plant failure hundreds of miles away results in an instantaneous loss of power. Perhaps you have held one or more of these misconceptions from fluid analogies.

What is the nature of an electric current? First, consider the nature of a metal conductor without a current. The atoms making up the metal have unattached electrons that are free to move about, much as the molecules of a gas in a container. They randomly move at high speed in all directions, often colliding with each other and with stationary positive ions of the metal. This motion is chaotic, and there is no net movement in any one direction, but the motion does increase with increases in the absolute temperature of the conductor.

When a potential difference is applied to the wire in a circuit, an electric field is established everywhere in the circuit. The *electric field* travels through the conductor at nearly the speed of light as it is established. A force is exerted on each electron by the field, which accelerates the free electrons in the direction of the force. The resulting increased velocity of the electrons is superimposed on their existing random, chaotic movement. This added motion is called the *drift velocity* of the electrons. The drift velocity of the electrons is a result of the imposed electric field. The electrons do not drift straight through the conductor, however, because they undergo countless collisions with other electrons and stationary positive ions. This results in a random zigzag motion with a net motion in one direction. *This net motion constitutes a current,* a flow of charge (Figure 6.14).

When the voltage across a conductor is zero, the drift velocity is zero, and there is no current. The current that occurs when there is a voltage depends on (1) the number of free electrons per unit volume of the conducting material, (2) the charge on each electron (the fundamental charge), (3) the drift velocity, which depends on the electronic structure of the conducting material and the temperature, and (4) the cross-sectional area of the conducting wire.

The relationship between the number of free electrons, charge, drift velocity, area, and current can be used to determine the drift velocity when a certain current flows in a certain size wire made of copper. A 1.0 amp current in copper bell wire (#18), for example, has an average drift velocity on the order of 0.01 cm/s. At that rate, it would take over 5 h for an electron to travel the 200 cm from your car battery to the brake light of your car (Figure 6.15). Thus, it seems clear that it is the *electric field*, not electrons, that causes your brake light to come on almost instantaneously when you apply the brake. The electric field accelerates the electrons already in the filament of the brake lightbulb. Collisions between the electrons in the filament cause the bulb to glow.

Conclusions about the nature of an electric current are that (1) an electric potential difference establishes, at near the speed of light, an electric field throughout a circuit, (2) the field causes

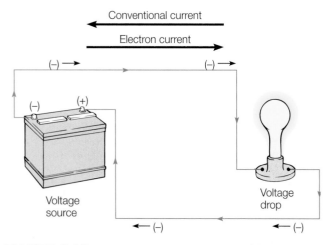

FIGURE 6.12 A conventional current describes positive charges moving from the positive terminal (+) to the negative terminal (−). An electron current describes negative charges (−) moving from the negative terminal (−) to the positive terminal (+).

FIGURE 6.13 What is the nature of the electric current carried by these conducting lines? It is an electric field that moves at near the speed of light. The field causes a net motion of electrons that constitutes a flow of charge, a current.

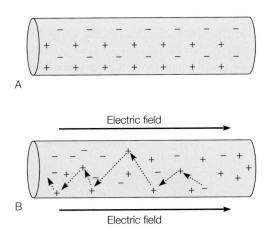

A

Electric field

B

Electric field

FIGURE 6.14 (A) A metal conductor without a current has immovable positive ions surrounded by a swarm of chaotically moving electrons. (B) An electric field causes the electrons to shift positions, creating a separation charge as the electrons move with a zigzag motion from collisions with stationary positive ions and other electrons.

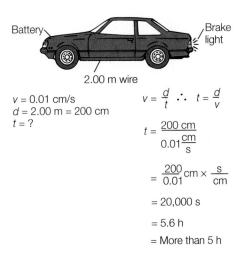

Battery

Brake light

2.00 m wire

$v = 0.01$ cm/s
$d = 2.00$ m $= 200$ cm
$t = ?$

$v = \dfrac{d}{t} \quad \therefore \quad t = \dfrac{d}{v}$

$t = \dfrac{200 \text{ cm}}{0.01\dfrac{\text{cm}}{\text{s}}}$

$= \dfrac{200}{0.01} \text{cm} \times \dfrac{\text{s}}{\text{cm}}$

$= 20{,}000$ s

$= 5.6$ h

$=$ More than 5 h

FIGURE 6.15 Electrons move very slowly in a direct current circuit. With a drift velocity of 0.01 cm/s, more than 5 h would be required for an electron to travel 200 cm from a car battery to the brake light. It is the electric field, not the electrons, that moves at near the speed of light in an electric circuit.

a net motion that constitutes a flow of charge, or current, and (3) the average velocity of the electrons moving as a current is very slow, even though the electric field that moves them travels with a speed close to the speed of light.

Another aspect of the nature of an electric current is the direction the charge is flowing. A circuit like the one described with your car battery has a current that always moves in one direction, a **direct current** (dc). Chemical batteries, fuel cells, and solar cells produce a direct current, and direct currents are utilized in electronic devices. Electric utilities and most of the electrical industry, on the other hand, use an **alternating current** (ac). An alternating current, as the name implies, moves the electrons alternately one way, then the other way. Since the electrons are simply moving back and forth, there is no electron drift along a conductor in an alternating current. Since household electric circuits use alternating current, there is no flow of electrons from the electrical outlets through the circuits. Instead, an electric field moves back and forth through the circuit at nearly the speed of light, causing electrons to jiggle back and forth. This constitutes a current that flows one way, then the other with the changing field. The current changes like this 120 times a second in a 60 hertz alternating current.

Myths, Mistakes, & Misunderstandings

What Is Electricity?

It is a misunderstanding that electricity is electrons moving through wires at near the speed of light. First, if electrons were moving that fast, they would fly out of the wires at every little turn. Electrons do move through wires in a dc circuit, but they do so slowly. In an ac circuit, electrons do not flow forward at all but rather jiggle back and forth in nearly the same place. Overall, electrons are neither gained nor lost in a circuit.

There is something about electricity that moves through a circuit at near the speed of light and is lost in a circuit. It is electromagnetic energy. If you are reading with a lightbulb, for example, consider that the energy lighting the bulb traveled from the power plant at near the speed of light. It was then changed to light and heat in the bulb.

Electricity is electrons moving slowly or jiggling back and forth in a circuit. It is also electromagnetic energy, and both the moving electrons and the moving energy are needed to answer the question, "What is electricity?"

ELECTRICAL RESISTANCE

Recall the natural random and chaotic motion of electrons in a conductor and their frequent collisions with each other and with the stationary positive ions. When these collisions occur, electrons lose energy that they gained from the electric field. The stationary positive ions gain this energy, and their increased energy of vibration results in a temperature increase. Thus, there is a resistance to the movement of electrons being accelerated by an electric field and a resulting energy loss. Materials have a

property of opposing or reducing a current, and this property is called **electrical resistance** (R).

Recall that the current (I) through a conductor is directly proportional to the potential difference (V) between two points in a circuit. If a conductor offers a small resistance, less voltage would be required to push an amp of current through the circuit. If a conductor offers more resistance, then more voltage will be required to push the same amp of current through the circuit. Resistance (R) is therefore a *ratio* between the potential difference (V) between two points and the resulting current (I). This ratio is

$$\text{resistance} = \frac{\text{electrical potential difference}}{\text{current}}$$

$$R = \frac{V}{I}$$

In units, this ratio is

$$1.00 \text{ ohm } (\Omega) = \frac{1.00 \text{ volt (V)}}{1.00 \text{ amp (A)}}$$

The ratio of volts/amps is the unit of resistance called an **ohm** (Ω) after G. S. Ohm (1789–1854), a German physicist who discovered the relationship. The resistance of a conductor is therefore 1.00 ohm if 1.00 volt is required to maintain a 1.00 amp current. The ratio of volt/amp is *defined as* an ohm. Therefore,

$$\text{ohm} = \frac{\text{volt}}{\text{amp}}$$

Another way to show the relationship between the voltage, current, and resistance is

$$V = IR$$

equation 6.5

which is known as **Ohm's law.** This is one of three ways to show the relationship, but this way (solved for V) is convenient for easily solving the equation for other unknowns.

The electrical resistance of a dc electrical conductor depends on four variables (Figure 6.16):

1. *Material.* Different materials have different resistances, as shown by the list of conductors in Table 6.1. Silver, for example, is at the top of the list because it offers the least resistance, followed by copper, gold, then aluminum. Of

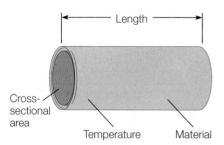

FIGURE 6.16 The four factors that influence the resistance of an electrical conductor are the length of the conductor, the cross-sectional area of the conductor, the material the conductor is made of, and the temperature of the conductor.

the materials listed in the table, nichrome is the conductor with the greatest resistance. By definition, conductors have less electrical resistance than insulators, which have a very large electrical resistance.

2. *Length.* The resistance of a conductor varies directly with the length; that is, a longer wire has more resistance and a shorter wire has less resistance. The longer the wire is, the greater the resistance.

3. *Diameter.* The resistance varies inversely with the cross-sectional area of a conductor. A thick wire has a greater cross-sectional area and therefore has less resistance than a thin wire. The thinner the wire is, the greater the resistance.

4. *Temperature.* For most materials, the resistance increases with increases in temperature. This is a consequence of the increased motion of electrons and ions at higher temperatures, which increases the number of collisions. At very low temperatures (100 K or less), the resistance of some materials approaches zero, and the materials are said to be *superconductors.*

EXAMPLE 6.3

A lightbulb in a 120 V circuit is switched on, and a current of 0.50 A flows through the filament. What is the resistance of the bulb?

SOLUTION

The current (I) of 0.50 A is given with a potential difference (V) of 120 V. The relationship to resistance (R) is given by Ohm's law (equation 6.5).

$$I = 0.50 \text{ A} \qquad V = IR \quad \therefore \quad R = \frac{V}{I}$$

$$V = 120 \text{ V}$$

$$R = ? \qquad\qquad = \frac{120}{0.50} \frac{\text{V}}{\text{A}}$$

$$= 240 \frac{\text{V}}{\text{A}}$$

$$= 240 \text{ ohm}$$

$$= \boxed{240 \ \Omega}$$

EXAMPLE 6.4

What current would flow through an electrical device in a circuit with a potential difference of 120 V and a resistance of 30 Ω? (Answer: 4 A)

ELECTRICAL POWER AND ELECTRICAL WORK

All electric circuits have three parts in common: (1) a *voltage source,* such as a battery or electric generator that uses some nonelectric source of energy to do work on electrons, moving them *against* an electric field to a higher potential; (2) an *electric device,* such as a lightbulb or electric motor where work is done *by* the electric field; and, (3) *conducting wires* that maintain the potential difference across the electrical device. In a direct current circuit, the electric field moves from one terminal of a battery to the electric device through one wire. The second

wire from the device carries the now low-potential field back to the other terminal, maintaining the potential difference. In an alternating current circuit, such as a household circuit, one wire supplies the alternating electric field from the electric generator of a utility company. The second wire from the device is connected to a pipe in the ground and is at the same potential as Earth. The observation that a bird can perch on a current-carrying wire without harm is explained by the fact that there is no potential difference across the bird's body. If the bird were to come into contact with Earth through a second, grounded wire, a potential difference would be established and there would be a current through it.

The work done by a voltage source (battery, electric generator) is equal to the work done by the electric field in an electric device (lightbulb, electric motor) *plus* the energy lost to resistance. Resistance is analogous to friction in a mechanical device, so low-resistance conducting wires are used to reduce this loss. Disregarding losses to resistance, electrical work can therefore be measured where the voltage source creates a potential difference by doing work (W) to move charges (q) to a higher potential (V). From equation 6.3, this relationship is

$$\text{work} = (\text{potential})(\text{charge})$$

or

$$W = (V)(q)$$

In units, the electrical potential is measured in joules/coulomb, and a quantity of charge is measured in coulombs. Therefore, the unit of electrical work is the *joule,*

$$W = (V)(q)$$

$$\text{joule} = \frac{\text{joules}}{\text{coulomb}} \times \text{coulomb}$$

Recall that a joule is a unit of work in mechanics (a newton-meter). In electricity, a joule is also a unit of work, but it is derived from moving a quantity of charge (coulomb) to higher potential difference (joules/coulomb). In mechanics, the work put into a simple machine equals the work output when you disregard *friction.* In electricity, the work put into an electric circuit equals the work output when you disregard *resistance.* Thus, the work done by a voltage source is ideally equal to the work done by electrical devices in the circuit.

Recall also that mechanical power (P) was defined as work (W) per unit time (t), or

$$P = \frac{W}{t}$$

Since electrical work is $W = Vq$, then electrical power must be

$$P = \frac{Vq}{t}$$

Equation 6.4 defined a quantity of charge (q) per unit time (t) as a current (I), or $I = q/t$. Therefore, electrical power is

$$P = \left(\frac{q}{t}\right)(V)$$

In units, you can see that multiplying the current A = C/s by the potential ($V = J/C$) yields

$$\frac{coulombs}{second} \times \frac{joules}{coulombs} = \frac{joules}{second}$$

A joule/second is a unit of power called the **watt** (W). Therefore, electrical power is measured in units of watts, and

$$W = A \cdot V$$

power (in watts) = current (in amps) $\times$ potential (in volts)

watts = amps $\times$ volts

This relationship is:

$$P = IV$$

equation 6.6

Household electrical devices are designed to operate on a particular voltage, usually 120 or 240 volts (Figure 6.17). They therefore draw a certain current to produce the designed power. Information about these requirements is usually found somewhere on the device. A lightbulb, for example, is usually stamped with the designed power, such as 100 W. Other electrical devices may be stamped with amp and volt requirements. You can determine the power produced in these devices by using equation 6.6, that is, amps $\times$ volts = watts. Another handy conversion factor to remember is that 746 watts are equivalent to 1.00 horsepower.

EXAMPLE 6.5

A 1,100 W hair dryer is designed to operate on 120 V. How much current does the dryer require?

SOLUTION

The power (P) produced is given in watts with a potential difference of 120 V across the dryer. The relationship between the units of amps, volts, and watts is found in equation 6.6, $P = IV$

$P = 1,100$ W $\qquad$ $P = IV$ $\;\; \therefore \;\; I = \dfrac{P}{V}$

$V = 120$ V

$I = ?$ A $\qquad\qquad = \dfrac{1,100\, \dfrac{\text{joule}}{\text{second}}}{120\, \dfrac{\text{joule}}{\text{coulomb}}}$

$$= \frac{1,100}{120}\; \frac{J}{s} \times \frac{C}{J}$$

$$= 9.2\; \frac{J \cdot C}{s \cdot J}$$

$$= 9.2\; \frac{C}{s}$$

$$= \boxed{9.2\ \text{A}}$$

EXAMPLE 6.6

An electric fan is designed to draw 0.5 A in a 120 V circuit. What is the power rating of the fan? (Answer: 60 W)

A

B

C

FIGURE 6.17 What do you suppose it would cost to run each of these appliances for one hour? (*A*) This lightbulb is designed to operate on a potential difference of 120 volts and will do work at the rate of 100 W. (*B*) The finishing sander does work at the rate of 1.6 amp $\times$ 120 volts, or 192 W. (*C*) The garden shredder does work at the rate of 8 amps $\times$ 120 volts, or 960 W.

Benjamin Franklin was the first great U.S. scientist. He made an important contribution to physics by arriving at an understanding of the nature of electric charge, introducing the terms *positive* and *negative* to describe charges. He also proved in a classic experiment that lightning is electrical in nature and went on to invent the lightning rod. In addition to being a scientist and inventor, Franklin is widely remembered as a statesman. He played a leading role in drafting the Declaration of Independence and the Constitution of the United States.

Franklin was born in Boston, Massachusetts, of British settlers on January 17, 1706. He started life with little formal instruction, and by the age of 10, he was helping his father in the tallow and soap business. Soon, apprenticed to his brother, a printer, he was launched into that trade, leaving home in 1724 to set himself up as a printer in Philadelphia. His business prospered, and he was soon active in journalism and publishing. He started the *Pennsylvania Gazette* but is better remembered for *Poor Richard's Almanac*. The almanac was a collection of articles and advice on a huge range of topics, "conveying instruction among the common people." Published in 1732, it was a great success and brought Franklin a considerable income.

In 1746, his business booming, Franklin turned his thoughts to electricity and spent the next seven years executing a remarkable series of experiments. Although he had little formal education, his voracious reading habits gave him the necessary background, and his practical skills, together with an analytical yet intuitive approach, enabled Franklin to put the whole topic on a very sound basis. It was said that he found electricity a curiosity and left it a science.

In 1752, Franklin carried out his famous experiments with kites. By flying a kite in a thunderstorm, he was able to produce sparks from the end of the wet string, which he held with a piece of insulating silk. The lightning rod used everywhere today owes its origin to these experiments. Furthermore, some of Franklin's last work in this area demonstrated that while most thunderclouds have negative charges, a few are positive—something confirmed in modern times.

Franklin also busied himself with such diverse topics as the first public library, bifocal lenses, population control, the rocking chair, and daylight-saving time.

Benjamin Franklin is arguably the most interesting figure in the history of science and not only because of his extraordinary range of interests, his central role in the establishment of the United States, and his amazing willingness to risk his life to perform a crucial experiment—a unique achievement in science. By conceiving of the fundamental nature of electricity, he began the process by which a most detailed understanding of the structure of matter has been achieved.

Source: Modified from the Hutchinson *Dictionary of Scientific Biography* © RM 2008. All rights reserved. Helicon Publishing is a division of RM.

An electric utility charges you for the electrical power used at a rate of cents per kilowatt-hour (typically 5–15 cents/kWh). The rate varies from place to place across the country, depending on the cost of producing the power. You can predict the cost of running a particular electric appliance with the following equation,

$$\text{cost} = \frac{(\text{watts})(\text{time})(\text{rate})}{1,000 \, \frac{\text{watt}}{\text{kilowatt}}}$$

equation 6.7

If the watt power rating is not given, it can be obtained by multiplying amps times volts. Also, since the time unit is in hours, the time must be converted to the decimal equivalent of an hour if you want to know the cost of running an appliance for a number of minutes (x min/60 min).

Table 6.2 provides a summary of the electrical quantities and units.

TABLE 6.2

Summary of electrical quantities and units

Quantity	Definition*	Units
Charge	$q = ne$	1.00 coulomb (C) = charge equivalent to 6.24×10^{18} particles such as the electron
Electric potential difference	$V = \frac{W}{q}$	$1.00 \text{ volt (V)} = \frac{1.00 \text{ joule (J)}}{1.00 \text{ coulomb (C)}}$
Electric current	$I = \frac{q}{t}$	$1.00 \text{ amp (A)} = \frac{1.00 \text{ coulomb (C)}}{1.00 \text{ second (s)}}$
Electrical resistance	$R = \frac{V}{I}$	$1.00 \text{ ohm } (\Omega) = \frac{1.00 \text{ volt (V)}}{1.00 \text{ amp (A)}}$
Electrical power	$P = IV$	$1.00 \text{ watt (W)} = \frac{C}{s} \times \frac{J}{C}$

*See Summary of Equations for more information.

CONCEPTS *Applied*

Shocking Costs

You can predict the cost of running an electric appliance with just a few calculations. For example, suppose you want to know the cost of using a 1,300-watt hair dryer for 20 minutes if the utility charges 10 cents per kilowatt-hour. The equation would look like this:

$$\text{cost} = \frac{(1{,}300 \text{ W})(0.33 \text{ h})(\$0.10 \text{ / kWh})}{1{,}000 \frac{\text{W}}{\text{kW}}}$$

Find answers to one or more of the following questions about the cost of running an electric appliance:

- What is your monthly electrical cost for watching television?
- What is the cost of drying your hair with a blow dryer?
- How much would you save by hanging out your clothes to dry rather than using an electric dryer?
- Compare the cost of using the following appliances: coffeemaker, toaster, can opener, vegetable steamer, microwave oven, and blender.
- How much does the electricity cost per month for the use of your desk lamp?
- Of all the electrical devices in a typical household, which three have the greatest monthly electrical cost?

EXAMPLE 6.7

What is the cost of operating a 100 W lightbulb for 1.00 h if the utility rate is $0.10 per kWh?

SOLUTION

The power rating is given as 100 W, so the volt and amp units are not needed. Therefore,

$$IV = P = 100 \text{ W}$$
$$t = 1.00 \text{ h}$$
$$\text{rate} = \$0.10/\text{kWh}$$
$$\text{cost} = ?$$

$$\text{cost} = \frac{(\text{watts})(\text{time})(\text{rate})}{1{,}000 \frac{\text{watts}}{\text{kilowatt}}}$$

$$= \frac{(100 \text{ W})(1.00 \text{ h})(\$0.10/\text{kWh})}{1{,}000 \frac{\text{W}}{\text{kW}}}$$

$$= \frac{(100)(.00)(0.10)}{1{,}000} \frac{\text{W}}{1} \times \frac{\text{h}}{1} \times \frac{\$}{\text{kWh}} \times \frac{\text{kW}}{\text{W}}$$

$$= \boxed{\$0.01}$$

The cost of operating a 100 W lightbulb at a rate of 10¢/kWh is 1¢/h.

EXAMPLE 6.8

An electric fan draws 0.5 A in a 120 V circuit. What is the cost of operating the fan if the rate is $0.10/kWh? (Answer: $0.006, which is 0.6 of a cent per hour)

MAGNETISM

The ability of a certain naturally occurring rock to attract iron has been known since at least 600 B.C. The early Greeks called this rock "Magnesian stone," since it came from the northern Greek county of Magnesia. Knowledge about the iron-attracting properties of the Magnesian stone grew slowly. About A.D. 100, the Chinese learned to magnetize a piece of iron with a Magnesian stone, and sometime before A.D. 1000, they learned to use the magnetized iron or stone as a direction finder (compass). Today, the rock that attracts iron is known to be the black iron oxide mineral named *magnetite*.

Magnetite is a natural magnet that strongly attracts iron and steel but also attracts cobalt and nickel. Such substances that are attracted to magnets are said to have *ferromagnetic properties*, or simply *magnetic* properties. Iron, cobalt, and nickel are considered to have magnetic properties, and most other common materials are considered not to have magnetic properties. Most of these nonmagnetic materials, however, are slightly attracted or slightly repelled by a strong magnet. In addition, certain rare Earth elements, as well as certain metal oxides, exhibit strong magnetic properties.

MAGNETIC POLES

Every magnet has two **magnetic poles,** or ends, about which the force of attraction seems to be concentrated. Iron filings or other small pieces of iron are attracted to the poles of a magnet, for example, revealing their location (Figure 6.18). A magnet suspended by a string will turn, aligning itself in a north-south direction. The north-seeking pole is called the *north pole* of the magnet. The south-seeking pole is likewise named the *south pole* of the magnet. All magnets have both a north pole and a south pole, and neither pole can exist by itself. You cannot separate a north pole from a south pole. If a magnet is broken into pieces, each new piece will have its own north and south poles (Figure 6.19).

You are probably familiar with the fact that two magnets exert forces on each other. For example, if you move the north pole of one magnet near the north pole of a second magnet, each will experience a repelling force. A repelling force also occurs if two south poles are moved close together. But if the north pole of one magnet is brought near the south pole of a second magnet, an attractive force occurs. The rule is "*like magnetic poles repel and unlike magnetic poles attract.*"

A similar rule of "like charges repel and unlike charges attract" was used for electrostatic charges, so you might wonder if there is some similarity between charges and poles. The answer is they are not related. A magnet has no effect on a charged glass rod, and the charged glass rod has no effect on either pole of a magnet.

MAGNETIC FIELDS

A magnet moved into the space near a second magnet experiences a magnetic force as it enters the **magnetic field** of the second magnet. A magnetic field can be represented by *magnetic field lines*. By convention, magnetic field lines are drawn to indicate how the *north pole* of a tiny imaginary magnet would point

FIGURE 6.18 Every magnet has ends, or poles, about which the magnetic properties seem to be concentrated. As this photo shows, more iron filings are attracted to the poles, revealing their location.

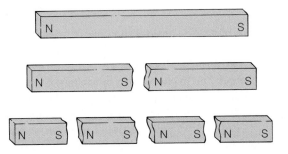

FIGURE 6.19 A bar magnet cut into halves always makes new, complete magnets with both a north and a south pole. The poles always come in pairs, and the separation of a pair into single poles, called monopoles, has never been accomplished.

when in various places in the magnetic field. Arrowheads indicate the direction that the north pole would point, thus defining the direction of the magnetic field. The strength of the magnetic field is greater where the lines are closer together and weaker where they are farther apart. Figure 6.20 shows the magnetic field lines around the familiar bar magnet. Note that magnetic field lines emerge from the magnet at the north pole and enter the magnet at the south pole. Magnetic field lines always form closed loops.

The north end of a magnetic compass needle points north because Earth has a magnetic field. Earth's magnetic field is

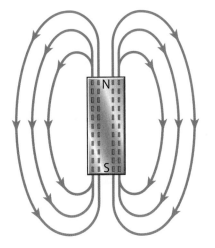

Magnetic compass

FIGURE 6.20 These lines are a map of the magnetic field around a bar magnet. The needle of a magnetic compass will follow the lines, with the north end showing the direction of the field.

shaped and oriented as if there were a huge bar magnet inside Earth (Figure 6.21). The geographic North Pole is the axis of Earth's rotation, and this pole is used to determine the direction of true north on maps. A magnetic compass does not point to true north because the north magnetic pole and the geographic North Pole are in two different places. The difference is called the *magnetic declination*. The map in Figure 6.22 shows approximately how many degrees east or west of true north a compass needle will point in different locations. Magnetic declination must be considered when navigating with a compass. If you are navigating with a compass, you might want to consider using an up-to-date declination map. The magnetic north pole is continuously moving, so any magnetic declination map is probably a snapshot of how it used to be in the past.

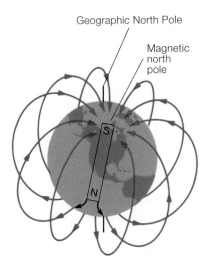

Geographic North Pole

Magnetic north pole

FIGURE 6.21 Earth's magnetic field. Note that the magnetic north pole and the geographic North Pole are not in the same place. Note also that the magnetic north pole acts as if the south pole of a huge bar magnet were inside the earth. You know that it must be a magnetic south pole, since the north end of a magnetic compass is attracted to it, and opposite poles attract.

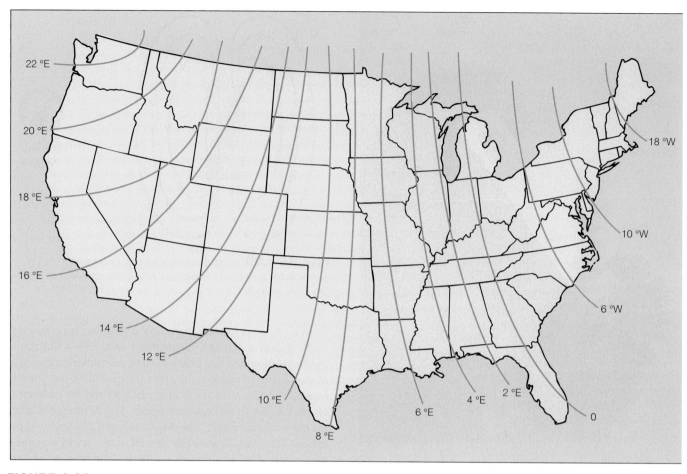

FIGURE 6.22 This magnetic declination map shows the approximate number of degrees east or west of the true geographic north that a magnetic compass will point in various locations.

Note in Figure 6.21 that Earth's magnetic field acts as if there were a huge bar magnet inside Earth with a *south* magnetic pole near Earth's geographic North Pole. This is not an error. The north pole of a magnet is attracted to the south pole of a second magnet, and the north pole of a compass needle points to the north. Therefore, the bar magnet must be arranged as shown. This apparent contradiction is a result of naming the magnetic poles after their "seeking" direction.

The typical compass needle pivots in a horizontal plane, moving to the left or right without up or down motion. Inspection of Figure 6.22, however, shows that Earth's magnetic field is horizontal to the surface only at the magnetic equator. A compass needle that is pivoted so that it moves only up and down will be horizontal only at the magnetic equator. Elsewhere, it shows the angle of the field from the horizontal, called the *magnetic dip*. The angle of dip is the vertical component of Earth's magnetic field. As you travel from the equator, the angle of magnetic dip increases from zero to a maximum of 90° at the magnetic poles.

THE SOURCE OF MAGNETIC FIELDS

The observation that like magnetic poles repel and unlike magnetic poles attract might remind you of the forces involved with like and unlike charges. Recall that electric charges exist as single isolated units of positive protons and units of negative electrons. An object becomes electrostatically charged when charges are separated, and the object acquires an excess or deficiency of negative charges. You might wonder, by analogy, if the poles of a magnet are similarly made up of an excess or deficiency of magnetic poles. The answer is no; magnetic poles are different from electric charges. Positive and negative charges *can* be separated and isolated. But suppose that you try to separate and isolate the poles of a magnet by cutting a magnet into two halves. Cutting a magnet in half will produce two new magnets, each with north and south poles. You could continue cutting each half into new halves, but each time the new half will have its own north and south poles (Figure 6.19). It seems that no subdivision will ever separate and isolate a single magnetic pole, called a *monopole*. Magnetic poles always come in matched pairs of north and south, and a monopole has never been found. The two poles are always found to come together, and as it is understood today, magnetism is thought to be produced by *electric currents,* not an excess of monopoles. The modern concept of magnetism is electric in origin, and magnetism is understood to be a property of electricity.

The key discovery about the source of magnetic fields was reported in 1820 by a Danish physics professor named Hans Christian Oersted. Oersted found that a wire conducting an electric current caused a magnetic compass needle below the

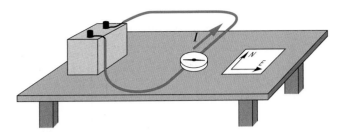

FIGURE 6.23 With the wire oriented along a north-south line, the compass needle deflects away from this line when there is a current in the wire.

wire to move. When the wire was not connected to a battery, the needle of the compass was lined up with the wire and pointed north as usual. But when the wire was connected to a battery, the compass needle moved perpendicular to the wire (Figure 6.23). Oersted had discovered that an electric current produces a magnetic field. An electric current is understood to be the movement of electric charges, so Oersted's discovery suggested that magnetism is a property of charges in motion.

Permanent Magnets

The magnetic fields of bar magnets, horseshoe magnets, and other so-called permanent magnets are explained by the relationship between magnetism and moving charges. Electrons in atoms are moving around the nucleus, so they produce a magnetic field. Electrons also have a magnetic field associated with their spin. In most materials, these magnetic fields cancel one another and neutralize the overall magnetic effect. In other materials, such as iron, cobalt, and nickel, the electrons are arranged and oriented in a complicated way that imparts a magnetic property to the atomic structure. These atoms are grouped in a tiny region called a **magnetic domain.** A magnetic domain is roughly 0.01 to 1 mm in length or width and does not have a fixed size (Figure 6.24). The atoms in each domain are magnetically aligned, contributing to the polarity of the

domain. Each domain becomes essentially a tiny magnet with a north and south pole. In an unmagnetized piece of iron, the domains are oriented in all possible directions and effectively cancel any overall magnetic effect. The net magnetism is therefore zero or near zero.

When an unmagnetized piece of iron is placed in a magnetic field, the orientation of the domain changes to align with the magnetic field, and the size of aligned domains may grow at the expense of unaligned domains. This explains why a "string" of iron paper clips is picked up by a magnet. Each paper clip has domains that become temporarily and slightly aligned by the magnetic field, and each paper clip thus acts as a temporary magnet while in the field of the magnet. In a strong magnetic field, the size of the aligned domains grows to such an extent that the paper clip becomes a "permanent magnet." The same result can be achieved by repeatedly stroking a paper clip with the pole of a magnet. The magnetic effect of a "permanent magnet" can be reduced or destroyed by striking, dropping, or heating the magnet to a sufficiently high temperature (770°C for iron). These actions randomize the direction of the magnetic domains, and the overall magnetic field disappears.

Earth's Magnetic Field

Earth's magnetic field is believed to originate deep within the earth. Like all other magnetic fields, Earth's magnetic field is believed to originate with moving charges. Earthquake waves and other evidence suggest that Earth has a solid inner core with a radius of about 1,200 km (about 750 mi), surrounded by a fluid outer core some 2,200 km (about 1,400 mi) thick. This core is probably composed of iron and nickel, which flows as Earth rotates, creating electric currents that result in Earth's magnetic field. How the electric currents are generated is not yet understood.

Other planets have magnetic fields, and there seems to be a relationship between the rate of rotation and the strength of the planet's magnetic field. Jupiter and Saturn rotate faster than Earth and have stronger magnetic fields than Earth. Venus and Mercury rotate more slowly than Earth and have weaker magnetic fields. This is indirect evidence that the rotation of a planet is associated with internal fluid movements, which somehow generate electric currents and produce a magnetic field.

In addition to questions about how the electric current is generated, there are puzzling questions from geologic evidence. Lava contains magnetic minerals that act like tiny compasses that are oriented to Earth's magnetic field when the lava is fluid but become frozen in place as the lava cools. Studies of these rocks by geologic dating and studies of the frozen magnetic mineral orientation show that Earth's magnetic field has undergone sudden reversals in polarity: the north magnetic pole becomes the south magnetic pole and vice versa. This has happened many times over the distant geologic past, and the most recent shift occurred about 780,000 years ago. The cause of such magnetic field reversals is unknown, but it must be related to changes in the flow patterns of Earth's fluid outer core of iron and nickel.

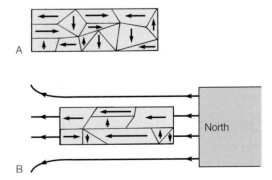

FIGURE 6.24 (*A*) In an unmagnetized piece of iron, the magnetic domains have a random arrangement that cancels any overall magnetic effect. (*B*) When an external magnetic field is applied to the iron, the magnetic domains are realigned, and those parallel to the field grow in size at the expense of the other domains, and the iron is magnetized.

As Oersted discovered, electric charges in motion produce a magnetic field. The direction of the magnetic field around a current-carrying wire can be determined by using a magnetic compass. The north-seeking pole of the compass needle will point in the direction of the magnetic field lines. If you move the compass around the wire, the needle will always move to a position that is tangent to a circle around the wire. Evidently, the magnetic field lines are closed concentric circles that are at right angles to the length of the wire (Figure 6.25).

CURRENT LOOPS

The magnetic field around a current-carrying wire will interact with another magnetic field, one formed around a permanent magnet or one from a second current-carrying wire. The two fields interact, exerting forces just like the forces between the fields of two permanent magnets. The force could be increased by increasing the current, but there is a more efficient way to obtain a larger force. A current-carrying wire that is formed into a loop has perpendicular, circular field lines that pass through the inside of the loop in the same direction. This has the effect of concentrating the field lines, which increases the magnetic field intensity. Since the field lines all pass through the loop in the same direction, one side of the loop will have a north pole and the other side a south pole (Figure 6.26).

Many loops of wire formed into a cylindrical coil are called a *solenoid.* When a current passes through the loops of wire in a solenoid, each loop contributes field lines along the length of the cylinder (Figure 6.27). The overall effect is a magnetic field around the solenoid that acts just like the magnetic field of a bar magnet. This magnet, called an **electromagnet,** can be turned on or off by turning the current on or off. In addition, the strength of the electromagnet depends on the magnitude of the current and the number of loops (ampere-turns). The strength of the electromagnet can also be increased by placing a piece of soft iron in the coil. The domains of the iron become aligned

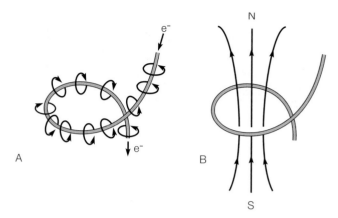

FIGURE 6.26 (*A*) Forming a wire into a loop causes the magnetic field to pass through the loop in the same direction. (*B*) This gives one side of the loop a north pole and the other side a south pole.

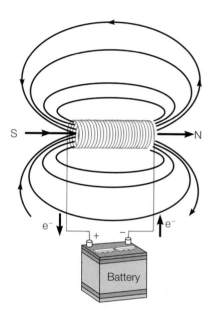

FIGURE 6.27 When a current is run through a cylindrical coil of wire, a solenoid, it produces a magnetic field like the magnetic field of a bar magnet.

by the influence of the magnetic field. This induced magnetism increases the overall magnetic field strength of the solenoid as the magnetic field lines are gathered into a smaller volume within the core.

APPLICATIONS OF ELECTROMAGNETS

The discovery of the relationship between an electric current, magnetism, and the resulting forces created much excitement in the 1820s and 1830s. This excitement was generated because it was now possible to explain some seemingly separate phenomena in terms of an interrelationship and because people began to see practical applications almost immediately. Within a year of Oersted's discovery, André Ampère had fully explored the magnetic effects of currents, combining experiments and theory to find the laws describing these effects.

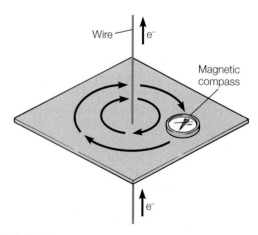

FIGURE 6.25 A magnetic compass shows the presence and direction of the magnetic field around a straight length of current-carrying wire.

Soon after Ampère's work, the possibility of doing mechanical work by sending currents through wires was explored. The electric motor, similar to motors in use today, was invented in 1834, only fourteen years after Oersted's momentous discovery.

The magnetic field produced by an electric current is used in many practical applications, including electrical meters, electromagnetic switches that make possible the remote or programmed control of moving mechanical parts, and electric motors. In each of these applications, an electric current is applied to an electromagnet.

Electric Meters

Since you cannot measure electricity directly, it must be measured indirectly through one of the effects that it produces. The strength of the magnetic field produced by an electromagnet is proportional to the electric current in the electromagnet. Thus, one way to measure a current is to measure the magnetic field that it produces. A device that measures currents from their magnetic fields is called a *galvanometer* (Figure 6.28). A galvanometer has a coil of wire that can rotate on pivots in the magnetic field of a permanent magnet. The coil has an attached pointer that moves across a scale and control springs that limit its motion and return the pointer to zero when there is no current. When there is a current in the coil, the electromagnetic field is attracted and repelled by the field of the permanent magnet. The larger the current, the greater the force and the more the coil will rotate until it reaches an equilibrium position with the control springs. The amount of movement of the coil (and thus the pointer) is proportional to the current in the coil. With certain modifications and applications, the galvanometer can be used to measure current (ammeter), potential difference (voltmeter), and resistance (ohmmeter).

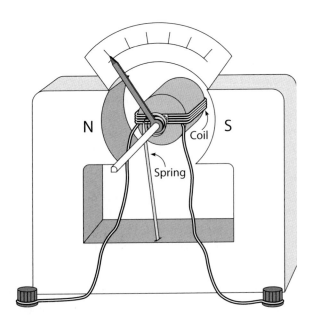

FIGURE 6.28 A galvanometer consists of a coil of wire, a permanent magnet, and a restoring spring to return the needle to zero when there is no current through the coil.

Lemon Battery

1. You can make a simple compass galvanometer that will detect a small electric current (Box Figure 6.1). All you need is a magnetic compass and some thin insulated wire (the thinner the better).

2. Wrap the thin insulated wire in parallel windings around the compass. Make as many parallel windings as you can, but leave enough room to see both ends of the compass needle. Leave the wire ends free for connections.

3. To use the galvanometer, first turn the compass so the needle is parallel to the wire windings. When a current passes through the coil of wire, the magnetic field produced will cause the needle to move from its north-south position, showing the presence of a current. The needle will deflect one way or the other depending on the direction of the current.

4. Test your galvanometer with a "lemon battery." Roll a soft lemon on a table while pressing on it with the palm of your hand. Cut two slits in the lemon about 1 cm apart. Insert a 8-cm (approximate) copper wire in one slit and a same-sized length of a straightened paper clip in the other slit, making sure the metals do not touch inside the lemon. Connect the galvanometer to the two metals. Try the two metals in other fruits, vegetables, and liquids. Can you find a pattern?

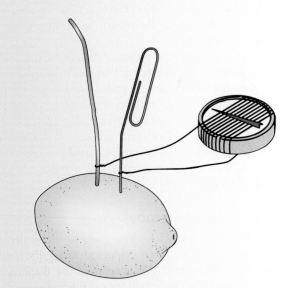

BOX FIGURE 6.1 You can use the materials shown here to create and detect an electric current.

Electromagnetic Switches

A *relay* is an electromagnetic switch device that makes possible the use of a low-voltage control current to switch a larger, high-voltage circuit on and off (Figure 6.29). A thermostat, for example, utilizes two thin, low-voltage wires in a glass tube of mercury. The glass tube of mercury is attached to a metal coil that expands and contracts with changes in temperature, tipping the attached

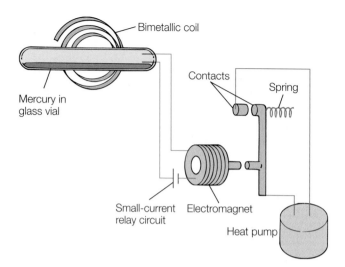

FIGURE 6.29 A schematic of a relay circuit. The mercury vial turns as changes in temperature expand or contract the coil, moving the mercury and making or breaking contact with the relay circuit. When the mercury moves to close the relay circuit, a small current activates the electromagnet, which closes the contacts on the large-current circuit.

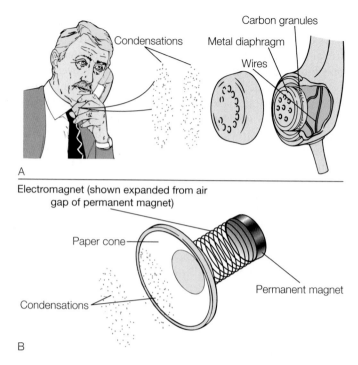

FIGURE 6.30 (*A*) Sound waves are converted into a changing electrical current in a telephone. (*B*) Changing electrical current can be changed to sound waves in a speaker by the action of an electromagnet pushing and pulling on a permanent magnet. The electromagnet is attached to a stiff paper cone or some other material that makes sound waves as it moves in and out.

glass tube. When the temperature changes enough to tip the glass tube, the mercury flows to the bottom end, which makes or breaks contact with the two wires, closing or opening the circuit. When contact is made, a weak current activates an electromagnetic switch, which closes the circuit on the large-current furnace or heat pump motor.

A solenoid is a coil of wire with a current. Some solenoids have a spring-loaded movable piece of iron inside. When a current flows in such a coil, the iron is pulled into the coil by the magnetic field, and the spring returns the iron when the current is turned off. This device could be utilized to open a water valve, turning the hot or cold water on in a washing machine or dishwasher, for example. Solenoids are also used as mechanical switches on VCRs, automobile starters, and signaling devices such as bells and buzzers.

Telephones and Loudspeakers

The mouthpiece of a typical telephone contains a cylinder of carbon granules with a thin metal diaphragm facing the front. When someone speaks into the telephone, the diaphragm moves in and out with the condensations and rarefactions of the sound wave (Figure 6.30). This movement alternately compacts and loosens the carbon granules, increasing and decreasing the electric current that increases and decreases with the condensations and rarefactions of the sound waves.

The moving electric current is fed to the earphone part of a telephone at another location. The current runs through a coil of wire that attracts and repels a permanent magnet attached to a speaker cone. When repelled forward, the speaker cone makes a condensation, and when attracted back, the cone makes a rarefaction. The overall result is a series of condensations and rarefactions that, through the changing electric current, accurately match the sounds made by the other person.

The loudspeaker in a radio or stereo system works from changes in an electric current in a similar way, attracting and repelling a permanent magnet attached to the speaker cone. You can see the speaker cone in a large speaker moving back and forth as it creates condensations and rarefactions.

Electric Motors

An electric motor is an electromagnetic device that converts electrical energy to mechanical energy. Basically, a motor has two working parts, a stationary electromagnet called a *field magnet* and a cylindrical, movable electromagnet called an *armature*. The armature is on an axle and rotates in the magnetic field of the field magnet. The axle turns fan blades, compressors, drills, pulleys, or other devices that do mechanical work.

Different designs of electric motors are used for various applications, but the simple demonstration motor shown in Figure 6.31 can be used as an example of the basic operating principle. Both the field coil and the armature are connected to an electric current. The armature turns, and it receives the current through a *commutator* and *brushes*. The brushes are contacts that brush against the commutator as it rotates, maintaining contact. When the current is turned on, the field coil and the armature become electromagnets, and the unlike poles attract, rotating the armature. If the current is dc, the armature would turn no farther, stopping as it does in a galvanometer. But the commutator has insulated segments so when it turns halfway, the commutator segments switch brushes and there is a current through the armature in the *opposite* direction. This switches the armature poles, which

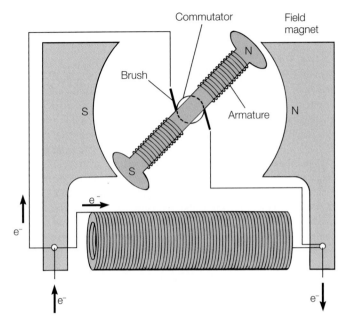

FIGURE 6.31 A schematic of a simple electric motor.

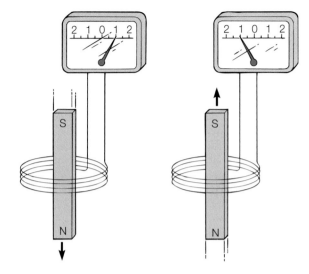

FIGURE 6.32 A current is induced in a coil of wire moved through a magnetic field. The direction of the current depends on the direction of motion.

are now repelled for another half-turn. The commutator again reverses the polarity, and the motion continues in one direction. An actual motor has many coils (called "windings") in the armature to obtain a useful force, and many commutator segments. This gives the motor a smoother operation with a greater turning force.

ELECTROMAGNETIC INDUCTION

So far, you have learned that (1) a moving charge and a current-carrying wire produce a magnetic field and (2) a second magnetic field exerts a force on a moving charge and exerts a force on a current-carrying wire as their magnetic fields interact.

Soon after the discovery of these relationships by Oersted and Ampère, people began to wonder if the opposite effect was possible; that is, would a magnetic field produce an electric current? The discovery was made independently in 1831 by Joseph Henry in the United States and by Michael Faraday in England. They found that *if a loop of wire is moved in a magnetic field, or if the magnetic field is changed, a voltage is induced in the wire.* The voltage is called an *induced voltage,* and the resulting current in the wire is called an *induced current.* The overall interaction is called **electromagnetic induction.**

One way to produce electromagnetic induction is to move a bar magnet into or out of a coil of wire (Figure 6.32). A galvanometer shows that the induced current flows one way when the bar magnet is moved toward the coil and flows the other way when the bar magnet is moved away from the coil. The same effect occurs if you move the coil back and forth over a stationary magnet. Furthermore, no current is detected when the magnetic field and the coil of wire are not moving. Thus, electromagnetic induction depends on the relative motion of the magnetic field and the coil of wire. It does not matter which moves or changes, but one must move or change relative to the other for electromagnetic induction to occur.

Electromagnetic induction occurs when the loop of wire moves across magnetic field lines or when magnetic field lines move across the loop. The magnitude of the induced voltage is proportional to (1) the number of wire loops passing through the magnetic field lines, (2) the strength of the magnetic field, and (3) the rate at which magnetic field lines pass through the wire.

 CONCEPTS *Applied*

Simple Generator

1. Make a coil of wire from insulated bell wire (#18 copper wire) by wrapping fifty windings around a cardboard tube from a roll of paper. Tape the coil at several places so it does not come apart, and discard the cardboard tube.
2. Make a current-detecting instrument from a magnetic compass and some thin insulated wire (the thinner the better). Wrap the thin insulated wire in parallel windings around the compass. Make as many parallel windings as you can, but leave enough room to see both ends of the compass needle. Connect the wire ends to the coil you made in step 1.
3. Orient the compass so the needle is parallel to the wire around the compass. When a current passes through the coil of wire, the magnetic field produced will cause the needle to move, showing the presence of a current.
4. First, move a bar magnet into and out of the stationary coil of wire and observe the compass needle. Second, move the coil of wire back and forth over a stationary bar magnet and observe the compass needle.
5. Experiment with a larger coil of wire, bar magnets of greater or weaker strengths, and by moving the coil at varying speeds. See how many generalizations you can make concerning electromagnetic induction.

A Closer Look

Current War

Thomas Edison (1847–1931) built the first electric generator and electrical distribution system to promote his new long-lasting lightbulbs. The dc generator and distribution system was built in lower Manhattan, New York City, and was switched on September 4, 1882. It supplied 110 V dc to fifty-nine customers. Edison studied both ac and dc systems and chose dc because of advantages it offered at the time. Direct current was used because batteries are dc, and batteries were used as a system backup. Also, dc worked fine with electric motors, and ac motors were not yet available.

George Westinghouse (1846–1914) was in the business of supplying gas for gas lighting, and he could see that electric lighting would soon be replacing all the gaslights. After studying the matter, he decided that Edison's low voltage system was not efficient enough. In 1885, he began experimenting with ac generators and transformers in Pittsburgh.

Nikola Tesla (1856–1943) was a Croatian-born U.S. physicist and electrical engineer who moved to the United States and worked for Thomas Edison in 1884. He then set up his own laboratory and workshop in 1887. His work led to a complicated set of patents covering the generation, transmission, and use of ac electricity. From 1888 on, Tesla was associated with George Westinghouse, who bought and successfully exploited Tesla's ideas, leading to the introduction of ac for power transmission.

Westinghouse's promotion of ac led to direct competition with Edison and his dc electrical systems. A "war of currents" resulted, with Edison claiming that transmission of such high voltage was dangerous. He emphasized this point by recommending the use of high voltage ac in an electric chair as the best way to execute prisoners.

The advantages of ac were greater since you could increase the voltage, transmit for long distances at a lower cost, and then decrease the voltage to a safe level. Eventually, even Edison's own General Electric company switched to producing ac equipment. Westinghouse turned his attention to the production of large steam turbines for producing ac power and was soon setting up ac distribution systems across the nation.

GENERATORS

Soon after the discovery of electromagnetic induction the **electric generator** was developed. The generator is essentially an axle with many wire loops that rotates in a magnetic field. The axle is turned by some form of mechanical energy, such as a water turbine or a steam turbine, which uses steam generated from fossil fuels or nuclear energy. As the coil rotates in a magnetic field, a current is induced in the coil (Figure 6.33).

TRANSFORMERS

Current from a power plant goes to a transformer to step up the voltage. A **transformer** is a device that steps up or steps down the ac voltage. It has two basic parts: (1) a *primary* or "input" coil and (2) a *secondary* or "output" coil, which is close by. Both coils are often wound on a single iron core but are always fully insulated from each other. When there is an alternating current through the primary coil, a magnetic field grows around the coil to a maximum size, collapses to zero, then grows to a maximum size with an opposite polarity. This happens 120 times a second as the alternating current oscillates at 60 hertz. The magnetic field is strengthened and directed by the iron core. The growing and collapsing magnetic field moves across the wires in the secondary coil, inducing a voltage in the secondary coil. The growing and collapsing magnetic field from the primary coil thus induces a voltage in the secondary coil, just as an induced voltage occurs in the wire loops of a generator.

The transformer increases or decreases the voltage in an alternating current because the magnetic field grows and collapses past the secondary coil inducing a voltage. If a direct current is applied to the primary coil, the magnetic field grows around the primary coil as the current is established but then becomes stationary. Recall that electromagnetic induction occurs when there is relative motion between the magnetic field

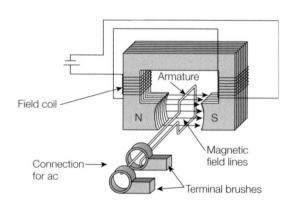

A

B

FIGURE 6.33 (*A*) Schematic of a simple alternator (ac generator) with one output loop. (*B*) Output of the single loop turning in a constant magnetic field, which alternates the induced current each half-cycle.

lines and a wire loop. Thus, an induced voltage occurs from a direct current (1) only for an instant when the current is established and the growing field moves across the secondary coil and (2) only for an instant when the current is turned off and the field collapses back across the secondary coil. In order to use dc in a transformer, the current must be continually interrupted to produce a changing magnetic field.

When an alternating current or a continually interrupted direct current is applied to the primary coil, the magnitude of the induced voltage in the secondary coil is proportional to the ratio of wire loops in the two coils. If they have the same number of loops, the primary coil produces just as many magnetic field lines as are intercepted by the secondary coil. In this case, the induced voltage in the secondary coil will be the same as the voltage in the primary coil. Suppose, however, that the secondary coil has one-tenth as many loops as the primary coil. This means that the secondary loops will move across one-tenth as many field lines as the primary coil produces. As a result, the induced voltage in the secondary coil will be one-tenth the voltage in the primary coil. This is called a *step-down transformer* because the voltage was stepped down in the secondary coil. On the other hand, more wire loops in the secondary coil will intercept more magnetic field lines. If the secondary coil has ten times *more* loops than the primary coil, then the voltage will be *increased* by a factor of 10. This is a *step-up transformer*. How much the voltage is stepped up or stepped down depends on the ratio of wire loops in the primary and secondary coils (Figure 6.34). Note that the *volts per wire loop* are the same in each coil. The relationship is

$$\frac{\text{volts}_{\text{primary}}}{(\text{number of loops})_{\text{primary}}} = \frac{\text{volts}_{\text{secondary}}}{(\text{number of loops})_{\text{secondary}}}$$

or

$$\frac{V_{\text{p}}}{N_{\text{p}}} = \frac{V_{\text{s}}}{N_{\text{s}}}$$

equation 6.8

EXAMPLE 6.9

A step-up transformer has five loops on its primary coil and twenty loops on its secondary coil. If the primary coil is supplied with an alternating current at 120 V, what is the voltage in the secondary coil?

SOLUTION

$N_{\text{p}} = 5$ loops

$N_{\text{s}} = 20$ loops

$V_{\text{p}} = 120$ V

$V_{\text{s}} = ?$

$$\frac{V_{\text{p}}}{N_{\text{p}}} = \frac{V_{\text{s}}}{N_{\text{s}}} \quad \therefore \quad V_{\text{s}} = \frac{V_{\text{p}}N_{\text{s}}}{N_{\text{p}}}$$

$$V_{\text{s}} = \frac{(120 \text{ V})(20 \text{ loops})}{5 \text{ loops}}$$

$$= \frac{120 \times 20}{5} \frac{\text{V·loops}}{\text{loops}}$$

$$= \boxed{480 \text{ V}}$$

A step-up or step-down transformer steps up or steps down the *voltage* of an alternating current according to the ratio of wire loops in the primary and secondary coils. Assuming no losses in the transformer, the *power input* on the primary coil equals the *power output* on the secondary coil. Since $P = IV$, you

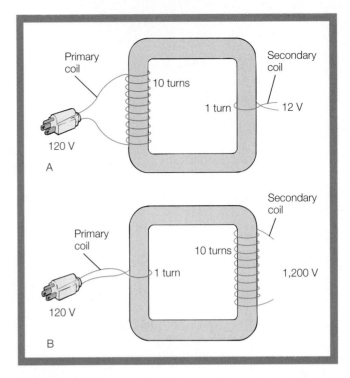

FIGURE 6.34 (*A*) This step-down transformer has ten turns on the primary for each turn on the secondary and reduces the voltage from 120 V to 12 V. (*B*) This step-up transformer increases the voltage from 120 V to 1,200 V, since there are ten turns on the secondary to each turn on the primary.

can see that when the voltage is stepped up the current is correspondingly decreased, as

$$\text{power input} = \text{power output}$$

$$\text{watts input} = \text{watts outpur}$$

$$(\text{amps} \times \text{volts})_{\text{in}} = (\text{amps} \times \text{volts})_{\text{out}}$$

or

$$V_{\text{p}}I_{\text{p}} = V_{\text{s}}I_{\text{s}}$$

equation 6.9

EXAMPLE 6.10

The step-up transformer in example 6.9 is supplied with an alternating current at 120 V and a current of 10.0 A in the primary coil. What current flows in the secondary circuit?

SOLUTION

$V_{\text{p}} = 120$ V

$I_{\text{p}} = 10.0$ A

$V_{\text{s}} = 480$ V

$I_{\text{s}} = ?$

$$V_{\text{p}}I_{\text{p}} = V_{\text{s}}I_{\text{s}} \quad \therefore \quad I_{\text{s}}\frac{V_{\text{p}}I_{\text{p}}}{V_{\text{s}}}$$

$$I_{\text{s}} = \frac{120 \text{ V} \times 10.0 \text{ A}}{480 \text{ V}}$$

$$= \frac{120 \times 10.0}{480} \frac{\text{V·A}}{\text{V}}$$

$$= \boxed{2.5 \text{ A}}$$

A

B

FIGURE 6.35 Energy losses in transmission are reduced by increasing the voltage, so the voltage of generated power is stepped up at the power plant. (*A*) These transformers, for example, might step up the voltage from tens to hundreds of thousands of volts. After a step-down transformer reduces the voltage at a substation, still another transformer (*B*) reduces the voltage to 120 for transmission to three or four houses.

CONCEPTS *Applied*

Swinging Coils

The interactions between moving magnets and moving charges can be easily demonstrated with two large magnets and two coils of wire.

1. Make a coil of wire from insulated bell wire (#18 copper wire) by wrapping fifty windings around a narrow jar. Tape the coil at several places so it does not come apart.
2. Now make a second coil of wire from insulated bell wire and tape the coil as before.
3. Suspend both coils of wire on separate ring stands or some other support on a table top. The coils should hang so they will swing with the broad circle of the coil moving back and forth. Place a large magnet on supports so it is near the center of each coil (Box Figure 6.2).
4. Connect the two coils of wire.
5. Move one of the coils of wire and observe what happens to the second coil. The second coil should move, mirroring the movements of the first coil (if it does not move, find some stronger magnets).
6. Explain what happens in terms of magnetic fields and currents at the first coil and at the second coil.

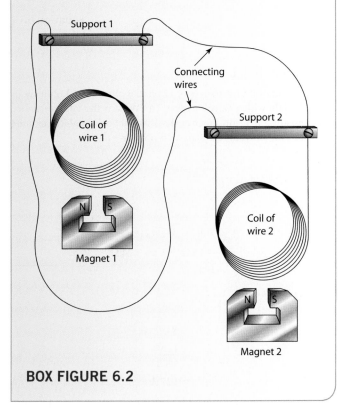

BOX FIGURE 6.2

Energy losses in transmission are reduced by stepping up the voltage. Recall that electrical resistance results in an energy loss and a corresponding absolute temperature increase in the conducting wire. If the current is large, there are many collisions between the moving electrons and positive ions of the wire, resulting in a large energy loss. Each collision takes energy from the electric

field, diverting it into increased kinetic energy of the positive ions and thus increased temperature of the conductor. The energy lost to resistance is therefore reduced by *lowering* the current, which is what a transformer does by increasing the voltage. Hence, electric power companies step up the voltage of generated power for economical transmission. A step-up transformer at a power plant, for example, might step up the voltage from 22,000 volts to 500,000 volts for transmission across the country to a city. This step up in voltage correspondingly reduces the current, lowering the resistance losses to a more acceptable 4 or 5 percent over long distances. A step-down transformer at a substation near the city reduces the voltage to several thousand volts for transmission around the city. Additional step-down transformers reduce this voltage to 120 volts for transmission to three or four houses (Figure 6.35).

CIRCUIT CONNECTIONS

Practically all of the electricity generated by power plants is *alternating current,* which is stepped up, transmitted over high lines, and stepped down for use in homes and industry. Electric circuits in automobiles, cell phones, MP3 players, and laptops, on the other hand, all have *direct current* circuits. Thus, most all industry and household circuits are ac circuits, and most movable or portable circuits are dc circuits. It works out that way because of the present need to use transformers for transmitting large currents, which can only be done economically with ac currents, and because chemical batteries are the main source of current for dc devices.

VOLTAGE SOURCES IN CIRCUITS

Most standard flashlights use two dry cells, each with a potential difference of 1.5 volts. All such dry cells are 1.5 volts, no matter how small or large they are, from penlight batteries up to much larger D cells. To increase the voltage above 1.5 volts, the cells must be arranged and connected in a **series circuit.** A series connection has the negative terminal of one cell connected to the positive terminal of another cell (see Figure 6.36A). The total voltage produced this way is equal to the sum of the single cell

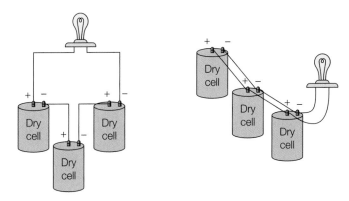

FIGURE 6.36 (*A*) A circuit connected with batteries in series will have the same current, and the voltages add. (*B*) A circuit connected with batteries in parallel will have the same voltage in the circuit of the largest battery, and each battery contributes a part of the total current.

In August 2003, problems in a huge electrical grid resulted in power plants shutting down across the Ohio Valley and a massive electric power blackout that affected some 50 million people. Scientists from the University of Maryland took advantage of the event to measure different levels of atmospheric air pollution while the fossil-fueled power plants were shut down. Scooping many air samples with a small airplane 24 hours after the blackout, they found 90 percent less sulfur dioxide, 50 percent less ozone, and 70 percent fewer light-scattering particles from the air in the same area than when the power plants were running. The scientists stated that the result could come from an underestimation of emissions from power plants or from unknown chemical reactions in the atmosphere.

QUESTIONS TO DISCUSS

1. Are there atmospheric factors that might have contributed to the findings?

2. Are there factors on the ground that might have contributed to the findings?

3. Do the results mean that power plants contribute that much pollution, or what else should be considered?

4. How would you conduct a pollution-measuring experiment that would leave no room for doubt about the results?

---

voltages. Thus, with three cells connected in a series circuit, the voltage of the circuit is 4.5 volts. A 9 volt battery is made up of six 1.5 volt cells connected in a series circuit. You can see the smaller cells of a 9 volt battery by removing the metal jacket. A 12 volt automobile battery works with different chemistry than the flashlight battery, and each cell in the automobile battery produces 2 volts. Each cell has its own water cap, and six cells are connected in a series circuit to produce a total of 12 volts.

Cells with all positive terminals connected and all negative terminals connected (see Figure 6.36B) are in a **parallel circuit.** A parallel circuit has a resultant voltage determined by the largest cell in the circuit. If the largest cell in the circuit is 1.5 volts, then the potential difference of the circuit is 1.5 volts. The purpose of connecting cells in parallel is to make a greater amount of electrical energy available. The electrical energy that can be furnished by dry cells in parallel is the sum of the energy that the individual cells can provide. A lantern battery, for example, is made up of four 1.5 volt dry cells in a parallel circuit. The total voltage of the battery is 1.5 volts. The four 1.5 volt cells in a parallel circuit will last much longer than they would if used individually.

RESISTANCES IN CIRCUITS

Electrical resistances, such as lightbulbs, can also be wired into a circuit in series (Figure 6.37). When several resistances are connected in series, the resistance (R) of the combination is equal to the sum of the resistances of each component. In symbols, this is written as

$$R_{total} = R_1 + R_2 + R_3 + \ldots$$

equation 6.10

EXAMPLE 6.11

Three resistors with resistances of 12 ohm, 8 ohm, and 24 ohm are in a series circuit with a 12 volt battery. (A) What is the total resistance of the resistors? (B) How much current can move through the circuit? (C) What is the current through each resistor?

SOLUTION

A.

$$R_{total} = R_1 + R_2 + R_3$$
$$= 12 \, \Omega + 8 \, \Omega + 24 \, \Omega$$
$$= 44 \, \Omega$$

B. Using Ohm's law,

$$I = \frac{V}{R}$$
$$= \frac{12 \, V}{44 \, \Omega}$$
$$= 0.27 \, A$$

C. Since the same current runs through one after the other, the current is the same in each.

In a parallel circuit, more than one path is available for the current, which divides and passes through each resistance independently (Figure 6.38). This lowers the overall resistance for the circuit, and the total resistance is less than any single

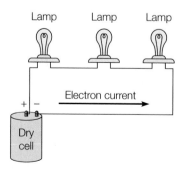

FIGURE 6.37 A series electric circuit.

You may be familiar with many solid-state devices such as calculators, computers, word processors, digital watches, VCRs, digital stereos, and camcorders. All of these are called solid-state devices because they use a solid material, such as the semiconductor silicon, in an electric circuit in place of vacuum tubes. Solid-state technology developed from breakthroughs in the use of semiconductors during the 1950s, and the use of thin pieces of silicon crystal is common in many electric circuits today.

A related technology also uses thin pieces of a semiconductor such as silicon but not as a replacement for a vacuum tube. This technology is concerned with photovoltaic devices, also called *solar cells,* that generate electricity when exposed to light (Box Figure 6.3). A solar cell is unique in generating electricity since it produces electricity directly, without moving parts or chemical reactions, and potentially has a very long lifetime. This reading is concerned with how a solar cell generates electricity.

The conducting properties of silicon can be changed by *doping,* that is, artificially forcing atoms of other elements into the crystal. Phosphorus, for example, has five electrons in its outermost shell compared to the four in a silicon atom. When phosphorus atoms replace silicon atoms in the crystal, there are extra electrons not tied up in the two electron bonds. The extra electrons move easily through the crystal, carrying a charge. Since the phosphorus-doped silicon carries a negative charge, it is called an *n-type* semiconductor. The n means negative charge carrier.

A silicon crystal doped with boron will have atoms with only three electrons in the outermost shell. This results in a deficiency, that is, electron "holes" that act as positive charges. A hole can move as an electron is attracted to it, but it leaves another hole elsewhere, where it moved from. Thus, a flow of electrons in one direction is equivalent to a flow of holes in the opposite direction. A hole, therefore, behaves as a positive charge. Since the boron-doped silicon carries a positive charge, it is called a *p-type* semiconductor. The p means positive charge carrier.

A B

BOX FIGURE 6.3 Solar cells are economical in remote uses such as (*A*) navigational aids and (*B*) communications. The solar panels in both of these examples are oriented toward the south.

The basic operating part of a silicon solar cell is typically an 8 cm wide and 3×10^{-1} mm (about one-hundredth of an inch) thick wafer cut from a silicon crystal. One side of the wafer is doped with boron to make p-silicon, and the other side is doped with phosphorus to make n-silicon. The place of contact between the two is called the p-n junction, which creates a *cell barrier*. The cell barrier forms as electrons are attracted from the n-silicon to the holes in the p-silicon. This creates a very thin zone of negatively charged p-silicon and positively charged n-silicon (Box Figure 6.4). Thus, an internal electric field is established at the p-n junction, and the field is the cell barrier.

The cell is thin, and light can penetrate through the p-n junction. Light strikes the p-silicon, freeing electrons. Low-energy free electrons might combine with a hole, but high-energy electrons cross the cell barrier into the n-silicon. The electron loses some of its energy, and the barrier prevents it from returning, creating an excess negative charge

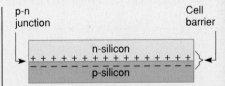

BOX FIGURE 6.4 The cell barrier forms at the p-n junction between the n-silicon and the p-silicon. The barrier creates a "one-way" door that accumulates negative charges in the n-silicon.

in the n-silicon and a positive charge in the p-silicon. This establishes a potential that will drive a current.

Today, solar cells are essentially hand-made and are economical only in remote power uses (navigational aids, communications, or irrigation pumps) and in consumer specialty items (solar-powered watches and calculators). Research continues on finding methods of producing highly efficient, highly reliable solar cells that are affordable.

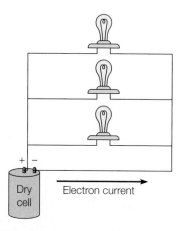

FIGURE 6.38 A parallel electric circuit.

resistance. In symbols, the effect of wiring resisters in parallel is

$$\frac{1}{R_{total}} = \frac{1}{R_1} + \frac{1}{R_2} + \frac{1}{R_3}$$

equation 6.11

EXAMPLE 6.12

Assume the three resistances in example 6.11 are now connected in parallel. (A) What is the combined resistance? (B) What is the current in the overall circuit? (C) What is the current through each resistance?

SOLUTION

A. The total resistance can be found from equation 6.11, and

$$\frac{1}{R_{total}} = \frac{1}{R_1} + \frac{1}{R_2} + \frac{1}{R_3}$$

$$= \frac{1}{12\,\Omega} + \frac{1}{8\,\Omega} + \frac{1}{24\,\Omega}$$

$$= \frac{2}{24\,\Omega} + \frac{3}{24\,\Omega} + \frac{1}{24\,\Omega}$$

$$\frac{1}{R_{total}} = \frac{6}{24\,\Omega}$$

$$6 \times R_{total} = 1 \times 24\,\Omega$$

$$R_{total} = \frac{24\,\Omega}{6}$$

$$= 4\,\Omega$$

B.

$$I = \frac{V}{R}$$

$$= \frac{12\,V}{4\,\Omega}$$

$$= 3\,A$$

C.

$$I_1 = \frac{V}{R_1} \qquad I_2 = \frac{V}{R_2} \qquad I_3 = \frac{V}{R_3}$$

$$= \frac{12\,V}{12\,\Omega} \qquad = \frac{12\,V}{8\,\Omega} \qquad = \frac{12\,V}{24\,\Omega}$$

$$= 1\,A \qquad = 1.5\,A \qquad = 0.5\,A$$

A series circuit has resistances connected one after the other, so the current passing through each resistance is the same through each resistance. Adding more resistances to a series circuit will cause a decrease in the current available in the circuit and a reduction of the voltage available for each individual resistance. Since power is determined from the product of the current and the voltage ($P = IV$), adding more lamps to a series circuit will result in dimmer lights. Perhaps you have observed such a dimming when you connected together two or more strings of decorating lights, which are often connected in a series circuit. Another disadvantage to a series circuit is that if one bulb burns out, the circuit is broken and all the lights go out.

In the parallel electric circuit, the current has alternate branches to follow, and the current in one branch does not affect the current in the other branches. The total current in the parallel circuit is therefore equal to the sum of the current flowing in each branch. Adding more resistances in a parallel circuit results in three major effects that are characteristic of all parallel circuits:

1. an increase in the current in the circuit;
2. the same voltage is maintained across each resistance; and
3. a lower total resistance of the entire circuit. The total resistance is lowered since additional branches provide more pathways for the current to move.

HOUSEHOLD CIRCUITS

Household wiring is a combination of series and parallel circuits (Figure 6.39). The light fixtures in a room, for example, are wired in parallel, and you can use just one light or all the lights at the same time. A switch, however, is wired in series with the light fixtures, as is the fuse or circuit breaker in the circuit.

Each appliance connected to such a parallel circuit has the same voltage available to do work, and each appliance draws current according to its resistance. This means that as additional appliances are turned on or plugged in, additional current flows through the circuit. This means you can turn on the lights and draw 1 amp, then start your 1,200 W blow dryer and draw 10 more amps for a total of 11 amps from the same circuit. Can you add still more appliances to the same circuit when it already has this load? The answer to this question will depend on what else you want to add (how many amps?) and on the preset load your fuse or circuit breaker will accept before breaking the circuit.

Without the "guard" role of fuses and circuit breakers, you could add more and more appliances to a circuit and pull more and more amps. The current could reach high enough levels to cause overheating and possibly a fire. A fuse or circuit breaker in the circuit is used to disconnect the circuit if it reaches the preset value, usually 15 or 20 amps. A fuse is a disposable, screw-in device that contains a short piece of metal with a high resistance and low melting point. When the current through the circuit reaches the preset rating, say 15 amps, the short metal strip melts. This gap opens the circuit just like a switch for the whole circuit. The circuit breaker has

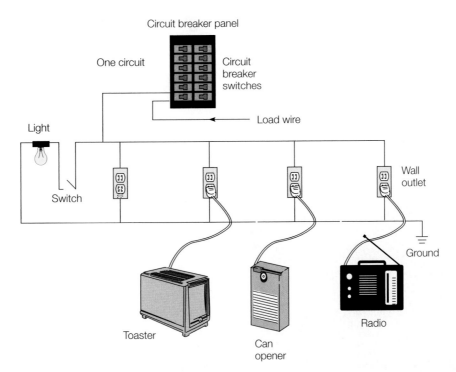

FIGURE 6.39 Each circuit has its own breaker switch (or fuse), and this simplified sketch shows one circuit and the wiring from the breaker panel to one light and four outlets. Household circuits are parallel circuits, and each appliance connected to such a circuit draws current. Each circuit has its own designed voltage and ampere rating based on the intended use.

the same purpose, but it uses a different procedure. A circuit breaker uses the directly proportional relationship between the magnitude of a current and the magnetic field around the conductor. When the current reaches the preset level, the associated magnetic field opens the circuit by the magnetic field attracting a piece of iron, thus opening a spring-loaded switch. The circuit breaker is reset by flipping the switch back to its original position.

In addition to overloads, fuses and circuit breakers are "blown" or "tripped" by a **short circuit,** a new path of lesser electrical resistance. A current always takes the path of least resistance. A short circuit occurs when some low resistance connection or "short cut" is accidentally established between two points in an electric circuit. Such a "short cut" could be provided by a frayed or broken wire that completes the circuit. The electrical resistance of copper is very low, which is why copper is used for a conducting wire in the first place. A short piece of copper wire involved in a short circuit might have a resistance of only 0.01 ohm. The current that could flow through this wire in a 120 volt circuit, according to Ohm's law, could reach 12,000 amp if a fuse or circuit breaker did not interrupt the circuit at the preset 15 amp load level.

A modern household electric circuit has many safety devices to protect people and property from electrical damage. Fuses or circuit breakers disconnect circuits before they become overloaded and thus overheated. People are also protected from electrical shock by three-pronged plugs, polarized plugs (Figure 6.40), and ground fault interrupters, which are discussed next.

A household circuit requires two wires to each electrical device: (1) an energized "load-carrying" wire that carries electrical energy from the electric utility and (2) a grounded or "neutral" wire that maintains a potential difference between the two wires. Suppose the load-carrying wire inside an appliance becomes frayed or broken or in some way makes contact with the metal housing of the appliance. If you touch the housing, you could become a part of the circuit as a current flows through you or parts of your body.

A *three-pronged plug* provides a third appliance-grounding wire through the grounded plug. The grounding wire connects the metal housing of an appliance directly to the ground. If there is a short circuit, the current will take the path of least resistance—through the grounding wire—rather than through you. A *polarized plug* has one prong larger than the other. The smaller prong is wired to the load-carrying wire, and the larger one is wired to the neutral or ground wire. The ordinary, nonpolarized plug can fit into an outlet either way, which means there is a 50-50 chance that one of the wires will be the one that carries the load. The polarized plug does not take that gamble since it always has the load-carrying wire on the same side of the circuit. Thus, the switch can be wired in so it is always on the load-carrying wire. The switch will function on either wire, but when it is on the ground wire, the appliance has an energy-carrying wire, just waiting for a potential difference to be established through a ground, perhaps through you. When the switch is on the load-carrying side, the appliance does not have this potential safety hazard.

FIGURE 6.40 This is the so-called "polarized" plug, with one prong larger than the other. The larger prong can fit only in the larger slot of the outlet, so the smaller prong (the current-carrying wire) always goes in the smaller slot. This is a safety feature that, when used correctly, results in the switch disconnecting the current-carrying wire rather than the ground wire.

Have you ever noticed a red push-button on electrical outlets where you live or where you have visited? The button is usually on outlets in bathrooms or outside a building, places where a person might become electrically grounded by standing in water. Usually, there is also a note on the outlet to "test monthly." This is another safety device called a *ground-fault interrupter* (GFI), which offers a different kind of protection from a fatal shock. Fuses and circuit breakers are designed to protect circuits from overheating that might occur from overloads or short circuits. It might occur to you that when a fuse or circuit breaker trips, there is already a problem somewhere in the circuit from a heavy load. If a person is in the circuit, they may have already received a fatal jolt. How does a GFI react fast enough to prevent a fatal shock? To answer that question, we must first review a few basics about household circuits.

A household circuit always has two wires, one that is carrying the load and one that is the neutral system ground. The load-carrying wire is usually black (or red), and the system ground is usually white. A third wire, usually bare or green, serves as an appliance ground. Normally, the currents in the load-carrying and system ground wires are the same. If a short occurs, some of the current is diverted to the appliance ground or, worse yet, through a person with wet feet. The GFI monitors the load-carrying and system ground wires. If any difference is detected, the GFI trips, opening the circuit within a fraction of a second. This is much quicker than the regular fuse or general circuit breaker can react, and the difference might be enough to prevent a fatal shock. The GFI, which can also be tripped by a line surge that might occur during a thunderstorm, is reset by pushing in the red button.

SUMMARY

The first electrical phenomenon recognized was the charge produced by friction, which today is called *static electricity*. By the early 1900s, the *electron theory of charge* was developed from studies of the *atomic nature of matter*. These studies led to the understanding that matter is made of *atoms*, which are composed of *negatively charged electrons* moving about a central *nucleus*, which contains *positively charged protons*. The two kinds of charges interact as *like charges produce a repellant force* and *unlike charges produce an attractive force*. An object acquires an *electric charge* when it has an excess or deficiency of electrons, which is called an *electrostatic charge.*

A *quantity of charge (q)* is measured in units of *coulombs* (C), the charge equivalent to the transfer of 6.24×10^{18} charged particles such as the electron. The *fundamental charge* of an electron or proton is 1.60×10^{-19} coulomb. The *electrical forces* between two charged objects can be calculated from the relationship between the quantity of charge and the distance between two charged objects. The relationship is known as *Coulomb's law.*

A charged object in an electric field has *electric potential energy* that is related to the charge on the object and the work done to move it into a field of like charge. The resulting *electric potential difference (V)* is a ratio of the work done (W) to move a quantity of charge (q). In units, a joule of work done to move a coulomb of charge is called a *volt.*

A flow of electric charge is called an *electric current (I).* A current requires some device, such as a generator or battery, to maintain a potential difference. The device is called a *voltage source.* An *electric circuit* contains (1) a voltage source, (2) a *continuous path* along which the current flows, and (3) a device such as a lamp or motor where work is done, called a *voltage drop.* Current (I) is measured as the *rate* of flow of charge, the quantity of charge (q) through a conductor in a period of time (t). The unit of current in coulomb/second is called an *ampere* or *amp* for short (A).

Current occurs in a conductor when a *potential difference* is applied and an *electric field* travels through the conductor at near the speed of light. The electrons drift *very slowly,* accelerated by the electric field. The field moves the electrons in one direction in a *direct current* (dc) and moves them back and forth in an *alternating current* (ac).

Materials have a property of opposing or reducing an electric current called *electrical resistance (R).* Resistance is a ratio between the potential difference (V) between two points and the resulting current (I), or $R = V/I$. The unit is called the *ohm* (Ω), and $1.00 \ \Omega = 1.00$ volt/1.00 amp. The relationship between voltage, current, and resistance is called *Ohm's law.*

Disregarding the energy lost to resistance, the *work* done by a voltage source is equal to the work accomplished in electrical devices in a circuit. The *rate* of doing work is *power,* or *work per unit time, P = W/t. Electrical power* can be calculated from the relationship of *P = IV,* which gives the power unit of *watts.*

Magnets have two poles about which their attraction is concentrated. When free to turn, one pole moves to the north and the other to the south. The north-seeking pole is called the *north pole* and the south-seeking pole is called the *south pole. Like poles repel* one another and *unlike poles attract.*

The property of magnetism is *electric in origin,* produced by charges in motion. *Permanent magnets* have tiny regions called *magnetic domains,* each with its own north and south poles. An *unmagnetized* piece of iron has randomly arranged domains. When *magnetized,* the domains become aligned and contribute to the overall magnetic effect.

A *current-carrying wire* has magnetic field lines of closed, *concentric circles* that are at right angles to the length of wire. The *direction* of the magnetic field depends on the direction of the current. A coil of many loops is called a *solenoid* or *electromagnet.* The electromagnet is the working part in electrical meters, electromagnetic switches, and the electric motor.

When a loop of wire is moved in a magnetic field, or if a magnetic field is moved past a wire loop, a voltage is induced in the wire loop. The interaction is called *electromagnetic induction.* An electric generator is a rotating coil of wire in a magnetic field. The coil is rotated by mechanical energy, and electromagnetic induction induces a voltage, thus converting mechanical energy to electrical energy. A *transformer* steps up or steps down the voltage of an alternating current. The ratio of input and output voltage is determined by the number of loops in the primary and secondary coils. Increasing the voltage decreases the current, which makes long-distance transmission of electrical energy economically feasible.

Batteries connected in *series* will have the *same current* and the *voltages add.* In *parallel,* the voltage in the circuit is *the same as each source,* and each battery contributes a part of the total current.

A *series circuit* has resistances connected one after the other so the same current flows through each resistance one after the other. A *parallel circuit* has comparable branches, separate pathways for the current to flow through. As more resistances are added to a parallel circuit, it has an *increase in the current, the same voltage* is maintained across each resistance, and *total resistance* of the entire circuit is lowered.

Household circuits are *parallel circuits,* so each appliance has the same voltage available to do work and each appliance draws current according to its resistance. *Fuses* and *circuit breakers* protect circuits from overheating from *overloads* or *short circuits.* A *short circuit* is a new path of lesser resistance. Other protective devices are *three-pronged plugs, polarized plugs,* and *ground fault interrupters.*

SUMMARY OF EQUATIONS

6.1

Quantity of charge = (number of electrons)(electron charge)

$$q = ne$$

6.2

Electrical force = (constant) $\times$ $\dfrac{\text{charge on one object} \times \text{charge on second object}}{\text{distance between objects squared}}$

$$F = k\frac{q_1 q_2}{d^2}$$

where $k = 9.00 \times 10^9$ newton·meters²/coulomb²

6.3

Electric potential = $\dfrac{\text{work to create potential}}{\text{charge moved}}$

$$V = \frac{W}{q}$$

6.4

Electric current = $\dfrac{\text{quantity of charge}}{\text{time}}$

$$I = \frac{q}{t}$$

6.5

Volts = current $\times$ resistance

$$V = IR$$

6.6

$$\text{Electrical power} = (\text{amps})(\text{volts})$$
$$P = IV$$

6.7

$$\text{Cost} = \frac{(\text{watts})(\text{time})(\text{rate})}{1{,}000 \text{ W/kW}}$$

6.8

$$\frac{\text{volts}_{\text{primary}}}{(\text{number of loops})_{\text{primary}}} = \frac{\text{volts}_{\text{secondary}}}{(\text{number of loops})_{\text{secondary}}}$$
$$\frac{V_p}{N_p} = \frac{V_s}{N_s}$$

6.9

$$(\text{volts}_{\text{primary}})(\text{current}_{\text{primary}}) = (\text{volts}_{\text{secondary}})(\text{current}_{\text{secondary}})$$
$$V_p I_p = V_s I_s$$

6.10 Resistances in series circuit

$$R_{\text{total}} = R_1 + R_2 + R_3 + \cdots$$

6.11 Resistances in parallel circuit

$$\frac{1}{R_{\text{total}}} = \frac{1}{R_1} + \frac{1}{R_2} + \frac{1}{R_3} + \cdots$$

KEY TERMS

alternating current (p. **150**)

amp (p. **147**)

ampere (p. **147**)

coulomb (p. **143**)

Coulomb's law (p. **144**)

direct current (p. **150**)

electric circuit (p. **147**)

electric current (p. **146**)

electric field (p. **144**)

electric generator (p. **162**)

electrical conductors (p. **142**)

electrical resistance (p. **150**)

electromagnet (p. **158**)

electromagnetic induction (p. **161**)

electrostatic charge (p. **141**)

force field (p. **144**)

fundamental charge (p. **143**)

magnetic domain (p. **157**)

magnetic field (p. **154**)

magnetic poles (p. **154**)

ohm (p. **150**)

Ohm's law (p. **150**)

parallel circuit (p. **166**)

series circuit (p. **165**)

short circuit (p. **169**)

transformer (p. **162**)

volt (p. **146**)

watt (p. **152**)

APPLYING THE CONCEPTS

1. Electrostatic charge results from
   a. transfer or redistribution of electrons.
   b. gain or loss of protons.
   c. separation of charge from electrons and protons.
   d. failure to keep the object clean of dust.

2. The unit of electric charge is the
   a. volt.
   b. amp.
   c. coulomb.
   d. watt.

3. An electric field describes the condition of space around a
   a. charged particle.
   b. magnetic pole.
   c. mass.
   d. All of the above are correct.

4. A material that has electrons that are free to move throughout the material is a (an)
   a. electrical conductor.
   b. electrical insulator.
   c. thermal insulator.
   d. thermal nonconductor.

5. An example of an electrical insulator is
   a. graphite.
   b. glass.
   c. aluminum.
   d. tungsten.

6. The electrical potential difference between two points in a circuit is measured in units of
   a. volt.
   b. amp.
   c. coulomb.
   d. watt.

7. The rate at which an electric current flows through a circuit is measured in units of
   a. volt.
   b. amp.
   c. coulomb.
   d. watt.

8. The law that predicts the behavior of electrostatic forces acting through space is
   a. law of universal gravitation.
   b. Watt's law.
   c. Coulomb's law.
   d. Ohm's law.

9. What type of electric current is produced by fuel cells and solar cells?
   a. ac
   b. dc
   c. 60 Hz
   d. 120 Hz

10. The electrical resistance of a conductor is measured in units of
    a. volt.
    b. amp.
    c. ohm.
    d. watt.

11. According to Ohm's law, what must be greater to maintain the same current in a conductor with more resistance?
    a. voltage
    b. current
    c. temperature
    d. cross-sectional area

12. A kilowatt-hour is a unit of
    a. power.
    b. work.
    c. current.
    d. potential difference.

13. If you multiply volts time amps, the answer will be in units of
    a. power.
    b. work.
    c. current.
    d. potential difference.

14. Units of joules per second is a measure called a
    a. volt.
    b. amp.
    c. ohm.
    d. watt.

15. A lodestone is a natural magnet that attracts
    a. iron.
    b. cobalt.
    c. nickel.
    d. All of the above are correct.

16. The north pole of a suspended or floating bar magnet currently points directly toward Earth's
    a. north magnetic pole.
    b. south magnetic pole.
    c. north geographic pole.
    d. south geographic pole.

17. A current-carrying wire always has
    a. a magnetic field with closed concentric field lines around the length of the wire.
    b. a magnetic field with field lines parallel to the length of the wire.
    c. an electric field but no magnetic field around the wire.
    d. nothing in the space around the wire.

18. Magnetism is produced by
    a. an excess of north monopoles.
    b. an excess of south monopoles.
    c. moving charges.
    d. separation of positive and negative charges.

19. Earth's magnetic field
    a. has undergone many reversals in polarity.
    b. has always been as it is now.
    c. is created beneath Earth's north geographic pole.
    d. is created beneath Earth's south geographic pole.

20. The strength of a magnetic field around a current-carrying wire varies directly with the
    a. amperage of the current.
    b. voltage of the current.
    c. resistance of the wire.
    d. temperature of the wire.

21. Reverse the direction of a current in a wire, and the magnetic field around the wire will
    a. have an inverse magnitude of strength.
    b. have a reversed north pole direction.
    c. become a conventional current.
    d. remain unchanged.

22. The operation of which of the following depends on the interaction between two magnetic fields?
    a. car stereo speakers
    b. telephone
    c. relay circuit
    d. All of the above are correct.

23. An electric meter measures
    a. actual number of charges moving through a conductor.
    b. current in packets of coulombs.
    c. the strength of a magnetic field.
    d. the difference in potential between two points in a conductor.

24. When a loop of wire cuts across magnetic field lines or when magnetic field lines move across a loop of wire
    a. electrons are pushed toward one end of the loop.
    b. an electrostatic charge is formed.
    c. the wire becomes a permanent magnet.
    d. a magnetic domain is created.

25. A step-up transformer steps up the
    a. voltage.
    b. current.
    c. power.
    d. energy.

26. Electromagnetic induction occurs when a coil of wire cuts across magnetic field lines. Which one of the following increases the voltage produced?
    a. fewer wire loops in the coil
    b. increased strength of the magnetic field
    c. slower speed of the moving coil of wire
    d. decreased strength of the magnetic field

27. Electric power companies step up the voltage of generated power for transmission across the country because higher voltage
    a. means more power is transmitted.
    b. reduces the current, which increases the resistance.
    c. means less power is transmitted.
    d. reduces the current, which lowers the energy lost to resistance.

28. A solar cell
    a. produces electricity directly.
    b. requires chemical reactions.
    c. has a very short lifetime.
    d. uses small moving parts.

29. Which of the following is most likely to acquire an electrostatic charge?
    a. electrical conductor
    b. electrical nonconductor
    c. Both are equally likely.
    d. None of the above is correct.

30. Which of the following units are measures of rates?
    a. amp and volt
    b. coulomb and joule
    c. volt and watt
    d. amp and watt

31. You are using which description of a current if you consider a current to be positive charges that flow from the positive to the negative terminal of a battery.
    a. electron current
    b. conventional current
    c. proton current
    d. alternating current

32. In an electric current, the electrons are moving
    a. at a very slow rate.
    b. at the speed of light.
    c. faster than the speed of light.
    d. at a speed described as supersonic.

33. In which of the following currents is there no electron movement from one end of a conducting wire to the other end?
    a. electron current
    b. direct current
    c. alternating current
    d. None of the above is correct.

34. If you multiply amps times volts, the answer will be in units of
    a. resistance.
    b. work.
    c. current.
    d. power.

35. A permanent magnet has magnetic properties because
    a. the magnetic fields of its electrons are balanced.
    b. of an accumulation of monopoles in the ends.
    c. the magnetic domains are aligned.
    d. All of the above are correct.

36. A current-carrying wire has a magnetic field around it because
    a. a moving charge produces a magnetic field of its own.
    b. the current aligns the magnetic domains in the metal of the wire.
    c. the metal was magnetic before the current was established, and the current enhanced the magnetic effect.
    d. None of the above is correct.

37. When an object acquires a negative charge, it actually
    a. gains mass.
    b. loses mass.
    c. has a constant mass.
    d. The answer is unknown.

38. A positive and a negative charge are initially 2 cm apart. What happens to the force on each as they are moved closer and closer together? The force
    a. increases while moving.
    b. decreases while moving.
    c. remains constant.
    d. The answer is unknown.

39. In order to be operational, a complete electric circuit must contain a source of energy, a device that does work, and
    a. a magnetic field.
    b. a conductor from the source to the working device and another conductor back to the source.
    c. connecting wires from the source to the working device.
    d. a magnetic field and a switch.

40. Which variable is inversely proportional to the resistance?
    a. length of conductor
    b. cross-sectional area of conductor
    c. temperature of conductor
    d. conductor material

41. Which of the following is not considered to have strong magnetic properties?
    a. iron
    b. nickel
    c. silver
    d. cobalt

42. A piece of iron can be magnetized or unmagnetized. This is explained by the idea that
    a. electrons in iron atoms are spinning and have magnetic fields around them.
    b. atoms of iron are grouped into tiny magnetic domains that may orient themselves in a particular direction or in a random direction.
    c. unmagnetized iron atoms can be magnetized by an external magnetic field.
    d. the north and south poles of iron can be segregated by the application of an external magnetic field.

43. Earth's magnetic field is believed to originate
    a. by a separation of north and south monopoles due to currents within Earth.
    b. with electric currents that are somehow generated in Earth's core.
    c. from a giant iron and cobalt bar magnet inside Earth.
    d. from processes that are not understood.

44. The speaker in a stereo system works by the action of
    a. a permanent magnet creating an electric current.
    b. an electromagnet pushing and pulling on a permanent magnet.
    c. sound waves pushing and pulling on an electromagnet.
    d. electrons creating sound waves.

45. Electromagnetic induction takes place because
    a. an electric current is measured by rate of movement of charges.
    b. the potential is determined by how much work is done.
    c. electrons have their own magnetic field, which interacts with an externally applied magnetic field.
    d. copper wire is magnetic, which induces magnetism.

46. The current in the secondary coil of a transformer is produced by a
    a. varying magnetic field.
    b. varying electric field.
    c. constant magnetic field.
    d. constant electric field.

47. An electromagnet uses
    a. a magnetic field to produce an electric current.
    b. an electric current to produce a magnetic field.
    c. a magnetic current to produce an electric field.
    d. an electric field to produce a magnetic current.

48. A transformer
    a. changes the voltage of a direct current.
    b. changes the power of a direct current.
    c. changes the voltage of an alternating current.
    d. changes the amperage of an alternating current.

49. A parallel circuit has
    a. wires that are lined up side by side.
    b. the same current flowing through one resistance after another.
    c. separate pathways for the current to flow through.
    d. None of the above is correct.

50. In which type of circuit would you expect a reduction of the available voltage as more and more resistances are added to the circuit?
    a. series circuit
    b. parallel circuit
    c. open circuit
    d. None of the above is correct.

51. In which type of circuit would you expect the same voltage with an increased current as more and more resistances are added to the circuit?
    a. series circuit
    b. parallel circuit
    c. open circuit
    d. None of the above is correct.

Answers

1. a **2.** c **3.** a **4.** a **5.** b **6.** a **7.** b **8.** c **9.** b **10.** c **11.** a **12.** b **13.** a **14.** d **15.** d **16.** a **17.** a **18.** c **19.** a **20.** a **21.** b **22.** d **23.** c **24.** a **25.** a **26.** b **27.** d **28.** a **29.** b **30.** d **31.** b **32.** a **33.** c **34.** d **35.** c **36.** a **37.** a **38.** a **39.** b **40.** b **41.** c **42.** b **43.** b **44.** b **45.** c **46.** a **47.** b **48.** c **49.** c **50.** a **51.** b

QUESTIONS FOR THOUGHT

1. Explain why a balloon that has been rubbed sticks to a wall for a while.
2. Explain what is happening when you walk across a carpet and receive a shock when you touch a metal object.
3. Why does a positively or negatively charged object have multiples of the fundamental charge?
4. Explain how you know that it is an electric field, not electrons, that moves rapidly through a circuit.
5. Is a kWh a unit of power or a unit of work? Explain.
6. What is the difference between ac and dc?
7. What is a magnetic pole? How are magnetic poles named?
8. How is an unmagnetized piece of iron different from the same piece of iron when it is magnetized?
9. Explain why the electric utility company increases the voltage of electricity for long-distance transmission.
10. Describe how an electric generator is able to generate an electric current.
11. Why does the north pole of a magnet point to the geographic North Pole if like poles repel?
12. Explain what causes an electron to move toward one end of a wire when the wire is moved across a magnetic field.

FOR FURTHER ANALYSIS

1. Explain how the model of electricity as electrons moving along a wire is an oversimplification that misrepresents the complex nature of an electric current.
2. What are the significant similarities and differences between ac and dc? What determines which is better for a particular application?
3. Transformers usually have signs warning, "Danger—High Voltage." Analyze if this is a contradiction since it is exposure to amps, not volts, that harms people.
4. Will a fuel cell be the automobile engine of the future? Identify the facts, beliefs, and theories that support or refute your answer.
5. Analyze the apparent contradiction in the statement that "solar energy is free" with the fact that solar cells are too expensive to use as a significant energy source.
6. What are the basic similarities and differences between an electric field and a magnetic field?
7. What are the advantages and disadvantages of using parallel circuits for household circuits?

INVITATION TO INQUIRY

Earth Power?

Investigate if you can use Earth's magnetic field to induce an electric current in a conductor. Connect the ends of a 10 m (about 33 ft) wire to a galvanometer. Have a partner hold the ends of the wire on the galvanometer while you hold the end of the wire loop and swing the double wire like a skip rope.

If you accept this invitation, try swinging the wire in different directions. Can you figure out a way to measure how much electricity you can generate?

PARALLEL EXERCISES

The exercises in groups A and B cover the same concepts. Solutions to group A exercises are located in appendix E.

Group A

1. A rubber balloon has become negatively charged from being rubbed with a wool cloth, and the charge is measured as 1.00×10^{-14} C. According to this charge, the balloon contains an excess of how many electrons?
2. One rubber balloon with a negative charge of 3.00×10^{-14} C is suspended by a string and hangs 2.00 cm from a second rubber balloon with a negative charge of 2.00×10^{-12} C. (a) What is the direction of the force between the balloons? (b) What is the magnitude of the force?
3. A dry cell does 7.50 J of work through chemical energy to transfer 5.00 C between the terminals of the cell. What is the electric potential between the two terminals?
4. An electric current through a wire is 6.00 C every 2.00 s. What is the magnitude of this current?

Group B

1. An inflated rubber balloon is rubbed with a wool cloth until an excess of a billion electrons is on the balloon. What is the magnitude of the charge on the balloon?
2. What is the force between two balloons with a negative charge of 1.6×10^{-10} C if the balloons are 5.0 cm apart?
3. How much energy is available from a 12 V storage battery that can transfer a total charge equivalent to 100,000 C?
4. A wire carries a current of 2.0 A. At what rate is the charge flowing?

5. A 1.00 A electric current corresponds to the charge of how many electrons flowing through a wire per second?

6. There is a current of 4.00 A through a toaster connected to a 120.0 V circuit. What is the resistance of the toaster?

7. What is the current in a 60.0 Ω resistor when the potential difference across it is 120.0 V?

8. A lightbulb with a resistance of 10.0 Ω allows a 1.20 A current to flow when connected to a battery. (a) What is the voltage of the battery? (b) What is the power of the lightbulb?

9. A small radio operates on 3.00 V and has a resistance of 15.0 Ω. At what rate does the radio use electric energy?

10. A 1,200 W hair dryer is operated on a 120 V circuit for 15 min. If electricity costs $0.10/kWh, what was the cost of using the blow dryer?

11. An automobile starter rated at 2.00 hp draws how many amps from a 12.0 V battery?

12. An average-sized home refrigeration unit has a 1/3 hp fan motor for blowing air over the inside cooling coils, a 1/3 hp fan motor for blowing air over the outside condenser coils, and a 3.70 hp compressor motor. (a) All three motors use electric energy at what rate? (b) If electricity costs $0.10/kWh, what is the cost of running the unit per hour? (c) What is the cost for running the unit 12 hours a day for a 30-day month?

13. A 15 ohm toaster is turned on in a circuit that already has a 0.20 hp motor, three 100 W lightbulbs, and a 600 W electric iron that are on. Will this trip a 15 A circuit breaker? Explain.

14. A power plant generator produces a 1,200 V, 40 A alternating current that is fed to a step-up transformer before transmission over the high lines. The transformer has a ratio of 200 to 1 wire loops. (a) What is the voltage of the transmitted power? (b) What is the current?

15. A step-down transformer has an output of 12 V and 0.5 A when connected to a 120 V line. Assuming no losses: (a) What is the ratio of primary to secondary loops? (b) What current does the transformer draw from the line? (c) What is the power output of the transformer?

16. A step-up transformer on a 120 V line has 50 loops on the primary and 150 loops on the secondary, and draws a 5.0 A current. Assuming no losses: (a) What is the voltage from the secondary? (b) What is the current from the secondary? (c) What is the power output?

17. Two 8.0 Ω lightbulbs are connected in a 12 V series circuit. What is the power of both glowing bulbs?

5. What is the magnitude of the least possible current that could theoretically exist?

6. There is a current of 0.83 A through a lightbulb in a 120 V circuit. What is the resistance of this lightbulb?

7. What is the voltage across a 60.0 Ω resistor with a current of 3 1/3 amp?

8. A 10.0 Ω lightbulb is connected to a 12.0 V battery. (a) What current flows through the bulb? (b) What is the power of the bulb?

9. A lightbulb designed to operate in a 120.0 V circuit has a resistance of 192 Ω. At what rate does the bulb use electric energy?

10. What is the monthly energy cost of leaving a 60 W bulb on continuously if electricity costs $0.10 per kWh?

11. An electric motor draws a current of 11.5 A in a 240 V circuit. What is the power of this motor in W?

12. A swimming pool requiring a 2.0 hp motor to filter and circulate the water runs for 18 hours a day. What is the monthly electrical cost for running this pool pump if electricity costs $0.10 per kWh?

13. Is it possible for two people to simultaneously operate 1,300 W hair dryers on the same 120 V circuit without tripping a 15 A circuit breaker? Explain.

14. A step-up transformer has a primary coil with 100 loops and a secondary coil with 1,500 loops. If the primary coil is supplied with a household current of 120 V and 15 A, (a) what voltage is produced in the secondary circuit? (b) What current flows in the secondary circuit?

15. The step-down transformer in a local neighborhood reduces the voltage from a 7,200 V line to 120 V. (a) If there are 125 loops on the secondary, how many are on the primary coil? (b) What current does the transformer draw from the line if the current in the secondary is 36 A? (c) What are the power input and output?

16. A step-down transformer connected to a 120 V electric generator has 30 loops on the primary for each loop in the secondary. (a) What is the voltage of the secondary? (b) If the transformer has a 90.0 A current in the primary, what is the current in the secondary? (c) What are the power input and output?

17. What is the power of an 8.0 ohm bulb when three such bulbs are connected in a 12 volt series circuit?

7
Light

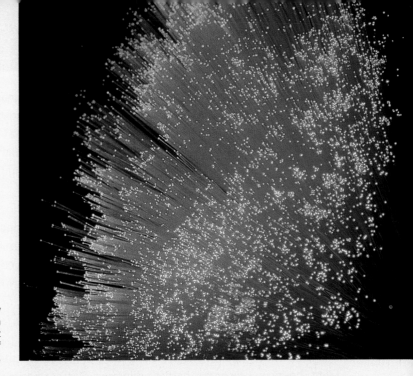

This fiber optics bundle carries pulses of light from an infrared laser to carry much more information than could be carried by electrons moving through wires. This is part of a dramatic change underway, a change that will first find a hybrid "optoelectronics" replacing the more familiar "electronics" of electrons and wires.

CORE **CONCEPT**
Light is electromagnetic radiation—energy—that interacts with matter.

OUTLINE

Light Interacts with Matter
Light that interacts with matter can be reflected, absorbed, or transmitted.

Refraction
Light moving from one transparent material to another undergoes a change of direction called refraction.

Evidence for Particles
The photoelectric effect and the quantization of energy provide evidence that light is a particle.

Reflection
The law of reflection states that the angle of an incoming ray of light and the angle of the reflected light are always equal.

Evidence for Waves
Interference and polarization provide evidence that light is a wave.

The Present Theory
Light is considered to have a dual nature, sometimes acting like a wave and sometimes acting like a particle.

You use light and your eyes more than any other sense to learn about your surroundings. All of your other senses—touch, taste, sound, and smell—involve matter, but the most information is provided by light. Yet light seems more mysterious than matter. You can study matter directly, measuring its dimensions, taking it apart, and putting it together to learn about it. Light, on the other hand, can only be studied indirectly in terms of how it behaves (Figure 7.1). Once you understand its behavior, you know everything there is to know about light. Anything else is thinking about what the behavior means.

The behavior of light has stimulated thinking, scientific investigations, and debate for hundreds of years. The investigations and debate have occurred because light cannot be directly observed, which makes the exact nature of light very difficult to pin down. For example, you know that light moves energy from one place to another place. You can feel energy from the sun as sunlight warms you, and you know that light has carried this energy across millions of miles of empty space. The ability of light to move energy like this could be explained (1) as energy transported by waves, just as sound waves carry energy from a source, or (2) as the kinetic energy of a stream of moving particles, which give up their energy when they strike a surface. The movement of energy from place to place could be explained equally well by a wave model of light or by a particle model of light. When two possibilities exist like this in science, experiments are designed and measurements are made to support one model and reject the other. Light, however, presents a baffling dilemma. Some experiments provide evidence that light consists of waves and not a stream of moving particles. Yet other experiments provide evidence of just the opposite, that light is a stream of particles and not a wave. Evidence for accepting a wave or particle model seems to depend on which experiments are considered.

The purpose of using a model is to make new things understandable in terms of what is already known. When these new things concern light, three models are useful in visualizing separate behaviors. Thus, the electromagnetic wave model will be used to describe how light is created at a source. Another model, a model of light as a ray, a small beam of light, will be used to discuss some common properties of light such as reflection and the refraction, or bending, of light. Finally, properties of light that provide evidence for a particle model will be discussed before ending with a discussion of the present understanding of light.

SOURCES OF LIGHT

The Sun and other stars, lightbulbs, and burning materials all give off light. When something produces light, it is said to be **luminous.** The Sun is a luminous object that provides almost all of the *natural* light on Earth. A small amount of light does reach Earth from the stars but not really enough to see by on a moonless night. The Moon and planets shine by reflected light and do not produce their own light, so they are not luminous.

Burning has been used as a source of *artificial* light for thousands of years. A wood fire and a candle flame are luminous because of their high temperatures. When visible light is given off as a result of high temperatures, the light source is said to be **incandescent.** A flame from any burning source, an ordinary lightbulb, and the Sun are all incandescent sources because of high temperatures.

How do incandescent objects produce light? One explanation is given by the electromagnetic wave model. This model describes a relationship between electricity, magnetism, and light. The model pictures an electromagnetic wave as forming whenever an electric charge is *accelerated* by some external force. Just as a rock thrown into a pond disturbs the water and generates a water wave that spreads out from the rock, an accelerating charge disturbs the electrical properties of the space around it, producing a wave consisting of electric and magnetic fields (Figure 7.2). This wave continues moving through space until it interacts with matter, giving up its energy.

The frequency of an electromagnetic wave depends on the acceleration of the charge; the greater the acceleration, the higher the frequency of the wave that is produced. The complete range of frequencies is called the *electromagnetic spectrum* (Figure 7.3). The spectrum ranges from radio waves at the low-frequency end of the spectrum to gamma rays at the high-frequency end. Visible light occupies only a small part of the middle portion of the complete spectrum.

Visible light is emitted from incandescent sources at high temperatures, but actually electromagnetic radiation is given off from matter at *any* temperature. This radiation is called **blackbody radiation,** which refers to an idealized material (the *blackbody*) that perfectly absorbs and perfectly emits electromagnetic radiation. From the electromagnetic wave model, the radiation originates from the acceleration of charged particles near the surface of an object. The frequency of the blackbody radiation is determined by the temperature of the object. Near absolute

FIGURE 7.1 Light, sounds, and odors can identify the pleasing environment of this garden, but light provides the most information. Sounds and odors can be identified and studied directly, but light can only be studied indirectly, that is, in terms of how it behaves. As a result, the behavior of light has stimulated scientific investigations and debate for hundreds of years. Perhaps you have wondered about light and its behaviors. What is light?

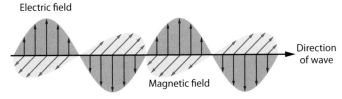

FIGURE 7.2 The electric and magnetic fields in an electromagnetic wave vary together. Here the fields are represented by arrows that indicate the strength and direction of the fields. Note the fields are perpendicular to each other and to the direction of the wave.

zero, there is little energy available and no radiation is given off. As the temperature of an object is increased, more energy is available, and this energy is distributed over a range of values, so more than one frequency of radiation is emitted. A graph of the frequencies emitted from the range of available energy is thus somewhat bell-shaped. The steepness of the curve and the position of the peak depend on the temperature (Figure 7.4). As the temperature of an object increases, there is an increase in the *amount* of radiation given off, and the peak radiation emitted progressively *shifts* toward higher and higher frequencies.

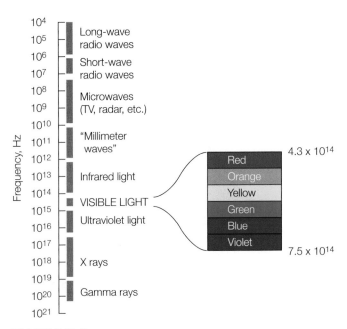

FIGURE 7.3 The electromagnetic spectrum. All electromagnetic waves have the same fundamental character and the same speed in a vacuum, but many aspects of their behavior depend on their frequency.

At room temperature the radiation given off from an object is in the infrared region, invisible to the human eye. When the temperature of the object reaches about 700°C (about 1,300°F), the peak radiation is still in the infrared region, but the peak has shifted enough toward the higher frequencies that a little visible light is emitted as a dull red glow. As the temperature of the object continues to increase, the amount of radiation increases, and the peak continues to shift toward shorter wavelengths. Thus, the object begins to glow brighter, and the color changes from red, to orange, to yellow, and eventually to white. The association of this color change with temperature is noted in the referent description of an object being "red hot," "white hot," and so forth.

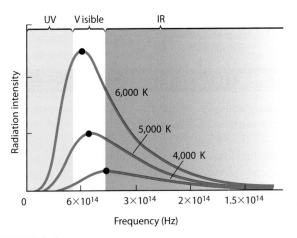

FIGURE 7.4 Three different objects emitting blackbody radiation at three different temperatures. The frequency of the peak of the curve (shown by dot) shifts to higher frequency at higher temperatures.

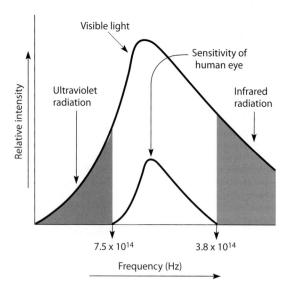

FIGURE 7.5 Sunlight is about 9 percent ultraviolet radiation, 40 percent visible light, and 51 percent infrared radiation before it travels through Earth's atmosphere.

The incandescent flame of a candle or fire results from the blackbody radiation of carbon particles in the flame. At a blackbody temperature of 1,500°C (about 2,700°F), the carbon particles emit visible light in the red to yellow frequency range. The tungsten filament of an incandescent lightbulb is heated to about 2,200°C (about 4,000°F) by an electric current. At this temperature, the visible light emitted is in the reddish, yellow-white range.

The radiation from the Sun, or sunlight, comes from the Sun's surface, which has a temperature of about 5,700°C (about 10,000°F). As shown in Figure 7.5, the Sun's radiation has a broad spectrum centered near the yellow-green frequency. Your eye is most sensitive to this frequency of sunlight. The spectrum of sunlight before it travels through Earth's atmosphere is infrared (about 51 percent), visible light (about 40 percent), and ultraviolet (about 9 percent). Sunlight originated as energy released from nuclear reactions in the Sun's core (see p. 359). This energy requires about a million years to work its way up to the surface. At the surface, the energy from the core accelerates charged particles, which then emit light. The sunlight requires about eight minutes to travel the distance from the Sun's surface to Earth.

PROPERTIES OF LIGHT

You can see luminous objects from the light they emit, and you can see nonluminous objects from the light they reflect, but you cannot see the path of the light itself. For example, you cannot see a flashlight beam unless you fill the air with chalk dust or smoke. The dust or smoke particles reflect light, revealing the path of the beam. This simple observation must be unknown to the makers of science fiction movies, since they always show visible laser beams zapping through the vacuum of space.

Some way to represent the invisible travels of light is needed in order to discuss some of its properties. Throughout history, a **light ray model** has been used to describe the travels of light. The meaning of this model has changed over time, but it has

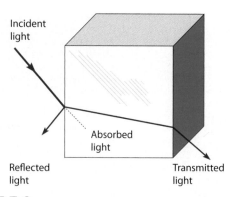

FIGURE 7.6 Light that interacts with matter can be reflected, absorbed, or transmitted through transparent materials. Any combination of these interactions can take place, but a particular substance is usually characterized by what it mostly does to light.

always been used to suggest that "something" travels in *straight-line paths*. The light ray is a line that is drawn to represent the straight-line travel of light. A line is drawn to represent this imaginary beam to illustrate the law of reflection (as from a mirror) and the law of refraction (as through a lens). There are limits to using a light ray for explaining some properties of light, but it works very well in explaining mirrors, prisms, and lenses.

LIGHT INTERACTS WITH MATTER

A ray of light travels in a straight line from a source until it encounters some object or particles of matter (Figure 7.6). What happens next depends on several factors, including (1) the smoothness of the surface, (2) the nature of the material, and (3) the angle at which the light ray strikes the surface.

The *smoothness* of the surface of an object can range from perfectly smooth to extremely rough. If the surface is perfectly smooth, rays of light undergo *reflection*, leaving the surface parallel to each other. A mirror is a good example of a very smooth surface that reflects light this way (Figure 7.7A). If a surface is not smooth, the light rays are reflected in many random directions as *diffuse reflection* takes place (Figure 7.7B). Rough and irregular surfaces and dust in the air make diffuse reflections. It is diffuse reflection that provides light in places not in direct lighting, such as under a table or under a tree. Such shaded areas would be very dark without the diffuse reflection of light.

Some materials allow much of the light that falls on them to move through the material without being reflected. Materials that allow transmission of light through them are called *transparent*. Glass and clear water are examples of transparent materials. Many materials do not allow transmission of any light and are called *opaque*. Opaque materials reflect light, absorb light, or some combination of partly absorbing and partly reflecting light (Figure 7.8). The light that is reflected varies with wavelength and gives rise to the perception of color, which will be discussed shortly. Absorbed light gives up its energy to the material and may be reemitted at a different wavelength, or it may simply show up as a temperature increase.

The *angle* of the light ray to the surface and the nature of the material determine if the light is absorbed, transmitted through

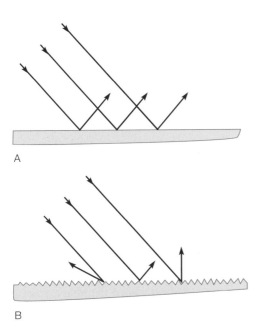

FIGURE 7.7 (*A*) Rays reflected from a perfectly smooth surface are parallel to each other. (*B*) Diffuse reflection from a rough surface causes rays to travel in many random directions.

a transparent material, or reflected. Vertical rays of light, for example, are mostly transmitted through a transparent material with some reflection and some absorption. If the rays strike the surface at some angle, however, much more of the light is reflected, bouncing off the surface. Thus, the glare of reflected sunlight is much greater around a body of water in the late afternoon than when the Sun is directly overhead.

Light that interacts with matter is reflected, transmitted, or absorbed, and all combinations of these interactions are possible.

FIGURE 7.9 (*A*) A one-way mirror reflects most of the light and transmits some light. You can see such a mirror around the top of the walls in this store. (*B*) Here is the view from behind the mirror.

Materials are usually characterized by which of these interactions they *mostly* do, but this does not mean that other interactions are not occurring too. For example, a window glass is usually characterized as a transmitter of light. Yet the glass always *reflects* about 4 percent of the light that strikes it. The reflected light usually goes unnoticed during the day because of the bright light that is transmitted from the outside. When it is dark outside, you notice the reflected light as the window glass now appears to act much like a mirror. A one-way mirror is another example of both reflection and transmission occurring (Figure 7.9). A mirror is usually characterized as a reflector of light. A one-way mirror, however, has a very thin silvering that reflects most of the light but still transmits a little. In a lighted room, a one-way mirror appears to reflect light just as any other mirror does. But a person behind the mirror in a dark room can see into the lighted room by means of the transmitted light. Thus, you know that this mirror transmits as well as reflects light. One-way mirrors are used to unobtrusively watch for shoplifters in many businesses.

FIGURE 7.8 Light travels in a straight line, and the color of an object depends on which wavelengths of light the object reflects. Each of these flowers absorbs most of the colors and reflects the color that you see.

Myths, Mistakes, & Misunderstandings

The Light Saber

The *Star Wars'* light saber is an impossible myth. Assuming that the light saber is a laser beam, we know that one laser beam will not stop another laser beam. Light beams simply pass through each other. Furthermore, a laser with a fixed length is not possible without a system of lenses that would also scatter the light, in addition to being cumbersome on a saber. Moreover, scattered laser light from reflective surfaces could result in injury to you.

REFLECTION

Most of the objects that you see are visible from diffuse reflection. For example, consider some object such as a tree that you see during a bright day. Each *point* on the tree must reflect light in all directions, since you can see any part of the tree from any angle (Figure 7.10). As a model, think of bundles of light rays entering your eye, which enable you to see the tree. This means that you can see any part of the tree from any angle because different bundles of reflected rays will enter your eye from different parts of the tree.

Light rays that are diffusely reflected move in all possible directions, but rays that are reflected from a smooth surface, such as a mirror, leave the mirror in a definite direction. Suppose you look at a tree in a mirror. There is only one place on the mirror where you look to see any one part of the tree. Light is reflecting off the mirror from all parts of the tree, but the only rays that reach your eyes are the rays that are reflected at a certain angle from the place where you look. The relationship between the light rays moving from the tree and the direction in

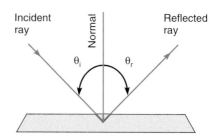

FIGURE 7.11 The law of reflection states that the angle of incidence (θ_i) is equal to the angle of reflection (θ_r). Both angles are measured from the *normal*, a reference line drawn perpendicular to the surface at the point of reflection.

which they are reflected from the mirror to reach your eyes can be understood by drawing three lines: (1) a line representing an original ray from the tree, called the *incident ray*, (2) a line representing a reflected ray, called the *reflected ray*, and (3) a reference line that is perpendicular to the reflecting surface and is located at the point where the incident ray struck the surface. This line is called the *normal*. The angle between the incident ray and the normal is called the *angle of incidence*, θ_i, and the angle between the reflected ray and the normal is called the *angle of reflection*, θ_r (Figure 7.11). The *law of reflection*, which was known to the ancient Greeks, is that the *angle of incidence equals the angle of reflection*, or

$$\theta_i = \theta_r$$

equation 7.1

Figure 7.12 shows how the law of reflection works when you look at a flat mirror. Light is reflected from all points on the block, and of course only the rays that reach your eyes are detected. These rays are reflected according to the law of reflection, with the angle of reflection equaling the angle of incidence. If you move your head slightly, then a different bundle of rays reaches your eyes. Of all the bundles of rays that reach your eyes, only two rays from a point are shown in the illustration. After these two rays are reflected, they continue to spread apart at the

FIGURE 7.10 Bundles of light rays are reflected diffusely in all directions from every point on an object. Only a few light rays are shown from only one point on the tree in this illustration. The light rays that move to your eyes enable you to see the particular point from which they were reflected.

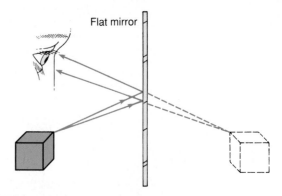

FIGURE 7.12 Light rays leaving a point on the block are reflected according to the law of reflection, and those reaching your eye are seen. After reflecting, the rays continue to spread apart at the same rate. You interpret this to be a block the same distance behind the mirror. You see a virtual image of the block, because light rays do not actually move from the image.

same rate that they were spreading before reflection. Your eyes and brain do not know that the rays have been reflected, and the diverging rays appear to come from behind the mirror, as the dashed lines show. The image, therefore, appears to be the same distance *behind* the mirror as the block is from the front of the mirror. Thus, a mirror image is formed where the rays of light *appear* to originate. This is called a **virtual image.** A virtual image is the result of your eyes' and brain's interpretations of light rays, not actual light rays originating from an image. Light rays that do originate from the other kind of image are called a **real image.** A real image is like the one displayed on a movie screen, with light originating from the image. A virtual image cannot be displayed on a screen, since it results from an interpretation.

Curved mirrors are either *concave,* with the center part curved inward, or *convex,* with the center part bulging outward. A concave mirror can be used to form an enlarged virtual image, such as a shaving or makeup mirror, or it can be used to form a real image, as in a reflecting telescope. Convex mirrors, such as the mirrors on the sides of trucks and vans, are often used to increase the field of vision. Convex mirrors are also used in a driveway to show a wide area (Figure 7.13).

FIGURE 7.13 This convex mirror increases your field of vision, enabling you to see merging cars from the left.

CONCEPTS *Applied*

Good Reflections

Do you need a full-length mirror to see your hair and your feet at the same time? Does how much you see in a mirror depend on how far you stand from the mirror? If you wink your right eye, does the right or left eye of your image wink back?

You can check your answers to the above questions with a large mirror on a wall, masking tape, a meter stick, and someone to help.

Stand facing the mirror and tell your helper exactly where to place a piece of masking tape on the mirror to mark the top of your head. Place a second piece of tape on the mirror to mark the top of your belt or some other part of your clothing.

Now we need some measurements on you and on the mirror. First, measure and record the actual distance from the top of your head to the top of your belt or the other clothing marked. Second, measure and record the distance between the two pieces of masking tape on the mirror.

Now step back, perhaps doubling your distance from the mirror, and repeat the procedure. What can you conclude about your distance from the mirror and the image you see? Is there a difference in how much you can see? How tall a mirror do you need to see yourself from head to toe? Finally, is there some way you can cause your image to wink the same eye as you?

REFRACTION

You may have observed that an object that is partly in the air and partly in water appears to be broken, or bent, where the air and water meet. When a light ray moves from one transparent material to another, such as from water through air, the ray undergoes a change in the direction of travel at the boundary between the two materials. This change of direction of a light ray at the boundary is called **refraction.** The amount of change can be measured as an angle from the normal, just as it was for the angle of reflection. The incoming ray is called the *incident ray* as before, and the new direction of travel is called the *refracted ray.* The angles of both rays are measured from the normal (Figure 7.14).

Refraction results from a *change in speed* when light passes from one transparent material into another. The speed of light in a vacuum is 3.00×10^8 m/s, but it is slower when moving through a transparent material. In water, for example, the speed of light is reduced to about 2.30×10^8 m/s. The speed of light has a magnitude that is specific for various transparent materials.

When light moves from one transparent material to another transparent material with a *slower* speed of light, the ray is refracted *toward* the normal (Figure 7.15A). For example, light travels through air faster than through water. Light traveling from air into water is therefore refracted toward the normal as it enters the water. On the other hand, if light has a *faster* speed in the new material, it is refracted *away* from the normal.

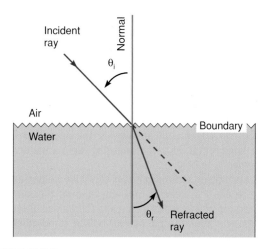

FIGURE 7.14 A ray diagram shows refraction at the boundary as a ray moves from air through water. Note that θ_i does not equal θ_r in refraction.

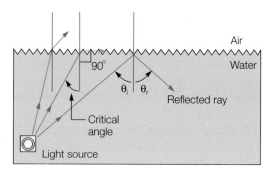

FIGURE 7.16 When the angle of incidence results in an angle of refraction of 90°, the refracted light ray is refracted along the water surface. The angle of incidence for a material that results in an angle of refraction of 90° is called the *critical angle*. When the incident ray is at this critical angle or greater, the ray is reflected internally. The critical angle for water is about 49°, and for a diamond it is about 25°.

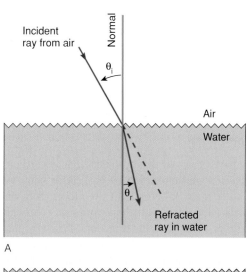

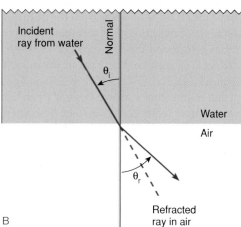

FIGURE 7.15 (*A*) A light ray moving to a new material with a slower speed of light is refracted toward the normal ($\theta_i > \theta_r$). (*B*) A light ray moving to a new material with a faster speed is refracted away from the normal ($\theta_i < \theta_r$).

Thus, light traveling from water into the air is refracted away from the normal as it enters the air (Figure 7.15B).

The magnitude of refraction depends on (1) the angle at which light strikes the surface and (2) the ratio of the speed of light in the two transparent materials. An incident ray that is perpendicular (90°) to the surface is not refracted at all. As the angle of incidence is increased, the angle of refraction is also increased. There is a limit, however, that occurs when the angle of refraction reaches 90°, or along the water surface. Figure 7.16 shows rays of light traveling from water to air at various angles. When the incident ray is about 49°, the angle of refraction that results is 90° along the water surface. This limit to the angle of incidence that results in an angle of refraction of 90° is called the *critical angle* for a water-to-air surface (Figure 7.16). At any incident angle greater than the critical angle, the light ray does not move from the water to the air but is *reflected* back from the surface as if it were a mirror. This is called **total internal reflection** and implies that the light is trapped inside if it arrived at the critical angle or beyond. Faceted transparent gemstones such as the diamond are brilliant because they have a small critical angle and thus reflect much light internally. Total internal reflection is also important in fiber optics.

CONCEPTS *Applied*

Internal Reflection

Seal a flashlight in a clear plastic bag to waterproof it, then investigate the critical angle and total internal reflection in a swimming pool, play pool, or large tub of water. In a darkened room or at night, shine the flashlight straight up from beneath the water, then at different angles until it shines almost horizontally beneath the surface. Report your observation of the critical angle for the water used.

TABLE 7.1

Index of refraction

Substance	$n = c/v$
Glass	1.50
Diamond	2.42
Ice	1.31
Water	1.33
Benzene	1.50
Carbon tetrachloride	1.46
Ethyl alcohol	1.36
Air (0°C)	1.00029
Air (30°C)	1.00026

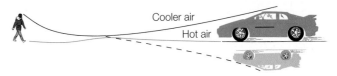

FIGURE 7.17 Mirages are caused by hot air near the ground refracting, or bending, light rays upward into the eyes of a distant observer. The observer believes he is seeing an upside-down image reflected from water on the highway.

As was stated earlier, refraction results from a change in speed when light passes from one transparent material into another. The ratio of the speeds of light in the two materials determines the magnitude of refraction at any given angle of incidence. The greatest speed of light possible, according to current theory, occurs when light is moving through a vacuum. The speed of light in a vacuum is accurately known to nine decimals but is usually rounded to 3.00×10^8 m/s for general discussion. The speed of light in a vacuum is a very important constant in physical science, so it is given a symbol of its own, c. The ratio of c to the speed of light in some transparent material, v, is called the **index of refraction,** n, of that material or

$$n = \frac{c}{v}$$

equation 7.2

The indexes of refraction for some substances are listed in Table 7.1. The values listed are constant physical properties and can be used to identify a specific substance. Note that a larger value means a greater refraction at a given angle. Of the materials listed, diamond refracts light the most and air the least. The index for air is nearly 1, which means that light is slowed only slightly in air.

EXAMPLE 7.1

What is the speed of light in a diamond?

SOLUTION

The relationship between the speed of light in a material (v), the speed of light in a vacuum ($c = 3.00 \times 10^8$ m/s), and the index of refraction is given in equation 7.2. The index of refraction of a diamond is found in Table 7.1 ($n = 2.42$).

$$n_{diamond} = 2.42 \qquad n = \frac{c}{v} \quad \therefore \quad v = \frac{c}{n}$$
$$c = 3.00 \times 10^8 \text{ m/s}$$
$$v = ? \qquad v = \frac{3.00 \times 10^8 \text{ m/s}}{2.42}$$
$$= \boxed{1.24 \times 10^8 \text{ m/s}}$$

Note that Table 7.1 shows that colder air at 0°C (32°F) has a higher index of refraction than warmer air at 30°C (86°F), which means that light travels faster in warmer air. This difference explains the "wet" highway that you sometimes see at a distance in the summer. The air near the road is hotter on a clear, calm day. Light rays traveling toward you in this hotter air are refracted upward as they enter the cooler air. Your brain interprets this refracted light as *reflected* light, but no reflection is taking place. Light traveling downward from other cars is also refracted upward toward you, and you think you are seeing cars "reflected" from the wet highway (Figure 7.17). When you reach the place where the "water" seemed to be, it disappears, only to appear again farther down the road.

Sometimes convection currents produce a mixing of warmer air near the road with the cooler air just above. This mixing refracts light one way, then the other, as the warmer and cooler air mix. This produces a shimmering or quivering that some people call "seeing heat." They are actually seeing changing refraction, which is a *result* of heating and convection. In addition to causing distant objects to quiver, the same effect causes the point source of light from stars to appear to twinkle. The light from closer planets does not twinkle because the many light rays from the disklike sources are not refracted together as easily as the fewer rays from the point sources of stars. The light from planets will appear to quiver, however, if the atmospheric turbulence is great.

CONCEPTS *Applied*

Seeing Around Corners

Place a coin in an empty cup. Position the cup so the coin appears to be below the rim, just out of your line of sight. Do not move from this position as your helper slowly pours water into the cup. Explain why the coin becomes visible, then appears to rise in the cup. Use a sketch such as one of those in Figure 7.15 to help with your explanation.

DISPERSION AND COLOR

Electromagnetic waves travel with the speed of light with a whole spectrum of waves of various frequencies and wavelengths. The

A Closer Look

Optics

Historians tell us there are many early stories and legends about the development of ancient optical devices. The first glass vessels were made about 1500 B.C., so it is possible that samples of clear, transparent glass were available soon after. One legend claimed that the ancient Chinese invented eyeglasses as early as 500 B.C. A burning glass (lens) was mentioned in an ancient Greek play written about 424 B.C. Several writers described how Archimedes saved his hometown of Syracuse with a burning glass in about 214 B.C. Syracuse was besieged by Roman ships when Archimedes supposedly used the burning glass to focus sunlight on the ships, setting them on fire. It is not known if this story is true or not, but it is known that the Romans indeed did have burning glasses. Glass spheres, which were probably used to start fires, have been found in Roman ruins, including a convex lens recovered from the ashes of Pompeii.

Today, lenses are no longer needed to start fires, but they are common in cameras, scanners, optical microscopes, eyeglasses, lasers, binoculars, and many other optical devices. Lenses are no longer just made from glass, and today many are made from a transparent, hard plastic that is shaped into a lens.

The primary function of a lens is to form an image of a real object by refracting incoming parallel light rays. Lenses have two basic shapes, with the center of a surface either bulging in or bulging out. The outward bulging shape is thicker at the center than around the outside edge, and this is called a *convex lens* (Box Figure 7.1A). The other basic lens shape is just the opposite, thicker around the outside edge than at the center, and is called a *concave lens* (Box Figure 7.1B).

Convex lenses are used to form images in magnifiers, cameras, eyeglasses, projectors, telescopes, and microscopes (Box Figure 7.2). Concave lenses are used in some eyeglasses and in combination with the convex lens to correct for defects. The convex lens is the most commonly used lens shape.

Your eyes are optical devices with convex lenses. Box Figure 7.3 shows the basic structure. First, a transparent hole called the *pupil* allows light to enter the eye. The size of the pupil is controlled by the *iris*, the colored part that is a muscular diaphragm. The *lens* focuses a sharp image

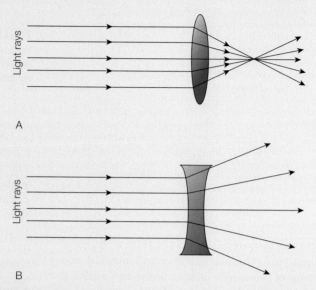

BOX FIGURE 7.1 (*A*) Convex lenses are called converging lenses because they bring together, or converge, parallel rays of light. (*B*) Concave lenses are called diverging lenses because they spread apart, or diverge, parallel rays of light.

on the back surface of the eye, the *retina*. The retina is made up of millions of light-sensitive structures, and nerves carry electrical signals from the retina to the optic nerve, then to the brain.

The lens is a convex, pliable material held in place and changed in shape by the attached *ciliary muscle*. When the eye is focused on a distant object, the ciliary muscle is completely relaxed. Looking at a closer object requires the contraction of the ciliary muscles to change the curvature of the lens. This adjustment of focus by the action of the ciliary muscle is called *accommodation*. The closest distance an object can be seen without a blurred image is called the *near point*, and this is the limit to accommodation.

The near point moves outward with age as the lens becomes less pliable. By middle age, the near point may be twice this distance or greater, creating the condition known as farsightedness. The condition of farsightedness, or *hyperopia*, is a problem associated with aging (called presbyopia). Hyperopia can be caused at an early age by an eye that is too short or by problems with the cornea or lens that focus the image behind the retina. Farsightedness can be corrected with a convex lens, as shown in Box Figure 7.4F.

Nearsightedness, or *myopia*, is a problem caused by an eye that is too long or problems with the cornea or lens that focus the image in front of the retina. Nearsightedness can be corrected with a concave lens, as shown in Box Figure 7.4D.

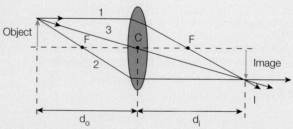

BOX FIGURE 7.2 A convex lens forms an inverted image from refracted light rays of an object outside the focal point. Convex lenses are mostly used to form images in cameras, file or overhead projectors, magnifying glasses, and eyeglasses.

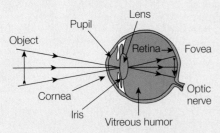

BOX FIGURE 7.3 Light rays from a distant object are focused by the lens onto the retina, a small area on the back of the eye.

The microscope is an optical device used to make things look larger. It is essentially a system of two lenses, one to produce an image of the object being studied, and the other to act as a magnifying glass and enlarge that image. The power of

the microscope is basically determined by the *objective lens,* which is placed close to the specimen on the stage of the microscope. Light is projected up through the specimen, and the objective lens makes an enlarged image of the specimen inside the tube between the two lenses. The *eyepiece lens* is adjusted up and down to make a sharp enlarged image of the image produced by the objective lens (Box Figure 7.5).

Telescopes are optical instruments used to provide enlarged images of near and distant objects. There are two major types of telescope: *refracting* telescopes that use two lenses and *reflecting* telescopes that use combinations of mirrors, or a mirror and a lens. The refracting telescope has two lenses, with the objective lens forming a reduced image, which is viewed with an eyepiece lens to enlarge that image. In reflecting telescopes,

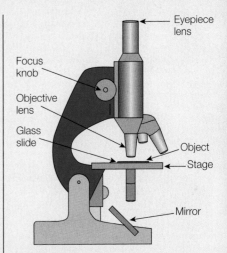

BOX FIGURE 7.5 A simple microscope uses a system of two lenses, which are an objective lens that makes an enlarged image of the specimen and an eyepiece lens that makes an enlarged image of that image.

mirrors are used instead of lenses to collect the light (Box Figure 7.6).

Finally, the *digital camera* is one of the more recently developed light-gathering and photograph-taking optical instruments. This camera has a group of small photocells, with perhaps thousands lined up on the focal plane behind a converging lens. An image falls on the array, and each photocell stores a charge that is proportional to the amount of light falling on the cell. A microprocessor measures the amount of charge registered by each photocell and considers it as a pixel, a small bit of the overall image. A shade of gray or a color is assigned to each pixel, and the image is ready to be enhanced, transmitted to a screen, printed, or magnetically stored for later use.

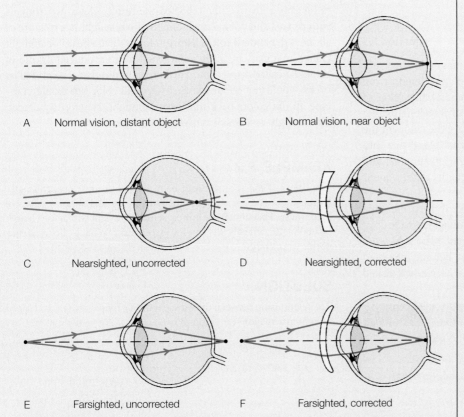

A   Normal vision, distant object   B   Normal vision, near object

C   Nearsighted, uncorrected   D   Nearsighted, corrected

E   Farsighted, uncorrected   F   Farsighted, corrected

BOX FIGURE 7.4 (*A*) The relaxed, normal eye forms an image of distant objects on the retina. (*B*) For close objects, the lens of the normal eye changes shape to focus the image on the retina. (*C*) In a nearsighted eye, the image of a distant object forms in front of the retina. (*D*) A diverging lens corrects for nearsightedness. (*E*) In a farsighted eye, the image of a nearby object forms beyond the retina. (*F*) A converging lens corrects for farsightedness.

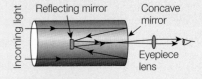

BOX FIGURE 7.6 This illustrates how the path of light moves through a simple reflecting astronomical telescope. Several different designs and mirror placements are possible.

FIGURE 7.18 The flowers appear to be red because they reflect light in the 7.9×10^{-7} m to 6.2×10^{-7} m range of wavelengths.

speed of electromagnetic waves (c) is related to the wavelength (λ) and the frequency (f) by

$$c = \lambda f$$

equation 7.3

Visible light is the part of the electromagnetic spectrum that your eyes can detect, a narrow range of wavelength from about 7.90×10^{-7} m to 3.90×10^{-7} m. In general, this range of visible light can be subdivided into ranges of wavelengths that you perceive as colors (Figure 7.18). These are the colors of the rainbow, and there are six distinct colors that blend one into another. These colors are *red, orange, yellow, green, blue,* and *violet*. The corresponding ranges of wavelengths and frequencies of these colors are given in Table 7.2.

In general, light is interpreted to be white if it has the same mixture of colors as the solar spectrum. That sunlight is made up of component colors was first investigated in detail by Isaac Newton (1642–1727). While a college student, Newton became interested in grinding lenses, light, and color. At the age of 23, Newton visited

a local fair and bought several triangular glass prisms. He then proceeded to conduct a series of experiments with a beam of sunlight in his room. In 1672, he reported the results of his experiments with prisms and color, concluding that white light is a mixture of all the independent colors. Newton found that a beam of sunlight falling on a glass prism in a darkened room produced a band of colors he called a *spectrum*. Further, he found that a second glass prism would not subdivide each separate color but would combine all the colors back into white sunlight. Newton concluded that sunlight consists of a mixture of the six colors.

EXAMPLE 7.2

The colors of the spectrum can be measured in units of wavelength, frequency, or energy, which are alternative ways of describing colors of light waves. The human eye is most sensitive to light with a wavelength of 5.60×10^{-7} m, which is a yellow-green color. What is the frequency of this wavelength?

SOLUTION

The relationship between the wavelength (λ), frequency (f), and speed of light in a vacuum (c), is found in equation 7.3, $c = \lambda f$.

$$c = 3.00 \times 10^8 \text{ m/s}$$
$$\lambda = 5.60 \times 10^{-7} \text{ m}$$
$$f = ?$$

$$c = \lambda f \quad \therefore \quad \frac{c}{\lambda}$$

$$f = \frac{3.00 \times 10^8 \frac{m}{s}}{5.60 \times 10^{-7} \text{ m}}$$

$$= \frac{3.00 \times 10^8}{5.60 \times 10^{-7}} \frac{m}{s} \times \frac{1}{m}$$

$$= 5.40 \times 10^{14} \frac{1}{s}$$

$$= \boxed{5.40 \times 10^{14} \text{ Hz}}$$

TABLE 7.2

Range of wavelengths and frequencies of the colors of visible light

Color	Wavelength (in meters)	Frequency (in hertz)
Red	7.9×10^{-7} to 6.2×10^{-7}	3.8×10^{14} to 4.8×10^{14}
Orange	6.2×10^{-7} to 6.0×10^{-7}	4.8×10^{14} to 5.0×10^{14}
Yellow	6.0×10^{-7} to 5.8×10^{-7}	5.0×10^{14} to 5.2×10^{14}
Green	5.8×10^{-7} to 4.9×10^{-7}	5.2×10^{14} to 6.1×10^{14}
Blue	4.9×10^{-7} to 4.6×10^{-7}	6.1×10^{14} to 6.6×10^{14}
Violet	4.6×10^{-7} to 3.9×10^{-7}	6.6×10^{14} to 7.7×10^{14}

Recall that the index of refraction is related to the speed of light in a transparent substance. A glass prism separates sunlight into a spectrum of colors because the index of refraction is different for different wavelengths of light. The same processes that slow the speed of light in a transparent substance have a greater effect on short wavelengths than they do on longer wavelengths. As a result, violet light is refracted most, red light is refracted least, and the other colors are refracted between these extremes. This results in a beam of white light being separated, or dispersed, into a spectrum when it is refracted. Any transparent material in which the index of refraction varies with wavelength has the property of *dispersion*. The dispersion of light by ice crystals sometimes produces a colored halo around the Sun and the Moon.

 CONCEPTS *Applied*

Colors and Refraction

A convex lens is able to magnify by forming an image with refracted light. This application is concerned with magnifying, but it is really more concerned with experimenting to find an explanation.

Here are three pairs of words:

SCIENCE **BOOK**
RAW **HIDE**
CARBON **DIOXIDE**

Hold a cylindrical solid glass rod over the three pairs of words, using it as a magnifying glass. A clear, solid, and transparent plastic rod or handle could also be used as a magnifying glass.

Notice that some words appear inverted but others do not. Does this occur because red letters are refracted differently than blue letters?

Make some words with red and blue letters to test your explanation. What is your explanation for what you observed?

EVIDENCE FOR WAVES

The nature of light became a topic of debate toward the end of the 1600s as Isaac Newton published his *particle theory* of light. He believed that the straight-line travel of light could be better explained as small particles of matter that traveled at great speed from a source of light. Particles, reasoned Newton, should follow a straight line according to the laws of motion. Waves, on the other hand, should bend as they move, much as water waves on a pond bend into circular shapes as they move away from a disturbance. About the same time that Newton developed his particle theory of light, Christian Huygens (pronounced "ni-ganz") (1629–1695) was concluding that light is not a stream of particles but rather a longitudinal wave.

Both theories had advocates during the 1700s, but the majority favored Newton's particle theory. By the beginning of the 1800s, new evidence was found that favored the wave theory, evidence that could not be explained in terms of anything but waves.

INTERFERENCE

In 1801, Thomas Young (1773–1829) published evidence of a behavior of light that could only be explained in terms of a wave model of light. Young's experiment is illustrated in Figure 7.19A. Light from a single source is used to produce two beams of light that are in phase, that is, having their crests and troughs together as they move away from the source. This light falls on a card with two slits, each less than a millimeter in width. The light moves out from each slit as an expanding arc. Beyond the card, the light from one slit crosses over the light from the other slit to produce a series of bright lines on a screen. Young had produced a phenomenon of light called **interference,** and interference can only be explained by waves.

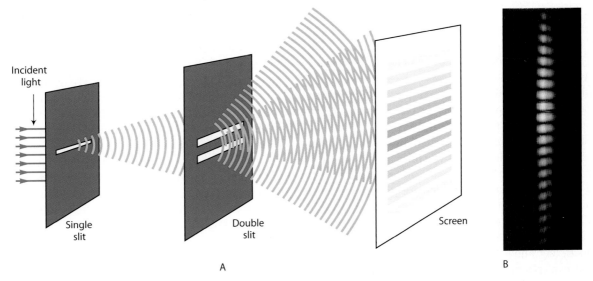

A   B

FIGURE 7.19 (*A*) The arrangement for Young's double-slit experiment. Sunlight passing through the first slit is coherent and falls on two slits close to each other. Light passing beyond the two slits produces an interference pattern on a screen. (*B*) The double-slit pattern of a small-diameter beam of light from a helium-neon laser.

A Closer Look

The Rainbow

A rainbow is a spectacular, natural display of color that is supposed to have a pot of gold under one end. Understanding the why and how of a rainbow requires information about water droplets and knowledge of how light is reflected and refracted. This information will also explain why the rainbow seems to move when you move—making it impossible to reach the end to obtain that mythical pot of gold.

First, note the pattern of conditions that occur when you see a rainbow. It usually appears when the Sun is shining low in one part of the sky and rain is falling in the opposite part. With your back to the Sun, you are looking at a zone of raindrops that are all showing red light, another zone that are all showing violet light, with zones of the other colors between (ROYGBV). For a rainbow to form like this requires a surface that refracts and reflects the sunlight, a condition met by spherical raindrops.

Water molecules are put together in such a way that they have a positive side and a negative side, and this results in strong molecular attractions. It is the strong attraction of water molecules for one another that results in the phenomenon of surface tension. Surface tension is the name given to the surface of water acting as if it is covered by an ultrathin elastic membrane that is contracting. It is surface tension that pulls

raindrops into a spherical shape as they fall through the air.

Box Figure 7.7 shows one thing that can happen when a ray of sunlight strikes a single spherical raindrop near the top of the drop. At this point, some of the sunlight is reflected, and some is refracted into the raindrop. The refraction disperses the light into its spectrum colors, with the violet light being refracted most and red the least. The refracted light travels through the drop to the opposite side, where some of it might be reflected back into the drop. The reflected part travels back through the drop again, leaving the front surface of the raindrop. As it leaves, the light is refracted for a second time. The combined refraction, reflection, and second refraction is the source of the zones of colors you see in a rainbow. This also explains why you see a rainbow in the part of the sky opposite from the sun.

The light from any one raindrop is one color, and that color comes from all drops on the arc of a circle that is a certain angle between the incoming sunlight and the refracted light. Thus, the raindrops in the red region refract red light toward your eyes at an angle of 42°, and all other colors are refracted over your head by these drops. Raindrops in the violet region refract violet light toward your eyes at an angle of 40° and the red and other colors toward your feet.

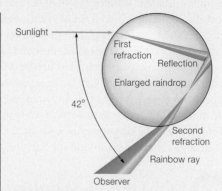

BOX FIGURE 7.7 Light is refracted when it enters a raindrop and when it leaves. The part that leaves the front surface of the raindrop is the source of the light in thousands upon thousands of raindrops from which you see zones of color—a rainbow.

Thus, the light from any one drop is seen as one color, and all drops showing this color are on the arc of a circle. An arc is formed because the angle between the sunlight and the refracted light of a color is the same for each of the spherical drops.

There is sometimes a fainter secondary rainbow, with colors reversed, that forms from sunlight entering the bottom of the drop, reflecting twice, and then refracting out the top. The double reflection reverses the colors, and the angles are 50° for the red and 54° for the violet. (See Figure 1.15 on p. 18.)

The pattern of bright lines and dark zones is called an *interference pattern* (Figure 7.19B). The light moved from each slit in phase, crest to crest and trough to trough. Light from both slits traveled the same distance directly across to the screen, so they arrived in phase. The crests from the two slits are superimposed here, and constructive interference produces a bright line in the center of the pattern. But for positions above and below the center, the light from the two slits must travel different distances to the screen. At a certain distance above and below the bright center line, light from one slit had to travel a greater distance and arrives one-half wavelength after light from the other slit. Destructive interference produces a zone of darkness at these positions. Continuing up and down the screen, a bright line of constructive interference will occur at each position where the distance traveled by light from the two slits differs by any whole number of wavelengths. A dark zone of destructive interference will occur at each position where the distance traveled by light from the two slits differs by any half-wavelength. Thus, bright lines occur above and below the center bright line at positions representing differences in paths of 1, 2,

3, 4, and so on wavelengths. Similarly, zones of darkness occur above and below the center bright line at positions representing differences in paths of ½, 1½, 2½, 3½, and so on wavelengths. Young found all of the experimental data such as this in full agreement with predictions from a wave theory of light. About fifteen years later, A. J. Fresnel (pronounced "fray-nel") (1788–1827) demonstrated mathematically that diffraction as well as other behaviors of light could be fully explained with the wave theory. In 1821, Fresnel determined that the wavelength of red light was about 8×10^{-7} m and of violet light about 4×10^{-7} m, with other colors in between these two extremes. The work of Young and Fresnel seemed to resolve the issue of considering light to be a stream of particles or a wave, and it was generally agreed that light must be waves.

POLARIZATION

Huygens' wave theory and Newton's particle theory could explain some behaviors of light satisfactorily, but there were some behaviors that neither (original) theory could explain.

The word *laser* is from *l*ight *a*mplification by *s*timulated *e*mission of *r*adiation. A laser is a device that produces a coherent beam of single-frequency, in-phase light. The beam comes from atoms that have been stimulated by electricity. Most ordinary light sources produce incoherent light; light that is emitted randomly and at different frequencies. The coherent light from a laser has the same frequency, phase, and direction, so it does not tend to spread out and it can be very intense. This has made possible a number of specialized applications, and the list of uses continues to grow.

There are different kinds of lasers in use and new ones are under development. One common type of laser is a gas-filled tube with mirrors at both ends. The mirror at one end is only partly silvered, which allows light to escape as the laser beam. The distance between the mirrors matches the resonant frequency of the light produced, so the trapped light will set up an optical standing wave. An electric discharge produces fast electrons that raise the energy level of the electrons of the specific gas atoms in the tube. The electrons of the energized gas atoms emit a particular frequency of light as they drop back to a lower level, and this emitted light sets up the standing wave. The standing wave stimulates other atoms of the gas, resulting in the emission of more light at the same frequency and phase.

Lasers are everywhere today and have connections with a wide variety of technologies. At the supermarket, a laser and detector unit reads the bar code on each grocery item. The laser sends the pattern to a computer, which sends a price to the register as well as tracks the store inventory. A low-powered laser and detector also reads your CD music or MP3 disc and can be used to make a three-dimensional image. Most laser printers use a laser, and a laser is the operational part of a fiber optics communication system. Stronger lasers are used for cutting, drilling, and welding. Lasers are used extensively in many different medical procedures, from welding a detached retina to bloodless surgery.

Both theories failed to explain some behaviors of light, such as light moving through certain transparent crystals. For example, a slice of the mineral tourmaline transmits what appears to be a low-intensity greenish light. But if a second slice of tourmaline is placed on the first and rotated, the transmitted light passing through both slices begins to dim. The transmitted light is practically zero when the second slice is rotated 90°. Newton suggested that this behavior had something to do with "sides" or "poles" and introduced the concept of what is now called the *polarization* of light.

The waves of Huygens' wave theory were longitudinal, moving like sound waves, with wave fronts moving in the direction of travel. A longitudinal wave could not explain the polarization behavior of light. In 1817, Young modified Huygens' theory by describing the waves as *transverse,* vibrating at right angles to the direction of travel. This modification helped explain the polarization behavior of light transmitted through the two crystals and provided firm evidence that light is a transverse wave. As shown in Figure 7.20A, **unpolarized light** is assumed to

consist of transverse waves vibrating in all conceivable random directions. Polarizing materials, such as the tourmaline crystal, transmit light that is vibrating in one direction only, such as the vertical direction in Figure 7.20B. Such a wave is said to be **polarized,** or *plane-polarized* since it vibrates only in one plane. The single crystal polarized light by transmitting only waves that vibrate parallel to a certain direction while selectively absorbing waves that vibrate in all other directions. Your eyes cannot tell the difference between unpolarized and polarized light, so the light transmitted through a single crystal looks just like any other light. When a second crystal is placed on the first, the amount of light transmitted depends on the alignment of the two crystals (Figure 7.21). When the two crystals are *aligned,* the polarized

FIGURE 7.20 (*A*) Unpolarized light has transverse waves vibrating in all possible directions perpendicular to the direction of travel. (*B*) Polarized light vibrates only in one plane. In this illustration, the wave is vibrating in a vertical direction only.

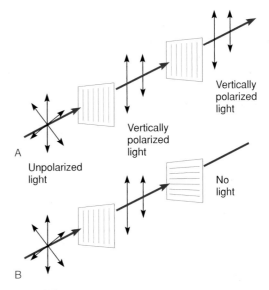

FIGURE 7.21 (*A*) Two crystals that are aligned both transmit vertically polarized light that looks like any other light. (*B*) When the crystals are crossed, no light is transmitted.

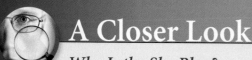
Why Is the Sky Blue?

Sunlight entering our atmosphere is scattered, or redirected, by interactions with air molecules. Sunlight appears to be white to the human eye but is actually a mixture of all the colors of the rainbow. The blue and violet part of the spectrum has shorter wavelengths than the red and orange part.

Shorter wavelength blue and violet light is scattered more strongly than red and orange light. When you look at the sky, you see the light that was redirected by the atmosphere into your line of sight. Since blue and violet light is scattered more efficiently than red and orange light, the sky appears

blue. When viewing a sunrise or sunset, you see only light that has not been scattered in other directions. The red and orange part of sunlight travels through a maximum length of the atmosphere, and the blue and violet has been scattered away, so a sunrise or sunset appears to be more orange and reddish.

---

light from the first crystal passes through the second with little absorption. When the crystals are *crossed* at 90°, the light transmitted by the first is vibrating in a plane that is absorbed by the second crystal, and practically all the light is absorbed. At some other angle, only a fraction of the polarized light from the first crystal is transmitted by the second.

You can verify whether or not a pair of sunglasses is made of polarizing material by rotating a lens of one pair over a lens of a second pair. Light is transmitted when the lenses are aligned but mostly absorbed at 90° when the lenses are crossed.

Light is completely polarized when all the waves are removed except those vibrating in a single direction. Light is partially polarized when some of the waves are in a particular orientation, and any amount of polarization is possible. There are several means of producing partially or completely polarized light, including (1) selective absorption, (2) reflection, and (3) scattering.

Selective absorption is the process that takes place in certain crystals, such as tourmaline, where light in one plane is transmitted and all the other planes are absorbed. A method of manufacturing a polarizing film was developed in the 1930s by Edwin H. Land (1909–1991). The film is called *Polaroid*. Today, Polaroid is made of long chains of hydrocarbon molecules that are aligned in a film. The long-chain molecules ideally absorb all light waves that are parallel to their lengths and transmit light that is perpendicular to their lengths. The direction that is *perpendicular* to the oriented molecular chains is thus called the polarization direction or the *transmission axis*.

Reflected light with an angle of incidence between 1° and 89° is partially polarized as the waves parallel to the reflecting surface are reflected more than other waves. Complete polarization, with all waves parallel to the surface, occurs at a particular angle of incidence. This angle depends on a number of variables, including the nature of the reflecting material. Figure 7.22 illustrates polarization by reflection. Polarizing sunglasses reduce the glare of reflected light because they have vertically oriented transmission axes. This absorbs the horizontally oriented reflected light. If you turn your head from side to side so as to rotate your sunglasses while looking at a reflected glare, you will see the intensity of the reflected light change. This means that the reflected light is partially polarized.

The phenomenon called *scattering* occurs when light is absorbed and reradiated by particles about the size of gas

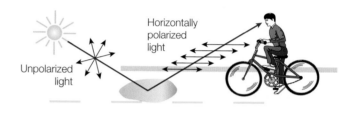

FIGURE 7.22 Light that is reflected becomes partially or fully polarized in a horizontal direction, depending on the incident angle and other variables.

molecules that make up the air. Sunlight is initially unpolarized. When it strikes a molecule, electrons are accelerated and vibrate horizontally and vertically. The vibrating charges reradiate polarized light. Thus, if you look at the blue sky with a pair of polarizing sunglasses and rotate them, you will observe that light from the sky is polarized. Bees are believed to be able to detect polarized skylight and use it to orient the direction of their flights. Violet and blue light have the shortest wavelengths of visible light, and red and orange light have the largest. The violet and blue rays of sunlight are scattered the most. At sunset the path of sunlight through the atmosphere is much longer than when the Sun is more directly overhead. Much of the blue and violet have been scattered away as a result of the longer path through the atmosphere at sunset. The remaining light that comes through is mostly red and orange, so these are the colors you see at sunset.

EVIDENCE FOR PARTICLES

The evidence from diffraction, interference, and polarization of light was very important in the acceptance of the wave theory because there was simply no way to explain these behaviors with a particle theory. Then, in 1850, J. L. Foucault (pronounced "Foo-co") (1819–1868) was able to prove that light travels much more slowly in transparent materials than it does in air. This was in complete agreement with the wave theory and completely opposed to the particle theory. By the end of the 1800s, James Maxwell's (1831–1879) theoretical concept of electric and

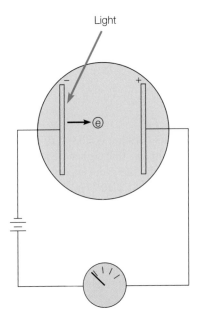

FIGURE 7.23 A setup for observing the photoelectric effect. Light strikes the negatively charged plate, and electrons are ejected. The ejected electrons move to the positively charged plate and can be measured as a current in the circuit.

magnetic fields changed the concept of light from mechanical waves to waves of changing electric and magnetic fields. Further evidence removed the necessity for ether, the material supposedly needed for waves to move through. Light was now seen as electromagnetic waves that could move through empty space. By this time it was possible to explain all behaviors of light moving through empty space or through matter with a wave theory. Yet there were nagging problems that the wave theory could not explain. In general, these problems concerned light that is absorbed by or emitted from matter.

PHOTOELECTRIC EFFECT

Light is a form of energy, and it gives its energy to matter when it is absorbed. Usually the energy of absorbed light results in a temperature increase, such as the warmth you feel from absorbed sunlight. Sometimes, however, the energy from absorbed light results in other effects. In some materials, the energy is acquired by electrons, and some of the electrons acquire sufficient energy to jump out of the material. The movement of electrons as a result of energy acquired from light is known as the **photoelectric effect.** The photoelectric effect is put to a practical use in a solar cell, which transforms the energy of light into an electric current (Figure 7.23).

The energy of light can be measured with great accuracy. The kinetic energy of electrons after they absorb light can also be measured with great accuracy. When these measurements were made of the light and electrons involved in the photoelectric effect, some unexpected results were observed. Monochromatic light, that is, light of a single, fixed frequency, was used to produce the photoelectric effect. First, a low-intensity, or dim, light was used, and the numbers and energy of the ejected electrons were measured. Then a high-intensity light was used, and the numbers and energy of the ejected electrons were again measured. Measurement showed that (1) low-intensity light caused

fewer electrons to be ejected, and high-intensity light caused many to be ejected, and (2) all electrons ejected from low- or high-intensity light ideally had the *same* kinetic energy. Surprisingly, the kinetic energy of the ejected electrons was found to be *independent* of the light intensity. This was contrary to what the wave theory of light would predict, since a stronger light should mean that waves with more energy have more energy to give to the electrons. Here is a behavior involving light that the wave theory could not explain.

QUANTIZATION OF ENERGY

In addition to the problem of the photoelectric effect, there were problems with blackbody radiation, light emitted from hot objects. The experimental measurements of light emitted through blackbody radiation did not match predictions made from theory. In 1900, Max Planck (pronounced "plonk") (1858–1947), a German physicist, found that he could fit the experimental measurements and theory together by assuming that the vibrating molecules that emitted the light could only have a *discrete amount* of energy. Instead of energy existing through a continuous range of amounts, Planck found that the vibrating molecules could only have energy in multiples of energy in certain amounts, or **quanta** (meaning "discrete amounts"; *quantum* is singular, and *quanta* plural).

Planck's discovery of quantized energy states was a radical, revolutionary development, and most scientists, including Planck, did not believe it at the time. Planck, in fact, spent considerable time and effort trying to disprove his own discovery. It was, however, the beginning of the quantum theory, which was eventually to revolutionize physics.

Five years later, in 1905, Albert Einstein (1879–1955) applied Planck's quantum concept to the problem of the photoelectric effect. Einstein described the energy in a light wave as quanta of energy called **photons.** Each photon has an energy E that is related to the frequency f of the light through Planck's constant h, or

$$E = hf$$

equation 7.4

The value of Planck's constant is 6.63×10^{-34} J·s.

This relationship says that higher-frequency light (e.g., blue light at 6.50×10^{14} Hz) has more energy than lower-frequency light (e.g., red light at 4.00×10^{14} Hz). The energy of such high- and low-frequency light can be verified by experiment.

The photon theory also explained the photoelectric effect. According to this theory, light is a stream of moving photons. It is the number of photons in this stream that determines if the light is dim or intense. A high-intensity light has many, many photons, and a low-intensity light has only a few photons. At any particular fixed frequency, all the photons would have the same energy, the product of the frequency and Planck's constant (hf). When a photon interacts with matter, it is absorbed and gives up all of its energy. In the photoelectric effect, this interaction takes place between photons and electrons. When an intense light is used, there are more photons to interact with the electrons, so more electrons are ejected. The energy given up by each photon is a function of the frequency of the light, so at a fixed frequency, the energy of each photon, hf, is the same, and the acquired kinetic energy of each ejected electron is the same. Thus, the photon theory explains the measured experimental results of the photoelectric effect.

EXAMPLE 7.3

What is the energy of a photon of red light with a frequency of 4.00×10^{14} Hz?

SOLUTION

The relationship between the energy of a photon (E) and its frequency (f) is found in equation 7.4. Planck's constant (h) is given as 6.63×10^{-34} J·s.

$$f = 4.00 \times 10^{14} \text{ Hz}$$
$$h = 6.63 \times 10^{-34} \text{ J·s}$$
$$E = ?$$

$$
\begin{aligned}
E &= hf \\
&= (6.63 \times 10^{-34} \text{ J·s}) \left(4.00 \times 10^{14} \frac{1}{s}\right) \\
&= (6.63 \times 10^{-34})(4.00 \times 10^{14}) \text{ J·s} \times \frac{1}{s} \\
&= 2.65 \times 10^{-19} \frac{\text{J·s}}{s} \\
&= \boxed{2.65 \times 10^{-19} \text{ J}}
\end{aligned}
$$

EXAMPLE 7.4

What is the energy of a photon of violet light with a frequency of 7.00×10^{14} Hz? (Answer: 4.64×10^{-19} J)

The photoelectric effect is explained by considering light to be photons with quanta of energy, not a wave of continuous energy. This is not the only evidence about the quantum nature of light, and more will be presented in chapter 8. But, as you can see, there is a dilemma. The electromagnetic wave theory and the photon theory seem incompatible. Some experiments cannot be explained by the wave theory and seem to support the photon theory. Other experiments are contradictions, providing seemingly equal evidence to reject the photon theory in support of the wave theory.

THE PRESENT THEORY

Today, light is considered to have a dual nature, sometimes acting like a wave and sometimes acting like a particle. A wave model is useful in explaining how light travels through space and how it exhibits such behaviors as refraction, interference, and diffraction. A particle model is useful in explaining how light is emitted from and absorbed by matter, exhibiting such behaviors as blackbody radiation and the photoelectric effect. Together, both of these models are part of a single theory of light, a theory that pictures light as having both particle and wave properties. Some properties are more useful when explaining some observed behaviors, and other properties are more useful when explaining other behaviors.

Frequency is a property of a wave, and the energy of a photon is a property of a particle. Both frequency and the energy of a photon are related in equation 7.4, $E = hf$. It is thus possible to describe light in terms of a frequency (or wavelength) or in terms of a quantity of energy. Any part of the electromagnetic spectrum can thus be described by units of frequency, wavelength, or energy, which are alternative means of describing light. The radio radiation parts of the spectrum are low-frequency, low-energy, and long-wavelength radiations. Radio radiations have more wave properties and practically no particle properties, since the energy levels are low. Gamma radiation, on the other hand, is high-frequency, high-energy, and short-wavelength radiation. Gamma radiation has more particle properties, since the extremely short wavelengths have very high energy levels. The more familiar part of the spectrum, visible light, is between these two extremes and exhibits both wave and particle properties, but it never exhibits both properties at the same time in the same experiment.

Part of the problem in forming a concept or mental image of the exact nature of light is understanding this nature in terms of what is already known. The things you already know about are observed to be particles, or objects, or they are observed to be waves. You can see objects that move through the air, such as baseballs or footballs, and you can see waves on water or in a field of grass. There is nothing that acts like a moving object in some situations but acts like a wave in other situations. Objects are objects, and waves are waves, but objects do not become waves, and waves do not become objects. If this dual nature did exist, it would seem very strange. Imagine, for example, hold-

A compact disc (CD) is a laser-read (also called *optically read*) data storage device. There are a number of different formats in use today, including music CDs, DVD movies, Blu-Ray DVD, and CDs for storing computer data. All of these utilize the general working principles described below, but some have different refinements. Some, for example, can fit much more data on a disc by utilizing smaller recording tracks.

The CD disc rotates between 200 and 500 revolutions per minute, but the drive changes speed to move the head at a constant linear velocity over the recording track, faster near the inner hub and slower near the outer edge of the disc. Furthermore, the drive reads from the inside out, so the disc will slow as it is played.

The CD disc is a 12 cm diameter, 1.3 mm thick sandwich of a hard plastic core, a mirrorlike layer of metallic aluminum, and a tough, clear plastic overcoating that protects the thin layer of aluminum. The CD records digitized data: music, video, or computer data that have been converted into a string of binary numbers. First, a master disc is made. The binary numbers are translated into a series of pulses that are fed to a laser. The laser is focused onto a photosensitive material on a spinning master disc. Whenever there is a pulse in the signal, the laser burns a small oval pit into the surface, making a pattern of pits and bumps on the track of the master disc. The laser beam is incredibly small, making marks about a micron or so in diameter. A micron is one-millionth of a meter, so you can fit a tremendous number of data tracks onto the disc, which has each track spaced 1.6 microns apart. Next, commercial CD discs are made by using the master disc as a mold.

Soft plastic is pressed against the master disc in a vacuum-forming machine so the small physical marks—the pits and bumps made by the laser—are pressed into the plastic. This makes a record of the strings of binary numbers that were etched into the master disc by the strong but tiny laser beam. During playback, a low-powered laser beam is reflected off the track to read the binary marks on it. The optical sensor head contains a tiny diode laser, a lens, mirrors, and tracking devices that can move the head in three directions. The head moves side to side to keep the head over a single track (within 1.6 micron), it moves up and down to keep the laser beam in focus, and it moves forward and backward as a fine adjustment to maintain a constant linear velocity.

The disadvantage of the commercial CD is the lack of ability to do writing or rewriting. Writing and rewritable optical media are available, and these are called CD-R and CD-RW.

A CD-R records data to a disc by using a laser to burn spots into an organic dye. Such a "burned" spot reflects less light than an area that was not heated by the laser. This is designed to mimic the way light reflects from pits and bumps of a commercial CD, except this time the string of binary numbers are burned (nonreflective) and not burned areas (reflective). Since this is similar to how data on a commerial CD is represented, a CD-R disc can generally be used in a CD player as if it were a commercial CD. The dyes in a CD-R disc are photosensitive organic compounds that are similar to those used in making photographs. The color of a CD-R disc is a result of the kind of dye that was used in the recording layer combined with the type of reflective coating used. Some of these dye and reflective coating combinations appear green, some appear blue, and others appear to be gold. Once a CD-R disc is burned, it cannot be rewritten or changed.

The CD-RW is designed to have the ability to do writing or rewriting. It uses a different technology but again mimics the way light reflects from the pits and bumps of a pressed commercial CD. Instead of a dye-based recording layer, the CD-RW uses a compound made from silver, indium, antimony, and tellurium. This layer has a property that permits rewriting the information on a disc. The nature of this property is that when it is heated to a certain temperature and cooled, it becomes crystalline. However, when it is heated to a higher temperature and cooled, it becomes noncrystalline. A crystalline surface reflects a laser beam while a noncrystalline surface absorbs the laser beam. The CD-RW is again designed to mimic the way light reflects from the pits and bumps of a commercial CD, except this time the string of binary numbers are noncrystalline (nonreflective) and crystalline areas (reflective). In order to write, erase, and read, the CD-RW recorder must have three different laser powers. It must have (1) a high power to heat spots to about 600°C, which cool rapidly and make noncrystalline spots that are less reflective. It must have (2) a medium power to erase data by heating the media to about 200°C, which allows the media to crystallize and have a uniform reflectivity. Finally, it must have (3) a low setting that is used for finding and reading nonreflective and the more reflective areas of a disc. The writing and rewriting of a CD-RW can be repeated hundreds of times.

---

ing a book at a certain height above a lake (Figure 7.24). You can make measurements and calculate the kinetic energy the book will have when dropped into the lake. When it hits the water, the book disappears, and water waves move away from the point of impact in a circular pattern that moves across the water. When the waves reach another person across the lake, a book identical to the one you dropped pops up out of the water as the waves disappear. As it leaves the water across the lake, the book has the same kinetic energy that your book had when it hit the water in front of you. You and the other person could measure things about either book, and you could measure things about the waves, but you could not measure both at the same time. You might say that this behavior is not only strange but impossible. Yet it is an analogy to the observed behavior of light.

As stated, light has a dual nature, sometimes exhibiting the properties of a wave and sometimes exhibiting the properties of moving particles but never exhibiting both properties at the same time. Both the wave and the particle nature are accepted as being part of one model today, with the understanding that the exact nature of light is not describable in terms of anything that is known to exist in the everyday-sized world. Light

FIGURE 7.24 It would seem very strange if there were not a sharp distinction between objects and waves in our everyday world. Yet this appears to be the nature of light.

is an extremely small-scale phenomenon that must be different, without a sharp distinction between a particle and a wave. Evidence about this strange nature of an extremely small-scale phenomenon will be considered again in chapter 8 as a basis for introducing the quantum theory of matter.

RELATIVITY

The electromagnetic wave model brought together and explained electric and magnetic phenomena, and explained that light can be thought of as an electromagnetic wave (see Figure 7.2). There remained questions, however, that would not be answered until Albert Einstein developed a revolutionary new theory. Even at the age of 17, Einstein was already thinking about ideas that would eventually lead to his new theory. For example, he wondered about chasing a beam of light if you were also moving at the speed of light. Would you see the light as an oscillating electric and magnetic field at rest? He realized there was no such thing, either on the basis of experience or according to Maxwell's theory of electromagnetic waves.

In 1905, at the age of 26, Einstein published an analysis of how space and time are affected by motion between an observer and what is being measured. This analysis is called the *special theory of relativity*. Eleven years later, Einstein published an interpretation of gravity as distortion of the structure of space and time. This analysis is called the *general theory of relativity*. A number of remarkable predictions have been made based on this theory, and all have been verified by many experiments.

SPECIAL RELATIVITY

The special theory of relativity is concerned with events as observed from different points of view, or different "reference frames." Here is an example. You are on a bus traveling straight down a highway at a constant 100 km/h. An insect is observed to fly from the back of the bus at 5 km/h. With respect to the

bus, the insect is flying at 5 km/h. To someone observing from the ground, however, the speed of the insect is 100 km/h plus 5 km/h, or 105 km/h. If the insect is flying toward the back of the bus, its speed is 100 km/h minus 5 km/h, or 95 km/h with respect to Earth. Generally, the reference frame is understood to be Earth, but this is not always stated. Nonetheless, we must specify a reference frame whenever a speed or velocity is measured.

Einstein's special theory is based on two principles. The first concerns frames of reference and the fact that all motion is relative to a chosen frame of reference. This principle could be called the **consistent law principle:**

> **The laws of physics are the same in all reference frames that are moving at a constant velocity with respect to each other.**

Ignoring vibrations, if you are in a windowless bus, you will not be able to tell if the bus is moving uniformly or if it is not moving at all. If you were to drop something—say, your keys—in a moving bus, they would fall straight down, just as they would in a stationary bus. The keys fall straight down with respect to the bus in either case. To an observer outside the bus, in a different frame of reference, the keys would appear to take a curved path because they have an initial velocity. Moving objects follow the same laws in a uniformly moving bus or any other uniformly moving frame of reference (Figure 7.25).

The second principle concerns the speed of light and could be called the **constancy of speed** principle:

> **The speed of light in empty space has the same value for all observers regardless of their velocity.**

FIGURE 7.25 All motion is relative to a chosen frame of reference. Here the photographer has turned the camera to keep pace with one of the cyclists. Relative to him, both the road and the other cyclists are moving. There is no fixed frame of reference in nature and, therefore, no such thing as "absolute motion"; all motion is relative.

James Clerk Maxwell (1831–1879)

James Maxwell was a British physicist who discovered that light consists of electromagnetic waves and established the kinetic theory of gases. He also proved the nature of Saturn's rings and demonstrated the principles governing color vision.

Maxwell was born at Edinburgh, Scotland, on November 13, 1831. He was educated at Edinburgh Academy from 1841 to 1847, then entered the University of Edinburgh. He then entered Cambridge University in 1850, graduating in 1854. He became professor of natural philosophy at Marischal College, Aberdeen, in 1856 and moved to London in 1860 to take up the post of professor of natural philosophy and astronomy at King's College. On the death of his father in 1865, Maxwell returned to his family home in Scotland and devoted himself to research. However, in 1871, he was persuaded to move to Cambridge, where he became the first professor of experimental physics and set up the Cavendish Laboratory, which opened in 1874. Maxwell continued in this position until 1879, when he contracted cancer. He died at Cambridge on November 5, 1879, at the age of 48.

Maxwell's development of the electromagnetic theory of light took many years. It began with the paper *On Faraday's Lines of Force*, in which Maxwell built on the views of Michael Faraday (1791–1867) that electric and magnetic effects result from fields of lines of force that surround conductors and magnets. Maxwell drew an analogy

between the behavior of the lines of force and the flow of an incompressible liquid, thereby deriving equations that represented known electric and magnetic effects. The next step toward the electromagnetic theory took place with the publication of the paper *On Physical Lines of Force* (1861–1862). In it, Maxwell developed a model for the medium in which electric and magnetic effects could occur.

In *A Dynamical Theory of the Electromagnetic Field* (1864), maxwell developed the fundamental equations that describe the electromagnetic field. These showed that light is propagated in two waves, one magnetic and the other electric, which vibrate perpendicular to each other and to the direction of propagation. This was confirmed in Maxwell's *Note on the Electromagnetic Theory of Light* (1868), which used an electrical derivation of the theory instead of the dynamical formulation, and Maxwell's whole work on the subject was summed up in *Treatise on Electricity and Magnetism* in 1873.

The treatise also established that light has a radiation pressure and suggested that a whole family of electromagnetic radiations must exist, of which light was only one. This was confirmed in 1888 with the sensational discovery of radio waves by Heinrich Hertz (1857–1894). Sadly, Maxwell did not live long enough to see this triumphant vindication of his work.

Maxwell is generally considered to be the greatest theoretical physicist of the 1800s, as his forebear Faraday was the greatest experimental physicist. His rigorous mathematical ability was combined with great insight to enable him to achieve brilliant syntheses of knowledge in the two most important areas of physics at that time. In building on Faraday's work to discover the electromagnetic nature of light, Maxwell not only explained electromagnetism but also paved the way for the discovery and application of the whole spectrum of electromagnetic radiation that has characterized modern physics.

Source: Modified from the Hutchinson *Dictionary of Scientific Biography* © RM 2008. All rights reserved. Helicon Publishing is a division of RM.

The speed of light in empty space is 3.00×10^8 m/s (186,000 mi/s). An observer traveling toward a source would measure the speed of light in empty space as 3.00×10^8 m/s. An observer not moving with respect to the source would measure this very same speed. This is not like the insect moving in a bus—you do not add or subtract the velocity of the source from the velocity of light. The velocity is always 3.00×10^8 m/s for all observers regardless of the velocity of the observers and regardless of the velocity of the source of light. Light behaves differently than anything in our everyday experience.

The special theory of relativity is based solely on the consistent law principle and the constancy of speed principle. Together, these principles result in some very interesting outcomes if you compare measurements from the ground of the length, time, and mass of a very fast airplane with measurements made by someone moving with the airplane. You, on the ground, would find that

- The length of an object is shorter when it is moving.
- Moving clocks run more slowly.
- Moving objects have increased mass.

The special theory of relativity shows that measurements of length, time, and mass are different in different moving reference frames. Einstein developed equations that describe each of the changes described above. These changes have been verified countless times with elementary particle experiments, and the data always fit Einstein's equations with predicted results.

GENERAL THEORY

Einstein's **general theory of relativity** could also be called Einstein's geometric theory of gravity. According to Einstein, a gravitational interaction does not come from some mysterious force called gravity. Instead, the interaction is between a mass and the geometry of space and time where the mass is located. Space and time can be combined into a fourth-dimensional "spacetime" structure. A mass is understood to interact with the spacetime, telling it how to curve. Spacetime also interacts with a mass, telling it how to move. A gravitational interaction is considered to be a local event of movement along a geodesic (shortest distance between two

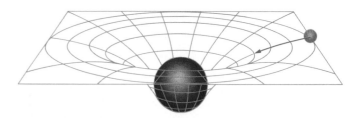

FIGURE 7.26 General relativity pictures gravity as a warping of the structure of space and time due to the presence of a body of matter. An object nearby experiences an attractive force as a result of this distortion in spacetime, much as a marble rolls toward the bottom of a saucer-shaped hole in the ground.

points on a curved surface) in curved spacetime (Figure 7.26). This different viewpoint has led to much more accurate measurements and has been tested by many events in astronomy (see p. 389 for one example).

SUMMARY

Electromagnetic radiation is emitted from all matter with a temperature above absolute zero, and as the temperature increases, more radiation and shorter wavelengths are emitted. Visible light is emitted from matter hotter than about 700°C, and this matter is said to be *incandescent*. The Sun, a fire, and the ordinary lightbulb are incandescent sources of light.

The behavior of light is shown by a light ray model that uses straight lines to show the straight-line path of light. Light that interacts with matter is *reflected* with parallel rays, moves in random directions by *diffuse reflection* from points, or is *absorbed*, resulting in a temperature increase. Matter is *opaque*, reflecting light, or *transparent*, transmitting light.

In reflection, the incoming light, or *incident ray*, has the same angle as the *reflected ray* when measured from a perpendicular from the point of reflection, called the *normal*. That the two angles are equal is called the *law of reflection*. The law of reflection explains how a flat mirror forms a *virtual image*, one from which light rays do not originate. Light rays do originate from the other kind of image, a *real image*.

Light rays are bent, or *refracted*, at the boundary when passing from one transparent medium to another. The amount of refraction depends on the *incident angle* and the *index of refraction*, a ratio of the speed of light in a vacuum to the speed of light in the medium. When the refracted angle is 90°, *total internal reflection* takes place. This limit to the angle of incidence is called the *critical angle*, and all light rays with an incident angle at or beyond this angle are reflected internally.

Each color of light has a range of wavelengths that forms the *spectrum* from red to violet. A glass prism has the property of *dispersion*, separating a beam of white light into a spectrum. Dispersion occurs because the index of refraction is different for each range of colors, with short wavelengths refracted more than larger ones.

A wave model of light can be used to explain interference and polarization. *Interference* occurs when light passes through two small slits or holes and produces an *interference pattern* of bright lines and dark zones. *Polarized light* vibrates in one direction only, in a plane. Light can be polarized by certain materials, by reflection, or by scattering. Polarization can only be explained by a transverse wave model.

A wave model fails to explain observations of light behaviors in the *photoelectric effect* and *blackbody radiation*. Max Planck found that he could modify the wave theory to explain blackbody radiation by assuming that vibrating molecules could only have discrete amounts, or *quanta*, of energy and found that the quantized energy is related to the frequency and a constant known today as *Planck's constant*. Albert Einstein applied Planck's quantum concept to the photoelectric effect and described a light wave in terms of quanta of energy called *photons*. Each photon has an energy that is related to the frequency and Planck's constant.

Today, the properties of light are explained by a model that incorporates both the wave and the particle nature of light. Light is considered to have both wave and particle properties and is not describable in terms of anything known in the everyday-sized world.

The *special theory of relativity* is an analysis of how space and time are affected by motion between an observer and what is being measured. The *general theory of relativity* relates gravity to the structure of space and time.

SUMMARY OF EQUATIONS

7.1

$$\text{angle of incidence} = \text{angle of reflection}$$
$$\theta_i = \theta_r$$

7.2

$$\text{index of refraction} = \frac{\text{speed of light in vacuum}}{\text{speed of light in material}}$$
$$n = \frac{c}{v}$$

7.3

$$\text{speed of light in vacuum} = (\text{wavelength})(\text{frequency})$$
$$c = \lambda f$$

7.4

$$\text{energy of photon} = \left(\text{Planck's constant}\right)(\text{frequency})$$
$$E = hf$$

KEY TERMS

blackbody radiation (p. **178**)
consistent law principle (p. **196**)
constancy of speed (p. **196**)
general theory of relativity (p. **198**)
incandescent (p. **178**)
index of refraction (p. **185**)
interference (p. **189**)
light ray model (p. **180**)
luminous (p. **178**)
photoelectric effect (p. **193**)
photon (p. **194**)
polarized (p. **191**)
quanta (p. **193**)
real image (p. **183**)
refraction (p. **183**)
total internal reflection (p. **184**)
unpolarized light (p. **191**)
virtual image (p. **183**)

APPLYING THE CONCEPTS

1. Which of the following is luminous?
   a. Moon
   b. Mars
   c. Sun
   d. All of the above are correct.

2. A source of light given off as a result of high temperatures is said to be
   a. luminous.
   b. blackbody radiation.
   c. incandescent.
   d. electromagnetic radiation.

3. An idealized material that absorbs and perfectly emits electromagnetic radiation is a (an)
   a. star.
   b. blackbody.
   c. electromagnetic wave.
   d. photon.

4. Electromagnetic radiation is given off from matter at any temperature. This radiation is called
   a. luminous.
   b. blackbody radiation.
   c. incandescent.
   d. electromagnetic radiation.

5. Light interacts with matter by which process?
   a. absorption
   b. reflection
   c. transmission
   d. All of the above are correct.

6. Materials that do not allow the transmission of any light are called
   a. transparent.
   b. colored.
   c. opaque.
   d. blackbody.

7. Light is said to travel in straight-line paths, as light rays, until it interacts with matter. A line representing the original ray before it interacts with matter is called
   a. incoming light ray.
   b. incident ray.
   c. reflected light ray.
   d. normal ray.

8. The image you see in a mirror is
   a. a real image.
   b. a virtual image.
   c. not really an image.

9. Refraction of light happens when light undergoes
   a. reflection from a surface.
   b. a change of speed between two transparent materials.
   c. movement through a critical angle.
   d. a 90° angle of incidence.

10. The ratio of the speed of light in a vacuum to the speed of light in a transparent material is called
    a. index of deflection.
    b. index of reflection.
    c. index of refraction.
    d. index of diffusion.

11. The part of the electromagnetic spectrum that our eyes can detect is
    a. ultraviolet.
    b. infrared.
    c. visible.
    d. All of the above are correct.

12. The component colors of sunlight were first studied by
    a. Joule.
    b. Galileo.
    c. Newton.
    d. Watt.

13. The color order of longer wavelength to smaller wavelength waves in the visible region is
    a. red, orange, yellow, green, blue, violet.
    b. red, violet, blue, yellow, green.
    c. violet, blue, green, yellow, orange, red.
    d. violet, red, blue, green, yellow, orange.

14. The separation of white light into its component colors is
    a. reflection.
    b. refraction.
    c. dispersion.
    d. transmission.

15. Polarization of light is best explained by considering light to be
    a. vibrating waves in one plane.
    b. moving particles in one plane.
    c. None of the above is correct.

16. Light in one plane is transmitted and light in all other planes is absorbed. This is
    a. selective absorption.
    b. polarized absorption.
    c. reflection.
    d. scattering.

17. The photoelectric effect is best explained by considering light to be
    a. vibrating waves.
    b. moving particles.
    c. None of the above is correct.

18. The concept that vibrating molecules emit light in discrete amounts of energy, called quanta, was proposed by
    a. Newton.
    b. Fresnel.
    c. Planck.
    d. Maxwell.

19. The photoelectric effect was explained, using Planck's work, by
    a. Planck.
    b. Einstein.
    c. Maxwell.
    d. Young.

20. Today, light is considered to be packets of energy with a frequency related to its energy. These packets are called
    a. gravitons.
    b. gluons.
    c. photons.
    d. quarks.

21. Fiber optics transmit information using
    a. sound.
    b. computers.
    c. light.
    d. All of the above are correct.

22. A luminous object is an object that
    a. reflects a dim blue-green light in the dark.
    b. produces light of its own by any method.
    c. shines by reflected light only, such as the moon.
    d. glows only in the absence of light.

23. An object is hot enough to emit a dull red glow. When this object is heated even more, it will emit
    a. shorter-wavelength, higher-frequency radiation.
    b. longer-wavelength, lower-frequency radiation.
    c. the same wavelengths as before with more energy.
    d. more of the same wavelengths.

24. The difference in the light emitted from a candle, an incandescent lightbulb, and the Sun is basically from differences in
    a. energy sources.
    b. materials.
    c. temperatures.
    d. phases of matter.

25. You are able to see in shaded areas, such as under a tree, because light has undergone
    a. refraction.
    b. incident bending.
    c. a change in speed.
    d. diffuse reflection.

26. An image that is not produced by light rays coming from the image but is the result of your brain's interpretations of light rays is called a (an)
    a. real image.
    b. imagined image.
    c. virtual image.
    d. phony image.

27. Any part of the electromagnetic spectrum, including the colors of visible light, can be measured in units of
    a. wavelength.
    b. frequency.
    c. energy.
    d. Any of the above are correct.

28. A prism separates the colors of sunlight into a spectrum because
    a. each wavelength of light has its own index of refraction.
    b. longer wavelengths are refracted more than shorter wavelengths.
    c. red light is refracted the most, violet the least.
    d. All of the above are correct.

29. Which of the following can only be explained by a wave model of light?
    a. reflection
    b. refraction
    c. interference
    d. photoelectric effect

30. The polarization behavior of light is best explained by considering light to be
    a. longitudinal waves.
    b. transverse waves.
    c. particles.
    d. particles with ends, or poles.

31. Max Planck made the revolutionary discovery that the energy of vibrating molecules involved in blackbody radiation existed only in
    a. multiples of certain fixed amounts.
    b. amounts that smoothly graded one into the next.
    c. the same, constant amount of energy in all situations.
    d. amounts that were never consistent from one experiment to the next.

32. Today, light is considered to be
    a. tiny particles of matter that move through space, having no wave properties.
    b. electromagnetic waves only, with no properties of particles.
    c. a small-scale phenomenon without a sharp distinction between particle and wave properties.
    d. something that is completely unknown.

33. As the temperature of an incandescent object is increased
    a. more infrared radiation is emitted with less UV.
    b. there is a decrease in the frequency of radiation emitted.
    c. the radiation emitted shifts toward infrared.
    d. more radiation is emitted with a shift to higher frequencies.

34. Is it possible to see light without the light interacting with matter?
    a. Yes.
    b. No.
    c. Only for opaque objects.
    d. Only for transparent objects.

35. The electromagnetic wave model defines an electromagnetic wave as having
    a. a velocity.
    b. a magnetic field.
    c. an electric field.
    d. All of the above are correct.

36. Of the following, the electromagnetic wave with the shortest wavelength is
    a. radio wave.
    b. infrared light.
    c. ultraviolet light.
    d. x rays.

37. Of the following, the electromagnetic wave with the lowest energy is
    a. radio wave.
    b. infrared light.
    c. ultraviolet light.
    d. x rays.

38. Green grass reflects
    a. yellow light.
    b. green light.
    c. blue light.
    d. white light.

39. Green grass absorbs
    a. yellow light.
    b. only green light.
    c. blue light.
    d. all light but green light.

40. We see a blue sky because
    a. air molecules absorb blue light.
    b. air molecules reflect red light.
    c. scattering of light by air molecules and dust is more efficient when its wavelength is longer.
    d. scattering of light by air molecules and dust is more efficient when its wavelength is shorter.

41. A pencil is placed in a glass of water. The pencil appears to be bent. This is an example of
    a. reflection.
    b. refraction.
    c. dispersion.
    d. polarization.

42. A one-way mirror works because it
    a. transmits all of the light falling on it.
    b. reflects all of the light falling on it.
    c. reflects and transmits light at the same time.
    d. neither reflects nor transmits light.

43. A mirage is the result of light being
    a. reflected.
    b. refracted.
    c. absorbed.
    d. bounced around a lot.

44. A glass prism separates sunlight into a spectrum of colors because
    a. shorter wavelengths are refracted the most.
    b. light separates into colors when reflected from crystal glass.
    c. light undergoes absorption in a prism.
    d. there are three surfaces that reflect light.

45. Polaroid sunglasses work best in eliminating glare because
    a. reflected light is refracted upward.
    b. unpolarized light vibrates in all possible directions.
    c. reflected light undergoes dispersion.
    d. reflected light is polarized in a horizontal direction only.

46. The condition of farsightedness, or hyperopia, can be corrected with a
    a. concave lens.
    b. convex lens.
    c. eyepiece lens.
    d. combination of convex and concave lenses.

47. Today, light is considered to be a stream of photons with a frequency related to its energy. This relationship finds that
    a. frequencies near the middle of the spectrum have more energy.
    b. more energetic light is always light with a lower frequency.
    c. higher-frequency light has more energy.
    d. lower-frequency light has more energy.

48. An instrument that produces a coherent beam of single frequency, in-phase light is a
    a. telescope.
    b. laser.
    c. camera.
    d. solar cell.

Answers

1. c 2. c 3. b 4. b 5. d 6. c 7. b 8. b 9. b 10. c 11. c 12. c 13. a 14. c 15. a 16. a 17. b 18. c 19. b 20. c 21. c 22. b 23. a 24. c 25. d 26. c 27. d 28. a 29. c 30. b 31. a 32. c 33. d 34. b 35. d 36. d 37. a 38. b 39. d 40. d 41. b 42. c 43. b 44. a 45. d 46. b 47. c 48. b

QUESTIONS FOR THOUGHT

1. What determines if an electromagnetic wave emitted from an object is a visible light wave or a wave of infrared radiation?

2. What model of light does the polarization of light support? Explain.

3. Which carries more energy, red light or blue light? Should this mean anything about the preferred color of warning and stop lights? Explain.

4. What model of light is supported by the photoelectric effect? Explain.

5. What happens to light that is absorbed by matter?

6. One star is reddish and another is bluish. Do you know anything about the relative temperatures of the two stars? Explain.

7. When does total internal reflection occur? Why does this occur in the diamond more than other gemstones?

8. Why does a highway sometimes appear wet on a hot summer day when it is not wet?

9. How can you tell if a pair of sunglasses is polarizing or not?

10. What conditions are necessary for two light waves to form an interference pattern of bright lines and dark areas?

11. Explain why the intensity of reflected light appears to change if you tilt your head from side to side while wearing polarizing sunglasses.

12. Why do astronauts in orbit around Earth see a black sky with stars that do not twinkle but see a blue Earth?

13. What was so unusual about Planck's findings about blackbody radiation? Why was this considered revolutionary?

14. Why are both the photon model and the electromagnetic wave model accepted today as a single theory? Why was this so difficult for people to accept at first?

FOR FURTHER ANALYSIS

1. Clarify the distinction between light reflection and light refraction by describing clear, obvious examples of each.

2. Describe how you would use questions alone to help someone understand that the shimmering they see above a hot pavement is not heat.

3. Use a dialogue as you "think aloud" considering the evidence that visible light is a wave, a particle, or both.

4. Compare and contrast the path of light through a convex and a concave lens. Give several uses for each lens and describe how the shape of the lens results in that particular use.

5. Analyze how the equation $E = hf$ could mean that visible light is a particle and a wave at the same time.

6. How are visible light and a radio wave different? How are they the same?

Obtain several different types of sunglasses. Design experiments to determine which combination of features will be found in the best pair of sunglasses. First, design an experiment to determine which reduces reflected glare the most. Find out how sunglasses are able to block ultraviolet radiation. According to your experiments and research, describe the "best" pair of sunglasses.

PARALLEL EXERCISES

The exercises in groups A and B cover the same concepts. Solutions to group A exercises are located in appendix E.

Group A

1. What is the speed of light while traveling through (a) water and (b) ice?

2. How many minutes are required for sunlight to reach Earth if the Sun is 1.50×10^8 km from Earth?

3. How many hours are required before a radio signal from a space probe near Pluto reaches Earth, 6.00×10^9 km away?

4. A light ray is reflected from a mirror with an angle 10° to the normal. What was the angle of incidence?

5. Light travels through a transparent substance at 2.20×10^8 m/s. What is the substance?

6. The wavelength of a monochromatic light source is measured to be 6.00×10^{-7} m in a diffraction experiment. (a) What is the frequency? (b) What is the energy of a photon of this light?

7. At a particular location and time, sunlight is measured on a 1 m² solar collector with a power of 1,000.0 W. If the peak intensity of this sunlight has a wavelength of 5.60×10^{-7} m, how many photons are arriving each second?

8. A light wave has a frequency of 4.90×10^{14} cycles per second. (a) What is the wavelength? (b) What color would you observe (see Table 7.2)?

9. What is the energy of a gamma photon of frequency 5.00×10^{20} Hz?

10. What is the energy of a microwave photon of wavelength 1.00 mm?

11. What is the speed of light traveling through glass?

12. What is the frequency of light with a wavelength of 5.00×10^{-7} m?

13. What is the energy of a photon of orange light with a frequency of 5.00×10^{14} Hz?

14. What is the energy of a photon of blue light with a frequency of 6.50×10^{14} Hz?

15. At a particular location and time, sunlight is measured on a 1 m² solar collector with an intensity of 1,000.0 watts. If the peak intensity of this sunlight has a wavelength of 5.60×10^{-7} m, how many photons are arriving each second?

Group B

1. (a) What is the speed of light while traveling through a vacuum? (b) While traveling through air at 30°C? (c) While traveling through air at 0°C?

2. How much time is required for reflected sunlight to travel from the Moon to Earth if the distance between Earth and the Moon is 3.85×10^5 km?

3. How many minutes are required for a radio signal to travel from Earth to a space station on Mars if the planet Mars is 7.83×10^7 km from Earth?

4. An incident light ray strikes a mirror with an angle of 30° to the surface of the mirror. What is the angle of the reflected ray?

5. The speed of light through a transparent substance is 2.00×10^8 m/s. What is the substance?

6. A monochromatic light source used in a diffraction experiment has a wavelength of 4.60×10^{-7} m. What is the energy of a photon of this light?

7. In black-and-white photography, a photon energy of about 4.00×10^{-19} J is needed to bring about the changes in the silver compounds used in the film. Explain why a red light used in a darkroom does not affect the film during developing.

8. The wavelength of light from a monochromatic source is measured to be 6.80×10^{-7} m. (a) What is the frequency of this light? (b) What color would you observe?

9. How much greater is the energy of a photon of ultraviolet radiation ($\lambda = 3.00 \times 10^{-7}$ m) than the energy of an average photon of sunlight ($\lambda = 5.60 \times 10^{-7}$ m)?

10. At what rate must electrons in a wire vibrate to emit microwaves with a wavelength of 1.00 mm?

11. What is the speed of light in ice?

12. What is the frequency of a monochromatic light used in a diffraction experiment that has a wavelength of 4.60×10^{-7} m?

13. What is the energy of a photon of red light with a frequency of 4.3×10^{14} Hz?

14. What is the energy of a photon of ultraviolet radiation with a wavelength of 3.00×10^{-7} m?

15. At a particular location and time, sunlight is measured on a 1 m² solar collector with an intensity of 500.0 watts. If the peak intensity of this sunlight has a wavelength of 5.60×10^{-7} m, how many photons are arriving each second?

8

Atoms and Periodic Properties

This is a picture of pure zinc, one of the eighty-nine naturally occurring elements found on the earth.

CORE CONCEPT

Different fields of study contributed to the development of a model of the atom.

OUTLINE

Discovery of the Electron
The electron was discovered from experiments with electricity.

Bohr's Theory
Experiments with light and line spectra and the application of the quantum concept led to the Bohr model of the atom.

The Periodic Table
The arrangement of elements in the periodic table has meaning about atomic structure and chemical behavior.

Atomic Structure Discovered
→ Discovery of the Electron
    The Nucleus ←
The Bohr Model
    The Quantum Concept
    Atomic Spectra
→ Bohr's Theory
Quantum Mechanics
    Matter Waves
    Wave Mechanics
    The Quantum Mechanics Model ←
Science and Society: Atomic Research
Electron Configuration
→ The Periodic Table
Metals, Nonmetals, and Semiconductors
A Closer Look: The Rare Earths
**People Behind the Science: Dmitri
    Ivanovich Mendeleyev**

The Nucleus
The nucleus and proton were discovered from experiments with radioactivity.

The Quantum Mechanics Model
Application of the wave properties of electrons led to the quantum mechanics model of the atom.

The development of the modern atomic model illustrates how modern scientific understanding comes from many different fields of study. For example, you will learn how studies of electricity led to the discovery that atoms have subatomic parts called *electrons*. The discovery of radioactivity led to the discovery of more parts, a central nucleus that contains protons and neutrons. Information from the absorption and emission of light was used to construct a model of how these parts are put together, a model resembling a miniature solar system with electrons circling the nucleus. The solar system model had initial, but limited, success and was inconsistent with other understandings about matter and energy. Modifications of this model were attempted, but none solved the problems. Then the discovery of wave properties of matter led to an entirely new model of the atom (Figure 8.1).

The atomic model will be put to use in later chapters to explain the countless varieties of matter and the changes that matter undergoes. In addition, you will learn how these changes can be manipulated to make new materials, from drugs to ceramics. In short, you will learn how understanding the atom and all the changes it undergoes not only touches your life directly but shapes and affects all parts of civilization.

ATOMIC STRUCTURE DISCOVERED

Did you ever wonder how scientists could know about something so tiny that you cannot see it, even with the most powerful optical microscope? The atom is a tiny unit of matter, so small that one gram of hydrogen contains about 600,000,000,000,000,000,000,000 (six-hundred-thousand-billion-billion or 6×10^{23}) atoms. Even more unbelievable is that atoms are not individual units but are made up of even smaller particles. How is it possible that scientists are able to tell you about the parts of something so small that it cannot be seen? The answer is that these things cannot be observed directly, but their existence can be inferred from experimental evidence. The following story describes the evidence and how scientists learned about the parts—electrons, the nucleus, protons, and neutrons—and how they are all arranged in the atom.

The atomic concept is very old, dating back to ancient Greek philosophers some 2,500 years ago. The ancient Greeks also reasoned about the way that pure substances are put together. A glass of water, for example, appears to be the same throughout. Is it the same? Two plausible, but conflicting, ideas were possible as an intellectual exercise. The water could have a continuous structure, that is, it could be completely homogeneous throughout. The other idea was that the water only appears to be continuous but is actually *discontinuous*. This means that if you continue to divide the water into smaller and smaller volumes, you would eventually reach a limit to this dividing, a particle that could not be further subdivided. The Greek philosopher Democritus (460–362 B.C.) developed this model in the fourth century B.C., and he called the indivisible particle an *atom,* from a Greek word meaning "uncuttable." However, neither Plato nor Aristotle accepted the atomic theory of matter, and it was not until about 2,000 years later that the atomic concept of matter was reintroduced. In the early 1800s, the English chemist John Dalton brought back the ancient Greek idea of hard, indivisible atoms to explain chemical reactions. Five statements will summarize his theory. As you will soon see, today we know that statement 2 is not strictly correct:

1. Indivisible minute particles called atoms make up all matter.
2. All the atoms of an element are exactly alike in shape and mass.
3. The atoms of different elements differ from one another in their masses.
4. Atoms chemically combine in definite whole-number ratios to form chemical compounds.
5. Atoms are neither created nor destroyed in chemical reactions.

During the 1800s, Dalton's concept of hard, indivisible atoms was familiar to most scientists. Yet the existence of atoms was not generally accepted by all scientists. There was skepticism about something that could not be observed directly. Strangely, full acceptance of the atom came in the early 1900s with the discovery that the atom was not indivisible after all. The atom has parts that give it an internal structure. The first part to be discovered was the *electron,* a part that was discovered through studies of electricity.

DISCOVERY OF THE ELECTRON

Scientists of the late 1800s were interested in understanding the nature of the recently discovered electric current. To observe a current directly, they tried to produce a current by itself, away from wires, by removing much of the air from a tube and then running a current through the rarefied air. When metal plates inside a tube were connected to the negative and positive terminals of a high-voltage source (Figure 8.2), a greenish beam was observed that seemed to move from the cathode (negative terminal) through the empty tube and collect at the anode

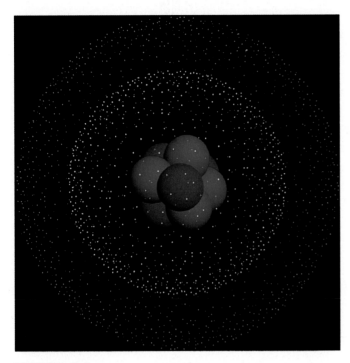

FIGURE 8.1 This is a computer-generated model of a beryllium atom, showing the nucleus and electron orbitals. This configuration can also be predicted from information on a periodic table (not to scale).

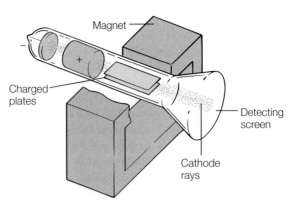

FIGURE 8.3 A cathode ray passed between two charged plates is deflected toward the positively charged plate. The ray is also deflected by a magnetic field. By measuring the deflection by both, J. J. Thomson was able to calculate the ratio of charge to mass. He was able to measure the deflection because the detecting screen was coated with zinc sulfide, a substance that produces a visible light when struck by a charged particle.

(positive terminal). Since this mysterious beam seemed to come out of the cathode, it was said to be a *cathode ray*.

The English physicist J. J. Thomson figured out what the cathode ray was in 1897. He placed charged metal plates on each side of the beam (Figure 8.3) and found that the beam was deflected away from the negative plate. Since it was known that like charges repel, this meant that the beam was composed of negatively charged particles.

The cathode ray was also deflected when caused to pass between the poles of a magnet. By balancing the deflections made by the magnet with the deflections made by the electric field, Thomson could determine the ratio of the charge to mass for an individual particle. Today, the charge-to-mass ratio is considered to be 1.7584×10^{11} coulomb/kilogram (see p. 143). A significant part of Thomson's experiments was that he found the charge-to-mass ratio was the same no matter what gas was in the tube or of what materials the electrodes were made. Thomson had discovered the **electron,** a fundamental particle of matter.

A method for measuring the charge and mass of the electron was worked out by an American physicist, Robert A. Millikan, around 1906. Millikan used an apparatus like the one illustrated in Figure 8.4 to measure the charge on tiny droplets of oil. Millikan found that none of the droplets had a charge less

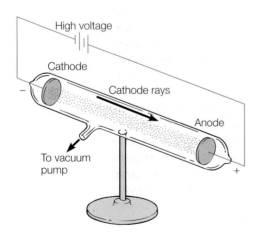

FIGURE 8.2 A vacuum tube with metal plates attached to a high-voltage source produces a greenish beam called *cathode rays*. These rays move from the cathode (negative charge) to the anode (positive charge).

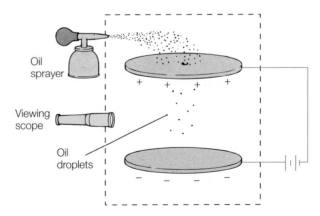

FIGURE 8.4 Millikan measured the charge of an electron by balancing the pull of gravity on oil droplets with an upward electrical force. Knowing the charge-to-mass ratio that Thomson had calculated, Millikan was able to calculate the charge on each droplet. He found that all the droplets had a charge of 1.60×10^{-19} coulomb or multiples of that charge. The conclusion was that this had to be the charge of an electron.

CHAPTER 8 Atoms and Periodic Properties **205**

than one particular value (1.60×10^{-19} coulomb) and that larger charges on various droplets were always multiples of this unit of charge. Since all of the droplets carried the single unit of charge or multiples of the single unit, the unit of charge was understood to be the charge of a single electron.

Knowing the charge of a single electron and knowing the charge-to-mass ratio that Thomson had measured now made it possible to calculate the mass of a single electron. The mass of an electron was thus determined to be about 9.11×10^{-31} kg, or about 1/1,840 of the mass of the lightest atom, hydrogen.

Thomson had discovered the negatively charged electron, and Millikan had measured the charge and mass of the electron. But atoms themselves are electrically neutral. If an electron is part of an atom, there must be something else that is positively charged, canceling the negative charge of the electron. The next step in the sequence of understanding atomic structure would be to find what is neutralizing the negative charge and to figure out how all the parts are put together.

Thomson had proposed a model for what was known about the atom at the time. He suggested that an atom could be a blob of massless, positively charged stuff in which electrons were stuck like "raisins in plum pudding." If the mass of a hydrogen atom is due to the electrons embedded in a massless, positively charged matrix, and since an electron was found to have 1/1,840 of the mass of a hydrogen atom, then 1,840 electrons would be needed together with sufficient positive stuff to make the atom electrically neutral. A different, better model of the atom was soon proposed by Ernest Rutherford, a British phycisist.

THE NUCLEUS

The nature of radioactivity and matter were the research interests of Rutherford. In 1907, Rutherford was studying the scattering of radiation particles directed toward a thin sheet of metal. As shown in Figure 8.5, the particles from a radioactive source were allowed to move through a small opening in a lead

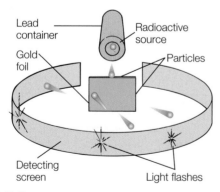

FIGURE 8.5 Rutherford and his coworkers studied alpha particle scattering from a thin metal foil. The alpha particles struck the detecting screen, producing a flash of visible light. Measurements of the angles between the flashes, the metal foil, and the source of the alpha particles showed that the particles were scattered in all directions, including straight back toward the source.

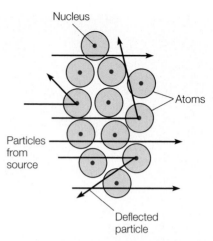

FIGURE 8.6 Rutherford's nuclear model of the atom explained the scattering results as positive particles experiencing a repulsive force from the positive nucleus. Measurements of the percentage of particles passing straight through and of the various angles of scattering of those coming close to the nuclei gave Rutherford a means of estimating the size of the nucleus.

container, so only a narrow beam of the massive, fast-moving particles would penetrate a very thin sheet of gold. The particles were detected by plates that produced small flashes of light when struck.

Rutherford found that most of the particles went straight through the foil. However, he was astounded to find that some were deflected at very large angles and some were even reflected backward. He could account for this only by assuming that the massive, positively charged particles were repelled by a massive positive charge concentrated in a small region of the atom (Figure 8.6). He concluded that an atom must have a tiny, massive, and positively charged **nucleus** surrounded by electrons.

From measurements of the scattering, Rutherford estimated electrons must be moving around the nucleus at a distance 100,000 times the radius of the nucleus. This means the volume of an atom is mostly empty space. A few years later Rutherford was able to identify the discrete unit of positive charge which we now call a **proton.** Rutherford also speculated about the existence of a neutral particle in the nucleus, a neutron. The **neutron** was eventually identified in 1932 by James Chadwick.

Today, the number of protons in the nucleus of an atom is called the **atomic number.** All of the atoms of a particular element have the same number of protons in their nuclei, so all atoms of an element have the same atomic number. Hydrogen has an atomic number of 1, so any atom that has one proton in its nucleus is an atom of the element hydrogen. In a neutral atom, the number of protons equals the number of electrons, so a neutral atom of hydrogen has one proton and one electron. Today, scientists have identified 117 different kinds of elements, each with a different number of protons.

The neutrons of the nucleus, along with the protons, contribute to the mass of an atom. Although all the atoms

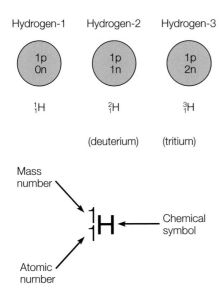

Hydrogen-1   Hydrogen-2   Hydrogen-3

1p   1p   1p
0n   1n   2n

$_{1}^{1}H$ $_{1}^{2}H$ $_{1}^{3}H$

(deuterium)   (tritium)

Mass
number

$_{1}^{1}H$ ← Chemical symbol

Atomic
number

FIGURE 8.7 The three isotopes of hydrogen have the same number of protons but different numbers of neutrons. Hydrogen-1 is the most common isotope. Hydrogen-2, with an additional neutron, is named *deuterium,* and hydrogen-3 is called *tritium.*

TABLE 8.1

Selected atomic weights calculated from mass and abundance of isotopes

Stable Isotopes	Mass of Isotope Compared to C-12	Abundance	Atomic Weight
$_{1}^{1}H$	1.007	99.985%	
$_{1}^{2}H$	2.0141	0.015%	1.0079
$_{4}^{9}Be$	9.01218	100.%	9.01218
$_{7}^{14}N$	14.00307	99.63%	
$_{7}^{15}N$	15.00011	0.37%	14.0067
$_{8}^{16}O$	15.99491	99.759%	
$_{8}^{17}O$	16.99914	0.037%	
$_{8}^{18}O$	17.00016	0.204%	15.9994
$_{9}^{19}F$	18.9984	100.%	18.9984
$_{10}^{20}Ne$	19.99244	90.92%	
$_{10}^{21}Ne$	20.99395	0.257%	
$_{10}^{22}Ne$	21.99138	8.82%	20.179
$_{13}^{22}Al$	26.9815	100.%	26.9815

of an element must have the same number of protons in their nuclei, the number of neutrons may vary. Atoms of an element that have different numbers of neutrons are called **isotopes.** There are three isotopes of hydrogen illustrated in Figure 8.7. All three isotopes have the same number of protons and electrons, but one isotope has no neutrons, one isotope has one neutron (deuterium), and one isotope has two neutrons (tritium).

An atom is very tiny, and it is impossible to find the mass of a given atom. It is possible, however, to compare the mass of one atom to another. The mass of any atom is compared to the mass of an atom of a particular isotope of carbon. This particular carbon isotope is assigned a mass of exactly 12.00 . . . units called **atomic mass units** (u). Since this isotope is *defined* to be exactly 12 u, it can have an infinite number of significant figures. This isotope, called *carbon-12,* provides the standard to which the masses of all other isotopes are compared. The relative mass of any isotope is based on the mass of a carbon-12 isotope.

The relative mass of the hydrogen isotope without a neutron is 1.007 when compared to carbon-12. The relative mass of the hydrogen isotope with 1 neutron is 2.0141 when compared to carbon-12. Elements occur in nature as a mixture of isotopes, and the contribution of each is calculated in the atomic weight. **Atomic weight** for the atoms of an element is an average of the isotopes based on their mass compared to carbon-12, and their relative abundance in nature. Of all the hydrogen isotopes, for example, 99.985 percent occur as the isotope without a neutron and 0.015 percent are the isotope with one neutron (the other isotope is not considered because it is radioactive). The fractional part of occurrence is multiplied by the relative atomic mass for each isotope and the results summed to obtain the atomic weight. Table 8.1 gives the atomic weight of hydrogen as 1.0079 as a result of this calculation.

The sum of the number of protons and neutrons in a nucleus of an atom is called the **mass number** of that atom. Mass numbers are used to identify isotopes. A hydrogen atom with 1 proton and 1 neutron has a mass number of 1 + 1, or 2, and is referred to as hydrogen-2. A hydrogen atom with 1 proton and 2 neutrons has a mass number of 1 + 2, or 3, and is referred to as hydrogen-3. Using symbols, hydrogen-3 is written as

$$_{1}^{3}H$$

where H is the chemical symbol for hydrogen, the subscript to the bottom left is the atomic number, and the superscript to the top left is the mass number.

How are the electrons moving around the nucleus? It might occur to you, as it did to Rutherford and others, that an atom might be similar to a miniature solar system. In this analogy, the nucleus is in the role of the Sun, electrons in the role of moving planets in their orbits, and electrical attractions between the nucleus and electrons in the role of gravitational attraction. There are, however, big problems with this idea. If electrons were moving in circular orbits, they would continually change their direction of travel and would therefore be accelerating. According to the Maxwell model of electromagnetic radiation, an accelerating electric charge emits electromagnetic radiation such as light. If an electron gave off light, it would lose energy. The energy loss would mean that the electron could not maintain its orbit, and it would be pulled into the oppositely charged nucleus. The atom would collapse as electrons spiraled into the nucleus. Since atoms do not collapse like this, there is a significant problem with the solar system model of the atom.

Atomic Parts

Identify the number of protons, neutrons, and electrons in an atom of $^{16}_{8}O$ Write your answer before reading the solution in the next paragraph.

The subscript to the bottom left is the atomic number. Atomic number is defined as the number of protons in the nucleus, so this number identifies the number of protons as 8. Any atom with 8 protons is an atom of oxygen, which is identified with the symbol O. The superscript to the top left identifies the mass number of this isotope of oxygen, which is 16. The mass number is defined as the sum of the number of protons and the number of neutrons in the nucleus. Since you already know the number of protons is 8 (from the atomic number), then the number of neutrons is 16 minus 8, or 8 neutrons. Since a neutral atom has the same number of electrons as protons, an atom of this oxygen isotope has 8 protons, 8 neutrons, and 8 electrons.

Now, can you describe how many protons, neutrons, and electrons are found in an atom of $^{17}_{8}O$? Compare your answer with a classmate's to check.

THE BOHR MODEL

Niels Bohr was a young Danish physicist who visited Rutherford's laboratory in 1912 and became very interested in questions about the solar system model of the atom. He wondered what determined the size of the electron orbits and the energies of the electrons. He wanted to know why orbiting electrons did not give off electromagnetic radiation. Seeking answers to questions such as these led Bohr to incorporate the *quantum concept* of Planck and Einstein with Rutherford's model to describe the electrons in the outer part of the atom. This quantum concept will be briefly reviewed before proceeding with the development of Bohr's model of the hydrogen atom.

THE QUANTUM CONCEPT

In the year 1900, Max Planck introduced the idea that matter emits and absorbs energy in discrete units that he called **quanta.** Planck had been trying to match data from spectroscopy experiments with data that could be predicted from the theory of electromagnetic radiation. In order to match the experimental findings with the theory, he had to assume that specific, discrete amounts of energy were associated with different frequencies of radiation. In 1905, Albert Einstein extended the quantum concept to light, stating that light consists of discrete units of energy that are now called **photons.** The energy of a photon is directly proportional to the frequency of vibration, and the higher the frequency of light, the greater the energy of the individual photons. In addition, the interaction of a photon

with matter is an "all-or-none" affair, that is, matter absorbs an entire photon or none of it. The relationship between frequency (f) and energy (E) is

$$E = hf$$

equation 8.1

where h is the proportionality constant known as *Planck's constant* (6.63×10^{-34} J·s). This relationship means that higher-frequency light, such as ultraviolet, has more energy than lower-frequency light, such as red light.

EXAMPLE 8.1

What is the energy of a photon of red light with a frequency of 4.60×10^{14} Hz?

SOLUTION

$$f = 4.60 \times 10^{14} \text{ Hz}$$
$$h = 6.63 \times 10^{-34} \text{ J·s}$$
$$E = ?$$

$$E = hf$$
$$= (6.63 \times 10^{-34} \text{ J·s}) \left(4.60 \times 10^{14} \frac{1}{\text{s}}\right)$$
$$= \left(6.63 \times 10^{-34}\right)\left(4.60 \times 10^{14}\right) \text{J·s} \times \frac{1}{\text{s}}$$
$$= \boxed{3.05 \times 10^{-19} \text{ J}}$$

EXAMPLE 8.2

What is the energy of a photon of violet light with a frequency of 7.30×10^{14} Hz? (Answer: 4.84×10^{-19} J)

ATOMIC SPECTRA

Planck was concerned with hot solids that emit electromagnetic radiation. The nature of this radiation, called *blackbody radiation,* depends on the temperature of the source. When this light is passed through a prism, it is dispersed into a *continuous spectrum,* with one color gradually blending into the next as in a rainbow. Today, it is understood that a continuous spectrum comes from solids, liquids, and dense gases because the atoms interact, and all frequencies within a temperature-determined range are emitted. Light from an incandescent gas, on the other hand, is dispersed into a **line spectrum,** narrow lines of colors with no light between the lines (Figure 8.8). The atoms in the incandescent gas are able to emit certain characteristic frequencies, and each frequency is a line of color that represents a definite value of energy. The line spectra are specific for a substance, and increased or decreased temperature changes only the intensity of the lines of colors. Thus, hydrogen always produces the same colors of lines in the same position. Helium has its own specific set of lines, as do other substances. Line spectra are a kind of fingerprint that can

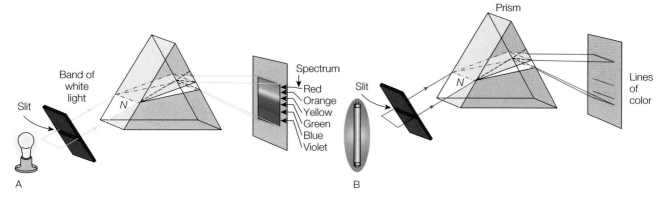

FIGURE 8.8 (*A*) Light from incandescent solids, liquids, or dense gases produces a continuous spectrum as atoms interact to emit all frequencies of visible light. (*B*) Light from an incandescent gas produces a line spectrum as atoms emit certain frequencies that are characteristic of each element.

be used to identify a gas. A line spectrum might also extend beyond visible light into ultraviolet, infrared, and other electromagnetic regions.

In 1885, a Swiss mathematics teacher named J. J. Balmer was studying the regularity of spacing of the hydrogen line spectra. Balmer was able to develop an equation that fit all the visible lines. By assigning values (*n*) of 3, 4, 5, and 6 to the four lines, he found the wavelengths fit the equation

$$\frac{1}{\lambda} = R\left(\frac{1}{2^2} - \frac{1}{n^2}\right)$$

equation 8.2

when *R* is a constant of 1.097×10^7 1/m.

Balmer's findings were:

Violet line	(*n* = 6)	$\lambda = 4.1 \times 10^{-7}$ m
Violet line	(*n* = 5)	$\lambda = 4.3 \times 10^{-7}$ m
Blue-green line	(*n* = 4)	$\lambda = 4.8 \times 10^{-7}$ m
Red line	(*n* = 3)	$\lambda = 6.6 \times 10^{-7}$ m

These four lines became known as the **Balmer series.** Other series were found later, outside the visible part of the spectrum

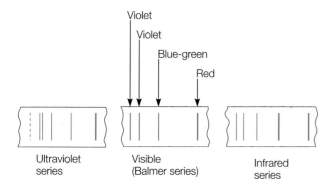

FIGURE 8.9 Atomic hydrogen produces a series of characteristic line spectra in the ultraviolet, visible, and infrared parts of the total spectrum. The visible light spectra always consist of two violet lines, a blue-green line, and a bright red line.

(Figure 8.9). The equations of the other series were different only in the value of *n* and the number in the other denominator.

Such regularity of observable spectral lines must reflect some unseen regularity in the atom. At this time, it was known that hydrogen had only one electron. How could one electron produce a series of spectral lines with such regularity?

EXAMPLE 8.3

Calculate the wavelength of the violet line (*n* = 6) in the hydrogen line spectra according to Balmer's equation.

SOLUTION

$$n = 6$$
$$R = 1.097 \times 10^7 \text{ 1/m}$$
$$\lambda = ?$$

$$\frac{1}{\lambda} = R\left(\frac{1}{2^2} - \frac{1}{n^2}\right)$$

$$= 1.097 \times 10^7 \frac{1}{\text{m}} \left(\frac{1}{2^2} - \frac{1}{6^2}\right)$$

$$= 1.097 \times 10^7 \left(\frac{1}{4} - \frac{1}{36}\right) \frac{1}{\text{m}}$$

$$= 1.097 \times 10^7 (0.222) \frac{1}{\text{m}}$$

$$\frac{1}{\lambda} = 2.44 \times 10^6 \frac{1}{\text{m}}$$

$$\lambda = \boxed{4.11 \times 10^{-7} \text{ m}}$$

BOHR'S THEORY

An acceptable model of the hydrogen atom would have to explain the characteristic line spectra and their regularity as described by Balmer. In fact, a successful model should be able to predict the occurrence of each color line as well as account for its origin. By 1913, Bohr was able to do this by applying the quantum concept to a solar system model of the atom. He began by considering the single hydrogen electron to be a single "planet" revolving in a circular orbit around the nucleus. There were three sets of rules that described this electron:

1. **Allowed Orbits.** An electron can revolve around an atom only in specific allowed orbits. Bohr considered the electron to be a particle with a known mass in motion around the nucleus and used Newtonian mechanics to calculate the distances of the allowed orbits. According to the Bohr model, electrons can exist only in one of these allowed orbits and nowhere else.

2. **Radiationless Orbits.** An electron in an allowed orbit does not emit radiant energy as long as it remains in the orbit. According to Maxwell's theory of electromagnetic radiation, an accelerating electron should emit an electromagnetic wave, such as light, which would move off into space from the electron. Bohr recognized that electrons moving in a circular orbit are accelerating, since they are changing direction continuously. Yet hydrogen atoms did not emit light in their normal state. Bohr decided that the situation must be different for orbiting electrons and that electrons could stay in their allowed orbits and not give off light. He postulated this rule as a way to make his theory consistent with other scientific theories.

3. **Quantum Leaps.** An electron gains or loses energy only by moving from one allowed orbit to another (Figure 8.10). In the Bohr model, the energy an electron has depends on which allowable orbit it occupies. The only way that an electron can change its energy is to jump from one allowed orbit to another in quantum "leaps." An electron must acquire energy to jump from a lower orbit to a higher one. Likewise, an electron gives up energy when jumping from a higher orbit to a lower one. Such jumps must be all at once, not part way and not gradual. An electron acquires energy from high temperatures or from electrical discharges to jump to a higher orbit. An electron jumping from a higher to a lower orbit gives up energy in the form of light. A single photon is emitted when a downward jump occurs, and the energy of the photon is *exactly* equal to the difference in the energy level of the two orbits.

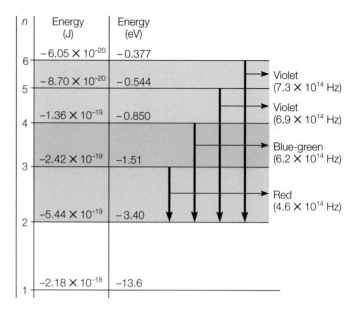

FIGURE 8.11 An energy level diagram for a hydrogen atom, not drawn to scale. The energy levels (*n*) are listed on the left side, followed by the energies of each level in J and eV. The color and frequency of the visible light photons emitted are listed on the right side, with the arrow showing the orbit moved from and to.

The energy level diagram in Figure 8.11 shows the energy states for the orbits of a hydrogen atom. The lowest energy state is the **ground state** (or normal state). The higher states are the **excited states.** The electron in a hydrogen atom would normally occupy the ground state, but high temperatures or electric discharge can give the electron sufficient energy to jump to one of the excited states. Once in an excited state, the electron immediately jumps back to a lower state, as shown by the arrows in the figure. The length of the arrow represents the frequency of the photon that the electron emits in the process. A hydrogen atom can give off only one photon at a time, and the many lines of a hydrogen line spectrum come from many atoms giving off many photons at the same time.

The reference level for the potential energy of an electron is considered to be zero when the electron is *removed* from an atom. The electron, therefore, has a lower and lower potential energy at closer and closer distances to the nucleus and has a negative value when it is in some allowed orbit. By way of analogy, you could consider ground level as a reference level where the potential energy of some object equals zero. But suppose there are two basement levels below the ground. An object on either basement level would have a gravitational potential energy less than zero, and work would have to be done on each object to bring it back to the zero level. Thus, each object would have a negative potential energy. The object on the lowest level would have the largest negative value of energy, since more work would have to be done on it to bring it back to the zero level. Therefore, the object on the lowest level would have the *least* potential energy, and this would be expressed as the *largest negative value.*

Just as the objects on different basement levels have negative potential energy, the electron has a definite negative potential energy in each of the allowed orbits. Bohr calculated the energy of an electron in the orbit closest to the nucleus to be -2.18×10^{-18} J,

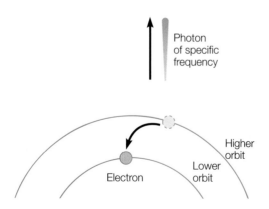

FIGURE 8.10 Each time an electron makes a "quantum leap," moving from a higher-energy orbit to a lower-energy orbit, it emits a photon of a specific frequency and energy value.

which is called the energy of the lowest state. The energy of electrons can be expressed in units of the **electron volt** (eV). An electron volt is defined as the energy of an electron moving through a potential of one volt. Since this energy is charge times voltage (from equation 6.3, $V = W/q$), 1.00 eV is equivalent to 1.60×10^{-19} J. Therefore, the energy of an electron in the innermost orbit is its energy in joules divided by 1.60×10^{-19} J/eV, or -13.6 eV.

Bohr found that the energy of each of the allowed orbits could be found from the simple relationship of

$$E_n = \frac{E_1}{n^2}$$

equation 8.3

where E_1 is the energy of the innermost orbit (-13.6 eV), and n is the quantum number for an orbit, or 1, 2, 3, and so on. Thus, the energy for the second orbit ($n = 2$) is $E_2 = -13.6$ eV$/4 = -3.40$ eV. The energy for the third orbit out ($n = 3$) is $E_3 = -13.6$ eV$/9 = -1.51$ eV, and so forth (Figure 8.11). Thus, the energy of each orbit is *quantized*, occurring only as a definite value.

In the Bohr model, the energy of the electron is determined by which allowable orbit it occupies. The only way that an electron can change its energy is to jump from one allowed orbit to another in quantum "jumps." An electron must *acquire* energy to jump from a lower orbit to a higher one. Likewise, an electron *gives up* energy when jumping from a higher orbit to a lower one. Such jumps must be all at once, not part way and not gradual. By way of analogy, this is very much like the gravitational potential energy that you have on the steps of a staircase. You have the lowest potential on the bottom step and the greatest amount on the top step. Your potential energy is quantized because you can increase or decrease it by going up or down a number of steps, but you cannot stop between the steps.

An electron acquires energy from high temperatures or from electrical discharges to jump to a higher orbit. An electron jumping from a higher to a lower orbit gives up energy in the form of light. A single photon is emitted when a downward jump occurs, and the *energy of the photon is exactly equal to the difference in the energy level* of the two orbits. If E_L represents the lower-energy level (closest to the nucleus) and E_H represents a higher-energy level (farthest from the nucleus), the energy of the emitted photon is

$$hf = E_H - E_L$$

equation 8.4

where h is Planck's constant, and f is the frequency of the emitted light (Figure 8.12).

As you can see, the energy level diagram in Figure 8.11 shows how the change of known energy levels from known orbits results in the exact energies of the color lines in the Balmer series. Bohr's theory did offer an explanation for the lines in the hydrogen spectrum with a remarkable degree of accuracy. However, the model did not have much success with larger atoms. Larger atoms had spectra lines that could not be explained by the Bohr model with its single quantum number. A German physicist, A. Sommerfeld, tried to modify Bohr's model by adding elliptical orbits in addition to Bohr's circular orbits. It soon became apparent that the "patched up" model, too, was not adequate. Bohr had made the rule that

FIGURE 8.12 These fluorescent lights emit light as electrons of mercury atoms inside the tubes gain energy from the electric current. As soon as they can, the electrons drop back to a lower-energy orbit, emitting photons with ultraviolet frequencies. Ultraviolet radiation strikes the fluorescent chemical coating inside the tube, stimulating the emission of visible light.

there were radiationless orbits without an explanation, and he did not have an explanation for the quantized orbits. There was something fundamentally incomplete about the model.

EXAMPLE 8.4

An electron in a hydrogen atom jumps from the excited energy level $n = 4$ to $n = 2$. What is the frequency of the emitted photon?

SOLUTION

The frequency of an emitted photon can be calculated from equation 8.4, $hf = E_H - E_L$. The values for the two energy levels can be obtained from Figure 8.11. (Note: E_H and E_L must be in joules. If the values are in electron volts, they can be converted to joules by multiplying by the ratio of joules per electron volt, or (eV)$(1.60 \times 10^{-19}$ J/eV$)$ = joules.)

$$E_H = -1.36 \times 10^{-19} \text{J}$$
$$E_L = -5.44 \times 10^{-19} \text{J}$$
$$h = 6.63 \times 10^{-34} \text{J·s}$$
$$f = ?$$

$$hf = E_H - E_L \quad \therefore \quad f = \frac{E_H - E_L}{h}$$

$$f = \frac{\left(-1.36 \times 10^{19} \text{ J}\right) - \left(-5.44 \times 10^{-19} \text{ J}\right)}{6.63 \times 10^{-34} \text{ J·s}}$$

$$= \frac{4.08 \times 10^{-19}}{6.63 \times 10^{-34}} \frac{\cancel{\text{J}}}{\cancel{\text{J}} \cdot \text{s}}$$

$$= 6.15 \times 10^{14} \frac{1}{\text{s}}$$

$$= \boxed{6.15 \times 10^{14} \text{Hz}}$$

This is approximately the blue-green line in the hydrogen line spectrum.

QUANTUM MECHANICS

The Bohr model of the atom successfully accounted for the line spectrum of hydrogen and provided an understandable mechanism for the emission of photons by atoms. However, the model did not predict the spectra of any atom larger than hydrogen, and there were other limitations. A new, better theory was needed. The roots of a new theory would again come from experiments with light. Experiments with light had established that sometimes light behaves like a stream of particles, and at other times it behaves like a wave (see chapter 7). Eventually, scientists began to accept that light has both wave properties and particle properties, which is now referred to as the *wave-particle duality of light*. This dual nature of light was recognized in 1905, when Einstein applied Planck's quantum concept to the energy of a photon with the relationship found in equation 8.1, $E = hf$, where E is the energy of a photon particle, f is the frequency of the associated wave, and h is Planck's constant.

MATTER WAVES

In 1923, Louis de Broglie, a French physicist, reasoned that symmetry is usually found in nature, so if a particle of light has a dual nature, then particles such as electrons should too. De Broglie reasoned further that if this is true, an electron in its circular path around the nucleus would have to have a particular wavelength that would fit into the circumference of the orbit (Figure 8.13). De Broglie derived a relationship from equations concerning light and energy, which was

$$\lambda = \frac{h}{mv}$$

equation 8.5

where λ is the wavelength, m is mass, v is velocity, and h is again Planck's constant. This equation means that any moving particle has a wavelength that is associated with its mass and velocity. In other words, de Broglie was proposing a wave-particle duality

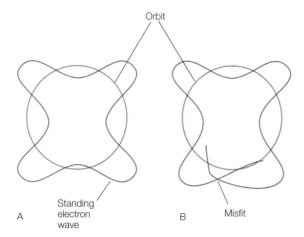

FIGURE 8.13 (*A*) Schematic of de Broglie wave, where the standing wave pattern will just fit in the circumference of an orbit. This is an allowed orbit. (*B*) This orbit does not have a circumference that will match a whole number of wavelengths; it is not an allowed orbit.

of matter, the existence of **matter waves.** According to equation 8.5, *any* moving object should exhibit wave properties. However, an ordinary-sized object would have wavelengths so small that they could not be observed. This is different for electrons because they have such a tiny mass.

EXAMPLE 8.5

What is the wavelength associated with a 0.150 kg baseball with a velocity of 50.0 m/s?

SOLUTION

$$m = 0.150 \text{ kg}$$
$$v = 50.0 \text{ m/s}$$
$$h = 6.63 \times 10^{-34} \text{ J·s}$$
$$\lambda = ?$$

$$\lambda = \frac{h}{mv}$$

$$= \frac{6.63 \times 10^{-34} \text{ J·s}}{(0.150 \text{ kg})\left(50.0 \frac{\text{m}}{\text{s}}\right)}$$

$$= \frac{6.63 \times 10^{-34}}{(0.150)(50.0)} \frac{\text{J·s}}{\text{kg} \times \frac{\text{m}}{\text{s}}}$$

$$= \frac{6.63 \times 10^{-34}}{7.50} \frac{\frac{\text{kg·m}^2}{\text{s}^2} \cdot \text{s}}{\frac{\text{kg·m}}{\text{s}}}$$

$$= \boxed{8.84 \times 10^{-35} \text{ m}}$$

What is the wavelength associated with an electron with a velocity of 6.00×10^6 m/s?

SOLUTION

$$m = 9.11 \times 10^{-31} \text{ kg}$$
$$v = 6.00 \times 10^6 \text{ m/s}$$
$$h = 6.63 \times 10^{-34} \text{ J·s}$$
$$\lambda = ?$$

$$\lambda = \frac{h}{mv}$$

$$= \frac{6.63 \times 10^{-34} \text{ J·s}}{(9.11 \times 10^{-31} \text{ kg})\left(6.00 \times 10^6 \frac{\text{m}}{\text{s}}\right)}$$

$$= \frac{6.63 \times 10^{-34}}{5.47 \times 10^{-24}} \frac{\text{J·s}}{\text{kg} \times \frac{\text{m}}{\text{s}}}$$

$$= 1.21 \times 10^{-10} \frac{\frac{\text{kg·m}^2}{\text{s}^2} \cdot \text{s}}{\frac{\text{kg·m}}{\text{s}}}$$

$$= \boxed{1.21 \times 10^{-10} \text{ m}}$$

The baseball wavelength of 8.84×10^{-35} m is much too small to be detected or measured. The electron wavelength of 1.21×10^{-10} m, on the other hand, is comparable to the distances between atoms in a crystal, so a beam of electrons through a crystal should produce diffraction.

The idea of matter waves was soon tested after de Broglie published his theory. Experiments with a beam of light passing by the edge of a sharp-edged obstacle produced interference patterns. This was part of the evidence for the wave nature of light, since such results could only be explained by waves, not particles. When similar experiments were performed with a beam of electrons, *identical* wave property behaviors were observed. This and many related experiments showed without doubt that electrons have both wave properties and particle properties. And, as was the case with light waves, measurements of the electron interference patterns provided a means to measure the wavelength of electron waves.

Recall that waves confined on a fixed string establish resonant modes of vibration called *standing waves* (see chapter 5). Only certain fundamental frequencies and harmonics can exist on a string, and the combination of the fundamental and overtones gives the stringed instrument its particular quality. The same result of resonant modes of vibrations is observed in *any* situation where waves are confined to a fixed space. Characteristic standing wave patterns depend on the wavelength and wave velocity for waves formed on strings, in enclosed columns of air, or for any kind of wave in a confined space. Electrons are confined to the space near a nucleus, and electrons have wave properties, so an electron in an atom must be a confined wave. Does an electron form a characteristic wave pattern? This was the question being asked in about 1925 when Heisenberg, Schrödinger, Dirac, and others applied the wave nature of the electron to develop a new model of the atom based on the mechanics of electron waves. The new theory is now called **wave mechanics,** or **quantum mechanics.**

WAVE MECHANICS

Erwin Schrödinger, an Austrian physicist, treated the atom as a three-dimensional system of waves to derive what is now called the *Schrödinger equation.* Instead of the simple circular planetary orbits of the Bohr model, solving the Schrödinger equation results in a description of three-dimensional shapes of the patterns that develop when electron waves are confined by a nucleus. Schrödinger first considered the hydrogen atom, calculating the states of vibration that would be possible for an electron wave confined by a nucleus. He found that the frequency of these vibrations, when multiplied by Planck's constant, matched exactly, to the last decimal point, the observed energies of the quantum states of the hydrogen atom ($E = hf$). The conclusion is that the wave nature of the electron is the important property to consider for a successful model of the atom.

The quantum mechanics theory of the atom proved to be very successful; it confirmed all the known experimental facts and predicted new discoveries. The theory does have some of the same quantum ideas as the Bohr model; for example, an electron emits a photon when jumping from a higher state to a lower one. The Bohr model, however, considered the particle nature of an electron moving in a circular orbit with a definitely assigned position at a given time. Quantum mechanics considers the wave nature, with the electron as a confined wave with well-defined shapes and frequencies. A wave is not localized like a particle and is spread out in space. The quantum mechanics model is, therefore, a series of orbitlike smears, or fuzzy statistical representations, of where the electron might be found.

THE QUANTUM MECHANICS MODEL

The quantum mechanics model is a highly mathematical treatment of the mechanics of matter waves. In addition, the wave properties are considered as three-dimensional problems, and three quantum numbers are needed to describe the fuzzy electron cloud. The mathematical detail will not be presented here. The following is a qualitative description of the main ideas in the quantum mechanics model. It will describe the results of the mathematics and will provide a mental visualization of what it all means.

First, understand that the quantum mechanical theory is not an extension or refinement of the Bohr model. The Bohr model considered electrons as particles in circular orbits that could be only certain distances from the nucleus. The quantum mechanical model, on the other hand, considers the electron as a wave and considers the energy of its harmonics, or modes, of standing waves. In the Bohr model, the location of an electron was certain—in an orbit. In the quantum mechanical model, the electron is a spread-out wave.

Quantum mechanics describes the energy state of an electron wave with four *quantum numbers:*

1. **Distance from the Nucleus.** The *principal quantum number* describes the *main energy level* of an electron in terms of its most probable distance from the nucleus. The lowest energy state possible is closest to the nucleus and is assigned the principal quantum number of 1 ($n = 1$). Higher states are assigned progressively higher positive whole numbers of $n = 2$, $n = 3$, $n = 4$, and so on. Electrons with higher principal quantum numbers have higher energies and are located farther from the nucleus.

2. **Energy Sublevel.** The *angular momentum quantum number* defines energy sublevels within the main energy levels. Each sublevel is identified with a letter. The first four of these letters, in order of increasing energy, are s, p, d, and f. The letter s represents the lowest sublevel, and the letter f represents the highest sublevel. A principal quantum number and a letter indicating the angular momentum quantum number are combined to identify the main energy state and energy sublevel of an electron. For an electron in the lowest main energy level, $n = 1$ and in the lowest sublevel, s, the number and letter are 1s (read as "one-s"). Thus, 1s indicates an electron that is as close to the nucleus as possible in the lowest energy sublevel possible.

There are limits to how many sublevels can occupy each of the main energy levels. Basically, the lowest main energy level can have only the lowest sublevel, and another sublevel is added as you move up through the main energy levels. Thus, the lowest main energy level, $n = 1$, can have only the

Science and Society

Atomic Research

There are two types of scientific research: basic and applied. Basic research is driven by a search for understanding and may or may not have practical applications. Applied research has a goal of solving some practical problem rather than just looking for answers.

Some people feel that all research should result in something practical, so all research should be applied. Hold that thought while considering if the following research discussed in this chapter is basic or applied:

1. J. J. Thomson investigates cathode rays.
2. Robert Millikan measures the charge of an electron.
3. Ernest Rutherford studies radioactive particles striking gold foil.
4. Niels Bohr proposes a solar system model of the atom by applying the quantum concept.
5. Erwin Schrödinger proposes a model of the atom based on the wave nature of the electron.

QUESTIONS TO DISCUSS

1. Were the five research topics basic or applied research?
2. Would we ever have developed a model of the atom if all research had to be practical?

---

s sublevel. The $n = 2$ can have s and p sublevels. The $n = 3$ main energy level can have the s, p, and d sublevels. Finally, the $n = 4$ main energy level can have all four sublevels, with s, p, d, and f. Therefore, the number of possible sublevels is the same as the principal quantum number.

The Bohr model considered the location of an electron as certain, like a tiny shrunken marble in an orbit. The quantum mechanical model considers the electron as a wave, and knowledge of its location is very uncertain. The **Heisenberg uncertainty principle** states that you cannot measure the exact position of a wave because a wave is spread out. One cannot specify the position and the momentum of a spread-out electron. The location of the electron can only be described in terms of *probabilities* of where it might be at a given instant. The probability of location is described by a fuzzy region of space called an **orbital**. An orbital defines the space where an electron is likely to be found. Orbitals have characteristic three-dimensional shapes and sizes and are identified with electrons of characteristic energy levels. An orbital shape represents where an electron could probably be located at any particular instant. This "probability cloud" could likewise have any particular orientation in space, and the direction of this orientation is uncertain.

3. **Orientation in Space.** An external magnetic field applied to an atom produces different energy levels that are related to the orientation of the orbital to the magnetic field. The orientation of an orbital in space is described by the *magnetic quantum number*. This number is related to the energies of orbitals as they are oriented in space relative to an external magnetic field, a kind of energy sub-sublevel. In general, the lowest-energy sublevel (s) has only one orbital orientation. The next higher-energy sublevel (p) can have three orbital orientations (Figure 8.14). The d sublevel can have five orbital orientations, and the highest sublevel, f, can have a total of seven different orientations (Table 8.2).

4. **Direction of Spin.** Detailed studies have shown that an electron spinning one way (say, clockwise) in an external

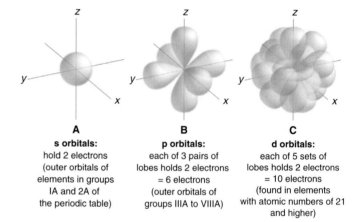

A
s orbitals:
hold 2 electrons
(outer orbitals of elements in groups IA and 2A of the periodic table)

B
p orbitals:
each of 3 pairs of lobes holds 2 electrons = 6 electrons
(outer orbitals of groups IIIA to VIIIA)

C
d orbitals:
each of 5 sets of lobes holds 2 electrons = 10 electrons
(found in elements with atomic numbers of 21 and higher)

FIGURE 8.14 The general shapes of s, p, and d orbitals, the regions of space around the nuclei of atoms in which electrons are likely to be found. (The f orbital is too difficult to depict.)

TABLE 8.2

Quantum numbers and electron distribution to $n = 4$

Main Energy Level	Energy Sublevels	Maximum Number of Electrons	Maximum Number of Electrons per Main Energy Level
$n = 1$	s	2	2
$n = 2$	s	2	
	p	6	8
$n = 3$	s	2	
	p	6	
	d	10	18
$n = 4$	s	2	
	p	6	
	d	10	
	f	14	32

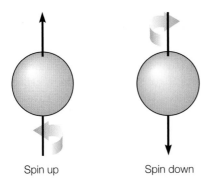

Spin up   Spin down

FIGURE 8.15 Experimental evidence supports the concept that electrons can be considered to spin one way or the other as they move about an orbital under an external magnetic field.

magnetic field would have a different energy than one spinning the other way (say, counterclockwise). The *spin quantum number* describes these two spin orientations (Figure 8.15).

Electron spin is an important property of electrons that helps determine the electronic structure of an atom. As it turns out, two electrons spinning in opposite directions, called an **electron pair,** can occupy the same orbital. This was summarized in 1924 by Wolfgang Pauli, an Austrian physicist. His summary, now known as the **Pauli exclusion principle,** states that *no two electrons in an atom can have the same four quantum numbers.* This provides the key for understanding the electron structure of atoms.

ELECTRON CONFIGURATION

The arrangement of electrons in orbitals is called the *electron configuration.* Before you can describe the electron arrangement, you need to know how many electrons are present in an atom. An atom is electrically neutral, so the number of protons (positive charge) must equal the number of electrons (negative charge). The atomic number therefore identifies the number of electrons as well as the number of protons. Now that you have a means of finding the number of electrons, consider the various energy levels to see how the electron configuration is determined.

According to the Pauli exclusion principle, no two electrons in an atom can have all four quantum numbers the same. As it works out, this means there can only be *a maximum of two electrons in any given orbital.* There are four things to consider: (1) the main energy level, (2) the energy sublevel, (3) the number of orbital orientations, and (4) the electron spin. Recall that the lowest-energy level is $n = 1$, and successive numbers identify progressively higher-energy levels. Recall also that the energy sublevels, in order of increasing energy, are s, p, d, and f. This electron configuration is written in shorthand, with 1s standing for the lowest-energy sublevel of the first energy level. A superscript gives the number of electrons present in a sublevel. Thus, the electron configuration for a helium atom, which has two electrons, is written as $1s^2$. This combination of symbols has the following meaning: The symbols mean an atom with two electrons in the s sublevel of the first main energy level.

TABLE 8.3

Electron configuration for the first twenty elements

Atomic Number	Element	Electron Configuration
1	Hydrogen	$1s^1$
2	Helium	$1s^2$
3	Lithium	$1s^2 2s^1$
4	Beryllium	$1s^2 2s^2$
5	Boron	$1s^2 2s^2 2p^1$
6	Carbon	$1s^2 2s^2 2p^2$
7	Nitrogen	$1s^2 2s^2 2p^3$
8	Oxygen	$1s^2 2s^2 2p^4$
9	Fluorine	$1s^2 2s^2 2p^5$
10	Neon	$1s^2 2s^2 2p^6$
11	Sodium	$1s^2 2s^2 2p^6 3s^1$
12	Magnesium	$1s^2 2s^2 2p^6 3s^2$
13	Aluminum	$1s^2 2s^2 2p^6 3s^2 3p^1$
14	Silicon	$1s^2 2s^2 2p^6 3s^2 3p^2$
15	Phosphorus	$1s^2 2s^2 2p^6 3s^2 3p^3$
16	Sulfur	$1s^2 2s^2 2p^6 3s^2 3p^4$
17	Chlorine	$1s^2 2s^2 2p^6 3s^2 3p^5$
18	Argon	$1s^2 2s^2 2p^6 3s^2 3p^6$
19	Potassium	$1s^2 2s^2 2p^6 3s^2 3p^6 4s^1$
20	Calcium	$1s^2 2s^2 2p^6 3s^2 3p^6 4s^2$

Table 8.3 gives the electron configurations for the first twenty elements. The configurations of the p energy sublevel have been condensed in this table. There are three possible orientations of the p orbital, each with two electrons. This is shown as p^6, which designates the number of electrons in all of the three possible p orientations. Note that the sum of the electrons in all the orbitals equals the atomic number. Note also that as you proceed from a lower atomic number to a higher one, the higher element has the same configuration as the element before it with the addition of one more electron. In general, it is then possible to begin with the simplest atom, hydrogen, and add one electron at a time to the order of energy sublevels and obtain the electron configuration for all the elements. The exclusion principle limits the number of electrons in any orbital, and allowances will need to be made for the more complex behavior of atoms with many electrons.

The energies of the orbital are different for each element, and there are several factors that influence their energies. The first orbitals are filled in a straightforward 1s, 2s, 2p, 3s, then 3p order. Then the order becomes contrary to what you might expect. One useful way of figuring out the order in which orbitals are filled is illustrated in Figure 8.16. Each row of this matrix represents a principal energy level with possible energy sublevels increasing from left to right. The order of filling is indicated by the diagonal arrows. There are exceptions to the order of filling shown by the matrix, but it works for most of the elements.

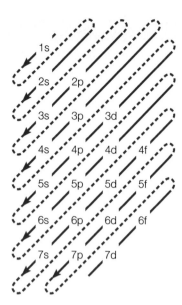

FIGURE 8.16 A matrix showing the order in which the orbitals are filled. Start at the top left, then move from the head of each arrow to the tail of the one immediately below it. This sequence moves from the lowest-energy level to the next higher level for each orbital.

 CONCEPTS *Applied*

Firework Configuration

Certain strontium (atomic number 38) chemicals are used to add the pure red color to flares and fireworks. Write the electron configuration of strontium and do this before looking at the solution that follows.

First, note that an atomic number of 38 means a total of thirty-eight electrons. Second, refer to the order of filling matrix in Figure 8.16. Remember that only two electrons can occupy an orbital, but there are three orientations of the p orbital, for a total of six electrons. There are likewise five possible orientations of the d orbital, for a total of ten electrons. Starting at the lowest energy level, two electrons go in 1s, making $1s^2$; then two go in 2s, making $2s^2$. That is a total of four electrons so far. Next $2p^6$ and $3s^2$ use eight more electrons, for a total of twelve so far. The $3p^6$, $4s^2$, $3d^{10}$, and $4p^6$ use up twenty-four more electrons, for a total of thirty-six. The remaining two go into the next sublevel, $5s^2$, and the complete answer is

Strontium: $1s^2\ 2s^2\ 2p^6\ 3s^2\ 3p^6\ 4s^2\ 3d^{10}\ 4p^6\ 5s^2$

THE PERIODIC TABLE

The periodic table is made up of rows and columns of cells, with each element having its own cell in a specific location. The cells are not arranged symmetrically. The arrangement has a meaning, both about atomic structure and about chemical behaviors. It will facilitate your understanding if you refer frequently to a periodic table during the following discussion (Figure 8.17).

An element is identified in each cell with its chemical symbol. The number above the symbol is the atomic number of the element, and the number below the symbol is the rounded atomic weight of the element. Horizontal rows of elements run from left to right with increasing atomic numbers. Each row is called a *period*. The periods are numbered from 1 to 7 on the left side. A vertical column of elements is called a *family* (or group) of elements. Elements in families have similar properties, but this is more true of some families than others. The table is subdivided into A and B groups. The members of the A-group families are called the *main group*, or **representative elements.** The members of the B group are called the **transition elements** (or metals). Some science organizations use a 1-to-18 designation for the A and B groups, as shown in figure 8.17. The A and B designations will be used throughout this text.

As shown in Table 8.4, all of the elements in the first column have an outside electron configuration of one electron. With the exception of hydrogen, the elements of the first column are shiny, low-density metals that are so soft you can cut them easily with a knife. These metals are called the *alkali metals* because they react violently with water to form an alkaline solution. The alkali metals do not occur in nature as free elements because they are so reactive. Hydrogen is a unique element in the periodic table. It is not an alkali metal and is placed in the group because it seems to fit there because it has one electron in its outer s orbital.

The elements in the second column all have an outside configuration of two electrons and are called the *alkaline earth metals*. The alkaline earth metals are soft, reactive metals but not as reactive or soft as the alkali metals. Calcium and magnesium, in the form of numerous compounds, are familiar examples of this group.

The elements in group VIIA all have an outside configuration of seven electrons, needing only one more electron to completely fill the outer (p) orbitals. These elements are called the *halogens*. The halogens are very reactive nonmetals. The halogens fluorine and chlorine are greenish-colored gases. Bromine is a reddish-brown liquid and iodine is a dark purple solid. Halogens are used as disinfectants, bleaches, and combined with a metal as a source of light in halogen lights. Halogens react with metals to form a group of chemicals called *salts*, such as sodium chloride. In fact, the word *halogen* is Greek, meaning "salt former."

TABLE 8.4

Electron structures of the alkali metal family

Element	Electron Configuration	Number of Electrons in Main Energy Level						
		1st	2nd	3rd	4th	5th	6th	7th
Lithium (Li)	[He]* $2s^1$	2	1	—	—	—	—	—
Sodium (Na)	[Ne] $3s^1$	2	8	1	—	—	—	—
Potassium (K)	[Ar] $4s^1$	2	8	8	1	—	—	—
Rubidium (Rb)	[Kr] $5s^1$	2	8	18	8	1	—	—
Cesium (Cs)	[Xe] $6s^1$	2	8	18	18	8	1	—
Francium (Fr)	[Rn] $7s^1$	2	8	18	32	18	8	1

*[He] is shorthand for the structure of He, which is $2s^2$. Lithium, therefore, is $2s^2\ 1s^1$. [Ne] means the structure of neon, and so on.

Periodic Table of the Elements

Metals
Semiconductors
Nonmetals

Transition Elements

Inner Transition Elements

Alkali Metals

Alkaline Earth Metals

Halogens

Noble Gases

Period	IA (1)	IIA (2)	IIIB (3)	IVB (4)	VB (5)	VIB (6)	VIIB (7)	VIIIB (8)	VIIIB (9)	VIIIB (10)	IB (11)	IIB (12)	IIIA (13)	IVA (14)	VA (15)	VIA (16)	VIIA (17)	VIIIA (18)
1	Hydrogen 1 **H** 1.008																	Helium 2 **He** 4.003
2	Lithium 3 **Li** 6.941	Beryllium 4 **Be** 9.012											Boron 5 **B** 10.81	Carbon 6 **C** 12.01	Nitrogen 7 **N** 14.01	Oxygen 8 **O** 16.00	Fluorine 9 **F** 19.00	Neon 10 **Ne** 20.18
3	Sodium 11 **Na** 22.99	Magnesium 12 **Mg** 24.31											Aluminum 13 **Al** 26.98	Silicon 14 **Si** 28.09	Phosphorus 15 **P** 30.97	Sulfur 16 **S** 32.07	Chlorine 17 **Cl** 35.45	Argon 18 **Ar** 39.95
4	Potassium 19 **K** 39.10	Calcium 20 **Ca** 40.08	Scandium 21 **Sc** 44.96	Titanium 22 **Ti** 47.88	Vanadium 23 **V** 50.94	Chromium 24 **Cr** 52.00	Manganese 25 **Mn** 54.94	Iron 26 **Fe** 55.85	Cobalt 27 **Co** 58.93	Nickel 28 **Ni** 58.69	Copper 29 **Cu** 63.55	Zinc 30 **Zn** 65.39	Gallium 31 **Ga** 69.72	Germanium 32 **Ge** 72.61	Arsenic 33 **As** 74.92	Selenium 34 **Se** 78.96	Bromine 35 **Br** 79.90	Krypton 36 **Kr** 83.80
5	Rubidium 37 **Rb** 85.47	Strontium 38 **Sr** 87.62	Yttrium 39 **Y** 88.91	Zirconium 40 **Zr** 91.22	Niobium 41 **Nb** 92.91	Molybdenum 42 **Mo** 95.94	Technetium 43 **Tc** (98)	Ruthenium 44 **Ru** 101.1	Rhodium 45 **Rh** 102.9	Palladium 46 **Pd** 106.4	Silver 47 **Ag** 107.9	Cadmium 48 **Cd** 112.4	Indium 49 **In** 114.8	Tin 50 **Sn** 118.7	Antimony 51 **Sb** 121.8	Tellurium 52 **Te** 127.6	Iodine 53 **I** 126.9	Xenon 54 **Xe** 131.3
6	Cesium 55 **Cs** 132.9	Barium 56 **Ba** 137.3	Lanthanum 57 **La** 138.9 †	Hafnium 72 **Hf** 178.5	Tantalum 73 **Ta** 180.9	Tungsten 74 **W** 183.8	Rhenium 75 **Re** 186.2	Osmium 76 **Os** 190.2	Iridium 77 **Ir** 192.2	Platinum 78 **Pt** 195.1	Gold 79 **Au** 197.0	Mercury 80 **Hg** 200.6	Thallium 81 **Tl** 204.4	Lead 82 **Pb** 207.2	Bismuth 83 **Bi** 209.0	Polonium 84 **Po** (209)	Astatine 85 **At** (210)	Radon 86 **Rn** (222)
7	Francium 87 **Fr** (223)	Radium 88 **Ra** (226)	Actinium 89 **Ac** (227) ‡	Rutherfordium 104 **Rf** (261)	Dubnium 105 **Db** (262)	Seaborgium 106 **Sg** (266)	Bohrium 107 **Bh** (264)	Hassium 108 **Hs** (277)	Meitnerium 109 **Mt** (268)	Ununnilium 110 **Uun** (281)	Unununium 111 **Uuu** (272)	Ununbium 112 **Uub** (285)	Ununtrium 113 **Uut** (284)	Ununquadium 114 **Uuq** (289)	Ununpentium 115 **Uup** (288)	Ununhexium 116 **Uuh** (292)		Ununoxtium 118 **Uuo** (294)

†Lanthanides 6

Cerium 58 **Ce** 140.1	Praseodymium 59 **Pr** 140.9	Neodymium 60 **Nd** 144.2	Promethium 61 **Pm** (145)	Samarium 62 **Sm** 150.4	Europium 63 **Eu** 152.0	Gadolinium 64 **Gd** 157.3	Terbium 65 **Tb** 158.9	Dysprosium 66 **Dy** 162.5	Holmium 67 **Ho** 164.9	Erbium 68 **Er** 167.3	Thulium 69 **Tm** 168.9	Ytterbium 70 **Yb** 173.0	Lutetium 71 **Lu** 175.0

‡Actinides 7

Thorium 90 **Th** 232.0	Protactinium 91 **Pa** 231.0	Uranium 92 **U** 238.0	Neptunium 93 **Np** (237)	Plutonium 94 **Pu** (244)	Americium 95 **Am** (243)	Curium 96 **Cm** (247)	Berkelium 97 **Bk** (247)	Californium 98 **Cf** (251)	Einsteinium 99 **Es** (252)	Fermium 100 **Fm** (257)	Mendelevium 101 **Md** (258)	Nobelium 102 **No** (259)	Lawrencium 103 **Lr** (262)

Values in parentheses are the mass numbers of the most stable or best-known isotopes.

Names and symbols for elements 110–118 are under review.

Key

element name → Hydrogen
symbol of element → **H**
1.008

atomic number
atomic weight

FIGURE 8.17 The periodic table of the elements.

TABLE 8.5

Electron structures of the noble gas family

Element	Electron Configuration	Number of Electrons in Main Energy Level						
		1st	2nd	3rd	4th	5th	6th	7th
Helium (He)	$1s^2$	2	—	—	—	—	—	—
Neon (Ne)	[He] $2s^2 2p^6$	2	8	—	—	—	—	—
Argon (Ar)	[Ne] $3s^2 3p^6$	2	8	8	—	—	—	—
Krypton (Kr)	[Ar] $4s^2 3d^{10} 4p^6$	2	8	18	8	—	—	—
Xenon (Xe)	[Kr] $5s^2 4d^{10} 5p^6$	2	8	18	18	8	—	—
Radon (Rn)	[Xe] $6s^2 4f^{14} 5d^{10} 6p^6$	2	8	18	32	18	8	—

As shown in Table 8.5, the elements in group VIIIA have orbitals that are filled to capacity. These elements are colorless, odorless gases that almost never react with other elements to form compounds. Sometimes they are called the noble gases because they are chemically inert, perhaps indicating they are above the other elements. They have also been called the *rare gases* because of their scarcity and *inert gases* because they are mostly chemically inert, not forming compounds. The noble gases are inert because they have filled outer electron configurations, a particularly stable condition.

Each period *begins* with a single electron in a new orbital. Second, each period *ends* with the filling of an orbital, completing the maximum number of electrons that can occupy that main energy level. Since the first A family is identified as IA, this means that all the atoms of elements in this family have one electron in their outer orbitals. All the atoms of elements in family IIA have two electrons in their outer orbitals. This pattern continues on to family VIIIA, in which all the atoms of elements have eight electrons in their outer orbitals except helium. Thus, the number identifying the A families *also identifies the number of electrons in the outer orbitals,* with the exception of helium. Helium is nonetheless similar to the other elements in this family, since all have filled outer orbitals. The electron theory of chemical bonding, which is discussed in chapter 9, states that only the electrons in the outermost orbitals of an atom are involved in chemical reactions. Thus, *the outer orbital electrons are mostly responsible for the chemical properties of an element.* Since the members of a family all have similar outer configurations, you would expect them to have similar chemical behaviors, and they do.

 CONCEPTS *Applied*

Periodic Practice

Identify the period and family of the element silicon. Write your answer before reading the solution in the next paragraph.

According to the list of elements on the inside back cover of this text, silicon has the symbol Si and an atomic number of 14. The square with the symbol Si and the atomic number 14 is located in the third period (third row) and in the column identified as IVA (14).

Now, can you identify the period and family of the element iron (Fe)? Compare your answer with a classmate's to check.

METALS, NONMETALS, AND SEMICONDUCTORS

As indicated earlier, chemical behavior is mostly concerned with the outer orbital electrons. The outer orbital electrons, that is, the highest energy level electrons, are conveniently represented with an **electron dot notation,** made by writing the chemical symbol with dots around it indicating the number of outer orbital electrons. Electron dot notations are shown for the representative elements in Figure 8.18. Again, note the pattern in Figure 8.18—all the noble gases are in group VIIIA, and all (except helium) have eight outer electrons. All the group IA elements (alkali metals) have one dot, all the IIA elements have two dots, and so on. This pattern will explain the difference in metals, nonmetals, and a third group of in-between elements called semiconductors.

One way to group substances is according to the physical properties of metals and nonmetals—luster, conductivity, malleability, and ductility. Metals and nonmetals also have certain chemical properties that are related to their positions in the periodic table. Figure 8.19 shows where the *metals, nonmetals,* and *semiconductors* are located. Note that about 80 percent of all the elements are metals.

The noble gases have completely filled outer orbitals in their highest energy levels, and this is a particularly stable arrangement. Other elements react chemically, either *gaining or losing electrons to attain a filled outermost energy level like the noble gases.* When an atom loses or gains electrons, it acquires an unbalanced electron charge and is called an **ion.** An atom of lithium, for example, has three protons (plus charges) and three electrons (negative charges). If it loses the outermost electron, it now has an outer filled orbital structure like helium, a noble gas. It is also now

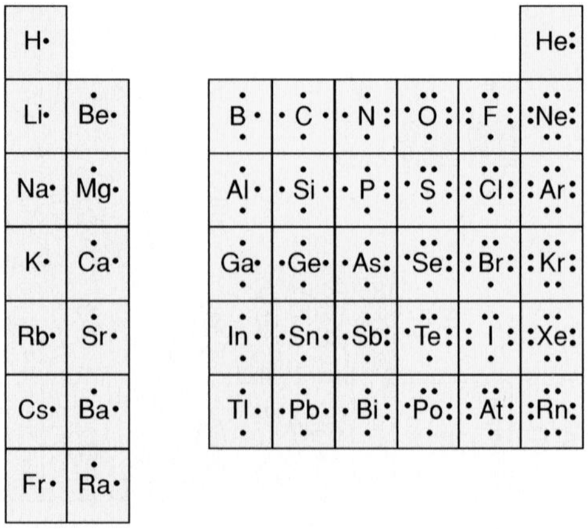

FIGURE 8.18 Electron dot notation for the representative elements.

Compounds of the rare earths were first identified when they were isolated from uncommon minerals in the late 1700s. The elements are very reactive and have similar chemical properties, so they were not recognized as elements until some fifty years later. Thus, they were first recognized as earths, that is, nonmetal substances, when in fact they are metallic elements. They were also considered to be rare since, at that time, they were known to occur only in uncommon minerals. Today, these metallic elements are known to be more abundant in the earth than gold, silver, mercury, or tungsten. The rarest of the rare earths, thulium, is twice as abundant as silver. The rare earth elements are neither rare nor earths, and they are important materials in glass, electronic, and metallurgical industries.

You can identify the rare earths in the two lowest rows of the periodic table. These rows contain two series of elements that actually belong in periods 6 and 7, but they are moved below so that the entire table is not so wide. Together, the two series are called the inner transition elements. The top series is fourteen elements wide from elements 58 through 71. Since this series belongs next to element 57, lanthanum, it is sometimes called the *lanthanide series*. This series is also known as the rare earths. The second series of fourteen elements is called the *actinide series*. These are mostly the artificially prepared elements that do not occur naturally.

You may never have heard of the rare earth elements, but they are key materials in many advanced or high-technology products. Lanthanum, for example, gives glass special refractive properties and is used in optic fibers and expensive camera lenses. Samarium, neodymium, and dysprosium are used to manufacture crystals used in lasers. Samarium, ytterbium, and terbium have special magnetic properties that have made possible new electric motor designs, magnetic-optical devices in computers, and the creation of a ceramic superconductor. Other rare earth metals are also being researched for use in possible high-temperature superconductivity materials. Many rare earths are also used in metal alloys; for example, an alloy of cerium is used to make heat-resistant jet-engine parts. Erbium is also used in high-performance metal alloys. Dysprosium and holmium have neutron-absorbing properties and are used in control rods to control nuclear fission. Europium should be mentioned because of its role in making the red color of color television screens. The rare earths are relatively abundant metallic elements that play a key role in many common and high-technology applications. They may also play a key role in superconductivity research.

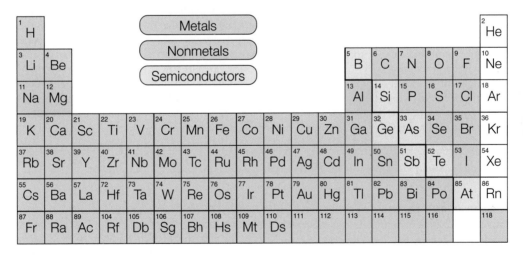

FIGURE 8.19 The location of metals, nonmetals, and semiconductors in the periodic table.

an ion, since it has three protons (3+) and two electrons (2−), for a net charge of 1+. A lithium ion thus has a 1+ charge.

 ## CONCEPTS *Applied*

Metals and Charge

Is strontium a metal, nonmetal, or semiconductor? What is the charge on a strontium ion?

The list of elements inside the back cover identifies the symbol for strontium as Sr (atomic number 38). In the periodic table, Sr is located in family IIA, which means that an atom of strontium has two electrons in its outer orbital. For several reasons, you know that strontium is a metal: (1) An atom of strontium has two electrons in its outer orbital and atoms with one, two, or three outer electrons are identified as metals; (2) strontium is located in the IIA family, the alkaline earth metals; and (3) strontium is located on the left side of the periodic table and, in general, elements located in the left two-thirds of the table are metals.

Elements with one, two, or three outer electrons tend to lose electrons to form positive ions. Since strontium has an atomic number of 38, you know that it has thirty-eight protons (38+) and thirty-eight electrons (38−). When it loses its two outer orbital electrons, it has 38+ and 36− for a charge of 2+.

People Behind the Science

Dmitri Ivanovich Mendeleyev (1834–1907)

Dmitri Mendeleyev was a Russian chemist whose name will always be linked with his outstanding achievement, the development of the periodic table. He was the first chemist to understand that all elements are related members of a single ordered system. He converted what had been a highly fragmented and speculative branch of chemistry into a true, logical science. The spelling of his name has been a source of confusion for students and frustration for editors for more than a century, and the forms Mendeléeff, Mendeléev, and even Mendelejeff can all be found in print.

Mendeleyev was born in Tobol'sk, Siberia, on February 7, 1834, the youngest of the seventeen children of the head of the local high school. His father went blind when Mendeleyev was a child, and the family had to rely increasingly on their mother for support. He was educated locally but could not gain admission to any Russian university because of prejudice toward the supposedly backward attainments of those educated in the provinces. In 1855, he finally qualified as a teacher at the Pedagogical Institute in St. Petersburg. He took an advanced-degree course in chemistry. In 1859, he was sent by the government for further study at the University of Heidelberg, and in 1861, he returned to St. Petersburg and became professor of general chemistry at the Technical Institute in 1864. He could find

no textbook adequate for his students' needs, so he decided to produce his own. The resulting *Principles of Chemistry* (1868–1870) won him international renown; it was translated into English in 1891 and 1897.

Before Mendeleyev produced his periodic law, understanding of the chemical elements had long been an elusive and frustrating task. According to Mendeleyev, the properties of the elements are periodic functions of their atomic weights. In 1869, he stated that "the elements arranged according to the magnitude of atomic weights show a periodic change of properties." Other chemists, notably Lothar Meyer in Germany, had meanwhile come to similar conclusions, with Meyer publishing his findings independently.

Mendeleyev compiled the first true periodic table, listing all the sixty-three elements then known. Not all elements would "fit" properly using the atomic weights of the time, so he altered indium from 76 to 114 (modern value 114.8) and beryllium from 13.8 to 9.2 (modern value 9.013). In 1871, he produced a revisionary paper showing the correct repositioning of seventeen elements.

To make the table work, Mendeleyev also had to leave gaps, and he predicted that further elements would eventually be discovered to fill them. These predictions provided the strongest endorsement

of the periodic law. Three were discovered in Mendeleyev's lifetime: gallium (1871), scandium (1879), and germanium (1886), all with properties that tallied closely with those he had assigned to them.

Farsighted though Mendeleyev was, he had no notion that the periodic recurrences of similar properties in the list of elements reflected anything in the structures of their atoms. It was not until the 1920s that it was realized that the key parameter in the periodic system is not the atomic weight but the atomic number of the elements—a measure of the number of protons in the atom. Since then, great progress has been made in explaining the periodic law in terms of the electronic structures of atoms and molecules.

Elements with one, two, or three outer electrons tend to lose electrons to form positive ions. The metals lose electrons like this, and the *metals are elements that lose electrons to form positive ions* (Figure 8.20). Nonmetals, on the other hand, are elements with five to seven outer electrons that tend to acquire electrons to fill their outer orbitals. *Nonmetals are elements that gain electrons to form negative ions.* In general, elements located in the left two-thirds or so of the periodic table are metals. The nonmetals are on the right side of the table (Figure 8.19).

CONCEPTS *Applied*

Outer Orbitals

How many outer orbital electrons are found in an atom of (a) oxygen, (b) calcium, and (c) aluminum? Write your answers before reading the answers in the next paragraph.

(a) According to the list of elements on the inside back cover of this text, oxygen has the symbol O and

an atomic number of 8. The square with the symbol O and the atomic number 8 is located in the column identified as VIA. Since the A family number is the same as the number of electrons in the outer orbital, oxygen has six outer orbital electrons. (b) Calcium has the symbol Ca (atomic number 20) and is located in column IIA, so a calcium atom has two outer orbital electrons. (c) Aluminum has the symbol Al (atomic number 13) and is located in column IIIA, so an aluminum atom has three outer orbital electrons.

The dividing line between the metals and nonmetals is a steplike line from the left top of group IIIA down to the bottom left of group VIIA. This is not a line of sharp separation between the metals and nonmetals, and elements *along* this line sometimes act like metals, sometimes like nonmetals, and sometimes like both. These hard-to-classify elements are called

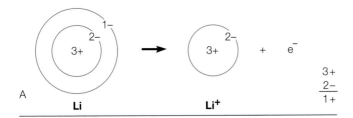

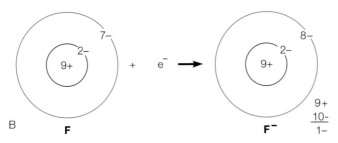

FIGURE 8.20 (*A*) Metals lose their outer electrons to acquire a noble gas structure and become positive ions. Lithium becomes a 1+ ion as it loses its one outer electron. (*B*) Nonmetals gain electrons to acquire an outer noble gas structure and become negative ions. Fluorine gains a single electron to become a 1– ion.

semiconductors (or *metalloids*). Silicon, germanium, and arsenic have physical properties of nonmetals; for example, they are brittle materials that cannot be hammered into a new shape. Yet these elements conduct electric currents under certain conditions. The ability to conduct an electric current is a property of a metal, and nonmalleability is a property of nonmetals, so as you can see, these semiconductors have the properties of both metals and nonmetals.

The transition elements, which are all metals, are located in the B-group families. Unlike the representative elements, which form vertical families of similar properties, the transition elements tend to form horizontal groups of elements with similar properties. Iron (Fe), cobalt (Co), and nickel (Ni) in group VIIIB, for example, are three horizontally arranged metallic elements that show magnetic properties.

A family of representative elements all form ions with the same charge. Alkali metals, for example, all lose an electron to form a 1+ ion. The transition elements have *variable charges*. Some transition elements, for example, lose their one outer electron to form 1+ ions (copper, silver). Copper, because of its special configuration, can also lose an additional electron to form a 2+ ion. Thus, copper can form either a 1+ ion or a 2+ ion. Most transition elements have two outer s orbital electrons and lose them both to form 2+ ions (iron, cobalt, nickel), but some of these elements also have special configurations that permit them to lose more of their electrons. Thus, iron and cobalt, for example, can form either a 2+ ion or a 3+ ion. Much more can be interpreted from the periodic table, and more generalizations will be made as the table is used in the following chapters.

SUMMARY

Attempts at understanding matter date back to ancient Greek philosophers, who viewed matter as being composed of *elements,* or simpler substances. Two models were developed that considered matter to be (1) *continuous,* or infinitely divisible, or (2) *discontinuous,* made up of particles called *atoms.*

In the early 1800s, Dalton published an *atomic theory,* reasoning that matter was composed of hard, indivisible atoms that were joined together or dissociated during chemical change.

Cathode rays were observed to move from the negative terminal in an evacuated glass tube. The nature of cathode rays was a mystery. The mystery was solved in 1897 when Thomson discovered they were negatively charged particles now known as *electrons.* Thomson had discovered the first elementary particle of which atoms are made and measured their charge-to-mass ratio.

Rutherford developed a solar system model based on experiments with alpha particles scattered from a thin sheet of metal. This model had a small, massive, and positively charged *nucleus* surrounded by moving electrons. These electrons were calculated to be at a distance from the nucleus of 100,000 times the radius of the nucleus, so the volume of an atom is mostly empty space. Later, Rutherford proposed that the nucleus contained two elementary particles: *protons* with a positive charge and *neutrons* with no charge. The *atomic number* is the number of protons in an atom. Atoms of elements with different numbers of neutrons are called *isotopes.* The mass of each isotope is compared to the mass of carbon-12, which is assigned a mass of exactly 12.00 *atomic mass units.* The mass contribution of the isotopes of an element according to their abundance is called the *atomic weight* of an element. Isotopes are identified by their *mass number,* which is the sum of the number of protons and neutrons in the nucleus. Isotopes are identified by their chemical symbol with the atomic number as a subscript and the mass number as a superscript.

Bohr developed a model of the hydrogen atom to explain the characteristic *line spectra* emitted by hydrogen. His model specified that (1) electrons can move only in allowed orbits, (2) electrons do not emit radiant energy when they remain in an orbit, and (3) electrons move from one allowed orbit to another when they gain or lose energy. When an electron jumps from a higher orbit to a lower one, it gives up energy in the form of a single photon. The energy of the photon corresponds to the difference in energy between the two levels. The Bohr model worked well for hydrogen but not for other atoms.

De Broglie proposed that moving particles of matter (electrons) should have wave properties like moving particles of light (photons). His derived equation, $\lambda = h/mv$, showed that these *matter waves* were only measurable for very small particles such as electrons. De Broglie's proposal was tested experimentally, and the experiments confirmed that electrons do have wave properties.

Schrödinger and others used the wave nature of the electron to develop a new model of the atom called *wave mechanics,* or *quantum mechanics*. This model was found to confirm exactly all the experimental data as well as predict new data. The quantum mechanical model describes the energy state of the electron in terms of quantum numbers based on the wave nature of the electron. The quantum numbers defined the *probability* of the location of an electron in terms of fuzzy regions of space called *orbitals*.

The *periodic table* has horizontal rows of elements called *periods* and vertical columns of elements called *families*. Members of a given family have the same outer orbital electron configurations, and it is the electron configuration that is mostly responsible for the chemical properties of an element.

SUMMARY OF EQUATIONS

8.1

$$energy = (Planck's\ constant)(frequency)$$

$$E = hf$$

where $h = 6.63 \times 10^{-34}$ J·s

8.2

$$\frac{1}{wavelength} = constsnt \left(\frac{1}{2^2} - \frac{1}{number^2} \right)$$

$$\frac{1}{\lambda} = R \left(\frac{1}{2^2} - \frac{1}{n^2} \right)$$

where $R = 1.097 \times 10^7$ l/m

8.3

$$energy\ state\ of\ orbit\ number = \frac{energy\ state\ of\ innnermost\ orbit}{number\ squared}$$

$$E_n = \frac{E_1}{n^2}$$

where $E_1 = -13.6$ eV, and $n = 1, 2, 3, \ldots$

8.4

$$\begin{matrix} energy \\ of \\ photon \end{matrix} = \begin{pmatrix} energy\ state \\ of \\ higher\ orbit \end{pmatrix} - \begin{pmatrix} energy\ state \\ of \\ lower\ orbit \end{pmatrix}$$

$$hf = E_H - E_L$$

where $h = 6.63 \times 10^{-34}$ J·s; E_H and E_L must be in joules

8.5

$$wavelength = \frac{Plank's\ constant}{(mass)(velocity)}$$

$$\lambda = \frac{h}{mv}$$

where $h = 6.63 \times 10^{-34}$ J·s

KEY TERMS

atomic mass unit (p. **207**)
atomic number (p. **206**)
atomic weight (p. **207**)
Balmer series (p. **209**)
electron (p. **205**)
electron dot notation (p. **218**)

electron pair (p. **215**)
electron volt (p. **211**)
excited states (p. **210**)
ground state (p. **210**)
Heisenberg uncertainty principle (p. **214**)
ion (p. **218**)
isotope (p. **207**)
line spectrum (p. **208**)
mass number (p. **207**)
matter waves (p. **212**)
neutron (p. **206**)
nucleus (p. **206**)
orbital (p. **214**)
Pauli exclusion principle (p. **215**)
photons (p. **208**)
proton (p. **206**)
quanta (p. **208**)
quantum mechanics (p. **213**)
representative elements (p. **216**)
semiconductors (p. **221**)
transition elements (p. **216**)
wave mechanics (p. **213**)

APPLYING THE CONCEPTS

1. Thomson was convinced that he had discovered a subatomic particle, the electron, from the evidence that
   a. the charge-to-mass ratio was the same for all materials.
   b. cathode rays could move through a vacuum.
   c. electrons were attracted toward a negatively charged plate.
   d. the charge was always 1.60×10^{-19} coulomb.

2. The existence of a tiny, massive, and positively charged nucleus was deduced from the observation that
   a. fast, massive, and positively charged radioactive particles all move straight through metal foil.
   b. radioactive particles were deflected by a magnetic field.
   c. some radioactive particles were deflected by metal foil.
   d. None of the above is correct.

3. According to Rutherford's calculations, the volume of an atom is mostly
   a. occupied by protons and neutrons.
   b. filled with electrons.
   c. occupied by tightly bound protons, electrons, and neutrons.
   d. empty space.

4. Millikan measured the charge on oil droplets and found that all the droplets had
   a. different charges.
   b. random charges, without any pattern.
   c. five groupings of different charges.
   d. the same or multiples of the same charge.

5. Rutherford's estimate of the radius of an atomic nucleus was based on
   a. the drift of oil droplets in an electric field.
   b. speculation about expected symmetry in gold foil.
   c. measurements of radioactive particle deflections from gold foil.
   d. measurements of the breakup of a nitrogen atom by collisions with radioactive particles.

6. The atomic number is the number of
   a. protons.
   b. protons plus neutrons.
   c. protons plus electrons.
   d. protons, neutrons, and electrons in an atom.

7. All neutral atoms of an element have the same
   a. atomic number.
   b. number of electrons.
   c. number of protons.
   d. All of the above are correct.

8. The main problem with a solar system model of the atom is that
   a. electrons move in circular, not elliptical orbits.
   b. the electrons should lose energy since they are accelerating.
   c. opposite charges should attract one another.
   d. the mass ratio of the nucleus to the electrons is wrong.

9. Atoms of an element that have different numbers of neutrons are called
   a. allotropes.
   b. isomers.
   c. isotopes.
   d. radioactive.

10. The sum of the numbers of protons and neutrons in the nucleus of an atom is the
    a. nucleon number.
    b. mass number.
    c. atomic weight.
    d. isotope number.

11. Atomic weight is
    a. determined by weighing individual atoms.
    b. an average weight of the isotopes of an element.
    c. the number of protons and neutrons in a nucleus.
    d. a weighted average of the masses of isotopes of an element based on abundance.

12. This isotope provides the standard to which the masses of all other isotopes are compared:
    a. carbon-12
    b. oxygen-16
    c. hydrogen-1
    d. gold-197

13. In 1910, Max Planck introduced the idea that matter emits and absorbs energy in
    a. light waves.
    b. discrete units called quanta.
    c. pulses with no particular pattern.
    d. pulses that vary in magnitude over time.

14. Energy of the electron is expressed in units of
    a. electron volts.
    b. electron watts.
    c. quantum leaps.
    d. orbit numbers.

15. The major success of the Bohr theory was in explaining
    a. how electrons move in circular orbits.
    b. why radiationless orbits existed.
    c. the colors in the hydrogen line spectrum.
    d. why the angular momentum of the electron should be by orbit quantum numbers.

16. Light from an incandescent gas is dispersed into narrow lines of colors with no light between the lines. This is called a (an)
    a. impossible spectrum.
    b. line spectrum.
    c. Balmer spectrum.
    d. Newton spectrum.

17. The lowest energy state or level of an atom is the
    a. bottom state.
    b. lowest level.
    c. ground state.
    d. basement state.

18. The basis of the quantum mechanics theory of the atom is
    a. spin and quantum leaps of electron masses.
    b. elliptical orbits of electrons.
    c. how electron particles move in orbits.
    d. the wave nature of electrons.

19. An electron moving from an excited state to the ground state
    a. emits a photon.
    b. gains a photon.
    c. gains a charge.
    d. loses a charge.

20. The existence of matter waves was proposed by
    a. Planck.
    b. Bohr.
    c. de Broglie.
    d. Einstein.

21. Any moving particle has a wavelength that is associated with its mass and velocity. This is a statement that proposed the existence of
    a. photoelectric effect.
    b. matter waves.
    c. quanta.
    d. photons.

22. The arrangement of electrons in orbitals is called
    a. electron configuration.
    b. periodic table.
    c. quantum numbers.
    d. energy levels.

23. Group IIA elements are called
    a. alkali metals.
    b. alkaline earth metals.
    c. alkaline salts.
    d. beryllium metals.

24. The elements in A groups are called
    a. alkali elements.
    b. transition elements.
    c. representative elements.
    d. metals.

25. The element chlorine belongs to which group?
    a. alkali metals
    b. lanthanides
    c. halogens
    d. noble gases

26. The gain or loss of electrons from an atom results in the formation of a (an)
    a. ion.
    b. metal.
    c. semiconductor.
    d. isotope.

27. Elements that have properties of both the metals and the nonmetals are
    a. semimetals.
    b. transition elements.
    c. semiconductors.
    d. noble gases.

28. Transition elements
    a. are metals.
    b. belong to the B group.
    c. have variable charges.
    d. All of the above are correct.

29. The energy of a photon
    a. varies inversely with the frequency.
    b. is directly proportional to the frequency.
    c. varies directly with the velocity.
    d. is inversely proportional to the velocity.

30. A photon of which of the following has the most energy?
    a. red light
    b. orange light
    c. green light
    d. blue light

31. The lines of color in a line spectrum from a given element
    a. change colors with changes in the temperature.
    b. are always the same, with a regular spacing pattern.
    c. are randomly spaced, having no particular pattern.
    d. have the same colors, with a spacing pattern that varies with the temperature.

32. Hydrogen, with its one electron, produces a line spectrum in the visible light range with
    a. one color line.
    b. two color lines.
    c. three color lines.
    d. four color lines.

33. According to the Bohr model, an electron gains or loses energy only by
    a. moving faster or slower in an allowed orbit.
    b. jumping from one allowed orbit to another.
    c. being completely removed from an atom.
    d. jumping from one atom to another atom.

34. According to the Bohr model, when an electron in a hydrogen atom jumps from an orbit farther from the nucleus to an orbit closer to the nucleus, it
    a. emits a single photon with energy equal to the energy difference of the two orbits.
    b. emits four photons, one for each of the color lines observed in the line spectrum of hydrogen.
    c. emits a number of photons dependent on the number of orbit levels jumped over.
    d. None of the above is correct.

35. The Bohr model of the atom
    a. explained the color lines in the hydrogen spectrum.
    b. could not explain the line spectrum of atoms larger than hydrogen.
    c. had some made-up rules without explanations.
    d. All of the above are correct.

36. The Bohr model of the atom described the energy state of electrons with one quantum number. The quantum mechanics model uses how many quantum numbers to describe the energy state of an electron?
    a. one
    b. two
    c. four
    d. ten

37. An electron in the second main energy level and the second sublevel is described by the symbols
    a. 1s.
    b. 2s.
    c. 1p.
    d. 2p.

38. The space in which it is probable that an electron will be found is described by a (an)
    a. circular orbit.
    b. elliptical orbit.
    c. orbital.
    d. geocentric orbit.

39. Two different isotopes of the same element have
    a. the same number of protons, neutrons, and electrons.
    b. the same number of protons and neutrons but different numbers of electrons.
    c. the same number of protons and electrons but different numbers of neutrons.
    d. the same number of neutrons and electrons but different numbers of protons.

40. The isotopes of a given element always have
    a. the same mass and the same chemical behavior.
    b. the same mass and a different chemical behavior.
    c. different masses and different chemical behaviors.
    d. different masses and the same chemical behavior.

41. If you want to know the number of protons in an atom of a given element, you would look up the
    a. mass number.
    b. atomic number.
    c. atomic weight.
    d. abundance of isotopes compared to the mass number.

42. If you want to know the number of neutrons in an atom of a given element, you would
    a. round the atomic weight to the nearest whole number.
    b. add the mass number and the atomic number.
    c. subtract the atomic number from the mass number.
    d. add the mass number and the atomic number, then divide by two.

43. Which of the following is always a whole number?
    a. atomic mass of an isotope
    b. mass number of an isotope
    c. atomic weight of an element
    d. None of the above is correct.

44. The quantum mechanics and Bohr models of the atom both *agree* on
    a. the significance of the de Broglie wavelength and the circumference of an orbit.
    b. the importance of momentum in determining the size of an orbit.
    c. how electrons are able to emit light.
    d. None of the above is correct.

45. Hydrogen, with its one electron, can produce a line spectrum with four visible colors because
    a. an isotope of hydrogen has four electrons.
    b. electrons occur naturally with four different colors.
    c. there are multiple energy levels that an electron can occupy.
    d. electrons are easily scattered.

46. A photon is emitted from the electronic structure of an atom when an electron
    a. jumps from a higher to a lower energy level.
    b. jumps from a lower to a lower higher level.
    c. reverses its spin by 180°.
    d. is removed from an atom by a high quantum of energy.

47. Which of the following represents a hydrogen isotope?
    a. $_1^1\text{H}$
    b. $_1^2\text{H}$
    c. $_1^3\text{H}$
    d. All of the above are correct.

48. In what are atoms of $_6^{12}\text{C}$ and $_6^{14}\text{C}$ different?
    a. number of protons
    b. number of neutrons
    c. number of electrons
    d. None of the above is correct.

49. An atom has 6 protons, 6 electrons, and 6 neutrons, so the *isotope symbol* is
    a. $_{12}^{18}\text{Mg}$
    b. $_{12}^{12}\text{Mg}$
    c. $_6^{12}\text{C}$
    d. $_6^{12}\text{C}$

Answers

1. a 2. c 3. d 4. d 5. c 6. a 7. d 8. b 9. c 10. b 11. d 12. a 13. b 14. a
15. c 16. b 17. c 18. d 19. a 20. c 21. b 22. a 23. b 24. c 25. c 26. a
27. c 28. d 29. b 30. d 31. b 32. d 33. b 34. a 35. d 36. c 37. d 38. c
39. c 40. d 41. b 42. c 43. b 44. c 45. c 46. a 47. d 48. b 49. c

QUESTIONS FOR THOUGHT

1. Describe the experimental evidence that led Rutherford to the concept of a nucleus in an atom.
2. What is the main problem with a solar system model of the atom?
3. Compare the size of an atom to the size of its nucleus.
4. An atom has 11 protons in the nucleus. What is the atomic number? What is the name of this element? What is the electron configuration of this atom?
5. Why do the energies of electrons in an atom have negative values? (*Hint:* It is *not* because of the charge of the electron.)
6. What is similar about the Bohr model of the atom and the quantum mechanical model? What are the fundamental differences?
7. What is the difference between a hydrogen atom in the ground state and one in the excited state?
8. Which of the following are whole numbers, and which are not whole numbers? Explain why for each.
    (a) atomic number
    (b) isotope mass
    (c) mass number
    (d) atomic weight

9. Why does the carbon-12 isotope have a whole-number mass but not the other isotopes?
10. What do the members of the noble gas family have in common? What are their differences?
11. How are the isotopes of an element similar? How are they different?
12. What patterns are noted in the electron structures of elements found in a period and in a family in the periodic table?

FOR FURTHER ANALYSIS

1. Evaluate Millikan's method for finding the charge of an electron. Are there any doubts about the results of using this technique?
2. What are the significant similarities and differences between the isotopes of a particular element?
3. Thomson's experiments led to the discovery of the electron. Analyze how you know for sure that he discovered the electron.
4. Describe a conversation between yourself and another person as you correct her belief that atomic weight has something to do with gravity.
5. Analyze the significance of the observation that matter only emits and absorbs energy in discrete units.
6. Describe several basic differences between the Bohr and quantum mechanics models of the atom.

INVITATION TO INQUIRY

Too Small to See?

As Rutherford knew when he conducted his famous experiment with radioactive particles and gold foil, the structure of an atom is too small to see, so it must be inferred from other observations. To illustrate this process, pour 50 mL of 95 percent ethyl alcohol (or some other almost pure alcohol) into a graduated cylinder. In a second graduated cylinder, measure 50 mL of water. Mix thoroughly and record the volume of the combined liquids. Assuming no evaporation took place, is the result contrary to your expectation? What question could be asked about the result?

Answers to questions about things that you cannot see often require a model that can be observed. For example, a model might represent water and alcohol molecules by using beans for alcohol molecules and sand for water molecules. Does mixing 50 mL of beans and 50 mL of sand result in 100 mL of mixed sand and beans? Explain the result of mixing alcohol and water based on your observation of mixing beans and sand.

To continue this inquiry, fill a water glass to the brim with water. Add a small amount of salt to the water. You know that two materials cannot take up the same space at the same time, so what happened to the salt? How can you test your ideas about things that are too tiny to see?

The exercises in groups A and B cover the same concepts. Solutions to group A exercises are located in appendix E.

Group A

1. A neutron with a mass of 1.68×10^{-27} kg moves from a nuclear reactor with a velocity of 3.22×10^3 m/s. What is the de Broglie wavelength of the neutron?

2. Calculate the energy (a) in eV and (b) in joules for the sixth energy level ($n = 6$) of a hydrogen atom.

3. How much energy is needed to move an electron in a hydrogen atom from $n = 2$ to $n = 6$? Give the answer (a) in joules and (b) in eV. (See Figure 8.11 for needed values.)

4. What frequency of light is emitted when an electron in a hydrogen atom jumps from $n = 6$ to $n = 2$? What color would you see?

5. How much energy is needed to completely remove the electron from a hydrogen atom in the ground state?

6. Thomson determined the charge-to-mass ratio of the electron to be -1.76×10^{11} coulomb/kilogram. Millikan determined the charge on the electron to be -1.60×10^{-19} coulomb. According to these findings, what is the mass of an electron?

7. Assume that an electron wave making a standing wave in a hydrogen atom has a wavelength of 1.67×10^{-10} m. Considering the mass of an electron to be 9.11×10^{-31} kg, use the de Broglie equation to calculate the velocity of an electron in this orbit.

8. Using any reference you wish, write the complete electron configurations for (a) boron, (b) aluminum, and (c) potassium.

9. Explain how you know that you have the correct *total* number of electrons in your answers for 8a, 8b, and 8c.

10. Refer to Figure 8.16 *only,* and write the complete electron configurations for (a) argon, (b) zinc, and (c) bromine.

11. Lithium has two naturally occurring isotopes: lithium-6 and lithium-7. Lithium-6 has a mass of 6.01512 relative to carbon-12 and makes up 7.42 percent of all naturally occurring lithium. Lithium-7 has a mass of 7.016 compared to carbon-12 and makes up the remaining 92.58 percent. According to this information, what is the atomic weight of lithium?

12. Identify the number of protons, neutrons, and electrons in the following isotopes:
    (a) $^{12}_{6}C$
    (b) $^{1}_{1}H$
    (c) $^{40}_{18}Ar$
    (d) $^{2}_{1}H$
    (e) $^{197}_{79}Au$
    (f) $^{235}_{92}U$

13. Identify the period and the family in the periodic table for the following elements:
    (a) Radon
    (b) Sodium
    (c) Copper
    (d) Neon
    (e) Iodine
    (f) Lead

Group B

1. An electron with a mass of 9.11×10^{-31} kg has a velocity of 4.3×10^6 m/s in the innermost orbit of a hydrogen atom. What is the de Broglie wavelength of the electron?

2. Calculate the energy (a) in eV and (b) in joules of the third energy level ($n = 3$) of a hydrogen atom.

3. How much energy is needed to move an electron in a hydrogen atom from the ground state ($n = 1$) to $n = 3$? Give the answer (a) in joules and (b) in eV.

4. What frequency of light is emitted when an electron in a hydrogen atom jumps from $n = 2$ to the ground state ($n = 1$)?

5. How much energy is needed to completely remove an electron from $n = 2$ in a hydrogen atom?

6. If the charge-to-mass ratio of a proton is 9.58×10^7 coulomb/kilogram and the charge is 1.60×10^{-19} coulomb, what is the mass of the proton?

7. An electron wave making a standing wave in a hydrogen atom has a wavelength of 8.33×10^{-11} m. If the mass of the electron is 9.11×10^{-31} kg, what is the velocity of the electron according to the de Broglie equation?

8. Using any reference you wish, write the complete electron configurations for (a) nitrogen, (b) phosphorus, and (c) chlorine.

9. Explain how you know that you have the correct *total* number of electrons in your answers for 8a, 8b, and 8c.

10. Referring to Figure 8.16 *only,* write the complete electron configuration for (a) neon, (b) sulfur, and (c) calcium.

11. Boron has two naturally occurring isotopes, boron-10 and boron-11. Boron-10 has a mass of 10.0129 relative to carbon-12 and makes up 19.78 percent of all naturally occurring boron. Boron-11 has a mass of 11.00931 compared to carbon-12 and makes up the remaining 80.22 percent. What is the atomic weight of boron?

12. Identify the number of protons, neutrons, and electrons in the following isotopes:
    (a) $^{14}_{7}N$
    (b) $^{7}_{3}Li$
    (c) $^{35}_{17}C$
    (d) $^{48}_{20}Ca$
    (e) $^{63}_{29}Cu$
    (f) $^{230}_{92}U$

13. Identify the period and the family in the periodic table for the following elements:
    (a) Xenon
    (b) Potassium
    (c) Chromium
    (d) Argon
    (e) Bromine
    (f) Barium

14. How many outer-orbital electrons are found in an atom of
    (a) Li?
    (b) N?
    (c) F?
    (d) Cl?
    (e) Ra?
    (f) Be?

15. Write electron dot notations for the following elements:
    (a) Boron
    (b) Bromine
    (c) Calcium
    (d) Potassium
    (e) Oxygen
    (f) Sulfur

16. Identify the charge on the following ions:
    (a) Boron
    (b) Bromine
    (c) Calcium
    (d) Potassium
    (e) Oxygen
    (f) Nitrogen

17. Use the periodic table to identify if the following are metals, nonmetals, or semiconductors:
    (a) Krypton
    (b) Cesium
    (c) Silicon
    (d) Sulfur
    (e) Molybdenum
    (f) Plutonium

18. From their charges, predict the periodic table family number for the following ions:
    (a) Br^{-1}
    (b) K^{+1}
    (c) Al^{+3}
    (d) S^{-2}
    (e) Ba^{+2}
    (f) O^{-2}

19. Use chemical symbols and numbers to identify the following isotopes:
    (a) Oxygen-16
    (b) Sodium-23
    (c) Hydrogen-3
    (d) Chlorine-35

14. How many outer-orbital electrons are found in an atom of
    (a) Na?
    (b) P?
    (c) Br?
    (d) I?
    (e) Te?
    (f) Sr?

15. Write electron dot notations for the following elements:
    (a) Aluminum
    (b) Fluorine
    (c) Magnesium
    (d) Sodium
    (e) Carbon
    (f) Chlorine

16. Identify the charge on the following ions:
    (a) Aluminum
    (b) Chlorine
    (c) Magnesium
    (d) Sodium
    (e) Sulfur
    (f) Hydrogen

17. Use the periodic table to identify if the following are metals, nonmetals, or semiconductors:
    (a) Radon
    (b) Francium
    (c) Arsenic
    (d) Phosphorus
    (e) Hafnium
    (f) Uranium

18. From their charges, predict the periodic table family number for the following ions:
    (a) F^{-1}
    (b) Li^{+1}
    (c) B^{+3}
    (d) O^{-2}
    (e) Be^{+2}
    (f) Si^{+4}

19. Use chemical symbols and numbers to identify the following isotopes:
    (a) Potassium-39
    (b) Neon-22
    (c) Tungsten-184
    (d) Iodine-127

9

Chemical Bonds

A chemical change occurs when iron rusts, and rust is a different substance with different physical and chemical properties than iron. This rusted anchor makes a colorful display on the bow of a grain ship.

CORE **CONCEPT**

Electron structure will explain how and why atoms join together in certain numbers.

OUTLINE

Compounds and Chemical Change
Chemical reactions are changes in matter in which different substances are created by forming or breaking chemical bonds.

Chemical Bonds
A chemical bond is an attractive force that holds atoms together in a compound.

Covalent Bonds
A covalent bond is a chemical bond formed by the sharing of electrons.

Valence Electrons and Ions
Atoms have a tendency to seek more stable half-filled or filled outer orbital arrangements of electrons.

Ionic Bonds
An ionic bond is a chemical bond of electrostatic attraction between ions.

Bond Polarity
If an ionic or covalent bond is formed, it is a result of the comparative ability of atoms of an element to attract bonding electrons.

In chapter 8, you learned how the modern atomic theory is used to describe the structures of atoms of different elements. The electron structures of different atoms successfully account for the position of elements in the periodic table as well as for groups of elements with similar properties. On a large scale, all metals were found to have a similarity in electron structure, as were nonmetals. On a smaller scale, chemical families such as the alkali metals were found to have the same outer electron configurations. Thus, the modern atomic theory accounts for observed similarities between elements in terms of atomic structure.

So far, only individual, isolated atoms have been discussed; we have not considered how atoms of elements join together to produce chemical compounds. There is a relationship between the electron structure of atoms and the reactions they undergo to produce specific compounds. Understanding this relationship will explain the changes that matter itself undergoes. For example, hydrogen is a highly flammable, gaseous element that burns with an explosive reaction. Oxygen, on the other hand, is a gaseous element that supports burning. As you know, hydrogen and oxygen combine to form water. Water is a liquid that neither burns nor supports burning. What happens when atoms of elements such as hydrogen and oxygen join to form molecules such as water? Why do such atoms join and why do they stay together? Why does water have different properties from the elements that combine to produce it? And finally, why is water H_2O and not H_3O or H_4O?

Answers to questions about why and how atoms join together in certain numbers are provided by considering the electronic structures of the atoms. Chemical substances are formed from the interactions of electrons as their structures merge, forming new patterns that result in molecules with new properties. It is the new electron pattern of the water molecule that gives water different properties than the oxygen or hydrogen from which it formed (Figure 9.1). Understanding how electron structures of atoms merge to form new patterns is understanding the changes that matter itself undergoes, the topic of this chapter.

COMPOUNDS AND CHEMICAL CHANGE

There are more than one hundred elements listed in the periodic table, and all matter on Earth is made of these elements. However, very few pure elements are found in your surroundings. The air you breathe, the liquids you drink, and all the other things around you are mostly *compounds,* substances made up of combinations of elements. Water, sugar, gasoline, and chalk are examples of compounds and each can be broken down into the elements that make it up. Examples of elements are hydrogen, carbon, and calcium. Why and how these elements join together in different ways to form the different compounds that make up your surroundings is the subject of this chapter.

You have already learned that elements are made up of atoms that can be described by the modern atomic theory. You can also consider an **atom** to be *the smallest unit of an element that can exist alone or in combination with other elements.* Compounds are formed when atoms are held together by an attractive force called a *chemical bond.* The chemical bond binds individual atoms together in a compound. A molecule is generally thought of as a tightly bound group of atoms that maintains its identity. More specifically, a **molecule** is defined *as the smallest particle of a compound, or a gaseous element, that can exist and still retain the characteristic chemical properties of a substance.* Compounds with one type of chemical bond, as you will see, have molecules that are electrically neutral groups of atoms held together strongly enough to be considered independent units.

For example, water is a compound. The smallest unit of water that can exist alone is an electrically neutral unit made up of two hydrogen atoms and one oxygen atom held together by chemical bonds. The concept of a molecule will be expanded as chemical bonds are discussed.

Compounds occur naturally as gases, liquids, and solids. Many common gases occur naturally as molecules made up of two or more atoms. For example, at ordinary temperatures, hydrogen gas occurs as molecules of two hydrogen atoms bound together. Oxygen gas also usually occurs as molecules of two oxygen atoms bound together. Both hydrogen and oxygen occur naturally as *diatomic molecules* (*di-* means "two"). Oxygen sometimes occurs as molecules of three oxygen atoms bound together. These *triatomic* oxygen molecules (*tri-* means "three") are called *ozone.* The noble gases are unique, occurring as single atoms called *monatomic* (*mon-* or *mono-* means "one") (Figure 9.2). These monatomic particles are sometimes called *monatomic molecules* since they are the smallest units of the noble gases that can exist alone. Helium and neon are examples of the monatomic noble gases.

When molecules of any size are formed or broken down into simpler substances, new materials with new properties are produced. This kind of a change in matter is called a chemical change, and the process is called a chemical reaction. A **chemical reaction** is defined as

a change in matter in which different chemical substances are created by breaking and/or forming chemical bonds.

FIGURE 9.1 Water is the most abundant liquid on Earth and is necessary for all life. Because of water's great dissolving properties, any sample is a solution containing solids, other liquids, and gases from the environment. This stream also carries suspended, ground-up rocks, called *rock flour,* from a nearby glacier.

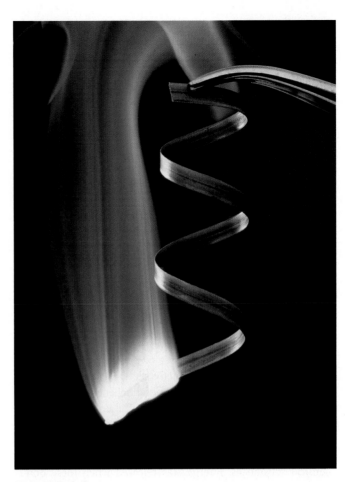

FIGURE 9.3 Magnesium is an alkaline earth metal that burns brightly in air, releasing heat and light. As chemical energy is released, new chemical substances are formed. One new chemical material produced here is magnesium oxide, a soft, powdery material that forms an alkaline solution in water (called *milk of magnesia*).

In general, chemical bonds are formed when atoms of elements are bound together to form compounds. Chemical bonds are broken when a compound is decomposed into simpler substances. Chemical bonds are electrical in nature, formed by electrical attractions, as discussed in chapter 6.

Chemical reactions happen all the time, all around you. A growing plant, burning fuels, and your body's utilization of food all involve chemical reactions. These reactions produce different chemical substances with greater or smaller amounts of internal potential energy (see chapter 4 for a discussion of internal potential energy). Energy is *absorbed* to produce new chemical substances with more internal potential energy. Energy is *released* when new chemical substances are produced with less

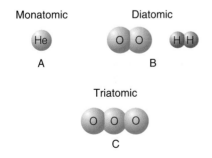

FIGURE 9.2 (*A*) The noble gases are monatomic, occurring as single atoms. (*B*) Many gases, such as hydrogen and oxygen, are diatomic, with two atoms per molecule. (*C*) Ozone is a form of oxygen that is triatomic, occurring with three atoms per molecule.

internal potential energy (Figure 9.3). In general, changes in internal potential energy are called **chemical energy.** For example, new chemical substances are produced in green plants through the process called *photosynthesis.* A green plant uses radiant energy (sunlight), carbon dioxide, and water to produce new chemical materials and oxygen. These new chemical materials, the stuff that leaves, roots, and wood are made of, contain more chemical energy than the carbon dioxide and water they were made from.

A **chemical equation** is a way of describing what happens in a chemical reaction. Later, you will learn how to use formulas in a chemical reaction. For now, the chemical reaction of photosynthesis will be described by using words in an equation:

$$\text{energy (sunlight)} + \text{carbon dioxide molecules} + \text{water molecules} \rightarrow \text{plant material} + \text{oxygen molecules}$$

The substances that are changed are on the left side of the word equation and are called *reactants.* The reactants are carbon dioxide molecules and water molecules. The equation also indicates that energy is absorbed, since the term *energy* appears on the left side. The arrow means *yields*. The new chemical substances are on the right side of the word equation and are called *products*. Reading the photosynthesis reaction as a sentence you would say, "Carbon dioxide and water use energy to react, yielding plant materials and oxygen."

The plant materials produced by the reaction have more internal potential energy, also known as *chemical energy,* than the reactants. You know this from the equation because the term *energy* appears on the left side but not the right. This means that the energy on the left went into internal potential energy on the right. You also know this because the reaction can be reversed to release the stored energy (Figure 9.4). When plant materials (such as wood) are burned, the materials react with oxygen, and chemical energy is released in the form of radiant energy (light) and high kinetic energy of the newly formed gases and vapors. In words,

plant oxygen carbon water
material + molecules → dioxide + molecules + energy
molecules molecules

If you compare the two equations, you will see that burning is the opposite of the process of photosynthesis! The energy released in burning is exactly the same amount of solar energy that was stored as internal potential energy by the plant. Such chemical changes, in which chemical energy is stored in one reaction and released by another reaction, are the result of the making, then the breaking, of chemical bonds. Chemical bonds were formed by utilizing energy to produce new chemical substances. Energy was released when these bonds were broken, then reformed to produce the original substances. In this example, chemical reactions and energy flow can be

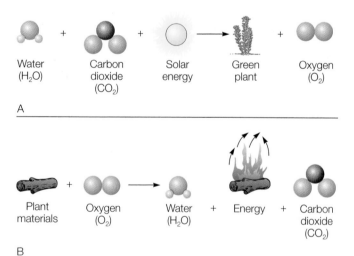

Water (H$_2$O) + Carbon dioxide (CO$_2$) + Solar energy → Green plant + Oxygen (O$_2$)

A

Plant materials + Oxygen (O$_2$) → Water (H$_2$O) + Energy + Carbon dioxide (CO$_2$)

B

FIGURE 9.4 (*A*) New chemical bonds are formed as a green plant makes new materials and stores solar energy through the photosynthesis process. (*B*) The chemical bonds are later broken, and the same amount of energy and the same original materials are released. The same energy and the same materials are released rapidly when the plant materials burn, and they are released slowly when the plant decomposes.

explained by the making and breaking of chemical bonds. Chemical bonds can be explained in terms of changes in the electron structures of atoms. Thus, the place to start in seeking understanding about chemical reactions is the electron structure of the atoms themselves.

VALENCE ELECTRONS AND IONS

As discussed in chapter 8, it is the number of electrons in the outermost orbital that usually determines the chemical properties of an atom. These outer electrons are called **valence electrons,** and it is the valence electrons that participate in chemical bonding. The inner electrons are in stable, fully occupied orbitals and do not participate in chemical bonds. The representative elements (the A-group families) have valence electrons in the outermost orbitals, which contain from one to eight valence electrons. Recall that you can easily find the number of valence electrons by referring to a periodic table. The number at the top of each representative family is the same as the number of outer orbital electrons (with the exception of helium).

The noble gases have filled outer orbitals and do not normally form compounds. Apparently, half-filled and filled orbitals are particularly stable arrangements. Atoms have a tendency to seek such a stable, filled outer orbital arrangement such as the one found in the noble gases. For the representative elements, this tendency is called the **octet rule.** The octet rule states that *atoms attempt to acquire an outer orbital with eight electrons* through chemical reactions. This rule is a generalization, and a few elements do not meet the requirement of eight electrons but do seek the same general trend of stability. There are a few other exceptions, and the octet rule should be considered a generalization that helps keep track of the valence electrons in most representative elements.

The family number of the representative element in the periodic table tells you the number of valence electrons and what the atom must do to reach the stability suggested by the octet rule. For example, consider sodium (Na). Sodium is in family IA, so it has one valence electron. If the sodium atom can get rid of this outer valence electron through a chemical reaction, it will have the same outer electron configuration as an atom of the noble gas neon (Ne) (compare Figures 9.5B and 9.5C).

When a sodium atom (Na) loses an electron to form a sodium ion (Na$^+$), it has the same, stable outer electron configuration as a neon atom (Ne). The sodium ion (Na$^+$) is still a form of sodium since it still has eleven protons. But it is now a sodium *ion,* not a sodium *atom,* since it has eleven protons (eleven positive charges) and now has ten electrons (ten negative charges) for a total of

$$11 + \text{(protons)}$$
$$\underline{10 - \text{(electrons)}}$$
$$1 + \text{(net charge on sodium ion)}$$

This charge is shown on the chemical symbol of Na$^+$ for the *sodium ion.* Note that the sodium nucleus and the inner orbitals

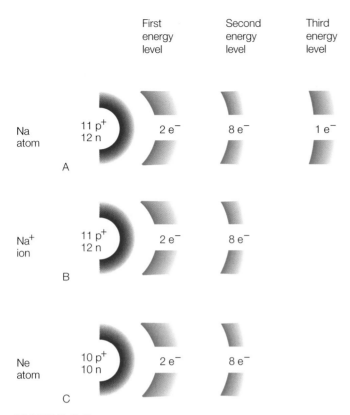

First energy level   Second energy level   Third energy level

Na atom — 11 p⁺ 12 n — 2 e⁻   8 e⁻   1 e⁻

A

Na⁺ ion — 11 p⁺ 12 n — 2 e⁻   8 e⁻

B

Ne atom — 10 p⁺ 10 n — 2 e⁻   8 e⁻

C

FIGURE 9.5 (*A*) A sodium atom has two electrons in the first energy level, eight in the second energy level, and one in the third level. (*B*) When it loses its one outer, or valence, electron, it becomes a sodium ion with the same electron structure as an atom of neon (*C*).

do not change when the sodium atom is ionized. The sodium ion is formed when a sodium atom loses its valence electron, and the process can be described by

$$\text{energy} + \text{Na} \cdot \longrightarrow \text{Na}^+ + e^-$$

where Na · is the electron dot symbol for sodium, and the e^- is the electron that has been pulled off the sodium atom.

CHEMICAL BONDS

Atoms gain or lose electrons through a chemical reaction to achieve a state of lower energy, the stable electron arrangement of the noble gas atoms. Such a reaction results in a **chemical bond,** an *attractive force that holds atoms together in a compound.* There are three general classes of chemical bonds: (1) ionic bonds, (2) covalent bonds, and (3) metallic bonds.

Ionic bonds are formed when atoms *transfer* electrons to achieve the noble gas electron arrangement. Electrons are given up or acquired in the transfer, forming positive and negative

ions. The electrostatic attraction between oppositely charged ions forms ionic bonds, and ionic compounds are the result. In general, ionic compounds are formed when a metal from the left side of the periodic table reacts with a nonmetal from the right side.

EXAMPLE 9.1

What is the symbol and charge for a calcium ion?

SOLUTION

From the list of elements on the inside back cover, the symbol for calcium is Ca, and the atomic number is 20. The periodic table tells you that Ca is in family IIA, which means that calcium has 2 valence electrons. According to the octet rule, the calcium ion must lose 2 electrons to acquire the stable outer arrangement of the noble gases. Since the atomic number is 20, a calcium atom has 20 protons (20+) and 20 electrons (20−). When it is ionized, the calcium ion will lose 2 electrons for a total charge of (20+) + (18−), or 2+. The calcium ion is represented by the chemical symbol for calcium and the charge shown as a superscript: Ca^{2+}.

EXAMPLE 9.2

What is the symbol and charge for an aluminum ion? (Answer: Al^{3+})

Covalent bonds result when atoms achieve the noble gas electron structure by *sharing* electrons. Covalent bonds are generally formed between the nonmetallic elements on the right side of the periodic table.

Metallic bonds are formed in solid metals such as iron, copper, and the other metallic elements that make up about 80 percent of all the elements. The atoms of metals are closely packed and share many electrons in a "sea" that is free to move throughout the metal, from one metal atom to the next. Metallic bonding accounts for metallic properties such as high electrical conductivity.

Ionic, covalent, and metallic bonds are attractive forces that hold atoms or ions together in molecules and crystals. There are two ways to describe what happens to the electrons when one of these bonds is formed: by considering (1) the new patterns formed when atomic orbitals overlap to form a combined orbital called a *molecular orbital* or (2) the atoms in a molecule as *isolated atoms* with changes in their outer shell arrangements. The molecular orbital description considers that the electrons belong to the whole molecule and form a molecular orbital with its own shape, orientation, and energy levels. The isolated atom description considers the electron energy levels as if the atoms in the molecule were alone, isolated from the molecule. The isolated atom description is less accurate than the molecular orbital description, but it is less complex and more easily understood. Thus, the following details about chemical bonding will mostly consider individual atoms and ions in compounds.

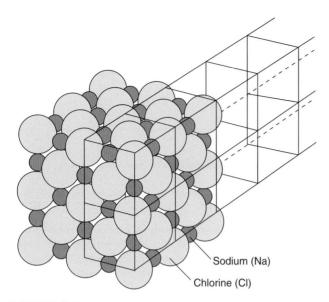

FIGURE 9.6 Sodium chloride crystals are composed of sodium and chlorine ions held together by electrostatic attraction. A crystal builds up, giving the sodium chloride crystal a cubic structure.

IONIC BONDS

An **ionic bond** is defined as the *chemical bond of electrostatic attraction* between negative and positive ions. Ionic bonding occurs when an atom of a metal reacts with an atom of a nonmetal. The reaction results in a transfer of one or more valence electrons from the metal atom to the valence shell of the nonmetal atom. The atom that loses electrons becomes a positive ion, and the atom that gains electrons becomes a negative ion. Oppositely charged ions attract one another, and when pulled together, they form an ionic solid with the ions arranged in an orderly geometric structure (Figure 9.6). This results in a crystalline solid that is typical of salts such as sodium chloride (Figure 9.7).

FIGURE 9.7 You can clearly see the cubic structure of these ordinary table salt crystals because they have been magnified about ten times.

As an example of ionic bonding, consider the reaction of sodium (a soft reactive metal) with chlorine (a pale yellow-green gas). When an atom of sodium and an atom of chlorine collide, they react violently as the valence electron is transferred from the sodium to the chlorine atom. This produces a sodium ion and a chlorine ion. The reaction can be illustrated with electron dot symbols as follows:

$$Na \cdot \ + \ \cdot \overset{..}{\underset{..}{Cl}} : \ \longrightarrow \ Na^+ \ (: \overset{..}{\underset{..}{Cl}} :)^-$$

As you can see, the sodium ion transferred its valence electron, and the resulting ion now has a stable electron configuration. The chlorine atom accepted the electron in its outer orbital to acquire a stable electron configuration. Thus, a stable positive ion and a stable negative ion are formed. Because of opposite electrical charges, the ions attract each other to produce an ionic bond. When many ions are involved, each Na^+ ion is surrounded by six Cl^- ions, and each Cl^- ion is surrounded by six Na^+ ions. This gives the resulting solid NaCl its crystalline cubic structure, as shown in Figure 9.7. In the solid state, all the sodium ions and all the chlorine ions are bound together in one giant unit. Thus, the term *molecule* is not really appropriate for ionic solids such as sodium chloride. But the term is sometimes used anyway, since any given sample will have the same number of Na^+ ions as Cl^- ions.

Energy and Electrons in Ionic Bonding

The sodium ions and chlorine ions in a crystal of sodium chloride can be formed from separated sodium and chlorine atoms. The energy involved in such a sodium-chlorine reaction can be assumed to consist of three separate reactions:

1. $energy \ + \ Na \cdot \ \longrightarrow \ Na^+ \ + \ e^-$

2. $\cdot \overset{..}{\underset{..}{Cl}} : \ + \ e^- \ \longrightarrow \ (: \overset{..}{\underset{..}{Cl}} :)^- + \ energy$

3. $Na^+ \ + \ (: \overset{..}{\underset{..}{Cl}} :)^- \ \longrightarrow \ Na^+ \ (: \overset{..}{\underset{..}{Cl}} :)^- + \ energy$

The overall effect is that energy is released and an ionic bond is formed. The energy released is called the **heat of formation.** It is also the amount of energy required to decompose the compound (sodium chloride) into its elements. The reaction does not take place in steps as described, however, but occurs all at once. Note again, as in the photosynthesis-burning reactions described earlier, that the total amount of chemical energy is conserved. The energy released by the formation of the sodium chloride compound is the *same* amount of energy needed to decompose the compound.

Ionic bonds are formed by electron transfer, and electrons are conserved in the process. This means that electrons are not created or destroyed in a chemical reaction. The same total number of electrons exists after a reaction that existed before the reaction. There are two rules you can use for keeping track of electrons in ionic bonding reactions:

1. Ions are formed as atoms gain or lose valence electrons to achieve the stable noble gas structure.
2. There must be a balance between the number of electrons lost and the number of electrons gained by atoms in the reaction.

The sodium-chlorine reaction follows these two rules. The loss of one valence electron from a sodium atom formed a stable sodium ion. The gain of one valence electron by the chlorine atom formed a stable chlorine ion. Thus, both ions have noble gas configurations (rule 1), and one electron was lost and one was gained, so there is a balance in the number of electrons lost and the number gained (rule 2).

Ionic Compounds and Formulas

The **formula** of a compound *describes what elements are in the compound and in what proportions.* Sodium chloride contains one positive sodium ion for each negative chlorine ion. The formula of the compound sodium chloride is NaCl. If there are no subscripts at the lower right part of each symbol, it is understood that the symbol has a number "1." Thus, NaCl indicates a compound made up of the elements sodium and chlorine, and there is one sodium atom for each chlorine atom.

Calcium (Ca) is an alkaline metal in family IIA, and fluorine (F) is a halogen in family VIIA. Since calcium is a metal and fluorine is a nonmetal, you would expect calcium and fluorine atoms to react, forming a compound with ionic bonds. Calcium must lose two valence electrons to acquire a noble gas configuration. Fluorine needs one valence electron to acquire a noble gas configuration. So calcium needs to lose two electrons and fluorine needs to gain one electron to achieve a stable configuration (rule 1). Two fluorine atoms, each acquiring one electron, are needed to balance the number of electrons lost and the number of electrons gained. The compound formed from the reaction, calcium fluoride, will therefore have a calcium ion with a charge of plus two for every two fluorine ions with a charge of minus one. Recalling that electron dot symbols show only the outer valence electrons, you can see that the reaction is

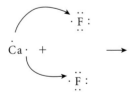

which shows that a calcium atom transfers two electrons, one each to two fluorine atoms. Now showing the results of the reaction, a calcium ion is formed from the loss of two electrons (charge 2+) and two fluorine ions are formed by gaining one electron each (charge 1−):

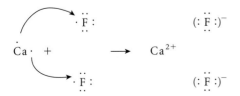

TABLE 9.1
Common ions of some representative elements

Element	Symbol	Ion
Lithium	Li	1+
Sodium	Na	1+
Potassium	K	1+
Magnesium	Mg	2+
Calcium	Ca	2+
Barium	Ba	2+
Aluminum	Al	3+
Oxygen	O	2−
Sulfur	S	2−
Hydrogen	H	1+, 1−
Fluorine	F	1−
Chlorine	Cl	1−
Bromine	Br	1−
Iodine	I	1−

The formula of the compound is therefore CaF_2, with the subscript 2 for fluorine and the understood subscript 1 for calcium. This means that there are two fluorine atoms for each calcium atom in the compound.

Sodium chloride (NaCl) and calcium fluoride (CaF_2) are examples of compounds held together by ionic bonds. Such compounds are called **ionic compounds.** Ionic compounds of the representative elements are generally white, crystalline solids that form colorless solutions. Sodium chloride, the most common example, is common table salt. Many of the transition elements form colored compounds that make colored solutions. Ionic compounds dissolve in water, producing a solution of ions that can conduct an electric current.

In general, the elements in families IA and IIA of the periodic table tend to form positive ions by losing electrons. The ion charge for these elements equals the family number of these elements. The elements in families VIA and VIIA tend to form negative ions by gaining electrons. The ion charge for these elements equals their family number minus 8. The elements in families IIIA and VA have less of a tendency to form ionic compounds, except for those in higher periods. Common ions of representative elements are given in Table 9.1. The transition elements form positive ions of several different charges. Some common ions of the transition elements are listed in Table 9.2.

The single-charge representative elements and the variable-charge transition elements form single, monatomic negative ions. There are also many polyatomic (*poly-* means "many") negative ions, charged groups of atoms that act like a single unit in ionic compounds. Polyatomic ions are listed in Table 9.3.

CHAPTER 9 Chemical Bonds **235**

TABLE 9.2

Common ions of some transition elements

Single-Charge Ions

Element	Symbol	Charge
Zinc	Zn	2+
Tungsten	W	6+
Silver	Ag	1+
Cadmium	Cd	2+

Variable-Charge Ions

Element	Symbol	Charge
Chromium	Cr	2+, 3+, 6+
Manganese	Mn	2+, 4+, 7+
Iron	Fe	2+, 3+
Cobalt	Co	2+, 3+
Nickel	Ni	2+, 3+
Copper	Cu	1+, 2+
Tin	Sn	2+, 4+
Gold	Au	1+, 3+
Mercury	Hg	1+, 2+
Lead	Pb	2+, 4+

TABLE 9.3

Some common polyatomic ions

Ion Name	Formula
Acetate	$(C_2H_3O_2)^-$
Ammonium	$(NH_4)^+$
Borate	$(BO_3)^{3-}$
Carbonate	$(CO_3)^{2-}$
Chlorate	$(ClO_3)^-$
Chromate	$(CrO_4)^{2-}$
Cyanide	$(CN)^-$
Dichromate	$(Cr_2O_7)^{2-}$
Hydrogen carbonate (or bicarbonate)	$(HCO_3)^-$
Hydrogen sulfate (or bisulfate)	$(HSO_4)^-$
Hydroxide	$(OH)^-$
Hypochlorite	$(ClO)^-$
Nitrate	$(NO_3)^-$
Nitrite	$(NO_2)^-$
Perchlorate	$(ClO_4)^-$
Permanganate	$(MnO_4)^-$
Phosphate	$(PO_4)^{3-}$
Phosphite	$(PO_3)^{3-}$
Sulfate	$(SO_4)^{2-}$
Sulfite	$(SO_3)^{2-}$

EXAMPLE 9.3

Use electron dot notation to predict the formula of a compound formed when aluminum (Al) combines with fluorine (F).

SOLUTION

Aluminum, atomic number 13, is in family IIIA, so it has three valence electrons and an electron dot notation of

According to the octet rule, the aluminum atom would need to lose three electrons to acquire the stable noble gas configuration. Fluorine, atomic number 9, is in family VIIA, so it has seven valence electrons and an electron dot notation of

Fluorine would acquire a noble gas configuration by accepting one electron. Three fluorine atoms, each acquiring one electron, are needed to balance the three electrons lost by aluminum. The reaction can be represented as

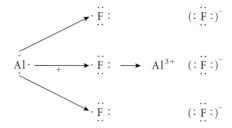

The ratio of aluminum atoms to fluorine atoms in the compound is 1:3. The formula for aluminum fluoride is therefore AlF_3.

EXAMPLE 9.4

Predict the formula of the compound formed between aluminum and oxygen using electron dot notation. (Answer: Al_2O_3).

COVALENT BONDS

Most substances do not have the properties of ionic compounds since they are not composed of ions. Most substances are molecular, composed of electrically neutral groups of atoms that are tightly bound together. As noted earlier, many gases are diatomic, occurring naturally as two atoms bound together as an electrically neutral molecule. Hydrogen, for example, occurs as molecules of H_2 and no ions are involved. The hydrogen atoms are held together by a covalent bond. A **covalent bond** is a *chemical bond formed by the sharing of at least a pair of electrons*. In the diatomic hydrogen molecule, each hydrogen atom contributes a single electron to the shared pair. Both hydrogen atoms count the shared pair of electrons in achieving their noble gas configuration. Hydrogen atoms both share one pair of electrons, but other elements might share more than one pair to achieve a noble gas structure.

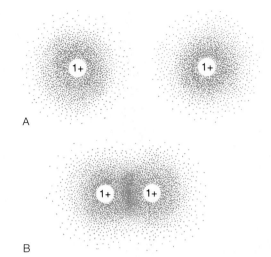

A

B

FIGURE 9.8 (*A*) Two hydrogen atoms, each with its own probability distribution of electrons about the nucleus. (*B*) When the hydrogen atoms bond, a new electron distribution pattern forms around the entire molecule, and both electrons occupy the molecular orbital.

Consider how the covalent bond forms between two hydrogen atoms by imagining two hydrogen atoms moving toward one another. Each atom has a single electron. As the atoms move closer and closer together, their orbitals begin to overlap. Each electron is attracted to the oppositely charged nucleus of the other atom and the overlap tightens. Then the repulsive forces from the like-charged nuclei will halt the merger. A state of stability is reached between the two nuclei and two electrons, and an H_2 molecule has been formed. The two electrons are now shared by both atoms, and the attraction of one nucleus for the other electron and vice versa holds the atoms together (Figure 9.8).

Covalent Compounds and Formulas

Electron dot notation can be used to represent the formation of covalent bonds. For example, the joining of two hydrogen atoms to form an H_2 molecule can be represented as

$$H \cdot \ + \ H \cdot \longrightarrow H : H$$

Since an electron pair is *shared* in a covalent bond, the two electrons move throughout the entire molecular orbital. Since each hydrogen atom now has both electrons on an equal basis, each can be considered to now have the noble gas configuration of helium. A dashed circle around each symbol shows that both atoms share two electrons:

$$H \cdot \ + \ H \cdot \longrightarrow \ (H : H)$$

Hydrogen and chlorine react to form a covalent molecule, and this bond can be represented with electron dots. Chlorine is in the VIIA family, so you know an atom of chlorine has seven valence electrons in the outermost energy level. The reaction is

$$H \cdot \ + \ \cdot \ddot{Cl} : \longrightarrow \ (H \ddot{:} \ddot{Cl} :)$$

Each atom shares a pair of electrons to achieve a noble gas configuration. Hydrogen achieves the helium configuration, and chlorine achieves the neon configuration. All the halogens have seven valence electrons, and all need to gain one electron (ionic bond) or share an electron pair (covalent bond) to achieve a noble gas configuration. This also explains why the halogen gases occur as diatomic molecules. Two chlorine atoms can achieve a noble gas configuration by sharing a pair of electrons:

$$\cdot \ddot{Cl} : \ + \ \cdot \ddot{Cl} : \longrightarrow \ (\ddot{Cl} \ddot{:} \ddot{Cl} :)$$

Each chlorine atom thus achieves the neon configuration by bonding together. Note that there are two types of electron pairs: (1) orbital pairs and (2) bonding pairs. Orbital pairs are not shared, since they are the two electrons in an orbital, each with a separate spin. Orbital pairs are also called *lone pairs*, since they are not shared. *Bonding pairs*, as the name implies, are the electron pairs shared between two atoms. Considering again the Cl_2 molecule,

Bonding pair

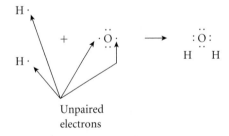

Lone pairs ⟵⟶ $: \ddot{Cl} : \ddot{Cl} :$ ⟵⟶ Lone pairs

Often, the number of bonding pairs that are formed by an atom is the same as the number of single, *unpaired* electrons in the atomic electron dot notation. For example, hydrogen has one unpaired electron, and oxygen has two unpaired electrons. Hydrogen and oxygen combine to form an H_2O molecule, as follows:

H ·

H ·

$+ \quad \cdot \ddot{O} \cdot \longrightarrow \quad : \ddot{O} :$
            H   H

Unpaired electrons

The diatomic hydrogen (H_2) and chlorine (Cl_2), hydrogen chloride (HCl), and water (H_2O) are examples of compounds held together by covalent bonds. A compound held together by covalent bonds is called a **covalent compound.** In general, covalent compounds form from nonmetallic elements on the right side of the periodic table. For elements in families IVA through VIIA, the number of unpaired electrons (and thus the number of covalent bonds formed) is eight minus the family number. You can get a lot of information from the periodic table from generalizations like this one. For another generalization, compare Table 9.4 with the periodic table. The table gives the structures of nonmetals combined with hydrogen and the resulting compounds.

TABLE 9.4

Structures and compounds of nonmetallic elements combined with hydrogen

Nonmetallic Elements	Element (E Represents Any Element of Family)	Compound
Family IVA: C, Si, Ge	·Ė·	H H:E:H H
Family VA: N, P, As, Sb	·Ë·	H:Ë:H H
Family VIA: O, S, Se, Te	·Ë:	H:Ë:H
Family VIIA: F, Cl, Br, I	·Ë:	H:Ë:

Multiple Bonds

Two dots can represent a lone pair of valence electrons, or they can represent a bonding pair, a single pair of electrons being shared by two atoms. Bonding pairs of electrons are often represented by a simple line between two atoms. For example,

H : H is shown as H — H

and

:Ö: is shown as (bent structure)
H   H   O
                       H   H

Note that the line between the two hydrogen atoms represents an electron pair, so each hydrogen atom has two electrons in the outer orbital, as does helium. In the water molecule, each hydrogen atom has two electrons as before. The oxygen atom has two lone pairs (a total of four electrons) and two bonding pairs (a total of four electrons) for a total of eight electrons. Thus, oxygen has acquired a stable octet of electrons.

A covalent bond in which a single pair of electrons is shared by two atoms is called a *single covalent bond* or simply a **single bond.** Some atoms have two unpaired electrons and can share more than one electron pair. A **double bond** is a covalent bond formed when *two pairs* of electrons are shared by two atoms. This happens mostly in compounds involving atoms of the elements C, N, O, and S. Ethylene, for example, is a gas given off from ripening fruit. The electron dot formula for ethylene is

H   H   H   H
:C :: C: or   \ C = C /
H   H   H   H

The ethylene molecule has a double bond between two carbon atoms. Since each line represents two electrons, you can simply count the lines around each symbol to see if the octet rule has been satisfied. Each H has one line, so each H atom is sharing two electrons. Each C has four lines so each C atom has eight electrons, satisfying the octet rule.

FIGURE 9.9 Acetylene is a hydrocarbon consisting of two carbon atoms and two hydrogen atoms held together by a triple covalent bond between the two carbon atoms. When mixed with oxygen gas, the resulting flame is hot enough to cut through most metals.

A **triple bond** is a covalent bond formed when *three pairs* of electrons are shared by two atoms. Triple bonds occur mostly in compounds with atoms of the elements C and N. Acetylene, for example, is a gas often used in welding torches (Figure 9.9). The electron dot formula for acetylene is

H : C ::: C : H   or   H — C ≡ C — H

The acetylene molecule has a triple bond between two carbon atoms. Again, note that each line represents two electrons. Each C atom has four lines, so the octet rule is satisfied.

BOND POLARITY

How do you know if a bond between two atoms will be ionic or covalent? In general, ionic bonds form between metal atoms and nonmetal atoms, especially those from the opposite sides of the periodic table. Also in general, covalent bonds form between the atoms of nonmetals. If an atom has a much greater electron-pulling ability than another atom, the electron is pulled completely away from the atom with lesser pulling ability, and an ionic bond is the result. If the electron-pulling ability is more even between the two atoms, the electron is shared, and a covalent bond results. As you can imagine, all kinds of reactions are possible between atoms with different combinations of electron-pulling abilities.

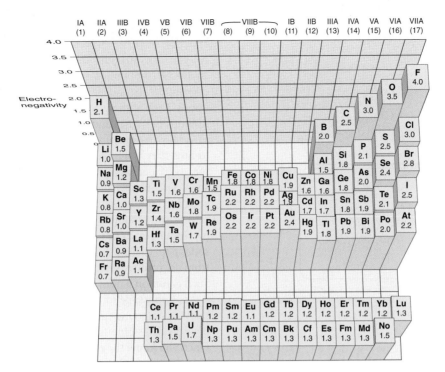

FIGURE 9.10 Elements with the highest electronegativity values have the strongest attraction for the electrons within a chemical bond. Note that the electronegativity of atoms in a group decreases moving down the periodic table, and the electronegativity of atoms in a period increases from left to right.

The result is that it is possible to form many gradations of bonding between completely ionic and completely covalent bonding. Which type of bonding will result can be found by comparing the electronegativity of the elements involved. **Electronegativity** is the *comparative ability of atoms of an element to attract bonding electrons.* The assigned numerical values for electronegativities are given in Figure 9.10. Elements with higher values have the greatest attraction for bonding electrons, and elements with the lowest values have the least attraction for bonding electrons.

The absolute ("absolute" means without plus or minus signs) difference in the electronegativity of two bonded atoms can be used to predict if a bond is ionic or covalent (Table 9.5). A large difference means that one element has a much greater attraction for bonding electrons than the other element. *If the absolute difference in electronegativity is 1.7 or more,* one atom pulls the bonding electron completely away and *an ionic bond results.* For example, sodium (Na) has an electronegativity of 0.9. Chlorine (Cl) has an electronegativity of 3.0. The difference is 2.1, so you can expect sodium and chloride to form ionic bonds. *If the absolute difference in electronegativity is 0.5 or less,* both atoms have about the same ability to attract bonding electrons. The result is that the electron is shared, and *a covalent bond results.* A given hydrogen atom (H) has an electronegativity of another hydrogen atom, so the difference is 0. Zero is less than 0.5 so you can expect a molecule of hydrogen gas to have a covalent bond.

An ionic bond can be expected when the difference in electronegativity is 1.7 or more, and a covalent bond can be

TABLE 9.5

The meaning of absolute differences in electronegativity

Absolute Difference	→	Type of Bond Expected
1.7 or greater	means	ionic bond
between 0.5 and 1.7	means	polar covalent bond
0.5 or less	means	covalent bond

expected when the difference is less than 0.5. What happens when the difference is between 0.5 and 1.7? A covalent bond is formed, but there is an inequality since one atom has a greater bonding electron attraction than the other atom. Thus, the bonding electrons are shared unequally. A **polar covalent bond** is *a covalent bond in which there is an unequal sharing of bonding electrons.* Thus, the bonding electrons spend more time around one atom than the other. The term *polar* means "poles," and that is what forms in a polar molecule. Since the bonding electrons spend more time around one atom than the other, one end of the molecule will have a negative pole, and the other end will have a positive pole. Since there are two poles, the molecule is sometimes called a *dipole.* Note that the molecule as a whole still contains an equal number of electrons and protons, so it is overall electrically neutral. The poles are created by an uneven charge distribution, not an imbalance of electrons and protons. Figure 9.11 shows this uneven charge distribution for a polar covalent compound. The bonding electrons spend more time

Electron distribution and kinds of bonding

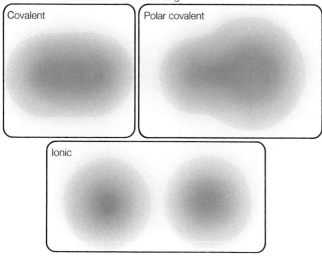

FIGURE 9.11 The absolute difference in electronegativities determines the kind of bond formed.

near the atom on the right, giving this side of the molecule a negative pole.

Figure 9.11 also shows a molecule that has an even charge distribution. The electron distribution around one atom is just like the charge distribution around the other. This molecule is thus a *covalent molecule.* Thus, a polar bond can be viewed as an intermediate type of bond between a covalent bond and an ionic bond. Many gradations are possible between the transition from a purely covalent bond and a purely ionic bond.

EXAMPLE 9.5

Predict if the following bonds are covalent, polar covalent, or ionic: (a) H-O; (b) C-Br; and (c) K-Cl

SOLUTION

From the electronegativity values in Figure 9.10, the absolute differences are

(a) H-O, 1.4
(b) C-Br, 0.3
(c) K-Cl, 2.2

Since an absolute difference of less than 0.5 means covalent, between 0.5 and 1.7 means polar covalent, and greater than 1.7 means ionic, then

(a) H-O, polar covalent
(b) C-Br, covalent
(c) K-Cl, ionic

EXAMPLE 9.6

Predict if the following bonds are covalent, polar covalent, or ionic: (a) Ca-O; (b) H-Cl; and, (c) C-O. Answer: (a) ionic; (b) polar covalent; (c) polar covalent.

COMPOSITION OF COMPOUNDS

As you can imagine, there are literally millions of different chemical compounds from all the possible combinations of over ninety natural elements held together by ionic or covalent bonds. Each of these compounds has its own name, so there are millions of names and formulas for all the compounds. In the early days, compounds were given *common names* according to how they were used, where they came from, or some other means of identifying them. Thus, sodium carbonate was called soda, and closely associated compounds were called baking soda (sodium bicarbonate), washing soda (sodium carbonate), caustic soda (sodium hydroxide), and the bubbly drink made by reacting soda with acid was called soda water, later called soda pop (Figure 9.12). Potassium carbonate was extracted from charcoal by soaking in water and came to be called potash. Such common names are colorful, and some are descriptive, but it was impossible to keep up with the names as the number of known compounds grew. So a systematic set of rules was developed to determine the name and formula of each compound. Once you know the rules, you can write the formula when you hear the name. Conversely, seeing the formula will tell you the systematic name of the compound. This can be an interesting intellectual activity and can also be important when reading the list of ingredients to understand the composition of a product.

There is a different set of systematic rules to be used with ionic compounds and covalent compounds, but there are a few rules in common. For example, a compound made of only two different elements always ends with the suffix *-ide.* So when you hear the name of a compound ending with *-ide,* you automatically know that the compound is made up of only two elements. Sodium chlor*ide* is an ionic compound made up of sodium and chlorine ions. Carbon diox*ide* is a covalent compound with carbon and oxygen atoms. Thus, the systematic name tells you what elements are present in a compound with an *-ide* ending.

FIGURE 9.12 These substances are made up of sodium and some form of a carbonate ion. All have common names with the term *soda* for this reason. Soda water (or "soda pop") was first made by reacting soda (sodium carbonate) with an acid, so it was called "soda water."

IONIC COMPOUND NAMES

Ionic compounds formed by representative metal ions are named by stating the name of the metal (positive ion) first, then the name of the nonmetal (negative ion). Ionic compounds formed by variable-charge ions of the transition elements have an additional rule to identify which variable-charge ion is involved. There was an old way of identifying the charge on the ion by adding either -*ic* or -*ous* to the name of the metal. The suffix -*ic* meant the higher of two possible charges, and the suffix -*ous* meant the lower of two possible charges. For example, iron has two possible charges, 2+ or 3+. The old system used the Latin name for the root. The Latin name for iron *is ferrum,* so a higher charged iron ion (3+) was named a ferric ion. The lower charged iron ion (2+) was called a ferrous ion.

You still hear the old names sometimes, but chemists now have a better way to identify the variable-charge ion. The newer system uses the English name of the metal with Roman numerals in parentheses to indicate the charge number. Thus, an iron ion with a charge of 2+ is called an iron(II) ion, and an iron ion with a charge of 3+ is an iron(III) ion. Table 9.6 gives some of the modern names for variable-charge ions. These names are used with the name of a nonmetal ending in -*ide*, just like the single-charge ions in ionic compounds made up of two different elements.

Some ionic compounds contain three or more elements, and so are more complex than a combination of a metal ion and a nonmetal ion. This is possible because they have *polyatomic ions,* groups of two or more atoms that are bound together tightly and behave very much like a single monatomic ion. For example, the OH^- ion is an oxygen atom bound to a hydrogen atom with a net charge of 1−. This polyatomic ion is called a *hydroxide ion.* The hydroxide compounds make up one of the main groups of ionic compounds, the *metal hydroxides.* A metal hydroxide is an ionic compound consisting of a metal with the hydroxide ion. Another main group consists of the salts with polyatomic ions.

The metal hydroxides are named by identifying the metal first and the term *hydroxide* second. Thus, NaOH is named sodium hydroxide and KOH is potassium hydroxide. The salts are similarly named, with the metal (or ammonium ion) identified first, then the name of the polyatomic ion. Thus, $NaNO_3$ is named sodium nitrate and $NaNO_2$ is sodium nitrite. Note that the suffix -*ate* means the polyatomic ion with one more oxygen atom than the -*ite* ion. For example, the chlor*ate* ion is $(ClO_3)^-$ and the chlor*ite* ion is $(ClO_2)^-$. Sometimes more than two possibilities exist, and more oxygen atoms are identified with the prefix *per-* and less with the prefix *hypo-*. Thus, the *per*chlor*ate* ion is $(ClO_4)^-$ and the *hypo*chlor*ite* ion is $(ClO)^-$.

IONIC COMPOUND FORMULAS

The formulas for ionic compounds are easy to write. There are two rules:

1. The symbols—write the symbol for the positive element first, followed by the symbol for the negative element (same order as in the name).
2. The subscripts—add subscripts to indicate the numbers of ions needed to produce an electrically neutral compound.

As an example, let us write the formula for the compound calcium chloride. The name tells you that this compound consists of positive calcium ions and negative chlorine ions. The suffix -*ide* tells you there are only two elements present. Following rule 1, the symbols would be CaCl.

For rule 2, note the calcium ion is Ca^{2+}, and the chlorine ion is Cl^-. You know the calcium is +2 and chlorine is −1 by applying the atomic theory, knowing their positions in the periodic table, or by using a table of ions and their charges. To be electrically neutral, the compound must have an equal number of pluses and minuses. Thus, you will need two negative chlorine ions for every calcium ion with its 2+ charge. Therefore, the formula is $CaCl_2$. The total charge of two chlorines is thus 2−, which balances the 2+ charge on the calcium ion.

One easy way to write a formula showing that a compound is electrically neutral is to cross over the absolute charge numbers (without plus or minus signs) and use them as subscripts. For example, the symbols for the calcium ion and the chlorine ion are

$$Ca^{2+}Cl^{1-}$$

Crossing the absolute numbers as subscripts, as follows

and then dropping the charge numbers gives

$$Ca_1Cl_2$$

No subscript is written for 1; it is understood. The formula for calcium chloride is thus

$$CaCl_2$$

TABLE 9.6
Modern names of some variable-charge ions

Ion	Name of Ion
Fe^{2+}	Iron(II) ion
Fe^{3+}	Iron(III) ion
Cu^+	Copper(I) ion
Cu^{2+}	Copper(II) ion
Pb^{2+}	Lead(II) ion
Pb^{4+}	Lead(IV) ion
Sn^{2+}	Tin(II) ion
Sn^{4+}	Tin(IV) ion
Cr^{2+}	Chromium(II) ion
Cr^{3+}	Chromium(III) ion
Cr^{6+}	Chromium(VI) ion

The crossover technique works because ionic bonding results from a transfer of electrons, and the net charge is conserved. A calcium ion has a 2+ charge because the atom lost two electrons and two chlorine atoms gain one electron each, for a total of two electrons gained. Two electrons lost equals two electrons gained, and the net charge on calcium chloride is zero, as it has to be. When using the crossover technique, it is sometimes necessary to reduce the ratio to the lowest common multiple. Thus, Mg_2O_2 means an equal ratio of magnesium and oxygen ions, so the correct formula is MgO.

The formulas for variable-charge ions are easy to write, since the Roman numeral tells you the charge number. The formula for tin(II) fluoride is written by crossing over the charge numbers (Sn^{2+}, F^{1-}), and the formula is SnF_2.

EXAMPLE 9.7

Name the following compounds: (a) LiF and (b) PbF_2. Write the formulas for the following compounds: (c) potassium bromide and (d) copper(I) sulfide.

SOLUTION

(a) The formula LiF means that the positive metal ions are lithium, the negative nonmetal ions are fluorine, and there are only two elements in the compound. Lithium ions are Li^{1+} (family IA), and fluorine ions are F^{1-} (family VIIA). The name is lithium fluoride.

(b) Lead is a variable-charge transition element (Table 9.6), and fluorine ions are F^{1-}. The lead ion must be Pb^{2+} because the compound PbF_2 is electrically neutral. Therefore, the name is lead(II) fluoride.

(c) The ions are K^{1+} and Br^{1-}. Crossing over the charge numbers and dropping the signs gives the formula KBr.

(d) The Roman numeral tells you the charge on the copper ion, so the ions are Cu^{1+} and S^{2-}. The formula is Cu_2S.

The formulas for ionic compounds with polyatomic ions are written from combinations of positive metal ions or the ammonium ion with the polyatomic ions, as listed in Table 9.3. Since the polyatomic ion is a group of atoms that has a charge and stays together in a unit, it is sometimes necessary to indicate this with parentheses. For example, magnesium hydroxide is composed of Mg^{2+} ions and $(OH)^{1-}$ ions. Using the crossover technique to write the formula, you get

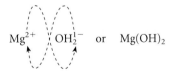

The parentheses are used and the subscript is written *outside* the parenthesis to show that the entire hydroxide unit is taken twice. The formula $Mg(OH)_2$ means

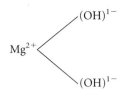

which shows that the pluses equal the minuses. Parentheses are not used, however, when only one polyatomic ion is present. Sodium hydroxide is NaOH, not $Na(OH)_1$.

EXAMPLE 9.8

Name the following compounds: (a) Na_2SO_4 and (b) $Cu(OH)_2$. Write formulas for the following compounds: (c) calcium carbonate and (d) calcium phosphate.

SOLUTION

(a) The ions are Na^+ (sodium ion) and $(SO_4)^{2-}$ (sulfate ion). The name of the compound is sodium sulfate.

(b) Copper is a variable-charge transition element (Table 9.6), and the hydroxide ion $(OH)^{1-}$ has a charge of 1−. Since the compound $Cu(OH)_2$ must be electrically neutral, the copper ion must be Cu^{2+}. The name is copper(II) hydroxide.

(c) The ions are Ca^{2+} and $(CO_3)^{2-}$. Crossing over the charge numbers and dropping the signs gives the formula $Ca_2(CO_3)_2$. Reducing the ratio to the lowest common multiple gives the correct formula of $CaCO_3$.

(d) The ions are Ca^{2+} and $(PO_4)^{3-}$ (from Table 9.3). Using the crossover technique gives the formula $Ca_3(PO_4)_2$. The parentheses indicate that the entire phosphate unit is taken twice.

COVALENT COMPOUND NAMES

Covalent compounds are molecular, and the molecules are composed of two *nonmetals,* as opposed to the metal and nonmetal elements that make up ionic compounds. The combinations of nonmetals alone do not present simple names as the ionic compounds did, so a different set of rules for naming and formula writing is needed.

Ionic compounds were named by stating the name of the positive metal ion, then the name of the negative nonmetal ion with an *-ide* ending. This system is not adequate for naming the covalent compounds. To begin, covalent compounds are composed of two or more nonmetal atoms that form a molecule. It is possible for some atoms to form single, double, or even triple bonds with other atoms, including atoms of the same element, and coordinate covalent bonding is also possible in some compounds. The net result is that the same two elements can form more than one kind of covalent compound. Carbon and oxygen, for example, can combine to form the gas released from burning and respiration, carbon dioxide (CO_2). Under certain conditions, the very same elements combine to produce a different gas, the poisonous carbon monoxide (CO). Similarly, sulfur and oxygen can combine differently to produce two different covalent compounds. A successful system for naming covalent

A microwave oven rapidly cooks foods that contain water, but paper, glass, and plastic products remain cool in the oven. If they are warmed at all, it is from the heat conducted from the food. The explanation of how the microwave oven heats water but not most other substances begins with the nature of the chemical bond.

A chemical bond acts much like a stiff spring, resisting both compression and stretching as it maintains an equilibrium distance between the atoms. As a result, a molecule tends to vibrate when energized or buffeted by other molecules. The rate of vibration depends on the "stiffness" of the spring, which is determined by the bond strength and the mass of the atoms making up the molecule. Each kind of molecule therefore has its own set of characteristic vibrations, a characteristic natural frequency.

Disturbances with a wide range of frequencies can impact a vibrating system. When the frequency of a disturbance matches the natural frequency, energy is transferred very efficiently, and the system undergoes a large increase in amplitude. Such a frequency match is called resonance. When the disturbance is visible light or some other form of radiant energy, a resonant match results in absorption of the radiant energy and an increase in the molecular kinetic energy of vibration. Thus, a resonant match results in a temperature increase.

The natural frequency of a water molecule matches the frequency of infrared radiation, so resonant heating occurs when infrared radiation strikes water molecules. It is the water molecules in your skin that absorb infrared radiation from the sun, a fire, or some hot object, resulting in the warmth that you feel. Because of this match between the frequency of infrared radiation and the natural frequency of a water molecule, infrared is often called "heat radiation."

The frequency ranges of visible light, infrared radiation, and microwave radiation are given in Box Table 9.1. Most microwave ovens operate at the lower end of the microwave frequency range, at 2.45 gigahertz. This frequency is too low for a resonant match with water molecules, so something else must transfer energy from the microwaves to heat the water. This something else is a result of another characteristic of

BOX TABLE 9.1

Approximate ranges of visible light, infrared radiation, and microwave radiation

Radiation	Frequency Range (Hz)
Visible light	4×10^{14} to 8×10^{14}
Infrared radiation	3×10^{11} to 4×10^{14}
Microwave radiation	1×10^{9} to 3×10^{11}

the water molecule, the type of covalent bond holding the molecule together.

The difference in electronegativity between a hydrogen and oxygen atom is 1.4, meaning the water molecule is held together by a polar covalent bond. The electrons are strongly shifted toward the oxygen end of the molecule, creating a negative pole at the oxygen end and a positive pole at the hydrogen end. The water molecule is thus a dipole, as shown in Box Figure 9.1A.

The dipole of water molecules has two effects: (1) the molecule can be rotated by the electric field of a microwave (see Box Figure 9.1B), and (2) groups of individual molecules are held together by an electrostatic attraction between the positive hydrogen ends of a water molecule and the negative oxygen end of another molecule (see Box Figure 9.1C).

One model to explain how microwaves heat water involves a particular group of three molecules, arranged so that the end molecules of the group are aligned with the microwave electric field, with the center molecule not aligned. The microwave torques the center molecule, breaking its hydrogen bond. The energy of the microwave goes into doing the work of breaking the hydrogen bond, and the molecule now has increased potential energy as a consequence. The detached water molecule reestablishes its hydrogen bond, giving up its potential energy, which goes into the vibration of the group of molecules. Thus, the energy of the microwaves is converted into a temperature increase of the water. The temperature increase is high enough to heat and cook most foods.

Microwave cooking is different from conventional cooking because the heating results from energy transfer in polar water molecules, not conduction and convection. The surface of the food never reaches a

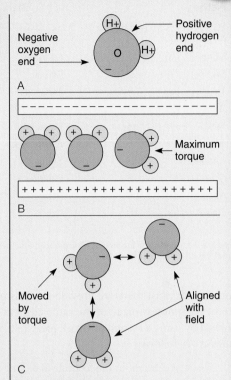

BOX FIGURE 9.1 (*A*) A water molecule is polar, with a negative pole on the oxygen end and positive poles on the hydrogen end. (*B*) An electric field aligns the water dipoles, applying a maximum torque at right angles to the dipole vector. (*C*) Electrostatic attraction between the dipoles holds groups of water molecules together.

temperature over the boiling point of water, so a microwave oven does not brown food (a conventional oven may reach temperatures almost twice as high). Large food items continue to cook for a period of time after being in a microwave oven as the energy is conducted from the water molecules to the food. Most recipes allow for this continued cooking by specifying a waiting period after removing the food from the oven.

Microwave ovens are able to defrost frozen foods because ice always has a thin layer of liquid water (which is what makes it slippery). To avoid "spot cooking" of small pockets of liquid water, many microwave ovens cycle on and off in the defrost cycle. The electrons in metals, like the dipole water molecules, are affected by the electric field of a microwave. A piece of metal near the wall of a microwave oven can result in sparking, which can ignite paper. Metals also reflect microwaves, which can damage the radio tube that produces the microwaves.

TABLE 9.7
Prefixes and element stem names

Prefixes		Stem Names	
Prefix	Meaning	Element	Stem
Mono-	1	Hydrogen	Hydr-
Di-	2	Carbon	Carb-
Tri-	3	Nitrogen	Nitr-
Tetra-	4	Oxygen	Ox-
Penta-	5	Fluorine	Fluor-
Hexa-	6	Phosphorus	Phosph-
Hepta-	7	Sulfur	Sulf-
Octa-	8	Chlorine	Chlor-
Nona-	9	Bromine	Brom-
Deca-	10	Iodine	Iod-

Note: The *a* or *o* ending on the prefix is often dropped if the stem name begins with a vowel, e.g., "tetroxide," not "tetraoxide."

compounds must therefore provide a means of identifying different compounds made of the same elements. This is accomplished by using a system of Greek prefixes (see Table 9.7). The rules are as follows:

1. The first element in the formula is named first with a prefix indicating the number of atoms if the number is greater than 1.
2. The stem name of the second element in the formula is next. A prefix is used with the stem if two elements form more than one compound. The suffix *-ide* is again used to indicate a compound of only two elements.

For example, CO is carbon monoxide and CO_2 is carbon dioxide. The compound BF_3 is boron trifluoride and N_2O_4 is dinitrogen tetroxide. Knowing the formula and the prefix and stem information in Table 9.7, you can write the name of any covalent compound made up of two elements by ending it with *-ide*. Conversely, the name will tell you the formula. However, there are a few polyatomic ions with *-ide* endings that are compounds made up of more than just two elements (hydroxide and cyanide). Compounds formed with ammonium will also have an *-ide* ending, and these are also made up of more than two elements.

Myths, Mistakes, & Misunderstandings

Ban DHMO?

"Dihydrogen monoxide (DHMO) is colorless, odorless, tasteless, and kills uncounted thousands of people every year. Most of these deaths are caused by accidental inhalation of DHMO, but the dangers do not end there. Prolonged exposure to its solid form causes severe tissue damage. Symptoms of DHMO ingestion can include excessive sweating and urination, and possibly a bloated feeling, nausea, vomiting, and body electrolyte imbalance. For those who have become dependent, DHMO withdrawal means certain death."

The above is part of a hoax that was recently circulated on the Internet. The truth is that dihydrogen monoxide is the chemical name of H_2O—water.

COVALENT COMPOUND FORMULAS

The systematic name tells you the formula for a covalent compound. The gas that dentists use as an anesthetic, for example, is dinitrogen monoxide. This tells you there are two nitrogen atoms and one oxygen atom in the molecule, so the formula is N_2O. A different molecule composed of the very same elements is nitrogen dioxide. Nitrogen dioxide is the pollutant responsible for the brownish haze of smog. The formula for nitrogen dioxide is NO_2. Other examples of formulas from systematic names are carbon dioxide (CO_2) and carbon tetrachloride (CCl_4).

Formulas of covalent compounds indicate a pattern of how many atoms of one element combine with atoms of another. Carbon, for example, combines with no more than two oxygen atoms to form carbon dioxide. Carbon combines with no more than four chlorine atoms to form carbon tetrachloride. Electron dot formulas show these two molecules as

$$\ddot{O} :: C :: \ddot{O} \qquad \begin{matrix} & \ddot{C}l : & \\ :\ddot{C}l : & C & :\ddot{C}l : \\ & \ddot{C}l : & \end{matrix}$$

Using a dash to represent bonding pairs, we have

$$O = C = O \qquad \begin{matrix} & Cl & \\ & | & \\ Cl - & C & - Cl \\ & | & \\ & Cl & \end{matrix}$$

In both of these compounds, the carbon atom forms four covalent bonds with another atom. The number of covalent bonds that an atom can form is called its **valence.** Carbon has a valence of four and can form single, double, or triple bonds. Here are the possibilities for a single carbon atom (combining elements not shown):

$$-\overset{|}{\underset{|}{C}}- \qquad -\overset{|}{C}= \qquad =C= \qquad -C\equiv$$

Hydrogen has only one unshared electron, so the hydrogen atom has a valence of one. Oxygen has a valence of two and nitrogen has a valence of three. Here are the possibilities for hydrogen, oxygen, and nitrogen:

$$H - \qquad -\ddot{O}- \qquad :\ddot{O}=$$

$$-\overset{..}{\underset{|}{N}}- \qquad -\overset{..}{N}= \qquad :N\equiv$$

Linus Carl Pauling (1901–1994)

Linus Pauling was a U.S. theoretical chemist and biologist whose achievements ranked among the most important of any in twentieth-century science. His main contribution was to the understanding of molecular structure and chemical bonding. He was one of the very few people to have been awarded two Nobel prizes: he received the 1954 Nobel Prize for chemistry (for his work on intermolecular forces) and the 1962 Peace Prize. Throughout his career, his work was noted for the application of intuition and inspiration, assisted by his phenomenal memory; he often carried over principles from one field of science and applied them to another.

Pauling was born in Portland, Oregon, on February 28, 1901, the son of a pharmacist. He began his scientific studies at Oregon State Agricultural College, from which he graduated in chemical engineering in 1922. He then began his research at the California Institute of Technology, Pasadena, gaining his Ph.D. in 1925. He became a full professor at Pasadena in 1931 and left there in 1936 to take up the post of director of the Gates and Crellin Laboratories, which he held for the next twenty-two years. He also held appointments at the University of California, San Diego, and Stanford University. His last appointment was as director of the Linus Pauling Institute of Science and Medicine at Menlo Park, California.

In 1931, Pauling published a classic paper, "The Nature of the Chemical Bond," in which he used quantum mechanics to explain that an electron-pair bond is formed by the interaction of two unpaired electrons, one from each of two atoms, and that once paired, these electrons cannot take part in the formation of other bonds. It was followed by the book *Introduction to Quantum Mechanics* (1935), of which he was coauthor. He was a pioneer in the application of quantum mechanical principles to the structures of molecules.

It was Pauling who introduced the concept of hybrid orbitals in molecules to explain the symmetry exhibited by carbon atoms in most of its compounds. Pauling also investigated electronegativity of atoms and polarization in chemical bonds. He assigned electronegativities on a scale up to 4.0. A pair of electrons in a bond is pulled preferentially toward an atom with a higher electronegativity. In hydrogen chloride (HCl), for example, hydrogen has an electronegativity of 2.1 and chlorine of 3.5. The bonding electrons are pulled toward the chlorine atom, giving it a small excess negative charge (and leaving the hydrogen atom with a small excess positive charge), polarizing the hydrogen-chlorine bond.

Pauling's ideas on chemical bonding are fundamental to modern theories of molecular structure. Much of this work was consolidated in his book *The Nature of the Chemical Bond, The Structure of Molecules and Crystals* (1939). In the 1940s, Pauling turned his attention to the chemistry of living tissues and systems. He applied his knowledge of molecular structure to the complexity of life, principally to proteins in blood. With Robert Corey, he worked on the structures of amino acids and polypeptides. They proposed that many proteins have structures held together with hydrogen bonds, giving them helical shapes. This concept assisted Francis Crick and James Watson in their search for the structure of DNA, which they eventually resolved as a double helix.

In his researches on blood, Pauling investigated immunology and sickle-cell anemia. Later work confirmed his hunch that the disease is genetic and that normal hemoglobin and the hemoglobin in abnormal "sickle" cells differ in electrical charge. Throughout the 1940s, he studied living materials; he also carried out research on anesthesia. At the end of this period, he published two textbooks, *General Chemistry* (1948) and *College Chemistry* (1950), which became best-sellers.

SUMMARY

Elements are basic substances that cannot be broken down into anything simpler, and an *atom* is the smallest unit of an element. *Compounds* are combinations of two or more elements and can be broken down into simpler substances. Compounds are formed when atoms are held together by an attractive force called a *chemical bond*. A *molecule* is the smallest unit of a compound, or a gaseous element, that can exist and still retain the characteristic properties of a substance.

A *chemical change* produces new substances with new properties, and the new materials are created by making or breaking chemical bonds. The process of chemical change in which different chemical substances are created by forming or breaking chemical bonds is called a *chemical reaction*. During a chemical reaction, different chemical substances with greater or lesser amounts of internal potential energy are produced. *Chemical energy* is the change of internal potential energy during a chemical reaction, and other reactions absorb energy. A *chemical equation* is a shorthand way of describing a chemical reaction. An equation shows the substances that are changed, the *reactants*, on the left side, and the new substances produced, the *products*, on the right side.

Chemical reactions involve *valence electrons*, the electrons in the outermost orbital of an atom. Atoms tend to lose or acquire electrons to achieve the configuration of the noble gases with stable, filled outer orbitals. This tendency is generalized as the *octet rule*, that atoms lose or gain electrons to acquire the noble gas structure of eight electrons in the outer orbital. Atoms form negative or positive *ions* in the process.

A chemical bond is an attractive force that holds atoms together in a compound. Chemical bonds that are formed when atoms transfer electrons to become ions are *ionic bonds*. An ionic bond is an electrostatic attraction between oppositely charged ions. Chemical bonds formed when atoms share electrons are *covalent bonds*.

Ionic bonds result in *ionic compounds* with a crystalline structure. The energy released when an ionic compound is formed is called the *heat of formation*. It is the same amount of energy that is required to decompose the compound into its elements. A *formula* of a compound uses symbols to tell what elements are in a compound and in what proportions. Ions of representative elements have a single, fixed charge, but many transition elements have variable charges. Electrons are conserved when ionic compounds are formed, and the ionic compound is electrically neutral. The formula shows this overall balance of charges.

Covalent compounds are molecular, composed of electrically neutral groups of atoms bound together by *covalent bonds*. A single covalent bond is formed by the sharing of a pair of electrons, with each atom contributing a single electron to the shared pair. Covalent bonds formed when two pairs of electrons are shared are called *double bonds*. A *triple bond* is the sharing of three pairs of electrons.

The electron-pulling ability of an atom in a bond is compared with arbitrary values of *electronegativity*. A high electronegative value means a greater attraction for bonding electrons. If the absolute difference in electronegativity of two bonded atoms is 1.7 or more, one atom pulls the bonding electron away, and an ionic bond results. If the difference is less than 0.5, the electrons are equally shared in a covalent bond. Between 0.5 and 1.7, the electrons are shared unequally in a *polar covalent bond*. A polar covalent bond results in electrons spending more time around the atom or atoms with the greater pulling ability, creating a negative pole at one end and a positive pole at the other. Such a molecule is called a *dipole*, since it has two poles, or centers, of charge.

Compounds are named with systematic rules for ionic and covalent compounds. Both ionic and covalent compounds that are made up of only two different elements always end with an *-ide* suffix, but there are a few *-ide* names for compounds that have more than just two elements.

The modern systematic system for naming variable-charge ions states the English name and gives the charge with Roman numerals in parentheses. Ionic compounds are electrically neutral, and formulas must show a balance of charge. The *crossover technique* is an easy way to write formulas that show a balance of charge.

Covalent compounds are molecules of two or more nonmetal atoms held together by a covalent bond. The system for naming covalent compounds uses Greek prefixes to identify the numbers of atoms, since more than one compound can form from the same two elements (CO and CO_2, for example).

KEY TERMS

atom (p. **230**)
chemical bond (p. **233**)
chemical energy (p. **231**)
chemical equation (p. **231**)
chemical reaction (p. **230**)
covalent bond (p. **236**)
covalent compound (p. **237**)
double bond (p. **238**)
electronegativity (p. **239**)
formula (p. **235**)
heat of formation (p. **234**)
ionic bond (p. **234**)
ionic compounds (p. **235**)
molecule (p. **230**)
octet rule (p. **232**)
polar covalent bond (p. **239**)
single bond (p. **238**)
triple bond (p. **238**)
valence (p. **244**)
valence electrons (p. **232**)

APPLYING THE CONCEPTS

1. Which of the following is **not** true of a compound? It is (a)
   a. pure substance.
   b. composed of combinations of atoms.
   c. held together by chemical bonds.
   d. substance that cannot be broken down into simpler units.

2. The smallest unit of an element that can exist alone or in combination with other elements is a (an)
   a. nucleus.
   b. atom.
   c. molecule.
   d. proton.

3. The smallest particle of a compound or a gaseous element that can exist and still retain the characteristic properties of that substance is a (an)
   a. molecule.
   b. element.
   c. atom.
   d. electron.

4. Which of the following is an example of a monatomic molecule?
   a. O_2
   b. N_2
   c. Ar
   d. O_3

5. What determines the chemical properties of an atom?
   a. valence electrons
   b. atomic weight
   c. neutrons and protons
   d. protons only

6. "Atoms attempt to acquire an outer orbital with eight electrons" is a statement of the
   a. Heisenberg certainty theory.
   b. atomic theory.
   c. octet rule.
   d. chemical energy balancing rule.

7. The family or group number of the representative elements represents the
   a. total number of electrons.
   b. total number of protons.
   c. total number of valence electrons.
   d. the atomic number.

8. What is the chemical symbol for a magnesium ion?
   a. Mg^+
   b. Mg^-
   c. Mg^{2+}
   d. Mg^{2-}

9. Atoms that achieve an octet by sharing electrons form
   a. covalent bonds.
   b. ionic bonds.
   c. metallic bonds.
   d. hydrogen bonds.

10. The electrostatic attraction between positive and negative ions is called a (an)
    a. ionic bond.
    b. metallic bond.
    c. covalent bond.
    d. electrostatic bond.

11. The energy released during the formation of an ionic bond is
    a. ionic energy.
    b. heat of formation.
    c. activation energy.
    d. heat of creation.

12. The comparative ability of atoms of an element to attract bonding electrons is called
    a. electron-attraction.
    b. electronegativity.
    c. bonding.
    d. electron transfer.

13. What kind of bond has unequal sharing of bonding electrons?
    a. polar covalent
    b. covalent
    c. ionic
    d. polar ionic

14. The correct name for the compound CS_2 is
    a. dicarbon disulfide.
    b. carbon disulfite.
    c. carbon disulfide.
    d. monocarbon trisulfide.

15. What is the formula for the compound zinc chloride?
    a. ZnCl
    b. Zn_2Cl
    c. $ZnCl_2$
    d. $ZnCl_4$

16. Pure substances that cannot be broken down into anything simpler are
    a. elements.
    b. compounds.
    c. molecules.
    d. mixtures.

17. Chemical energy is defined as
    a. change of internal potential energy during a chemical reaction.
    b. energy that is only absorbed during a chemical reaction.
    c. energy that is only released during a chemical reaction.
    d. energy added to a chemical reaction.

18. A metallic bond is defined by all of the following, except
    a. it is formed in solid metals.
    b. it restricts movement of electrons.
    c. metal atoms share "sea of electrons."
    d. it accounts for metallic properties such as conductivity and luster.

19. An ion can be described as an element or compound that
    a. gains electrons.
    b. loses electrons.
    c. gains or loses electrons.
    d. shares electrons.

20. You know that a chemical reaction is taking place if
    a. the temperature of a substance increases.
    b. electrons move in a steady current.
    c. chemical bonds are formed or broken.
    d. All of the above are correct.

21. What is the relationship between the energy released in burning materials produced by photosynthesis and the solar energy that was absorbed in making the materials? It is
    a. less than the solar energy absorbed.
    b. the same as the solar energy absorbed.
    c. more than the solar energy absorbed.
    d. variable, having no relationship to the energy absorbed.

22. The electrons that participate in chemical bonding are the
    a. valence electrons.
    b. electrons in fully occupied orbitals.
    c. stable inner electrons.
    d. All of the above are correct.

23. Atoms of the representative elements have a tendency to seek stability through
    a. acquiring the hydrogen gas structure.
    b. filling or emptying their outer orbitals.
    c. any situations that will fill all orbitals.
    d. All of the above are correct.

24. An ion is formed when an atom of a representative element
    a. gains or loses protons.
    b. shares electrons to achieve stability.
    c. loses or gains electrons to satisfy the octet rule.
    d. All of the above are correct.

25. An atom of an element in family VIA will have what charge when it is ionized?
    a. $+1$
    b. -1
    c. $+2$
    d. -2

26. Which type of chemical bond is formed by a transfer of electrons?
    a. ionic
    b. covalent
    c. metallic
    d. All of the above are correct.

27. Which type of chemical bond is formed between two atoms by the sharing of two electrons, with one electron from each atom?
    a. ionic
    b. covalent
    c. metallic
    d. All of the above are correct.

28. What type of compounds are salts, such as sodium chloride?
    a. ionic compounds
    b. covalent compounds
    c. polar compounds
    d. All of the above are correct.

29. What is the chemical formula if there are two bromide ions for each barium ion in a compound?
    a. $_2Br_1Ba$
    b. Ba_2Br
    c. $BaBr_2$
    d. None of the above is correct.

30. Which combination of elements forms crystalline solids that will dissolve in water, producing a solution of ions that can conduct an electric current?
    a. metal and metal
    b. metal and nonmetal
    c. nonmetal and nonmetal
    d. All of the above are correct.

31. In a single covalent bond between two atoms, a
    a. single electron from one of the atoms is shared.
    b. pair of electrons from one of the atoms is shared.
    c. pair of electrons, one from each atom, is shared.
    d. single electron is transferred from one atom.

32. Sulfur and oxygen are both in the VIA family of the periodic table. If element X combines with oxygen to form the compound X_2O, element X will combine with sulfur to form what compound?
    a. XS_2
    b. X_2S
    c. X_2S_2
    d. It is impossible to say without more information.

33. One element is in the IA family of the periodic table, and a second is in the VIIA family. What type of compound will the two elements form?
    a. ionic
    b. covalent
    c. They will not form a compound.
    d. More information is needed to answer this question.

34. One element is in the VA family of the periodic table, and a second is in the VIA family. What type of compound will these two elements form?
    a. ionic
    b. covalent
    c. They will not form a compound.
    d. More information is needed to answer this question.

35. A covalent bond in which there is an unequal sharing of bonding electrons is a
    a. single covalent bond.
    b. double covalent bond.
    c. triple covalent bond.
    d. polar covalent bond.

36. An inorganic compound made of only two different elements has a systematic name that always ends with the suffix
    a. -ite.
    b. -ate.
    c. -ide.
    d. -ous.

37. Dihydrogen monoxide is the systematic name for a compound that has the common name of
    a. laughing gas.
    b. water.
    c. smog.
    d. rocket fuel.

38. Which of the following is not an example of a chemical reaction?
    a. blending of a vanilla milkshake
    b. growing tomatoes
    c. burning logs in the fireplace
    d. digesting your dinner

39. Which of the following substances is not a compound?
    a. water
    b. table salt
    c. neon
    d. rust

40. Propane gas burns in oxygen to produce carbon dioxide, water, and energy. The reactants in this reaction are propane and
    a. energy.
    b. oxygen.
    c. carbon dioxide.
    d. water.

41. Propane gas burns in oxygen to produce carbon dioxide, water, and energy. For this reaction, energy is
    a. absorbed.
    b. released.
    c. equal on both sides of the reaction.
    d. completely consumed in the reaction.

42. Which is not true of an ionic bond? It involves
    a. a transfer of electrons.
    b. sharing of electrons.
    c. electrostatic attraction between ions.
    d. metals and nonmetals.

43. Which of the following does not have an ionic bond?
    a. $BaCl_2$
    b. Na_2O
    c. CS_2
    d. BaS

44. What is the correct formula for the magnesium chloride?
    a. MgCl
    b. Mg_2Cl
    c. $MgCl_2$
    d. Mg_2Cl_2

45. What is the formula for the sulfate ion?
    a. $(HSO_4)^-$
    b. $(SO_4)^{2-}$
    c. S^{2-}
    d. $(SO_3)^{2-}$

46. The correct name for the compound $NaNO_3$ is
    a. sodium nitrogen trioxygen.
    b. sodium nitrate.
    c. sodium nitrite.
    d. sodium nitrogen trioxide.

47. How many valence electrons does the element nitrogen (N) have?
    a. 3
    b. 7
    c. 5
    d. 14

48. A nonmetal atom combines with a second nonmetal atom. The first nonmetal atom
    a. gains electrons.
    b. loses electrons.
    c. remains neutral.
    d. shares electrons with the second nonmetal atom.

49. Polar covalent bonds result when the electronegativity difference between the two atoms in the bond is
    a. 1.7 and greater.
    b. between 0.5 and 1.7.
    c. less than 0.5.
    d. equal to 0.

50. What is the formula for the compound aluminum bromide?
    a. $AlBr_2$
    b. $AlBr_3$
    c. Al_2Br_3
    d. Al_3Br

51. The combination of nonmetals with nonmetals forms
    a. covalent compounds.
    b. metallic compounds.
    c. ionic compounds.
    d. no compounds.

Answers

1. d 2. b 3. a 4. c 5. a 6. c 7. c 8. c 9. a 10. a 11. b 12. b 13. a 14. c
15. c 16. a 17. a 18. b 19. c 20. c 21. b 22. a 23. b 24. c 25. d 26. a
27. b 28. a 29. c 30. b 31. c 32. b 33. a 34. b 35. d 36. c 37. b 38. a
39. c 40. b 41. b 42. b 43. c 44. c 45. b 46. b 47. c 48. d 49. b 50. b
51. a

QUESTIONS FOR THOUGHT

1. Describe how the following are alike and how they are different: (a) a sodium atom and a sodium ion, and (b) a sodium ion and a neon atom.

2. What is the difference between a polar covalent bond and a nonpolar covalent bond?

3. What is the difference between an ionic and covalent bond? What do atoms forming the two bond types have in common?

4. What is the octet rule?

5. Is there a relationship between the number of valence electrons and how many covalent bonds an atom can form? Explain.

6. Write electron dot formulas for molecules formed when hydrogen combines with (a) chlorine, (b) oxygen, and (c) carbon.

7. Sodium fluoride is often added to water supplies to strengthen teeth. Is sodium fluoride ionic, covalent, or polar covalent? Explain the basis of your answer.

8. What is the modern systematic name of a compound with the formula (a) SnF_2? (b) PbS?

9. What kinds of elements are found in (a) ionic compounds with a name ending with an -ide suffix? (b) Covalent compounds with a name ending with an -ide suffix?

10. Why is it necessary to use a system of Greek prefixes to name binary covalent compounds?

11. What are variable-charge ions? Explain how variable-charge ions are identified in the modern systematic system of naming compounds.

12. What is a polyatomic ion? Give the names and formulas for several common polyatomic ions.

13. Write the formula for magnesium hydroxide. Explain what the parentheses mean.

14. What is a double bond? A triple bond?

FOR FURTHER ANALYSIS

1. What are the significant similarities and differences between a physical change and a chemical change?

2. Analyze how you would know for sure that a pure substance you have is an ionic and not a covalent compound.

3. Make up an explanation for why ionic compounds are formed when a metal from the left side of the periodic table reacts with a nonmetal from the right side, but covalent bonds are formed between nonmetallic elements on the right side of the table.

4. Describe how you would teach a younger person how to name ionic and covalent compounds.

INVITATION TO INQUIRY

Seeing the Light

Dissolving an ionic compound in water results in ions being pulled from the compound to form free ions. A solution that contains ions will conduct an electric current, so electrical conductivity is one way to test a dissolved compound to see whether or not it is an ionic compound.

Make a conductivity tester from a 9 V battery and a miniature Christmas tree bulb with two terminal wires. Sand these wires to make a good electrical contact. Try testing (1) dry baking soda, (2) dry table salt, and (3) dry sugar. Test solutions of these substances dissolved in distilled water. Test pure distilled water and rubbing alcohol. Explain which contains ions and why.

The exercises in groups A and B cover the same concepts. Solutions to group A exercises are located in appendix E.

Group A

1. Use electron dot symbols in equations to predict the formula of the ionic compound formed from the following:
   (a) K and I
   (b) Sr and S
   (c) Na and O
   (d) Al and O

2. Name the following ionic compounds formed from variable-charge transition elements:
   (a) CuS
   (b) Fe_2O_3
   (c) CrO
   (d) PbS

3. Name the following polyatomic ions:
   (a) $(OH)^-$
   (b) $(SO_3)^{2-}$
   (c) $(ClO)^-$
   (d) $(NO_3)^-$
   (e) $(CO_3)^{2-}$
   (f) $(ClO_4)^-$

4. Use the crossover technique to write formulas for the following compounds:
   (a) Iron(III) hydroxide
   (b) Lead(II) phosphate
   (c) Zinc carbonate
   (d) Ammonium nitrate
   (e) Potassium hydrogen carbonate
   (f) Potassium sulfite

5. Write formulas for the following covalent compounds:
   (a) Carbon tetrachloride
   (b) Dihydrogen monoxide
   (c) Manganese dioxide
   (d) Sulfur trioxide
   (e) Dinitrogen pentoxide
   (f) Diarsenic pentasulfide

6. Name the following covalent compounds:
   (a) CO
   (b) CO_2
   (c) CS_2
   (d) N_2O
   (e) P_4S_3
   (f) N_2O_3

7. Predict if the bonds formed between the following pairs of elements will be ionic, polar covalent, or covalent:
   (a) Si and O
   (b) O and O
   (c) H and Te
   (d) C and H
   (e) Li and F
   (f) Ba and S

Group B

1. Use electron dot symbols in equations to predict the formulas of the ionic compounds formed between the following:
   (a) Li and F
   (b) Be and S
   (c) Li and O
   (d) Al and S

2. Name the following ionic compounds formed from variable-charge transition elements:
   (a) $PbCl_2$
   (b) FeO
   (c) Cr_2O_3
   (d) PbO

3. Name the following polyatomic ions:
   (a) $(C_2H_3O_2)^-$
   (b) $(HCO_3)^-$
   (c) $(SO_4)^{2-}$
   (d) $(NO_2)^-$
   (e) $(MnO_4)^-$
   (f) $(CO_3)^{2-}$

4. Use the crossover technique to write formulas for the following compounds:
   (a) Aluminum hydroxide
   (b) Sodium phosphate
   (c) Copper(II) chloride
   (d) Ammonium sulfate
   (e) Sodium hydrogen carbonate
   (f) Cobalt(II) chloride

5. Write formulas for the following covalent compounds:
   (a) Silicon dioxide
   (b) Dihydrogen sulfide
   (c) Boron trifluoride
   (d) Dihydrogen dioxide
   (e) Carbon tetrafluoride
   (f) Nitrogen trihydride

6. Name the following covalent compounds:
   (a) N_2O
   (b) SO_2
   (c) SiC
   (d) PF_5
   (e) $SeCl_6$
   (f) N_2O_4

7. Predict if the bonds formed between the following pairs of elements will be ionic, polar covalent, or covalent:
   (a) Si and C
   (b) Cl and Cl
   (c) S and O
   (d) Sr and F
   (e) O and H
   (f) K and F

10

Chemical Reactions

A clear solution of potassium iodide reacts with a clear solution of lead nitrate, producing a yellow precipitate of lead iodide. This is an example of one of the four types of chemical reactions.

CORE **CONCEPT**

Chemical symbols, formulas, and equations can be used to concisely represent elements, compounds, and what happens in a chemical reaction.

OUTLINE

Chemical Formulas
A chemical formula is a shorthand way of describing the elements or ions that make up a compound.

Types of Chemical Reactions
An oxidation-reduction reaction considers how electrons are transferred from one atom to another; oxidation is the part where there is a loss of electrons, and reduction is the part where there is a gain of electrons.

Chemical Formulas
   Molecular and Formula Weights
   Percent Composition of Compounds
Chemical Equations
   Balancing Equations
   Generalizing Equations
Types of Chemical Reactions
   Combination Reactions
   Decomposition Reactions
   Replacement Reactions
   Ion Exchange Reactions
Information from Chemical Equations
   Units of Measurement Used
     with Equations
   Quantitative Uses of Equations
Science and Society: The Catalytic Converter
People Behind the Science: Emma Perry Carr

Chemical Equations
A chemical equation is a shorthand way of describing what has happened before and after a chemical reaction.

Types of Chemical Reactions
Chemical reactions can be classified according to what happens to reactants and products, which leads to four basic categories: combination, decomposition, replacement, and ion exchange reactions.

251

OVERVIEW

We live in a chemical world that has been partly manufactured through controlled chemical change. Consider all of the synthetic fibers and plastics that are used in clothing, housing, and cars. Consider all the synthetic flavors and additives in foods, how these foods are packaged, and how they are preserved. Consider also the synthetic drugs and vitamins that keep you healthy. There are millions of such familiar products that are the direct result of chemical research. Most of these products simply did not exist sixty years ago.

Many of the products of chemical research have remarkably improved the human condition. For example, synthetic fertilizers have made it possible to supply food in quantities that would not otherwise be possible. Chemists learned how to take nitrogen from the air and convert it into fertilizers on an enormous scale. Other chemical research resulted in products such as weed killers, insecticides, and mold and fungus inhibitors. Fertilizers and these products have made it possible to supply food for millions of people who would otherwise have starved (Figure 10.1).

Yet we also live in a world with concerns about chemical pollutants, the greenhouse effect, acid rain, and a disappearing ozone shield. The very nitrogen fertilizers that have increased food supplies also wash into rivers, polluting the waterways and bays. Such dilemmas require an understanding of chemical products and the benefits and hazards of possible alternatives. Understanding requires a knowledge of chemistry, since the benefits, and risks, are chemical in nature.

Chapters 8 and 9 were about the modern atomic theory and how it explains elements and how compounds are formed in chemical change. This chapter is concerned with describing chemical changes and the different kinds of chemical reactions that occur. These reactions are explained with balanced chemical equations, which are concise descriptions of reactions that produce the products used in our chemical world.

CHEMICAL FORMULAS

In chapter 9, you learned how to name and write formulas for ionic and covalent compounds, including the ionic compound of table salt and the covalent compound of ordinary water. Recall that a formula is a shorthand way of describing the elements or ions that make up a compound. There are basically three kinds of formulas that describe compounds: (1) *empirical* formulas, (2) *molecular* formulas, and (3) *structural* formulas. Empirical and molecular formulas, and their use, will be considered in this chapter. Structural formulas will be considered in chapter 12.

An **empirical formula** identifies the elements present in a compound and describes the *simplest whole number ratio* of atoms of these elements with subscripts. For example, the empirical formula for ordinary table salt is NaCl. This tells you that the elements sodium and chlorine make up this compound, and there is one atom of sodium for each chlorine atom. The empirical formula for water is H_2O, meaning there are two atoms of hydrogen for each atom of oxygen.

Covalent compounds exist as molecules. A chemical formula that identifies the *actual numbers* of atoms in a molecule is known as a **molecular formula.** Figure 10.2 shows the structure of some common molecules and their molecular formulas. Note that each formula identifies the elements and numbers of atoms in each molecule. The figure also indicates how molecular

formulas can be written to show how the atoms are arranged in the molecule. Formulas that show the relative arrangements are called *structural formulas.* Compare the structural formulas in the illustration with the three-dimensional representations and the molecular formulas.

How do you know if a formula is empirical or molecular? First, you need to know if the compound is ionic or covalent. You know that ionic compounds are usually composed of metal and nonmetal atoms with an electronegativity difference greater than 1.7. Formulas for ionic compounds are *always* empirical formulas. Ionic compounds are composed of many positive and negative ions arranged in an electrically neutral array. There is no discrete unit, or molecule, in an ionic compound, so it is only possible to identify ratios of atoms with an empirical formula.

Covalent compounds are generally nonmetal atoms bonded to nonmetal atoms in a molecule. You could therefore assume that a formula for a covalent compound is a molecular formula unless it is specified otherwise. You can be certain it is a molecular formula if it is not the simplest whole-number ratio. Glucose, for example, is a simple sugar (also known as dextrose) with the formula $C_6H_{12}O_6$. This formula is divisible by six, yielding a formula with the simplest whole-number ratio of CH_2O. Therefore, CH_2O is the empirical formula for glucose, and $C_6H_{12}O_6$ is the molecular formula.

FIGURE 10.1 The products of chemical research have substantially increased food supplies but have also increased the possibilities of pollution. Balancing the benefits and hazards of the use of chemicals requires a knowledge of chemistry and a knowledge of the alternatives.

Name	Molecular formula	Space-filled	Structural formula
Water	H_2O		
Ammonia	NH_3		
Hydrogen peroxide	H_2O_2		
Carbon dioxide	CO_2		$O = C = O$

FIGURE 10.2 The name, molecular formula, sketch, and structural formula of some common molecules. Compare the kinds and numbers of atoms making up each molecule in the sketch to the molecular formula.

MOLECULAR AND FORMULA WEIGHTS

The **formula weight** of a compound is the sum of the atomic weights of all the atoms in a chemical formula. For example, the formula for water is H_2O. Hydrogen and oxygen are both nonmetals, so the formula means that one atom of oxygen is bound to two hydrogen atoms in a molecule. From the periodic table, you know that the approximate (rounded) atomic weight of hydrogen is 1.0 u and oxygen is 16.0 u. Adding the atomic weights for all the atoms,

Atoms	Atomic Weight		Totals
2 of H	2×1.0 u	=	2.0 u
1 of O	1×16.0 u	=	16.0 u
	Formula weight	=	18.0 u

Thus, the formula weight of a water molecule is 18.0 u.

The formula weight of an ionic compound is found in the same way, by adding the rounded atomic weights of atoms (or ions) making up the compound. Sodium chloride is NaCl, so the formula weight is 23.0 u plus 35.5 u, or 58.5 u. The *formula weight* can be calculated for an ionic or molecular substance. The **molecular weight** is the formula weight of a molecular substance. The term *molecular weight* is sometimes used for all substances, whether or not they have molecules. Since ionic substances such as NaCl do not occur as molecules, this is not strictly correct. Both molecular and formula weights are calculated in the same way, but *formula weight* is a more general term.

EXAMPLE 10.1

What is the molecular weight of table sugar (sucrose), which has the formula $C_{12}H_{22}O_{11}$?

SOLUTION

The formula identifies the numbers of each atom, and the atomic weights are from a periodic table:

Atoms	Atomic Weight		Totals
12 of C	12×12.0 u	=	144.0 u
22 of H	22×1.0 u	=	22.0 u
11 of O	11×16.0 u	=	176.0 u
	Formula weight	=	342.0 u

EXAMPLE 10.2

What is the molecular weight of ethyl alcohol, C_2H_5OH? (Answer: 46.0 u)

PERCENT COMPOSITION OF COMPOUNDS

The formula weight of a compound can provide useful information about the elements making up a compound (Figure 10.3). For example, suppose you want to know how much calcium is provided by a dietary supplement. The label lists the ingredient as calcium carbonate, $CaCO_3$. To find how much calcium is supplied by a pill with a certain mass, you need to find the *mass percentage* of calcium in the compound.

The mass percentage of an element in a compound can be found from

$$\frac{\left(\begin{array}{c}\text{atomic weight} \\ \text{of element}\end{array}\right)\left(\begin{array}{c}\text{number of atoms} \\ \text{of element}\end{array}\right)}{\text{Formula weight of compound}} \times \begin{array}{c}100\% \text{ of} \\ \text{compound}\end{array} = \begin{array}{c}\% \text{ of} \\ \text{element}\end{array}$$

equation 10.1

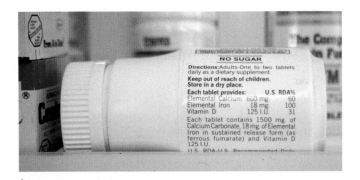

A

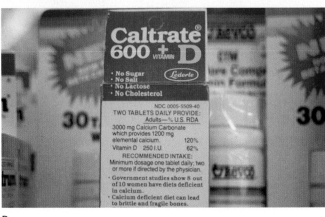

B

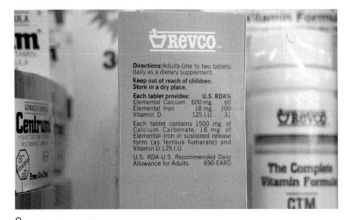

C

D

FIGURE 10.3 If you know the name of an ingredient, you can write a chemical formula, and the percent composition of a particular substance can be calculated from the formula. This can be useful information for consumer decisions.

The mass percentage of calcium in $CaCO_3$ can be found in two steps:

Step 1: Determine formula weight:

Atoms	Atomic Weight		Totals
1 of Ca	1×40.1 u	=	40.1 u
1 of C	1×12.0 u	=	12.0 u
3 of O	3×16.0 u	=	48.0 u
	Formula weight	=	100.1 u

Step 2: Determine percentage of Ca:

$$\frac{(40.1 \text{ u Ca})(1)}{100.1 \text{u } CaCO_3} \times 100\% \, CaCO_3 = 40.1\% \text{ Ca}$$

Knowing the percentage of the total mass contributed by the calcium, you can multiply this fractional part (as a decimal) by the mass of the supplement pill to find the calcium supplied. The mass percentage of the other elements can also be determined with equation 10.1.

EXAMPLE 10.3

Sodium fluoride is added to water supplies and to some toothpastes for fluoridation. What is the percentage composition of the elements in sodium fluoride?

SOLUTION

Step 1: Write the formula for sodium fluoride, NaF.

Step 2: Determine the formula weight.

Atoms	Atomic Weight		Totals
1 of Na	1×23.0 u	=	23.0 u
1 of F	1×19.0 u	=	19.0 u
	Formula weight	=	42.0 u

Step 3: Determine the percentage of Na and F.

For Na:

$$\frac{(23.0 \text{ u Na})(1)}{42.0 \text{ u NaF}} \times 100\% \text{ NaF} = \boxed{54.7\% \text{ Na}}$$

For F:

$$\frac{(19.0 \text{ u F})(1)}{42.0 \text{ u NaF}} \times 100\% \text{ NaF} = \boxed{45.2\% \text{ F}}$$

The percentage often does not total to exactly 100 percent because of rounding.

EXAMPLE 10.4

Calculate the percentage composition of carbon in table sugar, sucrose, which has a formula of $C_{12}H_{22}O_{11}$. (Answer: 42.1% C)

CONCEPTS *Applied*

Most for the Money

Chemical fertilizers are added to the soil when it does not contain sufficient elements essential for plant growth (see Figure 10.1). The three critical elements are nitrogen, phosphorus, and potassium, and these are the basic ingredients in most chemical fertilizers. In general, lawns require fertilizers high in nitrogen, and gardens require fertilizers high in phosphorus.

Read the labels on commercial packages of chemical fertilizers sold in a garden shop. Find the name of the chemical that supplies each of these critical elements; for example, nitrogen is sometimes supplied by ammonium nitrate NH_4NO_3. Calculate the mass percentage of each critical element supplied according to the label information. Compare these percentages to the grade number of the fertilizer, for example, 10–20–10. Determine which fertilizer brand gives you the most nutrients for the money.

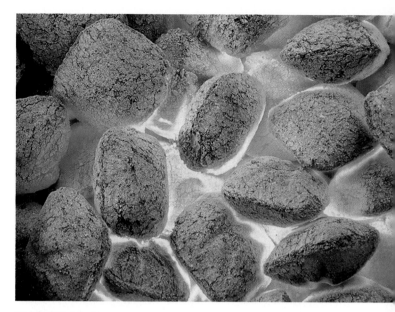

FIGURE 10.4 The charcoal used in a grill is basically carbon. The carbon reacts with oxygen to yield carbon dioxide. The chemical equation for this reaction, $C + O_2 \rightarrow CO_2$, contains the same information as the English sentence but has quantitative meaning as well.

CHEMICAL EQUATIONS

Chemical reactions occur when bonds between the outermost parts of atoms are formed or broken. Bonds are formed, for example, when a green plant uses sunlight—a form of energy—to create molecules of sugar, starch, and plant fibers. Bonds are broken and energy is released when you digest the sugars and starches or when plant fibers are burned. Chemical reactions thus involve changes in matter, the creation of new materials with new properties, and energy exchanges. So far, you have considered chemical symbols as a concise way to represent elements and formulas as a concise way to describe what a compound is made of. There is also a concise way to describe a chemical reaction, the **chemical equation.**

BALANCING EQUATIONS

Word equations are useful in identifying what has happened before and after a chemical reaction. The substances that existed before a reaction are called reactants, and the substances that exist after the reaction are called the products. The equation has a general form of

$$\text{reactants} \longrightarrow \text{products}$$

where the arrow signifies a separation in time; that is, it identifies what existed before the reaction and what exists after the reaction. For example, the charcoal used in a barbecue grill is carbon (Figure 10.4). The carbon reacts with oxygen while burning, and the reaction (1) releases energy and (2) forms carbon dioxide. The reactants and products for this reaction can be described as

$$\text{carbon} + \text{oxygen} \longrightarrow \text{carbon dioxide}$$

The arrow means *yields,* and the word equation is read as, "Carbon reacts with oxygen to yield carbon dioxide." This word equation describes what happens in the reaction but says nothing about the quantities of reactants or products.

Myths, Mistakes, & Misunderstandings

Chemical Reactions to Fool Breathalyzer Test?

Suppose someone consumes too much alcohol and then tries to drive a car. Police notice the erratic driving and stop the driver. Can the driver cause a chemical reaction to quickly fool the soon-to-be-administered breathalyzer test? Have you heard, for example, that sucking on a copper penny will cause a chemical reaction that will fool the breathalyzer, resulting in a low reading? This is a myth, of course, as the breathalyzer measures blood alcohol content by sampling air from the subject's lungs. Even if copper from a penny caused a chemical reaction in a person's saliva (which is doubtful), it would not react with or change the amount of alcohol in exhaled air from the lungs. Other variations on this myth that also have no scientific basis are sucking on a nickel; eating cough drops, garlic, or curry powder; or chewing vitamin C tablets. The bottom line is there is nothing you can do to change the breathalyzer reading—except for not consuming alcohol before driving.

Chemical symbols and formulas can be used in the place of words in an equation and the equation will have a whole new meaning. For example, the equation describing carbon reacting with oxygen to yield carbon dioxide becomes

$$C + O_2 \longrightarrow CO_2$$

(balanced)

The new, added meaning is that one atom of carbon (C) reacts with one molecule of oxygen (O_2) to yield one molecule of carbon dioxide (CO_2). Note that the equation also shows one atom of carbon and two atoms of oxygen (recall that oxygen occurs as a diatomic molecule) as reactants on the left side and one atom of carbon and two atoms of oxygen as products on the right side. Since the same number of each kind of atom appears on both sides of the equation, the equation is said to be *balanced*.

You would not want to use a charcoal grill in a closed room because there might not be enough oxygen. An insufficient supply of oxygen produces a completely different product, the poisonous gas carbon monoxide (CO). An equation for this reaction is

$$C + O_2 \longrightarrow CO$$

(not balanced)

As it stands, this equation describes a reaction that violates the **law of conservation of mass,** that matter is neither created nor destroyed in a chemical reaction. From the point of view of an equation, this law states that

mass of reactants = mass of products

Mass of reactants here means all that you start with, including some that might not react. Thus, elements are neither created nor destroyed, and this means the elements present and their mass. In any chemical reaction, the kind and mass of the reactive elements are identical to the kind and mass of the product elements.

From the point of view of atoms, the law of conservation of mass means that *atoms are neither created nor destroyed in the chemical reaction.* A chemical reaction is the making or breaking of chemical bonds between atoms or groups of atoms. Atoms are not lost or destroyed in the process, nor are they changed to a different kind. The equation for the formation of carbon monoxide has two oxygen atoms in the reactants (O_2) but only one in the product (in CO). An atom of oxygen has disappeared somewhere, and that violates the law of conservation of mass. You cannot fix the equation by changing the CO to a CO_2, because this would change the identity of the compounds. Carbon monoxide is a poisonous gas that is different from carbon dioxide, a relatively harmless product of burning and respiration. *You cannot change the subscript in a formula* because that would change the formula. A different formula means a different composition and thus a different compound.

You cannot change the subscripts of a formula, but you can place a number called a *coefficient* in *front* of the formula. Changing a coefficient changes the *amount* of a substance, not the identity. Thus, 2 CO means two molecules of carbon mon-

C	means		One atom of carbon
O	means		One atom of oxygen
O_2	means		One molecule of oxygen consisting of two atoms of oxygen
CO	means		One molecule of carbon monoxide consisting of one atom of carbon attached to one atom of oxygen
CO_2	means		One molecule of carbon dioxide consisting of one atom of carbon attached to two atoms of oxygen
3 CO_2	means		Three molecules of carbon dioxide, each consisting of one atom of carbon attached to two atoms of oxygen

FIGURE 10.5 The meaning of subscripts and coefficients used with a chemical formula. The subscripts tell you how many atoms of a particular element are in a compound. The coefficient tells you about the quantity, or number, of molecules of the compound.

oxide and 3 CO means three molecules of carbon monoxide. If there is no coefficient, 1 is understood as with subscripts. The meaning of coefficients and subscripts is illustrated in Figure 10.5.

Placing a coefficient of 2 in front of the C and a coefficient of 2 in front of the CO in the equation will result in the same numbers of each kind of atom on both sides:

$$2\,C + O2 \longrightarrow 2\,CO$$

Reactants: 2 C   Products: 2 C

2 O   2 O

The equation is now balanced.

Suppose your barbecue grill burns natural gas, not charcoal. Natural gas is mostly methane, CH_4. Methane burns by reacting with oxygen (O_2) to produce carbon dioxide (CO_2) and water vapor (H_2O). A balanced chemical equation for this reaction can be written by following a procedure of four steps.

Step 1: Write the correct formulas for the reactants and products in an unbalanced equation. The reactants and products could have been identified by chemical experiments, or they could have been predicted from what is known about chemical properties. This will be discussed in more detail later. For now, assume that the reactants and products are known and are given in words. For the burning of methane, the unbalanced, but otherwise correct, formula equation would be

$$CH_4 + O_2 \longrightarrow CO_2 + H_2O$$

(not balanced)

Step 2: Inventory the number of each kind of atom on both sides of the unbalanced equation. In the example there are

Reactants:	1 C	Products:	1 C
	4 H		2 H
	2 O		3 O

This shows that the H and O are unbalanced.

Step 3: Determine where to place coefficients in front of formulas to balance the equation. It is often best to focus on the simplest thing you can do with whole number ratios. The H and the O are unbalanced, for example, and there are 4 H atoms on the left and 2 H atoms on the right. Placing a coefficient 2 in front of H_2O will balance the H atoms:

$$CH_4 + O_2 \longrightarrow CO_2 + 2 H_2O$$

(not balanced)

Now take a second inventory:

Reactants:	1 C	Products:	1 C
	4 H		4 H
	2 O		4 O ($O_2 + 2$ O)

This shows the O atoms are still unbalanced with 2 on the left and 4 on the right. Placing a coefficient of 2 in front of O_2 will balance the O atoms.

$$CH_4 + 2 O_2 \longrightarrow CO_2 + 2 H_2O$$

(balanced)

Step 4: Take another inventory to determine if the numbers of atoms on both sides are now equal. If they are, determine if the coefficients are in the lowest possible whole-number ratio. The inventory is now

Reactants:	1 C	Products:	1 C
	4 H		4 H
	4 O		4 O

The number of each kind of atom on each side of the equation is the same, and the ratio of 1:2 → 1:2 is the lowest possible whole-number ratio. The equation is balanced, which is illustrated with sketches of molecules in Figure 10.6.

Balancing chemical equations is mostly a trial-and-error procedure. But with practice, you will find there are a few generalized "role models" that can be useful in balancing equations for many simple reactions. The key to success at balancing equations is to think it out step-by-step while remembering the following:

1. Atoms are neither lost nor gained, nor do they change their identity in a chemical reaction. The same kind and number of atoms in the reactants must appear in the products, meaning atoms are conserved.

Reaction:
   Methane reacts with oxygen to yield carbon dioxide and water

Balanced equation:
   $$CH_4 + 2 O_2 \longrightarrow CO_2 + 2 H_2O$$

Sketches representing molecules:

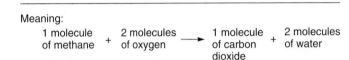

Meaning:

1 molecule of methane	+	2 molecules of oxygen	$\longrightarrow$	1 molecule of carbon dioxide	+	2 molecules of water

FIGURE 10.6 Compare the numbers of each kind of atom in the balanced equation with the numbers of each kind of atom in the sketched representation. Both the equation and the sketch have the same number of atoms in the reactants and in the products.

2. A correct formula of a compound cannot be changed by altering the number or placement of subscripts. Changing subscripts changes the identity of a compound and the meaning of the entire equation.
3. A coefficient in front of a formula multiplies everything in the formula by that number.

There are also a few generalizations that can be helpful for success in balancing equations:

1. Look first to formulas of compounds with the most atoms and try to balance the atoms or compounds they were formed from or decomposed to.
2. Polyatomic ions that appear on both sides of the equation should be treated as independent units with a charge. That is, consider the polyatomic ion as a unit while taking an inventory rather than the individual atoms making up the polyatomic ion. This will save time and simplify the procedure.
3. Both the "crossover technique" and the use of "fractional coefficients" can be useful in finding the least common multiple to balance an equation. All of these generalizations are illustrated in examples 10.5, 10.6, and 10.7.

The physical state of reactants and products in a reaction is often identified by the symbols (g) for gas, (l) for liquid, (s) for solid, and (aq) for an aqueous solution (*aqueous* means water). If a gas escapes, this is identified with an arrow pointing up (↑). A solid formed from a solution is identified with an arrow pointing down (↓). The Greek symbol delta (Δ) is often used under or over the yield sign to indicate a change of temperature or other physical values.

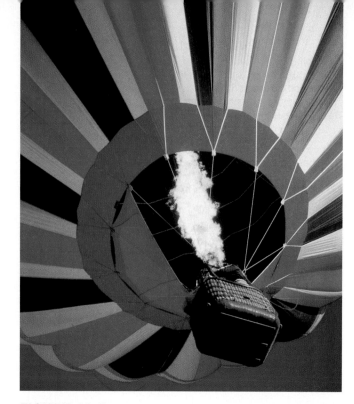

FIGURE 10.7 One of two burners is operating at the moment as this hot air balloon ascends. The burners are fueled by propane (C_3H_8), a liquified petroleum gas (LPG). Like other forms of petroleum, propane releases large amounts of heat during the chemical reaction of burning.

EXAMPLE 10.5

Propane is a liquified petroleum gas (LPG) that is often used as a bottled substitute for natural gas (Figure 10.7). Propane (C_3H_8) reacts with oxygen (O_2) to yield carbon dioxide (CO_2) and water vapor (H_2O). What is the balanced equation for this reaction?

SOLUTION

Step 1: Write the correct formulas of the reactants and products in an unbalanced equation.

$$C_3H_{8(g)} + O_{2(g)} \longrightarrow CO_{2(g)} + H_2O_{(g)}$$

(unbalanced)

Step 2: Inventory the numbers of each kind of atom.

Reactants:	3 C	Products:	1 C
	8 H		2 H
	2 O		3 O

Step 3: Determine where to place coefficients to balance the equation. Looking at the compound with the most atoms (generalization 1), you can **see** that a propane molecule has 3 C and 8 H. Placing a coefficient of 3 in front of CO_2 and a 4 in front of H_2O will balance these atoms (3 of C and $4 \times 2 = 8$ H atoms on the right has the same number of atoms as C_3H_8 on the left),

$$C_3H_{8(g)} + O_{2(g)} \longrightarrow 3\,CO_{2(g)} + 4\,H_2O_{(g)}$$

(not balanced)

A second inventory shows

Reactants:	3 C
	8 H
	2 O

Products:	3 C
	8 H ($4 \times 2 = 8$)
	10 O [$(3 \times 2) + (4 \times 1) = 10$]

The O atoms are still unbalanced. Place a 5 in front of O_2, and the equation is balanced ($5 \times 2 = 10$). Remember that you cannot change the subscripts and that oxygen occurs as a diatomic molecule of O_2.

$$C_3H_{8(g)} + 5\,O_{2(g)} \longrightarrow 3\,CO_{2(g)} + 4\,H_2O_{(g)}$$

(balanced)

Step 4: Another inventory shows (a) the number of atoms on both sides are now equal, and (b) the coefficients are 1:5 → 3:4, the lowest possible whole-number ratio. The equation is balanced.

EXAMPLE 10.6

One type of water hardness is caused by the presence of calcium bicarbonate in solution, $Ca(HCO_3)_2$. One way to remove the troublesome calcium ions from wash water is to add washing soda, which is sodium carbonate, Na_2CO_3. The reaction yields sodium bicarbonate ($NaHCO_3$) and calcium carbonate ($CaCO_3$), which is insoluble. Since $CaCO_3$ is insoluble, the reaction removes the calcium ions from solution. Write a balanced equation for the reaction.

SOLUTION

Step 1: Write the unbalanced equation

$$Ca(HCO_3)_{2(aq)} + Na_2CO_{3(aq)} \longrightarrow NaHCO_{3(aq)} + CaCO_3\downarrow$$

(not balanced)

Step 2: Inventory the numbers of each kind of atom. This reaction has polyatomic ions that appear on both sides, so they should be treated as independent units with a charge (generalization 2). The inventory is

Reactants:	1 Ca	Products:	1 Ca
	2 $(HCO_3)^{1-}$		1 $(HCO_3)^{1-}$
	2 Na		1 Na
	1 $(CO_3)^{2-}$		1 $(CO_3)^{2-}$

Step 3: Placing a coefficient of 2 in front of $NaHCO_3$ will balance the equation,

$$Ca(HCO_3)_{2(aq)} + Na_2CO_{3(aq)} \longrightarrow 2\,NaHCO_{3(aq)} + CaCO_3\downarrow$$

(balanced)

Step 4: An inventory shows

Reactants:	1 Ca	Products:	1 Ca
	2 $(HCO_3)^{1-}$		2 $(HCO_3)^{1-}$
	2 Na		2 Na
	1 $(CO_3)^{2-}$		1 $(CO_3)^{2-}$

The coefficient ratio of 1:1 → 2:1 is the lowest whole-number ratio. The equation is balanced.

EXAMPLE 10.7

Gasoline is a mixture of hydrocarbons, including octane (C_8H_{18}). Combustion of octane produces CO_2 and H_2O, with the release of energy. Write a balanced equation for this reaction.

SOLUTION

Step 1: Write the correct formulas in an unbalanced equation,

$$C_8H_{18(g)} + O_{2(g)} \longrightarrow CO_{2(g)} + H_2O_{(g)}$$

(not balanced)

Step 2: Take an inventory,

Reactants: 8 C   Products: 1 C

               18 H   2 H

               2 O   3 O

(not balanced)

Step 3: Start with the compound with the most atoms (generalization 1) and place coefficients to balance these atoms,

$$C_8H_{18(g)} + O_{2(g)} \longrightarrow 8\,CO_{2(g)} + 9\,H_2O_{(g)}$$

(not balanced)

Redo the inventory,

Reactants: 8 C   Products: 8 C

               18 H   18 H

               2 O   25 O

The O atoms are still unbalanced. There are 2 O atoms in the reactants but 25 O atoms in the products. Since the subscript cannot be changed, it will take 12.5 O_2 to produce 25 oxygen atoms (generalization 3).

$$C_8H_{18(g)} + 12.5\,O_{2(g)} \longrightarrow 8\,CO_{2(g)} + 9\,H_2O_{(g)}$$

(balanced)

Step 4: (a) An inventory will show that the atoms balance,

Reactants: 8 C   Products: 8 C

               18 H   18 H

               25 O   25 O

(b) The coefficients are not in the lowest whole-number ratio (one-half an O_2 does not exist). To make the lowest possible whole-number ratio, all coefficients are multiplied by 2. This results in a correct balanced equation of

$$2\,C_8H_{18(g)} + 25\,O_{2(g)} \longrightarrow 16\,CO_{2(g)} + 18\,H_2O_{(g)}$$

(balanced)

GENERALIZING EQUATIONS

In the previous chapters, you learned that the act of classifying, or grouping, something according to some property makes the study of a large body of information less difficult. Generalizing from groups of chemical reactions also makes it possible to

FIGURE 10.8 *Hydrocarbons* are composed of the elements hydrogen and carbon. Propane (C_3H_8) and gasoline, which contain octane (C_8H_{18}), are examples of hydrocarbons. *Carbohydrates* are composed of the elements hydrogen, carbon, and oxygen. Table sugar, for example, is the carbohydrate $C_{12}H_{22}O_{11}$. Generalizing, all hydrocarbons and carbohydrates react completely with oxygen to yield CO_2 and H_2O.

predict what will happen in similar reactions. For example, you have studied equations in the "Balancing Equations" section describing the combustion of methane (CH_4), propane (C_3H_8), and octane (C_8H_{18}). Each of these reactions involves a *hydrocarbon,* a compound of the elements hydrogen and carbon. Each hydrocarbon reacted with O_2, yielding CO_2 and releasing the energy of combustion. Generalizing from these reactions, you could predict that the combustion of any hydrocarbon would involve the combination of atoms of the hydrocarbon molecule with O_2 to produce CO_2 and H_2O with the release of energy. Such reactions could be analyzed by chemical experiments, and the products could be identified by their physical and chemical properties. You would find your predictions based on similar reactions would be correct, thus justifying predictions from such generalizations. Butane, for example, is a hydrocarbon with the formula C_4H_{10}. The balanced equation for the combustion of butane is

$$2\,C_4H_{10(g)} + 13\,O_{2(g)} \longrightarrow 8\,CO_{2(g)} + 10\,H_2O_{(g)}$$

You could extend the generalization further, noting that the combustion of compounds containing oxygen as well as carbon and hydrogen also produces CO_2 and H_2O (Figure 10.8). These compounds are *carbohydrates,* composed of carbon and water. Glucose, for example, was identified earlier as a compound with the formula $C_6H_{12}O_6$. Glucose combines with oxygen to produce CO_2 and H_2O, and the balanced equation is

$$C_6H_{12}O_{6(s)} + 6\,O_{2(g)} \longrightarrow 6\,CO_{2(g)} + 6\,H_2O_{(g)}$$

Note that three molecules of oxygen were not needed from the O_2 reactant because the other reactant, glucose, contains six oxygen atoms per molecule. An inventory of atoms will show that the equation is thus balanced.

Combustion is a rapid reaction with O_2 that releases energy, usually with a flame. A very similar, although much slower reaction takes place in plant and animal respiration. In respiration, carbohydrates combine with O_2 and release energy used for biological activities. This reaction is slow compared to combustion and requires enzymes to proceed at body temperature. Nonetheless, CO_2 and H_2O are the products.

TYPES OF CHEMICAL REACTIONS

The reactions involving hydrocarbons and carbohydrates with oxygen are examples of an important group of chemical reactions called *oxidation-reduction* reactions. Historically, when the term *oxidation* was first used, it specifically meant reactions involving the combination of oxygen with other atoms. But fluorine, chlorine, and other nonmetals were soon understood to have similar reactions to those of oxygen, so the definition was changed to one concerning the shifts of electrons in the reaction.

An **oxidation-reduction reaction** (or **redox reaction**) is broadly defined as a reaction in which electrons are transferred from one atom to another. As is implied by the name, such a reaction has two parts and each part tells you what happens to the electrons. *Oxidation* is the part of a redox reaction in which there is a loss of electrons by an atom. *Reduction* is the part of a redox reaction in which there is a gain of electrons by an atom. The name also implies that in any reaction in which oxidation occurs reduction must take place, too. One cannot take place without the other.

Substances that take electrons from other substances are called **oxidizing agents.** Oxidizing agents take electrons from the substances being oxidized. Oxygen is the most common oxidizing agent, and several examples have already been given about how it oxidizes foods and fuels. Chlorine is another commonly used oxidizing agent, often for the purposes of bleaching or killing bacteria (Figure 10.9).

A **reducing agent** supplies electrons to the substance being reduced. Hydrogen and carbon are commonly used reducing agents.

FIGURE 10.9 Oxidizing agents take electrons from other substances that are being oxidized. Oxygen and chlorine are commonly used, strong oxidizing agents.

Carbon is commonly used as a reducing agent to extract metals from their ores. For example, carbon (from coke, which is coal that has been baked) reduces Fe_2O_3, an iron ore, in the reaction

$$2\ Fe_2O_{3(s)} + 3\ C_{(s)} \longrightarrow 4\ Fe_{(s)} + 3\ CO_2\uparrow$$

The Fe in the ore gained electrons from the carbon, the reducing agent in this reaction.

CONCEPTS *Applied*

Silver Polish

Silverware and silver-plated objects often become tarnished when the silver is oxidized by sulfur, forming Ag_2S. Commercial silver polishes often act by removing the oxidized layer with an abrasive. The silver can also be polished by reducing the Ag_2S back to metallic silver without removing a layer. Place the tarnished silver in a clean aluminum pan with about 80 g sodium bicarbonate ($NaHCO_3$) and 80 g NaCl dissolved in each liter of near-boiling water. A sufficient amount should be prepared to cover the silver object or objects. The salts provide ions to help transfer electrons and facilitate the reaction. The reaction is

$$3\ Ag_2S + 2\ Al + 6\ H_2O \longrightarrow 6\ Ag + 2\ Al(OH)_3 + 3\ H_2S$$

(Note: H_2S has a rotten egg odor.)

Many chemical reactions can be classified as redox or nonredox reactions. Another way to classify chemical reactions is to consider what is happening to the reactants and products. This type of classification scheme leads to four basic categories of chemical reactions, which are (1) *combination,* (2) *decomposition,* (3) *replacement,* and (4) *ion exchange reactions.* The first three categories are subclasses of redox reactions. It is in the ion exchange reactions that you will find the first example of a reaction that is not a redox reaction.

COMBINATION REACTIONS

A **combination reaction** is a synthesis reaction in which two or more substances combine to form a single compound. The combining substances can be (1) elements, (2) compounds, or (3) combinations of elements and compounds. In generalized form, a combination reaction is

$$X + Y \longrightarrow XY$$

Many redox reactions are combination reactions. For example, metals are oxidized when they burn in air, forming a metal oxide. Consider magnesium, which gives off a bright white light as it burns:

$$2\ Mg_{(s)} + O_{2(g)} \longrightarrow 2\ MgO_{(s)}$$

Note how the magnesium-oxygen reaction follows the generalized form of $X + Y \rightarrow XY$.

The rusting of metals is oxidation that takes place at a slower pace than burning, but metals are nonetheless oxidized in the

FIGURE 10.10 Rusting iron is a common example of a combination reaction, where two or more substances combine to form a new compound. Rust is iron(III) oxide formed on these screws from the combination of iron and oxygen under moist conditions.

process (Figure 10.10). Again noting the generalized form of a combination reaction, consider the rusting of iron:

$$4\ Fe_{(s)} + 3\ O_{2(g)} \longrightarrow 2\ Fe_2O_{3(s)}$$

Nonmetals are also oxidized by burning in air, for example, when carbon burns with a sufficient supply of O_2:

$$C_{(s)} + O_{2(g)} \longrightarrow CO_{2(g)}$$

Note that all the combination reactions follow the generalized form of $X + Y \rightarrow XY$.

DECOMPOSITION REACTIONS

A **decomposition reaction**, as the term implies, is the opposite of a combination reaction. In decomposition reactions, a compound is broken down into (1) the elements that make up the compound, (2) simpler compounds, or (3) elements and simpler compounds. Decomposition reactions have a generalized form of

$$XY \longrightarrow X + Y$$

Decomposition reactions generally require some sort of energy, which is usually supplied in the form of heat or electrical energy. An electric current, for example, decomposes water into hydrogen and oxygen:

$$2\ H_2O_{(l)} \xrightarrow{\text{electricity}} 2\ H_{2(g)} + O_{2(g)}$$

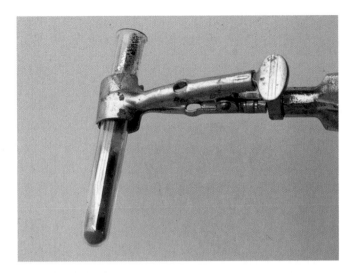

FIGURE 10.11 Mercury(II) oxide is decomposed by heat, leaving the silver-colored element mercury behind as oxygen is driven off. This is an example of a decomposition reaction, $2\ HgO \rightarrow 2\ Hg + O_2 \uparrow$. Compare this equation to the general form of a decomposition reaction.

Mercury(II) oxide is decomposed by heat, an observation that led to the discovery of oxygen (Figure 10.11):

$$2\ HgO_{(s)} \xrightarrow{\Delta} 2\ Hg_{(s)} + O_2\uparrow$$

Plaster is a building material made from a mixture of calcium hydroxide, $Ca(OH)_2$, and plaster of Paris, $CaSO_4$. The calcium hydroxide is prepared by adding water to calcium oxide (CaO), which is commonly called quicklime. Calcium oxide is made by heating limestone or chalk ($CaCO_3$), and

$$CaCO_{3(s)} \xrightarrow{\Delta} CaO_{(s)} + CO_2\uparrow$$

Note that all the decomposition reactions follow the generalized form of $XY \rightarrow X + Y$.

REPLACEMENT REACTIONS

In a **replacement reaction**, an atom or polyatomic ion is replaced in a compound by a different atom or polyatomic ion. The replaced part can be either the negative or positive part of the compound. In generalized form, a replacement reaction is

$$XY + Z \longrightarrow XZ + Y$$
(negative part replaced)

or

$$XY + A \longrightarrow AY + X$$
(positive part replaced)

Replacement reactions occur because some elements have a stronger electron-holding ability than other elements. Elements that have the least ability to hold on to their electrons are the most chemically active. Figure 10.12 shows a list of chemical activity of some metals, with the most chemically active at the top. Hydrogen is included because of its role in acids (see chapter 11). Take a few minutes to look over the generalizations listed in Figure 10.12. The generalizations apply to combination, decomposition, and replacement reactions.

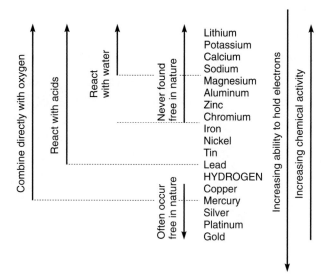

FIGURE 10.12 The activity series for common metals, together with some generalizations about the chemical activities of the metals. The series is used to predict which replacement reactions will take place and which reactions will not occur. (Note that hydrogen is not a metal and is placed in the series for reference to acid reactions.)

Replacement reactions take place as more active metals give up electrons to elements lower on the list with a greater electron-holding ability. For example, aluminum is higher on the activity series than copper. When aluminum foil is placed in a solution of copper(II) chloride, aluminum is oxidized, losing electrons to the copper. The loss of electrons from metallic aluminum forms aluminum ions in solution, and the copper comes out of solution as a solid metal (Figure 10.13).

$$2\ Al_{(s)} + 3\ CuCl_{2(aq)} \longrightarrow 2\ AlCl_{3(aq)} + 3\ Cu_{(s)}$$

FIGURE 10.13 This shows a reaction between metallic aluminum and the blue solution of copper(II) chloride. Aluminum is above copper in the activity series, and aluminum replaces the copper ions from the solution as copper is deposited as a metal. The aluminum loses electrons to the copper and forms aluminum ions in solution.

A metal will replace any metal ion in solution that it is above in the activity series. If the metal is listed below the metal ion in solution, no reaction occurs. For example, $Ag(s) + CuCl_{2(aq)} \rightarrow$ no reaction.

The very active metals (lithium, potassium, calcium, and sodium) react with water to yield metal hydroxides and hydrogen. For example,

$$2\ Na_{(s)} + 2\ H_2O_{(l)} \longrightarrow 2\ NaOH_{(aq)} + H_2\uparrow$$

Acids yield hydrogen ions in solution, and metals above hydrogen in the activity series will replace hydrogen to form a metal salt. For example,

$$Zn_{(s)} + H_2SO_{4(aq)} \longrightarrow ZnSO_{4(aq)} + H_2\uparrow$$

In general, the energy involved in replacement reactions is less than the energy involved in combination or decomposition reactions.

ION EXCHANGE REACTIONS

An **ion exchange reaction** is a reaction that takes place when the ions of one compound interact with the ions of another compound, forming (1) a solid that comes out of solution (a precipitate), (2) a gas, or (3) water.

A water solution of dissolved ionic compounds is a solution of ions. For example, solid sodium chloride dissolves in water to become ions in solution,

$$NaCl_{(s)} \longrightarrow Na^+_{(aq)} + Cl^-_{(aq)}$$

If a second ionic compound is dissolved with a solution of another, a mixture of ions results. The formation of a precipitate, a gas, or water, however, removes ions from the solution, and this must occur before you can say that an ionic exchange reaction has taken place. For example, water being treated for domestic use sometimes carries suspended matter that is removed by adding aluminum sulfate and calcium hydroxide to the water. The reaction is

$$3\ Ca(OH)_{2(aq)} + Al_2(SO_4)_{3(aq)} \longrightarrow 3\ CaSO_{4(aq)} + 2\ Al(OH)_3\downarrow$$

The aluminum hydroxide is a jellylike solid, which traps the suspended matter for sand filtration. The formation of the insoluble aluminum hydroxide removed the aluminum and hydroxide ions from the solution, so an ion exchange reaction took place.

In general, an ion exchange reaction has the form

$$AX + BY \longrightarrow AY + BX$$

where one of the products removes ions from the solution. The calcium hydroxide and aluminum sulfate reaction took place as the aluminum and calcium ions traded places. A solubility table such as the one in appendix B will tell you if an ionic exchange reaction has taken place. Aluminum hydroxide is insoluble, according to the table, so the reaction did take place. No ionic exchange reaction occurred if the new products are both soluble.

Another way for an ion exchange reaction to occur is if a gas or water molecule forms to remove ions from the solution.

When an acid reacts with a base (an alkaline compound), a salt and water are formed

$$HCl_{(aq)} + NaOH_{(aq)} \longrightarrow NaCl_{(aq)} + H_2O_{(l)}$$

The reactions of acids and bases are discussed in chapter 11.

 CONCEPTS *Applied*

Chemical Reactions

Look around your school and home for signs that a chemical reaction has taken place. Can you find evidence that a reaction has taken place with oxygen? Can you find new substances being made or decomposition taking place?

EXAMPLE 10.8

Write complete balanced equations for the following, and identify if each reaction is combination, decomposition, replacement, or ion exchange:

(a) silver$_{(s)}$ + sulfur$_{(g)}$ → silver sulfide$_{(s)}$
(b) aluminum$_{(s)}$ + iron(III) oxide$_{(s)}$ → aluminum oxide$_{(s)}$ + iron
(c) sodium chloride$_{(aq)}$ + silver nitrate$_{(aq)}$ → ?
(d) potassium chlorate$_{(s)}$ $\xrightarrow{\Delta}$ potassium chloride$_{(s)}$ + oxygen$_{(g)}$

SOLUTION

(a) The reactants are two elements, and the product is a compound, following the general form $X + Y \rightarrow XY$ of a combination reaction. Table 9.2 gives the charge on silver as Ag^{1+}, and sulfur (as the other nonmetals in family VIA) is S^{2-}. The balanced equation is

$$2 Ag_{(s)} + S_{(g)} \longrightarrow Ag_2S_{(s)}$$

Silver sulfide is the tarnish that appears on silverware.

(b) The reactants are an element and a compound that react to form a new compound and an element. The general form is $XY + Z \rightarrow XZ + Y$, which describes a replacement reaction. The balanced equation is

$$2 Al_{(s)} + Fe_2O_{3(s)} \longrightarrow Al_2O_{3(s)} + 2 Fe_{(s)}$$

This is known as a "thermite reaction," and in the reaction, aluminum reduces the iron oxide to metallic iron with the release of sufficient energy to melt the iron. The thermite reaction is sometimes used to weld large steel pieces, such as railroad rails.

(c) The reactants are water solutions of two compounds with the general form of $AX + BY \rightarrow$, so this must be the reactant part of an ion exchange reaction. Completing the products part of the equation by exchanging parts as shown in the general form and balancing,

$$NaCl_{(aq)} + AgNO_{3(aq)} \longrightarrow NaNO_{3(?)} + AgCl_{(?)}$$

Now consult the solubility chart in appendix B to find out if either of the products is insoluble. $NaNO_3$ is soluble and $AgCl$ is insoluble. Since at least one of the products is insoluble, the reaction did take place, and the equation is rewritten as

$$NaCl_{(aq)} + AgNO_{3(aq)} \longrightarrow NaNO_{3(aq)} + AgCl\downarrow$$

(d) The reactant is a compound, and the products are a simpler compound and an element, following the generalized form of a decomposition reaction, $XY \rightarrow X + Y$. The delta sign (Δ) also means that heat was added, which provides another clue that this is a decomposition reaction. The formula for the chlorate ion is in Table 9.3. The balanced equation is

$$2 KClO_{3(s)} \xrightarrow{\Delta} 2 KCl_{(s)} + 3 O_2\uparrow$$

INFORMATION FROM CHEMICAL EQUATIONS

A balanced chemical equation describes what happens in a chemical reaction in a concise, compact way. The balanced equation also carries information about (1) atoms, (2) molecules, and (3) atomic weights. The balanced equation for the combustion of hydrogen, for example, is

$$2 H_{2(g)} + O_{2(g)} \longrightarrow 2 H_2O_{(l)}$$

An inventory of each kind of atom in the reactants and products shows

Reactants:		Products:	
	4 hydrogen		4 hydrogen
	2 oxygen		2 oxygen
Total:	6 atoms	Total:	6 atoms

There are six atoms before the reaction and there are six atoms after the reaction, which is in accord with the law of conservation of mass.

In terms of molecules, the equation says that two diatomic molecules of hydrogen react with one (understood) diatomic molecule of oxygen to yield two molecules of water. The number of coefficients in the equation is the number of molecules involved in the reaction. If you are concerned how two molecules plus one molecule could yield two molecules, remember that *atoms* are conserved in a chemical reaction, not molecules.

Since atoms are conserved in a chemical reaction, their atomic weights should be conserved, too. One hydrogen atom has an atomic weight of 1.0 u, so the formula weight of a diatomic hydrogen molecule must be 2×1.0 u, or 2.0 u. The formula weight of O_2 is 2×16.0 u, or 32 u. If you consider the equation in terms of atomic weights, then

Equation

$$2 H_2 + O_2 \longrightarrow 2 H_2O$$

Formula weights

$$2 (1.0\, u + 1.0\, u) + (16.0\, u + 16.0\, u) \longrightarrow 2 (2 \times 1.0\, u + 16.0\, u)$$

$$4\, u + 32\, u \longrightarrow 36\, u$$

$$36\, u \longrightarrow 36\, u$$

The formula weight for H_2O is $(1.0\, u \times 2) + 16$ u, or 18 u. The coefficient of 2 in front of H_2O means there are two molecules of H_2O, so the mass of the products is 2×18 u, or 36 u. Thus, the reactants had a total mass of 4 u + 32 u, or 36 u, and the products had a total mass of 36 u. Again, this is in accord with the law of conservation of mass.

The equation says that 4 u of hydrogen will combine with 32 u of oxygen. Thus, hydrogen and oxygen combine in a mass ratio of 4:32, which reduces to 1:8. So 1 g of hydrogen will combine with 8 g of oxygen, and, in fact, they will combine in this ratio no matter what the measurement units are (gram, kilogram, pound, etc.). They always combine in this mass ratio because this is the mass of the individual reactants.

Back in the early 1800s, John Dalton (1766–1844) attempted to work out a table of atomic weights as he developed his atomic theory. Dalton made two major errors in determining the atomic weights, including (1) measurement errors about mass ratios of combining elements and (2) incorrect assumptions about the formula of the resulting compound. For water, for example, Dalton incorrectly measured that 5.5 g of oxygen combined with 1.0 g of hydrogen. He assumed that one atom of hydrogen combined with one atom of oxygen, resulting in a formula of HO. Thus, Dalton concluded that the atomic mass of oxygen was 5.5 u, and the atomic mass of hydrogen was 1.0 u. Incorrect atomic weights for hydrogen and oxygen led to conflicting formulas for other substances, and no one could show that the atomic theory worked.

The problem was solved during the first decade of the 1800s through the separate work of a French chemistry professor, Joseph Gay-Lussac (1778–1850), and an Italian physics professor, Amedeo Avogadro (1776–1856). In 1808, Gay-Lussac reported that reacting gases combined in small, whole number *volumes* when the temperature and pressure were constant. Two volumes of hydrogen, for example, combined with one volume of oxygen to form two volumes of water vapor. The term *volume* means any measurement unit, for example, a liter. Other reactions between gases were also observed to combine in small, whole number ratios, and the pattern became known as the *law of combining volumes* (Figure 10.14).

Avogadro proposed an explanation for the law of combining volumes in 1811. He proposed that equal volumes of all gases at the same temperature and pressure *contain the same number of molecules*. Avogadro's hypothesis had two important implications for the example of water. First, since two volumes of hydrogen combine with one volume of oxygen, it means that a molecule of water contains twice as many hydrogen atoms as oxygen atoms. The formula for water must be H_2O, not HO. Second, since *two* volumes of water vapor were produced, each molecule of hydrogen and each molecule of oxygen must be diatomic. Diatomic molecules of hydrogen and oxygen would double the number of hydrogen and oxygen atoms, thus producing twice as much water vapor. These two implications are illustrated in Figure 10.15, along with a balanced equation for the reaction. Note that the coefficients in the equation now have two meanings: (1) the number of molecules of each substance involved in the reaction and (2) the ratios of combining volumes. The coefficient of 2 in front of the H_2, for example, means two molecules of H_2. It also means two volumes of H_2 gas when all volumes are measured at the same temperature and pressure. Recall that equal volumes of any two gases at the same temperature and pressure contain the same number of molecules. Thus,

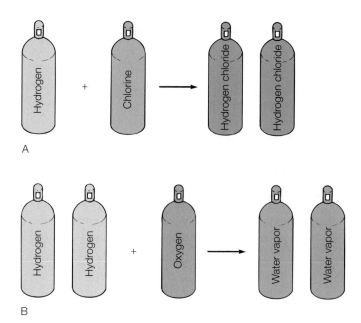

A

B

FIGURE 10.14 Reacting gases combine in ratios of small, whole-number volumes when the temperature and pressure are the same for each volume. (*A*) One volume of hydrogen gas combines with one volume of chlorine gas to yield two volumes of hydrogen chloride gas. (*B*) Two volumes of hydrogen gas combine with one volume of oxygen gas to yield two volumes of water vapor.

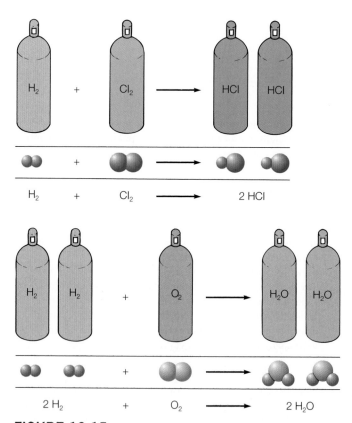

FIGURE 10.15 Avogadro's hypothesis of equal volumes of gas having equal numbers of molecules offered an explanation for the law of combining volumes.

the ratio of coefficients in a balanced equation means a ratio of *any number* of molecules, from 2 of H_2 and 1 of O_2, 20 of H_2 and 10 of O_2, 2,000 of H_2 and 1,000 of O_2, or however many are found in 2 L of H_2 and 1 L of O_2.

EXAMPLE 10.9

Propane is a hydrocarbon with the formula C_3H_8 that is used as a bottled gas. (a) How many liters of oxygen are needed to burn 1 L of propane gas? (b) How many liters of carbon dioxide are produced by the reaction? Assume all volumes to be measured at the same temperature and pressure.

SOLUTION

The balanced equation is

$$C_3H_{8(g)} + 5\,O_{2(g)} \longrightarrow 3\,CO_{2(g)} + 4\,H_2O_{(g)}$$

The coefficients tell you the relative number of molecules involved in the reaction, that 1 molecule of propane reacts with 5 molecules of oxygen to produce 3 molecules of carbon dioxide and 4 molecules of water. Since equal volumes of gases at the same temperature and pressure contain equal numbers of molecules, the coefficients also tell you the relative volumes of gases. Thus, 1 L of propane (a) requires 5 L of oxygen and (b) yields 3 L of carbon dioxide (and 4 L of water vapor) when reacted completely.

UNITS OF MEASUREMENT USED WITH EQUATIONS

The coefficients in a balanced equation represent a ratio of any *number* of molecules involved in a chemical reaction. The equation has meaning about the atomic *weights* and formula *weights* of reactants and products. The counting of numbers and the use of atomic weights are brought together in a very important measurement unit called a *mole* (from the Latin meaning "a mass"). Here are the important ideas in the mole concept:

1. Recall that the atomic weights of elements are average relative masses of the isotopes of an element. The weights are based on a comparison to carbon-12, with an assigned mass of exactly 12.00 (see chapter 8).

2. The *number* of C-12 atoms in exactly 12.00 g of C-12 has been measured experimentally to be 6.02×10^{23}. This number is called **Avogadro's number,** named after the scientist who reasoned that equal volumes of gases contain equal numbers of molecules.

3. An amount of a substance that contains Avogadro's number of atoms, ions, molecules, or any other chemical unit is defined as a **mole** of the substance. Thus, a mole is 6.02×10^{23} atoms, ions, etc., just as a dozen is 12 eggs, apples, etc. The mole is the chemist's measure of atoms, molecules, or other chemical units. A mole of Na^+ ions is 6.02×10^{23} Na^+ ions.

4. A mole of C-12 atoms is defined as having a mass of exactly 12.00 g, a mass that is numerically equal to its

atomic mass. So the mass of a mole of C-12 atoms is 12.00 g, or

$$\underset{(12.00\text{ u})}{\text{mass of one atom}} \times \underset{(6.02 \times 10^{23}) =}{\text{one mole}} = \underset{12.00\text{ g}}{\text{mass of a mole of C-12}}$$

The masses of all the other isotopes are *based* on a comparison to the C-12 atom. Thus, a He-4 atom has one-third the mass of a C-12 atom. An atom of Mg-24 is twice as massive as a C-12 atom. Thus

	1 Atom	$\times$	1 Mole	=	Mass of Mole
C-12:	12.00 u	$\times$	6.02×10^{23}	=	12.00 g
He-4:	4.00 u	$\times$	6.02×10^{23}	=	4.00 g
Mg-24:	24.00 u	$\times$	6.02×10^{23}	=	24.00 g

Therefore, the mass of a mole of any element is numerically equal to its atomic mass. Samples of elements with masses that are the same numerically as their atomic masses are 1 mole measures, and each will contain the same number of atoms (Figure 10.16).

This reasoning can be used to generalize about formula weights, molecular weights, and atomic weights since they are all based on atomic mass units relative to C-12. The **gram-atomic weight** is the mass in grams of one mole of an element that is numerically equal to its atomic weight. The atomic weight of carbon is 12.01 u; the gram-atomic weight of carbon is 12.01 g. The atomic weight of magnesium is 24.3 u; the gram-atomic weight of magnesium is 24.3 g. Any gram-atomic weight contains Avogadro's number of atoms. Therefore, the gram-atomic weights of the elements all contain the same number of atoms.

Similarly, the **gram-formula weight** of a compound is the mass in grams of one mole of the compound that is numerically equal to its formula weight. The **gram-molecular weight** is the gram-formula weight of a molecular compound. Note that one mole of Ne atoms (6.02×10^{23} neon atoms) has a gram-atomic weight of 20.2 g, but one mole of O_2 molecules (6.02×10^{23} oxygen molecules) has a gram-molecular weight of 32.0 g. Stated the other way around, 32.0 g of O_2 and 20.2 g of Ne both contain the same Avogadro's number of particles.

EXAMPLE 10.10

(a) A 100 percent silver chain has a mass of 107.9 g. How many silver atoms are in the chain? (b) What is the mass of one mole of sodium chloride, NaCl?

SOLUTION

The mole concept and Avogadro's number provide a relationship between numbers and masses. (a) The atomic weight of silver is 107.9 u, so the gram-atomic weight of silver is 107.9 g. A gram-atomic weight is one mole of an element, so the silver chain contains 6.02×10^{23} silver atoms. (b) The formula weight of NaCl is 58.5 u, so the gram-formula weight is 58.5 g. One mole of NaCl has a mass of 58.5 g.

A Each of the following represents one mole of an element:

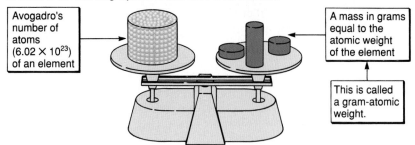

Avogadro's number of atoms (6.02×10^{23}) of an element

A mass in grams equal to the atomic weight of the element

This is called a gram-atomic weight.

B Each of the following represents one mole of a compound:

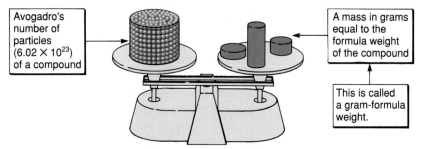

Avogadro's number of particles (6.02×10^{23}) of a compound

A mass in grams equal to the formula weight of the compound

This is called a gram-formula weight.

C Each of the following represents one mole of a molecular substance:

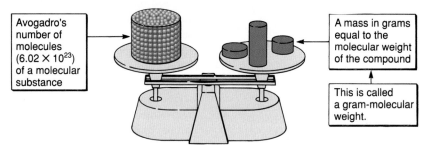

Avogadro's number of molecules (6.02×10^{23}) of a molecular substance

A mass in grams equal to the molecular weight of the compound

This is called a gram-molecular weight.

FIGURE 10.16 The mole concept for (*A*) elements, (*B*) compounds, and (*C*) molecular substances. A mole contains 6.02×10^{23} particles. Since every mole contains the same number of particles, the ratio of the mass of any two moles is the same as the ratio of the masses of individual particles making up the two moles.

QUANTITATIVE USES OF EQUATIONS

A balanced chemical equation can be interpreted in terms of (1) a *molecular ratio* of the reactants and products, (2) a *mole ratio* of the reactants and products, or (3) a *mass ratio* of the reactants and products. Consider, for example, the balanced equation for reacting hydrogen with nitrogen to produce ammonia,

$$3 \, H_{2(g)} + N_{2(g)} \longrightarrow 2 \, NH_{3(g)}$$

From a *molecular* point of view, the equation says that three molecules of hydrogen combine with one molecule of N_2 to form two molecules of NH_3. The coefficients of $3{:}1 \rightarrow 2$ thus express a molecular ratio of the reactants and the products.

The molecular ratio leads to the concept of a *mole ratio* since any number of molecules can react as long as they are in the ratio of $3{:}1 \rightarrow 2$. The number could be Avogadro's number, so $(3) \times (6.02 \times 10^{23})$ molecules of H_2 will combine with $(1) \times (6.02 \times 10^{23})$ molecules of N_2 to form $(2) \times (6.02 \times 10^{23})$

molecules of ammonia. Since 6.02×10^{23} molecules is the number of particles in a mole, the coefficients therefore represent the *numbers of moles* involved in the reaction. Thus, three moles of H_2 react with one mole of N_2 to produce two moles of NH_3.

The mole ratio of a balanced chemical equation leads to the concept of a *mass ratio* interpretation of a chemical equation. The gram-formula weight of a compound is the mass in grams of *one mole* that is numerically equal to its formula weight. Therefore, the equation also describes the mass ratios of the reactants and the products. The mass ratio can be calculated from the mole relationship described in the equation. The three interpretations are summarized in Table 10.1.

Thus, the coefficients in a balanced equation can be interpreted in terms of molecules, which leads to an interpretation of moles, mass, or any formula unit. The mole concept thus provides the basis for calculations about the quantities of reactants and products in a chemical reaction.

The modern automobile produces two troublesome products in the form of (1) nitrogen monoxide and (2) hydrocarbons from the incomplete combustion of gasoline. These products from the exhaust enter the air to react in sunlight, eventually producing an irritating haze known as photochemical smog. To reduce photochemical smog, modern automobiles are fitted with a catalytic converter as part of their exhaust system (Box Figure 10.1).

Molecules require a certain amount of energy to change chemical bonds. This certain amount of energy is called the *activation energy,* and it represents an energy barrier that must be overcome before a chemical reaction can take place. This explains why chemical reactions proceed at a faster rate at higher temperatures. At higher temperatures, molecules have greater average kinetic energies; thus, they already have part of the minimum energy needed for a reaction to take place.

The rate at which a chemical reaction proceeds is affected by a *catalyst,* a material that speeds up a chemical reaction without being permanently changed by the reaction. A catalyst appears to speed a chemical reaction by lowering the activation energy. Molecules become temporarily attached to the surface of the catalyst, which weakens the chemical bonds holding the molecule together. The weakened molecule is easier to break apart and the activation energy is lowered. Some catalysts do this better with some specific compounds than others, and extensive chemical research programs are devoted to finding new and more effective catalysts.

Automobile catalytic converters use metals such as platinum and transition metal oxides such as copper(II) oxide and chromium(III) oxide. Catalytic reactions that occur in the converter can reduce or oxidize about 90 percent of the hydrocarbons, 85 percent of the carbon monoxide, and 40 percent of the nitrogen monoxide from exhaust gases. Other controls, such as exhaust gas recirculation (EGR), are used to reduce further nitrogen monoxide formation.

A

B

BOX FIGURE 10.1 Catalytic converters speed up reactions in the exhaust gases from the automobile engine. This results in fewer pollutants being released into the air.

TABLE 10.1

Three interpretations of a chemical equation

Equation: $3\,H_2 + N_2 \rightarrow 2\,NH_3$

Molecular ratio:

$\quad$ 3 molecules H_2 + 1 molecule $N_2 \longrightarrow$ 2 molecules NH_3

Mole ratio:

$\quad$ 3 moles H_2 + 1 mole $N_2 \longrightarrow$ 2 moles NH_3

Mass ratio:

$\quad$ 6.0 g H_2 + 28.0 g $N_2 \longrightarrow$ 34.0 g NH_3

CONCEPTS *Applied*

Household Chemistry

Pick a household product that has a list of ingredients with names of covalent compounds or of ions you have met in this chapter. Write the brand name of the product and the type of product (example: Sani-Flush; toilet-bowl cleaner), then list the ingredients as given on the label, writing them one under the other (column 1). Beside each name put the formula, if you can figure out what it should be (column 2). Also, in a third column, put whatever you know or can guess about the function of that substance in the product (example: This is an acid; helps dissolve mineral deposits).

People Behind the Science

Emma Perry Carr (1880–1972)

Emma Perry Carr was a U.S. chemist, teacher, and researcher internationally renowned for her work in the field of spectroscopy.

Carr was born in Holmesville, Ohio, on July 23, 1880. After attending Coshocton High School, Ohio, she spent a year (1898–99) at Ohio State University and then transferred to Mount Holyoke College for a further two years' study. She continued as a chemistry assistant for three years and in 1905 finished her B.S. at the University of Chicago. Between 1905 and 1908, Carr worked as an instructor at Mount Holyoke, after which she returned to Chicago and completed her Ph.D. in physical chemistry in 1910. From 1910 to 1913, she was associate professor of chemistry at Mount Holyoke, after which she became professor of chemistry and head of the department, posts which she held until her retirement in 1946.

In 1913, Carr introduced a departmental research program to train students through collaborative research projects combining physical and organic chemistry. She achieved her renown in research into spectroscopy; under her leadership, she, her colleague Dorothy Hahn, and a team of students were among the first Americans to synthesize and analyze the structure of complex organic molecules using absorption spectroscopy. Her research into unsaturated hydrocarbons and far ultraviolet vacuum spectroscopy led to grants from the National Research Council and the Rockefeller Foundation in the 1930s. She served as a consultant on the spectra for the International Critical Tables, and during the 1920s and 1930s, she was three times a delegate to the International Union of Pure and Applied Chemistry. She was awarded four honorary degrees and was the first to receive the Garvan Medal, annually awarded to an American woman for achievement in chemistry.

Source: © Research Machines plc 2006. All Rights Reserved. Helicon Publishing is a division of Research Machines.

SUMMARY

A chemical formula is a shorthand way of describing the composition of a compound. An *empirical formula* identifies the simplest whole number ratio of atoms present in a compound. A *molecular formula* identifies the actual number of atoms in a molecule.

The sum of the atomic weights of all the atoms in any formula is called the *formula weight.* The *molecular weight* is the formula weight of a molecular substance. The formula weight of a compound can be used to determine the *mass percentage* of elements making up a compound.

A concise way to describe a chemical reaction is to use formulas in a *chemical equation.* A chemical equation with the same number of each kind of atom on both sides is called a *balanced equation.* A balanced equation is in accord with the *law of conservation of mass,* which states that atoms are neither created nor destroyed in a chemical reaction. To balance a chemical equation, *coefficients* are placed in front of chemical formulas. Subscripts of formulas may not be changed since this would change the formula, meaning a different compound.

One important group of chemical reactions is called *oxidation-reduction reactions,* or *redox reactions* for short. Redox reactions are reactions where shifts of electrons occur. The process of losing electrons is called *oxidation,* and the substance doing the losing is said to be *oxidized.* The process of gaining electrons is called *reduction,* and the substance doing the gaining is said to be *reduced.* Substances that take electrons from other substances are called *oxidizing agents.* Substances that supply electrons are called *reducing agents.*

Chemical reactions can also be classified as (1) *combination,* (2) *decomposition,* (3) *replacement,* or (4) *ion exchange.* The first three of these are redox reactions, but ion exchange is not.

A balanced chemical equation describes chemical reactions and has quantitative meaning about numbers of atoms, numbers of molecules, and conservation of atomic weights. The coefficients also describe the *volumes* of combining gases. At a constant temperature and pressure gases combine in small, whole number ratios that are given by the coefficients. Each volume at the same temperature and pressure contains the *same number of molecules.*

The number of atoms in exactly 12.00 g of C-12 is called *Avogadro's number,* which has a value of 6.02×10^{23}. Any substance that contains Avogadro's number of atoms, ions, molecules, or any chemical unit is called a *mole* of that substance. The mole is a measure of a number of atoms, molecules, or other chemical units. The mass of a mole of any substance is equal to the atomic mass of that substance.

The mass, number of atoms, and mole concepts are generalized to other units. The *gram-atomic weight* of an element is the mass in grams that is numerically equal to its atomic weight. The *gram-formula weight* of a compound is the mass in grams that is numerically equal to the formula weight of the compound. The *gram-molecular weight* is the gram-formula weight of a molecular compound. The relationships between the mole concept and the mass ratios can be used with a chemical equation for calculations about the quantities of reactants and products in a chemical reaction.

SUMMARY OF EQUATIONS

10.1

$$\frac{\left(\begin{array}{c}\text{atomic weight}\\ \text{of element}\end{array}\right)\left(\begin{array}{c}\text{number of atoms}\\ \text{of element}\end{array}\right)}{\text{formula weight of compound}} \times 100\% \text{ of} = \frac{\% \text{ of}}{\text{element}}$$

KEY TERMS

Avogadro's number (p. **265**)
chemical equation (p. **255**)
combination reaction (p. **260**)
decomposition reaction (p. **261**)
empirical formula (p. **252**)
formula weight (p. **253**)
gram-atomic weight (p. **265**)
gram-formula weight (p. **265**)
gram-molecular weight (p. **265**)
ion exchange reaction (p. **262**)
law of conservation of mass (p. **256**)
mole (p. **265**)
molecular formula (p. **252**)
molecular weight (p. **253**)
oxidation-reduction reaction (p. **260**)
oxidizing agents (p. **260**)
redox reaction (p. **260**)
reducing agent (p. **260**)
replacement reaction (p. **261**)

APPLYING THE CONCEPTS

1. An empirical formula does *not*
   a. identify elements in a compound.
   b. identify actual numbers of atoms in a molecule.
   c. provide the simplest whole number ratio of atoms of elements with subscripts.
   d. provide the formula for an ionic compound.

2. The relative arrangement of the atoms in a molecule is a (an)
   a. empirical formula.
   b. structural formula.
   c. molecular formula.
   d. covalent formula.

3. The sum of the atomic weights of all atoms in the ammonia molecule, NH_3, is called
   a. molecular weight.
   b. gravity weight.
   c. periodic weight.
   d. percent weight.

4. What is the meaning of the subscripts in a chemical formula?
   a. number of atoms of an element in a compound
   b. percent composition of that element in a compound
   c. number of molecules of that compound
   d. number of molecules in the reaction

5. "Atoms are neither created nor destroyed in a chemical reaction" is a statement of the
   a. law of definite proportions.
   b. law of conservation of mass.
   c. law of percent composition.
   d. atomic theory.

6. What products are formed from the complete reaction of hydrocarbons or carbohydrates with oxygen?
   a. $CO_2 + H_2O$
   b. $CO + H_2O$
   c. $CH_4 + CO_2$
   d. $CO_2 + H_2$

7. In a redox reaction, reduction is defined as the
   a. gain of electrons by an atom.
   b. loss of mass of an atom.
   c. loss of electrons by an atom.
   d. gain of mass of an atom.

8. In a redox reaction, oxidation is defined as the
   a. gain of electrons by an atom.
   b. loss of mass of an atom.
   c. loss of electrons by an atom.
   d. gain of mass of an atom.

9. To what chemical reaction class does this reaction belong? $CaCO_{3(s)} \longrightarrow CaO_{(s)} + CO_{2(g)}$
   a. combination
   b. ion exchange
   c. decomposition
   d. replacement

10. To what chemical reaction class does this reaction belong?
    $2\,Mg_{(s)} + O_{2(g)} \longrightarrow 2\,MgO_{(s)}$
    a. combination
    b. ion exchange
    c. decomposition
    d. replacement

11. Substances that take electrons from other substances are called
    a. oxidizing agents.
    b. reducing agents.
    c. ions.
    d. compounds.

12. Which is the more active metal?
    a. silver
    b. lead
    c. zinc
    d. sodium

13. A solid that comes out of solution as a result of an ion exchange reaction is called a
    a. precipitate.
    b. contaminant.
    c. base.
    d. metal.

14. A balanced chemical equation provides quantitative information about all of the following *except*
    a. time of reaction.
    b. atoms.
    c. molecules.
    d. atomic weights of reactants and products.

15. The number of atoms in exactly 12.00 g of C-12 is called
    a. atomic number.
    b. atomic weight.
    c. Avogadro's number.
    d. molecular number.

16. Any substance that contains Avogadro's number of particles is called a
    a. mole.
    b. atomic number.
    c. gopher.
    d. pound.

17. A material that speeds up a chemical reaction without being permanently changed by the reaction is a (an)
    a. catalyst.
    b. reactant.
    c. product.
    d. active metal.

18. The most active metals are found in Group
    a. IA.
    b. IIA.
    c. IB.
    d. IIB.

19. Potassium reacts with water to form potassium hydroxide and hydrogen gas. What kind of reaction is this?
    a. decomposition
    b. combination
    c. replacement
    d. ion exchange

20. The molecular weight of sulfuric acid, H_2SO_4, is
    a. 49 u.
    b. 50 u.
    c. 98 u.
    d. 194 u.

21. A balanced chemical equation has
    a. the same number of molecules on both sides of the equation.
    b. the same kinds of molecules on both sides of the equation.
    c. the same number of each kind of atom on both sides of the equation.
    d. All of the above are correct.

22. The law of conservation of mass means that
    a. atoms are not lost or destroyed in a chemical reaction.
    b. the mass of a newly formed compound cannot be changed.
    c. in burning, part of the mass must be converted into fire in order for mass to be conserved.
    d. molecules cannot be broken apart because this would result in less mass.

23. A chemical equation is balanced by changing
    a. subscripts.
    b. superscripts.
    c. coefficients.
    d. Any of the above is correct as necessary to achieve a balance.

24. Since wood is composed of carbohydrates, you should expect what gases to exhaust from a fireplace when complete combustion takes place?
    a. carbon dioxide, carbon monoxide, and pollutants
    b. carbon dioxide and water vapor
    c. carbon monoxide and smoke
    d. It depends on the type of wood being burned.

25. When carbon burns with an insufficient supply of oxygen, carbon monoxide is formed according to the following equation: $2\,C + O_2 \longrightarrow 2\,CO$. What category of chemical reaction is this?
    a. combination
    b. ion exchange
    c. replacement
    d. None of the above is correct because the reaction is incomplete.

26. According to the activity series for metals, adding metallic iron to a solution of aluminum chloride should result in
    a. a solution of iron chloride and metallic aluminum.
    b. a mixed solution of iron and aluminum chloride.
    c. the formation of iron hydroxide with hydrogen given off.
    d. no metal replacement reaction.

27. In a replacement reaction, elements that have the most ability to hold onto their electrons are
    a. the most chemically active.
    b. the least chemically active.
    c. not generally involved in replacement reactions.
    d. None of the above is correct.

28. Of the following elements, the one with the greatest electron-holding ability is
    a. sodium.
    b. zinc.
    c. copper.
    d. platinum.

29. Of the following elements, the one with the greatest chemical activity is
    a. aluminum.
    b. zinc.
    c. iron.
    d. mercury.

30. You know that an expected ion exchange reaction has taken place if the products include
    a. a precipitate.
    b. a gas.
    c. water.
    d. Any of the above is correct.

31. The incomplete equation of $2\,KClO_{3(s)} \longrightarrow$ probably represents which type of chemical reaction?
    a. combination
    b. decomposition
    c. replacement
    d. ion exchange

32. In the equation $2\,H_{2(g)} + O_{2(g)} \longrightarrow 2\,H_2O_{(l)}$
    a. the total mass of the gaseous reactants is less than the total mass of the liquid product.
    b. the total number of molecules in the reactants is equal to the total number of molecules in the products.
    c. one volume of oxygen combines with two volumes of hydrogen to produce two volumes of water.
    d. All of the above are correct.

33. If you have 6.02×10^{23} atoms of metallic iron, you will have how many grams of iron?
    a. 26
    b. 55.85
    c. 334.8
    d. 3.4×10^{25}

34. The molecular formula of benzene, C_6H_6, tells us that benzene is composed of six carbon atoms and six hydrogen atoms. What is the empirical formula of benzene?
   a. C_3H_3
   b. C_6H_6
   c. C_2H_2
   d. CH

35. What is the formula weight of acetic acid, CH_3COOH, which is found in vinegar?
   a. 29.0 u
   b. 48.0 u
   c. 58.0 u
   d. 60.0 u

36. What are the coefficients needed to balance this chemical equation?
   ___ $CaCl_2$ + ___ KOH → ___ $Ca(OH)_2$ + ___ KCl
   a. 1, 2, 1, 2
   b. 2, 1, 2, 1
   c. 1, 1, 1, 2
   d. 2, 1, 1, 1

37. What class of chemical reactions is not considered to belong to the class of oxidation-reduction reactions?
   a. combination
   b. decomposition
   c. ion exchange
   d. replacement

38. Which reaction is an example of a combination reaction?
   a. $2\ HgO_{(s)} → 2\ Hg_{(s)} + O_{2(g)}$
   b. $4\ Fe_{(s)} + 3\ O_{2(s)} → Fe_2O_{3(g)}$
   c. $CaCO_{3(s)} → CaO_{(s)} + CO_{2(g)}$
   d. $Na_{(s)} + KCl_{(aq)} → K_{(s)} + NaCl_{(aq)}$

39. To what chemical reaction class does this reaction belong?
   $Fe_{(s)} + CuSO_{4(aq)} → Cu_{(s)} + FeSO_{4(aq)}$
   a. ion exchange
   b. combination
   c. replacement
   d. decomposition

40. How many liters of propane gas are needed to produce 16 L of water vapor?
   $C_3H_{8(g)} + 5\ O_{2(g)} → 3\ CO_{2(g)} + 4\ H_2O_{(g)}$
   a. 2
   b. 3
   c. 4
   d. 16

41. How many sodium atoms are in two moles of sodium?
   a. 11
   b. 6.02×10^{23}
   c. 1.20×10^{24}
   d. 22

42. What is the gram-formula weight of sodium chloride?
   a. 23 u
   b. 23 g
   c. 58.5 g
   d. 58.5 u

43. A balanced chemical reaction provides all of the following information *except*
   a. molecular ratio of reactants and products.
   b. mole ratio of reactants and products.
   c. mass ratio of reactants and products.
   d. exchange rate of reactants and products.

44. The iron in a scouring pad will react with oxygen in the air to form iron oxide, Fe_2O_3, rust. The type of reaction that is described is
   a. replacement.
   b. combination.
   c. decomposition.
   d. ion exchange.

45. What is the formula weight of magnesium hydroxide, an ingredient found in antacids?
   a. 41.3 u
   b. 58.3 u
   c. 72.0 u
   d. 89.0 u

46. An ion exchange reaction can be identified by all of the following *except*
   a. formation of a precipitate.
   b. generation of a gas.
   c. formation of water.
   d. the required addition of heat.

47. What are the coefficients needed to balance this chemical equation?
   ___ $C_2H_{6(g)}$ + ___ $O_{2(g)}$ → ___ $CO_{2(g)}$ + ___ $H_2O_{(g)}$
   a. 1, 3, 2, 3
   b. 1, 5, 2, 3
   c. 2, 7, 4, 6
   d. 2, 5, 4, 3

48. What is the empirical formula of butane, C_4H_{10}?
   a. CH
   b. C_2H_5
   c. CH_3
   d. C_4H_{10}

49. How many moles of products are in this chemical equation?
   $CH_{4(g)} + 2\ O_{2(g)} → CO_{2(g)} + 2\ H_2O_{(g)}$
   a. 2
   b. 3
   c. 6
   d. 6.02×10^{23}

Answers

1. b 2. b 3. a 4. a 5. a 6. a 7. a 8. c 9. c 10. a 11. a 12. d 13. a 14. a 15. c 16. a 17. a 18. a 19. c 20. a 21. c 22. a 23. c 24. b 25. a 26. d 27. b 28. d 29. a 30. d 31. b 32. c 33. b 34. d 35. d 36. a 37. c 38. b 39. c 40. c 41. c 42. c 43. d 44. b 45. b 46. d 47. c 48. b 49. b

QUESTIONS FOR THOUGHT

1. How is an empirical formula like and unlike a molecular formula?

2. Describe the basic parts of a chemical equation. Identify how the physical state of elements and compounds is identified in an equation.

3. What is the law of conservation of mass? How do you know if a chemical equation is in accord with this law?

4. Describe in your own words how a chemical equation is balanced.

5. What is a hydrocarbon? What is a carbohydrate? In general, what are the products of complete combustion of hydrocarbons and carbohydrates?

6. Define and give an example in the form of a balanced equation of (a) a combination reaction, (b) a decomposition reaction, (c) a replacement reaction, and (d) an ion exchange reaction.

7. What must occur in order for an ion exchange reaction to take place? What is the result if this does not happen?

8. Predict the products for the following reactions: (a) The combustion of ethyl alcohol, C_2H_5OH, (b) the rusting of aluminum, and (c) the reaction between iron and sodium chloride.

9. The formula for butane is C_4H_{10}. Is this an empirical formula or a molecular formula? Explain the reason(s) for your answer.

10. How is the activity series for metals used to predict whether or not a replacement reaction will occur?

11. What is a gram-formula weight? How is it calculated?

12. What is the meaning and the value of Avogadro's number? What is a mole?

FOR FURTHER ANALYSIS

1. Analyze how you would know for sure that a pure substance you have is a compound and not an element.

2. What are the advantages and disadvantages of writing a chemical equation with chemical symbols and formulas rather than just words?

3. Provide several examples of each of the four basic categories of chemical reactions and describe how each illustrates a clear representation of the category.

4. Summarize for another person the steps needed to successfully write a balanced chemical equation.

INVITATION TO INQUIRY

Rate of Chemical Reactions

Temperature is one of the more important factors that influence the rate of a chemical reaction. You can use a "light stick" or "light tube" to study how temperature can influence a chemical reaction. Light sticks and tubes are devices that glow in the dark and have become very popular on July 4 and other times when people are outside after sunset. They work from a chemical reaction that is similar to the chemical reaction that produces light in a firefly. Design an experiment that uses light sticks to find out the effect of temperature on the brightness of light and how long the device will provide light. Perhaps you will be able to show by experimental evidence that use at a particular temperature produces the most light for the longest period of time.

PARALLEL EXERCISES

The exercises in groups A and B cover the same concepts. Solutions to group A exercises are located in appendix E.

Group A

1. Identify the following as empirical formulas or molecular formulas and indicate any uncertainty with (?):
   (a) $MgCl_2$
   (b) C_2H_2
   (c) BaF_2
   (d) C_8H_{18}
   (e) CH_4
   (f) S_8

2. What is the formula weight for each of the following compounds?
   (a) Copper(II) sulfate
   (b) Carbon disulfide
   (c) Calcium sulfate
   (d) Sodium carbonate

3. What is the mass percentage composition of the elements in the following compounds?
   (a) Fool's gold, FeS_2
   (b) Boric acid, H_3BO_3
   (c) Baking soda, $NaHCO_3$
   (d) Aspirin, $C_9H_8O_4$

4. Write balanced chemical equations for each of the following unbalanced reactions:
   (a) $SO_2 + O_2 \rightarrow SO_3$
   (b) $P + O_2 \rightarrow P_2O_5$
   (c) $Al + HCl \rightarrow AlCl_3 + H_2$
   (d) $NaOH + H_2SO_4 \rightarrow Na_2SO_4 + H_2O$
   (e) $Fe_2O_3 + CO \rightarrow Fe + CO_2$
   (f) $Mg(OH)_2 + H_3PO_4 \rightarrow Mg_3(PO_4)_2 + H_2O$

Group B

1. Identify the following as empirical formulas or molecular formulas and indicate any uncertainty with (?):
   (a) CH_2O
   (b) $C_6H_{12}O_6$
   (c) $NaCl$
   (d) CH_4
   (e) F_6
   (f) CaF_2

2. Calculate the formula weight for each of the following compounds:
   (a) Dinitrogen monoxide
   (b) Lead(II) sulfide
   (c) Magnesium sulfate
   (d) Mercury(II) chloride

3. What is the mass percentage composition of the elements in the following compounds?
   (a) Potash, K_2CO_3
   (b) Gypsum, $CaSO_4$
   (c) Saltpeter, KNO_3
   (d) Caffeine, $C_8H_{10}N_4O_2$

4. Write balanced chemical equations for each of the following unbalanced reactions:
   (a) $NO + O_2 \rightarrow NO_2$
   (b) $KClO_3 \rightarrow KCl + O_2$
   (c) $NH_4Cl + Ca(OH)_2 \rightarrow CaCl_2 + NH_3 + H_2O$
   (d) $NaNO_3 + H_2SO_4 \rightarrow Na_2SO_4 + HNO_3$
   (e) $PbS + H_2O_2 \rightarrow PbSO_4 + H_2O$
   (f) $Al_2(SO_4)_3 + BaCl_2 \rightarrow AlCl_3 + BaSO_4$

5. Identify the following as combination, decomposition, replacement, or ion exchange reactions:
   (a) $NaCl_{(aq)} + AgNO_{3(aq)} \rightarrow NaNO_{3(aq)} + AgCl\downarrow$
   (b) $H_2O_{(l)} + CO_{2(g)} \rightarrow H_2CO_{3(l)}$
   (c) $2 NaHCO_{3(s)} \rightarrow Na_2CO_{3(s)} + H_2O_{(g)} + CO_{2(g)}$
   (d) $2 Na_{(s)} + Cl_{2(g)} \rightarrow 2 NaCl_{(s)}$
   (e) $Cu_{(s)} + 2 AgNO_{3(aq)} \rightarrow Cu(NO_3)_{2(aq)} + 2 Ag_{(s)}$
   (f) $CaO_{(s)} + H_2O_{(l)} \rightarrow Ca(OH)_{2(aq)}$

6. Write complete, balanced equations for each of the following reactions:
   (a) $C_5H_{12(g)} + O_{2(g)} \rightarrow$
   (b) $HCl_{(aq)} + NaOH_{(aq)} \rightarrow$
   (c) $Al_{(s)} + Fe_2O_{3(s)} \rightarrow$
   (d) $Fe_{(s)} + CuSO_{4(aq)} \rightarrow$
   (e) $MgCl_{(aq)} + Fe(NO_3)_{2(aq)} \rightarrow$
   (f) $C_6H_{10}O_{5(s)} + O_{2(g)} \rightarrow$

7. Write complete, balanced equations for each of the following decomposition reactions. Include symbols for physical states, heating, and others as needed:
   (a) Solid potassium chloride and oxygen gas are formed when solid potassium chlorate is heated.
   (b) Upon electrolysis, molten bauxite (aluminum oxide) yields solid aluminum metal and oxygen gas.
   (c) Upon heating, solid calcium carbonate yields solid calcium oxide and carbon dioxide gas.

8. Write complete, balanced equations for each of the following replacement reactions. If no reaction is predicted, write "no reaction" as the product:
   (a) $Na_{(s)} + H_2O_{(l)} \rightarrow$
   (b) $Au_{(s)} + HCl_{(aq)} \rightarrow$
   (c) $Al_{(s)} + FeCl_{2(aq)} \rightarrow$
   (d) $Zn_{(s)} + CuCl_{2(aq)} \rightarrow$

9. Write complete, balanced equations for each of the following ion exchange reactions. If no reaction is predicted, write "no reaction" as the product:
   (a) $NaOH_{(aq)} + HNO_{3(aq)} \rightarrow$
   (b) $CaCl_{2(aq)} + KNO_{3(aq)} \rightarrow$
   (c) $Ba(NO_3)_{2(aq)} + Na_3PO_{4(aq)} \rightarrow$
   (d) $KOH_{(aq)} + ZnSO_{4(aq)} \rightarrow$

10. The gas welding torch is fueled by two tanks, one containing acetylene (C_2H_2) and the other pure oxygen (O_2). The very hot flame of the torch is produced as acetylene burns,
$$2 C_2H_2 + 5 O_2 \rightarrow 4 CO_2 + 2 H_2O$$
According to this equation, how many liters of oxygen are required to burn 1 L of acetylene?

5. Identify the following as combination, decomposition, replacement, or ion exchange reactions:
   (a) $ZnCO_{3(s)} \rightarrow ZnO_{(s)} + CO_2\uparrow$
   (b) $2 NaBr_{(aq)} + Cl_{2(g)} \rightarrow 2 NaCl_{(aq)} + Br_{2(g)}$
   (c) $2 Al_{(s)} + 3 Cl_{2(g)} \rightarrow 2 AlCl_{3(s)}$
   (d) $Ca(OH)_{2(aq)} + H_2SO_{4(aq)} \rightarrow CaSO_{4(aq)} + 2 H_2O_{(l)}$
   (e) $Pb(NO_3)_{2(aq)} + H_2S_{(g)} \rightarrow 2 HNO_{3(aq)} + PbS\downarrow$
   (f) $C_{(s)} + ZnO_{(s)} \rightarrow Zn_{(s)} + CO\uparrow$

6. Write complete, balanced equations for each of the following reactions:
   (a) $C_3H_{6(g)} + O_{2(g)} \rightarrow$
   (b) $H_2SO_{4(aq)} + KOH_{(aq)} \rightarrow$
   (c) $C_6H_{12}O_{6(s)} + O_{2(g)} \rightarrow$
   (d) $Na_3PO_{4(aq)} + AgNO_{3(aq)} \rightarrow$
   (e) $NaOH_{(aq)} + Al(NO_3)_{3(aq)} \rightarrow$
   (f) $Mg(OH)_{2(aq)} + H_3PO_{4(aq)} \rightarrow$

7. Write complete, balanced equations for each of the following decomposition reactions. Include symbols for physical states, heating, and others as needed:
   (a) When solid zinc carbonate is heated, solid zinc oxide and carbon dioxide gas are formed.
   (b) Liquid hydrogen peroxide decomposes to liquid water and oxygen gas.
   (c) Solid ammonium nitrite decomposes to liquid water and nitrogen gas.

8. Write complete, balanced equations for each of the following replacement reactions. If no reaction is predicted, write "no reaction" as the product:
   (a) $Zn_{(s)} + FeCl_{2(aq)} \rightarrow$
   (b) $Zn_{(s)} + AlCl_{3(aq)} \rightarrow$
   (c) $Cu_{(s)} + HgCl_{2(aq)} \rightarrow$
   (d) $Al_{(s)} + HCl_{(aq)} \rightarrow$

9. Write complete, balanced equations for each of the following ion exchange reactions. If no reaction is predicted, write "no reaction" as the product:
   (a) $Ca(OH)_{2(aq)} + H_2SO_{4(aq)} \rightarrow$
   (b) $NaCl_{(aq)} + AgNO_{3(aq)} \rightarrow$
   (c) $NH_4NO_{3(aq)} + Mg_3(PO_4)_{2(aq)} \rightarrow$
   (d) $Na_3PO_{4(aq)} + AgNO_{3(aq)} \rightarrow$

10. Iron(III) oxide, or hematite, is one mineral used as an iron ore. Other iron ores are magnetite (Fe_3O_4) and siderite ($FeCO_3$). Assume that you have pure samples of all three ores that will be reduced by reaction with carbon monoxide. Which of the three ores will have the highest yield of metallic iron?

11

Water and Solutions

Water is often referred to as the *universal solvent* because it makes so many different kinds of solutions. Eventually, moving water can dissolve solid rock, carrying it away in solution.

CORE **CONCEPT**

Water and solutions of water have unique properties.

OUTLINE

Properties of Water
Water is a universal solvent and has a high specific heat and a high latent heat of evaporation.

The Dissolving Process
Dissolving is the process of making a solution—a homogeneous mixture of ions or molecules of two or more substances.

Household Water
Properties of Water
Science and Society: Who Has the Right?
   Structure of Water Molecules
   The Dissolving Process
   Concentration of Solutions
   Solubility
A Closer Look: Decompression Sickness
Properties of Water Solutions
   Electrolytes
   Boiling Point
   Freezing Point
Acids, Bases, and Salts
   Properties of Acids and Bases
   Explaining Acid-Base Properties
   Strong and Weak Acids and Bases
   The pH Scale
   Properties of Salts
   Hard and Soft Water
A Closer Look: Acid Rain
People Behind the Science: Johannes Nicolaus Brönsted

Structure of Water Molecules
A water molecule is polar and able to establish hydrogen bonding.

Acids, Bases, and Salts
Solutions of acids, bases, and salts are evident in environmental quality, food, and everyday living.

What do you think about when you see a stream (Figure 11.1)? Do you wonder about the water quality and what might be dissolved in the water? Do you wonder where the stream comes from and if it will ever run out of water?

Many people can look at a stream, but they might think about different things. A farmer might think about how the water could be diverted and used for his crops. A city planner might wonder if the water is safe for domestic use, and if not, what it would cost to treat the water. Others might wonder if the stream has large fish they could catch. Many large streams can provide water for crops, domestic use, and recreation, and still meet the requirements for a number of other uses.

It is the specific properties of water that make it important for agriculture, domestic use, and recreation. Living things evolved in a watery environment, so water and its properties are essential to life on Earth. Some properties of water, such as the ability to dissolve almost anything, also make water very easy to pollute. This chapter is concerned with some of the unique properties of water, water solutions, and the household use of water.

HOUSEHOLD WATER

Water is an essential resource, not only because it is required for life processes but also because of its role in a modern society. Water is used in the home for drinking and cooking (2 percent), cleaning dishes (6 percent), laundry (11 percent), bathing (23 percent), toilets (29 percent), and for maintaining lawns and gardens (29 percent).

The water supply is obtained from streams, lakes, and reservoirs on the surface or from groundwater pumped from below the surface. Surface water contains more sediments, bacteria, and possible pollutants than water from a well because it is exposed to the atmosphere and water runs off the land into streams and rivers. Surface water requires filtering to remove suspended particles, treatment to kill bacteria, and sometimes processing to remove pollution. Well water is generally cleaner but still might require treatment to kill bacteria and remove pollution that has seeped through the ground from waste dumps, agricultural activities, or industrial sites.

Most pollutants are usually too dilute to be considered a significant health hazard, but there are exceptions. There are five types of contamination found in U.S. drinking water that are responsible for the most widespread danger, and these are listed in Table 11.1. In spite of these general concerns and other occasional local problems, the U.S. water supply is considered to be among the cleanest in the world.

The demand for domestic water sometimes exceeds the immediate supply in some metropolitan areas. This is most common during the summer, when water demand is high and rainfall is often low. Communities in these areas often have public education campaigns designed to help reduce the demand for water. For example, did you know that taking a tub bath can use up to 135 liters (about 36 gal) of water compared to only 95 liters (about 25 gal) for a regular shower? Even more water is saved by a shower that does not run continuously—wetting down, soaping up, and rinsing off uses only 15 liters (about 4 gal) of water. You can also save about 35 liters (about 9 gal) of

water by not letting the water run continuously while brushing your teeth.

It is often difficult to convince people to conserve water when it is viewed as an inexpensive, limitless resource. However, efforts to conserve water increase dramatically as the cost to the household consumer increases.

The issues involved in maintaining a safe water supply are better understood by considering some of the properties of water and water solutions. These are the topics of the following sections.

PROPERTIES OF WATER

Water is essential for life since living organisms are made up of cells filled with water and a great variety of dissolved substances. Foods are mostly water, with fruits and vegetables containing up to 95 percent water and meat consisting of about 50 percent water. Your body is over 70 percent water by weight. Since water is such a large component of living things, understanding the properties of water is important to understanding life. One important property is water's unusual ability to act as a solvent. Water is called a "universal solvent" because of its ability to dissolve most molecules. In living things, these dissolved molecules can be transported from one place to another by diffusion or by some kind of a circulatory system.

The usefulness of water does not end with its unique abilities as a solvent and transporter; it has many more properties that are useful, although unusual. For example, unlike other liquids, water in its liquid phase has a greater density than solid water (ice). This important property enables solid ice to float on the surface of liquid water, insulating the water below and permitting fish and other water organisms to survive the winter. If ice were denser than water, it would sink, freezing all lakes and rivers from the bottom up. Fish and most organisms that live in water would not be able to survive in a lake or river of solid ice.

FIGURE 11.1 A freshwater stream has many potential uses.

TABLE 11.1

Possible pollution problems in the U.S. water supply

Pollutant	Source	Risk
Lead	Lead pipes in older homes; solder in copper pipes; brass fixtures	Nerve damage, miscarriage, birth defects, high blood pressure, hearing problems
Chlorinated solvents	Industrial pollution	Cancer
Trihalomethanes	Chlorine disinfectant reacting with other pollutants	Liver damage, kidney damage, possible cancer
PCBs	Industrial waste, older transformers	Liver damage, possible cancer
Bacteria and viruses	Septic tanks, outhouses, overflowing sewer lines	Gastrointestinal problems, serious disease

unique solvent abilities, why solid water is less dense than liquid water, its high specific heat, its high latent heat of vaporization, and perhaps why no two snowflakes seem to be alike.

STRUCTURE OF WATER MOLECULES

In chapter 9, you learned that atoms combine in two ways. Atoms from opposite sides of the periodic table form ionic bonds after transferring one or more electrons. Atoms from the right side of the periodic table form covalent bonds by sharing one or more pairs of electrons. This distinction is clear-cut in many compounds but not in water. The way atoms share electrons in a water molecule is not exactly covalent, but it is not ionic, either. As you learned in chapter 9, the bond that is not exactly covalent or ionic is called a *polar covalent bond.*

In a water molecule, an oxygen atom shares a pair of electrons with each of two hydrogen atoms with polar covalent bonds. Oxygen has six outer electrons and needs two more to satisfy the octet rule, achieving the noble gas structure of eight. Each hydrogen atom needs one more electron to fill its outer orbital with two. Therefore, one oxygen atom bonds with two hydrogen atoms, forming H_2O. Both oxygen and hydrogen are more stable with the outer orbital configuration of the noble gases (neon and helium in this case).

Electrons are shared in a water molecule but not equally. Oxygen, with its eight positive protons, has a greater attraction for the shared electrons than do either of the hydrogens with a single proton. Therefore, the shared electrons spend more time around the oxygen part of the molecule than they do around the hydrogen part. This results in the oxygen end of the molecule being more negative than the hydrogen end. When electrons in a covalent bond are not equally shared, the molecule is said to be polar. A **polar molecule** has a *dipole* (*di* = two; *pole* = side or end), meaning it has a positive end and a negative end.

As described in chapter 4, water is also unusual because it has a high specific heat. The same amount of sunlight falling on equal masses of soil and water will warm the soil 5°C for each 1°C increase in water temperature. Thus, it will take five times more sunlight to increase the temperature of the water as much as the soil temperature change. This enables large bodies of water to moderate the atmospheric temperature, making it more even.

A high latent heat of vaporization is yet another unusual property of water. This property enables people to dissipate large amounts of heat by evaporating a small amount of water. Since people carry this evaporative cooling system with them, they can survive some very warm desert temperatures, for example.

Finally, other properties of water are not crucial for life but are interesting nonetheless. For example, why do all snowflakes have six sides? Is it true that no two snowflakes are alike? The unique structure of the water molecule will explain water's

Science and Society

Who Has the Right?

As the population grows and new industries develop, more and more demands are placed on the water supply. This raises some issues about how water should be divided among agriculture, industries, and urban domestic use. Agricultural interests claim they should have the water because they produce the food and fibers that people must have. Industrial interests claim they should have the water because they create jobs and products that people must have. Cities, on the other hand, claim that domestic consumption is the most important because people cannot survive without water. Yet others claim that no group has a right to use water when it is needed to maintain habitats.

QUESTIONS TO DISCUSS

1. Who should have the first priority for water use?

2. Who should have the last priority for water use?

3. What determined your answers to questions 1 and 2?

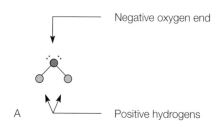

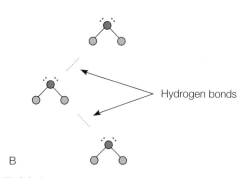

FIGURE 11.2 (*A*) The water molecule is polar, with centers of positive and negative charges. (*B*) Attractions between these positive and negative centers establish hydrogen bonds between adjacent molecules.

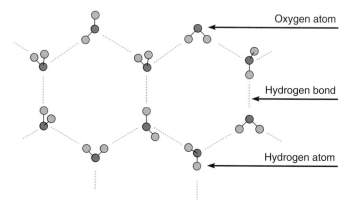

FIGURE 11.3 The hexagonal structure of ice. Hydrogen bonding between the oxygen atom and two hydrogen atoms of other water molecules results in a tetrahedral arrangement, which forms the open, hexagonal structure of ice. Note the angles of the water molecules do not change but have different orientations.

A water molecule has a negative center at the oxygen end and a positive center at the hydrogen end. The positive charges on the hydrogen end are separated, giving the molecule a bent rather than straight-line arrangement. Figure 11.2A shows a model of a water molecule showing its polar nature.

It is the polar structure of the water molecule that is responsible for many of the unique properties of water. Polar molecules of any substances have attractions between the positive end of a molecule and the negative end of another molecule. When the polar molecule has hydrogen at one end and fluorine, oxygen, or nitrogen on the other, the attractions are strong enough to make a type of bonding called **hydrogen bonding.** Hydrogen bonding is a strong bond that occurs between the hydrogen end of a molecule and the fluorine, oxygen, or nitrogen end of similar molecules. A better name for this would be a hydrogen-fluorine bond, a hydrogen-oxygen bond, or a hydrogen-nitrogen bond. However, for brevity the second part of the bond is not named and all the hydrogen-something bonds are simply known as "hydrogen" bonds. The dotted line between the hydrogen and oxygen molecules in Figure 11.2B represents a hydrogen bond. A dotted line is used to represent a bond that is not as strong as the bond represented by the solid line of a covalent compound.

Hydrogen bonding accounts for the physical properties of water, including its unusual density changes with changes in temperature. Figure 11.3 shows the hydrogen-bonded structure of ice. Water molecules form a six-sided hexagonal structure that extends out for billions of molecules. The large channels, or holes, in the structure result in ice being less dense than water. The shape of the hexagonal arrangement also suggests why snowflakes always have six sides. Why does it seem like no two snowflakes are alike? Perhaps the answer can be found in the almost infinite variety of shapes that can be built from billions and billions of tiny hexagons of ice crystals.

When ice is warmed, the increased vibrations of the molecules begin to expand and stretch the hydrogen bond structure. When ice melts, about 15 percent of the hydrogen bonds break and the open structure collapses into the more compact

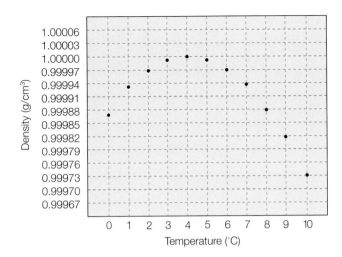

FIGURE 11.4 The density of water from 0°C to 10°C. The density of water is at a maximum at 4°C, becoming less dense as it is cooled or warmed from this temperature. Hydrogen bonding explains this unusual behavior.

arrangement of liquid water. As the liquid water is warmed from 0°C still more, hydrogen bonds break down and the density of the water steadily increases. At 4°C, the expansion of water from the increased molecular vibrations begins to predominate, and the density decreases steadily with further warming (Figure 11.4). Thus, water has its greatest density at a temperature of 4°C.

The heat of fusion, specific heat, and heat of vaporization of water are unusually high when compared to other, chemically similar substances. These high values are accounted for by the additional energy needed to break hydrogen bonds.

THE DISSOLVING PROCESS

A **solution** is a homogeneous mixture of ions or molecules of two or more substances. *Dissolving* is the process of making a solution. During dissolving, the different components that make up the solution become mixed. For example, when sugar dissolves in water, the molecules of sugar become uniformly dispersed throughout the molecules of water. The uniform taste of sweetness of any part of the sugar solution is a result of this uniform mixing.

The general terms *solvent* and *solute* identify the components of a solution. The solvent is the component present in the larger amount. The solute is the component that dissolves in the solvent. Atmospheric air, for example, is about 78 percent nitrogen, so nitrogen is considered the solvent. Oxygen (about 21 percent), argon (about 0.9 percent), and other gases make up the solutes. If one of the components of a solution is a liquid, it is usually identified as the solvent. An *aqueous solution* is a solution of a solid, a liquid, or a gas in water.

A solution is formed when the molecules or ions of two or more substances become homogeneously mixed. But the process of dissolving must be more complicated than the simple mixing together of particles because (1) solutions become saturated, meaning there is a limit on solubility, and (2) some substances

are *insoluble,* not dissolving at all or at least not noticeably. In general, the forces of attraction between molecules or ions of the solvent and solute determine if something will dissolve and if there are any limits on the solubility. These forces of attraction and their role in the dissolving process will be considered in the following examples.

First, consider the dissolving process in gaseous and liquid solutions. In a gas, the intermolecular forces are small, so gases can mix in any proportion. Fluids that can mix in any proportion without separating into phases are called **miscible fluids.** Fluids that do not mix are called *immiscible fluids.* Air is a mixture of gases, so gases (including vapors) are miscible.

Liquid solutions can dissolve a gas, another liquid, or a solid. Gases are miscible in liquids, and a carbonated beverage (your favorite cola) is the common example, consisting of carbon dioxide dissolved in water. Whether or not two given liquids form solutions depends on some similarities in their molecular structures. The water molecule, for example, is a polar molecule with a negative end and a positive end. On the other hand, carbon tetrachloride (CCl_4) is a molecule with covalent bonds that are symmetrically arranged. Because of the symmetry, CCl_4 has no negative or positive ends, so it is nonpolar. Thus, some liquids have polar molecules and some have nonpolar molecules. The general rule for forming solutions is *like dissolves like.* A nonpolar compound, such as carbon tetrachloride, will dissolve oils and greases because they are nonpolar compounds. Water, a polar compound, will not dissolve the nonpolar oils and greases. Carbon tetrachloride was at one time used as a cleaning solvent because of its oil and grease dissolving abilities. Its use is no longer recommended because it causes liver damage.

Some molecules, such as soap, have a part of the molecule that is polar and a part that is nonpolar. Washing with water alone will not dissolve oils because water and oil are immiscible. When soap is added to the water, however, the polar end of the soap molecule is attracted to the polar water molecules, and the nonpolar end is absorbed into the oil. A particle (larger than a molecule) is formed, and the oil is washed away with the water (Figure 11.5).

CONCEPTS *Applied*

How to Mix

Obtain a small, clear water bottle with a screw-on cap. Fill the bottle halfway with water, then add some food coloring and swirl to mix. Now add enough mineral or cooking oil to almost fill the bottle. Seal the bottle tightly with the cap.

Describe the oil and water in the bottle. Shake the bottle vigorously for about 30 seconds, then observe what happens when you stop shaking. Does any of the oil and water mix?

Try mixing the oil and water again, this time after adding a squirt of liquid dishwashing soap. Describe what happens before and after adding the soap. What does this tell you about the structure of the oil and water molecules? How did soap overcome these differences?

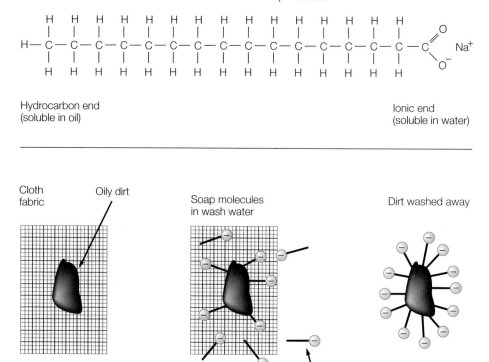

Structural formula of a soap molecule

Hydrocarbon end
(soluble in oil)

Ionic end
(soluble in water)

Cloth fabric

Oily dirt

Soap molecules in wash water

Dirt washed away

A soap molecule

FIGURE 11.5 Soap cleans oil and grease because one end of the soap molecule is soluble in water and the other end is soluble in oil and grease. Thus, the soap molecule provides a link between two substances that would otherwise be immiscible.

The "like dissolves like" rule applies to solids and liquid solvents as well as liquids and liquid solvents. Polar solids, such as salt, will readily dissolve in water, which has polar molecules, but do not dissolve readily in oil, grease, or other nonpolar solvents. Polar water readily dissolves salt because the charged polar water molecules are able to exert an attraction on the ions, pulling them away from the crystal structure. Thus, ionic compounds dissolve in water.

Ionic compounds vary in their solubilities in water. This difference is explained by the existence of two different forces involved in an ongoing "tug of war." One force is the attraction between an ion on the surface of the crystal and a water molecule, an *ion-polar molecule force*. When solid sodium chloride and water are mixed together, the negative ends of the water molecules (the oxygen ends) become oriented toward the positive sodium ions on the crystal. Likewise, the positive ends of water molecules (the hydrogen ends) become oriented toward the negative chlorine ions. The attraction of water molecules for ions is called *hydration*. If the force of hydration is greater than the attraction between the ions in the solid, they are pulled away from the solid, and dissolving occurs (Figure 11.6). Considering sodium chloride only, the equation is

$$Na^+Cl^-_{(s)} \longrightarrow Na^+_{(aq)} + Cl^-_{(aq)}$$

which shows that the ions were separated from the solid to become a solution of ions. In other compounds, the attraction between the ions in the solid might be greater than the energy of hydration. In this case, the ions of the solid would win the "tug-of-war," and the ionic solid is insoluble.

The saturation of soluble compounds is explained in terms of hydration eventually occupying a large number of the polar water molecules. Fewer available water molecules means less attraction on the ionic solid, with more solute ions being pulled back to the surface of the solid. The tug-of-war continues back and forth as an equilibrium condition is established.

CONCENTRATION OF SOLUTIONS

The relative amounts of solute and solvent are described by the **concentration** of a solution. In general, a solution with a large amount of solute is *concentrated,* and a solution with much less solute is *dilute.* The terms *dilute* and *concentrated* are somewhat arbitrary, and it is sometimes difficult to know the difference between a solution that is "weakly concentrated" and one that is "not very diluted." More meaningful information is provided by measurement of the *amount of solute in a solution.* There are different ways to express concentration measurements, each lending itself to a particular kind of solution or to how the information will be used. For example, you read about concentrations of parts per million in an article about pollution, but most of the concentrations of solutions sold in stores are reported in percent by volume or percent by weight (Figure 11.7). Each of these concentrations is concerned with the amount of *solute* in the *solution.*

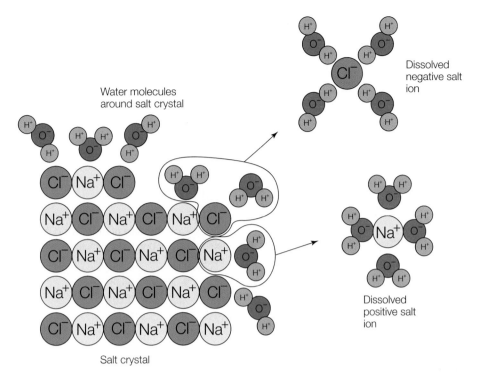

Water molecules
around salt crystal

Dissolved
negative salt
ion

Dissolved
positive salt
ion

Salt crystal

FIGURE 11.6 An ionic solid dissolves in water because the number of water molecules around the surface is greater than the number of other ions of the solid. The attraction between polar water molecules and a charged ion enables the water molecules to pull ions away from the crystal, so the salt crystals dissolve in the water.

Concentration ratios that describe small concentrations of solute are sometimes reported as a ratio of *parts per million* (ppm) or *parts per billion* (ppb). This ratio could mean ppm by volume or ppm by weight, depending on whether the solution is a gas or a liquid. For example, a drinking water sample with 1 ppm Na^+ by weight has 1 weight measure of solute, sodium ions, *in* every 1,000,000 weight measures of the total solution. By way of analogy, 1 ppm expressed in money means 1 cent in every $10,000 (which is 1 million cents). A concentration of 1 ppb means 1 cent in $10,000,000. Thus, the concentrations of very dilute solutions, such as certain salts in seawater, minerals in drinking water, and pollutants in water or in the atmosphere are often reported in ppm or ppb.

 CONCEPTS *Applied*

ppm or ppb to Percent

Sometimes it is useful to know the conversion factors between ppm or ppb and the more familiar percent concentration by weight. These factors are ppm ÷ (1 × 10^4) = percent concentration and ppb ÷ (1 × 10^7) = percent concentration. For example, very hard water (water containing Ca^{2+} or Mg^{2+} ions), by definition, contains more than 300 ppm of the ions. This is a percent concentration of 300 ÷ 1 × 10^4, or 0.03 percent. To be suitable for agricultural purposes, irrigation water must not contain more than 700 ppm of total dissolved salts, which means a concentration no greater than 0.07 percent salts.

The concentration term of *percent by volume* is defined as the *volume of solute in 100 volumes of solution*. This concentration term is just like any other percentage ratio, that is, "part" divided by the "whole" times 100 percent. The distinction is that the part and the whole are concerned with a volume of solute and a volume of solution. Knowing the meaning of percent by volume can be useful in consumer decisions. Rubbing alcohol, for example, can be purchased at a wide range of prices. The various brands range from a concentration, according to the labels, of "12% by volume" to "70% by volume." If the volume unit is mL, a "12% by volume" concentration contains 12 mL of pure isopropyl (rubbing) alcohol in every 100 mL of solution. The "70% by volume" contains 70 mL of isopropyl alcohol in every 100 mL of solution. The relationship for % by volume is

$$\frac{\text{volume solute}}{\text{volume solution}} \times 100\% \text{ solution} = \% \text{ solute}$$

or

$$\frac{V_{\text{solute}}}{V_{\text{solution}}} \times 100\% \text{ solution} = \% \text{ solute}$$

equation 11.1

The concentration term of *percent by weight* is defined as the *weight of solute in 100 weight units of solution*. This concentration term is just like any other percentage composition, the difference being that it is concerned with the weight of solute (the part) in a weight of solution (the whole). Hydrogen peroxide, for example, is usually sold in a concentration of "3% by weight." This means that 3 oz (or other weight units) of pure

A Solution strength by parts

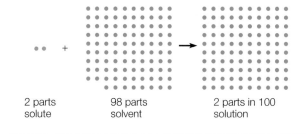

| 2 parts solute | 98 parts solvent | 2 parts in 100 solution |

B Solution strength by percent (volume)

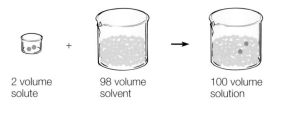

| 2 volume solute | 98 volume solvent | 100 volume solution |

C Solution strength by percent (weight)

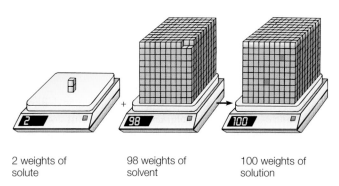

| 2 weights of solute | 98 weights of solvent | 100 weights of solution |

FIGURE 11.7 Three ways to express the amount of solute in a solution: (*A*) as parts (e.g., parts per million), this is 2 parts per 100; (*B*) as a percent by volume, this is 2 percent by volume; (*C*) as percent by weight, this is 2 percent by weight.

hydrogen peroxide are in 100 oz of solution. Since weight is proportional to mass in a given location, mass units such as grams are sometimes used to calculate a percent by weight. The relationship for percent by weight (using mass units) is

$$\frac{\text{mass of solute}}{\text{mass of solution}} \times 100\% \text{ solution} = \% \text{ solute}$$

or

$$\frac{m_{\text{solute}}}{m_{\text{solution}}} \times 100\% \text{ solution} = \% \text{ solute}$$

equation 11.2

Both percent by volume and percent by weight are defined as the volume or weight per 100 units of solution because percent *means* parts per hundred. The measure of dissolved salts in seawater is called *salinity*. **Salinity** is defined as the mass of salts dissolved in 1,000 g of solution. As illustrated in Figure 11.8, evaporation of 965 g of water from 1,000 g of seawater will leave an average of 35 g salts. Thus, the average salinity of the sea-

water is 35‰. Note the ‰, which means parts per thousand just as % means parts per hundred. The equivalent percent measure for salinity is 3.5%, which equals 35‰.

EXAMPLE 11.1

Vinegar that is prepared for table use is a mixture of acetic acid in water, usually 5.00% by weight. How many grams of pure acetic acid are in 25.0 g of vinegar?

SOLUTION

The percent by weight is given (5.00%), the mass of the solution is given (25.0 g), and the mass of the solute (CH_3COOH) is the unknown. The relationship between these quantities is found in equation 11.2, which can be solved for the mass of the solute:

$$\% \text{ solute} = 5.00\%$$
$$m_{\text{solution}} = 25.0 \text{ g}$$
$$m_{\text{solute}} = ?$$

$$\frac{m_{\text{solute}}}{m_{\text{solution}}} \times 100\% \text{ solution} = \% \text{ solute}$$

$$\therefore$$

$$m_{\text{solute}} = \frac{(m_{\text{solution}})(\% \text{ solute})}{100\% \text{ solution}}$$

$$= \frac{(m_{\text{solution}})(\% \text{ solute})}{100\% \text{ solution}}$$

$$= \frac{(25.0 \text{ g})(5.00)}{100} \text{ solute}$$

$$= \boxed{1.25 \text{ g solute}}$$

EXAMPLE 11.2

A solution used to clean contact lenses contains 0.002% by volume of thimerosal as a preservative. How many L of this preservative are needed to make 100,000 L of the cleaning solution? (Answer: 2L)

Recall from chapter 10 that a *mole* is a measure of amount used in chemistry. One mole is defined as the amount of a substance that contains the same number of elementary units as there are atoms in exactly 12 grams of the carbon-12 isotope. The number of units in this case is called *Avogadro's number*, which is 6.02×10^{23}—a very large number. This measure can be

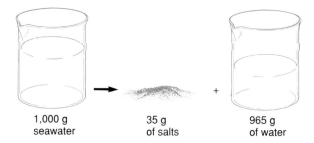

| 1,000 g seawater | 35 g of salts | 965 g of water |

FIGURE 11.8 Salinity is a measure of the amount of salts dissolved in 1 kg of solution. If 1,000 g of seawater were evaporated, 35.0 g of salts would remain as 965.0 g of water leave.

Decompression sickness (DCS) is a condition caused by the formation of nitrogen bubbles in the blood and tissues of a scuba diver who surfaces too quickly, causing a rapid drop in pressure. This condition is usually marked by joint pain but can include chest pain, skin irritation, and muscle cramps. These symptoms may be barely noticed in mild cases, but severe cases can be fatal. The joint pain is often called *bends*.

DCS can occur in any situation in which a person is subjected to a higher air pressure for some period of time and then experiences a rapid decompression. Here is what happens: Air is about 79 percent nitrogen and 21 percent oxygen. Nitrogen is inert to the human body, and what we inhale is exhaled, but some becomes dissolved in the blood. Dissolved nitrogen is a normal occurrence and does not present a problem. However, when a person is breathing air under higher pressure, more nitrogen is breathed in and more becomes dissolved in the blood and other tissues. If the person returns to normal pressure slowly, the extra dissolved nitrogen is expelled by the lungs and there is no problem. If the return to normal pressure is too rapid, however, nitrogen bubbles form in the blood and other tissues, and this causes DCS. The nitrogen bubbles can cause pressure on nerves, block circulation, and cause joint pain. These symptoms usually appear when the diver returns to the surface or within eight hours of returning to normal air pressure.

One way to prevent DCS is for divers to make "decompression stops" as they return to the surface. These stops allow the dissolved nitrogen to diffuse into the lungs rather than making bubbles in tissues and the bloodstream. Also, it is not a good idea for a diver to fly on a commercial airliner for a day or so after diving. Commercial airliners are pressurized to an altitude of about 2,500 m (about 8,200 ft). This is safe for the normal passenger but can cause more decompression problems for the recent diver.

DCS can be treated by placing the affected person in a specially sealed chamber. Pressure is slowly increased in the chamber to cause nitrogen bubbles to go back into solution, then slowly decreased to allow the dissolved nitrogen to be expelled by the lungs.

---

compared with identifying amounts in the grocery store by the dozen. You know that a dozen is twelve of something. Now you know that a mole is 6.02×10^{23} of whatever you are measuring.

Chemists use a measure of concentration that is convenient for considering chemical reactions of solutions. The measure is based on moles of solute since a mole is a known number of particles (atoms, molecules, or ions). The concentration term of **molarity** (M) is defined as the number of moles of solute dissolved in 1 L of solution. Thus,

$$\text{Molarity (M)} = \frac{\text{moles of solute}}{\text{liters of solution}}$$

equation 11.3

An aqueous solution of NaCl that has a molarity of 1.0 contains 1.0 mole NaCl per liter of solution. To make such a solution, you would place 58.5 g (1.0 mole) NaCl in a beaker, then add water to make 1 L of solution.

SOLUBILITY

There is a limit to how much solid can be dissolved in a liquid. You may have noticed that a cup of hot tea will dissolve several teaspoons of sugar, but the limit of solubility is reached quickly in a glass of iced tea. The limit of how much sugar will dissolve seems to depend on the temperature of the tea. More sugar added to the cold tea after the limit is reached will not dissolve, and solid sugar granules begin to accumulate at the bottom of the glass. At this limit, the sugar and tea solution is said to be *saturated*. Dissolving does not actually stop when a solution becomes saturated, and undissolved sugar continues to enter the solution. However, dissolved sugar is now returning to the undissolved state at the same rate as it is dissolving. The overall equilibrium condition of sugar dissolving as sugar is coming out of solution is called a *saturated solution*. A saturated solution is a *state of equilibrium that exists between dissolving solute and solute coming out of solution.* You actually cannot see the dissolving and coming out of solution that occurs in a saturated solution because the exchanges are taking place with particles the size of molecules or ions.

Not all compounds dissolve as sugar does, and more or less of a given compound may be required to produce a saturated solution at a particular temperature. In general, the difficulty of dissolving a given compound is referred to as *solubility*. More specifically, the **solubility** of a solute is defined as the *concentration that is reached in a saturated solution at a particular temperature.* Solubility varies with the temperature, as the sodium and potassium salt examples show in Figure 11.9. These solubility curves describe the amount of solute required to reach the saturation equilibrium at a particular temperature. In general, the solubilities of most ionic solids increase with temperature, but there are exceptions. In addition, some salts release heat when dissolved in water, and other salts absorb heat when dissolved. The "instant cold pack" used for first aid is a bag of water containing a second bag of ammonium nitrate (NH_4NO_3). When the bag of ammonium nitrate is broken, the compound dissolves and absorbs heat.

You can usually dissolve more of a solid, such as salt or sugar, as the temperature of the water is increased. Contrary to what you might expect, gases usually become *less* soluble in water as the temperature increases. As a glass of water warms, small bubbles collect on the sides of the glass as dissolved air comes out of solution. The first bubbles that appear when warming a pot of water to boiling are also bubbles of dissolved air coming out of solution. This is why water that has been boiled usually tastes "flat." The dissolved air has been removed by the heating. The "normal" taste of water can be restored by pouring the boiled water back and forth between two glasses. The water dissolves more air during this process, restoring the usual taste.

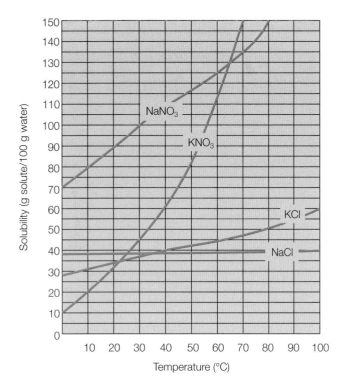

FIGURE 11.9 Approximate solubility curves for sodium nitrate, potassium nitrate, potassium chloride, and sodium chloride.

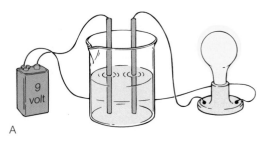

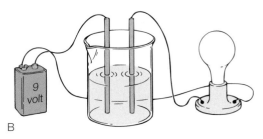

FIGURE 11.10 (*A*) Water solutions that conduct an electric current are called *electrolytes*. (*B*) Water solutions that do not conduct electricity are called *nonelectrolytes*.

Changes in pressure have no effect on the solubility of solids in liquids but greatly affect the solubility of gases. The release of bubbles (fizzing) when a bottle or can of soda is opened occurs because pressure is reduced on the beverage and dissolved carbon dioxide comes out of solution. In general, *gas solubility decreases with temperature and increases with pressure.* As usual, there are exceptions to this generalization.

PROPERTIES OF WATER SOLUTIONS

Pure solvents have characteristic physical and chemical properties that are changed by the presence of the solute. Following are some of the more interesting changes.

ELECTROLYTES

Water solutions of ionic substances will conduct an electric current, so they are called **electrolytes.** Ions must be present and free to move in a solution to carry the charge, so electrolytes are solutions containing ions. Pure water will not conduct an electric current because it is a covalent compound, which ionizes only very slightly. Water solutions of sugar, alcohol, and most other covalent compounds are nonconductors, so they are called *nonelectrolytes.* Nonelectrolytes are covalent compounds that form molecular solutions, so they cannot conduct an electric current (Figure 11.10).

Some covalent compounds are nonelectrolytes as pure liquids but become electrolytes when dissolved in water. Pure hydrogen chloride (HCl), for example, does not conduct an electric

current, so you can assume that it is a molecular substance. When dissolved in water, hydrogen chloride does conduct a current, so it must now contain ions. Evidently, the hydrogen chloride has become *ionized* by the water. The process of forming ions from molecules is called *ionization.* Hydrogen chloride, like water, has polar molecules. The positive hydrogen atom on the HCl molecule is attracted to the negative oxygen end of a water molecule, and the force of attraction is strong enough to break the hydrogen-chlorine bond, forming charged particles (Figure 11.11). The reaction is

$$HCl_{(l)} + H_2O_{(l)} \longrightarrow H_3O^+{}_{(aq)} + Cl^-{}_{(aq)}$$

The H_3O^+ ion is called a **hydronium ion.** A hydronium ion is basically a molecule of water with an attached hydrogen ion.

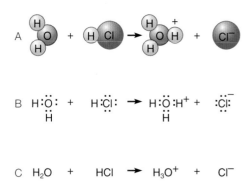

FIGURE 11.11 Three representations of water and hydrogen chloride in an ionizing reaction. (*A*) Sketches of molecules involved in the reaction. (*B*) Electron dot equation of the reaction. (*C*) The chemical equation for the reaction. Each of these representations shows the hydrogen being pulled away from the chlorine atom to form H_3O^+, the hydronium ion.

The presence of the hydronium ion gives the solution new chemical properties; the solution is no longer hydrogen chloride but is *hydrochloric acid*. Hydrochloric acid, and other acids, will be discussed shortly.

BOILING POINT

Boiling occurs when the pressure of the vapor escaping from a liquid is equal to the atmospheric pressure on the liquid. The *normal* boiling point is defined as the temperature at which the vapor pressure is equal to the average atmospheric pressure at sea level. For pure water, this temperature is 100°C (212°F). It is important to remember that boiling is a purely physical process. No bonds within water molecules are broken during boiling.

The vapor pressure over a solution is *less* than the vapor pressure over the pure solvent at the same temperature. Molecules of a liquid can escape into the air only at the surface of the liquid, and the presence of molecules of a solute means that fewer solvent molecules can be at the surface to escape. Thus, the vapor pressure over a solution is less than the vapor pressure over a pure solvent (Figure 11.12).

Because the vapor pressure over a solution is less than that over the pure solvent, the solution boils at a higher temperature. A higher temperature is required to increase the vapor pressure to that of the atmospheric pressure. Some cooks have been observed to add a pinch of salt to a pot of water before boiling. Is this to increase the boiling point and therefore cook the food more quickly? How much does a pinch of salt increase the boiling temperature? The answers are found in the relationship between the concentration of a solute and the boiling point of the solution.

It is the number of solute particles (ions or molecules) at the surface of a solution that increases the boiling point. Recall that a mole is a measure that can be defined as a number of particles called Avogadro's number. Since the number of particles at the surface is proportional to the ratio of particles in the solution, the concentration of the solute will directly influence the increase in the boiling point. In other words, the boiling point of any dilute solution is increased proportional to the concentration of the solute. For water, the boiling point is increased 0.521°C for every mole of solute dissolved in 1,000 g of water. Thus, any water solution will boil at a higher temperature than pure water. Since it boils at a higher temperature, it also takes a longer time to reach the boiling point.

It makes no difference what substance is dissolved in the water; one mole of solute in 1,000 g of water will elevate the boiling point by 0.521°C. A mole contains Avogadro's number of particles, so a mole of any solute will lower the vapor pressure by the same amount. Sucrose, or table sugar, for example, is $C_{12}H_{22}O_{11}$ and has a gram-formula weight of 342 g. Thus, 342 g of sugar in 1,000 g of water (about a liter) will increase the boiling point by 0.521°C. Therefore, if you measure the boiling point of a sugar solution, you can determine the concentration of sugar in the solution. For example, pancake syrup that boils at 100.261°C (sea-level pressure) must contain 171 g of sugar dissolved in 1,000 g of water. You know this because the increase of 0.261°C over 100°C is one-half of 0.521°C. If the boiling point were increased by 0.521°C over 100°C, the syrup would have the full gram-formula weight (342 g) dissolved in a kg of water.

Since it is the number of particles of solute in a specific sample of water that elevates the boiling point, different effects are observed in dissolved covalent and dissolved ionic compounds (Figure 11.13). Sugar is a covalent compound, and the solute is molecules of sugar moving between the water molecules. Sodium chloride, on the other hand, is an ionic compound and dissolves by the separation of ions, or

$$Na^+Cl^-_{(s)} \longrightarrow Na^+_{(aq)} + Cl^-_{(aq)}$$

This equation tells you that one mole of NaCl separates into one mole of sodium ions and one mole of chlorine ions for a total of *two* moles of solute. The boiling point elevation of a solution made from one mole of NaCl (58.5 g) is therefore multiplied by two, or $2 \times 0.521°C = 1.04°C$. The boiling point of a solution made by adding 58.5 g of NaCl to 1,000 g of water is therefore 101.04°C at normal sea-level pressure.

Now back to the question of how much a pinch of salt increases the boiling point of a pot of water. Assuming the pot contains about a liter of water (about a quart), and assuming that a pinch of salt has a mass of about 0.2, the boiling point will be increased by 0.0037°C. Thus, there must be some reason other than increasing the boiling point that a cook adds a pinch of salt to a pot of boiling water. Perhaps the salt is for seasoning?

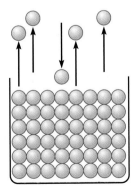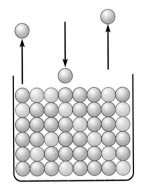

FIGURE 11.12 The rate of evaporation, and thus the vapor pressure, is less for a solution than for a solvent in the pure state. The greater the solute concentration, the less the vapor pressure.

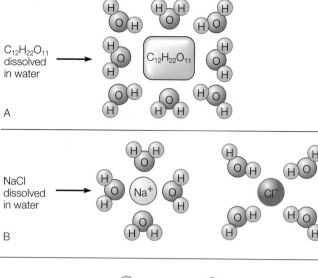

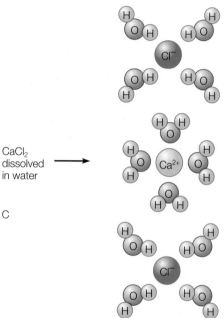

FIGURE 11.13 Since ionic compounds dissolve by the separation of ions, they provide more particles in solution than molecular compounds. (*A*) A mole of sugar provides Avogadro's number of particles. (*B*) A mole of NaCl provides two times Avogadro's number of particles. (*C*) A mole of $CaCl_2$ provides three times Avogadro's number of particles.

FREEZING POINT

Freezing occurs when the kinetic energy of molecules has been reduced sufficiently so the molecules can come together, forming the crystal structure of the solid. Reduced kinetic energy of the molecules, that is, reduced temperature, results in a specific freezing point for each pure liquid. The *normal* freezing point for pure water, for example, is 0°C (32°F) under normal pressure. The presence of solute particles in a solution interferes with the water molecules as they attempt to form the six-sided hexagonal structure. The water molecules cannot get by the solute particles until the kinetic energy of the solute particles is reduced, that is, until the temperature is below the normal freezing point. Thus,

the presence of solute particles lowers the freezing point, and solutions freeze at a lower temperature than the pure solvent.

The freezing-point depression of a solution has a number of interesting implications for solutions such as seawater. When seawater freezes, the water molecules must work their way around the salt particles, as was described in the section on the structure of water molecules. Thus, the solute particles are *not* normally included in the hexagonal structure of ice. Ice formed in seawater is practically pure water. Since the solute was *excluded* when the ice formed, the freezing of seawater increases the salinity. Increased salinity means increased concentration, so the freezing point of seawater is further depressed and more ice forms only at a lower temperature. When this additional ice forms, more pure water is removed, and the process goes on. Thus, seawater does not have a fixed freezing point but has a lower and lower freezing point as more and more ice freezes.

The depression of the freezing point by a solute has a number of interesting applications in colder climates. Salt, for example, is spread on icy roads to lower the freezing point (and thus the melting point) of the ice. Calcium chloride, $CaCl_2$, is a salt that is often used for this purpose. Water in a car radiator would also freeze in colder climates if a solute, called antifreeze, were not added to the radiator water. Methyl alcohol has been used as an antifreeze because it is soluble in water and does not damage the cooling system. Methyl alcohol, however, has a low boiling point and tends to boil away. Ethylene glycol has a higher boiling point, so it is called a "permanent" antifreeze. Like other solutes, ethylene glycol also raises the boiling point, which is an added benefit for summer driving.

ACIDS, BASES, AND SALTS

The electrolytes known as *acids, bases,* and *salts* are evident in environmental quality, foods, and everyday living. Environmental quality includes the hardness of water, which is determined by the presence of certain salts; the acidity of soils, which determines how well plants grow; and acid rain, which is a by-product of industry and automobiles. Many concerns about air and water pollution are often related to the chemistry concepts of acids, bases, and salts. These concepts, and uses of acids, bases, and salts, will be considered in this section.

PROPERTIES OF ACIDS AND BASES

Acids and bases are classes of chemical compounds that have certain characteristic properties. These properties can be used to identify if a substance is an acid or a base (Tables 11.2 and 11.3). The following are the properties of *acids* dissolved in water:

1. Acids have a sour taste, such as the taste of citrus fruits.
2. Acids change the color of certain substances; for example, litmus changes from blue to red when placed in an acid solution (Figure 11.14A).
3. Acids react with active metals, such as magnesium or zinc, releasing hydrogen gas.
4. Acids *neutralize* bases, forming water and salts from the reaction.

TABLE 11.2

Some common acids

Name	Formula	Comment
Acetic acid	CH_3COOH	A weak acid found in vinegar
Boric acid	H_3BO_3	A weak acid used in eyedrops
Carbonic acid	H_2CO_3	The weak acid of carbonated beverages
Formic acid	$HCOOH$	Makes the sting of some insects and certain plants
Hydrochloric acid	HCl	Also called muriatic acid; used in swimming pools, soil acidifiers, and stain removers
Lactic acid	$CH_3CHOHCOOH$	Found in sour milk, sauerkraut, and pickles; gives tart taste to yogurt
Nitric acid	HNO_3	A strong acid
Phosphoric acid	H_3PO_4	Used in cleaning solutions; added to carbonated beverages for tartness
Sulfuric acid	H_2SO_4	Also called oil of vitriol; used as battery acid and in swimming pools

TABLE 11.3

Some common bases

Name	Formula	Comment
Sodium hydroxide	$NaOH$	Also called lye or caustic soda; a strong base used in oven cleaners and drain cleaners
Potassium hydroxide	KOH	Also called caustic potash; a strong base used in drain cleaners
Ammonia	NH_3	A weak base used in household cleaning solutions
Calcium hydroxide	$Ca(OH)_2$	Also called slaked lime; used to make brick mortar
Magnesium hydroxide	$Mg(OH)_2$	Solution is called milk of magnesia; used as antacid and laxative

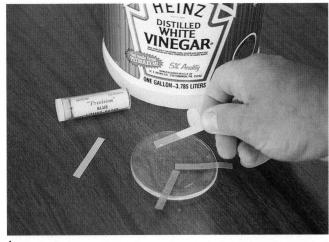

A

B

FIGURE 11.14 (*A*) Acid solutions will change the color of blue litmus to red. (*B*) Solutions of bases will change the color of red litmus to blue.

3. Basic solutions feel slippery on the skin. They have a *caustic* action on plant and animal tissue, converting tissue into soluble materials. A strong base, for example, reacts with fat to make soap and glycerine. This accounts for the slippery feeling on the skin.
4. Bases *neutralize* acids, forming water and salts from the reaction.

Tasting an acid or base to see if it is sour or bitter can be hazardous, since some are highly corrosive or caustic. Many organic acids are not as corrosive and occur naturally in foods. Citrus fruit, for example, contains citric acid, vinegar is a solution of acetic acid, and sour milk contains lactic acid. The stings or bites of some insects (bees, wasps, and ants) and some plants (stinging nettles) are painful because an organic acid, formic acid, is injected by the insect or plant. Your stomach contains a solution of hydrochloric acid. In terms of relative strength, the hydrochloric acid in your stomach is about ten times stronger than the carbonic acid (H_2CO_3) of carbonated beverages.

Likewise, *bases* have their own characteristic properties. Bases are also called alkaline substances, and the following are the properties of bases dissolved in water:

1. Bases have a bitter taste, for example, the taste of caffeine.
2. Bases reverse the color changes that were caused by acids. Red litmus is changed back to blue when placed in a solution containing a base (Figure 11.14B).

CHAPTER 11 Water and Solutions

Examples of bases include solutions of sodium hydroxide (NaOH), which has a common name of lye or caustic soda, and potassium hydroxide (KOH), which has a common name of caustic potash. These two bases are used in products known as drain cleaners. They open plugged drains because of their caustic action, turning grease, hair, and other organic "plugs" into soap and other soluble substances that are washed away. A weaker base is a solution of ammonia (NH_3), which is often used as a household cleaner. A solution of magnesium hydroxide, $Mg(OH)_2$, has a common name of milk of magnesia and is sold as an antacid and laxative.

Many natural substances change color when mixed with acids or bases. You may have noticed that tea changes color slightly, becoming lighter, when lemon juice (which contains citric acid) is added. Some plants have flowers of one color when grown in acidic soil and flowers of another color when grown in basic soil. A vegetable dye that changes color in the presence of acids or bases can be used as an **acid-base indicator**. An indicator is simply a vegetable dye that is used to distinguish between acid and base solutions by a color change. Litmus, for example, is an acid-base indicator made from a dye extracted from certain species of lichens. The dye is applied to paper strips, which turn red in acidic solutions and blue in basic solutions.

 CONCEPTS *Applied*

Cabbage Indicator

To see how acids and bases change the color of certain vegetable dyes, consider the dye that gives red cabbage its color. Shred several leaves of red cabbage and boil them in a pan of water to extract the dye. After you have a purple solution, squeeze the juice from the cabbage into the pan and allow the solution to cool. Add vinegar in small amounts as you stir the solution, continuing until the color changes. Add ammonia in small amounts, again stirring until the color changes again. Reverse the color change again by adding vinegar in small amounts. Will this purple cabbage acid-base indicator tell you if other substances are acids or bases?

EXPLAINING ACID-BASE PROPERTIES

Comparing the lists in Tables 11.2 and 11.3, you can see that acids and bases appear to be chemical opposites. Notice in Table 11.2 that the acids all have an H, or hydrogen atom, in their formulas. In Table 11.3, most of the bases have a hydroxide ion, OH^-, in their formulas. Could this be the key to acid-base properties?

The modern concept of an acid considers the properties of acids in terms of the hydronium ion, H_3O^+. As was mentioned earlier, the hydronium ion is a water molecule to which an H^+ ion is attached. Since a hydrogen ion is a hydrogen atom without its single electron, it could be considered as an ion consisting of a single proton. Thus, the H^+ ion can be called a *proton*. An **acid** is defined as any substance that is a *proton donor* when dissolved in water, increasing the hydronium ion concentration.

For example, hydrogen chloride dissolved in water has the following reaction:

$$H{-}Cl_{(l)} \ + \ H_2O_{(l)} \ \longrightarrow \ H_3O^+_{(aq)} \ + \ Cl^-_{(aq)}$$

The dotted circle and arrow were added to show that the hydrogen chloride donated a proton to a water molecule. The resulting solution contains H_3O^+ ions and has acid properties, so the solution is called hydrochloric acid. It is the H_3O^+ ion that is responsible for the properties of an acid.

The bases listed in Table 11.3 all appear to have a hydroxide ion, OH^-. Water solutions of these bases do contain OH^- ions, but the definition of a base is much broader. A **base** is defined as any substance that is a *proton acceptor* when dissolved in water, increasing the hydroxide ion concentration. For example, ammonia dissolved in water has the following reaction:

$$NH_{3(g)} \ + \ H{-}OH_{(l)} \ \longrightarrow \ (NH_4)^+ \ + \ OH^-$$

The dotted circle and arrow show that the ammonia molecule accepted a proton from a water molecule, providing a hydroxide ion. The resulting solution contains OH^- ions and has basic properties, so a solution of ammonium hydroxide is a base.

Carbonates, such as sodium carbonate (Na_2CO_3), form basic solutions because the carbonate ion reacts with water to produce hydroxide ions.

$$(CO_3)^{2-}_{(aq)} + H_2O_{(l)} \longrightarrow (HCO_3)^-_{(aq)} + OH^-_{(aq)}$$

Thus, sodium carbonate produces a basic solution.

Acids could be thought of as simply solutions of hydronium ions in water, and bases could be considered solutions of hydroxide ions in water. The proton donor and proton acceptor definition is much broader, and it does include the definition of acids and bases as hydronium and hydroxide compounds. The broader, more general definition covers a wider variety of reactions and is therefore more useful.

The modern concept of acids and bases explains why the properties of acids and bases are **neutralized,** or lost, when acids and bases are mixed together. For example, consider the hydronium ion produced in the hydrochloric acid solution and the hydroxide ion produced in the ammonia solution. When these solutions are mixed together, the hydronium ion reacts with the hydroxide ion, and

$$H_3O^+_{(aq)} + OH^-_{(aq)} \longrightarrow H_2O_{(l)} + H_2O_{(l)}$$

Thus, a proton is transferred from the hydronium ion (an acid), and the proton is accepted by the hydroxide ion (a base). Water is produced, and both the acid and base properties disappear or are neutralized.

STRONG AND WEAK ACIDS AND BASES

Acids and bases are classified according to their degree of ionization when placed in water. *Strong acids* ionize completely in water, with all molecules dissociating into ions.

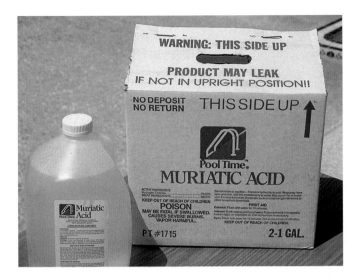

FIGURE 11.15 Hydrochloric acid (HCl) has the common name of *muriatic* acid. Hydrochloric acid is a strong acid used in swimming pools, soil acidifiers, and stain removers.

Nitric acid, for example, reacts completely in the following equation:

$$HNO_{3(aq)} + H_2O_{(l)} \longrightarrow H_3O^+_{(aq)} + (NO_3)^-_{(aq)}$$

Nitric acid, hydrochloric acid (Figure 11.15), and sulfuric acid are common strong acids.

Acids that ionize only partially and produce fewer hydronium ions are weaker acids. *Weak acids* are only partially ionized. Vinegar, for example, contains acetic acid that reacts with water in the following reaction:

$$HC_2H_3O_2 + H_2O \longrightarrow H_3O^+ + (C_2H_3O_2)^-$$

Only about 1 percent or less of the acetic acid molecules ionize, depending on the concentration.

Bases are also classified as strong or weak. A *strong base* is completely ionic in solution and has hydroxide ions. Sodium hydroxide, or lye, is the most common example of a strong base. It dissolves in water to form a solution of sodium and hydroxide ions:

$$Na^+OH^-_{(s)} \longrightarrow Na^+_{(aq)} + OH^-_{(aq)}$$

A *weak base* is only partially ionized. Ammonia, magnesium hydroxide, and calcium hydroxide are examples of weak bases. Magnesium and calcium hydroxide are only slightly soluble in water, and this reduces the *concentration* of hydroxide ions in a solution. It would appear that $Ca(OH)_2$ would produce two moles of hydroxide ions. It would, if it were completely soluble and reacted completely. It is the concentration of hydroxide ions in solution that determines if a base is weak or strong, not the number of ions per mole.

THE pH SCALE

The strength of an acid or a base is usually expressed in terms of a range of values called a **pH scale.** The pH scale is based on the concentration of the hydronium ion (in moles/L) in an acidic or a basic solution. To understand how the scale is able to express both acid and base strength in terms of the hydronium ion, first note that pure water is very slightly ionized in the reaction:

$$H_2O_{(l)} + H_2O_{(l)} \longrightarrow H_3O^+_{(aq)} + OH^-_{(aq)}$$

The amount of self-ionization by water has been determined through measurements. In pure water at 25°C or any neutral water solution at that temperature, the H_3O^+ concentration is 1×10^{-7} moles/L, and the OH^- concentration is also 1×10^{-7} moles/L. Since both ions are produced in equal numbers, then the H_3O^+ concentration equals the OH^- concentration, and pure water is neutral, neither acidic nor basic.

In general, adding an acid substance to pure water increases the H_3O^+ concentration. Adding a base substance to pure water increases the OH^- concentration. Adding a base also *reduces* the H_3O^+ concentration as the additional OH^- ions are able to combine with more of the hydronium ions to produce un-ionized water. Thus, at a given temperature, an increase in OH^- concentration is matched by a *decrease* in H_3O^+ concentration. The concentration of the hydronium ion can be used as a measure of acidic, neutral, and basic solutions. In general, (1) acidic solutions have H_3O^+ concentrations above 1×10^{-7} moles/L, (2) neutral solutions have H_3O^+ concentrations equal to 1×10^{-7} moles/L, and (3) basic solutions have H_3O^+ concentrations less than 1×10^{-7} moles/L. These three statements lead directly to the pH scale, which is named from the French *pouvoir hydrogene*, meaning "hydrogen power." Power refers to the exponent of the hydronium ion concentration, and the pH is a *power of ten notation that expresses the H_3O^+ concentration* (Table 11.4).

TABLE 11.4

The pH and hydronium ion concentration (moles/L)

Hydronium Ion Concentration (moles/L)	Reciprocal of Hydronium Ion Concentration	pH	Meaning
10^0	10^0	0	
10^{-1}	10^1	1	
10^{-2}	10^2	2	Increasing acidity
10^{-3}	10^3	3	
10^{-4}	10^4	4	
10^{-5}	10^5	5	
10^{-6}	10^6	w	
10^{-7}	10^7	7	neutral
10^{-8}	10^8	8	
10^{-9}	10^9	9	
10^{-10}	10^{10}	10	Increasing basicity
10^{-11}	10^{11}	11	
10^{-12}	10^{12}	12	
10^{-13}	10^{13}	13	
10^{-14}	10^{14}	14	

TABLE 11.5

The approximate pH of some common substances

Substance	pH (or pH Range)
Hydrochloric acid (4%)	0
Gastric (stomach) solution	1.6–1.8
Lemon juice	2.2–2.4
Vinegar	2.4–3.4
Carbonated soft drinks	2.0–4.0
Grapefruit	3.0–3.2
Oranges	3.2–3.6
Acid rain	4.0–5.5
Tomatoes	4.2–4.4
Potatoes	5.7–5.8
Natural rainwater	5.6–6.2
Milk	6.3–6.7
Pure water	7.0
Seawater	7.0–8.3
Blood	7.4
Sodium bicarbonate solution	8.4
Milk of magnesia	10.5
Ammonia cleaning solution	11.9
Sodium hydroxide solution	13.0

FIGURE 11.16 The pH increases as the acidic strength of these substances decreases from left to right. Did you know that lemon juice is more acidic than vinegar? That a soft drink is more acidic than orange juice or grapefruit juice?

A neutral solution has a pH of 7.0. Acidic solutions have pH values below 7, and smaller numbers mean greater acidic properties. Increasing the OH^- concentration decreases the H_3O^+ concentration, so the strength of a base is indicated on the same scale with values greater than 7. Note that the pH scale is logarithmic, so a pH of 2 is ten times as acidic as a pH of 3. Likewise, a pH of 10 is one hundred times as basic as a pH of 8. Table 11.5 compares the pH of some common substances (Figure 11.16).

CONCEPTS *Applied*

Acid or Base?

Pick some household product that probably has an acid or base character (example: pH increaser for aquariums). Write down the listed ingredients and identify any you believe would be distinctly acidic or basic in a water solution. Tell whether you expect the product to be an acid or a base. Describe your findings from a litmus paper test.

PROPERTIES OF SALTS

Salt is produced by a neutralization reaction between an acid and a base. A **salt** is defined as any ionic compound except those with hydroxide or oxide ions. Table salt, NaCl, is but one example of this large group of ionic compounds. As an example of a salt produced by a neutralization reaction, consider the reaction of HCl (an acid in solution) with $Ca(OH)_2$ (a base in solution). The reaction is

$$2\ HCl_{(aq)} + Ca(OH)_{2(aq)} \longrightarrow CaCl_{2(aq)} + 2\ H_2O_{(l)}$$

This is an ionic exchange reaction that forms molecular water, leaving Ca^{2+} and Cl^- in solution. As the water is evaporated, these ions begin forming ionic crystal structures as the solution concentration increases. When the water is all evaporated, the white crystalline salt of $CaCl_2$ remains.

If sodium hydroxide had been used as the base instead of calcium hydroxide, a different salt would have been produced:

$$HCl_{(aq)} + NaOH_{(aq)} \longrightarrow NaCl_{(aq)} + H_2O_{(l)}$$

Salts are also produced when elements combine directly, when an acid reacts with a metal, and by other reactions.

Salts are essential in the diet both as electrolytes and as a source of certain elements, usually called *minerals* in this context. Plants must have certain elements that are derived from water-soluble salts. Potassium, nitrates, and phosphate salts are often used to supply the needed elements. There is no scientific evidence that plants prefer to obtain these elements from natural sources, as compost, or from chemical fertilizers. After all, a nitrate ion is a nitrate ion, no matter what its source. Table 11.6 lists some common salts and their uses.

HARD AND SOFT WATER

Salts vary in their solubility in water, and a solubility chart appears in appendix B. Table 11.7 lists some generalizations concerning the various common salts. Some of the salts are dissolved by water that will eventually be used for domestic supply. When the salts are soluble calcium or magnesium compounds, the water will contain calcium or magnesium ions in solution. A solution of Ca^{2+} of Mg^{2+} ions is said to be *hard water* because it is hard to make soap lather in the water. "Soft" water, on the other hand, makes a soap lather easily. The difficulty occurs because soap is a sodium or potassium compound that is soluble in water. The calcium or magnesium ions, when present, replace the sodium or potassium ions in the soap compound, forming an insoluble compound. It is this insoluble compound that forms a "bathtub ring" and also collects on clothes being washed, preventing cleansing.

TABLE 11.6

Some common salts and their uses

Common Name	Formula	Use
Alum	$KAl(SO_4)_2$	Medicine, canning, baking powder
Baking soda	$NaHCO_3$	Fire extinguisher, antacid, deodorizer, baking powder
Bleaching powder (chlorine tablets)	$CaOCl_2$	Bleaching, deodorizer, disinfectant in swimming pools
Borax	$Na_2B_4O_7$	Water softener
Chalk	$CaCO_3$	Antacid tablets, scouring powder
Cobalt chloride	$CoCl_2$	Hygrometer (pink in damp weather, blue in dry weather)
Chile saltpeter	$NaNO_3$	Fertilizer
Epsom salt	$MgSO_4 \cdot 7\ H_2O$	Laxative
Fluorspar	CaF_2	Metallurgy flux
Gypsum	$CaSO_4 \cdot 2\ H_2O$	Plaster of Paris, soil conditioner
Lunar caustic	$AgNO_3$	Germicide and cauterizing agent
Niter (or saltpeter)	KNO_3	Meat preservative, makes black gunpowder (75 parts KNO_3, 15 of carbon, 10 of sulfur)
Potash	K_2CO_3	Makes soap, glass
Rochelle salt	$KNaC_4H_4O_6$	Baking powder ingredient
TSP	Na_3PO_4	Water softener, fertilizer

TABLE 11.7

Generalizations about salt solubilities

Salts	Solubility	Exceptions
Sodium	Soluble	None
Potassium		
Ammonium		
Nitrate	Soluble	None
Acetate		
Chlorate		
Chlorides	Soluble	Ag and Hg (I) are insoluble
Sulfates	Soluble	Ba, Sr, and Pb are insoluble
Carbonates	Insoluble	Na, K, and NH_4 are soluble
Phosphates		
Silicates		
Sulfides	Insoluble	Na, K, and NH_4 are soluble; Mg, Ca, Sr, and Ba decompose

The key to "softening" hard water is to remove the troublesome calcium and magnesium ions. If the hardness is caused by magnesium or calcium *bicarbonates,* the removal is accomplished by simply heating the water. Upon heating, they decompose, forming an insoluble compound that effectively removes the ions from solution. The decomposition reaction for calcium bicarbonate is

$$Ca^{2+}(HCO_3)_{2(aq)} \longrightarrow CaCO_{3(s)} + H_2O_{(l)} + CO_2 \uparrow$$

The reaction is the same for magnesium bicarbonate. As the solubility chart in appendix B shows, magnesium and calcium carbonates are insoluble, so the ions are removed from solution in the solid that is formed. Perhaps you have noticed such a white compound forming around faucets if you live where bicarbonates are a problem. Commercial products to remove such deposits usually contain an acid, which reacts with the carbonate to make a new, soluble salt that can be washed away.

Water hardness is also caused by magnesium or calcium *sulfate,* which requires a different removal method. Certain chemicals such as sodium carbonate (washing soda), trisodium phosphate (TSP), and borax will react with the troublesome ions, forming an insoluble solid that removes them from solution. For example, washing soda and calcium sulfate react as follows:

$$Na_2CO_{3(aq)} + CaSO_{4(aq)} \longrightarrow Na_2SO_{4(aq)} + CaCO_3 \downarrow$$

Calcium carbonate is insoluble; thus, the calcium ions are removed from solution before they can react with the soap. Many laundry detergents have Na_2CO_3, TSP, or borax ($Na_2B_4O_7$) added to soften the water. TSP causes problems, however, because the additional phosphates in the waste water can act as a fertilizer, stimulating the growth of algae to such an extent that other organisms in the water die.

A water softener unit is an ion exchanger (Figure 11.17). The unit contains a mineral that exchanges sodium ions for calcium

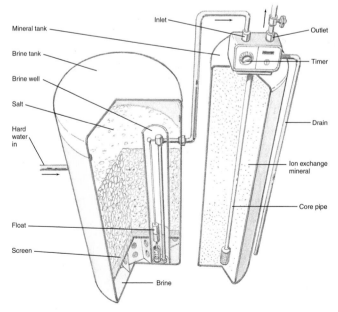

FIGURE 11.17 A water softener exchanges sodium ions for the calcium and magnesium ions of hard water.

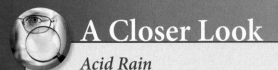
Acid Rain

Acid rain is a general term used to describe any acidic substances, wet or dry, that fall from the atmosphere. Wet acidic deposition could be in the form of rain, but snow, sleet, and fog could also be involved. Dry acidic deposition could include gases, dust, or any solid particles that settle out of the atmosphere to produce an acid condition.

Pure, unpolluted rain is naturally acidic. Carbon dioxide in the atmosphere is absorbed by rainfall, forming carbonic acid (H_2CO_3). Carbonic acid lowers the pH of pure rainfall to a range of 5.6 to 6.2. Decaying vegetation in local areas can provide more CO_2, making the pH even lower. A pH range of 4.5 to 5.0, for example, has been measured in remote areas of the Amazon jungle. Human-produced exhaust emissions of sulfur and nitrogen oxides can lower the pH of rainfall even more, to a 4.0 to 5.5 range. This is the pH range of acid rain.

The sulfur and nitrogen oxides that produce acid rain come from exhaust emissions of industries and electric utilities that burn coal and from the exhaust of cars, trucks, and buses (Box Figure 11.1). The emissions are sometimes called "SO_x" and "NO_x," which is read as "socks" and "knox." The x subscript implies the variable presence of any or all of the oxides, for example, nitrogen monoxide (NO), nitrogen dioxide (NO_2), and dinitrogen tetroxide (N_2O_4) for NO_x.

SO_x and NO_x are the raw materials of acid rain and are not themselves acidic. They react with other atmospheric chemicals to form sulfates and nitrates, which combine with water vapor to form sulfuric acid (H_2SO_4) and nitric acid (HNO_3). These are the chemicals of concern in acid rain.

Many variables influence how much and how far SO_x and NO_x are carried in the atmosphere and if they are converted to acid rain or simply return to the surface as a dry gas or particles. During the 1960s and 1970s, concerns about local levels of pollution led to the replacement of short smokestacks of about 60 m (about 200 ft) with taller smokestacks of about 200 m (about 650 ft). This did reduce the local levels of pollution by dumping the exhaust higher in the atmosphere where winds could carry it away. It also set the stage for longer-range transport of SO_x and NO_x and their eventual conversion into acids.

There are two main reaction pathways by which SO_x and NO_x are converted to acids: (1) reactions in the gas phase and (2) reactions in the liquid phase, such as in water droplets in clouds and fog. In the gas phase, SO_x and NO_x are oxidized to acids, mainly by hydroxyl ions and ozone, and the acid is absorbed by cloud droplets and precipitated as rain or snow. Most of the nitric acid in acid rain and about one-fourth of the sulfuric acid is formed in gas-phase reactions. Most of the liquid-phase reactions that produce sulfuric acid involve the absorbed SO_x and hydrogen peroxide (H_2O_2), ozone, oxygen, and particles of carbon, iron oxide, and manganese oxide particles. These particles also come from the exhaust of fossil fuel combustion.

Acid rain falls on the land, bodies of water, forests, crops, buildings, and people. The concerns about acid rain center on its environmental impact on lakes, forests, crops, materials, and human health. Lakes in different parts of the world, for example, have been increasing in acidity over the past fifty years. Lakes in northern New England, the Adirondacks, and parts of Canada now have a pH of less than 5.0, and correlations have been established between lake acidity and decreased fish populations. Trees, mostly conifers, are dying at unusually rapid rates in the northeastern United States. Red spruce in Vermont's Green Mountains and the mountains of New York and New Hampshire have been affected by acid rain, as have pines in New Jersey's Pine Barrens. It is believed that acid rain leaches essential nutrients, such as calcium, from the soil and also mobilizes aluminum ions. The

BOX FIGURE 11.1 Natural rainwater has a pH of 5.6 to 6.2. Exhaust emissions of sulfur and nitrogen oxides can lower the pH of rainfall to a range of 4.0 to 5.5. The exhaust emissions come from industries, electric utilities, and automobiles. Not all emissions are as visible as those pictured in this illustration.

aluminum ions disrupt the water equilibrium of fine root hairs, and when the root hairs die, so do the trees.

Human-produced emissions of sulfur and nitrogen oxides from burning fossil fuels are the cause of acid rain. The heavily industrialized northeastern part of the United States, from the Midwest through New England, release sulfur and nitrogen emissions that result in a precipitation pH of 4.0 to 4.5. This region is the geographic center of the nation's acid rain problem. The solution to the problem is found in (1) using fuels other than fossil fuels and (2) reducing the thousands of tons of SO_x and NO_x that are dumped into the atmosphere per day when fossil fuels are used.

and magnesium ions as water is run through it. The softener is regenerated periodically by flushing with a concentrated sodium chloride solution. The sodium ions replace the calcium and magnesium ions, which are carried away in the rinse water. The softener is then ready for use again. The frequency of renewal cycles depends on the water hardness, and each cycle can consume from 4 to 20 lb of sodium chloride per renewal cycle. In general, water with less than 75 ppm calcium and magnesium ions is called soft water; with greater concentrations, it is called hard water. The greater the concentration above 75 ppm, the harder the water.

Johannes Brönsted was a Danish physical chemist whose work in solution chemistry, particularly electrolytes, resulted in a new theory of acids and bases.

Brönsted was born on February 22, 1879, in Varde, Jutland, the son of a civil engineer. He was educated at local schools before going to study chemical engineering at the Technical Institute of the University of Copenhagen in 1897. He graduated two years later and then turned to chemistry, in which he qualified in 1902. After a short time in industry, he was appointed an assistant in the university's chemical laboratory in 1905, becoming professor of physical and inorganic chemistry in 1908.

Brönsted's early work was wide ranging, particularly in the fields of electrochemistry, the measurement of hydrogen ion concentrations, amphoteric electrolytes, and the behavior of indicators. He discovered a method of eliminating potentials in the measurement of hydrogen ion concentrations and devised a simple equation that connects the activity and osmotic coefficients of an electrolyte, as well as another that relates activity coefficients to reaction velocities. From the absorption spectra of chromic—chromium(III)—salts, he concluded that strong electrolytes are completely dissociated and that the changes of molecular conductivity and freezing point that accompany changes in concentration are caused by the electrical forces between ions in solution.

In 1887, Svante Arrhenius had proposed a theory of acidity that explained its nature on an atomic level. He defined an acid as a compound that could generate hydrogen ions in aqueous solution and an alkali as a compound that could generate hydroxyl ions. A strong acid is completely ionized (dissociated) and produces many hydrogen ions, whereas a weak acid is only partly dissociated and produces few hydrogen ions. Conductivity measurements confirm the theory, as long as the solutions are not too concentrated.

In 1923, Brönsted published (simultaneously with Thomas Lowry in Britain) a new theory of acidity, which has certain important advantages over that of Arrhenius. Brönsted defined an acid as a proton donor and a base as a proton acceptor. The definition applies to all solvents, not just water. It also explains the different behavior of pure acids in solution. Pure dry liquid sulfuric acid or acetic (ethanoic) acid does not change the color of indicators nor does it react with carbonates or metals. But as soon as water is added, all of these reactions occur.

Source: From the Hutchinson *Dictionary of Scientific Biography.* © Research Machines plc 2003. All Rights Reserved. Helicon Publishing is a division of Research Machines.

SUMMARY

A water molecule consists of two hydrogen atoms and an oxygen atom with covalent bonding. Oxygen has more positive protons than either of the hydrogens, so electrons spend more time around the oxygen, producing a *polar molecule,* with centers of negative and positive charge. Polar water molecules interact with an attractive force between the negative center of one molecule and the positive center of another. This force is called a *hydrogen bond.* The hydrogen bond accounts for the decreased density of ice, the high heat of fusion, and the high heat of vaporization of water. The hydrogen bond is also involved in the *dissolving* process.

A *solution* is a homogeneous mixture of ions or molecules of two or more substances. The substance present in the large amount is the *solvent,* and the *solute* is dissolved in the solvent. If one of the components is a liquid, however, it is called the solvent. Fluids that mix in any proportion are called *miscible fluids,* and *immiscible fluids* do not mix. Polar substances dissolve in polar solvents but not nonpolar solvents, and the general rule is *like dissolves like.* Thus oil, a nonpolar substance, is immiscible in water, a polar substance.

The relative amount of solute in a solvent is called the *concentration* of a solution. Concentrations are measured (1) in *parts per million* (ppm) or *parts per billion* (ppb), (2) *percent by volume,* the volume of a solute per 100 volumes of solution, (3) *percent by weight,* the weight of solute per 100 weight units of solution, and (4) *salinity,* the mass of salts in 1 kg of solution.

A limit to dissolving solids in a liquid occurs when the solution is *saturated.* A *saturated solution* is one with equilibrium between solute dissolving and solute coming out of solution. The *solubility* of a solid is the concentration of a saturated solution at a particular temperature.

Water solutions that carry an electric current are called *electrolytes,* and nonconductors are called *nonelectrolytes.* In general, ionic substances make electrolyte solutions, and molecular substances make nonelectrolyte solutions. Polar molecular substances may be *ionized* by polar water molecules, however, making an electrolyte from a molecular solution.

The *boiling point of a solution* is greater than the boiling point of the pure solvent, and the increase depends only on the concentration of the solute (at a constant pressure). For water, the boiling point is increased $0.521°C$ for each mole of solute in each kg of water. The *freezing point of a solution* is lower than the freezing point of the pure solvent, and the depression also depends on the concentration of the solute.

Acids, bases, and salts are chemicals that form ionic solutions in water, and each can be identified by simple properties. These properties are accounted for by the modern concepts of each. *Acids* are *proton donors* that form *hydronium ions* (H_3O^+) in water solutions. *Bases* are *proton acceptors* that form *hydroxide ions* (OH^-) in water solutions. *Strong acids* and *strong bases* ionize completely in water, and *weak acids* and *weak bases* are only partially ionized. The strength of an acid or base is measured on the *pH scale,* a power of ten notation of the hydronium ion concentration. On the scale, numbers from 0 up to 7 are acids, 7 is neutral, and numbers above 7 and up to 14 are bases. Each unit represents a tenfold increase or decrease in acid or base properties.

A *salt* is any ionic compound except those with hydroxide or oxide ions. Salts provide plants and animals with essential elements. The solubility of salts varies with the ions that make up the compound. Solutions of magnesium or calcium produce *hard water,* water in which it is hard to make soap lather. Hard water is softened by removing the magnesium and calcium ions.

SUMMARY OF EQUATIONS

11.1

Percent by volume

$$\frac{V_{solute}}{V_{solution}} \times 100\% \text{ solution} = \% \text{ solute}$$

11.2

Percent by weight (mass)

$$\frac{m_{solute}}{m_{solution}} \times 100\% \text{ solution} = \% \text{ solute}$$

11.3

$$\text{Molarity (M)} = \frac{\text{moles of solute}}{\text{liters of solution}}$$

KEY TERMS

acid (p. **288**)
acid-base indicator (p. **288**)
base (p. **288**)
concentration (p. **280**)
electrolyte (p. **284**)
hydrogen bonding (p. **278**)
hydronium ion (p. **284**)
miscible fluids (p. **279**)
molarity (p. **283**)
neutralized (p. **288**)
pH scale (p. **289**)
polar molecule (p. **277**)
salinity (p. **282**)
salt (p. **290**)
solubility (p. **283**)
solution (p. **279**)

APPLYING THE CONCEPTS

1. A major use for water in the average home is for
   a. drinking and cooking.
   b. bathing.
   c. toilets.
   d. laundry.

2. Freshwater is obtained from all of the following *except*
   a. oceans.
   b. streams.
   c. lakes.
   d. rivers.

3. A molecule with a positive end and a negative end is called a (an)
   a. polar molecule.
   b. nonpolar molecule.
   c. neutral molecule.
   d. ionic molecule.

4. The material present in a solution in the largest amount is the
   a. solvent.
   b. solute.
   c. salt.
   d. molecules.

5. Fluids that mix in any proportion without separating into phases are said to be
   a. miscible.
   b. concentrated.
   c. immiscible.
   d. solvated.

6. The relative amount of solute and solvent in a solution is defined as the
   a. solubility.
   b. miscibility.
   c. concentration.
   d. polarity.

7. A solution with a state of equilibrium between the dissolving solute and solute coming out of solution is
   a. unsaturated.
   b. saturated.
   c. supersaturated.
   d. undersaturated.

8. The solubility of most ionic salts in water
   a. increases with temperature.
   b. decreases with temperature.
   c. depends on the amount of salt.
   d. increase with stirring.

9. Ionic substances that dissolve in water and conduct an electric current are called
   a. salts.
   b. compounds.
   c. molecules.
   d. electrolytes.

10. The temperature at which the vapor pressure is equal to the average atmospheric pressure at sea level is called the
    a. normal pressure point.
    b. normal boiling point.
    c. normal liquid point.
    d. normal temperature point.

11. The temperature at which a liquid undergoes a phase change to the solid state at normal pressure is the
    a. solidification temperature.
    b. freezing point.
    c. condensation point.
    d. compression temperature.

12. All of the following are electrolytes *except*
    a. acids.
    b. salts.
    c. bases.
    d. sugars.

13. Some covalent compounds, such as HCl, become electrolytes through the process of
    a. ionization.
    b. oxidation-reduction.
    c. decomposition.
    d. combination.

14. Which of the following is *not* a property of an acid?
    a. sour taste
    b. changes litmus from red to blue
    c. reacts with active metals to generate hydrogen gas
    d. changes litmus from blue to red

15. What is an acid when it is dissolved in water?
    a. proton acceptor
    b. proton donor
    c. indicator
    d. salt

16. Which of the following is *not* a property of a base?
    a. changes red litmus to blue
    b. changes blue litmus to red
    c. slippery to touch
    d. bitter taste

17. The pH scale is based on the concentration of what in solution?
    a. hydroxide ions
    b. hydronium ions
    c. electrolytes
    d. solute

18. The products produced in a neutralization reaction are
    a. acid and bases.
    b. salt and water.
    c. molecules and water.
    d. water only.

19. Which of the following is *not* a solution?
    a. seawater
    b. carbonated water
    c. sand
    d. brass

20. Atmospheric air is a homogeneous mixture of gases that is mostly nitrogen gas. The nitrogen is therefore the
    a. solvent.
    b. solution.
    c. solute.
    d. None of the above is correct.

21. A homogeneous mixture is made up of 95 percent alcohol and 5 percent water. In this case, the water is the
    a. solvent.
    b. solution.
    c. solute.
    d. None of the above is correct.

22. The solution concentration terms of *parts per million, percent by volume*, and *percent by weight* are concerned with the amount of
    a. solvent in the solution.
    b. solute in the solution.
    c. solute compared to solvent.
    d. solvent compared to solute.

23. A concentration of 500 ppm is reported in a news article. This is the same concentration as
    a. 0.005 percent.
    b. 0.05 percent.
    c. 5 percent.
    d. 50 percent.

24. According to the label, a bottle of vodka has a 40 percent by volume concentration. This means the vodka contains 40 mL of pure alcohol
    a. in each 140 mL of vodka.
    b. to every 100 mL of water.
    c. to every 60 mL of vodka.
    d. mixed with water to make 100 mL vodka.

25. A bottle of vinegar is 4 percent by weight, so you know that the solution contains 4 weight units of pure vinegar with
    a. 96 weight units of water.
    b. 99.6 weight units of water.
    c. 100 weight units of water.
    d. 104 weight units of water.

26. If a salt solution has a salinity of 40 parts per thousand (‰), what is the equivalent percentage measure?
    a. 400 percent
    b. 40 percent
    c. 4 percent
    d. 0.4 percent

27. A salt solution has solid salt on the bottom of the container and salt is dissolving at the same rate that it is coming out of solution. You know the solution is
    a. an electrolyte.
    b. a nonelectrolyte.
    c. a buffered solution.
    d. a saturated solution.

28. As the temperature of water *decreases*, the solubility of carbon dioxide gas in the water
    a. increases.
    b. decreases.
    c. remains the same.
    d. increases or decreases, depending on the specific temperature.

29. At what temperature does water have the greatest density?
    a. 100°C
    b. 20°C
    c. 4°C
    d. 0°C

30. An example of a hydrogen bond is a weak-to-moderate bond between
    a. any two hydrogen atoms.
    b. a hydrogen atom of one polar molecule and an oxygen atom of another polar molecule.
    c. two hydrogen atoms on two nonpolar molecules.
    d. a hydrogen atom and any nonmetal atom.

31. A certain solid salt is insoluble in water, so the strongest force must be the
    a. ion-water molecule force.
    b. ion-ion force.
    c. force of hydration.
    d. polar molecule force.

32. Which of the following will conduct an electric current?
    a. pure water
    b. a water solution of a covalent compound
    c. a water solution of an ionic compound
    d. All of the above are correct.

33. Ionization occurs upon solution of
    a. ionic compounds.
    b. some polar molecules.
    c. nonpolar molecules.
    d. None of the above is correct.

34. Adding sodium chloride to water raises the boiling point of water because
    a. sodium chloride has a higher boiling point.
    b. sodium chloride ions occupy space at the water surface.
    c. sodium chloride ions have stronger ion-ion bonds than water.
    d. the energy of hydration is higher.

35. The ice that forms in freezing seawater is
   a. pure water.
   b. the same salinity as liquid seawater.
   c. more salty than liquid seawater.
   d. denser than liquid seawater.

36. Salt solutions freeze at a lower temperature than pure water because
   a. more ionic bonds are present.
   b. salt solutions have a higher vapor pressure.
   c. ions get in the way of water molecules trying to form ice.
   d. salt naturally has a lower freezing point than water.

37. Which of the following would have a pH of less than 7?
   a. a solution of ammonia
   b. a solution of sodium chloride
   c. pure water
   d. carbonic acid

38. Which of the following would have a pH of more than 7?
   a. a solution of ammonia
   b. a solution of sodium chloride
   c. pure water
   d. carbonic acid

39. Solutions of acids, bases, and salts have what in common? All have
   a. proton acceptors.
   b. proton donors.
   c. ions.
   d. polar molecules.

40. When a solution of an acid and a base are mixed together,
   a. a salt and water are formed.
   b. they lose their acid and base properties.
   c. both are neutralized.
   d. All of the above are correct.

41. A substance that ionizes completely into hydronium ions is known as a
   a. strong acid.
   b. weak acid.
   c. strong base.
   d. weak base.

42. A scale of values that expresses the hydronium ion concentration of a solution is known as
   a. an acid-base indicator.
   b. the pH scale.
   c. the solubility scale.
   d. the electrolyte scale.

43. Substance A has a pH of 2 and substance B has a pH of 3. This means that
   a. substance A has more basic properties than substance B.
   b. substance B has more acidic properties than substance A.
   c. substance A is ten times more acidic than substance B.
   d. substance B is ten times more acidic than substance A.

44. Water is a (an)
   a. ionic compound.
   b. covalent compound.
   c. polar covalent compound.
   d. triatomic molecule.

45. The heat of fusion, specific heat, and heat of vaporization of water are high compared to similar substances such as hydrogen sulfide, H_2S, because
   a. ionic bonds form in water molecules.
   b. hydrogen bonds form between water molecules.
   c. covalent bonds form between water molecules.
   d. covalent bonds form in water molecules.

46. The substance that is a nonelectrolyte is
   a. sodium chloride, NaCl.
   b. copper sulfate, $CuSO_4$.
   c. glucose, $C_6H_{12}O_6$.
   d. aluminum chloride, $AlCl_3$.

47. Salt is spread on icy roads because it
   a. lowers the freezing point of water.
   b. increases the freezing point of water.
   c. increases the boiling point of water.
   d. increases the melting point of water.

48. Hard water is a solution of
   a. Na^+ and Cl^- ions.
   b. Na^+ and K^+ ions.
   c. Ca^{2+} and Mg^{2+} ions.
   d. Ba^{2+} and Cl^- ions.

49. A solution has a pH of 9. This means that the solution is
   a. acidic.
   b. neutral.
   c. basic.
   d. unknown.

Answers

1. c **2.** a **3.** a **4.** a **5.** a **6.** c **7.** b **8.** a **9.** d **10.** b **11.** b **12.** d **13.** a **14.** b **15.** b **16.** b **17.** b **18.** b **19.** c **20.** a **21.** c **22.** b **23.** b **24.** d **25.** a **26.** c **27.** d **28.** a **29.** c **30.** b **31.** b **32.** c **33.** b **34.** b **35.** a **36.** c **37.** d **38.** a **39.** c **40.** d **41.** a **42.** b **43.** c **44.** c **45.** b **46.** c **47.** a **48.** c **49.** c

QUESTIONS FOR THOUGHT

1. How is a solution different from other mixtures?
2. Explain why some ionic compounds are soluble while others are insoluble in water.
3. Explain why adding salt to water increases the boiling point.
4. A deep lake in Minnesota is covered with ice. What is the water temperature at the bottom of the lake? Explain your reasoning.
5. Explain why water has a greater density at 4°C than at 0°C.
6. What is hard water? How is it softened?
7. According to the definition of an acid and the definition of a base, would the pH increase, decrease, or remain the same when NaCl is added to pure water? Explain.
8. What is a hydrogen bond? Explain how a hydrogen bond forms.
9. What feature of a soap molecule gives it cleaning ability?
10. What ion is responsible for (a) acidic properties? (b) For basic properties?
11. Explain why a pH of 7 indicates a neutral solution—why not some other number?

FOR FURTHER ANALYSIS

1. What are the basic differences and similarities between the concentration measures of salinity and percent by weight?
2. Compare and contrast the situations where you would express concentration in (1) parts per million, (2) parts per billion, (3) percent (volume or weight), and (4) salinity.

3. Analyze the basic reason that water is a universal solvent, becomes less dense when it freezes, has a high heat of fusion, has a high specific heat, and has a high heat of vaporization.

4. What is the same and what is different between a salt that will dissolve in water and one that is insoluble?

5. There are at least three ways to change the boiling point of water, so describe how you know for sure that 100°C (212°F) is the boiling point?

6. What are the significant similarities and differences between an acid, a base, and a salt?

7. Describe how you would teach someone why the pH of an acid is a low number (less than 7), while the pH of a base is a larger number (greater than 7).

8. Describe at least four different examples of how you could make hard water soft.

INVITATION TO INQUIRY

Water Temperature and Dissolving Gas

What relationship exists, if any, between the temperature of water and how much gas will dissolve in the water? You can find out by experimenting with water temperature and Alka-Seltzer tablets. Set up a flask on a sensitive balance, with two tablets wrapped in a tissue and lodged in the neck of the flask. Record the weight, and then gently push the tissue so it falls into the water. The difference in mass will be a result of carbon dioxide leaving the flask. Compare how much carbon dioxide is dissolved in water of different temperatures. Should you expect the same dissolving rate for oxygen?

PARALLEL EXERCISES

The exercises in groups A and B cover the same concepts. Solutions to group A exercises are located in appendix E.

Group A

1. A 50.0 g sample of a saline solution contains 1.75 g NaCl. What is the percentage by weight concentration?

2. A student attempts to prepare a 3.50 percent by weight saline solution by dissolving 3.50 g NaCl in 100 g of water. Since equation 11.2 calls for 100 g of solution, the correct amount of solvent should have been 96.5 g water ($100 - 3.5 = 96.5$). What percent by weight solution did the student actually prepare?

3. Seawater contains 30,113 ppm by weight dissolved sodium and chlorine ions. What is the percent by weight concentration of sodium chloride in seawater?

4. What is the mass of hydrogen peroxide, H_2O_2, in 250 grams of a 3.0 percent by weight solution?

5. How many mL of pure alcohol are in a 200 mL glass of wine that is 12 percent alcohol by volume?

6. How many mL of pure alcohol are in a single cocktail made with 50 mL of 40 percent vodka? (Note: "Proof" is twice the percent, so 80 proof is 40 percent.)

7. If fish in a certain lake are reported to contain 5 ppm by weight DDT, (a) what percentage of the fish meat is DDT? (b) How much of this fish would have to be consumed to reach a poisoning accumulation of 17.0 g of DDT?

8. For each of the following reactants, draw a circle around the proton donor and a box around the proton acceptor. Label which acts as an acid and which acts as a base.
   (a) $HC_2H_3O_{2(aq)} + H_2O_{(l)} \rightarrow H_3O^+_{(aq)} + C_2H_3O_2^-_{(aq)}$
   (b) $C_6H_6NH_{2(l)} + H_2O_{(l)} \rightarrow C_6H_6NH_3^+_{(aq)} + OH^-_{(aq)}$
   (c) $HClO_{4(aq)} + HC_2H_3O_{2(aq)} \rightarrow H_2C_2H_3O_2^+_{(aq)} + ClO_4^-_{(aq)}$
   (d) $H_2O_{(l)} + H_2O_{(l)} \rightarrow H_3O^+_{(aq)} + OH^-_{(aq)}$

Group B

1. What is the percent by weight of a solution containing 2.19 g NaCl in 75 g of the solution?

2. What is the percent by weight of a solution prepared by dissolving 10 g of NaCl in 100 g of H_2O?

3. A concentration of 0.5 ppm by volume SO_2 in air is harmful to plant life. What is the percent by volume of this concentration?

4. What is the volume of water in a 500 mL bottle of rubbing alcohol that has a concentration of 70 percent by volume?

5. If a definition of intoxication is an alcohol concentration of 0.05 percent by volume in blood, how much alcohol would be present in the average (155 lb) person's 6,300 mL of blood if that person was intoxicated?

6. How much pure alcohol is in a 355 mL bottle of a "wine cooler" that is 5.0 percent alcohol by volume?

7. In the 1970s, when lead was widely used in "ethyl" gasoline, the blood level of the average American contained 0.25 ppm lead. The danger level of lead poisoning is 0.80 ppm. (a) What percent of the average person was lead? (b) How much lead would be in an average 80 kg person? (c) How much more lead would the average person need to accumulate to reach the danger level?

8. Draw a circle around the proton donor and a box around the proton acceptor for each of the reactants and label which acts as an acid and which acts as a base.
   (a) $H_3PO_{4(aq)} + H_2O_{(l)} \rightarrow H_3O^+_{(aq)} + H_2PO_4^-_{(aq)}$
   (b) $N_2H_{4(l)} + H_2O_{(l)} \rightarrow N_2H_5^+_{(aq)} + OH^-_{(aq)}$
   (c) $HNO_{3(aq)} + HC_2H_3O_{2(aq)} \rightarrow H_2C_2H_3O_2^+_{(aq)} + NO_3^-_{(aq)}$
   (d) $2\,NH_4^+_{(aq)} + Mg_{(s)} \rightarrow Mg^{2+}_{(aq)} + 2\,NH_3^+_{(aq)} + H_{2(g)}$

12

Organic Chemistry

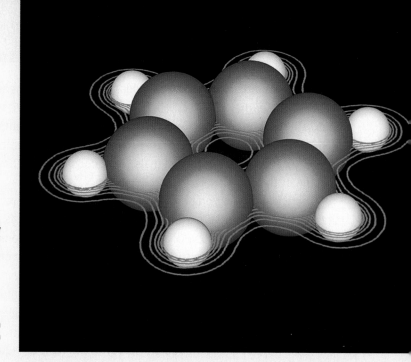

This is a computer-generated model of a benzene molecule showing six carbon atoms (gold) and six hydrogen atoms (white). Benzene is a hydrocarbon, an organic compound made up of the elements carbon and hydrogen.

CORE **CONCEPT**

The nature of the carbon atom allows for a great variety of organic compounds, many of which play vital roles in living.

Hydrocarbons
A great variety of organic compounds can be formed.

Hydrocarbon Derivatives
There are eight hydrocarbon functional groups, which include alcohols, ethers, aldehydes, ketones, esters, and organic acids.

OUTLINE

Petroleum
Petroleum is a mixture of hydrocarbons that are distilled into a variety of products.

Organic Compounds of Life
Living things build or break down large molecules of proteins, carbohydrates, and fats and oils.

The impact of ancient Aristotelian ideas on the development of understandings of motion, elements, and matter was discussed in earlier chapters. Historians also trace the "vitalist theory" back to Aristotle. According to Aristotle's idea, all living organisms are composed of the four elements (earth, air, fire, and water) and have in addition an *actuating force*, the life or soul that makes the organism different from nonliving things made of the same four elements. Plants, as well as animals, were considered to have this actuating, or vital, force in the Aristotelian scheme of things.

There were strong proponents of the vitalist theory as recently as the early 1800s. Their basic argument was that organic matter, the materials and chemical compounds recognized as being associated with life, could not be produced in the laboratory. Organic matter could only be produced in a living organism, they argued, because the organism had a vital force that is not present in laboratory chemicals. Then, in 1828, a German chemist named Friedrich Wöhler decomposed a chemical that was *not organic* to produce urea (N_2H_4CO), a known *organic* compound that occurs in urine. Wöhler's production of an organic compound was soon followed by the production of other organic substances by other chemists. The vitalist theory gradually disappeared with each new reaction, and a new field of study, organic chemistry, emerged.

This chapter is an introductory survey of the field of organic chemistry, which is concerned with compounds and reactions of compounds that contain carbon. You will find this an interesting, informative introduction, particularly if you have ever wondered about synthetic materials, natural foods and food products, or any of the thousands of carbon-based chemicals you use every day. The survey begins with the simplest of organic compounds, those consisting of only carbon and hydrogen atoms, compounds known as hydrocarbons. Hydrocarbons are the compounds of crude oil, which is the source of hundreds of petroleum products (Figure 12.1).

Most common organic compounds can be considered derivatives of the hydrocarbons, such as alcohols, ethers, fatty acids, and esters. Some of these are the organic compounds that give flavors to foods, and others are used to make hundreds of commercial products, from face cream to oleo. The main groups, or classes, of derivatives will be briefly introduced, along with some interesting examples of each group. Some of the important organic compounds of life, including proteins, carbohydrates, and fats, are discussed next. The chapter concludes with an introduction to synthetic polymers, what they are, and how they are related to the fossil fuel supply.

ORGANIC COMPOUNDS

Today, **organic chemistry** is defined as the study of compounds in which carbon is the principal element, whether the compound was formed by living things or not. The study of compounds that do not contain carbon as a central element is called **inorganic chemistry.** An *organic compound* is thus a compound that contains carbon as the principal element, and an *inorganic compound* is any other compound.

Organic compounds, by definition, must contain carbon, whereas all the inorganic compounds can contain all the other elements. Yet the majority of known compounds are organic. Several million organic compounds are known and thousands of new ones are discovered every year. You use organic compounds every day, including gasoline, plastics, grain alcohol, foods, flavorings, and many others.

It is the unique properties of carbon that allow it to form so many different compounds. A carbon atom has a valence of four and can combine with one, two, three, or four *other carbon atoms,* in addition to a wide range of other kinds of atoms (Figure 12.2). The number of possible molecular combinations is almost limitless, which explains why there are so many organic compounds. Fortunately, there are patterns of groups of carbon atoms and groups of other atoms that lead to similar chemical characteristics, making the study of organic chemistry less difficult. The key to success in studying organic chemistry is to recognize patterns and to understand the code and meaning of organic chemical names.

HYDROCARBONS

A **hydrocarbon** is an organic compound consisting of only two elements. As the name implies, these elements are hydrogen and carbon. The simplest hydrocarbon has one carbon atom and four hydrogen atoms (Figure 12.3), but since carbon atoms

FIGURE 12.1 Refinery and tank storage facilities, like this one in Texas, are needed to change the hydrocarbons of crude oil to many different petroleum products. The classes and properties of hydrocarbons form one topic of study in organic chemistry.

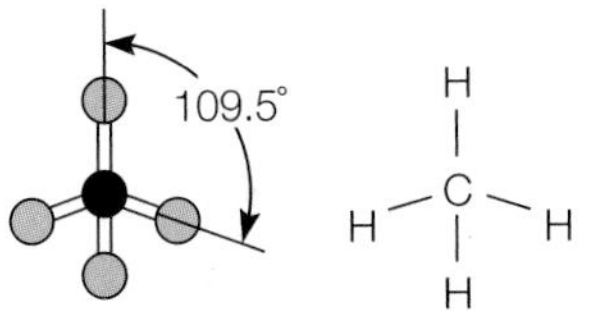

A Three-dimensional model

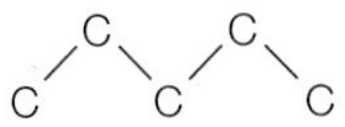

B An unbranched chain

$$C—C—C—C—C$$

C Simplified unbranched chain

FIGURE 12.2 (*A*) The carbon atom forms bonds in a tetrahedral structure with a bond angle of 109.5°. (*B*) Carbon-to-carbon bond angles are 109.5°, so a chain of carbon atoms makes a zigzag pattern. (*C*) The unbranched chain of carbon atoms is usually simplified in a way that looks like a straight chain, but it is actually a zigzag, as shown in (*B*).

CH_4

A Molecular formula

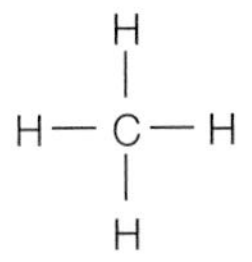

B Structural formula

FIGURE 12.3 A molecular formula (*A*) describes the numbers of different kinds of atoms in a molecule, and a structural formula (*B*) represents a two-dimensional model of how the atoms are bonded to each other. Each dash represents a bonding pair of electrons.

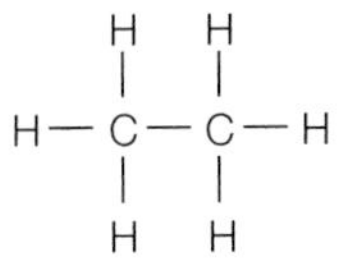

A Ethane

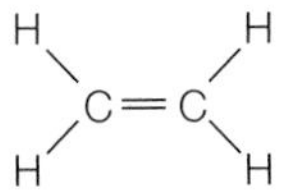

B Ethene

$$H—C≡C—H$$

C Ethyne

FIGURE 12.4 Carbon-to-carbon bonds can be single (*A*), double (*B*), or triple (*C*). Note that in each example, each carbon atom has four dashes, which represent four bonding pairs of electrons, satisfying the octet rule.

can combine with one another, there are thousands of possible structures and arrangements. The carbon-to-carbon bonds are covalent and can be single, double, or triple (Figure 12.4). Recall that the dash in a structural formula means one shared electron pair, a covalent bond. To satisfy the octet rule, this means that each carbon atom must have a total of four dashes around it, no more and no less. Note that when the carbon atom has double or triple bonds, fewer hydrogen atoms can be attached as the octet rule is satisfied. There are four groups of hydrocarbons that are classified according to how the carbon atoms are put together: (1) *alkanes,* (2) *alkenes,* (3) *alkynes,* and (4) *aromatic hydrocarbons.*

The **alkanes** are *hydrocarbons with single covalent bonds* between the carbon atoms. Alkanes that are large enough to form chains of carbon atoms occur with a straight structure, a branched structure, or a ring structure, as shown in Figure 12.5. (The "straight" structure is actually a zigzag, as shown in Figure 12.2.) You are familiar with many alkanes, for they make up the bulk of petroleum and petroleum products, which will be discussed shortly. The clues and codes in the names of the alkanes will be considered first.

The alkanes are also called the *paraffin series.* The alkanes are not as chemically reactive as the other hydrocarbons, and the term *paraffin* means "little affinity." They are called a series because *each higher molecular weight alkane has an additional CH_2.* The simplest alkane is methane, CH_4, and the next highest molecular weight alkane is ethane, C_2H_6. As you can see, C_2H_6 is CH_4 with an additional CH_2. If you compare the first ten alkanes

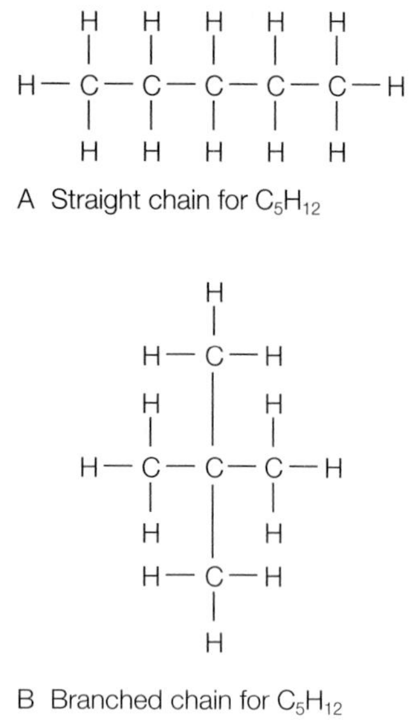

A Straight chain for C_5H_{12}

B Branched chain for C_5H_{12}

C Ring chain for C_5H_{10}

FIGURE 12.5 Carbon-to-carbon chains can be (A) straight, (B) branched, or (C) in a closed ring. (Some carbon bonds are drawn longer but are actually the same length.)

A n-butane, C_4H_{10}

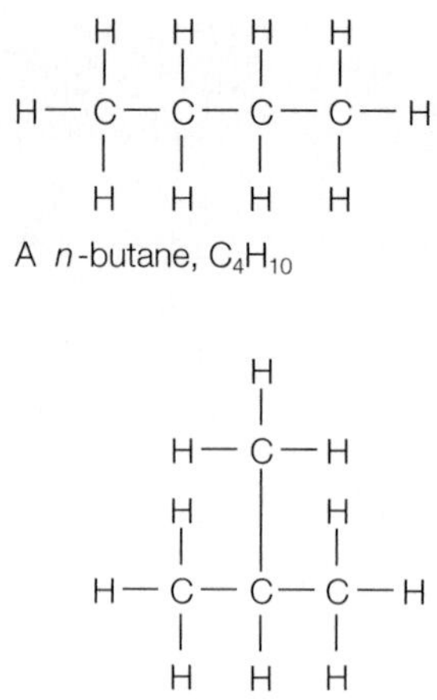

B Isobutane (2-methylpropane), C_4H_{10}

FIGURE 12.6 (A) A straight-chain alkane is identified by the prefix n- for "normal" in the common naming system. (B) A branched-chain alkane isomer is identified by the prefix iso- for "isomer" in the common naming system. In the IUPAC name, isobutane is 2-methylpropane. (Carbon bonds are actually the same length.)

in Table 12.1, you will find that each successive compound in the series always has an additional CH_2.

Note the names of the alkanes listed in Table 12.1. From pentane on, the names have a consistent prefix and suffix pattern. The prefix and suffix pattern is a code that provides a clue about the compound. The Greek prefix tells you the *number of carbon atoms* in the molecule, for example, *oct-* means eight, so octane has eight carbon atoms. The suffix *-ane* tells you this hydrocarbon is a member of the alk*ane* series, so it has single bonds only. With the general alkane formula of C_nH_{2n+2}, you can now write the formula when you hear the name. Octane has eight carbon atoms with single bonds and $n = 8$. Two times 8 plus 2 ($2n + 2$) is 18, so the formula for octane is C_8H_{18}. Most organic chemical names provide clues like this.

The alkanes in Table 12.1 all have straight chains. A straight, continuous chain is identified with the term *normal,* which is abbreviated n. Figure 12.6A shows n-butane with a straight chain and a molecular formula of C_4H_{10}. Figure 12.6B shows a different branched structural formula that has the same C_4H_{10} molecular formula. Compounds with the same molecular formulas with different structures are called

isomers. Since the straight-chained isomer is called n-butane, the branched isomer is called *isobutane.* The isomers of a particular alkane, such as butane, have different physical and chemical properties because they have different structures. Isobutane, for example, has a boiling point of $-10°C$. The boiling point of n-butane, on the other hand, is $-0.5°C$. In the "Petroleum" section, you will learn that the various isomers of the octane hydrocarbon perform differently in automobile engines, requiring the "reforming" of n-octane to *iso-octane* before it can be used.

Methane, ethane, and propane can have only one structure each, and butane has two isomers. The number of possible isomers for a particular molecular formula increases rapidly as the number of carbon atoms increases. After butane, hexane has five isomers, octane eighteen isomers, and decane seventy-five isomers. Because they have different structures, each isomer has different physical properties. A different naming system is needed because there are just too many isomers to keep track of. The system of naming the branched-chain alkanes is described by rules agreed upon by a professional organization, the International Union of Pure and Applied Chemistry, or IUPAC. Here are the steps in naming the alkane isomers.

Step 1: The longest continuous chain of carbon atoms determines the *base name* of the molecule. The longest continuous chain is not necessarily straight and can take any number of right-angle turns as long as the continuity is not broken. The base name corresponds to the number of carbon atoms in this chain as in Table 12.1. For example, the structure has six carbon atoms in the longest chain, so the base name is *hexane.*

TABLE 12.1

The first ten straight-chain alkanes

Name	Molecular Formula	Structural Formula	Name	Molecular Formula	Structural Formula
Methane	CH_4	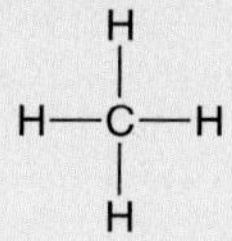	Hexane	C_6H_{14}	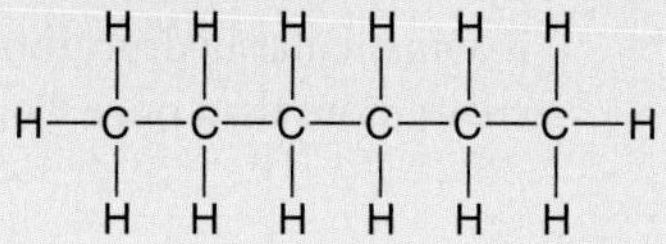
Ethane	C_2H_6	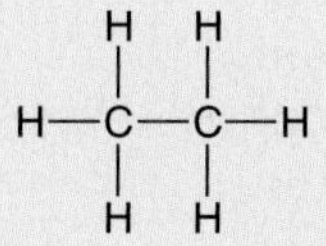	Heptane	C_7H_{16}	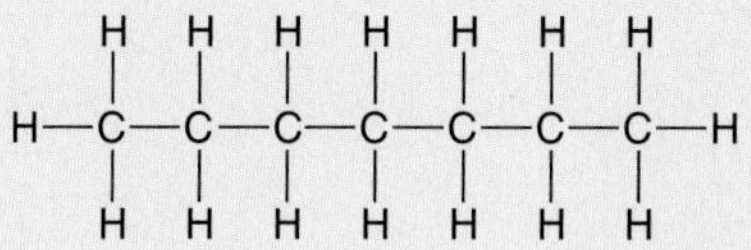
Propane	C_3H_8	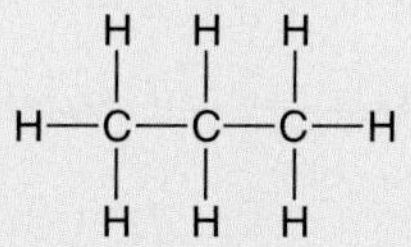	Octane	C_8H_{18}	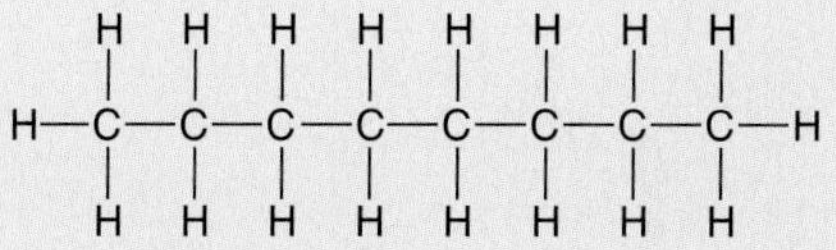
Butane	C_4H_{10}	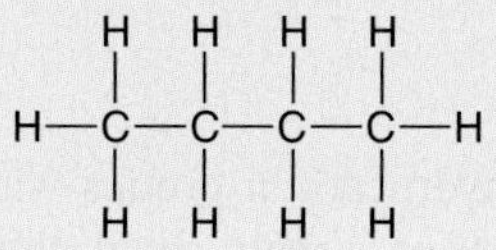	Nonane	C_9H_{20}	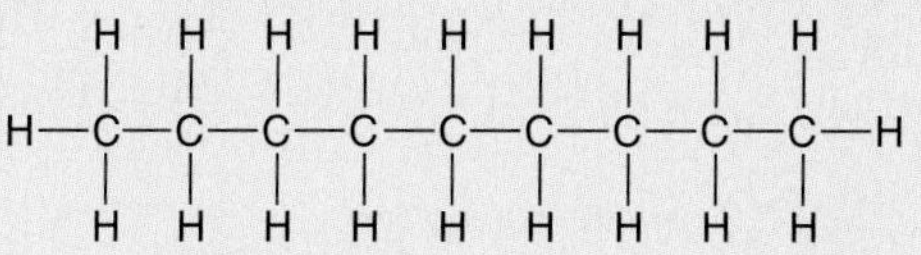
Pentane	C_5H_{12}	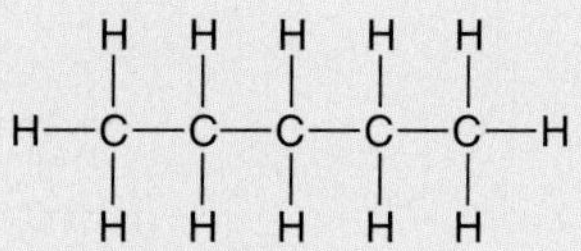	Decane	$C_{10}H_{22}$	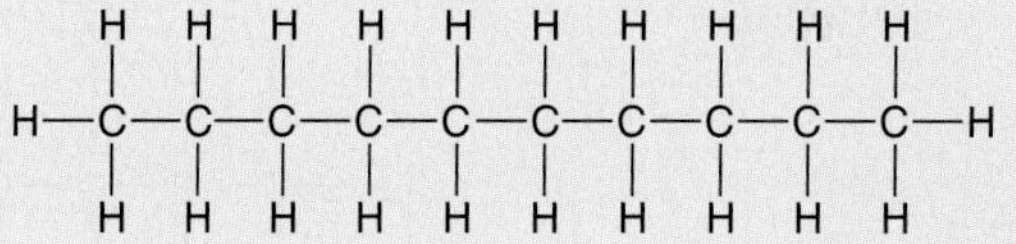

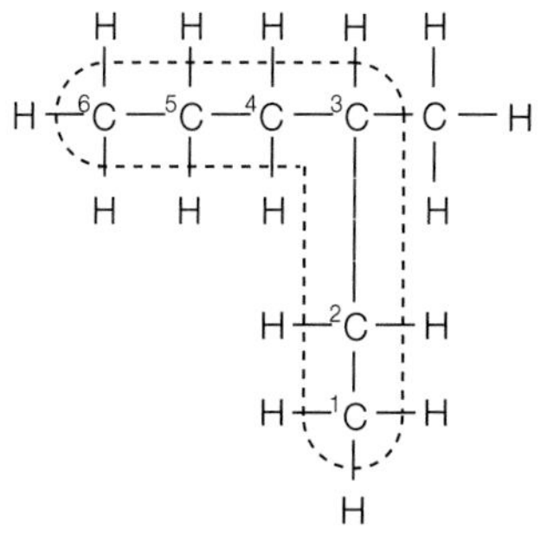

Step 2: The locations of other groups of atoms attached to the base chain are identified by counting carbon atoms from either the left or the right. The direction selected is the one that results in the *smallest* numbers for attachment locations. For example, the hexane chain has a CH_3 attached to the third or the fourth carbon atom, depending on which way you count. The third atom direction is chosen since it results in a smaller number.

Step 3: The hydrocarbon groups attached to the base chain are named from the number of carbons in the group by changing the alkane suffix -*ane* to -*yl*. Thus, a hydrocarbon group attached to a base chain that has one carbon atom is called meth*yl*. Note that the -*yl* hydrocarbon groups have one less hydrogen than the corresponding alkane. Therefore, methane is CH_4, and a *methyl group* is CH_3. The first ten alkanes and their corresponding hydrocarbon group names are listed in Table 12.2. In the example, a methyl group is attached to the third carbon atom of the base hexane chain. The name and address of this hydrocarbon group is 3-methyl. The compound is named 3-methylhexane.

Step 4: The prefixes *di-*, *tri-*, and so on are used to indicate if a particular hydrocarbon group appears on the main chain more than once. For example,

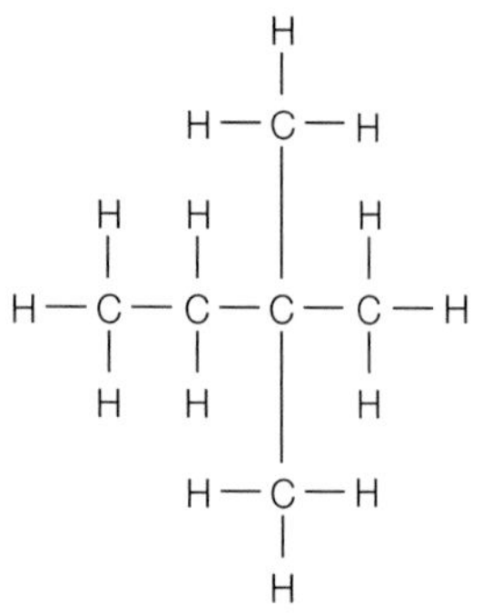

(or)

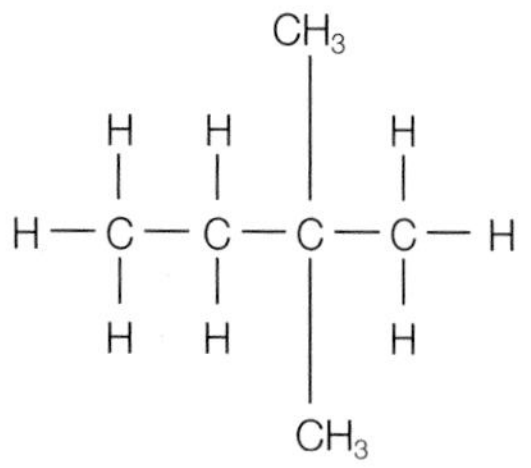

is 2,2-dimethylbutane and

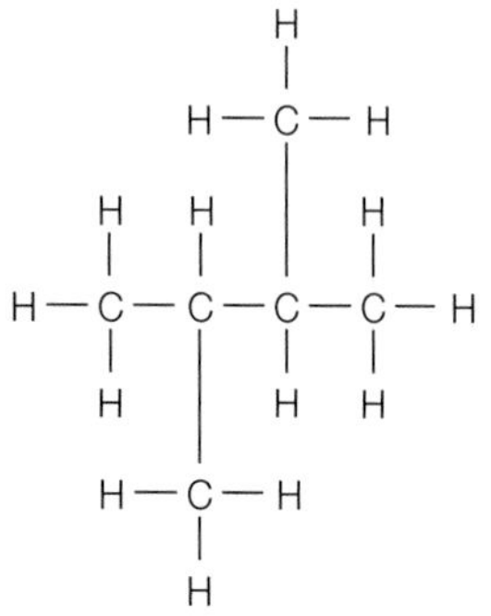

(or)

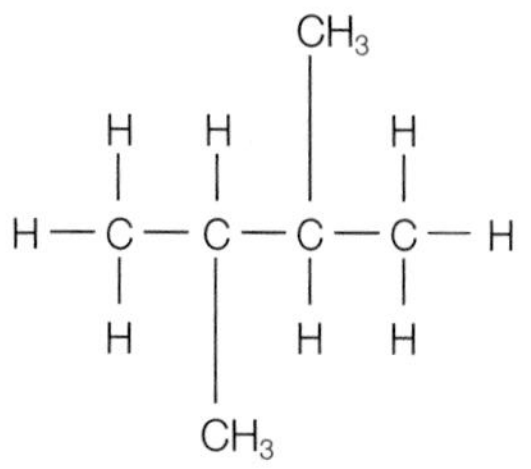

is 2,3-dimethylbutane.

TABLE 12.2

Alkane hydrocarbons and corresponding hydrocarbon groups

Alkane Name	Molecular Formula	Hydrocarbon Group	Molecular Formula
Methane	CH_4	Methyl	$-CH_3$
Ethane	C_2H_6	Ethyl	$-C_2H_5$
Propane	C_3H_8	Propyl	$-C_3H_7$
Butane	C_4H_{10}	Butyl	$-C_4H_9$
Pentane	C_5H_{12}	Amyl	$-C_5H_{11}$
Hexane	C_6H_{14}	Hexyl	$-C_6H_{13}$
Heptane	C_7H_{16}	Heptyl	$-C_7H_{15}$
Octane	C_8H_{18}	Octyl	$-C_8H_{17}$
Nonane	C_9H_{20}	Nonyl	$-C_9H_{19}$
Decane	$C_{10}H_{22}$	Decyl	$-C_{10}H_{21}$

Note: $-CH_3$ means $*-C-H$ where * denotes unattached. The attachment takes place on a base chain or functional group.

If hydrocarbon groups with different numbers of carbon atoms are on a main chain, they are listed in alphabetic order. For example,

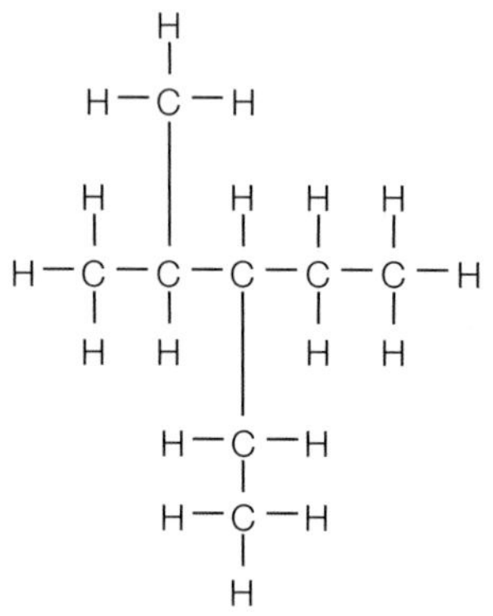

(or)

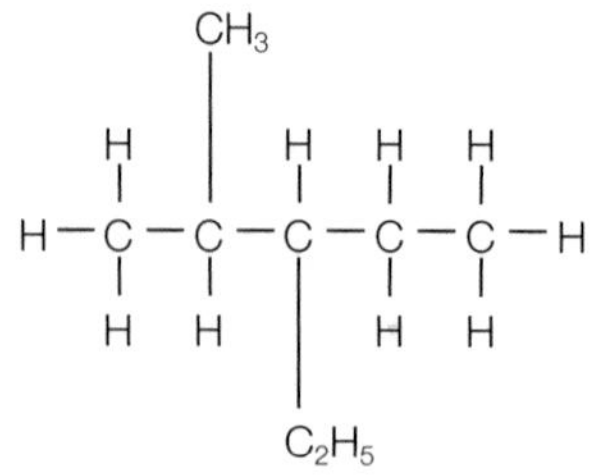

is named 3-ethyl-2-methylpentane. Note how numbers are separated from names by hyphens.

EXAMPLE 12.1

What is the name of an alkane with the following formula?

$$
\begin{array}{c}
\text{H} \\
| \\
\text{H}-\text{C}-\text{H} \qquad\qquad \text{H} \\
\qquad | \qquad\qquad\quad | \\
\text{H}-\text{C}-\text{H} \quad \text{H}-\text{C}-\text{H} \\
\end{array}
$$

H—C—C—C—C—C—C—C—H

H—C—H

SOLUTION

The longest continuous chain has seven carbon atoms, so the base name is heptane. The smallest numbers are obtained by counting from right to left and counting the carbons on this chain; there is a methyl group in carbon atom 2, a second methyl group on atom 4, and an ethyl group on atom 5. There are two methyl groups, so the prefix *di-* is needed, and the "e" of the ethyl group comes first in the alphabet, so ethyl is listed first. The name of the compound is 5-ethyl-2,4-dimethylheptane.

EXAMPLE 12.2

Write the structural formula for 2,2-dichloro-3-methyloctane. Answer:

H—C—C—C—C—C—C—C—C—H

ALKENES AND ALKYNES

The **alkenes** are *hydrocarbons with a double covalent carbon-to-carbon bond.* To denote the presence of a double bond, the *-ane* suffix of the alkanes is changed to *-ene* as in alk*ene* (Table 12.3). Figure 12.4 shows the structural formula for (A) ethane, C_2H_6, and (B) ethene, C_2H_4. Alkenes have room for two fewer hydrogen atoms because of the double bond, so the general alkene formula is C_nH_{2n}. Note the simplest alkene is called ethene but is commonly known as ethylene.

Ethylene is an important raw material in the chemical industry. Obtained from the processing of petroleum, about half of the commercial ethylene is used to produce the familiar polyethylene plastic. It is also produced by plants to ripen fruit, which

TABLE 12.3

The general molecular formulas and molecular structures of the alkanes, alkenes, and alkynes

Group	General Molecular Formula	Example Compound	Molecular Structure
Alkanes	C_nH_{2n+2}	Ethane	H—C—C—H
Alkenes	C_nH_{2n}	Ethene	C=C
Alkynes	C_nH_{2n-2}	Ethyne	H—C≡C—H

explains why unripe fruit enclosed in a sealed plastic bag with ripe fruit will ripen more quickly (Figure 12.7). The ethylene produced by the ripe fruit acts on the unripe fruit. Commercial fruit packers sometimes use small quantities of ethylene gas to quickly ripen fruit that was picked while green.

Perhaps you have heard the terms *saturated* and *unsaturated* in advertisements for cooking oil and margarine. An organic molecule, such as a hydrocarbon, that does not contain the maximum number of hydrogen atoms is an *unsaturated* hydrocarbon. For example, ethylene can add more hydrogen atoms by reacting with hydrogen gas to form ethane:

C=C + H_2 → H—C—C—H

Ethylene + Hydrogen → Ethane

FIGURE 12.7 Ethylene is the gas that ripens fruit, and a ripe fruit emits the gas, which will act on unripe fruit. Thus, a ripe tomato placed in a sealed bag with green tomatoes will help ripen them.

The ethane molecule has all the hydrogen atoms possible, so ethane is a *saturated* hydrocarbon. Unsaturated molecules are less stable, which means that they are more chemically reactive than saturated molecules.

Alkenes are named as the alkanes are, except (1) the longest chain of carbon atoms must contain the double bond, (2) the base name now ends in *-ene*, (3) the carbon atoms are numbered from the end nearest the double bond, and (4) the base name is given a number of its own, which identifies the address of the double bond. For example,

$$H-C=C-C-C-C-H$$

is named 4-methyl-1-pentene. The 1-pentene tells you there is a double bond (*-ene*), and the 1 tells you the double bond is after the first carbon atom in the longest chain containing the double bond. The methyl group is on the fourth carbon atom in this chain.

An **alkyne** is a *hydrocarbon with a carbon-to-carbon triple bond* and the general formula of C_nH_{2n-2}. The alkynes are highly reactive, and the simplest one, ethyne, has a common name of acetylene. Acetylene is commonly burned with oxygen gas in a welding torch because the flame reaches a temperature of about 3,000°C. Acetylene is also an important raw material in the production of plastics. The alkynes are named as the alkenes are, except the longest chain must contain the triple bond, and the base name suffix is changed to *-yne*.

CYCLOALKANES AND AROMATIC HYDROCARBONS

The hydrocarbons discussed up until now have been straight or branched open-ended chains of carbon atoms. Carbon atoms can also bond to each other to form a ring, or cyclic, structure. Figure 12.8 shows the structural formulas for some of these cyclic structures. Note that the cycloalkanes have the same molecular formulas as the alkenes; thus, they are isomers of the alkenes. They are, of course, very different compounds, with different physical and chemical properties. This shows the importance of structural, rather than simply molecular formulas in referring to organic compounds.

The six-carbon ring structure shown in Figure 12.9A has three double bonds that do not behave like the double bonds in the alkenes. In this six-carbon ring, the double bonds are not localized in one place but are spread over the whole molecule. Instead of alternating single and double bonds, all the bonds are something in between. This gives the C_6H_6 molecule increased stability. As a result, the molecule does not behave like other unsaturated compounds, that is, it does not readily react in order to add hydrogen to the ring. The C_6H_6 molecule is the organic compound named *benzene*. Organic compounds that are based on the benzene ring structure are called **aromatic**

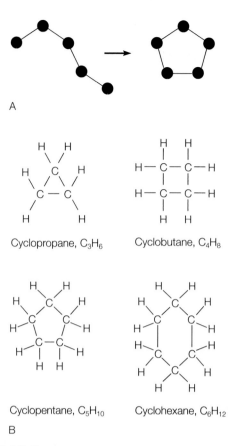

FIGURE 12.8 (*A*) The "straight" chain has carbon atoms that are able to rotate freely around their single bonds, sometimes linking up in a closed ring. (*B*) Ring compounds of the first four cycloalkanes.

hydrocarbons. To denote the six-carbon ring with delocalized electrons, benzene is represented by the symbol shown in Figure 12.9B.

The circle in the six-sided benzene symbol represents the delocalized electrons. Figure 12.9B illustrates how this benzene ring symbol is used to show the structural formula of some aromatic hydrocarbons. You may have noticed some of the names on labels of paints, paint thinners, and lacquers. Toluene and the xylenes are commonly used in these products as solvents. A benzene ring attached to another molecule or functional group is given the name *phenyl*.

PETROLEUM

Petroleum is a mixture of alkanes, cycloalkanes, and some aromatic hydrocarbons. The origin of petroleum is uncertain, but it is believed to have formed from the slow decomposition of buried marine life, primarily microscopic plankton and algae in the absence of oxygen (i.e., anaerobic). Time, temperature, pressure, and perhaps bacteria are considered important in the formation of petroleum. As the petroleum formed, it was forced through porous rock until it reached a rock type or rock structure that stopped it. Here, it accumulated to saturate the porous

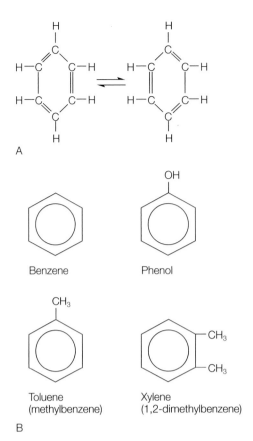

A

Benzene

Phenol

Toluene
(methylbenzene)

Xylene
(1,2-dimethylbenzene)

B

FIGURE 12.9 (A) The bonds in C_6H_6 are something between single and double, which gives it different chemical properties than double-bonded hydrocarbons. (B) The six-sided symbol with a circle represents the benzene ring. Organic compounds based on the benzene ring are called *aromatic hydrocarbons* because of their aromatic character.

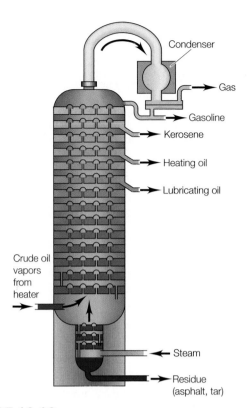

FIGURE 12.10 Fractional distillation is used to separate petroleum into many products. This simplified illustration shows how the about 9 m (30 ft) tower is used to separate the different "fractions" by differences in their boiling points.

rock, forming an accumulation called an *oil field*. The composition of petroleum varies from one oil field to the next. The oil from a given field might be dark or light in color, and it might have an asphalt or paraffin base. Some oil fields contain oil with a high quantity of sulfur, referred to as "sour crude." Because of such variations, some fields have oil with more desirable qualities than oil from other fields.

Early settlers found oil seeps in the eastern United States and collected the oil for medicinal purposes. One enterprising oil peddler tried to improve the taste by running the petroleum through a whiskey still. He obtained a clear liquid by distilling the petroleum and, by accident, found that the liquid made an excellent lamp oil. This was fortunate timing, for the lamp oil used at that time was whale oil, and whale oil production was declining. This clear liquid obtained by distilling petroleum is today known as *kerosene*.

Wells were drilled, and crude oil refineries were built to produce the newly discovered lamp oil. Gasoline was a by-product of the distillation process and was used primarily as a spot remover. With Henry Ford's automobile production and Edison's electric light invention, the demand for gasoline increased, and the demand for kerosene decreased. The refineries

were converted to produce gasoline, and the petroleum industry grew to become one of the world's largest industries.

Crude oil is petroleum that is pumped from the ground, a complex and variable mixture of hydrocarbons with an upper limit of about fifty carbon atoms. This thick, smelly black mixture is not usable until it is refined, that is, separated into usable groups of hydrocarbons called petroleum products. Petroleum products are separated from crude oil by distillation, and any particular product has a boiling point range, or "cut" of the distilled vapors (Figure 12.10). Thus, each product, such as gasoline, heating oil, and so forth is made up of hydrocarbons within a range of carbon atoms per molecule (Figure 12.11). The products, their boiling ranges, and ranges of carbon atoms per molecule are listed in Table 12.4.

The hydrocarbons that have one to four carbon atoms (CH_4 to C_4H_{10}) are gases at room temperature. They can be pumped from certain wells as a gas, but they also occur dissolved in crude oil. *Natural gas* is a mixture of hydrocarbon gases, but it is about 95 percent methane (CH_4). Propane (C_3H_8) and butane (C_4H_{10}) are liquified by compression and cooling and are sold as liquified petroleum gas, or *LPG*. LPG is used where natural gas is not available for cooking or heating and is widely used as a fuel in barbecue grills and camp stoves.

Gasoline is a mixture of hydrocarbons that may have five to twelve carbon atoms per molecule. Gasoline distilled from crude oil consists mostly of straight-chain molecules not suitable for use as an automotive fuel. Straight-chain molecules

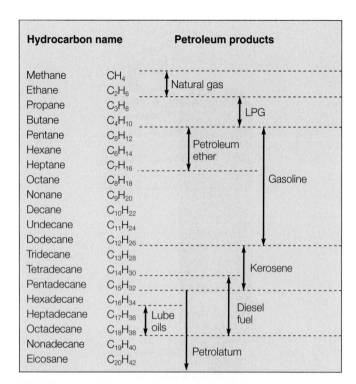

FIGURE 12.11 Petroleum products and the ranges of hydrocarbons in each product.

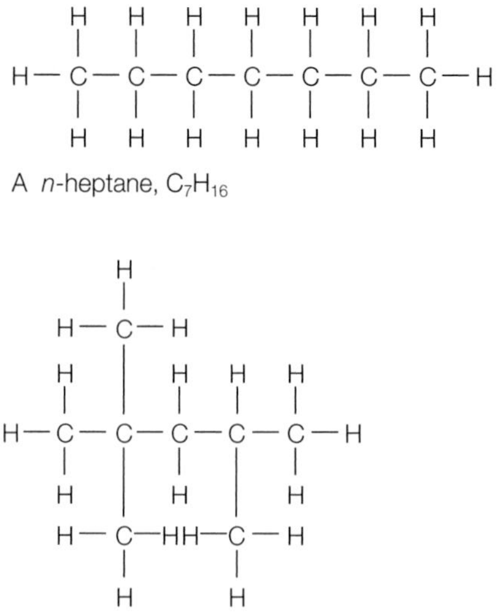

A n-heptane, C_7H_{16}

B 2,2,4-trimethylpentane (or iso-octane), C_8H_{18}

FIGURE 12.12 The octane rating scale is a description of how rapidly gasoline burns. It is based on (A) n-heptane, with an assigned octane number of 0, and (B) 2,2,4-trimethylpentane, with an assigned number of 100.

TABLE 12.4
Petroleum products

Name	Boiling Range (°C)	Carbon Atoms per Molecule
Natural gas	Less than 0	C_1 to C_4
Petroleum ether	35–100	C_5 to C_7
Gasoline	35–215	C_5 to C_{12}
Kerosene	35–300	C_{12} to C_{15}
Diesel fuel	300–400	C_{15} to C_{18}
Motor oil, grease	350–400	C_{16} to C_{18}
Paraffin	Solid, melts at about 55	C_{20}
Asphalt	Boiler residue	C_{40} or more

burn too rapidly in an automobile engine, producing more of an explosion than a smooth burn. You hear these explosions as a knocking or pinging in the engine, and they indicate poor efficiency and could damage the engine. On the other hand, branched-chain molecules burn slower by comparison, without the pinging or knocking explosions. The burning rate of gasoline is described by the *octane number* scale. The scale is based on pure *n*-heptane, straight-chain molecules that are assigned an octane number of 0, and a multiple branched isomer of octane, 2,2,4-trimethylpentane, which is assigned an octane number of 100 (Figure 12.12). Most gasolines have an octane rating of 87, which could be obtained with a mixture that is 87 percent 2,2,4-trimethylpentane and 13 percent *n*-heptane. Gasoline, however, is a much more complex mixture.

It is expensive to process gasoline because some of the straight-chain hydrocarbon molecules must be converted into branched molecules. The process is one of "cracking and reforming" some of the straight-chain molecules. First, the gasoline is passed through metal tubes heated to 500°C to 800°C (932°F to 1,470°F). At this high temperature, and in the absence of oxygen, the hydrocarbon molecules decompose by breaking into smaller carbon-chain units. These smaller hydrocarbons are then passed through tubes containing a catalyst, which causes them to reform into branched-chain molecules. Unleaded gasoline is produced by the process. Without the reforming that produces unleaded gasoline, low-numbered hydrocarbons (such as ethylene) can be produced. Ethylene is used as a raw material for many plastic materials, antifreeze, and other products. Cracking is also used to convert higher-numbered hydrocarbons, such as heating oil, into gasoline.

Kerosene is a mixture of hydrocarbons that have from twelve to fifteen carbon atoms. The petroleum product called kerosene is also known by other names, depending on its use. Some of these names are lamp oil (with coloring and odorants added), jet fuel (with a flash flame retardant added), heating oil, #1 fuel oil, and in some parts of the country, "coal oil."

Diesel fuel is a mixture of a group of hydrocarbons that have from fifteen to eighteen carbon atoms per molecule. Diesel fuel also goes by other names, again depending on its use; for example, diesel fuel, distillate fuel oil, heating oil, or #2 fuel oil. During the summer season, there is a greater demand for gasoline than for heating oil, so some of the supply is converted to gasoline by the cracking process.

Motor oil and *lubricating oils* have sixteen to eighteen carbon atoms per molecule. Lubricating grease is heavy oil that is thickened with soap. *Petroleum jelly*, also called petrolatum (or Vaseline), is a mixture of hydrocarbons with sixteen to thirty-two carbon atoms per molecule. *Mineral oil* is a light lubricating oil that has been decolorized and purified.

Depending on the source of the crude oil, varying amounts of *paraffin* wax (C_{20} or greater) or *asphalt* (C_{36} or more) may be present. Paraffin is used for candles, waxed paper, and home canning. Asphalt is mixed with gravel and used to surface roads. It is also mixed with refinery residues and lighter oils to make a fuel called #6 fuel oil or residual fuel oil. Industries and utilities often use this semisolid material that must be heated before it will pour. Number 6 fuel oil is used as a boiler fuel, costing about half as much as #2 fuel oil.

HYDROCARBON DERIVATIVES

The hydrocarbons account for only about 5 percent of the known organic compounds, but the other 95 percent can be considered hydrocarbon derivatives. **Hydrocarbon derivatives** are formed when *one or more hydrogen atoms on a hydrocarbon have been replaced by some element or group of elements other than hydrogen.* For example, the halogens (F_2, Cl_2, Br_2) react with an alkane in sunlight or when heated, replacing a hydrogen:

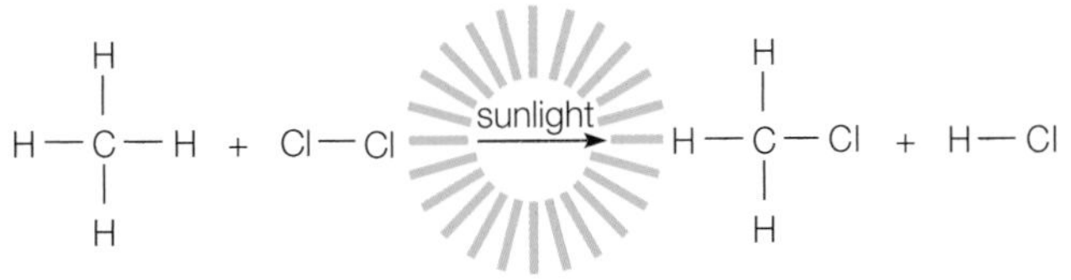

In this particular *substitution reaction,* a hydrogen atom on methane is replaced by a chlorine atom to form methyl chloride. Replacement of any number of hydrogen atoms is possible, and a few *organic halides* are illustrated in Figure 12.13.

If a hydrocarbon molecule is unsaturated (has a multiple bond), a hydrocarbon derivative can be formed by an *addition reaction:*

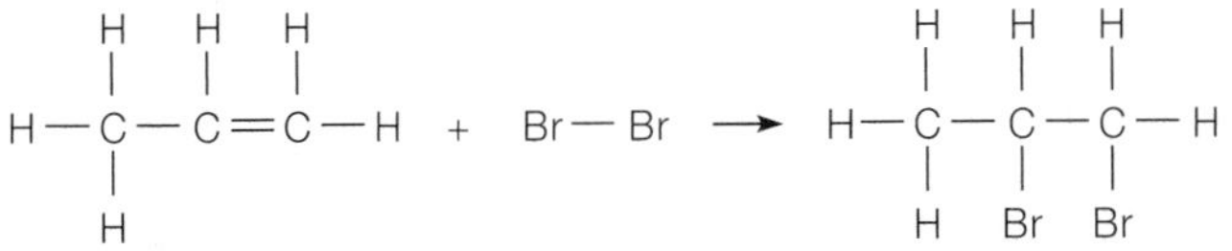

The bromine atoms add to the double bond on propene, forming 1,2-dibromopropane.

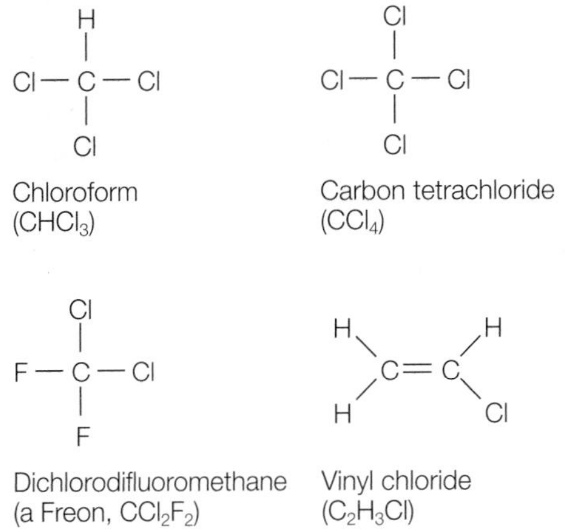

Chloroform
(CHCl₃)

Carbon tetrachloride
(CCl₄)

Dichlorodifluoromethane
(a Freon, CCl₂F₂)

Vinyl chloride
(C₂H₃Cl)

FIGURE 12.13 Common examples of organic halides.

Alkene molecules can also add to each other in an addition reaction to form a very long chain consisting of hundreds of molecules. A long chain of repeating units is called a **polymer** (*poly-* = many; *-mer* = segment), and the reaction is called *addition polymerization.* Ethylene, for example, is heated under pressure with a catalyst to form *polyethylene.* Heating breaks the double bond,

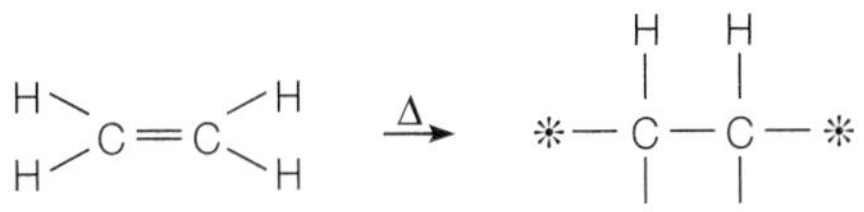

which provides sites for single covalent bonds to join the ethylene units together,

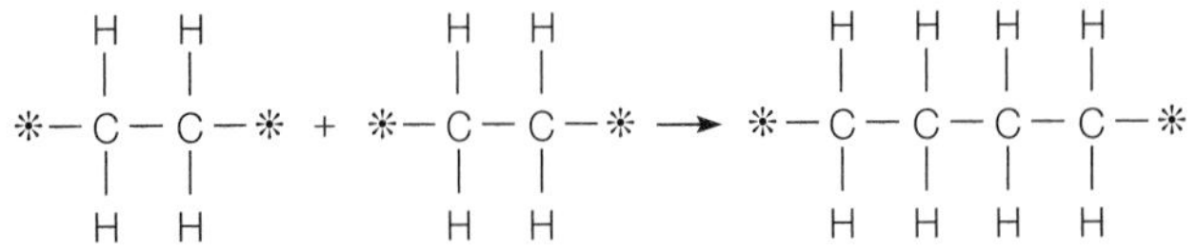

which continues the addition polymerization until the chain is hundreds of units long. Synthetic polymers such as polyethylene are discussed in the "Synthetic Polymers" section.

The addition reaction and the addition polymerization reaction can take place because of the double bond of the alkenes, and, in fact, the double bond is the site of most alkene reactions. The atom or group of atoms in an organic molecule that is the site of a chemical reaction is identified as a **functional group.** *It is the functional group that is responsible for the chemical properties of an organic compound.* Functional groups usually have (1) multiple bonds or (2) lone pairs of electrons that cause them to be sites of reactions. Table 12.5 lists some of the common hydrocarbon functional groups. Look over this list, comparing the structure of the functional group with the group name. Some of the more interesting examples from a few of these groups will be considered next. Note that the R and R′ (pronounced, "R prime") stand for one or more of the hydrocarbon groups from Table 12.2. For example, in the reaction between methane and chlorine, the product is methyl chloride. In this case, the R in RCl stands for methyl, but it could represent any hydrocarbon group.

ALCOHOLS

An *alcohol* is an organic compound formed by replacing one or more hydrogens on an alkane with a hydroxyl functional group (−OH). The hydroxyl group should not be confused with the hydroxide ion, OH⁻. The hydroxyl group is attached to an organic compound and does not form ions in solution as the hydroxide ion does. It remains attached to a hydrocarbon group (R), giving the compound its set of properties that are associated with alcohols.

The name of the hydrocarbon group (Table 12.2) determines the name of the alcohol. If the hydrocarbon group in ROH is methyl, for example, the alcohol is called *methyl alcohol.* Using the IUPAC naming rules, the name of an alcohol

TABLE 12.5

Selected organic functional groups

Name of Functional Group	General Formula	General Structure
Organic halide	RCl	R—C̈l:
Alcohol	ROH	R—Ö—H
Ether	ROR'	R—Ö—R'
Aldehyde	RCHO	R—C—H, ‖ :O:
Ketone	RCOR'	R—C—R', ‖ :O:
Organic acid	RCOOH	R—C—Ö—H, ‖ :O:
Ester	RCOOR'	R—C—Ö—R', ‖ :O:
Amine	RNH₂	R—N̈—H, │ H

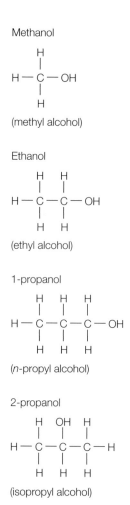

Methanol

(methyl alcohol)

Ethanol

(ethyl alcohol)

1-propanol

(n-propyl alcohol)

2-propanol

(isopropyl alcohol)

FIGURE 12.14 Four different alcohols. The IUPAC name is given above each structural formula, and the common name is given below.

has the suffix *-ol*. Thus, the IUPAC name of methyl alcohol is *methanol*.

All alcohols have the hydroxyl functional group, and all are chemically similar (Figure 12.14). Alcohols are toxic to humans, except that ethanol can be consumed in limited quantities. Consumption of other alcohols such as 2-propanol (isopropyl alcohol, or "rubbing alcohol") can result in serious gastric distress. Consumption of methanol can result in blindness and death. Ethanol, C_2H_5OH, is produced by the action of yeast or by a chemical reaction of ethylene derived from petroleum refining. Yeast acts on sugars to produce ethanol and CO_2. When beer, wine, and other such beverages are the desired products, the CO_2 escapes during fermentation, and the alcohol remains in solution. In baking, the same reaction utilizes the CO_2 to make the dough rise, and the alcohol is evaporated during baking. Most alcoholic beverages are produced by the yeast fermentation reaction, but some are made from ethanol derived from petroleum refining.

Alcohols with six or fewer carbon atoms per molecule are soluble in both alkanes and water. A solution of ethanol and gasoline is called *gasohol* (Figure 12.15). Alcoholic beverages are a solution of ethanol and water. The *proof* of such a beverage is double the ethanol concentration by volume. Therefore, a solution of 40 percent ethanol by volume in water is 80 proof, and wine that is 12 percent alcohol by volume is 24 proof. Distillation alone will produce a 190 proof

FIGURE 12.15 Gasoline is a mixture of hydrocarbons (C_8H_{18}, for example) that contain no atoms of oxygen. Gasohol contains ethyl alcohol, C_2H_5OH, which does contain oxygen. The addition of alcohol to gasoline, therefore, adds oxygen to the fuel. Since carbon monoxide forms when there is an insufficient supply of oxygen, the addition of alcohol to gasoline helps cut down on carbon monoxide emissions.

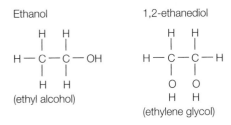

Ethanol
1,2-ethanediol

```
     H   H
     |   |
 H — C — C — OH
     |   |
     H   H
 (ethyl alcohol)
```

```
     H   H
     |   |
 H — C — C — H
     |   |
     O   O
     H   H
 (ethylene glycol)
```

1,2,3-propanetriol

```
     H   H   H
     |   |   |
 H — C — C — C — H
     |   |   |
     O   O   O
     H   H   H
 (glycerol or glycerin)
```

FIGURE 12.16 Common examples of alcohols with one, two, and three hydroxyl groups per molecule. The IUPAC name is given above each structural formula, and the common name is given below.

Carbonyl group

```
 — C —
    ||
    O
 A
```

An aldehyde

```
 R(H) — C — H
         ||
         O
 B
```

A ketone

```
 R — C — R'
     ||
     O
```

Methanal

```
 H — C — H
     ||
     O
 (formaldehyde)
 C
```

Propanone

```
 CH₃ — C — CH₃
       ||
       O
 (acetone)
```

FIGURE 12.17 The carbonyl group (*A*) is present in both aldehydes and ketones, as shown in (*B*). (*C*) The simplest example of each, with the IUPAC name above and the common name below each formula.

concentration, but other techniques are necessary to obtain 200 proof absolute alcohol. *Denatured alcohol* is ethanol with acetone, formaldehyde, and other chemicals in solution that are difficult to separate by distillation. Since these denaturants make consumption impossible, denatured alcohol is sold without the consumption tax.

Methanol, ethanol, and isopropyl alcohol all have one hydroxyl group per molecule. An alcohol with two hydroxyl groups per molecule is called a *glycol*. Ethylene glycol is perhaps the best-known glycol since it is used as an antifreeze. An alcohol with three hydroxyl groups per molecule is called *glycerol* (or *glycerin*). Glycerol is a by-product in the making of soap. It is added to toothpastes, lotions, and some candies to retain moisture and softness. Ethanol, ethylene glycol, and glycerol are compared in Figure 12.16.

Glycerol reacts with nitric acid in the presence of sulfuric acid to produce glyceryl trinitrate, commonly known as *nitroglycerine*. Nitroglycerine is a clear oil that is violently explosive, and when warmed, it is extremely unstable. In 1867, Alfred Nobel discovered that a mixture of nitroglycerine and siliceous earth was more stable than pure nitroglycerine but was nonetheless explosive. The mixture is packed in a tube and is called *dynamite*. Old dynamite tubes, however, leak pure nitroglycerine that is again sensitive to a slight shock.

ETHERS, ALDEHYDES, AND KETONES

An *ether* has a general formula of ROR′, and the best-known ether is diethylether. In a molecule of diethylether, both the R and the R′ are ethyl groups. Diethylether is a volatile, highly flammable liquid that was used as an anesthetic in the past. Today, it is used as an industrial and laboratory solvent.

Aldehydes and *ketones* both have a functional group of a carbon atom doubly bonded to an oxygen atom called a *carbonyl group*. The *aldehyde* has a hydrocarbon group, R (or a hydrogen in one case), and a hydrogen attached to the carbonyl group.

A *ketone* has a carbonyl group with two hydrocarbon groups attached (Figure 12.17).

The simplest aldehyde is *formaldehyde*. Formaldehyde is soluble in water, and a 40 percent concentration called *formalin* has been used as an embalming agent and to preserve biological specimens. Formaldehyde is also a raw material used to make plastics such as Bakelite. All the aldehydes have odors, and the odors of some aromatic hydrocarbons include the odors of almonds, cinnamon, and vanilla. The simplest ketone is *acetone*. Acetone has a fragrant odor and is used as a solvent in paint removers and nail polish removers. By sketching the structural formulas, you can see that ethers are isomers of alcohols, while aldehydes and ketones are isomers of each other. Again, the physical and chemical properties are quite different.

ORGANIC ACIDS AND ESTERS

Mineral acids, such as hydrochloric and sulfuric acid, are made of inorganic materials. Acids that were derived from organisms are called *organic acids*. Because many of these organic acids can be formed from fats, they are sometimes called *fatty acids*. Chemically, they are known as the *carboxylic acids* because they contain the carboxyl functional group, −COOH, and have a general formula of RCOOH.

The simplest carboxylic acid has been known since the Middle Ages, when it was isolated by the distillation of ants. The Latin word *formica* means "ant," so this acid was given the name *formic acid* (Figure 12.18). Formic acid is

```
 H — C — O H
     ||
     O
```

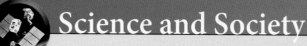

Science and Society

Aspirin, a Common Organic Compound

Aspirin went on the market as a prescription drug in 1897 after being discovered only two years earlier! However, even primitive humans were familiar with its value as a pain reliever. The bark of the willow tree was known for its "magical" pain relief power thousands of years ago. People in many cultures stripped and chewed the bark for its medicinal effect. It is estimated that more than 80 million tablets of aspirin are consumed in the United States daily. It is the most widely used drug in the world. Just what does it do? Aspirin is really acetylsalicylic acid, which inhibits the body's production of compounds known as prostaglandins, the cause of the pain.

QUESTIONS TO DISCUSS

1. What structural parts of aspirin would make it an organic acid?
2. Go to the Internet and find out just how this organic compound acts as a pain reliever.

FIGURE 12.18 These red ants, like other ants, make the simplest of the organic acids, formic acid. The sting of bees, ants, and some plants contains formic acid, along with some other irritating materials. Formic acid is HCOOH.

It is formic acid, along with other irritating materials, that causes the sting of bees, ants, and certain plants such as the stinging nettle.

Acetic acid, the acid of vinegar, has been known since antiquity. Acetic acid forms from the oxidation of ethanol. An oxidized bottle of wine contains acetic acid in place of the alcohol, which gives the wine a vinegar taste. Before wine is served in a restaurant, the person ordering is customarily handed the bottle cork and a glass with a small amount of wine. You first break the cork in half to make sure it is dry, which tells you that the wine has been sealed from oxygen. The small sip is to taste for vinegar before the wine is served. If the wine has been oxidized, the reaction is

$$\underset{\text{Ethanol}}{H-\overset{\overset{\displaystyle H}{|}}{\underset{\underset{\displaystyle H}{|}}{C}}-\overset{\overset{\displaystyle H}{|}}{\underset{\underset{\displaystyle H}{|}}{C}}-OH} \xrightarrow{\text{oxidation}} \underset{\text{Acetic acid}}{H-\overset{\overset{\displaystyle H}{|}}{\underset{\underset{\displaystyle H}{|}}{C}}-\overset{\overset{\displaystyle O}{\|}}{C}-OH}$$

Organic acids are common in many foods. The juice of citrus fruit, for example, contains citric acid, which relieves a thirsty feeling by stimulating the flow of saliva. Lactic acid is found in sour milk, buttermilk, sauerkraut, and pickles. Lactic acid also forms in your muscles as a product of carbohydrate metabolism, causing a feeling of fatigue. Citric and lactic acids are small molecules compared to some of the carboxylic acids that are formed from fats. Palmitic acid, for example, is $C_{16}H_{32}O_2$ and comes from palm oil. The structure of palmitic acid is a chain of fourteen CH_2 groups with CH_3- at one end and $-COOH$ at the other. Again, it is the functional carboxyl group, $-COOH$, that gives the molecule its acid properties. Organic acids are also raw materials used in the making of polymers of fabric, film, and paint.

Esters are common in both plants and animals, giving fruits and flowers their characteristic odor and taste. Esters are also used in perfumes and artificial flavorings. A few of the flavors for which particular esters are responsible are listed in Table 12.6. These liquid esters can be obtained from natural sources, or they can be chemically synthesized. Whatever the source, amyl acetate, for example, is the chemical responsible for what you identify as

TABLE 12.6

Flavors and esters

Ester Name	Formula	Flavor
Amyl acetate	$CH_3-\underset{\underset{\displaystyle O}{\|}}{C}-O-C_5H_{11}$	Banana
Octyl acetate	$CH_3-\underset{\underset{\displaystyle O}{\|}}{C}-O-C_8H_{17}$	Orange
Ethyl butyrate	$C_3H_7-\underset{\underset{\displaystyle O}{\|}}{C}-O-C_2H_5$	Pineapple
Amyl butyrate	$C_3H_7-\underset{\underset{\displaystyle O}{\|}}{C}-O-C_5H_{11}$	Apricot
Ethyl formate	$H-\underset{\underset{\displaystyle O}{\|}}{C}-O-C_2H_5$	Rum

the flavor of banana. Natural flavors, however, are complex mixtures of these esters along with other organic compounds. Lower molecular weight esters are fragrant-smelling liquids, but higher molecular weight esters are odorless oils and fats.

 CONCEPTS *Applied*

Organic Products

Pick a household product that has ingredients that sound like they could be organic compounds. On a separate sheet of paper, write the brand name of the product and the type of product (example: Oil of Olay; skin moisturizer), then list the ingredients one under the other (column 1). In a second column beside each name, put the type of compound if you can figure it out from its name or find it in any reference (example: cetyl palmitate—an ester of cetyl alcohol and palmitic acid). In a third column, put the structural formula if you can figure it out or find it in any reference such as a CRC handbook or the Merck Index. Finally, in a fourth column, put whatever you know or can find out about the function of that substance in the product.

ORGANIC COMPOUNDS OF LIFE

The chemical processes regulated by living organisms begin with relatively small organic molecules and water. The organism uses energy and matter from the surroundings to build large *macromolecules*. A **macromolecule** is a very large molecule that is a combination of many smaller, similar molecules joined together in a chainlike structure. Macromolecules have molecular weights of thousands or millions of atomic mass units. There are four main types of macromolecules: (1) proteins, (2) carbohydrates, (3) fats and oils, and (4) nucleic acids. A living organism, even a single-cell organism such as a bacterium, contains six thousand or so different kinds of macromolecules. The basic unit of an organism is called a *cell*. Cells are made of macromolecules that are formed inside the cell. The cell decomposes organic molecules taken in as food and uses energy from the food molecules to build more macromolecules. The process of breaking down organic molecules and building up macromolecules is called *metabolism*. Through metabolism, the cell grows, then divides into two cells. Each cell is a generic duplicate of the other, containing the same number and kinds of macromolecules. Each new cell continues the process of growth, then reproduces again, making more cells. This is the basic process of life. The complete process is complicated and very involved, easily filling a textbook in itself, so the details will not be presented here. The following discussion will be limited to three groups of organic molecules involved in metabolic processes: proteins, carbohydrates, and fats and oils.

PROTEINS

Proteins are macromolecular polymers made up of smaller molecules called amino acids. These very large macromolecules have molecular weights that vary from about six thousand to fifty million. Some proteins are simple straight-chain polymers of amino acids, but others contain metal ions such as Fe^{2+} or parts of organic molecules derived from vitamins. Proteins serve as major structural and functional materials in living things. *Structurally,* proteins are major components of muscles, connective tissue, and the skin, hair, and nails. *Functionally,* some proteins are enzymes, which catalyze metabolic reactions; hormones, which regulate body activities; hemoglobin, which carries oxygen to cells; and antibodies, which protect the body.

Proteins are formed from twenty **amino acids,** which are organic molecules with acid and amino functional groups with the general formula of

$$
R - \underset{\underset{NH_2}{|}}{\overset{\overset{H}{|}}{C}} - \overset{\overset{O}{||}}{C} - OH
$$

alpha

Note the carbon atom labeled "alpha" in the general formula. The amino functional group (NH_2) is attached to this carbon atom, which is next to the carboxylic group (COOH). This arrangement is called an *alpha-amino acid,* and the building blocks of proteins are all alpha-amino acids. The twenty amino acids differ in the nature of the R group, also called the *side chain*. It is the linear arrangements of amino acids and their side chains that determine the properties of a protein.

Amino acids are linked to form a protein by a peptide bond between the amino group of one amino acid and the carboxyl group of a second amino acid. A polypeptide is a polymer formed from linking many amino acid molecules. If the polypeptide is involved in a biological structure or function, it is called a *protein*. A protein chain can consist of different combinations of the twenty amino acids with hundreds or even thousands of amino acid molecules held together with peptide bonds (Figure 12.19). The arrangement or sequence of these amino acid molecules determines the structure that gives the protein its unique set of biochemical properties. Insulin, for example, is a protein hor-

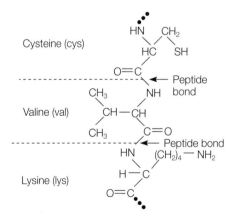

FIGURE 12.19 Part of a protein polypeptide made up of the amino acids cysteine (cys), valine (val), and lysine (lys). A protein can have from fifty to a thousand of these amino acid units; each protein has its own unique sequence.

mone that biochemically regulates the blood sugar level. Insulin contains 86 amino acid molecules in the chain. Hemoglobin is the protein that carries oxygen in the bloodstream, and its biochemical characteristics are determined by its chain of 146 amino acid molecules.

 CONCEPTS *Applied*

Fats Foods

Pick a food product and write the number of grams of fats, proteins, and carbohydrates per serving according to the information on the label. Multiply the number of grams of proteins and carbohydrates each by 4 Cal/g and the number of grams of fat by 9 Cal/g. Add the total Calories per serving. Does your total agree with the number of Calories per serving given on the label? Also examine the given serving size. Is this a reasonable amount to be consumed at one time, or would you probably eat two or three times this amount? Write the rest of the nutrition information (vitamins, minerals, sodium content, etc.), and then write the list of ingredients. Tell what ingredient you think is providing which nutrient. (Example: vegetable oil—source of fat; milk—provides calcium and vitamin A; MSG—source of sodium.)

CARBOHYDRATES

Carbohydrates are an important group of organic compounds that includes sugars, starches, and cellulose, and they are important in plants and animals for structure, protection, and food. Cellulose is the skeletal substance of plants and plant materials, and chitin is a similar material that forms the hard, protective covering of insects and shellfish such as crabs and lobsters. *Glucose*, $C_6H_{12}O_6$, is the most abundant carbohydrate and serves as a food and a basic building block for other carbohydrates.

Carbohydrates were named when early studies found that water vapor was given off and carbon remained when sugar was heated. The name *carbohydrate* literally means "watered carbon," and the empirical formulas for most carbohydrates indeed indicate carbon (C) and water (H_2O). Glucose, for example, could be considered to be six carbons with six waters, or $C_6(H_2O)_6$. However, carbohydrate molecules are more complex than just water attached to a carbon atom. They are polyhydroxyl aldehydes and ketones, two of which are illustrated in Figure 12.20. The two carbohydrates in this illustration belong to a group of carbohydrates known as **monosaccharides,** or *simple sugars.* They are called simple sugars because they are the smallest units that have the characteristics of carbohydrates, and they can be combined to make larger complex carbohydrates. There are many kinds of simple sugars, but they are mostly 6-carbon molecules such as glucose and fructose. Glucose (also called dextrose) is found in the sap of plants, and in the human bloodstream, it is called *blood sugar.* Corn syrup, which is often used as a sweetener, is mostly glucose. Fructose, as its name implies, is the sugar that occurs in fruits, and it is sometimes called *fruit sugar.* Both glucose and fructose have the same molecular formula,

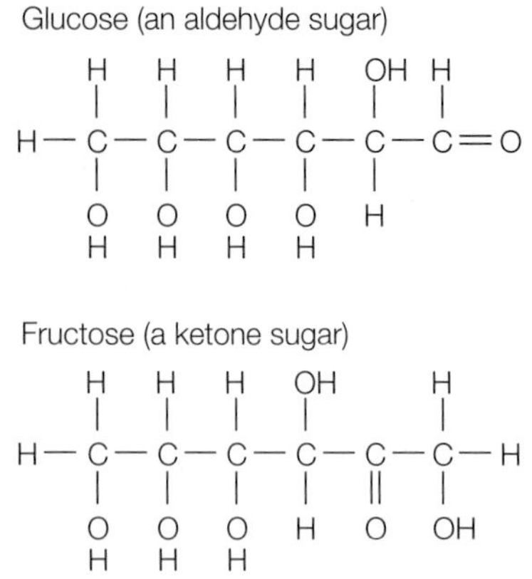

Glucose (an aldehyde sugar)

Fructose (a ketone sugar)

FIGURE 12.20 Glucose (blood sugar) is an aldehyde, and fructose (fruit sugar) is a ketone. Both have a molecular formula of $C_6H_{12}O_6$.

but glucose is an aldehyde sugar and fructose is a ketone sugar (Figure 12.20). A mixture of glucose and fructose is found in honey. This mixture also is formed when table sugar (sucrose) is reacted with water in the presence of an acid, a reaction that takes place in the preparation of canned fruit and candies. The mixture of glucose and fructose is called *invert sugar.* Thanks to fructose, invert sugar is about twice as sweet to the taste as the same amount of sucrose.

Two monosaccharides are joined together to form **disaccharides** with the loss of a water molecule, for example,

$$\underset{\text{glucose}}{C_6H_{12}O_6} + \underset{\text{fructose}}{C_6H_{12}O_6} \longrightarrow \underset{\text{sucrose}}{C_{12}H_{22}O_{11}} + H_2O$$

The most common disaccharide is *sucrose,* or ordinary table sugar. Sucrose occurs in high concentrations in sugarcane and sugar beets. It is extracted by crushing the plant materials, then dissolving the sucrose from the materials with water. The water is evaporated and the crystallized sugar is decolorized with charcoal to produce white sugar. Other common disaccharides include *lactose* (milk sugar) and *maltose* (malt sugar). All three disaccharides have similar properties, but maltose tastes only about one-third as sweet as sucrose. Lactose tastes only about one-sixth as sweet as sucrose. No matter which disaccharide sugar is consumed (sucrose, lactose, or maltose), it is converted into glucose and transported by the bloodstream for use by the body.

Polysaccharides are polymers consisting of monosaccharide units joined together in straight or branched chains. Polysaccharides are the energy-storage molecules of plants and animals (starch and glycogen) and the structural molecules of plants (cellulose). **Starches** are a group of complex carbohydrates composed of many glucose units that plants use as a stored food source. Potatoes, rice, corn, and wheat store starch granules and serve as an important source of food for humans. The human body breaks down the starch molecules to glucose, which is transported by the bloodstream and utilized just like any other glucose. This digestive process begins with enzymes secreted with saliva in the mouth. You may have noticed a result of this enzyme-catalyzed reaction as you eat bread. If you chew the bread for a while, it begins to taste sweet.

A starch

Cellulose

FIGURE 12.22 Starch and cellulose are both polymers of glucose, but humans cannot digest cellulose. The difference in the bonding arrangement might seem minor, but enzymes must fit a molecule very precisely. Thus, enzymes that break down starch do nothing to cellulose.

FIGURE 12.21 These plants and their flowers are made up of a mixture of carbohydrates that were manufactured from carbon dioxide and water, with the energy of sunlight. The simplest of the carbohydrates are the monosaccharides, simple sugars (fruit sugar) that the plant synthesizes. Food is stored as starches, which are polysaccharides made from the simpler monosaccharide glucose. The plant structure is held upright by fibers of cellulose, another form of a polysaccharide composed of glucose.

Plants store sugars in the form of starch polysaccharides, and animals store sugars in the form of the polysaccharide *glycogen.* Glycogen is a starchlike polysaccharide that is synthesized by the human body and stored in the muscles and liver. Glycogen, like starch, is a very high molecular weight polysaccharide, but it is more highly branched. These highly branched polysaccharides serve as a direct reserve source of energy in the muscles. In the liver, they serve as a reserve source to maintain the blood sugar level.

Cellulose is a polysaccharide that is abundant in plants, forming the fibers in cell walls that preserve the structure of plant materials (Figure 12.21). Cellulose molecules are straight chains, consisting of large numbers of glucose units. These glucose units are arranged in a way that is very similar to the arrangement of the glucose units of starch but with differences in the bonding arrangement that holds the glucose units together (Figure 12.22). This difference turns out to be an important one where humans are concerned because enzymes that break down starches do not affect cellulose. Humans do not have the necessary enzymes to break down the cellulose chain (digest it), so humans receive no food value from cellulose. Cattle and termites that do utilize cellulose as a source of food have protozoa and bacteria (with the necessary enzymes) in their digestive systems. Cellulose is still needed in the human diet, however, for fiber and bulk.

Myths, Mistakes, & Misunderstandings

Fruit Juice Versus Cola

Misconception: Fruit juices have fewer calories from sugar (monosaccharides) than nondiet soft drinks.

In fact, apple juice has more sugar than the same amount of cola. Did you also know that it is estimated that people in the United States consume more calories in the form of simple sugar than of meat, chicken, vegetables, and breads combined?

FATS AND OILS

Cereal grains and other plants also provide carbohydrates, the human body's preferred food for energy. When an excess amount of carbohydrates is consumed, the body begins to store some of its energy source in the form of glycogen in the

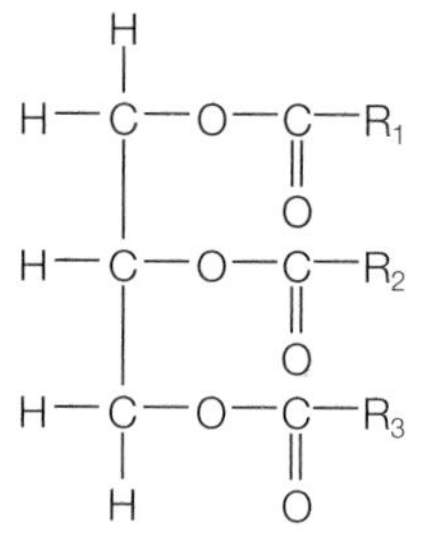

FIGURE 12.23 The triglyceride structure of fats and oils. Note the glycerol structure on the left and the ester structure on the right. Also notice that R_1, R_2, and R_3 are long-chained molecules of 12, 14, 16, 18, 20, 22, or 24 carbons that might be saturated or unsaturated.

muscles and liver. Beyond this storage for short-term needs, the body begins to store energy in a different chemical form for longer-term storage. This chemical form is called **fat** in animals and **oil** in plants. Fats and oils are esters formed from glycerol (1,2,3-trihydroxypropane) and three long-chain carboxylic acids (fatty acids). This ester is called a **triglyceride,** and its structural formula is shown in Figure 12.23. Fats are solids and oils are liquids at room temperature, but they both have this same general structure.

Fats and oils usually have two or three different fatty acids, and several are listed in Table 12.7. Animal fats can be either saturated or unsaturated, but most are saturated. Oils are liquids at room temperature because they contain a higher number of unsaturated units. These unsaturated oils (called "polyunsaturated" in news and advertisements), such as safflower and corn oils, are used as liquid cooking oils because unsaturated oils are believed to lead to lower cholesterol levels in the bloodstream. Saturated fats, along with cholesterol, are believed to contribute to hardening of the arteries over time.

Cooking oils from plants, such as corn and soybean oil, are hydrogenated to convert the double bonds of the unsaturated

oil to the single bonds of a saturated one. As a result, the liquid oils are converted to solids at room temperature. For example, one brand of margarine lists ingredients as "liquid soybean oil (nonhydrogenated) and partially hydrogenated cottonseed oil with water, salt, preservatives, and coloring." Complete hydrogenation would result in a hard solid, so the cottonseed oil is partially hydrogenated and then mixed with liquid soybean oil. Coloring is added because oleo is white, not the color of butter. Vegetable shortening is the very same product without added coloring. Reaction of a triglyceride with a strong base such as KOH or NaOH yields a fatty acid of salt and glycerol. A sodium or potassium fatty acid is commonly known as *soap.*

Excess food from carbohydrate, protein, or fat and oil sources is converted to fat for long-term energy storage in *adipose tissue,* which also serves to insulate and form a protective padding. In terms of energy storage, fats yield more than twice the energy per gram oxidized as carbohydrates or proteins.

CONCEPTS *Applied*

Brand News

Pick two competing brands of a product you use (example: Tylenol and a store-brand acetaminophen) and write the following information: All ingredients, amount of each ingredient (if this is not given, remember that by law, ingredients have to be listed in order of their percent by weight), and the cost per serving or dose. Comment on whether each of the listed ingredients is a single substance and thus something with the same properties wherever it is found (example: salt, as a label ingredient, means sodium chloride no matter in what product it appears) or a mixture and thus possibly different in different products (example: tomatoes, as a ketchup ingredient, might be of better quality in one brand of ketchup than another). Then draw a reasonably informed conclusion as to whether there is any significant difference between the two brands or whether the more expensive one is worth the difference in price. Finally, do your own consumer test to check your prediction.

SYNTHETIC POLYMERS

Polymers are huge, chainlike molecules made of hundreds or thousands of smaller, repeating molecular units called *monomers.* Polymers occur naturally in plants and animals. Cellulose, for example, is a natural plant polymer made of glucose monomers. Wool and silk are natural animal polymers made of amino acid monomers. *Synthetic polymers* are manufactured from a wide variety of substances, and you are familiar with these polymers as synthetic fibers such as nylon and the inexpensive light plastic used for wrappings and containers (Figure 12.24).

The first synthetic polymer was a modification of the naturally existing cellulose polymer. Cellulose was chemically modified in 1862 to produce celluloid, the first *plastic.* The term *plastic* means that celluloid could be molded to any desired shape. Celluloid was produced by first reacting cotton with a

TABLE 12.7		
Some fatty acids occurring in fats		
Common Name	Condensed Structure	Source
Lauric acid	$CH_3(CH_2)_{10}COOH$	Coconuts
Palmitic acid	$CH_3(CH_2)_{14}COOH$	Palm oil
Stearic acid	$CH_3(CH_2)_{16}COOH$	Animal fats
Oleic acid	$CH_3(CH_2)_7CH{=}CH(CH_2)_7COOH$	Corn oil
Linoleic acid	$CH_3(CH_2)_4CH{=}CHCH_2{=}CH(CH_2)_7COOH$	Soybean oil
Linolenic acid	$CH_3CH_2(CH{=}CHCH_2)_3(CH_2)_6COOH$	Fish oils

Name	Chemical unit	Uses	Name	Chemical unit	Uses
Polyethylene	$\left[\begin{array}{c} H\ H \\ -C-C- \\ H\ H \end{array}\right]_n$	Squeeze bottles, containers, laundry and trash bags, packaging	Polyvinyl acetate	$\left[\begin{array}{c} H\ CH_3 \\ -C-C- \\ H\ O \\ C-CH_3 \\ O \end{array}\right]_n$	Mixed with vinyl chloride to make vinylite; used as an adhesive and resin in paint
Polypropylene	$\left[\begin{array}{c} H\ H \\ -C-C- \\ H\ CH_3 \end{array}\right]_n$	Indoor-outdoor carpet, pipe valves, bottles	Styrene-butadiene rubber	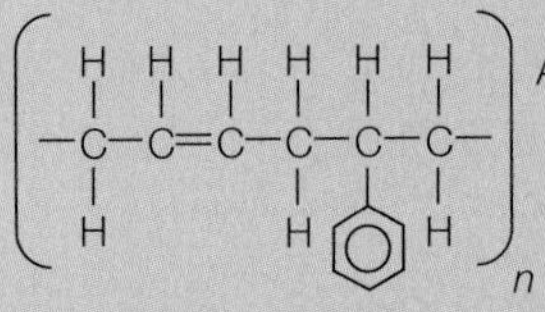 $\left[\begin{array}{c} H\ H\ H\ H\ H\ H \\ -C-C=C-C-C-C- \\ H\ \ \ \ \ H\ \bigcirc\ H \end{array}\right]_n$	Automobile tires
Polyvinyl chloride (PVC)	$\left[\begin{array}{c} H\ H \\ -C-C- \\ H\ Cl \end{array}\right]_n$	Plumbing pipes, synthetic leather, plastic tablecloths, phonograph records, vinyl tile	Polychloroprene (Neoprene)	$\left[\begin{array}{c} H\ Cl\ H\ H \\ -C-C=C-C- \\ H\ \ \ \ \ \ H \end{array}\right]_n$	Shoe soles, heels
Polyvinylidene chloride (Saran)	$\left[\begin{array}{c} H\ Cl \\ -C-C- \\ H\ Cl \end{array}\right]_n$	Flexible food wrap	Polymethyl methacrylate (Plexiglas, Lucite)	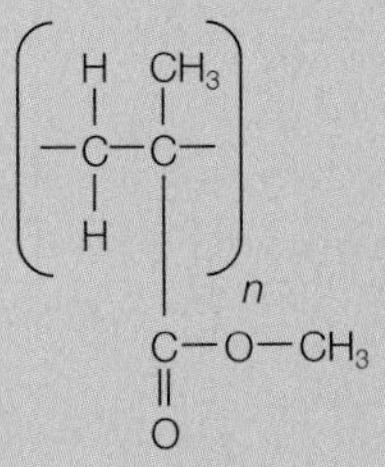 $\left[\begin{array}{c} H\ CH_3 \\ -C-C- \\ H\ \ \ \\ C-O-CH_3 \\ O \end{array}\right]_n$	Moldings, transparent surfaces on furniture, lenses, jewelry, transparent plastic "glass"
Polystyrene (Styrofoam)	$\left[\begin{array}{c} H\ H \\ -C-C- \\ H\ \bigcirc \end{array}\right]_n$	Coolers, cups, insulating foam, shock-resistant packing material, simulated wood furniture	Polycarbonate (Lexan)	$\left[\bigcirc-\begin{array}{c} CH_3 \\ C \\ CH_3 \end{array}-\bigcirc-O-\begin{array}{c} C \\ O \end{array}-O-\right]_n$	Tough, molded articles such as motorcycle helmets
Polytetrafluoroethylene (Teflon)	$\left[\begin{array}{c} F\ F \\ -C-C- \\ F\ F \end{array}\right]_n$	Gears, bearings, coating for nonstick surface of cooking utensils	Polyacrylonitrile (Orlon, Acrilan, Creslan)	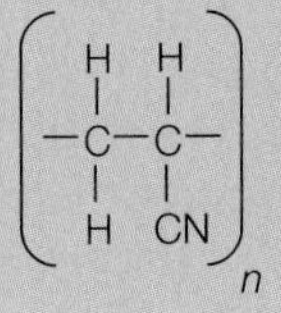 $\left[\begin{array}{c} H\ H \\ -C-C- \\ H\ CN \end{array}\right]_n$	Textile fibers

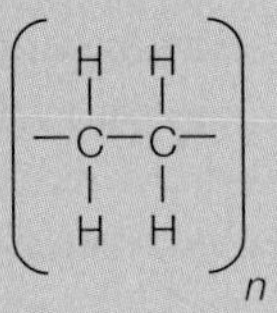

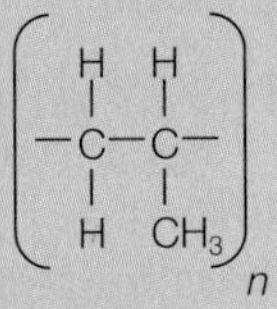

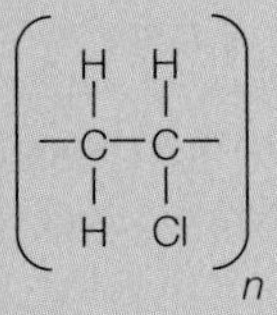

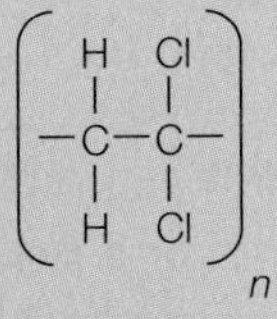

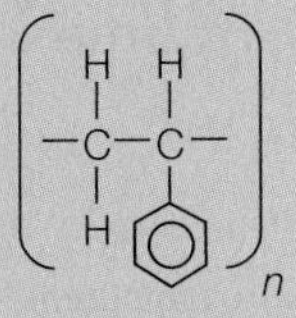

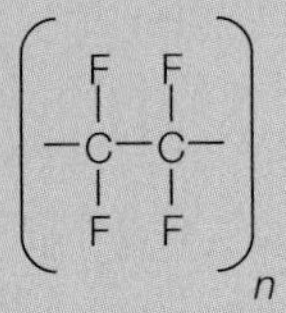

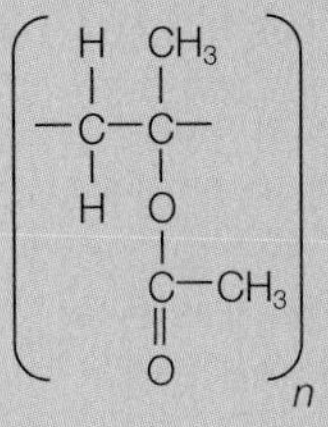

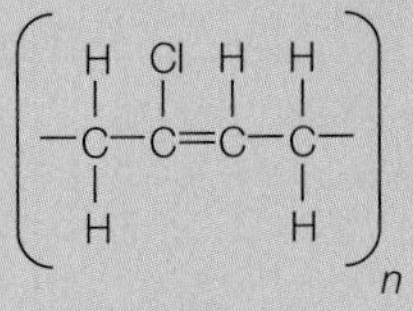

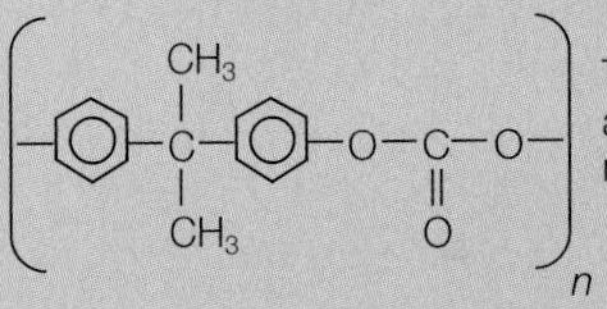

FIGURE 12.24 Synthetic polymers, the polymer unit, and some uses of each polymer.

mixture of nitric and sulfuric acids, which produced an ester of cellulose nitrate. This ester is an explosive compound known as "guncotton," or smokeless gunpowder. When made with ethanol and camphor, the product is less explosive and can be formed and molded into useful articles. This first plastic, celluloid, was used to make dentures, combs, eyeglass frames, and photographic film. Before the discovery of celluloid, many of these articles, including dentures, were made from wood. Today, only Ping-Pong balls are made from cellulose nitrate.

Cotton or other sources of cellulose reacted with acetic acid and sulfuric acid produce a cellulose acetate ester. This polymer, through a series of chemical reactions, produces viscose rayon filaments when forced through small holes. The filaments are twisted together to form viscose rayon thread. When the polymer is forced through a thin slit, a sheet is formed rather than filaments, and the transparent sheet is called *cellophane*. Both rayon and cellophane, like celluloid, are manufactured by modifying the natural polymer of cellulose.

The first truly synthetic polymer was produced in the early 1900s by reacting two chemicals with relatively small molecules rather than modifying a natural polymer. Phenol, an aromatic hydrocarbon, was reacted with formaldehyde, the simplest aldehyde, to produce the polymer named *Bakelite*. Bakelite is a *thermosetting* material that forms cross-links between the polymer chains. Once the links are formed during production, the plastic becomes permanently hardened and cannot be softened or made to flow. Some plastics are *thermoplastic* polymers and soften during heating and harden during cooling because they do not have cross-links.

A Closer Look

How to Sort Plastic Bottles for Recycling

Plastic containers are made of different types of plastic resins; some are suitable for recycling and some are not. How do you know which are suitable and how to sort them? Most plastic containers have a code stamped on the bottom. The code is a number in the recycling arrow logo, sometimes appearing with some letters. Here is what the numbers and letters mean in terms of (a) the plastic, (b) how it is used, and (c) if it is usually recycled or not.

a. Polyethylene terephthalate (PET)
b. Large soft-drink bottles, salad dressing bottles
c. Frequently recycled

a. High-density polyethylene (HDPE)
b. Milk jugs, detergent and bleach bottles, others
c. Frequently recycled

a. Polyvinyl chloride (PVC or PV)
b. Shampoos, hair conditioners, others
c. Rarely recycled

a. Low-density polyethylene (LDPE)
b. Plastic wrap, laundry and trash bags
c. Rarely recycled but often reused

a. Polypropylene (PP)
b. Food containers
c. Rarely recycled

a. Polystyrene (PS)
b. Styrofoam cups, burger boxes, plates
c. Occasionally recycled

a. Mixed resins
b. Catsup squeeze bottles, other squeeze bottles
c. Rarely recycled

Polyethylene is a familiar thermoplastic polymer used for vegetable bags, dry cleaning bags, grocery bags, and plastic squeeze bottles. Polyethylene is a polymer produced by a polymerization reaction of ethylene, which is derived from petroleum. Polyethylene was invented just before World War II and was used as an electrical insulating material during the war. Today, there are many variations of polyethylene that are produced by different reaction conditions or by the substitution of one or more hydrogen atoms in the ethylene molecule. When soft polyethylene near the melting point is rolled in alternating perpendicular directions or expanded and compressed as it is cooled, the polyethylene molecules become ordered in a way that improves the rigidity and tensile strength. This change in the microstructure produces *high-density polyethylene* with a superior rigidity and tensile strength compared to *low-density polyethylene*. High-density polyethylene is used in milk jugs, as liners in screw-on jar tops, in bottle caps, and as a material for toys.

Alfred Nobel was a Swedish industrial chemist and philanthropist who invented dynamite and endowed the Nobel Foundation, which after 1901 awarded the annual Nobel Prizes.

Nobel was born in Stockholm, Sweden, on October 21, 1833, the son of a builder and industrialist. His father, Immanuel Nobel, was also something of an inventor, and his grandfather had been one of the most important Swedish scientists of the seventeenth century. Alfred Nobel attended St. Jakob's Higher Apologist School in Stockholm before the family moved to St. Petersburg, Russia, where he and his brothers were taught privately by Russian and Swedish tutors, always being encouraged to be inventive by their father. From 1850 to 1852, Nobel made a study trip to Germany, France, Italy, and North America, improving his knowledge of chemistry and mastering all the necessary languages. During the Crimean War (1853–1856), Nobel worked for his father's munitions company in St. Petersburg, whose output during the war was large. After the fighting ended, Nobel's father went bankrupt,

and in 1859, the family returned to Sweden. During the next few years, Nobel developed several new explosives and factories for making them, and became a rich man.

Guncotton, a more powerful explosive than gunpowder, had been discovered in 1846 by a German chemist. It was made by nitrating cotton fiber with a mixture of concentrated nitric and sulfuric acids. A year later, an Italian discovered nitroglycerin, made by nitrating glycerin (glycerol). This extremely powerful explosive gives off twelve hundred times its own volume of gas when it explodes, but for many years, it was too dangerous to use because it can be set off much too easily by rough handling or shaking. Alfred and his father worked independently on both explosives when they returned to Sweden, and in 1862, Immanuel Nobel devised a comparatively simple way of manufacturing nitroglycerin on a factory scale. In 1863, Alfred Nobel invented a mercury fulminate detonator for use with nitroglycerin in blasting.

In 1864, their nitroglycerin factory blew up, killing Nobel's younger brother

and four other people. Nobel turned his attention to devising a safer method of handling the sensitive liquid nitroglycerin. His many experiments with nitroglycerin led to the development of the explosive dynamite, which he patented in Sweden, Britain, and the United States in 1867. Consisting of nitroglycerin absorbed by a porous diatomite mineral, dynamite was thought to be convenient to handle and safer to use.

Nobel was a prolific inventor whose projects touched on the fields of electrochemistry, optics, biology, and physiology. He was also involved in problem solving in the synthetic silk, leather, and rubber industries, and in the manufacture of artificial semiprecious stones from fused alumina. In his will, made in 1895, he left almost all his fortune to set up a foundation that would bestow annual awards on "those who, during the preceding year, shall have conferred the greatest benefit on mankind." In 1958, the new element number 102 was named nobelium in his honor.

Source: Modified from the Hutchinson *Dictionary of Scientific Biography* © RM 2008. All rights reserved. Helicon Publishing is a division of R M.

The properties of polyethylene are changed by replacing one of the hydrogen atoms in a molecule of ethylene. If the hydrogen is replaced by a chlorine atom, the compound is called vinyl chloride, and the polymer formed from vinyl chloride is

$$\underset{H}{\overset{H}{\diagdown}} C = C \underset{Cl}{\overset{H}{\diagup}}$$

polyvinyl chloride (PVC). Polyvinyl chloride is used to make plastic water pipes, synthetic leather, and other vinyl products. It differs from the waxy plastic of polyethylene because of the chlorine atom that replaces hydrogen on each monomer.

The replacement of a hydrogen atom with a benzene ring makes a monomer called *styrene*. Styrene is

$$\underset{H}{\overset{H}{\diagdown}} C = C \underset{\bigcirc}{\overset{H}{\diagup}}$$

and polymerization of styrene produces *polystyrene*. Polystyrene is puffed full of air bubbles to produce the familiar Styrofoam coolers, cups, and insulating materials.

If all hydrogens of an ethylene molecule are replaced with atoms of fluorine, the product is polytetrafluoroethylene, a tough plastic that resists high temperatures and acts more like a metal than a plastic. Since it has a low friction, it is used for bearings, gears, and as a nonsticking coating on frying pans. You probably know of this plastic by its trade name of *Teflon.*

There are many different polymers in addition to PVC, Styrofoam, and Teflon, and the monomers of some of these are shown in Figure 12.24. There are also polymers of isoprene, or synthetic rubber, in wide use. Fibers and fabrics may be polyamides (such as nylon), polyesters (such as Dacron), or polyacrylonitriles (Orlon, Acrilan, Creslan), which have a CN in place of a hydrogen atom on an ethylene molecule and are called acrylic materials. All of these synthetic polymers have added much to practically every part of your life. It would be impossible to list all of their uses here; however, they present problems because (1) they are manufactured from raw materials obtained from coal and a dwindling petroleum supply, and (2) they do not readily decompose when dumped into rivers, oceans, or other parts of the environment. However, research in the polymer sciences is beginning to reflect new understandings learned from research on biological tissues. This could lead to whole new molecular designs for synthetic polymers that will be more compatible with the ecosystems.

SUMMARY

Organic chemistry is the study of compounds that have carbon as the principal element. Such compounds are called *organic compounds,* and all the rest are *inorganic compounds.* There are millions of organic compounds because a carbon atom can link with other carbon atoms as well as atoms of other elements.

A *hydrocarbon* is an organic compound consisting of hydrogen and carbon atoms. The simplest hydrocarbon is one carbon atom and four hydrogen atoms, or CH_4. All hydrocarbons larger than CH_4 have one or more carbon atoms bonded to another carbon atom. The bond can be single, double, or triple, and this forms a basis for classifying hydrocarbons. A second basis is whether the carbons are in a ring or not. The *alkanes* are hydrocarbons with single carbon-to-carbon bonds, the *alkenes* have a double carbon-to-carbon bond, and the *alkynes* have a triple carbon-to-carbon bond. The alkanes, alkenes, and alkynes can have straight- or branched-chain molecules. When the number of carbon atoms is greater than three, there are different arrangements that can occur for a particular number of carbon atoms. The different arrangements with the same molecular formula are called isomers. *Isomers* have different physical properties, so each isomer is given its own name. The name is determined by (1) identifying the longest continuous carbon chain as the base name, (2) locating the attachment of other atoms or hydrocarbon groups by counting from the direction that results in the smallest numbers, (3) identifying attached hydrocarbon groups by changing the *-ane* suffix of alkanes to *-yl*, (4) identifying the number of these hydrocarbon groups with prefixes, and (5) identifying the location of the groups with the carbon atom number.

The alkanes have all the hydrogen atoms possible, so they are *saturated* hydrocarbons. The alkenes and the alkynes can add more hydrogens to the molecule, so they are *unsaturated* hydrocarbons. Unsaturated hydrocarbons are more chemically reactive than saturated molecules.

Hydrocarbons that occur in a ring or cycle structure are cyclohydrocarbons. A six-carbon cyclohydrocarbon with three double bonds has different properties than the other cyclohydrocarbons because the double bonds are not localized. This six-carbon molecule is *benzene,* the basic unit of the *aromatic hydrocarbons.*

Petroleum is a mixture of alkanes, cycloalkanes, and a few aromatic hydrocarbons that formed from the slow decomposition of buried marine plankton and algae. Petroleum from the ground, or *crude oil,* is distilled into petroleum products of *natural gas, LPG, petroleum ether, gasoline, kerosene, diesel fuel,* and *motor oils.* Each group contains a range of hydrocarbons and is processed according to use.

In addition to oxidation, hydrocarbons react by *substitution, addition,* and *polymerization* reactions. Reactions take place at sites of multiple bonds or lone pairs of electrons on the *functional groups.* The functional group determines the chemical properties of organic compounds. Functional group results in the *hydrocarbon derivatives* of *alcohols, ethers, aldehydes, ketones, organic acids, esters,* and *amines.*

Living organisms have an incredible number of highly organized chemical reactions that are catalyzed by *enzymes,* using food and energy to grow and reproduce. The process involves building large *macromolecules* from smaller molecules and units. The organic molecules involved in the process are proteins, carbohydrates, and fats and oils.

Proteins are macromolecular polymers of *amino acids* held together by *peptide bonds.* There are twenty amino acids that are used

in various polymer combinations to build structural and functional proteins. *Structural proteins* are muscles, connective tissue, and the skin, hair, and nails of animals. *Functional proteins* are enzymes, hormones, and antibodies.

Carbohydrates are polyhydroxyl aldehydes and ketones that form three groups: the monosaccharides, disaccharides, and polysaccharides. The *monosaccharides* are simple sugars such as *glucose* and *fructose.* Glucose is *blood sugar,* a source of energy. The disaccharides are *sucrose* (table sugar), *lactose* (milk sugar), and *maltose* (malt sugar). The disaccharides are broken down (digested) to glucose for use by the body. The polysaccharides are polymers of glucose in straight or branched chains used as a near-term source of stored energy. Plants store the energy in the form of *starch,* and animals store it in the form of *glycogen. Cellulose* is a polymer similar to starch that humans cannot digest.

Fats and oils are esters formed from three fatty acids and glycerol into a *triglyceride. Fats* are usually solid triglycerides associated with animals, and *oils* are liquid triglycerides associated with plant life, but both represent a high-energy storage material.

Polymers are huge, chainlike molecules of hundreds or thousands of smaller, repeating molecular units called *monomers.* Polymers occur naturally in plants and animals, and many *synthetic polymers* are made today from variations of the ethylene-derived monomers. Among the more widely used synthetic polymers derived from ethylene are polyethylene, polyvinyl chloride, polystyrene, and Teflon. Problems with the synthetic polymers include that (1) they are manufactured from fossil fuels that are also used as the primary energy supply, and (2) they do not readily decompose and tend to accumulate in the environment.

KEY TERMS

alkane (p. **301**)

alkene (p. **305**)

alkyne (p. **306**)

amino acids (p. **313**)

aromatic hydrocarbons (p. **306**)

carbohydrates (p. **314**)

cellulose (p. **315**)

disaccharides (p. **314**)

fat (p. **316**)

functional group (p. **309**)

hydrocarbon (p. **300**)

hydrocarbon derivatives (p. **309**)

inorganic chemistry (p. **300**)

isomers (p. **302**)

macromolecule (p. **313**)

monosaccharides (p. **314**)

oil (p. **316**)

organic chemistry (p. **300**)

petroleum (p. **306**)

polymer (p. **309**)

polysaccharides (p. **314**)

proteins (p. **313**)

starches (p. **314**)

triglyceride (p. **316**)

APPLYING THE CONCEPTS

1. An organic compound consisting only of carbon and hydrogen is called a (an)
   a. hydrogen carbide.
   b. hydrocarbon.
   c. organic compound.
   d. hydrocarbide.

2. Carbon-to-carbon bonds are
   a. ionic.
   b. polar covalent.
   c. covalent.
   d. metallic.

3. All of the following are hydrocarbon groups *except*
   a. alkanes.
   b. alkenes.
   c. alkynes.
   d. alkalines.

4. The group $-C_2H_5$ is called
   a. methyl.
   b. ethyl.
   c. dimethyl.
   d. propyl.

5. Compounds with the same molecular formulas but with different structures are called
   a. isotopes.
   b. allotropes.
   c. isomers.
   d. polymers.

6. Hydrocarbons with double bonds are the
   a. alkanes.
   b. alkenes.
   c. alkynes.
   d. alkines.

7. A hydrocarbon with a carbon-to-carbon triple bond is an
   a. alkane.
   b. alkene.
   c. alkyne.
   d. alkaline.

8. Hydrocarbons that have one to four carbon atoms are
   a. gases.
   b. liquids.
   c. solids.
   d. semisolids.

9. The process where large hydrocarbons are converted into smaller ones is called
   a. snapping.
   b. distillation.
   c. cracking.
   d. fractionation.

10. The atom or group of atoms in an organic molecule that is the site of a chemical reaction is called the
    a. reaction site group.
    b. functional group.
    c. addition site.
    d. reactant group.

11. A hydroxyl, $-OH$, functional group is found in
    a. ketone.
    b. alkanes.
    c. aldehydes.
    d. alcohols.

12. The oxidation of ethanol, C_2H_5OH, produces
    a. acetone.
    b. formaldehyde.
    c. acetic acid.
    d. formic acid.

13. An organic compound is a compound that
    a. contains carbon and was formed by a living organism.
    b. is a natural compound.
    c. contains carbon, whether it was formed by a living thing or not.
    d. was formed by a plant.

14. There are millions of organic compounds but only thousands of inorganic compounds because
    a. organic compounds were formed by living things.
    b. there is more carbon on Earth's surface than any other element.
    c. atoms of elements other than carbon never combine with themselves.
    d. carbon atoms can combine with up to four other atoms, including other carbon atoms.

15. You know for sure that the compound named decane has
    a. more than ten isomers.
    b. ten carbon atoms in each molecule.
    c. only single bonds.
    d. All of the above are correct.

16. An alkane with four carbon atoms would have how many hydrogen atoms in each molecule?
    a. four
    b. eight
    c. ten
    d. sixteen

17. Isomers are compounds with the same
    a. molecular formula with different structures.
    b. molecular formula with different atomic masses.
    c. atoms but different molecular formulas.
    d. structures but different formulas.

18. Isomers have
    a. the same chemical and physical properties.
    b. the same chemical but different physical properties.
    c. the same physical but different chemical properties.
    d. different physical and chemical properties.

19. The organic compound 2,2,4-trimethylpentane is an isomer of
    a. propane.
    b. pentane.
    c. heptane.
    d. octane.

20. Which of the following would not occur as an unsaturated hydrocarbon?
    a. alkane
    b. alkene
    c. alkyne
    d. None of the above is correct.

21. Petroleum is believed to have formed mostly from the anaerobic decomposition of buried
    a. dinosaurs.
    b. fish.
    c. pine trees.
    d. plankton and algae.

22. The label on a container states that the product contains "petroleum distillates." Which of the following hydrocarbons is probably present?
    a. CH_4
    b. C_5H_{12}
    c. $C_{16}H_{34}$
    d. $C_{40}H_{82}$

23. The reaction of $C_2H_2 + Br_2 \rightarrow C_2H_2Br_2$ is a (an)
    a. substitution reaction.
    b. addition reaction.
    c. addition polymerization reaction.
    d. substitution polymerization reaction.

24. Ethylene molecules can add to each other in a reaction to form a long chain called a
    a. monomer.
    b. dimer.
    c. trimer.
    d. polymer.

25. Chemical reactions usually take place on an organic compound at the site of a
    a. double bond.
    b. lone pair of electrons.
    c. functional group.
    d. Any of the above is correct.

26. The R in ROH represents
    a. a functional group.
    b. a hydrocarbon group with a name ending in -yl.
    c. an atom of an inorganic element.
    d. a polyatomic ion that does not contain carbon.

27. The OH in ROH represents
    a. a functional group.
    b. a hydrocarbon group with a name ending in -yl.
    c. the hydroxide ion, which ionizes to form a base.
    d. the site of chemical activity in a strong base.

28. What is the proof of a "wine cooler" that is 5 percent alcohol by volume?
    a. 2.5 proof
    b. 5 proof
    c. 10 proof
    d. 50 proof

29. An alcohol with two hydroxyl groups per molecule is called
    a. ethanol.
    b. glycerol.
    c. glycerin.
    d. glycol.

30. A bottle of wine that has "gone bad" now contains
    a. CH_3OH.
    b. CH_3OCH_3.
    c. CH_3COOH.
    d. CH_3COOCH_3.

31. A protein is a polymer formed from the linking of many
    a. glucose units.
    b. DNA molecules.
    c. amino acid molecules.
    d. monosaccharides.

32. Which of the following is not converted to blood sugar by the human body?
    a. lactose
    b. dextrose
    c. cellulose
    d. None of the above is correct.

33. Fats from animals and oils from plants have the general structure of a (an)
    a. aldehyde.
    b. ester.
    c. amine.
    d. ketone.

34. Liquid oils from plants can be converted to solids by adding what to the molecule?
    a. metal ions
    b. carbon
    c. polyatomic ions
    d. hydrogen

35. The basic difference between a monomer of polyethylene and a monomer of polyvinyl chloride is
    a. the replacement of a hydrogen by a chlorine.
    b. the addition of four fluorines.
    c. the elimination of double bonds.
    d. the removal of all hydrogens.

36. Many synthetic polymers become a problem in the environment because they
    a. decompose to nutrients, which accelerates plant growth.
    b. do not readily decompose and tend to accumulate.
    c. do not contain vitamins as natural materials do.
    d. become a source of food for fish but ruin the flavor of fish meat.

37. Which of the following terms does not describe an alkane?
    a. paraffin
    b. single bonds
    c. double bonds
    d. straight chain

38. A straight chain alkane with four carbons is called
    a. butane.
    b. tetraalkane.
    c. quatrane.
    d. propane.

39. What is the general formula for alkenes?
    a. C_nH_{2n+2}
    b. C_nH_{2n+4}
    c. C_nH_{2n}
    d. C_nH_{2n-2}

40. An organic molecule that does not contain the maximum number of hydrogen atoms is called
    a. deficient.
    b. unsaturated.
    c. saturated.
    d. incomplete.

41. What is the general formula for an alkyne?
   a. C_nH_{2n+2}
   b. C_nH_{2n+4}
   c. C_nH_{2n}
   d. C_nH_{2n-2}

42. Acetylene is a fuel used in welding and an important reactant in the synthesis of plastics. Acetylene is an
   a. alkane.
   b. alkene.
   c. alkyne.
   d. alkaline.

43. Aromatic hydrocarbons are often used as
   a. perfumes.
   b. soaps.
   c. solvents.
   d. odorants.

44. Fractional distillation separates different fractions by differences in their
   a. melting point.
   b. freezing point.
   c. boiling point.
   d. origin.

45. The rate of combustion of gasoline is quantified by
   a. NASCAR rating.
   b. octane number.
   c. kern rating.
   d. leading number.

46. A petroleum product that is composed of hydrocarbons with sixteen to eighteen carbons atoms per molecules is
   a. gasoline.
   b. asphalt.
   c. paraffin.
   d. motor oil.

47. The common name of 2-propanol is
   a. isopropyl alcohol.
   b. wood alcohol.
   c. methyl alcohol.
   d. moonshine.

48. The organic acid that is found in sour milk, pickles, and in your muscles as a product of carbohydrate metabolism is
   a. acetic acid.
   b. palmitic acid.
   c. lactic acid.
   d. citric acid.

49. What are the organic compounds that are used in perfumes and for artificial flavors?
   a. carboxylic acids
   b. alcohols
   c. ethers
   d. esters

50. Sugars, starches, and cellulose belong to which class of organic compounds?
   a. proteins
   b. fatty acids
   c. carbohydrates
   d. enzymes

Answers

1. b 2. c 3. d 4. b 5. c 6. b 7. c 8. a 9. c 10. b 11. d 12. c 13. c 14. d 15. d 16. c 17. a 18. d 19. d 20. a 21. d 22. b 23. b 24. d 25. d 26. b 27. a 28. c 29. d 30. c 31. c 32. c 33. b 34. d 35. a 36. b 37. c 38. a 39. c 40. b 41. d 42. c 43. c 44. c 45. b 46. d 47. a 48. c 49. d 50. c

QUESTIONS FOR THOUGHT

1. What is an organic compound?

2. There are millions of organic compounds but only thousands of inorganic compounds. Explain why this is the case.

3. What is cracking and reforming? For what purposes are either or both used by the petroleum industry?

4. Is it possible to have an isomer of ethane? Explain.

5. Suggest a reason that ethylene is an important raw material used in the production of plastics but ethane is not.

6. What are (a) natural gas, (b) LPG, and (c) petroleum ether?

7. What does the octane number of gasoline describe? On what is the number based?

8. What is a functional group? What is it about the nature of a functional group that makes it the site of chemical reactions?

9. Draw a structural formula for alcohol. Describe how alcohols are named.

10. What are fats and oils? What are saturated and unsaturated fats and oils?

11. What is a polymer? Give an example of a naturally occurring plant polymer. Give an example of a synthetic polymer.

12. Explain why a small portion of wine is customarily poured before a bottle of wine is served. Sometimes the cork is handed to the person doing the ordering with the small portion of wine. What is the person supposed to do with the cork and why?

FOR FURTHER ANALYSIS

1. Many people feel that they will get better performance by using a higher-octane gasoline in their vehicles. What is the difference between high-octane and regular gasoline? Does scientific evidence bear this out? What octane gasoline is recommended for your vehicle?

2. There have been some health concerns about the additives used in gasoline. What are these additives? What purpose do they serve? What do opponents of using such additives propose are their negative impact?

3. The so-called "birth control pill," or "the pill," has been around since the early 1960s. This medication is composed of a variety of organic molecules. What is the nature of these compounds? How do they work to control conception since they do not control birth? What are some of the negative side effects and some of the benefits of taking "the pill"?

4. Many communities throughout the world are involved in recycling. One of the most important classes of materials recycled is plastic. There has been concern that the cost of recycling plastics is higher than the cost of making new plastic products. Go to the Internet and
   a. Identify the kinds of plastics that are commonly recycled.
   b. Explain how a plastic such as HDPE is recycled.
   c. Present an argument for eliminating the recycling of HDP and for expanding the recycling of HDPE.

Alcohol: What Do You Really Know?

Archaeologists, anthropologists, chemists, biologists, and health care professionals agree that the drinking of alcohol dates back thousands of years. Evidence also exists that this practice has occurred in most cultures around the world. Use the Internet to search out answers to the following questions:

1. What is the earliest date for which there is evidence for the production of ethyl alcohol?

2. In which culture did this occur?

3. What is the molecular formula and structure of ethanol?

4. Do alcohol and water mix?

5. How much ethanol is consumed in the form of beverages in the United States each year?

6. What is the legal blood alcohol limit in your state?

7. How is this level measured?

8. Why is there an extra tax on alcoholic beverages?

9. How do the negative effects of drinking alcohol compare between men and women?

10. Have researchers demonstrated any beneficial effects of drinking alcohol?

Compare what you thought you knew to what is now supported with scientific evidence.

PARALLEL EXERCISES

The exercises in groups A and B cover the same concepts. Solutions to group A exercises are located in appendix E.

Group A

1. Draw the structural formulas for (a) *n*-pentane and (b) an isomer of pentane with the maximum possible branching. (c) Give the IUPAC name of this isomer.

2. Write structural formulas for all the hexane isomers you can identify. Write the IUPAC name for each isomer.

3. Write structural formulas for
   a. 3,3,4-trimethyloctane.
   b. 2-methyl-1-pentene.
   c. 5,5-dimethyl-3-heptyne.

4. Write the IUPAC name for each of the following:

Group B

1. Write structural formulas for (a) *n*-octane, (b) an isomer of octane with the maximum possible branching. (c) Give the IUPAC name of this isomer.

2. Write the structural formulas for all the heptane isomers you can identify. Write the IUPAC name for each isomer.

3. Write structural formulas for
   a. 2,3-dimethylpentane.
   b. 1-butene.
   c. 3-ethyl-2-methyl-3-hexene.

4. Write the IUPAC name for each of the following:

C

```
              H
              |
          H — C — H
              |
     H        |        H
     |        |        |
H  — C  —  C  —  C  =  C  —  C — H
     |     |              |   |
     H     H              H   H
              |
          H — C — H
              |
          H — C — H
              |
              H
```

C
```
     H        H        H
     |        |        |
H  — C  —  C  =  C  —  C  =  C  —  C — H
     |     |              |  |  |
     H     Cl             H  H  H
```

D
```
     H     H     H     H
     |     |     |     |
Br — C  —  C  =  C  —  C — H
     |                 |
     H                 H
```

5. Which would have the higher octane rating, 2,2,3-trimethyl-butane or 2,2-dimethylpentane? Explain with an illustration.

6. Use the information in Table 12.5 to classify each of the following as an alcohol, ether, organic acid, ester, or amide.

A
```
     H   H   H
     |   |   |
H  — C — C — C — OH
     |   |   |
     H   H   H
```

B
```
     H   H   H
     |   |   |
H  — C — C — C — NH₂
     |   |   |
     H   H   H
```

C
```
     H   H   H       H   H   H
     |   |   |       |   |   |
H  — C — C — C — O — C — C — C — H
     |   |   |       |   |   |
     H   H   H       H   H   H
```

D
```
     H   H       H   H   H
     |   |       |   |   |
H  — C — C — O — C — C — C — C — H
     |   |       ‖   |   |   |
     H   H       O   H   H   H
```

E
```
     H   H   H   H
     |   |   |   |
H  — C — C — C — C — C — OH
     |   |   |   |   ‖
     H   H   H   H   O
```

5. Which would have the higher octane rating, 2-methyl-butane or dimethylpropane? Explain with an illustration.

6. Classify each of the following as an alcohol, ether, organic acid, ester, or amide.

A
```
     H   H   H
     |   |   |
H  — C — C — C — H
     |   |   |
     O   O   O
     |   |   |
     H   H   H
```

B
```
     H   H
     |   |
H  — C — C — C — OH
     |   |   ‖
     H   H   O
```

C
```
     H   H       H   H
     |   |       |   |
H  — C — C — O — C — C — H
     |   |       |   |
     H   H       H   H
```

D
```
     H   H   H
     |   |   |
H  — C — C — C — NH₂
     |   |   |
     H   H   H
```

E
```
     H   H       H   H   H   H   H   H   H
     |   |       |   |   |   |   |   |   |
H  — C — C — O — C — C — C — C — C — C — C — H
     |   |       ‖   |   |   |   |   |   |
     H   H       O   H   H   H   H   H   H
```

13
Nuclear Reactions

With the top half of the steel vessel and control rods removed, fuel rod bundles can be replaced in the water-flooded nuclear reactor.

CORE CONCEPT
Nuclear reactions involve changes in the nucleus of the atom.

OUTLINE

Natural Radioactivity
Natural radioactivity is the spontaneous emission of particles or energy from a disintegrating nucleus.

Types of Radioactive Decay
An unstable nucleus becomes more stable by emitting alpha, beta, or gamma radiation from the nucleus.

Nuclear Power Plants
A nuclear reactor in a power plant can only release energy at a comparatively slow rate, and it is impossible for a nuclear power plant to produce a nuclear explosion.

Natural Radioactivity
   Nuclear Equations
   The Nature of the Nucleus
   Types of Radioactive Decay
   Radioactive Decay Series
Measurement of Radiation
   Measurement Methods
A Closer Look: How Is Half-Life Determined?
A Closer Look: Carbon Dating
   Radiation Units
   Radiation Exposure
A Closer Look: Radiation and Food Preservation
Nuclear Energy
A Closer Look: Nuclear Medicine
   Nuclear Fission
   Nuclear Power Plants
A Closer Look: Three Mile Island and Chernobyl
   Nuclear Fusion
A Closer Look: Nuclear Waste
Science and Society: High-Level Nuclear Waste
   The Source of Nuclear Energy
People Behind the Science: Marie Curie

The Nature of the Nucleus
Nuclear instability results from an imbalance between the attractive nuclear force and the repulsive electromagnetic force.

Nuclear Energy
The relationship between energy and mass changes is $E = mc^2$.

327

The ancient alchemist dreamed of changing one element into another, such as lead into gold. The alchemist was never successful, however, because such changes were attempted with chemical reactions. Chemical reactions are reactions that involve only the electrons of atoms. Electrons are shared or transferred in chemical reactions, and the internal nucleus of the atom is unchanged. Elements thus retain their identity during the sharing or transferring of electrons. This chapter is concerned with a different kind of reaction, one that involves the *nucleus* of the atom. In nuclear reactions, the nucleus of the atom is often altered, changing the identity of the elements involved. The ancient alchemist's dream of changing one element into another was actually a dream of achieving a nuclear change, that is, a nuclear reaction.

Understanding nuclear reactions is important because although fossil fuels are the major source of energy today, there are growing concerns about (1) air pollution from fossil fuel combustion, (2) increasing levels of CO_2 from fossil fuel combustion, which may be warming Earth (the greenhouse effect), and (3) the dwindling fossil fuel supply itself, which cannot last forever. Energy experts see nuclear energy as a means of meeting rising energy demands in an environmentally acceptable way. However, the topic of nuclear energy is controversial, and discussions of it often result in strong emotional responses. Decisions about the use of nuclear energy require some understandings about nuclear reactions and some facts about radioactivity and radioactive materials (Figure 13.1). These understandings and facts are the topics of this chapter.

NATURAL RADIOACTIVITY

Natural **radioactivity** is the spontaneous emission of particles or energy from an atomic nucleus as it disintegrates. It was discovered in 1896 by Henri Becquerel, a French scientist who was very interested in the recent discovery of X rays. Becquerel was experimenting with fluorescent minerals, minerals that give off visible light after being exposed to sunlight. He wondered if fluorescent minerals emitted X rays in addition to visible light. From previous work with X rays, Becquerel knew that they would penetrate a wrapped, light-tight photographic plate, exposing it as visible light exposes an unprotected plate. Thus, Becquerel decided to place a fluorescent uranium mineral on a protected photographic plate while the mineral was exposed to sunlight. Sure enough, he found a silhouette of the mineral on the plate when it was developed. Believing the uranium mineral emitted X rays, he continued his studies until the weather turned cloudy. Storing a wrapped, protected photographic plate and the uranium mineral together during the cloudy weather, Becquerel returned to the materials later and developed the photographic plate to again find an image of the mineral (Figure 13.2). He concluded that the mineral was emitting an "invisible radiation" that was not induced by sunlight. The emission of invisible radiation was later named *radioactivity*. Materials that have the property of radioactivity are called *radioactive* materials.

Becquerel's discovery led to the beginnings of the modern atomic theory and to the discovery of new elements. Ernest Rutherford studied the nature of radioactivity and found that there are three kinds, which are today known by the first three letters of the Greek alphabet—alpha (α), beta (β), and gamma (γ).

FIGURE 13.1 Decisions about nuclear energy require some understanding of nuclear reactions and the nature of radioactivity. This is one of the three units of the Palo Verde Nuclear Generating Station in Arizona. With all three units running, enough power is generated to meet the electrical needs of nearly 4 million people.

A

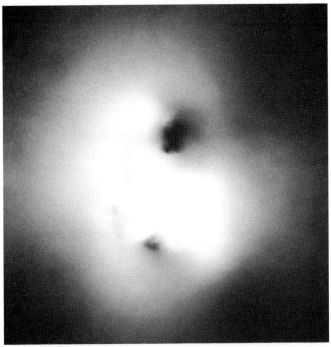

B

FIGURE 13.2 Radioactivity was discovered by Henri Becquerel when he exposed a light-tight photographic plate to a radioactive mineral, then developed the plate. (*A*) A photographic film is exposed to a uraninite ore sample. (*B*) The film, developed normally after a four-day exposure to uraninite. Becquerel found an image like this one and deduced that the mineral gave off invisible radiation.

These Greek letters were used at first before the nature of the radiation was known. Today, an **alpha particle** (sometimes called an alpha ray) is known to be the nucleus of a helium atom, that is, two protons and two neutrons. A **beta particle** (or beta ray) is a high-energy electron. A **gamma ray** is electromagnetic radiation, as is light, but of very short wavelength (Figure 13.3).

At Becquerel's suggestion, Marie Curie searched for other radioactive materials and in the process discovered two new

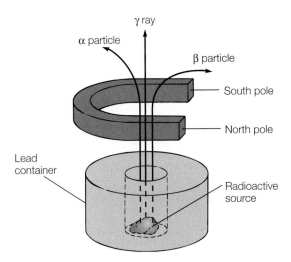

FIGURE 13.3 Radiation passing through a magnetic field shows that massive, positively charged alpha particles are deflected one way, and less massive beta particles with their negative charge are greatly deflected in the opposite direction. Gamma rays, like light, are not deflected.

elements: polonium and radium. More radioactive elements have been discovered since that time, and, in fact, all the isotopes of all the elements with an atomic number greater than 83 (bismuth) are radioactive. As a result of radioactive disintegration, the nucleus of an atom often undergoes a change of identity. The spontaneous disintegration of a given nucleus is a purely natural process and cannot be controlled or influenced. The natural spontaneous disintegration or decomposition of a nucleus is also called **radioactive decay.** Although it is impossible to know *when* a given nucleus will undergo radioactive decay, as you will see later, it is possible to deal with the *rate* of decay for a given radioactive material with precision.

NUCLEAR EQUATIONS

There are two main subatomic particles in the nucleus: the proton and the neutron. The proton and neutron are called **nucleons.** Recall that the number of protons, the *atomic number,* determines what element an atom is and that all atoms of a given element have the same number of protons. The number of neutrons varies in *isotopes,* which are atoms with the same atomic number but different numbers of neutrons (Figure 13.4). The number of protons and neutrons together determines the *mass number,* so different isotopes of the same element are identified with their mass numbers. Thus, the two most common, naturally occurring isotopes of uranium are referred to as uranium-238 and uranium-235, and the 238 and 235 are the mass numbers of these isotopes. Isotopes are also represented by the following symbol:

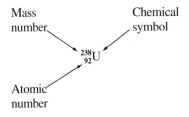

Mass number

Chemical symbol

$$^{238}_{92}\text{U}$$

Atomic number

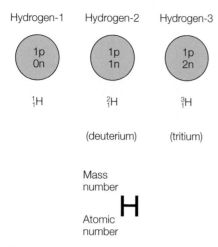

Hydrogen-1   Hydrogen-2   Hydrogen-3

1p   1p   1p
0n   1n   2n

$^1_1$H $^2_1$H $^3_1$H

(deuterium)   (tritium)

Mass
number

H

Atomic
number

FIGURE 13.4 The three isotopes of hydrogen have the same number of protons but different numbers of neutrons. Hydrogen-1 is the most common isotope. Hydrogen-2, with an additional neutron, is named *deuterium,* and hydrogen-3 is called *tritium.*

Subatomic particles involved in nuclear reactions are represented by symbols with the following form:

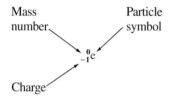

Mass
number

Particle
symbol

$^0_{-1}$e

Charge

Symbols for these particles are illustrated in Table 13.1.

Symbols are used in an equation for a nuclear reaction that is written much like a chemical reaction with reactants and products. When a uranium-238 nucleus emits an alpha particle ($^4_2$He), for example, it loses two protons and two neutrons. The nuclear reaction is written in equation form as

$$^{238}_{92}U \rightarrow {}^{234}_{90}Th + {}^4_2He$$

The *products* of this nuclear reaction from the decay of a uranium-238 nucleus are (1) the alpha particle ($^4_2$He) given off and (2) the nucleus, which remains after the alpha particle leaves the original nucleus. What remains is easily determined since all nuclear equations must show conservation of charge and conservation of the total number of nucleons. In an alpha emission reaction, (1) the number of protons (positive charge) remains

the same, and the sum of the subscripts (atomic number, or numbers of protons) in the reactants must equal the sum of the subscripts in the products; and (2) the total number of nucleons remains the same, and the sum of the superscripts (atomic mass, or number of protons plus neutrons) in the reactants must equal the sum of the superscripts in the products. The new nucleus remaining after the emission of an alpha particle, therefore, has an atomic number of 90 (92 − 2 = 90). According to the table of atomic numbers on the inside back cover of this text, this new nucleus is thorium (Th). The mass of the thorium isotope is 238 minus 4, or 234. The emission of an alpha particle thus decreases the number of protons by 2 and the mass number by 4. From the subscripts, you can see that the total charge is conserved (92 = 90 + 2). From the superscripts, you can see that the total number of nucleons is also conserved (238 = 234 + 4). The mass numbers (superscripts) and the atomic numbers (subscripts) are *balanced* in a correctly written nuclear equation. Such nuclear equations are considered to be independent of any chemical form or chemical reaction. Nuclear reactions are independent and separate from chemical reactions, whether or not the atom is in the pure element or in a compound. Each particle that is involved in nuclear reactions has its own symbol with a superscript indicating mass number and a subscript indicating the charge. These symbols, names, and numbers are given in Table 13.1.

EXAMPLE 13.1

A plutonium-242 nucleus undergoes radioactive decay, emitting an alpha particle. Write the nuclear equation for this nuclear reaction.

SOLUTION

Step 1: The table of atomic weights on the inside back cover gives the atomic number of plutonium as 94. Plutonium-242 therefore has a symbol of, $^{242}_{94}$Pu. The symbol for an alpha particle is ($^4_2$He) so the nuclear equation so far is

$$^{242}_{94}Pu \rightarrow {}^4_2He + ?$$

Step 2: From the subscripts, you can see that 94 = 2 + 92, so the new nucleus has an atomic number of 92. The table of atomic weights identifies element 92 as uranium with a symbol of U.

Step 3: From the superscripts, you can see that the mass number of the uranium isotope formed is 242 − 4 = 238, so the product nucleus is $^{238}_{92}$U and the complete nuclear equation is

$$^{242}_{94}Pu \rightarrow {}^4_2He + {}^{238}_{92}U$$

Step 4: Checking the subscripts (94 = 2 + 92) and the superscripts (242 = 4 + 238), you can see that the nuclear equation is balanced.

EXAMPLE 13.2

What is the product nucleus formed when radium emits an alpha particle? (Answer: Radon-222, a chemically inert, radioactive gas)

TABLE 13.1

Names, symbols, and properties of subatomic particles in nuclear equations

Name	Symbol	Mass Number	Charge
Proton	$^1_1$H (or $^1_1$p)	1	1+
Electron	$^0_{-1}$e (or $^0_{-1}\beta$)	0	1−
Neutron	$^1_0$n	1	0
Gamma photon	$^0_0\gamma$	0	0

THE NATURE OF THE NUCLEUS

The modern atomic theory does not picture the nucleus as a group of stationary protons and neutrons clumped together by some "nuclear glue." The protons and neutrons are understood to be held together by a **nuclear force,** a strong fundamental force of attraction that is functional only at very short distances, on the order of 10^{-15} m or less. At distances greater than about 10^{-15} m, the nuclear force is negligible, and the weaker **electromagnetic force,** the force of repulsion between like charges, is the operational force. Thus, like-charged protons experience a repulsive force when they are farther apart than about 10^{-15} m. When closer together than 10^{-15} m, the short-range, stronger nuclear force predominates, and the protons experience a strong attractive force. This explains why the like-charged protons of the nucleus are not repelled by their like electric charges.

Observations of radioactive decay reactants and products, and experiments with nuclear stability have led to a **shell model of the nucleus.** This model considers the protons and neutrons moving in energy levels, or shells, in the nucleus analogous to the orbital structure of electrons in the outermost part of the atom. As in the electron orbitals, there are certain configurations of nuclear shells that have a greater stability than others. Considering electrons, filled and half-filled orbitals are more stable than other arrangements, and maximum stability occurs with the noble gases and their 2, 10, 18, 36, 54, and 86 electrons. Considering the nucleus, atoms with 2, 8, 20, 28, 50, 82, or 126 protons or neutrons have a maximum nuclear stability. The stable numbers are not the same for electrons and nucleons because of differences in nuclear and electromagnetic forces.

Isotopes of uranium, radium, and plutonium, as well as other isotopes, emit an alpha particle during radioactive decay to a simpler nucleus. The alpha particle is a helium nucleus, $_2^4$He. The alpha particle contains two protons as well as two neutrons, which is one of the nucleon numbers of stability, so you would expect the helium nucleus (or alpha particle) to have a stable nucleus, and it does. *Stable* means it does not undergo radioactive decay. Pairs of protons and pairs of neutrons have increased stability, just as pairs of electrons in a molecule do. As a result, nuclei with an *even number* of both protons and neutrons are, in general, more stable than nuclei with odd numbers of protons and neutrons. There are a little more than 150 stable isotopes with an even number of protons and an even number of neutrons, but there are only four stable isotopes with odd numbers of each. Just as in the case of electrons, there are other factors that come into play as the nucleus becomes larger and larger with increased numbers of nucleons.

The results of some of these factors are shown in Figure 13.5, which is a graph of the number of neutrons versus the number of protons in nuclei. As the number of protons increases, the neutron-to-proton ratio of the *stable nuclei* also increases in a **band of stability.** Within the band, the neutron-to-proton ratio increases from about 1:1 at the bottom left to about 1½:1 at the top right. The increased ratio of neutrons is needed to produce a stable nucleus as the number of protons increases. Neutrons provide additional attractive *nuclear*

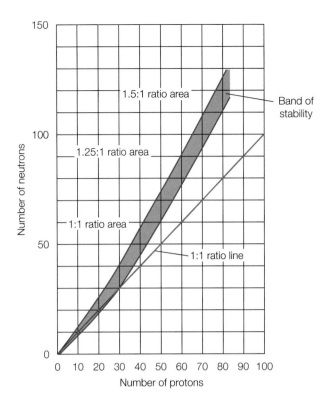

FIGURE 13.5 The shaded area indicates stable nuclei, which group in a band of stability according to their neutron-to-proton ratio. As the size of nuclei increases, so does the neutron-to-proton ratio that represents stability. Nuclei outside this band of stability are radioactive.

(not electrical) forces, which counter the increased electrical repulsion from a larger number of positively charged protons. Thus, more neutrons are required in larger nuclei to produce a stable nucleus. However, there is a limit to the additional attractive forces that can be provided by more and more neutrons, and all isotopes of all elements with more than 83 protons are unstable and thus undergo radioactive decay.

The generalizations about nuclear stability provide a means of predicting if a particular nucleus is radioactive. The generalizations are as follows:

1. All isotopes with an atomic number greater than 83 have an unstable nucleus.
2. Isotopes that contain 2, 8, 20, 28, 50, 82, or 126 protons or neutrons in their nuclei are more stable than those with other numbers of protons or neutrons.
3. Pairs of protons and pairs of neutrons have increased stability, so isotopes that have nuclei with even numbers of both protons and neutrons are generally more stable than those that have nuclei with odd numbers of both protons and neutrons.
4. Isotopes with an atomic number less than 83 are stable when the ratio of neutrons to protons in the nucleus is about 1:1 in isotopes with up to 20 protons, but the ratio increases in larger nuclei in a band of stability (see Figure 13.5). Isotopes with a ratio to the left or right of this band are unstable and thus will undergo radioactive decay.

EXAMPLE 13.3

Would you predict the following isotopes to be radioactive or stable?

(a) $^{60}_{27}\text{Co}$

(b) $^{222}_{86}\text{Rn}$

(c) $^{3}_{1}\text{H}$

(d) $^{40}_{20}\text{Ca}$

SOLUTION

(a) Cobalt-60 has 27 protons and 33 neutrons, both odd numbers, so you might expect $^{60}_{27}\text{Co}$ to be radioactive.

(b) Radon has an atomic number of 86, and all isotopes of all elements beyond atomic number 83 are radioactive. Radon-222 is therefore radioactive.

(c) Hydrogen-3 has an odd number of protons and an even number of neutrons, but its 2:1 neutron-to-proton ratio places it outside the band of stability. Hydrogen-3 is radioactive.

(d) Calcium-40 has an even number of protons and an even number of neutrons, containing 20 of each. The number 20 is a particularly stable number of protons or neutrons, and calcium-40 has 20 of each. In addition, the neutron-to-proton ratio is 1:1, placing it within the band of stability. All indications are that calcium-40 is stable, not radioactive.

TYPES OF RADIOACTIVE DECAY

Through the process of radioactive decay, an unstable nucleus becomes a more stable one with less energy. The three more familiar types of radiation emitted—alpha, beta, and gamma—were introduced earlier. There are five common types of radioactive decay, and three of these involve alpha, beta, and gamma radiation.

1. *Alpha emission.* Alpha (α) emission is the expulsion of an alpha particle ($^{4}_{2}\text{He}$) from an unstable, disintegrating nucleus. The alpha particle, a helium nucleus, travels from 2 to 12 cm through the air, depending on the energy of emission from the source. An alpha particle is easily stopped by a sheet of paper close to the nucleus. As an example of alpha emission, consider the decay of a radon-222 nucleus:

$$^{222}_{86}\text{Rn} \rightarrow {}^{218}_{84}\text{Po} + {}^{4}_{2}\text{He}$$

The spent alpha particle eventually acquires two electrons and becomes an ordinary helium atom.

2. *Beta emission.* Beta (β^-) emission is the expulsion of a different particle, a beta particle, from an unstable disintegrating nucleus. A beta particle is simply an electron ($^{0}_{-1}\text{e}$) ejected from the nucleus at a high speed. The emission of a beta particle *increases the number of protons* in a nucleus. It is as if a neutron changed to a proton by emitting an electron, or

$$^{1}_{0}\text{n} \rightarrow {}^{1}_{1}\text{p} + {}^{0}_{-1}\text{e}$$

Carbon-14 is a carbon isotope that decays by beta emission:

$$^{14}_{6}\text{C} \rightarrow {}^{14}_{7}\text{N} + {}^{0}_{-1}\text{e}$$

Note that the number of protons increased from six to seven, but the mass number remained the same. The mass number is unchanged because the mass of the expelled electron (beta particle) is negligible.

Beta particles are more penetrating than alpha particles and may travel several hundred centimeters through the air. They can be stopped by a thin layer of metal close to the emitting nucleus, such as a 1 cm thick piece of aluminum. A spent beta particle may eventually join an ion to become part of an atom, or it may remain a free electron.

3. *Gamma emission.* Gamma (γ) emission is a high-energy burst of electromagnetic radiation from an excited nucleus. It is a burst of light (photon) of a wavelength much too short to be detected by the eye. Other types of radioactive decay, such as alpha or beta emission, sometimes leave the nucleus with an excess of energy, a condition called an *excited state*. As in the case of excited electrons, the nucleus returns to a lower energy state by emitting electromagnetic radiation. From a nucleus, this radiation is in the high-energy portion of the electromagnetic spectrum. Gamma is the most penetrating of the three common types of nuclear radiation. Like X rays, gamma rays can pass completely through a person, but all gamma radiation can be stopped by a 5 cm thick piece of lead close to the source. As with other types of electromagnetic radiation, gamma radiation is absorbed by and gives its energy to materials. Since the product nucleus changed from an excited state to a lower energy state, there is no change in the number of nucleons. For example, radon-222 is an isotope that emits gamma radiation:

$$^{222}_{86}\text{Rn}^{\star} \rightarrow {}^{222}_{86}\text{Rn} + {}^{0}_{0}\gamma$$

(*denotes excited state)

Radioactive decay by alpha, beta, and gamma emission is summarized in Table 13.2, which also lists the unstable nuclear conditions that lead to the particular type of emission. Just as electrons seek a state of greater stability, a nucleus undergoes radioactive decay to achieve a balance between nuclear attractions, electromagnetic repulsions, and a low quantum of nuclear shell energy. The key to understanding the types of reactions that occur is found in the band of stable nuclei illustrated in Figure 13.5. The isotopes within this band have achieved the

TABLE 13.2
Radioactive decay

Unstable Condition	Type of Decay	Emitted	Product Nucleus
More than 83 protons	Alpha emission	$^{4}_{2}\text{H}$	Lost 2 protons and 2 neutrons
Neutron-to-proton ratio too large	Beta emission	$^{0}_{-1}\text{e}$	Gained 1 proton, no mass change
Excited nucleus	Gamma emission	$^{0}_{0}\gamma$	No change
Neutron-to-proton ratio too small	Other emission	$^{0}_{1}\text{e}$	Lost 1 proton, no mass change

state of stability, and other isotopes above, below, or beyond the band are unstable and thus radioactive.

Nuclei that have a neutron-to-proton ratio beyond the upper right part of the band are unstable because of an imbalance between the proton-proton electromagnetic repulsions and all the combined proton and neutron nuclear attractions. Recall that the neutron-to-proton ratio increases from about 1:1 to about 1½:1 in the larger nuclei. The additional neutron provided additional nuclear attractions to hold the nucleus together, but atomic number 83 appears to be the upper limit to this additional stabilizing contribution. Thus, all nuclei with an atomic number greater than 83 are outside the upper right limit of the band of stability. Emission of an alpha particle reduces the number of protons by two and the number of neutrons by two, moving the nucleus more toward the band of stability. Thus, you can expect a nucleus that lies beyond the upper right part of the band of stability to be an alpha emitter (Figure 13.6).

A nucleus with a neutron-to-proton ratio that is too large will be on the left side of the band of stability. Emission of a beta particle decreases the number of neutrons and increases the number of protons, so a beta emission will lower the neutron-to-proton ratio. Thus, you can expect a nucleus with a large neutron-to-proton ratio, that is, one to the left of the band of stability, to be a beta emitter.

A nucleus that has a neutron-to-proton ratio that is too small will be on the right side of the band of stability. These nuclei can increase the number of neutrons and reduce the number of protons in the nucleus by other types of radioactive decay. As is usual when dealing with broad generalizations and trends, there are exceptions to the summarized relationships between neutron-to-proton ratios and radioactive decay.

EXAMPLE 13.4

Refer to Figure 13.6 and predict the type of radioactive decay for each of the following unstable nuclei:

(a) $^{131}_{53}\text{I}$

(b) $^{242}_{94}\text{Pu}$

SOLUTION

(a) Iodine-131 has a nucleus with 53 protons and 131 minus 53, or 78 neutrons, so it has a neutron-to-proton ratio of 1.47:1. This places iodine-131 on the left side of the band of stability, with a high neutron-to-proton ratio that can be reduced by beta emission. The nuclear equation is

$$^{131}_{53}\text{I} \rightarrow {}^{131}_{54}\text{Xe} + {}^{0}_{-1}\text{e}$$

(b) Plutonium-242 has 94 protons and 242 minus 94, or 148 neutrons, in the nucleus. This nucleus is to the upper right, beyond the band of stability. It can move back toward stability by emitting an alpha particle, losing 2 protons and 2 neutrons from the nucleus. The nuclear equation is

$$^{242}_{94}\text{Pu} \rightarrow {}^{238}_{92}\text{U} + {}^{4}_{2}\text{He}$$

RADIOACTIVE DECAY SERIES

A radioactive decay reaction produces a simpler and eventually more stable nucleus than the reactant nucleus. As discussed in the section on types of radioactive decay, large nuclei with an atomic number greater than 83 decay by alpha emission, giving up two protons and two neutrons with each alpha particle. A nucleus with an atomic number greater than 86, however, will emit an alpha particle and *still* have an atomic number greater than 83, which means the product nucleus will also be radioactive. This nucleus will also undergo radioactive decay, and the process will continue through a series of decay reactions until a stable nucleus is achieved. Such a series of decay reactions that (1) begins with one radioactive nucleus, which (2) decays to a second nucleus, which (3) then decays to a third nucleus, and so on until (4) a stable nucleus is reached is called a **radioactive decay series.** There are three naturally occurring radioactive decay series. One begins with thorium-232 and ends with lead-208, another begins with uranium-235 and ends with lead-207, and the third series begins with uranium-238 and ends with lead-206. Figure 13.7 shows the uranium-238 radioactive decay series.

As Figure 13.7 illustrates, the uranium-238 begins with uranium-238 decaying to thorium-234 by alpha emission. Thorium has a new position on the graph because it now has a new atomic number and a new mass number. Thorium-234 is unstable and decays to protactinium-234 by beta emission, which is also unstable and decays by beta emission to uranium-234. The process continues with five sequential alpha emissions, then two beta-beta-alpha decay steps before the series terminates with the stable lead-206 nucleus.

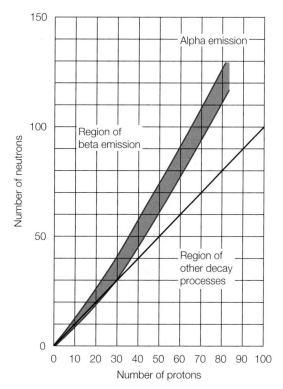

FIGURE 13.6 Unstable nuclei undergo different types of radioactive decay to obtain a more stable nucleus. The type of decay depends, in general, on the neutron-to-proton ratio, as shown.

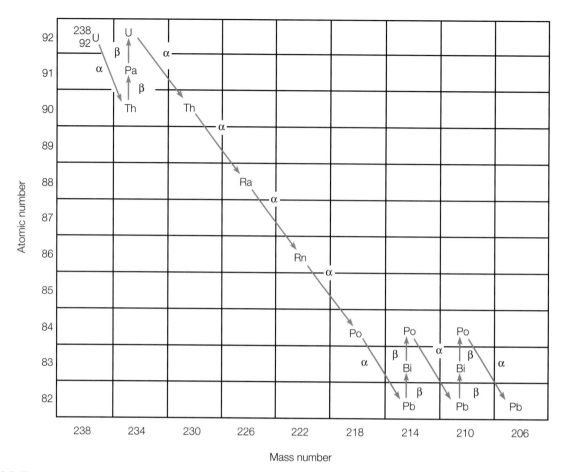

FIGURE 13.7 The radioactive decay series for uranium-238. This is one of three naturally occurring series.

The rate of radioactive decay is usually described in terms of its *half-life*. The **half-life** is the time required for one-half of the unstable nuclei to decay. Since each isotope has a characteristic decay constant, each isotope has its own characteristic half-life. Half-lives of some highly unstable isotopes are measured in fractions of seconds, and other isotopes have half-lives measured in seconds, minutes, hours, days, months, years, or billions of years. Table 13.3 lists half-lives of some of the isotopes, and the process is illustrated in Figure 13.8.

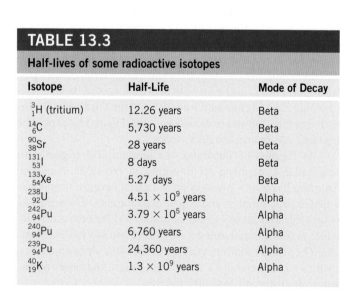

TABLE 13.3

Half-lives of some radioactive isotopes

Isotope	Half-Life	Mode of Decay
^3_1H (tritium)	12.26 years	Beta
$^{14}_6C$	5,730 years	Beta
$^{90}_{38}Sr$	28 years	Beta
$^{131}_{53}I$	8 days	Beta
$^{133}_{54}Xe$	5.27 days	Beta
$^{238}_{92}U$	4.51×10^9 years	Alpha
$^{242}_{94}Pu$	3.79×10^5 years	Alpha
$^{240}_{94}Pu$	6,760 years	Alpha
$^{239}_{94}Pu$	24,360 years	Alpha
$^{40}_{19}K$	1.3×10^9 years	Alpha

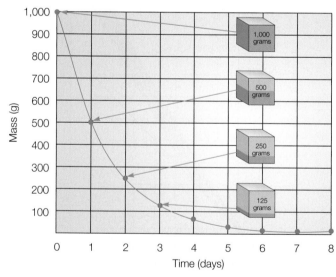

FIGURE 13.8 Radioactive decay of a hypothetical isotope with a half-life of one day. The sample decays each day by one-half to some other element. Actual half-lives may be in seconds, minutes, or any time unit up to billions of years.

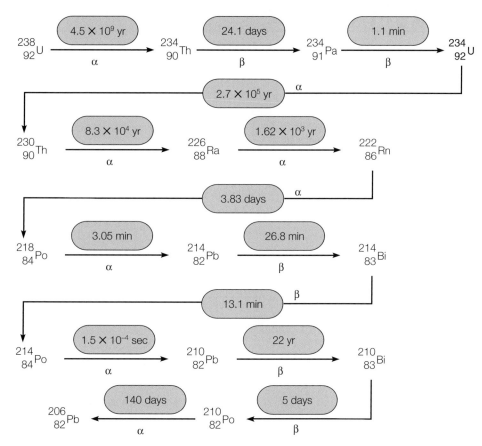

FIGURE 13.9 The half-life of each step in the uranium-238 radioactive decay series.

MEASUREMENT OF RADIATION

The measurement of radiation is important in determining the half-life of radioactive isotopes. Radiation measurement is also important in considering biological effects, which will be discussed in the section on radiation units. As is the case with electricity, it is not possible to make direct measurements on things as small as electrons and other parts of atoms. Indirect measurement methods are possible, however, by considering the effects of the radiation.

As an example of the half-life measure, consider a hypothetical isotope that has a half-life of one day. The half-life is independent of the amount of the isotope being considered, but suppose you start with a 1.0 kg sample of this element with a half-life of one day. One day later, you will have half of the original sample, or 500 g. The other half did not disappear, but it is now the decay product, that is, some new element. During the next day, half of the remaining nuclei will disintegrate, and only 250 g of the initial sample is still the original element. One-half of the remaining sample will disintegrate each day until the original sample no longer exists.

The half-life of uranium-238 is 4.5 billion years. Figure 13.9 gives the half-life for each step in the uranium-238 decay series.

MEASUREMENT METHODS

As Becquerel discovered, radiation affects photographic film, exposing it as visible light does. Since the amount of film exposure is proportional to the amount of radiation, photographic film can be used as an indirect measure of radiation. Today, people who work around radioactive materials or X rays carry light-tight film badges. The film is replaced periodically and developed. The optical density of the developed film provides a record of the worker's exposure to radiation because the darkness of the developed film is proportional to the exposure.

There are also devices that indirectly measure radiation by measuring an effect of the radiation. An **ionization counter** is one type of device that measures ions produced by radiation. A second type of device is called a **scintillation counter.** *Scintillate* is a word meaning "sparks or flashes," and a scintillation counter measures the flashes of light produced when radiation strikes a phosphor.

The most common example of an ionization counter is known as a **Geiger counter** (Figure 13.10). The working components of a Geiger counter are illustrated in Figure 13.11. Radiation is received in a metal tube filled with an inert gas, such as argon. An insulated wire inside the tube is connected to the positive terminal of a direct current source. The metal cylinder around the insulated wire is connected to the negative terminal. There is no current between the center wire and the metal cylinder because the gas acts as an insulator. When radiation passes

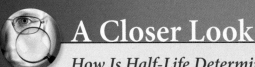

A Closer Look

How Is Half-Life Determined?

It is not possible to predict when a radioactive nucleus will decay because it is a random process. It is possible, however, to deal with nuclear disintegration statistically, since the rate of decay is not changed by any external conditions of temperature or pressure, or any chemical state. When dealing with a large number of nuclei, the ratio of the rate of nuclear disintegration per unit of time to the total number of radioactive nuclei is a constant, or

$$\frac{\text{radioactive decay}}{\text{constant}} = \frac{\text{decay rate}}{\text{number of nuclei}}$$

The radioactive decay constant is a specific constant for a particular isotope, and each isotope has its own decay constant that can be measured. For example, a 238 g sample of uranium-238 (1 mole) that has 2.93×10^6 disintegrations per second would have a decay constant of

$$\begin{aligned}\frac{\text{radioactive decay}}{\text{constant}} &= \frac{\text{decay rate}}{\text{number of nuclei}} \\ &= \frac{2.93 \times 10^6 \text{ nuclei/s}}{6.02 \times 10^{23} \text{ nuclei}} \\ &= 4.87 \times 10^{-18} \text{ l/s}\end{aligned}$$

The half-life of a radioactive nucleus is related to its radioactive decay constant by

$$\text{half-life} = \frac{\text{a mathematical constant}}{\text{decay costant}}$$

The half-life of uranium-238 is therefore

$$\begin{aligned}\text{half-life} &= \frac{\text{a mathematical constant}}{\text{decay costant}} \\ &= \frac{0.693}{4.87 \times 10^{-18} \text{ l/s}} \\ &= 1.42 \times 10^{17} \text{ s}\end{aligned}$$

This is the half-life of uranium-238 in seconds. There are $60 \times 60 \times 24 \times 365$, or 3.15×10^7 s in a year, so

$$\frac{1.42 \times 10^{17} \text{ s}}{3.15 \times 10^7 \text{ s/yr}} = 4.5 \times 10^9 \text{ yr}$$

The half-life of uranium-238 is thus 4.5 billion years.

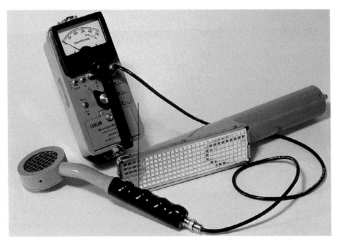

FIGURE 13.10 This is a beta-gamma probe, which can measure beta and gamma radiation in millirems per unit of time.

through the window, however, it ionizes some of the gas atoms, releasing free electrons. These electrons are accelerated by the field between the wire and cylinder, and the accelerated electrons ionize more gas molecules, which results in an *avalanche* of free electrons. The avalanche creates a pulse of current that is amplified and then measured. More radiation means more avalanches, so the pulses are an indirect means of measuring radiation. When connected to a speaker or earphone, each avalanche produces a "pop" or "click."

Some materials are *phosphors,* substances that emit a flash of light when excited by radiation. Zinc sulfide, for example, is used in television screens and luminous watches, and it was used by Rutherford to detect alpha particles. A zinc sulfide atom gives off a tiny flash of light when struck by radiation. A scintillation counter measures the flashes of light through the photoelectric effect, producing free electrons that are accelerated to produce a pulse of current. Again, the pulses of current are used as an indirect means to measure radiation.

RADIATION UNITS

You have learned that *radioactivity* is a property of isotopes with unstable, disintegrating nuclei and *radiation* is emitted particles (alpha or beta) or energy traveling in the form of photons (gamma). Radiation can be measured (1) at the source of radioactivity or (2) at a place of reception, where the radiation is absorbed.

The *activity* of a radioactive source is a measure of the number of nuclear disintegrations per unit of time. The unit of activity at the source is called a **curie** (Ci), which is defined as 3.70×10^{10} nuclear disintegrations per second. Activities are usually expressed in terms of fractions of curies, for example, a *picocurie* (pCi), which is a millionth of a millionth of a curie. Activities are sometimes expressed in terms of so many picocuries per liter (pCi/L).

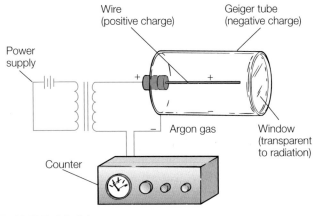

FIGURE 13.11 The working parts of a Geiger counter.

Carbon is an element that occurs naturally as several isotopes. The most common form is carbon-12. A second, heavier radioactive isotope, carbon-14, is constantly being produced in the atmosphere by cosmic rays. Radioactive elements are unstable and break down into other forms of matter. Hence, radioactive carbon-14 naturally decays. The rate at which carbon-14 is formed and the rate at which it decays is about the same; therefore, the concentration of carbon-14 on the earth stays relatively constant. All living things contain large quantities of the element carbon. Plants take in carbon in the form of carbon dioxide from the atmosphere, and animals obtain carbon from the food they eat. While an organism is alive, the proportion of carbon-14 to carbon-12 within its body is equal to its surroundings. When an organism dies, the carbon-14 within its tissues disintegrates, but no new carbon-14 is added. Therefore, the age of plant and animal remains can be determined by the ratio of carbon-14 to carbon-12 in the tissues. The older the specimen, the less carbon-14 present. Radioactive decay rates are measured in half-life. One half-life is the amount of time it takes for one-half of a radioactive sample to decay.

The half-life of carbon-14 is 5,730 years. Therefore, a bone containing one-half the normal proportion of carbon-14 is 5,730 years old. If the bone contains one-quarter of the normal proportion of carbon-14, it is $2 \times 5,730 = 11,460$ years old, and if it contains one-eighth of the naturally occurring proportion of carbon-14, it is $3 \times 5,730 = 17,190$ years old. As the amount of carbon-14 in a sample becomes smaller, it becomes more difficult to measure the amount remaining. Therefore, carbon-14 dating is generally only useful for dating things that are less than 50,000 years old.

The International System of Units (SI) unit for radioactivity is the *Becquerel* (Bq), which is defined as one nuclear disintegration per second. The unit for reporting radiation in the United States is the curie, but the Becquerel is the internationally accepted unit. Table 13.4 gives the names, symbols, and conversion factors for units of radioactivity.

As radiation from a source moves out and strikes a material, it gives the material energy. The amount of energy released by radiation striking living tissue is usually very small, but it can cause biological damage nonetheless because chemical bonds are broken and free polyatomic ions are produced by radiation.

The amount of radiation received by a human is expressed in terms of radiological dose. Radiation dose is usually written in units of a **rem,** which takes into account the possible biological damage produced by different types of radiation. Doses are usually expressed in terms of fractions of the rem, for example, a *millirem* (mrem). The SI unit for radiation dose is the *millisievert* (mSv). Both the millirem and the millisievert relate ionizing radiation and biological effect to humans. The natural radiation that people receive from nature in one day is about 1 millirem (0.01 millisievert). A single dose of 100,000 to 200,000 millirems (1,000 to 2,000 millisieverts) can cause radiation sickness in humans (Table 13.5). A single dose of 500,000 millirems (5,000 millisieverts) results in death about 50 percent of the time.

Another measure of radiation received by a material is the **rad.** The term *rad* is from *rad*iation *a*bsorbed *d*ose. The SI unit for radiation received by a material is the *gray* (Gy). One gray is equivalent to an exposure of 100 rad.

Overall, there are many factors and variables that affect the possible damage from radiation, including the distance from

TABLE 13.4

Names, symbols, and conversion factors for radioactivity

Name	Symbol	To Obtain	Multiply by
Becquerel	Bq	Ci	2.7×10^{-11}
gray	Gy	rad	100
sievert	Sv	rem	100
curie	Ci	Bq	3.7×10^{10}
rem	rem	Sv	0.01
millirem	mrem	rem	0.001
rem	rem	millirem	1,000

TABLE 13.5

Approximate single dose, whole-body effects of radiation exposure

Level	Comment
0.130 rem	Average annual exposure to natural background radiation
0.500 rem	Upper limit of annual exposure to general public
25.0 rem	Threshold for observable effects such as reduced blood cell count
100.0 rem	Fatigue and other symptoms of radiation sickness
200.0 rem	Definite radiation sickness, bone marrow damage, possibility of developing leukemia
500.0 rem	Lethal dose for 50 percent of individuals
1,000.0 rem	Lethal dose for all

the source and what shielding materials are between a person and a source. A *millirem* is the unit of choice when low levels of radiation are discussed.

RADIATION EXPOSURE

Natural radioactivity is a part of your environment, and you receive between 100 and 500 millirems each year from natural sources. This radiation from natural sources is called **background radiation.** Background radiation comes from outer space in the form of cosmic rays and from unstable isotopes in the ground, building materials, and foods. Many activities and situations will increase your yearly exposure to radiation. For example, the atmosphere absorbs some of the cosmic rays from space, so the less atmosphere above you, the more radiation you will receive. You are exposed to 1 additional millirem per year for each 100 feet you live above sea level. You receive approximately 0.3 millirem for each hour spent on a jet flight. Airline crews receive an additional 300 to 400 millirems per year because they spend so much time high in the atmosphere. Additional radiation exposure comes from medical X rays and television sets. In general, the background radiation exposure for the average person is about 130 millirems per year.

What are the consequences of radiation exposure? Radiation can be a hazard to living organisms because it produces ionization along its path of travel. This ionization can (1) disrupt chemical bonds in essential macromolecules such as DNA and (2) produce molecular fragments, which are free polyatomic ions that can interfere with enzyme action and other essential cell functions. Tissues with rapidly dividing cells, such as blood-forming tissue, are more vulnerable to radiation damage than others. Thus, one of the symptoms of an excessive radiation exposure is a lowered red and white blood cell count. Table 13.5 compares the estimated results of various levels of acute radiation exposure.

Radiation is not a mysterious, unique health hazard. It is a hazard that should be understood and neither ignored nor exaggerated. Excessive radiation exposure should be avoided, just as you avoid excessive exposure to other hazards such as certain chemicals, electricity, or even sunlight. Everyone agrees that *excessive* radiation exposure should be avoided, but there is some controversy about long-term, low-level exposure and its possible role in cancer. Some claim that tolerable low-level exposure does not exist because that is not possible. Others point to many studies comparing high and low background radioactivity with cancer mortality data. For example, no cancer mortality differences could be found between people receiving 500 or more millirems a year and those receiving less than 100 millirems a year. The controversy continues, however, because of lack of knowledge about long-term exposure. Two models of long-term, low-level radiation exposure have been proposed: (1) a linear model and (2) a threshold model. The *linear model* proposes that any radiation exposure above zero is damaging and can produce cancer and genetic damage. The *threshold model* proposes that the human body can repair damage and get rid of damaging free polyatomic ions up to a certain exposure level called the threshold (Figure 13.12). The controversy over

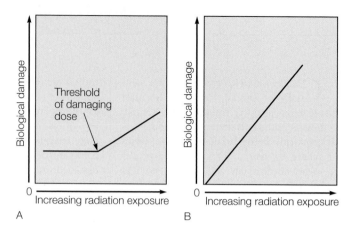

FIGURE 13.12 Graphic representation of the (*A*) threshold model and (*B*) the linear model of low-level radiation exposure. The threshold model proposes that the human body can repair damage up to a threshold. The linear model proposes that any radiation exposure is damaging.

long-term, low-level radiation exposure will probably continue until there is clear evidence about which model is correct. Whichever is correct will not lessen the need for rational risks versus cost-benefit analyses of all energy alternatives.

 Myths, Mistakes, & Misunderstandings

Antiradiation Pill?

It is a myth that there is an antiradiation pill that will protect you from ionizing radiation. There is a pill, an iodine supplement, that when taken, saturates your thyroid with a nonradioactive isotope of iodine. Once saturated, your thyroid will not absorb radioactive isotopes for storage in the gland, which could be dangerous.

NUCLEAR ENERGY

Some nuclei are unstable because they are too large or because they have an unstable neutron-to-proton ratio. These unstable nuclei undergo radioactive decay, eventually forming products of greater stability. An example of this radioactive decay is the alpha emission reaction of uranium-238 to thorium-234,

$$^{238}_{92}\text{U} \rightarrow {}^{234}_{90}\text{Th} + {}^{4}_{2}\text{He}$$

$$238.0003 \text{ u} \rightarrow 233.9942 \text{ u} + 4.00150 \text{ u}$$

The numbers below the nuclear equation are the *nuclear* masses (u) of the reactant and products. As you can see, there seems to be a loss of mass in the reaction,

$$233.9942 + 4.00150 - 238.0003 = -0.0046 \text{ u}$$

A Closer Look

Radiation and Food Preservation

Radiation can be used to delay food spoilage and preserve foods by killing bacteria and other pathogens, just as heat is used to pasteurize milk. Foods such as wheat, flour, fruits, vegetables, pork, chicken, turkey, ground beef, and other uncooked meats are exposed to gamma radiation from cobalt-60 or cesium-137 isotopes, X rays, and electron beams. This kills insects, parasites such as *Trichinella spiralis* and tapeworms, and bacteria such as *E. coli*, *Listeria*, *Salmonellae*, and *Staphylococcus*. The overall effect is that many food-borne causes of human disease are eliminated and it is possible to store foods longer.

Food in the raw state, processed, or frozen is passed through a machine where it is irradiated while cold or frozen. This process does not make the food radioactive because the food does not touch any radioactive substance. In addition, the radiation used in the process is not strong enough to disrupt the nuclei of atoms in food molecules, so it does not produce radioactivity, either.

In addition to killing the parasites and bacteria, the process might result in some nutritional loss but no more than that which normally occurs in canning. Some new chemical products may be formed by the exposure to radiation, but studies in several countries have not been able to identify any health problems or ill effects from these compounds.

Treatment with radiation works better for some foods than others. Dairy products undergo some flavor changes that are undesirable, and some fruits such as peaches become soft. Irradiated strawberries, on the other hand, remain firm and last for weeks instead of a few days in the refrigerator. Foods that are sterilized with a stronger dose of radiation can be stored for years without refrigeration just like canned foods that have undergone heat pasteurization.

In the United States, the Food and Drug Administration (FDA) regulates which products can be treated by radiation and the dosages used in the treatment. The U.S. Department of Agriculture (USDA) is responsible for the inspection of irradiated meat and poultry products. All foods that have undergone a radiation treatment must show the international logo for this and a statement. The logo, called a radura, is a stylized flower inside a circle with five openings on the top part (Box Figure 13.1).

BOX FIGURE 13.1

This change in mass is related to the energy change according to the relationship that was formulated by Albert Einstein in 1905. The relationship is

$$E = mc^2$$

equation 13.1

where E is a quantity of energy, m is a quantity of mass, and c is a constant equal to the speed of light in a vacuum, 3.00×10^8 m/s. According to this relationship, matter and energy are the same thing, and energy can be changed to matter and vice versa.

The products of a mole of uranium-238 decaying to more stable products (1) have a lower energy of 4.14×10^{11} J and (2) lost a mass of 4.6×10^{-6} kg. As you can see, a very small amount of matter was converted into a large amount of energy in the process, forming products of lower energy.

The relationship between mass and energy explains why the mass of a nucleus is always *less* than the sum of the masses of the individual particles of which it is made. The difference between (1) the mass of the individual nucleons making up a nucleus and (2) the actual mass of the nucleus is called the **mass defect** of the nucleus. The explanation for the mass defect is again found in $E = mc^2$. When nucleons join to make a nucleus, energy is released as the more stable nucleus is formed.

The energy equivalent released when a nucleus is formed is the same as the **binding energy,** the energy required to break the nucleus into individual protons and neutrons. The binding energy of the nucleus of any isotope can be calculated from the mass defect of the nucleus.

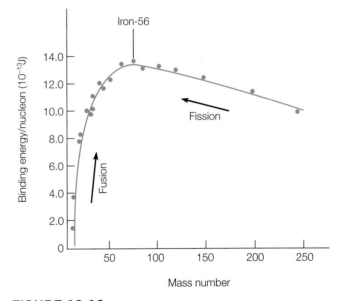

FIGURE 13.13 The maximum binding energy per nucleon occurs around mass number 56, then decreases in both directions. As a result, fission of massive nuclei and fusion of less massive nuclei both release energy.

The ratio of binding energy to nucleon number is a reflection of the stability of a nucleus (Figure 13.13). The greatest binding energy per nucleon occurs near mass number 56, with about 1.4×10^{-12} J per nucleon, then decreases for both more massive and less massive nuclei. This means that more massive

A Closer Look

Nuclear Medicine

Nuclear medicine had its beginnings in 1946 when radioactive iodine was first successfully used to treat thyroid cancer patients. Then physicians learned that radioactive iodine could also be used as a diagnostic tool, providing a way to measure the function of the thyroid and to diagnose thyroid disease. More and more physicians began to use nuclear medicine to diagnose thyroid disease as well as to treat hyperthyroidism and other thyroid problems. Nuclear medicine is a branch of medicine using radiation or radioactive materials to diagnose as well as treat diseases.

The development of new nuclear medicine technologies, such as cameras, detection instruments, and computers, has led to a remarkable increase in the use of nuclear medicine as a diagnostic tool. Today, there are nearly one hundred different nuclear medicine imaging procedures. These provide unique, detailed information about virtually every major organ system within the body, information that was unknown just years ago. Treatment of disease with radioactive materials continues to be a valuable part of nuclear medicine, too. The material that follows will consider some techniques of using nuclear medicine as a diagnostic tool, followed by a short discussion of the use of radioactive materials in the treatment of disease.

Nuclear medicine provides diagnostic information about organ function, compared to conventional radiology, which provides images about the structure. For example, a conventional X-ray image will show if a bone is broken or not, while a bone

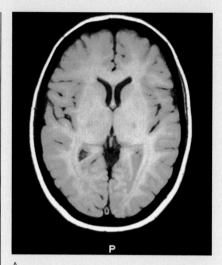
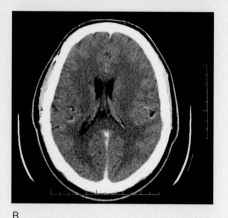

BOX FIGURE 13.2 (*A*) MRI scan of brain; (*B*) CAT scan of brain.

imaging nuclear scan will show changes caused by tumors, hairline fractures, or arthritis. There are procedures for making detailed structural X-ray pictures of internal organs such as the liver, kidney, or heart, but these images often cannot provide diagnostic information, showing only the structure. Nuclear medicine scans, on the other hand, can provide information about how much heart tissue is still alive after a heart attack or if a kidney is working, even when there are no detectable changes in organ appearance.

An X-ray image is produced when X rays pass through the body and expose photographic film on the other side. Some X-ray exams improve photographic contrast

by introducing certain substances. A barium sulfate "milk shake," for example, can be swallowed to highlight the esophagus, stomach, and intestine. More information is provided if X rays are used in a CAT scan (CAT stands for "computed axial tomography"). The CAT-scan is a diagnostic test that combines the use of X rays with computer technology. The CAT scan shows organs of interest by making X-ray images from many different angles as the source of the X-rays moves around the patient. Contrast-improving substances, such as barium sulfate, might also be used with a CAT scan. In any case, CAT scan images are assembled by a computer into a three-dimensional picture that can show organs, bones, and tissues in great detail.

---

nuclei can gain stability by splitting into smaller nuclei with the release of energy. It also means that less massive nuclei can gain stability by joining together with the release of energy. The slope also shows that more energy is released in the coming together process than in the splitting process.

The nuclear reaction of splitting a massive nucleus into more stable, less massive nuclei with the release of energy is **nuclear fission** (Figure 13.14). Nuclear fission occurs rapidly in an atomic bomb explosion and occurs relatively slowly in a nuclear reactor. The nuclear reaction of less massive nuclei coming together to form more stable, and more massive, nuclei with the release of energy is **nuclear fusion.** Nuclear fusion occurs rapidly in a hydrogen bomb explosion and occurs continually in the sun, releasing the energy essential for the continuation of life on Earth.

NUCLEAR FISSION

Nuclear fission was first accomplished in the late 1930s when researchers were attempting to produce isotopes by bombarding massive nuclei with neutrons. In 1938, two German scientists, Otto Hahn and Fritz Strassman, identified the element barium in a uranium sample that had been bombarded with neutrons. Where the barium came from was a puzzle at the time, but soon afterward Lise Meitner deduced that uranium nuclei had split, producing barium. The reaction might have been

$$_{0}^{1}n + _{92}^{235}U \rightarrow _{56}^{141}Ba + _{36}^{92}Kr + 3_{0}^{1}n$$

The phrase "might have been" is used because a massive nucleus can split in many different ways, producing different products. About thirty-five different, less massive elements have been

The gamma camera is a key diagnostic imaging tool used in nuclear medicine. Its use requires a radioactive material, called a radiopharmaceutical, to be injected into or swallowed by the patient. A given radiopharmaceutical tends to go to a specific organ of the body; for example, radioactive iodine tends to go to the thyroid gland, and others go to other organs. Gamma-emitting radiopharmaceuticals are used with the gamma camera, and the gamma camera collects and processes these gamma rays to produce images. These images provide a way of studying the structure as well as measuring the function of the selected organ. Together, the structure and function provide a way of identifying tumors, areas of infection, or other problems. The patient experiences little or no discomfort, and the radiation dose is small.

A SPECT scan (single photon emission computerized tomography) is an imaging technique employing a gamma camera that rotates around the patient, measuring gamma rays and computing their point of origin. Cross-sectional images of a three-dimensional internal organ can be obtained from such data, resulting in images that have higher resolution and thus more diagnostic information than a simple gamma camera image. A gallium radiopharmaceutical is often used in a scan to diagnose and follow the progression of tumors or infections. Gallium scans also can be used to evaluate the heart, lungs, or any other organ that may be involved with inflammatory disease.

Use of MRI (magnetic resonance imaging) also produces images as an infinite number of projections through the body. Unlike CAT, gamma, or SPECT scans, MRI does not use any form of ionizing radiation. MRI uses magnetic fields, radio waves, and a computer to produce detailed images. As the patient enters an MRI scanner, his or her body is surrounded by a large magnet. The technique requires a very strong magnetic field, a field so strong that it aligns the nuclei of the person's atoms. The scanner sends a strong radio signal, temporarily knocking the nuclei out of alignment. When the radio signal stops, the nuclei return to the aligned position, releasing their own faint radio frequencies. These radio signals are read by the scanner, which uses them in a computer program to produce very detailed images of the human anatomy (Box Figure 13.2).

The PET scan (positron emission tomography) produces 3D images superior to gamma camera images. This technique is built around a radiopharmaceutical that emits positrons (like an electron with a positive charge). Positrons collide with electrons, releasing a burst of energy in the form of photons. Detectors track the emissions and feed the information into a computer. The computer has a program to plot the source of radiation and translates the data into an image. Positron-emitting radiopharmaceuticals used in a PET scan can be low atomic weight elements such as carbon, nitrogen, and oxygen. This is important for certain purposes since these are the same elements found in many biological substances such as sugar, urea, or carbon dioxide. Thus, a PET scan can be used to study processes in such organs as the brain and heart where glucose is being broken down or oxygen is being consumed. This diagnostic method can be used to detect epilepsy or brain tumors, among other problems.

Radiopharmaceuticals used for diagnostic examinations are selected for their affinity for certain organs, if they emit sufficient radiation to be easily detectable in the body, and if they have a rather short half-life, preferably no longer than a few hours. Useful radioisotopes that meet these criteria for diagnostic purposes are technetium-99, gallium-67, indium-111, iodine-123, iodine-131, thallium-201, and krypton-81.

The goal of therapy in nuclear medicine is to use radiation to destroy diseased or cancerous tissue while sparing adjacent healthy tissue. Few radioactive therapeutic agents are injected or swallowed, with the exception of radioactive iodine—mentioned earlier as a treatment for cancer of the thyroid. Useful radioisotopes for therapeutical purposes are iodine-131, phosphorus-32, iridium-192, and gold-198. The radioactive source placed in the body for local irradiation of a tumor is normally iridium-192. A nuclear pharmaceutical is a physiologically active carrier to which a radioisotope is attached. Today, it is possible to manufacture chemical or biological carriers that migrate to a particular part of the human body, and this is the subject of much ongoing medical research.

identified among the fission products of uranium-235. Some of these products are fission fragments, and some are produced by unstable fragments that undergo radioactive decay. Selected fission fragments are listed in Table 13.6, together with their major modes of radioactive decay and half-lives. Some of the isotopes are the focus of concern about nuclear wastes, the topic of A Closer Look reading at the end of this chapter.

The fission of a uranium-235 nucleus produces two or three neutrons along with other products. These neutrons can each move to other uranium-235 nuclei where they are absorbed, causing fission with the release of more neutrons, which move to other uranium-235 nuclei to continue the process. A reaction where the products are able to produce more reactions in a self-sustaining series is called a **chain reaction.** A chain reaction is self-sustaining until all the uranium-235 nuclei have fissioned or until the neutrons fail to strike a uranium-235 nucleus (Figure 13.15).

You might wonder why all the uranium in the universe does not fission in a chain reaction. Natural uranium is mostly uranium-238, an isotope that does not fission easily. Only about 0.7 percent of natural uranium is the highly fissionable uranium-235. This low ratio of readily fissionable uranium-235 nuclei makes it unlikely that a stray neutron would be able to achieve a chain reaction.

To achieve a chain reaction, there must be (1) a sufficient mass with (2) a sufficient concentration of fissionable nuclei. When the mass and concentration are sufficient to sustain a chain reaction, the amount is called a **critical mass.** Likewise, a

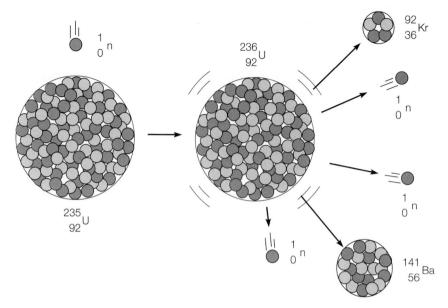

FIGURE 13.14 The fission reaction occurring when a neutron is absorbed by a uranium-235 nucleus. The deformed nucleus splits any number of ways into lighter nuclei, releasing neutrons in the process.

TABLE 13.6

Fragments and products from nuclear reactors using fission of uranium-235

Isotope	Major Mode of Decay	Half-Life	Isotope	Major Mode of Decay	Half-Life
Tritium	Beta	12.26 years	Cerium-144	Beta, gamma	285 days
Carbon-14	Beta	5,730 years	Promethium-147	Beta	2.6 years
Argon-41	Beta, gamma	1.83 hours	Samarium-151	Beta	90 years
Iron-55	Electron capture	2.7 years	Europium-154	Beta, gamma	16 years
Cobalt-58	Beta, gamma	71 days	Lead-210	Beta	22 years
Cobalt-60	Beta, gamma	5.26 years	Radon-222	Alpha	3.8 days
Nickel-63	Beta	92 years	Radium-226	Alpha, gamma	1,620 years
Krypton-85	Beta, gamma	10.76 years	Thorium-229	Alpha	7,300 years
Strontium-89	Beta	5.4 days	Thorium-230	Alpha	26,000 years
Strontium-90	Beta	28 years	Uranium-234	Alpha	2.48×10^5 years
Yttrium-91	Beta	59 days	Uranium-235	Alpha, gamma	7.13×10^8 years
Zirconium-93	Beta	9.5×10^5 years	Uranium-238	Alpha	4.51×10^9 years
Zirconium-95	Beta, gamma	65 days	Neptunium-237	Alpha	2.14×10^6 years
Niobium-95	Beta, gamma	35 days	Plutonium-238	Alpha	89 years
Technetium-99	Beta	2.1×10^5 years	Plutonium-239	Alpha	24,360 years
Ruthenium-106	Beta	1 year	Plutonium-240	Alpha	6,760 years
Iodine-129	Beta	1.6×10^7 years	Plutonium-241	Beta	13 years
Iodine-131	Beta, gamma	8 days	Plutonium-242	Alpha	3.79×105 years
Xenon-133	Beta, gamma	5.27 days	Americium-241	Alpha	458 years
Cesium-134	Beta, gamma	2.1 years	Americium-243	Alpha	7,650 years
Cesium-135	Beta	2×10^6 years	Curium-242	Alpha	163 days
Cesium-137	Beta	30 years	Curium-244	Alpha	18 years
Cerium-141	Beta	32.5 days			

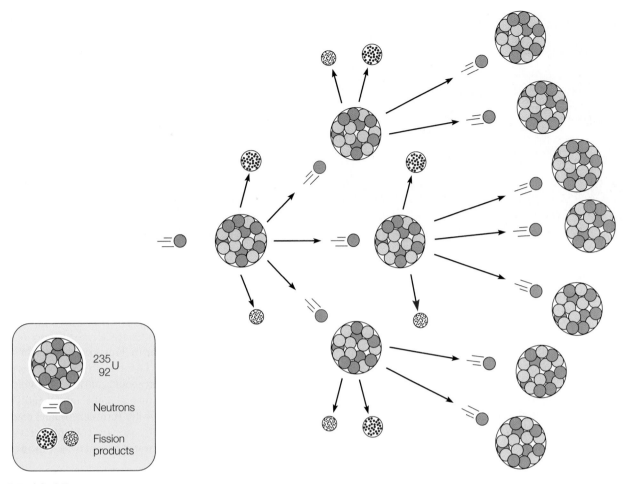

235
92 U

Neutrons

Fission products

FIGURE 13.15 A schematic representation of a chain reaction. Each fissioned nucleus releases neutrons, which move out to fission other nuclei. The number of neutrons can increase quickly with each series.

mass too small to sustain a chain reaction is called a *subcritical mass.* A mass of sufficiently pure uranium-235 (or plutonium-239) that is large enough to produce a rapidly accelerating chain reaction is called a *supercritical mass.* An atomic bomb is simply a device that uses a small, conventional explosive to push sub-critical masses of fissionable material into a supercritical mass. Fission occurs almost instantaneously in the supercritical mass, and tremendous energy is released in a violent explosion.

NUCLEAR POWER PLANTS

The nuclear part of a nuclear power plant is the **nuclear reactor,** a steel vessel in which a controlled chain reaction of fissionable material releases energy (Figure 13.16). In the most popular design, called a pressurized light-water reactor, the fissionable material is enriched 3 percent uranium-235 and 97 percent uranium-238 that has been fabricated in the form of small ceramic pellets (Figure 13.17A). The pellets are encased in a long zirconium alloy tube called a **fuel rod.** The fuel rods are locked into a *fuel rod assembly* by locking collars, arranged to permit pressurized water to flow around each fuel rod (Figure 13.17B) and to allow the insertion of *control rods* between the fuel rods. **Control rods** are constructed of materials, such as cadmium, that absorb neutrons. The lowering or raising of control rods

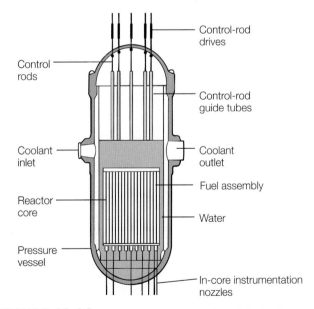

FIGURE 13.16 A schematic representation of the basic parts of a nuclear reactor. The largest commercial nuclear power plant reactors are 23 to 28 cm (9–11in) thick steel vessels with a stainless steel liner, standing about 12m (40 ft) high with a diameter of about 5 m (16 ft). Such a reactor has four pumps, which move 1.67 million L (440, 000 gal) of water per minute through the primary loop.

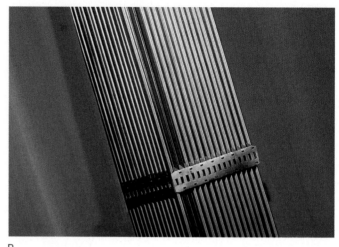

A   B

FIGURE 13.17 (*A*) These are uranium oxide fuel pellets that are stacked inside fuel rods, which are then locked together in a fuel rod assembly. (*B*) A fuel rod assembly. See also Figure 13.20, which shows a fuel rod assembly being loaded into a reactor.

within the fuel rod assemblies slows or increases the chain reaction by varying the amount of neutrons absorbed. When they are lowered completely into the assembly, enough neutrons are absorbed to stop the chain reaction.

It is physically impossible for the low-concentration fuel pellets to form a supercritical mass. A nuclear reactor in a power plant can only release energy at a comparatively slow rate, and it is impossible for a nuclear power plant to produce a nuclear

explosion. In a pressurized water reactor, the energy released is carried away from the reactor by pressurized water in a closed pipe called the **primary loop** (Figure 13.18). The water is pressurized at about 150 atmospheres (about 2,200 lb/in^2) to keep it from boiling, since its temperature may be 350°C (about 660°F).

In the pressurized light-water (ordinary water) reactor, the circulating pressurized water acts as a coolant, carrying heat away from the reactor. The water also acts as a **moderator,** a

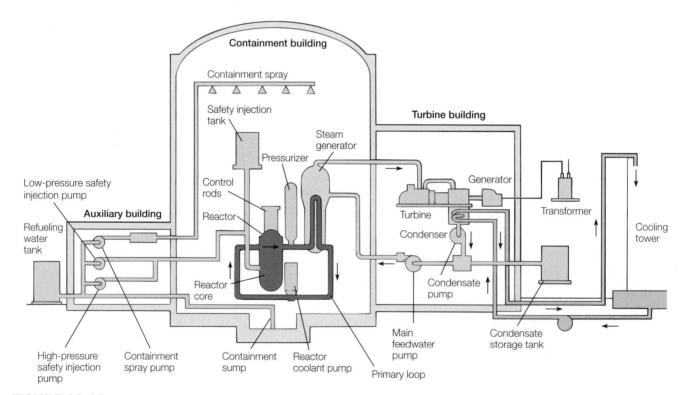

FIGURE 13.18 A schematic general system diagram of a pressurized water nuclear power plant, not to scale. The containment building is designed to withstand an internal temperature of 149°C (300°F) at a pressure of 4 atmospheres (60 lb/in^2) and still maintain its leak-tight integrity.

FIGURE 13.19 The turbine deck of a nuclear generating station. There is one large generator in line with four steam turbines in this non-nuclear part of the plant. The large silver tanks are separators that remove water from the steam after it has left the high-pressure turbine and before it is recycled back into the low-pressure turbines.

FIGURE 13.20 Spent fuel rod assemblies are removed and new ones are added to a reactor head during refueling. This shows an initial fuel load to a reactor, which has the upper part removed and set aside for the loading.

substance that slows neutrons so they are more readily absorbed by uranium-235 nuclei. Other reactor designs use heavy water (dideuterium monoxide) or graphite as a moderator.

Water from the closed primary loop is circulated through a heat exchanger called a **steam generator** (Figure 13.18). The pressurized high-temperature water from the reactor moves through hundreds of small tubes inside the generator as *feedwater* from the **secondary loop** flows over the tubes. The water in the primary loop heats feedwater in the steam generator and then returns to the nuclear reactor to become heated again. The feedwater is heated to steam at about 235°C (455°F) with a pressure of about 68 atmospheres (1,000 lb/in^2). This steam is piped to the turbines, which turn an electric generator (Figure 13.19).

After leaving the turbines, the spent steam is condensed back to liquid water in a second heat exchanger receiving water from the cooling towers. Again, the cooling water does not mix with the closed secondary loop water. The cooling-tower water enters the condensing heat exchanger at about 32°C (90°F) and leaves at about 50°C (about 120°F) before returning to a cooling tower, where it is cooled by evaporation. The feedwater is preheated, then recirculated to the steam generator to start the cycle over again. The steam is condensed back to liquid water because of the difficulty of pumping and reheating steam.

After a period of time, the production of fission products in the fuel rods begins to interfere with effective neutron transmission, so the reactor is shut down annually for refueling. During refueling, about one-third of the fuel that had the longest exposure in the reactor is removed as "spent" fuel. New fuel rod assemblies are inserted to make up for the part removed (Figure 13.20). However, only about 4 percent of the "spent" fuel is unusable waste, about 94 percent is uranium-238, 0.8 percent is uranium-235, and about 0.9 percent is plutonium (Figure 13.21). Thus, "spent" fuel rods contain an appreciable amount of usable

uranium and plutonium. For now, spent reactor fuel rods are mostly stored in cooling pools at the nuclear plant sites. In the future, a decision will be made either to reprocess the spent fuel, recovering the uranium and plutonium through chemical reprocessing, or to put the fuel in terminal storage. Concerns about reprocessing are based on the fact that plutonium-239 and uranium-235 are fissionable and could possibly be used by terrorist groups to construct nuclear explosive devices. Six other countries do have reprocessing plants, however, and the spent fuel rods represent an energy source equivalent to more than 25 billion barrels of petroleum. Some energy experts say that it would be inappropriate to dispose of such an energy source.

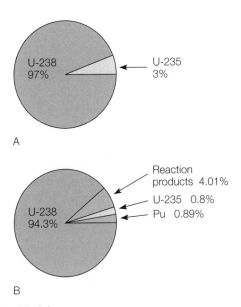

FIGURE 13.21 The composition of the nuclear fuel in a fuel rod (*A*) before and (*B*) after use over a three-year period in a nuclear reactor.

A Closer Look

Three Mile Island and Chernobyl

Three Mile Island (downriver from Harrisburg, Pennsylvania, U.S.A.) and Chernobyl (former U.S.S.R., now Ukraine) are two nuclear power plants that became famous because of accidents. Here is a brief accounting of what happened.

THREE MILE ISLAND

It was March 28, 1979, and the 880-megawatt Three Mile Island Nuclear Plant, operated by Metropolitan Edison Company, was going full blast. At 4 A.M., the main feedwater pump that pumps water to the steam generator failed for some unexplained reason (follow this description in Figure 13.18). Backup feedwater pumps kicked in, but the valves that should have been open were closed for maintenance, blocking the backup source of water for the steam generator. All of a sudden, there was not a source of water for the steam generator that removes heat from the primary loop. Events began to happen quickly at this point.

The computer sensors registered that the steam generator was not receiving water, and it began to follow a shutdown procedure. First, the turbine was shut down as steam was vented from the steam line out through the turbine building. It sounded much like a large jet plane. Within six seconds, the reactor was "scrammed," shut down with control rods dropped between the fuel rods in the reactor vessel. Fissioning began to slow, but the reactor was still hot.

Between three and six seconds, a pressure relief valve opened on the primary loop, reducing the excess pressure that was generated because feedwater was not entering the steam generator to remove heat from the primary loop. The valve should have closed when the excess pressure was released, but it did not. It was stuck in an open position. Pressurized water and steam were pouring from the primary loop into the containment building. As water was lost from the primary loop, temperatures inside the reactor began to climb. The loss of pressure resulted in high-temperature water flashing into steam. If an operator had pressed the right button in the control room, it would have closed the open valve, but this did not happen until thirty-two minutes later.

At this point, the reactor could have recovered from the events. Two minutes after the initial shutdown, the computer sensors noted the loss of pressure from the open valve and kicked in the emergency core cooling system, which pumps more water into the reactor vessel. However, for some unknown reason, the control room operators shut down one pump four and a half minutes after the initial event and the second pump six minutes later.

Water continued to move through the open pressure relief valve into the containment building. At seven and a half minutes after the accident began, the radioactive water on the floor was two feet deep and the sump pumps started pumping water into tanks in the auxiliary building. This water would become the source of the radioactivity that did escape the plant. It escaped because the pump seals leaked and spilled radioactive water. The filters had been removed from the auxiliary building air vents, allowing radioactive gases to escape.

Eleven minutes after the start of the accident, an operator restarted the emergency core cooling system that had been turned off. With the cooling water flowing again, the pressure in the reactor stopped falling. The fuel rods, some thirty-six thousand in this reactor, had not yet suffered any appreciable damage. This would be taken care of by the next incredible event.

The next incredible event was that operators began turning off the emergency cooling pumps, perhaps because they were vibrating too much. In any case, with the pumps off, the water level in the reactor fell again, this time uncovering the fuel rods. Within minutes, the temperature was high enough to rupture fuel rods, dropping radioactive oxides into the bottom of the reactor vessel. The operators now had a general emergency.

It was eleven hours later that the operators decided to start the main reactor coolant pump. This pump had shut down at the beginning of the series of events. Water again covered the fuel rods, and the pressure and temperature stabilized.

The consequences of this series of events were as follows:

1. Local residences did receive some radiation from the release of gases. They received 10 millirems (0.1 millisievert) in a low-exposure area and up to 25 millirems (0.25 millisievert) in a high-exposure area.

The technology to dispose of fuel rods exists if the decision is made to do so. The longer half-life waste products are mostly alpha emitters. These metals could be converted to oxides, mixed with powdered glass (or a ceramic), melted, and then poured into stainless steel containers. The solidified canisters would then be buried in a stable geologic depository. The glass technology is used in France for disposal of high-level wastes. Buried at 610 to 914 m (2,000–3,000 ft) depths in solid granite, the only significant means of the radioactive wastes reaching the surface would be through groundwater dissolving the stainless steel, glass, and waste products and then transporting them back to the surface. Many experts believe that if such groundwater dissolving were to take place, it would require thousands of years. The radioactive isotopes would thus undergo natural radioactive decay by the time they could reach the surface. Nonetheless, research is continuing on nuclear waste and its disposal. In the meantime, the question of whether it is best to reprocess fuel rods or place them in permanent storage remains unanswered.

NUCLEAR FUSION

As the graph of nuclear binding energy versus mass numbers shows (see Figure 13.13), nuclear energy is released when (1) massive nuclei such as uranium-235 undergo fission and (2) when less massive nuclei come together to form more massive nuclei through nuclear fusion. Nuclear fusion is responsible for the energy released by the Sun and other stars. At the

2. Cleaning up the damaged reactor vessel and core required more than ten years to cut it up, pack it into canisters, and ship everything to the Federal Nuclear Reservation at Idaho Falls.

3. The cost of the cleanup was more than $1 billion.

4. Changes at other nuclear power plants as a consequence: Pressure relief valves have been removed; operators can no longer turn off the emergency cooling system; and, operators must now spend about one-fourth of their time in training.

CHERNOBYL

The Soviet-designed Chernobyl reactor was a pressurized water reactor with individual fuel channels, which is very different from the pressurized water reactors used in the United States. The Chernobyl reactor was constructed with each fuel assembly in an individual pressure tube with circulating pressurized water. This heated water was circulated to a steam separator, and the steam was directed to turbines, which turned a generator to produce electricity. Graphite blocks surrounded the pressure tubes, serving as moderators to slow the neutrons involved in the chain reaction. The graphite was cooled by a mixture of helium and nitrogen. The reactor core was located in a concrete bunker that acted as a radiation shield. The top part was a steel cap and shield that supported the fuel assemblies. There were no containment buildings around the Soviet reactors as there are in the United States.

The Chernobyl accident was the result of a combination of a poorly engineered reactor design, poorly trained reactor operators, and serious mistakes made by the operators on the day of the accident.

The reactor design was flawed because at low power, steam tended to form pockets in the water-filled fuel channels, creating a condition of instability. Instability occurred because (1) steam is not as efficient at cooling as is liquid water, and (2) liquid water acts as a moderator and neutron absorber while steam does not. Excess steam therefore leads to overheating and increased power generation. Increased power can lead to increased steam generation, which leads to further increases in power. This coupled response is very difficult to control because it feeds itself.

On April 25, 1986, the operators of Chernobyl unit 4 started a test to find out how long the turbines would spin and supply power following the loss of electrical power. The operators disabled the automatic shutdown mechanisms and then started the test early on April 26, 1986. The plan was to stabilize the reactor at 1,000 MW, but an error was made; the power fell to about 30 MW and pockets of steam became a problem. Operators tried to increase the power by removing all the control rods. At 1 A.M., they were able to stabilize the reactor at 200 MW. Then instability returned, and the operators were making continuous adjustments to maintain a constant power. They reduced the feedwater to maintain steam pressure, and this created even more steam voids in the fuel channels. Power surged to

a very high level and fuel elements ruptured. A steam explosion moved the reactor cap, exposing individual fuel channels and releasing fission products to the environment. A second steam explosion knocked a hole in the roof, exposing more of the reactor core, and the graphite, which served as a moderator, burst into flames. The graphite burned for nine days, releasing about 324 million Ci (12×10^{18} Bq) into the environment.

The fiery release of radioactivity was finally stopped by using a helicopter to drop sand, boron, lead, and other materials onto the burning graphite reactor. After the fire was out, the remains of the reactor were covered with a large concrete shelter.

In addition to destroying the reactor, the accident killed 30 people, with 28 of these dying from radiation exposure. Another 134 people were treated for acute radiation poisoning, and all recovered from the immediate effects.

Cleanup crews over the next year received about 10 rem (100 millisieverts) to 25 rem (250 millisieverts) and some received as much as 50 rem (500 millisieverts). In addition to this direct exposure, large expanses of Belarus, Ukraine, and Russia were contaminated by radioactive fallout from the reactor fire. Hundreds of thousands of people have been resettled into less contaminated areas. The World Health Organization and other international agencies have studied the data to understand the impact of radiation-related disease. These studies do confirm a rising incidence of thyroid cancer but no increases in leukemia so far.

present halfway point in the Sun's life—with about 5 billion years to go—the core is now 35 percent hydrogen and 65 percent helium. Through fusion, the Sun converts about 650 million tons of hydrogen to 645 million tons of helium every second. The other roughly 5 million tons of matter are converted into energy. Even at this rate, the Sun has enough hydrogen to continue the process for an estimated 5 billion years. There are several fusion reactions that take place between hydrogen and helium isotopes, including the following:

$$^1_1H + {}^1_1H \rightarrow {}^2_1H + {}^0_1e$$

$$^2_1H + {}^2_1H \rightarrow {}^3_2He + {}^0_1n$$

$$^3_2He + {}^3_2He \rightarrow {}^4_2He + 2{}^1_1H$$

The fusion process would seem to be a desirable energy source on earth because (1) two isotopes of hydrogen, deuterium (^2_1H) and tritium (^3_1H), undergo fusion at a relatively low temperature; (2) the supply of deuterium is practically unlimited, with each gallon of seawater containing about a teaspoonful of heavy water; and (3) enormous amounts of energy are released with no radioactive by-products.

The oceans contain enough deuterium to generate electricity for the entire world for millions of years, and tritium can be constantly produced by a fusion device. Researchers know what needs to be done to tap this tremendous energy source. The problem is *how* to do it in an economical, continuous energy-producing fusion reactor. The problem, one of the most difficult engineering tasks ever attempted, is meeting three basic

Nuclear Waste

There are two general categories of nuclear wastes: (1) low-level wastes and (2) high-level wastes. The *low-level wastes* are produced by hospitals, universities, and other facilities. They are also produced by the normal operation of a nuclear reactor. Radioactive isotopes sometimes escape from fuel rods in the reactor and in the spent fuel storage pools. These isotopes are removed from the water by ion-exchange resins and from the air by filters. The used resins and filters will contain the radioactive isotopes and will become low-level wastes. In addition, any contaminated protective clothing, tools, and discarded equipment also become low-level wastes.

Low-level liquid wastes are evaporated, mixed with cement, then poured into 55-gal steel drums. Solid wastes are compressed and placed in similar drums. The drums are currently disposed of by burial in government-licensed facilities. In general, low-level waste has an activity of less than 1.0 curie per cubic foot. Contact with the low-level waste could expose a person to up to 20 millirems per hour of contact.

High-level wastes from nuclear power plants are spent nuclear fuel rods. At the present time, most of the commercial nuclear power plants have these rods in temporary storage at the plant sites. These rods are "hot" in the radioactive sense, producing about 100,000 curies per cubic foot. They are also hot in the thermal sense, continuing to generate heat for months after removal from the reactor. The rods are cooled by heat exchangers connected to storage pools; they could otherwise achieve an internal temperature as high as 800°C for several decades. In the future, these spent fuel rods will be reprocessed or disposed of through terminal storage.

Agencies of the U.S. government have also accumulated millions of gallons of high-level wastes from the manufacture of nuclear weapons and nuclear research programs. These liquid wastes are stored in million-gallon stainless steel containers that are surrounded by concrete. The future of this large amount of high-level wastes may be evaporation to a solid form or mixture with a glass or ceramic matrix, which is melted and poured into stainless steel containers. These containers would be buried in solid granite rock in a stable geologic depository. Such high-level wastes must be contained for thousands of years as they undergo natural radioactive decay (Box Figure 13.3). Burial at a depth of 610 to 914 m (2,000–3,00 ft) in solid granite

BOX FIGURE 13.3 This is a standard warning sign for a possible radioactive hazard. Such warning signs would have to be maintained around a nuclear waste depository for thousands of years.

would provide protection from exposure by explosives, meteorite impact, or erosion. One major concern about this plan is that a hundred generations later, people might lose track of what is buried in the nuclear garbage dump.

---

fusion reaction requirements of (1) temperature, (2) density, and (3) time (Figure 13.22):

1. *Temperature.* Nuclei contain protons and are positively charged, so they experience the electromagnetic repulsion of like charges. This force of repulsion can be overcome, moving the nuclei close enough to fuse together, by giving the nuclei sufficient kinetic energy. The fusion reaction of deuterium and tritium, which has the lowest temperature requirements of any fusion reaction known at the present time, requires temperatures on the order of 100 million°C.
2. *Density.* There must be a sufficiently dense concentration of heavy hydrogen nuclei, on the order of $10^{14}/cm^3$, so many reactions occur in a short time.
3. *Time.* The nuclei must be confined at the appropriate density up to a second or longer at pressures of at least 10 atmospheres to permit a sufficient number of reactions to take place.

The temperature, density, and time requirements of a fusion reaction are interrelated. A short time of confinement, for example,

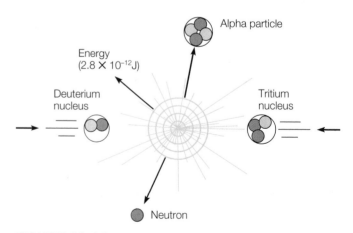

FIGURE 13.22 A fusion reaction between a tritium nucleus and a deuterium nucleus requires a certain temperature, density, and time of containment to take place.

In 1982, the U.S. Congress established a national policy to solve the problem of nuclear waste disposal. This policy is a federal law called the Nuclear Waste Policy Act. Congress based this policy on what most scientists worldwide agreed is the best way to dispose of nuclear waste.

The Nuclear Waste Policy Act made the U.S. Department of Energy responsible for finding a site, building, and operating an underground disposal facility called a geologic respository.

In 1983, the Energy Department selected nine locations in six states for consideration as potential respository sites. This was based on data collected for nearly ten years. The nine sites were studied, and results of these preliminary studies were reported in 1985. Based on these reports, President Ronald Reagan approved three sites for intensive scientific study called site characterization. The three sites were Handford, Washington; Deaf Smith County, Texas; and Yucca Mountain, Nevada.

In 1987, Congress amended the Nuclear Waste Policy Act and directed the Energy Department to study only Yucca Mountain. On July 9, 2002, the U.S. Senate cast the final legislative vote approving the development of a repository at Yucca Mountain. On July 23, 2002, President George W. Bush signed House Joint Resolution 87, allowing the Energy Department to take the next step in establishing a safe repository in which to store our nation's nuclear waste.

In July 2004, the Energy Department issued a report that describes how commercial spent nuclear fuel and high-level waste will be acquired, transported to federal facilities, and disposed of in the Yucca Mountain Geologic Repository. The plan is to accept scheduled allocations from nuclear power plant owners beginning in 2010. In 2010, 400 metric tons of uranium will be accepted. This number will grow to 3,000 metric tons per year in 2014 and for each of the next five years.

The Department of Energy is preparing an application to obtain a license from the U.S. Nuclear Regulatory Commission to construct a repository.

Source: www.ocrwm.doe.gov/ymp/about/history.shtml

requires an increased density, and a longer confinement time requires less density. The primary problems of fusion research are the high-temperature requirements and confinement. No material in the world can stand up to a temperature of 100 million°C, and any material container would be instantly vaporized. Thus, research has centered on meeting the fusion reaction requirements without a material container. Two approaches are being tested: *magnetic confinement* and *inertial confinement.*

Magnetic confinement utilizes a very hot **plasma,** a gas consisting of atoms that have been stripped of their electrons because of the high kinetic energies. The resulting positively and negatively charged particles respond to electrical and magnetic forces, enabling researchers to develop a "magnetic bottle," that is, magnetic fields that confine the plasma and avoid the problems of material containers that would vaporize. A magnetically confined plasma is very unstable, however, and researchers have compared the problem to trying to carry a block of jello on a pair of rubber bands. Different magnetic field geometries and magnetic "mirrors" are the topics of research in attempts to stabilize the hot, wobbly plasma. Electric currents, injection of fast ions, and radio frequency (microwave) heating methods are also being studied.

Inertial confinement is an attempt to heat and compress small frozen pellets of deuterium and tritium with energetic laser beams or particle beams, producing fusion. The focus of this research is new and powerful lasers, light ion and heavy ion beams. If successful, magnetic or inertial confinement will provide a long-term solution for future energy requirements.

THE SOURCE OF NUCLEAR ENERGY

When elements undergo the natural radioactive decay process, energy is released, and the decay products have less energy than the original reactant nucleus. When massive nuclei undergo fission, much energy is rapidly released along with fission products that continue to release energy through radioactive decay. What is the source of all this nuclear energy? The answer to this question is found in current theories about how the universe started and in theories about the life cycle of the stars. Theories about the life cycle of stars are discussed in chapters 14 and 15. For now, consider just a brief introduction to the life cycle of a star in order to understand the ultimate source of nuclear energy.

The current universe is believed to have started with a "big bang" of energy, which created a plasma of protons and neutrons. This primordial plasma cooled rapidly and, after several minutes, began to form hydrogen nuclei. Throughout the newly formed universe, massive numbers of hydrogen atoms—on the order of 10^{57} nuclei—were gradually pulled together by gravity into masses that would become the stars. As the hydrogen atoms fell toward the center of each mass of gas, they accelerated, just like any other falling object. As they accelerated, the contracting mass began to heat up because the average kinetic energy of the atoms increased from acceleration. Eventually, after say 10 million years or so of collapsing and heating, the mass of hydrogen condensed to a sphere with a diameter of 1.5 million miles or so, or about twice the size of the Sun today. At the same time, the interior temperature increased to millions of degrees, reaching the critical points of density, temperature, and containment for a fusion reaction to begin. Thus, a star was born as hydrogen nuclei fused into helium nuclei, releasing enough energy that the star began to shine.

Hydrogen nuclei in the newborn star had a higher energy per nucleon than helium nuclei, and helium nuclei had more energy per nucleon than other nuclei up to around iron. The fusion process continued for billions of years, releasing energy as heavier and heavier nuclei were formed. Eventually, the star

People Behind the Science

Marie Curie (1867–1934)

Marie Curie was a Polish-born French scientist who, with her husband, Pierre Curie (1859–1906), was an early investigator of radioactivity. From 1896, the Curies worked together, building on the results of Henri Becquerel, who had discovered radioactivity from uranium salts. Marie Curie discovered that thorium also emits radiation and found that the mineral pitchblende was even more radioactive than could be accounted for by any uranium and thorium content. The Curies then carried out an exhaustive search and in July 1898 announced the discovery of polonium, followed in December of that year with the discovery of radium. They shared the 1903 Nobel Prize for physics with Henri Becquerel for the discovery of radioactivity. The Curies did not participate in Becquerel's discovery but investigated radioactivity and gave the phenomenon its name. Marie Curie went on to study the chemistry and medical applications of radium, and was awarded the 1911 Nobel Prize for chemistry in recognition of her work in isolating the pure metal.

At the outbreak of World War I in 1914, Marie Curie helped to equip ambulances with X-ray equipment and drove the ambulances to the front lines. The International Red Cross made her head of its Radiological Service. She taught medical orderlies and doctors how to use the new technique. By the late 1920s, her health began to deteriorate: continued exposure to high-energy radiation had given her leukemia. She entered a sanatorium and died on July 4, 1934.

Throughout much of her life, Marie Curie was poor, and the painstaking radium extractions were carried out in primitive conditions. The Curies refused to patent any of their discoveries, wanting them to benefit everyone freely. They used the Nobel Prize money and other financial rewards to finance further research. One of the outstanding applications of their work has been the use of radiation to treat cancer, one form of which cost Marie Curie her life.

Source: Modified from the Hutchinson *Dictionary of Scientific Biography.* © Research Machines plc 2003. All Rights Reserved. Helicon Publishing is a division of Research Machines.

materials were fused into nuclei around iron, the element with the lowest amount of energy per nucleon, and the star used up its energy source. Larger, more massive dying stars explode into supernovas (discussed in chapter 14). Such an explosion releases a flood of neutrons, which bombard medium-weight nuclei and build them up to more massive nuclei, all the way from iron up to uranium. Thus, the more massive elements were born from an exploding supernova, then spread into space as dust. In a process to be discussed in chapter 15, this dust became the materials of which planets were made, including Earth. The point for the present discussion, however, is that the energy of naturally radioactive elements, and the energy released during fission, can be traced back to the force of gravitational attraction, which provided the initial energy for the whole process.

SUMMARY

Radioactivity is the spontaneous emission of particles or energy from an unstable atomic nucleus. The modern atomic theory pictures the nucleus as protons and neutrons held together by a short-range *nuclear force* that has moving *nucleons* (protons and neutrons) in *energy shells* analogous to the shell structure of electrons. A graph of the number of neutrons to the number of protons in a nucleus reveals that stable nuclei have a certain neutron-to-proton ratio in a *band of stability.* Nuclei that are above or below the band of stability, and nuclei that are beyond atomic number 83, are radioactive and undergo *radioactive decay.*

Three common examples of radioactive decay involve the emission of an *alpha particle,* a *beta particle,* and a *gamma ray.* An alpha particle is a helium nucleus, consisting of two protons and two neutrons. A beta particle is a high-speed electron that is ejected from the nucleus. A gamma ray is a short-wavelength electromagnetic radiation from an excited nucleus. In general, nuclei with an atomic number of 83 or larger become more stable by alpha emission. Nuclei with a neutron-to-proton ratio that is too large become more stable by beta emission. Gamma ray emission occurs from a nucleus that was left in a high-energy state by the emission of an alpha or beta particle.

Each radioactive isotope has its own specific rate of nuclear disintegration. The rate is usually described in terms of *half-life,* the time required for one-half the unstable nuclei to decay.

Radiation is measured by (1) its effects on photographic film, (2) the number of ions it produces, or (3) the flashes of light produced on a phosphor. It is measured at a source in units of a *curie,* defined as 3.70×10^{10} nuclear disintegrations per second. It is measured where received in units of a *rad.* A *rem* is a measure of radiation that takes into account the biological effectiveness of different types of radiation damage. In general, the natural environment exposes everyone to 100 to 500 millirems per year, an exposure called *background radiation.* Lifestyle and location influence the background radiation received, but the average is 130 millirems per year.

Energy and mass are related by Einstein's famous equation of $E = mc^2$, which means that *matter can be converted to energy and energy to matter.* The mass of a nucleus is always less than the sum of the masses of the individual particles of which it is made. This *mass defect* of a nucleus is equivalent to the energy released when the nucleus was formed according to $E = mc^2$. It is also the *binding energy,* the energy required to break the nucleus apart into nucleons.

When the binding energy is plotted against the mass number, the greatest binding energy per nucleon is seen to occur for an atomic number near that of iron. More massive nuclei therefore release energy by *fission,* or splitting to more stable nuclei. Less massive nuclei release energy by *fusion,* the joining of less massive nuclei to produce a more stable, more massive nucleus. Nuclear fission provides the energy for atomic explosions and nuclear power plants. Nuclear fusion is the energy source of the Sun and other stars and also holds promise as a future energy source for humans. The source of the energy of a nucleus can be traced back to the gravitational attraction that formed a star.

SUMMARY OF EQUATIONS

13.1

energy = mass × the speed of light squared

$$E = mc^2$$

KEY TERMS

alpha particle (p. **329**)

background radiation (p. **338**)

band of stability (p. **331**)

beta particle (p. **329**)

binding energy (p. **339**)

chain reaction (p. **341**)

control rods (p. **343**)

critical mass (p. **341**)

curie (p. **336**)

electromagnetic force (p. **331**)

fuel rod (p. **343**)

gamma ray (p. **329**)

Geiger counter (p. **335**)

half-life (p. **334**)

ionization counter (p. **335**)

mass defect (p. **339**)

moderator (p. **344**)

nuclear fission (p. **340**)

nuclear force (p. **331**)

nuclear fusion (p. **340**)

nuclear reactor (p. **343**)

nucleons (p. **329**)

plasma (p. **349**)

primary loop (p. **344**)

rad (p. **337**)

radioactive decay (p. **329**)

radioactive decay series (p. **333**)

radioactivity (p. **328**)

rem (p. **337**)

scintillation counter (p. **335**)

secondary loop (p. **345**)

shell model of the nucleus (p. **331**)

steam generator (p. **345**)

APPLYING THE CONCEPTS

1. Natural radioactivity is a result of
   a. adjustments to balance nuclear attractions and repulsions.
   b. experiments with human-made elements.
   c. absorption of any type of radiation by very heavy elements.
   d. a mass defect of the nucleus.

2. Which one of the following is an electron emitted by a nucleus as it undergoes radioactive decay?
   a. alpha particle
   b. beta particle
   c. gamma ray
   d. Z ray

3. Which one of the following is an alpha particle emitted by a nucleus as it undergoes radioactive decay?
   a. electron
   b. helium nucleus
   c. photon of very short wavelength
   d. Z ray

4. Protons and neutrons collectively are called
   a. particles.
   b. nucleons.
   c. heavy particles.
   d. alpha particles.

5. The number of protons and neutrons in a nucleus defines the
   a. atomic number.
   b. neutron number.
   c. nucleon number.
   d. mass number.

6. Atoms with the same atomic number but with different numbers of neutrons are
   a. allotropes.
   b. isomers.
   c. isotopes.
   d. allomers.

7. Which of the following types of radiation will not penetrate clothing?
   a. alpha particles
   b. beta particles
   c. gamma rays
   d. X rays

8. The half-life of a radioactive isotope is the time required for half of
   a. any size sample to sublimate.
   b. a molar mass to disappear through radioactivity.
   c. the total radioactivity to be given off.
   d. the nuclei of any size sample to decay.

9. The activity of a radioactive source is measured in units of
   a. curie.
   b. rad.
   c. rem.
   d. roentgen.

10. The measure of radiation that takes into account damage done by alpha, beta, and gamma is the
    a. curie.
    b. Becquerel.
    c. rem.
    d. rad.

11. The "threshold model" of long-term, low-level radiation exposure to the human body proposes that
    a. any radiation exposure is damaging.
    b. radiation damage will be repaired up to a certain level.
    c. any radiation exposure will result in cancer and other damage.
    d. the door should be shut on nuclear energy.

12. Radiation from sources in your everyday life is called
    a. safe radiation.
    b. good radiation.
    c. background radiation.
    d. cosmic radiation.

13. The relationship between mass changes and energy changes in a nuclear reaction is described by the
    a. law of conservation of mass.
    b. law of conservation of matter.
    c. formula $E = mc^2$.
    d. molar mass.

14. When protons and neutrons join to make a nucleus, energy is released and
    a. mass is lost.
    b. mass is gained.
    c. the mass is constant.
    d. mass is gained or lost, depending on the atomic weight.

15. The energy required to break a nucleus into its constituent protons and neutrons is the same as the
    a. nuclear energy.
    b. breaking energy.
    c. binding energy.
    d. splitting energy

16. The nuclear reaction that takes place when a nucleus splits into more stable, less massive nuclei with the release of energy is called nuclear
    a. chaos.
    b. fission.
    c. fusion.
    d. disintegration.

17. When the thing that starts a reaction is produced by the reaction, you have a
    a. chain reaction.
    b. nuclear series.
    c. decay series.
    d. nuclear sequence.

18. There must be sufficient mass and concentration of a fissionable material to support a chain reaction. This is called the
    a. vital mass.
    b. critical mass.
    c. essential concentration.
    d. crucial mass.

19. The control rods in a nuclear reactor are made of special materials that
    a. spin, slowing the reaction.
    b. absorb energy, cooling the reaction rate.
    c. absorb neutrons.
    d. interfere with the effective transmission of electrons.

20. The amount of unusable waste in a used nuclear fuel rod is
    a. 100 percent.
    b. 67 percent.
    c. 20 percent.
    d. 4 percent.

21. One of the major problems with a fusion reactor would be the
    a. production of radioactive waste.
    b. high temperature requirement.
    c. lack of fusion fuel on Earth.
    d. lack of understanding about requirements for fusion reaction.

22. Water is used in a pressurized water reactor as a coolant, but it also has the critical job of
    a. absorbing radioactivity.
    b. keeping waste products wet.
    c. slowing neutrons.
    d. preventing a rapid cooldown.

23. The ejection of a beta particle from a nucleus results in
    a. an increase in the atomic number by one.
    b. an increase in the atomic mass by four.
    c. a decrease in the atomic number by two.
    d. None of the above is correct.

24. The ejection of an alpha particle from a nucleus results in
    a. an increase in the atomic number by one.
    b. an increase in the atomic mass by four.
    c. a decrease in the atomic number by two.
    d. None of the above is correct.

25. The emission of a gamma ray from a nucleus results in
    a. an increase in the atomic number by one.
    b. an increase in the atomic mass by four.
    c. a decrease in the atomic number by two.
    d. None of the above is correct.

26. An atom of radon-222 loses an alpha particle to become a more stable atom of
    a. radium.
    b. bismuth.
    c. polonium.
    d. radon.

27. The nuclear force is
    a. attractive when nucleons are closer than 10^{-15} m.
    b. repulsive when nucleons are closer than 10^{-15} m.
    c. attractive when nucleons are farther than 10^{-15} m.
    d. repulsive when nucleons are farther than 10^{-15} m.

28. Which of the following is most likely to be radioactive?
    a. nuclei with an even number of protons and neutrons
    b. nuclei with an odd number of protons and neutrons
    c. nuclei with the same number of protons and neutrons
    d. The number of protons and neutrons has nothing to do with radioactivity.

29. Hydrogen-3 is a radioactive isotope of hydrogen. Which type of radiation would you expect an atom of this isotope to emit?
    a. an alpha particle
    b. a beta particle
    c. either of the above
    d. neither of the above

30. A sheet of paper will stop a (an)
    a. alpha particle.
    b. beta particle.
    c. gamma ray.
    d. None of the above is correct.

31. The most penetrating of the three common types of nuclear radiation is the
    a. alpha particle.
    b. beta particle.
    c. gamma ray.
    d. All have equal penetrating ability.

32. An atom of an isotope with an atomic number greater than 83 will probably emit a (an)
    a. alpha particle.
    b. beta particle.
    c. gamma ray.
    d. None of the above is correct.

33. An atom of an isotope with a large neutron-to-proton ratio will probably emit a (an)
    a. alpha particle.
    b. beta particle.
    c. gamma ray.
    d. None of the above is correct.

34. All of the naturally occurring radioactive decay series end when the radioactive elements have decayed to
    a. lead.
    b. bismuth.
    c. uranium.
    d. hydrogen.

35. The rate of radioactive decay can be increased by increasing the
    a. temperature.
    b. pressure.
    c. surface area.
    d. None of the above is correct.

36. Isotope A has a half-life of seconds, and isotope B has a half-life of millions of years. Which isotope is more radioactive?
    a. isotope A
    b. isotope B
    c. It depends on the temperature.
    d. The answer is unknown from the information given.

37. A Geiger counter indirectly measures radiation by measuring
    a. ions produced.
    b. flashes of light.
    c. alpha, beta, and gamma.
    d. fog on photographic film.

38. A measure of radiation received that considers the biological effect resulting from the radiation is the
    a. curie.
    b. rad.
    c. rem.
    d. Any of the above is correct.

39. Used fuel rods from a nuclear reactor contain about
    a. 96 percent usable uranium and plutonium.
    b. 33 percent usable uranium and plutonium.
    c. 4 percent usable uranium and plutonium.
    d. 0 percent usable uranium and plutonium.

40. The source of energy from the Sun is
    a. chemical (burning).
    b. fission.
    c. fusion.
    d. radioactive decay.

41. The isotope of hydrogen that has one proton and one neutron is:
    a. ordinary hydrogen.
    b. deuterium.
    c. tritium.
    d. hydrogen-1.

42. Hydrogen-3 emits a beta particle and the new nucleus formed is
    a. helium-3.
    b. helium-4.
    c. hydrogen-2.
    d. hydrogen-3.

43. Which of the following is most likely to happen to an alpha particle after it is emitted?
    a. It escapes to space, where it is known as a cosmic ray.
    b. After being absorbed by matter, it is dissipated as radiant energy.
    c. After joining with subatomic particles, it finds itself in a child's balloon.
    d. It joins others of its kind to make lightning and thunder.

44. Which of the following is most likely to happen to a beta particle after it is emitted?
    a. It escapes to space, where it is known as a cosmic ray.
    b. After being absorbed by matter, it is dissipated as radiant energy.
    c. After joining with subatomic particles, it finds itself in a child's balloon.
    d. It joins others of its kind to make lightning and thunder.

45. A nucleus emits a beta particle and the number of nucleons it contains is now
    a. more.
    b. less.
    c. the same.

46. Enriched uranium has more
    a. energy.
    b. radioactivity.
    c. deuterium.
    d. uranium-235.

Answers

1. a 2. b 3. b 4. b 5. d 6. c 7. a 8. d 9. a 10. c 11. b 12. c 13. c 14. a 15. c 16. b 17. a 18. b 19. c 20. d 21. b 22. c 23. a 24. c 25. d 26. c 27. a 28. b 29. b 30. a 31. c 32. a 33. b 34. a 35. d 36. a 37. a 38. c 39. a 40. c 41. b 42. a 43. c 44. d 45. c 46. d

QUESTIONS FOR THOUGHT

1. How is a radioactive material different from a material that is not radioactive?

2. What is radioactive decay? Describe how the radioactive decay rate can be changed if this is possible.

3. Describe three kinds of radiation emitted by radioactive materials. Describe what eventually happens to each kind of radiation after it is emitted.

4. How are positively charged protons able to stay together in a nucleus since like charges repel?

5. What is half-life? Give an example of the half-life of an isotope, describing the amount remaining and the time elapsed after five half-life periods.

6. Would you expect an isotope with a long half-life to be more, the same, or less radioactive than an isotope with a short half-life? Explain.

7. What is (a) a curie? (b) a rad? (c) a rem?

8. What is meant by background radiation? What is the normal radiation dose for the average person from background radiation?

9. Why is there controversy about the effects of long-term, low levels of radiation exposure?

10. What is a mass defect? How is it related to the binding energy of a nucleus? How can both be calculated?

11. Compare and contrast nuclear fission and nuclear fusion.

FOR FURTHER ANALYSIS

1. What are the significant differences between a radioactive isotope and an isotope that is not radioactive?

2. Analyze the different types of radioactive decay to explain how each is a hazard to living organisms.

3. Make up a feasible explanation for why some isotopes have half-lives of seconds, while other kinds of isotopes have half-lives in the billions of years.

4. Suppose you believe the threshold model of radiation exposure is correct. Describe a conversation between yourself and another person who feels strongly that the linear model of radiation exposure is correct.

5. Explain how the fission of heavy elements and the fusion of light elements both release energy.

6. Write a letter to your congressional representative describing why used nuclear fuel rods should be reprocessed rather than buried as nuclear waste.

7. What are the similarities and differences between a nuclear fission power plant and a nuclear fusion power plant?

INVITATION TO INQUIRY

How Much Radiation?

Ionizing radiation is understood to be potentially harmful if certain doses are exceeded. How much ionizing radiation do you acquire from the surroundings where you live, from your lifestyle, and from medical procedures? Investigate radiation from cosmic sources, the Sun, television sets, time spent in jet airplanes, and dental or other X-ray machines. What are other sources of ionizing radiation in your community? How difficult is it to find relevant information and make recommendations? Does any agency monitor the amount of radiation that people receive? What are the problems and issues with such monitoring?

PARALLEL EXERCISES

The exercises in groups A and B cover the same concepts. Solutions to group A exercises are located in appendix E.

Group A

Note: *You will need the table of atomic weights inside the back cover of this text.*

1. Give the number of protons and the number of neutrons in the nucleus of each of the following isotopes:
   (a) cobalt-60
   (b) potassium-40
   (c) neon-24
   (d) lead-208

2. Write the nuclear symbols for each of the nuclei in exercise 1.

3. Predict if the nuclei in exercise 1 are radioactive or stable, giving your reasoning behind each prediction.

4. Write a nuclear equation for the decay of the following nuclei as they give off a beta particle:
   (a) $^{56}_{26}\text{Fe}$
   (b) $^{7}_{4}\text{Be}$
   (c) $^{64}_{29}\text{Cu}$
   (d) $^{24}_{11}\text{Na}$
   (e) $^{214}_{82}\text{Pb}$
   (f) $^{32}_{15}\text{P}$

5. Write a nuclear equation for the decay of the following nuclei as they undergo alpha emission:
   (a) $^{235}_{92}\text{U}$
   (b) $^{226}_{88}\text{Ra}$
   (c) $^{239}_{94}\text{Pu}$
   (d) $^{214}_{83}\text{Bi}$
   (e) $^{230}_{90}\text{Th}$
   (f) $^{210}_{84}\text{Po}$

6. The half-life of iodine-131 is 8 days. How much of a 1.0 oz sample of iodine-131 will remain after 32 days?

Group B

Note: *You will need the table of atomic weights inside the back cover of this text.*

1. Give the number of protons and the number of neutrons in the nucleus of each of the following isotopes:
   (a) aluminum-25
   (b) technetium-95
   (c) tin-120
   (d) mercury-200

2. Write the nuclear symbols for each of the nuclei in exercise 1.

3. Predict if the nuclei in exercise 1 are radioactive or stable, giving your reasoning behind each prediction.

4. Write a nuclear equation for the beta emission decay of each of the following:
   (a) $^{14}_{6}\text{C}$
   (b) $^{60}_{27}\text{Co}$
   (c) $^{24}_{11}\text{Na}$
   (d) $^{214}_{94}\text{Pu}$
   (e) $^{131}_{53}\text{I}$
   (f) $^{210}_{82}\text{Pb}$

5. Write a nuclear equation for each of the following alpha emission decay reactions:
   (a) $^{241}_{95}\text{Am}$
   (b) $^{223}_{90}\text{Th}$
   (c) $^{223}_{88}\text{Ra}$
   (d) $^{234}_{92}\text{U}$
   (e) $^{242}_{96}\text{Cm}$
   (f) $^{237}_{93}\text{Np}$

6. If the half-life of cesium-137 is 30 years, approximately how much time will be required to reduce a 1 kg sample to about 1 g?

14
The Universe

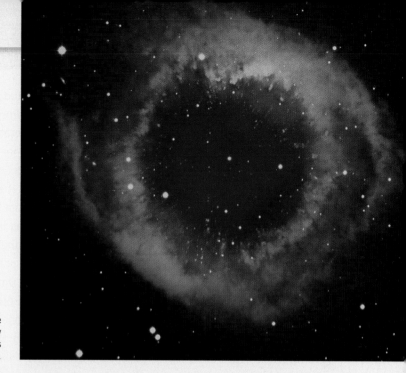

This is a planetary nebula in the constellation Aquarius. Planetary nebulae are clouds of ionized gases with no relationship to any planet. They were named long ago, when they appeared similar to the planets Neptune and Uranus when viewed through early telescopes.

CORE **CONCEPT**

The night sky is filled with billions of stars, and the sun is an ordinary star with an average brightness.

OUTLINE

Origin of Stars
Stars are very large accumulations of gases that release energy from nuclear fusion reactions.

Star Temperature
Stars appear to be different colors because they have different temperatures and chemical compositions.

Galaxies
Stars are organized in galaxies, the basic unit of the universe.

The Night Sky
Stars
  Origin of Stars
  Brightness of Stars
  Star Temperature
  Star Types
  The Life of a Star
Science and Society: Light Pollution
Galaxies
  The Milky Way Galaxy
  Other Galaxies
A Closer Look: Extraterrestrials?
  The Life of a Galaxy
A Closer Look: Redshift and Hubble's Law
A Closer Look: Dark Energy
A Closer Look: Dark Matter
People Behind the Science:
  Jocelyn (Susan) Bell Burnell

Brightness of Stars
The brightness of a star varies with the amount of light produced, the size of the star, and the distance to the star. The brightness you see is compared to an apparent magnitude scale of brightness.

Star Types
The Hertzsprung-Russell diagram is a plot of temperature versus brightness of stars.

Astronomy is an exciting field of science that has fascinated people since the beginnings of recorded history. Ancient civilizations searched the heavens in wonder, some recording on clay tablets what they observed. Many religious and philosophical beliefs were originally based on interpretations of these ancient observations. Today, we are still awed by space, but now we are fascinated with ideas of space travel, black holes, and the search for extraterrestrial life. Throughout history, people have speculated about the universe and their place in it and watched the sky and wondered (Figure 14.1). What is out there and what does it all mean? Are there other people on other planets, looking at the star in their sky that is our Sun, wondering if we exist?

Until about thirty years ago, progress in astronomy was limited to what could be observed and photographed. Today, our understanding has expanded by the use of technology and spacecraft. It is now understood that our Sun is but one of billions of stars that move in large groups of stars called galaxies. Where did the Sun come from? Will it always be here? Our understanding has now developed to the point where we can answer such questions, but we can also calculate the conditions in the center of our Sun and other stars. We can also describe what the universe looked like billions of years ago, and we can predict what will probably happen to it billions of years in the future. How stars and galaxies form and what eventually will happen are two big ideas in this chapter.

Theoretical ideas about stars, galaxies, and how the universe formed will lead to some understanding of how stars are arranged in space and how our Sun fits in the big picture. This will provide a framework of understanding for moving closer to home, the local part of our galaxy—the solar system—which will be considered in chapter 15. This will lead to more understanding of our Earth and Moon.

THE NIGHT SKY

Early civilizations had a much better view of the night sky before city lights, dust, and pollution obscured much of it. Today, you must travel far from the cities, perhaps to a remote mountaintop, to see a clear night sky as early people observed it. Back then, people could clearly see the motion of the Moon and stars night after night, observing recurring cycles of motion. These cycles became important as people associated them with certain events. Thus, watching the Sun, Moon, and star movements became a way to identify when to plant crops, when to harvest, and when it was time to plan for other events. Observing the sky was an important activity, and many early civilizations built observatories with sighting devices to track and record astronomical events. Stonehenge, for example, was an ancient observatory built in England around 2600 B.C. by Neolithic people (Figure 14.2).

Light from the stars and planets must pass through Earth's atmosphere to reach you, and this affects the light. Stars appear as point *sources* of light, and each star generates its own light. The stars seem to twinkle because density differences in the atmosphere refract the point of starlight one way, then the other, as the air moves. The result is the slight dancing about and change in intensity called twinkling. The points of starlight are much steadier when viewed on a calm night or when viewed from high in the mountains where there is less atmosphere for the starlight to pass through. Astronauts outside the atmosphere see no twinkling, and the stars appear as steady point sources of light.

Back at ground level, within the atmosphere, the *reflected* light from a planet does not seem to twinkle. A planet appears as a disk of light rather than a point source, so refraction from moving air of different densities does not affect the image as much. Sufficient air movement can cause planets to appear to shimmer, however, just as a road appears to shimmer on a hot summer day.

How far away is a star? When you look at the sky, it appears that all the stars are at the same distance. It seems impossible to know anything about the actual distance to any given star. Standard referent units of length such as kilometers or miles have little meaning in astronomy since there are no referent points of comparison. Distance without a referent point can be measured in terms of angles or time. The unit of astronomical distance that uses time is the **light-year** (ly). A light-year is the distance that light travels in one year, about 9.5×10^{12} km (about 6×10^{12} mi).

To locate planets, or anything else in the sky, you need something to refer to, a referent system. A referent system is easily established by first imagining the sky to be a *celestial sphere* (Figure 14.3). A coordinate system of lines can be visualized on this celestial sphere just as you think of the coordinate system of latitude and longitude lines on Earth's surface (see p. 411). Imagine that you could inflate Earth until its surface touched the celestial sphere. If you now transfer the latitude and longitude lines to the celestial sphere, you will have a system of sky coordinates. The line of the equator of Earth on the celestial sphere is called the **celestial equator.** The North Pole of Earth touches the celestial sphere at a point called the **north celestial pole.** From the surface of Earth, you can see that the celestial equator is a line on the celestial sphere directly above Earth's equator, and the north celestial pole is a point directly above the North Pole of Earth. Likewise, the **south celestial pole** is a point directly above the South Pole of Earth.

A

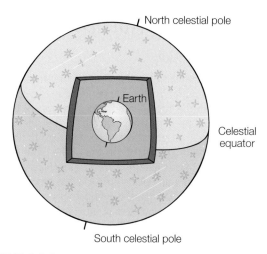

FIGURE 14.3 The celestial sphere with the celestial equator directly above Earth's equator, and the celestial poles directly above Earth's poles.

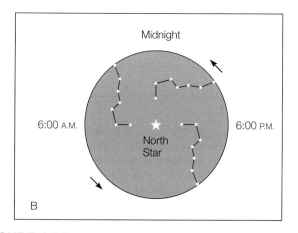

B

FIGURE 14.1 Ancient civilizations used celestial cycles of motion as clocks and calendars. (*A*) This photograph shows the path of stars around the North Star. (*B*) A "snapshot" of the position of the Big Dipper over a period of 24 hours as it turns around the North Star one night. This shows how the Big Dipper can be used to help you keep track of time.

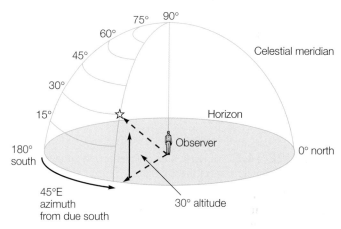

FIGURE 14.4 Once you have established the celestial equator, the celestial poles, and the celestial meridian, you can use a two-coordinate horizon system to locate positions in the sky. One popular method of using this system identifies the altitude angle (in degrees) from the horizon up to an object on the celestial sphere and the azimuth angle (again in degrees) of the object on the celestial sphere is east or west of due south, where the celestial meridian meets the horizon. The illustration shows an altitude of 30° and an azimuth of 45° east of due south.

FIGURE 14.2 The stone pillars of Stonehenge were positioned so they could be used to follow the movement of the Sun and Moon with the seasons of the year.

You can only see half of the overall celestial sphere from any one place on the surface of Earth. Imagine a point on the celestial sphere directly above where you are located. An imaginary line that passes through this point, then passes north through the north celestial pole, continuing all the way around through the south celestial pole and back to the point directly above you makes a big circle called the **celestial meridian** (Figure 14.4). Note that the celestial meridian location is determined by where *you* are on Earth. The celestial equator and the celestial poles, on the other hand, are always in the same place no matter where you are.

CONCEPTS *Applied*

The Night Sky

Become acquainted with the night sky by first locating the *Big Dipper*. In autumn, the Big Dipper is close to the northern horizon, with the cup open upward. The stars of the far end of the cup, to the east, are called the pointer stars. Trace an imaginary line through these two stars upward to find the Little Dipper. The bright star at the end of the Little Dipper handle is *Polaris*, also called the *North Star* (Figure 14.5). Observe the stars around Polaris until you can describe their apparent movement. Compare the brightness and color of different stars. Can you find stars of different colors?

If you have a camera that has a "time" or "bulb" exposure setting and can be attached to a tripod, try a twenty-minute exposure of Polaris and nearby stars. To find the direction of apparent movement, try a fifteen-minute exposure, cover the lens for five minutes, then remove the cover for an additional two minutes.

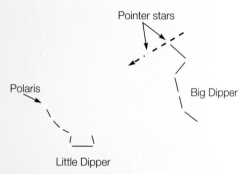

FIGURE 14.5 The North Star, or Polaris, is located by using the pointer stars of the Big Dipper.

Myths, Mistakes, & Misunderstandings

Stars Visible from the Bottom of a Well?

Perhaps you have heard that stars are visible from the bottom of a well, even in full daylight. This is a myth, as you cannot see any stars from the bottom of a well in full daylight. This idea is based on the belief that the brightness of the Sun, which obliterates the light from stars during the day, is somehow less in the bottom of a well. This idea might be based on the observation that looking at a star that you cannot see well is aided by looking at it through a long, dark tube (a telescope). This overlooks the role of lenses or mirrors in a telescope.

STARS

If you could travel by spaceship a few hundred light-years from Earth, you would observe the Sun shrink to a bright point of light among the billions and billions of other stars. The Sun is just an ordinary star with an average brightness. Like the other stars, the Sun is a massive, dense ball of gases with a surface heated to incandescence by energy released from fusion reactions deep within. Since the Sun is an average star, it can be used as a reference for understanding all the other stars.

ORIGIN OF STARS

Theoretically, stars are born from swirling clouds of hydrogen gas in the deep space between other stars. Such interstellar (between stars) clouds are called **nebulae.** These clouds consist of random, swirling atoms of gases that have little gravitational attraction for one another because they have little mass. Complex motions of stars, however, can produce a shock wave that causes particles to move closer together and collide, making local compressions. Their mutual gravitational attraction then begins to pull them together into a cluster. The cluster grows as more atoms are pulled into it, which increases the mass and thus the gravitational attraction, and still more atoms are pulled in from farther away. Theoretical calculations indicate that on the order of 1×10^{57} atoms are necessary, all within a distance of 3 trillion km (about 1.9 trillion mi). When these conditions occur, the cloud of gas atoms begins to condense by gravitational attraction to a **protostar,** an accumulation of gases that will become a star.

EXAMPLE 14.1

Compared to the 10^{19} molecules/cm^3 of air on Earth, an average concentration of 1,000 hydrogen atoms/cm^3 in the Orion Nebula does not seem very dense. However, considering that the Orion Nebula is about 20 light-years across (20×10^{18} cm), a sphere with a volume of 4.19×10^{57} cm^3 would enclose the Orion Nebula, and it would contain

$$\frac{1,000 \text{ atoms}}{\text{cm}^3} \times (4.19 \times 10^{57} \text{cm}^3) = 4.19 \times 10^{60} \text{ atoms}$$

This is a sufficient number of hydrogen atoms to produce

$$\frac{4.19 \times 10^{60} \text{ atoms}}{1 \times 10^{57} \text{ atoms/star}} = 4,190 \text{ stars}$$

Thus, there is a sufficient number of hydrogen atoms in the Orion Nebula to produce 4,190 average stars like the Sun.

Gravitational attraction pulls the average protostar from a cloud with a diameter of trillions of kilometers (trillions of miles) down to a dense sphere with a diameter of 2.5 million km (1.6 million mi) or so. As gravitational attraction accelerates the atoms toward the center, they gain kinetic energy, and the interior temperature increases. Over a period of some 10 million years of contracting and heating, the temperature and density conditions at the center of the protostar are sufficient to start nuclear fusion reactions. Pressure from hot gases and energy from increasing fusion reactions begin to balance the gravitational attraction over the next 17 million years, and the newborn, average star begins its stable life, which will continue for the next 10 billion years.

The interior of an average star, such as the Sun, is modeled after the theoretical pressure, temperature, and density conditions

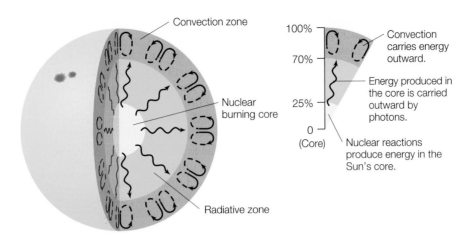

FIGURE 14.6 Energy-producing nuclear reactions occur only within the inner 25 percent of the Sun's radius. The energy produced by these reactions is carried outward by photons to 70 percent of the Sun's radius. From that distance outward, convection carries most of the Sun's energy.

that would be necessary to produce the observed energy and light from the surface. This model describes the interior as a set of three shells: (1) the core, (2) a radiation zone, and (3) the convection zone (Figure 14.6).

Our model describes the *core* as a dense, very hot region where nuclear fusion reactions release gamma and X-ray radiation. The density of the core is about twelve times that of solid lead. Because of the plasma conditions, however, the core remains in a gaseous state even at this density.

Our model describes the *radiation zone* as less dense than the core, having a density about the same as that of water. Energy in the form of gamma and X rays from the core is absorbed and reemitted by collisions with atoms in this zone. The radiation slowly diffuses outward because of the countless collisions over a distance comparable to the distance between Earth and the Moon. It could take millions of years before this radiation finally escapes the radiation zone.

The model *convection zone* begins about seven-tenths of the way to the surface, where the density of the gases is about 1 percent the density of water. Gases at the bottom of this zone are heated by radiation from the radiation zone below, expand from the heating, and rise to the surface by convection. At the surface, the gases emit energy in the form of visible light, ultraviolet radiation, and infrared radiation, which moves out into space. As they lose energy, the gases contract in volume and sink back to the radiation zone to become heated again, continuously carrying energy from the radiation zone to the surface in convection cells. The surface is continuously heated by the convection cells as it gives off energy to space, maintaining a temperature of about 5,800 K (about 5,500°C).

As an average star, the Sun converts about 1.4×10^{17} kg of matter to energy every year as hydrogen nuclei are fused to produce helium. The Sun was born about 5 billion years ago and has sufficient hydrogen in the core to continue shining for another 4 or 5 billion years. Other stars, however, have masses that are much greater or much less than the mass of the Sun, so they have different life spans. More massive stars generate higher temperatures in the core because they have a greater

gravitational contraction from their greater masses. Higher temperatures mean increased kinetic energy, which results in increased numbers of collisions between hydrogen nuclei with the end result an increased number of fusion reactions. Thus, a more massive star uses up its hydrogen more rapidly than a less massive star. On the other hand, stars that are less massive than the Sun use their hydrogen at a slower rate so they have longer life spans. The life spans of the stars range from a few million years for large, massive stars, to 10 billion years for average stars like the Sun, to trillions of years for small, less massive stars.

BRIGHTNESS OF STARS

Stars generate their own light, but some stars appear brighter than others in the night sky. As you can imagine, this difference in brightness could be related to (1) the amount of light produced by the stars, (2) the size of each star, or (3) the distance to a particular star. A combination of these factors is responsible for the brightness of a star as it appears to you in the night sky. A classification scheme for different levels of brightness that you see is called the **apparent magnitude** scale (Table 14.1). The apparent magnitude scale is based on a system established by a Greek astronomer over two thousand years ago. Hipparchus made a catalog of the stars he could see and assigned a numerical value to each to identify its relative brightness. The brightness values ranged from 1 to 6, with the number 1 assigned to the brightest star and the number 6 assigned to the faintest star that could be seen. Stars assigned the number 1 came to be known as first-magnitude stars, those a little dimmer as second-magnitude stars, and so on to the faintest stars visible, the sixth-magnitude stars.

When technological developments in the nineteenth century made it possible to measure the brightness of a star, Hipparchus's system of brightness values acquired a precise, quantitative meaning. Today, a first-magnitude star is defined as one that is 100 times brighter than a sixth-magnitude star, with five uniform multiples of decreasing brightness on a scale from the first magnitude to the sixth magnitude.

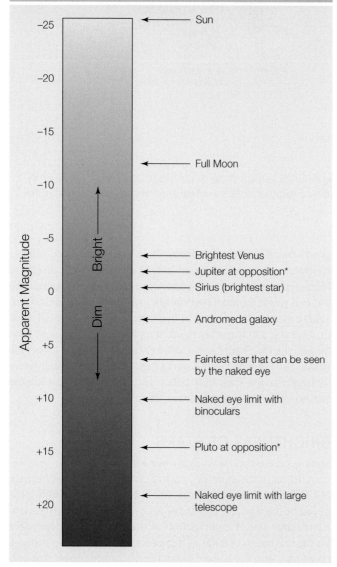

TABLE 14.1

The apparent magnitude scale comparing some familiar objects and some observable limits

Apparent Magnitude scale (Bright/Dim):

- Sun — near −25
- Full Moon — about −12
- Brightest Venus — about −4
- Jupiter at opposition* — about −3
- Sirius (brightest star) — about −1.5
- Andromeda galaxy — about +4
- Faintest star that can be seen by the naked eye — about +6
- Naked eye limit with binoculars — about +10
- Pluto at opposition* — about +15
- Naked eye limit with large telescope — about +20

*When the body is opposite the Sun in the sky.

The apparent magnitude of a star depends on how far away stars are in addition to differences in the stars themselves. Stars at a farther distance will appear fainter, and those closer will appear brighter, just like any other source of light. To compensate for distance differences, astronomers calculate the brightness that stars would appear to have if they were all at a defined, standard distance (32.6 light-years). The brightness of a star at this distance is called the **absolute magnitude.** The Sun, for example, is the closest star and has an apparent magnitude of −26.7 at an average distance from Earth. When viewed from the standard distance, the Sun would have an absolute magnitude of +4.8, which is about the brightness of a faint star.

The absolute magnitude is an expression of **luminosity,** the total amount of energy radiated into space each second from the surface of a star. The Sun, for example, radiates 4×10^{26} joules per second from its surface. The luminosity of stars is often compared to the Sun's luminosity, with the Sun considered to have a luminosity of 1 unit. When this is done, the luminosity of the stars ranges from a low of 10^{-6} sun units for the dimmest stars up to a high of 10^5 sun units. Thus, the Sun is somewhere in the middle of the range of star luminosity.

STAR TEMPERATURE

If you observe the stars on a clear night, you will notice that some are brighter than others, but you will also notice some color differences. Some stars have a reddish color, some have a bluish white color, and others have a yellowish color. This color difference is understood to be a result of the relationship that exists between the color and the temperature of an incandescent object. The colors of the various stars are a result of the temperatures of the stars. You see a cooler star as reddish in color and comparatively hotter stars as bluish white. Stars with in-between temperatures, such as the Sun, appear to have a yellowish color (Figure 14.7).

Astronomers analyze starlight to measure the temperature and luminosity as well as the chemical composition of a star. When the starlight is analyzed in a spectroscope, specific elements can be identified from the unique set of spectral lines that each element emits. Temperature and spectra are used as the basis for a star classification scheme. Originally, the classification scheme was based on sixteen categories according to the strength of the hydrogen line spectra. The groups were identified alphabetically with A for the group with the strongest hydrogen line spectrum, B for slightly weaker lines, and on to the last group with the faintest lines. Later, astronomers realized that the star temperature was the important variable, so they rearranged the categories according to decreasing temperatures. The original letter categories were retained, however, resulting in classes of stars with the hottest temperature first and the coolest last with the sequence O B A F G K M. Table 14.2 compares the color, temperature ranges, and other features of the stellar spectra classification scheme.

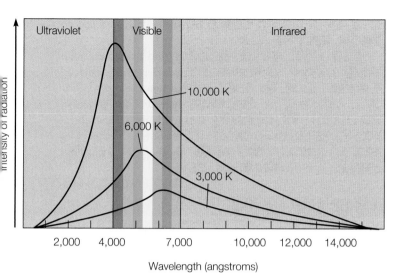

FIGURE 14.7 The distribution of radiant energy emitted is different for stars with different surface temperatures. Note that the peak radiation of a cooler star is more toward the red part of the spectrum, and the peak radiation of a hotter star is more toward the blue part of the spectrum.

STAR TYPES

Henry Russell in the United States and Ejnar Hertzsprung in Denmark independently developed a scheme to classify stars with a temperature-luminosity graph. The graph is called the **Hertzsprung-Russell diagram,** or the *H-R diagram* for short. The diagram is a plot with temperature indicated by spectral types, and the true brightness indicated by absolute magnitude. The diagram, as shown in Figure 14.8, plots temperature by spectral types sequenced O through M, so the temperature

TABLE 14.2

Major stellar spectral types and temperatures

Type	Color	Temperature (K)	Comment
O	Bluish	30,000–80,000	Spectrum with ionized helium and hydrogen but little else; short-lived and rare stars
B	Bluish	10,000–30,000	Spectrum with neutral helium, none ionized
A	Bluish	7,500–10,000	Spectrum with no helium, strongest hydrogen, some magnesium and calcium
F	White	6,000–7,500	Spectrum with ionized calcium, magnesium, neutral atoms of iron
G	Yellow	5,000–6,000	The spectral type of the sun. Spectrum shows sixty-seven elements in the sun
K	Orange-red	3,500–5,000	Spectrum packed with lines from neutral metals
M	Reddish	2,000–3,500	Band spectrum of molecules, e.g., titanium oxide; other related spectral types (R, N, and S) are based on other molecules present in each spectral type

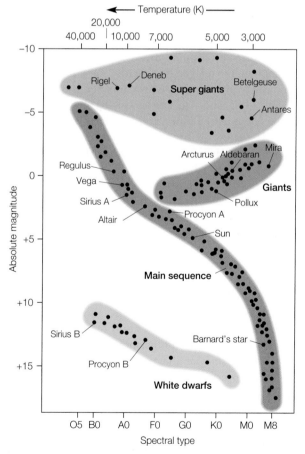

FIGURE 14.8 The Hertzsprung-Russell diagram. The main sequence and giant regions contain most of the stars, whereas hot underluminous stars, the white dwarfs, lie below and to the left of the main sequence.

decreases from left to right. The hottest, brightest stars are thus located at the top left of the diagram, and the coolest, faintest stars are located at the bottom right.

Each dot is a data point representing the surface temperature and brightness of a particular star. The Sun, for example, is a type G star with an absolute magnitude of about +5, which places the data point for the Sun almost in the center of the diagram. This means that the Sun is an ordinary, average star with respect to both surface temperature and true brightness.

Most of the stars plotted on an H-R diagram fall in or close to a narrow band that runs from the top left to the lower right. This band is made up of **main sequence stars.** Stars along the main sequence band are normal, mature stars that are using their nuclear fuel at a steady rate. Those stars on the upper left of the main sequence are the brightest, bluest, and most massive stars on the sequence. Those at the lower right are the faintest, reddest, and least massive of the stars on the main sequence. In general, most of the main sequence stars have masses that fall between a range from ten times greater than the mass of the Sun (upper left) to one-tenth the mass of the Sun (lower right). The extremes, or ends, of the main sequence range from about sixty times more massive than the Sun to one-twenty-fifth of the Sun's mass. It is the *mass* of a main sequence star that determines its brightness, its temperature, and its location on the H-R diagram. High-mass stars on the main sequence are brighter, hotter, and have shorter lives than low-mass stars. These relationships do not apply to the other types of stars in the H-R diagram.

There are two groups of stars that have a different set of properties than the main sequence stars. The **red giant stars** are bright, but low-temperature, giants. These reddish stars are enormously bright for their temperature because they are very large, with an enormous surface area giving off light. A red giant might be one hundred times larger but have the same mass as the Sun. These low-density red giants are located in the upper right part of the H-R diagram. The **white dwarf stars,** on the other hand, are located at the lower left because they are faint, white-hot stars. A white dwarf is faint because it is small, perhaps twice the size of Earth. It is also very dense, with a mass approximately equal to the Sun's. During its lifetime, a star will be found in different places on the H-R diagram as it undergoes changes. Red giants and white dwarfs are believed to be evolutionary stages that aging stars pass through, and the path a star takes across the diagram is called an evolutionary track. During the lifetime of the Sun, it will be a main sequence star, a red giant, and then a white dwarf.

Stars such as the Sun emit a steady light because the force of gravitational contraction is balanced by the outward flow of energy. *Variable stars,* on the other hand, are stars that change in brightness over a period of time. A **Cepheid variable** is a bright variable star that is used to measure distances. There is a general relationship between the period and the brightness: the longer the time needed for one pulse, the greater the apparent brightness of that star. The period-brightness relationship to distance was calibrated by comparing the apparent brightness with the absolute magnitude (true brightness) of a Cepheid at a known distance with a known period. Using the period to predict how bright the star would appear at various distances

allowed astronomers to calculate the distance to a Cepheid given its apparent brightness.

Edwin Hubble used the Cepheid period-brightness relationship to find the distances to other galaxies and discovered yet another relationship: the greater the distance to a galaxy, the greater a shift in spectral lines toward the red end of the spectrum (redshift). This relationship is called Hubble's law, and it forms the foundation for understandings about our expanding universe. Measuring redshift provides another means of establishing distances to other, far-out galaxies.

THE LIFE OF A STAR

A star is born in a gigantic cloud of gas and dust in interstellar space, then spends billions of years calmly shining while it fuses hydrogen nuclei in the core. How long a star shines and what happens to it when it uses up the hydrogen in the core depends on the mass of the star. Of course, no one has observed a star's life cycle over billions of years. The life cycle of a star is a theoretical outcome based on what is known about nuclear reactions. The predicted outcomes seem to agree with observations of stars today, with different groups of stars that can be plotted on the H-R diagram. Thus, the groups of stars on the diagram—main sequence, red giants, and white dwarfs, for example—are understood to be stars in various stages of their lives.

Protostar Stage

The first stage in the theoretical model of the life cycle of a star is the formation of the protostar. As gravity pulls the gas of a protostar together, the density, pressure, and temperature increase from the surface down to the center. Eventually, the conditions are right for nuclear fusion reactions to begin in the core, which requires a temperature of 10 million kelvins. The initial fusion reaction essentially combines four hydrogen nuclei to form a helium nucleus with the release of much energy. This energy heats the core beyond the temperature reached by gravitational contraction, eventually to 16 million kelvins. Since the star is a gas, the increased temperature expands the volume of the star. The outward pressure of expansion balances the inward pressure from gravitational collapse, and the star settles down to a balanced condition of calmly converting hydrogen to helium in the core, radiating the energy released into space (Figure 14.9). The theoretical time elapsed from the initial formation and collapse of the protostar to the main sequence is about 50 million years for a star of a solar mass (a star with the mass of our Sun).

Main Sequence Stage

Where the star is located on the main sequence and what happens to it next depend only on how massive it is. The more massive stars have higher core temperatures and use up their hydrogen more rapidly as they shine at higher surface temperatures (O type stars). Less massive stars shine at lower surface temperatures (M type stars) as they use their fuel at a slower rate. The overall life span on the main sequence ranges from millions of years for O type stars to trillions of years for M type stars. An average one-solar-mass star will last about 10 billion years.

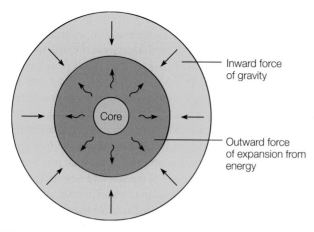

FIGURE 14.9 A star becomes stable when the outward forces of expansion from the energy released in nuclear fusion reactions balance the inward forces of gravity.

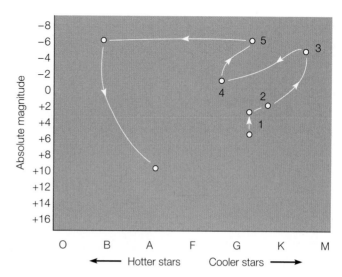

FIGURE 14.10 The evolution of a star of solar mass as it depletes hydrogen in the core (1), fuses hydrogen in the shell to become a red giant (2 to 3), becomes hot enough to produce helium fusion in the core (3 to 4), then expands to a red giant again as helium and hydrogen fusion reactions move out into the shells (4 to 5). It eventually becomes unstable and blows off the outer shells to become a white dwarf star.

Red Giant Stage

The next stage in the theoretical life of a star begins when the hydrogen in the core has been fused into helium. With fewer hydrogen fusion reactions, less energy is released and less outward balancing pressure is produced, so the star begins to collapse. The collapse heats the core, which now is composed primarily of helium, and the surrounding shell where hydrogen still exists. The increased temperature causes the hydrogen in the shell to undergo fusion, and the increased release of energy causes the outer layers of the star to expand. With an increased surface area, the amount of radiation emitted per unit area is less, and the star acquires the properties of a brilliant red giant. Its position on the H-R diagram changes since it now has different luminosity and temperature properties. (The star has not physically *moved*. The changing properties move its temperature-luminosity data point, not the star, to a new position.)

Back Toward Main Sequence

After about 500 million years as a red giant, the star now has a surface temperature of about 4,000 kelvins compared to its main sequence surface temperature of 6,000 kelvins. The radius of the red giant is now a thousand times greater, a distance that will engulf Earth when the Sun reaches this stage, assuming Earth is in the same position as today. Even though the surface temperature has decreased from the expansion, the helium core is continually heating and eventually reaches a temperature of 100 million kelvins, the critical temperature necessary for the helium nuclei to undergo fusion to produce carbon. The red giant now has helium fusion reactions in the core and hydrogen fusion reactions in a shell around the core. This changes the radius, the surface temperature, and the luminosity, with the overall result depending on the composition of the star. In general, the radius and luminosity decrease when this stage is reached, moving the star back toward the main sequence (Figure 14.10).

Beginning of the End for Less Massive Stars

After millions of years of helium fusion reactions, the core is gradually converted to a carbon core, and helium fusion begins in the shell surrounding the core. The core reactions decrease as the star now has a helium fusing shell surrounded by a second hydrogen fusing shell. This releases additional energy, and the star again expands to a red giant for the second time. A star the size of the Sun or less massive may cool enough at this point that nuclei at the surface become neutral atoms rather than a plasma. As neutral atoms, they can absorb radiant energy coming from within the star, heating the outer layers. Changes in temperature produce changes in pressure, which change the balance between the temperature, pressure, and the internal energy generation rate. The star begins to expand outward from heating. The expanded gases are cooled by the expansion process, however, and are pulled back to the star by gravity, only to be heated and expand outward again. In other words, the outer layers of the star begin to pulsate in and out. Finally, a violent expansion blows off the outer layers of the star, leaving the hot core. Such blown-off outer layers of a star form circular nebulae called *planetary nebulae* (Figure 14.11). The nebulae continue moving away from the core, eventually adding to the dust and gases between the stars. The remaining carbon core and helium-fusing shell begin gravitationally to contract to a small, dense *white dwarf* star. A star with the original mass of the Sun or less slowly cools from white, to red, then to a black lump of carbon in space (Figure 14.12).

Beginning of the End for Massive Stars

A more massive star will have a different theoretical ending than the slow cooling of a white dwarf. A massive star will contract, just like the less massive stars, after blowing off its outer

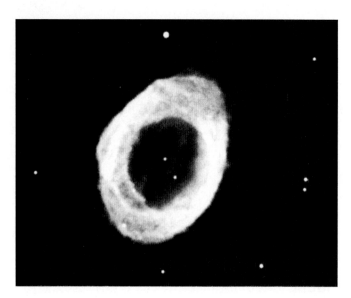

FIGURE 14.11 The blown-off outer layers of stars form ringlike structures called *planetary nebulae.*

shells. In a more massive star, however, heat from the contraction may reach the critical temperature of 600 million kelvins to begin carbon fusion reactions. Thus, a more massive star may go through a carbon fusing stage and other fusion reaction stages that will continue to produce new elements until the element iron is reached. After iron, energy is no longer released by the fusion process (see chapter 13), and the star has used up all of its energy sources. Lacking an energy source, the star is no longer able to maintain its internal temperature. The star loses the outward pressure of expansion from the high temperature, which had previously balanced the inward pressure from gravitational attraction. The star thus collapses, then rebounds like a compressed spring into a catastrophic explosion called a **supernova.** A supernova produces a brilliant light in the sky that may last for months before it begins to dim as the new elements that were created during the life of the star diffuse into space. These include all the elements up to iron that were produced by fusion reactions during the life of the star and heavier elements that were created during the instant of the explosion. All the elements heavier than iron were created as some less massive nuclei disintegrated in the explosion, joining with each other and with lighter nuclei to produce the nuclei of the elements from iron to uranium. As you will see in chapter 15, these newly produced, scattered elements will later become the building blocks for new stars and planets such as the Sun and Earth.

If the core of a supernova has a remaining mass greater than 1.4 solar masses, the gravitational forces on the remaining matter, together with the compressional forces of the supernova explosion, are great enough to collapse nuclei, forcing protons and electrons together into neutrons, forming the core of a **neutron star.** A neutron star is the very small (10 to 20 km diameter), superdense (10^{11} kg/cm^3 or greater) remains of a supernova with a center core of pure neutrons.

Because it is a superdense form of matter, the neutron star also has an extremely powerful magnetic field, capable of becoming a pulsar. A **pulsar** is a very strongly magnetized

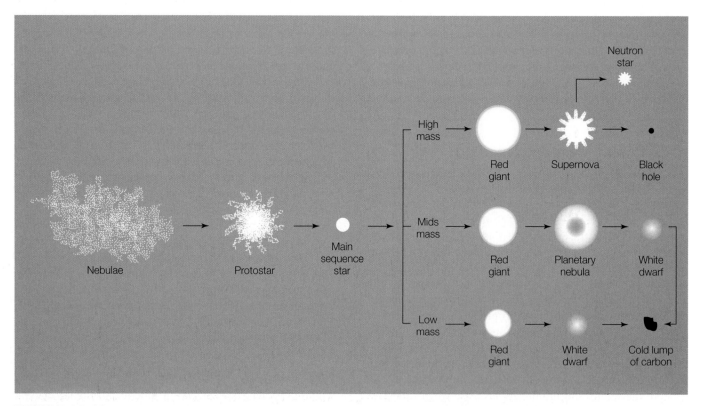

FIGURE 14.12 This flowchart shows some of the possible stages in the birth and aging of a star. The differences are determined by the mass of the star.

Light pollution is an adverse effect of light including sky glow, light clutter, and glare. In general, light pollution can destroy our view of the night sky, creating problems for astronomers and others who might want to view or study the stars.

Go with your group to a dark area on a clear, moonless night. Agree upon a plan to consider a part of the sky, such as a fourth of the sky—a quadrant. Everyone should count and record all the stars they can see in the quadrant. Average everyone's count to obtain an overall estimate of the number of visible stars in the quadrant. Repeat this exact same exercise when there is a full moon, which creates light pollution conditions. Compare the result and discuss how light pollution destroys the number of stars you can see. Compare what you would see near cities with light pollution and far from cities where there is no light pollution.

QUESTIONS TO DISCUSS

1. What quantitative evidence illustrates that light pollution limits viewing a quadrant of the night sky?

2. How could photographs of the night sky provide qualitative evidence of light pollution?

3. Is qualitative evidence as good as quantitative evidence? Explain.

4. How could quantitative and qualitative evidence be used to motivate people to consider light pollution when installing outdoor lighting?

---

neutron star that emits a uniform series of equally spaced electromagnetic pulses. Evidently, the magnetic field of a rotating neutron star makes it a powerful electric generator, capable of accelerating charged particles to very high energies. These accelerated charges are responsible for emitting a beam of electromagnetic radiation, which sweeps through space with amazing regularity (Figure 14.13). The pulsating radio signals from a pulsar were a big mystery when first discovered. For a time, extraterrestrial life was considered as the source of the signals, so they were jokingly identified as LGM (for "little green men"). Over three hundred pulsars have been identified, and most emit radiation in the form of radio waves. Two, however, emit visible light, two emit beams of gamma radiation, and one emits X-ray pulses.

Another theoretical limit occurs if the remaining core has a mass of about 3 solar masses or more. At this limit, the force of gravity overwhelms *all* nucleon forces, including the repulsive forces between like charged particles. If this theoretical limit is reached, nothing can stop the collapse, and the collapsed star will become so dense that even light cannot escape. The star is now a **black hole** in space. Since nothing can stop the collapsing star, theoretically a black hole would continue to collapse to a pinpoint and then to a zero radius called a *singularity.* This event seems contrary to anything that can be directly observed in the physical universe, but it does agree with the general theory of relativity and concepts about the curvature of space produced by such massively dense objects. Black holes are theoretical and none has been seen, of course, because a black hole theoretically pulls in radiation of all wavelengths and emits nothing. Evidence for the existence of a black hole is sought by studying X rays that would be given off by matter as it is accelerated into a black hole.

Evidence of the existence of a black hole has been provided by photographs from the Hubble Space Telescope. Hubble pictured a disk of gas only about 60 light-years out from the center of a galaxy (M87), moving at more than 1.6 million km/h (about 1 million mi/h). The only known possible explanation for such a massive disk of gas moving with this velocity at the distance observed would require the presence of a 1–2 billion solar-mass black hole. This gas disk could only be resolved by the Hubble Space Telescope, so this telescope has provided the first observational *evidence* of a black hole.

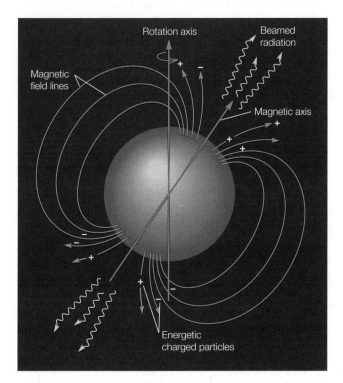

FIGURE 14.13 The magnetic axis of the pulsar is inclined with respect to the rotation axis. Rapidly moving electrons in the regions near the magnetic poles emit radiation in a beam pointed outward. When the beam sweeps past Earth, a pulse is detected.

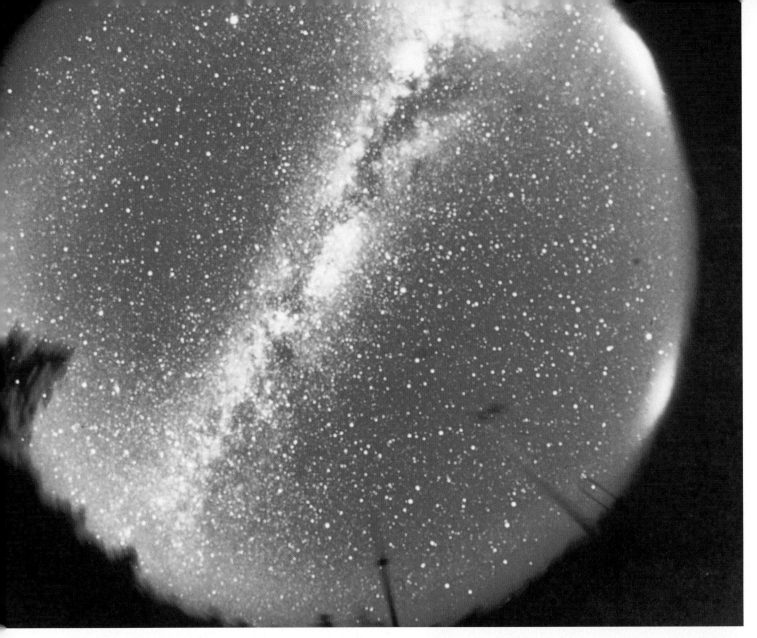

FIGURE 14.14 A wide-angle view toward the center of the Milky Way galaxy. Parts of the white, milky band are obscured from sight by gas and dust clouds in the galaxy.

GALAXIES

Stars are associated with other stars on many different levels, from double stars that orbit a common center of mass, to groups of tens or hundreds of stars that have gravitational links and a common origin, to the billions and billions of stars that form the basic unit of the universe, a **galaxy.** The Sun is but one of an estimated 100 billion stars that are held together by gravitational attraction in the Milky Way galaxy. The numbers of stars and vastness of the Milky Way galaxy alone seem almost beyond comprehension, but there is more to come. The Milky Way is but one of *billions* of galaxies that are associated with other galaxies in clusters, and these clusters are associated with one another in superclusters. Through a large telescope, you can see more galaxies than individual stars in any direction, each galaxy with its own structure of billions of stars. Yet there are similarities

that point to a common origin. Some of the similarities and associations of stars will be introduced in this section along with the Milky Way galaxy, the vast, flat, spiraling arms of stars, gas, and dust where the Sun is located (Figure 14.14).

THE MILKY WAY GALAXY

Away from city lights, you can clearly see the faint, luminous band of the Milky Way galaxy on a moonless night. Through a telescope or a good pair of binoculars, you can see that the luminous band is made up of countless numbers of stars. You may also be able to see the faint glow of nebulae: concentrations of gas and dust. There are dark regions in the Milky Way that also give an impression of something blocking starlight, such as dust. You can also see small groups of stars called *galactic clusters.* Galactic clusters are gravitationally bound subgroups

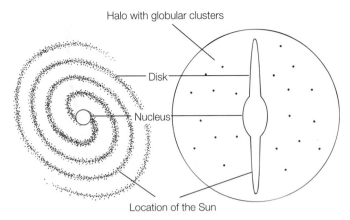

FIGURE 14.15 The structure of the Milky Way galaxy.

of as many as one thousand stars that move together within the Milky Way. Other clusters are more symmetrical and tightly packed, containing as many as a million stars, and are known as *globular clusters.*

Viewed from a distance in space, the Milky Way would appear to be a huge, flattened cloud of spiral arms radiating out from the center. There are three distinct parts: (1) the spherical concentration of stars at the center of the disk called the *galactic nucleus;* (2) the rotating *galactic disk,* which contains most of the bright, blue stars along with much dust and gas; and (3) a spherical *galactic halo,* which contains some 150 globular clusters located outside the galactic disk (Figure 14.15). The Sun is located in one of the arms of the galactic disk, some 25,000 to 30,000 light-years from the center. The galactic disk rotates, and the Sun completes one full rotation every 200 million years.

The diameter of the *galactic disk* is about 100,000 light-years. Yet in spite of the 100 billion stars in the Milky Way, it is mostly full of emptiness. By way of analogy, imagine reducing the size of the Milky Way disk until stars like the Sun were reduced to the size of tennis balls. The distance between two of these tennis-ball-sized stars would now compare to the distance across the state of Texas. The space between the stars is not actually empty since it contains a thin concentration of gas, dust, and molecules of chemical compounds. The gas particles outnumber the dust particles about 10^{12} to 1. The gas is mostly hydrogen, and the dust is mostly solid iron, carbon, and silicon compounds. Over forty different chemical molecules have been discovered in the space between the stars, including many organic molecules. Some nebulae consist of clouds of molecules with a maximum density of about 10^6 molecules/cm^3. The gas, dust, and chemical compounds make up part of the mass of the galactic disk, and the stars make up the remainder. The gas plays an important role in the formation of new stars, and the dust and chemical compounds play an important role in the formation of planets.

OTHER GALAXIES

Outside the Milky Way is a vast expanse of emptiness, lacking even the few molecules of gas and dust spread thinly through the galactic nucleus. There is only the light from faraway galaxies

and the time that it takes for this light to travel across the vast vacuum of intergalactic space. How far away is the nearest galaxy? Recall that the Milky Way is so large that it takes light 100,000 years to travel the length of its diameter.

The nearest galactic neighbor to the Milky Way is a dwarf spherical galaxy only 80,000 light-years from our solar system. The nearby galaxy is called a dwarf because it has a diameter of only about 1,000 light-years. It is apparently in the process of being pulled apart by the gravitational pull of the Milky Way, which now is known to have eleven satellite galaxies.

The nearest galactic neighbor similar to the Milky Way is Andromeda, about 2 million light-years away. Andromeda is similar to the Milky Way in size and shape, with about 100 billion stars, gas, and dust turning in a giant spiral pinwheel (Figure 14.16). Other galaxies have other shapes and other characteristics. The American astronomer Edwin Hubble developed a classification scheme for the structure of galaxies based on his study of some six hundred different galaxies. The basic galactic structures were identified as elliptical, spiral, barred, and irregular.

FIGURE 14.16 The Andromeda galaxy, which is believed to be similar in size, shape, and structure to the Milky Way galaxy.

A Closer Look

Extraterrestrials?

Extraterrestrial is a descriptive term, meaning a thing or event outside Earth or its atmosphere. The term is also used to describe a being, a life form that originated away from Earth. This reading is concerned with the search for extraterrestrials, intelligent life that might exist beyond Earth and outside the solar system.

Why do people believe that extraterrestrials might exist? The affirmative answer comes from a mixture of current theories about the origin and development of stars, statistical odds, and faith. Considering the statistical odds, note that our Sun is one of the some 300 billion stars that make up our Milky Way galaxy. The Milky Way galaxy is one of some 10 billion galaxies in the observable universe. Assuming an average of about 300 billion stars per galaxy, this means there are some 300 billion times 10 billion, or 3×10^{21}, stars in the observable universe. There is nothing special or unusual about our Sun, and astronomers believe it to be quite ordinary among all the other stars (all 3,000,000,000,000,000,000,000 or so).

So the Sun is an ordinary star, but what about our planet? Not too long ago, most people, including astronomers, thought our solar system with its life-supporting planet (Earth) to be unique. Evidence collected over the past decade or so, however, has strongly suggested that this is not so. Planets are now believed to be formed as a natural part of the star-forming process. Evidence of planets around other stars has also been detected by astronomers. One of the stars with planets is "only" 53 light-years from Earth.

Even with a very low probability of planetary systems forming with the development of stars, a population of 3×10^{21} stars means there are plenty of planetary systems in existence, some with the conditions necessary to support life (Note: If 1 percent have planetary systems, this means 3×10^{19} stars have planets). Thus, it is a statistical observation that suitable planets for life are very likely to exist. In addition, radio astronomers have found that many organic molecules exist, even in the space between the stars. Based on statistics alone, there should be life on other planets, life that may have achieved intelligence and developed into a technological civilization.

If extraterrestrials exist, why have we not detected them or why have they not contacted us? The answer to this question is found in the unbelievable distances involved in interstellar space. For example, a logical way to search for extraterrestrial intelligence is to send radio signals to space and analyze those coming from space. Modern radio telescopes can send powerful radio beams, and present-day computers and data processing techniques can now search through incoming radio signals for the patterns of artificially generated radio signals (Box Figure 14.1). Radio signals, however, travel through space at the speed of light. The diameter of our Milky Way galaxy is about 100,000 light-years, which means 100,000 years would be required for a radio transmission to travel across the galaxy. If we were to transmit a super, super strong radio beam from Earth, it would travel at the speed of light and cross the distance of our galaxy in 100,000 years. If some extraterrestrials on the other side of the Milky Way galaxy did detect the message and send a reply, it could not arrive at Earth until 200,000 years after the message was sent. Now consider the fact that of all the 10 billion other galaxies in the observable universe, our nearest galactic neighbor similar to the Milky Way is Andromeda. Andromeda is 2 million light-years from the Milky Way galaxy.

BOX FIGURE 14.1 The 300 m (984 ft) diameter radio telescope at Arecibo, Puerto Rico, is the largest fixed-dish radio telescope in the world.

In addition to problems with distance and time, there are questions about in which part of the sky you should send and look for radio messages, questions about which radio frequency to use, and problems with the power of present-day radio transmitters and detectors. Realistically, the hope for any exchange of radio-transmitted messages would be restricted to within several hundred light-years of Earth.

Considering all the limitations, what sort of signals should we expect to receive from extraterrestrials? Probably a series of pulses that somehow indicate counting, such as 1, 2, 3, 4, and so on, repeated at regular intervals. This is the most abstract, while at the same time the simplest, concept that an intelligent being anywhere would have. It could provide the foundation for communications between the stars.

THE LIFE OF A GALAXY

Hubble's classification of galaxies into distinctly different categories of shape was an exciting accomplishment because it suggested that some relationship or hidden underlying order might exist in the shapes. Finding underlying order is important because it leads to the discovery of the physical laws that govern the universe. Soon after Hubble published his classification results in 1926, two models of galactic evolution were proposed. One model, which was suggested by Hubble, had extremely slowly spinning spherical galaxies forming first, which gradually flattened out as their rate of spin increased while they condensed. This is a model of spherical galaxies flattening out to increasingly elliptical shapes, eventually spinning off spirals until they finally broke up into irregular shapes over a long period of time.

Among its many uses, the Hubble Space Telescope is used to study young galaxies and galaxies that are on collision courses. Based on these studies, astronomers today recognize that the

As described in chapter 5, the Doppler effect tells us that the frequency of a wave depends on the relative motion of the source and observer. When the source and observer are moving toward each other, the frequencies appear to be higher. If the source and observer are moving apart, the frequency appears to be lower.

Light from a star or galaxy is changed by the Doppler effect, and the frequency of the observed spectral lines depends on the relative motion. The Doppler effect changes the frequency from what it would be if the star or galaxy were motionless relative to the observer. If the star or galaxy is moving toward the observer, a shift occurs in the spectral lines toward a higher frequency (blueshift). If the star or galaxy is moving away from the observer, a shift occurs in the spectral lines toward a lower frequency (redshift). Thus, a redshift or blueshift in the spectral lines will tell you if a star or galaxy is moving toward or away from you.

One of the first accurate measurements of the distance to other galaxies was made by Edwin Hubble at Mount Wilson Observatory in California. When Hubble compared the distance figures with the observed redshifts, he found that the recession speeds were proportional to the distance. Farther-away galaxies were moving away from the Milky Way, but galaxies that are more distant are moving away faster than closer galaxies. This proportional relationship between galactic speed and distances was discovered in 1929 by Hubble and today is know as *Hubble's law*. The conclusion was that all the galaxies are moving away from one another, and an observer on any given galaxy would have the impression that all galaxies were moving away in all directions. In other words, the universe is expanding with component galaxies moving farther and farther apart.

different shapes of galaxies do not represent an evolutionary sequence. The different shapes of galaxies are understood to be a result of the different conditions under which the galaxies were formed.

The current model of how galaxies form is based on the **big bang theory** of the creation of the universe. The big bang theory considers the universe to have had an explosive beginning. According to this theory, all matter in the universe was located together in an arbitrarily dense state from which it began to expand, an expansion that continues today. Evidence that supports the big bang theory comes from (1) present-day microwave radiation from outer space, (2) current data on the expansion of the universe, (3) the relative abundance of elements that were altered in the core of older stars—this agrees with predictions based on analysis of the big bang, and (4) the *Cosmic Background Explorer (COBE)* spacecraft, which studied diffuse cosmic background radiation to help answer such questions as how matter is distributed in the universe, whether the universe is uniformly expanding, and how and when galaxies first formed.

The 2003 results from NASA's orbiting *Wilkinson Microwave Anisotropy Probe (WMAP)* produced a precision map of the remaining cosmic microwave background from the big bang. *WMAP* surveyed the entire sky for a whole year with a resolution some forty times greater than that of *COBE*. Analysis of *WMAP* data revealed that the universe is 13.7 billion years old, with 1 percent margin of error. The *WMAP* data found strong support for the big bang and expanding universe theories. It also revealed that the contents of the universe includes 4 percent ordinary matter, 23 percent of an unknown type of dark matter, and 73 percent of a mysterious dark energy.

The initial evidence for the big bang theory came from Edwin Hubble and his earlier work with galaxies. Hubble had determined the distances to some of the galaxies that had redshifted spectra. From this expansive redshift, it was known that these galaxies were moving away from the Milky Way. Hubble found a relationship between the distance to a galaxy and the velocity with which it was moving away. He found the velocity to be directly proportional to the distance; that is, the greater the distance to a galaxy, the greater the velocity. This means that a galaxy twice as far from the Milky Way as a second galaxy is moving away from the Milky Way at twice the speed as the second galaxy. Since this relationship was seen in all directions, it meant that the universe is expanding uniformly. The same effect would be viewed from any particular galaxy; that is, all the other galaxies are moving away with a velocity proportional to the distance to the other galaxies. This points to a common beginning, a time when all matter in the universe was together.

Myths, Mistakes, & Misunderstandings

Hello, *Enterprise?*

It is a myth that spaceships could have instant two-way communication. Radio waves travel at the speed of light, but distances in space are huge. The table below gives the approximate time required for a radio signal to travel from Earth to the planet or place. Two-way communication would require twice the time listed.

Mars	4.3 min
Jupiter	35 min
Saturn	1.2 h
Uranus	2.6 h
Neptune	4 h
Pluto	5.3 h
Across Milky Way galaxy	100,000 yr
To Andromeda galaxy	2,000,000 yr

How old is the universe? As mentioned earlier, astronomical and physical "clocks" indicate that the universe was created in a "big bang" some 13.7 billion years ago expanding as an intense

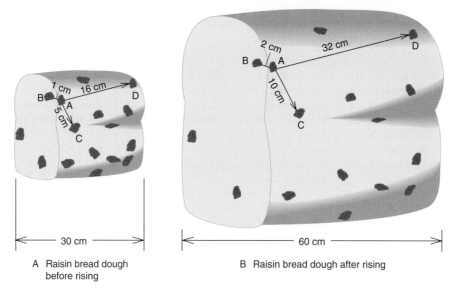

A   Raisin bread dough
    before rising

B   Raisin bread dough after rising

FIGURE 14.17 As the universe (here represented by a loaf of raisin bread) expands, the expansion carries galaxies (represented by the raisins) away from each other at speeds that are proportional to their distances from each other. It doesn't matter within which galaxy an astronomer resides; the other galaxies all appear to be moving away.

and brilliant explosion from a primeval fireball with a temperature of some 10^{12} kelvins. This estimate of the age is based on precise measurements of the rate at which galaxies are moving apart, an expansion that started with the big bang. Astronomers used data on the expansion to back-calculate the age, much like running a movie backward, to arrive at the age estimate. At first, this technique found that the universe was about 18 to 20 billion years old. But then astronomers found that the universe is not expanding at a constant rate. Instead, the separation of galaxies is actually accelerating, pushed by a mysterious force known as "dark energy." By adding in calculations for this poorly understood force, the estimate of 13 to 14 billion years was developed.

Other age-estimating techniques agree with this ballpark age. Studies of white dwarfs, for example, established their rate of cooling. By looking at the very faintest and oldest white dwarfs with the Hubble Space Telescope, astronomers were able to use the cooling rate to estimate the age of the universe. The result found the dimmest of the white dwarfs to be about 13 billion years old, plus or minus about half a billion years. More recently, data from the orbiting *WMAP* spacecraft has produced a precision map of the cosmic microwave background that shows the universe to be 13.7 billion years old. An age of 13.7 billion years may not be the final answer for the age of the universe, but when three independent measurement techniques agree closely, the answer becomes more believable.

An often-used analogy for the movement of galaxies after the big bang is a loaf of rising raisin bread. Consider galaxies as raisins in a loaf of raisin bread dough as it rises (Figure 14.17). As

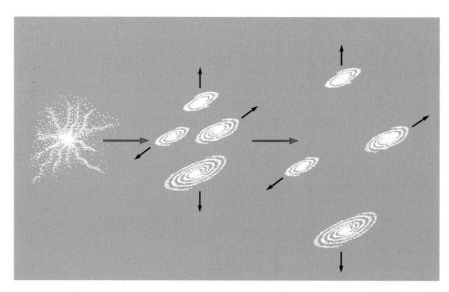

FIGURE 14.18 Will the universe continue expanding as the dust and gas in galaxies become locked up in white dwarf stars, neutron stars, and black holes?

The first evidence came from astronomers trying to measure how fast the expanding universe is slowing. At the time, the generally accepted model of the universe was that it was still expanding but had been slowing from gravitational attraction since the big bang. The astronomers intended to measure the slowing rate to determine the average density of matter in the universe, providing a clue about the extent of dark matter. Their idea was to compare light traveling from a supernova in the distant universe with light traveling from a much closer supernova. Light from the distant universe is just now reaching us because it was emitted when the universe was very young. The brightness of the older and younger supernova would provide information about distance, and expansive redshift data would provide information about their speed of expansion.

By comparing these two light sources, astronomers would be able to calculate the expansion rate for now and in the distant past.

To their surprise, the astronomers did not find a rate of slowing expansion. Rather, they found that the expansion is speeding up. There was no explanation for this speeding up other than to assume that some unknown antigravity force was at work. This unknown force must be pushing the galaxies farther apart at the same time that gravity is trying to pull them together. The unknown repulsive force became known as "dark energy."

The idea of an unknown repulsive force is not new. It was a new idea when Einstein added a "cosmological constant" to the equations in the theory of general relativity. This constant represents a force that opposes gravity, and the force grows as a function of space. This means there was not much—shall we say, dark energy—when the universe was smaller, and gravity slowed the expansion. As the distance between galaxies increased, there was an increase of dark energy, and the expansion accelerated. Einstein removed this constant when Edwin Hubble reported evidence that the universe is expanding. Removing the constant may have been a mistake. One of the problems in understanding what was happening to the early universe compared to now was a lack of information about the past as represented by the distant, faraway universe. Today, this is changing as more new technology is helping astronomers study the faraway universe. Will they find the meaning of dark energy? What *is* dark energy? The answer is that no one knows. Stay tuned . . . this story is to be continued.

the dough expands in size, it carries along the raisins, which are moved farther and farther apart by the expansion. If you were on a raisin, you would see all the other raisins moving away from you, and the speed of this movement would be directly proportional to their distance away. It would not matter which raisin you were on since all the raisins would appear to be moving away to all the other raisins. In other words, the galaxies are not expanding into space that is already there. It is space itself that is expanding, and the galaxies move with the expanding space (Figure 14.18).

The ultimate fate of the universe will strongly depend on the mass of the universe and if the expansion is slowing or continuing. All the evidence tells us it began by expanding with a big bang. Will it continue to expand, becoming increasingly cold and diffuse? Will it slow to a halt, then start to contract to a fiery finale of a big crunch (Figure 14.19)? Researchers continue searching for answers with experimental data and theoretical models, looking for clues in matches between the data and models.

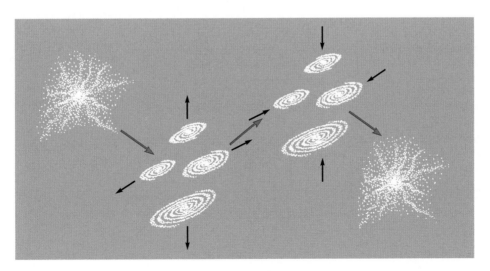

FIGURE 14.19 The oscillating theory of the universe assumes that the space between the galaxies is expanding, as does the big bang theory, but in the oscillating theory, the galaxies gradually come back together to begin all over in another big bang.

Dark Matter

Will the universe continue to expand forever, or will it be pulled back together into a really big crunch? Whether the universe will continue expanding or gravity will pull it back together again depends on dark energy and the *critical density* of the universe. If there is enough matter, the actual density of the universe will be above the critical density, and gravity will stop the current expansion and pull everything back together again. On the other hand, if the actual density of the universe is less than or equal to the critical density, the universe will be able to escape its own gravity and continue expanding. Is the actual density small enough for the universe to escape itself?

All the detailed calculations of astrophysicists point to an actual density that is less than the critical density, meaning the universe will continue to expand forever. However, these calculations also show that there is more matter in the universe than can be accounted for in the stars and galaxies. This means there must be matter in the universe that is not visible, or not shining at least, so we cannot see it. Three examples follow that show why astrophysicists believe more matter exists than you can see.

1. There is a relationship between the light emitted from a star and the mass of that star. Thus, you can indirectly measure the amount of matter in a galaxy by measuring the light output of that galaxy. Clusters of galaxies have been observed moving, and the motion of galaxies within a cluster does not agree with matter information from the light. The motion suggests that galaxies are attracted by a gravitational force from about ten times more matter than can be accounted for by the light coming from the galaxies.

2. There are mysterious variations in the movement of stars within an individual galaxy. The rate of rotation of stars about the center of rotation of a galaxy is related to the distribution of matter in that galaxy. The outer stars are observed to rotate too fast for the amount of matter that can be seen in the galaxy. Again, measurements indicate there is about ten times more matter in each galaxy than can be accounted for by the light coming from the galaxies.

3. Finally, there are estimates of the total matter in the universe that can be made by measuring ratios such as deuterium to ordinary hydrogen, lithium-7 to helium-3, and helium to hydrogen. In general, theoretical calculations all seem to account for only about 4 percent of all the matter in the universe.

Calculations from a variety of sources all seem to agree that 23 percent or more of the matter that makes up the universe is missing. The missing matter and the mysterious variations in the orbits of galaxies and stars can be accounted for by the presence of *dark matter,* which is invisible and unseen. What is the nature of this dark matter? You could speculate that dark matter is simply normal matter that has been overlooked, such as dark galaxies, brown dwarfs, or planetary material such as rock, dust, and the like. You could also speculate that dark matter is dark because it is in the form of subatomic particles, too small to be seen. As you can see, scientists have reasons to believe that dark matter exists, but they can only speculate about its nature.

The nature of dark matter represents one of the major unsolved problems in astronomy today. There are really two dark matter problems: (1) the nature of the dark matter and (2) how much dark matter contributes to the actual density of the universe. Not everyone believes that dark matter is simply normal matter that has been overlooked. There are at least two schools of thought on the nature of dark matter and what should be the focus of research. One school of thought focuses on *particle physics* and contends that dark matter consists primarily of exotic undiscovered particles or known particles such as the neutrino. Neutrinos are electrically neutral, stable subatomic particles. This school of thought is concerned with forms of dark matter called WIMPs (after "weakly interacting massive particles"). In spite of the name, particles under study are not always massive. It is entirely possible that some low-mass species of neutrino—or some as of yet unidentified WIMP—will be experimentally discovered and found to have the mass needed to account for the dark matter and thus close the universe.

The other school of thought about dark matter focuses on *astrophysics* and contends that dark matter is ordinary matter in the form of brown dwarfs, unseen planets outside the solar system, and galactic halos. This school of thought is concerned with forms of dark matter called MACHOs (after "massive astrophysical compact halo objects"). Dark matter considerations in this line of reasoning consider massive objects, but ordinary matter (protons and neutrons) is also considered as the material making up galactic halos. Protons and neutrons belong to a group of subatomic particles called the *baryons,* so they are sometimes referred to as *baryonic dark matter.* Some astronomers feel that baryonic dark matter is the most likely candidate for halo dark matter.

In general, astronomers can calculate the probable cosmological abundance of WIMPs, but there is no proof of their existence. By contrast, MACHOs dark matter candidates are known to exist, but there is no way to calculate their abundance. MACHOs astronomers assert that halo dark matter and probably all dark matter may be baryonic. Do you believe WIMPs or MACHOs will provide an answer about the future of the universe? What is the nature of dark matter, how much dark matter exists, and what is the fate of the universe? Answers to these and more questions await further research.

Jocelyn Bell is a British astronomer who discovered pulsating radio stars—pulsars—an important astronomical discovery of the 1960s.

Bell was born in Belfast, Ireland, on July 15, 1943. The Armagh Observatory, of which her father was architect, was sited near her home, and the staff there were particularly helpful and offered encouragement when they learned of her early interest in astronomy. From 1956 to 1961, she attended the Mount School in York, England. She then went to the University of Glasgow, receiving her B.Sc. degree in 1965. In the summer of 1965, she began to work for her Ph.D. under the supervision of Anthony Hewish at the University of Cambridge. It was during the course of this work that the discovery of pulsars was made. Having completed her doctorate at Cambridge, she went on to work in gamma-ray astronomy at the University of Southampton, and from 1974 to 1982, she worked at the Mullard Space Science Laboratory in X-ray astronomy. In 1982, she was appointed a senior research fellow at the Royal Observatory, Edinburgh, where she worked on infrared and optical astronomy. She was head of the James Clark Maxwell Telescope section, responsible for the British end of the telescope project based in Hawaii. In 1991, she was appointed professor of physics and departmental chair at the Open University, Milton Keynes, U.K. A winner of the Royal Astronomical Society's prestigious Herschel medal (1989), she has made significant contributions in the fields of X-ray and gamma-ray astronomy.

She spent her first two years in Cambridge building a radio telescope that was specially designed to track quasars—her Ph.D. research topic. The telescope that she and her team built had the ability to record rapid variations in signals. It was also nearly 2 hectares (about 5 acres) in area, equivalent to a dish of 150 m (about 500 ft) in diameter, making it an extremely sensitive instrument. The sky survey began when the telescope was finally completed in 1967, and Bell was given the task of analyzing the signals received. One day, while scanning the charts of recorded signals, she noticed a rather unusual radio source that had occurred during the night and had been picked up in a part of the sky that was opposite in direction to the sun. This was curious because strong variations in the signals from quasars are caused by solar wind and are usually weak during the night. At first, she thought that the signal might be due to a local interference. After a month of further observations, it became clear that the position of the peculiar signals remained fixed with respect to the stars, indicating that it was neither terrestrial nor solar in origin. A more detailed examination of the signal showed that it was in fact composed of a rapid set of pulses that occurred precisely every 1.337 seconds. The pulsed signal was as regular as the most regular clock on the earth.

One attempted explanation of this curious phenomenon was that it represented an interstellar beacon sent out by extraterrestrial life on another star, so initially it was nicknamed LGM, for "little green men." Within a few months of noticing this signal, however, Bell located three other similar sources. They too pulsed at an extremely regular rate, but their periods varied over a few fractions of a second, and they all originated from widely spaced locations in our galaxy. Thus, it seemed that a more likely explanation of the signals was that they were being emitted by a special kind of star—a pulsar.

Since the astonishing discovery was announced, other observatories have searched the heavens for new pulsars. Some three hundred are now known to exist, their periods ranging from hundredths of a second to four seconds. It is thought that neutron stars are responsible for the signal. These are tiny stars, only about 7 km (about 4.3 mi) in diameter, but they are incredibly massive. The whole star and its associated magnetic field are spinning at a rapid rate, and the rotation produces the pulsed signal.

Source: From the Hutchinson *Dictionary of Scientific Biography.* © Research Machines plc 2003. All Rights Reserved. Helicon Publishisearch Machines.

SUMMARY

Theoretically, stars are born in clouds of hydrogen gas and dust in the space between other stars. Gravity pulls huge masses of hydrogen gas together into a *protostar,* a mass of gases that will become a star. The protostar contracts, becoming increasingly hot at the center, eventually reaching a temperature high enough to start *nuclear fusion* reactions between hydrogen atoms. Pressure from hot gases balances the gravitational contraction, and the average newborn star will shine quietly for billions of years. The average star has a dense, hot *core* where nuclear fusion releases radiation, a less dense *radiation zone* where radiation moves outward, and a thin *convection zone* that is heated by the radiation at the bottom, then moves to the surface to emit light to space.

The brightness of a star is related to the amount of energy and light it is producing, the size of the star, and the distance to the star. The *apparent magnitude* is the brightness of a star as it appears to you. To compensate for differences in brightness due to distance, astronomers calculate the brightness that stars would have at a standard distance called the *absolute magnitude.* Absolute magnitude is an expression of *luminosity,* the total amount of energy radiated into space each second from the surface of a star.

Stars appear to have different colors because they have different surface temperatures. A graph of temperature by spectral types and brightness by absolute magnitude is called the *Hertzsprung-Russell diagram,* or H-R diagram for short. Such a graph shows that normal, mature stars fall on a narrow band called the *main sequence* of stars. Where a star falls on the main sequence is determined by its brightness and temperature, which in turn are determined by the mass of the star. Other groups of stars on the H-R diagram have different sets of properties that are determined by where they are in their evolution.

The life of a star consists of several stages, the longest of which is the *main sequence* stage after a relatively short time as a *protostar.* After using up the hydrogen in the core, a star with an average mass expands to a *red giant,* then blows off the outer shell to become a *white dwarf star,* which slowly cools to a black dwarf. The blown-off outer shell forms a *planetary nebula,* which disperses over time to become the gas and dust of interstellar space. More massive stars collapse into *neutron stars* or *black holes* after a violent *supernova* explosion.

Galaxies are the basic units of the universe. The Milky Way galaxy has three distinct parts: (1) the *galactic nucleus,* (2) a rotating *galactic disk,* and (3) a *galactic halo.* The galactic disk contains subgroups of stars that move together as *galactic clusters.* The halo contains symmetrical and tightly packed clusters of millions of stars called *globular clusters.*

All the billions of galaxies can be classified into groups of four structures: *elliptical, spiral, barred,* and *irregular.* Evidence from astronomical and physical "clocks" indicates that the galaxies formed some 13.7 billion years ago, expanding ever since from a common origin in a *big bang.* The *big bang theory* describes how the universe began by expanding.

KEY TERMS

absolute magnitude (p. **360**)

apparent magnitude (p. **359**)

big bang theory (p. **369**)

black hole (p. **365**)

celestial equator (p. **356**)

celestial meridian (p. **357**)

Cepheid variable (p. **362**)

galaxy (p. **366**)

Hertzsprung-Russell diagram (p. **361**)

light-year (p. **356**)

luminosity (p. **360**)

main sequence stars (p. **362**)

nebulae (p. **358**)

neutron star (p. **364**)

north celestial pole (p. **356**)

protostar (p. **358**)

pulsar (p. **364**)

red giant stars (p. **362**)

south celestial pole (p. **356**)

supernova (p. **364**)

white dwarf stars (p. **362**)

APPLYING THE CONCEPTS

1. A referent system that can be used to locate objects in the sky is the
   a. celestial globe.
   b. celestial coordinates.
   c. celestial sphere.
   d. celestial maps.

2. Our Sun is
   a. one of several stars in the solar system.
   b. the brightest star in the night sky.
   c. an average star, considering mass and age.
   d. a protostar.

3. The brightness of a star at a defined standard distance is
   a. apparent magnitude.
   b. apparent longitude.
   c. absolute magnitude.
   d. absolute latitude.

4. The total amount of energy radiated from the surface of a star into space each second is
   a. stellar radiation.
   b. luminosity.
   c. convection.
   d. dispersion.

5. Which of the following colors of starlight is from the hottest star?
   a. blue-white
   b. red
   c. orange-red
   d. yellow

6. The color of a star is related to its
   a. composition.
   b. apparent magnitude.
   c. absolute magnitude.
   d. surface temperature.

7. The temperature-luminosity graph used to classify stars was developed by
   a. Russell.
   b. Hubble.
   c. Hubble and Russell.
   d. Hertzsprung and Russell.

8. The period-brightness relationship of a Cepheid variable star allows astronomers to measure
   a. distance.
   b. age of stars.
   c. luminosity.
   d. size.

9. Where does our Sun belong based on the Hertzsprung-Russell classification of stars?
   a. main sequence
   b. red giant
   c. white dwarf
   d. Cepheid variable

10. The lifetime of a star depends on its
    a. composition.
    b. mass.
    c. temperature.
    d. location.

11. The stages in the life of a star are
    a. protostar, main sequence, white dwarf, and red giant.
    b. protostar, main sequence, red giant, and red dwarf.
    c. protostar, main sequence, red giant, and white dwarf.
    d. protostar, main sequence, red giant, and nebulae.

12. In the process of a less massive star growing old, its outer layer is sometimes blown off into space, forming a
    a. white dwarf.
    b. black hole.
    c. Cepheid variable.
    d. planetary nebula.

13. The collapse of a massive star
    a. results in a tremendous explosion called a supernova.
    b. forms a white dwarf.
    c. forms planetary nebulae.
    d. results in a tremendous explosion called a pulsar.

14. A rapidly rotating neutron star with a strong magnetic field is a
    a. neutron star.
    b. pulsar.
    c. supernova.
    d. black hole.

15. The product of the collapse of a massive star with a core three times the mass of our Sun is a
    a. neutron star.
    b. pulsar.
    c. red giant.
    d. black hole.

16. The basic unit of the universe is a
    a. star.
    b. solar system.
    c. galaxy.
    d. constellation.

17. A galaxy is held together by
    a. stellar dust.
    b. gravitational attraction.
    c. electronic attractions.
    d. pulsating stars.

18. The diameter of the Milky Way is about
    a. 500 light-years.
    b. 50,000 light-years.
    c. 100,000 light-years.
    d. 200 million light-years.

19. Approximately how old is the universe?
    a. 5 billion years
    b. 10 billion years
    c. 14 billion years
    d. 25 billion years

20. Stars twinkle and planets do not twinkle because
    a. planets shine by reflected light, and stars produce their own light.
    b. all stars are pulsing light sources.
    c. stars appear as point sources of light, and planets are disk sources.
    d. All of the above are correct.

21. How much of the celestial meridian can you see from any given point on the surface of Earth?
    a. one-fourth
    b. one-half
    c. three-fourths
    d. all of it

22. Which of the following of the coordinate system of lines depends on where you are on the surface of Earth?
    a. celestial meridian
    b. celestial equator
    c. north celestial pole
    d. None of the above is correct.

23. The angle that you see Polaris, the North Star, above the horizon is about the same as your approximate location on
    a. the celestial meridian.
    b. the celestial equator.
    c. a northern longitude.
    d. a northern latitude.

24. If you were at the north celestial pole looking down on Earth, how would it appear to be moving? (Use a globe if you wish.)
    a. clockwise
    b. counterclockwise
    c. one way, then the other as a pendulum
    d. It would not appear to move from this location.

25. Your answer to question 24 means that Earth turns
    a. from the west toward the east.
    b. from the east toward the west.
    c. at the same rate it is moving in its orbit.
    d. not at all.

26. Your answer to question 25 means that the Moon, Sun, and stars that are not circumpolar appear to rise in the
    a. west, move in an arc, then set in the east.
    b. north, move in an arc, then set in the south.
    c. east, move in an arc, then set in the west.
    d. south, move in an arc, then set in the north.

27. How many degrees of arc above the horizon is a star located halfway between directly over your head and the horizon?
    a. 45°
    b. 90°
    c. 135°
    d. 180°

28. In which part of a newborn star does the nuclear fusion take place?
    a. convection zone
    b. radiation zone
    c. core
    d. All of the above are correct.

29. Which of the following stars would have the longer life spans?
    a. the less massive
    b. between the more massive and the less massive
    c. the more massive
    d. All have the same life span.

30. A bright blue star on the main sequence is probably
    a. very massive.
    b. less massive.
    c. between the more massive and the less massive.
    d. None of the above is correct.

31. The brightest of the stars listed are the
    a. first magnitude.
    b. second magnitude.
    c. fifth magnitude.
    d. sixth magnitude.

32. The basic property of a main sequence star that determines most of its other properties, including its location on the H-R diagram, is
    a. brightness.
    b. color.
    c. temperature.
    d. mass.

33. All of the elements that are more massive than the element iron were formed in a
    a. nova.
    b. white dwarf.
    c. supernova.
    d. black hole.

34. If the core remaining after a supernova has a mass between 1.5 and 3 solar masses, it collapses to form a
    a. white dwarf.
    b. neutron star.
    c. red giant.
    d. black hole.

35. The relationship between the different shapes of galaxies is:
    a. spherical galaxies form first, which flatten out to elliptical galaxies, then spin off spirals until they break up in irregular shapes.
    b. irregular shapes form first, which collapse to spiral galaxies, then condense to spherical shapes.
    c. There is no relationship as the different shapes probably resulted from different rates of swirling gas clouds.
    d. None of the above is correct.

36. Microwave radiation from space, measurements of the expansion of the universe, the age of the oldest stars in the Milky Way galaxy, and ratios of radioactive decay products all indicate that the universe is about how old?
    a. 6,000 years
    b. 4.5 billion years
    c. 13.7 billion years
    d. 100,000 billion years

37. Whether the universe will continue to expand or will collapse back into another big bang seems to depend on what property of the universe?
    a. the density of matter in the universe
    b. the age of galaxies compared to the age of their stars
    c. the availability of gases and dust between the galaxies
    d. the number of black holes

38. Distance in outer space is measured in terms of
    a. kilometers.
    b. megameters.
    c. light-years.
    d. miles.

39. You are studying physical science in Tennessee. Where are you located with respect to the celestial equator?
    a. north of the celestial equator
    b. south of the celestial equator
    c. on the celestial equator
    d. east of the celestial equator

40. The Crab Nebula is what remains of a
    a. pulsar.
    b. supernova.
    c. black hole.
    d. constellation.

41. The interior of an average star is proposed to contain all of the following *except* a
    a. conduction zone.
    b. core.
    c. radiation zone.
    d. convection zone.

42. Approximately 1.4×10^{17} kg of matter is converted to energy each year by the Sun. How much matter is converted to energy in twenty years?
    a. 4.2×10^{25} kg
    b. 1.3×10^{34} kg
    c. 2.8×10^{21} kg
    d. 2.8×10^{18} kg

43. Red giant stars are
    a. bright, low temperature giants.
    b. dim, low temperature giants.
    c. bright, high temperature giants.
    d. dim, high temperature giants.

44. Stars that are faint, very dense, white hot, and close to the end of their lifetime are
    a. red giant stars.
    b. novas.
    c. white dwarf stars.
    d. Cepheid variable stars.

45. Which of the following elements forms in a supernova explosion of a dying star?
    a. hydrogen
    b. carbon
    c. nitrogen
    d. nickel

46. The greater the distance to a galaxy, the greater a redshift in its spectral lines. This is known as
    a. Doppler's law.
    b. Cepheid's law.
    c. Hubble's law.
    d. Kepler's law.

47. Evidence that points to the existence of black holes was provided by
    a. Mauna Lao Observatory.
    b. Hubble Space Telescope.
    c. radio astronomy.
    d. LGM.

48. The name of our galaxy is the
    a. solar system.
    b. main sequence.
    c. Milky Way.
    d. Polaris.

49. What is the obstacle to finding extraterrestrial life?
    a. money
    b. number of stars
    c. distance and time
    d. language

50. Outer space is mostly
    a. galaxies.
    b. stars.
    c. planets.
    d. empty space.

51. Initial evidence that supports the big bang theory of the creation of the universe came from
    a. experiments conducted by shuttle astronauts.
    b. unmanned space probes.
    c. redshift calculations by Edwin Hubble.
    d. data from the Hubble Space Telescope.

Answers

1. c 2. c 3. c 4. b 5. a 6. d 7. d 8. a 9. a 10. b 11. c 12. d 13. a 14. b 15. d 16. c 17. b 18. c 19. c 20. c 21. b 22. a 23. d 24. b 25. a 26. c 27. a 28. c 29. a 30. a 31. a 32. d 33. c 34. b 35. c 36. c 37. a 38. c 39. a 40. b 41. a 42. d 43. a 44. c 45. d 46. c 47. b 48. c 49. c 50. d 51. c

1. Would you ever observe the Sun to move along the celestial meridian? Explain.

2. What is a light-year and how is it defined?

3. Why are astronomical distances not measured with standard referent units of distance such as kilometers or miles?

4. Explain why a protostar heats up internally as it gravitationally contracts.

5. Describe in general the structure and interior density, pressure, and temperature conditions of an average star such as the Sun.

6. Which size of star has the longest life span: a star sixty times more massive than the Sun, one just as massive as the Sun, or a star that has a mass of one-twenty-fifth that of the Sun? Explain.

7. What is the difference between apparent magnitude and absolute magnitude?

8. What does the color of a star indicate about the surface temperature of the star? What is the relationship between the temperature of a star and the spectrum of the star? Describe in general the spectral classification scheme based on temperature and stellar spectra.

9. What is the Hertzsprung-Russell diagram? What is the significance of the diagram?

10. What is meant by the main sequence of the H-R diagram? What one thing determines where a star is plotted on the main sequence?

11. Describe in general the life history of a star with an average mass like the Sun.

12. What is a nova? What is a supernova?

13. Describe the theoretical physical circumstances that lead to the creation of (a) a white dwarf star, (b) a red giant, (c) a neutron star, (d) a black hole, and (e) a supernova.

14. Describe the two forces that keep a star in a balanced, stable condition while it is on the main sequence. Explain how these forces are able to stay balanced for a period of billions of years or longer.

15. What is the source of all the elements in the universe that are more massive than helium but less massive than iron? What is the source of all the elements in the universe that are more massive than iron?

16. Why must the internal temperature of a star be hotter for helium fusion reactions than for hydrogen fusion reactions?

17. When does a protostar become a star? Explain.

18. What is a red giant star? Explain the conditions that lead to the formation of a red giant. How can a red giant become brighter than it was as a main sequence star if it now has a lower surface temperature?

19. Why is an average star like the Sun unable to have carbon fusion reactions in its core?

20. If the universe is expanding, are the galaxies becoming larger? Explain.

21. What is the evidence that supports a big bang theory of the universe?

1. A star is 517 light-years from Earth. During what event in history did the light now arriving at Earth leave the star?

2. What are the significant differences between the life and eventual fate of a massive star and an average-sized star such as the Sun?

3. Analyze when apparent magnitude is a better scale of star brightness and when absolute magnitude is a better scale of star brightness.

4. What is the significance of the Hertzsprung-Russell diagram?

5. The Milky Way galaxy is a huge, flattened cloud of spiral arms radiating out from the center. Describe several ideas that explain why it has this shape. Identify which idea you favor and explain why.

It Keeps Going, and Going, and . . .

Pioneer 10 was the first space probe to visit an outer planet of our solar system. It was launched March 2, 1972, and successfully visited Jupiter on June 13, 1983. After transmitting information and relatively close-up pictures of Jupiter, *Pioneer 10* continued on its trajectory, eventually becoming the first space probe to leave the solar system. Since then, it continued to move silently into deep space and sent the last signal on January 22, 2003, when it was 12.2 billion km (7.6 billion mi) from Earth. It will now continue to drift for the next 2 million years toward the star Aldebaran in the constellation Taurus.

As the first human-made object to go out of the solar system, *Pioneer 10* carries a gold-plated plaque with the image shown in Box Figure 14.2. Perhaps intelligent life will find the plaque and decipher the image to learn about us. What information is in the image? Try to do your own deciphering to reveal the information. When you have exhausted your efforts, see grin.hq.nasa.gov/ABSTRACTS/GPN-2000-001623.html For more on the *Pioneer 10* mission, see nssdc.gsfc.nasa.gov/nwc/tmp/1972-012A.html

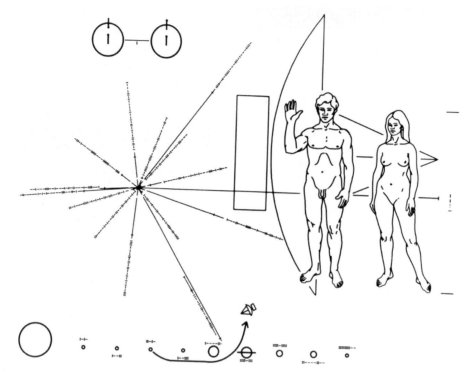

BOX FIGURE 14.2 *Pioneer* plaque symbology.

15

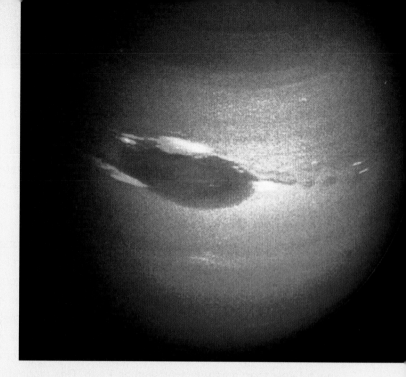

The Solar System

Neptune, the most distant and smallest of the gas giant planets, is a cold and interesting place. It has a Great Dark Spot, as you can see in this photograph taken by the *Voyager* probe. This spot is about the size of Earth and is similar to the Great Red Spot on Jupiter. Neptune has the strongest winds of any planet of the solar system—up to 2,000 km/h (1,200 mi/h). Clouds were observed by *Voyager* to be "scooting" around Neptune every 16 hours or so. *Voyager* scientists called these fast-moving clouds "scooters."

CORE **CONCEPT**

The solar system is composed of the Sun and orbiting planets, dwarf planets, and small solar system bodies.

OUTLINE

Planets, Moons, and Other Bodies
The interior planets of Mercury, Venus, Earth, and Mars are composed of rocky materials with a metallic nickel and iron core.

Origin of the Solar System
The Sun, planets, dwarf planets, moons, asteroids, and comets are believed to have formed from gas, dust, and chemical elements created by massive, previously existing stars.

Planets, Moons, and Other Bodies
   Mercury
   Venus
   Mars
Science and Society: Worth the Cost?
   Jupiter
   Saturn
   Uranus and Neptune
Small Bodies of the Solar System
   Comets
   Asteroids
   Meteors and Meteorites
Origin of the Solar System
   Stage A
   Stage B
   Stage C
Ideas About the Solar System
   The Geocentric Model
   The Heliocentric Model
People Behind the Science: Percival Lowell

Planets, Moons, and Other Bodies
The outer planets of Jupiter, Saturn, Uranus, and Neptune are mostly composed of hydrogen, helium, and methane.

For generations, people have observed the sky in awe, wondering about the bright planets moving across the background of stars, but they could do no more than wonder. You are among the first generations on Earth to see close-up photographs of the planets, comets, and asteroids, and to see Earth as it appears from space. Spacecrafts have now made thousands of photographs of the other planets and their moons, measured properties of the planets, and, in some cases, studied their surfaces with landers. Astronauts have left Earth and visited the Moon, bringing back rock samples, data, and photographs of Earth as seen from the Moon (Figure 15.1). All of these photographs and findings have given us a new perspective of Earth, the planets, and the moons, comets, and asteroids that make up the solar system.

Viewed from the Moon, Earth is a spectacular blue globe with expanses of land and water covered by huge, changing patterns of white clouds. Viewed from a spacecraft, other planets present a very different picture, each unique in its own way. Mercury has a surface that is covered with craters, looking very much like the surface of Earth's Moon. Venus is covered with clouds of sulfuric acid over an atmosphere of mostly carbon dioxide, which is under great pressures with surface temperatures hot enough to melt lead. The surface of Mars has great systems of canyons, inactive volcanoes, dry riverbeds and tributaries, and ice beneath the surface. The giant planets Jupiter and Saturn have orange, red, and white bands of organic and sulfur compounds and storms with gigantic lightning discharges much larger than anything ever seen on Earth. One moon of Jupiter has active volcanoes spewing out liquid sulfur and gaseous sulfur dioxide. The outer giant planets Uranus and Neptune have moons and particles in rings that appear to be covered with powdery, black carbon.

These and many more findings, some fascinating surprises and some expected, have stimulated the imagination as well as added to our comprehension of the frontiers of space. The new information about the Sun's impressive system of planets, moons, comets, and asteroids has also added to speculations and theories about the planets and how they evolved over time in space. This information, along with the theories and speculations, will be presented in this chapter to give you a picture of the solar system.

PLANETS, MOONS, AND OTHER BODIES

The International Astronomical Union (IAU) is the governing authority over names of celestial bodies. At the August 24, 2006, meeting of the IAU, the definitions of planets, dwarf planets, and small solar system bodies were clarified and approved. To be a classical **planet,** an object must be orbiting the Sun, nearly spherical, and large enough to clear all matter from its orbital zone. A **dwarf planet** is defined as an object that is orbiting the Sun, is nearly spherical, but has *not* cleared matter from its orbital zone and is *not* a satellite. All other objects orbiting the Sun are referred to collectively as **small solar system bodies.**

Some astronomers had dismissed Pluto as a true planet for years because it has properties that do not fit with the other planets. The old definition of a planet was anything spherical that orbits the Sun, which resulted in nine planets. But then in 2003, a new astronomical body was discovered. This body was named "Eris" after the Greek goddess of discord. Eris is larger than Pluto, round, and circles the Sun. Is it a planet? If so, the asteroid Ceres would also be a planet, as would the fifty or so large, icy bodies believed to be orbiting the Sun far beyond Pluto. The idea of fifty or sixty planets in the solar system spurred astronomers to clarify the definition of a planet. Because Pluto does not clear its orbital zone, it was downgraded to a dwarf planet.

Today, there are eight planets, three dwarf planets (Pluto, Eris, and the giant asteroid Ceres), and many, many small solar system bodies. These definitions may change again in the future as more is learned about our solar system.

In this chapter, we will visit each of the planets and other bodies of the solar system (Figure 15.2).

The Sun has seven hundred times the mass of all the planets, moons, and minor members of the solar system together. It is the force of gravitational attraction between the comparatively massive Sun and the rest of our solar system that holds it all together. The distance from Earth to the Sun is known as one **astronomical unit** (AU). One AU is about 1.5×10^8 km (about 9.3×10^7 mi). The astronomical unit is used to describe distances in the solar system, for example, Earth is 1 AU from the Sun.

Table 15.1 compares the basic properties of the eight planets. From this table, you can see that the planets can be classified into two major groups based on size, density, and nature of the atmosphere. The interior planets Mercury, Venus, and Mars have densities and compositions similar to those of Earth, so these planets, along with Earth, are known as the **terrestrial planets.**

Outside the orbit of Mars are four **giant planets,** which are similar in density and chemical composition. The terrestrial planets are mostly composed of rocky materials and metallic

nickel and iron. The giant planets Jupiter, Saturn, Uranus, and Neptune, on the other hand, are massive giants mostly composed of hydrogen, helium, and methane. The density of the giant planets suggests the presence of rocky materials and iron as a core surrounded by a deep layer of compressed gases beneath a deep atmosphere of vapors and gases. Note that the terrestrial planets are separated from the giant planets by the asteroid belt.

We will start with the planet closest to the Sun and work our way outward, moving farther and farther from the Sun as we learn about our solar system.

MERCURY

Mercury is the innermost planet, moving rapidly in a highly elliptical orbit that averages about 0.4 astronomical unit, or about 0.4 of the average distance of Earth from the Sun. Mercury is the smallest planet and is slightly larger than Earth's Moon. Mercury is very bright because it is so close to the Sun, but it is difficult to observe because it only appears briefly for a few hours immediately after sunset or before sunrise. This appearance, low on the horizon, means that Mercury must be viewed through more of Earth's atmosphere, making the study of such a small object difficult at best (Figure 15.3).

Mercury moves around the Sun in about three Earth months, giving Mercury the shortest "year" of all the planets. With the highest orbital velocity of all the planets, Mercury was appropriately named after the mythical Roman messenger of speed. Oddly, however, this speedy planet has a rather long day in spite of its very short year. With respect to the stars, Mercury rotates once every fifty-nine days. This means that Mercury rotates on its axis three times every two orbits.

The long Mercury day with a nearby large, hot Sun means high temperatures on the surface facing the Sun. High

FIGURE 15.1 This view of the rising Earth was seen by the *Apollo 11* astronauts after they entered orbit around the Moon. Earth is just above the lunar horizon in this photograph.

FIGURE 15.2 The order of the planets out from the Sun. The planets are (1) Mercury, (2) Venus, (3) Earth, (4) Mars, (5) Jupiter, (6) Saturn, (7) Uranus, and (8) Neptune. The orbits and the planet sizes are not drawn to scale, and not all rings or moons are shown. Also, the planets are not in a line as shown.

TABLE 15.1

Properties of the planets

	Mercury	Venus	Earth	Mars	Jupiter	Saturn	Uranus	Neptune
Average distance from the Sun:								
in 10^6 km	58	108	150	228	778	1,400	3,000	4,497
in AU	0.38	0.72	1.0	1.5	5.2	9.5	19.2	30.1
Inclination to ecliptic	7°	3.4°	0°	1.9°	1.3°	2.5°	0.8°	1.8°
Revolution period (Earth years)	0.24	0.62	1.00	1.88	11.86	29.46	84.01	164.8
Rotation period (Earth days, h, min, and s)	59 days	−243 days*	23 h 56 min 4 s	24 h 37 min 23 s	9 h 50 min 30 s	10 h 39 min	−17 h* 14 min	16 h 6.7 min
Mass (Earth = 1)	0.05	0.82	1.00	0.11	317.9	95.2	14.6	17.2
Equatorial dimensions:								
diameter in km	4,880	12,104	12,756	6,787	142,984	120,536	57,118	49,528
in Earth radius = 1	0.38	0.95	1.00	0.53	11	9	4	4
Density (g/cm^3)	5.43	5.25	5.52	3.95	1.33	0.69	1.29	1.64
Atmosphere (major compounds)	None	CO_2	N_2,O_2	CO_2	H_2,He	H_2,He	H_2,He,CH_4	H_2,He,CH_4
Solar energy received (cal/cm^2/s)	13.4	3.8	2.0	0.86	0.08	0.02	0.006	0.002

*Negative means spin is opposite to motion in orbit.

temperatures mean higher gas kinetic energies, and with a low gravity, gases easily escape from Mercury so it has only trace gases for an atmosphere. The lack of an atmosphere to even the heat gains from the long days and heat losses from the long nights results in some very large temperature differences. The temperature of the surface of Mercury ranges from above the melting point of lead on the sunny side to below the temperature of liquid oxygen on the dark side.

Mercury has been visited by *Mariner 10,* which flew by three times in 1973 and 1974. The photographs transmitted by *Mariner 10* revealed that the surface of Mercury is covered with craters and very much resembles the surface of Earth's Moon. There are large craters, small craters, superimposed craters, and craters with lighter colored rays coming from them just like the craters on the Moon. Also as on Earth's Moon, there are hills and smooth areas with light and dark colors that were covered by lava in the past, some time after most of the impact craters were formed (Figure 15.4).

A spacecraft named *MESSENGER* is now on its way to investigate the planet Mercury. Its name is an acronym for *ME*rcury Surface, Space *EN*vironment, *GE*ochemistry, and *R*ang*ing*. It was launched August 3, 2004, on a 7.9 billion km (about 4.9 billion mi) trip designed to slow the spacecraft as it falls toward the Sun. Overall, it looped by Earth in August 2005, then twice by Venus in October 2006 and June 2007, and then three times by Mercury in January 2008, October 2008, and September 2009. Eventually, it will be slow enough to be captured by the planet Mercury as it flies by in March 2011. It will then become the first spacecraft to orbit Mercury. Mercury has a high surface temperature, and the spacecraft instruments will be protected from high radiant energy from Mercury and the Sun by a sunshade of heat-resistant ceramic fabric. There is a NASA mission page at www.nasa.gov/mission_pages/messenger/main/index.html.

Mercury has no natural satellites, or moons, it has a weak magnetic field, and it has an average density more similar to Venus or Earth than to the Moon. The presence of the magnetic field and the relatively high density for such a small body must mean that Mercury probably has a relatively large core of iron with at least part of the core molten. Because of its high density, it is thought that Mercury lost much of its less dense, outer layer of rock materials sometime during its formation.

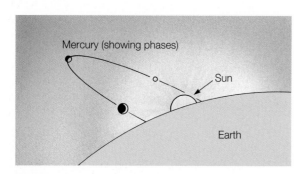

FIGURE 15.3 Mercury is close to the Sun and is visible only briefly before or after sunrise or sunset, showing phases. Mercury actually appears much smaller and is in an orbit that is not tilted as much as shown in this figure.

FIGURE 15.4 A photomosaic of Mercury made from pictures taken by the *Mariner 10* spacecraft. The surface of Mercury is heavily cratered, looking much like the surface of Earth's Moon. All the interior planets and the Moon were bombarded early in the life of the solar system.

VENUS

Venus is the brilliant evening and morning "star" that appears near sunrise or sunset, sometimes shining so brightly that you can see it while it is still daylight. Venus orbits the Sun at an average distance of about 0.7 AU. Venus is sometimes to the left of the Sun, appearing as the evening star, and sometimes to the right of the Sun, appearing as the morning star. Venus also has phases just as the Moon does. When Venus is in the full phase, it is small and farthest away from Earth. A crescent Venus appears much larger and thus the brightest because Venus is closer to Earth when it is in its crescent phase. You can see the phases of Venus with a good pair of binoculars.

Venus shines brightly because it is covered with clouds that reflect about 80 percent of the sunlight, making it the brightest object in the sky after the Sun and Moon. These same clouds prevented any observations of the surface of Venus until the early 1960s, when astronomers using radar were able to penetrate the clouds and measure the planet's rate of movement. Venus was found to spin slowly, so slowly that each day on Venus is longer than a Venus year! Also a surprise, Venus was found to spin in the *opposite* direction to its direction of movement in its orbit. On Venus, you would observe the Sun to rise in the west and set in the east, if you could see it, that is, through all the clouds.

In addition to early studies by radio astronomers, Venus has been the target of many American and Soviet spacecraft probes (Table 15.2). Venus has long been called Earth's sister planet because its mass, size, and density are very similar. That is where the similarities with Earth end, however, as Venus has been found to have a very hostile environment. Spacecraft probes found a hot, dry surface under tremendous atmospheric pressure (Figure 15.5). The atmosphere consists mostly of carbon dioxide, a few percent of nitrogen, and traces of water vapor and other gases. The atmospheric pressure at the surface of Venus is almost one hundred times the pressure at the surface of Earth, a pressure many times beyond what a human could tolerate. The average surface temperature is comparable to the surface temperature on Mercury, which is hot enough to melt lead. The hot temperature on Venus, which is nearly twice the distance from the Sun as Mercury, is a result of the greenhouse effect. Sunlight filters through the atmosphere of Venus, warming the surface. The surface reemits the energy in the form of infrared radiation, which is absorbed by the almost pure carbon dioxide atmosphere. Carbon dioxide molecules absorb the infrared radiation, increasing their kinetic energy and the temperature.

The surface of Venus is mostly a flat, rolling plain but there are several raised areas, or "continents," on about 5 percent of

TABLE 15.2

Spacecraft missions to Venus

Date	Name	Owner	Remark
Feb 12, 1961	*Venera 1*	U.S.S.R.	Flyby
Aug 27, 1962	*Mariner 2*	U.S.	Flyby
Apr 2, 1964	*Zond 1*	U.S.S.R.	Flyby
Nov 12, 1965	*Venera 2*	U.S.S.R.	Flyby
Nov 16, 1965	*Venera 3*	U.S.S.R.	Crashed on Venus
Jun 12, 1967	*Venera 4*	U.S.S.R.	Impacted Venus
Jun 14, 1967	*Mariner 5*	U.S.	Flyby
Jan 5, 1969	*Venera 5*	U.S.S.R.	Impacted Venus
Jan 10, 1969	*Venera 6*	U.S.S.R.	Impacted Venus
Aug 17, 1970	*Venera 7*	U.S.S.R.	Venus landing
Mar 27, 1972	*Venera 8*	U.S.S.R.	Venus landing
Nov 3, 1973	*Mariner 10*	U.S.	Venus, Mercury flyby photos
Jun 8, 1975	*Venera 9*	U.S.S.R.	Lander/orbiter
Jun 14, 1975	*Venera 10*	U.S.S.R.	Lander/orbiter
May 20, 1978	*Pioneer 12* (also called *Pioneer Venus 1* or *Pioneer Venus*)	U.S.	Orbital studies of Venus
Aug 8, 1978	*Pioneer 13* (also called *Pioneer Venus 2* or *Pioneer Venus*)	U.S.	Orbital studies of Venus
Sep 9, 1978	*Venera 11*	U.S.S.R.	Lander; sent photos
Sep 14, 1978	*Venera 12*	U.S.S.R.	Lander; sent photos
Oct 30, 1981	*Venera 13*	U.S.S.R.	Lander; sent photos
Nov 4, 1981	*Venera 14*	U.S.S.R.	Lander; sent photos
Jun 2, 1983	*Venera 15*	U.S.S.R.	Radar mapper
Jun 7, 1983	*Venera 16*	U.S.S.R.	Radar mapper
Dec 15, 1984	*Vega 1*	U.S.S.R.	Venus/Comet Halley probe
Dec 21, 1984	*Vega 2*	U.S.S.R.	Venus/Comet Halley probe
May 4, 1989	*Magellan*	U.S.	Orbital radar mapper
Oct 18, 1989	*Galileo*	U.S.	Flyby measurements and photos
Apr 11, 2006	*Venus Express*	ESA	Orbital spacecraft
Jun 5, 2007	*MESSENGER*	U.S.	Flyby photos

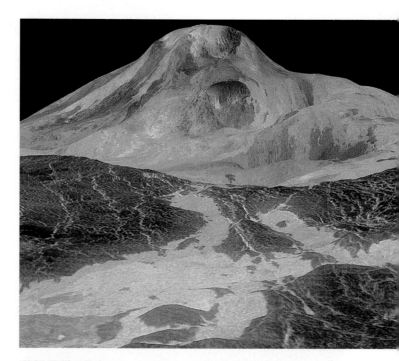

FIGURE 15.5 This is an image of an 8 km (5 mi) high volcano on the surface of Venus. The image was created by a computer using *Magellan* radar data, simulating a viewpoint elevation of 1.7 kilometers (1 mi) above the surface. The lava flows extend for hundreds of kilometers across the fractured plains shown in the foreground. The simulated colors are based on color images recorded by the Soviet *Venera 13* and *14* spacecraft.

Venus, like Mercury, has no satellites. Venus also does not have a magnetic field, as might be expected. Two conditions seem to be necessary in order for a planet to generate a magnetic field: a molten center part and a relatively rapid rate of rotation. Since Venus takes 243 days to complete one rotation, it is the slowest of all the planets, and it does not have a magnetic field even if some of the interior of Venus is still liquid as in Earth.

The European Space Agency's (ESA) orbiting *Venus Express* is conducting an in-depth observation of the structure and chemistry of the atmosphere of Venus. One of the early findings was hot, extensive, 21 km (13 mi) deep clouds of sulfuric acid moved by 334 km/h (220 mph) winds. The mission also plans to study the Venus greenhouse effect and volcanic activity of the past and present, if any.

MARS

Mars has always attracted attention because of its unique, bright reddish color. The properties and surface characteristics have also attracted attention, particularly since Mars seems to have similarities to Earth. It orbits the Sun at an average distance of about 1.5 AU. It makes a complete orbit every 687 days, about twice the time that Earth takes. Mars rotates in twenty-four hours, thirty-seven minutes, so the length of a day on Mars is about the same as the length of a day on Earth. The observations that Mars has an atmosphere, light and dark regions that appear to be greenish and change colors with the seasons, and white polar caps that grow and shrink with the seasons led to

the surface, a mountain larger than Mount Everest, a great valley deeper and wider than the Grand Canyon, and many large, old impact craters. In general, the surface of Venus appears to have evolved much as did the surface of Earth but without the erosion caused by ice, rain, and running water.

early speculations (and many fantasies!) about the possibilities of life on Mars. These speculations increased dramatically in 1877, when Schiaparelli, an Italian astronomer, reported seeing "channels" on the Martian surface. Other astronomers began interpreting the dark greenish regions as vegetation and the white polar caps as ice caps as Earth has. In the early part of the twentieth century, the American astronomer who founded the Lowell Observatory in Arizona, Percival Lowell, published a series of popular books showing a network of hundreds of canals on Mars. Lowell and other respectable astronomers interpreted what they believed to be canals as evidence of intelligent life on Mars. Other astronomers, however, interpreted the greenish colors and the canals to be illusions, imagined features of astronomers working with the limited telescopes of that time. Since canals never appeared in photographs, said the skeptics, the canals were the result of the human tendency to see patterns in random markings where no patterns actually exist.

This speculation ended in the late 1960s and early 1970s with extensive studies and probes by spacecraft (Table 15.3).

TABLE 15.3

Completed spacecraft missions to Mars

Date	Name	Owner	Remark
Nov 5, 1964	*Mariner 3*	U.S.	Flyby
Nov 28, 1964	*Mariner 4*	U.S.	First photos
Feb 24, 1969	*Mariner 6*	U.S.	Flyby
Mar 27, 1969	*Mariner 7*	U.S.	Flyby
May 19, 1971	*Mars 2*	U.S.S.R.	Lander
May 28, 1971	*Mars 3*	U.S.S.R.	Orbiter/lander
May 30, 1971	*Mariner 9*	U.S.	Orbiter
Jul 21, 1973	*Mars 4*	U.S.S.R.	Probe
Jul 25, 1973	*Mars 5*	U.S.S.R.	Orbiter
Aug 5, 1973	*Mars 6*	U.S.S.R.	Lander
Aug 9, 1973	*Mars 7*	U.S.S.R.	Flyby/lander
Aug 20, 1975	*Viking 1*	U.S.	Lander/orbiter
Sep 9, 1975	*Viking 2*	U.S.	Lander/orbiter
Jul 7, 1988	*Phobos 1*	U.S.S.R.	Orbiter/Phobos lander
Jul 12, 1988	*Phobos 2*	U.S.S.R.	Orbiter/Phobos lander
Nov 7, 1996	*Global Surveyor*	U.S.	Orbiter
Dec 4, 1996	*Pathfinder*	U.S.	Lander/Surface rover
Oct 23, 2001	*2001 Mars Odyssey*	U.S.	Orbiter
Dec 25, 2003	*Mars Express*	ESA	Orbiter/Lander
Jan 4, 2004	*Spirit*	U.S.	Lander/Surface rover
Jan 25, 2004	*Opportunity*	U.S.	Lander/Surface rover
Mar 10, 2006	*Mars Reconnaissance Orbiter*	U.S.	Orbiter
May 25, 2008	*Phoenix Mars Lander*	U.S.	Lander

Limited photographs by *Mariner* flybys in 1965 and 1969 had provided some evidence that the surface of Mars was much like the Moon, with no canals, vegetation, or much of anything else. Then in 1971, *Mariner 9* became the first spacecraft to orbit Mars, photographing the entire surface as well as making extensive measurements of the Martian atmosphere, temperature ranges, and chemistry. For about a year, *Mariner 9* sent a flood of new and surprising information about Mars back to Earth.

Mariner 9 found the surface of Mars not to be a crater-pitted surface as is found on the Moon. Mars has had a geologically active past and has four provinces, or regions, of related surface features. There are (1) volcanic regions with inactive volcanoes, one larger than any found on Earth, (2) regions with systems of canyons, some larger than any found on Earth, (3) regions of terraced plateaus near the poles, and (4) flat regions pitted with impact craters. Surprisingly, dry channels suggesting former water erosion were discovered near the cratered regions. These are sinuous, dry riverbed features with dry tributaries. At one time, Mars must have had an abundance of liquid water. Liquid water may have been present on Mars in the past, but none is to be found today. However, scientists using instruments on NASA's *Mars Odyssey* spacecraft found strong signals that ice—perhaps enough to twice fill Lake Michigan—lies just beneath the surface.

The atmosphere of Mars is very thin, exerting an average pressure at the surface that is only 0.6 percent of the average atmospheric pressure on Earth's surface. Moreover, this thin Martian atmosphere is about 95 percent carbon dioxide, and 20 percent of this freezes as dry ice at the Martian South Pole every winter.

Does life exist on Mars? Two *Viking* spacecraft were sent to Mars in 1975 to search for signs of life. The two *Viking* spacecraft were identical, each consisting of an orbiter and a lander. After eleven months of travel time, *Viking 1* entered an orbit around Mars in June 1976 and spent a month sending high-resolution images of the surface back to Earth. From these images, a landing site was selected for the *Viking 1* lander. Using retrorockets, parachutes, and descent rockets, the *Viking 1* lander arrived on a dusty, rocky slope in the southern hemisphere on July 20, 1976. The *Viking 2* lander arrived forty-five days later but farther to the north. The *Viking* lander contained a mechanical soil-retrieving arm and a miniature computerized lab to analyze the soil for evidence of metabolism, respiration, and other life processes. Neither lander detected any evidence of life processes or any organic compounds that would indicate life now or in the past.

The *Viking* spacecraft continued sending images and weather data back to Earth until 1982. During their six-year life, the orbiters sent about fifty-two thousand images and mapped about 97 percent of the Martian surface. The landers sent an additional forty-five hundred images, recorded a major "Marsquake," and recorded data about regular dust storms that occur on Mars with seasonal changes.

Some answers about the geology and history of Mars were provided by Mars *Pathfinder*. On July 4, 1997, a *Pathfinder* lander and rover started sending images and data from the Ares Vallis area of Mars. The lander served as a base communications station, but it also had cameras and measurement instruments of

People are fascinated with ideas of space travel as well as new discoveries about the universe and how it formed. Information from unmanned space probes, new kinds of instruments, and new kinds of telescopes have resulted in new information. We are learning about what is happening in outer space away from Earth, as well as the existence of other planets.

Few would deny that the space program has provided valuable information. Some people wonder, however, if it is worth the cost. They point to many problems here on Earth, such as growing energy and water needs, pollution problems, and on-going health problems. They say that the money spent on exploring space could be better used for helping resolve problems on Earth.

In addition to new information and understanding, supporters of the space program point to new technology that helps people living on Earth. Satellites now provide valuable information for agriculture, including land use and weather monitoring. Untold numbers of lives have been saved thanks to storm warnings provided by weather satellites. There are many other spin-offs from the space program, including improvements in communication systems.

QUESTIONS TO DISCUSS

1. Have the gain of knowledge and the spin-offs merited the expense?
2. Alternatively, would the funds be better spent on solving other problems?

---

its own. The first vehicle to roam the surface of another planet, a skateboard-sized rover named *Sojourner,* was designed to move about and analyze the chemical makeup of the surface of Mars. It was programmed to move from rock to rock by instructions relayed to it through the lander, then send analysis information back to Earth—again through the lander. This provided data of the geochemistry and petrology of soils and rocks, which in turn provided data about the early environments and conditions that have existed on Mars. After the *Pathfinder* lander completed its primary thirty-day mission and fell silent, it was renamed the Carl Sagan Memorial Station.

The Mars Exploration Rovers, named *Spirit* and *Opportunity,* landed on Mars on January 4 and January 25, 2004, to answer questions about the history of water on Mars. The spacecraft were sent to sites on opposite sides of Mars that appear to have been affected by liquid water in the past. After parachute and airbag landings, the 185 kg (408 lb) rovers charged their solar-powered batteries. They then began driving to different locations to perform on-site scientific investigations over the course of their mission (Figure 15.6).

What did the rovers find? They found that Mars was made of basalt rock (see chapter 17) and groundwater that was dilute sulfuric acid. The acid interacted with the rock, dissolving things out of it, and then evaporated, leaving sulfur-rich salts. The *Spirit* and *Opportunity* rover results confirmed that sufficient amounts of water to alter the rocks have been present in the past (Figure 15.7). The results also confirm the premission interpretations of remote sensing data. This provides evidence that other present and future remote sensing data is accurate.

The presence of past or present life on Mars remains an open question. Scientists already knew there was liquid water in the past, and water and life go together. Beyond that, the Rover mission has really not changed the prospect of finding evidence of past or present life, so the search goes on.

Mars was named for a mythical god of war, so the two satellites that circle Mars were named Deimos and Phobos after the

FIGURE 15.6 Researchers used the rover *Spirit*'s rock abrasion tool to help them study a rock dubbed "Uchben" in the "Colombia Hills" of Mars. The tool ground into the rock, creating a shallow hole 4.5 cm (1.8 in) in diameter in the central upper portion of this image. It also used wire bristles to brush a portion of the surface below and to the right of the hole. *Spirit* used its panoramic camera during the rover's 293rd martian day (October 29, 2004) to take the frames combined into this approximately true-color image. *Source:* NASA/1PL/Cornell.

FIGURE 15.7 A rock dubbed "Palenque" in the "Colombia Hills" of Mars has contrasting textures in upper and lower portions. This view of the rock combines two frames taken by the panoramic camera on NASA's Mars Exploration Rover *Spirit* during the rover's 278th martian day (October 14, 2004). The layers meet each other at an angular unconformity that may mark a change in environmental conditions between the formation of the two portions of the rock. Scientists would have liked the rover to take a closer look, but Palenque is not on a north-tilted slope, which is the type of terrain needed to keep the rover's solar panels tilted toward the winter sun. The exposed portion of the rock is about 100 cm (39 in) long. *Source:* NASA/1PL/Cornell.

two companions of the Roman god. Both satellites are small, irregularly shaped, and highly cratered. Phobos is the larger of the two, about 22 km (about 14 mi) across the longest dimension, and Deimos is about 13 km (about 8 mi) across. Both satellites reflect light poorly and have a much lower density than Mars. They are assumed to be captured asteroids rather than naturally occurring moons.

JUPITER

Jupiter is the largest of all the planets, with a mass equivalent to some 318 Earths and, in fact, is more than twice as massive as all the other planets combined. This massive planet is located an average 5 AU from the Sun in an orbit that takes about twelve Earth years for one complete trip around the Sun. The internal heating from gravitational contraction was tremendous when this giant formed, and today, it still radiates twice the energy that it receives from the Sun. The source of this heat is the slow gravitational compression of the planet, not nuclear reactions as in the Sun. Jupiter would have to be about eighty times as massive to create the internal temperatures needed to start nuclear fusion reactions or, in other words, to become a star itself. Nonetheless, the giant Jupiter and its system of satellites seem almost like a smaller version of a planetary system within the solar system.

Jupiter has an average density that is about a quarter of the density of Earth. This low density indicates that Jupiter is mostly made of light elements, such as hydrogen and helium, but it does contain a percentage of heavier rocky substances. The model of Jupiter's interior (Figure 15.8) is derived from this and other information from spectral studies, studies of spin rates, and measurements of heat flow. The model indicates a solid, rocky core that is more than twice the size of Earth. Surrounding this core is a thick layer of liquid hydrogen, compressed so

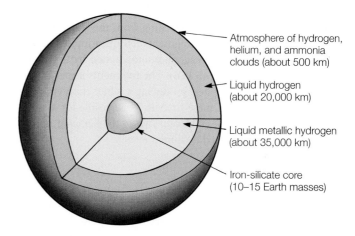

Atmosphere of hydrogen, helium, and ammonia clouds (about 500 km)

Liquid hydrogen (about 20,000 km)

Liquid metallic hydrogen (about 35,000 km)

Iron-silicate core (10–15 Earth masses)

FIGURE 15.8 The structure of Jupiter.

tightly by millions of atmospheres of pressure that it is able to conduct electric currents. Liquid hydrogen with this property is called *metallic hydrogen* because it has the conductive ability of metals. Above the layer of metallic hydrogen is a thick layer of ordinary liquid hydrogen, which is under less pressure. The outer layer, or atmosphere, of Jupiter is a zone with hydrogen, helium, ammonia gas, crystalline compounds, and a mixture of ice and water. It is the uppermost ammonia clouds, perhaps mixed with sulfur and organic compounds, that form the bright orange, white, and yellow bands around the planet. The banding is believed to be produced by atmospheric convection, in which bright, hot gases are forced to the top where they cool, darken in color, and sink back to the surface.

Jupiter's famous Great Red Spot is located near the equator. This permanent, deep, red oval feature was first observed by

Robert Hooke in the 1600s and has generated much speculation over the years. The red oval, some 40,000 km (about 25,000 mi) long has been identified by infrared observations to be a high pressure region, with higher and colder clouds, that has lasted for at least three hundred years. The energy source for such a huge, long-lasting feature is unknown (Figure 15.9).

Jupiter has many satellites, and the four brightest and largest can be seen from Earth with a good pair of binoculars. These four are called the *Galilean moons* because they were discovered by Galileo in 1610. The Galilean moons are named Io, Europa, Ganymede, and Callisto (Figure 15.10). Observations by the *Pioneer* and *Voyager* spacecrafts revealed some fascinating and intriguing information about the moons of Jupiter. Io, for example, was discovered to have active volcanoes that eject enormous plumes of molten sulfur and sulfur dioxide gas. Europa is covered with a 19.3 km (about 12 mi) thick layer of smooth water ice, which has a network of long, straight, dark cracks. Ganymede has valleys, ridges, folded mountains, and other evidence of an active geologic history. Callisto, the most distant of the Galilean moons, was found to be the most heavily cratered object in the solar system.

Impact events such as those that marked Callisto are still occurring. In 1994, the Comet Shoemaker-Levy broke apart into a "string of pearls" (Figure 15.11A) and then produced a once-in-a-lifetime spectacle as it proceeded to leave its imprint on Jupiter as well as on people of Earth watching from the sidelines. The string of twenty-two comet fragments fell onto Jupiter during July 1994, creating a show eagerly photographed by telescopes around the world (Figure 15.11B). The fragments impacted the upper atmosphere of Jupiter, producing visible, energetic fireballs. The aftereffects of these fireballs were visible for about a year. There are chains of craters on two of the Galilean moons that may have been formed by similar events.

SATURN

Saturn is slightly smaller and substantially less massive than Jupiter and has similar features (see Figure 15.8) to those of Jupiter, but it is readily identified by its unique, beautiful system of rings. Saturn's rings consist of thousands of narrow bands of particles. Some rings are composed of particles large enough to be measured in meters, while others are composed of particles that are dust-sized (Figure 15.12). Saturn is about 9.5 AU from the Sun, but its system of rings is easily spotted with a good pair of binoculars. Saturn also has the lowest average density of any of the planets, about 0.7 times the density of water.

The surface of Saturn, like Jupiter's surface, has bright and dark bands that circle the planet parallel to the equator. Saturn also has a smaller version of Jupiter's Great Red Spot, but in general, the bands and spot are not as highly contrasted or brightly colored as they are on Jupiter.

The international *Cassini-Huygens* mission entered orbit around Saturn on July 1, 2004, after a 3.5 billion km (2.2 billion mi), seven-year voyage from Earth. Establishing orbit was the first step of a four-year study of Saturn and its rings and moons. So far, the mission has found that Saturn's largest moon, Titan, has a surface shaped by rock fracturing, winds, and erosion.

A

B

FIGURE 15.9 Photos of Jupiter taken by *Voyager 1*. (*A*) From a distance of about 36 million km (about 22 million mi). (*B*) A closer view, from the Great Red Spot to the South Pole, showing organized cloud patterns. In general, dark features are warmer, and light features are colder. The Great Red Spot soars about 25 km (about 15 mi) above the surrounding clouds and is the coldest place on the planet.

FIGURE 15.10 The four Galilean moons pictured by *Voyager 1*. Clockwise from upper left, Io, Europa, Ganymede, and Callisto. Io and Europa are about the size of Earth's Moon; Ganymede and Callisto are larger than Mercury.

Among the new discoveries is a 1,500 km (930 mi) long river of liquid methane. Titan's atmosphere is rich in organic (meaning carbon-containing) molecules. Most of the clouds on Titan occur over the south pole, and scientists believe this is where a cycle of methane rain, runoff with channel carving, and liquid methane evaporation is most active.

On January 14, 2005, *Cassini* delivered a detachable probe, called *Huygens,* to the moon Titan. The probe dropped by parachute, sampling the chemical composition of the atmosphere for the 2.5 hour descent and for 90 minutes after landing. It landed with more of a "splat" than a thud or splash, indicating that even though the surface temperature was −180°C, the surface was more like a thick organic stew than a solid or liquid.

Cassini has also discovered that Saturn's icy moon Enceladus has a significant atmosphere. Before this discovery, Titan was the only moon in the solar system known to have an atmosphere. Enceladus is too small to hold onto an atmosphere, so it must have a continuous source that may be volcanic, geysers, or gases escaping from the interior.

Cassini also provided a test of Einstein's general theory of relativity (p. 198). According to this theory, a massive object such as the Sun should cause spacetime to curve. This curve means that it should take longer for anything to travel by the Sun since a curved path is a greater distance. The general theory of relativity was tested by analyzing radio waves that traveled by the Sun on their way from *Cassini* to Earth. The researchers found that the radio waves traveling by the Sun were indeed delayed, and the amount of delay agreed precisely with predictions made from Einstein's theory.

The *Cassini-Huygens* mission is a cooperative project of NASA, the European Space Agency, and the Italian Space Agency.

URANUS AND NEPTUNE

Uranus and Neptune are two more giant planets that are far, far away from Earth. Uranus revolves around the Sun at an average distance of over 19 AU, taking about 84 years to circle the Sun once. Neptune is an average 30 AU from the Sun and takes

CHAPTER 15 The Solar System **389**

A

B

FIGURE 15.11 (*A*) This image, made by the Hubble Space Telescope, clearly shows the large impact site made by fragment G of former Comet Shoemaker-Levy 9 when it collided with Jupiter. (*B*) This is a picture of Comet Shoemaker-Levy 9 after it broke into twenty-two pieces, lined up in this row, then proceeded to plummet into Jupiter during July 1994. The picture was made by the Hubble Space Telescope.

about 165 years for one complete orbit. Thus, Uranus is about twice as far away from the Sun as Saturn, and Neptune is three times as far away. To give you an idea of these tremendous distances, consider that the time required for radio signals to travel from Uranus to Earth is more than 2.5 hours! It would be most difficult to carry on a conversation by radio with someone such a distance away. Even farther away, a radio signal from a transmitter near Neptune would require over 4 hours to reach

Earth, which means 8 hours would be required for two people to just say "Hello" to each other!

Uranus and Neptune are more similar to each other than Saturn is to Jupiter (Figure 15.13). Both are the smallest of the giant planets, with a diameter of about 50,000 km (about 30,000 mi), which is about a third the size of Jupiter. Both planets are thought to have similar interior structures (Figure 15.14), which consist of water and water ice surrounding a rocky core and an atmosphere of hydrogen and helium. Because of their great distances from the Sun, both have very low average surface temperatures.

SMALL BODIES OF THE SOLAR SYSTEM

Comets, asteroids, and meteorites are the leftovers from the formation of the Sun and planets. Presently, the total mass of all these leftovers in and around the solar system may account for a significant fraction of the mass of the solar system, perhaps as much as two-thirds of the total mass. It must have been much greater in the past, however, as evidenced by the intense bombardment that took place on the Moon and other planets up to some four billion years ago.

COMETS

A **comet** is known to be a relatively small, solid body of frozen water, carbon dioxide, ammonia, and methane, along with dusty and rocky bits of materials mixed in. Until the 1950s, most astronomers believed that comet bodies were mixtures of sand and gravel. Fred Whipple proposed what became known as the *dirty-snowball cometary model*, which was recently verified when spacecraft probes observed Halley's comet in 1986 (Table 15.4).

Based on calculation of their observed paths, comets are estimated to originate some 30 AU to a light-year or more from the Sun. Here, according to other calculations and estimates, is a region of space containing billions and billions of objects. There is a spherical "cloud" of the objects beyond the orbit of Pluto from about 30,000 AU out to a light-year or more from the Sun, called the **Oort cloud** (Figure 15.15). The icy aggregates of the Oort cloud are understood to be the source of long-period comets, with orbital periods of more than two hundred years.

There is also a disk-shaped region of small icy bodies, which ranges from about 30 to 100 AU from the Sun, called the **Kuiper Belt.** The small icy bodies in the Kuiper Belt are understood to be the source of short-period comets, with orbital periods of less than two hundred years. There are thousands of Kuiper Belt objects that are larger than 100 km in diameter, and six are known to be orbiting between Jupiter and Neptune. Called *Centaurs,* these objects are believed to have escaped the Kuiper Belt. Centaurs might be small, icy bodies similar to Pluto. Indeed, some speculate that Pluto is a large Kuiper Belt object.

The current theory of the origin of comets was developed by the Dutch astronomer Jan Oort in 1950. According to the

FIGURE 15.12 A part of Saturn's system of rings, pictured by *Voyager 2* from a distance of about 3 million km (about 2 million mi). More than sixty bright and dark ringlets are seen here; different colors indicate different surface compositions.

theory, the huge cloud and belt of icy, dusty aggregates are leftovers from the formation of the solar system and have been slowly circling the solar system ever since it formed. Something, perhaps a gravitational nudge from a passing star, moves one of the icy bodies enough that it is pulled toward the Sun in what will become an extremely elongated elliptical orbit. The icy, dusty body forms the only substantial part of a comet, and the body is called the comet *nucleus.*

Observations by the *Vega* and *Giotto* spacecrafts found the nucleus of Halley's comet to be an elongated mass of about 8 by 11 km (about 5 by 7 mi) with an overall density less than one-fourth that of solid water ice. As the comet nucleus moves toward the Sun, it warms from the increasingly intense solar radiation. Somewhere between Jupiter and Mars, the ice and frozen gases begin to vaporize, releasing both grains of dust and evaporated ices. These materials form a large, hazy head around the comet called a *coma.* The coma grows larger with increased vaporization, perhaps several hundred or thousands of km across. The coma reflects sunlight as well as producing its own light, making it visible from Earth. The coma generally appears when a comet is within about 3 AU of the Sun. It reaches its maximum diameter about 1.5 AU from the Sun. The nucleus and coma together are called the *head* of the comet. In addition, a large cloud of invisible hydrogen gas surrounds the head, and this hydrogen *halo* may be hundreds of thousands of km across.

FIGURE 15.13 This is a photo image of Neptune taken by *Voyager*. Neptune has a turbulent atmosphere over a very cold surface of frozen hydrogen and helium.

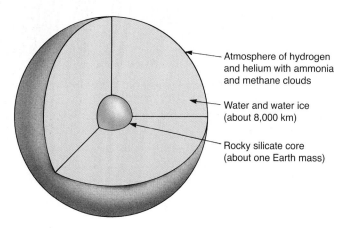

Atmosphere of hydrogen and helium with ammonia and methane clouds

Water and water ice (about 8,000 km)

Rocky silicate core (about one Earth mass)

FIGURE 15.14 The structure of Uranus and Neptune.

TABLE 15.4

Spacecraft missions to study comets and asteroids

Date	Name	Owner	Remark
Sep 11, 1985	*ISSE 3* (or *ICE*)	U.S.	Studies of electric and magnetic fields around Giacobini-Zinner comet from 7,860 km (4,880 mi)
Mar 6, 1986	*Vega 1*	U.S.S.R.	Photos and studies of nucleus of Halley's comet from 8,892 km (5,525 mi)
Mar 8, 1986	*Suisei*	Japan	Studied hydrogen halo of Halley's comet from 151,000 km (93,800 mi)
Mar 9, 1986	*Vega 2*	U.S.S.R.	Photos and studies of nucleus of Halley's comet from 8,034 km (4,992 mi)
Mar 11, 1986	*Sakigake*	Japan	Studied solar wind in front of Halley's comet 7.1 million km (4.4 million mi)
Mar 28, 1986	*Giotto*	ESA	Photos and studies of Halley's comet from 541 km (336 mi)
Mar 28, 1986	*ISSE 3* (or *ICE*)	U.S.	Studies of electric and magnetic fields around Halley's comet from 32 million km (20 million mi)
Feb 17, 1996	*NEAR Shoe-maker*	U.S.	Studied the asteroid Eros
Oct 24, 1998	*Deep Space 1*	U.S.	Flyby of asteroid Braille and comet Borrelly
Feb 7, 1999	*Stardust*	U.S.	Studied comet Wild 2 and returned a sample of cosmic dust to Earth
Mar 2, 2004	*Rosetta*	ESA	Two probes will be launched into comet 67PChruyumov-Gerasimenko on Nov 2014
Jan 12, 2005	*Deep Impact*	U.S.	Studied comet Tempel 1 by sending impact probe on Jul 4, 2005
Sep 27, 2007	*Dawn*	U.S.	Studied the two most massive asteroids Ceres and Vesta

As the comet nears the Sun, the solar wind and solar radiation ionize gases and push particles from the coma, pushing both into the familiar visible *tail* of the comet. Comets may have two types of tails: (1) ionized gases and (2) dust. The dust is pushed from the coma by the pressure from sunlight. It is visible because of reflected sunlight. The ionized gases are pushed into the tail by magnetic fields carried by the solar wind. The ionized gases of the tail are fluorescent, emitting visible light because they are excited by ultraviolet radiation from the Sun. The tail generally points away from the Sun, so it follows the comet as it approaches the Sun but leads the comet as it moves away from the Sun (Figure 15.16).

Comets are not very massive or solid, and the porous, snowlike mass has a composition more similar to the giant planets than to the terrestrial planets in comparison. Each time a comet passes near the Sun, it loses some of its mass through evaporation of gases and loss of dust to the solar wind. After passing the Sun, the surface forms a thin, fragile crust covered with carbon and other dust particles. Each pass by the Sun means a loss of matter, and the coma and tail are dimmer with each succeeding pass. About 20 percent of the approximately six hundred comets that are known have orbits

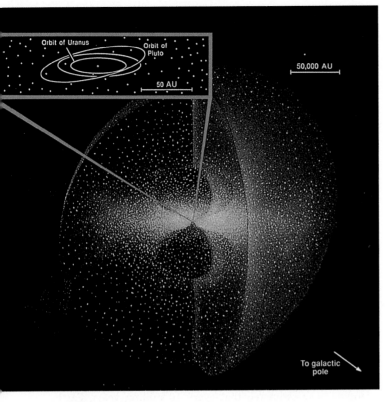

FIGURE 15.15 Although there may be as many as 1 trillion comets in the Oort cloud, the volume of space that the Oort cloud occupies is so immense that the comets are separated from one another by distances that are typically about 10 AU.

that return them to the Sun within a two-hundred-year period, some of which return as often as every five or ten years. The other 80 percent have long elliptical orbits that return them at intervals exceeding two hundred years. The famous Halley's comet has a smaller elliptical orbit and returns about every seventy-six years. Halley's comet, like all other comets, may eventually break up into a trail of gas and dust particles that orbit the Sun.

The *Stardust* spacecraft had a mission to fly to a comet named P/Wild 2 and collect samples of dust and volatiles from the coma of the comet. It returned these samples to Earth, along with interstellar dust samples collected on the way to the comet. These samples represent primitive substances from the early formation of the solar system and continue undergoing detailed analysis.

ASTEROIDS

Between the orbits of Mars and Jupiter is a belt, or circular region, of thousands of small, rocky bodies called **asteroids** (Figure 15.17). This belt contains thousands of asteroids that range in size from 1 km or less up to the largest asteroid, named Ceres, which has a diameter of about 1,000 km (over 600 mi). The asteroids are thinly distributed in the belt, 1 million km or so apart (about 600,000 mi), but there is evidence of collisions occurring in the past. Most asteroids larger than 50 km (about 30 mi) have

FIGURE 15.16 As a comet nears the Sun, it grows brighter, with the tail always pointing away from the Sun.

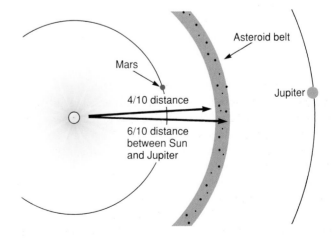

FIGURE 15.17 Most of the asteroids in the asteroid belt are about halfway between the Sun and Jupiter.

been studied by analyzing the sunlight reflected from their surfaces. These spectra provide information about the composition of the asteroids. Asteroids on the inside of the belt, toward the Sun, are made of stony materials, and those on the outside of the belt are dark with carbon minerals. Still other asteroids are metallic, containing iron and nickel. These spectral composition studies, analyses of the orbits of asteroids, and studies of meteorites that have fallen to Earth all indicate that the asteroids are not the remains of a planet or planets that were broken up. The asteroids are now believed to have formed some 4.6 billion years ago from the original solar nebula. During their formation, or shortly thereafter, their interiors were partly melted, perhaps from the heat of short-lived radioactive decay reactions. Their location close to Jupiter, with its gigantic gravitational field, prevented the slow gravitational clumping together process that would have formed a planet.

Jupiter's gigantic gravitational field also captured some of the asteroids, pulling them into its orbit. Today, there are two groups of asteroids, called the *Trojan asteroids,* that lead and follow Jupiter in its orbit. They lead and follow at a distance where the gravitational forces of Jupiter and the Sun balance to keep them in the orbit. A third group of asteroids, called the *Apollo asteroids,* has orbits that cross the orbit of Earth.

METEORS AND METEORITES

Comets leave trails of dust and rock particles after encountering the heat of the Sun, and collisions between asteroids in the past have ejected fragments of rock particles into space. In space, the remnants of comets and asteroids are called **meteoroids.** When a meteoroid encounters Earth moving through space, it accelerates toward the surface with a speed that depends on its direction of travel and the relative direction that Earth is moving. It soon begins to heat from air compression in the upper atmosphere, melting into a visible trail of light and smoke. The streak of light and smoke in the sky is called a **meteor.** The "falling star" or "shooting star" is a meteor. Most meteors burn up or evaporate completely within seconds after reaching an altitude of about 100 km (about 60 mi) because they are nothing more than specks of dust. A **meteor shower** occurs when Earth passes through a stream of particles left by a comet in its orbit. Earth might meet the stream of particles concentrated in such an orbit on a regular basis as it travels around the Sun, resulting in predictable meteor showers (Table 15.5). In the third week of October, for example, Earth crosses the orbital path of Halley's comet, resulting in a shower of some ten to fifteen meteors per hour. Meteor showers are named for the constellation in which they appear to originate. The October meteor shower resulting from an encounter with the orbit of Halley's comet, for example, is called the Orionid shower because it appears to come from the constellation Orion.

Did you know that atom-bomb-sized meteoroid explosions often occur high in Earth's atmosphere? Most smaller meteors melt into the familiar trail of light and smoke. Larger meteors may fragment upon entering the atmosphere, and

TABLE 15.5

Some annual meteor showers

Name	Date of Maximum	Hour Rate
Quadrantid	January 3	30
Aquarid	May 4	5
Perseid	August 12	40
Orionid	October 22	15
Taurids	November 1, 16	5
Leonid	November 17	5
Geminid	December 12	55

the smaller fragments will melt into multiple light trails. Still larger meteors may actually explode at altitudes of about 32 km (about 20 mi) or so. Military satellites that watch Earth for signs of rockets blasting off or nuclear explosions record an average of eight meteor explosions a year. These are big explosions, with an energy equivalent estimated to be similar to a small nuclear bomb. Actual explosions, however, may be ten times larger than the estimation. Based on statistical data, scientists have estimated that every 10 million years, Earth should be hit by a very, very large meteor. The catastrophic explosion and aftermath would devastate life over much of the planet, much like the theoretical dinosaur-killing impact of 65 million years ago.

If a meteoroid survives its fiery trip through the atmosphere to strike the surface of Earth, it is called a **meteorite.** Most meteors are from fragments of comets, but most meteorites generally come from particles that resulted from collisions between asteroids that occurred long ago. Meteorites are classified into three basic groups according to their composition: (1) **iron meteorites,** (2) **stony meteorites,** and (3) **stony-iron meteorites** (Figure 15.18). The most common meteorites are stony, composed of the same minerals that make up rocks on Earth. The stony meteorites are further subdivided into two groups according to their structure, the **chondrites** and the **achondrites.** Chondrites have a structure of small, spherical lumps of silicate minerals or glass, called **chondrules,** held together by a fine-grained cement. The achondrites do not have the chondrules, as their name implies, but have a homogeneous texture more like volcanic rocks such as basalt that cooled from molten rock.

The iron meteorites are about half as abundant as the stony meteorites. They consist of variable amounts of iron and nickel, with traces of other elements. In general, there is proportionally much more nickel than is found in the rocks of Earth. When cut, polished, and etched, beautiful crystal patterns are observed on the surface of the iron meteorite. The patterns mean that the iron was originally molten, then cooled very slowly over millions of years as the crystal patterns formed.

A meteorite is not, as is commonly believed, a ball of fire that burns up the landscape where it lands. The iron or

A   B

FIGURE 15.18 (*A*) A stony meteorite. The smooth, black surface was melted by friction with the atmosphere. (*B*) An iron meteorite that has been cut, polished, and etched with acid. The pattern indicates that the original material cooled from a molten material over millions of years.

rock has been in the deep freeze of space for some time, and it travels rapidly through Earth's atmosphere. The outer layers become hot enough to melt, but there is insufficient time for this heat to be conducted to the inside. Thus, a newly fallen iron meteorite will be hot since metals are good heat conductors, but it will not be hot enough to start a fire. A stone meteorite is a poor conductor of heat, so it will be merely warm.

ORIGIN OF THE SOLAR SYSTEM

Any model of how the solar system originated presents a problem in testing or verification. This problem is that the solar system originated a long time ago, some 5 billion years ago according to a number of different independent sources of evidence, and that there are no other planetary systems either in existence or in the process of being formed that can be directly observed. From the distance at which they occur, even the Hubble Space Telescope would not be able to directly observe planets around their suns. Astronomers have identified about one hundred extra-solar planets. They identify the presence of planets by measuring the very slight wobble of a central star and then using the magnitude of this motion to determine the presence of orbiting planets, the size and shape of their orbits, and their mass. The technique works only for larger planets and cannot detect those much smaller than about half the mass of Saturn. The technique does not provide a visual image of the planets but only measures the gravitational effect of the planets on the star.

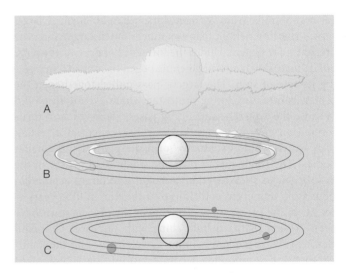

FIGURE 15.19 Formation of the solar system according to the protoplanet nebular model, not drawn to scale. (*A*) The process starts with a nebula of gas, dust, and chemical elements from previously existing stars. (*B*) The nebula is pulled together by gravity, collapsing into the protosun and protoplanets. (*C*) As the planets form, they revolve around the Sun in orbits.

The most widely accepted theory of the origin of the solar system is called the **protoplanet nebular model.** A *protoplanet* is the earliest stage in the formation of a planet. The model can be considered in stages, which are not really a part of the model but are simply a convenient way to organize the total picture (Figure 15.19).

STAGE A

The first important event in the formation of our solar system involves stars that disappeared billions of years ago, long before the Sun was born. Earth, the other planets, and all the members of the solar system are composed of elements that were manufactured by these former stars. In a sequence of nuclear reactions, hydrogen fusion in the core of large stars results in the formation of the elements up to iron. Elements heavier than iron are formed in rare supernova explosions of dying massive stars. Thus, *stage A* of the formation of the solar system consisted of the formation of elements heavier than hydrogen in many, many previously existing stars, including the supernovas of more massive stars. Many stars had to live out their life cycles to provide the raw materials of the solar system. The death of each star, including supernovas, added newly formed elements to the accumulating gases and dust in interstellar space. Over a long period of time, these elements began to concentrate in one region of space as dust, gases, and chemical compounds, but hydrogen was still the most abundant element in the nebula that was destined to become the solar system.

STAGE B

During *stage B,* the hydrogen gas, dust, elements, and chemical compounds from former stars began to form a large, slowly rotating nebula that was much, much larger than the present solar system. Under the influence of gravity, the large but diffuse, slowly rotating nebula began to contract, increasing its rate of spin. The largest mass pulled together in the center, contracting to the protostar, which eventually would become the Sun. The remaining gases, elements, and dusts formed an enormous, fat, bulging disk called an *accretion disk,* which would eventually form the planets and smaller bodies. The fragments of dust and other solid matter in the disk began to stick together in larger and larger accumulations from numerous collisions over the first million years or so. All of the present-day elements of the planets must have been present in the nebula along with the most abundant elements hydrogen and helium. The elements and the familiar chemical compounds accumulated into basketball-sized or larger chunks of matter.

Did the planets have an icy slush beginning? Over a period of time, perhaps 100 million years or so, huge accumulations of frozen water, frozen ammonia, and frozen crystals of methane began to collect, together with silicon, aluminum, and iron oxide plus other metals in the form of rock and mineral grains. Such a slushy mixture would no doubt have been surrounded by an atmosphere of hydrogen, helium, and other vapors thinly interspersed with smaller rocky grains of dust. Evidence for this icy beginning is found today in the Oort cloud and Kuiper Belt. Local concentrations of certain minerals might have occurred throughout the whole accretion disk, with a greater concentration of iron, for example, in the disk where the protoplanet Mars was forming compared to where the protoplanet Earth was forming. Evidence

for this part of the model is found in Mars today, with its greater abundance of iron. It is the abundant iron oxides that make Mars the red planet.

All of the protoplanets might have started out somewhat similarly as huge accumulations of a slushy mixture with an atmosphere of hydrogen and helium gases. Gravitational attraction must have compressed the protoplanets as well as the protosun. During this period of contraction and heating, gravitational adjustments continued, and about a fifth of the disk nearest to the protosun must have been pulled into the central body of the protosun, leaving a larger accumulation of matter in the outer part of the accretion disk.

STAGE C

During *stage C,* the warming protosun became established as a star, perhaps undergoing an initial flare-up that has been observed today in other newly forming stars. Such a flare-up might have been of such a magnitude that it blasted away the hydrogen and helium atmospheres of the interior planets (Mercury, Venus, Earth, and Mars) out past Mars, but it did not reach far enough out to disturb the hydrogen and helium atmospheres of the outer planets. The innermost of the outer planets, Jupiter and Saturn, might have acquired some of the matter blasted away from the inner planets, becoming the giants of the solar system by comparison. This is just speculation, however, and the two giants may have simply formed from greater concentrations of matter in that part of the accretion disk.

The evidence, such as separation of heavy and light mineral matter, shows that the protoplanets underwent heating early in their formation. Much of the heating may have been provided by gravitational contraction, the same process that gave the protosun sufficient heat to begin its internal nuclear fusion reactions. Heat was also provided from radioactive decay processes inside the protoplanets, and the initial greater heating from the Sun may have played a role in the protoplanet heating process. Larger bodies were able to retain this heat better than smaller ones, which radiated it to space more readily. Thus, the larger bodies underwent a more thorough heating and melting, perhaps becoming completely molten early in their history. In the larger bodies, the heavier elements, such as iron, were pulled to the center of the now-molten mass, leaving the lighter elements near the surface. The overall heating and cooling process took millions of years as the planets and smaller bodies were formed. Gases from the hot interiors formed secondary atmospheres of water vapor, carbon dioxide, and nitrogen on the larger interior planets.

Interestingly, the belt of asteroids was discovered from a prediction made by the German astronomer Bode at the end of the eighteenth century. Bode had noticed a pattern of regularity in the spacing of the planets that were known at the time. He found that by expressing the distances of the planets from the Sun in astronomical units, these distances could be approximated by the relationship $(n + 4)/10$, where n is a number in the sequence 0, 3, 6, 12, and so on where each number (except the first) is doubled in succession. When these calculations were done, the distances turned out to be very close to the distances of all the planets known at that time, but

TABLE 15.6

Distances from the Sun to planets known in the 1790s

Planet	n	Distance Predicted by (n + 4)/10 (AU)	Actual Distance (AU)
Mercury	0	0.4	0.39
Venus	3	0.7	0.72
Earth	6	1.0	1.0
Mars	12	1.6	1.5
(Asteroid belt)	24	2.8	—
Jupiter	48	5.2	5.2
Saturn	96	10.0	9.5
Uranus	192	19.6	19.2

the numbers also predicted a planet between Mars and Jupiter where there was none. Later, a belt of asteroids was found where the Bode numbers predicted there should be a planet. This suggested to some people that a planet had existed between Mars and Jupiter in the past and somehow this planet was broken into pieces, perhaps by a collision with another large body (Table 15.6).

Myths, Mistakes, & Misunderstandings

UFOs and You

UFOs (unidentified flying objects) are observed around the world. UFOs are generally sighted near small towns and out in the country, often near a military installation. Statistically, most sightings occur during the month of July at about 9 P.M., then at 3 A.M. UFOs can be grouped into three categories:

1. **Natural Phenomena.** Most sightings of UFOs can be explained as natural phenomena. For example, a bright light that flashes red and green on the horizon could be a star viewed through atmospheric refraction. The vast majority of all UFO sightings are natural phenomena such as atmospheric refraction, ball lighting, swamp gas, or something simple such as a drifting weather balloon or military flares on parachutes, high in the atmosphere and drifting across the sky. This category of UFOs should also include exaggeration and fraud—such as balloons released with burning candles as a student prank.
2. **Aliens.** The idea that UFOs are alien spacecraft from other planets is very popular. No authentic, unambiguous evidence exists, however, that would prove the existence of aliens. In fact, most unidentified flying objects are not objects at all. They are lights that can eventually be identified.
3. **Psychological Factors.** This group of sightings includes misperceptions of natural phenomena resulting from unusual conditions that influence a person's perception and psychological health. This includes people who claim to receive information from aliens by "channeling" messages.

Divide your group into three subgroups, with each subgroup selecting one of the three categories above. After preparing for a few minutes, have each group present reasons why we should understand UFOs to be natural phenomena, aliens from other planets, or psychological misperceptions. Then have the entire group discuss the three categories and try to come to a consensus.

Such patterns of apparent regularity in the spacing of planetary orbits were of great interest because if a true pattern existed, it could hold meaning about the mechanism that determined the location of planets at various distances from the Sun. Many attempts have been made to explain the mechanism of planetary spacing and why a belt of asteroids exists where the Bode numbers predict there should be a planet. The most successful explanations concern Jupiter and the influence of its gigantic gravitational field on the formation of clumps of matter at certain distances from the Sun. In other words, a planet does not exist today between Mars and Jupiter because there never was a planet there. The gravitational influence of Jupiter prevented the clumps of matter from joining together to form a planet, and a belt of asteroids formed instead.

IDEAS ABOUT THE SOLAR SYSTEM

If you observe the daily motion of the Sun and Moon and the nightly motion of the Moon and stars over a period of time, you can easily convince yourself that all the heavenly bodies revolve around a fixed, motionless Earth. The Sun, the Moon, and the planets do appear to move from east to west across the sky, and the stars seem to be fixed on a turning sphere, maintaining the same positions relative to one another as they move as units on the turning sphere. To consider that Earth moves rather than the Sun, the Moon, and the planets seems contrary to all these observations.

There is a problem with the observed motion of the planets, however, one troublesome observation that spoils this model of a motionless Earth with everything moving around it. When the planets are observed over a period of a year or more, they do not hold the same relative positions as would be expected of this model. The stars and the planets both rise in the east and set in the west, but the planets do not maintain the same positions relative to the background of stars. Over time, the planets appear to move across the background of stars, sometimes slowing, then reversing their direction in a loop before resuming their normal motion. This *retrograde*, or reverse, motion for the planet Mars is shown in Figure 15.20.

THE GEOCENTRIC MODEL

Early Greek astronomers and philosophers had attempted to explain the observed motions of the Sun, Moon, and stars with a geometric model, a model of perfect geometrical spheres with attached celestial bodies rotating around a fixed Earth in perfect circles. To explain the occasional retrograde motion of the planets with this model, later Greek astronomers had to

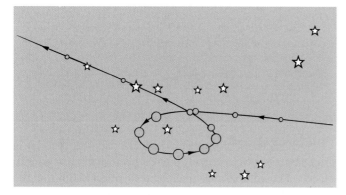

FIGURE 15.20 The apparent position of Mars against the background of stars as it goes through retrograde motion. Each position is observed approximately two weeks after the previous position.

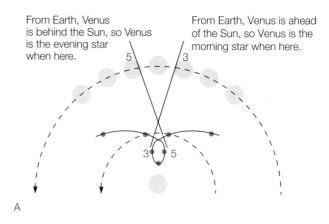

A

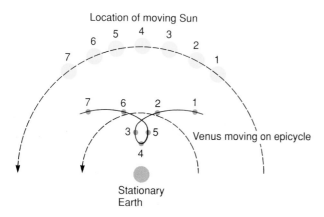

FIGURE 15.21 The paths of Venus and the Sun at equal time intervals according to the Ptolemaic system. The combination of epicycle and Sun movement explains retrograde motion with a stationary Earth.

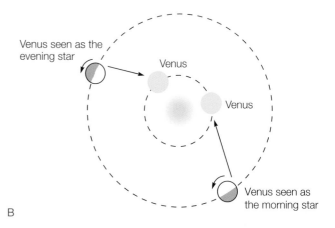

B

FIGURE 15.22 (*A*) The Ptolemaic system explanation of Venus as a morning star and evening star. (*B*) The heliocentric system explanation of Venus as a morning star and evening star.

modify it by assuming a secondary motion of the planets. The modification required each planet to move in a secondary circular orbit as it moved along with the turning sphere. The small circular orbit, or *epicycle*, was centered on the surface of the sphere as it turned around Earth. The planet was understood to move around the epicycle once as the sphere turned around Earth. The combined motion of the movement of the planet around the epicycle as the sphere turned resulted in a loop with retrograde motion (Figure 15.21). Thus, the earlier model of the solar system, modified with epicycles, was able to explain all that was known about the movement of the stars and planets at that time.

A version of the explanation of retrograde motion—perfectly circular epicycle motion on perfectly spherical turning spheres—was published by Ptolemy in the second century A.D. and came to be known as the **Ptolemaic system.** This system did account for the facts that were known about the solar system at that time. Not only did the system describe retrograde motion with complex paths, as shown in Figure 15.21, but it

also explained other observations such as why Venus is sometimes seen for a short time near the Sun at sunrise (the morning "star") and other times seen near the Sun for a short time at sunset (the evening "star") (Figure 15.22).

Over the years, inconsistencies of observed and predicted positions of the planets were discovered, and epicycles were added to epicycles in an attempt to fit the system with observations. The system became increasingly complicated, but it did seem to agree with other ideas of that time. Those ideas were that (1) humans were at the center of a universe that was created for them, and (2) heaven was a perfect place, so it would naturally be a place of perfectly circular epicycles moving on perfect spheres, which were in turn moving in perfect circles. Thus, the Ptolemaic system of a geocentric, or Earth-centered, solar system came to be supported by the church, and this was the accepted understanding of the solar system for the next fourteen centuries.

THE HELIOCENTRIC MODEL

The idea that Earth revolves around the Sun rather than the Sun moving around Earth was proposed by a Polish astronomer, Nicolas Copernicus, in a book published in 1543. In his book,

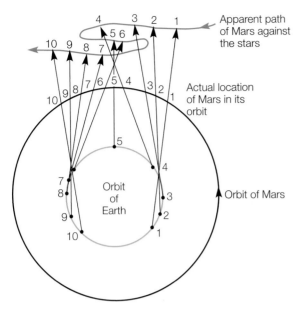

FIGURE 15.23 The heliocentric system explanation of retrograde motion.

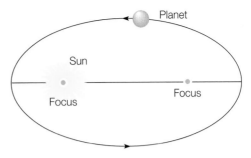

FIGURE 15.24 Kepler's first law describes the shape of a planetary orbit as an ellipse, which is exaggerated in this figure. The Sun is located at one focus of the ellipse.

Copernicus pointed out that the observed motions of the planets could be explained by a model of Earth and the other planets revolving around the Sun as well as by the Ptolemaic system. In Copernicus' model, each planet moved around the Sun in perfect circles at different distances, moving at faster speeds in orbits closer to the Sun. When viewed from a moving Earth, the other planets would appear to undergo retrograde motion because of the combined motions of Earth and the planets. Earth, for example, moves along its inner orbit with about twice the angular speed as Mars in its outer orbit, which is about one and one-half times farther from the Sun. Since Earth is moving faster in the inside orbit, it will move even with, then pass, Mars in the outer orbit (Figure 15.23). As this happens, Mars will appear to slow to a stop, move backward, stop again, then move on. This combined motion is similar to what you observe when you pass a slower-moving car, which appears to move backward against the background of the landscape as you pass it. In an outer circular path, the car would appear to slow, move backward, then move forward again as you pass it.

The **Copernican system** of a heliocentric, or Sun-centered, solar system provided a simpler explanation for retrograde motion than the Ptolemaic system, but it was only an alternative way to consider the solar system. The Copernican system offered no compelling reasons why the alternative Ptolemaic system should be rejected. Furthermore, Copernicus had retained the old Greek idea of planets moving in perfect circles, so there were inconsistencies in predicted and observed motions with this model, too. Clear-cut evidence for rejecting the Ptolemaic system would have to await the detailed measurements of planetary motions made by Tycho Brahe and analysis of those measurements by Johannes Kepler.

Tycho Brahe was a Danish nobleman who constructed highly accurate observatories for his time, which was before the telescope. From his observatory on a little island about 32 km (20 mi) from Copenhagen, Brahe spent about twenty years (1576–1597) making systematic, uninterrupted measurements of the Sun, Moon, planets, and stars. His skilled observations resulted in the first precise, continuous record of planetary position. In 1600, Brahe hired a young German, Johannes Kepler, as an assistant. When Brahe died in 1601, Kepler was promoted to Brahe's position and was given access to the vast collection of observation records. Kepler spent the next twenty-five years analyzing the data to find if planets followed circular paths or if they followed the paths of epicycles. Using the careful observations of Tycho Brahe, Kepler found that the planets did not move in epicycles nor did they move in perfect circles. Planets move in the path of an ellipse of certain dimensions. He published his findings in 1609 and 1619, establishing the actual paths of planetary movement. Kepler had found the first evidence that the Ptolemaic system of complicated epicycles was unnecessary and unacceptable as a model of the solar system. His findings also required adjustments of the heliocentric Copernican system, which are described by his three laws of planetary motion. Today, his findings are called **Kepler's laws of planetary motion.**

Kepler's first law states that each planet moves in an orbit that has the shape of an ellipse, with the Sun located at one focus (Figure 15.24). **Kepler's second law** states that an imaginary line between the Sun and a planet moves over equal areas of the ellipse during equal time intervals (Figure 15.25). This means that the orbital velocity of a planet varies with where the planet is in the orbit, since the distance from the focus to a given position varies around the ellipse. The point at which an orbit comes closest to the Sun is called the *perihelion*, and the point at which an orbit is farthest from the Sun is called the *aphelion*. The shortest line from a planet to the Sun at perihelion means that the planet moves most rapidly when here. The short line and rapidly moving planet would sweep out a certain area in a certain time period, for example, one day. The longest line from a planet to the Sun at aphelion means that

People Behind the Science

Percival Lowell (1855–1916)

Percival Lowell was an American astronomer, mathematician, and the founder of an important observatory in the United States whose main field of research was the planets of the solar system. Responsible for the popularization in his time of the theory of intelligent life on Mars, he also predicted the existence of a body beyond Neptune that was later discovered and named Pluto.

Lowell was born in Boston, Massachusetts, on March 13, 1855. His interest in astronomy began to develop during his early school years. In 1876, he graduated from Harvard University, where he had concentrated on mathematics, and then traveled for a year before entering his father's cotton business. Six years later, Lowell left the business and went to Japan. He spent most of the next ten years traveling around the Far East, partly for pleasure, partly to serve business interests, but also holding a number of minor diplomatic posts.

Lowell returned to the United States in 1893 and soon afterwards decided to concentrate on astronomy. He set up an observatory at Flagstaff, Arizona, at an altitude more than 2,000 m (6,564 ft) above sea level, on a site chosen for the clarity of its air and its favorable atmospheric conditions. He first used borrowed telescopes of 12- and 18-in (30 and 45 cm) diameters to study Mars, which at that time was in a particularly suitable position. In 1896, he acquired a larger telescope and studied Mars by night and Mercury and Venus during the day. Overwork led to a deterioration in Lowell's health, and from 1897 to 1901, he could do little research, although he was able to participate in an expedition to Tripoli in 1900 to study a solar eclipse.

He was made nonresident professor of astronomy at the Massachusetts Institute of Technology in 1902 and gave several lecture series in that capacity. He led an expedition to the Chilean Andes in 1907 that produced the first high-quality photographs of Mars. The author of many books and the holder of several honorary degrees, Lowell died in Flagstaff on November 12, 1916.

The planet Mars was a source of fascination for Lowell. Influenced strongly by the work of Giovanni Schiaparelli (1835–1910)—and possibly misled by the current English translation of "canals" for the Italian *canali* ("channels")—Lowell set up his observatory at Flagstaff originally with the sole intention of confirming the presence of advanced life forms on the planet. Thirteen years later, the expedition to South America was devoted to the study and photography of Mars. Lowell "observed" a complex and regular network of canals and believed that he detected regular seasonal variations that strongly indicated agricultural activity. He found darker waves that seemed to flow from the poles to the equator and suggested that the polar caps were made of frozen water. (The waves were later attributed to dust storms, and the polar caps are now known to consist not of ice but mainly of frozen carbon dioxide. Lowell's canal system also seems to have arisen mostly out of wishful thinking; part of the system does indeed exist, but it is not artificial and is apparent only because of the chance apposition of dark patches on the Martian surface.)

Lowell also made observations at Flagstaff of all the other planets of the solar system. He studied Saturn's rings, Jupiter's atmosphere, and Uranus's rotation period. Finding that the perturbations in the orbit of Uranus were not fully accounted for by the presence of Neptune, Lowell predicted the position and brightness of a planet that he called Planet X but was unable to discover. (Nearly fourteen years after Lowell's death, Clyde Tombaugh found the body—Pluto—on March 12, 1930; the discovery was made at Lowell's observatory and announced on the seventy-fifth anniversary of Lowell's birth.)

Lowell is remembered as a scientist of great patience and originality. He contributed to the advancement of astronomy through his observations and his establishment of a fine research center, and he did much to bring the excitement of the subject to the general public.

Source: From the Hutchinson *Dictionary of Scientific Biography.* © Research Machines plc 2003. All Rights Reserved. Helicon Publishing is a division of Research Machines.

the planet moves most slowly at aphelion. The long line and slowly moving planet would sweep out the same area in one day as was swept out at perihelion. Earth travels fastest in its orbit at perihelion on about January 3 and slowest at aphelion on about July 1.

Kepler's third law states that the square of the period of a planet's orbit is proportional to the cube of that planet's semimajor axis, or $t^2 \propto d^3$. When the time is expressed in Earth units of one year for a revolution and a radius of astronomical unit, the distance to a planet can be determined by observing the period of revolution and comparing the orbit of the planet with that of Earth. For example, suppose a planet is observed to require eight Earth years to complete one orbit. Then,

$$\frac{t\,(\text{planet})^2}{t\,(\text{Earth})^2} = \frac{d\,(\text{planet})^3}{d\,(\text{Earth})^3}$$

$$\frac{(8)^2}{1} = \frac{(\text{distance})^3}{1}$$

$$64 = (\text{distance})^3$$

$$\text{distance} = \sqrt[3]{64}\ \text{AU}$$

$$\text{distance} = 4\ \text{AU}$$

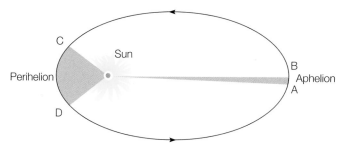

FIGURE 15.25 Kepler's second law. A line from the Sun to a planet at point A sweeps over a certain area as the planet moves to point B in a given time interval. A line from the Sun to a planet at point C will sweep over the same area as the planet moves to point D during the same time interval. The time required to move from point A to point B is the same as the time required to move from point C to point D, so the planet moves faster in its orbit at perihelion.

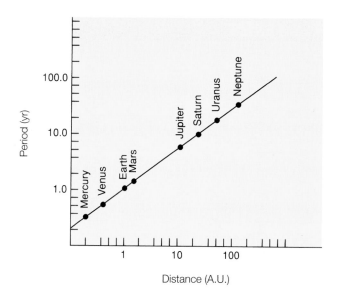

Distance (A.U.)

FIGURE 15.26 Kepler's third law describes a relationship between the time required for a planet to move around the Sun and its average distance from the Sun. The relationship is that the time squared is proportional to the distance cubed.

Thus, a planet that takes eight times as long to complete an orbit is four times as far from the Sun as Earth. In general, Kepler's third law means that the more distant a planet is from the Sun, the longer the time required to complete one orbit. Figure 15.26 shows this relationship for the planets of the solar system. Kepler's third law applies to moons, satellites, and comets in addition to the planets. In the case of moons and satellites, the distance is to the planet the moon or satellite is orbiting, not to the Sun.

Kepler's laws were empirically derived from the data collected by Tycho Brahe, and the reason planets followed these relationships would not be known or understood until Isaac Newton published the law of gravitation some sixty years later (see chapter 2). In the meantime, Galileo constructed his telescope and added observational support to the heliocentric theory. By the time Newton derived the law of gravitation and then improved the accuracy of Kepler's third law, the solar system was understood to consist of planets that move around the Sun in elliptical orbits, paths that could be predicted by applying the law of gravitation. The heliocentric model of the solar system had evolved to a conceptual model that both explained and predicted what was observed.

SUMMARY

The planets can be classified into two major groups: (1) the *terrestrial planets* of Mercury, Venus, Mars, and Earth, and (2) the *giant planets* of Jupiter, Saturn, Uranus, and Neptune. Table 15.1 summarizes the characteristics of the various planets.

Comets are porous aggregates of water ice, frozen methane, frozen ammonia, dry ice, and dust. The solar system is surrounded by the *Kuiper Belt* and the *Oort cloud* of these objects. Something nudges one of the icy bodies, and it falls into a long, elliptical orbit around the Sun. As it approaches the Sun, increased radiation evaporates ices and pushes ions and dust into a long, visible tail. *Asteroids* are rocky or metallic bodies that are mostly located in a belt between Mars and Jupiter. The remnants of comets, fragments of asteroids, and dust are called *meteoroids*. A meteoroid that falls through Earth's atmosphere and melts to a visible trail of light and smoke is called a *meteor*. A meteoroid that survives the trip through the atmosphere to strike the surface of Earth is called a *meteorite*. Most meteors are fragments and pieces of dust from comets. Most meteorites are fragments that resulted from collisions between asteroids.

The *protoplanet nebular model* is the most widely accepted theory of the origin of the solar system, and this theory can be considered as a series of events, or stages. *Stage A* is the creation of all the elements heavier than hydrogen in previously existing stars. *Stage B* is the formation of a nebula from the raw materials created in stage A. The nebula contracts from gravitational attraction, forming the *protosun* in the center with a fat, bulging *accretion disk* around it. The Sun will form from the protosun, and the planets will form in the accretion disk. *Stage C* begins as the protosun becomes established as a star. The icy remains of the original nebula are the birthplace of *comets*. Asteroids are other remains that underwent some melting.

KEY TERMS

achondrites (p. **394**)

asteroids (p. **393**)

astronomical unit (p. **380**)

chondrites (p. **394**)

chondrules (p. **394**)

comet (p. **390**)

Copernican system (p. **399**)

dwarf planet (p. **380**)

giant planets (p. **380**)

iron meteorites (p. **394**)

Kepler's first law (p. **399**)

Kepler's laws of planetery motion (p. **399**)

Kepler's second law (p. **399**)

Kepler's third law (p. **400**)

Kuiper Belt (p. **390**)

meteor (p. **394**)

meteorite (p. **394**)

meteoroids (p. **394**)

meteor shower (p. **394**)

Oort cloud (p. **390**)

planet (p. **380**)

protoplanet nebular model (p. **395**)

Ptolemaic system (p. **398**)

small solar system bodies (p. **380**)

stony-iron meteorites (p. **394**)

stony meteorites (p. **394**)

terrestrial planets (p. **380**)

APPLYING THE CONCEPTS

1. The mass of the Sun is how much larger than all of the other planets, moons, asteroids, and other bodies in the solar system?
   a. one hundred times larger
   b. two hundred times larger
   c. five hundred times larger
   d. seven hundred times larger

2. The distance from Earth to the Sun is called a (an)
   a. light-year.
   b. solar year.
   c. astronomical unit.
   d. astronomical year.

3. What type of planets are Mercury, Venus, Earth, and Mars?
   a. early planets
   b. terrestrial planets
   c. small planets
   d. hot planets

4. Which of the following is most likely found on Jupiter?
   a. hydrogen
   b. argon
   c. nickel
   d. carbon dioxide

5. What is the outermost planet?
   a. Mercury
   b. Mars
   c. Neptune
   d. Jupiter

6. The planet that was named after the mythical Roman messenger of speed is
   a. Jupiter.
   b. Mercury.
   c. Saturn.
   d. Mars.

7. A day on which planet is longer than a year on that planet?
   a. Mercury
   b. Venus
   c. Neptune
   d. Jupiter

8. The day on which planet is about the same time period as a day on Earth?
   a. Mercury
   b. Venus
   c. Mars
   d. Jupiter

9. Mars has distinct surface feature-related regions. Which of the following is not such a region?
   a. volcanic
   b. systems of canyons
   c. plateaus near the poles
   d. ocean floors

10. How many moons orbit Mars?
    a. one
    b. two
    c. four
    d. none

11. What is the largest planet in our solar system?
    a. Mars
    b. Jupiter
    c. Saturn
    d. Earth

12. Callisto, Europa, Gannymede, and Io are
    a. Galilean moons.
    b. volcanoes on Jupiter.
    c. red spots.
    d. comets that struck Jupiter.

13. The density of Jupiter is
    a. 50 percent greater than that of Earth.
    b. 50 percent that of the density of Earth.
    c. 25 percent that of the density of Earth.
    d. five hundred times the density of Earth.

14. The only moon in the solar system with a substantial atmosphere is
    a. Io.
    b. Moon.
    c. Phobos.
    d. Titan.

15. Saturn's rings are thought to be
    a. composed of thousands of particles of various sizes.
    b. remains of a collision with Jupiter.
    c. preventing the atmosphere on Saturn from escaping.
    d. made up of captured asteroids.

16. The planet with the lowest average density, which is less than that of liquid water, is
    a. Uranus.
    b. Neptune.
    c. Saturn.
    d. None of the above is correct.

17. The planet that is not a giant is
    a. Saturn.
    b. Naptune.
    c. Mercury.
    d. Uranus.
18. What planets are considered "twins"?
    a. Saturn and Jupiter
    b. Mercury and Pluto
    c. Neptune and Uranus
    d. Earth and Mars
19. Area of the solar system where long period comets originate is the
    a. asteroid belt.
    b. Oort cloud.
    c. Kuiper belt.
    d. region between Jupiter and Neptune.
20. Short period comets have orbital periods of
    a. less than two hundred years.
    b. more than two hundred years.
    c. variable duration but between ten and one hundred years.
    d. variable duration but between one hundred and two hundred years.
21. Remnants of comets and asteroids found in space are called
    a. meteoroids.
    b. meteors.
    c. meteorites.
    d. minor planets.
22. Meteorites are classified into the following groups *except*
    a. iron.
    b. stony.
    c. lead.
    d. stony-iron.
23. The most widely accepted theory on the origin of the solar system is
    a. big bang.
    b. conservation of matter.
    c. protoplanet nebular.
    d. collision of planets.
24. The belt of asteroids between Mars and Jupiter is probably
    a. the remains of a planet that exploded.
    b. clumps of matter that condensed from the accretion disk but never got together as a planet.
    c. the remains of two planets that collided.
    d. the remains of a planet that collided with an asteroid or comet.
25. Which of the following planets would be mostly composed of hydrogen, helium, and methane and have a density of less than 2 g/cm^3?
    a. Uranus
    b. Mercury
    c. Mars
    d. Venus
26. Which of the following planets probably still has its original atmosphere?
    a. Mercury
    b. Venus
    c. Mars
    d. Jupiter

27. Venus appears the brightest when it is in the
    a. full phase.
    b. half phase.
    c. quarter phase.
    d. crescent phase.
28. The small body with a composition and structure closest to the materials that condensed from the accretion disk is a (an)
    a. asteroid.
    b. meteorite.
    c. comet.
    d. None of the above is correct.
29. A small body from space that falls on the surface of Earth is a
    a. meteoroid.
    b. meteor.
    c. meteor shower.
    d. meteorite.
30. Planets in our solar system are classified according to all of the following characteristics *except*
    a. density.
    b. size.
    c. nature of the atmosphere.
    d. inhabitants.
31. What separates the terrestrial planets from the giant planets?
    a. Mars
    b. asteroid belt
    c. empty space
    d. the Moon
32. The planet that has the shortest "year" among the eight planets is
    a. Jupiter.
    b. Neptune.
    c. Mercury.
    d. Earth.
33. What planet is called the "morning star" and the "evening star," depending on its position with respect to the Sun?
    a. Mercury
    b. Venus
    c. Mars
    d. Saturn
34. Venus "shines" because it is
    a. composed of rocky materials.
    b. covered with metallic iron and nickel.
    c. powered by fusion reactions.
    d. covered with clouds that reflect sunlight.
35. On Venus, the sun rises in the west. This is because
    a. the Venus day is very long.
    b. Venus rotates in the opposite direction of its revolution.
    c. Venus is the second planet from the Sun.
    d. Venus is covered with clouds and the Sun cannot be seen until later in its day.
36. The "sister" planet to Earth is
    a. Mercury.
    b. Venus.
    c. Mars.
    d. Moon.

37. What feature on Mars was considered by some to be evidence of life on Mars?
    a. ice caps
    b. red color
    c. channels or canals
    d. radio signals

38. Jupiter radiates twice as much energy as it receives from the Sun because of
    a. internal fusion reactions.
    b. rapid gravitational expansion of the planet.
    c. slow gravitational compression of the planet.
    d. internal fission reactions.

39. The Great Red Spot is thought to be
    a. a low pressure region with stagnant clouds.
    b. a high pressure region with high cold clouds.
    c. hydrogen and helium storms.
    d. ammonia and sulfur clouds.

40. The metallic hydrogen that surrounds the core of Jupiter is not
    a. a liquid.
    b. under tremendous pressure.
    c. a gas.
    d. a conductor of electric currents.

41. A shooting star is a (an)
    a. meteoroid.
    b. meteor.
    c. meteorite.
    d. asteroid.

Answers

1. d 2. c 3. b 4. a 5. c 6. b 7. b 8. c 9. d 10. b 11. b 12. a 13. c 14. d
15. a 16. c 17. c 18. c 19. b 20. a 21. a 22. c 23. c 24. b 25. a 26. d
27. d 28. c 29. d 30. d 31. b 32. c 33. b 34. d 35. b 36. b 37. c 38. c
39. b 40. c 41. b

QUESTIONS FOR THOUGHT

1. Describe the protoplanet nebular model of the origin of the solar system. Which part or parts of this model seem least credible to you? Explain. What information could you look for today that would cause you to accept or modify this least credible part of the model?

2. What are the basic differences between the terrestrial planets and the giant planets? Describe how the protoplanet nebular model accounts for these differences.

3. Describe the surface and atmospheric conditions on Mars.

4. What evidence exists that Mars at one time had abundant liquid water? If Mars did have liquid water at one time, what happened to it and why?

5. Describe the internal structure of Jupiter and Saturn.

6. What are the rings of Saturn?

7. Describe some of the unusual features found on the moons of Jupiter.

8. What are the similarities and the differences between the Sun and Jupiter?

9. Give one idea about why the Great Red Spot exists on Jupiter. Does the existence of a similar spot on Saturn support or not support this idea? Explain.

10. What is so unusual about the motions and orbits of Venus and Uranus?

11. What evidence exists today that the number of rocks and rock particles floating around in the solar system was much greater in the past soon after the planets formed?

12. Using the properties of the planets other than Earth, discuss the possibilities of life on each of the other planets.

13. What are "shooting stars"? Where do they come from? Where do they go?

14. What is an asteroid? What evidence indicates that asteroids are parts of a broken-up planet? What evidence indicates that asteroids are not parts of a broken-up planet?

15. Where do comets come from? Why are astronomers so interested in studying the physical and chemical structure of a comet?

16. What is a meteor? What is the most likely source of meteors?

17. What is a meteorite? What is the most likely source of meteorites?

18. Technically speaking, what is wrong with calling a rock that strikes the surface of the Moon a meteorite? Again speaking technically, what should you call a rock that strikes the surface of the Moon (or any planet other than Earth)?

19. If a comet is an icy, dusty body, explain why it appears brightly in the night sky.

FOR FURTHER ANALYSIS

1. What are the significant similarities and differences between the terrestrial and giant planets? Speculate why these similarities and differences exit.

2. Draw a sketch showing the positions of Earth, the Sun, and Venus when it appears as the morning star. Draw a second sketch showing the positions when it appears as the evening star.

3. Evaluate the statement that Venus is Earth's sister planet.

4. Describe the possibility and probability of life on each of the other planets.

5. Provide arguments that Pluto should be considered a planet. Counter this argument with evidence that it should not be classified as a planet.

6. Describe and analyze why it would be important to study the nucleus of a comet.

INVITATION TO INQUIRY

What's Your Sign?

Form a team to investigate horoscope forecasts in a newspaper or Internet site. Each team member should select one birthday and track what is forecast to happen and what actually happens each day for a week. Analyze the ways in which the forecasts are written that may make them "come true." Compare the prediction, actual results, and analysis of each team member.

16

Earth in Space

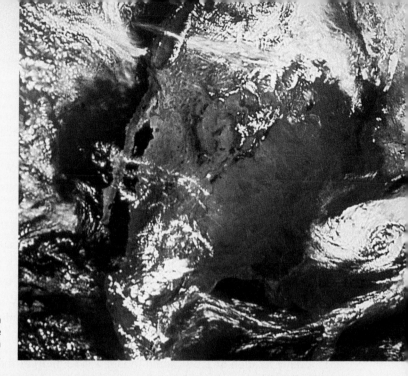

Earth as seen from space. Do you see the United States, with a storm over the East Coast? Do you see Denver and Los Angeles? One topic of this chapter is identifying places on Earth, which should help you find places in the United States.

CORE **CONCEPT**

The way Earth moves in space is used to define time and describe location on the surface, and causes recurrent phenomena such as seasons.

OUTLINE

Revolution
Earth moves around the Sun in a yearly revolution with the same orientation of its axis relative to the background of stars.

Identifying Place
Earth's axis and equator serve as references for finding locations on the surface with parallels and meridians.

Phases of the Moon
The Moon phases are caused by the relative positions of the Earth, Moon, and Sun, and our viewing different parts of the Moon that are in sunlight and not in sunlight.

Shape and Size of Earth
Motions of Earth
   Revolution
   Rotation
   Precession
Place and Time
   Identifying Place
   Measuring Time
Science and Society: Saving Time?
The Moon
   Composition and Features
   History of the Moon
The Earth-Moon System
   Phases of the Moon
   Eclipses of the Sun and Moon
   Tides
People Behind the Science: Carl Edward Sagan

Rotation
Earth spins on its axis in a daily rotation.

Measuring Time
Recurring cycles associated with the rotation of Earth on its axis and its revolution around the Sun are used to define measures of time.

Tides
A tide is the periodic rise and fall of sea level, and the timing and size of a given tide is influenced by the relative positions of Earth, Moon, and Sun, the orbit of the Moon, and the size, shape, and depth of the basin holding the water.

Earth is a common object in the solar system, one of eight planets that goes around the Sun once a year in an almost circular orbit. Earth is the third planet out from the Sun, it is fifth in mass and diameter, and it has the greatest density of all the planets (Figure 16.1). Earth is unique because of its combination of an abundant supply of liquid water, a strong magnetic field, and a particular atmospheric composition. In addition to these physical properties, Earth has a unique set of natural motions that humans have used for thousands of years as a frame of reference to mark time and to identify the events of their lives. These references to Earth's motions are called the day, the month, and the year.

Eventually, about three hundred years ago, people began to understand that their references for time came from an Earth that spins like a top as it circles the Sun. It was still difficult, however, for them to understand Earth's place in the universe. The problem was not unlike that of a person trying to comprehend the motion of a distant object while riding a moving merry-go-round being pulled by a cart. Actually, the combined motions of Earth are much more complex than a simple moving merry-go-round being pulled by a cart. Imagine trying to comprehend the motion of a distant object while undergoing a combination of Earth's more conspicuous motions, which are as follows:

1. A daily rotation of 1,670 km/h (about 1,040 mi/h) at the equator and less at higher latitudes.
2. A monthly revolution of Earth around the Earth-Moon center of gravity at about 50 km/h (about 30 mi/h).
3. A yearly revolution around the Sun at about an average 106,000 km/h (about 66,000 mi/h).
4. A motion of the solar system around the core of the Milky Way at about 370,000 km/h (about 230,000 mi/h).
5. A motion of the local star group that contains the Sun as compared to other star clusters of about 1,000,000 km/h (about 700,000 mi/h).
6. Movement of the entire Milky Way galaxy relative to other, remote galaxies at about 580,000 km/h (about 360,000 mi/h).
7. Minor motions such as cycles of change in the size and shape of Earth's orbit and the tilt of Earth's axis. In addition to these slow changes, there is a gradual slowing of the rate of Earth's daily rotation.

Basically, Earth is moving through space at fantastic speeds, following the Sun in a spiral path of a giant helix as it spins like a top (Figure 16.2). This ceaseless and complex motion in space is relative to various frames of reference, however, and the limited perspective from Earth's surface can result in some very different ideas about Earth and its motions. This chapter is about the more basic, or fundamental, motions of Earth and its moon. In addition to conceptual understandings and evidences for the motions, some practical human uses of the motions will be discussed.

SHAPE AND SIZE OF EARTH

The most widely accepted theory about how the solar system formed pictures the planets forming in a disk-shaped nebula with a turning, swirling motion. The planets formed from separate accumulations of materials within this disk-shaped, turning nebula, so the orbit of each planet was established along with its rate of rotation as it formed. Thus, all the planets move around the Sun in the same direction in elliptical orbits that are nearly circular. The flatness of the solar system results in the observable planets moving in, or near, the plane of Earth's orbit, which is called the **plane of the ecliptic.**

When viewed from Earth, the planets appear to move only within a narrow band across the sky as they move in the plane of the ecliptic. The Sun also appears to move in the center of this band, which is called the *ecliptic*. As viewed from Earth, the Sun appears to move across the background of stars, completely around the ecliptic each year.

Today, almost everyone has seen pictures of Earth from space, and it is difficult to deny that it has a rounded shape (Figure 16.3). During the fifth and sixth centuries B.C., the ancient Greeks decided that Earth must be round because (1) philosophically, they considered the sphere to be the perfect shape and they considered Earth to be perfect, so therefore Earth must be a sphere, (2) Earth was observed to cast a circular shadow on the Moon during a lunar eclipse, and (3) ships were observed to slowly disappear below the horizon as they sailed off into the distance. More abstract evidence of a round Earth was found in the observation that the altitude of the North Star above the horizon appeared to increase as a person traveled

FIGURE 16.1 Artist's concept of the solar system. Shown are the orbits of the planets, Earth being the third planet from the Sun, and the other planets and their relative sizes and distances from each other and to the Sun. Also shown is the solar system as seen looking toward Earth from the Moon (not to scale).

northward. This established that Earth's surface was curved, at least, which seemed to fit with other evidence.

The shape and size of Earth have been precisely measured by artificial satellites circling Earth. These measurements have found that Earth is not a perfectly round sphere as believed by the ancient Greeks. It is flattened at the poles and has an equatorial bulge, as do many other planets. In fact, you can observe through a telescope that both Jupiter and Saturn are considerably

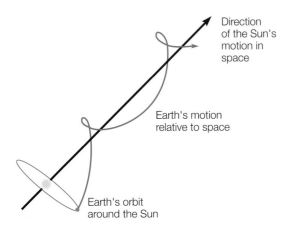

FIGURE 16.2 Earth undergoes many different motions as it moves through space. There are seven more conspicuous motions, three of which are more obvious on the surface. Earth follows the path of a gigantic helix, moving at fantastic speeds as it follows the Sun and the galaxy through space.

FIGURE 16.3 Earth as seen from space.

flattened at the poles. A shape that is flattened at the poles has a greater distance through the equator than through the poles, which is described as an *oblate* shape. Earth, like a water-filled, round balloon resting on a table, has an oblate shape. It is not perfectly symmetrically oblate, however, since the North Pole is slightly higher and the South Pole is slightly lower than the average surface. In addition, it is not perfectly circular around the equator, with a lump in the Pacific and a depression in the Indian

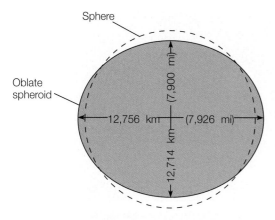

FIGURE 16.4 Earth has an irregular, slightly lopsided, slightly pear-shaped form. In general, it is considered to have the shape of an oblate spheroid, departing from a perfect sphere as shown here.

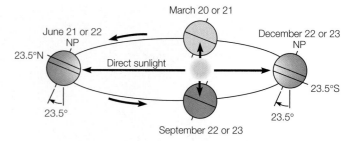

FIGURE 16.5 The consistent tilt and orientation of Earth's axis as it moves around its orbit is the cause of the seasons. The North Pole is pointing toward the Sun during the summer solstice and away from the Sun during the winter solstice.

Oceans. The shape of Earth is a slightly pear-shaped, slightly lopsided *oblate spheroid* (Figure 16.4). All the elevations and depressions are less than 85 m (about 280 ft), however, which is practically negligible compared to the size of Earth. Thus, Earth is very close to, but not exactly, an oblate spheroid. The significance of this shape will become apparent when Earth's motions are discussed next.

MOTIONS OF EARTH

Ancient civilizations had a fairly accurate understanding of the size and shape of Earth but had difficulty accepting the idea that Earth moves. The geocentric theory of a motionless Earth with the Sun, Moon, planets, and stars circling it was discussed in chapter 15. Ancient people had difficulty with anything but a motionless Earth for at least two reasons: (1) they could not sense any motion of Earth, and (2) they had ideas about being at the center of a universe that was created for them. Thus, it was not until the 1700s that the concept of an Earth in motion became generally accepted. Today, Earth is understood to move a number of different ways, seven of which were identified in the introduction to this chapter. Three of these motions are independent of motions of the Sun and the galaxy. These are (1) a yearly revolution around the Sun, (2) a daily rotation on its axis, and (3) a slow, clockwise wobble of its axis.

REVOLUTION

Earth moves constantly around the Sun in a slightly elliptical orbit that requires an average of one year for one complete circuit. The movement around the Sun is called a **revolution,** and all points of Earth's orbit lie in the plane of the ecliptic. The average distance between Earth and the Sun is about 150 million km (about 93 million mi).

Earth's orbit is slightly elliptical, so it moves with a speed that varies. It moves fastest when it is closer to the Sun in January and slowest when it is farthest away from the Sun in early July. Earth is about 2.5 million km (about 1.5 million mi) closer to

the Sun in January and about the same distance farther away in July than it would be if the orbit were a circle. This total difference of about 5 million km (about 3 million mi) results in a January Sun with an apparent diameter that is 3 percent larger than the July Sun, and Earth as a whole receives about 6 percent more solar energy in January. The effect of being closer to the Sun is much less than the effect of some other relationships, and winter occurs in the Northern Hemisphere when Earth is closest to the Sun. Likewise, summer occurs in the Northern Hemisphere when the Sun is at its greatest distance from Earth (Figure 16.5).

The important directional relationships that override the effect of Earth's distance from the Sun involve the daily **rotation,** or spinning, of Earth around an imaginary line through the geographic poles called Earth's *axis.* The important directional relationships are a constant inclination of Earth's axis to the plane of the ecliptic and a constant orientation of the axis to the stars. The *inclination of Earth's axis* to the plane of the ecliptic is about 66.5° (or 23.5° from a line perpendicular to the plane). This relationship between the plane of Earth's orbit and the tilt of its axis is considered to be the same day after day throughout the year, even though small changes do occur in the inclination over time. Likewise, the *orientation of Earth's axis* to the stars is considered to be the same throughout the year as Earth moves through its orbit. Again, small changes do occur in the orientation over time. Thus, in general, the axis points in the same direction, remaining essentially parallel to its position during any day of the year. The essentially constant orientation and inclination of the axis result in the axis pointing toward the Sun as Earth moves in one part of its orbit, then pointing away from the Sun six months later. The generally constant inclination and orientation of the axis, together with Earth's rotation and revolution, combine to produce three related effects: (1) days and nights that vary in length, (2) changing seasons, and (3) climates that vary with latitude.

Figure 16.5 shows how the North Pole points toward the Sun on June 21 or 22, then away from the Sun on December 22 or 23 as it maintains its orientation to the stars. When the North Pole is pointed toward the Sun, it receives sunlight for a full twenty-four hours, and the South Pole is in Earth's shadow for a full twenty-four hours. This is summer in the Northern Hemisphere with the longest daylight periods and the Sun

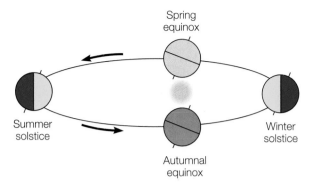

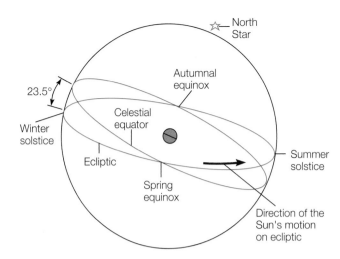

FIGURE 16.6 The length of daylight during each season is determined by the relationship of Earth's shadow to the tilt of the axis. At the equinoxes, the shadow is perpendicular to the latitudes, and day and night are of equal length everywhere. At the summer solstice, the North Pole points toward the Sun and is completely out of the shadow for a twenty-four-hour day. At the winter solstice, the North Pole is in the shadow for a twenty-four-hour night. The situation is reversed for the South Pole.

FIGURE 16.7 The position of the Sun on the celestial sphere at the solstices and the equinoxes.

at its maximum noon height in the sky. Six months later, on December 22 or 23, the orientation is reversed with winter in the Northern Hemisphere, the shortest daylight periods, and the Sun at its lowest noon height in the sky.

The beginning of a season can be recognized from any one of the three related observations: (1) the length of the daylight period, (2) the altitude of the Sun in the sky at noon, or (3) the length of a shadow from a vertical stick at noon. All of these observations vary with changes in the direction of Earth's axis of rotation relative to the Sun (Figure 16.6). On about June 22 and December 22, the Sun reaches its highest and lowest noon altitudes as Earth moves to point the North Pole directly toward the Sun (June 21 or 22) and directly away from the Sun (December 22 or 23). Thus, the Sun appears to stop increasing or decreasing its altitude in the sky, stop, then reverse its movement twice a year. These times are known as **solstices** after the Latin meaning "Sun stand still." The Northern Hemisphere's **summer solstice** occurs on about June 22 and identifies the beginning of the summer season. At the summer solstice, the Sun at noon has the highest altitude, and the shadow from a vertical stick is shorter than on any other day of the year. The Northern Hemisphere's **winter solstice** occurs on about December 22 and identifies the beginning of the winter season. At the winter solstice, the Sun at noon has the lowest altitude, and the shadow from a vertical stick is longer than on any other day of the year.

As Earth moves in its orbit between pointing its North Pole toward the Sun on about June 22 and pointing it away on about December 22, there are two times when it is halfway. At these times, Earth's axis is perpendicular to a line between the center of the Sun and Earth, and daylight and night are of equal length. These are called the **equinoxes** after the Latin meaning "equal nights." The **spring equinox** (also called the **vernal equinox**) occurs on about March 21 and identifies the beginning of the spring season. The **autumnal equinox** occurs on about September 23 and identifies the beginning of the fall season.

The relationship between the apparent path of the Sun on the celestial sphere and the seasons is shown in Figure 16.7. Recall that the celestial equator is a line on the celestial sphere directly above Earth's equator. The equinoxes are the points on the celestial sphere where the ecliptic, the path of the Sun, crosses the celestial equator. Note also that the summer solstice occurs when the ecliptic is 23.5° north of the celestial equator, and the winter solstice occurs when it is 23.5° south of the celestial equator.

CONCEPTS *Applied*

Sunrise, Sunset

Make a chart to show the time of sunrise and sunset for a month. Calculate the amount of daylight and darkness. Does the sunrise change in step with the sunset, or do they change differently? What models can you think of that would explain all your findings?

ROTATION

Observing the apparent turning of the celestial sphere once a day and seeing the east-to-west movement of the Sun, Moon, and stars, it certainly seems as if it is the heavenly bodies and not Earth doing the moving. You cannot sense any movement, and there is little apparent evidence that Earth indeed moves. Evidence of a moving Earth comes from at least three different observations: (1) the observation that the other planets and the Sun rotate, (2) the observation of the changing plane of a long, heavy pendulum at different latitudes on Earth, and (3) the observation of the direction of travel of something moving across, but above, Earth's surface, such as a rocket.

Other planets, such as Jupiter, and the Sun can be observed to rotate by keeping track of features on the surface such as the Great Red Spot on Jupiter and sunspots on the Sun. While such observations are not direct evidence that Earth also rotates, they

FIGURE 16.8 As is being demonstrated in this old woodcut, Foucault's insight helped people understand that Earth turns. The pendulum moves back and forth without changing its direction of movement, and we know this is true because no forces are involved. We turn with Earth, and this makes the pendulum appear to change its plane of rotation. Thus, we know Earth rotates.

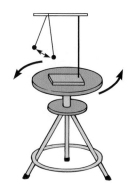

FIGURE 16.9 The Foucault pendulum swings back and forth in the same plane while a stool is turned beneath it. Likewise, a Foucault pendulum on Earth's surface swings back and forth in the same plane while Earth turns beneath it. The amount of turning observed depends on the latitude of the pendulum.

do show that other members of the solar system spin on their axes. As described earlier, Jupiter is also observed to be oblate, flattened at its poles with an equatorial bulge. Since Earth is also oblate, this is again indirect evidence that it rotates, too.

The most easily obtained and convincing evidence about Earth's rotation comes from a *Foucault pendulum,* a heavy mass swinging from a long wire. This pendulum is named after the French physicist Jean Foucault, who first used a long pendulum in 1851 to prove that Earth rotates. Foucault started a long, heavy pendulum moving just above the floor, marking the plane of its back-and-forth movement. Over some period of time, the pendulum appeared to slowly change its position, smoothly shifting its plane of rotation. Science museums often show this shifting plane of movement by setting up small objects for the pendulum to knock down. Foucault demonstrated that the pendulum actually maintains its plane of movement in space (inertia) while Earth rotates eastward (counterclockwise) under the pendulum. It is Earth that turns under the pendulum, causing the pendulum to appear to change its plane of rotation. It is difficult to imagine the pendulum continuing to move in a fixed direction in space while Earth, and everyone on it, turns under the swinging pendulum (Figure 16.8).

Figure 16.9 illustrates the concept of the Foucault pendulum. A pendulum is attached to a support on a stool that is free to rotate. If the stool is slowly turned while the pendulum is swinging, you will observe that the pendulum maintains its plane of rotation while the stool turns under it. If you were much smaller and looking from below the pendulum, it would appear to turn as you rotate with the turning stool. This is what happens on Earth. Such a pendulum at the North Pole would make a complete turn in about twenty-four hours. Moving south from the North Pole, the change decreases with latitude until, at the equator, the pendulum would not appear to turn at all. At higher latitudes, the plane of the pendulum appears to move clockwise in the Northern Hemisphere and counterclockwise in the Southern Hemisphere.

More evidence that Earth rotates is provided by objects that move above and across Earth's surface. As shown in Figure 16.10, Earth has a greater rotational velocity at the equator than at the poles. As an object leaves the surface and moves north or south, the surface has a different rotational velocity, so it rotates beneath the object as it proceeds in a straight line. This gives the moving object an apparent deflection to the right of the direction of movement in the Northern Hemisphere and to the left in the Southern Hemisphere. The apparent deflection caused by Earth's rotation is called the **Coriolis effect.** The Coriolis effect will explain Earth's prevailing wind systems as well as the characteristic direction of wind in areas of high pressure and areas of low pressure (see chapter 23).

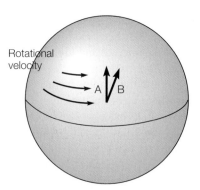

FIGURE 16.10 Earth has a greater rotational velocity at the equator and less toward the poles. As an object moves north or south (*A*), it passes over land with a different rotational velocity, which produces a deviation to the right in the Northern Hemisphere (*B*) and to the left in the Southern Hemisphere.

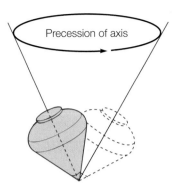

FIGURE 16.11 A spinning top wobbles as it spins, and the axis of the top traces out a small circle. The wobbling of the axis is called *precession.*

PRECESSION

If Earth were a perfect spherically shaped ball, its axis would always point to the same reference point among the stars. The reaction of Earth to the gravitational pull of the Moon and the Sun on its equatorial bulge, however, results in a slow wobbling of Earth as it turns on its axis. This slow wobble of Earth's axis, called **precession,** causes it to swing in a slow circle like the wobble of a spinning top (Figure 16.11). It takes Earth's axis about twenty-six thousand years to complete one turn, or wobble. Today, the axis points very close to the North Star, Polaris, but is slowly moving away to point to another star. In about twelve thousand years, the star Vega will appear to be in the position above the North Pole, and Vega will be the new North Star. The moving pole also causes changes over time in which particular signs of the zodiac appear with the spring equinox. Because of precession, the occurrence of the spring equinox has been moving backward (westward) through the zodiac constellations at about 1 degree every seventy-two years. Thus, after about twenty-six thousand years, the spring equinox will have moved through all the constellations and will again approach the constellation of Aquarius for the next "age of Aquarius" (Figure 16.12).

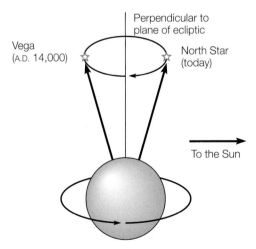

FIGURE 16.12 The slow, continuous precession of Earth's axis results in the North Pole pointing around a small circle over a period of about 26,000 years.

PLACE AND TIME

The continuous rotation and revolution of Earth establish an objective way to determine direction, location, and time on Earth. If Earth were an unmoving sphere, there would be no side, end, or point to provide a referent for direction and location. Earth's rotation, however, defines an axis of rotation, which serves as a reference point for determination of direction and location on the entire surface. Earth's rotation and revolution together define cycles, which define standards of time. The following describes how Earth's movements are used to identify both place and time.

IDENTIFYING PLACE

A system of two straight lines can be used to identify a point, or position, on a flat, two-dimensional surface. The position of the letter X on this page, for example, can be identified by making a line a certain number of measurement units from the top of the page and a second line a certain number of measurement units from the left side of the page. Where the two lines intersect will identify the position of the letter *X,* which can be recorded or communicated to another person (Figure 16.13).

A system of two straight lines can also be used to identify a point, or position, on a sphere, except this time the lines are circles. The reference point for a sphere is not as simple as in the flat, two-dimensional case, however, since a sphere does not have a top or side edge. Earth's axis provides the north-south reference point. The equator is a big circle around Earth that is exactly halfway between the two ends, or poles, of the rotational axis. An infinite number of circles are imagined to run around Earth parallel to the equator as shown in Figure 16.14. The east- and west-running parallel circles are called **parallels.** Each parallel is the same distance between the equator and one of the poles all the way around Earth. The distance from the equator to a point on a parallel is called the **latitude** of that point. Latitude tells you how far north or south a point is from the equator by telling you the parallel the point is located on. The distance is measured northward from the equator (which is 0°) to the

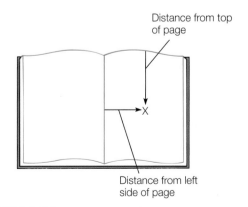

FIGURE 16.13 Any location on a flat, two-dimensional surface is easily identified with two references from two edges. This technique does not work on a motionless sphere because there are no reference points.

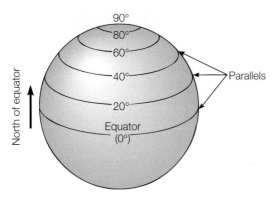

FIGURE 16.14 A circle that is parallel to the equator is used to specify a position north or south of the equator. A few of the possibilities are illustrated here.

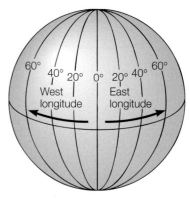

FIGURE 16.16 Meridians run pole to pole and perpendicular to the parallels, providing a reference for specifying east and west directions.

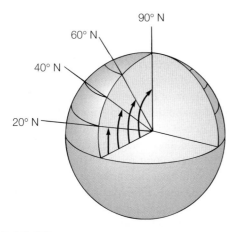

FIGURE 16.15 If you could see to Earth's center, you would see that latitudes run from 0° at the equator north to 90° at the North Pole (or to 90° south at the South Pole).

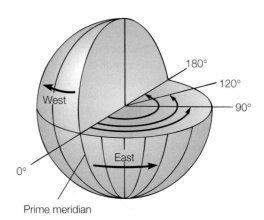

FIGURE 16.17 If you could see inside Earth, you would see 360° around the equator and 180° of longitude east and west of the prime meridian.

North Pole (90° north) or southward from the equator (0°) to the South Pole (90° south) (Figure 16.15). If you are somewhere at a latitude of 35° north, you are somewhere on Earth on the 35° latitude line north of the equator.

Since a parallel is a circle, a location of 40°N latitude could be anyplace on that circle around Earth. To identify a location, you need another line, this time one that runs pole to pole and perpendicular to the parallels. These north-south running arcs that intersect at both poles are called **meridians** (Figure 16.16). There is no naturally occurring, identifiable meridian that can be used as a point of reference such as the equator serves for parallels, so one is identified as the referent by international agreement. The referent meridian is the one that passes through the Greenwich Observatory near London, England, and this meridian is called the **prime meridian.** The distance from the prime meridian east or west is called the **longitude.** The degrees of longitude of a point on a parallel are measured to the east or to the west from the prime meridian up to 180° (Figure 16.17). New Orleans, Louisiana, for example, has a latitude of about 30°N of the equator and a longitude of about 90°W of the prime meridian. The location of New Orleans is therefore described as 30°N, 90°W.

Locations identified with degrees of latitude north or south of the equator and degrees of longitude east or west of the prime meridian are more precisely identified by dividing each degree of latitude into subdivisions of 60 minutes (60′) per degree, and each minute into 60 seconds (60″). On the other hand, latitudes near the equator are sometimes referred to in general as the *low atitudes,* and those near the poles are sometimes called the *high latitudes.*

In addition to the equator (0°) and the poles (90°), the parallels of 23.5°N and 23.5°S from the equator are important references for climatic consideration. The parallel of 23.5°N is called the **tropic of Cancer,** and 23.5°S is called the **tropic of Capricorn.** These two parallels identify the limits toward the poles within which the Sun appears directly overhead during the course of a year. The parallel of 66.5°N is called the **Arctic Circle,** and the parallel of 66.5°S is called the **Antarctic Circle.** These two parallels identify the limits toward the equator within which the Sun appears above the horizon all day during the summer (Figure 16.18). This starts with six months of daylight every day at the pole, then decreases as you get fewer days of full light until reaching the limit of one day of twenty-four-hour daylight at the 66.5° limit.

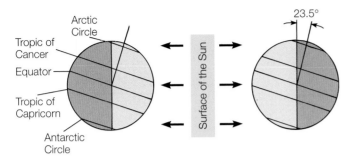

Summer solstice   Winter solstice

FIGURE 16.18 At the summer solstice, the noon Sun appears directly overhead at the tropic of Cancer (23.5°N) and twenty-four hours of daylight occurs north of the Arctic Circle (66.5°N). At the winter solstice, the noon Sun appears overhead at the tropic of Capricorn (23.5°S) and twenty-four hours of daylight occurs south of the Antarctic Circle (66.5°S).

MEASURING TIME

Standards of time are determined by intervals between two successive events that repeat themselves in a regular way. Since ancient civilizations, many of the repeating events used to mark time have been recurring cycles associated with the rotation of Earth on its axis and its revolution around the Sun. Thus, the day, month, season, and year are all measures of time based on recurring natural motions of Earth. All other measures of time are based on other events or definitions of events. There are, however, several different ways to describe the day, month, and year, and each depends on a different set of events. These events are described in the following section.

Daily Time

The technique of using astronomical motions for keeping time originated some four thousand years ago with the Babylonian culture. The Babylonians marked the yearly journey of the Sun against the background of the stars, which was divided into twelve periods, or months, after the signs of the zodiac. Based on this system, the Babylonian year was divided into twelve months with a total of 360 days. In addition, the Babylonians invented the week and divided the day into hours, minutes, and seconds. The week was identified as a group of seven days, each based on one of the seven heavenly bodies that were known at the time. The hours, minutes, and seconds of a day were determined from the movement of the shadow around a straight, vertical rod.

As seen from a place in space above the North Pole, Earth rotates counterclockwise turning toward the east. On Earth, this motion causes the Sun to appear to rise in the east, travel across the sky, and set in the west. The changing angle between the tilt of Earth's axis and the Sun produces an apparent shift of the Sun's path across the sky, northward in the summer season and southward in the winter season. The apparent movement of the Sun across the sky was the basis for the ancient as well as the modern standard of time known as the day.

Today, everyone knows that Earth turns as it moves around the Sun, but it is often convenient to regard space and astronomical motions as the ancient Greeks did, as a celestial sphere that turns around a motionless Earth. Recall that the celestial meridian is a great circle on the celestial sphere that passes directly overhead where you are and continues around Earth through both celestial poles. The movement of the Sun across the celestial meridian identifies an event of time called **noon.** As the Sun appears to travel west, it crosses meridians that are farther and farther west, so the instant identified as noon moves west with the Sun. The instant of noon at any particular longitude is called the **apparent local noon** for that longitude because it identifies noon from the apparent position of the Sun in the sky. The morning hours before the Sun crosses the meridian are identified as *ante meridiem* (A.M.) hours, which is Latin for "before meridian." Afternoon hours are identified as *post meridiem* (P.M.) hours, which is Latin for "after the meridian."

There are several ways to measure the movement of the Sun across the sky. The ancient Babylonians, for example, used a vertical rod called a *gnomon* to make and measure a shadow that moved as a result of the apparent changes of the Sun's position. The gnomon eventually evolved into a *sundial*, a vertical or slanted gnomon with divisions of time marked on a horizontal plate beneath the gnomon. The shadow from the gnomon indicates the **apparent local solar time** at a given place and a given instant from the apparent position of the Sun in the sky. If you have ever read the time from a sundial, you know that it usually does not show the same time as a clock or a watch (Figure 16.19). In addition, sundial time is nonuniform, fluctuating throughout the course of a year, sometimes running ahead of clock time and sometimes running behind clock time.

A sundial shows the apparent local solar time, but clocks are set to measure a uniform standard time based on **mean solar time.** Mean solar time is a uniform time averaged from the apparent solar time. The apparent solar time is nonuniform, fluctuating because (1) Earth moves sometimes faster and sometimes slower in its elliptical orbit around the Sun and (2) the equator of Earth is inclined to the ecliptic. The

FIGURE 16.19 A sundial indicates the apparent local solar time at a given instant in a given location. The time read from a sundial, which is usually different from the time read from a clock, is based on an average solar time.

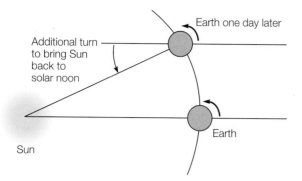

FIGURE 16.20 Because Earth is moving in orbit around the Sun, it must rotate an additional distance each day, requiring about four minutes to bring the Sun back across the celestial meridian (local solar noon). This explains why the stars and constellations rise about four minutes earlier every night.

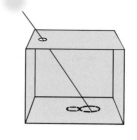

FIGURE 16.21 (A) During a year, a beam of sunlight traces out a lopsided figure eight on the floor if the position of the light is marked at noon every day. (B) The location of the point of light on the figure eight during each month.

combined consequence of these two effects is a variable, non-uniform sundial time as compared to the uniform mean solar time, otherwise known as clock time.

A day is defined as the length of time required for Earth to rotate once on its axis. There are different ways to measure this rotation, however, which result in different definitions of the day. A **sidereal day** is the interval between two consecutive crossings of the celestial meridian by a particular star (sidereal means "star"). This interval of time depends only on the time Earth takes to rotate 360° on its axis. One sidereal day is practically the same length as any other sidereal day because Earth's rate of rotation is constant for all practical purposes.

An **apparent solar day** is the interval between two consecutive crossings of the celestial meridian by the Sun, for example, from one local solar noon to the next solar noon. Since Earth is moving in orbit around the Sun, it must turn a little bit farther to compensate for its orbital movement, bringing the Sun back to local solar noon (Figure 16.20). As a consequence, the apparent solar day is about four minutes longer than the sidereal day. This additional time accounts for the observation that the stars and constellations of the zodiac rise about four minutes earlier every night, appearing higher in the sky at the same clock time until they complete a yearly cycle. A sidereal day is twenty-three hours, fifty-six minutes, and four seconds long. A **mean solar day** is twenty-four hours long, averaged from the mean solar time to keep clocks in closer step with the Sun than would be possible using the variable apparent solar day. Just how out of synchronization the apparent solar day can become with a clock can be illustrated with another ancient way of keeping track of the Sun's motions in the sky, the "hole in the wall" sun calendar and clock.

Variations of the "hole in the wall" sun calendar were used all over the world by many different ancient civilizations, including the early Native Americans of the American Southwest. More than one ancient Native American ruin has small holes in the western wall aligned in such a way as to permit sunlight to enter a chamber only on the longest and shortest days of the year. This established a basis for identifying the turning points in the yearly cycle of seasons.

A hole in the roof can be used as a sun clock, but it will require a whole year to establish the meaning of a beam of sunlight shining on the floor. Imagine a beam of sunlight passing through a small hole to make a small spot of light on the floor. For a year, you mark the position of the spot of light on the floor *each day* when your clock tells you the *mean solar time is noon.* You trace out an elongated, lopsided figure eight with the small end pointing south and the larger end pointing north (Figure 16.21A). Note by following the monthly markings shown in Figure 16.21B that the figure-eight shape is actually traced out by the spot of sunlight making two S shapes as the Sun changes its apparent position in the sky. Together, the two S shapes make the shape of the figure eight.

Why did the sunbeam trace out a figure eight over a year? The two extreme north-south positions of the figure are easy to understand because by December, Earth is in its orbit with the North Pole tilted away from the Sun. At this time, the direct rays of the Sun fall on the tropic of Capricorn (23.5° south of the equator), and the Sun appears low in the sky as seen from the Northern Hemisphere. Thus, on this date, the winter solstice, a beam of sunlight strikes the floor at its northernmost position beneath the hole. By June, Earth has moved halfway around its orbit, and the North Pole is now tilted toward the Sun. The direct rays of the Sun now fall on the tropic of Cancer (23.5° north of the equator), and the Sun appears high in the sky as seen from the Northern Hemisphere (Figure 16.22). Thus, on this date, the summer solstice, a beam of sunlight strikes the floor at its southernmost position beneath the hole.

If everything else were constant, the path of the spot would trace out a straight line between the northernmost and southernmost positions beneath the hole. The east and west movements of the point of light as it makes an S shape on the floor must mean, however, that the Sun crosses the celestial meridian (noon) earlier one part of the year and later the other part. This early and late arrival is explained in part by Earth moving at different speeds in its orbit.

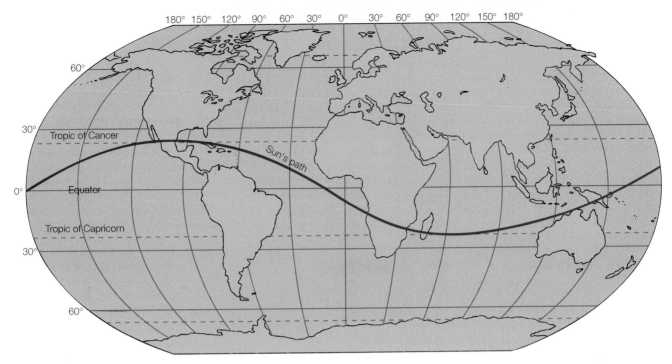

FIGURE 16.22 The path of the Sun's direct rays during a year. The Sun is directly over the tropic of Cancer at the summer solstice and high in the Northern Hemisphere sky. At the winter solstice, the Sun is directly over the tropic of Capricorn and low in the Northern Hemisphere sky.

If changes in orbital speed were the only reason that the Sun does not cross the sky at the same rate during the year, the spot of sunlight on the floor would trace out an oval rather than a figure eight. The plane of the ecliptic, however, does not coincide with the plane of Earth's equator, so the Sun appears at different angles in the sky, and this makes it appear to change its speed during different times of the year. This effect changes the length of the apparent solar day by making the Sun up to ten minutes later or earlier than the mean solar time four times a year between the solstices and equinoxes.

The two effects add up to a cumulative variation between the apparent local solar time (sundial time) and the mean solar time (clock time) (Figure 16.23). This cumulative variation is known as the **equation of time,** which shows how many minutes sundial time is faster or slower than clock time during different days of the year. The equation of time is often shown on

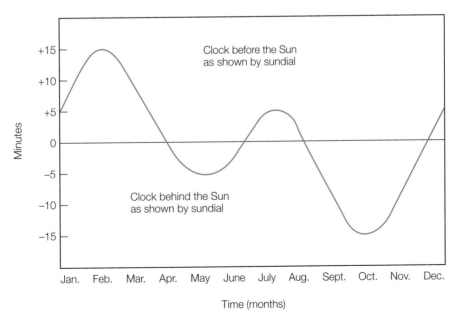

FIGURE 16.23 The equation of time, which shows how many minutes sundial time is faster or slower than clock time during different months of the year.

The purpose of daylight saving time is to make better use of daylight during the summer by moving an hour of daylight from the morning to the evening. In the United States, daylight saving time is observed from the second Sunday in March to the first Sunday in November. Clocks are changed on these Sundays according to the saying, "Spring ahead, fall back." Arizona and Hawaii choose not to participate and stay on standard time all year.

Americans who say they like daylight saving time say they like it is because it gives them more light in the evenings, and it saves energy. Some people do not like daylight saving time because it requires them to reset all their clocks and adjust their sleep schedule twice a year. They also complain that the act of changing the clock is not saving daylight at all, but it is sending them to bed an hour earlier. Farmers also complain that plants and animals are regulated by the Sun, not the clock, so they have to plan all their nonfarm interactions on a different schedule.

QUESTIONS TO DISCUSS

Divide your group into two subgroups, one representing those who like daylight saving time and the other representing those who do not. After a few minutes of preparation, have a short debate about the advantages and disadvantages of daylight saving time.

globes in the figure-eight shape called an *analemma,* which also can be used to determine the latitude of direct solar radiation for any day of the year.

Since the local mean time varies with longitude, every place on an east-west line around Earth could possibly have clocks that were a few minutes ahead of those to the west and a few minutes behind those to the east. To avoid the confusion that would result from many clocks set to local mean solar time, Earth's surface is arbitrarily divided into one-hour **standard time zones** (Figure 16.24). Since there are 360° around Earth and 24 hours in a day, this means that each time zone is 360° divided by 24, or 15° wide. These 15° zones are adjusted so that whole states are in the same time zone, or the zones are adjusted for other political reasons. The time for each zone is defined as the mean solar time at the middle of each zone. When you cross a boundary between two zones, the clock is set ahead one hour if you are traveling east and back one hour if you are traveling west. Most states adopt **daylight saving time** during the summer, setting clocks ahead one hour in the spring and back one hour in the fall ("spring ahead and fall back"). Daylight saving time results in an extra hour of daylight during summer evenings.

The 180° meridian is arbitrarily called the **international date line,** an imaginary line established to compensate for cumulative time zone changes (Figure 16.25). A traveler crossing the date line gains or loses a day just as crossing a time zone boundary results in the gain or loss of an hour. A person moving across the line while traveling westward gains a day; for example, the day after June 2 would be June 4. A person crossing the line while traveling eastward repeats a day; for example, the day after June 6 would be June 6. Note that the date line is curved around land masses to avoid local confusion.

Yearly Time

A *year* is generally defined as the interval of time required for Earth to make one complete revolution in its orbit. As was the case for definitions of a day, there are different definitions of what is meant by a year. The most common definition of a year is the interval between two consecutive spring equinoxes, which is known as the **tropical year** (*trope* is Greek for "turning"). The tropical year is 365 days, 5 hours, 48 minutes, and 46 seconds, or 365.24220 mean solar days.

A **sidereal year** is defined as the interval of time required for Earth to move around its orbit so the Sun is again in the same position relative to the stars. The sidereal year is slightly longer than the tropical year because Earth rotates more than 365.25 times during one revolution. Thus, the sidereal year is 365.25636 mean solar days, which is about 20 minutes longer than the tropical year.

The tropical and sidereal years would be the same interval of time if Earth's axis pointed in a consistent direction. The precession of the axis, however, results in the axis pointing in a slightly different direction with time. This shift of direction over the course of a year moves the position of the spring equinox westward, and the equinox is observed twenty minutes before the orbit has been completely circled. The position of the spring equinox against the background of the stars thus moves westward by some 50 seconds of arc per year.

It is the *tropical year* that is used as a standard time interval to determine the calendar year. Earth does not complete an exact number of turns on its axis while completing one trip around the Sun, so it becomes necessary to periodically adjust the calendar so it stays in step with the seasons. The calendar system that was first designed to stay in step with the seasons was devised by the ancient Romans. Julius Caesar reformed the calendar, beginning in 46 B.C., to have a 365-day year with a 366-day year (leap year) every fourth year. Since the tropical year of 365.24220 mean solar days is very close to 365¼ days, the system, called a *Julian calendar,* accounted for the ¼ day by adding a full day to the calendar every fourth year. The Julian calendar was very similar to the one now used, except the year began in March, the month of the spring equinox. The month of July was named in honor of Julius Caesar, and the following month was later named after his successor, Augustus.

There was a slight problem with the Julian calendar because it was longer than the tropical year by 365.25 minus 365.24220, or 0.0078 day per year. This small interval (which is 11 minutes,

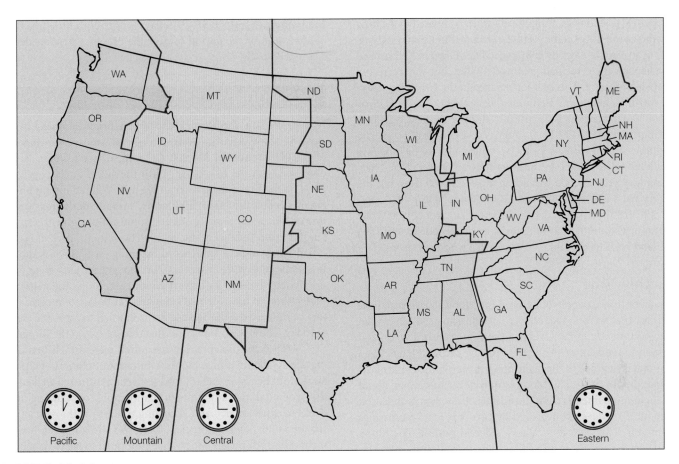

FIGURE 16.24 The standard time zones. Hawaii and most of Alaska are two hours earlier than Pacific Standard Time.

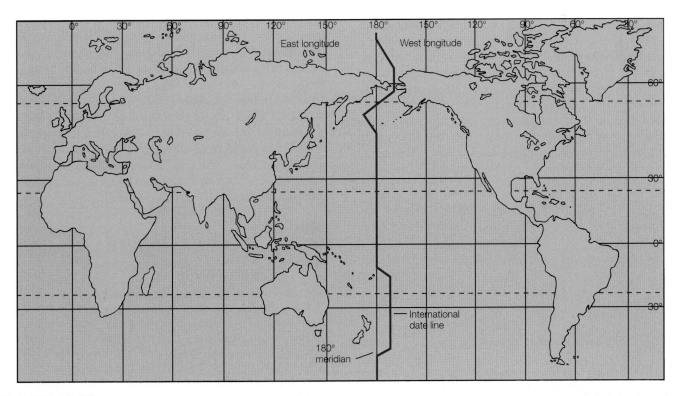

FIGURE 16.25 The international date line follows the 180° meridian but is arranged in a way that land areas and island chains have the same date.

14 seconds) does not seem significant when compared to the time in a whole year. But over the years, the error of minutes and seconds grew to an error of days. By 1582, when Pope Gregory XIII revised the calendar, the error had grown to 13 days but was corrected for 10 days of error. This revision resulted in the *Gregorian calendar,* which is the system used today. Since the accumulated error of 0.0078 day per year is almost 0.75 day per century, it follows that four centuries will have 0.75 times 4, or 3 days of error. The Gregorian system corrects for the accumulated error by dropping the additional leap year day three centuries out of every four. Thus, the century year of 2000 was a leap year with 366 days, but the century years of 2100, 2200, and 2300 will not be leap years. You will note that this approximation still leaves an error of 0.0003 day per century, so another calendar revision will be necessary in a few thousand years to keep the calendar in step with the seasons.

Monthly Time

In ancient times, people often used the Moon to measure time intervals that were longer than a day but shorter than a year. The word *month,* in fact, has its origins in the word *moon* and its period of revolution. The Moon revolves around Earth in an orbit that is inclined to the plane of Earth's orbit, the plane of the ecliptic, by about 5°. The Moon is thus never more than about ten apparent diameters from the ecliptic. It revolves in this orbit in about 27⅓ days as measured by two consecutive crossings of any star. This period is called a **sidereal month.** The Moon rotates in the same period as the time of revolution, so the sidereal month is also the time required for one rotation. Because the rotation and revolution rates are the same, you always see the same side of the Moon from Earth.

The ancient concept of a month was based on the **synodic month,** the interval of time from new moon to new moon (or any two consecutive identical phases). The synodic month is longer than a sidereal month at a little more than 29½ days. The Moon's phases (see the section on the Moon) are determined by the relative positions of Earth, Moon, and Sun. As shown in Figure 16.26, the Moon moves with Earth in its orbit around the Sun. During one sidereal, month, the Moon has to revolve a greater distance before the same phase is observed on Earth, and this greater distance requires 2.2 days. This makes

the synodic month about 29½ days long, only a little less than ¹⁄₁₂ of a year, or the period of time the present calendar identifies as a "month."

THE MOON

Next to the Sun, the Moon is the largest, brightest object in the sky. The Moon is Earth's nearest neighbor at an average distance of 380,000 km (about 238,000 mi), and surface features can be observed with the naked eye. With the aid of a telescope or a good pair of binoculars, you can see light-colored mountainous regions called the **lunar highlands,** smooth, dark areas called **maria,** and many sizes of craters, some with bright streaks extending from them (Figure 16.27). The smooth, dark areas are called maria after a Latin word meaning "sea." They acquired this name from early observers who thought the dark areas were oceans and the light areas were continents. Today, the maria are understood to have formed from ancient floods of molten lava that poured across the surface and solidified to form the "seas" of today. There is no water and no atmosphere on the Moon.

Many facts known about the Moon were established during the *Apollo* missions, the first human exploration of a place away from Earth. A total of twelve *Apollo* astronauts walked on the Moon, taking thousands of photographs, conducting hundreds of experiments, and returning to Earth with over 380 kg (about 840 lb) of moon rocks (Table 16.1). In addition, instruments were left on the Moon that continued to radio data back to Earth after the *Apollo* program ended in 1972. As a result of the *Apollo* missions, many questions were answered about the Moon, but unanswered questions still remain.

Unmanned missions to the Moon have returned data about mineral composition, topography, the presence of water, and other important information. In 1994, the spacecraft *Clementine* measured the vertical topography of the Moon, which has resulted in the best global map of the Moon to date. The *Lunar Prospector* was launched in January 1998 and later found evidence of extensive deposits of water ice in permanently shadowed areas of deep craters. Deposits of ice on the Moon could be important for future manned lunar exploration. Shipping water to the Moon would be expensive at $2,000 to $20,000 per liter, according to NASA. Water on the Moon could serve as drinking water, of course, but it could also serve as a source of oxygen and hydrogen, which could be used as rocket fuel.

CONCEPTS *Applied*

A Bigger Moon?

Why does the Moon appear so large when it is on the horizon? The truth is that the Moon is actually *smaller* when it is on the horizon than when it is overhead. It is smaller because when it is on the horizon, it is farther away by one Earth radius. It is an optical illusion that the Moon is larger on the horizon. You can test this by bending a paperclip so that it is "one moon" wide when held at arm's length. Repeat the paperclip measurement when the Moon is overhead.

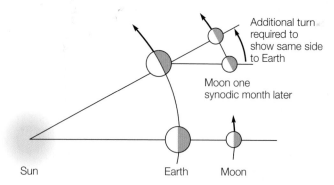

FIGURE 16.26 As the Moon moves in its orbit around Earth, it must revolve a greater distance to bring the same part to face Earth. The additional turning requires about 2.2 days, making the synodic month longer than the sidereal month.

FIGURE 16.27 You can easily see the light-colored lunar highlands, smooth and dark maria, and many craters on the surface of Earth's nearest neighbor in space.

TABLE 16.1

The Apollo missions

Mission	Date	Crew	Comments
Apollo 1	Jan 27, 1967	Gus Grissom, Ed White, Roger Chaffee	The crew died in their spacecraft during a test three weeks before they would have flown in space.
Apollo 7	Oct 11–22, 1968	Wally Schirra, Donn F. Eisele, Walter Cunningham	This was the first *Apollo* mission in space following the *Apollo 1* launchpad fire. It was an eleven-day mission to validate *Apollo* hardware in low Earth orbit.
Apollo 8	Dec 21–27, 1968	Frank Borman, Jim Lovell, Bill Anders	First space mission to orbit the Moon. The first picture of Earth taken from deep space.
Apollo 9	Mar 3–13, 1969	James A. McDivitt, David R. Scott, Russel L. Schweickhart	First test of the lunar module (LM) in space. *Apollo 9* was an Earth orbital mission. *Apollo* type rendezvous and docking was tested after a 6-hour, 113-mile separation.
Apollo 10	May 18–26, 1969	Tom Stafford, John Young, Gene Cernan	Trial rehearsal of Moon landing. The LM was taken to the Moon and separated from the command module, but it did not land on the Moon. The LM was tested in lunar orbit.
Apollo 11	Jul 16–24, 1969	Neil Armstrong, Mike Collins, Buzz Aldrin	The lunar module *Eagle* landed the first man, Neil Armstrong, on the Moon on July 20, 1969, and established the first manned Moon base.
Apollo 12	Nov 14–24, 1969	Peter Conrad, Dick Gordon, Al Bean	Landing was very accurate, only 163 m (535 ft) from *Surveyor III*. The crew conducted two moon walks and put up both a geophysical station and a nuclear power station.
Apollo 13	Apr 11–17, 1970	Jim Lovell, Jack Swigert, Fred Haise	This was the first abort in deep space (321,860 km or 200,000 mi from Earth). The lunar module was used as a lifeboat to return the crew safely.
Apollo 14	Jan 31–Feb 9, 1971	Alan Shepard, Stuart A. Roosa, Edgar D. Mitchell	This was the third mission to land on the Moon, landing in the Fra Mauro region. Al Shepard hit two golf balls on the Moon. Lunar specimens (43 kg, or 95 lb) were collected.
Apollo 15	Jul 26–Aug 7, 1971	Dave Scott, Alfred M. Worden, James B. Irwin	During their record time on the Moon (66 h 54 min), the crew placed a subsatellite in lunar orbit and were the first to use the lunar rover.
Apollo 16	Apr 16–27, 1972	John Young, Thomas K. Mattingly II, Charles M. Duke	Highest landing on the Moon (elevation 7,830 m or 25,688 ft); lunar rover land speed record of 18.0 km/h (11.2 mi/h) and distance record of 36 m (22.4 mi) covered. The crew returned 97 kg (213 lb) of lunar samples.
Apollo 17	Dec 7–19, 1972	Gene Cernan, Ronald E. Evans, Harrison H. Schmitt	This, the last of the *Apollo* flights, was the first time an *Apollo* flight was launched at night. A record of 75 hours was set for time spent on the Moon, and 114 kg (250 lb) of lunar samples were returned.
Apollo-Soyuz Mission	Jul 15–24, 1975	Tom Stafford, Deke Slayton, Vance Brand	First international space rendezvous. This was the first coordinated launch of two spacecraft from different countries.

Source: Data from NASA.

COMPOSITION AND FEATURES

The *Apollo* astronauts found that the surface of the Moon is covered by a 3 m (about 10 ft) layer of fine gray dust that contains microscopic glass beads. The dust and beads were formed from millions of years of continuous bombardment of micrometeorites. These very small meteorites generally burn up in Earth's atmosphere. The Moon does not have an atmosphere, so it is continually fragmented and pulverized. The glass beads are believed to have formed when larger meteorite impacts melted part of the surface, which was immediately forced into a fine spray that cooled rapidly while above the surface.

The rocks on the surface of the Moon were found to be mostly *basalts*, a type of rock formed on Earth from the cooling and solidification of molten lava. The dark-colored rocks from the maria are similar to Earth's basalts but contain greater amounts of titanium and iron oxides. The light-colored rocks from the highlands are mostly *brecchias*, a kind of rock made up of rock fragments that have been compacted together. On the Moon, the compacting was done by meteorite impacts. The rocks from the highlands contain more aluminum and less iron and thus have a lower density (2.9 g/cm^3) than the darker maria rocks (3.3 g/cm^3).

All the moon rocks contain a substantial amount of radioactive elements, which made it possible to precisely measure their age. The light-colored rocks from the highlands were formed some 4 billion years ago. The dark-colored rocks from the maria were much younger, with ages ranging from 3.1 to 3.8 billion years. This indicates a period of repeated volcanic eruptions and lava flooding over a 700-million-year period that ended about 3 billion years ago.

Seismometers left on the Moon by *Apollo* astronauts detected only very weak moonquakes, so weak that they would not be felt by a person. These moonquakes are thought to be produced by the nearby impact of larger meteoroids or by a slight cracking of the crust from gravitational interactions with Earth and the Sun. The movement of these seismic waves through the Moon's interior suggests that the Moon has an internal structure. The outer layer of solid rock, or crust, is about 65 km (about 40 mi) thick on the side that always faces Earth and is about twice as thick on the far side. The data also suggest a small, partly molten iron core at a depth of about 900 km (about 600 mi) beneath the surface. A small core would account for the Moon's low density (3.34 g/cm^3) as compared to Earth's average (5.5 g/cm^3) and for the observation that the Moon has no general magnetic field.

HISTORY OF THE MOON

The moon rocks brought back to Earth, the results of the lunar seismographs, and all the other data gathered through the *Apollo* missions have increased our knowledge about the Moon, leading to new understandings of how it formed. This model pictures the present Moon developing through four distinct stages of events.

Stage 1. The *origin stage* describes how the Moon originally formed. The Moon is believed to have originated from the impact of Earth with a very large object, perhaps as large as Mars or larger. The Moon formed from ejected material produced by this collision. The collision is believed to have vaporized the colliding body as well as part of Earth. Some of the debris condensed away from Earth to form the Moon.

The collision that resulted in the Moon took place after Earth's iron core formed, so there is not much iron in moon rocks. The difference in Moon and Earth rocks can be accounted for by the presence of materials from the impacting body.

Stage 2. The *molten surface stage* occurred during the first 200 million years after the collision. Heating from a number of sources could have been involved in the melting of the entire lunar surface 100 km (about 60 mi) or so deep. The heating required to melt the surface is believed to have resulted from the impacts of rock fragments, which were leftover debris from the formation of the solar system that intensely bombarded the Moon. After a time, there were fewer fragments left to bombard the Moon, and the molten outer layer cooled and solidified to solid rock. The craters we see on the Moon today are the result of meteorite bombardment that occured between 3.9 and 4.2 billion years ago after the crust had formed.

Stage 3. The *molten interior stage* involved the melting of the interior of the Moon. Radioactive decay had been slowly heating the interior, and 3.8 billion years ago, or about a billion years after the Moon formed, sufficient heat accumulated to melt the interior. The light and heavier rock materials separated during this period, perhaps producing a small iron core. Molten lava flowed into basins on the surface during this period, forming the smooth, darker maria seen today. The lava flooding continued for about 700 million years, ending about 3.1 billion years ago.

Stage 4. The *cold and quiet stage* began 3.1 billion years ago as the last lava flow cooled and solidified. Since that time, the surface of the Moon has been continually bombarded by micrometeorites and a few larger meteorites. With the exception of a few new craters, the surface of the Moon has changed little in the last 3 billion years.

THE EARTH-MOON SYSTEM

Earth and its Moon are unique in the solar system because of the size of the Moon. It is not the largest satellite, but the ratio of its mass to Earth's mass is greater than the mass ratio of any other moon to its planet. The Moon has a diameter of 3,476 km (about 2,159 mi), which is about one-fourth the diameter of Earth, and a mass of about 1/81 of Earth's mass. This is a small fraction of Earth's mass, but it is enough to affect Earth's motion as it revolves around the Sun.

If the Moon had a negligible mass, it would circle Earth with a center of rotation (center of mass) located at the center of Earth. In this situation, the center of Earth would follow a smooth path around the Sun (Figure 16.28A). The mass of the Moon, however, is great enough to move the center of rotation away from Earth's center toward the Moon. As a result, both bodies act as a system, moving around a center of mass. The center of mass between Earth and the Moon follows a smooth orbit around the Sun. Earth follows a slightly wavy path around the Sun as it slowly revolves around the common center of mass (Figure 16.28B).

PHASES OF THE MOON

The phases of the Moon are a result of the changing relative positions of Earth, the Moon, and the Sun as the Earth-Moon system moves around the Sun. Sunlight always illuminates

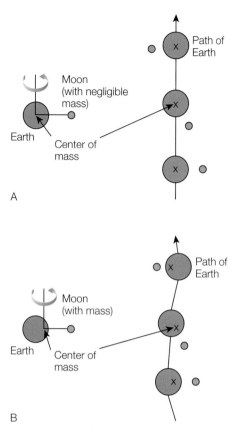

A

B

FIGURE 16.28 (*A*) If the Moon had a negligible mass, the center of gravity between the Moon and Earth would be Earth's center, and Earth would follow a smooth orbit around the Sun. (*B*) The actual location of the center of mass between Earth and Moon (identified with X) results in a slightly in and out, or wavy, path around the Sun.

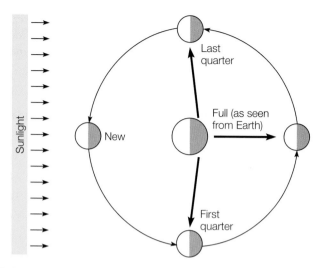

FIGURE 16.29 Half of the Moon is always lighted by the Sun, and half is always in the shadow. The moon phases result from the view of the lighted and dark parts as the Moon revolves around Earth.

FIGURE 16.30 The cusps, or horns, of the Moon always point away from the Sun. A line drawn from the tip of one cusp to the other is perpendicular to a straight line between the Moon and the Sun.

half of the Moon, and half is always in shadow. As the Moon's path takes it between Earth and the Sun, then to the dark side of Earth, you see different parts of the illuminated half called *phases* (Figure 16.29). When the Moon is on the dark side of Earth, you see the entire illuminated half of the Moon called the **full moon** (or the full phase). Halfway around the orbit, the lighted side of the Moon now faces away from Earth, and the unlighted side now faces Earth. This dark appearance is called the **new moon** (or the new phase). In the new phase, the Moon is not *directly* between Earth and the Sun, so it does not produce an eclipse (see the section on eclipses).

As the Moon moves from the new phase in its orbit around Earth, you will eventually see half the lighted surface, which is known as the **first quarter.** Often the unlighted part of the Moon shines with a dim light of reflected sunlight from Earth called *earthshine*. Note that the division between the lighted and unlighted part of the Moon's surface is curved in an arc. A straight line connecting the ends of the arc is perpendicular to the direction of the Sun (Figure 16.30). After the first quarter, the Moon moves to its full phase, then to the **last quarter** (see Figure 16.29). The period of time between two consecutive phases, such as new moon to new moon, is the synodic month, or about 29.5 days.

Myths, Mistakes, & Misunderstandings

Moon Mistakes

1. It is a common misunderstanding that Earth's shadow creates the moon phases. In fact, the moon phases are caused by viewing different parts of the Moon that are in sunlight and not in sunlight.

2. The phrase "blue moon" is not a reference to the color of the Moon. Instead, it is a reference to a second full moon that occurs in a calendar month. This doesn't happen very often, and the phrase "once in a blue moon" means a very long period of time.

3. A popular myth has more accidents happening during a full moon phase. However, studies have found no statistically significant relationships between the accident rate and the moon phase. This myth is probably a result of people remembering the full moon phase but not the other moon phases during an accident.

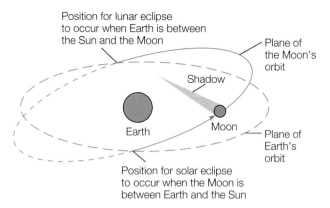

Position for lunar eclipse to occur when Earth is between the Sun and the Moon

Plane of the Moon's orbit

Shadow

Earth   Moon

Plane of Earth's orbit

Position for solar eclipse to occur when the Moon is between Earth and the Sun

FIGURE 16.31 The plane of the Moon's orbit is inclined to the plane of Earth's orbit by about 5°. An eclipse occurs only where the two planes intersect, and Earth, the Moon, and the Sun are in a line.

ECLIPSES OF THE SUN AND MOON

Sunlight is not visible in the emptiness of space because there is nothing to reflect the light, so the long conical shadow behind each spherical body is not visible, either. One side of Earth and one side of the Moon are always visible because they reflect sunlight. The shadow from Earth or from the Moon becomes noticeable only when it falls on the illuminated surface of the other body. This event of Earth's or the Moon's shadow falling on the other body is called an **eclipse.** Most of the time eclipses do not occur because the plane of the Moon's orbit is inclined to Earth's orbit about 5° (Figure 16.31). As a result, the shadow from the Moon or the shadow from Earth usually falls above or below the other body, too high or too low to produce an eclipse. An eclipse occurs only when the Sun, Moon, and Earth are in a line with each other (Table 16.2).

The shadow from Earth and the shadow from the Moon are long cones that point away from the Sun. Both cones have two parts, an inner cone of a complete shadow called the **umbra** and an outer cone of partial shadow called the **penumbra.** When and where the umbra of the Moon's shadow falls on Earth, people see a **total solar eclipse.** During a total solar eclipse, the new Moon completely covers the disk of the Sun. The total solar eclipse is preceded and followed by a partial eclipse, which is seen when the observer is in the penumbra. If the observer is in a location where only the penumbra passes, then only a partial eclipse will be observed (Figure 16.32). More people see partial than full solar eclipses because the penumbra covers a larger area. The occurrence of a total solar eclipse is a rare event in a given location, occurring once every several hundred years and then lasting for less than seven minutes.

The Moon's cone-shaped shadow averages a length of 375,000 km (about 233,000 mi), which is less than the average distance between Earth and the Moon. The Moon's elliptical orbit brings it sometimes closer to and sometimes farther from Earth. A total solar eclipse occurs only when the Moon is close enough so at least the tip of its umbra reaches the surface of Earth. If the Moon's umbra fails to reach Earth, an **annular eclipse** occurs. *Annular* means "ring-shaped," and during this eclipse, the edge of the Sun is seen to form a bright ring around

TABLE 16.2

Total Eclipses in the United States—2008–2025*

Total Solar Eclipses

Aug 21, 2017	Path: Oregon to South Carolina
Apr 8, 2024	Path: Texas to Maine

Total Lunar Eclipses

Feb 21, 2008	All visible eastern; moonrise western
Dec 21, 2010	All visible
Dec 10, 2011	Visible at moonset
Apr 15, 2014	Visible at moonset
Oct 8, 2014	All visible western; at moonset eastern
Apr 4, 2015	Visible at moonset
Sept 28, 2015	Western visible at moonrise; all eastern
Jan 31, 2018	Visible at moonset
Jan 21, 2019	All visible
May 26, 2021	Visible at moonset
May 16, 2022	Visible moonrise western; all eastern
Mar 14, 2025	All visible

*Eclipse predictions by Fred Espenak, NASA/GSFC. See http://eclipse.gsfc.nasa.gov/eclipse.html.

the Moon. As before, people located in the area where the penumbra falls will see a partial eclipse. The annular eclipse occurs more frequently than the total solar eclipse.

When the Moon is full and the Sun, Moon, and Earth are lined up so Earth's shadow falls on the Moon, a **lunar eclipse** occurs. Earth's shadow is much larger than the Moon's diameter, so a lunar eclipse is visible to everyone on the night side of Earth. This larger shadow also means a longer eclipse that may last for hours. As the umbra moves over the Moon, the darkened part takes on a reddish, somewhat copper-colored glow from light refracted and scattered into the umbra by Earth's atmosphere. This light passes through the thickness of Earth's atmosphere on its way to the eclipsed Moon, and it acquires the reddish color for the same reason that a sunset is red: much of the blue light has been removed by scattering in Earth's atmosphere.

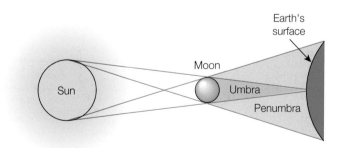

Earth's surface

Sun   Moon   Umbra   Penumbra

FIGURE 16.32 People in a location where the tip of the umbra falls on the surface of Earth see a total solar eclipse. People in locations where the penumbra falls on the Earth's surface see a partial solar eclipse.

TIDES

If you live near or have ever visited a coastal area of the ocean, you are familiar with the periodic rise and fall of sea level known as **tides.** The relationship between the motions of the Moon and the magnitude and timing of tides has been known and studied since very early times. These relationships are that (1) the greatest range of tides occurs at full and new moon phases, (2) the least range of tides occurs at quarter moon phases, and (3) in most oceans, the time between two high tides or between two low tides is an average of twelve hours and twenty-five minutes. The period of twelve hours and twenty-five minutes is half the average time interval between consecutive passages of the Moon across the celestial meridian. A location on the surface of Earth is directly under the Moon when it crosses the meridian and directly opposite it on the far side of Earth an average twelve hours and twenty-five minutes later. There are two *tidal bulges* that follow the Moon as it moves around Earth, one on the side facing the Moon and one on the opposite side. In general, tides are a result of these bulges moving westward around Earth.

A simplified explanation of the two tidal bulges involves two basic factors: the gravitational attraction of the Moon and the motion of the Earth-Moon system (Figure 16.33). Water on Earth's surface is free to move, and the Moon's gravitational attraction pulls the water to the tidal bulge on the side of Earth facing the Moon. This tide-raising force directed toward the Moon bulges the water in mid-ocean some .75 m (about 2.5 ft), but it also bulges the land, producing a land tide. Since land is much more rigid than water, the land tide is much smaller at about 12 cm (about 4.5 in). Since all parts of the land bulge together, this movement is not evident without measurement by sensitive instruments.

The tidal bulge on the side of Earth opposite the Moon occurs as Earth is pulled away from the ocean by the Earth-Moon gravitational interaction. Between the tidal bulges facing the Moon and the tidal bulge on the opposite side, sea level is depressed across the broad surface. The depression is called a *tidal trough,* even though it does not actually have the shape of a trough. The two tidal bulges, with the trough between, move slowly eastward following the Moon. Earth turns more rapidly on its axis, however, which forces the tidal bulge to stay in front of the Moon moving through its orbit. Thus, the tidal axis is not aligned with the Earth-Moon gravitational axis.

The tides do not actually appear as alternating bulges that move around Earth. There are a number of factors that influence the making and moving of the bulges in complex interactions that determine the timing and size of a tide at a given time in a given location. Some of these factors include (1) the relative positions of Earth, Moon, and Sun, (2) the elliptical orbit of the Moon, which sometimes brings it closer to Earth, and (3) the size, shape, and depth of the basin holding the water.

The relative positions of Earth, Moon, and Sun determine the size of a given tide because the Sun, as well as the Moon, produces a tide-raising force. The Sun is much more massive than the Moon, but it is so far away that its tide-raising force is about half that of the closer Moon. Thus, the Sun basically modifies lunar tides rather than producing distinct tides of its own. For example, Earth, Moon, and Sun are nearly in line during the full and new moon phases. At these times, the lunar and solar tide-producing forces act together, producing tides that are unusually high and corresponding low tides that are unusually low. The periods of these unusually high and low tides are called **spring tides.** Spring tides occur every two weeks and have nothing to do with the spring season. When the Moon is in its quarter phases, the Sun and Moon are at right angles to one another, and the solar tides occur between the lunar tides, causing unusually less pronounced high and low tides called **neap tides.** The period of neap tides also occurs every two weeks.

The size of the lunar-produced tidal bulge varies as the Moon's distance from Earth changes. The Moon's elliptical orbit brings it closest to Earth at a point called **perigee** and farthest from Earth at a point called **apogee.** At perigee, the Moon is about 44,800 km (about 28,000 mi) closer to Earth than at apogee, so its gravitational attraction is much greater. When perigee coincides with a new or full moon, especially high spring tides result.

The open basins of oceans, gulfs, and bays are all connected but have different shapes and sizes and have bordering landmasses in all possible orientations to the westward-moving tidal bulges. Water in each basin responds differently to the tidal forces, responding as periodic resonant oscillations that move back and forth much like the water in a bowl shifts when carried. Thus, coastal regions on open seas may experience tides that range between about 1 and 3 m (about 3 to 10 ft), but mostly enclosed basins such as the Gulf of Mexico have tides of less than about 1/3 m (about 1 ft). The Gulf of Mexico, because of its size, depth, and limited connections with the open ocean, responds only to the stronger tidal attractions and has only one high and one low tide per day. Even lakes and ponds respond to tidal attractions, but the result is too small to be noticed. Other basins, such as the Bay of Fundy in Nova Scotia, are funnel-shaped and undergo an unusually high tidal range. The Bay of Fundy has experienced as much as a 15 m (about 50 ft) tidal range.

As the tidal bulges are pulled against a rotating Earth, friction between the moving water and the ocean basin tends to slow Earth's rotation over time. This is a very small slowing effect that is increasing the length of each day by about 1.5 seconds per year.

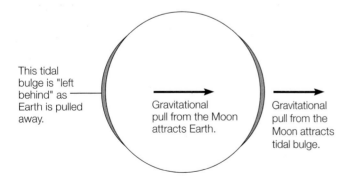

This tidal bulge is "left behind" as Earth is pulled away.

Gravitational pull from the Moon attracts Earth.

Gravitational pull from the Moon attracts tidal bulge.

FIGURE 16.33 Gravitational attraction pulls on Earth's waters on the side of Earth facing the Moon, producing a tidal bulge. A second tidal bulge on the side of Earth opposite the Moon is produced when Earth, which is closer to the Moon, is pulled away from the waters.

Carl Edward Sagan (1934–1996)

Carl Edward Sagan was an American astronomer and popularizer of astronomy whose main research was on planetary atmospheres, including that of the primordial Earth. His most remarkable achievement was to provide valuable insights into the origin of life on our planet.

Sagan was born on November 19, 1934, in New York City. Completing his education at the University of Chicago, he obtained his bachelor's degree in 1955 and his doctorate in 1960. Then, for two years, he was a research fellow at the University of California in Berkeley before he transferred to the Smithsonian Astrophysical Observatory in Cambridge, Massachusetts, lecturing also at Harvard University, where he became assistant professor. Finally, in 1968, Sagan moved to Cornell University in Ithaca (New York) and took a position as director of the Laboratory for Planetary Studies; in 1970, he became professor of astronomy and space science there. He died on December 20, 1996.

In the early 1960s, Sagan's first major research was into the planetary surface and atmosphere of Venus. At the time, although intense emission of radiation had shown that the dark-side temperature of Venus was nearly 600K, it was thought that the surface itself remained relatively cool—leaving open the possibility that there was some form of life on the planet. Various hypotheses were put forward to account for the strong emission actually observed: perhaps it was due

to interactions between charged particles in Venus's dense upper atmosphere; perhaps it was glow discharge between positive and negative charges in the atmosphere; or perhaps emission was due to a particular radiation from charged particles trapped in the Venusian equivalent of a Van Allen Belt. Sagan showed that each of these hypotheses was incompatible with other observed characteristics or with implications of these characteristics. The positive part of Sagan's proposal was to show that all the observed characteristics were compatible with the straightforward hypothesis that the surface of Venus was very hot. On the basis of radar and optical observations, the distance between surface and clouds was calculated to be between 44 km (27 mi) and 65 km (40 mi); given the cloud-top temperature and Sagan's expectation of a "greenhouse effect" in the atmosphere, surface temperature on Venus was computed to be between 500K (227°C/440°F) and 800K

(527°C/980°F)—the range that would also be expected on the basis of emission rate.

Sagan then turned his attention to the early planetary atmosphere of Earth, with regard to the origins of life. One way of understanding how life began is to try to form the compounds essential to life in conditions analogous to those of the primeval atmosphere. Before Sagan, Stanley Miller and Harold Urey had used a mixture of methane, ammonia, water vapor, and hydrogen, sparked by a corona discharge that simulated the effect of lightning, to produce amino and hydroxy acids of the sort found in life forms. Later experiments used ultraviolet light or heat as sources of energy, and even these had less energy than would have been available in Earth's primordial state. Sagan followed a similar method and, by irradiating a mixture of methane, ammonia, water, and hydrogen sulfide, was able to produce amino acids—and, in addition, glucose, fructose, and nucleic acids. Sugars can be made from formaldehyde under alkaline conditions and in the presence of inorganic catalysts. These sugars include five-carbon sugars, which are essential to the formation of nucleic acids, glucose, and fructose—all common metabolites found as constituents of present-day life forms. Sagan's simulated primordial atmosphere not only showed the presence of those metabolites, it also contained traces of adenosine triphosphate (ATP)—the foremost agent used by living cells to store energy.

Source: From the Hutchinson *Dictionary of Scientific Biography.* © Research Machines plc 2003. All Rights Reserved. Helicon Publishing is a division of Research Machines.

Evidence for this slowing comes from a number of sources, including records of ancient solar eclipses. The solar eclipses of two thousand years ago occurred three hours earlier than would be expected by using today's time but were on the mark if a lengthening day is considered. Fossils of a certain species of coral still living today provide further evidence of a lengthening day. This particular coral adds daily growth rings, and 500-million-year-old fossils

show that the day was about twenty-one hours long at that time. Finally, the Moon is moving away from Earth at a rate of about 4 cm (about 1.5 in) per year. This movement out to a larger orbit is a necessary condition to conserve angular momentum as Earth slows. As the Moon moves away from Earth, the length of the month increases. Some time in the distant future, both the day and the month will be equal, about fifty of the present days long.

SUMMARY

Earth is an *oblate spheroid* that undergoes three basic motions: (1) a yearly *revolution* around the Sun, (2) a daily *rotation* on its axis, and (3) a slow wobble of its axis called *precession.*

As Earth makes its yearly *revolution* around the Sun, it maintains a generally *constant inclination of its axis* to the *plane of the ecliptic* of

66.5°, or 23.5° from a line perpendicular to the plane. In addition, Earth maintains a generally constant *orientation of its axis* to the stars, which always points in the same direction. The constant inclination and orientation of the axis, together with Earth's rotation and revolution, produce three effects: (1) days and nights that vary in length, (2) seasons that change during the course of a year, and (3) climates that vary with latitude. When Earth is at a place in its orbit so the axis points toward the Sun, the Northern Hemisphere experiences the longest days and the summer season. This begins on June 21 or 22, which is called the *summer solstice*. Six months later, the axis points away from the Sun, and the Northern Hemisphere experiences the shortest days and the winter season. This begins on December 22 or 23 and is called the *winter solstice*. On March 20 or 21, Earth is halfway between the solstices and has days and nights of equal length, which is called the *spring* (or *vernal*) *equinox*. On September 22 or 23, the *autumnal equinox*, another period of equal nights and days, identifies the beginning of the fall season.

Precession is a slow wobbling of the axis as Earth spins. Precession is produced by the gravitational tugs of the Sun and Moon on Earth's equatorial bulge.

Lines around Earth that are parallel to the equator are circles called *parallels*. The distance from the equator to a point on a parallel is called the *latitude* of that point. North and south arcs that intersect at the poles are called *meridians*. The meridian that runs through the Greenwich Observatory is a reference line called the *prime meridian*. The distance of a point east or west of the prime meridian is called the *longitude* of that point.

The event of time called *noon* is the instant the Sun appears to move across the celestial meridian. The instant of noon at a particular location is called the *apparent local noon*. The time at a given place that is determined by a sundial is called the *apparent local solar time*. It is the basis for an averaged, uniform standard time called the *mean solar time*. Mean solar time is the time used to set clocks.

A *sidereal day* is the interval between two consecutive crossings of the celestial meridian by a star. An *apparent solar day* is the interval between two consecutive crossings of the celestial meridian by the Sun, from one apparent solar noon to the next. A *mean solar day* is twenty-four hours as determined from mean solar time. The *equation of time* shows how the local solar time is faster or slower than the clock time during different days of the year.

Earth's surface is divided into one-hour *standard time zones* that are about 15° of meridian wide. The *international date line* is the 180° meridian; you gain a day if you cross this line while traveling westward and repeat a day if you are traveling eastward.

A *tropical year* is the interval between two consecutive spring equinoxes. A *sidereal year* is the interval of time between two consecutive crossings of a star by the Sun. It is the tropical year that is used as a standard time interval for the calendar year. A *sidereal month* is the interval of time between two consecutive crossings of a star by the Moon. The *synodic month* is the interval of time from a new moon to the next new moon. The *synodic month* is about 29 1/2 days long, which is about 1/12 of a year.

The surface of the Moon has light-colored mountainous regions called *highlands*, smooth dark areas called *maria*, many sizes of *craters*, and is covered by a layer of fine *dust*. Samples of rocks returned to Earth by *Apollo* astronauts revealed that the highlands are composed of basalt breccias that were formed some 4 billion years ago. The maria are basalts that formed from solidified lava some 3.1 to 3.8 billion years ago. This, and other data, indicate that the Moon developed through four stages.

Earth and the Moon act as a system, with both bodies revolving around a common center of mass located under Earth's surface. This combined motion around the Sun produces three phenomena: (1) as the Earth-Moon system revolves around the Sun, different parts of the illuminated lunar surface, called *phases,* are visible from Earth; (2) a *solar eclipse* is observed where the Moon's shadow falls on Earth, and a *lunar eclipse* is observed where Earth's shadow falls on the Moon; and (3) the *tides,* a periodic rising and falling of sea level, are produced by gravitational attractions of the Moon and Sun and by the movement of the Earth-Moon system.

KEY TERMS

annular eclipse (p. **423**)

Antarctic Circle (p. **412**)

apogee (p. **424**)

apparent local noon (p. **413**)

apparent local solar time (p. **413**)

apparent solar day (p. **414**)

Arctic Circle (p. **412**)

autumnal equinox (p. **409**)

Coriolis effect (p. **410**)

daylight saving time (p. **416**)

eclipse (p. **423**)

equation of time (p. **415**)

equinoxes (p. **409**)

first quarter (p. **422**)

full moon (p. **422**)

international date line (p. **416**)

last quarter (p. **422**)

latitude (p. **411**)

longitude (p. **412**)

lunar eclipse (p. **423**)

lunar highlands (p. **418**)

maria (p. **418**)

mean solar day (p. **414**)

mean solar time (p. **413**)

meridians (p. **412**)

neap tide (p. **424**)

new moon (p. **422**)

noon (p. **413**)

parallels (p. **411**)

penumbra (p. **423**)

perigee (p. **424**)

plane of the ecliptic (p. **406**)

precession (p. **411**)

prime meridian (p. **412**)

revolution (p. **408**)

rotation (p. **408**)

sidereal day (p. **414**)

sidereal month (p. **418**)

sidereal year (p. **416**)

solstices (p. **409**)

spring equinox (p. **409**)

spring tides (p. **424**)

standard time zones (p. **416**)

summer solstice (p. **409**)

synodic month (p. **418**)

tides (p. **424**)

total solar eclipse (p. **423**)

tropical year (p. **416**)
tropic of Cancer (p. **412**)
tropic of Capricorn (p. **412**)
umbra (p. **423**)
vernal equinox (p. **409**)
winter solstice (p. **409**)

APPLYING THE CONCEPTS

1. The plane of Earth's orbit is called the
   a. plane of Earth.
   b. plane of the solar system.
   c. plane of the ecliptic.
   d. plane of the Sun.

2. The spinning of a planet on its axis, an imaginary line through its poles, is called
   a. twenty-four hour day.
   b. rotation.
   c. revolution.
   d. retrograde motion.

3. The consistent tilt and the orientation of its axis as Earth moves around its orbit is responsible for
   a. tides.
   b. seasons.
   c. volcanic eruptions.
   d. earthquakes.

4. In the Northern Hemisphere, the North Pole points toward the Sun during the
   a. summer solstice.
   b. spring solstice.
   c. winter equinox.
   d. winter solstice.

5. The referent meridian is the
   a. prime meridian.
   b. principal meridian.
   c. first meridian.
   d. primary meridian.

6. The parallel at 66.5° S is called the
   a. Arctic Circle.
   b. Antarctic Circle.
   c. Tropic of Cancer.
   d. Tropic of Capricorn.

7. The movement of the Sun across the celestial meridian is defined as
   a. midnight.
   b. 6 A.M.
   c. 6 P.M.
   d. noon.

8. Clocks and watches are set to measure a uniform standard time based on
   a. mean solar time.
   b. apparent solar day.
   c. sidereal day.
   d. apparent solar time.

9. How many standard time zones are there?
   a. ten
   b. twelve
   c. twenty
   d. twenty-four

10. The 180° meridian is called the
    a. Tropic of Cancer.
    b. prime meridian.
    c. international date line.
    d. Arctic Circle.

11. The time period from one new moon to the next new moon is called the
    a. sidereal month.
    b. synodic month.
    c. lunar month.
    d. apparent month.

12. Maria are
    a. craters on the Moon.
    b. mountains on the Moon.
    c. ancient molten lava seas on the Moon.
    d. dried-up oceans on the Moon.

13. Unmanned missions to the Moon did not find or identify
    a. active volcanoes.
    b. water.
    c. minerals.
    d. lunar topography.

14. Rocks on the surface of the Moon are primarily
    a. sedimentary.
    b. basaltic.
    c. limestone.
    d. metamorphic.

15. The atmosphere of the Moon is
    a. primarily H_2.
    b. primarily N_2.
    c. primarily CO_2.
    d. It does not have an atmosphere.

16. The approximate age of the Moon was determined by
    a. radioactive dating of moon rocks.
    b. exploration by *Apollo* astronauts.
    c. measurement of the depth of the surface dust.
    d. laser measurement of the distance from Earth to the Moon.

17. What is the accepted theory about the origin of the Moon?
    a. It formed from particles revolving around Earth.
    b. A very large object hit Earth and the Moon was formed from ejected material.
    c. Earth captured an incoming asteroid and it became the Moon.
    d. Earth and the Moon were created at the same time.

18. The Moon is positioned between Earth and the Sun. This phase is the
    a. full moon.
    b. new moon.
    c. first quarter moon.
    d. last quarter moon.

19. Tides that occur at the full and new moon phases are called
    a. new tides.
    b. lunar tides.
    c. spring tides.
    d. tidal waves.

20. Friction between the tides and the ocean basin is
    a. wearing away the ocean basin.
    b. moving Earth closer to the Moon.
    c. increasing the rotation of Earth.
    d. slowing the rotation of Earth.

21. Earth is undergoing a combination of how many different major motions?
    a. zero
    b. one
    c. three
    d. seven

22. In the Northern Hemisphere, city A is located a number of miles north of city B. At 12 noon in city B, the Sun appears directly overhead. At this very same time, the Sun over city A will appear
    a. to the north of overhead.
    b. directly overhead.
    c. to the south of overhead.

23. Earth as a whole receives the most solar energy during what month?
    a. January
    b. March
    c. July
    d. September

24. During the course of a year and relative to the Sun, Earth's axis points
    a. always toward the Sun.
    b. toward the Sun half the year and away from the Sun the other half.
    c. always away from the Sun.
    d. toward the Sun for half a day and away from the Sun the other half.

25. If you are located at 20°N latitude, when will the Sun appear directly overhead?
    a. never
    b. once a year
    c. twice a year
    d. four times a year

26. If you are located on the equator (0° latitude), when will the Sun appear directly overhead?
    a. never
    b. once a year
    c. twice a year
    d. four times a year

27. If you are located at 40°N latitude, when will the Sun appear directly overhead?
    a. never
    b. once a year
    c. twice a year
    d. four times a year

28. During the equinoxes
    a. a vertical stick in the equator will not cast a shadow at noon.
    b. at noon the Sun is directly overhead at 0° latitude.
    c. daylight and night are of equal length.
    d. All of the above are correct.

29. Evidence that Earth is rotating is provided by
    a. varying length of night and day during a year.
    b. seasonal climatic changes.
    c. stellar parallax.
    d. a pendulum.

30. In about twelve thousand years, the star Vega will be the North Star, not Polaris, because of Earth's
    a. uneven equinox.
    b. tilted axis.
    c. precession.
    d. recession.

31. The significance of the tropic of Cancer (23.5°N latitude) is that
    a. the Sun appears directly overhead north of this latitude some time during a year.
    b. the Sun appears directly overhead south of this latitude some time during a year.
    c. the Sun appears above the horizon all day for six months during the summer north of this latitude.
    d. the Sun appears above the horizon all day for six months during the summer south of this latitude.

32. The significance of the Arctic Circle (66.5°N latitude) is that
    a. the Sun appears directly overhead north of this latitude some time during a year.
    b. the Sun appears directly overhead south of this latitude some time during a year.
    c. the Sun appears above the horizon all day at least one day during the summer.
    d. the Sun appears above the horizon all day for six months during the summer.

33. In the time 1 P.M., the P.M. means
    a. "past morning."
    b. "past midnight."
    c. "before the meridian."
    d. "after the meridian."

34. Clock time is based on
    a. sundial time.
    b. an averaged apparent solar time.
    c. the apparent local solar time.
    d. the apparent local noon.

35. An apparent solar day is
    a. the interval between two consecutive local solar noons.
    b. about four minutes longer than the sidereal day.
    c. of variable length throughout the year.
    d. All of the above are correct.

36. The time as read from a sundial is the same as the time read from a clock
    a. all the time.
    b. only once a year.
    c. twice a year.
    d. four times a year.

37. You are traveling west by jet and cross three time zone boundaries. If your watch reads 3 P.M. when you arrive, you should reset it to
    a. 12 noon.
    b. 6 P.M.
    c. 12 midnight.
    d. 6 A.M.

38. If it is Sunday when you cross the international date line while traveling westward, the next day is
    a. Wednesday.
    b. Sunday.
    c. Tuesday.
    d. Saturday.

39. What has happened to the surface of the Moon during the last 3 billion years?
    a. heavy meteorite bombardment, producing craters
    b. widespread lava flooding from the interior
    c. both widespread lava flooding and meteorite bombardment
    d. not much

40. If you see a full moon, an astronaut on the Moon looking back at Earth at the same time would see a
    a. full Earth.
    b. new Earth.
    c. first quarter Earth.
    d. last quarter Earth.

41. A lunar eclipse can occur only during the moon phase of
    a. full moon.
    b. new moon.
    c. first quarter.
    d. last quarter.

42. A total solar eclipse can occur only during the moon phase of
    a. full moon.
    b. new moon.
    c. first quarter.
    d. last quarter.

43. A lunar eclipse does not occur every month because
    a. the plane of the Moon's orbit is inclined to the ecliptic.
    b. of precession.
    c. Earth moves faster in its orbit when closest to the Sun.
    d. Earth's axis is tilted with respect to the Sun.

44. The smallest range between high and low tides occurs during
    a. full moon.
    b. new moon.
    c. quarter moon phases.
    d. an eclipse.

45. Earth's axis points toward the
    a. constellation Aquarius.
    b. constellation Leo.
    c. North Star.
    d. Vega.

46. At the summer solstice, the Sun is
    a. low in the Northern Hemisphere sky.
    b. high in the Northern Hemisphere sky.
    c. directly over the Tropic of Capricorn.
    d. directly over the equator.

47. Earth is positioned between the Sun and the Moon in a nearly straight line. This positioning results in a (an)
    a. lunar eclipse.
    b. solar eclipse.
    c. annular eclipse.
    d. new moon.

Answers:

1. c 2. b 3. b 4. a 5. a 6. b 7. d 8. a 9. d 10. c 11. b 12. c 13. a 14. b 15. d 16. a 17. b 18. b 19. c 20. d 21. d 22. c 23. a 24. b 25. c 26. c 27. a 28. d 29. d 30. c 31. b 32. c 33. d 34. b 35. d 36. d 37. a 38. c 39. d 40. b 41. a 42. b 43. a 44. c 45. c 46. b 47. a

QUESTIONS FOR THOUGHT

1. Briefly describe the more conspicuous of Earth's motions. Identify which of these motions are independent of the Sun and the galaxy.

2. Describe some evidences that (a) Earth is shaped like a sphere and (b) Earth moves.

3. Use sketches with brief explanations to describe how the constant inclination and constant orientation of Earth's axis produces (a) a variation in the number of daylight hours and (b) a variation in seasons throughout a year.

4. Where on Earth are you if you observe the following at the instant of apparent local noon on September 23? (a) The shadow from a vertical stick points northward. (b) There is no shadow on a clear day. (c) The shadow from a vertical stick points southward.

5. What is the meaning of the word *solstice*? What causes solstices? On about what dates do solstices occur?

6. What is the meaning of *equinox*? What causes equinoxes? On about what dates do equinoxes occur?

7. What is precession?

8. Briefly describe how Earth's axis is used as a reference for a system that identifies locations on Earth's surface.

9. Use a map or a globe to identify the latitude and longitude of your present location.

10. The tropic of Cancer, tropic of Capricorn, Arctic Circle, and Antarctic Circle are parallels that are identified with specific names. What parallels do the names represent? What is the significance of each?

11. What is the meaning of (a) noon, (b) A.M., and (c) P.M.?

12. Explain why standard time zones were established. In terms of longitude, how wide is a standard time zone? Why was this width chosen?

13. When it is 12 noon in Texas, what time is it (a) in Jacksonville, Florida? (b) in Bakersfield, California? (c) at the North Pole?

14. Explain why a lunar eclipse is not observed once a month.

15. Use a sketch and briefly describe the conditions necessary for a total eclipse of the Sun.

16. Using sketches, briefly describe the positions of Earth, Moon, and Sun during each of the major moon phases.

17. If you were on the Moon as people on Earth observed a full moon, in what phase would you observe Earth?

18. What are the smooth, dark areas that can be observed on the face of the Moon? When did they form?

19. What made all the craters that can be observed on the Moon? When did this happen?

20. What phase is the Moon in if it rises at sunset? Explain your reasoning.

21. Why doesn't an eclipse of the Sun occur at each new moon when the Moon is between Earth and the Sun?

22. Is the length of time required for the Moon to make one complete revolution around Earth the same length of time required for a complete cycle of moon phases? Explain.

23. What is an annular eclipse? Which is more common, an annular eclipse or a total solar eclipse? Why?

24. Does an eclipse of the Sun occur during any particular moon phase? Explain.

25. Identify the moon phases that occur with (a) a spring tide and (b) a neap tide.

26. What was the basic problem with the Julian calendar? How does the Gregorian calendar correct this problem?

27. What is the source of the dust found on the Moon?

28. Describe the four stages in the Moon's history.

29. Explain why everyone on the dark side of Earth can see a lunar eclipse but only a limited few ever see a solar eclipse on the lighted side.

30. Explain why there are two tidal bulges on opposite sides of Earth.

31. Describe how the Foucault pendulum provides evidence that Earth turns on its axis.

FOR FURTHER ANALYSIS

1. What is the significance of the special designation of the latitudes called the tropic of Cancer and the Arctic Circle?

2. What are the significant similarities and differences between a solstice and an equinox?

3. On what date is Earth closest to the Sun? What season is occurring in the Northern Hemisphere at this time? Explain this apparent contradiction.

4. Explain why an eclipse of the Sun does not occur at each new moon phase when the Moon is between Earth and the Sun.

5. Explain why sundial time is often different than the time as shown by a clock.

6. Analyze why the time between two consecutive tides is twelve hours and twenty-five minutes rather than twelve hours. Explore as many different explanations as you can imagine, then select the best and explain how it could be tested.

INVITATION TO INQUIRY

Hello, Moon!

Observe where the Moon is located relative to the landscape where you live each day or night that it is visible. Make a sketch of the outline of the landscape. For each date, draw an accurate sketch of where the Moon is located and note the time. Also, sketch the moon phase for each date.

Continue your observations until you have enough information for analysis. Analyze the data for trends that would enable you to make predictions. For each day, draw circles to represent the relative positions of Earth, the Moon, and the Sun.

17
Rocks and Minerals

These rose-red rhodochrosite crystals are a naturally occurring form of manganese carbonate. Rhodochrosite is but one of about twenty-five hundred minerals that are known to exist, making up the solid materials of Earth's crust.

CORE **CONCEPT**

Earth is a dynamic body that cycles rocks and minerals through ongoing changes.

OUTLINE

Minerals
Most of Earth's surface is composed of minerals made of oxygen and silicon.

Igneous Rocks
Igneous rocks formed from a hot, molten mass or from melted rock materials.

Metamorphic Rocks
Metamorphic rocks formed by the action of heat, pressure, or hot solutions acting on previously existing different rocks.

Solid Earth Materials
Minerals
    Crystal Structures
    Silicates and Nonsilicates
    Physical Properties of Minerals
Mineral-Forming Processes
Rocks
  Igneous Rocks
Science and Society: Costs of Mining Mineral Resources
A Closer Look: Asbestos
  Sedimentary Rocks
Science and Society: Using Mineral Resources
  Metamorphic Rocks
The Rock Cycle
People Behind the Science: Victor Moritz Goldschmidt

Rocks
A rock is formed from one or more minerals that have been brought together as a cohesive solid.

Sedimentary Rocks
Sedimentary rocks formed from particles or dissolved materials derived from previously existing rocks.

The Rock Cycle
Igneous, sedimentary, and metamorphic rocks are continuously changed through cycles of one rock type to another.

OVERVIEW

Some of the most recent and revolutionary ideas of science have been developed in the earth sciences. The revolution began with the understanding that parts of Earth's surface are continually moving. This led to a new theory about Earth's surface consisting of moving plates. The new theory brought together the dynamics of moving continents, ocean basins that are constantly being created and destroyed, and a crust that is subjected to volcanoes and earthquakes. The creation of new ideas continued as scientists began to understand that the separate parts of Earth—the interior, the surface, the atmosphere, and the water on the surface—are all engaged in an ongoing cycling of materials (Figure 17.1).

The fascinating story of Earth's unique, unified cycles continues over the ensuing chapters. The story begins with this chapter on rocks and minerals, the basic materials that make up Earth. This is followed by evidence from a number of sources that is used to describe how Earth's interior is arranged, as well as how rocks and minerals cycle into and out of the interior. Evidence from the ocean then leads to a theory about plates that move on the surface. Next, you will learn how rocks are folded, faulted, and sculptured to produce the landscape you see on Earth today. Understanding the sculpting process will enable you to interpret the landscape where you live for meaning about Earth's past and perhaps its future.

As you can see, Earth is not made of isolated, independent parts. All parts of this dynamic Earth are related, and changes in one part affect the other parts. Yet, in spite of all the changes and cycles, each part of Earth has remained amazingly stable for millions of years. The study of earth science is thus viewed in a broad context, a context that will help you to understand Earth where you live, as well as present-day environmental concerns. Understanding all this starts with the basic materials of Earth—rocks and minerals—the subject of this chapter.

SOLID EARTH MATERIALS

Earth, like all other solid matter in the universe, is made up of chemical elements. The different elements are not distributed equally throughout the mass of Earth, however, nor are they equally abundant. As you shall see in chapter 18, there is evidence that Earth was molten during an early stage in its development. During this molten stage, most of the heavier abundant elements, such as iron and nickel, apparently sank to the deep interior of Earth, leaving a thin layer of lighter elements on the surface. This thin layer is called the *crust*. The rocks and rock materials that you see on the surface and the materials sampled in even the deepest mines and well holes are all materials of Earth's crust. The bulk of Earth's mass lies below the crust and has not been directly sampled.

Chemical analysis of thousands of rocks from Earth's surface found that only eight elements make up about 98.6 percent of the crust. All the other elements make up the remaining 1.4 percent of the crust. Oxygen is the most abundant element, making up about 50 percent of the weight of the crust. Silicon makes up over 25 percent, so oxygen and silicon alone make up about 75 percent of Earth's solid surface.

Figure 17.2 shows the eight most abundant elements that occur as elements or combine to form the chemical compounds of Earth's crust. They make up the solid materials of Earth's crust that are known as *minerals* and *rocks*. For now, consider a mineral to be a solid material of Earth that has both a known chemical composition and a crystalline structure that is unique to that mineral. About twenty-five hundred minerals are known to exist, but only about twenty are common in the crust. Examples of these common minerals are quartz, calcite, and gypsum.

Minerals are the fundamental building blocks of the rocks making up Earth's crust. A rock is a solid aggregation of one or more minerals that have been cohesively brought together by a rock-forming process. There are many possibilities of different kinds of rocks that could exist from many different variations of mineral mixtures. Within defined ranges of composition, however, there are only about twenty common rocks making up the crust. Examples of common rocks are sandstone, limestone, and granite.

How the atoms of elements combine depends on the number and arrangement of electrons around the atoms. How they form a mineral, with specific properties and a specific crystalline structure, depends on the electrons and the size of the ions as well. A discussion of the chemical principles that determine the structure and properties of chemical compounds is found in chapters 8 and 9 of this book. You may wish to review these principles before continuing with the discussion of common minerals and rocks.

FIGURE 17.1 No other planet in the solar system has the unique combination of fluids of Earth. Earth has a surface that is mostly covered with liquid water, water vapor in the atmosphere, and both frozen and liquid water on the land.

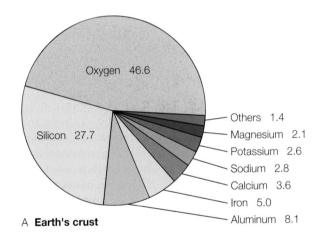

A **Earth's crust**

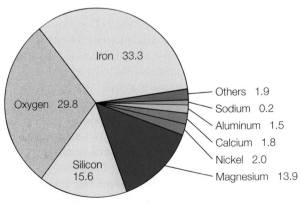

B **Whole Earth**

FIGURE 17.2 (*A*) The percentage by mass of the elements that make up Earth's crust. (*B*) The percentage by mass of the elements that make up the whole Earth.

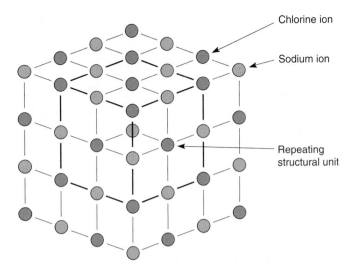

FIGURE 17.3 A crystal is composed of a structural unit that is repeated in three dimensions. This is the basic structural unit of a crystal of sodium chloride, the mineral halite.

MINERALS

In everyday usage, the word *mineral* can have several different meanings. It can mean something your body should have (vitamins and minerals), something a fertilizer furnishes for a plant (nitrogen, potassium, and phosphorus), or sand, rock, and coal taken from Earth for human use (mineral resources). In the earth sciences, a **mineral** is defined as a naturally occurring, inorganic solid element or compound with a crystalline structure (Figure 17.3). This definition means that the element or compound cannot be synthetic (must be naturally occurring), cannot be made of organic molecules (see chapter 12) (must be inorganic), and must have atoms arranged in a regular, repeating pattern (a crystal structure). Note that the crystal structure of a mineral can be present on the microscopic scale, and it is not necessarily obvious to the unaided eye. Even crystals that could be observed with the unaided eye are sometimes not noticed (Figure 17.4).

CRYSTAL STRUCTURES

The crystal structure of a mineral can be made up of atoms of one or more kinds of elements. Diamond, for example, is a mineral with only carbon atoms in a strong crystal structure. Quartz, on the other hand, is a mineral with atoms of silicon and oxygen in a different crystal structure (Figure 17.5). No matter how many kinds of atoms are present, each mineral has its own defined chemical composition or range of chemical compositions. A range of chemical compositions is possible because the composition of some minerals can vary with the substitution of chemically similar elements. For example, some atoms of magnesium might be substituted for some chemically similar atoms of calcium. Such substitutions might slightly alter some properties but not enough to make a different mineral.

Crystals can be classified and identified on the basis of the symmetry of their surfaces. This symmetry is an outward

FIGURE 17.4 The structural unit for a crystal of table salt, sodium chloride, is cubic, as you can see in the individual grains.

FIGURE 17.5 These quartz crystals are hexagonal prisms (compare with Figure 17.6).

 Myths, Mistakes, & Misunderstandings

Healing Crystals?

Practitioners of crystal healing have long held that certain crystals aid physical and mental healing. Quartz crystals, for example, are believed to "draw out" pain and "align" healing energy. Other varieties of quartz are believed to have other healing abilities. Those who wear amethyst, for example, are "helped" as the amethyst relaxes the mind and nervous system. Citrine is believed to be a self-esteem crystal that helps increase knowledge of self and overcomes fear, worry, or depression.

No scientific studies have found that crystals are beneficial. Healing crystals seem to belong in the same category of folk medicine as wearing copper bracelets to treat arthritis or amber beads to protect against colds. Such folk medicine remedies appear to serve as placebos, nothing more.

expression of the internal symmetry in the arrangement of the atoms making up the crystal. Thus, on the basis of symmetry, crystalline substances are classified into six major systems, which, in turn, are subdivided into smaller groups. Some of the more common examples of these forms are illustrated in Figure 17.6.

SILICATES AND NONSILICATES

Silicon and oxygen are the most abundant elements in Earth's crust, and, as you would expect, the most common minerals contain these two elements. All minerals are classified on the basis of whether the mineral structure contains these two elements or not. The two main groups are thus called the *silicates* and the *nonsilicates* (Table 17.1). Note, however, that the silicates can contain some other elements in addition to silicon and oxygen. The silicate minerals are by far the most abundant, making up about 92 percent of Earth's crust. When an atom of silicon (Si^{+4}) combines with four oxygen atoms (O^{-2}), a tetrahedral ionic structure of $(SiO_4)^{-4}$ forms (see Figure 17.7). All **silicates** have a basic silicon-oxygen tetrahedral unit either isolated or joined together in the crystal structure. The structure has four unattached electrons on the oxygen atoms that can combine with metallic ions such as iron or magnesium. They can also combine with the silicon atoms of *other* tetrahedral units. Some silicate minerals are thus made up of single tetrahedral units combined with metallic ions. Other silicate minerals are combinations of tetrahedral units combined in single chains, double chains, or sheets (Figure 17.8).

The silicate minerals can be conveniently subdivided into two groups based on the presence of iron and magnesium. The basic tetrahedral structure joins with ions of iron, magnesium, calcium, and other elements in the *ferromagnesian silicates*. Examples of ferromagnesian silicates are *olivine, augite, hornblende,* and *biotite* (Figure 17.9). They have a greater density and

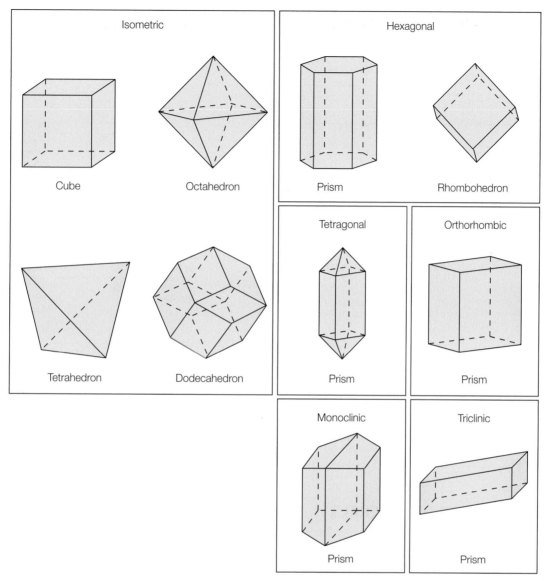

FIGURE 17.6 Crystalline substances are classified into six major systems: isometric, hexagonal, tetragonal, orthorhombic, monoclinic, and triclinic. The six systems are based on the arrangement of crystal axes, which reflect how the atoms or molecules are arranged inside. For example, crystals in the orthorhombic system have three axes of different lengths intersecting at 90° angles, while crystals in the hexagonal system have three horizontal axes intersecting at 60° and one vertical axis. Some common examples in each system are illustrated here.

a darker color than the other silicates because of the presence of the metal ions. Augite, hornblende, and biotite are very dark in color, practically black, and olivine is light green.

The *nonferromagnesian silicates* have a light color and a low density compared to the ferromagnesians. This group includes the minerals *muscovite* (*white mica*), the *feldspars,* and *quartz* (Figure 17.10).

The silicate minerals can also be classified according to the structural differences, or arrangements of the basic tetrahedral structures. There are four major arrangements of the units: (1) isolated tetrahedrons, (2) chain silicates, (3) sheet silicates, and (4) framework silicates (see Figure 17.8). Other structures are possible but not as common. The impact of meteorites, for example, creates sufficiently high temperatures and pressures to form a silicate structure with six oxygens

around each silicon atom rather than the typical four. A similar structure is believed to exist deep within Earth. Another interesting, less common structure is found in the asbestos minerals. These minerals have a sheet silicate structure that is rolled into fiberlike strands, resulting in a wide variety of silicates that are fibrous.

The remaining 8 percent of minerals making up Earth's crust that do not have silicon-oxygen tetrahedrons in their crystal structure are called *nonsilicates*. There are eight subgroups of nonsilicates: (1) carbonates, (2) sulfates, (3) oxides, (4) sulfides, (5) halides, (6) phosphates, (7) hydroxides, and (8) native elements. Some of these are identified in Table 17.1. The carbonates are the most abundant of the nonsilicates, but others are important as fertilizers, sources of metals, and sources of industrial chemicals.

TABLE 17.1

Classification scheme of some common minerals

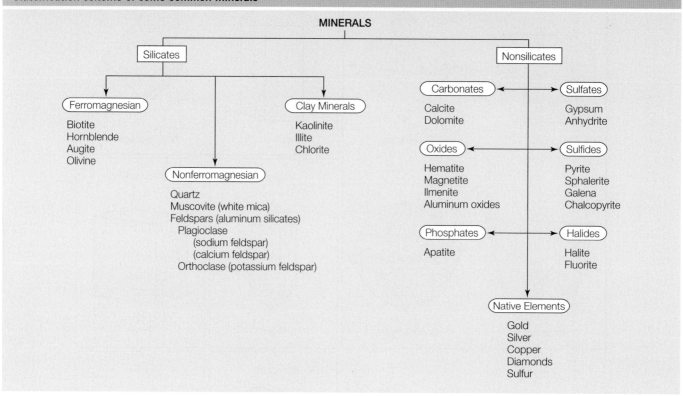

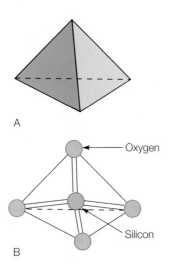

FIGURE 17.7 (*A*) The geometric shape of a tetrahedron with four equal sides. (*B*) A silicon and four oxygen atoms are arranged in the shape of a tetrahedron with the silicon in the center. This is the basic building block of all silicate minerals.

PHYSICAL PROPERTIES OF MINERALS

Each mineral has its own set of physical properties because it has a unique chemical composition and crystal structure. If a particular mineral sample happens to have formed large crystals with well-developed shapes, it is often possible to tell one mineral from another through identifying characteristics. There are about eight characteristics, or physical properties, that are useful in identifying minerals. These are the characteristics of color, streak, hardness, crystal form, cleavage, fracture, luster, and density.

The *color* of a mineral is an obvious characteristic, but it is often not very useful for identification. While some minerals always seem to appear the same color, many will vary from one specimen to the next. Variation in color is usually caused by the presence of small amounts of chemical impurities in the mineral that have nothing to do with its basic composition. The mineral quartz, for example, is colorless in its pure form, but other samples may appear milky white, rose pink, golden yellow, or purple.

A more consistent characteristic of a mineral is *streak*, the color of the mineral when it is finely powdered. Streak is tested by rubbing a mineral across a piece of unglazed tile or porcelain, which leaves a line of powdered mineral on the tile. Surprisingly, the streak of a mineral is more consistent than is the color of the overall sample. The streak of the same mineral usually shows the same color even though different samples of the mineral may have different colors. Hematite, for example, always leaves a red-brown streak even though the colors of different samples may range from reddish brown to black, depending on variations in grain size.

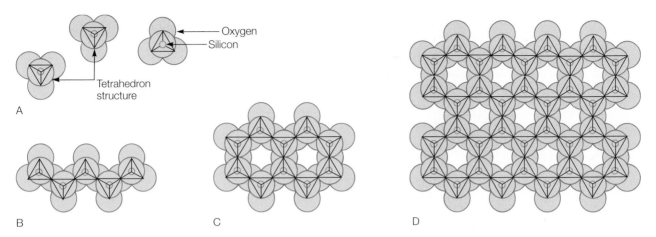

FIGURE 17.8 (*A*) Isolated silicon-oxygen tetrahedrons do not share oxygens. This structure occurs in the mineral olivine. (*B*) Single chains of tetrahedra are formed by each silicon ion having two oxygens all to itself and sharing two with other silicons at the same time. This structure occurs in augite. (*C*) Double chains of tetrahedra are formed by silicon ions sharing either two or three oxygens. This structure occurs in hornblende. (*D*) The sheet structure in which each silicon shares three oxygens occurs in the micas, resulting in layers that pull off easily because of cleavage between the sheets.

FIGURE 17.9 Compare the dark colors of the ferromagnesian silicates augite (right), hornblende (left), and biotite to the light-colored nonferromagnesian silicates in Figure 17.10.

FIGURE 17.10 Compare the light colors of the nonferromagnesian silicates mica (front center), white and pink orthoclase (top and center), and quartz, to the dark-colored ferromagnesian silicates in Figure 17.9.

Hardness is the resistance of a mineral to being scratched (Figure 17.11). Classically, hardness is measured by using the Mohs' hardness scale, which is a list of ten minerals in order of hardness (Table 17.2). The softest mineral is talc, which is assigned a hardness of 1. The hardest mineral is diamond, which is assigned a hardness of 10. A hardness test is made by trying to scratch an unknown mineral or by using the unknown mineral to try to scratch one of the test minerals. If the unknown mineral scratches a test mineral, the unknown mineral is harder than the test mineral. If the unknown mineral is scratched by the test mineral, the unknown mineral is not as hard as the test mineral. If both minerals are scratched by each other, they have the same hardness. However, the hardness test yields only approximate findings since there are many minerals of a particular hardness.

A

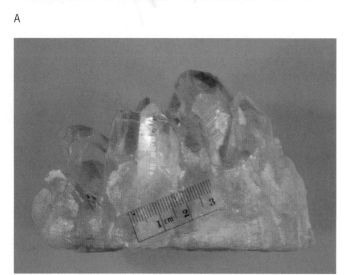

B

FIGURE 17.11 (*A*) Gypsum, with a hardness of 2, is easily scratched by a fingernail. (*B*) Quartz, with a hardness of 7, is so hard that even a metal file will not scratch it.

CONCEPTS *Applied*

Grow Your Own

Experiment with growing crystals from solutions of alum, copper sulfate, salt, or potassium permanganate. Write a procedure that will tell others what the important variables are for growing large, well-formed crystals.

The *crystal form,* or shape of a well-developed crystal of a mineral, is often a useful clue to its identity. The crystal form is related to the internal geometric arrangement of the atoms making up the crystal structure. The ions of sodium chloride, for example, are arranged in a cubic structure and table salt tends to crystallize in the shape of cubes (Figure 17.4). Thus, halite, a mineral composed of sodium chloride, occurs with a cubic structure. There are six basic groups of crystal forms (see Figure 17.6), each with a characteristic symmetry of the flat surfaces, or faces of the crystal. A crystal with a cubic structure, for example, belongs to the isometric group, which has three equal-length crystal axes at right angles to each other.

Another property that is controlled by the internal crystal structure is *cleavage,* the tendency of minerals to break along smooth planes. Where the cleavage occurs depends on zones of weakness in the crystal structure. Mica, for example, will break along zones of weakness into very thin sheets. Calcite and halite will break in three directions, and if you hit either mineral with a hammer, it will shatter into little pieces with angles consistent with the original specimen (assuming it was cleaved).

If a mineral does not have a well-defined zone of weakness, it may show *fracture* rather than cleavage. In fracture, the broken surface is irregular and not in the flat plane of a cleavage. A distinctive type of fracture is the conchoidal fracture of volcanic glass, quartz, and a few other minerals. Conchoidal fracture breaks along smooth surfaces like a shell (see Figure 17.15).

The *luster* of a mineral describes the surface sheen, that is, the way the mineral reflects light. Minerals that have the surface sheen of a metal are described as being *metallic.* Other descriptions of luster include *pearly* (like a pearl), *vitreous* (like glass), and *earthy.*

Density is a ratio of the mass of a mineral to its volume, or the compactness of the matter making up the mineral. Often mineral density is expressed as *specific gravity,* which is a ratio of the mineral density to the density of water. In the metric system, the density of water is 1 g/cm^3, so specific gravity will have the same numerical value as its density. The specific gravity of a mineral will depend on two factors: (1) the kinds of atoms of which it is composed and (2) the way the atoms are packed in the crystal lattice. A diamond, for example, has carbon atoms arranged in a close-packed structure and has a specific gravity of 3.5. Graphite, however, has carbon atoms in a loose-packed structure and has a specific gravity of 2.2. To obtain an exact specific gravity, a mineral sample must be pure and without cracks, bubbles, or substitutions of chemically similar elements. These conditions are difficult to meet, and a range of specific gravities is sometimes specified for certain minerals.

There are a few other properties and tests that can be used for a few minerals such as taste, feel, melting point, reaction to a magnet, and so forth. Some of these special properties, such as the double image seen through a calcite crystal, might identify an unknown mineral in an instant. Otherwise, an analysis of the other, more general properties will be needed. In general, the properties of minerals are used to find out what an unknown mineral is not, that is, what possible minerals can be ruled out by certain tests.

MINERAL-FORMING PROCESSES

Mineral crystals usually form in a liquid environment, but they can also form from gases or in solids under the right conditions. Two liquid environments where minerals usually form are (1) water solutions and (2) a solution of a hot, molten mass of melted rock materials known as **magma.** Magma may cool and crystallize to solid minerals either below or on the surface of Earth. Magma that is forced out to the surface is also called **lava.** Lava is the familiar molten material associated with an erupting volcano. High kinetic energy prevents the forming of bonds

between atoms when water solutions and magma are very hot. As cooling occurs, the kinetic energy is reduced enough that bonds will begin to form. Atoms then combine into small groups called *nucleation centers* that grow into crystals. Crystals can also form from cooler water solution, but dissolved ions must be highly concentrated for this to happen. In a concentrated solution, the ions are close enough together that their charges can pull them together. When this happens, crystals form and are precipitated from the solution.

Mineral-forming processes are influenced by *temperature, pressure, time,* and *the availability and concentration of ions in solution.* Different minerals are stable under different conditions of temperature and pressure. Clearly, a given mineral can form only if the ions of which it is made are present in the environment. Ion concentrations in solution must be high enough to permit the formation of the embryonic nucleation centers required for crystal growth. The amount of time available for crystal formation and the amount of water present determine the size of the crystals. Large crystals result from long, slow cooling of magma or from high water content in the magma, both of which favor the free migration of ions. Rapid cooling and low water content, on the other hand, suppress ion migration and result in small or even microscopic crystals. Sudden chilling can even prevent crystal growth altogether, resulting in *glass,* a solid that cooled too quickly for its atoms to move into ordered crystal structures.

Early in the twentieth century, N.L. Bowen conducted a series of experiments concerning the sequence that minerals crystallized in a cooling magma. The sequence became known as **Bowen's reaction series.** The series, as shown in Figure 17.12, is arranged with minerals at the top that crystallize at higher temperatures and minerals at the bottom that crystallize at lower temperatures. Bowen's idea was that different crystals separated

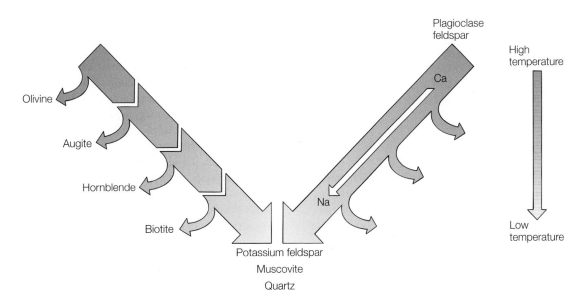

FIGURE 17.12 Bowen's reaction series. Minerals at the top of the series (olivine, augite, and calcium-rich plagioclase) crystallize at higher temperatures, leaving the magma enriched in silica. Later, the residual magma cools and lighter-colored, less dense minerals (orthoclase feldspar, quartz, and white mica) crystallize.

from a homogenous mixture, and the separation caused a change in the remaining materials available to form crystals. In more detail, consider that the minerals at the top of the series are ferromagnesian silicates, minerals that are rich in iron and magnesium. Minerals at the bottom of the series are nonferromagnesian silicates, minerals that are rich in silicon and generally lack iron and magnesium. Since the ferromagnesian silicates crystallize at a higher temperature, these minerals are the first to form in a cooling magma. If magma crystallizes directly, it will contain the ferromagnesian minerals listed at the top of the series. On the other hand, if it cools slowly, perhaps far below the surface, the minerals containing the iron and magnesium will form crystals that sink toward the bottom of the liquid magma. Thus, the remaining magma, and the minerals that crystallize later, will become progressively richer in silicon as more and more iron and magnesium are removed. Mineral separation does not take place in nature to the extent that Bowen envisioned from his experiments, but his work did lead to other theories on the behavior of magmas.

ROCKS

Elements are *chemically* combined to make minerals. Minerals are *physically* combined to make rocks. A **rock** is defined as an aggregation of one or more minerals and perhaps other materials that have been brought together into a cohesive solid. These materials include volcanic glass, a silicate that is not considered a mineral because it lacks a crystalline structure. Thus, a rock can consist of one or more kinds of minerals that are somewhat "glued" together by other materials such as glass. Most rocks are composed of silicate minerals, as you might expect since most minerals are silicates. Granite, for example, is a rock that is primarily three silicate minerals: quartz, mica, and feldspar. You can see the grains of these three minerals in a freshly broken surface of most samples of granite (Figure 17.13).

There is a classification scheme that is based on the way the rocks were formed. There are three main groups: (1) *igneous rocks* formed as a hot, molten mass of rock materials cooled and solidified; (2) *sedimentary rocks* formed from particles or dissolved materials from previously existing rocks; and, (3) *metamorphic rocks* formed from rocks that were subjected to high temperatures and pressures that deformed or recrystallized the rock without complete melting.

IGNEOUS ROCKS

The word *igneous* comes from the Latin *ignis,* which means "fire." This is an appropriate name for **igneous rocks,** which are rocks formed from a hot, molten mass of melted rock materials. The first step in forming igneous rocks is the creation of some very high temperature, hot enough to melt rocks. Recall that a mass of melted rock materials is called *magma.* Magma may cool and crystallize to solid igneous rock either below or on the surface of Earth. Earth has had a history of molten materials, and all rocks of Earth were at one time igneous rocks. Today, about two-thirds of the outer layer, or crust, is made up of igneous rocks. This is not apparent in many locations because the

FIGURE 17.13 Granite is a coarse-grained igneous rock composed mostly of light-colored, light-density, nonferromagnesian minerals. Earth's continental areas are dominated by granite and by rocks with the same mineral composition as granite.

surface is covered by other kinds of rocks and rock materials (sand, soil, etc.).

As a magma cools, atoms in the melt begin to lose kinetic energy and come together to form the orderly array of a crystal structure. How rapidly the cooling takes place determines the *texture* of the igneous rock being formed. In general, a *coarse-grained* texture means that you can see mineral crystals with the unaided eye. The texture is said to be fine-grained if you need a lens or a microscope to see the crystals. The presence of a fine-grained or coarse-grained texture tells you something about the cooling history of a particular igneous rock.

How rapidly a magma cools and hardens is generally determined by its location. Magma that cools slowly deep below the surface produces coarse-grained **intrusive igneous rocks.** Below the surface, the magma loses heat slowly and the atoms have sufficient time to produce large crystals. Lava that cools rapidly above the surface produces fine-grained **extrusive igneous rocks.** Rapid cooling does not result in sufficient time for large crystals to form so extrusive rocks are fine-grained. Very rapid cooling results in no time for *any* crystals to form, and a volcanic glass is produced as a result. Glass does not have an orderly arrangement of atoms and is therefore not a crystal.

A general classification scheme for igneous rocks is given in Figure 17.14. Igneous rocks are various mixtures of minerals, and this scheme names the rocks according to (1) their mineral composition and (2) their texture. Note that the mineral composition changes continuously from one side of the chart to the other. There are many intermediate types of igneous rocks possible, but this chart identifies only the most important ones.

Ancient humans exploited mineral resources as they mined copper minerals for the making of tools. They also used salt, clay, and other mineral materials for nutrients and pot making. These early people were few in number, and their simple tools made little impact on the environment as they mined what they needed. As the numbers of people grew and technology advanced, more and more mineral resources were utilized to build machines and provide energy. With advances in population and technology came increasing impacts on the environment in both size and scope. In addition to copper minerals and clay, the metal ores of iron, chromium, aluminum, nickel, tin, uranium, manganese, platinum, cobalt, zinc, and many others were now in high demand.

Today, there are three categories of costs recognized with the mining of any mineral resource. First, the *economic cost* is the money needed to lease or buy land, acquire equipment, and pay for labor to run the equipment. A second category is the *resource cost* of mining. It takes energy to concentrate the ore and transport it to smelters or refineries. Sometimes other resources are needed, such as large quantities of water for the extraction or concentration of a mineral resource. If the energy and water are not readily available, the resource cost might be converted to economic cost, which could ultimately determine if the operation will be profitable or not. Finally, the *environmental cost* of mining the resource must be considered. Environmental cost is converted to economic cost as controls on pollution are enforced. It is expensive to clean pollution from the land and to restore the ecosystem that was changed by mining operations. Consideration of the conversion of environmental cost to economic cost can also determine if a mining operation is feasible or not.

All mining operations start by making a mineral resource accessible so it can be removed. This might take place by strip mining, which begins with the removal of the top layers of soil and rock overlying a resource deposit. This overburden is placed somewhere else, to the side, so the mineral deposit can be easily removed. Access to a smaller, deeper mineral deposit might be gained by building a tunnel to the resource. The debris from building such a tunnel is usually piled outside the entrance. The rock debris from both strip and tunnel mining is an eyesore, and it is difficult for vegetation to grow on the barren rock. Since plants are not present, water may wash away small rock particles, causing erosion of the land and silting of the streams. The debris might also contain arsenic, lead, and other minerals that can pollute the water supply.

Today, regulations on the mining industry require less environmental damage than had been previously tolerated. The cost of finding and processing the minerals is also increasing as the easiest to use, less expensive resources have been utilized first. As current mineral resource deposits become exhausted, pressure will increase to use the minerals in protected areas. The environmental costs for utilization of these areas will indeed be large.

QUESTIONS TO DISCUSS

Divide your group into three subgroups: one representing economic cost, one representing resource cost, and one representing environmental cost. After a few minutes of preparation, have a short debate about the necessity of having mineral resources at the lowest cost possible versus the need to protect our environment no matter what the cost.

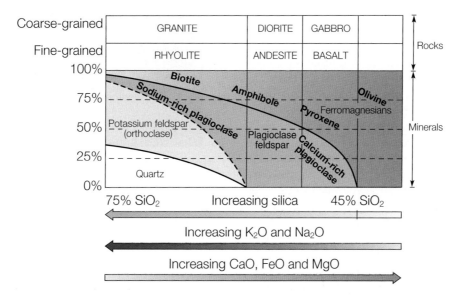

FIGURE 17.14 Igneous rock classification scheme based on mineral composition and texture. There are other blends of minerals with various textures, many of which have specific names.

A Closer Look

Asbestos

Asbestos is a common name for any of several minerals that can be separated into fireproof fibers, that will not melt or ignite. The fibers can be woven into a fireproof cloth, used directly as fireproof insulation material, or mixed with plaster or other materials. People now have a fear of all asbestos because it is presumed to be a health hazard (Box Figure 17.1). However, there are about six commercial varieties of asbestos. Five of these varieties are made from an amphibole mineral and are commercially called "brown" and "blue" asbestos. The other variety is made from *chrysotile*, a *serpentine* family of minerals and is commercially called "white" asbestos. White asbestos is the asbestos mined and most commonly used in North America. It is only the amphibole asbestos (brown and blue asbestos) that has been linked to cancer, even for a short exposure time. There is, however, no evidence that exposure to white asbestos results in an increased health hazard. It makes sense to ban the use of and remove all the existing amphibole asbestos from public buildings. It does not make sense to ban or remove the serpentine asbestos since it is not a proven health hazard.

BOX FIGURE 17.1 Did you know there are different kinds of asbestos? Are all kinds of asbestos a health hazard?

Igneous rocks on the left side of Figure 17.14 are blends of nonferromagnesian minerals. Thus, these rocks are comparatively light in density and color, appearing to be light gray, white, or ivory colored. The most common igneous rock with the minerals on the left side of the chart is **granite.** If you look closely at the surface of a freshly broken piece of granite, you will note that it is *coarse-grained* with noticeable particles of different size, shape, and color. The vitreous, white particles are probably orthoclase feldspar, which makes up about 45 percent of the particles. The clear, glassy-looking particles are probably quartz crystals, which make up about 25 percent of the total sample. The remaining particles of black specks are ferromagnesian minerals.

Rocks with the chemical composition of granite make up the bulk of Earth's continents, and granite is the most common intrusive rock in the continental crust. As shown in the chart, *rhyolite* and *obsidian* are the chemical equivalents of granite, except they have a different texture (Figure 17.15). Rhyolite is fine-grained and obsidian is a translucent volcanic glass.

Igneous rocks on the right side of the chart usually have a greater density than rocks with the granite chemical composition, and they are very dark in color. The most common example of these dark, relatively high-density igneous rocks is *basalt*. Basalt is the dark, *fine-grained* igneous rock that you probably associate with cooled and hardened lava.

FIGURE 17.15 This is a piece of obsidian, which has the same chemical composition as the granite shown in Figure 17.13. Obsidian has a different texture because it does not have crystals and is a volcanic glass. The curved fracture surface is common in noncrystalline substances such as glass.

Basalt is fine-grained, so you cannot see any mineral particles, and a freshly broken surface looks sugary. As shown in the chart, basalt is about half plagioclase feldspars and about half ferromagnesian minerals.

Basaltic rocks, meaning rocks with the chemical composition of basalt, make up the ocean basins and much of Earth's interior. Basalt is the most common extrusive rock that is found on Earth's surface. The coarse-grained chemical equivalent of basalt is called *gabbro*.

CONCEPTS *Applied*

Collect Your Own

Make a collection of rocks and minerals that can be found in your location, showing the name and location found for each. What will determine the total number of rocks and minerals it is possible to collect in your particular location?

SEDIMENTARY ROCKS

Sedimentary rocks are rocks that formed from particles or dissolved materials from previously existing rocks. Chemical reactions with air and water tend to break down and dissolve the less chemically stable parts of rocks, freeing more stable particles and grains in the process. The remaining particles are transported by moving water and are deposited as sediments. **Sediments** are accumulations of silt, sand, or other materials that settle out of water. Weathered rock fragments and dissolved rock materials both contribute to sediment deposits (Table 17.3).

Weathered rock fragments are called **clastic sediments** after a Greek word meaning "broken." Clastic sediments accumulated from rocks that are in various stages of being broken down, so there is a wide range of sizes of clastic sediments (Table 17.4). The largest of the clastic sediments, boulders and gravel, are the raw materials for the sedimentary rock that is called *conglomerate* or *breccia*, depending on if the fragments are well-rounded or angular (Figure 17.16). *Sandstone*, as the name implies, is a sedimentary rock formed from sand that has been consolidated into solid rock (Figure 17.17). The smallest clastic sediments, silt and clay, are consolidated into solid *siltstone* and *claystone*. If either of these sedimentary rocks tends to break along planes into flat

TABLE 17.4

A simplified classification scheme for clastic sediments and rocks

Sediment Name	Size Range	Rock
Boulder	Over 256 mm (10 in)	
Gravel	2 to 256 mm (0.08–10 in)	Conglomerate or breccia*
Sand	1/16 to 2 mm (0.025–0.08 in)	Sandstone
Silt (or dust)	1/256 to 1/16 mm (0.00015–0.025 in)	Siltstone**
Clay (or dust)	Less than 1/256 mm (less than 0.00015 in)	Claystone**

*Conglomerate has a rounded fragment; breccia has an angular fragment.
**Both also known as mudstone; called shale if it splits along parallel planes.

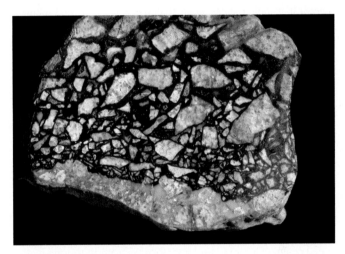

FIGURE 17.16 This is a piece of breccia, a sedimentary rock that formed from the consolidation of large angular fragments into a solid rock. This sample with gravel-sized and smaller fragments is from Colorado.

FIGURE 17.17 This is a piece of sandstone, a sedimentary rock that formed from the consolidation of sand grains into a solid rock. This sample with iron-oxide banding is from Tasmania, Australia.

TABLE 17.3

A classification scheme for sedimentary rocks

Sediment Type	Particle or Composition	Rock
Clastic	Larger than sand	Conglomerate or breccia
Clastic	Sand	Sandstone
Clastic	Silt and clay	Siltstone, claystone, or shale
Chemical	Calcite	Limestone
Chemical	Dolomite	Dolomite
Chemical	Gypsum	Gypsum
Chemical	Halite (sodium chloride)	Salt

FIGURE 17.18 This is a piece of limestone, a sedimentary rock that formed under water—sometimes with the remains of marine organisms. Can you find the mold of a brachiopod fossil on this sample?

pieces, it is called *shale*. Note that when a clastic sediment is referred to as "clay," it means a sediment size (less than 1/256 mm) and not the name of the clay mineral. When deposited from the air, clay- and silt-sized particles are commonly called "dust."

Dissolved rock materials form **chemical sediments** that are removed from solution to form sedimentary rocks. The dissolved materials are ions from minerals and rocks that have been completely broken down. Once they are transported to lakes or oceans, the dissolved ions are available to make sediments through one of three paths. These are (1) chemical precipitation from solution, (2) crystallization from evaporating water, or (3) biological sediments. The most abundant chemical sedimentary rocks are the carbonates and evaporates. The carbonates are *limestone* and *dolomite* (Figure 17.18). Limestone is composed of calcium carbonate, which is also the composition of the mineral called calcite. Dolomite probably formed from limestone by the replacement of calcium ions with magnesium ions (both belong to the same chemical family). Limestone is precipitated directly from freshwater or salt water or indirectly by the actions of plants and animals that form shells of calcium carbonate.

Most sediments are deposited as many separate particles that accumulate in certain environments as loose sediments.

Such accumulations of rock fragments, chemical deposits, or animal shells must become consolidated into a solid, coherent mass to become sedimentary rock. There are two main parts to this *lithification,* or rock-forming process: (1) compaction and (2) cementation (Figure 17.19).

The weight of an increasing depth of overlying sediments causes an increasing pressure on the sediments below. This pressure squeezes the deeper sediments together, gradually reducing the pore space between the individual grains. This **compaction** of the grains reduces the thickness of a sediment deposit, squeezing out water as the grains are packed more tightly together. Compaction alone is usually not enough to make loose sediment into solid rock. Cementation is needed to hold the compacted grains together.

In **cementation,** the spaces between the sediment particles are filled with a chemical deposit. As underground water moves through the remaining spaces, solid chemical deposits can precipitate and bind the loose grains together. The chemical deposit binds the particles together into the rigid, cohesive mass of a sedimentary rock. Compaction and cementation may occur at the same time, but the cementing agent must have been introduced before compaction restricts the movement of the fluid through the open spaces. Many soluble materials can serve as cementing agents and calcite (calcium carbonate) and silica (silicon dioxide) are common.

 CONCEPTS *Applied*

Minerals in Sand

Collect dry sand from several different locations. Use a magnifying glass to determine the minerals found in each sample.

METAMORPHIC ROCKS

The third group of rocks are called metamorphic. **Metamorphic rocks** are previously existing rocks that have been changed by heat, pressure, or hot solutions into a distinctly different rock. The heat, pressure, or hot solutions that produced the changes

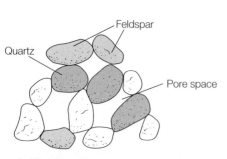 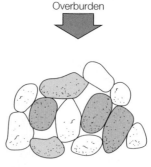 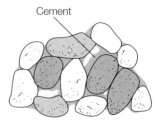

FIGURE 17.19 Lithification of sand grains to become sandstone. (*A*) Loose sand grains are deposited with open pore space between the grains. (*B*) The weight of overburden compacts the sand into a tighter arrangement, reducing pore space. (*C*) Precipitation of cement in the pores by groundwater binds the sand into the rock sandstone, which has a clastic texture.

Most people understand that our mineral resources are limited and that when we use them, they are gone. Of course, some mineral resources can be recycled, reducing the need to mine more minerals. For example, aluminum can be recycled and used over and over again. Glass, copper, iron, and other metals can similarly be recycled repeatedly. Other critical resources, however, cannot be recycled and cannot be replaced. Crude oil, for example, is a dwindling resource that will eventually become depleted. Oil is not recyclable once it is burned, and no new supplies are being created, at least not at a rate that would make them available in the immediate future. Even if Earth were a hollow vessel completely filled with oil, it would eventually become depleted, perhaps sooner than you might think.

There is also another of our mineral resources that is critically needed for our survival but will eventually be depleted. That resource is phosphorus derived from phosphate rock. Phosphorus is an essential nutrient required for plant growth, and if its concentration in soils is too low, plants grow poorly, if at all. Most agricultural soils are artificially fertilized with phosphate minerals. Without this amendment, plant productivity would decline and, in some cases, cease altogether.

Phosphate occurs naturally as the mineral apatite. Deposits of apatite were formed where ocean currents carried water rich in dissolved phosphate ions to the continental shelf. Here, phosphate ions replaced the carbonate ions in limestone, forming the mineral apatite. Apatite also occurs as a minor accessory mineral in most igneous, sedimentary, and metamorphic rocks. Some igneous rocks serve as a source of phosphate fertilizer, but most phosphate is mined from formerly submerged coastal areas of limestone, such as those found in Florida.

Trends in phosphate production and use suggest that the world reserves of phosphate rock will eventually be exhausted. New sources might be discovered, but eventually, phosphate rock will no longer be available for use as a fertilizer. When this happens, the food supply will have to be grown on lands that already have adequate phosphate minerals. Estimates are that the worldwide existing land area with adequate phosphate minerals will supply food for only 2 billion people on the entire Earth. Phosphate is an essential element for all life on Earth, and no other element can function in its place.

QUESTIONS TO DISCUSS

Discuss with your group the following questions concerning the use of mineral resources:

1. Should the mining industry be permitted to exhaust an important mineral resource? Provide reasons with your answer.

2. What are the advantage and disadvantages of a controlled mining industry?

3. If phosphate minerals supplies become exhausted, who should be responsible for developing new supplies or substitutes, the mining industry or governments?

---

are associated with geologic events of (1) movement of the crust, which will be discussed in chapter 18, and with (2) heating and hot solutions from the intrusion of a magma. Pressures from movement of the crust can change the rock texture by flattening, deforming, or realigning mineral grains. Temperatures from an intruded magma must be just right to produce a metamorphic rock. They must be high enough to disrupt the crystal structures to cause them to recrystallize but not high enough to melt the rocks and form igneous rocks (Figure 17.20).

The exact changes caused by heat and pressure depend on the mineral composition of the parent rock and the extent of the pressure, temperature, and hot solutions that may or may not be present to induce chemical changes. Pressure on parent rocks with flat crystal flakes (such as clays and mica) tends to align the flakes in parallel sheets. This new crystal alignment is called **foliation** after the Latin for "leaf" (as in the leaves, or pages, of a closed book). Foliation gives a metamorphic rock the property of breaking along the planes between the aligned mineral grains, a characteristic known as *rock cleavage*. The extent of foliation is determined by the extent of the metamorphic changes (Table 17.5). For example, *slate* is a metamorphic rock formed from the sedimentary rock shale. Slate is fine-grained with no crystals visible to the unaided eye. Alignment of the microscopic crystals results in a tendency of slate to split into flat sheets. Greater heat and pressure can cause more metamorphic change, resulting in larger crystals and increased foliation. The metamorphic rock called *schist* can be produced from slate by

further metamorphism. In schist, the cleavage surfaces are now visible, and coarser mica crystals are visible to the unaided eye. Still further metamorphism of schist may break down the mica crystals and produce alternating bands of light and dark minerals. These bands are characteristic of the metamorphic rock *gneiss*

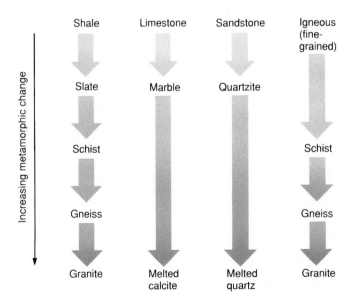

FIGURE 17.20 Increasing metamorphic change occurs with increasing temperatures and pressures. If the melting point is reached, the change is no longer metamorphic, and igneous rocks are formed.

TABLE 17.5

A classification scheme for metamorphic rocks

Metamorphic Texture	Metamorphic Rock
Nonfoliated	Quartzite and marble
Very finely foliated	Slate
Finely foliated	Schist
Coarsely foliated	Gneiss

(pronounced "nice") (Figure 17.21). Gneiss can also be produced by strong metamorphism of other rock types such as granite. Slate, schist, and gneiss are but three examples of a continuous transition that can take place from the metamorphism of shale all the way until it is completely melted to become an igneous rock.

Some metamorphic rocks are nonfoliated because they consist mainly of one mineral, and the grains are not aligned into sheets. When a quartz-rich sandstone is metamorphosed, the new rock has recrystallized, tightly locking grains. The resulting metamorphic rock is the tough, hard rock called *quartzite*. *Marble* is another nonfoliated metamorphic rock that forms from recrystallized limestone (Figure 17.22).

THE ROCK CYCLE

Earth is a dynamic planet with a constantly changing surface and interior. As you will see in the next chapters, internal changes alter Earth's surface by moving the continents and, for

FIGURE 17.21 This banded metamorphic rock is very old; at an age of 3.8 billion years, it is probably among the oldest rocks on the surface of Earth.

FIGURE 17.22 This is a sample of marble, a coarse-grained metamorphic rock with interlocking calcite crystals. The calcite crystals were recrystallized from limestone during metamorphism.

example, building mountains that are eventually worn away by weathering and erosion. Seas advance and retreat over the continents as materials are cycled from the atmosphere to the land and from the surface to the interior of Earth and then back again. Rocks are transformed from one type to another through this continual change. There is not a single rock on Earth's surface today that has remained unchanged through Earth's long history. The concept of continually changing rocks through time is called the **rock cycle** (Figure 17.23). The rock cycle concept views an igneous, a sedimentary, or a metamorphic rock as the present but temporary stage in the ongoing transformation of rocks to new types. Any particular rock sample today has gone through countless transformations in the 4.6-billion-year history of Earth and will continue to do so in the future.

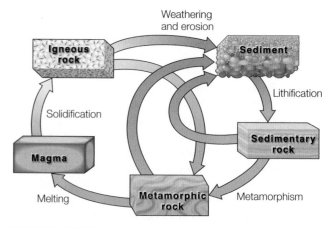

FIGURE 17.23 A schematic diagram of the rock cycle concept, which states that geologic processes act continuously to produce new rocks from old ones.

Victor Moritz Goldschmidt (1888–1947)

NORSKE GEOLOGER

V·M·GOLDSCHMIDT 1888–1947

NORGE 85

K.LØKKE·S 1974

Victor Goldschmidt was a Swiss-born Norwegian chemist who has been called the founder of modern geochemistry.

Goldschmidt's doctoral thesis on contact metamorphism in rocks is recognized as a fundamental work in geochemistry. It set the scene for a huge program of research on the elements, their origins, and their relationships, which was to occupy him for the next thirty years. He broke new ground when he applied the concepts of Josiah Gibbs's phase rule to the colossal chemical processes of geological time, which he considered to be interpretable in terms of the laws of chemical equilibrium. The evidence of geological change over millions of years represents a series of chemical processes on a scarcely imaginable scale, and even an imperceptibly slow reaction can yield megatons of product over the time scale involved.

A shortage of materials during World War I led Goldschmidt to speculate further on the distribution of elements in Earth's crust. In the next few years, he and his coworkers studied two hundred compounds of seventy-five elements and produced the first tables of ionic and atomic radii. The new science of X-ray crystallography, developed by William and Lawrence Bragg after Max von Laue's original discovery in 1912, could hardly have been more opportune. Goldschmidt was able to show that, given an electrical balance between positive and negative ions, the most important factor in crystal structure is ionic size. He suggested furthermore that complex natural minerals, such as hornblende $(OH)_2Ca_2Mg_5Si_8O_{22}$, can be explained by the balancing of charge by means of substitution based primarily on size. This led to the relationships between close-packing of identical spheres and the various interstitial sites available for the formation of crystal lattices. He also established the relation of hardness to interionic distances.

At Göttingen (1929), Goldschmidt pursued his general researches and extended them to include meteorites, pioneering spectrographic methods for the rapid determination of small amounts of elements. Exhaustive analysis of results from geochemistry, astrophysics, and nuclear physics led to his work on the cosmic abundance of the elements and the important links between isotopic stability and abundance. Studies of terrestrial abundance reveal about eight predominant elements. Recalculation of atom and volume percentages led to the remarkable notion that Earth's crust is composed largely of oxygen anions (90 percent of the volume), with silicon and the common metals filling up the rest of the space. Goldschmidt was a brilliant scientist with the rare ability to arrive at broad generalizations that draw together many apparently unconnected pieces of information.

Source: From the Hutchinson *Dictionary of Scientific Biography.* © Research Machines plc 2003. All Rights Reserved. Helicon Publishing is a division of Research Machines.

SUMMARY

The elements silicon and oxygen make up 75 percent of all the elements in the outer layer, or *crust,* of Earth. The elements combine to make crystalline chemical compounds called minerals. A *mineral* is defined as a naturally occurring, inorganic solid element or compound with a crystalline structure.

About 92 percent of the minerals of Earth's crust are composed of silicon and oxygen, the *silicate minerals.* The basic unit of the silicates is a *tetrahedral structure* that combines with positive metallic ions or with other tetrahedral units to form chains, sheets, or an interlocking framework.

The *ferromagnesian silicates* are tetrahedral structures combined with ions of iron, magnesium, calcium, and other elements. The ferromagnesian silicates are darker in color and more dense than other silicates. The *nonferromagnesian silicates* do not have iron or magnesium ions, and they are lighter in color and less dense than the ferromagnesians. The *nonsilicate minerals* do not contain silicon and are carbonates, sulfates, oxides, halides, sulfides, and native elements.

A *rock* is defined as an aggregation of one or more minerals that have been brought together into a cohesive solid. *Igneous rocks* formed as hot, molten *magma* cooled and crystallized to firm, hard rocks. Magma that cools slowly produces coarse-grained *intrusive igneous rocks.* Magma that cools rapidly produces fine-grained *extrusive igneous rocks.* The most abundant igneous rocks are the ferromagnesian-rich *basalt* and the silicon-rich and ferromagnesian-poor *granite.*

Sedimentary rocks are formed from *sediments,* accumulations of weathered rock materials that settle out of the atmosphere or out of water. Sedimentary rocks from *clastic sediments,* or rock fragments, are named according to the size of the sediments making up the rock: *conglomerate, sandstone,* and *shale,* in decreasing sediment size. Chemical sediments form from precipitation, crystallization, or the action of plants and animals. *Limestone* is the most common sedimentary rock from chemical sediments. Sediments become sedimentary rocks through *lithification,* a rock-forming process that involves both the *compaction* and *cementation* of the sediments.

Metamorphic rocks are previously existing rocks that have been changed by heat, pressure, or hot solution into a different kind of rock without melting. Increasing metamorphism can change the sedimentary rock *shale* to *slate,* which is then changed to *schist,* which can then be changed to *gneiss.* Each of these stages has a characteristic crystal size and alignment known as *foliation. Quartzite* and *marble* are examples of two nonfoliated metamorphic rocks.

The *rock cycle* is a concept that an igneous, a sedimentary, or a metamorphic rock is a temporary stage in the ongoing transformation of rocks to new types.

KEY TERMS

Bowen's reaction series (p. **439**)
cementation (p. **444**)
chemical sediments (p. **444**)
clastic sediments (p. **443**)
compaction (p. **444**)
extrusive igneous rocks (p. **440**)
foliation (p. **445**)
granite (p. **442**)
igneous rocks (p. **440**)
intrusive igneous rocks (p. **440**)
lava (p. **439**)
magma (p. **439**)
metamorphic rocks (p. **444**)
mineral (p. **433**)
rock (p. **440**)
rock cycle (p. **446**)
sedimentary rocks (p. **443**)
sediments (p. **443**)
silicates (p. **434**)

APPLYING THE CONCEPTS

1. A naturally occurring inorganic solid element or compound with a crystalline structure is a
   a. mineral.
   b. crystal.
   c. rock.
   d. stone.

2. A structural unit that is repeated in three dimensions is called a
   a. mineral.
   b. rock.
   c. crystal.
   d. glass.

3. Which element is the most abundant in Earth's crust?
   a. oxygen
   b. silicon
   c. sodium
   d. aluminum

4. Minerals are classified as
   a. silicates.
   b. nonsilicates.
   c. silicates or nonsilicates.
   d. silicates and oxides.

5. The most abundant class of nonsilicates is the
   a. oxides.
   b. sulfates.
   c. halides.
   d. carbonates.

6. Silicates are classified into two groups based on the presence of
   a. iron and manganese.
   b. iron and magnesium.
   c. aluminum and iron.
   d. calcium and potassium.

7. The color of a mineral when it is finely powdered is defined as the
   a. streak.
   b. spot.
   c. luster.
   d. color.

8. The hardness of a mineral is rated using the
   a. hardness scale.
   b. Richter scale.
   c. Mohs' scale.
   d. cleavage scale.

9. The ratio of the mineral's density to the density of water is
   a. 1 g/mL.
   b. specific gravity.
   c. 1 g/cm^3.
   d. specific density.

10. Molten rock material from which minerals crystallize is called
    a. rock salts.
    b. liquid crystals.
    c. magma.
    d. the mother lode.

11. An aggregation of one or more minerals that have been brought together into a cohesive solid is a
    a. silicate.
    b. rock.
    c. clay mineral.
    d. magma.

12. Rocks that are formed from molten minerals are
    a. sedimentary.
    b. igneous.
    c. volcanic.
    d. metamorphic.

13. Igneous rock that slowly cooled deep below Earth's crust is
    a. intrusive.
    b. extrusive.
    c. magma.
    d. sedimentary.

14. The rocks that make up the bulk of Earth's continents are
    a. basalt.
    b. olivine.
    c. granite.
    d. clay minerals.

15. The rocks that make up the ocean basins and much of Earth's interior are
    a. basalt.
    b. sandstone.
    c. granite.
    d. clay minerals.

16. Rocks that are formed from particles of other rocks or from dissolved materials from previously existing rocks are classified as
    a. sedimentary.
    b. igneous.
    c. silicates.
    d. metamorphic.

17. Accumulations of silt, sand, or other materials that settle out of water are called
    a. precipitates.
    b. solutes.
    c. clay.
    d. sediments.

18. Limestone and dolomite are
    a. sandstone.
    b. carbonates.
    c. evaporates.
    d. silicates.

19. Heat and pressure change rocks into
    a. igneous rocks.
    b. sedimentary rocks.
    c. metamorphic rocks.
    d. clastic rocks.

20. The relationship between rocks that are continually changing over long periods of time is called
    a. geology.
    b. mineralogy.
    c. rock cycle.
    d. weathering.

21. The thin layer that covers Earth's surface is the
    a. continental shelf.
    b. crust.
    c. mantle.
    d. core.

22. Based on its abundance in Earth's crust, most rocks will contain a mineral composed of oxygen and the element
    a. sulfur.
    b. carbon.
    c. silicon.
    d. iron.

23. The most common rock in Earth's crust is
    a. igneous.
    b. sedimentary.
    c. metamorphic.
    d. None of the above is correct.

24. An intrusive igneous rock will have which type of texture?
    a. fine-grained
    b. coarse-grained
    c. medium-grained
    d. no grains

25. Which igneous rock would have the greatest density?
    a. fine-grained granite (rhyolite)
    b. coarse-grained granite
    c. one composed of nonferromagnesian silicates
    d. basalt

26. Which of the following formed from previously existing rocks?
    a. sedimentary rocks
    b. igneous rocks
    c. metamorphic rocks
    d. All of the above are correct.

27. Sedimentary rocks are formed by the processes of compaction and
    a. pressurization.
    b. melting.
    c. cementation.
    d. heating but not melting.

28. The greatest extent of metamorphic changes has occurred in
    a. gneiss.
    b. schist.
    c. slate.
    d. shale.

29. Which type of rock probably existed first, starting the rock cycle?
    a. metamorphic
    b. igneous
    c. sedimentary
    d. All of the above are correct.

30. Earth is unique because it has
    a. CO_2 in its atmosphere.
    b. water covering most of its surface.
    c. fusion-powered energy in its core.
    d. "land" quakes.

31. The common structural feature of all silicates is
    a. silicon-silicon chains.
    b. silicon-oxygen tetrahedral unit.
    c. silicon-silicon rings.
    d. silicon-oxygen chains.

32. The one group that is not a subgroup of the silicate minerals is
    a. ferromagnesian.
    b. nonferromagnesian.
    c. plastic silicates.
    d. clay minerals.

33. The property that is not considered useful in identifying minerals is
    a. hardness.
    b. luster.
    c. density.
    d. melting point.

34. The specific gravity of a mineral depends on the
    a. kinds of atoms of which it is composed.
    b. manner in which the atoms are packed in the crystal lattice.
    c. amount of the mineral.
    d. a and b

35. Fluorite is a mineral that floats in liquid mercury. The specific gravity of fluorite must be
    a. equal to the specific gravity of mercury.
    b. less than the specific gravity of mercury.
    c. greater than the specific gravity of mercury.

36. The group that is not a class of rocks is
    a. sedimentary.
    b. igneous.
    c. volcanic.
    d. metamorphic.

37. The classification of rocks is based on
    a. location.
    b. how they were formed.
    c. age.
    d. composition.

38. An example of a sedimentary rock is
    a. basalt.
    b. granite.
    c. shale.
    d. slate.
39. The term that does not describe a size of clastic sediment is
    a. gravel.
    b. sandstone.
    c. silt.
    d. boulder.
40. Dissolved rock materials form
    a. chemical sediments.
    b. clastic sediments.
    c. basaltic rocks.
    d. igneous rocks.
41. An example of a metamorphic rock is
    a. marble.
    b. granite.
    c. limestone.
    d. sandstone.
42. Extrusive igneous rocks are formed on Earth's surface from
    a. magma.
    b. lava.
    c. slowly cooling hot solutions.
    d. sediments.
43. Foliation is found in
    a. sedimentary rocks.
    b. igneous rocks.
    c. metamorphic rocks.
    d. all rocks.

Answers

1. a 2. c 3. a 4. c 5. d 6. b 7. a 8. c 9. b 10. c 11. b 12. b 13. a 14. c
15. a 16. a 17. d 18. b 19. c 20. c 21. b 22. c 23. a 24. b 25. d 26. d
27. c 28. a 29. b 30. b 31. b 32. c 33. d 34. d 35. b 36. c 37. b 38. c
39. b 40. a 41. a 42. b 43. c

QUESTIONS FOR THOUGHT

1. What are the characteristics that make a mineral different from other solid materials of Earth?
2. Describe the silicate minerals in terms of structural arrangement; in terms of composition.
3. Explain why each mineral has its own unique set of physical properties.
4. Identify at least eight physical properties that are useful in identifying minerals. From this list, identify two properties that are probably the most useful and two that are probably the least useful in identifying an unknown mineral. Give reasons for your choices.
5. Explain how the identity of an unknown mineral is determined by finding out what the mineral is not.
6. What is a rock?

7. Describe the concept of the rock cycle.
8. Briefly explain the basic differences among the three major kinds of rocks based on the way they were formed.
9. Which major kind of rock, based on the way it is formed, would you expect to find most of in Earth's crust? Explain.
10. What is the difference between magma and lava?
11. What is meant by the "texture" of an igneous rock? What does the texture of an igneous rock tell you about its cooling history?
12. What are the basic differences between basalt and granite, the two most common igneous rocks of Earth's crust? In what part of Earth's crust are basalt and granite most common? Explain.
13. Explain why a cooled and crystallized magma might have ferromagnesian silicates in the lower part and nonferromagnesian silicates in the upper part.
14. Is the igneous rock basalt *always* fine-grained? Explain.
15. What are clastic sediments? How are they classified and named?
16. Briefly describe the rock-forming process that changes sediments into solid rock.
17. What are metamorphic rocks? What limits the maximum temperatures possible in metamorphism? Explain.
18. Describe what happens to the minerals as shale is metamorphosed to slate, then schist, then gneiss. Is it possible to metamorphose shale directly to gneiss, or must it go through the slate and schist sequences first? Explain.
19. What is the rock cycle? Why is it unique to the planet Earth?

FOR FURTHER ANALYSIS

1. What are the significant similarities and differences between igneous, sedimentary, and metamorphic rocks?
2. Is ice a mineral? Describe reasons to agree that ice is a mineral. Describe reasons to argue that ice is not a mineral.
3. If ice is a mineral, is a glacier a rock? Describe reasons to support or argue against calling a glacier a rock according to the definition of a rock.
4. The rock cycle describes how igneous, metamorphic, and sedimentary rocks are changed into each other. If this is true, analyze why most of the rocks on Earth's surface are sedimentary.

INVITATION TO INQUIRY

Building Rocks

Survey the use of rocks used in building construction in your community. Compare the type of rocks that are used for building interiors and those that are used for building exteriors. Where were the rocks quarried? Are any trends apparent for buildings constructed in the past and those built more recently? If so, are there reasons (cost, shipping, other limitations) underlying a trend, or is it simply a matter of style?

18

Plate Tectonics

All of the rocks that you can see on Earth's surface are part of a very thin veneer, barely covering the surface of Earth.

CORE **CONCEPT**
Earth has an internal structure and cycles materials between the surface and the interior.

451

OVERVIEW

In chapter 17, you learned about rocks and minerals of Earth's surface. What is below the rocks and minerals that you see on the surface? What is deep inside Earth? What you can see—the rocks, minerals, and soil on the surface—is a thin veneer. Nothing has ever been directly observed below this veneer, however (Figure 18.1). The deepest mine has penetrated to depths of about 3 km (about 2 mi), and the deepest oil wells may penetrate down to about 8 km (about 5 mi). But Earth has a radius of about 6,370 km (about 3,960 mi). How far have the mines and wells penetrated into Earth? By way of analogy, consider the radius of Earth to be the length of a football field, from one goal line to the other. The deep mine represents progress of 4.3 cm (1.7 in) from one goal line. The deep oil well represents progress of about 11.5 cm (about 4.5 in). It should be obvious that human efforts have only sampled materials directly beneath the surface. What is known about Earth's interior was learned indirectly, from measurements of earthquake waves, how heat moves through rocks, and Earth's magnetic field.

Indirect evidence suggests that Earth is divided into three main parts—the crust on the surface, a rocky mantle beneath the crust, and a metallic core. The crust and the uppermost mantle can be classified on a different basis, as a rigid layer made up of the crust and part of the upper mantle, and as a plastic, movable layer below in the upper mantle.

Understanding that Earth has a rigid upper layer on top of a plastic, movable layer is important in understanding the concepts of plate tectonics. Plate tectonics describes how the continents and the seafloor are moving on giant, rigid plates over the plastic layer below. This movement can be measured directly. In some places, the movement of a continent is about as fast as your fingernail grows, but movement does occur.

Understanding that Earth's surface is made up of moving plates is important in understanding a number of Earth phenomena. These include earthquakes, volcanoes, why deep sea trenches exist where they do, and why mountains exist where they do. This chapter is the "whole Earth" chapter, describing all of Earth's interior and the theory of plate tectonics. A bit of indirect information about Earth's interior, a theory of plate tectonics, and the observation of a number of related Earth phenomena all fit together. You can use this concept to explain many things that happen on the surface of Earth.

HISTORY OF EARTH'S INTERIOR

Many of the properties and characteristics of Earth, including the structure of its interior, can be explained from current theories of how it formed and evolved. Theories and ideas about how Earth and the rest of the solar system formed were discussed in detail in chapter 15. Here is the theoretical summary of how Earth's interior was formed, discussed as if it were a fact. Keep in mind, however, that the following is all conjecture, even if it is conjecture based on facts.

In brief, Earth is considered to have formed about 4.6 billion years ago in a rotating disk of particles and grains that had condensed around a central protosun. The condensed rock, iron, and mineral grains were pulled together by gravity, growing eventually to a planet-sized mass. Not all the bits and pieces of matter in the original solar nebula were incorporated into the newly formed planets. They were soon being pulled by gravity to the newly born planets and their satellites. All sizes of these leftover bits and pieces of matter thus began bombarding the planets and their moons. Evidently, the bombardment was so intense that the heat generated by impact after impact increased

the surface temperature to the melting point. Evidence visible on the Moon and other planets today indicates that the bombardment was substantial as well as lengthy, continuing for several hundred million years. Calculations of the heating resulting from this tremendous bombardment indicate that sufficient heat was liberated to melt the entire surface of Earth to a layer of glowing, molten lava. Thus, the early Earth had a surface of molten lava that eventually cooled and crystallized to solid igneous rocks as the bombardment gradually subsided, then stopped.

Then Earth began to undergo a second melting, this time from the inside. The interior slowly accumulated heat from the radioactive decay of uranium, thorium, and other radioactive isotopes (see chapter 13). Heat conducts slowly through great thicknesses of rock and rock materials. After about a 100 million years or so of accumulating heat, parts of the interior became hot enough to melt to pockets of magma. Iron and other metals were pulled from the magma toward the center of Earth, leaving less dense rocks toward the surface. The melting probably did not occur all at one time throughout the interior but rather in local pockets of magma. Each magma pocket became molten, cooled to a solid, and perhaps

FIGURE 18.1 This drilling ship samples sediment and rock from the deep ocean floor. It can only sample materials well within the upper crust of Earth, however, barely scratching the surface of Earth's interior.

repeated the cycle numerous times. With each cyclic melting, the heavier abundant elements were pulled by gravity toward the center of Earth, and additional heat was generated by the release of gravitational energy. Today, Earth's interior still contains an outer core of molten material that is predominantly iron. The environment of the center of Earth today is extreme, with estimates of pressures up to 3.5 million atmospheres (3.5 million times the pressure of the atmosphere at the surface). Recent estimates of the temperatures at Earth's core are about the same as the temperature of the surface of the Sun, about 6,000°C (11,000°F).

The melting and flowing of iron to Earth's center were the beginnings of *differentiation,* the separation of materials that gave Earth its present-day stratified or layered interior. The different crystallization temperatures of the basic minerals, as illustrated in Bowen's reaction series, further differentiated the materials in Earth's interior.

EARTH'S INTERNAL STRUCTURE

The theoretical formation of Earth and the layered structure of its interior are supported by indirect evidence from measurements of vibrations in Earth, Earth's magnetic field, gravity, and heat flow. First, we will consider how vibrations tell us about Earth's interior.

If you have ever felt vibrations in Earth from a passing train, an explosion, or an earthquake, you know that Earth can vibrate. In fact, a large disturbance such as a nuclear explosion or really big earthquake can generate waves that pass through the entire Earth. A vibration that moves through any part of Earth is called a **seismic wave.** Geologists use seismic waves to learn about Earth's interior.

Seismic waves radiate outward from an earthquake, spreading in all directions through the solid Earth's interior like sound waves from an explosion. There are basically three kinds of waves:

1. A longitudinal (compressional) wave called a *P-wave* (Figure 18.2A). P-waves are the fastest and move through

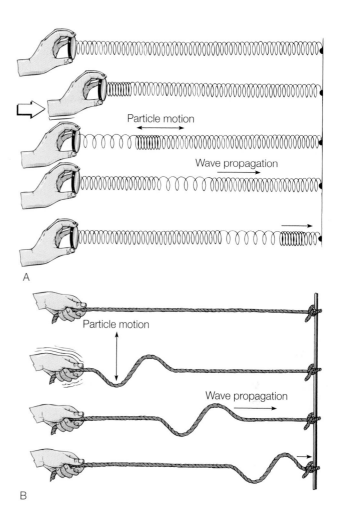

FIGURE 18.2 (*A*) A P-wave is illustrated by a sudden push on a stretched spring. The pushed-together section (compression) moves in the direction of the wave movement, left to right in the example. (*B*) An S-wave is illustrated by a sudden shake of a stretched rope. The looped section (sideways) moves perpendicular to the direction of wave movement, again left to right in the illustration.

surface rocks and solid and liquid materials below the surface. The *P* stands for primary.

2. A transverse (shear) wave called an *S-wave* (Figure 18.2B). The *S* stands for secondary. S-waves are second fastest after the P-waves. S-waves do not travel through liquids because liquids do not have the cohesion necessary to transmit a shear, or side-to-side, motion.

3. Up-and-down (crest and trough) and side-to-side waves that travel across the surface are called *surface waves* that are much like waves on water that move across the solid surface of Earth. Surface waves are the slowest and occur where S- or P-waves reach the surface. There are two important types of surface waves: Love waves and Rayleigh waves. Love waves are horizontal S waves that move from side to side. This motion knocks buildings off their foundations and can destroy bridges and overpasses. Rayleigh waves are more like rolling water waves. Rayleigh waves are more destructive because they produce more up, down, and sideways ground movement for a longer time.

Using data from seismic waves, scientists were able to determine that the interior of Earth can be broken down into three zones (Figure 18.3). The *crust* is the outer layer of rock that forms a thin shell around Earth. Below the crust is the *mantle,* a much thicker shell than the crust. The mantle separates the crust from the center part, which is called the *core.* The following discussion starts on Earth's surface, at the crust, and then digs deeper and deeper into Earth's interior.

THE CRUST

Seismic studies have found that Earth's **crust** is a thin skin that covers the entire Earth, existing below the oceans as well as making up the continents. According to seismic waves, there are differences in the crust making up the continents and the crust beneath the oceans (Table 18.1). These differences are (1) the oceanic crust is much thinner than the continental crust and (2) seismic waves move through the oceanic crust faster than they do through continental crust. The two types of crust vary because they are made up of different kinds of rock.

The boundary between the crust and the mantle is marked by a sharp increase in the velocity of seismic waves as they pass from the crust to the mantle. Today, this boundary is called the *Mohorovicic discontinuity,* or the "Moho" for short. The boundary is a zone where seismic P-waves increase in velocity because of changes in the composition of the materials. The increase occurs because the composition on both sides of the boundary is different. The mantle is richer in ferromagnesian minerals and poorer in silicon than the crust.

TABLE 18.1

Comparison of oceanic crust and continental crust

	Oceanic Crust	Continental Crust
Age	Less than 200 million years	Up to 3.8 billion years old
Thickness	5 to 8 km (3 to 5 mi)	10 to 75 km (6 to 47 mi)
Density	3.0 g/cm^3	2.7 g/cm^3
Composition	Basalt	Granite, schist, gneiss

Studies of the Moho show that the crust varies in thickness around Earth's surface. It is thicker under the continents and thinner under the oceans.

The age of rock samples from Earth's continents has been compared with the age samples of rocks taken from the seafloor by oceanographic ships. This sampling has found the continental crust to be much older, with parts up to 3.8 billion years old. By comparison, the oldest oceanic crust is less than 200 million years old.

Comparative sampling also found that continental crust is a less dense, granite-type rock with a density of about 2.7 g/cm^3. Oceanic crust, on the other hand, is made up of basaltic rock with a density of about 3.0 g/cm^3. The less dense crust behaves as if it were floating on the mantle, much as less dense ice floats on water. There are exceptions, but in general, the thicker, less dense continental crust "floats" in the mantle above sea level, and the thin, dense oceanic crust "floats" in the mantle far below sea level (Figure 18.4).

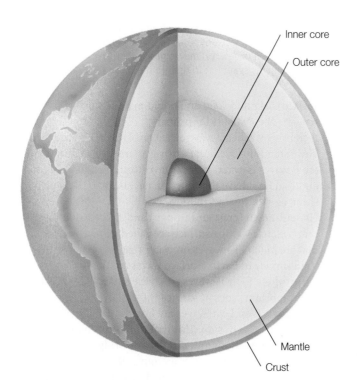

FIGURE 18.3 The structure of Earth's interior.

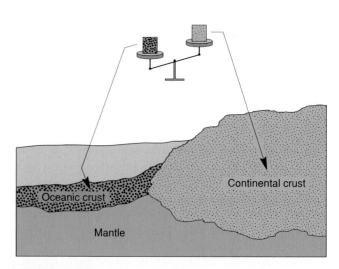

FIGURE 18.4 Continental crust is less dense, granite-type rock, while the oceanic crust is more dense, basaltic rock. Both types of crust behave as if they were floating on the mantle, which is more dense than either type of crust.

THE MANTLE

The middle part of Earth's interior is called the **mantle.** The mantle is a thick shell between the core and the crust. This shell takes up about 80 percent of the total volume of Earth and accounts for about two-thirds of Earth's total mass. Information about the composition and nature of the mantle comes from (1) studies of seismological data, (2) studies of the nature of meteorites, and (3) studies of materials from the mantle that have been ejected to Earth's surface by volcanoes. The evidence from these separate sources all indicates that the mantle is composed of silicates, predominantly the ferromagnesian silicate *olivine.* Meteorites, as discussed in chapter 15, are basically either iron meteorites or stony meteorites. Most of the stony meteorites are silicates with a composition that would produce the chemical composition of olivine if they were melted and the heavier elements separated by gravity. This chemical composition also agrees closely with the composition of basalt, the most common volcanic rock found on the surface of Earth.

THE CORE

Information about the nature of the **core,** the center part of Earth, comes from studies of three sources of information, (1) seismological data, (2) the nature of meteorites, and (3) geological data at the surface of Earth. Seismological data provides the primary evidence for the structure of the core of Earth. Seismic P-waves spread through Earth from a large earthquake. Figure 18.5 shows how the P-waves spread out, soon arriving at seismic measuring stations all around the world. However, there are places between 103° and 142° of arc from the earthquake that do not receive P-waves. This region is called the *P-wave shadow zone,* since no P-waves are received here. The P-wave shadow zone is explained by P-waves being refracted by

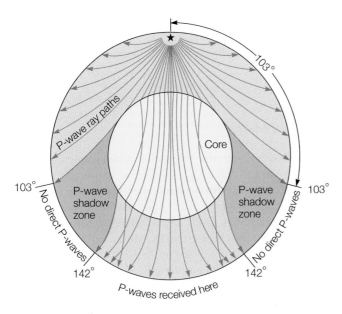

FIGURE 18.5 The P-wave shadow zone, caused by refraction of P-waves within Earth's core.

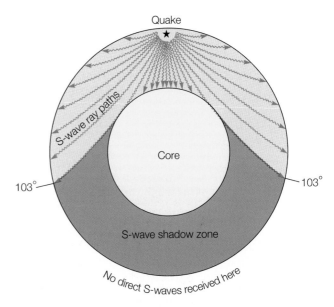

FIGURE 18.6 The S-wave shadow zone. Since S-waves cannot pass through a liquid, at least part of the core is either a liquid or has some of the same physical properties as a liquid.

the core, leaving a shadow (Figure 18.5). The paths of P-waves can be accurately calculated, so the size and shape of Earth's core can also be accurately calculated.

Seismic S-waves leave a different pattern at seismic receiving stations around Earth. Recall that S- (sideways or transverse) waves can travel only through solid materials. An *S-wave shadow zone* also exists and is larger than the P-wave shadow zone (Figure 18.6). S-waves are not recorded in the entire region more than 103° away from the epicenter. The S-wave shadow zone seems to indicate that S-waves do not travel through the core at all. If this is true, it implies that the core of Earth is a liquid or at least acts like a liquid.

Analysis of P-wave data suggests that the core has two parts: a *liquid outer core* and a *solid inner core* (Figure 18.7). Both the P-wave and S-wave data support this conclusion. Overall, the core makes up about 15 percent of Earth's total volume and about one-third of its mass.

Evidence from the nature of meteorites indicates that Earth's core is mostly iron. Earth has a strong magnetic field that has its sources in the turbulent flow of the liquid part of Earth's core. To produce such a field, the material of the core would have to be an electrical conductor, that is, a metal such as iron. There are two general kinds of meteorites that fall to Earth: (1) stony meteorites that are made of silicate minerals and (2) iron meteorites that are made of iron or of a nickel-iron alloy. Since Earth has a silicate-rich crust and mantle, by analogy Earth's core must consist of iron or a nickel and iron alloy.

A MORE DETAILED STRUCTURE

There is strong evidence that Earth has a layered structure with a core, mantle, and crust. This description of the structure is important for historical reasons and for understanding how Earth

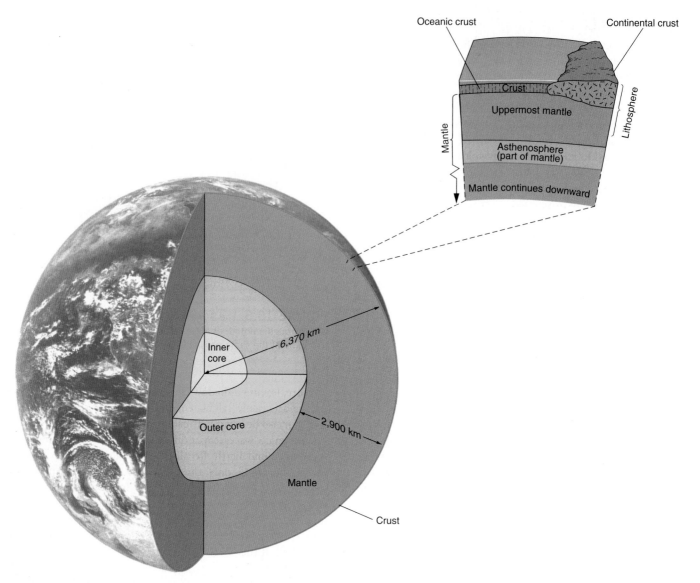

FIGURE 18.7 Earth's interior, showing the weak, plastic layer called the *asthenosphere*. The rigid, solid layer above the asthenosphere is called the *lithosphere*. The lithosphere is broken into plates that move on the upper mantle like giant ice sheets floating on water. This arrangement is the foundation for plate tectonics, which explains many changes that occur on Earth's surface such as earthquakes, volcanoes, and mountain building.

evolved over time. There is also another, more detailed structure that can be described. This structure is far more important in understanding the history and present appearance of Earth's surface, including the phenomena of earthquakes and volcanoes.

The important part of this different structural description of Earth's interior was first identified from seismic data. There is a thin zone in the mantle where seismic waves undergo a sharp *decrease* in velocity. This low-velocity zone is evidently a hot, elastic semiliquid layer that extends around the entire Earth. It is called the **asthenosphere** after the Greek for "weak shell (Figure 18.7)." The asthenosphere is weak because it is plastic, mobile, and yields to stresses. In some regions, the asthenosphere is completely liquid, containing pockets of magma.

The rocks above and below the asthenosphere are rigid, lacking a partial melt. The solid layer above the asthenosphere is called the **lithosphere** after the Greek for "stone shell." The lithosphere is also known as the "strong layer" in contrast to the "weak layer" of the asthenosphere. The lithosphere includes the entire crust, the Moho, and the upper part of the mantle. As you will see in the section on the theory of plate tectonics, the asthenosphere is one important source of magma that reaches Earth's surface. It is also a necessary part of the mechanism involved in the movement of the crust. The lithosphere is made up of comparatively rigid plates that are moving, floating in the upper mantle like giant ice sheets floating in the ocean.

The CAT scan is a common diagnostic imaging procedure that combines the use of X rays with computer technology. CAT stands for "computed axial tomography," and the word *tomography* means "drawing slices." The CAT scan shows organs of interest by making X-ray images from many different angles as the source of the X rays moves around the patient. CAT scan images are assembled by a computer into a three-dimensional picture that can show organs, bones, and tissues in great detail.

The CAT scan is applied to Earth's interior in a *seismic tomography* procedure. It works somewhat like the medical CAT scan, but seismic tomography uses seismic waves instead of X rays. The velocities of S and P seismic waves vary with depth, changing with density, temperature, pressure, and the composition of Earth's interior. Interior differences in temperature, pressure, and composition cause the motion of tectonic plates

on Earth's surface. Thus, a picture of Earth's interior can be made by mapping variations in seismic wave speeds.

Suppose an earthquake occurs and a number of seismic stations record when the S- and P-waves arrive. From these records, you could compare the seismic velocity data and identify if there were low seismic velocities between the source and some of the receivers. The late arrival of seismic waves could mean some difference in the structure of Earth between the source and the observing station. When an earthquake occurs in a new location, more data will be collected at observing stations, and this provides additional information about the shape and structure of whatever is slowing the seismic waves. Now, imagine repeating these measurements for many new earthquakes until you have enough data to paint a picture of what is beneath the surface.

Huge amounts of earthquake data—perhaps 10 million data points in 5 million

groups—are needed to construct a picture of Earth's interior. A really fast computer may take days to process this much data and construct cross sections through some interesting place, such as a subduction zone where an oceanic plate dives into the mantle.

Seismic tomography has also identified massive plumes of molten rock rising toward the surface from deep within Earth. These *superplumes* originate from the base of the mantle, rising to the lithosphere. The hot material was observed to spread out horizontally under the lithosphere toward mid-ocean ridges. This may contribute to tectonic plate movement. Regions above the superplumes tend to bulge upward, and other indications of superplumes, such as variations in gravity, have been measured. Scientists continue to work with new seismic tomography data with more precise images and higher resolutions to better describe the interior structure of Earth.

THEORY OF PLATE TECTONICS

If you observe the shape of the continents on a world map or a globe, you will notice that some of the shapes look as if they would fit together like the pieces of a puzzle. The most obvious is the eastern edge of North and South America, which seem to fit the western edge of Europe and Africa in a slight S-shaped curve. Such patterns between continental shapes seem to suggest that the continents were at one time together, breaking apart and moving to their present positions some time in the past (Figure 18.8).

In the early 1900s, a German geologist named Alfred Wegener became enamored with the idea that the continents had shifted positions and published papers on the subject for nearly two decades. Wegener supposed that at one time there was a single large landmass that he called "Pangaea," which is from the Greek meaning "all lands." He pointed out that similar fossils found in landmasses on both sides of the Atlantic Ocean today must be from animals and plants that lived in Pangaea, which later broke up and split into smaller continents. Wegener's concept came to be known as *continental drift*, the idea that individual continents could shift positions on Earth's surface. Some people found the idea of continental drift plausible, but most had difficulty imagining how a huge and massive continent could "drift" around on a solid Earth. Since Wegener had provided no good explanation of why or how continents

might do this, most scientists found the concept unacceptable. The concept of continental drift was dismissed as an interesting but odd idea. Then new evidence discovered in the 1950s and 1960s began to indicate that the continents have indeed moved. The first of this evidence would come from the bottom of the ocean and would lead to a new, broader theory about the movement of Earth's crust.

EVIDENCE FROM EARTH'S MAGNETIC FIELD

Earth's magnetic field is probably created by electric currents within the slowly circulating liquid part of the iron core. However, there is nothing static about Earth's magnetic poles. Geophysical studies have found that the magnetic poles are moving slowly around the geographic poles. Studies have also found that Earth's magnetic field occasionally undergoes magnetic reversal. Magnetic reversal is the flipping of polarity of Earth's magnetic field. During a magnetic reversal, the north magnetic pole and the south magnetic pole exchange positions. The present magnetic field orientation has persisted for the past seven hundred thousand years and, according to the evidence, is now preparing for another reversal. The evidence, such as the magnetized iron particles found in certain Roman ceramic artifacts, shows that the magnetic field was 40 percent stronger two thousand years ago than it is today. If the present decay rate were to continue, Earth's magnetic field would be near zero by the end of the next

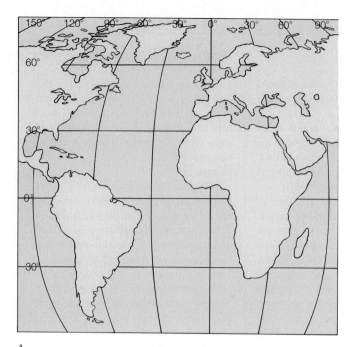

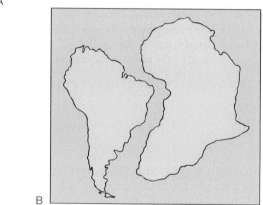

FIGURE 18.8 (A) Normal position of the continents on a world map. (B) A sketch of South America and Africa, suggesting that they once might have been joined together and subsequently separated by continental drift.

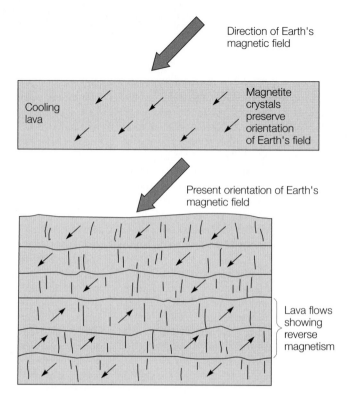

FIGURE 18.9 Magnetite mineral grains align with Earth's magnetic field and are frozen into position as the magma solidifies. This magnetic record shows Earth's magnetic field has reversed itself in the past.

two thousand years—if it decays that far before reversing orientation, then increasing to its usual value.

Many igneous rocks contain a record of the strength and direction of Earth's magnetic field at the time the rocks formed. Iron minerals, such as magnetite (Fe_3O_4), crystallize in a cooling magma and become magnetized and oriented to Earth's magnetic field at the time like tiny compass needles. When the rock crystallizes to a solid, these tiny compass needles become frozen in the orientations they had at the time. Such rocks thus provide evidence of the direction and distance to Earth's ancient magnetic poles. The study of ancient magnetism, called *paleomagnetics,* provides the information that Earth's magnetic field has undergone twenty-two magnetic reversals during the past 4.5 million years (Figure 18.9).

The record shows the time between pole flips is not consistent, sometimes reversing in as little as ten thousand years and

sometimes taking as long as 25 million years. Once a reversal starts, however, it takes about five thousand years to complete the process.

EVIDENCE FROM THE OCEAN

The first important studies concerning the movement of continents came from studies of the ocean basin, the bottom of the ocean floor. The basins are covered by 4 to 6 km (about 3 to 4 mi) of water and were not easily observed during Wegener's time. It was not until the development and refinement of sonar and other new technologies that scientists began to learn about the nature of the ocean basin. They found that it was not the flat, featureless plain that many had imagined. There are valleys, hills, mountains, and mountain ranges. Long, high, and continuous chains of mountains that seem to run clear around Earth were discovered, and these chains are called **oceanic ridges.** The *Mid-Atlantic Ridge* is one such oceanic ridge that is located in the center of the Atlantic Ocean basin. The Mid-Atlantic Ridge divides the Atlantic Ocean into two nearly equal parts. Where it is high enough to reach sea level, it makes oceanic islands such as Iceland (Figure 18.10). The basins also contain **oceanic trenches.** These trenches are long, narrow, and deep troughs with steep sides. Oceanic trenches always run parallel to the edges of continents.

Studies of the Mid-Atlantic Ridge found at least three related groups of data and observations: (1) Submarine earthquakes

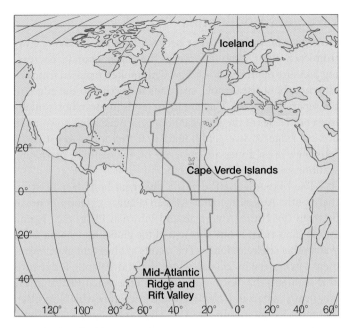

FIGURE 18.10 The Mid-Atlantic Ridge divides the Atlantic Ocean into two nearly equal parts. Where the ridge reaches above sea level, it makes oceanic islands, such as Iceland.

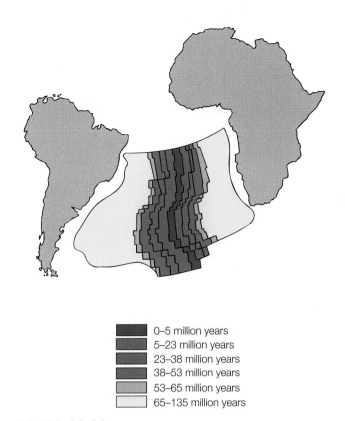

■	0–5 million years
■	5–23 million years
■	23–38 million years
■	38–53 million years
■	53–65 million years
□	65–135 million years

FIGURE 18.11 The pattern of seafloor ages on both sides of the Mid-Atlantic Ridge reflects seafloor spreading activity. Younger rocks are found closer to the ridge.

were discovered and measured, but the earthquakes were all observed to occur mostly in a narrow band under the crest of the Mid-Atlantic Ridge. (2) A long, continuous **rift**, or valley, was observed to run along the crest of the Mid-Atlantic Ridge for its length. (3) A large amount of heat was found to be escaping from the rift. One explanation of the related groups of findings is that the rift might be a crack in Earth's crust, a fracture through which basaltic lava flowed to build up the ridge. The evidence of excessive heat flow, earthquakes along the crest of the ridge, and the very presence of the ridge all led to a **seafloor spreading** hypothesis (see the reading on Harry Hess on p. 464). This hypothesis explained that hot, molten rock moved up from the interior of Earth to emerge along the rift, flowing out in both directions to create new rocks along the ridge. The creation of new rock like this would tend to spread the seafloor in both directions, thus the name. The test of this hypothesis would come from further studies, this time on the ages and magnetic properties of the seafloor along the ridge (Figure 18.11).

Evidence of the age of sections of the seafloor was obtained by drilling into the ocean floor from a research ship. From these drillings, scientists were able to obtain samples of fossils and sediments at progressive distances outward from the Mid-Atlantic Ridge. They found thin layers of sediments near the ridge that became progressively thicker toward the continents. This is a pattern you would expect if the seafloor were spreading, because older layers would have more time to accumulate greater depths of sediments. The fossils and sediments in the bottom of the layer were also progressively older at increasing distances from the ridge. The oldest, which were about 150 million years old, were near the continents. This would seem to indicate that the Atlantic Ocean did not exist until 150 million years ago. At that time, a fissure formed

between Africa and South America, and new materials have been continuously flowing, adding new crust to the edges of the fissure.

More convincing evidence for the support of seafloor spreading came from the paleomagnetic discovery of patterns of magnetic strips in the basaltic rocks of the ocean floor. Earth's magnetic field has been reversed many times in the last 150 million years. The periods of time between each reversal were not equal, ranging from thousands to millions of years. Since iron minerals in molten basalt formed, became magnetized, then froze in the orientation they had when the rock cooled, they made a record of reversals in Earth's ancient magnetic field (Figure 18.12). Analysis of the magnetic pattern in the rocks along the Mid-Atlantic Ridge found identical patterns of magnetic bands on both sides of the ridge. This is just what you would expect if molten rock flowed out of the rift, cooled to solid basalt, then moved away from the rift on both sides. The pattern of magnetic bands also matched patterns of reversals measured elsewhere, providing a means of determining the age of the basalt. This showed that the oceanic crust is like a giant conveyer belt that is moving away from the Mid-Atlantic Ridge in both directions. It is moving at an average 5 cm (about 2 in) a year, which is about how fast your fingernails grow. This means that in fifty years, the seafloor will have moved 5 cm/yr × 50 yr, or 2.5 m (about 8 ft). This slow rate is why most people do not recognize that the seafloor—and the continents—move.

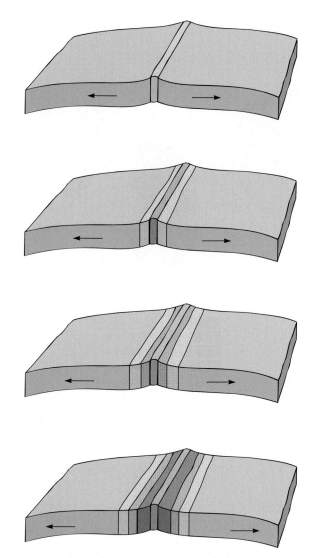

FIGURE 18.12 Formation of magnetic strips on the seafloor. As each new section of seafloor forms at the ridge, iron minerals become magnetized in a direction that depends on the orientation of Earth's field at that time. This makes a permanent record of reversals of Earth's magnetic field.

LITHOSPHERE PLATES AND BOUNDARIES

The strong evidence for seafloor spreading soon led to the development of a new theory called **plate tectonics.** According to plate tectonics, the lithosphere is broken into a number of fairly rigid plates that move on the asthenosphere. Some plates, as shown in Figure 18.13, contain continents and part of an ocean basin, while other plates contain only ocean basins. The plates move, and the movement is helping to explain why mountains form where they do, the occurrence of earthquakes and volcanoes, and in general, the entire changing surface of Earth.

Earthquakes, volcanoes, and most rapid changes in Earth's crust occur at the edge of a plate, which is called a *plate boundary.* There are three general kinds of plate boundaries that describe how one plate moves relative to another: divergent, convergent, and transform.

Divergent Boundaries

Divergent boundaries occur between two plates moving away from each other. Magma forms as the plates separate, decreasing pressure on the mantle below. This molten material from the asthenosphere rises, cools, and adds new crust to the edges of the separating plates. The new crust tends to move horizontally from both sides of the divergent boundary, usually known as an oceanic ridge. A divergent boundary is thus a **new crust zone.** Most new crust zones are presently on the seafloor, producing seafloor spreading (Figure 18.14).

The Mid-Atlantic Ridge is a divergent boundary between the South American and African Plates, extending north between the North American and Eurasian Plates (see Figure 18.13). This ridge is one segment of the global mid-ocean ridge system that encircles Earth. The results of divergent plate movement can be seen in Iceland, where the Mid-Atlantic Ridge runs as it separates the North American and Eurasian Plates. In the northeastern part of Iceland, ground cracks are widening, often accompanied by volcanic activity. The movement was measured extensively between 1975 and 1984, when displacements caused a total separation of about 7 m (about 23 ft).

The measured rate of spreading along the Mid-Atlantic Ridge ranges from 1 to 6 cm per year. This may seem slow, but the process has been going on for millions of years and has caused a tiny inlet of water between the continents of Europe, Africa, and the Americas to grow into the vast Atlantic Ocean that exists today.

Another major ocean may be in the making in East Africa, where a divergent boundary has already moved Saudi Arabia away from the African continent, forming the Red Sea. If this spreading between the African Plate and the Arabian Plate continues, the Indian Ocean will flood the area and the easternmost corner of Africa will become a large island.

Convergent Boundaries

Convergent boundaries occur between two plates moving toward each other. The creation of new crust at a divergent boundary means that old crust must be destroyed somewhere else at the same rate, or else Earth would have a continuously expanding diameter. Old crust is destroyed by returning to the asthenosphere at convergent boundaries. The collision produces an elongated belt of down-bending called a **subduction zone.** The lithosphere of one plate, which contains the crust, is subducted beneath the second plate and partially melts, then becoming part of the mantle. The more dense components of this may become igneous materials that remain in the mantle. Some of it may eventually migrate to a spreading ridge to make new crust again. The less dense components may return to the surface as a silicon, potassium, and sodium-rich lava, forming volcanoes on the upper plate, or they may cool below the surface to form a body of granite. Thus, the oceanic lithosphere is being recycled through this process, which explains why ancient seafloor rocks do not exist. Convergent boundaries produce related characteristic geologic features depending on the nature of the materials in the plates, and there are three general possibilities: (1) converging continental and oceanic plates; (2) converging oceanic plates; and (3) converging continental plates.

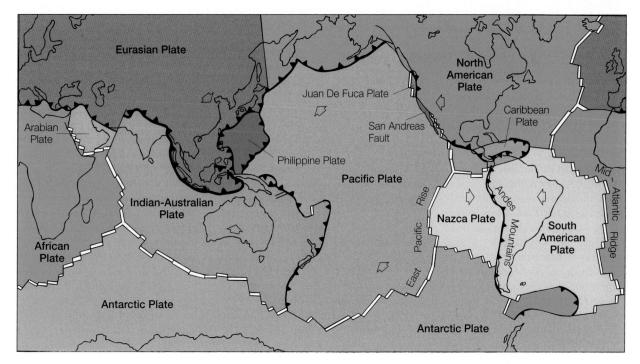

FIGURE 18.13 The major plates of the lithosphere that move on the asthenosphere. *Source:* After W. Hamilton, U.S. Geological Survey.

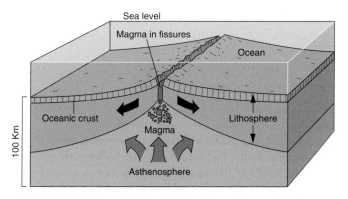

FIGURE 18.14 A diverging boundary at a mid-oceanic ridge. Hot asthenosphere wells upward beneath the ridge crest. Magma forms and squirts into fissures. Solid material that does not melt remains as mantle in the lower part of the lithosphere. As the lithosphere moves away from the spreading axis, it cools, becomes denser, and sinks to a lower level.

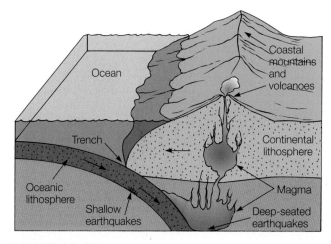

FIGURE 18.15 Ocean-continent plate convergence. This type of plate boundary accounts for shallow and deep-seated earthquakes, an oceanic trench, volcanic activity, and mountains along the coast.

As an example of *ocean-continent plate convergence,* consider the plate containing the South American continent (the South American Plate) and its convergent boundary with an oceanic plate (the Nazca Plate) along its western edge. Continent-oceanic plate convergence produces a characteristic set of geologic features as the oceanic plate of denser basaltic material is subducted beneath the less dense granite-type continental plate (Figure 18.15). The subduction zone is marked by an oceanic trench (the Peru-Chile Trench), deep-seated earthquakes, and volcanic

mountains on the continent (the Andes Mountains). The trench is formed from the down-bending associated with subduction and the volcanic mountains from subducted and melted crust that rise up through the overlying plate to the surface. The earthquakes are associated with the movement of the subducted crust under the overlying crust.

Ocean-ocean plate convergence produces another set of characteristics and related geologic features (Figure 18.16). The northern boundary of the oceanic Pacific Plate, for example, converges

A Closer Look

Measuring Plate Movement

According to the theory of plate tectonics, Earth's outer shell is made up of moving plates. The plates making up the continents are about 100 km thick and are gradually drifting at a rate of about 0.5 to 10 cm (0.2 to 4 in) per year. This reading is about one way that scientists know that Earth's plates are moving and how this movement is measured.

The very first human lunar landing mission took place in July 1969. Astronaut Neil Armstrong stepped onto the lunar surface, stating, "That's one small step for a man, one giant leap for mankind." In addition to fulfilling a dream, the *Apollo* project carried out a program of scientific experiments. The *Apollo 11* astronauts placed a number of experiments on the lunar surface in the Sea of Tranquility. Among the experiments was the first Laser Ranging Retro-reflector

Experiment, which was designed to reflect pulses of laser light from Earth. Three more reflectors were later placed on the Moon, including two by other *Apollo* astronauts and one by an unmanned Soviet *Lunakhod 2* lander.

The McDonald Observatory in Texas, the Lure Observatory on the island of Maui, Hawaii, and a third observatory in southern France have regularly sent laser beams through optical telescopes to the reflectors. The return signals, which are too weak to be seen with the unaided eye, are detected and measured by sensitive detection equipment at the observatories. The accuracy of these measurements, according to NASA reports, is equivalent to determining the distance between a point on the east and a point on the west coast of the United States

to an accuracy of 0.5 mm, about the size of the period at the end of this sentence.

Reflected laser light experiments have found that the Moon is pulling away from Earth at about 4 cm/yr (about 1.6 in/yr), that the shape of Earth is slowly changing, undergoing adjustment from the compression by the glaciers during the last ice age, and that the observatory in Hawaii is slowly moving away from the one in Texas. This provides a direct measurement of the relative drift of two of Earth's tectonic plates. Thus, one way that changes on the surface of Earth are measured is through lunar ranging experiments. Results from lunar ranging, together with laser ranging to artificial satellites in Earth orbit, have revealed the small but constant drift rate of the plates making up Earth's dynamic surface.

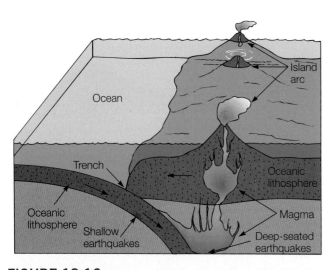

FIGURE 18.16 Ocean-ocean plate convergence. This type of plate convergence accounts for shallow and deep-focused earthquakes, an oceanic trench, and a volcanic arc above the subducted plate.

with the oceanic part of the North American Plate near the Bering Sea. The Pacific Plate is subducted, forming the Aleutian oceanic trench with a zone of earthquakes that are shallow near the trench and progressively more deep-seated toward the continent. The deeper earthquakes are associated with the movement of more deeply subducted crust into the mantle. The Aleutian Islands are typical **island arcs,** curving chains of volcanic islands that occur over the belt of deep-seated earthquakes. These islands form where the melted subducted material rises up through the overriding plate above sea level. The Japanese, Marianas, and Indonesians are similar groups of arc islands associated with converging oceanic-oceanic plate boundaries.

During *continent-continent plate convergence,* subduction does not occur as the less dense, granite-type materials tend to resist subduction (Figure 18.17). Instead, the colliding plates pile up into a deformed and thicker crust of the lighter material. Such a collision produced the thick, elevated crust known as the Tibetan Plateau and the Himalayan Mountains.

Transform Boundaries

Transform boundaries occur between two plates sliding by each other. Crust is neither created nor destroyed at transform boundaries as one plate slides horizontally past another along a long, vertical fault. The movement is neither smooth nor equal along the length of the fault, however, as short segments move independently with sudden jerks that are separated by periods without motion. The Pacific Plate, for example, is moving slowly to the northwest, sliding past the North American Plate. The

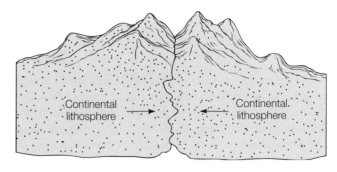

FIGURE 18.17 Continent-continent plate convergence. Rocks are deformed and some lithosphere thickening occurs, but neither plate is subducted to any great extent.

San Andreas fault is one boundary along the California coastline. Vibrations from plate movements along this boundary are the famous California earthquakes.

Myths, Mistakes, & Misunderstandings

Bye Bye California?

It is a myth that California will eventually fall off the continent into the ocean. The San Andreas fault is the boundary between the Pacific and North American Plates. The Pacific Plate is moving northwest along the North American Plate at 45 mm per year (about the rate your fingernails grow). The plates are moving horizontally by each other, so there is no reason to believe California will fall into the ocean. However, some 15 million years and millions of earthquakes from now, Los Angeles might be across the bay from San Francisco. See "Earthquakes, Mega Quakes, and the Movies" at

http://earthquake.usgs.gov/learning/topics/?topicID= 36

PRESENT-DAY UNDERSTANDINGS

The theory of plate tectonics, developed during the late 1960s and early 1970s, is new compared to most major scientific theories. Measurements are still being made, evidence is being gathered and evaluated, and the exact number of plates and their boundaries are yet to be determined with certainty. The major question that remains to be answered is what drives the plates, moving them apart, together, and by each other? One explanation is that slowly turning *convective cells* in the plastic asthenosphere drive the plates (Figure 18.18). According to this hypothesis, hot mantle materials rise at the diverging boundaries. Some of the material escapes to form new crust, but most of it spreads out beneath the lithosphere. As it moves beneath the lithosphere, it drags the overlying plate with it. Eventually, it cools and sinks back inward under a subduction zone.

There is uncertainty about the existence of convective cells in the asthenosphere and their possible role because of a lack of clear evidence. Seismic data is not refined enough to show convective cell movement beneath the lithosphere. In addition,

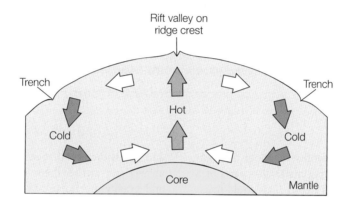

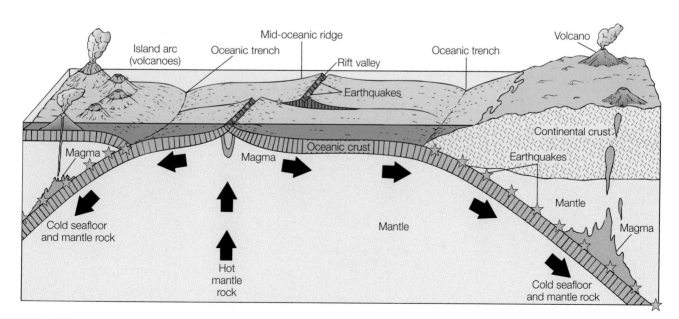

FIGURE 18.18 Not to scale. One idea about convection in the mantle has a convection cell circulating from the core to the lithosphere, dragging the overlying lithosphere laterally away from the oceanic ridge.

People Behind the Science

Harry Hammond Hess (1906-1969)

Harry Hess was the U.S. geologist who played the key part in the plate tectonics revolution of the 1960s.

Born in New York, Hess first studied electrical engineering and then trained in geology at Yale and later Princeton universities. From 1931, he began geophysical researches into the oceans, accompanying F. A. Vening Meinesz on a Caribbean submarine expedition to measure gravity and take soundings. During World War II, Hess continued his oceanographic investigations, undertaking extended echo-soundings while captain of the assault transport *USS Cape Johnson*. During the war years, he made studies of the flat-topped sea mounts that he called *guyots* (after Arnold Guyot, an earlier Princeton geologist). In the postwar years, he was one of the main advocates of the Mohole project, whose aim was to drill down through Earth's crust to gain access to the upper mantle.

The vast increase in seabed knowledge (deriving in part from wartime naval activities) led to the recognition that certain parts of the ocean floor were anomalously young.

Building on Maurice Ewing's discovery of the global distribution of mid-ocean ridges and their central rift valleys, Hess enunciated in

1962 the notion of deep-sea spreading. This contended that convection within Earth was continually creating new ocean floor at mid-ocean ridges. Material, Hess claimed, was incessantly rising from Earth's mantle to create the mid-ocean ridges, which then flowed horizontally to constitute new oceanic crust. It would follow that the further from the mid-ocean ridge, the older would be the crust—an expectation confirmed by research in 1963 by D. H. Matthews and his student F. J. Vine into the magnetic anomalies of the seafloor. Hess envisioned that the process of seafloor spreading would continue as far as the continental margins, where the oceanic crust would slide down beneath the lighter continental crust into a subduction zone, the entire operation thus constituting a kind of terrestrial conveyor belt. Within a few years, the plate tectonics revolution Hess had spearheaded had proved entirely successful. Hess's role in this "revolution in the earth sciences" was largely due to his remarkable breadth as geophysicist, geologist, and oceanographer.

Source: From the Hutchinson *Dictionary of Scientific Biography.* © Research Machines plc 2003. All Rights Reserved. Helicon Publishing is a division of Research Machines.

deep-seated earthquakes occur to depths of about 700 km (435 mi), which means that descending materials—parts of a subducted plate—must extend to that depth. This could mean that a convective cell might operate all the way down to the core-mantle boundary some 2,900 km (1,802 mi) below the surface. This presents another kind of problem because little is known about the lower mantle and how it interacts with the upper mantle. Theorizing without information is called speculation,

and that is the best that can be done with existing data. The full answer may include the role of heat and the role of gravity. Heat and gravity are important in a proposed mechanism of plate motion called "*ridge-push*" (Figure 18.19). This idea has a plate cooling and thickening as it moves away from a divergent boundary. As it subsides, it cools asthenospheric mantle to lithospheric mantle, forming a sloping boundary between the lithosphere and the asthenosphere. The plate slides down this boundary.

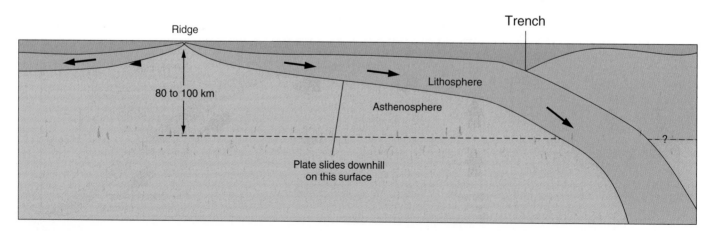

FIGURE 18.19 Ridge-push. A plate may slide downhill on the sloping boundary between the lithosphere and the asthenosphere at the base of the plate.

Geothermal energy means earth (*geo*) heat (*thermal*) energy, or energy in the form of heat from Earth. Beneath the surface of Earth is a very large energy resource in the form of hot water and steam that can be used directly for heat or converted to electricity.

Most of the U.S. geothermal resources are located in the western part of the nation where the Juan de Fuca Plate is subducted beneath the continental lithosphere. The Juan de Fuca partially melts, forming magma that is buoyed toward Earth's surface, erupting as volcanoes (see Figure 19.25). This subduction is the source of heating for the hot water and steam resources in ten western states. There are also geothermal resources in Hawaii from a different geothermal source.

One use of geothermal energy is to generate electricity. Geothermal power plants are located in California (ten sites), Hawaii, Nevada (ten sites), Oregon, and Utah (two sites). These sites have a total generating capacity of 2,700 megawatts (MW), which is enough electricity to supply the needs of 3.5 million people. The world's largest geothermal power plant is located at the Geysers Power Plant in northern California. Dry steam provides the energy for twenty-three units at this site, which generate more than 1,700 MW of electrical power. All of the other sites use hot water rather than steam to generate electricity, with a total generating capacity of 1,000 MW.

In addition to producing electricity, geothermal hot water is used directly for space heating. Space heating in individual houses is accomplished by piping hot water from one geothermal well. District systems, on the other hand, pipe hot water from one or more geothermal wells to several buildings, houses, or blocks of houses. Currently, geothermal hot water is used in individual and district space-heating systems at more than 120 locations. There are more than 1,200 potential geothermal sites that could be developed to provide hot water to more than 370 cities in eight states. The creation of such geothermal districts could result in a savings of up to 50 percent over the cost of natural gas heating.

Geothermal hot water is also used directly in greenhouses and aquaculture facilities. There are more than thirty-five large geothermal-energized greenhouses raising vegetables and flowers and more than twenty-five geothermal-energized aquaculture facilities raising fish in Arizona, California, Colorado, Idaho, Montana, Nevada, New York, Oregon, South Dakota, Utah, and Wyoming (see http://geoheat.oit.edu/dusys.htm. A food dehydration facility in Nevada, for example, uses geothermal energy to process more than 15 million pounds of dried onions and garlic per year. Other uses of geothermal energy include laundries, swimming pools, spas, and resorts. Over two hundred resorts are using geothermal hot water in the United States.

Geothermal energy is considered to be one of the renewable energy resources since the energy supply is maintained by plate tectonics. Currently, geothermal energy production is ranked third behind hydroelectricity and biomass but ahead of solar and wind. It has been estimated that known geothermal resources could supply thousands of megawatts more power beyond current production, and development of the potential direct-use applications could displace the use—and greenhouse gas emissions—of 18 million barrels of oil per year.

QUESTIONS TO DISCUSS

Discuss with your group the following questions concerning the use of geothermal energy:

1. Why is the development of geothermal energy not proceeding more rapidly?
2. Should the government provide incentives for developing geothermal resources? Give reasons for your answer.
3. What are the advantages and disadvantages of a government-controlled geothermal energy industry?
4. As other energy supplies become depleted, who should be responsible for developing new energy supplies, investor-owned industry or government agencies?

Another proposed mechanism of plate motion is called "*slab-pull*" (Figure 18.20). In this mechanism, the subducting plate is colder and therefore denser than the surrounding hot mantle, so it pulls the surface part of the plate downward. Density of the slab may also be increased by loss of water and reforming of minerals into more dense forms.

What is generally accepted about the plate tectonic theory is the understanding that the solid materials of Earth are engaged in a continual cycle of change. Oceanic crust is subducted, melted, then partly returned to the crust as volcanic igneous rocks in island arcs and along continental plate boundaries. Other parts of the subducted crust become mixed with the upper mantle, returning as new crust at diverging boundaries. The materials of the crust and the mantle are thus cycled back and forth in a mixing that may include the

deep mantle and the core as well. There is more to this story of a dynamic Earth that undergoes a constant change. The story continues in the next chapters with different cycles to consider.

CONCEPTS *Applied*

New Earthquakes and Volcanoes

Locate and label the major plates of the lithosphere on an outline map of the world according to the most recent findings in plate tectonics. Show all types of boundaries and associated areas of volcanoes and earthquakes.

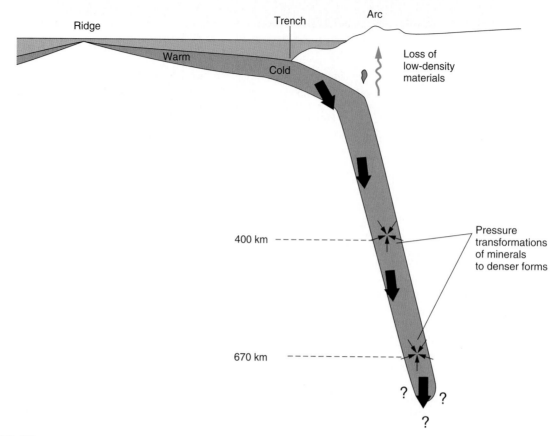

FIGURE 18.20 Slab-pull. The dense, leading edge of a subducting plate pulls the rest of the plate along. Plate density increases due to cooling, loss of low-density material, and pressure transformation of minerals to denser forms.

SUMMARY

Earth has a layered interior that formed as Earth's materials underwent *differentiation,* the separation of materials while in the molten state. The center part, or *core,* is predominantly iron with a solid inner part and a liquid outer part. The core makes up about 15 percent of Earth's total volume and about a third of its total mass. The *mantle* is the middle part of Earth's interior that accounts for about two-thirds of Earth's total mass and about 80 percent of its total volume. The mantle is predominantly composed of the ferromagnesian silicate *olivine,* which undergoes structural changes at two depths from the increasing heat and pressure. The outer layer, or *crust,* of Earth is separated from the mantle by the *Mohorovicic discontinuity.* The crust of the *continents* is composed mostly of less dense granite-type rock. The crust of the *ocean basins* is composed mostly of the more dense basaltic rocks.

Another way to consider Earth's interior structure is to consider the weak layer in the upper mantle, the *asthenosphere* that extends around the entire Earth. The rigid, solid, and brittle layer above the asthenosphere is called the *lithosphere.* The lithosphere includes the entire crust, the Moho, and the upper part of the mantle.

The shapes of the continents suggested to some that the continents were together at one time and have shifted positions to their present locations. This idea, first developed in the early 1900s, came to be known as *continental drift* and was generally dismissed by most scientists. Evidence from the ocean floor that was gathered in the 1950s and 1960s revived interest in the idea that continents could move. The evidence for *seafloor spreading* came from related observations concerning oceanic ridge systems, sediment and fossil dating of materials outward from the ridge, and from magnetic patterns of seafloor rocks. Confirmation of seafloor spreading led to the *plate tectonic theory.* According to plate tectonics, new basaltic crust is added at *diverging boundaries* of plates, and old crust is *subducted* at *converging boundaries.* Mountain building, volcanoes, and earthquakes are seen as *related geologic features* that are caused by plate movements. The force behind the movement of plates is uncertain, but it may involve *convection* in the deep mantle.

KEY TERMS

asthenosphere (p. **456**)

convergent boundaries (p. **460**)

core (p. **455**)

crust (p. **454**)

divergent boundaries (p. **460**)

island arcs (p. **462**)

lithosphere (p. **456**)

mantle (p. **455**)

new crust zone (p. **460**)

oceanic ridges (p. **458**)

oceanic trenches (p. **458**)

plate tectonics (p. **460**)

rift (p. **459**)

seafloor spreading (p. **459**)

seismic wave (p. **453**)

subduction zone (p. **460**)

transform boundaries (p. **462**)

APPLYING THE CONCEPTS

1. The core of Earth is composed of
   a. iron and nickel.
   b. silicon and oxygen.
   c. iron and lead.
   d. nickel and copper.

2. The middle part of Earth's interior is
   a. oceanic crust.
   b. mantle.
   c. inner crust.
   d. outer core.

3. The separation of materials that gave Earth its layered interior is called
   a. stratification.
   b. separation.
   c. differentiation.
   d. partition.

4. A vibration that moves through any part of Earth is called a
   a. seismic wave.
   b. gamma wave.
   c. Z wave.
   d. radio wave.

5. The S wave is a
   a. longitudinal wave.
   b. transverse wave.
   c. radio wave.
   d. surface wave.

6. Waves that occur where S or P waves reach the surface are
   a. secondary waves.
   b. surface waves.
   c. T waves.
   d. minor waves.

7. The three main areas of Earth's interior are
   a. core, mantle, and crust.
   b. core, secondary layer, and surface.
   c. nucleus, mantle, and crust.
   d. core, mantle, and shell.

8. The boundary between the crust and the mantle is called
   a. crust to mantle discontinuity.
   b. continental to oceanic boundary.
   c. Litho discontinuity.
   d. Moho discontinuity.

9. The mantle is composed of
   a. sulfides.
   b. silicates.
   c. clay minerals.
   d. oxides.

10. Seismological studies suggests that the core
    a. is solid.
    b. is liquid.
    c. has a liquid outer core and solid inner core.
    d. has a solid outer core and liquid inner core.

11. Evidence from meteorite studies proposes that the core is composed of
    a. nickel.
    b. iron.
    c. lead and iron.
    d. nickel and iron.

12. The layer in Earth where seismic waves sharply decrease in velocity is called the
    a. crust.
    b. mantle.
    c. lithosphere.
    d. asthenosphere.

13. The layer that is broken up into plates that move in the upper mantle is the
    a. crust.
    b. lithosphere.
    c. upper mantle.
    d. asthenosphere.

14. The name of the single large continent suggested by Wegener is
    a. Pan-continental.
    b. Atlantis.
    c. Pangaea.
    d. Asia Major.

15. Records of the strength and directions of Earth's magnetic field are found in
    a. igneous rocks.
    b. meteorites.
    c. sedimentary rocks.
    d. minerals.

16. The chain of mountains found in the center of the Atlantic Ocean basin is called the
    a. Atlantic Mountains.
    b. Mid-Atlantic Ridge.
    c. Continental Ridge.
    d. Icelandic Ridge.

17. Long, deep, and narrow oceanic trenches are located
    a. in the middle of oceanic ridges.
    b. on either side of the oceanic ridges.
    c. parallel to the edges of continents.
    d. perpendicular to the edges of continents.

18. The theory that the lithosphere is composed of several rigid plates that "float" in the asthenosphere is called
    a. continental drift.
    b. plate tectonics.
    c. plate boundaries.
    d. continental tectonics.

19. The plate boundary associated with the formation of new crust is called
    a. divergent.
    b. convergent.
    c. transform.
    d. None of the above is correct.

20. The movement of one plate under another plate creates a
    a. divergent boundary.
    b. ridge.
    c. subduction zone.
    d. boundary zone.

21. Transform boundaries occur when
    a. two plates slide by each other without the formation or loss of crust.
    b. two plates come together and "transform" each other.
    c. one plate slides under another plate.
    d. ridges are formed.

22. What is the current theory about why the plates move?
    a. Plates follow the rotation of Earth.
    b. Convective cells move from the core to the lithosphere.
    c. gravity
    d. Conduction cells move from the core to the lithosphere.

23. The seismic waves that cause the most damage during an earthquake are
    a. P waves.
    b. S waves.
    c. surface waves.
    d. P and S waves.

24. Earth's mantle has a chemical composition that agrees closely with the composition of
    a. basalt.
    b. iron and nickel.
    c. granite.
    d. gneiss.

25. From seismological data, Earth's shadow zone indicates that part of Earth's interior must be
    a. liquid.
    b. solid throughout.
    c. plastic.
    d. hollow.

26. The Mohorovicic discontinuity is a change in seismic wave velocity that is believed to take place because of
    a. structural changes in minerals of the same composition.
    b. changes in the composition on both sides of the boundary.
    c. a shift in the density of minerals of the same composition.
    d. changes in the temperature with depth.

27. The oldest rocks are found in
    a. continental crust.
    b. oceanic crust.
    c. neither, since both are the same age.

28. The least dense rocks are found in
    a. continental crust.
    b. oceanic crust.
    c. neither, since both are the same density.

29. The idea of seafloor spreading along the Mid-Atlantic Ridge was supported by evidence from
    a. changes in magnetic patterns and ages of rocks moving away from the ridge.
    b. faulting and volcanoes on the continents.
    c. the observations that there was no relationship between one continent and another.
    d. All of the above are correct.

30. According to the plate tectonics theory, seafloor spreading takes place at a
    a. convergent boundary.
    b. subduction zone.
    c. divergent boundary.
    d. transform boundary.

31. The presence of an oceanic trench, a chain of volcanic mountains along the continental edge, and deep-seated earthquakes is characteristic of a (an)
    a. ocean-ocean plate convergence.
    b. ocean-continent plate convergence.
    c. continent-continent plate convergence.
    d. None of the above is correct.

32. The presence of an oceanic trench with shallow earthquakes and island arcs with deep-seated earthquakes is characteristic of a (an)
    a. ocean-ocean plate convergence.
    b. ocean-continent plate convergence.
    c. continent-continent plate convergence.
    d. None of the above is correct.

33. The ongoing occurrence of earthquakes without seafloor spreading, oceanic trenches, or volcanoes is most characteristic of a
    a. convergent boundary between plates.
    b. subduction zone.
    c. divergent boundary between plates.
    d. transform boundary between plates.

34. The evidence that Earth's core is part liquid or acts like a liquid comes from
    a. the P-wave shadow zone.
    b. the S-wave shadow zone.
    c. meteorites.
    d. All of the above are correct.

35. The surfaces of early planets in our solar system were thought to be formed from
    a. melting by the Sun.
    b. bombardment by dust and debris remaining from initial planet formation.
    c. nuclear decay processes.
    d. volcanic eruptions.

36. The early Earth's core is thought to have formed from the
    a. exterior melting from radioactive decay.
    b. gravitational pull of heavier materials to the center of Earth.
    c. exterior melting from fusion reactions.
    d. interior melting from lava.

37. Indirect evidence that supports the theory of how Earth formed is not supported by the study of
    a. vibrations in Earth.
    b. Earth's magnetic field.
    c. heat flow.
    d. samples from Earth's core.

38. The oceanic crust is
    a. thicker than the continental crust.
    b. found under the continental crust.
    c. thinner than the continental crust.
    d. the same in its properties as the continental crust.

39. Seismic waves that do not travel through liquid are
    a. P waves.
    b. S waves.
    c. both P and S waves.
    d. surface waves.

40. The fastest seismic wave is the
    a. P wave.
    b. S wave.
    c. P and S waves are equally fast.
    d. surface wave.

41. Information about the composition and nature of the mantle does not come from
    a. seismological data.
    b. meteorite studies.
    c. debris ejected from volcanoes.
    d. samples obtained from drilling.

42. Primary information about the nature of the core is provided by
    a. seismological data.
    b. meteorite studies.
    c. debris ejected from volcanoes.
    d. samples obtained from drilling.

43. The asthenosphere is not defined as
    a. rigid.
    b. plastic.
    c. mobile.
    d. elastic.

44. Earth's magnetic field is thought to be generated by
    a. seismic waves throughout Earth.
    b. electrical currents in the core.
    c. electrical currents in the crust.
    d. electrical currents in the asthenosphere.

45. Studies of the Mid-Atlantic Ridge provided evidence of
    a. seafloor spreading.
    b. seafloor closing.
    c. ancient shipwrecks.
    d. early man.

46. Evidence that supports seafloor spreading does *not* include the
    a. earthquakes occurring along the crest of the Mid-Atlantic Ridge.
    b. presence of rift along the crest of the Mid-Atlantic Ridge.
    c. younger sediments near the continents.
    d. discovery of identical magnetic patterns in rocks on both sides of the ridge.

47. A geologic feature that was produced by divergent boundaries is
    a. the Alps.
    b. the Mid-Atlantic Ridge.
    c. the Marianas Trench.
    d. Japan.

48. Which type of plate boundary accounts for the formation of the Appalachian Mountains?
    a. divergent
    b. ocean to continental convergence
    c. continental to continental convergence
    d. transform

49. Which type of plate boundary was responsible for the formation of the Japanese Islands?
    a. divergent
    b. ocean to ocean convergence
    c. continental to continental convergence
    d. transform

50. A famous transform boundary in the United States is the
    a. Rocky Mountains fault.
    b. Mississippi fault.
    c. San Andreas fault.
    d. San Jose fault.

51. Plate movement is measured by
    a. reflected laser light experiments.
    b. age of sediments and fossils.
    c. analysis of magnetic patterns in rocks.
    d. All of the above are correct.

52. Islands that form when melted subducted material rises through the plates above sea level are called
    a. Pacific islands.
    b. arc islands.
    c. convergence islands.
    d. plate islands.

Answers

1. a 2. b 3. c 4. a 5. b 6. b 7. a 8. d 9. b 10. c 11. d 12. d 13. b
14. c 15. a 16. b 17. c 18. b 19. a 20. c 21. a 22. b 23. c 24. a
25. a 26. b 27. a 28. a 29. a 30. c 31. b 32. a 33. d 34. b 35. b
36. b 37. d 38. c 39. b 40. a 41. d 42. a 43. a 44. b 45. a 46. c
47. b 48. c 49. b 50. c 51. d 52. b

QUESTIONS FOR THOUGHT

1. Describe one theory of how Earth came to have a core composed mostly of iron. What evidence provides information about the nature of Earth's core?

2. Briefly describe the internal composition and structure of the (a) core, (b) mantle, and (c) crust of Earth.

3. What is the asthenosphere? Why is it important in modern understandings of Earth?

4. Describe the parts of Earth included in the (a) lithosphere, (b) asthenosphere, (c) crust, and (d) mantle.

5. What is continental drift? How is it different from plate tectonics?

6. Rocks, sediments, and fossils around an oceanic ridge have a pattern concerning their ages. What is the pattern? Explain what the pattern means.

7. Describe the origin of the magnetic strip patterns found in the rocks along an oceanic ridge.

8. Explain why ancient rocks are not found on the seafloor.

9. Describe the three major types of plate boundaries and what happens at each.

10. What is an island arc? Where are they found? Explain why they are found at this location.

11. Briefly describe a model that explains how Earth developed a layered internal structure.

12. Briefly describe the theory of plate tectonics and how it accounts for the existence of certain geologic features.

13. What is an oceanic trench? What is the relationship of trenches to major plate boundaries? Explain this relationship.

14. Describe the probable source of all the earthquakes that occur in southern California.

15. The northwestern coast of the United States has a string of volcanoes running along the coast. According to plate tectonics, what does this mean about this part of the North American Plate? What geologic feature would you expect to find on the seafloor off the northwestern coast? Explain.

16. Explain how the crust of Earth is involved in a dynamic, ongoing recycling process.

FOR FURTHER ANALYSIS

1. Why are there no active volcanoes in the eastern United States or Canada? Explain why you would or would not expect volcanoes there in the future.

2. Describe cycles that occur on Earth's surface and cycles that occur between the surface and the interior. Explain why these cycles do not exist on other planets of the solar system.

3. Discuss evidence that would explain why plate tectonics occurs on Earth but not on other planets.

4. Analyze why you would expect most earthquakes to be localized, shallow occurrences near a plate boundary.

INVITATION TO INQUIRY

Measuring Plate Motion

Tectonic plate motion can be measured with several relatively new technologies, including satellite laser ranging (SLR) and use of the Global Positioning System (GPS). Start your inquiry by visiting the Tectonic Plate Motion website at cddisa.gsfc.nasa.gov/926/slrtecto.html. Study the regional plate motion of the North American Plate, for example. Note the scale of 50 mm/yr, then measure to find the rate of movement at the different stations shown. Is this recent data on plate motion consistent with other available information on plate motion? How can you account for the different rates of motion at adjacent stations?

19

Building Earth's Surface

Folding, faulting, and lava flows, such as the one you see here, tend to build up, or elevate, Earth's surface.

CORE **CONCEPT**

The surface of Earth is involved in plate tectonic processes that result in an ongoing building up of the surface.

Interpreting Earth's Surface
The surface of Earth undergoes slow, uniform change, and the geologic processes responsible for the change are the same today as they were in the distant past.

Folding
Stress on deeply buried layers of horizontal rocks can result in a wrinkling of the layers into folds.

Earthquakes
Stress on deeply buried rock can result in fracture that produces vibrations, and the resulting quaking, shaking, and upheaval of the ground is an earthquake.

Stress and Strain
Rocks are subjected to forces associated with plate tectonics and other forces.

Faulting
Stress on cooler, less plastic layers of horizontal rocks can result in breaking of the layers with relative movement, producing a fault.

Origin of Mountains
Mountain ranges are features of folding and faulting on a very large scale; a volcano is a hill or mountain formed by lava or rock fragments from magma below.

The central idea of plate tectonics, which was discussed in chapter 18, is that Earth's surface is made up of rigid plates that are moving slowly across the surface. Since the plates and the continents riding on them are in constant motion, any given map of the world is only a snapshot that shows the relative positions of the continents at a given time. The continents occupied different positions in the distant past. They will occupy different positions in the distant future. The surface of Earth, which seems so solid and stationary, is in fact mobile.

Plate tectonics has changed the accepted way of thinking about the solid, stationary nature of Earth's surface and ideas about the permanence of the surface as well. The surface of Earth is no longer viewed as having a permanent nature but is understood to be involved in an ongoing cycle of destruction and renewal. Old crust is destroyed as it is plowed back into the mantle through subduction, becoming mixed with the mantle. New crust is created as molten materials move from the mantle through seafloor spreading and volcanoes. Over time, much of the crust must cycle into and out of the mantle.

The movement of plates, the crust-mantle cycle, and the rock cycle all combine to produce a constantly changing surface. There are basically two types of surface changes: (1) changes that originate within Earth, resulting in a building up of the surface (Figure 19.1), and (2) changes that result from rocks being exposed to the atmosphere and water, resulting in a sculpturing and tearing down of the surface. This chapter is about the building up of the land. The concepts of this chapter will provide you with something far more interesting about Earth's surface than the scenic aspect. The existence of different features (such as mountains, folded hills, islands) and the occurrence of certain events (such as earthquakes, volcanoes, faulting) are all related. The related features and events also have a story to tell about Earth's past, a story about the here and now, and yet another story about the future.

INTERPRETING EARTH'S SURFACE

Because many geologic changes take place slowly, it is difficult for a person to see significant change occur to mountains, canyons, and shorelines in the brief span of a lifetime. Given a mental framework based on a lack of appreciation of change over geologic time, how do you suppose people interpreted the existence of features such as mountains and canyons? Some believed, as they had observed in their lifetimes, that the mountains and canyons had "always" been there. Statements such as "unchanging as the hills" or "old as the hills" illustrate this lack of appreciation of change over geologic time. Others did not believe the features had always been there but believed they were formed by a sudden, single catastrophic event (Figure 19.2). A catastrophe created a feature of Earth's surface all at once, with little or no change occurring since that time. The Grand Canyon, for example, was not interpreted as the result of incomprehensibly slow river erosion but as the result of a giant crack or rip that appeared in the surface. The canyon that you see today was interpreted as forming when Earth split open and the Colorado River fell into the split. This interpretation was used to explain the formation of major geologic features based on the lack of change that could be observed during a person's lifetime.

About two hundred years ago, the idea of unchanging, catastrophically formed landscapes was challenged by James Hutton, a Scottish physician. Hutton, who is known today as the founder of modern geology, traveled widely throughout the British Isles. Hutton was a keen observer of rocks, rock structures, and other features of the landscape. He noted that sandstone, for example, was made up of rock fragments that appeared to be (1) similar to the sand being carried by rivers and (2) similar to the sand making up the beaches next to the sea. He also noted fossil shells of sea animals in sandstone on the land, while the living relatives of these animals were found in the shallow waters of the sea. This and other evidence led Hutton to realize that rocks were being ground into fragments, then carried by rivers to the sea. He surmised that these particles would be reformed into rocks later, then lifted and shaped into the hills and mountains of the land. He saw all this as quiet, orderly change that required only *time* and the ongoing work of the water and some forces to make the sediments back into rocks. With Hutton's logical conclusion came the understanding that Earth's history could be interpreted by tracing it backward, from the present to the past. This tracing required a frame of reference of slow, uniform change, not the catastrophic frame of reference of previous thinkers. The frame of reference of uniform changes is today called the **principle of uniformity** (also called *uniformitarianism*). The principle of uniformity is often represented by a statement that "the present is the key to the past." This statement means that the geologic processes you see changing rocks today are the very same processes that changed them in the ancient past, although not necessarily at the same rate. The principle of uniformity does not *exclude* the happening of sudden or catastrophic events on the surface of Earth.

FIGURE 19.1 An aerial view from the south of the eruption of Mount St. Helens volcano on May 18, 1980.

FIGURE 19.2 Would you believe that this rock island has "always" existed where it is? Would you believe it was formed by a sudden, single event? What evidence would it take to convince you that the rock island formed ever so slowly, starting as a part of southern California and moving very slowly, at a rate of centimeters per year, to its present location near the coast of Alaska?

A violent volcanic explosion, for example, is a catastrophic event that most certainly modifies the surface of Earth. What the principle of uniformity does state is that the physical and chemical laws we observe today operated exactly the same way in the past. The rates of operation may or may not have been the same in the past, but the events you see occurring today are the same events that occurred in the past. Given enough time, you can explain the formation of the structures of Earth's surface with known events and concepts.

The principle of uniformity has been used by geologists since the time of Hutton. The concept of how the constant changes occur has evolved with the development of plate tectonics, but the basic frame of reference is the same. You will see how the principle of uniformity is applied by first considering what can happen to rocks and rock layers that are deeply buried.

DIASTROPHISM

All the possible movements of Earth's plates, including drift toward or away from other plates and any process that deforms or changes Earth's surface, are included in the term *diastrophism*. Diastrophism is the process of deformation that changes Earth's surface. It produces many of the basic structures you see on the surface, such as plateaus, mountains, and folds in the crust. The movement of magma is called *vulcanism* or *volcanism*. Diastrophism, volcanism, and earthquakes are closely related, and their occurrence can usually be explained by events involving plate tectonics (chapter 18). The results of diastrophism are discussed in the section on stress and strain, which is followed by a discussion of earthquakes, volcanoes, and mountain chains. Again, remember that diastrophism, volcanism, earthquakes, and the movement of Earth's plates are very closely related. All are involved with the shapes, arrangements, and interrelationships of different parts of Earth's crust and the forces that change it. We will begin with a discussion of some of these forces before discussing what the forces can do.

STRESS AND STRAIN

Any solid material responds to a force in a way that depends on the extent of coverage (force per unit area, or pressure), the nature of the material, and other variables such as the temperature. Consider, for example, what happens if you place the point of a ballpoint pen on the side of an aluminum pop (soda) can and apply an increasing pressure. With increasing pressure, you can observe at least four different and separate responses:

1. At first, the metal successfully resists a slight pressure and *nothing happens*.
2. At a somewhat greater pressure, you will be able to deform or bend the metal into a concave surface. The metal will return to its original shape, however, when the pressure is removed. This is called an *elastic deformation* since the metal was able to spring back into its original shape.
3. At a still greater pressure, the metal is deformed to a concave surface, but this time the metal does not return to

its original shape. This means the *elastic limit* of the metal has been exceeded, and it has now undergone a *plastic deformation.* Plastic deformation permanently alters the shape of a material.

4. Finally, at some great pressure, the metal will rupture, resulting in a *break* in the material.

Many materials, including rocks, respond to increasing pressures in this way, showing (1) no change, (2) an elastic change with recovery, (3) a plastic change with no recovery, and (4) finally breaking from the pressure.

A **stress** is a force that tends to compress, pull apart, or deform a rock. Rocks in Earth's solid outer crust are subjected to forces as Earth's plates move into, away from, or alongside each other. However, not all stresses are generated directly by plate interaction. Thus, there are three types of forces that cause rock stress:

1. *Compressive stress* is caused by two plates moving together or by one plate pushing against another plate that is not moving.
2. *Tensional stress* is the opposite of compressional stress. It occurs when one part of a plate moves away, for example, and another part does not move.
3. *Shear stress* is produced when two plates slide past one another or by one plate sliding past another plate that is not moving.

Just like the metal in the soda can, a rock is able to withstand stress up to a limit. Then it might undergo elastic deformation, plastic deformation, or breaking with progressively greater pressures. The adjustment to stress is called **strain.** A rock unit might respond to stress by changes in volume, changes in shape, or by breaking. Thus, there are three types of strain: elastic, plastic, and fracture.

1. In *elastic strain,* rock units recover their original shape after the stress is released.
2. In *plastic strain,* rock units are molded or bent under stress and do not return to their original shape after the stress is released.
3. In *fracture strain,* rock units crack or break, as the name suggests.

The relationship between stress and strain, that is, exactly how the rock responds depends on at least four variables. They are (1) the nature of the rock, (2) the temperature of the rock, (3) how slowly or quickly the stress is applied, and (4) the confining pressure on the rock. The temperature and confining pressure are generally a function of how deeply the rock is buried. In general, rocks are better able to withstand compressional than pulling-apart stresses. Cold rocks are more likely to break than warm rocks, which tend to undergo plastic deformation. In addition, a stress that is applied quickly tends to break the rock, whereas stress applied more slowly over time, perhaps thousands of years, tends to result in plastic strain.

In general, rocks at great depths are under great pressure at higher temperatures. These rocks tend to undergo plastic deformation, then plastic flow, so rocks at great depths are bent and

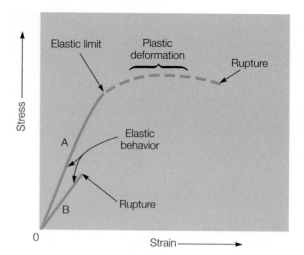

FIGURE 19.3 Stress and deformation relationships for deeply buried, warm rocks under high pressure (*A*) and cooler rocks near the surface (*B*). Breaking occurs when stress exceeds rupture strength.

deformed extensively. Rocks closer to the surface can also bend, but they have a lower elastic limit and break more readily (Figure 19.3). Rock deformation often results in recognizable surface features called folds and faults, the topics of the next sections.

FOLDING

Sediments that form most sedimentary rocks are deposited in nearly flat, horizontal layers at the bottom of a body of water. Conditions on the land change over time, and different mixtures of sediments are deposited in distinct layers of varying thickness. Thus, most sedimentary rocks occur naturally as structures of horizontal layers or beds (Figure 19.4).

A sedimentary rock layer that is not horizontal may have been subjected to some kind of compressive stress. The source of such a stress could be from colliding plates, from the intrusion of magma, or from a plate moving over a hot spot. Stress on buried layers of horizontal rocks can result in plastic strain, resulting in a wrinkling of the layers into *folds.* **Folds** are bends in layered bedrock (Figure 19.5). They are analogous to layers of rugs or blankets that were stacked horizontally, then pushed into a series of arches and troughs. Folds in layered bedrock of all shapes and sizes can occur from plastic strain, depending generally on the regional or local nature of the stress and other factors. Of course, when the folding occurred, the rock layers were in a plastic condition, probably under considerable confining pressure from deep burial. However, you see the results of the folding when the rocks are under very different conditions at the surface.

Widespread, regional horizontal stress on deeply buried sedimentary rock layers can produce symmetrical up and down folds shaped like waves on water. A vertical, upward stress, on the other hand, can produce a large, upwardly bulging fold called a *dome.* A corresponding downward bulging fold is called a *basin.* When the stress is great and extensive, complex overturned folds can result.

A

B

FIGURE 19.4 (*A*) Rock bedding on a grand scale in the Grand Canyon. (*B*) A closer example of rock bedding can be seen in this roadcut.

FIGURE 19.5 These folded rock layers are in the Calico Hills, California. Can you figure out what might have happened to flat rock layers to make folds like this?

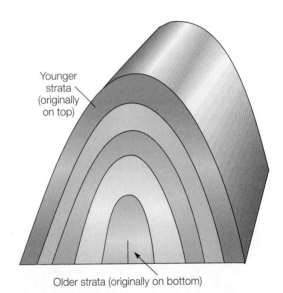

Younger strata (originally on top)

Older strata (originally on bottom)

FIGURE 19.6 An anticline, or arching fold, in layered sediments. Note that the oldest strata are at the center.

The most common regional structures from deep plastic deformation are arch-shaped and trough-shaped folds. In general, an arch-shaped fold is called an **anticline** (Figure 19.6). The corresponding trough-shaped fold is called a **syncline** (Figure 19.7). Anticlines and synclines sometimes alternate across the land like waves on water. You can imagine that a great compressional stress must have been involved over a wide region to wrinkle the land like this.

Anticlines, synclines, and other types of folds are not always visible as such on Earth's surface. The ridges of anticlines are constantly being weathered into sediments. The sediments, in turn, tend to collect in the troughs of synclines, filling them in. The Appalachian Mountains have ridges of rocks that are more resistant to weathering, forming hills and mountains. The San Joaquin Valley, on the other hand, is a very large syncline in California.

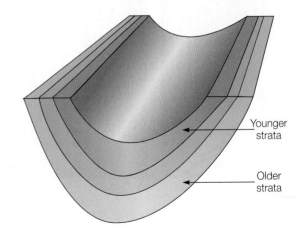

Younger strata

Older strata

FIGURE 19.7 A syncline, showing the reverse age pattern.

Note that any kind of rock can be folded. Sedimentary rocks are usually the best example of folding, however, since the fold structures of rock layers are easy to see and describe. Folding is much harder to see in igneous or metamorphic rocks that are blends of minerals without a layered structure.

FAULTING

Rock layers do not always respond to stress by folding. Rocks near the surface are cooler and under less pressure, so they tend to be more brittle. A sudden stress on these rocks may reach the rupture point, resulting in a cracking and breaking of the rock structure. If there is breaking of rock without a relative displacement on either side of the break, the crack is called a *joint*. Joints are common in rocks exposed at the surface of Earth. They can be produced from compressional stresses, but they are also formed by other processes such as the contraction of an igneous rock while cooling. Basalt often develops *columnar jointing* from the contraction of cooling, solidified magma. The joints are parallel and evenly spaced, resulting in the appearance of hexagonal columns (Figure 19.8). The Devil's Post Pile in California and Devil's Tower in Wyoming are classic examples of columnar jointing.

When there is relative movement between the rocks on either side of a fracture, the crack is called a **fault**. When faulting occurs, the rocks on one side move relative to the rocks on the other side along the surface of the fault, which is called the *fault plane*. Faults are generally described in terms of (1) the steepness of the fault plane, that is, the angle between the plane and an imaginary horizontal plane, and (2) the direction of relative movement. There are basically three ways that rocks on one side of a fault can move relative to the rocks on the other side: (1) up and down (called "dip"), (2) horizontally, or sideways

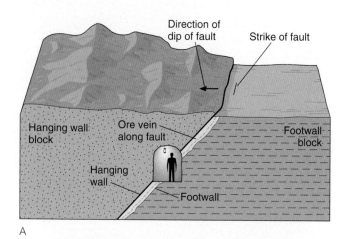

A

B

FIGURE 19.9 (*A*) The relationship between the hanging wall block and footwall block of fault. (*B*) A photo of a fault near Kingman, Arizona, showing how the hanging wall has moved relative to the footwall.

A

B

FIGURE 19.8 Columnar jointing forms at right angles to the surface as basalt cools. (*A*) Devil's Post Pile, San Joaquin River, California. (*B*) The Devil's Tower, Wyoming.

(called "strike"), and (3) with elements of both directions of movement (called "oblique").

One classification scheme for faults is based on an orientation referent borrowed from mining (many ore veins are associated with fault planes). Imagine a mine with a fault plane running across a horizontal shaft. Unless the plane is perfectly vertical, a miner would stand on the mass of rock below the fault plane and look up at the mass of rock above. Therefore, the mass of rock below is called the *footwall* and the mass of rock above is called the *hanging wall* (Figure 19.9). How the

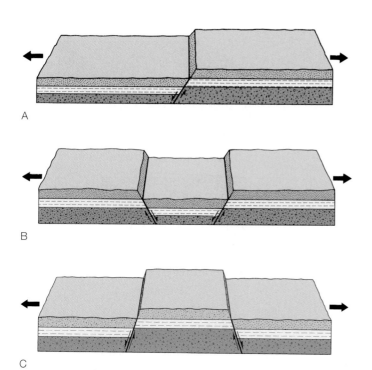

A

B

C

FIGURE 19.10 How tensional stress could produce (*A*) a normal fault, (*B*) a graben, and (*C*) a horst.

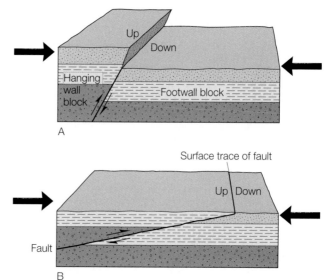

A

B

FIGURE 19.11 How compressive stress could produce (*A*) a reverse fault and (*B*) a thrust fault.

footwall and hanging wall have moved relative to one another describes three basic classes of faults: (1) normal, (2) reverse, and (3) thrust. A **normal fault** is one in which the hanging wall has moved downward relative to the footwall. This seems "normal" in the sense that you would expect an upper block to slide *down* a lower block along a slope (Figure 19.10A). Sometimes a huge block of rock bounded by normal faults will drop down, creating a *graben* (Figure 19.10B). The opposite of a graben is a *horst,* which is a block bounded by normal faults that is uplifted (Figure 19.10C). A very large block lifted sufficiently becomes a fault-block mountain. Many parts of the western United States are characterized by numerous fault-block mountains separated by adjoining valleys.

In a **reverse fault,** the hanging wall block has moved upward relative to the footwall block. As illustrated in Figure 19.11A, a reverse fault is probably the result of horizontal compressive stress.

A reverse fault with a low-angle fault plane is also called a *thrust fault* (Figure 19.11B). In some thrust faults, the hanging wall block has completely overridden the lower footwall for 10 to 20 km (6 to 12 mi). This is sometimes referred to as an "overthrust."

As shown in Figures 19.10 and 19.11, the relative movement of blocks of rocks along a fault plane provides information about the stresses that produced the movement. Reverse and thrust faulting result from compressional stress in the direction of the movement. Normal faulting, on the other hand, results from a pulling-apart stress that might be associated with diverging plates. It might also be associated with the stretching and bulging up of the crust over a hot spot.

CONCEPTS *Applied*

Fold and Fault Models

Make clay, plaster, or papier-mâché models to illustrate folding and faulting. Use arrows stuck on pins to show the forces that would produce the fold or fault.

EARTHQUAKES

This section is concerned with the nature and origin of earthquakes. The use of seismic waves to determine Earth's internal structure was discussed in chapter 18. In this section, seismic waves of earthquakes will be considered in more detail, along with how the quakes are measured and located by measuring the waves. The effects of earthquakes, such as ground motion, ground displacement, damage, and tsunamis ("tidal waves") will also be described.

CAUSES OF EARTHQUAKES

What is an earthquake? An **earthquake** is a quaking, shaking, vibrating, or upheaval of the ground. Earthquakes are the result of the sudden release of energy that comes from *stress* on rock beneath Earth's surface. In the section on diastrophism, you learned that rock units can bend and become deformed in response to stress, but there are limits as to how much stress rock can take before it fractures. When it does fracture, the sudden movement of blocks of rock produces vibrations that move out as waves throughout Earth. These vibrations are called *seismic waves.* It is strong seismic waves that people feel as a shaking, quaking, or vibrating during an earthquake.

Seismic waves are generated when a huge mass of rock breaks and slides into a different position. As you learned in

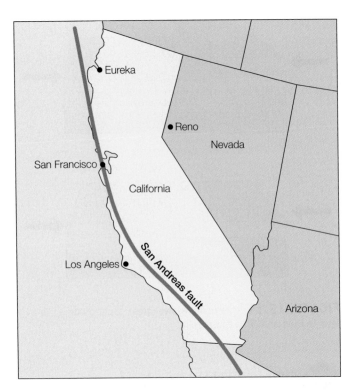

FIGURE 19.12 The San Andreas fault, with the Pacific Plate moving on one side and the North American Plate moving on the other.

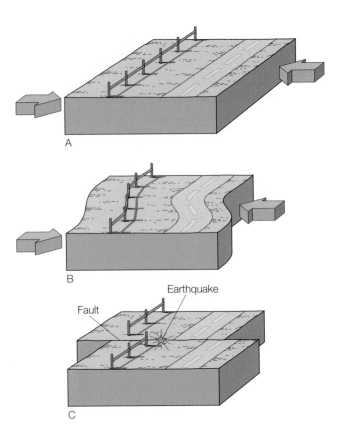

FIGURE 19.13 The elastic rebound theory of the cause of earthquakes. (*A*) Rock with stress acting on it. (*B*) Stress has caused strain in the rock. Strain builds up over a long period of time. (*C*) Rock breaks suddenly, releasing energy, with rock movement along a fault. Horizontal motion is shown; rocks can also move vertically. (*D*) Horizontal offset of rows in a lettuce field, 1979, El Centro, California. (*D*) Photo by University of Colorado; courtesy National Geophysical Data Center, Boulder, Colorado.

the diastrophism section, the plane between two rock masses that have moved into new relative positions is called a *fault.* Major earthquakes occur along existing fault planes or when a new fault is formed by the fracturing of rock. In either case, most earthquakes occur along a fault plane when there is displacement of one side relative to the other.

Most earthquakes occur along a fault plane, and they occur near Earth's surface. You might expect this to happen since the rocks near the surface are brittle, and those deeper are more ductile from increased temperature and pressure. Shallow-focus earthquakes are typical of those that occur at the boundary of the North American Plate, which is moving against the Pacific Plate. In California, the boundary between these two plates is known as the *San Andreas fault* (Figure 19.12). The San Andreas fault runs north-south for some 1,300 km (800 mi) through California, with the Pacific Plate moving on one side and the North American Plate moving on the other. The two plates are tightly pressed against each other, and friction between the rocks along the fault prevents them from moving easily. Stress continues to build along the entire fault as one plate attempts to move along the other. Some elastic deformation does occur from the stress, but eventually, the rupture strength of the rock (or the friction) is overcome. The stressed rock, now released of the strain, snaps suddenly into new positions in the phenomenon known as **elastic rebound** (Figure 19.13). The rocks are displaced to new positions on either side of the fault, and the vibrations from the sudden movement are felt as an earthquake. The elastic rebound and movement tend to occur along short segments of the fault at different times rather

than along long lengths. Thus, the resulting earthquake tends to be a localized phenomenon rather than a regional one.

Most earthquakes are explained by the movement of rock blocks along faults, but there are also other causes. Earthquakes have occurred in the eastern United States and without any apparent relationship to known faults at or near the surface. One

478   CHAPTER 19   Building Earth's Surface

of the largest earthquakes on record in the United States occurred not in California but in the region of New Madrid, Missouri, in 1811. This quake toppled chimneys four hundred miles away in Ohio and was felt from the Rocky Mountains to the Atlantic Coast and from Canada to the Gulf of Mexico. The New Madrid fault zone is theorized to represent a failed rift that is being reactivated by compressional stress from the west and east.

Some of the few earthquakes that are difficult to explain seem to be associated with deeply buried anticlines or other deeply buried structures. Earthquakes are also associated with the movement of magma that occurs beneath a volcano before an eruption. Earthquakes also occur during an explosive volcanic eruption. Earthquakes associated with volcanic activity, however, are always relatively feeble compared to those associated with faulting.

LOCATING AND MEASURING EARTHQUAKES

Most earthquakes occur near plate boundaries. Occurrences do happen elsewhere, but they are rare. The actual place where seismic waves originate beneath the surface is called the **focus** of the earthquake. The focus is considered to be the center of earthquake and the place of initial rock movement on a fault. The point on Earth's surface directly above the focus is called the earthquake *epicenter* (Figure 19.14).

Seismic waves from an earthquake are detected and measured by an instrument called a *seismometer* (Figure 19.15). Seismic waves were introduced in chapter 18. These waves radiate outward from the earthquake focus, spreading in all directions through the solid Earth's interior like the sound waves from an explosion. The seismograph detects three kinds of waves: (1) the longitudinal (compressional) wave called a *P-wave*, (2) the transverse (shear) wave called an *S-wave,* and (3) up-and-down (crests and troughs) waves that travel across the surface called *surface waves,* which are much like a water wave that moves across the solid surface of Earth. S- and P-waves provide information about the location and magnitude of an earthquake, and they also provide information about Earth's interior.

Seismic S- and P-waves leave the focus of an earthquake at essentially the same time. As they travel away from the focus,

they gradually separate because the P-waves travel faster than the S-waves. To locate an epicenter, at least three recording stations measure the time lag between the arrival of the first P-waves and the first slower S-waves. The difference in the speed between the two waves is a constant. Therefore, the farther they travel, the greater the time lag between the arrival of the faster P-waves and the slower S-waves (Figure 19.16A). By measuring the time lag and knowing the speed of the two waves, it is possible to calculate the distance to their source. However, the calculated distance provides no information about the direction or location of the source of the waves. The location is found by first using the calculated distance as the radius of a circle drawn on a map. The place where the circles from the three recording stations intersect is the location of the source of the waves (Figure 19.16B).

Seismographic data can also be used to calculate the depth of the earthquake focus. Earthquakes are classified into three groups according to their depth of focus:

1. *Shallow-focus* earthquakes occur within the depth of the continental crust, which is from the surface down to 70 km (about 45 mi) deep.
2. *Intermediate-focus* earthquakes occur in the upper part of the mantle. This is defined as between the bottom of the crust down to where the seismic wave velocities increase because of a change in the character of the mantle materials. Based on this definition, the upper mantle is 70 to 350 km (45 and 220 mi) deep.
3. *Deep-focus* earthquakes occur in the lower part of the upper mantle. This is defined as the boundary where a deeper change in the seismic wave velocities indicates still another change in the character of mantle materials. This boundary occurs 350 to 700 km (about 220 to 430 mi) deep.

About 85 percent of all earthquakes are of a shallow focus, occurring in the top 70 km (about 45 mi) of the surface. Only about 3 percent, on the other hand, are earthquakes of a deep focus. The reasons that most earthquakes are shallow ones are (1) the nature of the rocks near the surface (brittle) differs from that of those under great pressure and higher temperatures (plastic) and (2) the mechanism that results in earthquakes (plate tectonics) has the most resistance to movement near the surface between the plates.

MEASURING EARTHQUAKE STRENGTH

The intensities of earthquakes vary widely, from the many that are barely detectable to the few that cause widespread destruction. Destruction is caused by the seismic waves, which cause the land and buildings to vibrate. Vibrations during small quakes can crack windows and walls, but vibrations in strong quakes can topple bridges and buildings. Injuries and death are usually the result of falling debris, crumbling buildings, or falling bridges. Fire from broken gas pipes was a problem in the 1906 and 1989 earthquakes in San Francisco and the 1994 earthquake in Los Angeles. Broken water mains made it difficult to fight the 1906 San Francisco fires, but in 1989, fireboats and fire hoses using water pumped from the bay were able to extinguish the fires.

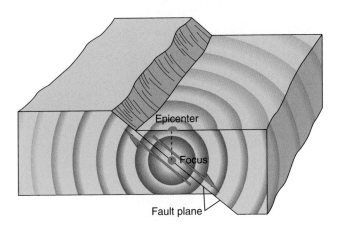

FIGURE 19.14 Simplified diagram of a fault, illustrating component parts and associated earthquake terminology.

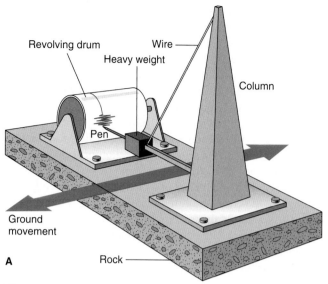

Revolving drum

Wire

Heavy weight

Column

Pen

Ground movement

A

Rock

FIGURE 19.15 (*A*) A seismometer for horizontal motion. Some seismometers record earth motion on moving strips of paper. The mass is suspended by a wire from the column and swings like a pendulum when the ground moves horizontally. A pen attached to the mass records the motion on a moving strip of paper. (*B*) A seismogram of a 1967 earthquake in Taiwan, magnitude 6.2, recorded in Berkeley, California, 6,300 miles away. First arrivals of P-, S-, and surface waves are shown. *Source:* Courtesy of Berkeley Seismological Laboratory University of California Berkeley.

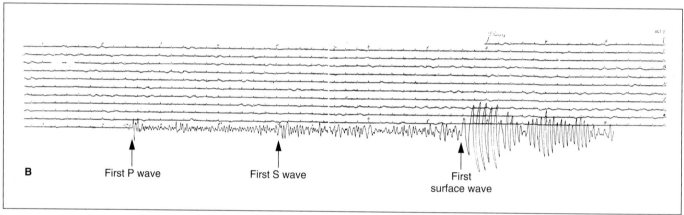

B

First P wave   First S wave   First surface wave

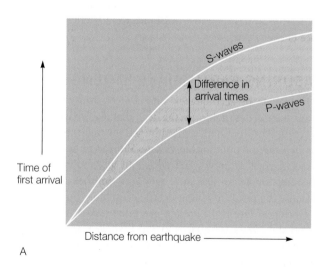

S-waves

Difference in arrival times

P-waves

Time of first arrival

Distance from earthquake ⟶

A

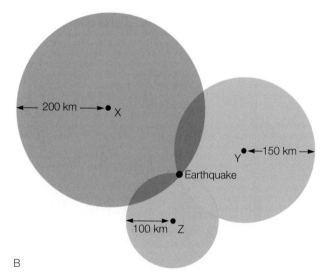

⟵ 200 km ⟶ • X

Y • ⟵ 150 km ⟶

• Earthquake

⟵ 100 km ⟶ • Z

B

FIGURE 19.16 Use of seismic waves in locating earthquakes. (*A*) Difference in times of first arrival of P-waves and S-waves is a function of the distance from the focus. (*B*) Triangulation using data from several seismograph stations allows location of the earthquake.

Other effects of earthquakes include landslides and displacement of the land surface. Vertical, horizontal, or both vertical and horizontal displacement of the land can occur during a quake. People sometimes confuse cause and effect when they see a land displacement, saying things like, "Look what the earthquake did!" The fact is that the movement of the land probably produced the seismic waves (the earthquake). The seismic waves did not produce the land displacement. Such displacements from a single earthquake can be up to 10 to 15 m (about 30 to 50 ft), but such displacements rarely happen.

The effect of an earthquake on people and buildings can be used to determine the *relative intensity* of the earthquake. The modified *Mercalli scale* expresses such intensities with Roman numerals that range from I to XII. In general, intensities of I through VI are concerned with increasing levels of awareness by people. Intensity level I is "not felt," and level VI is "felt by all." Intensity levels V through VII are concerned with increasing levels of damage to plaster, dishes, furniture, and so forth. Level V, for example, describes "some dishes broken." Levels VIII through X are concerned with increasing levels of damage to structures. Finally, an XI on the Mercalli scale means that few, if any, buildings remain standing. XII means total destruction with visible waves moving across the ground surface. As you can see, the Mercalli measure of relative earthquake intensity has its advantages since it requires no instruments. It could also be misleading since the intensity could vary with the type of construction, the type of material (clay or sand, for example) under the buildings, the distance from the epicenter, and various combinations of these factors.

Myths, Mistakes, & Misunderstandings

Do Earthquakes Swallow People?

It is a myth that the ground opens up, swallows people, then closes during an earthquake. There might be shallow crevasses that form during an earthquake, but faults do not open up and close during an earthquake. Movement occurs along a fault, not perpendicular to it. If faults opened up, no earthquake would occur because there would be no friction to lock the sides of the fault together.

The size of an earthquake can be measured in terms of vibrations, in terms of displacement, or in terms of the amount of energy released at the site of the earthquake. The larger the quake, the larger the waves recorded on a seismometer. From these recorded waves, scientists assign a number called the *magnitude*. Magnitude is a measure of the energy released during an earthquake. Earthquake magnitude is often reported by the media using the **Richter scale** (Table 19.1). The Richter scale was developed by Charles Richter, a seismologist at the California Institute of Technology in the early 1930s. The scale was based on the widest swing in the back-and-forth line traces of a seismograph recording. The higher the magnitude of an earthquake, as measured by the Richter scale, the greater the

TABLE 19.1

Effects of earthquakes of various magnitudes

Richter Magnitudes	Description
0–2	Smallest detectable earthquake
2–3	Detected and measured but not generally felt
3–4	Felt as small earthquake but no damage occurs
4–5	Minor earthquake with local damage
5–6	Moderate earthquake with structural damage
6–7	Strong earthquake with destruction
7–8	Major earthquake with extensive damage and destruction
8–9	Great earthquake with total destruction

(1) severity of the ground-shaking vibrations and (2) energy released by the earthquake. An increase of one on the Richter scale means that the amount of movement of the ground increased by a factor of 10 and the amount of energy released increased by a factor of 30. An earthquake measuring below 3 on the scale is usually not felt by people near the epicenter. The largest earthquake measured so far had a magnitude over 9, but there is actually no upper limit to the scale. Today, professional seismologists rate the size of earthquakes in different ways, depending on what they are comparing and why, but each way results in logarithmic scales similar to the Richter scale.

Tsunami is a Japanese term now used to describe the very large ocean waves that can be generated by an earthquake, landslide, or volcanic explosion. Such large waves were formerly called "tidal waves." Since the large, fast waves were not associated with tides or tidal forces in any way, the term *tsunami* or *seismic sea wave* is preferred.

A tsunami, like other ocean waves, is produced by a disturbance. Most common ocean waves are produced by winds, travel at speeds of 90 km/h (55 mi/h), and produce a wave height of 0.6 to 3 m (2 to 10 ft) when they break on the shore. A tsunami, on the other hand, is produced by some strong disturbance in the seafloor, travels at speeds of 725 km/h (450 mi/h), and produces a wave height of over 8 m (about 25 ft) when it breaks on the shore. A tsunami may have a very long wavelength of 200 km (120 mi) compared to the wavelength of ordinary wind-generated waves of 400 m (1,300 ft). Because of its great wavelength, a tsunami does not just break on the shore, then withdraw. Depending on the seafloor topography, the water from a tsunami may continue to rise for five to ten minutes, flooding the coastal region before the wave withdraws. A gently sloping seafloor and a funnel-shaped bay can force tsunamis to great heights as they break on the shore. The current record height was established in 1971 when a tsunami broke at 85 m (278 ft) on an island south of Japan. Waves of such great height and long wavelength can be very destructive, sweeping away trees, lighthouses, and buildings up to 30 m (100 ft) above sea level. The size of a particular

A Closer Look

Earthquake Safety

tsunami depends on the nature of how the seafloor was disturbed. Generally, a "big" earthquake that causes the seafloor suddenly to rise or fall favors the generation of a large tsunami.

On December 26, 2004, an earthquake measuring 9.0 on the Richter scale occurred west of the Indonesian island of Sumatra. This was the largest quake worldwide in four decades. The focus was about 10 km (about 6 mi) beneath the ocean floor at the interface of the India and Burma Plates and was caused by the release of stresses that develop as the India Plate subducts beneath the overriding Burma Plate. A 100 km (62 mi) wide rupture occurred in the ocean floor, about 1,200 km (746 mi) long and parallel to the Sunda trench. The ocean floor was suddenly uplifted more than 2 m (about 7 ft), with a movement about 10 m (about 33 ft) to the west-southwest. This displacement acted like a huge paddle at the bottom of the ocean, vertically displacing billions of tons of water and triggering a tsunami. The tsunami created a path of destruction across the 4,500 km (about 2,800 mi) wide Indian Ocean over the next seven hours. Series of very large waves struck the coast of the Indian Ocean, with an estimated death toll of about 295,000 and over a million left homeless.

The world's largest recorded earthquakes have all occurred where one tectonic plate subducts beneath another. These include the magnitude-9.5 1960 Chile earthquake, the magnitude-9.2 1964 Prince William Sound, Alaska, earthquake, the magnitude-9.1 1957 Andreanof Islands, Alaska, earthquake, and the magnitude-9.0 1952 Kamchatka, Russia, earthquake. Such earthquakes often create large tsunamis that result in death and destruction over a wide area.

ORIGIN OF MOUNTAINS

Most of the interesting features of Earth's surface have been created by folding and faulting, and the most prominent of these features are **mountains.** Mountains are elevated parts of Earth's crust that rise abruptly above the surrounding surface. Most mountains do not occur in isolation but in groups that are associated in chains or belts. These long, thin belts are generally found along the edges of continents rather than in the continental interior. There are a number of complex processes involved in the origin of mountains and mountain chains, and no two mountains are exactly alike. For convenience, however, mountains can be classified according to three basic origins: (1) folding, (2) faulting, and (3) volcanic activity.

FOLDED AND FAULTED MOUNTAINS

The major mountain ranges of Earth—the Appalachian, Rocky, and Himalayan mountains, for example—have a great vertical relief that involves complex folding on a very large scale. The crust was thickened in these places as compressional forces produced tight, almost vertical folds. Thus, folding is a major feature of the major mountain ranges, but faulting and igneous intrusions are invariably also present. Differential weathering of different rock types produced the parallel features of the Appalachian Mountains that are so prominent in satellite photographs (Figure 19.17). The folded sedimentary rocks of the Rockies are evident in the almost upright beds along the flanks of the front range.

A broad arching fold, which is called a dome, produced the Black Hills of South Dakota. The sedimentary rocks from the top of the dome have been weathered away, leaving a somewhat circular area of more resistant granite hills surrounded by upward-tilting sedimentary beds (Figure 19.18). The Adirondack Mountains of New York are another example of this type of mountain formed from folding, called domed mountains.

Compressional forces on a regional scale can produce large-scale faults, shifting large crustal blocks up or down relative to one another. Huge blocks of rocks can be thrust to mountainous heights, creating a series of fault block

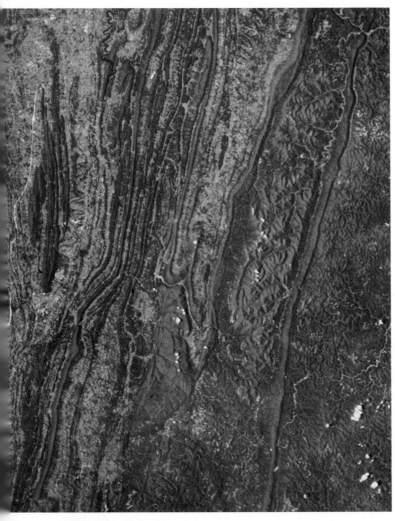

FIGURE 19.17 The folded structure of the Appalachian Mountains, revealed by weathering and erosion, is obvious in this *Skylab* photograph of the Virginia-Tennessee-Kentucky boundary area. The clouds are over the Blue Ridge Mountains.

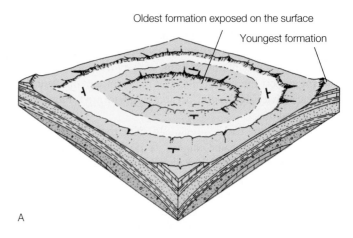

Oldest formation exposed on the surface
Youngest formation

A

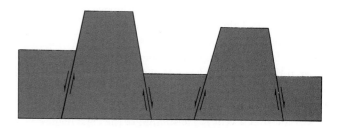

B

FIGURE 19.18 (*A*) A sketch of an eroded structural dome where all the rock layers dip away from the center.(*B*) A photo of a dome named Little Sundance Mountain (Wyoming),showing the more resistant sedimentary layers that dip away from the center.

mountains. Fault block mountains rise sharply from the surrounding land along the steeply inclined fault plane. The mountains are not in the shape of blocks, however, as weathering has carved them into their familiar mountainlike shapes (Figure 19.19). The Teton Mountains of Wyoming and the Sierra Nevadas of California are classic examples of fault block mountains that rise abruptly from the surrounding land. The various mountain ranges of Nevada, Arizona, Utah, and southeastern California have large numbers of fault block mountains that generally trend north and south.

VOLCANIC MOUNTAINS

Lava and other materials from volcanic vents can pile up to mountainous heights on the surface. These accumulations can form local volcano-formed mountains near mountains produced by folding or faulting. Such mixed-origin mountains are

Original sharply bounded
fault blocks softened by
erosion and sedimentation

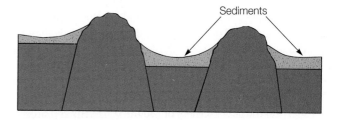

Sediments

FIGURE 19.19 Fault block mountains are weathered and eroded as they are elevated, resulting in a rounded shape and sedimentation rather than sharply edged fault blocks.

common in northern Arizona, New Mexico, and western Texas. The Cascade Mountains of Washington and Oregon are a series of towering volcanic peaks, most of which are not active today. As a source of mountains, volcanic activity has an overall limited impact on the continents. The major mountains built by volcanic activity are the mid-oceanic ridges formed at diverging plate boundaries.

Deep within Earth, previously solid rock melts at high temperatures to become *magma,* a pocket of molten rock. Magma is not just melted rock alone, however, as the melt contains variable mixtures of minerals (resulting in different types of lava flows). It also includes gases such as water vapor, sulfur dioxide, hydrogen sulfide, carbon dioxide, and hydrochloric acid. You can often smell some of these gases around volcanic vents and hot springs. Hydrogen sulfide smells like rotten eggs or sewer gas. The sulfur smells like a wooden match that has just been struck.

The gases dissolved in magma play a major role in forcing magma out of the ground. As magma nears the surface, it comes under less pressure, and this releases some of the dissolved gases from the magma. The gases help push the magma out of the ground. This process is similar to releasing the pressure on a can of warm soda, which releases dissolved carbon dioxide.

Magma works its way upward from its source below to Earth's surface, here to erupt into a lava flow or a volcano. A **volcano** is a hill or mountain formed by the extrusion of lava or rock fragments from magma below. Some lavas have a lower viscosity than others, are more fluid, and flow out over the land rather than forming a volcano. Such *lava flows* can accumulate into a plateau of basalt, the rock that the lava formed as it cooled and solidified. The Columbia Plateau of the states of Washington, Idaho, and Oregon is made up of layer after layer of basalt that accumulated from lava flows. Individual flows of lava formed basalt layers up to 100 m (about 330 ft) thick, covering an area of hundreds of square kilometers. In places, the Columbia Plateau is up to 3 km (about 2 mi) thick from the accumulation of many individual lava flows.

The hill or mountain of a volcano is formed by ejected material that is deposited in a conical shape. The materials are deposited around a central *vent,* an opening through which an eruption takes place. The *crater* of a volcano is a basinlike depression over a vent at the summit of the cone. Figure 19.20 is an aerial view of Mount St. Helens, looking down into the crater at a volcanic dome that formed as magma periodically welled upward into the floor of the crater. This photo was taken several years after Mount St. Helens erupted in May 1980. The volcano was quiet between that time and late 2004, when thousands of small earthquakes preceded the renewed growth of the lava dome inside the crater. According to the U.S. Geological Survey, the new dome grew to more than 152 m (500 ft) above the old dome before the volcano entered a relatively quiet state. Go to http://vulcan.wr.usgs.gov/News/framework.html for a fact sheet on the latest activity of Mount St. Helens and links to current information.

Mount Garibaldi, Mount Rainier, Mount Hood, Mount Baker, and about ten other volcanic cones of the Cascade Range

FIGURE 19.20 This is the top of Mount St. Helens several years after the 1980 explosive eruption.

in Washington and Oregon could erupt next year, in the next decade, or in the next century. History and plate movement (see Figure 19.25) make it very likely that any one of the volcanic cones of the Cascade Range will indeed erupt again.

Volcanic materials do not always come out from a central vent. In a *flank eruption,* lava pours from a vent on the side of a volcano. The crater on Mount St. Helens was formed by a flank explosion. Magma was working its way toward the summit of Mount St. Helens, along with the localized earthquakes that usually accompany the movement of magma beneath the surface. One of the earthquakes was fairly strong and caused a landslide, removing the rock layers from over the slope. This reduced the pressure on the magma, and gases were suddenly released in a huge explosion, which ripped away the north flank of the volcano. The exploding gases continued to propel volcanic ash into the atmosphere for the next thirty hours.

There are three major types of volcanoes: (1) shield, (2) cinder cone, and (3) composite. The *shield volcanoes* are broad, gently sloping cones constructed of solidified lava flows. The lava that forms this type of volcano has a low viscosity, spreading widely from a vent. The islands of Hawaii are essentially a series of shield volcanoes built upward from the ocean floor. These enormous Hawaiian volcanoes range up to elevations of 8.9 km (5.5 mi) above the ocean floor, typically with slopes that range from 2° to 20° (Figure 19.21).

The Hawaiian volcanoes form from a magma about 60 km (about 40 mi) deep, probably formed by a mantle plume. The magma rises buoyantly to the volcanic cones, where it erupts by pouring from vents. Hawaiian eruptions are rarely explosive. Shield volcanoes also occur on the oceanic ridge (Iceland) and

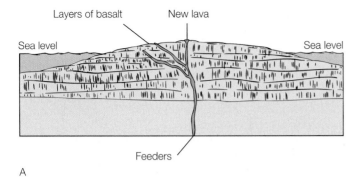

Layers of basalt   New lava

Sea level   Sea level

Feeders

A

FIGURE 19.22 Sunset Crater, a cinder cone volcano near Flagstaff, Arizona.

B

FIGURE 19.21 (*A*) A schematic cross section of an idealized shield volcano. (*B*) A photo of a shield volcano, Mauna Loa in Hawaii.

occasionally on the continents (Columbia Plateau).

A *cinder cone volcano* is constructed of, as the name states, cinders. The cinders are rock fragments, usually with sharp edges since they formed from frothy blobs of lava that cooled as they were thrown into the air. The cinder cone volcano is a pile or piles of loose cinders, usually red or black in color, that have been ejected from a vent. Cinder cones can have steep sides, with slope angles of 35° to 40°. This is the maximum steepness at which the sharp-edged, unconsolidated cinders will stay without falling. The cinder cone volcanoes are much steeper than the gentle slopes of the shield volcanoes. They also tend to be much smaller, with cones that are usually no more than about 500 m (1,600 ft) high (Figure 19.22).

A *composite volcano* (also called stratovolcano) is built up of alternating layers of cinders, ash, and lava flows (Figure 19.23), forming what many people believe is the most imposing and majestic of Earth's mountains. The steepness of the sides, as you might expect, is somewhere between the steepness of the low shield volcanoes and the steep cinder cone volcanoes. The Cascade volcanoes are composite volcanoes, but the mixture of lava flows and cinders seems to vary from one volcano to the next. In addition to the alternating layers of solid basaltic rock and deposits of cinders and ash, a third type of layer is formed by

volcanic mudflow. The volcanic mud can be hot or cold, and it rolls down the volcano cone like wet concrete, depositing layers of volcanic conglomerate. Volcanic mudflows often do as much or more damage to the countryside than the other volcanic hazards.

Shield, cinder cone, and composite volcanoes form when magma breaks out at Earth's surface. Only a small fraction of all the magma generated actually reaches Earth's surface. Most of it remains below the ground, cooling and solidifying to form *intrusive rocks,* igneous rocks that were described in chapter 17. A large amount of magma that has crystallized below the surface is known as a *batholith.* A small protrusion from a batholith is called a *stock.* By definition, a stock has less than 100 km^2 (40 mi^2) of exposed surface area, and a batholith is larger. Both batholiths and stocks become exposed at the surface through erosion of the overlying rocks and rock materials, but not much is known about their shape below. The sides seem to angle away with depth, suggesting that they become larger with depth. The intrusion of a batholith sometimes tilts rock layers upward, forming a *hogback,* a ridge with equal slopes on its sides (Figure 19.24). Other forms of intruded rock were formed as moving magma took the paths of least resistance, flowing into joints, faults, and planes between sedimentary bodies of rock. An intrusion that has flowed into a joint or fault that cuts across rock bodies is called a *dike.* A dike is usually tabular in shape, sometimes appearing as a great wall when exposed at the surface. One dike can occur by itself, but frequently dikes occur in great numbers, sometimes radiating out from a batholith like spokes around a wheel. If the intrusion flowed into the plane of contact between sedimentary rock layers, it is called a *sill.* A *laccolith* is similar to a sill but has an arched top where the intrusion has raised the overlying rock into a blisterlike uplift (see Figure 19.24).

Where does the magma that forms volcanoes and other volcanic features come from? It is produced within the outer 100 km (about 60 mi) or so of Earth's surface, presumably from a partial melting of the rocks within the crust or the uppermost part of the mantle. Basalt, the same rock type that makes up the oceanic crust, is the most abundant extrusive rock, both on

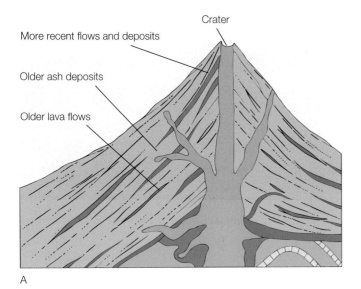

A

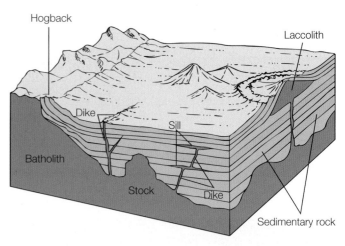

FIGURE 19.24 Here are the basic intrusive igneous bodies that form from volcanic activity.

B

FIGURE 19.23 (*A*) A schematic cross section of an idealized composite volcano, which is built up of alternating layers of cinders, ash, and lava flows. (*B*) A photo of Mount Shasta, a composite volcano in California. You can still see the shapes of former lava flows from Mount Shasta.

the continents and along the mid-oceanic ridges. Volcanoes that rim the Pacific Ocean and Mediterranean extrude lava with a slightly different chemistry. This may be from a partial melting of the oceanic crust as it is subducted beneath the continental crust. The Cascade volcanoes are typical of those that rim the Pacific Ocean. The source of magma for Mount St. Helens and the other Cascade volcanoes is the Juan de Fuca Plate, a small plate whose spreading center is in the Pacific Ocean a little west of the Washington and Oregon coastline (Figure 19.25). The Juan de Fuca Plate is subducted beneath the continental lithosphere, and as it descends, it comes under higher and higher temperatures. Partial melting takes place, forming magma. The magma is less dense than the surrounding rock and is buoyed toward Earth's surface, erupting as a volcano.

Overall, the origin of mountain systems and belts of mountains such as the Cascades involves a complex mixture of volcanic activity as well as folding and faulting. An individual mountain, such as Mount St. Helens, can be identified as having a volcanic origin. The overall picture is best seen, however, from generalizations about how the mountains have grown along the edges of plates that are converging. Such converging boundaries are the places of folding, faulting, and associated earthquakes. They are also the places of volcanic activities, events that build and thicken Earth's crust. Thus, plate tectonics explains that mountains are built as the crust thickens at a convergent boundary between two plates. These mountains are slowly weathered and worn down as the next belt of mountains begins to build at the new continental edge. How long does it take to build a mountain and then wear it completely down to sea level? How would you ever find an answer to this question? These questions will be discussed in chapter 21. First, we must consider how the land is worn down, the topic of chapter 20.

Volcanoes Change the World

A volcanic eruption changes the local landscape, that much is obvious. What is not so obvious are the worldwide changes that can happen just because of the eruption of a single volcano. Perhaps the most discussed change brought about by a volcano occurred back in 1815–16 after the eruption of Tambora in Indonesia. The Tambora eruption was massive, blasting huge amounts of volcanic dust, ash, and gas high into the atmosphere. Most of the ash and dust fell back to Earth around the volcano, but some dust particles and sulfur dioxide gas were pushed high into the stratosphere.

It is known today that the sulfur dioxide from explosive volcanic eruptions reacts with water vapor in the stratosphere, forming tiny droplets of diluted sulfuric acid. In the stratosphere, there is no convection, so the droplets of acid and dust from the volcano eventually form a layer around the entire globe. This forms a haze that remains in the stratosphere for years, reflecting and scattering sunlight.

What were the effects of volcanic haze from Tambora around the entire world? There were fantastic, brightly colored sunsets from the added haze in the stratosphere. On the other hand, it was also cooler than usual, presumably because of the reflected sunlight that did not reach Earth's surface. It snowed in New England in June 1816, and the cold continued into July. Crops failed, and 1816 became known as the "year without summer."

More information is available about the worldwide effects of present-day volcanic eruptions because there are now instruments to make more observations in more places. However, it is still necessary to do a great deal of estimating because of the relative inaccessibility of the worldwide stratosphere. It was estimated, for example, that the 1982 eruption of El Chichon in Mexico created enough haze in the stratosphere to reflect 5 percent of the solar radiation away from Earth. Researchers also estimated that the El Chichon eruption cooled the

global temperatures by a few tenths of a degree for two or three years. The cooling did take place, but the actual El Chichon contribution to the cooling is not clear because of other interactions. Earth may have been undergoing global warming from the greenhouse effect, for example, so the El Chichon cooling effect could have actually been much greater. Other complicating factors such as the effects of El Niño or La Niña (see chapter 23) make changes difficult to predict.

In June 1991, the Philippine volcano Mount Pinatubo erupted, blasting twice as much gas and dust into the stratosphere as El Chichon had about a decade earlier. Dust from Mount Pinatubo remained in Earth's atmosphere for the next ten years. The haze from such eruptions has the potential to cool the climate about 0.5°C (1°F). The overall result, however, will always depend on a possible greenhouse effect, a possible El Niño or La Niña effect, and other complications.

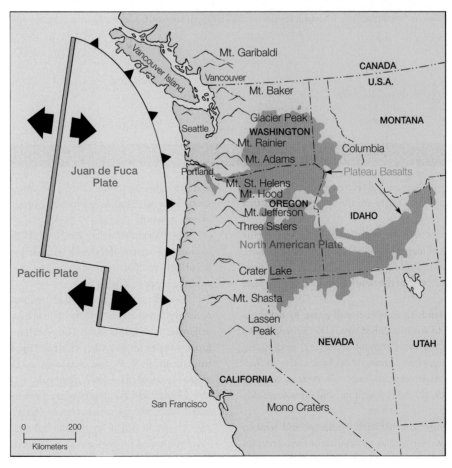

FIGURE 19.25 The Juan de Fuca Plate, the Cascade volcanoes, and the Columbia Plateau Basalts.

People Behind the Science

James Hutton (1726–1797)

James Hutton was a Scottish natural philosopher who pioneered uniformitarian geology. The son of an Edinburgh merchant, Hutton studied at Edinburgh University, Paris, and Leiden, training first for the law but taking his doctorate in medicine in 1749 (though he never practiced). He spent the next two decades traveling and farming in the southeast of Scotland. During this time, he cultivated a love of science and philosophy, developing a special taste for geology. About 1768, he returned to his native Edinburgh. A friend of Joseph Black, William Cullen, and James Watt, Hutton shone as a leading member of the scientific and literary establishment, playing a large role in the early history of the Royal Society of Edinburgh and in the Scottish Enlightenment.

Hutton wrote widely on many areas of natural science, including chemistry, but he is best known for his geology, set out in his *Theory of the Earth,* of which a short version appeared in 1788, followed by the definitive statement in 1795. In that work, Hutton attempted (on the basis of both theoretical considerations and personal fieldwork) to demonstrate that Earth formed a steady-state system in which terrestrial causes had always been of the same kind as at present, acting with comparable intensity (the principle later known as uniformitarianism). In Earth's economy, in the imperceptible creation and devastation of landforms, there was no vestige of a beginning nor prospect of an end. Continents were continually being gradually eroded by rivers and weather. Denuded debris accumulated on the seabed, to be consolidated into strata and subsequently thrust upward to form new continents thanks to the action of Earth's central heat. Nonstratified rocks such as granite were of igneous origin. All Earth's processes were exceptionally leisurely, and hence, Earth must be incalculably old.

Though supported by the experimental findings of Sir James Hall, Hutton's theory was vehemently attacked in its day, partly because it appeared to point to an eternal Earth and hence to atheism. It found more favor when popularized by Hutton's friend, John Playfair, and later by Charles Lyell. The notion of uniformitarianism still forms the groundwork for much geological reasoning.

Source: From the Hutchinson *Dictionary of Scientific Biography.* © Research Machines plc 2003. All Rights Reserved. Helicon Publishing is a division of Research Machines.

SUMMARY

The *principle of uniformity* is the frame of reference that the geologic processes you see changing rocks today are the same processes that changed them in the past.

Diastrophism is the process of deformation that changes Earth's surface, and the movement of magma is called *vulcanism* or *volcanism. Diastrophism, volcanism,* and *earthquakes* are closely related, and their occurrence can be explained most of the time by events involving *plate tectonics.*

Stress is a force that tends to compress, pull apart, or deform a rock, and the adjustment to stress is called *strain.* Rocks respond to stress by (1) withstanding the stress without change, (2) undergoing *elastic strain,* (3) undergoing *plastic strain,* or (4) by *breaking in fracture strain.* Exactly how a particular rock responds to stress depends on (1) the nature of the rock, (2) the temperature, and (3) how quickly the stress is applied.

Deeply buried rocks are at a higher temperature and tend to undergo *plastic deformation,* resulting in a wrinkling of the layers into *folds.* The most common are an arch-shaped fold called an *anticline* and a trough-shaped fold called a *syncline.* Anticlines and synclines are most easily observed in sedimentary rocks because they have bedding planes, or layers.

Rocks near the surface tend to break from a sudden stress. A break without movement of the rock is called a *joint.* When movement does occur between the rocks on one side of a break relative to the other side, the break is called a *fault.*

The vibrations that move out as waves from the movement of rocks are called an *earthquake.* The actual place where an earthquake originates is called its *focus.* The place on the surface directly above a focus is called an *epicenter.* There are three kinds of waves that travel from the focus: *S-, P-,* and *surface waves.* The magnitude of earthquake waves is measured on the *Richter scale.*

Folding and faulting produce prominent features on the surface called *mountains.* Mountains can be classified as having a *folding, faulting,* or *volcanic* origin. In general, mountains that occur in long, narrow belts called *ranges* have an origin that can be explained by *plate tectonics.*

anticline (p. **475**)
earthquake (p. **477**)
elastic rebound (p. **478**)
fault (p. **476**)
focus (p. **479**)
folds (p. **474**)
mountains (p. **482**)
normal fault (p. **477**)
principle of uniformity (p. **472**)
reverse fault (p. **477**)
Richter scale (p. **481**)
strain (p. **474**)
stress (p. **474**)
syncline (p. **475**)
tsunami (p. **481**)
volcano (p. **484**)

APPLYING THE CONCEPTS

1. The premise that the present is the key to understanding the past is called
   a. history.
   b. principle of uniformity.
   c. principles of geology.
   d. philosophy.

2. The process of deformation that changes Earth's surface is called
   a. continental drift.
   b. plate tectonics.
   c. diastrophism.
   d. volcanism.

3. A force that compresses, pulls apart, or deforms a rock is called
   a. stress.
   b. strain.
   c. pressure.
   d. tension.

4. Rock stress caused by two plates moving together is
   a. compressive stress.
   b. tensional stress.
   c. shear stress.
   d. transform stress.

5. Adjustment to stress is defined as
   a. release.
   b. strain.
   c. pressure.
   d. relief.

6. Rocks at great depths are under
   a. lower temperature and higher pressure.
   b. higher temperature and higher pressure.
   c. higher temperature and lower pressure.
   d. lower temperature and lower pressure.

7. A bend in layered bedrock that resulted from stress is called a
   a. fracture.
   b. fold.
   c. fault.
   d. twist.

8. Folds that resemble an arch are called
   a. inverted basins.
   b. clines.
   c. anticlines.
   d. synclines.

9. A fold that forms a trough is called a (an)
   a. syncline.
   b. basin.
   c. inverted arch.
   d. semicline.

10. Movement between rocks on one side of a fracture relative to the rocks on the other side of the fracture is called a
    a. fracture.
    b. transformation.
    c. fault.
    d. displacement.

11. The actual place where seismic waves originate is the earthquake
    a. epicenter.
    b. focus.
    c. root.
    d. source.

12. The point on Earth's surface directly above the focus of an earthquake is called the
    a. fault line.
    b. epicenter.
    c. source.
    d. quake starting point.

13. An earthquake that occurs in the upper part of the mantle is called
    a. shallow focus.
    b. deep focus.
    c. lithosphere focus.
    d. intermediate focus.

14. The majority of earthquakes (85 percent) are
    a. shallow focus.
    b. deep focus.
    c. lithosphere focus.
    d. intermediate focus.

15. The size of an earthquake is measured by
    a. how many buildings are damaged
    b. the amount of energy released at the focus.
    c. the height of the tsunami it generated.
    d. the time of the earthquake.

16. The energy of the vibrations or the magnitude of an earthquake is reported using the
    a. seismograph numbers.
    b. Richter scale.
    c. magnitude scale.
    d. bigone scale.

17. Earthquakes are detected and measured by (a)
    a. seismometer.
    b. Geiger counter.
    c. Doppler radar.
    d. oil well monitors.

18. Elevated parts of Earth's crust that rise above the surrounding surface are called
    a. ridges.
    b. hills.
    c. plateaus.
    d. mountains.

19. Which of the following is not a classification of mountain origin?
    a. folding
    b. faulting
    c. parallel
    d. volcanic

20. Mountains that rise sharply from surrounding land along a steeply inclined fault plane are
    a. weathered mountains.
    b. domed mountains.
    c. fault block mountains.
    d. folded mountains.

21. A large amount of magma that has crystallized beneath Earth's surface is a
    a. batholith.
    b. monolith.
    c. sill.
    d. laccolith.

22. The most abundant extrusive rock is
    a. granite.
    b. silicate.
    c. basalt.
    d. limestone.

23. The basic difference between the frame of reference called the principle of uniformity and the catastrophic frame of reference used by previous thinkers is
    a. the energy requirements for catastrophic changes are much less.
    b. the principle of uniformity requires more time for changes to take place.
    c. catastrophic changes have a greater probability of occurring.
    d. None of the above is correct.

24. The difference between elastic deformation and plastic deformation of rocks is that plastic deformation
    a. permanently alters the shape of a rock layer.
    b. always occurs just before a rock layer breaks.
    c. results in the rock returning to its original shape after the pressure is removed.
    d. All of the above are correct.

25. Whether a rock layer subjected to stress undergoes elastic deformation, plastic deformation, or rupture depends on
    a. the temperature of the rock.
    b. the confining pressure on the rock.
    c. how quickly or how slowly the stress is applied over time.
    d. All of the above are correct.

26. When subjected to stress, rocks buried at great depths are under great pressure at high temperatures, so they tend to undergo
    a. no change because of the pressure.
    b. elastic deformation because of the high temperature.
    c. plastic deformation.
    d. breaking or rupture.

27. A sedimentary rock layer that has not been subjected to stress occurs naturally as
    a. a basin, or large downward bulging fold.
    b. a dome, or a large upwardly bulging fold.
    c. a series of anticlines and synclines.
    d. beds, or horizontal layers.

28. The difference between a joint and a fault is
    a. the fault is larger.
    b. the fault is a long, far-ranging fracture, and a joint is short.
    c. relative movement has occurred on either side of a fault.
    d. All of the above are correct.

29. A fault where the footwall has moved upward relative to the hanging wall is called a
    a. normal fault.
    b. reverse fault.
    c. thrust fault.
    d. None of the above is correct.

30. Reverse faulting probably resulted from which type of stress?
    a. compressional stress
    b. pulling-apart stress
    c. a twisting stress
    d. stress associated with diverging tectonic plates

31. Earthquakes that occur at the boundary between two tectonic plates moving against each other occur along
    a. the entire length of the boundary at once.
    b. short segments of the boundary at different times.
    c. the entire length of the boundary at different times.
    d. None of the above is correct.

32. Each higher number of the Richter scale
    a. increases with the magnitude of an earthquake.
    b. means ten times more ground movement.
    c. indicates about thirty times more energy is released.
    d. All of the above are correct.

33. The removal of "older" crust from the surface of Earth is accomplished by
    a. erosion.
    b. subduction.
    c. transform strain.
    d. volcanoes.

34. Hutton observed that rocks, rock structures, and features of Earth are all related. This relationship is called
    a. history.
    b. principle of uniformity.
    c. principles of geology.
    d. philosophy.

35. The principle of uniformity has a basic frame of reference. This frame of reference is
    a. plate tectonics.
    b. continental drift.
    c. changes and deformations of rocks today and in the past.
    d. compressive strain.

36. What is not considered a type of strain?
    a. elastic
    b. plastic
    c. rigidity
    d. fracture

37. How a rock responds to stress and strain does not depend on the
    a. nature of the rock.
    b. temperature of the rock.
    c. pressure on the rock.
    d. mass of the rock.

38. Which rock is more likely to break under stress?
    a. cold
    b. hot
    c. deep in the earth
    d. under great pressure

39. Rocks near or on the surface
    a. are not cooler than those below the surface.
    b. are not hotter that those below the surface.
    c. are less brittle.
    d. are under more pressure.

40. Rocks recover their original shape after strain is released. This type of strain is called
    a. plastic strain.
    b. elastic strain.
    c. fracture strain.
    d. pop-up strain.

41. Which is not a type of fault?
    a. normal.
    b. reverse.
    c. thrust.
    d. forward.

42. Where do most earthquakes occur?
    a. along plate boundaries
    b. in the oceans
    c. on ridges
    d. along the Moho discontinuity

43. The name of the fault that is of concern to people living along the Mississippi River is the
    a. Mississippi fault.
    b. Memphis fault.
    c. New Madrid fault.
    d. Missouri fault.

44. P-waves travel _____ S-waves.
    a. faster than
    b. slower than
    c. farther than
    d. at the same rate as

45. The epicenter is located by
    a. measuring how long the earthquake lasts.
    b. measuring the time difference between P and S waves.
    c. using time data from several seismographs to triangulate the location.
    d. b and c

46. An earthquake is
    a. the result of the sudden release of energy that comes from stress on rock.
    b. ground displacement and motion.
    c. the cause of tsunamis.
    d. All of the above are correct.

47. The Black Hills in South Dakota and the Adirondack Mountains in New York are
    a. arched mountains.
    b. domed mountains.
    c. volcanic mountains.
    d. compressed mountains.

48. The Appalachian Mountains were formed when
    a. North America split from South America.
    b. North America collided with Europe and Africa.
    c. North America collided with South America.
    d. North America split from Europe and Africa.

49. Mountains that were formed as a result of volcanic eruptions are the
    a. Alps.
    b. Cascades.
    c. Rockies.
    d. Appalachians.

50. The source of magma for Mount St. Helens volcano is the
    a. Cascade Mountains.
    b. subduction of the continental lithosphere under the Juan de Fuca Plate.
    c. subduction of the Juan de Fuca Plate under the continental lithosphere.
    d. continental lithosphere to Juan de Fuca Plate divergence.

Answers

1. b 2. c 3. a 4. a 5. b 6. b 7. b 8. c 9. a 10. c 11. b 12. b 13. d 14. a 15. b 16. b 17. a 18. d 19. c 20. c 21. a 22. c 23. b 24. a 25. d 26. c 27. d 28. c 29. a 30. a 31. b 32. d 33. b 34. b 35. c 36. c 37. d 38. a 39. b 40. b 41. d 42. a 43. c 44. a 45. d 46. d 47. b 48. b 49. b 50. c

QUESTIONS FOR THOUGHT

1. What is the principle of uniformity? What are the underlying assumptions of this principle?

2. Describe the responses of rock layers to increasing compressional stress when it (a) increases slowly on deeply buried, warm layers, (b) increases slowly on cold rock layers, and (c) is applied quickly to rock layers of any temperature.

3. Describe the difference between a syncline and an anticline, using sketches as necessary.

4. What does the presence of folded sedimentary rock layers mean about the geologic history of an area?

5. Describe the conditions that would lead to faulting as opposed to folding of rock layers.

6. How would plate tectonics explain the occurrence of normal faulting? Reverse faulting?

7. What is an earthquake? What produces an earthquake?

8. Where would the theory of plate tectonics predict that earthquakes would occur?

9. Describe how the location of an earthquake is identified by a seismic recording station.

10. Briefly explain how and where folded mountains form and how fault block mountains form.

11. The magnitude of an earthquake is measured on the Richter scale. What does each higher number mean about an earthquake?

12. Identify three areas of probable volcanic activity today in the United States. Explain your reasoning for selecting these three areas.

13. Discuss the basic source of energy that produces the earthquakes in southern California.

14. Describe any possible relationships between volcanic activity and changes in the weather.

15. What is the source of magma that forms volcanoes? Explain how the magma is generated.

16. Describe how the nature of the lava produced results in the three major classification types of volcanoes.

17. What are mountains? Why do they tend to form in long, thin belts?

1. Evaluate the statement "the present is the key to the past" as it represents the principle of uniformity. What evidence supports this principle?
2. Does the theory of plate tectonics support or not support the principle of uniformity? Provide evidence to support your answer.
3. What are the significant similarities and differences between elastic deformation and plastic deformation?
4. Explain the combination of variables that results in solid rock layers folding rather than faulting.

INVITATION TO INQUIRY

Earthquake Patterns?

Click on http://earthquake.usgs.gov/eqcenter/index.php for a list of magnitude-1 earthquakes catalogued in the last week for the United States. Earthquakes magnitude 3 and greater are in a separate list. The most recent earthquakes are at the top of each list. Click on the column header for information about that column. Click on the word *map* to see a map displaying the location. Click on an event's *date* or *depth* for additional information about the earthquake. Explore the location and magnitude of earthquakes until you can answer the following questions:

1. Do earthquakes appear to occur anywhere or mostly in specific regions?
2. Is there a recognizable pattern to where earthquakes seem to occur?
3. Is there any relationship between the earthquake locations and a map showing tectonic plates?
4. Examine a map that shows the types of tectonic plate boundaries (convergent, divergent, transform). Do the *deep* earthquakes seem to correlate to any particular type of plate boundary?
5. Can you explain any patterns or correlations that you found?

20
Shaping Earth's Surface

Weathering and erosion tend to sculpt, or wear down, the wland, sometimes leaving behind rock structures of more resistant rocks such as those you can see here.

CORE **CONCEPT**

The surface of Earth is involved in an ongoing process of destruction and tearing down of higher elevations.

OUTLINE

Weathering
Physical, chemical, and biological processes that break up or crumble solid rock are called weathering.

Erosion
Erosion is the removal of fragments, clays, and solutions that have been produced from solid rock by weathering.

Weathering, Erosion, and Transportation
Weathering
Soils
Erosion
  Mass Movement
  Running Water
  Glaciers
  Wind
Science and Society: Acid Rain
Development of Landscapes
People Behind the Science:
  John Wesley Powell
  Rock Structure
  Weathering and Erosion Processes
  State of Development

Soils Soil is a mixture of the products of weathering—sand, silt, and clay—and humus.

Erosion
Gravity, running water, glaciers, and wind are agents of erosion.

As discussed in chapter 19, the movement of Earth's plates produces mountains, folded structures, and other surface features. These are produced through volcanic activity and as a result of stress, which results in folded and faulted structures. Vulcanism and diastrophism are constructive forces of Earth that result in a building up of the surface, an increase in the elevation of the land. There is also a destructive side, however, a side that goes mostly unseen. The elevated land is now subjected to a sculpturing and tearing down of its surface.

Sculpturing of Earth's surface takes place through agents and processes acting so gradually that humans are usually not aware that it is happening. Sure, some events such as a landslide or the movement of a big part of a beach by a storm are noticed. But the continual, slow, downhill drift of all the soil on a slope or the constant shift of grains of sand along a beach are outside the awareness of most people. People do notice the muddy water moving rapidly downstream in the swollen river after a storm, but few are conscious of the slow, steady dissolution of limestone by acid rain percolating through it. Yet it is the processes of slow moving, shifting grains and bits of rocks, and slow dissolving that will wear down the mountains, removing all the features of the landscape that you can see.

Each of the agents that wear down Earth's surface—gravity, moving water, glaciers, and the wind—has its own way of removing and redepositing the fragments of the land. Each produces a set of characteristic sculpturing and depositional features. Thus, it is possible to recognize how a particular landscape formed in the past, even though different agents may be working today. This chapter is about the decomposition and sculpturing changes that occur in the landscape. Knowledge of these changes can be used to account for some of the varied and interesting scenery that you can observe across the countryside, telling you something about the history of the region (Figure 20.1).

WEATHERING, EROSION, AND TRANSPORTATION

A mountain of solid granite on the surface of Earth might appear to be a very solid, substantial structure, but it is always undergoing slow and gradual changes. Granite on Earth's surface is exposed to and constantly altered by air, water, and other agents of change. It is altered both in appearance and in composition, slowly crumbling, and then dissolving in water. Smaller rocks and rock fragments are moved downhill by gravity or streams, exposing more granite that was previously deeply buried. The process continues, and ultimately—over much time—a mountain of solid granite is reduced to a mass of loose rock fragments and dissolved materials. The photograph in Figure 20.2 is a snapshot of a mountain-sized rock mass in a stage somewhere between its formation and its eventual destruction to rock fragments. Can you imagine the length of time that such a process requires?

WEATHERING

The slow changes that result in the breaking up, crumbling, and destruction of any kind of solid rock are called **weathering.** The term implies changes in rocks from the action of the weather, but it actually includes chemical, physical, and biological processes. These weathering processes are important and necessary in (1) the rock cycle, (2) the formation of soils, and (3) the movement of rock materials over Earth's surface. Weathering is important in the rock cycle because it produces sediment, the raw materials for new rocks. It is important in the formation of soils because soil is an accumulation of rock fragments and organic matter. Weathering is also important because it reduces the size of rock particles, preparing the rock materials for transport by wind or moving water. Before the process of weathering, the rock is mostly confined to one location as a solid mass.

Weathering breaks down rocks physically and chemically, and this breaking down can occur while the rocks are stationary or while they are moving. The process of physically removing weathered materials is called **erosion.** *Weathering* prepares the way for erosion by breaking solid rock into fragments. The fragments are then *eroded,* physically picked up by an agent such as a stream or a glacier. After they are eroded, the materials are then removed by *transportation.* Transportation is the movement of eroded materials by agents such as rivers, glaciers, wind, or waves. The weathering process continues during transportation. A rock being tumbled downstream, for example, is physically worn down as it bounces from rock to rock. It may be chemically altered as well as it is bounced along by the moving water. Overall, the combined action of weathering and erosion

FIGURE 20.1 This famous natural bridge is an example of a landform created by the sculpturing power of weathering and erosion. It is Rainbow Bridge in the Rainbow Bridge National Monument, Utah.

wears away and lowers the elevated parts of Earth and sculpts their surfaces.

There are two basic kinds of weathering that act to break down rocks: *mechanical weathering* and *chemical weathering*. *Mechanical weathering* is the physical breaking up of rocks without any changes in their chemical composition. Mechanical weathering results in the breaking up of rocks into smaller and smaller pieces, so it is also called *disintegration*. If you smash a sample of granite into smaller and smaller pieces, you are mechanically weathering the granite. *Chemical weathering* is the alteration of minerals by chemical reactions with water, gases of the atmosphere, or solutions. Chemical weathering results in the dissolving or breaking down of the minerals in rocks, so it is also called *decomposition*. If you dissolve a sample of limestone in a container of acid, you are chemically weathering the limestone.

Examples of mechanical weathering in nature include the disintegration of rocks caused by (1) *wedging effects* and (2) the *effects of reduced pressure*. Wedging effects are often caused by the repeated freezing and thawing of water in the pores and small cracks of otherwise solid rock. If you have ever seen what happens when water in a container freezes, you know that freezing water expands and exerts a pressure on the sides of its

container. As water in a pore or a crack of a rock freezes, it also expands, exerting a pressure on the walls of the pore or crack, making it slightly larger. The ice melts and the enlarged pore or crack again becomes filled with water for another cycle of freezing and thawing. As the process is repeated many times, small pores and cracks become larger and larger, eventually forcing pieces of rock to break off. This process is called *frost wedging* (Figure 20.3A). It is an important cause of mechanical weathering in mountains and other locations where repeated cycles of water freezing and thawing occur. The roots of trees and shrubs can also mechanically wedge rocks apart as they grow into cracks. You may have noticed the work of roots when trees or shrubs have grown next to a sidewalk for some period of time (Figure 20.4).

The other example of mechanical weathering is believed to be caused by the reduction of pressure on rocks. As more and more weathered materials are removed from the surface, the downward pressure from the weight of the material on the rock below becomes less and less. The rock below begins to expand upward, fracturing into concentric sheets from the effect of reduced pressure. These curved, sheetlike plates fall away later in the mechanical weathering process called *exfoliation* (Figure 20.3B). *Exfoliation* is the term given to the process

FIGURE 20.2 The piles of rocks and rock fragments around a mass of solid rock is evidence that the solid rock is slowly crumbling away. This solid rock that is crumbling to rock fragments is in the Grand Canyon, Arizona.

of spalling off of layers of rock, somewhat analogous to peeling layers from an onion. Granite commonly weathers by exfoliation, producing characteristic dome-shaped hills and rounded boulders. Stone Mountain in Georgia is a well-known example of an exfoliation-shaped dome. The onionlike structure of exfoliated granite is a common sight in the Sierras, Adirondacks, and any mountain range where older granite is exposed at the surface (Figure 20.5).

Examples of chemical weathering include (1) oxidation, (2) carbonation, and (3) hydration. *Oxidation* is a reaction between oxygen and the minerals making up rocks. The ferromagnesian minerals contain iron, magnesium, and other metal ions in a silicate structure. Iron can react with oxygen to produce several different iron oxides, each with its own characteristic color. The most common iron oxide (hematite) has a deep red color. Other oxides of iron are brownish to yellow-brownish. It is the presence of such iron oxides that color many sedimentary rocks and soils. The red soils of Oklahoma, Georgia, and many other places are colored by the presence of iron oxides produced by chemical weathering.

Carbonation is a reaction between carbonic acid and the minerals making up rocks. Rainwater is naturally somewhat acidic because it dissolves carbon dioxide from the air. This forms a weak acid known as carbonic acid (H_2CO_3), the same acid that is found in carbonated soda. Carbonic acid rain falls on the land, seeping into cracks and crevices where it reacts with minerals. Limestone, for example, is easily weathered to a soluble form by carbonic acid. The limestone caves of Missouri, Kentucky, New Mexico, and elsewhere were produced by the chemical weathering of limestone by carbonation (Figure 20.6). Minerals containing calcium, magnesium, sodium, potassium, and iron are chemically weathered by carbonation to produce salts that are soluble in water.

Hydration is a reaction between water and the minerals of rocks. The process of hydration includes (1) the dissolving of a mineral and (2) the combining of water directly with a mineral. Some minerals, such as halite (which is sodium chloride), dissolve in water to form a solution. The carbonates formed from carbonation are mostly soluble, so they are easily leached from a rock by dissolving. Water also combines directly with some minerals to form new, different minerals. The feldspars, for example, undergo hydration and carbonation to produce (1) water-soluble potassium carbonate, (2) a chemical product that combines with water to produce a clay mineral, and (3) silica. The silica, which is silicon dioxide (SiO_2), may appear as a suspension of finely divided particles or in solution.

Mechanical and chemical weathering are interrelated, working together in breaking up and decomposing solid rocks of Earth's surface. In general, mechanical weathering results in cracks in solid rocks and broken-off coarse fragments. Chemical weathering results in finely pulverized materials and ions in solution, the ultimate decomposition of a solid rock. Consider, for example, a mountain of solid granite, the most common rock found on continents. In general, granite is made up of 65 percent feldspars, 25 percent quartz, and about 10 percent ferromagnesian minerals. Mechanical weathering begins the destruction process as exfoliation and frost wedging create cracks in the solid mass of granite. Rainwater, with dissolved oxygen and carbon dioxide, flows and seeps into the cracks and reacts with ferromagnesian minerals to form soluble carbonates and metal oxides. Feldspars undergo carbonation and hydration, forming clay minerals and soluble salts, which are washed away. Quartz is less susceptible to chemical weathering and remains mostly unchanged to form sand grains. The end products of the complete weathering of granite are quartz sand, clay minerals, metal oxides, and soluble salts.

Water freezing in crack expands, widening crack.

Eventually, rock is broken free by progressive fracturing.

A

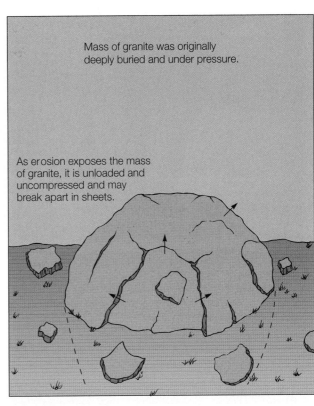

Mass of granite was originally deeply buried and under pressure.

As erosion exposes the mass of granite, it is unloaded and uncompressed and may break apart in sheets.

B

FIGURE 20.3 (*A*) Frost wedging and (*B*) exfoliation are two examples of mechanical weathering, or disintegration, of solid rock.

A

B

FIGURE 20.4 Growing trees can break, separate, and move solid rock. (*A*) Note how this tree has raised the sidewalk. (*B*) This tree is surviving by growing roots into tiny joints and cracks, which become larger as the tree grows.

FIGURE 20.5 Spheroidal weathering of granite. The edges and corners of an angular rock are attacked by weathering from more than one side and retreat faster than flat rock faces. The result is rounded granite boulders, which often shed partially weathered minerals in onion-like layers.

A

SOILS

Accumulations of the products of weathering—sand, silt, and clay—result in a layer of unconsolidated earth materials known as *soil*. Soil is, however, usually understood to be more than just loose weathered materials such as sand or clay. **Soil** is a mixture of unconsolidated weathered earth materials and *humus,* which is altered, decay-resistant organic matter. A mature, fertile soil is the result of centuries of mechanical and chemical weathering of rock, combined with years of accumulated decayed plants and other organic matter. Soil forms over the solid rock below, which is generally known as *bedrock.*

There are thousands of different soil types, depending on such things as the parent rock type, climate, time of accumulation, topographic relief, elevation, rainfall, percentage of clay, sand, or silt, amount of humus, and a number of other environmental variables. In general, soils formed in cold and dry climates are shallower with less humus than soils produced in wet and warm climates. This is because chemical reactions occur at a faster pace in warmer, wetter soil than they do in dry, cooler soil. In addition, the wet and warm climate would be more conducive to plant growth, which would provide more organic matter for the formation of humus.

A soil that has varying proportions of sand, silt, and clay mixed with an abundance of humus is called *loam.* Loam is a great soil for gardening because it is fertile and drains well but holds enough moisture for sustained plant growth. Loam is usually found in the topmost layers of soil, so it is also referred to as *topsoil.* Topsoil is usually more fertile because it is closer to the source of humus. The soil beneath, referred to as *subsoil,* often contains more rocks and mineral accumulations and lacks humus.

B

FIGURE 20.6 Limestone caves develop when slightly acidic groundwater dissolves limestone along joints and bedding planes, carrying away rock components in solution. (*A*) Joints and bedding planes in a limestone bluff. (*B*) This stream has carried away less-resistant rock components, forming a cave under the ledge.

As described in the section on weathering, quartz grains of sand and clay minerals are the two minerals that usually remain after rock has weathered completely. The sand grains help keep the soil loose and aerated, allowing good water drainage. Clay minerals, on the other hand, help hold water in a soil. A good, fertile loam contains some clay and some sand. The balanced mixture of clay minerals and sand provides plants with both the air and water that they need for optimum root growth.

 CONCEPTS *Applied*

City Weathering

Prepare a display of photographs that show weathering and erosion processes at work in a *city* environment. Provide location, agent of weathering or erosion, and a description of what could be done to slow or stop each process illustrated.

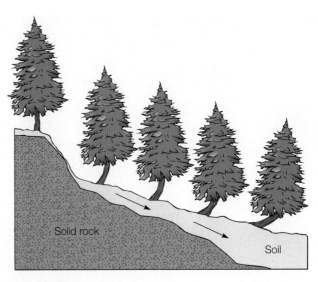

FIGURE 20.7 The slow creep of soil is evidenced by the strange growth pattern of these trees.

EROSION

Weathering has prepared the way for erosion and for some agent of transportation to move or carry away the fragments, clays, and solutions that have been produced from solid rock. The weathered materials can be moved to a lower elevation by the direct result of gravity acting alone. They can also be moved to a lower elevation by gravity acting through some intermediate agent, such as running water, wind, or glaciers. The erosion of weathered materials as a result of gravity alone will be considered first.

MASS MOVEMENT

Gravity constantly acts on every mass of materials on the surface of Earth, pulling parts of elevated regions toward lower levels. Rocks in the elevated regions are able to temporarily resist this constant pull through their cohesiveness with a main rock mass or by the friction of the rock on a slope. Whenever anything happens to reduce the cohesiveness or to reduce the friction, gravity pulls the freed material to a lower elevation. Thus, gravity acts directly on individual rock fragments and on large amounts of surface materials as a mass, pulling all to a lower elevation. Erosion caused by gravity acting directly is called **mass movement** (also called mass wasting). Mass movement can be so slow that it is practically imperceptible. *Creep,* the slow downhill movement of soil down a steep slope, for example, is detectable only from the peculiar curved growth patterns of trees growing in the slowly moving soil (Figure 20.7). At the other extreme, mass movement can be as sudden and swift as a single rock bounding and clattering down a slope from a cliff. A *landslide* is a generic term used to describe any slow to rapid movement of any type or mass of materials, from the short slump of a hillside to the slide of a whole mountainside. Either slow or sudden, mass movement is a small victory for gravity in the ongoing process of leveling the landmass of Earth.

 CONCEPTS *Applied*

Creeping Trees

Investigate how the creep of soil has changed trees on a steep hillside. Hold a protractor with the curved side down and attach a string to the top center. Tie a weight to the other end of the string. Use the protractor and weighted string to find the angle of trees growing on the steep hillside. Make a histogram of the number of trees versus leaning angle. Compare this to a graph of trees growing on level ground.

RUNNING WATER

Running water is the most important of all the erosional agents of gravity that remove rock materials to lower levels.

Erosion by running water begins with rainfall. Each raindrop impacting the soil moves small rock fragments about, but it also begins to dissolve some of the soluble products of weathering. If the rainfall is heavy enough, a shallow layer or sheet of water forms on the surface, transporting small fragments and dissolved materials across the surface. This *sheet erosion* picks up fragments and dissolved material, then transports them to small streams at lower levels (Figure 20.8). The small streams move to larger channels, and the running water transports materials three different ways: (1) as dissolved rock materials carried in solution, (2) as clay minerals and small grains carried in suspension, and (3) as sand and larger rock fragments that are rolled, bounced, and slid along the bottom of the streambed. Just how much material is eroded and transported by the stream depends on the volume of water, its velocity, and the load that it is already carrying.

Streams and major rivers are at work, for the most part, twenty-four hours a day every day of the year moving rock

FIGURE 20.8 Moving streams of water carry away dissolved materials and sediments as they slowly erode the land.

FIGURE 20.9 A river usually stays in its channel, but during a flood, it spills over and onto the adjacent flat land called the *floodplain.*

fragments and dissolved materials from elevated landmasses to the oceans. Any time you see mud, clay, and sand being transported by a river, you know that the river is at work moving mountains, bit by bit, to the ocean. It has been estimated that rivers remove enough dissolved materials and sediments to make the whole surface of the United States flat in a little over 20 million years, a very short time compared to the 4.6 billion-year age of Earth.

In addition to transporting materials that were weathered and eroded by other agents of erosion, streams do their own erosive work. Streams can dissolve soluble materials directly from rocks and sediments. They also quarry and pluck fragments and pieces of rocks from beds of solid rock by hydraulic action. Most of the erosion accomplished directly by streams, however, is done by the more massive fragments that are rolled, bounced, and slid along the streambed and against each other. This results in a grinding and filing action on the fragments and a wearing away of the streambed.

As a stream cuts downward into its bed, other agents of erosion such as mass movement begin to widen the channel as

materials slump into the moving water. The load that the stream carries is increased by this slumping, which slows the stream. As the stream slows, it begins to develop bends, or *meanders,* along the channel. Meanders have a dramatic effect on stream erosion because the water moves faster around an outside bank than it does around the inside bank downstream. This difference in stream velocity means that the stream has a greater erosion ability on the outside, downstream side and less on the sheltered area inside of curves. The stream begins to widen the floor of the valley through which it runs by eroding on the outside of the meander, then depositing the eroded material on the inside of another bend downstream. The stream thus begins to erode laterally, slowly working its way across the land. Sometimes two bends in the stream meet, forming a cut-off meander called an *oxbow lake.*

A stream, along with mass movement, develops a valley on a widening floodplain. A **floodplain** is the wide, level floor of a valley built by a stream (Figure 20.9). It is called a floodplain because this is where the stream floods when it spills out of its channel. The development of a stream channel into a widening floodplain seems to follow a general, idealized aging pattern (Figure 20.10). When a stream is on a recently uplifted landmass, it has a steep gradient, a vigorous, energetic ability to erode the land, and characteristic features known as the stage of youth. *Youth* is characterized by a steep gradient, a V-shaped valley without a floodplain, and the presence of features that interrupt its smooth flow such as boulders in the streambed, rapids, and waterfalls (Figure 20.11). Stream erosion during youth is predominantly downward. The stream eventually erodes its way into *maturity* by eroding away the boulders, rapids, and waterfalls, and in general smoothing and lowering the stream gradient. During maturity, meanders form over a wide floodplain that now occupies the valley floor. The higher elevations are now more sloping hills at the edge of the wide floodplain rather than steep-sided walls close to the river channel. *Old age* is marked by a very low gradient in extremely broad, gently sloping valleys. The stream now flows slowly in broad meanders over the wide floodplain. Floods are more common in old age

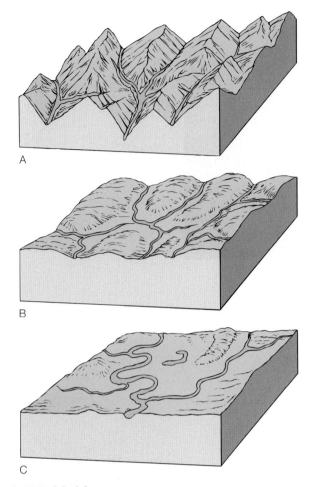

FIGURE 20.10 Three stages in the aging and development of a stream valley: (A) youth, (B) maturity, and (C) old age.

FIGURE 20.11 The waterfall and rapids on the Yellowstone River in Wyoming indicate that the river is actively downcutting. Note the V-shaped cross-profile and lack of floodplain, characteristics of a young stream valley.

since the stream is carrying a full load of sediments and flows sluggishly.

Many assumptions are made in any generalized scheme of the erosional aging of a stream. Streams and rivers are dynamic systems that respond to local conditions, so it is possible to find an "old age feature" such as meanders in an otherwise youthful valley. This is not unlike finding a gray hair on an 18-year-old youth, and in this case, the presence of the gray hair does not mean old age. In general, old age characteristics are observed near the *mouth* of a stream, where it flows into an ocean, lake, or another stream. Youthful characteristics are observed at the *source*, where the water collects to first form the stream channel. As the stream slowly lowers the land, the old age characteristics will move slowly but surely toward the source.

When the stream flows into the ocean or a lake, it loses all of its sediment-carrying ability. It drops the sediments, forming a deposit at the mouth called a **delta** (Figure 20.12). Large rivers such as the Mississippi River have large and extensive deltas that actually extend the landmass more and more over time. In a way, you could think of the Mississippi River delta as being formed from pieces and parts of the Rocky Mountains, the Ozark Mountains, and other elevated landmasses that the Mississippi has carried there over time.

 CONCEPTS *Applied*

Stream Relationships

Measure and record the speed of a river or stream every day for a month. The speed can be calculated from the time required for a floating object to cover a measured distance. Define the clarity of the water by shining a beam of light through a sample, then comparing to a beam of light through clear water. Use a scale, such as 1 for perfectly clear to 10 for no light coming through, to indicate clarity. Graph your findings to see if there is a relationship between clarity and speed of flow of the stream.

GLACIERS

Glaciers presently cover only about 10 percent of Earth's continental land area, and much of this is at higher latitudes, so it might seem that glaciers would not have much of an overall effect in eroding the land. However, ice has sculptured much

A

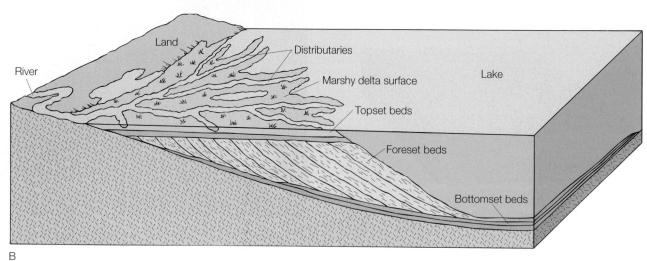

B

FIGURE 20.12 (*A*) Delta of Nooksack River, Washington. Note the sediment-laden water and how the land is being built outward by river sedimentation. (*B*) Cross section showing how a small delta might form. Large deltas are more complicated than this.

of the present landscape, and features attributed to glacial episodes are found over about three-quarters of the continental surface. Only a few tens of thousands of years ago, sheets of ice covered major portions of North America, Europe, and Asia. Today, the most extensive glaciers in the United States are those of Alaska, which cover about 3 percent of the state's land area. Less extensive glacier ice is found in the mountainous regions of Washington, Montana, California, Colorado, and Wyoming.

A **glacier** is a mass of ice on land that moves under its own weight. Glacier ice forms gradually from snow, but the quantity of snow needed to form a glacier does not fall in a single winter. Glaciers form in cold climates where some snow and ice persist throughout the year. The amount of winter snowfall must exceed the summer melting to accumulate a sufficient mass of snow to form a glacier. As the snow accumulates, it is gradually transformed into ice. The weight of the overlying snow packs it down, driving out much of the air, and causing it to recrystallize into a coarser, denser mass of interlocking ice crystals that appears to have a blue to deep blue color. Complete conver-

sion of snow into glacial ice may take from five to thirty-five hundred years, depending on such factors as climate and rate of snow accumulation at the top of the pile. Eventually, the mass of ice will become large enough that it begins to flow, spreading out from the accumulated mass. Glaciers that form at high elevations in mountainous regions are called *alpine glaciers*. If these glaciers flow down into a valley, they are also called *valley glaciers* (Figure 20.13). Glaciers that cover a large area of a continent are called *continental glaciers*. Continental glaciers can cover whole continents and reach a thickness of 1 km (3,295 ft) or more. Today, the remaining continental glaciers are found on Greenland and the Antarctic.

Glaciers move slowly and unpredictably, spreading like a huge blob of putty under the influence of gravity. As an alpine glacier moves downhill through a V-shaped valley, the sides and bottom of the valley are eroded wider and deeper. When the glacier later melts, the V-shaped valley is now a U-shaped valley that has been straightened and deepened by the glacial erosion. The glacier does its erosional work using three different techniques: (1) by bulldozing, (2) by abrasion, and (3) by plucking.

FIGURE 20.13 Valley glacier on Mount Logan, Yukon Territory.

Bulldozing, as the term implies, is the pushing along of rocks, soil, and sediments by the leading edge of an advancing glacier. Deposits of bulldozed rocks and other materials that remain after the ice melts are called *moraines. Plucking* occurs as water seeps into cracked rocks and freezes, becoming a part of the solid glacial mass. As the glacier moves on, it pulls the fractured rock apart and plucks away chunks of it. The process is accelerated by the frost-wedging action of the freezing water. Plucking at the uppermost level of an alpine glacier, combined with weathering of the surrounding rocks, produces a rounded or bowl-like depression known as a *cirque* (Figure 20.14). *Abrasion* occurs as the rock fragments frozen in the moving glacial ice scratch, polish, and grind against surrounding rocks at the base and along the valley walls. The result of this abrasion is the pulverizing of rock into ever finer fragments, eventually producing a powdery, silt-sized sediment called *rock flour.* Suspended rock flour in meltwater from a glacier gives the water a distinctive gray to blue-gray color.

Glaciation is continuously at work eroding the landscape in Alaska and many mountainous regions today. The glaciation that formed the landscape features in the Rockies, the Sierras, and across the northeastern United States took place thousands of years ago.

Myths, Mistakes, & Misunderstandings

Unchanging as the Hills?

It is a mistake to say something is "unchanging as the hills." The hills may appear to be tranquil and unchanging, but they are actually under constant attack, weathered and eroded bit by bit. As more and more of the hills are carried away over time, they are slowly changing, eventually to cease to exist as hills.

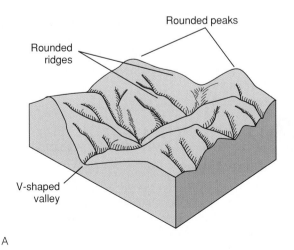

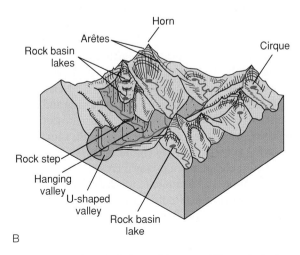

FIGURE 20.14 (*A*) A stream-carved mountainside before glaciation. (*B*) The same area after glaciation, with some of the main features of mountain glaciation labeled.

WIND

Like running water and ice, wind also acts as an agent shaping the surface of the land. It can erode, transport, and deposit materials. However, wind is considerably less efficient than ice or water in modifying the surface. Wind is much less dense and does not have the eroding or carrying power of water or ice. In addition, a stream generally flows most of the time, but the wind blows only occasionally in most locations. Thus, on a worldwide average, winds move only a few percent as much material as do streams. Wind also lacks the ability to attack rocks chemically as water does through carbonation and other processes, and wind cannot carry dissolved sediments in solution. Even in many deserts, more sediment is moved during the brief periods of intense surface runoff following the occasional rainstorms than is moved by wind during the prolonged dry periods.

Flowing air and moving water do have much in common as agents of erosion since both are fluids. Both can move larger particles by rolling them along the surface and can move finer particles by carrying them in suspension. Both can move larger and more massive particles with increased velocities. Water is denser and more viscous than air, so it is more efficient at transporting quantities of material than is the wind, but the processes are quite similar.

Two major processes of wind erosion are called (1) abrasion and (2) deflation. *Wind abrasion* is a natural sandblasting process that occurs when the particles carried along by the wind break off small particles and polish what they strike. Generally, the harder mineral grains such as quartz sand accomplish this best near the ground where the wind is bouncing them along. Wind abrasion can strip paint from a car exposed to the moving particles of a dust storm, eroding the paint just as it erodes rocks on the surface. Rocks and boulders exposed to repeated wind storms where the wind blows consistently from one or a few directions may be planed off from the repeated action of this natural sandblasting. Rocks sculptured by wind abrasion are called *ventifacts,* after the Latin meaning "wind-made" (Figure 20.15).

Deflation, after the Latin meaning "to blow away," is the widespread picking up of loose materials from the surface. Deflation is naturally most active where winds are unobstructed and the materials are exposed and not protected by vegetation. These conditions are often found on deserts, beaches, and unplanted farmland between crops. During the 1930s, many farmers in the Plains states replaced the native grassland vegetation when they established farms. A series of drought years occurred and the crops died, leaving the soil exposed. Unusually strong winds eroded the unprotected surface, removing and transporting hundreds of millions of tons of soil. This period of prolonged drought, dust storms, and general economic disaster for farmers in the area is known as the Dust Bowl episode.

The most common wind-blown deposits are (1) dunes and (2) loess. A **dune** is a low mound or ridge of sand or other sediments. Dunes form when sediment-bearing wind encounters an obstacle that reduces the wind velocity. With a slower velocity, the wind cannot carry as large a load, so sediments are deposited on the surface. This creates a larger windbreak, which results in a growing obstacle, a dune. Once formed, a dune tends to migrate,

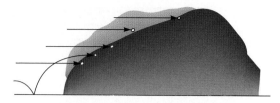

If wind is predominantly from one direction, rocks will be planed off or flattened on the upwind side.

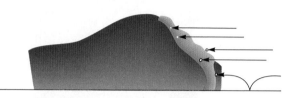

With a persistent shift in wind direction, additional facets are cut in the rock.

FIGURE 20.15 Ventifact formation by abrasion from one or several directions.

particularly if the winds blow predominantly from one direction. Dunes are commonly found in semiarid areas or near beaches.

Another common wind deposit is called *loess* (pronounced "luss"). Loess is a very fine dust, or silt, that has been deposited over a large area. One such area is located in the central part of the United States, particularly to the east sides of the major rivers of the Mississippi basin. Apparently, this deposit originated from the rock flour produced during the last great ice age. The rock flour was probably deposited along the major river valleys and later moved eastward by the prevailing westerly winds. Since rock flour is produced by the mechanical grinding action of glaciers, it has not been chemically broken down. Thus, the loess deposit contains many minerals that were not leached out of the deposit as typically occurs with chemical weathering. It also has an open, porous structure since it does not have as much of the chemically produced clay minerals. The good moisture-holding capacity from this open structure, together with the presence of minerals that serve as plant nutrients, makes farming on the soils developed from these deposits particularly productive.

CONCEPTS *Applied*

Dune Angles

Use a protractor to measure the slopes in a sand dune. Compare the angle of the slope made by sand moving upslope to the angle of the slope where sand moves downslope. Investigate what factors seem to influence the steepness of the slopes.

Science and Society

Acid Rain

*A*cid rain is a general term used to describe any acidic substances, wet or dry, that fall from the atmosphere. Wet acidic deposition could be in the form of rain, but snow, sleet, and fog could also be involved. Dry acidic deposition could include gases, dust, or any solid particles that settle out of the atmosphere to produce an acid condition.

Pure, unpolluted rain is naturally acidic. Carbon dioxide in the atmosphere is absorbed by rainfall, forming carbonic acid (H_2CO_3). Carbonic acid lowers the pH of pure rainfall to a range of 5.6–6.2. Decaying vegetation in local areas can provide more CO_2, making the pH even lower. A pH range of 4.5–5.0, for example, has been measured in rainfall of the Amazon jungle. Human-produced exhaust emissions of sulfur and nitrogen oxides can lower the pH of rainfall even more, to a 4.0–5.5 range. This is the pH range of acid rain (Box Table 20.1).

The sulfur and nitrogen oxides that produce acid rain come from exhaust emissions of industries and electric utilities that burn fossil fuels and of cars, trucks, and buses. The oxides are the raw materials of acid rain and are not themselves acidic. They react with other atmospheric chemicals to form sulfates and nitrates, which combine with water vapor to form sulfuric acid and nitric acid. These are the chemicals of concern in acid rain.

Acid rain falls on the land, bodies of water, forests, crops, buildings, and people, so the concerns about acid rain center on its environmental impact on lakes, forest, crops, materials, and human health. Acid rain accelerates chemical weathering, leaches essential nutrients from the soil, and acidifies lakes and streams. All of these processes affect plants, animals and microbes. Chemical weathering by acid rain also can cause deterioration of buildings and other structures.

The type of rocks making up the local landscape can either moderate or aggravate the problems of acid rain. Limestone and the soils of arid climates tend to neutralize acid, while waters in granite rocks and soils cannot neutralize acid and tend already to be somewhat acidic.

Although natural phenomena, such as volcanoes, contribute acids to the atmosphere, human-produced emissions of sulfur and nitrogen oxides from burning fossil fuels are the primary cause of acid rain. The heavily industrialized part of the United States, from the Midwest through New England, releases sulfur and nitrogen emissions that result in a precipitation pH of 4.0–4.5. Unfortunately, the area of New England and adjacent Canada downwind of major acid rain sources has granite bedrock, which means that the effects of acid rain will not be moderated as they would in the West or Midwest. This region is the geographic center of the acid rain problem in North America. A solution to the problem is being sought by (1) using fuels other than fossil when possible and (2) reducing the thousands of tons of sulfur and nitrogen oxides that are dumped into the atmosphere each day when fossil fuels are used.

BOX TABLE 20.1

The approximate pH of some common acidic substances

Substance	pH (or pH Range)
Hydrochloric acid (4%)	0
Gastric (stomach) solution	1.6–1.8
Lemon juice	2.2–2.4
Vinegar	2.4–3.4
Carbonated soft drinks	2.0–4.0
Grapefruit	3.0–3.2
Oranges	3.2–3.6
Acid rain	**4.0–5.5**
Tomatoes	4.2–4.4
Potatoes	5.7–5.8
Natural rainwater	**5.6–6.2**
Milk	6.3–6.7
Pure water	7.0

QUESTIONS TO DISCUSS

Discuss with your group the following questions concerning acid rain:

1. Should the use of fossil fuels be taxed to cut the source of acid rain and fund solutions? Provide reasons with your answer.
2. Electric utilities are required to remove sulfur dioxide from power plant exhaust according to the best technology that was available when the plant was constructed. Should they retrofit all plants with more expensive technology to reduce the amount released even more? What if doing so would increase your electric bill by 50 percent? Would you still support further reductions in the amount of sulfur dioxide released?
3. What are the advantages and disadvantages of a total ban on the use of fossil fuels?
4. Brainstorm with your group to see how many solutions you can think of to stop acid rain damage.

DEVELOPMENT OF LANDSCAPES

The landscape provides interesting scenery with a variety of features such as mountains, valleys, and broad, rolling hills. The features of Earth's surface are called **landforms.** Landforms include (1) *broad features* such as a mountain, plain, or plateau and (2) *minor features* such as a hill, valley, or canyon. Broad or minor, all landforms are temporary expressions between the forces that elevate the land and the weathering and erosion that level it.

No two landforms are identical because each has been produced and sculptured by a variety of processes. Thus, there is no exact way to describe how a particular landform came to be, but generalizations are possible. Generalizations are based on three factors: (1) rock structure, (2) weathering and erosion processes, and (3) stages of erosion.

People Behind the Science

John Wesley Powell (1834–1902)

John Wesley Powell was the most romantic figure in nineteenth-century U.S. geology. The son of intensely pious Methodist immigrants, Powell was intended by his farmer father for the Methodist ministry, but early on he developed a love for natural history. In the 1850s, he became secretary of the Illinois Society of Natural History, traveling widely and building up his natural history collections and his geological expertise. While he was fighting in the Civil War, his right arm was shot off, but he continued in the service, rising to the rank of colonel.

After the end of the war, Powell occupied various chairs in geology in Illinois, while continuing with intrepid fieldwork (he was one of the first to steer a way down the Grand Canyon). In 1870, Congress appointed him to lead an official survey of the natural resources of the Utah, Colorado, and Arizona area, the findings of which were published in his *The Exploration of the Colorado River* (1875) and *The Geology of the Eastern Portion of the Uinta Mountains* (1876).

Powell's enormous and original studies produced lasting insights on fluvial erosion, volcanism, isostasy, and orogeny. His greatness as a geologist and geomorphologist stemmed from his capacity to grasp the interconnections of geological and climatic causes. In 1881, he was appointed director of the U.S. Geological Survey. He encouraged most of the great U.S. geologists of the next generation, including Grove Karl Gilbert, Clarence E. Dutton, and W. H. Holmes.

Powell drew attention to the aridity of the American southwest and for a couple of decades campaigned for massive funds for irrigation projects and dams and for the geological surveys necessary to implement adequate water strategies. He also asserted the need in the drylands for changes in land policy and farming techniques. Failing to win political support on such matters, he resigned in 1894 from the Geological Survey.

Source: From the Hutchinson *Dictionary of Scientific Biography.* © Research Machines plc 2003. All Rights Reserved. Helicon Publishing is a division of Research Machines.

ROCK STRUCTURE

The *structure* of the rocks determines the shape of the minor landforms. Structure refers to the (1) *type* of rock (igneous, metamorphic, or sedimentary) and (2) their *attitude,* that is, if they have been disturbed by faulting or folding. The type of rock determines how well a rock resists weathering. The sedimentary rock limestone, for example, is highly susceptible to chemical weathering, while the metamorphic rock quartzite is highly resistant to chemical weathering. The attitude of the rock also determines how well a rock resists weathering and erosion. Limestone beds that have been faulted and folded, for example, are more easily eroded than flat-lying beds of limestone.

WEATHERING AND EROSION PROCESSES

The processes of weathering and erosion that attack the rock structure are influenced and controlled by other factors such as climate and elevation. Chemical weathering, for example, is more dominant in warm, moist climates, and mechanical weathering is more dominant in dry climates. Thus, landforms in warm, moist climates tend to have softer, rounded outlines from the accumulations of clay minerals, sand, and other finely divided products of chemical weathering. The landforms in dry climates, on the other hand, tend to have sharp, angular outlines from the mass movement of rock material from vertical cliffs. Lacking as much chemical weathering, the landscapes in dry climate regions tend to have sharper outlines.

STATE OF DEVELOPMENT

The *stage* of landform development describes how effective the processes have been in attacking the rock structure. Stage describes the extent to which the processes have completed their work, that is, the amount of the original surface that remains. Mountains, for example, are said to be *youthful* when the processes of weathering and erosion have not had time to do much of their work. Youthful mountains are characterized by prominent relief of steep peaks and narrow, steep valleys. The steep peaks may be from cirques produced by glaciers, and the steep valleys have been cut by streams, but neither process has yet greatly altered the original structure (Figure 20.16). By *maturity,* the original structure has been worn down to rounded forms and slopes. Eventually, even the mightiest mountain is worn down to nearly flat, rolling plains during *old age*. The nearly flat surface is called a *peneplain,* which means a region that is "almost" a plain. Often hills of resistant rock called *monadnocks* exist on the peneplain during the last stages of old age. Theoretically, the monadnocks and peneplain will be reduced to the lowest level possible, which approximates sea level. More than likely, however, the land will be uplifted before this happens, causing *rejuvenation* of the erosion processes. Rejuvenation renews the effectiveness of the weathering and erosion processes, and the cycle begins again with youthful landform structures being superimposed on the old age structures.

A

B

FIGURE 20.16 This melting glacier (*A*) is the source for a stream (*B*) that flows through a valley in the youth stage.

SUMMARY

Weathering is the breaking up, crumbling, and destruction of any kind of solid rock. The process of physically picking up weathered rock materials is called *erosion*. After the eroded materials are picked up, they are removed by *transportation* agents. The combined action of weathering, erosion, and transportation wears away and lowers the surface of Earth.

The physical breaking up of rocks is called *mechanical weathering*. Mechanical weathering occurs by *wedging effects* and the *effects of reduced pressure*. *Frost wedging* is a wedging effect that occurs from repeated cycles of water freezing and thawing. The process of spalling off of curved layers of rock from reduced pressure is called *exfoliation*.

The breakdown of minerals by chemical reactions is called *chemical weathering.* Examples include *oxidation,* a reaction between oxygen and the minerals making up rocks; *carbonation,* a reaction between carbonic acid (carbon dioxide dissolved in water) and minerals making up rocks; and *hydration,* the dissolving or combining of a mineral with water. Granite is composed of feldspars, quartz, and some ferromagnesian minerals. The end products of complete mechanical and chemical weathering of granite are quartz sand, clay minerals, metal oxides, and soluble salts.

Soil is a mixture of weathered earth materials and *humus.* When the end products of complete weathering of rocks are removed directly by gravity, the erosion is called *mass movement. Landslide* is a generic term meaning any type of movement by any type of material. Erosion and transportation also occur through the agents of *running water, glaciers,* or *wind.* Each creates its own characteristic features of erosion and deposition.

KEY TERMS

delta (p. **501**)
dune (p. **504**)
erosion (p. **494**)
floodplain (p. **500**)
glacier (p. **502**)
landforms (p. **505**)
mass movement (p. **499**)
soil (p. **498**)
weathering (p. **494**)

APPLYING THE CONCEPTS

1. Small changes that result in the breaking up, crumbling, and destruction of any kind of rock are
   a. decomposition.
   b. weathering.
   c. corrosion.
   d. erosion.
2. The process of physically removing weathered materials is called
   a. weathering.
   b. transportation.
   c. erosion.
   d. corrosion.
3. Muddy water rushing downstream after a heavy rain is an example of
   a. weathering.
   b. washing.
   c. erosion.
   d. transportation.
4. The physical breakup of rocks without any changes in their chemical composition is
   a. oxidation.
   b. mechanical weathering.
   c. erosion.
   d. transportation.
5. Chemical weathering, the dissolving or breaking down of minerals in rocks, is also called
   a. oxidation.
   b. reduction.
   c. disintegration.
   d. decomposition.

6. The process of peeling off layers of rock, reducing the pressure on the rock's surface, is called
   a. chemical peel.
   b. exfoliation.
   c. wedging.
   d. disintegration.
7. The weak acid formed by the reaction of water with carbon dioxide is
   a. carbonic acid.
   b. carbonate acid.
   c. hydrocarbonic acid.
   d. dihydrocarbonate acid.
8. A mixture of unconsolidated weathered earth materials and humus is
   a. gravel.
   b. soil.
   c. dirt.
   d. mud.
9. Decay-resistant, altered organic material is
   a. soil.
   b. clay.
   c. topsoil.
   d. humus.
10. Two minerals that usually remain after granite has completely weathered are
   a. quartz and clay.
   b. quartz and hematite.
   c. olivine and granite.
   d. quartz and basalt.
11. Weathered materials move to lower elevations due to only
   a. wind.
   b. rain.
   c. gravity.
   d. erosion.
12. The slow movement downhill of soil on the side of a mountain is called
   a. rockslide.
   b. avalanche.
   c. creep.
   d. crawl.
13. The wide, level floor of a valley built by a stream is called a
   a. channel.
   b. valley.
   c. floodplain.
   d. island.
14. The deposit at the mouth of a river where sediments are dropped is called a
   a. riverbank.
   b. delta.
   c. mouth.
   d. source.
15. Rock fragments frozen in moving glacier ice polish and scratch rocks at the base and on the walls of the glacier. This process is
   a. bulldozing.
   b. polishing.
   c. plucking.
   d. abrasion.
16. The agent that has the least ability to erode is
   a. wind.
   b. gravity.
   c. streams.
   d. glaciers.

17. The major processes of wind erosion are
    a. abrasion and bulldozing.
    b. polishing and plucking.
    c. abrasion and deflation.
    d. deflation and polishing.

18. The picking up of loose materials from the surface by wind is
    a. abrasion.
    b. deflation.
    c. inflation.
    d. polishing.

19. What is the pH of natural rainwater?
    a. 5.0–5.5
    b. 5.6–6.2
    c. 6.3–6.7
    d. 7.0

20. Freezing water exerts pressure on the wall of a crack in a rock mass, making the crack larger. This is an example of
    a. mechanical weathering.
    b. chemical weathering.
    c. exfoliation.
    d. hydration.

21. Of the following rock weathering events, the last one to occur would probably be
    a. exfoliation.
    b. frost wedging.
    c. carbonation.
    d. disintegration.

22. Which of the following would have the greatest overall effect in lowering the elevation of a continent such as North America?
    a. continental glaciers
    b. alpine glaciers
    c. wind
    d. running water

23. Broad meanders on a very wide, gently sloping floodplain with oxbow lakes are characteristics you would expect to find in a river valley during what stage?
    a. newborn
    b. youth
    c. maturity
    d. old age

24. A glacier forms when
    a. the temperature does not rise above freezing.
    b. snow accumulates to form ice, which begins to flow.
    c. a summer climate does not occur.
    d. a solid mass of snow moves downhill under the influence of gravity.

25. A likely source of loess is
    a. rock flour.
    b. a cirque.
    c. a terminal moraine.
    d. an accumulation of ventifacts.

26. The landscape in a dry climate tends to be more angular because the dry climate
    a. has more winds.
    b. lacks as much chemical weathering.
    c. has less rainfall.
    d. has stronger rock types.

27. Peneplains and monadnocks are prevented from forming by
    a. mass movement.
    b. running water.
    c. deflation.
    d. rejuvenation.

28. The phrase "weathering of rocks" means
    a. able to resist any changes, as in "weathers the storm."
    b. a discoloration caused by the action of the weather.
    c. physical or chemical destruction.
    d. the same thing as rusting.

29. What are you are doing to a rock if you pick up the small pieces of a smashed rock?
    a. mechanical weathering
    b. chemical weathering
    c. erosion
    d. transportation

30. What are you doing to the fragments of a smashed rock if you carry the fragments to a new location?
    a. mechanical weathering
    b. chemical weathering
    c. erosion
    d. transportation

31. What are you doing to a rock if you dissolve it in acid?
    a. mechanical weathering
    b. chemical weathering
    c. erosion
    d. transportation

32. A deeper, richer layer of soil would be expected where the climate is
    a. wet and warm.
    b. dry and cold.
    c. tropical.
    d. arctic.

33. The soil called loam is
    a. all sand and humus.
    b. mostly humus with some sand.
    c. equal amounts of sand, clay, and humus.
    d. from the C horizon of a soil profile.

34. A moraine is a
    a. wind deposit.
    b. glacier deposit.
    c. river deposit.
    d. Any of the above is correct.

35. The breaking up, crumbling, chemical decomposition, and destruction of rocks at or near Earth's surface is called
    a. rockslide.
    b. mining.
    c. weathering.
    d. erosion.

36. Crushing of rock at a quarry to make smaller-sized gravel is an example of
    a. physical weathering.
    b. chemical weathering.
    c. mechanical weathering.
    d. reduction weathering.

37. Fragments of rocks fall into a mountain stream and are carried into the valley. This is an example of
    a. weathering.
    b. erosion.
    c. transportation.
    d. decomposition.

38. Tree roots grow and expand, and eventually break though a sidewalk. This is an example of
   a. corrosion.
   b. erosion.
   c. wedging.
   d. disintegration.

39. Damage to the Lincoln Memorial by rain and smog is an example of
   a. physical weathering.
   b. chemical weathering.
   c. exfoliation.
   d. erosion.

40. Ferromagnesian minerals will react with oxygen to produce deeply colored iron oxides. This is an example of
   a. hydration.
   b. carbonation.
   c. oxidation.
   d. combination.

41. You are planning a garden and need a soil that will hold moisture for plant growth but also drain well. The soil that you want to purchase is
   a. humus.
   b. loam.
   c. clay-based.
   d. sandy.

42. The formation of a shallow layer of water by rain on the surface, which dissolves materials and carries fragments away, is called
   a. rain erosion.
   b. sheet erosion.
   c. waterfall.
   d. stream erosion.

43. The most extensive glaciers in the United States are found in
   a. Montana.
   b. Wyoming.
   c. Washington.
   d. Alaska.

44. Continental glaciers are found in
   a. Iceland.
   b. Antarctic.
   c. Greenland.
   d. Greenland and Antarctic.

45. An example of a chemical weathering process that is a major concern in the Northeastern United States is
   a. erosion.
   b. acid rain.
   c. hydration.
   d. oxidation.

Answers

1. b 2. c 3. d 4. b 5. d 6. b 7. a 8. b 9. d 10. a 11. c 12. c 13. c 14. b
15. d 16. a 17. c 18. b 19. b 20. a 21. c 22. d 23. d 24. b 25. a 26. b
27. d 28. c 29. c 30. d 31. b 32. a 33. c 34. b 35. c 36. c 37. c 38. c
39. b 40. c 41. b 42. b 43. d 44. d 45. b

QUESTIONS FOR THOUGHT

1. Compare and contrast mechanical and chemical weathering.
2. Granite is the most common rock found on continents. What are the end products after granite has been completely weathered? What happens to these weathering products?

3. What other erosion processes are important as a stream of running water carves a valley in the mountains? Explain.
4. Describe three ways in which a river erodes its channel.
5. What is a floodplain?
6. Describe the characteristic features associated with stream erosion as the stream valley passes through the stages of youth, maturity, and old age.
7. What is a glacier? How does a glacier erode the land?
8. What is rock flour and how is it produced?
9. Could a glacier erode the land lower than sea level? Explain.
10. Explain why glacial erosion produces a U-shaped valley, but stream erosion produces a V-shaped valley.
11. Name and describe as many ways as you can think of that mechanical weathering occurs in nature. Do not restrict your thinking to those discussed in this chapter.
12. What characteristics of a soil make it a good soil?
13. What essential condition must be met before mass wasting can occur?
14. Compare the features caused by stream erosion, wind erosion, and glacial erosion.
15. Compare the materials deposited by streams, wind, and glaciers.
16. Why do certain stone buildings tend to weather more rapidly in cities than they do in rural areas?
17. Why would mechanical weathering speed up chemical weathering in a humid climate but not in a dry climate?
18. Discuss all the reasons you can in favor of and in opposition to clearing away and burning tropical rainforests for agricultural purposes.

FOR FURTHER ANALYSIS

1. What are the significant similarities and difference between weathering and erosion?
2. Speculate if the continents will ever be weathered and eroded flat, at sea level. Provide evidence to support your speculation.
3. Is it possible for running water to erode below sea level? Provide evidence or some observation to support your answer. Is it possible for any agent of erosion to erode the land to below sea level?

INVITATION TO INQUIRY

Frost Wedging

Rocks undergo mechanical weathering as freezing water expands, exerting pressure on both side of a crack in a rock, making it slightly larger. How much does water expand upon freezing? To investigate, pour water into a plastic cylinder, mark the water level, and place it in a freezer. How much did the ice expand?

21
Geologic Time

These are modern-day stromatolites from Hamlin Pool in Western Australia. The dome-shaped structures shown in the photograph are composed of layers of cyanobacteria and materials they secrete. They grow up to 60 cm (about 2 ft) tall. Some of the oldest fossils are of ancient stromatolites that developed in shallow marine environments about 3.5 billion years ago. When samples from fossil stromatolites are cut into slices, microscopic images can be produced that show the fossil remains of some of the world's oldest cells.

CORE **CONCEPT**

Earth and its kinds of living things have changed greatly over billions of years.

OUTLINE

Fossils
Fossils provide evidence for past life and show that the kinds of organisms on Earth have changed.

The Geologic Time Scale
In the past, the positions of the continents changed, the climate changed, and sea level changed.

Fossils
    Early Ideas About Fossils
    Types of Fossilization
Reading Rocks
    Arranging Events in Order
    Correlation
Geologic Time
    Early Attempts at Earth Dating
    Modern Techniques
The Geologic Time Scale
    Geologic Periods and Typical Fossils
    Mass Extinctions
People Behind the Science: Eduard Suess
    Interpreting Geologic History—A Summary

Reading Rocks
The same geologic processes operate today as they did in the past.

The Geologic Time Scale
Living things can be organized into logical evolutionary sequences.

OVERVIEW

Geology is the study of Earth and the processes that shape it. *Physical geology* is a branch of geology concerned with the materials of Earth, processes that bring about changes in the materials and structures they make up, and the physical features of Earth formed as a result. *Historical geology,* on the other hand, is a branch of geology concerned with the development of Earth and the organisms on it over time. Physical and historical geology together provide a basis for understanding much about Earth and how it has developed.

One reason to study geology is to satisfy an intellectual curiosity about how Earth works. Piecing together the history of how a mountain range formed or inferring the history of an individual rock can be exciting as well as satisfying. As a result, you appreciate the beauty of our Earth from a different perspective. Distinctive features such as the granite domes of Yosemite, the geysers of Yellowstone, and the rocks exposed in a roadcut take on a whole new meaning. Often, part of the new meaning is a story that tells the history of that distinctive feature and how it came to be.

There are many different and fascinating stories that can be read from a given landscape. The structures in the landscape, such as hills and valleys, tell a story about folding, faulting, and other building-up mechanisms, or processes, that were described in chapter 19. There is also a story about the present stage of weathering, erosion, and sculpturing, the processes that were described in chapter 20. Thus, the landscape has a story about the building and sculpturing of surface features and what this must mean about the history of the region (Figure 21.1).

The story reaching back the farthest in time is told by individual rocks. Each rock was formed by processes that were described in chapter 17. Each rock has its own combination of minerals that began to change the moment the rock was created.

Altogether, the story of the individual rock and the landscape features describes the history of the region and how it came to be what it is today. The resulting knowledge of geologic processes and events can also have a practical aspect. Certain earth materials are used for energy or in the manufacture of technological devices. Knowing how, where, and when such resources are formed can be very useful information to modern society.

FOSSILS

A **fossil** is any evidence of former life, so the term means more than fossilized remains, such as those pictured in Figure 21.2. Evidence can include actual or altered remains of plants and animals. It could also be just simple evidence of former life, such as the imprint of a leaf, the footprint of a dinosaur, or droppings from bats in a cave.

What would you think if you were on the top of a mountain, broke open a rock, and discovered the fossil fish pictured in Figure 21.2? How could you explain what you found? There are several ways that a fish could end up on a mountaintop as a fossil. For example, perhaps the ocean was once much deeper and covered the mountaintop. On the other hand, maybe the mountaintop was once below sea level and pushed its way up to its present high altitude. Another explanation might be that someone left the fossil on the mountain as a practical joke. What would you look for to help you figure out what actually happened? In every rock and fossil, there are fascinating clues that help you read what happened in the past, including clues that tell you if an ocean had covered the area or if a mountain pushed its way up from lower levels. There are even clues that

tell you if a rock has been brought in from another place. This chapter is about some of the clues found in fossils and rocks, and what the clues mean.

EARLY IDEAS ABOUT FOSSILS

The story about finding a fossil fish on a mountain is not as far-fetched as it might seem, and in fact, one of the first recorded evidences of understanding the meaning of fossils took place in a similar setting. The ancient Greek historian Herodotus was among the first to realize that fossil shells found in rocks far from any ocean were remnants of organisms left by a bygone sea. Other Greek philosophers were not convinced that this conclusion was as obvious as it might seem today. Aristotle, for example, could see no connection between the shells of organisms of his time and the fossils, which he also believed to have formed inside the rocks. Note that he also believed that living organisms could arise by spontaneous generation from mud. A belief that the fossils must have "grown" in place in rocks would seem to be consistent with a belief in spontaneous generation.

It was a long time before it was generally recognized that fossils had anything to do with living things. Even when

FIGURE 21.1 The Painted Desert, Arizona. This landscape has a story to tell, and each individual rock and even the colors mean something about the past.

FIGURE 21.2 The fossil record of the hard parts is beautifully preserved, along with a carbon film, showing a detailed outline of the fish and some of its internal structure.

people started to recognize some similarities between living organisms and certain fossils, they did not make a connection. Fossils were considered to be the same as quartz crystals, or any other mineral crystals, meaning they were either formed with Earth or grew there later (depending on the philosophical view of the interpreter). Fossils of marine organisms that were well preserved and very similar to living organisms were finally accepted as remains of once-living organisms that were buried in Noah's flood. By the time of the Renaissance, some people were starting to think of other fossils, too, as the remains of former life forms. Leonardo da Vinci, like other Renaissance scholars, argued that fossils were the remains of organisms that had lived in the past.

By the early 1800s, the true nature of fossils was becoming widely accepted. William "Strata" Smith, an English surveyor, discovered at this time that sedimentary rock strata could be identified by the fossils they contained. Smith grew up in a region of England where fossils were particularly plentiful. He became a collector, keeping careful notes on where he found each fossil and in which type of sedimentary rock layers. During his travels, he discovered that the succession of rock layers on the south coast of England was the same as the succession of rock layers on the east coast. Through his keen observations, Smith found that each kind of sedimentary rock had a distinctive group of fossils that was unlike the group of fossils in other rock layers. Smith amazed his friends by telling them where and in what type of rock they had found their fossils.

Today, the science of discovering fossils, studying the fossil record, and deciphering the history of life from fossils is known as *paleontology*. The word *paleontology* was invented in 1838 by the British geologist Charles Lyell to describe this newly established branch of geology. It is derived from classic Greek roots and means "study of ancient life," and this requires a study of fossils.

People sometimes blur the distinction between *paleontology* and *archaeology*. Archaeology is the study of past human life and culture from material evidence of artifacts, such as graves, buildings, tools, pottery, landfills, and so on (*artifact* literally means "something made"). The artifacts studied in archaeology can be of any age, from the garbage added to the city landfill yesterday to the pot shards of an ancient tribe that disappeared hundreds of years ago. The word *fossil* originally meant "anything dug up" but today carries the meaning of any *evidence of ancient organisms in the history of life*. Artifacts are, therefore, not fossils, as you can see from the definitions.

TYPES OF FOSSILIZATION

Considering all the different things that can happen to the remains of an organism and considering the conditions needed to form a fossil, it seems amazing that any fossils are formed and then found. Consider, for example, the animals you see killed beside the road or the dead trees that fall over in a forest. Rarely do they become fossils, because scavengers eat the remains of dead animals and decay organisms break down the organic remains of plants and animals. As a result, very little digestible organic matter escapes destruction, but indigestible skeletal material, such as shells, bones, and teeth, has a much better chance of not being destroyed. Thus, a fossil is not likely to form unless there is rapid burial of a recently

TABLE 21.1

Summary of the types of fossil preservation

I. Preservation of all or part of the organism
  A. Unaltered
    1. Soft parts
    2. Hard parts
  B. Altered
    1. Mineralization
    2. Replacement
    3. Carbon films
II. Preservation of the organism's shape
  A. Cast
  B. Mold
III. Signs of activity
  A. Tracks
  B. Trails
  C. Burrows
  D. Borings
  E. Coprolites

FIGURE 21.3 This fly was stuck in and covered over with plant pitch. When the pitch fossilized to form amber, the fly was preserved as well. The entire fly is preserved in the amber.

deceased organism. The presence of hard parts, such as a shell or a skeleton, will also favor the formation of a fossil if there is rapid burial.

There are three broad ways in which fossils are commonly formed (Table 21.1):

1. preservation or alteration of hard parts.
2. preservation of the shape.
3. preservation of signs of activity.

Occasionally, entire organisms are preserved, and in rare cases, even the unaltered remains of an organism's *soft parts* are found. In order for this to occur, the organism must be quickly protected from scavengers and decomposers following death. There are several conditions that allow this to occur. The best examples of this uncommon method of fossilization include protection by freezing, entombing in tree resin, or embalming in tar. Mammoths, for example, have been found frozen and preserved by natural refrigeration in the ices of Alaska and Siberia. The body of a human from the Bronze Age has also been discovered frozen in ice. Insects and spiders, complete with delicate appendages, have been found preserved in amber, which is fossilized tree resin (Figure 21.3). The bones of saber-toothed tigers and other vertebrates were found embalmed in the tars of the La Brea tar pit in Los Angeles, California. In each case—ice, resin, and tar—the remains were protected from scavengers, insects, and bacteria.

Fossils are more commonly formed from *remains of hard parts* such as shells, bones, and teeth of animals or the pollen and spores of plants (Figure 21.4). Such parts are composed of calcium carbonate, calcium phosphate, silica, chitin, or other tough organic coverings. The fish fossil pictured in

Figure 21.2 is from Wyoming's Green River Formation. This freshwater fish died in its natural environment and was soon covered by fine-grained sediment. The sediment preserved the complete articulated skeleton, along with some carbon traces of soft tissue. Plant fossils are often found as carbon traces, sometimes looking like a photograph of a leaf on a slab of shale or limestone.

Shells and other hard parts of invertebrates are sometimes preserved without alteration or with changes in the chemical composition. Some protozoans and most corals, mollusks, and other shelled invertebrates have calcium carbonate shells, but some do have calcium phosphate shells. Silica makes up the hard shells of some protozoans, sponges, and diatoms. Diatoms

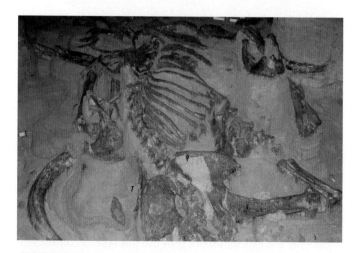

FIGURE 21.4 Skeletons of recent organisms are often preserved as fossils in sediments. These are the bones of a mammoth.

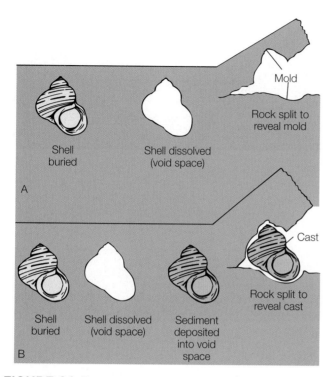

FIGURE 21.5 Origin of molds and casts. (*A*) Formation of a mold. (*B*) Formation of a cast.

are among the most abundant marine microorganisms. Silica is the most resistant common substance found in fossils. Chitin is the tough material that makes up the exoskeletons of insects, crabs, and lobsters.

Calcium carbonate shell material may be dissolved by groundwater in certain buried environments, leaving an empty *mold* in the rock. Sediment or groundwater deposits may fill the mold and make a *cast* of the organism (Figure 21.5).

Figure 21.6 is a photograph of part of the Petrified Forest National Park in Arizona. There are two processes involved in

FIGURE 21.6 Petrified fossils are the result of an ion-for-ion replacement of the buried organisms with mineral material. These logs of petrified wood are in the Petrified Forest National Park, Arizona.

the making of petrified fossils, and they are not restricted to just wood. The processes involve (1) *mineralization,* which is the filling of pore spaces with deposits of calcium carbonate, silica, or pyrite, (2) *replacement,* which is the dissolving of the original material and depositing of new material an ion at a time, or (3) both mineralization and replacement. Petrified wood is formed by both processes over a long period of time. As it decayed, the original wood was replaced by mineral matter an ion at a time. Over time, the "mix" of minerals being deposited changed and the various resulting colors appear to preserve the texture of the wood. Since the mineralization or replacement processes take place an ion at a time, a great deal of detail can be preserved in such fossils. The size and shape of the cells and growth rings in petrified wood are preserved well enough that they can be compared to modern plants. The skeletons or shells of many extinct organisms are also typically preserved in this way. The "fossilized bones" of dinosaurs or the "fossilized shells" of many invertebrates are examples.

Finally, there are many kinds of fossils that are not the preserved remains of an organism but are preserved indications of the activities of organisms. Some of the most interesting of such fossils are the footprints of various kinds of animals. There are examples of footprints of dinosaurs and various kinds of extinct mammals, including ancestors of humans. If an animal walked through mud or other soft substrates and the substrates were covered with silt or volcanic ash, the pattern of the footprints can be preserved. The tunnels of burrowing animals, the indentations left by crawling worms, and the nests of dinosaurs have been preserved in a similar manner. Even the eggs of dinosaurs and other animals have been preserved in a mineralized form.

As you can see, there are many different ways in which a fossil can be formed, but it must be *found and studied* if it is to reveal its part in the history of life. This means the rocks in which the fossil formed must now somehow make it back to the surface of Earth. This usually involves movement and uplift of the rock and weathering and erosion of the surrounding rock to release or reveal the fossil. Most fossils are found in recently eroded sedimentary rocks—sometimes atop mountains that were under the ocean a long, long time ago. The complete record of what has happened in the past is not found in the fossil alone but requires an understanding of the layers of rocks present, the relationship of the layers to each other, and their age.

CONCEPTS *Applied*

Find Fossils

Collect fossils from roadcuts and old quarries in your area. Make an exhibit showing the fossils with sketches of what the animals or plants were like. What do these particular animals and plants tell you about the history of your area?

READING ROCKS

Reading history from the rocks of Earth's crust requires both a feel for the immensity of geologic time and an understanding of geologic processes. By "geologic time," we mean the age of Earth, the very long span of Earth's history. This span of time is difficult for most of us to comprehend. Human history is measured in units of hundreds and thousands of years, and even the events that can take place in a thousand years are hard to imagine. Geologic time, on the other hand, is measured in units of millions and billions of years.

The understanding of geologic processes has been made possible through the development of various means of measuring ages and time spans in geologic systems. An understanding of geologic time leads to an understanding of geologic processes, which then leads to an understanding of the environmental conditions that must have existed in the past. Thus, the mineral composition, texture, and sedimentary structure of rocks are clues to past events, events that make up the history of Earth.

ARRANGING EVENTS IN ORDER

The clues provided by thinking about geologic processes that must have occurred in the past are interpreted within a logical frame of reference that can be described by several basic principles. The following is a summary of these basic guiding principles that are used to read a story of geologic events from the rocks.

Recall that the cornerstone of the logic used to guide thinking about geologic time is the **principle of uniformity.** As described in chapter 19, this principle is sometimes stated as "the present is the key to the past." This means that the geologic features that you see today have been formed in the past by the same processes of crustal movement, erosion, and deposition that are observed today. By studying the processes now shaping Earth, you can understand how it has evolved through time. This principle establishes the understanding that the surface of Earth has been continuously and gradually modified over the immense span of geologic time.

The **principle of original horizontality** is a basic principle that is applied to sedimentary rocks. It is based on the observation that, on a large scale, sediments are commonly deposited in flat-lying layers. Old rocks are continually being changed to new ones in the continuous processes of crustal movement, erosion, and deposition. As sediments are deposited in a basin of deposition, such as a lake or ocean, they accumulate in essentially

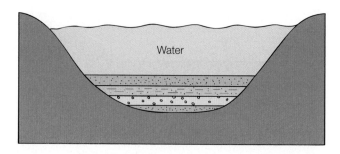

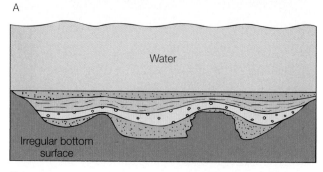

FIGURE 21.7 The principle of original horizontality. (*A*) Sediments tend to be deposited in horizontal layers. (*B*) Even where the sediments are draped over an irregular surface, they tend toward the horizontal.

flat-lying, approximately horizontal layers (Figure 21.7). Thus, any layer of sedimentary rocks that is not horizontal has been subjected to forces that have deformed Earth's surface.

The **principle of superposition** is another logical and obvious principle that is applied to sedimentary rocks. Layers of sediments are usually deposited in succession in horizontal layers, which later are compacted and cemented into layers of sedimentary rock. An undisturbed sequence of horizontal layers is thus arranged in chronological order with the oldest layers at the bottom. Each consecutive layer will be younger than the one below it (Figure 21.8). This is true, of course,

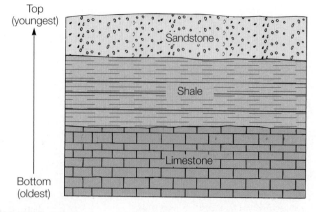

FIGURE 21.8 The principle of superposition. In an undisturbed sedimentary sequence, the rocks on the bottom were deposited first, and the depositional ages decrease as you progress to the top of the pile.

FIGURE 21.9 The Grand Canyon, Arizona, provides a majestic cross section of horizontal sedimentary rocks. According to the principle of superposition, traveling deeper and deeper into the Grand Canyon means that you are moving into older and older rocks.

only if the layers have not been turned over by deforming forces (Figure 21.9).

The **principle of crosscutting relationships** is concerned with igneous and metamorphic rock, in addition to sedimentary rock layers. Any geologic feature that cuts across or is intruded into a rock mass must be younger than the rock mass. Thus, if a fault cuts across a layer of sedimentary rocks, the fault is the youngest feature. Faults, folds, and igneous intrusions are always younger than the rocks they originally occur in. Often, there is a further clue to the correct sequence: The hotter igneous rock may have "baked," or metamorphosed, the surrounding rock immediately adjacent to it, so again the igneous rock must have come second (Figure 21.10).

Shifting Sites of Erosion and Deposition

The principle of uniformity states that Earth processes going on today have always been occurring. This does not mean,

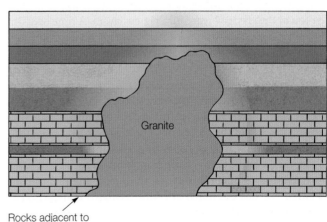

Rocks adjacent to intruding magma may also be metamorphosed by its heat.

FIGURE 21.10 A granite intrusion cutting across older rocks.

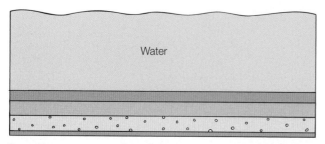

Deposition

Erosional surface

Rocks tilted, eroded

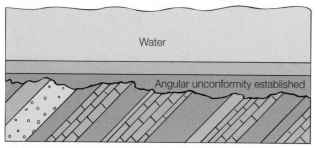

Water

Angular unconformity established

Subsequent deposition

FIGURE 21.11 Angular unconformity. Development involves some deformation and erosion before sedimentation is resumed.

FIGURE 21.12 A time break in the rock record in the Grand Canyon, Arizona. The horizontal sedimentary rock layers overlie almost vertically foliated metamorphic rocks. Metamorphic rocks form deep in Earth, so they must have been uplifted, and the overlying material eroded away before being buried again.

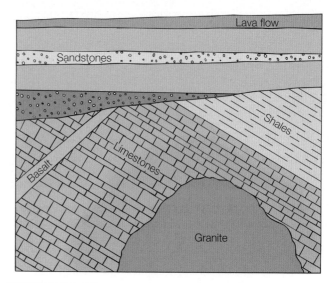

FIGURE 21.13 Deciphering a complex rock sequence. The limestones must be oldest (law of superposition), followed by the shales. The granite and basalt must both be younger than the limestone they crosscut (note the metamorphosed zone around the granite). It is not possible to tell whether the igneous rocks predate or postdate the shales or to determine whether the sedimentary rocks were tilted before or after the igneous rocks were emplaced. After the limestones and shales were tilted, they were eroded, and then the sandstones were deposited on top. Finally, the lava flow covered the entire sequence.

however, that they always occur in the same place. As erosion wears away the rock layers at a site, the sediments produced are deposited someplace else. Later, the sites of erosion and deposition may shift, and the sediments are deposited on top of the eroded area. When the new sediments later are formed into new sedimentary rocks, there will be a time lapse between the top of the eroded layer and the new layers. A time break in the rock record is called an **unconformity.** The unconformity is usually shown by a surface within a sedimentary sequence on which there was a lack of sediment deposition or where active erosion may even have occurred for some period of time. When the rocks are later examined, that time span will not be represented in the record, and if the unconformity is erosional, some of the record once present will have been lost. An unconformity may occur within a sedimentary sequence of the same kind or between different kinds of rocks. The most obvious kind of unconformity to spot is an **angular unconformity.** An angular unconformity, as illustrated in Figure 21.11, is one in which the bedding planes above and below the unconformity are not parallel. An angular unconformity usually implies some kind of tilting or folding, followed by a significant period of erosion, which in turn was followed by a period of deposition (Figure 21.12).

CORRELATION

The principle of superposition, the principle of crosscutting relationships, and the presence of an unconformity all have meaning about the order of geologic events that have occurred in the past. This order can be used to unravel a complex sequence of events such as the one shown in Figure 21.13. The presence of fossils can help, too. The **principle of faunal succession**

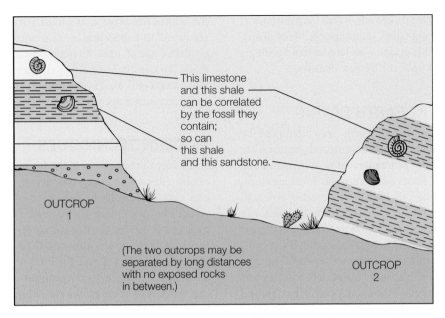

This limestone
and this shale
can be correlated
by the fossil they
contain;
so can
this shale
and this sandstone.

OUTCROP
1

(The two outcrops may be
separated by long distances
with no exposed rocks
in between.)

OUTCROP
2

FIGURE 21.14 Similarity of fossils suggests similarity of ages, even in different rocks widely separated in space.

recognizes that life-forms have changed through time. Old life-forms disappear from the fossil record and new ones appear, but the same form is never exactly duplicated independently at two different times in history. This principle implies that the same type of fossil organisms that lived only a brief geologic time should occur only in rocks that are the same age. According to the principle of faunal succession, then, once the basic sequence of fossil forms in the rock record is determined, rocks can be placed in their correct relative chronological position on the basis of the fossils contained in them. The principle also means that if the same type of fossil organism is preserved in two different rocks, the rocks should be the same age. This is logical even if the two rocks have very different compositions and are from places far, far apart (Figure 21.14).

Distinctive fossils of plant or animal species that were distributed widely over Earth but lived only a brief time with a common extinction time are called **index fossils.** Index fossils, together with the other principles used in reading rocks, make it possible to compare the ages of rocks exposed in two different locations. This is called age **correlation** between rock units. Correlations of exposed rock units separated by a few kilometers are easier to do, but correlations have been done with exposed rock units that are separated by an ocean. Correlation allows the ordering of geologic events according to age. Since this process is only able to determine the age of a rock unit or geologic event relative to some other unit or event, it is called *relative dating* (Figure 21.15). Dates with numerical ages are determined by means different from correlation.

The usefulness of correlation and relative dating through the concept of faunal succession is limited because the principles can be applied only to rocks in which fossils are well preserved, which are almost exclusively sedimentary rocks. Correlation also can be based on the occurrence of unusual rock types,

FIGURE 21.15 This dinosaur footprint is in shale near Tuba City, Arizona. It tells you something about the relative age of the shale, since it must have been soft mud when the dinosaur stepped here.

distinctive rock sequences, or other geologic similarities. All this is useful in clarifying relative age relationships among rock units. It is not useful in answering questions about the age of rocks or the time required for certain events, such as the eruption of a volcano, to occur. Questions requiring numerical answers went unanswered until the twentieth century.

GEOLOGIC TIME

How do you measure and track time intervals for something as old as Earth? First, you would need to know the age of Earth; then you would need some consistent, measurable events to divide the overall age into intervals. Questions about the age of

Earth have puzzled people for thousands of years, dating back at least to the time of the ancient Greek philosophers. Many people have attempted to answer this question and understand geologic time but with little success until the last few decades.

EARLY ATTEMPTS AT EARTH DATING

One early estimate of the age of Earth was attempted by Archbishop Ussher of Ireland in the seventeenth century. He painstakingly counted up the generations of people mentioned in biblical history, added some numerological considerations, and arrived at the conclusion that Earth was created at 9 A.M. on Tuesday, October 26, in the year 4004 B.C. On the authority of biblical scholars, this date was generally accepted for the next century or so, even though some people thought that the geology of Earth seemed to require far longer to develop. The date of 4004 B.C. meant that Earth and all of the surface features had formed over a period of about six thousand years. This required a model of great cataclysmic catastrophes to explain how all Earth's features could possibly have formed over a span of six thousand years.

Near the end of the eighteenth century, James Hutton reasoned out the principle of uniformity and people began to assume a much older Earth. The problem then became one of finding some uniform change or process that could serve as a geologic clock to measure the age of Earth. To serve as a geologic clock, a process or change would need to meet three criteria: (1) the process must have been operating since Earth began, (2) the process must be uniform or at least subject to averaging, and (3) the process must be measurable.

During the nineteenth century, many attempts were made to find Earth processes that would meet the criteria to serve as a geologic clock. Among others, the processes explored were (1) the rate that salt is being added to the ocean, (2) the rate that sediments are being deposited, and (3) the rate that Earth is cooling. Comparing the load of salts being delivered to the ocean by all the rivers, and assuming the ocean was initially pure water, it was calculated that about 100 million years would be required for the present salinity to be reached. The calculations did not consider the amount of materials being removed from the ocean by organisms and by chemical sedimentation, however, so this technique was considered to be unacceptable. Even if the amount of materials removed were known, it would actually result in the age of the ocean, not the age of Earth.

A number of separate and independent attempts were made to measure the rate of sediment deposition, then compare that rate to the thickness of sedimentary rocks found on Earth. Dividing the total thickness by the rate of deposition resulted in estimates of an Earth age that ranged from about 20 to 1,500 million years. The wide differences occurred because there are gaps in many sedimentary rock sequences, periods when sedimentary rocks were being eroded away to be deposited elsewhere as sediments again. There were just too many unknowns for this technique to be considered acceptable.

The idea of measuring the rate that Earth is cooling for use as a geologic clock assumed that Earth was initially a molten mass that has been cooling ever since. Calculations estimating

the temperature that Earth must have been to be molten were compared to Earth's present rate of cooling. This resulted in an estimated age of 20 to 40 million years. These calculations were made back in the nineteenth century before it was understood that natural radioactivity is adding heat to Earth's interior, so it has required much longer to cool down to its present temperature.

MODERN TECHNIQUES

Soon after the beginning of the twentieth century, the discovery of the radioactive decay process in the elements of minerals and rocks led to the development of a new, accurate geologic clock. This clock finds the *radiometric age* of rocks in years by measuring the radioactive decay of unstable elements within the crystals of certain minerals. Since radioactive decay occurs at a constant, known rate, the ratio of the remaining amount of an unstable element to the amount of decay products present can be used to calculate the time that the unstable element has been a part of that crystal (see chapter 13). Certain radioactive isotopes of potassium, uranium, and thorium are often included in the minerals of rocks, so they are often used as "radioactive clocks." By using radiometric aging techniques along with other information, we arrive at a generally accepted age for Earth of about 4.5 billion years. It should be noted that radiometric aging is only useful in aging igneous rocks, since sedimentary rocks are the result of weathering and deposition of other rocky materials. Table 21.2 lists several radioactive isotopes, their decay products, and half-lives. Often two or more isotopes are used to determine an age for a rock. Agreement between them increases the scientist's confidence in the estimates of the age of the rock. Because there are great differences in the half-lives, some are useful for dating things back to several billion years, while others, such as carbon-14, are only useful for dating things to perhaps fifty thousand years. Carbon-14 is not used to age rocks, but is very useful in aging materials that are of relatively recent biological origin, since carbon is an important part of all living things. Also, a slightly different method is used to determine carbon-14 dating. (See "A Closer Look: Carbon Dating?" on p.337).

A recently developed geologic clock is based on the magnetic orientation of magnetic minerals. These minerals become

TABLE 21.2

Radioactive isotopes and half-lives

Radioactive Isotope	Stable Daughter Product	Half-Life
Samarium-147	Neodymium-143	106 billion years
Rubidium-87	Strontium-87	48.8 billion years
Thorium-232	Lead-208	14.0 billion years
Uranium-238	Lead-206	4.5 billion years
Potassium-40	Argon-40	1.25 billion years
Uranium-235	Lead-207	704 million years
Carbon-14	Nitrogen-14	5,730 years

aligned with Earth's magnetic field when the igneous rock crystallizes, making a record of the magnetic field at that time. Earth's magnetic field is global and has undergone a number of reversals in the past. A *geomagnetic time scale* has been established from the number and duration of magnetic field reversals occurring during the past 6 million years. Combined with radiometric age dating, the geomagnetic time scale is making possible a worldwide geologic clock that can be used to determine local chronologies.

THE GEOLOGIC TIME SCALE

A yearly calendar helps you keep track of events over periods of time by dividing the year into months, weeks, and days. In a similar way, the **geologic time scale** helps you keep track of events that have occurred in Earth's geologic history. The first development of this scale came from the work of William "Strata" Smith, the English surveyor described in the section on fossils earlier in this chapter. Recall that Smith discovered that certain rock layers in England occurred in the same order, top to bottom, wherever they were located. He also found that he could correlate and identify each layer by the kinds of fossils in the rocks of the layers. In 1815, he published a geologic map of England, identifying the rock layers in a sequence from oldest to youngest. Smith's work was followed by extensive geological studies of the rock layers in other countries. Soon it was realized that similar, distinctive index fossils appeared in rocks of the same age when the principle of superposition was applied. For example, the layers at the bottom contained fossils of trilobites (Figure 21.16A), but trilobites were not found in the upper levels. (Trilobites are extinct marine arthropods that may be closely related to living scorpions, spiders, and horseshoe crabs.) On the other hand, fossil shells of ammonites (Figure 21.16B) appeared in the middle levels but not the lower nor upper levels of the rocks. The topmost layer was found to contain the fossils of animals identified as still living today. The early appearance and later disappearance of fossils in progressively younger rocks is explained by organic evolution and extinction, events that could be used to mark the time boundaries of Earth's geologic history.

The major blocks of time in Earth's geologic history are called **eons,** and there have been four eons since the origin of Earth. The **Phanerozoic** (Greek for "visible life") eon is the geologic time of the abundant fossil record and living organisms. The eon before this time of visible life is called the **Proterozoic,** which is Greek for "beginning life." The other two eons are the **Prearchean** and the **Archean** (Figure 21.17).

The Prearchean, Archean, and Proterozoic eons together are know as the **Precambrian,** which refers to the time before the time of life. The Precambrian denotes the largest block of time in Earth's history—more than 85 percent of the total time.

The Phanerozoic eon—the time of visible life—is divided into blocks of time called **eras,** and each era is identified by the appearance and disappearance of particular fossils in the sedimentary rock record. There are three main eras, and the pie chart in Figure 21.18 shows how long each era lasted compared to the Precambrian. The eras are: (1) **Cenozoic,** which refers to the time of recent life. Recent life means that the fossils for

A Trilobites

B Ammonites

FIGURE 21.16 Since both trilobites (*A*) and ammonites (*B*) were common in the oceans throughout the world, they have been used as index fossils. Trilobites are always found in older rocks than are ammonites. Trilobites were common in the Paleozoic era, and ammonites were common in the earlier parts of the Mesozoic era.

this time period are similar to the life found on Earth today. (2) **Mesozoic,** which refers to the time of middle life. Middle life means that some of the fossils for this time period are similar to the life found on Earth today, but many are different from anything living today. (3) **Paleozoic,** which refers to the time of ancient life. Ancient life means that the fossils for this time period are very different from anything living on Earth today. The eras are divided into blocks of time called **periods** (Figure 21.19), and the periods are further subdivided into smaller blocks of time called **epochs.**

GEOLOGIC PERIODS AND TYPICAL FOSSILS

The Precambrian period of time contains the earliest fossils. The Precambrian fossils that have been found are chiefly those of deposits from bacteria, algae, a few fungi, unusual

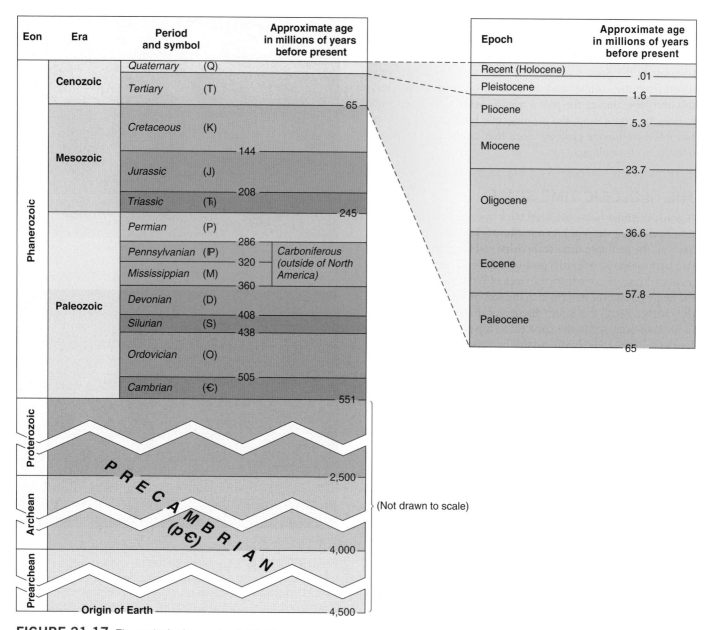

Eon	Era	Period and symbol		Approximate age in millions of years before present
Phanerozoic	Cenozoic	Quaternary	(Q)	
		Tertiary	(T)	65
	Mesozoic	Cretaceous	(K)	144
		Jurassic	(J)	208
		Triassic	(Ŧ)	245
	Paleozoic	Permian	(P)	286
		Pennsylvanian	(IP)	320
		Mississippian	(M)	360
		Devonian	(D)	408
		Silurian	(S)	438
		Ordovician	(O)	505
		Cambrian	(€)	551

Carboniferous (outside of North America)

Epoch	Approximate age in millions of years before present
Recent (Holocene)	.01
Pleistocene	1.6
Pliocene	5.3
Miocene	23.7
Oligocene	36.6
Eocene	57.8
Paleocene	65

	Precambrian (p€)	
Proterozoic		2,500
Archean		4,000
Prearchean		

Origin of Earth — 4,500

(Not drawn to scale)

FIGURE 21.17 The geologic time scale. Modified from "Decade of North American Geology," 1983 Geologic Time Scale—Geological Society of America.

soft-bodied animals, and the burrow holes of worms. It appears that there were no animals with hard parts; thus, the fossil record is incomplete since it is the hard parts of animals or plants that form fossils, usually after rapid burial. Another problem in finding fossils of soft-bodied or extremely small life-forms in these extremely old rocks is that heat and pressure have altered many of the ancient rocks over time, destroying any fossil evidence that may have been present.

The Paleozoic era was a time when there was great change in the kinds of plants and animals present. In general, the earliest abundant fossils are found in rocks from the Cambrian period at the beginning of the Paleozoic era (see Figure 21.17). These rocks show an abundance of oceanic life that represents all the major groups of marine animals found today. There is no fossil evidence of life of any kind living on the land during the Cambrian. The dominant life-forms of the Cambrian ocean were echinoderms, mollusks, trilobites, and brachiopods. The trilobites, now extinct, made up more than half of the kinds of living things during the Cambrian.

During the Ordovician and Silurian periods, most living things were still marine organisms, with various kinds of jawless fish becoming prominent. Also, by the end of the Silurian, some primitive plants were found on land. The Devonian period saw the further development of different kinds of fish, including those that had jaws, and many kinds of land plants and animals. Coral reefs were also common in

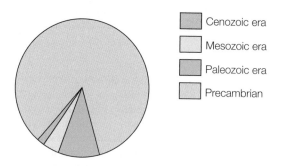

FIGURE 21.18 Geologic history is divided into four main eras. The Precambrian denotes the first 4 billion years, or about 85 percent of the total 4.5 billion years of geologic time. The Paleozoic lasted about 10 percent of geologic time, the Mesozoic about 4 percent, and the Cenozoic only about 1.5 percent of all geologic time.

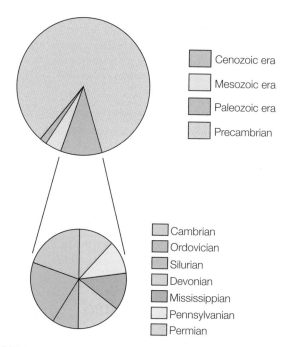

FIGURE 21.19 The periods of the Paleozoic era, which refers to the time of ancient life. Ancient life means that the fossils for this time period are very different from anything living on Earth today. Each period is characterized by specific kinds of plants and animals. This pie chart compares the relative time that each period lasted.

the Devonian. The Carboniferous period was a time of vast swamps of ferns, horsetails, and other primitive nonseed plants that would form great coal deposits. Fossils of the first reptiles and the first winged insects are found in rocks from this age. The Paleozoic era closed with the extinction at the end of the Permian period of about 90 percent of plant and animal life of that time.

The Mesozoic era was a time when the development of life on land flourished. The dinosaurs first appeared in the Triassic period, outnumbering all the other reptiles until the close of the Mesozoic. Fossils of the first mammals and modern forms

of gymnosperms (cone-bearing plants) developed in the Triassic. The first flowering plants, the first deciduous trees, and the first birds appeared in the Cretaceous period. The Cretaceous is the final period of the Mesozoic era and is characterized by the dominance of dinosaurs and extensive evolution of flowering plants and the insects that pollinate them. Birds and mammals also increased in variety. Like the close of the Paleozoic, the Mesozoic era ended with a great dying of land and marine life that resulted in the extinction of many species, including the dinosaurs.

As the Cenozoic era opened, the dinosaurs were extinct and the mammals became the dominant vertebrate life-form. The Cenozoic is thus called the "Age of the Mammals." However, there were also major increases in the kinds of insects, flowering plants (particularly grasses), and birds. Finally, toward the end of this period of time, humans arrived on the scene, and many other kinds of large mammals such as mammoths, mastodons, giant ground sloths, and saber-toothed tigers went extinct.

MASS EXTINCTIONS

When we look at the fossil record, there is evidence of several mass extinctions. Five are recognized for causing the extinction of 50 percent or more of the species present. It is important to understand that although these extinctions were "sudden" in geologic time, they occurred over millions of years. Each resulted in a change in the kinds of organisms present with major groups going extinct and the evolution of new kinds of organisms. The boundaries between many of the geologic periods are defined by major extinction events. Geologists have developed theories about the causes of each of these mass extinctions. Many of these theories involve changes in the size and location of continents as a result of plate tectonics.

The mass extinction at the end of the Ordovician period resulted in the extinction of 60 percent of genera of organisms. At that time, most organisms lived in the oceans. It is thought that the large continent of Pangaea migrated to the South Pole and this resulted in the development of large glaciers and a drastic drop in sea level along with a cooling of the waters.

At the end of the Devonian period, there was a mass extinction that affected primarily marine organisms. Approximately 60 percent of genera went extinct. Since many of the organisms that went extinct were warm-water, marine organisms, glaciation along with a cooling of the oceans is a widely held theory for the cause of this extinction.

The mass extinction at the end of the Permian period is unusual in several ways. It resulted in the extinction of about 90 percent of organisms and took place over a very short time— less than a million years. Because of this, it is often referred to as the "Great Dying." Both marine and terrestrial organisms were affected. Because this extinction event occurred over a short time and affected all species, it is assumed that a major, worldwide event was responsible. However, at this time, there is no clearly identifiable cause. Suggestions include a meteorite impact, massive volcanic activity, or a combination of factors.

Eduard Suess was an Austrian geologist who helped pave the way for modern theories of the continents. Born in London, though of Bohemian ancestry, Suess was educated in Vienna and at the University of Prague. He moved to Vienna in 1856 and became professor of geology there in 1861. In addition to his geological interests, he occupied himself with public affairs, serving as a member of the Reichstag for twenty-five years. His geological researches took several directions. As a paleontologist, he investigated graptolites, brachiopods, ammonites, and the fossil mammals of the Danube Basin. He wrote an original text on economic geology. He undertook important research on the structure of the Alps, the tectonic geology of Italy, and seismology. The possibility of a former land bridge between North Africa and southern Europe caught his attention.

North Africa and southern Europe caught his attention.

The outcome of these interests was *The Face of Earth* (1885–1909), a massive work devoted to analyzing the physical agencies contributing to Earth's geographical evolution. Suess offered an encyclopedic view of crustal movement, the structure and grouping of mountain chains, sunken continents, and the history of the oceans. He also made significant contributions to rewriting the structural geology of each continent. In many respects, Suess cleared the path for the new views associated with the theory of continental drift in the twentieth century. In view of geological similarities among parts of the southern continents, Suess suggested that there had once been a great supercontinent, made up of the present southern continents; this he named Gondwanaland, after a region of India. Wegener's work was later to establish the soundness and penetration of such speculations.

Source: From the Hutchinson *Dictionary of Scientific Biography.* © Research Machines plc 2003. All Rights Reserved. Helicon Publishing is a division of Research Machines.

The extinction at the end of the Triassic period was relatively mild compared to others. About 50 percent of species appear to have gone extinct. There is no clear cause for this extinction.

The mass extinction at the end of the Cretaceous period resulted in the extinction of about 60 percent of species. Based on evidence of a thin clay layer marking the boundary between the Cretaceous and Tertiary periods, one theory proposes that a huge (16 km, or 9.9 mi, diameter and 10^{15} kg, or 1.1×10^{12} tons, mass) meteorite struck Earth. The impact would have thrown a tremendous amount of dust into the atmosphere, obscuring the Sun and significantly changing the climate and thus the conditions of life on Earth. The resulting colder climate may have led to the extinction of many plant and animal species, including the dinosaurs. This theory is based on the clay layer, which theoretically formed as the dust settled, and its location in the rock record at the time of the extinctions. The layer is enriched with a rare metal, iridium, which is not found on Earth in abundance but occurs in certain meteorites in greater abundance.

As we approach current times, the extinction of the many species of large mammals during the Quaternary period is thought to be due to either a major change in climate at the end of the last ice age or hunting by humans as they expanded their range from Africa to Europe, Asia, and the Americas. Many people are convinced that we are currently experiencing a mass extinction because of our ability to alter the face of Earth and destroy the habitats needed by plants and animals.

CONCEPTS *Applied*

Fossil Animals

You have no doubt seen many illustrations of various kinds of extinct animals, for example, the different kinds of dinosaurs. Have you ever wondered how anyone can know what these animals looked like? Certainly the general size and shape can be determined by assembling fossils of individual skeletal parts, but most of the shape of an animal is determined by the size and distribution of its muscles. An understanding of physics provides part of the answer. The skeleton of an animal is basically arranged like a large number of levers. For example, a leg and foot consists of several bones linked end to end. The size of a bone gives a clue to the amount of mass that particular bone had to support. Various bumps on the bones are places where muscles and tendons attached. By knowing where the muscles are attached and analyzing the kinds of levers they represent, we can calculate the size of the muscle. Thus, the general shape of the animal can be estimated. However, this doesn't provide any information about the color or surface texture of the animal. That information must come from other sources.

INTERPRETING GEOLOGIC HISTORY— A SUMMARY

When interpreting the fossil record in any part of the world, there are several things to keep in mind:

1. *We are dealing with long periods of time.* The history of life on Earth goes back to 3.5 billion years ago, and the evolution of humans took place over a period of several million years. This is important because many of the processes of sedimentation, continental drift, and climate change took place slowly over many millions of years.

2. *Earth has changed greatly over its history.* There have been repeated periods of warming and cooling, and some cooling periods resulted in the formation of glaciers. Since glaciers tie up water, they had the effect of lowering sea level, which in turn exposed more land and changed the climate of continents. In addition, the continents were not fixed in position. Changes in position affected the climate that the continent experienced. For example, at one time what is now North America was attached to Antarctica near the south pole.

3. *There have been many periods in the history of Earth when most of the organisms went extinct.* Cooling climates, changes in sea level, and meteorite impacts are all suspected of causing mass extinctions. However, it should not be thought that these were sudden extinctions. Most took place over millions of years. Some of the extinctions affected as much as 90 percent of the things living at the time.

4. *New forms of life evolved that replaced those that went extinct.* The earliest organisms we see in the fossil record were marine organisms similar to present-day bacteria. The oldest fossils of these organisms date to about 3.5 billion years ago. The first multicellular organisms were present by about 1 billion years ago. The development of multicellular organisms ultimately led to the colonization of land by plants and animals, with plants colonizing about 500 million years ago and animals at about 450 million years ago.

5. *Although there were massive extinctions, there are many examples of the descendants of early life-forms present today.* Bacteria and many kinds of simple organisms are extremely common today, as are various kinds of algae and primitive forms of plants. In the oceans, many kinds of marine animals such as starfish, jellyfish, and clams are descendants of earlier forms.

6. *The kinds of organisms present have changed the nature of Earth.* The oxygen in the atmosphere is the result of the process of photosynthesis. Its presence has altered the amount of ultraviolet radiation reaching Earth. Plants tend to reduce the erosive effects of running water, and humans have significantly changed the surface of Earth.

SUMMARY

A *fossil* is any evidence of former life. This evidence could be in the form of *actual remains, altered remains, preservation of the shape of an organism,* or *any sign of activity (trace fossils).* Actual remains of former organisms are rare, occurring usually from protection of remains by freezing, entombing in tree resin, or embalming in tar. Remains of organisms are sometimes altered by groundwater in the process of *mineralization,* deposition of mineral matter in the pore spaces of an object. *Replacement* of original materials occurs by dissolution and deposition of mineral matter. *Petrified wood* is an example of mineralization and replacement of wood. Removal of an organism may leave a *mold,* a void where an organism was buried. A *cast* is formed if the void becomes filled with mineral matter.

Clues provided by geologic processes are interpreted within a logical framework of references to read the story of geologic events from the rocks. These clues are interpreted within a frame of reference based on (1) the *principle of uniformity,* (2) the *principle of original horizontality,* (3) the *principle of superposition,* (4) the *principle of crosscutting relationships,* (5) the fact that *sites of past erosion and deposition have shifted* over time (shifting sites produce an *unconformity,* or break in the rock record when erosion removes part of the rocks), and (6) the *principle of faunal succession.*

Geologic time is measured through the radioactive decay process, determining the *radiometric age* of rocks in years. Correlation and the determination of the numerical ages of rocks and events has led to the development of a *geologic time scale.* The major blocks of time on this calendar are called *eras.* The eras are the (1) *Cenozoic,* the time of recent life, (2) *Mesozoic,* the time of middle life, (3) *Paleozoic,* the time of ancient life, and (4) *Precambrian,* the time before the time of ancient life. The eras are divided into smaller blocks of time called *periods,* and the periods are further subdivided into *epochs.* The fossil record is seen to change during each era, ending with *great extinctions* of plant and animal life.

KEY TERMS

APPLYING THE CONCEPTS

1. Evidence of former life is called a
   a. ghost.
   b. fossil.
   c. history.
   d. relic.

2. In the early 1800s, William Smith noted that sedimentary rock layers could be identified by
   a. location.
   b. color.
   c. fossils.
   d. size.

3. The science of discovering fossils and their history is
   a. history.
   b. paleontology.
   c. archeology.
   d. anthropology.

4. The meaning of the word *fossil* is
   a. petrified rock.
   b. ancient artifact.
   c. petrified bones and the study of these bones.
   d. evidence of ancient organisms.

5. There are three methods of fossil formation. The method that is *not* a method of fossil formation is
   a. preservation of soft organic materials.
   b. preservation of hard parts.
   c. preservation of signs of activity.
   d. replacement.

6. Hard parts of organisms that form fossils do not include
   a. shells.
   b. teeth.
   c. plants.
   d. bones.

7. Fossils are most often found in what type of rock?
   a. sedimentary
   b. metamorphic
   c. igneous
   d. volcanic

8. "Geologic time" means
   a. the relative age of humankind.
   b. the relative age of fossils.
   c. the relative age of Earth.
   d. the relative age of the universe.

9. Clues to the past that we get from rocks do not include
   a. mineral composition.
   b. place of origin.
   c. rock texture.
   d. sedimentary structure.

10. A time break in the rock record is called a (an)
    a. unconformity.
    b. time lag.
    c. crosscutting relationship.
    d. time warp.

11. Distinctive fossils of plants or animals that were distributed widely but lived only a short time with a common extinction time are called
    a. short-lived fossils.
    b. correlation fossils.
    c. index fossils.
    d. faunal successive fossils.

12. The ability to use index fossils along with the ages of rocks in different locations is
    a. age correlation.
    b. fossil association.
    c. age comparison.
    d. relative dating.

13. The process that is adding heat to the interior of Earth is
    a. conduction.
    b. revolution.
    c. radioactive decay.
    d. solar energy.

14. Modern geologic clocks include
    a. radiometric dating.
    b. geomagnetic measurements.
    c. both a and b.
    d. neither a nor b.

15. Major blocks of time in the geologic time scale are called
    a. decades.
    b. eons.
    c. centuries.
    d. millennia.

16. The time before the time of life is called
    a. Precambrian.
    b. Postcambrian.
    c. Phanerozoic.
    d. Post-Phanerozoic.

17. The smallest block of time assigned to the geologic time period is the
    a. eon.
    b. period.
    c. era.
    d. epoch.

18. Which of the following does not describe a unit of geologic time?
    a. eon
    b. epoch
    c. period
    d. century

19. The time of recent life era is the
    a. Cenozoic.
    b. Mesozoic.
    c. Paleozoic.
    d. Phanerozoic.

20. The earliest fossils represented life from
    a. land.
    b. oceans.
    c. air.
    d. All of the above are correct.

21. Mammals first appeared in what era?
    a. Cenozoic
    b. Mesozoic
    c. Paleozoic
    d. Phanerozoic

22. Pangaea formed approximately
    a. 25 million years ago.
    b. 65 million years ago.
    c. 225 million years ago.
    d. 3.5 billion years ago.

23. A fossil is
    a. actual remains of plants or animals such as shells or wood.
    b. remains that have been removed and replaced by mineral matter.
    c. any sign of former life older than ten thousand years.
    d. Any of the above is correct.

24. Some of the oldest fossils are about how many years old?
    a. 4.55 billion
    b. 3.5 billion
    c. 250 million
    d. ten thousand

25. According to the evidence, a human footprint found preserved in baked clay is about five thousand years old. According to the definitions used by paleontologists, this footprint is a (an)
    a. fossil since it is an indication of former life.
    b. mold since it preserves the shape.
    c. actual fossil since it is preserved.
    d. None of the above is correct.

26. A fossil of a fly, if found, would most likely be
    a. a cast.
    b. formed by mineralization.
    c. preserved in amber.
    d. Any of the above is correct.

27. Which of the basic guiding principles used to read a story of geologic events tells you that layers of undisturbed sedimentary rocks have progressively older layers as you move toward the bottom?
    a. superposition
    b. horizontality
    c. crosscutting relationships
    d. None of the above is correct.

28. In any sequence of sedimentary rock layers that has not been subjected to stresses, you would expect to find
    a. essentially horizontal stratified layers.
    b. the oldest layers at the bottom and the youngest at the top.
    c. no faults, folds, or intrusions in the rock layers.
    d. All of the above are correct.

29. An unconformity is a
    a. rock bed that is not horizontal.
    b. rock bed that has been folded.
    c. rock sequence with rocks missing from the sequence.
    d. All of the above are correct.

30. Correlation and relative dating of rock units are made possible by application of the
    a. principle of crosscutting relationships.
    b. principle of faunal succession.
    c. principle of superposition.
    d. All of the above are correct.

31. The geologic time scale identified major blocks of time in Earth's past by
    a. major worldwide extinctions of life on Earth.
    b. the beginning of radioactive decay of certain unstable elements.
    c. measuring reversals in Earth's magnetic field.
    d. None of the above is correct.

32. You would expect to find the least number of fossils in rocks from which era?
    a. Cenozoic
    b. Mesozoic
    c. Paleozoic
    d. Precambrian

33. You would expect to find fossils of life very different from anything living today in rocks from which era?
    a. Cenozoic
    b. Mesozoic
    c. Paleozoic
    d. Precambrian

34. The numerical dates associated with events on the geologic time scale were determined by
    a. relative dating of the rate of sediment deposition.
    b. radiometric dating using radioactive decay.
    c. the temperature of Earth.
    d. the rate that salt is being added to the ocean.

35. Which of the following is not a fossil?
    a. dinosaur footprints
    b. petrified wood
    c. saber-toothed tiger bones
    d. pieces of ancient Native American pottery

36. An early record about the discovery of fossils and connection of fossils to living organisms comes from the
    a. ancient Greeks.
    b. ancient Romans.
    c. Babylonians.
    d. Vikings.

37. Signs of activity that are a type of fossil preservation include all but
    a. tracks.
    b. footprints.
    c. burrows.
    d. shells.

38. The observation that sediments are commonly deposited in flat-lying layers is called the
    a. principle of uniformity.
    b. principle of superposition.
    c. principle of original horizontality.
    d. principle of crosscutting relationships.

39. A geologic feature that cuts across or is intruded into a rock mass must be younger than the rock mass. This is based on the
    a. principle of uniformity.
    b. principle of superposition.
    c. principle of original horizontality.
    d. principle of crosscutting relationships.

40. Criteria for a geologic clock process do not include that the
    a. process must have been operating since Earth began.
    b. process must be measurable.
    c. process must be uniform or subject to averaging.
    d. process must include humans.

41. How many eons has there been since the birth of Earth?
    a. two
    b. four
    c. sixteen
    d. over one hundred

42. The geologic eon that is known for abundant fossil records and living organisms is called
    a. Precambrian.
    b. Postcambrian.
    c. Phanerozoic.
    d. Post-Phanerozoic.

43. The earliest abundant fossils are found in what era?
    a. Cenozoic
    b. Mesozoic
    c. Paleozoic
    d. Phanerozoic

44. Dinosaurs first appeared in what era?
    a. Cenozoic
    b. Mesozoic
    c. Paleozoic
    d. Phanerozoic

45. What era is also called the "Age of Mammals"?
    a. Cenozoic
    b. Mesozoic
    c. Paleozoic
    d. Phanerozoic

46. The oxygen in our atmosphere is the result of
    a. ultraviolet radiation.
    b. photosynthesis.
    c. evolution of mammals.
    d. cooling climates.

47. Earth is though to be
    a. 225 million years old.
    b. 2.5 billion years old.
    c. 3.8 billion years old.
    d. 4.5 billion years old.

Answers

1. b 2. c 3. b 4. d 5. a 6. c 7. a 8. c 9. b 10. a 11. c 12. a 13. c 14. c
15. b 16. a 17. d 18. d 19. a 20. b 21. b 22. c 23. d 24. b 25. a 26. c
27. a 28. d 29. c 30. d 31. a 32. d 33. c 34. b 35. d 36. a 37. d 38. c
39. d 40. d 41. b 42. c 43. c 44. a 45. a 46. b 47. d

QUESTIONS FOR THOUGHT

1. What is the principle of uniformity? What are the underlying assumptions of this principle?
2. What is the geologic time scale? What is the meaning of the eras?
3. Why does the rock record go back only 3.8 billion years? If this missing record were available, what do you think it would show? Explain.
4. Do igneous, metamorphic, or sedimentary rocks provide the most information about Earth's history? Explain.
5. What major event marked the end of the Paleozoic and Mesozoic eras according to the fossil record? Describe one theory that proposes to account for this.
6. Briefly describe the principles and assumptions that form the basis of interpreting Earth's history from the rocks.
7. Describe the sequence of geologic events represented by an angular unconformity. Begin with the deposition of the sediments that formed the oldest rocks represented.
8. Describe how the principles of superposition, horizontality, and faunal succession are used in the relative dating of sedimentary rock layers.
9. How are the numbers of the ages of eras and other divisions of the geologic time scale determined?
10. Describe the three basic categories of fossilization methods.
11. Describe some of the things that fossils can tell you about the earth's history.

FOR FURTHER ANALYSIS

1. Analyze the significant reasons that a pot chard of an ancient tribe is *not* considered to be a fossil.
2. Considering the ways fossils are formed, describe what percent you would expect to be (1) preservation or alteration of hard parts, (2) preservation of shape, and (3) preservation of signs of activity. Give reasons for your selections.
3. Suppose you believe the principle of uniformity to be correct. Describe a conversation between yourself and another person who feels strongly that all geologic features were formed at the same time in the recent past.
4. What are the significant differences between the yearly time scale and the geologic time scale?
5. Summarize why the different blocks of time (eon, era, periods) exist in the geologic time scale.
6. The history of Earth has been one of many changes over a long period of time, so why do some people today expect no change in Earth and its atmosphere?

INVITATION TO INQUIRY

Look for Clues

Visit a cliff or place where sedimentary rock has been exposed in a hill by the building of a road. Make a drawing of the layers and measure their thickness. If it took a hundred years to form a millimeter of sediment, how many years are represented by each different layer? Look for clues about the type of sediment and how it was deposited. Look for fossils and other signs about past events. Make up a description about conditions in the region at the time the sediments were being deposited.

22

The Atmosphere of Earth

This cloud forms a thin covering over the mountaintop. Likewise, Earth's atmosphere forms a thin shell around Earth, with 99 percent of the mass within 32 km (about 20 mi) of the surface.

CORE **CONCEPT**

The atmosphere is a thin shell of gases that surround the solid Earth, cycling materials back and forth.

OUTLINE

Composition of the Atmosphere
The atmosphere is made up of relatively fixed amounts of nitrogen, oxygen, and argon plus variable amounts of water vapor and carbon dioxide.

Structure of the Atmosphere
Almost all weather occurs in the troposphere, which is the lower 11 km (about 6.7 mi) of the atmosphere.

Evaporation and Condensation
Relative humidity is a ratio of the amount of water vapor to the maximum at that temperature.

The Atmosphere
  Composition of the Atmosphere
  Atmospheric Pressure
  Warming the Atmosphere
A Closer Look: Hole in the Ozone Layer?
  Structure of the Atmosphere
The Winds
  Local Wind Patterns
  Global Wind Patterns
A Closer Look: The Windchill Factor
Science and Society: Use Wind Energy?
Water and the Atmosphere
  Evaporation and Condensation
  Fog and Clouds
People Behind the Science: James Ephraim Lovelock

Warming the Atmosphere
The process of heating the lower part of the atmosphere by the absorption and reemission of infrared radiation is called the greenhouse effect.

The Winds
Horizontal movement of air is called wind, and the direction of a wind is the direction from which it blows.

Evaporation and Condensation
Cooling, saturation, and condensation nuclei are needed to form the tiny droplets of fog or clouds.

Earth's atmosphere has a unique composition because of the cyclic flow of materials that takes place between different parts. These cycles involve the movement of materials between the surface and the interior (see chapter 18) and the building-up and tearing-down cycles on the surface (see chapters 19–20). Materials also cycle in and out of Earth's atmosphere. Carbon dioxide, for example, is the major component of the atmospheres around Venus and Mars, and the early Earth had a similar atmosphere. Today, carbon dioxide is a very minor part of Earth's atmosphere. It has been maintained as a minor component in a mostly balanced state for about the past 570 million years, cycling into and out of the atmosphere.

Water is also involved in a global cyclic flow between the atmosphere and the surface. Water on the surface is mostly in the ocean, with lesser amounts in lakes, streams, and underground. Not much water is found in the atmosphere at any one time on a worldwide basis, but billions of tons are constantly evaporating into the atmosphere each year and returning as precipitation in an ongoing cycle.

The cycling of carbon dioxide and water to and from the atmosphere takes place in a dynamic system that is energized by the Sun. Radiant energy from the Sun heats some parts of Earth more than others. Winds redistribute this energy with temperature changes, rain, snow, and other changes that are generally referred to as the *weather*.

Understanding and predicting the weather is the subject of **meteorology.** Meteorology is the science of the atmosphere and weather phenomena, from understanding everyday rain and snow to predicting not-so-common storms and tornadoes (Figure 22.1). Understanding weather phenomena depends on a knowledge of the atmosphere and the role of radiant energy on a rotating Earth that is revolving around the Sun. This chapter is concerned with understanding the atmosphere of Earth, its cycles, and the influence of radiant energy on the atmosphere. This understanding will be put to use in chapter 23, which is concerned with weather and climate.

THE ATMOSPHERE

The atmosphere is a relatively thin shell of gases that surrounds the solid Earth. If you could see the molecules making up the atmosphere, you would see countless numbers of rapidly moving particles, all undergoing a terrific jostling from the billions of collisions occurring every second. Since this jostling mass of tiny particles is pulled toward Earth by gravity, more are found near the surface than higher up. Thus, the atmosphere thins rapidly with increasing distance above the surface, gradually merging with the very diffuse medium of outer space.

To understand how rapidly the atmosphere thins with altitude, imagine a very tall stack of open boxes. At any given instant, each consecutively higher box would contain fewer of the jostling molecules than the box below it. Molecules in the lowest box on the surface, at sea level, might be able to move a distance of only 1×10^{-8} m (about 3×10^{-6} in) before colliding with another molecule. A box moved to an altitude of 80 km (about 50 mi) above sea level would have molecules that could move perhaps 10^{-2} m (about 1/2 in) before colliding with another molecule. At 160 km (about 100 mi), the distance traveled would be about 2 m (about 7 ft). As you can see, the distance between molecules increases rapidly with increasing altitude. Since air density is defined by the number of molecules in a unit volume, the density of the atmosphere decreases rapidly with increasing altitude (Figure 22.2).

It is often difficult to imagine a distance above the surface of Earth because there is nothing visible in the atmosphere for comparison. Imagine our stack of boxes from the previous

FIGURE 22.1 The probability of a storm can be predicted, but nothing can be done to stop or slow a storm. Understanding the atmosphere may help in predicting weather changes, but it is doubtful that weather will ever be controlled on a large scale.

example, each box with progressively fewer molecules per unit volume. Imagine that this stack of boxes is so tall that it reaches from the surface past the top of the atmosphere. Now imagine that this tremendously tall stack of boxes is tipped over and carefully laid out horizontally on the surface of Earth. How far

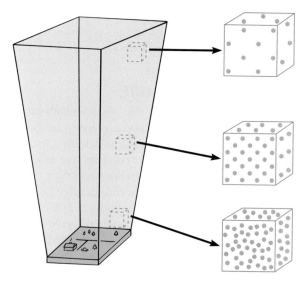

FIGURE 22.2 At greater altitudes, the same volume contains fewer molecules of the gases that make up the air. This means that the density of air decreases with increasing altitude.

would you have to move along these boxes to reach the box that was in outer space, outside of the atmosphere? From the bottom box, you would cover a distance of only 5.6 km (about 3.5 mi) to reach the box that was above 50 percent of the mass of Earth's atmosphere. At 12 km (about 7 mi), you would reach the box that was above 75 percent of Earth's atmosphere. At 16 km (about 10 mi), you would reach the box that was above about 90 percent of the atmosphere. And, after only 32 km (about 20 mi), you would reach the box that was above 99 percent of Earth's atmosphere. The significance of these distances might be better appreciated if you can imagine the distances to some familiar locations; for example, from your campus to a store 16 km (about 10 mi) away would place you above 90 percent of the atmosphere if you were to travel this same distance straight up.

Since the average radius of the solid Earth is about 6,373 km (3,960 mi), you can see that the atmosphere is a very thin shell with 99 percent of the mass within 32 km (about 20 mi) by comparison. The outer edge of the atmosphere is much closer to Earth than most people realize (Figure 22.3).

COMPOSITION OF THE ATMOSPHERE

A sample of pure, dry air is colorless, odorless, and composed mostly of the molecules of just three gases: nitrogen (N_2), oxygen (O_2), and argon (Ar). Nitrogen is the most abundant (about 78 percent of the total volume), followed by oxygen (about 21 percent), then argon (about 1 percent). The molecules of these three gases are well mixed, and this composition is nearly constant everywhere near Earth's surface (Figure 22.4). Nitrogen does not readily enter into chemical reactions with rocks, so it has accumulated in the atmosphere. Some nitrogen is removed from the atmosphere by certain bacteria in the soil and by lightning. The compounds that are formed are absorbed by plants and consequently utilized throughout the food chain. Eventually, the nitrogen is returned to the atmosphere through the decay of plant and animal matter. Overall, these processes of nitrogen removal and release must be in balance since the amount of nitrogen in the atmosphere is essentially constant over time.

Oxygen gas also cycles into and out of the atmosphere in balanced processes of removal and release. Oxygen is removed by (1) living organisms as food is oxidized to carbon dioxide and water and by (2) chemical weathering of rocks as metals and other elements combine with oxygen to form oxides. Oxygen is released by green plants as a result of photosynthesis, and the amount released balances the amount removed by organisms and weathering. So oxygen, as well as nitrogen, is maintained in a state of constant composition through balanced chemical reactions.

The third major component of the atmosphere, argon, is inert and does not enter into any chemical reactions or cycles. It is produced as a product of radioactive decay and, once released, remains in the atmosphere as an inactive filler.

In addition to the relatively fixed amounts of nitrogen, oxygen, and argon, the atmosphere contains variable amounts of water vapor. Water vapor is the invisible, molecular form of water in the gaseous state, which should not be confused with fog or clouds. Fog and clouds are tiny droplets of liquid water, not water in the single molecular form of water vapor. The amount of

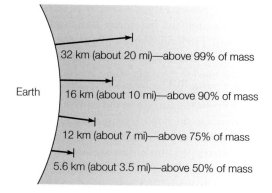

FIGURE 22.3 Earth's atmosphere thins rapidly with increasing altitude and is much closer to Earth's surface than most people realize.

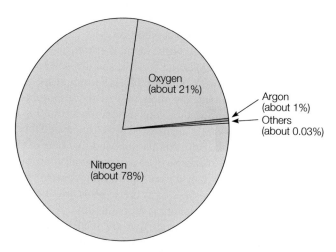

FIGURE 22.4 Earth's atmosphere has a unique composition of gases when compared to that of the other planets in the solar system.

water vapor in the atmosphere can vary from a small fraction of a percent composition by volume in cold, dry air to about 4 percent in warm, humid air. This small, variable percentage of water vapor is essential in maintaining life on Earth. It enters the atmosphere by evaporation, mostly from the ocean, and leaves the atmosphere as rain or snow. The continuous cycle of evaporation and precipitation is called the **hydrologic cycle.** The hydrologic cycle is considered in detail in chapters 23 and 24.

Apart from the variable amounts of water vapor, the relatively fixed amounts of nitrogen, oxygen, and argon make up about 99.97 percent of the volume of a sample of dry air. The remaining 0.03 percent is mostly carbon dioxide (CO_2) and traces of the inert gases neon, helium, krypton, and xenon, along with less than 5 parts per million of free hydrogen, methane, and nitrous oxide. The carbon dioxide content varies locally near cities from the combustion of fossil fuels and from the respiration and decay of organisms and materials produced by organisms. The overall atmospheric concentration of carbon dioxide is regulated (1) by removal from the atmosphere through the photosynthesis process of green plants, (2) by massive exchanges of carbon dioxide between the ocean and the atmosphere, and (3) by chemical reactions between the atmosphere and rocks of the surface, primarily limestone.

The ocean contains some fifty times more carbon dioxide than the atmosphere in the form of carbonate ions and in the form of a dissolved gas. The ocean seems to serve as an equilibrium buffer, absorbing more if the atmospheric concentration increases and releasing more if the atmospheric concentration decreases. Limestone rocks contain an amount of carbon dioxide that is equal to about twenty times the mass of all of Earth's present atmosphere. If all this chemically locked-up carbon dioxide were released, the atmosphere would have a concentration of carbon dioxide similar to the present atmosphere of Venus. This amount of carbon dioxide would result in a tremendous increase in the atmospheric pressure and temperatures on Earth. Overall, however, equilibrium exchange processes with the ocean, rocks, and living things regulate the amount of carbon dioxide in the atmosphere. Measurements have indicated a yearly increase of about 1 part per million of carbon dioxide in the atmosphere over the last several decades. This increase is believed to be a result of the destruction of tropical rainforests along with increased fossil fuel combustion.

In addition to gases and water vapor, the atmosphere contains particles of dust, smoke, salt crystals, and tiny solid or liquid particles called *aerosols.* These particles become suspended and are dispersed among the molecules of the atmospheric gases. Aerosols are produced by combustion, often resulting in air pollution. Aerosols are also produced by volcanoes and forest fires. Volcanoes, smoke from combustion, and the force of the wind lifting soil and mineral particles into the air all contribute to dust particles larger than aerosols in the atmosphere. These larger particles are not suspended as the aerosols are, and they soon settle out of the atmosphere as dust and soot.

Tiny particles of salt crystals that are suspended in the atmosphere come from the mist created by ocean waves and the surf. This mist forms an atmospheric aerosol of seawater that evaporates, leaving the solid salt crystals suspended in the air. The aerosol of salt crystals and dust becomes well mixed in the lower atmosphere around the globe, playing a large and important role in the formation of clouds.

ATMOSPHERIC PRESSURE

The atmosphere exerts a pressure that decreases with increasing altitude above the surface. Atmospheric pressure can be understood in terms of two different frames of reference that will be useful for different purposes. These two frames of reference are called (1) a hydrostatic frame of reference and (2) a molecular frame of reference. *Hydrostatics* is a consideration of the pressure exerted by a fluid at rest. In this frame of reference, atmospheric pressure is understood to be produced by the mass of the atmosphere being pulled to Earth's surface by gravity. In other words, atmospheric pressure is the pressure from the weight of the atmosphere above you. The atmosphere is deepest at Earth's surface, so the greatest pressure is found at the surface. Pressure is less at higher altitudes because there is less air above you and the air is thinner. The *molecular* frame of reference is a consideration of the number of molecules (nitrogen, oxygen, etc.) and the force with which they strike a surface. Air pressure in this frame of reference is understood to be the result of the composite bombardment of air molecules. Atmospheric pressure is greatest at Earth's surface because there are more molecules at lower levels. At higher altitudes in the atmosphere, fewer molecules are present per unit volume, so the pressure they exert is less.

At Earth's surface (sea level), the atmosphere exerts a force of about 10.0 newtons on each square centimeter (14.7 lb/sq in). As you go to higher altitudes above sea level, the pressure rapidly decreases with increasing altitude. At an altitude of about 5.6 km (about 3.5 mi), the air pressure is about half of what it is at sea level, about 5.0 newtons/cm^2 (7.4 lb/in^2). At 12 km (about 7 mi), the air pressure is about 2.5 newtons/cm^2 (3.7 lb/in^2). Compare this decreasing air pressure at greater elevations to Figure 22.3. Again, you can see that most of the atmosphere is very close to Earth, and it thins rapidly with increasing altitude. Even a short elevator ride takes you high enough that the atmospheric pressure on your eardrum is reduced. You equalize the pressure by opening your mouth, allowing the air under greater pressure inside the eardrum to move through the eustachian tube. This makes a "pop" sound that most people associate with changes in air pressure.

Atmospheric pressure is measured by an instrument called a **barometer.** The mercury barometer was invented in 1643 by an Italian named Torricelli. He closed one end of a glass tube and then filled it with mercury. The tube was then placed, open end down, in a bowl of mercury while holding the mercury in the tube with a finger. When Torricelli removed his finger with the open end below the surface in the bowl, a small amount of mercury moved into the bowl, leaving a vacuum at the top end of the tube. The mercury remaining in the tube was supported by the atmospheric pressure on the surface of the mercury in the bowl. The pressure exerted by the weight of the mercury in the tube thus balanced the pressure exerted by the atmosphere. At sea level, Torricelli found that atmospheric pressure balanced a column of mercury about 76.00 cm (29.92 in) tall (Figure 22.5).

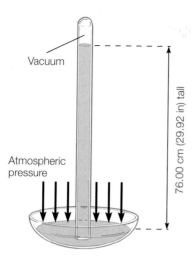

FIGURE 22.5 The mercury barometer measures the atmospheric pressure from the balance between the pressure exerted by the weight of the mercury in a tube and the pressure exerted by the atmosphere. As the atmospheric pressure increases and decreases, the mercury rises and falls. This sketch shows the average height of the column at sea level.

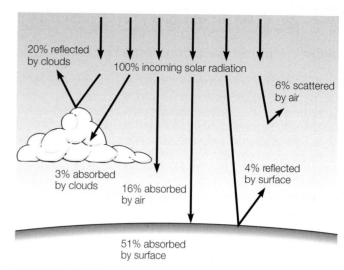

FIGURE 22.6 On the average, Earth's surface absorbs only 51 percent of the incoming solar radiation after it is filtered, absorbed, and reflected. This does not include the radiation emitted back to the surface from the greenhouse effect.

As the atmospheric pressure increases and decreases, the height of the supported mercury column moves up and down. Atmospheric pressure can be expressed in terms of the height of such a column of mercury. Public weather reports give the pressure by referring to such a mercury column, for example, "The pressure is 30 inches (about 76 cm) and rising." If the atmospheric pressure at sea level is measured many times over long periods of time, an average value of 76.00 cm (29.92 in) of mercury is obtained. This average measurement is called the **standard atmospheric pressure** and is sometimes referred to as the *normal pressure*. It is also called *one atmosphere of pressure*.

WARMING THE ATMOSPHERE

Temperature, heat, and the heat transfer processes were discussed in chapter 4. The temperature of a substance is related to the kinetic energy of the molecules making up that substance. The higher the temperature, the greater the kinetic energy. Thus, a high temperature means high molecular kinetic energy, and low temperature means low molecular kinetic energy. In general, all objects with a temperature above absolute zero emit radiant energy. The higher the temperature, the greater the amount of all wavelengths emitted but with greater proportions of shorter wavelengths. You know, for example, that a "red hot" piece of metal is not as hot as an "orange hot" piece of metal and that the bluish-white light from a welder's torch means a very hot temperature. The wavelengths are progressively shorter from red to orange to blue, indicating a progressively higher temperature in each case. Since the surface of the Sun is very hot (6,000 K), it radiates much of its energy at short wavelengths. The cooler surface of Earth (300 K) does more of its radiating at longer wavelengths, wavelengths too long to be detected by the human eye. The wavelengths that Earth emits are in the infrared range of the

electromagnetic spectrum. Humans cannot see infrared radiation, but they can feel warmth when this radiation is absorbed. Thus, it is often referred to as "heat radiation."

Radiation from the Sun must pass through the atmosphere before reaching Earth's surface. The atmosphere filters, absorbs, and reflects incoming solar radiation, as shown in Figure 22.6. On the average, Earth as a whole reflects about 30 percent of the total radiation back into space, with two-thirds of the reflection occurring from clouds. The amount reflected at any one time depends on the extent of cloud cover, the amount of dust in the atmosphere, and the extent of snow and vegetation on the surface. Substantial changes in any of these influencing variables could increase or decrease the reflectivity, leading to increased heating or cooling of the atmosphere.

As Figure 22.6 shows, only about one-half of the incoming solar radiation reaches Earth's surface. The reflection and selective filtering by the atmosphere allow a global average of about 240 watts per square meter to reach the surface. Wide variations from the average occur with latitude as well as with the season.

The incoming sunlight that does reach Earth's surface is mostly absorbed (Figure 22.6). Rocks, soil, water, and anything on the ground become warmer as a result. These materials emit the absorbed solar energy as infrared radiation, wavelengths longer than the visible part of the electromagnetic spectrum. This longer-wavelength infrared radiation has a frequency that matches some of the natural frequencies of vibration of carbon dioxide and water molecules. This match means that carbon dioxide and water molecules readily absorb infrared radiation that is emitted from the surface of Earth. The absorbed infrared energy shows up as an increased kinetic energy of the molecules, which is indicated by an increase in temperature. Carbon dioxide and water vapor molecules in the atmosphere now emit infrared radiation of their own, this time in all directions. Some

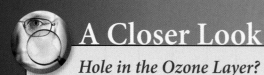
Hole in the Ozone Layer?

Ozone is triatomic oxygen (O_3) that is concentrated mainly in the upper portions of the stratosphere. Diatomic molecules of oxygen (O_2) are concentrated in the troposphere, and monatomic molecules of oxygen (O) are found in the outer edges of the atmosphere. Although the amount of ozone present in the stratosphere is not great, its presence is vital to life on Earth's surface. Ultraviolet radiation (UV) causes mutations to the DNA of organisms. Without the stratospheric ozone, much more ultraviolet radiation would reach the surface of Earth, causing mutations in all kinds of organisms. In humans, ultraviolet radiation is known to cause skin cancer and damage the cornea of eyes. It is believed that the incidence of skin cancer would rise dramatically without the protection offered by the ozone.

Here is how stratospheric ozone shields Earth from ultraviolet radiation. The ozone concentration is not static because there is an ongoing process of ozone formation and destruction. For ozone to form, diatomic oxygen (O_2) must first be broken down into the monatomic form (O). Short-wave ultraviolet (UV) radiation is absorbed by diatomic oxygen (O_2), which breaks it down into the single-atom (O) form. This reaction is significant because of (1) the high-energy ultraviolet radiation that is removed from the sunlight and (2) the monatomic oxygen that is formed, which will combine with diatomic oxygen to make triatomic ozone, which will absorb even more ultraviolet radiation. This initial reaction is

$$O_2 + UV \rightarrow O + O$$

When the O molecule collides with an O_2 molecule and any third, neutral molecule (NM), the following reaction takes place:

$$O_2 + O + NM \rightarrow O_3 + NM$$

When O_3 is exposed to ultraviolet, the ozone absorbs the UV radiation and breaks down to two forms of oxygen in the following reaction:

$$O_3 + UV \rightarrow O_2 + O$$

The monatomic molecule that is produced combines with an ozone molecule to produce two diatomic molecules,

$$O + O_3 \rightarrow 2O_2$$

and the process starts all over again.

There is much concern about Freon (CF_2Cl_2) and other similar chemicals that make their way into the stratosphere. These chemicals are broken down by UV radiation, releasing chlorine (Cl), which reacts with ozone. The reaction might be

$$CF_2Cl_2 + UV \rightarrow CF_2Cl-* + Cl-*$$
$$Cl-* + O_3 \rightarrow *-ClO + O_2$$
$$*-ClO + O \rightarrow O_2 + Cl-*$$

(−* is an unattached bond)

Note that ozone is decomposed in the second step, so it is removed from the UV-radiation-absorbing process. Furthermore, the second and third steps repeat, so the destruction of one molecule of Freon can result in the decomposition of many molecules of ozone.

A regional zone of decreased ozone availability in the stratosphere is referred to as a "hole in the ozone layer" since much high-energy UV can now reach the surface below. Concerns about the impact of chlorine-containing chemicals such as Freon on the ozone layer led to an international agreement known as the Montreal Protocol. The agreement has resulted in a phaseout of the use of Freon and similar compounds in most of the world. Subsequently, it appears that the sizes of the "holes in the ozone layer" are getting smaller.

---

of this reemitted radiation is again absorbed by other molecules in the atmosphere, some is emitted to space, and significantly, some is absorbed by the surface to start the process all over again. The net result is that less of the energy from the Sun escapes immediately to space after being absorbed and emitted as infrared. It is retained through the process of being redirected to the surface, increasing the surface temperature more than it would have otherwise been. The more carbon dioxide that is present in the atmosphere, the more energy that will be bounced around and redirected back toward the surface, increasing the temperature near the surface. The process of heating the atmosphere in the lower parts by the absorption of solar radiation and reemission of infrared radiation is called the **greenhouse effect.** It is called the greenhouse effect because greenhouse glass allows the short wavelengths of solar radiation to pass into the greenhouse but does not allow all of the longer infrared radiation to leave. This analogy is misleading, however, because carbon dioxide and water vapor molecules do not "trap" infrared radiation, but they are involved in a dynamic absorption and downward reemission process that increases the surface temperature. The more carbon dioxide molecules that are involved in this dynamic process, the more infrared radiation that will be redirected back to Earth and the more the temperature will increase. More layers of glass on a greenhouse will not increase the temperature significantly. The significant heating factor in a real greenhouse is the blockage of convection by the glass, a process that does not occur from the presence of carbon dioxide and water vapor in the atmosphere.

STRUCTURE OF THE ATMOSPHERE

Convection currents and the repeating absorption and reemission processes of the greenhouse effect tend to heat the atmosphere from the ground up. In addition, the higher-altitude parts of the atmosphere lose radiation to space more readily than the lower-altitude parts. Thus, the lowest part of the atmosphere is warmer, and the temperature decreases with increasing altitude. On the average, the temperature decreases about 6.5°C for each kilometer of altitude (3.5°F/1,000 ft). This change of temperature with altitude is called the **observed lapse rate.** The observed lapse rate applies only to air that is not rising or sinking, and the actual change with altitude can be very different from this average value. For example, a stagnant mass of very cold air may settle over an area, producing colder temperatures near the surface than in the air layers above. Such a

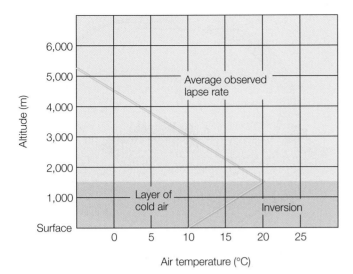

FIGURE 22.7 On the average, the temperature decreases about 6.5°C/1,000 m, which is known as the *observed lapse rate*. An inversion is a layer of air in which the temperature increases with height.

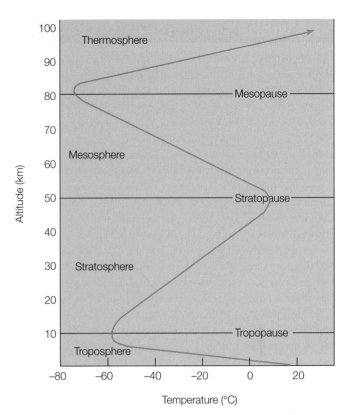

FIGURE 22.8 The structure of the atmosphere based on temperature differences. Note that the "pauses" are actually not lines but broad regions that merge.

layer where the temperature *increases* with height is called an **inversion** (Figure 22.7). Inversions often result in a "cap" of cooler, more dense air overlying the warmer air beneath. This often leads to an increase of air pollution because the inversion prevents dispersion of the pollutants.

Temperature decreases with height at the observed lapse rate until an average altitude of about 11 km (about 6.7 mi), where it then begins to remain more or less constant with increasing altitude. The layer of the atmosphere from the surface up to where the temperature stops decreasing with height is called the **troposphere.** Almost all weather occurs in the troposphere, which is derived from Greek words meaning "turning layer." The upper boundary of the troposphere is called the **tropopause.** The tropopause is identified by the altitude where the temperature stops decreasing and remains constant with increasing altitude. This altitude varies with latitude and with the season. In general, the tropopause is nearly one and one-half times higher than the average over the equator and about half the average altitude over the poles. It is also higher in the summer than in the winter at a given latitude. Whatever its altitude, the tropopause marks the upper boundary of the atmospheric turbulence and the weather that occurs in the troposphere. The average temperature at the tropopause is about −60°C (about −80°F).

Above the tropopause is the second layer of the atmosphere called the **stratosphere.** This layer's name is derived from the Greek for "stratified layer." It is stratified, or layered, because the temperature increases with height. Cooler air below means that consecutive layers of air are denser on the bottom, which leads to a stable situation rather than the turning turbulence of the troposphere below. The stratosphere contains little moisture or dust and lacks convective turbulence, making it a desirable altitude at which to fly. Temperature in the lower stratosphere increases gradually with increasing altitude to a height of about 48 km (about 30 mi), where it reaches a maximum

of about 10°C (about 50°F). This altitude marks the upper boundary of the stratosphere, the **stratopause** (Figure 22.8).

Above the stratopause, the temperature decreases again, just as in the stratosphere, then increases with altitude. The rising temperature is caused by the absorption of solar radiation by molecular fragments present at this altitude.

Layers above the stratopause are the **mesosphere** (Greek for "middle layer") and the **thermosphere** (Greek for "warm layer"). The name *thermosphere* and the high temperature readings of the thermosphere would seem to indicate an environment that is actually not found at this altitude. The gas molecules here do have a high kinetic energy, but the air here is very thin and the molecules are far apart. Thus, the average kinetic energy is very high, but the few molecules do not transfer much energy to a thermometer. A thermometer here would show a temperature far below zero for this reason, even though the same average kinetic energy back at the surface would result in a temperature beyond any temperature ever recorded in the hottest climates.

The **exosphere** (Greek for "outer layer") is the outermost layer where the molecules merge with the diffuse vacuum of space. Molecules of this layer that have sufficient kinetic energy are able to escape and move off into space. The thermosphere and upper mesosphere are sometimes called the **ionosphere** because of the free electrons and ions at this altitude. The electrons and ions here are responsible for reflecting radio waves around Earth and for the northern lights.

CHAPTER 22 The Atmosphere of Earth **535**

THE WINDS

The troposphere is heated from the bottom up as the surface of Earth absorbs sunlight and emits infrared radiation. The infrared radiation is absorbed and reemitted numerous times by carbon dioxide and water molecules as the energy works its way back to space. The overall result of this ongoing absorption and reemission process is the observed lapse rate, the decrease of temperature upward in the troposphere. The observed lapse rate is an *average* value, which means that the actual rate at a given place and time is probably higher or lower than this value. The composition of the surface varies from place to place, consisting of many different types and forms of rock, soil, water, ice, snow, and so forth. These various materials absorb and emit energy at various rates, which results in an uneven heating of the surface. You may have noticed that different materials vary in their abilities to absorb and emit energy if you have ever walked barefooted across some combination of grass, concrete, asphalt, and dry sand on a hot, sunny day.

Uneven heating of Earth's surface sets the stage for *convection* (see chapter 4). As a local region of air becomes heated, the increased kinetic energy of the molecules expands the mass of air, reducing its density. This less dense air is buoyant and is pushed upward by nearby cooler, more dense air. This results in three general motions of air: (1) the upward movement of air over a region of greater heating, (2) the sinking of air over a cooler region, and (3) a horizontal air movement between the cooler and warmer regions. In general, a horizontal movement of air is called **wind**, and the direction of a wind is defined as the direction from which it blows.

Air in the troposphere rises, moves as the wind, and sinks. All three of these movements are related, and all occur at the same time in different places. During a day with gentle breezes on the surface, the individual, fluffy clouds you see are forming over areas where the air is moving upward. The clear air between the clouds is over areas where the air is moving downward. On a smaller scale, air can be observed moving from a field of cool grass toward an adjacent asphalt parking lot on a calm, sunlit day. Soap bubbles or smoke will often reveal the gentle air movement of this localized convection.

Depending on local surface conditions, which are in the section on local wind patterns, the wind usually averages about 16 km/h (about 10 mi/h) and has an average rising and sinking velocity of about 2 km/h (about 1 mi/h). These normal, average values are greatly exceeded during storms and severe weather events. A hurricane has winds that exceed 120 km/h (about 75 mi/h), and a thunderstorm can have updrafts and downdrafts between 50 and 100 km/h (about 30 to 60 mi/h). The force exerted by such winds can be very destructive to structures on the surface. An airplane unfortunate enough to be caught in a thunderstorm can be severely damaged as it is tossed about by the updrafts and downdrafts.

LOCAL WIND PATTERNS

Considering average conditions, there are two factors that are important for a generalized model to help you understand local wind patterns. These factors are (1) the relationship between

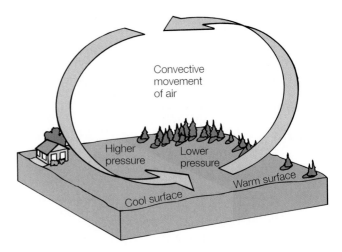

FIGURE 22.9 A model of the relationships between differential heating, the movement of air, and pressure difference in a convective cell. Cool air pushes the less dense, warm air upward, reducing the surface pressure. As the uplifted air cools and becomes more dense, it sinks, increasing the surface pressure.

air temperature and air density and (2) the relationship between air pressure and the movement of air.

The upward and downward movement of air leads to the second part of the generalized model, that (1) the upward movement produces a "lifting" effect on the surface that results in an area of lower atmospheric pressure and (2) the downward movement produces a "piling up" effect on the surface that results in an area of higher atmospheric pressure. On the surface, air is seen to move from the "piled up" area of higher pressure horizontally to the "lifted" area of lower pressure (Figure 22.9). In other words, air moves from an area of higher pressure to an area of lower pressure. The movement of air and the pressure differences occur together, and neither is the cause of the other. This is an important relationship in a working model of air movement that can be observed and measured on a very small scale, such as between an asphalt parking lot and a grass field. It can also be observed and measured for local, regional wind patterns and for worldwide wind systems.

A local wind pattern may result from the temperature differences between a body of water and adjacent landmasses. If you have ever spent some time along a coast, you may have observed that a cool, refreshing gentle breeze blows from the water toward the land during the summer. During the day, the temperature of the land increases more rapidly than the water temperature. The air over the land is therefore heated more, expands, and becomes less dense. Cool, dense air from over the water moves inland under the air over the land, buoying it up. The air moving from the sea to the land is called a **sea breeze.** The sea breeze along a coast may extend inland several miles during the hottest part of the day in the summer. The same pattern is sometimes observed around the Great Lakes during the summer, but this breeze usually does not reach more than several city blocks inland. During the night, the land surface cools more rapidly than the water, and the air moves from the land to the sea (Figure 22.10).

CONCEPTS *Applied*

Why the Beach Is Cooler

Adjacent areas of the surface can have different temperatures because of different heating or cooling rates. The difference is very pronounced between adjacent areas of land and water. Under identical conditions of incoming solar radiation, the temperature changes experienced by the water will be much less than the changes experienced by the adjacent land. There are three principal reasons for this difference: (1) The specific heat of water is about five times the specific heat of soil. This means that it takes more energy to increase the temperature of water than it does for soil. Equal masses of soil and water exposed to sunlight will result in the soil heating about 5°C while the water heats 1°C from absorbing the same amount of solar radiation. (2) Water is a transparent fluid that is easily mixed, so the incoming solar radiation warms a body of water throughout, spreading out the heating effect. Incoming solar radiation on land, on the other hand, warms a relatively thin layer on the top, concentrating the heating effect. (3) The water is cooled by evaporation, which helps keep a body of water at a lower temperature than an adjacent landmass under identical conditions of incoming solar radiation.

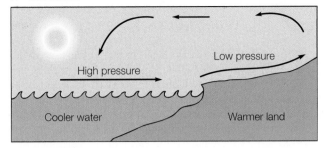

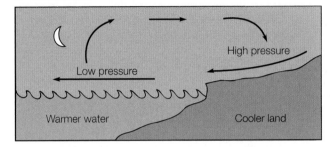

FIGURE 22.10 The land warms and cools more rapidly than an adjacent large body of water. (*A*) During the day, the land is warmer, and air over the land expands and is buoyed up by cooler, more dense air from over the water. (*B*) During the night, the land cools more rapidly than the water, and the direction of the breeze is reversed.

Another pattern of local winds develops in mountainous regions. If you have ever visited a mountain in the summer, you may have noticed that there is usually a breeze or wind blowing up the mountain slope during the afternoon. This wind pattern develops because the air over the mountain slope is heated more than the air in a valley. As shown in Figure 22.11, the air over the slope becomes warmer because it receives more direct sunlight than the valley floor. Sometimes this air movement is so gentle that it would be unknown except for the evidence of clouds that form over the peaks during the day and evaporate at night. During the night, the air on the slope cools as the land loses radiant energy to space. As the air cools, it becomes denser and flows downslope, forming a reverse wind pattern to the one observed during the day.

During cooler seasons, cold, dense air may collect in valleys or over plateaus, forming a layer or "puddle" of cold air. Such an accumulation of cold air often results in some very cold nighttime temperatures for cities located in valleys, temperatures that are much colder than anywhere in the surrounding region. Some weather disturbance, such as an approaching front, can disturb such an accumulation of cold air and cause it to pour out of its resting place and through canyons or lower valleys. Air moving from a higher altitude like this becomes compressed as it moves to lower elevations under increasing atmospheric pressure. Compression of air increases the temperature by increasing the kinetic energy of the molecules. This creates a wind called a **Chinook,** which is common to mountainous and adjacent regions. A Chinook is a wind of

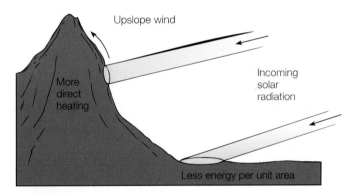

FIGURE 22.11 Incoming solar radiation falls more directly on the side of a mountain, which results in differential heating. The same amount of sunlight falls on the areas shown in this illustration, with the valley floor receiving a more spread-out distribution of energy per unit area. The overall result is an upslope mountain breeze during the day. During the night, dense, cool air flows downslope for a reverse wind pattern.

compressed air with sharp temperature increases that can sublimate or melt away any existing snow cover in a single day. The *Santa Ana* is a well-known compressional wind that occurs in southern California.

GLOBAL WIND PATTERNS

Local wind patterns tend to mask the existence of the overall global wind pattern that is also present. The global wind pattern is not apparent if the winds are observed and measured

The Windchill Factor

The term *windchill* is attributed to the Antarctic explorer Paul A. Siple. During the 1940s, Siple and Charles F. Passel conducted experiments on how long it took a can of water to freeze at various temperatures and wind speeds at a height of 10 m (33 ft) above the ground (typical height of an anemometer). They found that the time depended on the air temperature and the wind speed. From these data, an equation was developed to calculate the windchill factor for humans.

In 2001, the National Weather Service changed to a new method of computing windchill temperatures. The new windchill formula was developed by a year-long cooperative effort between the U.S. and Canadian governments and university scientists. The new standard is based on wind speeds at an average height of 1.5 m (5 ft) above the ground, which is closer to the height of a human face than the height of an anemometer and takes advantage of advances in science, technology, and computers. The new chart also highlights the danger of frostbite.

Here is the reason why the windchill factor is an important consideration. The human body constantly produces heat to maintain a core temperature, and some of this heat is radiated to the surroundings. When the wind is not blowing (and you are not moving), your body heat is also able to warm some of the air next to your body. This warm blanket of air provides some insulation, protecting your skin from the colder air farther away. If the wind blows, however, it moves this air away from your body and you feel cooler. How much cooler depends on how fast the air is moving and the outside temperature—which is what the windchill factor tells you. Thus, windchill is an attempt to measure the combined effect of low temperature and wind on humans (Box Figure 22.1). It is just one of the many factors that can affect winter comfort. Others include the type of clothes, level of physical exertion, amount of sunshine, humidity, age, and body type.

There is a windchill calculator at the National Weather Service website, www.nws.noaa.gov/om/windchill/. All you need to do is enter the air temperature (in degrees Fahrenheit) and the wind speed (in miles per hour), and the calculator will give you the windchill using both the old and the new formulas.

Wind (mph)	Temperature (°F)																	
	40	35	30	25	20	15	10	5	0	−5	−10	−15	−20	−25	−30	−35	−40	−45
5	36	31	25	19	13	7	1	−5	−11	−16	−22	−28	−34	−40	−46	−52	−57	−63
10	34	27	21	15	9	3	−4	−10	−16	−22	−28	−35	−41	−47	−53	−59	−66	−72
15	32	25	19	13	6	0	−7	−13	−19	−26	−32	−39	−45	−51	−58	−64	−71	−77
20	30	24	17	11	4	−2	−9	−15	−22	−29	−35	−42	−48	−55	−61	−68	−74	−81
25	29	23	16	9	3	−4	−11	−17	−24	−31	−37	−44	−51	−58	−64	−71	−78	−84
30	28	22	15	8	1	−5	−12	−19	−26	−33	−39	−46	−53	−60	−67	−73	−80	−87
35	28	21	14	7	0	−7	−14	−21	−27	−34	−41	−48	−55	−62	−69	−76	−82	−89
40	27	20	13	6	−1	−8	−15	−22	−29	−36	−43	−50	−57	−64	−71	−78	−84	−91
45	26	19	12	5	−2	−9	−16	−23	−30	−37	−44	−51	−58	−65	−72	−79	−86	−93
50	26	19	12	4	−3	−10	−17	−24	−31	−38	−45	−52	−60	−67	−74	−81	−88	−95
55	25	18	11	4	−3	−11	−18	−25	−32	−39	−46	−54	−61	−68	−75	−82	−89	−97
60	25	17	10	3	−4	−11	−19	−26	−33	−40	−48	−55	−62	−69	−76	−84	−91	−98

Frostbite times ☐ 30 minutes ☐ 10 minutes ☐ 5 minutes

BOX FIGURE 22.1 Windchill chart.

for a particular day, week, or month. It does become apparent when the records for a long period of time are analyzed. These records show that Earth has a large-scale pattern of atmospheric circulation that varies with latitude. There are belts in which the winds average an overall circulation in one direction, belts of higher atmospheric pressure averages, and belts of lower atmospheric pressure averages. This has led to a generalized pattern of atmospheric circulation and a global atmospheric model. This model, as you will see, today provides the basis for the daily weather forecast for local and regional areas.

As with local wind patterns, it is temperature imbalances that drive the global circulation of the atmosphere. Earth receives more direct solar radiation in the equatorial region than it does at higher latitudes (Figure 22.12). As a result, the temperatures of the lower troposphere are generally higher in the equatorial region, decreasing with latitude toward both poles. The lower troposphere from 10°N to 10°S of the equator is heated, expands, and becomes less dense. Hot air rises in this belt around the equator, known as the **intertropical convergence zone.** The rising air cools because it expands as it rises,

Millions of windmills were installed in rural areas of the United States between the late 1800s and the late 1940s. These windmills used wind energy to pump water, grind grain, or generate electricity. Some are still in use today, but most were no longer needed after inexpensive electric power became generally available in rural areas of the country.

In the 1970s, wind energy began making a comeback as a clean, renewable energy alternative to fossil fuels. The windmills of the past were replaced by wind turbines of today. A wind turbine is usually mounted on a tower, with blades that are rotated by the wind. This rotary motion drives a generator that produces electricity. A location should have yearly wind speeds of at least 19 km/h (12 mi/h) to provide enough wind energy for a turbine, and a greater yearly average means more energy is available. Farms, homes, and businesses in these locations can use smaller turbines, which are generally 50 kW or less. Large turbines of 500 kW or more are used in "wind farms," which are large clusters of interconnected wind turbines connected to a utility power grid.

Many areas of the United States have a high potential for wind-power use. North Dakota, South Dakota, and Texas have enough wind resources to provide electricity for the entire nation. Today, only California has extensively developed wind farms, with more than thirteen thousand wind turbines on three wind farms in the Altamont Pass region (east of San Francisco), Techachapi (southeast of Bakersfield), and San Gorgonio (near Palm Springs). With a total of 2,361 MW of installed capacity in California, wind energy generates enough electricity to more than meet the needs of a city the size of San Francisco. The wind farms in California have a rated capacity that is comparable to two large coal-fired power plants but without the pollution and limits of this nonrenewable energy source. Wind energy makes economic as well as environmental sense, and new wind farms are being developed in Minnesota, Oregon, and Wyoming. Other states with a strong wind-power potential include Kansas, Montana, Nebraska, Oklahoma, Iowa, Colorado, Michigan, and New York. All of these states, in fact, have a greater wind-energy potential than California.

QUESTIONS TO DISCUSS

Discuss with your group the following questions concerning wind power:

1. Why have electric utilities not used much wind power as an energy source?

2. Should governments provide a tax break to encourage people to use wind power? Why or why not?

3. What are the advantages and disadvantages of using wind power in place of fossil fuels?

4. What are the advantages and disadvantages of the government building huge wind farms in North Dakota, South Dakota, and Texas to supply electricity for the entire nation?

---

resulting in heavy average precipitation. The tropical rainforests of Earth occur in this zone of high temperatures and heavy rainfall. As the now dry, rising air reaches the upper parts of the troposphere, it begins to spread toward the north and toward the south, sinking back toward Earth's surface (Figure 22.13). The descending air reaches the surface to form a high-pressure belt that is centered about 30°N and 30°S of the equator. Air moving on the surface away from this high-pressure belt produces

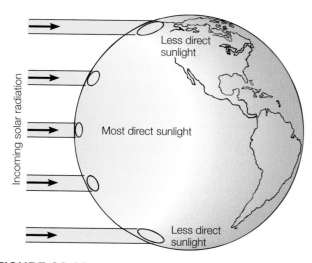

FIGURE 22.12 On a global, yearly basis, the equatorial region of Earth receives more direct incoming solar radiation than the higher latitudes. As a result, average temperatures are higher in the equatorial region and decrease with latitude toward both poles. This sets the stage for worldwide patterns of prevailing winds, high and low areas of atmospheric pressure, and climatic patterns.

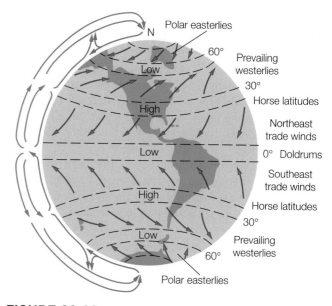

FIGURE 22.13 Simplified pattern of horizontal and vertical circulation in the actual atmosphere. Regions of high and low pressure are indicated.

the prevailing northeast trade winds and the prevailing westerly winds of the Northern Hemisphere. The great deserts of Earth are also located in this high-pressure belt of descending dry air.

Poleward of the belt of high pressure, the atmospheric circulation is controlled by a powerful belt of wind near the top of the troposphere called a **jet stream.** Jet streams are meandering loops of winds that tend to extend all the way around Earth, moving generally from the west in both hemispheres at speeds of 160 km/h (about 100 mi/h) or more. A jet stream may occur as a single belt, or loop, of wind, but sometimes it divides into two or more parts. The jet stream develops north and south loops of waves much like the waves you might make on a very long rope. These waves vary in size, sometimes beginning as a small ripple but then growing slowly as the wave moves eastward. Waves that form on the jet stream bulge toward the poles (called a crest) or toward the equator (called a trough). Warm air masses move toward the poles ahead of a trough, and cool air masses move toward the equator behind a trough as it moves eastward. The development of a wave in the jet stream is understood to be one of the factors that influences the movement of warm and cool air masses, a movement that results in weather changes on the surface.

The intertropical convergence zone, the 30° belt of high pressure, and the northward and southward migration of a meandering jet stream all shift toward or away from the equator during the different seasons of the year. The troughs of the jet stream influence the movement of alternating cool and warm air masses over the belt of the prevailing westerlies, resulting in frequent shifts of fair weather to stormy weather, then back again. The average shift during the year is about 6° of latitude, which is sufficient to control the overall climate in some locations. The influence of this shift of the global circulation of Earth's atmosphere will be considered as a climatic influence after considering the roles of water and air masses in frequent weather changes.

WATER AND THE ATMOSPHERE

Water exists on Earth in all three states: (1) as a liquid when the temperature is generally above the freezing point of 0°C (32°F), (2) as a solid in the form of ice, snow, or hail when the temperature is generally below the freezing point, and (3) as the invisible, molecular form of water in the gaseous state, which is called *water vapor.*

Over 98 percent of all the water on Earth exists in the liquid state, mostly in the ocean, and only a small, variable amount of water vapor is in the atmosphere at any given time. Since so much water seems to fall as rain or snow at times, it may be a surprise that the overall atmosphere really does not contain very much water vapor. If the average amount of water vapor in Earth's atmosphere were condensed to liquid form, the vapor *and* all the droplets present in clouds would form a uniform layer around Earth only 3 cm (about 1 in) thick. Nonetheless, it is this small amount of water vapor that is eventually responsible for (1) contributing to the greenhouse effect, which helps make Earth a warmer planet, (2) serving as one of the principal

agents in the weathering and erosion of the land, which creates soils and sculptures the landscape, and (3) maintaining life, for almost all plants and animals cannot survive without water. It is the ongoing cycling of water vapor into and out of the atmosphere that makes all this possible. Understanding this cycling process and the energy exchanges involved is also closely related to understanding Earth's weather patterns.

EVAPORATION AND CONDENSATION

Water tends to undergo a liquid-to-gas or a gas-to-liquid phase change at any temperature. The phase change can occur in either direction at any temperature. The temperature of liquid water and the temperature of water vapor are associated with the *average* kinetic energy of the water molecules. The word *average* implies that some of the molecules have a greater kinetic energy and some have less. If a molecule of water that has an exceptionally high kinetic energy is near the surface and is headed in the right direction, it may overcome the attractive forces of the other water molecules and escape the liquid to become a gas. This is the process of evaporation. A supply of energy must be present to maintain the process of evaporation, and the water robs this energy from the surroundings. This explains why water at a higher temperature evaporates more rapidly than water at a lower temperature. More energy is available at higher temperatures to maintain the process at a faster rate.

Water molecules that evaporate move about in all directions, and some will strike the liquid surface. The same forces that it escaped from earlier now capture the molecule, returning it to the liquid state. This is called the process of condensation. Condensation is the opposite of evaporation. In *evaporation,* more molecules are leaving the liquid state than are returning. In *condensation,* more molecules are returning to the liquid state than are leaving. This is a dynamic, ongoing process with molecules leaving and returning continuously (Figure 22.14). If the air were perfectly dry and still, more

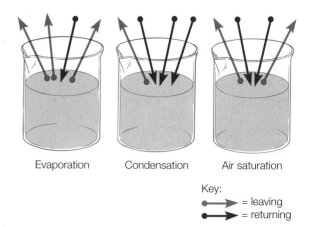

Evaporation   Condensation   Air saturation

Key:
•——→ = leaving
•——→ = returning

FIGURE 22.14 Evaporation and condensation are occurring all the time. If the number of molecules leaving the liquid state exceeds the number returning, the water is evaporating. If the number of molecules returning to the liquid state exceeds the number leaving, the water vapor is condensing. If both rates are equal, saturation has occured; that is, the relative humidity is 100 percent.

molecules would leave (evaporate) the liquid state than would return (condense). Eventually, however, an equilibrium would be reached with as many molecules returning to the liquid state per unit of time as are leaving. An equilibrium condition between evaporation and condensation occurs at **saturation.** Saturation occurs when the processes of evaporation and condensation are in balance.

Air will remain saturated as long as (1) the temperature remains constant and (2) the processes of evaporation and condensation remain balanced. Temperature influences the equilibrium condition of saturated air because increases or decreases in the temperature mean increases or decreases in the kinetic energy of water vapor molecules. Water vapor molecules usually undergo condensation when attractive forces between the molecules can pull them together into the liquid state. Lower temperature means lower kinetic energies, and slow-moving water vapor molecules spend more time close to one another and close to the surface of liquid water. Spending more time close together means an increased likelihood of attractive forces pulling the molecules together. On the other hand, higher temperature means higher kinetic energies, and molecules with higher kinetic energy are less likely to be pulled together. As the temperature increases, there is therefore less tendency for water molecules to return to the liquid state. If the temperature is increased in an equilibrium condition, more water vapor must be added to the air to maintain the saturated condition. A warm atmosphere can therefore hold more water vapor than a cooler atmosphere. In fact, a warm atmosphere on a typical summer day can hold five times as much water vapor as a cold atmosphere on a cold winter day.

The rate of evaporation or condensation depends on the temperature and some variables concerning the surface of the solid, liquid, or gaseous water vapor. Temperature, however, is the important variable, with evaporation increasing with warming and condensation increasing with cooling. It is the temperature of the water or water vapor that is important, but weather broadcasters often refer to the *air* temperature when discussing humidity. The air temperature is nearly the same as the water vapor temperature, and the important thing to remember is that you are actually referring to the waver vapor (not air) temperature when discussing humidity.

Also, broadcasters sometimes refer to the air's ability "to hold water vapor." Here we understand that air has nothing to do with how much water vapor can be present in the atmosphere. Saying that "air can hold more water vapor" is an easy way to say that evaporation exceeds condensation, therefore increasing the amount of water vapor in the atmosphere. Likewise, saying that "air has a lower capacity to hold water vapor" is an easy way to say that condensation exceeds evaporation, therefore decreasing the amount of water vapor in the atmosphere.

Humidity

The amount of water vapor is referred to generally as **humidity.** Damp, moist conditions are more likely to have condensation than evaporation, and this is said to be a *high humidity.* Dry conditions are more likely to have evaporation than condensation,

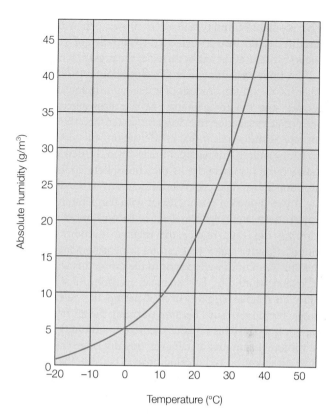

FIGURE 22.15 The maximum amount of water vapor that can be in the air at different temperatures. The amount of water vapor in the air at a particular temperature is called the *absolute humidity.*

on the other hand, and this is said to be a *low humidity.* A measurement of the amount of water vapor in the atmosphere at a particular time is called the **absolute humidity** (Figure 22.15). At room temperature, for example, humid air might contain 15 grams of water vapor in each cubic meter. At the same temperature, air of low humidity might have an absolute humidity of only 2 g/cm^3. The absolute humidity can range from near zero at temperatures well below freezing, up to a maximum that is determined by the temperature at the time.

The relationship between the *actual* absolute humidity at a particular temperature and the maximum absolute humidity that can occur at that temperature is called the **relative humidity.** Relative humidity is a ratio between (1) the amount of water vapor and (2) the amount of water vapor needed to reach saturation at that temperature. The relationship is

$$\frac{\text{absolute humidity at present temperature}}{\text{maximum absolute humidity at present temperature}} \times 100\%$$

$$= \frac{\text{realtive}}{\text{humidity}}$$

For example, suppose a measurement of the water vapor at 10°C (50°F) finds an absolute humidity of 5.0 g/m^3. According to Figure 22.15, the maximum amount of water vapor when the temperature is 10°C is about 10 g/m^3. The relative humidity is then

$$\frac{5.0 \text{ g/m}^3}{10 \text{ g/m}^3} \times 100\% = 50\%$$

CHAPTER 22 The Atmosphere of Earth   **541**

If the absolute humidity had been 10 g/m³, then the relative humidity would be 100 percent. A humidity of 100 percent means that saturation has occurred at the present temperature.

The important thing to understand about relative humidity is that the maximum amount of water vapor in the atmosphere changes with the temperature. As the atmosphere becomes colder, the evaporation rate decreases faster than the condensation rate, with the result that condensation will occur at a certain temperature. This temperature is called the **dew point.** The relative humidity increases as the atmosphere cools, reaching saturation and 100 percent relative humidity at the dew point.

The evaporation rate increases with temperature increases because the water vapor molecules become more energized. Warming can thus reduce the relative humidity since the amount of water vapor that can be present is increased. With the same amount of water vapor in the room, this will decrease the relative humidity. For example, warming of a room can reduce the relative humidity from 50 percent to 3 percent. Lower relative humidity results because warming the air increases the evaporation rate. This explains the need to humidify a home in the winter. Evaporation occurs very rapidly when the humidity is low. Evaporation is a cooling process because the molecules with higher kinetic energy are the ones to escape, lowering the average kinetic energy as they evaporate. Dry air will therefore cause you to feel cool even though the air temperature is fairly high. Adding moisture to the air will enable you to feel warmer at lower air temperatures and thus lower your fuel bill.

You have probably heard the expression, "It's not the heat, it's the humidity." A higher humidity will lessen the tendency for evaporation to exceed condensation, slowing the cooling process of evaporation. You often fan your face or use an electric fan to move evaporated water vapor molecules away from the surface of your skin, increasing the net rate of evaporation. The use of a fan will not help cool you at all if the relative humidity is 100 percent. At 100 percent relative humidity, the air is saturated, and the rate of evaporation is equal to the rate of condensation, so there can be no net cooling.

Evaporation occurs at a rate that is proportional to the absolute humidity, ranging from a maximum rate when the air is driest to no net evaporation when the air is saturated. Since evaporation is a cooling process, it is possible to use a thermometer to measure humidity. An instrument called a **psychrometer** has two thermometers, one of which has a damp cloth wick around its bulb end. As air moves past the two thermometer bulbs, the ordinary thermometer (the dry bulb) will measure the present air temperature. Water will evaporate from the wet wick (the wet bulb) until an equilibrium is reached between water vapor leaving the wick and water vapor returning to the wick from the air. Since evaporation lowers the temperature, the depression of the temperature of the wet bulb thermometer is an indirect measure of the water vapor present in the air. The relative humidity can be determined by obtaining the dry and wet bulb temperature readings and referring to a relative humidity table such as the one found in appendix C. If the humidity is 100 percent, no net evaporation will take place from the wet bulb, and both wet and dry bulb temperatures will be the same. The lower the humidity, the greater the difference in the temperature reading of the two thermometers.

You may have noticed that your hair becomes more or less curly in humid weather. The human hair absorbs moisture from the air, becoming longer in humid air and shorter in dry air. Since the change of length is proportional to changes in humidity, it is possible to use human hair to measure the humidity. A **hair hygrometer** is an instrument that measures the humidity from changes in the length of hair. A bundle of the hair is held under tension by a spring, and changes in the length move a pointer on a dial that is calibrated to read the relative humidity.

 CONCEPTS *Applied*

Humidity Factors

Compare the relative humidity in a classroom, over a grass lawn, over a paved parking lot, and other places you might be during a particular day. What do the findings mean?

The Condensation Process

In still air under a constant pressure, the rate of evaporation depends primarily on three factors: (1) the surface area of the liquid that is exposed to the atmosphere, (2) the air and water temperature, and (3) the amount of water vapor in the air at the time, that is, the relative humidity. The opposite process, condensation, depends primarily on two factors: (1) the relative humidity and (2) the temperature of the air, or more directly, the kinetic energy of the water vapor molecules. During condensation, molecules of water vapor join together to produce liquid water on the surface as dew or in the air as the droplets of water making up fog or clouds. Water molecules may also join together to produce solid water in the form of frost or snow. Before condensation can occur, however, the air must be saturated, which means that the relative humidity must be 100 percent. A parcel of air can become saturated as a result of (1) water vapor being added to the air from evaporation, (2) cooling, which reduces the evaporation rate faster than the condensation rate and therefore increases the relative humidity, or (3) a combination of additional water vapor with cooling.

The process of condensation of water vapor explains a number of common observations. You are able to "see your breath" on a cold day, for example, because the high moisture content of your exhaled breath is condensed into tiny water droplets by cold air. The small fog of water droplets evaporates as it spreads into the surrounding air with a lower moisture content. The white trail behind a high-flying jet aircraft is also a result of condensation of water vapor. Water is one of the products of combustion, and the white trail is condensed water vapor, a trail of tiny droplets of water in the cold upper atmosphere. The trail of water droplets is called a *contrail* after "condensation trail." Back on the surface, a cold glass of beverage seems to "sweat" as water vapor molecules near the outside of the glass are cooled, moving more slowly. Slowly moving water vapor molecules spend more time closer together, and the molecular forces between the molecules pull them together, forming a thin layer of liquid water on the outside of the cold glass. This same condensation process sometimes results in a small stream of water from the cold air conditioning coils of an automobile or home mechanical air conditioner.

As air is cooled, the evaporation rate decreases faster than the condensation rate. Even without water vapor being added to the air, a temperature will eventually be reached at which saturation, 100 percent humidity, occurs. Further cooling below this temperature will result in condensation. The temperature at which condensation begins is called the dew point temperature. If the dew point is above 0°C (32°F) the water vapor will condense on surfaces as a liquid called **dew.** If the temperature is at or below 0°C, the vapor will condense on surfaces as a solid called **frost.** Note that dew and frost form on the tops, sides, and bottoms of objects. Dew and frost condense directly on objects and do not "fall out" of the air. Note also that the temperature that determines if dew or frost forms is the temperature of the object where they condense. This temperature near the open surface can be very different from the reported air temperature, which is measured more at eye level in a sheltered instrument enclosure.

Observations of where and when dew and frost form can lead to some interesting things to think about. Dew and frost, for example, seem to form on "C" nights, nights that can be described by the three "C" words of clear, calm, and cool. Dew and frost also seem to form more (1) in open areas than under trees or other shelters, (2) on objects such as grass than on the flat, bare ground, and (3) in low-lying areas before they form on slopes or the sides of hills. What is the meaning of these observations?

Dew and frost are related to clear nights and open areas because these are the conditions best suited for the loss of infrared radiation. Air near the surface becomes cooler as infrared radiation is emitted from the grass, buildings, streets, and everything else that absorbed the shorter-wavelength radiation of incoming solar radiation during the day. Clouds serve as a blanket, keeping the radiation from escaping to space so readily. So a clear night is more conducive to the loss of infrared radiation and therefore to cooling. On a smaller scale, a tree serves the same purpose, holding in radiation and therefore retarding the cooling effect. Thus, an open area on a clear, calm night would have cooler air near the surface than would be the case on a cloudy night or under the shelter of a tree.

The observation that dew and frost form on objects such as grass before forming on flat, bare ground is also related to loss of infrared radiation. Grass has a greater exposed surface area than the flat, bare ground. A greater surface area means a greater area from which infrared radiation can escape, so grass blades cool more rapidly than the flat ground. Other variables, such as specific heat, may be involved, but overall, frost and dew are more likely to form on grass and low-lying shrubs before they form on the flat, bare ground.

Dew and frost form in low-lying areas before forming on slopes and the sides of hills because of the density differences of cool and warm air. Cool air is more dense than warm air and is moved downhill by gravity, pooling in low-lying areas. You may have noticed the different temperatures of low-lying areas if you have ever driven across hills and valleys on a clear, calm, and cool evening. Citrus and other orchards are often located on

FIGURE 22.16 Fans like this one are used to mix the warmer, upper layers of air with the cooling air in the orchard on nights when frost is likely to form.

slopes of hills rather than on valley floors because of the gravity drainage of cold air.

It is air near the surface that is cooled first by the loss of radiation from the surface. Calm nights favor dew or frost formation because the wind mixes the air near the surface that is being cooled with warmer air above the surface. If you have ever driven near a citrus orchard, you may have noticed the huge, propellerlike fans situated throughout the orchard on poles. These fans are used on "C" nights when frost is likely to form to mix the warmer, upper layers of air with the cooling air in the orchard (Figure 22.16).

Condensation occurs on the surface as frost or dew when the dew point is reached. When does condensation occur in the air? Water vapor molecules in the air are constantly colliding and banging into each other, but they do not just join together to form water droplets, even if the air is saturated. The water molecules need something to condense upon. Condensation of water vapor into fog or cloud droplets takes place on tiny particles present in the air. The particles are called **condensation nuclei.** There are hundreds of tiny dust, smoke, soot, and salt crystals suspended in each cubic centimeter of the air that serve as condensation nuclei. Tiny salt crystals, however, are particularly effective condensation nuclei because salt crystals attract water molecules. You may have noticed that salt in a salt shaker becomes moist on a humid day because of the way it attracts water molecules. Tiny salt crystals suspended in the air act the same way, serving as nuclei that attract water vapor into tiny droplets of liquid water.

After water vapor molecules begin to condense on a condensation nucleus, other water molecules will join the liquid water already formed, and the tiny droplet begins to increase in volume. The water droplets that make up a cloud are about

CHAPTER 22 The Atmosphere of Earth   **543**

Condensation nucleus
(0.2 micron)

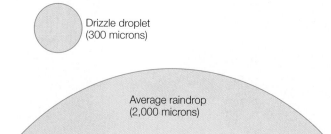
Average cloud droplet
(20 microns)

Large cloud droplet
(100 microns)

Drizzle droplet
(300 microns)

Average raindrop
(2,000 microns)

FIGURE 22.17 This figure compares the size of the condensation nuclei to the size of typical condensation droplets. Note that 1 micron is 1/1,000 mm.

A

B

FIGURE 22.18 (*A*) An early morning aerial view of fog between mountain at top and river below that developed close to the ground in cool, moist air on a clear, calm night. (*B*) Fog forms over the ocean where air moves from over a warm current to over a cool current, and the fog often moves inland.

fifteen hundred times larger than a condensation nucleus, and these droplets can condense out of the air in a matter of minutes. As the volume increases, however, the process slows, and hours and days are required to form the even larger droplets and drops. For comparison to the sizes shown in Figure 22.17, consider that the average human hair is about 100 microns in diameter. This is about the same diameter as the large cloud droplet of water. Large raindrops have been observed falling from clouds that formed only a few hours previously, so it must be some process or processes other than the direct condensation of raindrops that form precipitation. These processes are discussed in chapter 23.

Myths, Mistakes, & Misunderstandings

Moon Weather Forecasts?

The cusps, or horns, of the Moon always point away from the Sun (see Figure 16.30). You can imagine that the cusps are the edges of a cup. Sometimes this "cup" points down, toward the surface of Earth. Other times it points up, away from Earth. It is a myth that Earth has wet weather when the cusps point down ("spilling water") and dry weather when the cusps point up ("holding water").

Another weather-related saying about the Moon concerns the presence of a white ring, or "halo," around the Moon. It is not a myth that this ring is usually followed by a change in the weather. The halo is formed when light from the Moon passes through thin upper clouds composed of ice crystals, which are cirrus clouds.

FOG AND CLOUDS

Fog and clouds are both accumulations of tiny droplets of water that have been condensed from the air. These water droplets are very small, and a very slight upward movement of the air will keep them from falling. If they do fall, they usually evaporate. Fog is sometimes described as a cloud that forms at or near the surface. A fog, as a cloud, forms because air containing water vapor and condensation nuclei has been cooled to the dew point. Some types of fog form under the same "C" night conditions favorable for dew or frost to form, that is, on clear, cool, and calm nights when the relative humidity is high. Sometimes this type of fog forms only in valleys and low-lying areas where cool air accumulates (Figure 22.18). This type of fog is typical of inland

James Lovelock is an English scientist, specializing in the atmospheric sciences, who began the study of chlorofluorocarbons (CFCs) in the 1960s and who invented the concept of Earth as a single organism—the Gaia hypothesis.

Born in London on July 26, 1919, Lovelock was educated at London and Manchester Universities in the early years of World War II. Graduating in 1941, he worked at the National Institute for Medical Research (NIMR) in London on wartime problems such as the measurement of blood pressure under water, the freezing of viable cells, and the design of an acoustic anemometer. Twenty years later, feeling stifled by the security of his position at the institute, Lovelock gave up his job and worked briefly for NASA on the first lunar *Surveyor* mission in California.

Determined not to become part of an institution—a government department, university, or multinational company—which he believes compromises the freedom of scientists to express themselves, he left NASA in 1964. Income from his inventions, especially from Hewlett-Packard, helped him to support himself and his family while he developed his Gaia hypothesis. In 1966,

he discovered chlorofluorocarbons (CFCs) by monitoring what had been thought to be clean Atlantic air on the west coast of Ireland, using the electron capture detector he had invented at NIMR (which measures minute traces of atmospheric gases). However, it was not until he accumulated the money to travel to the Antarctic in 1971 that he was able to corroborate this finding by detecting more CFCs. His discovery sparked research by U.S. chemist F. Sherwood Rowland (1927–) and Mexican chemist Mario Molina (1943–), who predicted the destruction of the ozone layer by human use of CFCs in 1974.

Lovelock presented his Gaia hypothesis in print first in *Gaia: A New Look at Life on Earth* (1979). Dismissed by some scientists as pseudoscientific but seen as a workable hypothesis by others, it has been widely accepted by conservationists, ecologists, Greens, and "New Age" thinkers. (The name *Gaia* was suggested by English novelist William Golding, Lovelock's neighbor in Wiltshire, after the Greek earth goddess who drew the living world forth from Chaos.) Existing theories held that the evolution of plants and animals is distinct from the evolution of the inanimate planet.

According to Lovelock's theory, life and the environment are inextricably linked and mutually dependent. The rocks, oceans and atmosphere, and all living things are part of one great organism that has evolved through the aeons. Life regulates the atmosphere, and the atmosphere provides the conditions necessary for life. Thus, Earth has maintained a more or less constant temperature (unlike stars, which get hotter as they age). The theory has not been properly accepted in the formal traditional sciences but has provoked debate and speculation on the planet's reaction to the greenhouse effect—will it be able to look after itself and adapt to these new conditions? In *The Ages of Gaia: A Biography of Our Living Earth* (1988), Lovelock evolves and refines the nature of Gaia and discusses the greenhouse effect, acid rain, the depletion of the ozone layer, and other topics in detail, demonstrating the geophysical interaction of atmosphere, oceans, climate, and Earth's crust that are comfortably regulated by the use of the Sun's energy by living organisms.

Lovelock is president of the British Marine Biology Association, a fellow of the Royal Society, and visiting professor of cybernetics at Reading University, England.

Source: From the Hutchinson *Dictionary of Scientific Biography.* © Research Machines plc 2003. All Rights Reserved. Helicon Publishing is a division of Research Machines.

fogs, those that form away from bodies of water. Other types of fog may form somewhere else, such as in the humid air over an ocean, and then move inland. Many fogs that occur along coastal regions were formed over the ocean and then carried inland by breezes. A third type of fog looks much like mist rising from melting snow on a street, mist rising over a body of water into cold air, or mist rising over streets after a summer rain shower. These are examples of a temporary fog that forms as a lot of water vapor is added to cool air. This is a cool fog, like other fogs, and is not hot as the mistlike appearance may lead you to believe.

Sometimes a news report states something about the Sun "burning off" a fog. A fog does not burn, of course, because it is made up of droplets of water. What the reporter really means is that the Sun's radiation will increase the temperature, which increases the air's capacity to hold water vapor. With an increased capacity to hold water, the relative humidity drops, and the

fog simply evaporates back to the state of invisible water vapor molecules.

Clouds, like fogs, are made up of tiny droplets of water that have been condensed from the air. Luke Howard, an English weather observer, made one of the first cloud classification schemes. He used the Latin terms *cirrus* (curly), *cumulus* (piled up), and *stratus* (spread out) to identify the basic shapes of clouds (Figure 22.19). The clouds usually do not occur just in these basic cloud shapes but in combinations of the different shapes. Later, Howard's system was modified by expanding the different shapes of clouds into ten classes by using the basic cloud shapes and altitude as criteria. Clouds give practical hints about the approaching weather. The relationship between the different cloud shapes and atmospheric conditions and what clouds can mean about the coming weather are discussed in chapter 23.

FIGURE 22.19 (*A*) Cumulus clouds. (*B*) Stratus and stratocumulus. Note the small stratocumulus clouds forming from increased convection over each of the three small islands. (*C*) An aerial view between the patchy cumulus clouds below and the cirrus and cirrostratus above (the patches on the ground are clear-cut forests). (*D*) Altocumulus. (*E*) A rain shower at the base of a cumulonimbus. (*F*) Stratocumulus.

Earth's *atmosphere* thins rapidly with increasing altitude. Pure, dry air is mostly *nitrogen, oxygen,* and *argon,* with traces of *carbon dioxide* and other gases. Atmospheric air also contains a variable amount of *water vapor.* Water vapor cycles into and out of the atmosphere through the *hydrologic cycle* of evaporation and precipitation.

Atmospheric pressure is measured with a *mercury barometer.* At sea level, the atmospheric pressure will support a column of mercury about 76.00 cm (about 29.92 in) tall. This is the average pressure at sea level, and it is called the *standard atmospheric pressure, normal pressure,* or *one atmosphere of pressure.*

Sunlight is absorbed by the materials on Earth's surface, which then emit infrared radiation. Infrared is absorbed by carbon dioxide and water molecules in the atmosphere, which then reemit the energy many times before it reaches outer space again. The overall effect warms the lower atmosphere from the bottom up in a process called the *greenhouse effect.*

The layer of the atmosphere from the surface up to where the temperature stops decreasing with height is called the *troposphere.* The *stratosphere* is the layer above the troposphere. Temperatures in the stratosphere increase because of the interaction between ozone (O_3) and ultraviolet radiation from the Sun. Other layers of the atmosphere are the *mesosphere, thermosphere, exosphere,* and *ionosphere.*

The surface of Earth is not heated uniformly by sunlight. This results in a *differential heating,* which sets the stage for *convection.* The horizontal movement of air on the surface from convection is called *wind.* A generalized model for understanding why the wind blows involves (1) the relationship between *air temperature and air density* and (2) the relationship between *air pressure and the movement of air.* This model explains local wind patterns and wind patterns observed on a global scale.

The amount of water vapor in the air at a particular time is called the *absolute humidity.* The *relative humidity* is a ratio between the amount of water vapor that is in the air and the amount needed to saturate the air at the present temperature.

When the air is saturated, condensation can take place. The temperature at which this occurs is called the *dew point temperature.* If the dew point temperature is above freezing, *dew* will form. If the temperature is below freezing, *frost* will form. Both dew and frost form directly on objects and do not fall from the air.

Water vapor condenses in the air on *condensation nuclei.* If this happens near the ground, the accumulation of tiny water droplets is called a *fog.* Clouds are accumulations of tiny water droplets in the air above the ground. In general, there are three basic shapes of clouds: *cirrus, cumulus,* and *stratus.* These basic cloud shapes have meaning about the atmospheric conditions and about the coming weather conditions.

KEY TERMS

absolute humidity (p. **541**)
barometer (p. **532**)
Chinook (p. **537**)
condensation nuclei (p. **543**)
dew (p. **543**)
dew point (p. **542**)
exosphere (p. **535**)
frost (p. **543**)
greenhouse effect (p. **534**)
hair hygrometer (p. **542**)
humidity (p. **541**)
hydrologic cycle (p. **532**)
intertropical convergence zone (p. **538**)
inversion (p. **535**)
ionosphere (p. **535**)
jet stream (p. **540**)
mesosphere (p. **535**)
meteorology (p. **530**)
observed lapse rate (p. **534**)
psychrometer (p. **542**)
relative humidity (p. **541**)
saturation (p. **541**)
sea breeze (p. **536**)
standard atmospheric pressure (p. **533**)
stratopause (p. **535**)
stratosphere (p. **535**)
thermosphere (p. **535**)
tropopause (p. **535**)
troposphere (p. **535**)
wind (p. **536**)

APPLYING THE CONCEPTS

1. The science that studies the atmosphere and weather phenomena is
   a. astronomy.
   b. astrology.
   c. meteorology.
   d. space science.

2. Up from the surface, 99 percent of the mass of Earth's atmosphere is found within
   a. 12 km (7 mi).
   b. 16 km (10 mi).
   c. 24 km (15 mi).
   d. 32 km (20 mi).

3. The most abundant gas in the atmosphere is
   a. oxygen.
   b. nitrogen.
   c. argon.
   d. carbon dioxide.

4. Approximately how much of the total volume of the atmosphere is oxygen?
   a. 21 percent
   b. 32 percent
   c. 78 percent
   d. 93 percent

5. Fog and clouds are composed of
   a. water vapor.
   b. tiny droplets of liquid water.
   c. pockets of rain or snow.
   d. large crystals of water vapor.

6. The continuous cycle of water precipitation and evaporation is called
   a. rain.
   b. weather.
   c. climate.
   d. hydrologic cycle.

7. Tiny solid or liquid smoke, soot, dust, and salt crystals found in the atmosphere are
   a. of no concern.
   b. harmful, reducing rainfall.
   c. aerosols.
   d. clouds.

8. Atmospheric pressure is measured using a
   a. barometer.
   b. psychrometer.
   c. thermometer.
   d. hydrometer.

9. Which molecules in the atmosphere absorb infrared radiation?
   a. water
   b. oxygen
   c. carbon dioxide
   d. a and c

10. What is the process of heating the atmosphere by the absorption of solar radiation and reemission of infrared radiation?
    a. solar energy
    b. greenhouse effect
    c. ozone depletion
    d. global emissions

11. Near the surface, what happens to the temperature of the atmosphere with increasing altitude?
    a. increases
    b. decreases
    c. stays the same
    d. depends on the season

12. What is the layer of the atmosphere where we live?
    a. mesosphere
    b. stratosphere
    c. troposphere
    d. ozone layer

13. A layer of the atmosphere where the temperature increases with height is a (an)
    a. conversion.
    b. inversion.
    c. diversion.
    d. clouds.

14. What is the boundary between the troposphere and the stratosphere?
    a. thermal plane
    b. tropopause
    c. stratopause
    d. inversion layer

15. In what lower layer of the atmosphere would transcontinental aircraft escape convective turbulence?
    a. mesosphere
    b. stratosphere
    c. troposphere
    d. ionosphere

16. Ultraviolet radiation is filtered by the
    a. ozone shield.
    b. inversion layer.
    c. Earth's magnetic field.
    d. greenhouse.

17. The term that does not describe a layer of the atmosphere is
    a. thermosphere.
    b. stratosphere.
    c. mesosphere.
    d. temposphere.

18. Uneven heating of Earth's surface directly leads to
    a. rain.
    b. condensation.
    c. convection.
    d. thermal radiation.

19. A general horizontal movement of air is called
    a. wind.
    b. unusual.
    c. hurricane.
    d. storm.

20. Air moving from the ocean to the land is called a
    a. tsunami.
    b. sea breeze.
    c. wind.
    d. storm.

21. A rapidly moving "stream" of air near the top of the troposphere is
    a. an unusual wind pattern.
    b. a jet steam.
    c. El Niño.
    d. impossible.

22. Water vapor in the atmosphere does not
    a. participate in weather.
    b. act as a greenhouse gas.
    c. maintain life on Earth.
    d. filter ultraviolet radiation out of the atmosphere.

23. The amount of water vapor at a particular temperature is defined as
    a. effective humidity.
    b. actual humidity.
    c. absolute humidity.
    d. humidity.

24. The temperature at which condensation begins is the
    a. dew point.
    b. boiling point.
    c. frost point.
    d. melting point.

25. The basic shapes of clouds do not include
    a. cirrus.
    b. cotton tuffs.
    c. cumulus.
    d. stratus.

26. An airplane flying at about 6 km (20,000 ft) is above how much of Earth's atmosphere?
    a. 99 percent
    b. 90 percent
    c. 75 percent
    d. 50 percent

27. Earth's atmosphere is mostly composed of
    a. oxygen, carbon dioxide, and water vapor.
    b. nitrogen, oxygen, and argon.
    c. oxygen, carbon dioxide, and nitrogen.
    d. oxygen, nitrogen, and water vapor.

28. Which of the following gases cycle into and out of the atmosphere?
    a. nitrogen
    b. carbon dioxide
    c. oxygen
    d. All of the above are correct.

29. If it were not for the ocean, Earth's atmosphere would probably be mostly
    a. nitrogen.
    b. carbon dioxide.
    c. oxygen.
    d. argon.

30. Your ear makes a "pop" sound as you descend in an elevator because
    a. air is moving from the atmosphere into your eardrum.
    b. air is moving from your eardrum to the atmosphere.
    c. air is not moving into or out of your eardrum.
    d. None of the above is correct.

31. Most of the total energy radiated by the Sun is
    a. visible light.
    b. ultraviolet radiation.
    c. infrared radiation.
    d. gamma radiation.

32. How much of the total amount of solar radiation reaching the outermost part of Earth's atmosphere reaches the surface?
    a. all of it
    b. about 99 percent
    c. about 75 percent
    d. about half

33. The solar radiation that does reach Earth's surface
    a. is eventually radiated back to space.
    b. shows up as an increase in temperature.
    c. is reradiated at different wavelengths.
    d. All of the above are correct.

34. The greenhouse effect results in warmer temperatures near the surface because
    a. clouds trap infrared radiation near the surface.
    b. some of the energy is reradiated back toward the surface.
    c. carbon dioxide molecules do not permit the radiation to leave.
    d. carbon dioxide and water vapor both trap infrared radiation.

35. The temperature increases with altitude in the stratosphere because
    a. it is closer to the Sun than the troposphere.
    b. heated air rises to the stratosphere.
    c. of a concentration of ozone.
    d. the air is less dense in the stratosphere.

36. Ozone is able to protect Earth from harmful amounts of ultraviolet radiation by
    a. reflecting it back to space.
    b. absorbing it and decomposing, then reforming.
    c. refracting it to a lower altitude.
    d. All of the above are correct.

37. Summertime breezes would not blow if Earth did not experience
    a. cumulus clouds.
    b. differential heating.
    c. the ozone layer.
    d. a lapse rate in the troposphere.

38. On a clear, calm, cool night, you would expect the air temperature over a valley floor to be what compared to the air temperature over a slope to the valley?
    a. cooler
    b. warmer
    c. the same temperature
    d. sometimes warmer and sometimes cooler

39. Air moving down a mountain slope is often warm because
    a. it has been closer to the Sun.
    b. cool air is more dense and settles to lower elevations.
    c. it is compressed as it moves to lower elevations.
    d. this occurs only during the summertime.

40. Considering Earth's overall atmosphere, you would expect more rainfall to occur in a zone of
    a. high atmospheric pressure.
    b. low atmospheric pressure.
    c. prevailing westerly winds.
    d. prevailing trade winds.

41. Considering Earth's overall atmosphere, you would expect to find a desert located in a zone of
    a. high atmospheric pressure.
    b. low atmospheric pressure.
    c. prevailing westerly winds.
    d. prevailing trade winds.

42. Water molecules can go (1) from the liquid state to the vapor state and (2) from the vapor state to the liquid state. When is the movement from the liquid to the vapor state only?
    a. evaporation
    b. condensation
    c. saturation
    d. This usually does not occur alone.

43. What condition means a balance between the number of water molecules moving to and from the liquid state?
    a. evaporation
    b. condensation
    c. saturation
    d. None of the above is correct.

44. Without adding or removing any water vapor, a sample of the atmosphere experiencing an increase in temperature will have
    a. a higher relative humidity.
    b. a lower relative humidity.
    c. the same relative humidity.
    d. a changed absolute humidity.

45. Cooling a sample of water vapor results in a (an)
    a. increased capacity.
    b. decreased capacity.
    c. unchanged capacity.

46. On a clear, calm, and cool night, dew or frost is most likely to form
    a. under trees or other shelters.
    b. on bare ground on the side of a hill.
    c. under a tree on the side of a hill.
    d. on grass in an open, low-lying area.

47. The density of the atmosphere
   a. increases with increasing altitude.
   b. decreases with increasing altitude.
   c. remains the same, regardless of altitude.
   d. decreases with decreasing altitude.
48. Condensation nuclei provide a surface for fog or cloud formation. These particles include
   a. salt crystals.
   b. soot.
   c. dust.
   d. All of the above are correct.

Answers

1. c 2. d 3. b 4. a 5. b 6. d 7. c 8. a 9. d 10. b 11. b 12. c 13. b 14. b
15. b 16. a 17. d 18. c 19. a 20. b 21. b 22. d 23. c 24. a 25. b 26. d
27. b 28. d 29. b 30. a 31. c 32. d 33. d 34. b 35. c 36. b 37. b 38. a
39. c 40. b 41. a 42. d 43. c 44. b 45. b 46. d 47. b 48. d

QUESTIONS FOR THOUGHT

1. What is the hydrologic cycle? Why is it important?
2. What is the meaning of "normal" atmospheric pressure?
3. Explain the "greenhouse effect." Is a greenhouse a good analogy for Earth's atmosphere? Explain.
4. What is a temperature inversion? Why does it increase air pollution?
5. Describe how the ozone layer protects living things on Earth's surface. Why is there some concern about this ozone layer?
6. What is wind? What is the energy source for wind? Explain.
7. Explain the relationship between air temperature and air density.
8. Why does heated air rise?
9. Provide an explanation for the observation that an airplane flying at the top of the troposphere takes several hours longer to fly from the east coast to the west coast than it does to make the return trip.
10. If evaporation cools the surroundings, does condensation warm the surroundings? Explain.
11. Explain why a cooler temperature can result in a higher relative humidity when no water vapor was added to the air.
12. What is the meaning of the expression, "It's not the heat, it's the humidity"?
13. What is the meaning of the dew point temperature?
14. Explain why frost is more likely to form on a clear, calm, and cool night than on nights with other conditions.

FOR FURTHER ANALYSIS

1. Describe how you could use a garden hose and a bucket of water to make a barometer. How high a column of water would standard atmospheric pressure balance in a water barometer?
2. If heated air rises, why is there snow on top of a mountain and not at the bottom?
3. According to the U.S. National Oceanic and Atmospheric Administration, the atmospheric concentration of CO_2 has been increasing, global surface temperatures have increased 0.2 to 0.3°C (0.4°F) over the past twenty-five years, and sea level has been rising 1 to 2 mm/yr since the late nineteenth century. Describe what evidence you would look for to confirm these increases are due to human activity rather than changes in the Sun's output or energy, or changes in Earth's orbit.
4. Evaluate the requirement that differential heating must take place before wind will blow. Do any winds exist without differential heating?
5. Given the current air temperature and relative humidity, explain how you could use the graph in Figure 22.15 to find the dew point temperature.

INVITATION TO INQUIRY

What Is Your Angle?

A radiometer is a device that can be used to measure radiant energy. It has four vanes that spin in a partial vacuum, and the rate of spinning is related to the amount of radiant energy received. Point a flashlight beam so it shines directly on the vanes. Move far enough away so you can count the revolutions per minute. Investigate the relationship between the angle of radiation received and the amount of energy received as shown by the revolutions per minute. Vary the angle so it is 0° (straight in), 30°, 60°, and 90° (straight down). What is the relationship between the angle and energy received? How many examples can you find that involve this relationship between the angle of sunlight received and the amount of radiant energy received?

23

Weather and Climate

You cannot control the weather, but the weather certainly can control you. As you can imagine, this weather change over Miami Beach, Florida, probably altered many plans for golf, swimming, and many other outdoor activities, as well as how people dressed to go outside.

CORE **CONCEPT**

Solar radiation drives cycles in Earth's atmosphere, and some of these cycles determine weather and climate.

OUTLINE

Clouds and Precipitation
Water cycles into Earth's atmosphere as water vapor and out as condensation and precipitation.

Weather Fronts
Movement of air masses brings about rapid changes in the weather.

Climate
Climate is the general pattern of weather.

Air Masses
A uniform body of air is called an air mass.

Major Storms
Rapid and violent changes of weather occur with three kinds of major storms: thunderstorms, tornadoes, and hurricanes.

551

The condition of the atmosphere can be described by the appearance of the sky, direction of the wind, and humidity ("feel" of the air) (Figure 23.1). This is what you do when you step outside and decide to carry an umbrella or not. You know that the probability of rain or other weather phenomena can be predicted by observing certain patterns in atmospheric conditions.

Weather proverbs are statements relating to atmospheric conditions by the appearance of the sky, wind direction, and humidity. They usually are meaningful as forecasting tools, as long as they were not transported from the part of the world where they were developed. Here are some such proverbs:

Mare's tails and mackerel scales make tall ships carry low sails.

A rainbow in the morning gives you fair warning.

A ring around the Sun or Moon means rain or snow coming soon.

If the salt is sticky and gains weight, it will rain before too late.

Can you figure out the atmospheric conditions described by the proverbs? Assuming that you live in the mid-latitudes, where the wind normally moves from west to east, can you determine what the proverbs mean about weather predictions? Return to these proverbs after you finish the chapter to confirm your skill at forecasting the weather from observations and weather proverbs.

CLOUDS AND PRECIPITATION

Water cycles continuously into and out of the atmosphere through the processes of evaporation, condensation, and precipitation. When water evaporates, individual water molecules leave the liquid and enter the atmosphere as the gas called *water vapor*. While in the liquid state, water molecules are held together by attractive molecular forces. Water molecules have a wide range of kinetic energies, and occasionally the more energetic ones are able to overcome the attractive forces, breaking away. When they do escape, water vapor molecules carry the latent heat of vaporization with them, as discussed in chapter 4. Because water vapor molecules take energy with them, evaporation is a cooling process. If incoming solar radiation did not supply energy, Earth's surface and the ocean would soon become cooler and cooler from the continuous evaporation that takes place. The Sun supplies the energy that maintains surface temperatures, which allows the ongoing evaporation of water. Thus, it is the Sun that supplies the energy required to evaporate water.

Water vapor in the atmosphere does not remain for more than several weeks, but during this time, it is transported by the winds of Earth. Eventually, the air becomes cooled and the relative humidity increases to 100 percent. The water vapor in the saturated air now condenses to form the tiny droplets of clouds. The water returns to the surface as precipitation that falls from the clouds. Each year, on the average, about 97 cm (about 38 in) of water evaporate from Earth's oceans, but only 90 cm (about 35 in) are returned by precipitation. The deficit is made up, on the average, by the return of 7 cm (about 3 in) per year by streams flowing from the continents into the oceans. If rivers and streams did not cycle water back to the ocean, it would be lowered each year by a depth of 7 cm (about 3 in). On the other hand,

if the atmosphere did not cycle water vapor back over the land, there would eventually be no water on the land. Both streams and precipitation are part of a never-ending series of events involving the ocean and lands of Earth. The series of events is called the **hydrologic cycle.** Overall, the hydrologic cycle can be considered to have four main events: (1) *evaporation* of water from the ocean, (2) *transport* of water vapor through the atmosphere, (3) *condensation and precipitation* of water on the lands, and (4) the *return of water* to the ocean by rivers and streams (Figure 23.2). This definition of the hydrologic cycle involves only the ocean and the lands, but water vapor also evaporates from the land and may condense and precipitate back to the land without ever returning to the ocean. This can be considered a small subcycle within the overall hydrologic cycle. The ocean-land exchange is the major cycle, and many small subcycles also exist. The section on cloud-forming processes is about the part of the hydrologic cycle that returns water to Earth's surface. The cloud-forming condensation processes will be considered first, followed by a discussion of the processes that result in precipitation falling from the clouds.

CLOUD-FORMING PROCESSES

Clouds form when a mass of air above the surface is cooled to its dew point temperature. In general, the mass of air is cooled because something has given it an upward push, moving it to higher levels in the atmosphere. There are three major causes of upward air movement: (1) *convection* resulting from differential heating, (2) mountain ranges that serve as *barriers* to moving air masses, and (3) the meeting of *moving air masses* with different densities; for example, a cold, dense mass of air meeting a warm, less dense mass of air.

FIGURE 23.1 Weather is a description of the changeable aspects of the atmosphere, the temperature, rainfall, pressure, and so forth, at a particular time. These changes usually affect your daily life one way or another, but some of them seem more inconvenient than others.

The three major causes of uplifted air sometimes result in clouds, but just as often they do not. Whether clouds form or not depends on the condition of the atmosphere at the time. As a parcel of warm air is moved upward, it tends to stay together, mixing very little with the surrounding air. As it is forced upward, it expands and becomes cooler because it is expanding. Similarly, the temperature of a gas increases when it is compressed. So rising air is cooled and descending air is warmed.

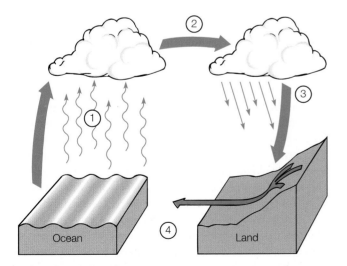

FIGURE 23.2 The main events of the hydrologic cycle are: (1) The evaporation of water from the ocean, (2) the transport of water vapor through the atmosphere, (3) the condensation and precipitation of water on the land, and (4) the return of water to the ocean by rivers and streams.

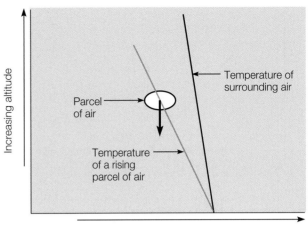

FIGURE 23.3 In a state of atmospheric stability, the parcel of air will always be cooler, and therefore more dense, than the surrounding air at any altitude. It will, therefore, return to the original level when the upward force is removed.

What happens to a parcel of air that is pushed upward depends on the difference in density between the parcel and the surrounding air. Air temperature will tell you about air density since the density of air is determined by its temperature. Instruments attached to a weather balloon can measure the change of temperature with altitude. By comparing this change with the rate of cooling by expansion, the state of atmospheric stability can be determined. There are many different states of atmospheric stability, and the following is a simplified description of just a few of the possible states, first considering dry air only.

The atmosphere is in a state of *stability* when a lifted parcel of air is cooler than the surrounding air (Figure 23.3). Being cooler, the parcel of air will be more dense than the surroundings. If it is moved up to a higher level and released, it will move back to its former level. A lifted parcel of air always returns to its original level when the atmosphere is stable. Any clouds that do develop in a stable atmosphere are usually arranged in the horizontal layers of stratus-type clouds.

The atmosphere is in a state of *instability* when a lifted parcel of air is warmer than the surrounding air (Figure 23.4). Being warmer, the parcel of air will be less dense than the surroundings. If it is moved up to a higher level, it will continue moving after the uplifting force is removed. Cumulus clouds usually develop in an unstable atmosphere, and the rising parcels of air, called thermals, can result in a very bumpy airplane ride.

So far, only dry air has been considered. As air moves upward and cools from expansion, sooner or later the dew point is reached and the air becomes saturated. As some of the water vapor in the rising parcel condenses to droplets, the latent heat of vaporization is released. The release of latent heat warms the air in the parcel and decreases the density even more, accelerating the ascent. This leads to further condensation and the formation of towering cumulus clouds, often leading to rain.

The state of the atmosphere and the moisture content of an uplifted parcel of air are not the only variables affecting whether or not clouds form. When the dew point temperature is reached,

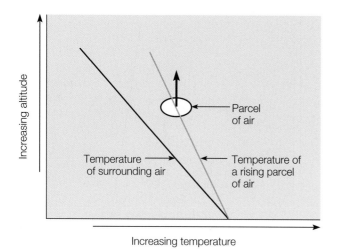

FIGURE 23.4 In a state of atmospheric instability, a parcel of air will always be warmer, and therefore less dense, than the surrounding air at any altitude. The parcel will, therefore, continue on in the direction pushed when the upward force is removed.

the water vapor in the air tends to condense, but now it requires the help of tiny microscopic particles called *condensation nuclei*. Without condensation nuclei, water vapor condenses to tiny droplets that are soon torn apart by collisions with other water vapor molecules. Thus, without condensation nuclei, further cooling can result in the air parcel becoming *supersaturated*, containing more than its normal saturation amount of water vapor. Condensation nuclei promote the condensation of water vapor into tiny droplets that are not torn apart by molecular collisions. Soon the stable droplets grow on a condensation nucleus to a diameter of about one-tenth the diameter of a human hair, but a wide range of sizes may be present. This accumulation of tiny droplets can easily remain suspended in the air from the slightest air movement. An average cloud may contain hundreds of such droplets per cubic centimeter, with an average density of 1 g/m^3. So, a big cloud that has a volume of 1 km^3 would contain about a million liters (about 264,000 gal) of water.

A single water droplet in a cloud is so tiny that it would be difficult to see with the unaided eye. When water droplets accumulate in huge numbers in clouds, what you see depends on the size of the droplets making up the cloud and the relative position of you, the cloud, and the Sun. You will see a white cloud, for example, if you are between the cloud and the Sun so that you see reflected sunlight from the cloud. The same cloud will appear to be gray if it is between you and the Sun, positioned so that it filters the sunlight coming toward you.

ORIGIN OF PRECIPITATION

Water that returns to the surface of Earth, in either the liquid or solid form, is called **precipitation** (Figure 23.5). Note that dew and frost are not classified as precipitation because they form directly on the surface and do not fall through the air. Precipitation seems to form in clouds by one of two processes: (1) the *coalescence* of cloud droplets or (2) the *growth of ice crystals.*

FIGURE 23.5 Precipitation is water in the liquid or solid form that returns to the surface of Earth. The precipitation falling on the hills to the left is liquid, and each raindrop is made from billions of the tiny droplets that make up the clouds. The tiny droplets of clouds become precipitation by merging to form larger droplets or by the growth of ice crystals that melt while falling.

It would appear difficult for cloud droplets to merge, or coalesce, with one another since any air movement moves them all at the same time, not bringing them together. Condensation nuclei come in different sizes, however, and cloud droplets of many different sizes form on these different-sized nuclei. Larger cloud droplets are slowed less by air friction as they drift downward, and they collide and merge with smaller droplets as they fall. They may merge, or coalesce, with a million other droplets before they fall from the cloud as raindrops. This *coalescence process* of forming precipitation is thought to take place in warm cumulus clouds that form near the ocean in the tropics. These clouds contain giant salt condensation nuclei and have been observed to produce rain within about twenty minutes after forming.

Clouds at middle latitudes, away from the ocean, also produce precipitation, so there must be a second way that precipitation forms. The *ice-crystal process* of forming precipitation is important in clouds that extend high enough in the atmosphere to be above the freezing point of water. Water molecules are more strongly bonded to each other in an ice crystal than in liquid water. Thus, an ice crystal can capture water molecules and grow to a larger size while neighboring water droplets are evaporating. As they grow larger and begin to drift toward the surface, they may coalesce with other ice crystals or droplets of water, soon falling from the cloud. During the summer, they fall through warmer air below and reach the ground as raindrops. During the winter, they fall through cooler air below and reach the ground as snow.

Tiny water droplets do not freeze as readily as a larger mass of liquid water, and many droplets do not freeze until the temperature is below about −40°C (−40°F). Water that is still in the liquid state when the temperature is below the freezing temperature is said to be *supercooled*. Supercooled clouds of water droplets are common between the temperatures of −40°C and 0°C (−40°F and 32°F), a range of temperatures that is often found in the upper atmosphere. The liquid droplets at these temperatures need solid particles called *ice-forming nuclei* to freeze upon. Generally, dust from the ground serves as ice-forming nuclei that start the ice-crystal process of forming precipitation. Artificial rainmaking has been successful by (1) dropping crushed dry ice, which is cooler than −40°C, on top of a supercooled cloud and (2) by introducing "seeds" of ice-forming nuclei in supercooled clouds. Tiny crystals from the burning of silver iodide are effective ice-forming nuclei, producing ice crystals at temperatures as high as −4.0°C (about 25°F). Attempts at ground-based cloud seeding with silver iodide in the mountains of the western United States have suggested up to 15 percent more snowfall, but it is difficult to know how much snowfall would have resulted without the seeding.

The basic form of a cloud has meaning about the general type of precipitation that can occur as well as the coming weather. Cumulus clouds usually produce showers or thunderstorms that last only brief periods of time. Longer periods of drizzle, rain, or snow usually occur from stratus clouds. Cirrus clouds do not produce precipitation of any kind, but they may tell us about the coming weather.

WEATHER PRODUCERS

The idealized model of the general atmospheric circulation starts with the poleward movement of warm air from the tropics. The region between 10°N and 10°S of the equator receives more direct radiation, on the average, than other regions of Earth's surface. The air over this region is heated more, expands, and becomes less dense as a consequence of the heating. This less dense air is buoyed up by convection to heights up to 20 km (about 12 mi) as it is cooled by radiation to less than −73°C (about −110°F). This accumulating mass of cool, dry air spreads north and south toward both poles (see Figure 22.13 on p. 541), then sinks back toward the surface at about 30°N and 30°S. The descending air is warm and dry by the time it reaches the surface. Part of the sinking air then moves back toward the equator across the surface, completing a large convective cell. This giant cell has a low-pressure belt over the equator and high-pressure belts over the subtropics near latitudes of 30°N and 30°S. The other part of the sinking air moves poleward across the surface, producing belts of westerly winds in both hemispheres to latitudes of about 60°. On an Earth without landmasses next to bodies of water, a belt of low pressure would probably form around 60° in both hemispheres, and a high-pressure region would form at both poles.

The overall pattern of pressure belts and belts of prevailing winds is seen to shift north and south with the seasons, resulting in a seasonal shift in the types of weather experienced at a location. This shift of weather is related to three related weather producers: (1) the movement of large bodies of air, called *air*

masses, that have acquired the temperature and moisture conditions where they have been located, (2) the leading *fronts* of air masses when they move, and (3) the local *high- and low-pressure* patterns that are associated with air masses and fronts. These are the features shown on almost all daily weather maps, and they are the topics of this section.

AIR MASSES

An **air mass** is defined as a large, horizontally uniform body of air with nearly the same temperature and moisture conditions. An air mass forms when a large body of air, perhaps covering millions of square kilometers, remains over a large area of land or water for an extended period of time. While it is stationary, it acquires the temperature and moisture characteristics of the land or water through the heat transfer processes of conduction, convection, and radiation and through the moisture transfer processes of evaporation and condensation. For example, a large body of air that remains over the cold, dry, snow-covered surface of Siberia for some time will become cold and dry. A large body of air that remains over a warm tropical ocean, on the other hand, will become warm and moist. Knowledge about the condition of air masses is important because they tend to retain the acquired temperature and moisture characteristics when they finally break away, sometimes moving long distances. An air mass that formed over Siberia can bring cold, dry air to your location, while an air mass that formed over a tropical ocean will bring warm, moist air.

Air masses are classified according to the temperature and moisture conditions where they originate. There are two temperature extreme possibilities, a **polar air mass** from a cold region and a **tropical air mass** from a warm region. There are also two moisture extreme possibilities, a moist **maritime air mass** from over the ocean and a generally dry **continental air mass** from over the land. Thus, there are four main types of air masses that can influence the weather (1) continental polar, (2) maritime polar, (3) continental tropical, and (4) maritime tropical. Figure 23.6 shows the general direction in which these air masses usually move over the mainland United States.

Once an air mass leaves its source region, it can move at speeds of up to 800 km (about 500 mi) per day while mostly retaining the temperature and moisture characteristics of the source region (Figure 23.7). If it slows and stagnates over a new location, however, the air may again begin to acquire a new temperature and moisture equilibrium with the surface. When a location is under the influence of an air mass, the location is having a period of *air mass weather.* This means that the weather conditions will generally remain the same from day to day with slow, gradual changes. Air mass weather will remain the same until a new air mass moves in or until the air mass acquires the conditions of the new location. This process may take days or several weeks, and the weather conditions during this time depend on the conditions of the air mass and conditions at the new location. For example, a polar continental air mass arriving over a cool, dry land area may produce a temperature inversion with the air colder near the surface than higher up. When the temperature increases with height, the air

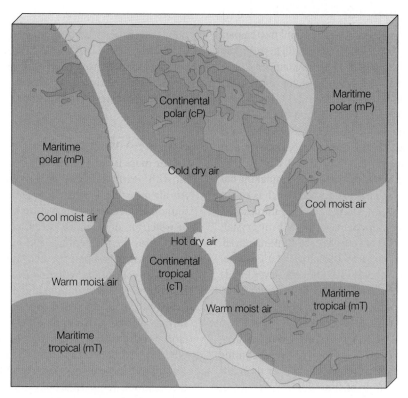

FIGURE 23.6 The air masses that affect weather in North America. The importance of the various air masses depends on the season. In winter, for instance, the continental tropical air mass disappears and the continental polar air mass exerts its greatest influence.

is stable and cloudless, and cold weather continues with slow, gradual warming. The temperature inversion may also result in hazy periods of air pollution in some locations. A continental air mass arriving over a generally warmer land area, on the other hand, results in a condition of instability. In this situation, each day will start clear and cold, but differential heating during the day develops cumulus clouds in the unstable air. After sunset, the clouds evaporate, and a clear night results because the thermals during the day carried away the dust and air pollution. Thus, a dry, cold air mass can bring different weather conditions, each depending on the properties of the air mass and the land it moves over.

WEATHER FRONTS

The boundary between air masses of different temperatures is called a **front.** A front is actually a thin transition zone between two air masses that ranges from about 5 to 30 km (about 3 to 20 mi) wide, and the air masses do not mix other than in this narrow zone. The density differences between the two air masses prevent any general mixing since the warm, less-dense air mass is forced upward by the cooler, more dense air moving under it. You may have noticed on a daily weather map that fronts are usually represented with a line bulging outward in the direction of cold air mass movement (Figure 23.8). A cold air mass is much like a huge, flattened bubble of air that moves across the land (Figure 23.9). The line on a weather map represents the

place where the leading edge of this huge, flattened bubble of air touches the surface of Earth.

A **cold front** is formed when a cold air mass moves into warmer air, displacing it in the process. A cold front is generally steep, and when it runs into the warmer air, it forces the warmer air to rise quickly. If the warm air is moist, it is quickly cooled by expansion to the dew point temperature, resulting in large, towering cumulus clouds and thunderclouds along the front (Figure 23.10). You may have observed that thunderstorms created by an advancing cold front often form in a line along the front. These thunderstorms can be intense but are usually over quickly, soon followed by a rapid drop in temperature from the cold air mass. The passage of the cold front is also marked by a rapid shift in the wind direction and a rapid increase in the barometric pressure. Before the cold front arrives, the wind is generally moving toward the front as warm, less dense air is forced upward by the cold, more dense air. The lowest barometric pressure reading is associated with the lifting of the warm air at the front. After the front passes your location, you are in the cooler, more dense air that is settling outward, so the barometric pressure increases and the wind shifts with the movement of the cold air mass.

A **warm front** forms when a warm air mass advances over a mass of cooler air. Since the advancing warm air is less dense than the cooler air it is displacing, it generally overrides the cooler air, forming a long, gently sloping front. Because of this, the overriding warm air may form clouds far in advance of the

FIGURE 23.7 This satellite photograph shows the result of a polar air mass moving southeast over the southern United States. Clouds form over the warmer waters of the Gulf of Mexico and the Atlantic Ocean, showing the state of atmospheric instability from the temperature differences.

ground-level base of the front (Figure 23.11). This may produce high cirrus clouds a day or more in advance of the front, which are followed by thicker and lower stratus clouds as the front advances. Usually these clouds result in a broad band of drizzle, fog, and the continuous light rain usually associated with stratus clouds. This light rain (and snow in the winter) may last for days as the warm front passes.

Sometimes the forces influencing the movement of a cold or warm air mass lessen or become balanced, and the front stops advancing. When this happens, a stream of cold air moves along the north side of the front, and a stream of warm air moves along the south side in an opposite direction. This is called a **stationary front** because the edge of the front is not advancing (Figure 23.12A). A stationary front may sound as if it is a mild frontal weather maker because it is not moving. Actually, a stationary front represents an unstable situation that can result in a major atmospheric storm. This type of storm is discussed in the section on waves and cyclones.

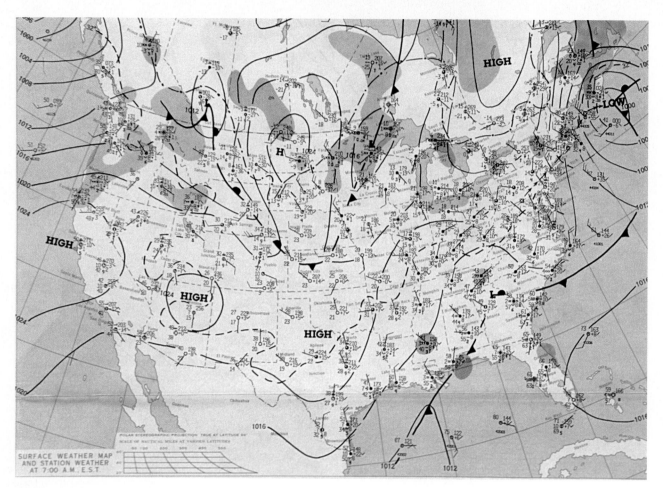

FIGURE 23.8 This weather map of the United States shows two fronts with associated low-pressure areas and five areas with high pressure.

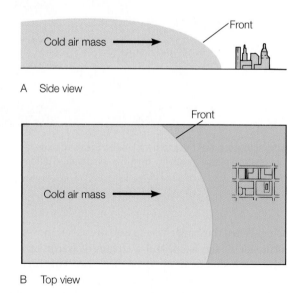

A Side view

B Top view

FIGURE 23.9 (*A*) A cold air mass is similar to a huge, flattened bubble of cold air that moves across the land. The front is the boundary between two air masses, a narrow transition zone of mixing. (*B*) A front is represented by a line on a weather map, which shows the location of the front at ground level.

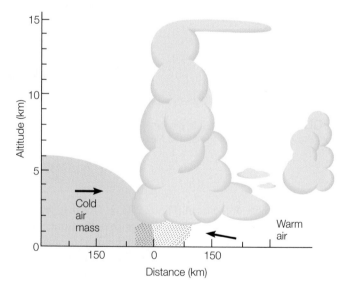

FIGURE 23.10 An idealized cold front, showing the types of clouds that might occur when an unstable cold air mass moves through unstable warm air. Stable air would result in more stratus clouds rather than cumulus clouds.

Urban areas are often 2 to 10 degrees warmer at night than the surrounding rural areas. Climatologists call this difference the "urban heat island effect." This effect is created by all the bricks, concrete, and asphalt of buildings, parking lots, and streets. These structures are warmed by sunlight during the day, then reradiate stored heat at night. Building surfaces also disrupt the normal cooling by infrared radiation. Surrounding rural areas are cooler because less heat is absorbed during the day, and more is released by radiation at night.

It is possible to reduce the heat island effect by using alternatives to conventional asphalt and concrete. Resins and rubberized asphalt, for example, do not absorb as much heat during the day. Landscaping to create islands of "urban forests" also helps.

QUESTIONS TO DISCUSS

1. Developers are slow to consider taking steps to diminish heat island effects. Should governments provide a tax break to encourage developers to diminish the heat island effect? Why or why not?

2. What are the advantages and disadvantages of required urban planning, including the creation of forest islands in urban areas?

3. What solutions can you think of to diminish the urban heat island effect?

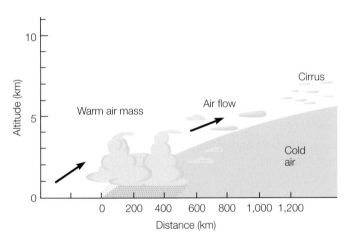

FIGURE 23.11 An idealized warm front, showing a warm air mass overriding and pushing cold air in front of it. Notice that the overriding warm air produces a predictable sequence of clouds far in advance of the moving front.

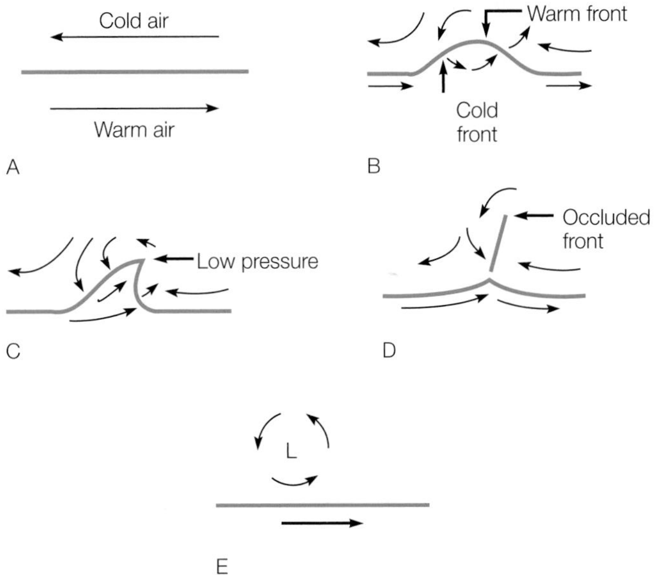

FIGURE 23.12 The development of a low-pressure center, or cyclonic storm, along a stationary front as seen from above. (*A*) A stationary front with cold air on the north side and warm air on the south side. (*B*) A wave develops, producing a warm front moving northward on the right side and a cold front moving southward on the left side. (*C*) The cold front lifts the warm front off the surface at the apex, forming a low-pressure center. (*D*) When the warm front is completely lifted off the surface, an occluded front is formed. (*E*) The cyclonic storm is now a fully developed low-pressure center.

WAVES AND CYCLONES

A stationary front often develops a bulge, or *wave,* in the boundary between cool and warm air moving in opposite directions (Figure 23.12B). The wave grows as the moving air is deflected, forming a warm front moving northward on the right side and a cold front moving southward on the left side. Cold air is more dense than warm air, and the cold air moves faster than the slowly moving warm front. As the faster-moving cold air catches up with the slower-moving warm air, the cold air underrides the warm air, lifting it upward. This lifting action produces a low-pressure area at the point where the two fronts come together (Figure 23.12C). The lifted air expands, cools by expansion, and reaches the dew point. Clouds form and precipitation begins from the lifting and cooling action. Within days after the wave first appears, the cold front completely overtakes the warm front, forming an occlusion (Figure 23.12D). An **occluded front** is one that has been lifted completely off the ground into the atmosphere. The disturbance is now a *cyclonic storm* with a fully developed low-pressure center. After forming, the low-pressure cyclonic

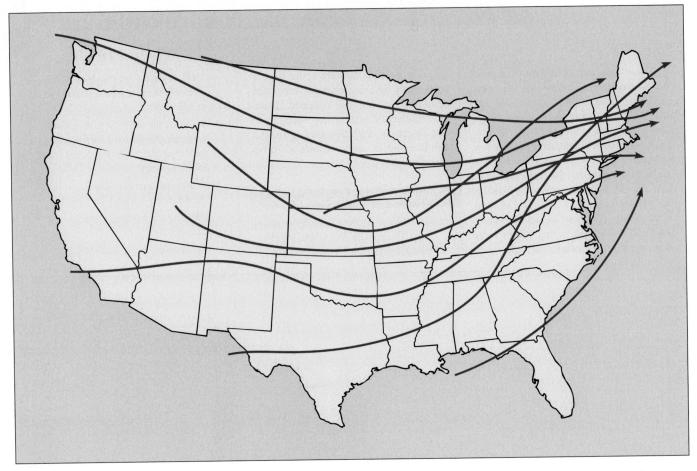

FIGURE 23.13 Cyclonic storms usually follow principal storm tracks across the continental United States in a generally easterly direction. This makes it possible to predict where the low-pressure storm might move next.

storm continues moving, taking the associated stormy weather with it in a generally easterly direction. Such cyclonic storms usually follow principal tracks along a front (Figure 23.13). Since they are observed generally to follow these same tracks, it is possible to predict where the storm might move next.

A **cyclone** is defined as a low-pressure center where the winds move into the low-pressure center and are forced upward. As air moves in toward the center, the Coriolis effect (see chapter 16) and friction with the ground cause the moving air to veer. In the Northern Hemisphere, this produces a counterclockwise circulation pattern (Figure 23.14). The upward movement associated with the low-pressure center of a cyclone cools the air, resulting in clouds, precipitation, and stormy conditions.

Air is sinking in the center of a region of high pressure, producing winds that move outward. In the Northern Hemisphere, the Coriolis effect and frictional forces deflect this wind to the right, producing a clockwise circulation (Figure 23.14). A high-pressure center is called an **anticyclone,** or simply a **high.** Since air in a high-pressure zone sinks, it is warmed, and the relative humidity is lowered. Thus, clear, fair weather is usually associated with a high. By observing the barometric pressure, you can watch for decreasing pressure, which can mean the coming of a cyclone and its associated stormy weather. You can also watch

for increasing pressure, which means a high and its associated fair weather are coming. Consulting a daily weather map makes such projections a much easier job, however.

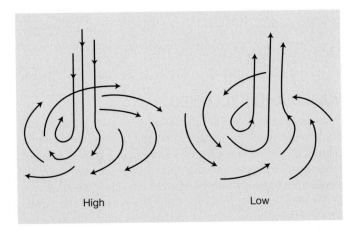

High   Low

FIGURE 23.14 Air sinks over a high-pressure center and moves away from the center on the surface, veering to the right in the Northern Hemisphere to create a clockwise circulation pattern. Air moves toward a low-pressure center on the surface, rising over the center.

MAJOR STORMS

A wide range of weather changes can take place as a front passes because there is a wide range of possible temperature, moisture, stability, and other conditions between the new air mass and the air mass that it is displacing. The changes that accompany some fronts may be so mild that they go unnoticed. Others are noticed only as a day with breezes or gusty winds. Still other fronts are accompanied by a rapid and violent weather change called a **storm**. A snowstorm, for example, is a rapid weather change that may happen as a cyclonic storm moves over a location. The most rapid and violent changes occur with three kinds of major storms: (1) thunderstorms, (2) tornadoes, and (3) hurricanes.

Thunderstorms

A **thunderstorm** is a brief but intense storm with rain, lightning and thunder, gusty and often strong winds, and sometimes hail. Thunderstorms usually develop in warm, very moist, and unstable air. These conditions set the stage for a thunderstorm to develop when something lifts a parcel of air, starting it moving upward. This is usually accomplished by the same three general causes that produce cumulus clouds: (1) differential heating, (2) mountain barriers, or (3) an occluded or cold front. Thunderstorms that occur from differential heating usually occur during warm, humid afternoons after the Sun has had time to establish convective thermals. In the Northern Hemisphere, most of these convective thunderstorms occur during the month of July. Frontal thunderstorms, on the other hand, can occur any month and any time of the day or night that a front moves through warm, moist, and unstable air.

Frontal thunderstorms generally move with the front that produced them. Thunderstorms that developed in mountains or over flat lands from differential heating can move miles after they form, sometimes appearing to wander aimlessly across the land. These storms are not just one big rain cloud but are sometimes made up of cells that are born, grow to maturity, then die out in less than an hour. The thunderstorm, however, may last longer than an hour because new cells are formed as old ones die out. Each cell is about 2 to 8 km (about 1 to 5 mi) in diameter and goes through three main stages in its life: (1) cumulus, (2) mature, and (3) final (Figure 23.15).

Damage from a thunderstorm is usually caused by the associated lightning, strong winds, or hail. As illustrated in Figure 23.15, the first stage of a thunderstorm begins as convection, mountains, or a dense air mass slightly lifts a mass of warm, moist air in an unstable atmosphere. The lifted air mass expands and cools to the dew point temperature, and a cumulus cloud forms. The latent heat of vaporization released by the condensation process accelerates the upward air motion, called an *updraft,* and the cumulus cloud continues to grow to towering heights. Soon the upward-moving, saturated air reaches the freezing level and ice crystals and snowflakes begin to form. When they become too large to be supported by the updraft, they begin to fall toward the surface, melting into raindrops in the warmer air they fall through. When they reach the surface, this marks the beginning of the mature stage. As the raindrops fall through the air, friction between the falling drops and the

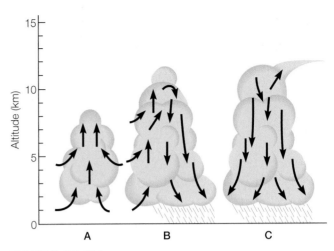

FIGURE 23.15 Three stages in the life of a thunderstorm cell. (*A*) The cumulus stage begins as warm, moist air is lifted in an unstable atmosphere. All the air movement is upward in this stage. (*B*) The mature stage begins when precipitation reaches the ground. This stage has updrafts and downdrafts side by side, which create violent turbulence. (*C*) The final stage begins when all the updrafts have been cut off and only downdrafts exist. This cuts off the supply of moisture, and the rain decreases as the thunderstorm dissipates. The anvil-shaped top is a characteristic sign of this stage.

cool air produces a downdraft in the region of the precipitation. The cool air accelerates toward the surface at speeds up to 90 km/h (about 55 mi/h), spreading out on the ground when it reaches the surface. In regions where dust is raised by the winds, this spreading mass of cold air from the thunderstorm has the appearance of a small cold front with a steep, bulging leading edge. This miniature cold front may play a role in lifting other masses of warm, moist air in front of the thunderstorm, leading to the development of new cells. This stage in the life of a thunderstorm has the most intense rainfall, winds, and possibly hail. As the downdraft spreads throughout the cloud, the supply of new moisture from the updrafts is cut off and the thunderstorm enters the final, dissipating stage. The entire life cycle, from cumulus cloud to the final stage, lasts for about an hour as the thunderstorm moves across the surface. During the mature stage of powerful updrafts, the top of the thunderstorm may reach all the way to the top of the troposphere, forming a cirrus cloud that is spread into an anvil shape by the strong winds at this high altitude.

The updrafts, downdrafts, and falling precipitation separate tremendous amounts of electric charges that accumulate in different parts of the thundercloud. Large drops of water tend to carry negative charges, and cloud droplets tend to lose them. The upper part of the thunderstorm develops an accumulation of positive charges as cloud droplets are uplifted, and the middle portion develops an accumulation of negative charges from larger drops that fall. There are many other charging processes at work, and the lower part of the thundercloud develops both negative and positive charges. The voltage of these charge centers builds to the point that the electrical insulating ability of the air between them is overcome and a giant electrical

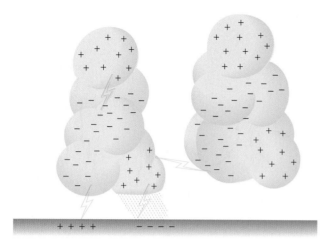

FIGURE 23.16 Different parts of a thunderstorm cloud develop centers of electric charge. Lightning is a giant electric spark that discharges the accumulated charges.

FIGURE 23.17 These hailstones fell from a thunderstorm in Iowa, damaging automobiles, structures, and crops.

discharge called *lightning* occurs (Figure 23.16). Lightning discharges occur from the cloud to the ground, from the ground to a cloud, from one part of the cloud to another part, or between two different clouds. The discharge takes place in a fraction of a second and may actually consist of a number of strokes rather than one big discharge. An extremely high temperature is produced around the path of the discharge, which may be only 6 cm (about 2 in) or so wide. The air it travels through is heated quickly, expanding into a sudden pressure wave that you hear as *thunder*. A nearby lightning strike produces a single, loud crack. Farther away strikes sound more like a rumbling boom as the sounds from the separate strokes become separated over distance. Echoing of the thunder produced at farther distances also adds to the rumbling sounds. The technique of calculating the distance to a lightning stroke by measuring the interval between the flash of the lightning and the boom of the thunder is discussed in chapter 5. Lightning can present a risk for people in the open, near bodies of water, or under a single, isolated tree during a thunderstorm. The safest place to be during a thunderstorm is inside a car or a building with a metal frame.

Updrafts are also responsible for *hail,* a frozen form of precipitation that can be very destructive to crops, automobiles, and other property. Hailstones can be irregular, somewhat spherical, or flattened forms of ice that range from the size of a BB to the size of a softball (Figure 23.17). Most hailstones, however, are less than 2 cm (about 1 in) in diameter. The larger hailstones have alternating layers of clear and opaque, cloudy ice. These layers are believed to form as the hailstone goes through cycles of falling then being returned to the upper parts of the thundercloud by updrafts. The clear layers are believed to form as the hailstone moves through heavy layers of supercooled water droplets, which accumulate quickly on the hailstone but freeze slowly because of the release of the latent heat of fusion. The cloudy layers are believed to form as the hailstone accumulates snow crystals or moves through a part of the cloud with less supercooled water droplets. In either case, rapid freezing traps air bubbles, which result in the opaque,

cloudy layer. Thunderstorms with hail are most common during the month of May in Colorado, Kansas, and Nebraska.

Tornadoes

A **tornado** is the smallest, most violent weather disturbance that occurs on Earth (Figure 23.18). Tornadoes occur with intense thunderstorms; they resemble long, narrow funnel or ropelike structures that drop down from a thundercloud and may or may not touch the ground. This ropelike structure is a rapidly whirling column of air, usually 100 to 400 m (about 330 to 1,300 ft) in diameter. An average tornado will travel 6 to 8 km (about 4 to 5 mi) on the ground, sometimes skipping into the air, then back down again. The bottom of the column moves across the ground at speeds that average about 50 km/h (about 30 mi/h). The speed of the whirling air in the column has been estimated to be up to about 480 km/h (about 300 mi/h), but most tornadoes have winds of less than 180 km/h (112 mi/h). The destruction is produced by the powerful winds, the sudden drop in atmospheric pressure that occurs at the center of the funnel, and the debris that is flung through the air like projectiles. A passing tornado sounds like very loud, continuous rumbling thunder with cracking and hissing noises that are punctuated by the crashing of debris projectiles.

On average, several hundred tornadoes are reported in the United States every year. These occur mostly during spring and early summer afternoons over the Great Plains states. Texas, Oklahoma, Kansas, and Iowa have such a high occurrence of tornadoes that the region is called "tornado alley." During the spring and early summer, this region has maritime tropical air from the Gulf of Mexico at the surface. Above this warm, moist layer is a layer of dry, unstable air that has just crossed the Rocky Mountains, moved along rapidly by the jet stream. The stage is now set for some event, such as a cold air mass moving in from the north, to shove the warm, moist air upward, and the result will be violent thunderstorms with tornadoes.

FIGURE 23.18 A tornado might be small, but it is the most violent storm that occurs on Earth. This tornado, moving across an open road, eventually struck Dallas, Texas.

CONCEPTS *Applied*

Tornado Damage

Tornadoes are rated on wind speed and damage. Here is the scale with approximate wind speeds:

0: Light damage, winds under 120 km/h (75 mi/h). Damage to chimneys, tree limbs broken, small trees pushed over, signs damaged.

1: Moderate damage, winds 121–180 km/h (76–112 mi/h). Roofing materials removed, mobile homes and moving autos pushed around or overturned.

2: Considerable damage, winds 181–250 km/h (113–155 mi/h). Roofs torn off homes, mobile homes demolished, boxcars overturned, large trees snapped or uprooted, light objects become missiles.

3: Severe damage, winds 251–330 km/h (156–205 mi/h). Roofs and some walls torn off homes, whole trains overturned, most trees uprooted, cars lifted off the ground and thrown.

4: Devastating damage, winds 331–418 km/h (206–260 mi/h). Homes leveled, cars thrown, large missiles generated.

5: Incredible damage, winds 419–512 km/h (261–318 mi/h). Homes demolished and swept away, automobile-sized missiles fly though the air more than 100 yards, trees debarked.

Hurricanes

What is the difference between a tropical depression, a tropical storm, and a hurricane? In general, they are all storms with strong upward atmospheric motion and a cyclonic surface wind circulation (Figure 23.19). They are born over tropical or subtropical waters and are not associated with a weather front. The varieties of storm intensities are classified according to the *speed* of the maximum sustained surface winds.

A *tropical depression* is an area of low pressure around which the winds are generally moving 55 km/h (about 35 mi/h) or less. The tropical depression might dissolve into nothing, or it might develop into a more intense disturbance. A *tropical storm* is a more intense low-pressure area with winds between 56 and 120 km/h (about 35 to 75 mi/h). A **hurricane** is a very intense low-pressure area with winds greater than 120 km/h (about 75 mi/h). A strong storm of this type is called a "hurricane" if it occurs over the Atlantic Ocean or the Pacific Ocean east of the international date line. It is called a "typhoon" if it occurs over the North Pacific Ocean west of the international date line.

A tropical cyclone is similar to the wave cyclone of the mid-latitudes because both have low-pressure centers with a counterclockwise circulation in the Northern Hemisphere. They are different because a wave cyclone is usually about 2,500 km (about 1,500 mi) wide, has moderate winds, and receives its energy from the temperature differences between two air masses. A tropical cyclone, on the other hand, is often less than 200 km (about 125 mi) wide, has very strong winds, and receives its energy from the latent heat of vaporization released during condensation.

A fully developed hurricane has heavy bands of clouds, showers, and thunderstorms that rapidly rotate around a relatively clear, calm eye (Figure 23.20). As a hurricane approaches a location, the air seems unusually calm as a few clouds appear, then thicken as the wind begins to gust. Over the next six hours or so, the overall wind speed increases as strong gusts and intense rain showers occur. Thunderstorms, perhaps with tornadoes, and the strongest winds occur just before the winds suddenly die down and the sky clears with the arrival of the eye

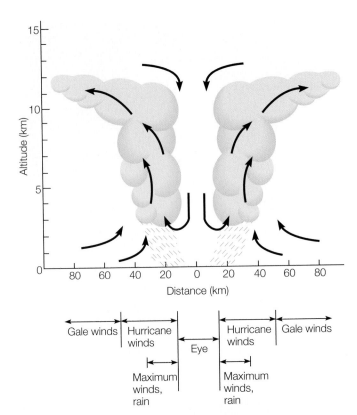

FIGURE 23.20 Cross section of a hurricane.

FIGURE 23.19 This is a satellite photo of hurricane John, showing the eye and counterclockwise motion.

of the hurricane. The eye is an average of 10 to 15 km (about 6 to 9 mi) across, and it takes about an hour or so to cross a location. When the eye passes, the intense rain showers, thunderstorms, and hurricane-speed winds begin again, this time blowing from the opposite direction. The whole sequence of events may be over in a day or two, but hurricanes are unpredictable and sometimes stall in one location for days. In general, they move at a rate of 15 to 50 km/h (about 10 to 30 mi/h).

Most of the damage from hurricanes results from strong winds, flooding, and the occasional tornado. Flooding occurs from the intense, heavy rainfall but also from the increased sea level that results from the strong, constant winds blowing seawater toward the shore. The sea level can be raised some 5 m (about 16 ft) above normal, with storm waves up to 15 m (about 50 ft) high on top of this elevated sea level. Overall, large inland areas can be flooded, resulting in extensive property damage. A single hurricane moving into a populated coastal region has

caused billions of dollars of damage and the loss of hundreds of lives in the past. Today, the National Weather Service tracks hurricanes by weather satellites. Warnings of hurricanes, tornadoes, and severe thunderstorms are broadcast locally over special weather alert stations located across the country.

In August 2005, hurricane Katrina initially struck near Miami, Florida, as a category 1 hurricane. It then moved into the Gulf of Mexico and grew to a strong category 3 hurricane that moved up the Gulf to the eastern Louisiana and western Mississippi coast. This massive storm had winds that extended outward 190 km (120 mi) from the center, resulting in severe storm damage over a wide area as it struck the coast on August 29. Damage resulted from a storm surge that exceeded 8m (25 ft), heavy rainfall, wind damage, and the failure of the levee system in New Orleans. Overall, this resulted in an estimated $81 billion in damages and more than 1,836 fatalities.

Katrina had sustained winds of 200 km/h (125 mi/h) as it stuck the shore, but other, smaller hurricanes have had stronger sustained winds when they struck the shore. These include:

- hurricane Camille, which hit Mississippi with 306 km/h (190 mi/h) sustained winds on August 17, 1969;
- hurricane Andrew, which hit south Florida with 266 km/h (165 mi/h) sustained winds on August 24, 1992; and
- hurricane Charley, which hit Punta Gorda, Florida, with 240 km/h (150 mi/h) sustained winds on August 13, 2009.*

*Hurricane data from National Climatic Data Center: www.ncdc.noaa.gov/oa/climate/research/2005/katrina.html.

CONCEPTS *Applied*

Hurricane Damage

Hurricanes are classified according to category and damage to be expected. Here is the classification scheme:

Category	Damage	Winds
1	Minimal	120–153 km/h (75–95 mi/h)
2	Moderate	154–177 km/h (96–110 mi/h)
3	Extensive	178–210 km/h (111–130 mi/h)
4	Extreme	211–250 km/h (131–155 mi/h)
5	Catastrophic	over 250 km/h (155 mi/h)

WEATHER FORECASTING

Today, weather predictions are based on information about the characteristics, location, and rate of movement of air masses and associated fronts and pressure systems. This information is summarized as average values, then fed into a computer model of the atmosphere. The model is a scaled-down replica of the real atmosphere, and changes in one part of the model result in changes in another part of the model, just as they do in the real atmosphere. Underlying the computer model are the basic scientific laws concerning solar radiation, heat, motion, and the gas laws. All these laws are written as a series of mathematical equations, which are applied to thousands of data points in a three-dimensional grid that represents the atmosphere. The computer is given instructions about the starting conditions at each data point, that is, the average values of temperature, atmospheric pressure, humidity, wind speed, and so forth. The computer is then instructed to calculate the changes that will take place at each data point, according to the scientific laws, within a very short period of time. This requires billions of mathematical calculations when the program is run on a worldwide basis. The new calculated values are then used to start the process all over again, and it is repeated some 150 times to obtain a one-day forecast (Figure 23.21).

A problem with the computer model of the atmosphere is that small-scale events are inadequately treated, and this introduces errors that grow when predictions are attempted for further and further into the future. Small eddies of air, for example, or gusts of wind in a region have an impact on larger-scale atmospheric motions such as those larger than a cumulus cloud. But all of the small eddies and gusts cannot be observed without filling the atmosphere with measuring instruments. This lack of ability to observe small events that can change the large-scale events introduces uncertainties in the data, which, over time, will increasingly affect the validity of a forecast.

To find information about the accuracy of a forecast, the computer model can be run several different times, with each run having slightly different initial conditions. If the results of all the runs are close to each other, the forecasters can feel confident that the atmosphere is in a predictable condition, and this

FIGURE 23.21 Supercomputers make routine weather forecasts possible by solving mathematical equations that describe changes in a mathematical model of the atmosphere. This "fish-eye" view was necessary to show all of this Cray supercomputer at CERN, the European Center of Particle Physics.

means the forecast is probably accurate. In addition, multiple computer runs can provide forecasts in the form of probabilities. For example, if eight out of ten forecasts indicate rain, the "averaged" forecast might call for an 80 percent chance of rain.

The use of new computer technology has improved the accuracy of next-day forecasts tremendously, and the forecasts up to three days are fairly accurate, too. For forecasts of more than five days, however, the number of calculations and the effect of uncertainties increase greatly. It has been estimated that the reductions of observational errors could increase the range of accurate forecasting up to two weeks. The ultimate range of accurate forecasting will require a better understanding—and thus an improved model—of patterns of changes that occur in the ocean as well as in the atmosphere. All of this increased understanding and reduction of errors leads to an estimated ultimate future forecast of three weeks, beyond which any pinpoint forecast would be only slightly better than a wild guess. In the meantime, regional and local daily weather forecasts are fairly accurate, and computer models of the atmosphere now provide the basis for extending the forecasts for up to about a week.

CLIMATE

Changes in the atmospheric condition over a brief period of time, such as a day or a week, are referred to as changes in the *weather*. Weather changes follow a yearly pattern of seasons that are referred to as winter weather, summer weather, and so on. All of these changes are part of a composite, larger pattern called **climate**. Climate is the general pattern of the weather that occurs for a region over a number of years. Among other things, the climate determines what types of vegetation grow in a particular region, resulting in characteristic groups of plants associated

FIGURE 23.22 The climate determines what types of plants and animals live in a location, the types of houses that people build, and the lifestyles of people. This orange tree, for example, requires a climate that is relatively frost-free, yet it requires some cool winter nights to produce a sweet fruit.

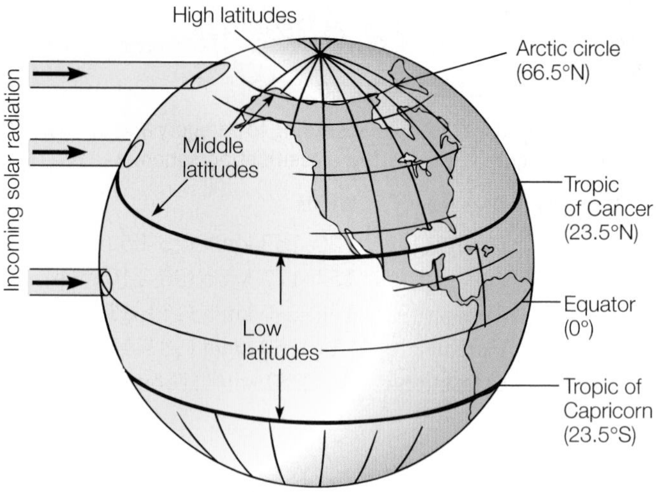

FIGURE 23.23 Latitude groups based on incoming solar radiation. The low latitudes receive vertical solar radiation at noon some time of the year, the high latitudes receive no solar radiation at noon during some time of the year, and the middle latitudes are in between.

with the region (Figure 23.22). For example, orange, grapefruit, and palm trees grow in a region that has a climate with warm monthly temperatures throughout the year. On the other hand, blueberries, aspen, and birch trees grow in a region that has cool temperature patterns throughout the year. Climate determines what types of plants and animals live in a location, the types of houses that people build, and the lifestyles of people. Climate also influences the processes that shape the landscape, the type of soils that form, the suitability of the region for different types of agriculture, and how productive the agriculture will be in a region. This section is about climate, what determines the climate of a region, and how climate patterns are classified.

MAJOR CLIMATE GROUPS

Earth's atmosphere is heated directly by incoming solar radiation and by absorption of infrared radiation from the surface. The amount of heating at any particular latitude on the surface depends primarily on two factors: (1) the *intensity* of the incoming solar radiation, which is determined by the angle at which the radiation strikes the surface, and (2) the *time* that the radiation is received at the surface, that is, the number of daylight hours compared to the number of hours of night.

Earth is so far from the Sun that all rays of incoming solar radiation reaching Earth are essentially parallel. Earth, however, has a mostly constant orientation of its axis with respect to the stars as it moves around the Sun in its orbit. Since the inclined axis points toward the Sun part of the year and away from the Sun the other part, radiation reaches different latitudes at different angles during different parts of the year. The orientation of Earth's axis to the Sun during different parts of the year also results in days and nights of nearly equal length in the equatorial region but increasing differences at increasing latitudes to

the poles. During the polar winter months, the night is twenty-four hours long, which means no solar radiation is received at all. The equatorial region receives more solar radiation during a year, and the amount received decreases toward the poles as a result of (1) yearly changes in intensity and (2) yearly changes in the number of daylight hours.

In order to generalize about the amount of radiation received at different latitudes, some means of organizing, or grouping, the latitudes is needed (Figure 23.23). For this purpose, the latitudes are organized into three groups:

1. the *low latitudes,* those that some time of the year receive *vertical* solar radiation at noon,
2. the *high latitudes,* those that some time of the year receive *no* solar radiation at noon, and
3. the *middle latitudes,* which are between the low and high latitudes.

This definition of low, middle, and high latitudes means that the low latitudes are between the tropics of Cancer and Capricorn (between 23.5°N and 23.5°S latitudes) and that the high latitudes are above the Arctic and Antarctic Circles (above 66.5°N and above 66.5°S latitudes).

In general,

1. the *low latitudes* receive a high amount of incoming solar radiation that varies little during a year. Temperatures are high throughout the year, varying little from month to month.
2. The *middle latitudes* receive a higher amount of incoming radiation during one part of the year and a lower amount during the other part. Overall temperatures are cooler than in the low latitudes and have a wide seasonal variation.

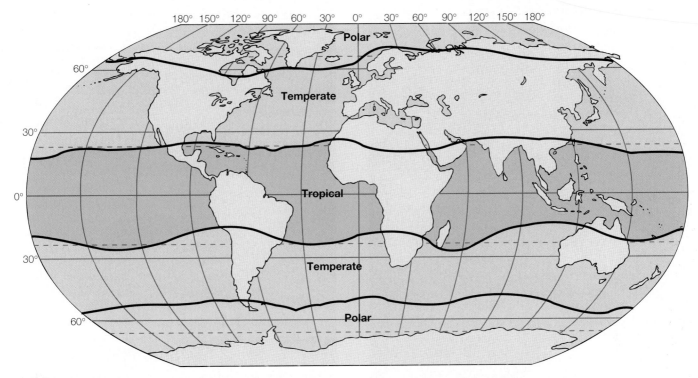

FIGURE 23.24 The principal climate zones are defined in terms of yearly temperature averages, which are determined by the amount of solar radiation received at the different latitude groups.

3. The *high latitudes* receive a maximum amount of radiation during one part of the year and none during another part. Overall temperatures are low, with the highest range of annual temperatures.

The low, middle, and high latitudes provide a basic framework for describing Earth's climates. These climates are associated with the low, middle, and high latitudes illustrated in Figure 23.23, but they are defined in terms of yearly temperature averages. It is necessary to define the basic climates in terms of temperature because land and water surfaces react differently to incoming solar radiation, creating a different temperature. Temperature and moisture are the two most important climate factors, and temperature will be considered first.

The principal climate zones are defined in terms of yearly temperature averages, which occur in broad regions (Figure 23.24). They are

1. the *tropical climate zone* of the low latitudes (Figure 23.25),
2. the *polar climate zone* of the high latitudes (Figure 23.26), and
3. the *temperate climate zone* of the middle latitudes (Figure 23.27).

The tropical climate zone is near the equator and receives the greatest amount of sunlight throughout the year. Overall, the tropical climate zone is hot. Average monthly temperatures stay above 18°C (64°F), even during the coldest month of the year.

The other extreme is found in the polar climate zone, where the Sun never sets during some summer days and never rises during some winter days. Overall, the polar climate zone is cold. Average monthly temperatures stay below 10°C (50°F), even during the warmest month of the year.

The temperate climate zone is between the polar and tropical zones, with average temperatures that are neither very cold nor very hot. Average monthly temperatures stay between 10°C and 18°C (50°F and 64°F) throughout the year.

General patterns of precipitation and winds are also associated with the low, middle, and high latitudes. An idealized model of the global atmospheric circulation and pressure patterns was described in chapter 22. Recall that this model described a huge convective movement of air in the low latitudes, with air being forced upward over the equatorial region. This air expands, cools to the dew point, and produces abundant rainfall throughout most of the year. On the other hand, air is slowly sinking over 30°N and 30°S of the equator, becoming warm and dry as it is compressed. Most of the great deserts of the world are near 30°N or 30°S latitude for this reason. There is another wet zone near 60° latitudes and another dry zone near the poles. These wet and dry zones are shifted north and south during the year with the changing seasons. This results in different precipitation patterns in each season. Figure 23.28 shows where the wet and dry zones are in winter and in summer seasons.

FIGURE 23.25 A wide variety of plant life can grow in a tropical climate, as you can see here.

FIGURE 23.26 Polar climates occur at high elevations as well as high latitudes. This mountain location has a highland polar climate and tundra vegetation but little else.

REGIONAL CLIMATIC INFLUENCE

Latitude determines the basic tropical, temperate, and polar climatic zones, and the wet and dry zones move back and forth over the latitudes with the seasons. If these were the only factors influencing the climate, you would expect to find the same climatic conditions at all locations with the same latitude. This is not what is found, however, because there are four major factors that affect a regional climate. These are (1) altitude, (2) mountains, (3) large bodies of water, and (4) ocean currents. The following describes how these four factors modify the climate of a region.

The first of the four regional climate factors is *altitude.* The atmosphere is warmed mostly by the greenhouse effect from the surface upward, and air at higher altitudes increasingly radiates more and more of its energy to space. Average air temperatures, therefore, decrease with altitude, and locations with higher altitudes will have lower average temperatures. This is why the tops of mountains are often covered with snow when none is found at lower elevations. St. Louis, Missouri, and Denver, Colorado, are located almost at the same latitude (within 1° of 39°N), so you might expect the two cities to have about the same average temperature. Denver, however, has an altitude of 1,609 m (5,280 ft), and the altitude of St. Louis is 141 m (465 ft). The yearly average temperature for Denver is about 10°C (about 50°F), and for St. Louis it is about 14°C

(about 57°F). In general, higher altitude means lower average temperature.

The second of the regional climate factors is *mountains.* In addition to the temperature change caused by the altitude of the mountain, mountains also affect the conditions of a passing air mass. The western United States has mountainous regions along the coast. When a moist air mass from the Pacific meets these mountains, it is forced upward and cools. Water vapor in the moist air mass condenses, clouds form, and the air mass loses much of its moisture as precipitation falls on the western side of the mountains. Air moving down the eastern slope is compressed and becomes warm and dry. As a result, the western slopes of these mountains are moist and have forests of spruce, redwood, and fir trees. The eastern slopes are dry and have grassland or desert vegetation.

The third of the regional climate factors is a large body of *water.* Water, as discussed previously, has a higher specific heat than land material, is transparent, and loses energy through evaporation. All of these affect the temperature of a landmass located near a large body of water, making the temperatures more even from day to night and from summer to winter. San Diego, California, and Dallas, Texas, for example, are at about the same latitude (both almost 33°N), but San Diego is at a seacoast and Dallas is inland. Because of its nearness to water, San Diego has an average summer temperature about 7°C

FIGURE 23.27 This temperate-climate deciduous forest responds to seasonal changes in autumn with a show of color.

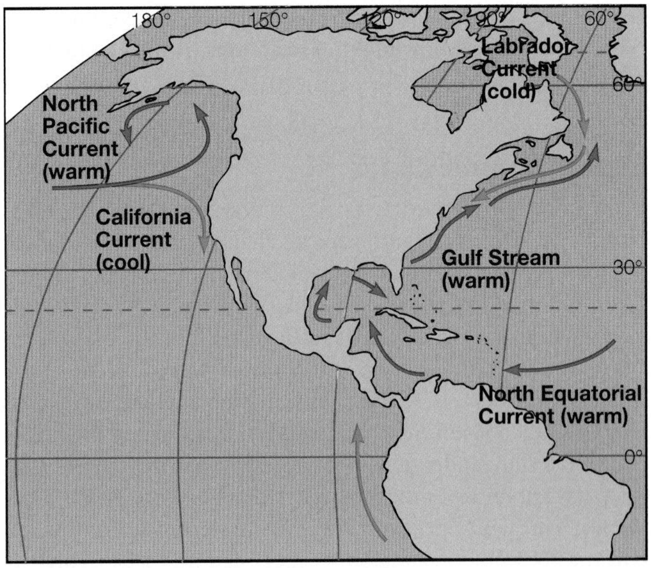

FIGURE 23.29 Ocean currents can move large quantities of warm or cool water to influence the air temperatures of nearby landmasses.

(about 13°F) cooler and an average winter temperature about 5°C (about 9°F) warmer than the average temperatures in Dallas. Nearness to a large body of water keeps the temperature more even at San Diego. This relationship is observed to occur for Earth as a whole as well as locally. The Northern Hemisphere is about 39 percent land and 61 percent water and has an average yearly temperature range of about 14°C (about 25°F). On the other hand, the Southern Hemisphere is about 19 percent land and 81 percent water and has an average yearly

temperature range of about 7°C (about 13°F). There is little doubt about the extent to which nearness to a large body of water influences the climate.

The fourth of the regional climate factors is *ocean currents*. In addition to the evenness brought about by being near the ocean, currents in the ocean can bring water that has a different temperature than the land. For example, currents can move warm water northward or they can move cool water southward (Figure 23.29; see also Figure 24.22 on p. 601). This can influence the temperatures of air masses that move from the water to the land and, thus, the temperatures of the land. For example, the North Pacific current brings warm waters to the western coast of North America, which results in warmer temperatures for cities near the coast.

DESCRIBING CLIMATES

Describing Earth's climates presents a problem because there are no sharp boundaries that exist naturally between two adjacent regions with different climates. Even if two adjacent climates are very different, one still blends gradually into the other. For example, if you are driving from one climate zone to another, you might drive for miles before becoming aware that the vegetation is now different than it was an hour ago. Since the vegetation is very different from what it was before, you know that you have driven from one regional climate zone to another.

Actually, no two places on Earth have exactly the same climate. Some plants will grow on the north or south side of a building, for example, but not on the other side. The two sides of the building could be considered as small, local climate zones within a larger, major climate zone. The

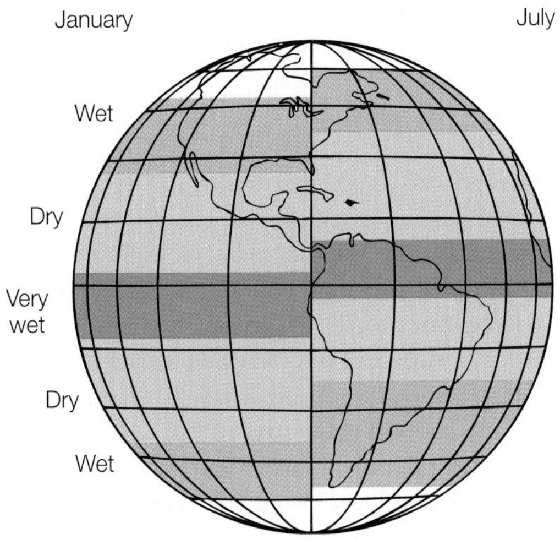

FIGURE 23.28 The idealized general rainfall patterns over Earth change with seasonal shifts in the wind and pressure areas of the planet's general atmospheric circulation patterns.

following is a general description of the regional climates of North America, followed by descriptions of how some local climates develop.

Major Climate Zones

Major climate zones are described by first considering the principal polar, temperate, and tropical climate zones, then looking at subdivisions within each that result from differences in moisture. Within one of the major zones, for example, there may be an area near the ocean that is influenced by air masses from the ocean. If so, the area has a **marine climate.** Because of the influence of the ocean, areas with marine climates have mild winters and cool summers compared to areas farther inland. In addition to the moderate temperatures, areas with a marine climate have abundant precipitation, with an average 50 to 75 cm (20 to 30 in) yearly precipitation. The western coast of Canada and the coasts of Washington, Oregon, and northern California have a marine climate. These regions are covered with forests of spruce, fir, and other conifers.

Also within a major polar, temperate, or tropical zone may be an area that is far from an ocean, influenced mostly by air masses from large land areas. If so, the area has a **continental climate.** A continental climate does not have an even temperature as does the marine climate because the land heats and cools rapidly. Thus, summers are hot and winters are cold.

Climates can be further classified as being **arid,** which means dry, or **humid,** which means moist. An area with an arid climate is defined as one that receives less than 25 cm (10 in) of precipitation per year. An area with a humid climate is defined as one that receives 50 cm (20 in) or more precipitation per year. An area that receives between 25 and 50 cm (10 and 20 in) precipitation per year is defined as **semiarid.**

Table 23.1 describes the principal climate types of North America, and Figure 23.30 gives the general location of these climates. Both are based on average temperatures and amounts of precipitation as described in the table. Recall that the actual climate in a given location may not agree with the general description because a local climate factor may change the climate. Also, recall that the climates blend gradually from one location to the next and do not change suddenly as you move across one of the lines.

Local Climates

The spread of cities, construction of high-rise buildings, paving of roads, and changes in the natural vegetation and landscape can change the local climate. Concrete, metal, stone, and glass react differently to incoming solar radiation than the natural vegetation and soils they replaced. High-rise buildings not only have a greater area exposed to solar radiation but also are capable of slowing and channeling the wind. Concrete and asphalt streets, roads, and parking lots also change the local climate because they are better absorbers of incoming solar radiation than natural vegetation and become heat sources for

TABLE 23.1
Principal climate types

Climate Type	Description
Polar	Long, very cold winters, cold summers, dry
Humid continental (subarctic)	Long, cold winters, cool summers, moderate precipitation
Humid continental (middle latitudes)	Cold winters, moderate summers, moderate precipitation
Humid continental (low latitudes)	Mild winters, hot summers, moderate precipitation
Humid subtropical	Short, mild winters, humid summers, moderate precipitation
Tropical wet	Hot and humid all year with heavy precipitation
Tropical wet/dry (subtropical)	Hot all year with alternating wet and dry seasons
Semiarid	Varying temperatures with low precipitation
Desert	Hot days and cool nights, arid
Marine	Moderate, rainy summers, mild winters
Mediterranean	Hot, dry summers with short, mild, and wet winters
Highland	Conditions vary with altitude

increased convection. They also make it impossible for precipitation to soak into the ground, increasing the likelihood of flooding. Large cities make a greater contribution to what has been called the "heat island" or the "heat dome" effect. In addition to causing the small changes in the actual air temperature, the buildings, concrete, and asphalt in large cities emit much more infrared radiation than is given off from the landscape in the country. This increased radiation causes people to feel much warmer in the city during the day and especially at night. The overall feeling of being warmer is also influenced by the decreased wind speed that occurs because large buildings block the wind, as well as the increased amounts of humidity in the city air.

Changes in the local climate are not restricted to large cities. A local pattern of climate is called a **microclimate.** A large city creates a new microclimate. Certain plants will grow within one microclimate but not another. For example, in some locations, lichens or mosses will grow on one side of a tree but not the other. A different microclimate exists on each side of the tree. The planting or cutting down of trees around a house can change the microclimate around the house. Air pollution also creates a new microclimate. Dust, particulates, and smog also contribute to the "heat island" of a large city by holding in radiation at night and reflecting incoming solar radiation during the day. This reduces convection, making the air pollution and smog last longer.

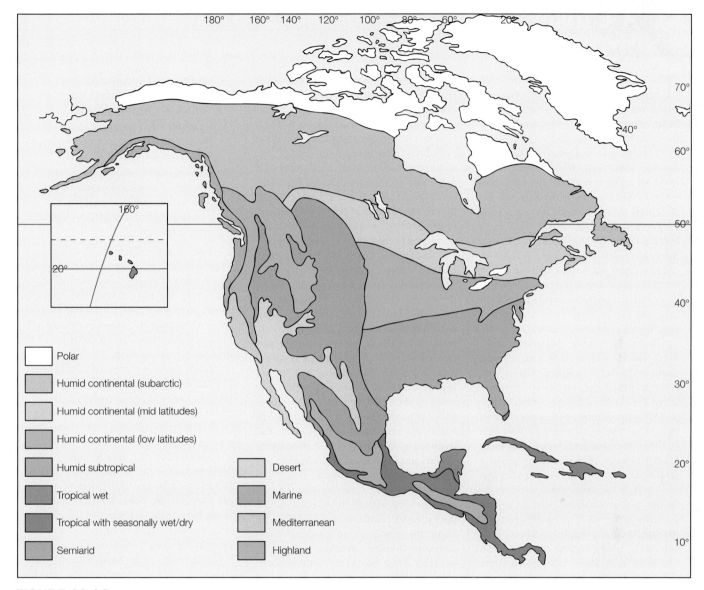

Polar

Humid continental (subarctic)

Humid continental (mid latitudes)

Humid continental (low latitudes)

Humid subtropical

Tropical wet

Tropical with seasonally wet/dry

Semiarid

Desert

Marine

Mediterranean

Highland

FIGURE 23.30 This map highlights the approximate location of the major types of climates in North America. See Table 23.1 for a description of these climates.

CONCEPTS *Applied*

Cities Change the Weather

Using satellite measurements, NASA researchers have confirmed that the presence of a city does influence the weather. First, because of all the concrete, asphalt, and buildings, cities are 0.6 to 5.6 Celsius degrees (about 1 to 10 Fahrenheit degrees) warmer than the surrounding country. The added heat of this "urban heat island" tends to make the air less stable. Less stable air brings about the second influence of more rain falling downwind of a city. Evidently, the less stable air and rougher surfaces cause air

moving toward a city to rise, and the turbulence mixes in city pollutants that add nuclei for water vapor to condense upon. The result is an average 28 percent increase in the amount of rainfall 30 to 60 km (about 19 to 37 mi) downwind of cities. The NASA research team's satellite measurements were verified by data from a large array of ground-based thermometers and rain gauges. The verification of these satellite-based findings are important for urban planning, water resource management, and decisions about where to farm the land. It may also show that local surface environments are more important in computer weather forecast models than had previously been believed.

A Closer Look

El Niño and La Niña

The term *El Niño* was originally used to describe an occurrence of warm, above-normal ocean temperatures off the South American coast. Fishermen along this coast learned long ago to expect associated changes in fishing patterns about every three to seven years, which usually lasted for about eighteen months. They called this event El Niño, which is Spanish for "the boy child" or "Christ child," because it typically occurred near Christmas. The El Niño event occurs when the trade winds along the equatorial Pacific become reduced or calm, allowing sea surface temperatures to increase much above normal. The warm water and changes in the circulation pattern drive the fish to deeper waters or farther away from usual fishing locations.

Today, El Niño is understood to be much more involved than just a warm ocean in the Pacific. It is more than a local event, and the bigger picture is sometimes called the "El Niño-Southern Oscillation," or ENSO. In addition to the warmer ocean of El Niño, "the boy," the term *La Niña*, "the girl," has been used to refer to the times when the water of the tropical Pacific is *colder* than normal. The "Southern Oscillation" part of the name comes from observations that atmospheric pressure around Australia seems to be inversely linked to the atmospheric pressure in Tahiti. They seem to be linked because when the pressure is low in Australia, it is high in Tahiti. Conversely, when the atmospheric pressure is high in Australia, it is low in Tahiti. The strength of this Southern Oscillation is measured by the Southern Oscillation Index (SOI), which is defined as the pressure at Darwin, Australia, subtracted from that at Tahiti. Negative values of SOI are usually associated with El Niño events, so the Southern Oscillation and the El Niño are obviously linked. How ENSO can impact the weather in other parts of the world has only recently become better understood.

The atmosphere is a system that responds to incoming solar radiation, the spinning Earth, and other factors, such as the amount of water vapor present. The ocean and atmospheric systems undergo changes by interacting with each other, most visibly in the tropical cyclone. The ocean system supplies water vapor, latent heat, and condensation nuclei, which are the essential elements of a tropical cyclone as well as everyday weather changes and climate. The atmosphere, on the other hand, drives the ocean with prevailing winds, moving warm or cool water to locations where they affect the climate on the land. There is a complex, interdependent relationship between the ocean and the atmosphere, and it is probable that even small changes in one system can lead to bigger changes in the other.

Normally, during non–El Niño times, the Pacific Ocean along the equator has established systems of prevailing wind belts, pressure systems, and ocean currents. In July, these systems push the surface seawater offshore from South America, westward along the equator and toward Indonesia. During El Niño times, the trade winds weaken and the warm water moves back eastward, across the Pacific to South America, where it then spreads north and south along the coast. Why the trade winds weaken and become calm is unknown and the subject of ongoing research.

Warmer waters along the coast of South America bring warmer, more humid air and the increased possibility of thunderstorms. Thus, the possibility of towering thunderstorms, tropical storms, or hurricanes increases along the Pacific coast of South America as the warmer waters move north and south. This creates the possibility of weather changes not only along the western coast but elsewhere, too. The towering thunderstorms reach high into the atmosphere, adding tropical moisture and creating changes in prevailing wind belts. These wind belts carry or steer weather systems across the middle latitudes of North America, so typical storm paths are shifted. This shifting can result in

- increased precipitation in California during the fall through spring season;
- a wet winter with more and stronger storms along regions of the southern United States and Mexico;

- a warmer and drier-than-average winter across the northern regions of Canada and the United States;
- a variable effect on central regions of the United States, ranging from reduced snowfall to no effect at all; and
- other changes in the worldwide global complex of ocean and weather events, such as droughts in normally wet climates and heavy rainfall in normally dry climates.

One major problem in these predictions is a lack of understanding of what causes many of the links and a lack of consistency in the links themselves. For example, southern California did not always have an unusually wet season every time an El Niño occurred and in fact experienced a drought during one event.

Scientists have continued to study the El Niño since the mid-1980s, searching for patterns that will reveal consistent cause-and-effect links. Part of the problem may be that other factors, such as a volcanic eruption, may influence part of the linkage but not another part. Another part of the problem may be the influence of unknown factors, such as the circulation of water deep beneath the ocean surface, the track taken by tropical cyclones, or the energy released by tropical cyclones one year compared to the next.

The results so far have indicated that atmosphere-ocean interactions are much more complex than early theoretical models had predicted. Sometimes a new model will predict some weather changes that occur with El Niño, but no model is yet consistently correct in predicting the conditions that lead to the event and the weather patterns that result. All this may someday lead to a better understanding of how the ocean and the atmosphere interact on this dynamic planet.

Recent years in which El Niño events have occurred are 1951, 1953, 1957–1958, 1965, 1969, 1972–1973, 1976, 1982–1983, 1986–1987, 1991–1992, 1994, 1997–1998, 2002–2003, and 2005–2006.

CLIMATE CHANGE

As stated earlier, *weather* describes changes in the atmospheric condition (rain, wind, etc.) over a brief period of time (day, week). *Climate* describes the general pattern of weather that occurs over a region for a number of months or years. **Climate change** is a departure from the expected average pattern of climate for a region over time. Scientists have measured climate patterns during the past 120 years or so by using thermometers, rain gauges, and other instruments. Climate patterns before 120 years ago are inferred by analyzing evidence found in tree rings, lake-bottom sediments, ice cores drilled from a glacier, and other sources. A natural source used to infer temperature change, rainfall, or some other past climate condition is called **proxy data.**

Proxy data indicates that Earth's climate has undergone major changes in the past, with cold *ice ages* and glaciers dominating the climate for the past several million years (Figure 23.31). The most recent ice age covered almost a third of Earth's land surface with ice sheets up to 4 km (2.5 mi) thick during its maximum extent. About half of the states in the United States were covered by ice, some completely and others partially. A large amount of water was locked up in glaciers, and this caused the sea level to drop some 90 m (about 300 ft), exposing a land bridge between Siberia and Alaska, among other things. The ice sheets of the most recent ice age advanced and retreated at least four times, with the sea level fluctuating with each advance and retreat.

About every one hundred thousand years, Earth enters an *interglacial warming period* before returning to another ice age. We are currently in such a warming period, and the current period began about eighteen thousand years ago. About fifteen thousand years ago, Earth's climate had warmed enough to stop the advance of glaciers, and since that time, the glaciers have been retreating. By about 4,000 B.C., the average temperatures had risen to a few degrees warmer that those of today. The proxy record indicates that such interglacial periods last from fifteen thousand to twenty thousand years before beginning a new glacial period. If the cycle continues as it has in the past, we

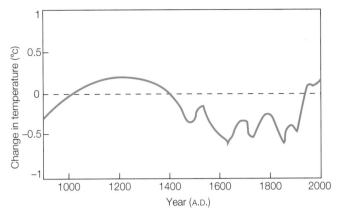

FIGURE 23.32 According to proxy data, Earth has experienced a period of warming and a period of cooling over the past one thousand years. (*Source:* www.geocraft.com/WVFossils/ice_ages.html.)

are near the end of the current interglacial period, nearing the beginning of the next ice age.

Data about ice ages and interglacial warming show that periods of warming and cooling occur in large cycles. During the past one thousand years, Earth has also undergone a number of smaller warming and cooling cycles (Figure 23.32). For example, it was warm—about like today—from 1000 up to 1400. Then from 1400 to 1860, there was a period of cooling called the "little ice age." The little ice age is why the Vikings left Greenland after successfully farming there from the tenth until the thirteenth century.

During the past one hundred years, Earth has had two warming cycles separated by one period of cooling. The temperature increased from the early 1800s—before the Industrial Revolution—because we were coming out of the little ice age. Then the temperature decreased from about 1940 to the late 1970s (Figure 23.33). And then a slight warming cycle began in the 1980s, continuing through today. Recording stations began reading small but steady increases in temperatures near the surface, averaging about 0.25°C (0.4°F) in the past twenty-five

FIGURE 23.31 Much of Earth's surface is covered with ice sheets and glaciers during an ice age. Both are formed from accumulations of compacted snow.

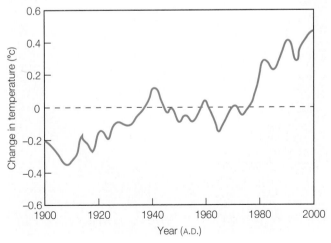

FIGURE 23.33 According to land-based weather stations, Earth has experienced a period of warming, a period of cooling, and most recently, a period of warming over the past one hundred years. (*Source:* www.ncdc.noaa.gov.)

years. Sea level has increased 1 to 2 mm per year for the past one hundred years, and this increase is due to thermal expansion of seawater in addition to melting ice.

The recent temperature increase is not uniform around the globe since the southeastern United States has cooled over the past one hundred years. Arctic sea ice has decreased since the early 1970s, but sea ice in the Antarctic has *increased* during the same period. Nonetheless, some people became concerned that the temperature increase was caused by the release of greenhouse gases—mostly carbon dioxide—and would result in a runway greenhouse effect (see p. 534 for information on the greenhouse effect).

CAUSES OF GLOBAL CLIMATE CHANGE

Climate change is brought about by a complex interaction of a number of factors. Some of these factors could be astronomical, and some could be occurring in the atmosphere. Solar energy is fundamentally responsible for weather and climate, and changes in the Sun's energy output can change the climate. The Sun's output of energy changes with *sunspots,* dark spots that appear to move around the surface of the Sun (Figure 23.34). The number of sunspots varies from year to year, and there are years when sunspots are rare or absent and years when a peak is reached. The maximum number seems to occur in a cycle that averages 11.1 years. Evidently, the amount of solar energy increases as the number of

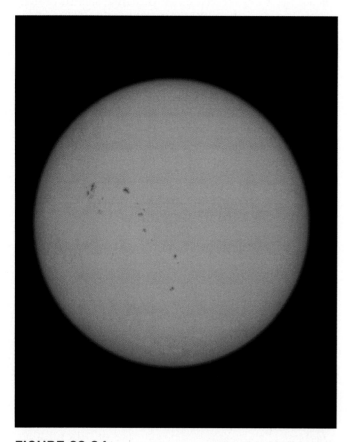

FIGURE 23.34 Sunspots appear on the Sun's surface as dark spots because they have a much cooler temperature than their surroundings. Sunspots appear in cycles that vary in number, and cycle 24 began to increase from a minimum on January 4, 2008.

sunspots increases. More sunspots deliver more energy to Earth's surface, which increases the temperature. Estimates are that the Sun's energy output varies by up to 0.2 percent with each sunspot cycle. If this increase continues into the mid-twenty-first century, Earth's surface temperature will increase by about 0.5°C (about 1°F). However, a relationship between the sunspot cycle *length* and temperatures also comes into play. Higher-than-normal temperatures tend to occur with shorter cycles, and lower-than-normal temperatures occur with longer cycles.

In addition to changes in the Sun, changes in the orientation of Earth's tilt, orbital shape, and axis wobble change our orientation to the Sun in predictable cycles. These are called Milankovitch cycles after Milutin Milankovitch (1879–1958), a Yugoslav physicist who calculated how the cycles would affect the climate by altering the amount of solar energy received by Earth (Figure 23.35). As it works out, the amount of energy received by high latitudes varies up to 20 percent. A shorter summer allows ice to accumulate, making an ice age. The timing of the ice age and cycles of warmer or cooler average temperatures fit with periods of Earth's orbital variations.

There are also atmospheric factors that can cause climatic changes. The greenhouse gases, for example, can reradiate heat in the atmosphere, producing a warmer climate. The most abundant greenhouse gas is water vapor, which is also the dominant gas in terms of increasing the temperature. Water vapor is followed by carbon dioxide, methane, and then some trace gases. The natural greenhouse effect is a good thing, for without it, the average Earth surface temperature would be −18°C (0.4°F). Thanks to the greenhouse effect, the global average is 14°C (57°F).

Carbon dioxide has been increasing in the atmosphere since the late 1800s (Figure 23.36), and some believe the burning of fossil fuels is responsible. Precise measurements of atmosphere carbon dioxide concentration made since 1958 found that the year-to-year concentration varied, but had an average increase of 1.5 parts per million by volume. Carbon dioxide is naturally released and absorbed by plants and animals as well as by Earth's oceans. The overall concentration varies because green plants convert carbon dioxide to plant materials. Carbon dioxide is also absorbed by ocean waters, used by marine organisms to make shells, and converted to mineral deposits such as limestone. Also, note that carbon dioxide can go into solution, and warm ocean water dissolves less carbon dioxide than cool ocean water. Just like a glass of soda, warmer water will release carbon dioxide. Tropical deforestation is reducing one means of removing carbon dioxide from the atmosphere, and this adds up to an estimated 40 percent as much carbon dioxide as the burning of fossil fuels. How much land, the ocean, and plant and animals remove and add carbon dioxide to the atmosphere is still highly uncertain.

GLOBAL WARMING

How do you predict the climate of the future? Scientists use mathematical models to calculate the evolving state of the atmosphere in response to changes in factors. The model is run on a large computer, which uses current climatic data from sunlight, land and atmosphere interactions, and interactions with the ocean. A change in some factor is then introduced to

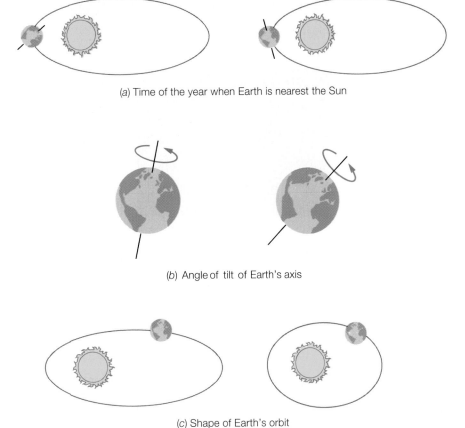

(a) Time of the year when Earth is nearest the Sun

(b) Angle of tilt of Earth's axis

(c) Shape of Earth's orbit

FIGURE 23.35 Three variations in Earth's motion that may be responsible for causing ice ages. (*A*) The time of year when Earth is nearest the Sun varies with a period of about twenty-three thousand years. (*B*) The angle of tilt of Earth's axis of rotation varies with a period of about forty-one thousand years. (*C*) The shape of Earth's elliptical orbit varies with a period of about one hundred thousand years. These variations have relatively little effect on the total sunlight reaching Earth but a considerable effect on the sunlight reaching the polar regions in summer. The ellipticity of Earth's orbit is vastly exaggerated here; the orbit is actually less than 2 percent away from a perfect circle.

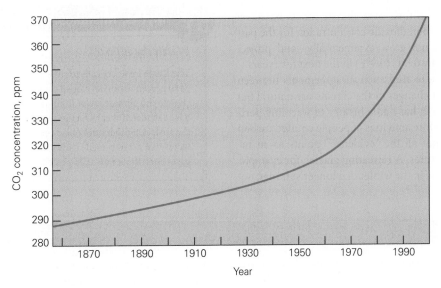

FIGURE 23.36 Carbon dioxide concentration in the atmosphere since 1860, in parts per million (ppm).

People Behind the Science

Vilhelm Firman Koren Bjerknes (1862–1951)

Vilhelm Bjerknes was the Norwegian scientist who created modern meteorology. Bjerknes came from a talented family. His father was professor of mathematics at the Christiania (now Oslo) University and a highly influential geophysicist who clearly shaped his son's studies. Bjerknes held chairs at Stockholm and Leipzig before founding the Bergen Geophysical Institute in 1917. By developing hydrodynamic models of the oceans and the atmosphere, Bjerknes made momentous contributions that transformed meteorology into an accepted science. Not least, he showed how weather prediction could be put on a statistical basis, dependent on the use of mathematical models.

During World War I, Bjerknes instituted a network of weather stations throughout Norway; coordination of the findings from such stations led him and his coworkers to develop the highly influential theory of polar fronts, on the basis of the discovery that the atmosphere is made up of discrete air masses displaying dissimilar features. Bjerknes coined the word *front* to delineate the boundaries between such air masses. One of the many contributions of the "Bergen frontal theory" was to explain the generation of cyclones over the Atlantic, at the junction of warm and cold air wedges. Bjerknes's work gave modern meteorology its theoretical tools and methods of investigation.

Source: From the Hutchinson *Dictionary of Scientific Biography.* © Research Machines plc 2003. All Rights Reserved. Helicon Publishing is a division of Research Machines.

calculate a hypothetical future. For example, the changing factor could be an increased concentration of carbon dioxide in the atmosphere. Climate models are imperfect because they do not include all the important factors; for example, they leave out the cycling of carbon dioxide into and out of the atmosphere.

The United Nations established the Intergovernmental Panel on Climate Change (IPCC) in 1988 to address climate change. The IPCC releases reports on the state of the global climate and makes projections about climate change based on changes in climate models. For example, a 2002 report predicts increased global temperatures during the next one hundred years, somewhere between 1.4°C (2.5°F) to 5.8°C (10.4°F), as a result of increased carbon dioxide in the atmosphere. Not all the experts agree with this prediction because it is based on an estimated carbon dioxide concentration that grows exponentially, increasing 1 percent per year. These experts run their own climate model based on the actual increases in carbon dioxide concentration for the past thirty years—which did not grow exponentially—and project additional warming of only 0.5°C (0.9°F) in the next fifty years.

It may be a surprise to find such disagreements between experts. Some argue that changes of the climate are natural but not well understood. Earth has had a history of warming periods, and this was long before activities of people emitted carbon dioxide. Furthermore, not all the "evidence" points to an increase in global temperatures. A retreating glacier, for example, can be a result of less snow in the winter or a longer summer, not

necessarily warmer temperatures. Do we have increased temperatures and retreating glaciers because of increased carbon dioxide, or are these natural variations in climate cycles that will continue to change? The answer will not come from computer simulation of the global climate because the computer model will always be simpler than the climate itself. Scientists can only weigh the evidence and make a professional judgment, and this can result in disagreement. Today, there is no consensus about the cause of the supposed slight warming observed during the past forty years. At least, the underlying science of climate change should be understood. Then a critical analysis of climate change should be completed before any future policies about human-made carbon dioxide are implemented.

Myths, Mistakes, & Misunderstandings

Worth the Effort?

The Kyoto Protocol calls for mandatory carbon dioxide reductions of 30 percent from developed countries. It is a myth that this would change the climate, even if followed by all participants. Scientists have calculated that reducing emissions as proposed would reduce the surface temperature one-twentieth of one degree by 2050. The fact is that human-made carbon dioxide, methane, and other trace gases contribute only 0.25 percent of the total greenhouse effect.

SUMMARY

The *cloud-forming process* begins when something gives a parcel of air an upward push. The three major causes of upward air movement are (1) *convection*, (2) *barriers* to moving air masses, and (3) the

meeting of moving air masses. As a parcel of air is pushed upward, it comes under less atmospheric pressure and expands. An expanding parcel of air does work on the surroundings and becomes cooler.

The rate of cooling as a parcel of dry air is lifted in the atmosphere is 10°C/km (about 5.5°F/1,000 ft). The relationship between the temperature of the parcel of air cooled by expansion and the temperature of the surrounding air determines whether a state of *atmospheric stability* or *atmospheric instability* exists.

Condensation nuclei act as centers of condensation as water vapor forms tiny droplets around the microscopic particles. The accumulation of large numbers of tiny droplets is what you see as a cloud.

Water that returns to Earth in liquid or solid form falls from the clouds as *precipitation*. Precipitation forms in clouds through two processes: (1) the *coalescence* of cloud droplets or (2) the *growth of ice crystals* at the expense of water droplets.

Weather changes are associated with the movement of large bodies of air called *air masses,* the leading *fronts* of air masses when they move, and local *high-* and *low-pressure* patterns that accompany air masses or fronts. Examples of air masses include (1) *continental polar,* (2) *maritime polar,* (3) *continental tropical,* and (4) *maritime tropical.*

When a location is under the influence of an air mass, the location is having *air mass weather* with slow, gradual changes. More rapid changes take place when the *front,* a thin transition zone between two air masses, passes a location.

A *stationary front* often develops a bulge, or *wave,* that forms into a moving cold front and a moving warm front. The faster-moving cold front overtakes the warm front, lifting it into the air to form an *occluded front.* The lifting process forms a low-pressure center called a *cyclone.* Cyclones are associated with heavy clouds, precipitation, and stormy conditions because of the lifting action.

A *thunderstorm* is a brief, intense storm with rain, lightning and thunder, gusty and strong winds, and sometimes hail. A *tornado* is the smallest, most violent weather disturbance that occurs on Earth. A *hurricane* is a *tropical cyclone,* a large, violent circular storm that is born over warm tropical waters near the equator.

The general pattern of the weather that occurs for a region over a number of years is called *climate.* The three principal climate zones are (1) the *tropical climate zone,* (2) the *polar climate zone,* and (3) the *temperate climate zone.* The climate in these zones is influenced by four factors that determine the local climate: (1) *altitude,* (2) *mountains,* (3) *large bodies of water,* and (4) *ocean currents.* The climate for a given location is described by first considering the principal climate zone, then looking at subdivisions within each that result from local influences.

Earth's climate has undergone *major changes* in the past, caused by *changes in the Sun, changes in Earth's orbit,* and *changes in the atmosphere.*

KEY TERMS

air mass (p. **555**)

anticyclone (p. **560**)

arid (p. **570**)

climate (p. **565**)

climate change (p. **573**)

cold front (p. **556**)

continental air mass (p. **555**)

continental climate (p. **570**)

cyclone (p. **560**)

front (p. **556**)

high (p. **560**)

humid (p. **570**)

hurricane (p. **563**)

hydrologic cycle (p. **552**)

marine climate (p. **570**)

maritime air mass (p. **555**)

microclimate (p. **570**)

occluded front (p. **559**)

polar air mass (p. **555**)

precipitation (p. **554**)

proxy data (p. **573**)

semiarid (p. **570**)

stationary front (p. **557**)

storm (p. **561**)

thunderstorm (p. **561**)

tornado (p. **562**)

tropical air mass (p. **555**)

warm front (p. **556**)

APPLYING THE CONCEPTS

1. Condensation of water vapor into clouds or fog requires
   a. high temperatures.
   b. storms.
   c. condensation nuclei.
   d. no wind.
2. Clouds that are between you and the Sun will appear gray because
   a. the clouds filter sunlight.
   b. the clouds reflect sunlight.
   c. the clouds refract sunlight.
   d. the Sun's shadow makes the clouds dark.
3. Which is not an example of precipitation?
   a. rain
   b. dew
   c. snow
   d. ice
4. Cloud droplets merge and fuse with millions of other droplets to form large raindrops. This process is called
   a. seeding.
   b. precipitation.
   c. coalescence.
   d. combination.
5. What type of clouds will usually produce a long, cold winter storm with drizzle, rain, ice, and snow?
   a. cirrus
   b. cumulus
   c. stratus
   d. storm

6. The transfer of heat from a region of higher temperature to a region of lower temperature by the displacement of high-energy molecules is
   a. thermal energy.
   b. conduction.
   c. convection.
   d. radiation.

7. The one term that does not describe an air mass is
   a. maritime polar.
   b. continental southern.
   c. maritime tropical.
   d. continental polar.

8. A boundary between air masses is called a
   a. boundary.
   b. front.
   c. dividing line.
   d. barrier.

9. A low-pressure center where the winds move counterclockwise into the low-pressure center is called a (an)
   a. occluded front.
   b. stationary front.
   c. cyclone.
   d. high.

10. Clear, fair weather is associated with a
    a. high.
    b. low.
    c. stationary front.
    d. low-pressure center.

11. A thunderstorm usually does not develop under what conditions?
    a. warm air
    b. cold air
    c. moist air
    d. humid air

12. Upward air motion that leads to the growth of cumulus clouds to tremendous heights is called a (an)
    a. lift.
    b. flight.
    c. updraft.
    d. rise.

13. The separation of charge associated with the movement of water droplets as they fall or as they are lifted by an updraft can result in
    a. lightning.
    b. hail.
    c. thunder.
    d. a rainbow.

14. An intense low-pressure area with widespread winds greater than 120 km/h is a
    a. tornado.
    b. tropical storm.
    c. hurricane.
    d. cyclone.

15. A hurricane does not have
    a. gale winds.
    b. rain and thunderstorms.
    c. a high-pressure area.
    d. an "eye" at its center.

16. Weather prediction is not based on
    a. use of computer models.
    b. the *Farmer's Almanac.*
    c. the study of the movement of air masses.
    d. knowledge of basic scientific laws.

17. The general pattern of weather that occurs for a region over a number of years is called
    a. a weather history.
    b. the climate.
    c. a forecast.
    d. the environment.

18. The source of energy that drives the hydrologic cycle is
    a. the ocean.
    b. latent heat from evaporating water.
    c. the Sun.
    d. Earth's interior.

19. Considering the average amount of water that evaporates from the Earth's oceans each year and the average amount that returns by precipitation,
    a. evaporation is greater than precipitation.
    b. precipitation is greater than evaporation.
    c. precipitation balances evaporation.
    d. there is no pattern that can be generalized.

20. A thunderstorm that occurs at 3 A.M. over a flat region of the country was probably formed by
    a. convection.
    b. differential heating.
    c. the meeting of moving air masses.
    d. Any of the above is correct.

21. White, puffy cumulus clouds that form over a flat region of the country during the late afternoon of a clear, warm day are probably the result of
    a. convection.
    b. a barrier to moving air.
    c. the meeting of moving air masses.
    d. None of the above is correct.

22. Without any heat being added or removed, a parcel of air that is expanding is becoming
    a. neither warmer nor cooler.
    b. warmer.
    c. cooler.
    d. the temperature of the surrounding air.

23. A parcel of air shoved upward into atmospheric air in a state of instability will expand and become cooler,
    a. but not as cool as the surrounding air.
    b. and thus colder than the surrounding air.
    c. reaching the same temperature as the surrounding air.
    d. then warmer than the surrounding air.

24. A parcel of air shoved upward into atmospheric air in a state of stability will expand and become cooler,
    a. but not as cool as the surrounding air.
    b. and thus colder than the surrounding air.
    c. reaching the same temperature as the surrounding air.
    d. then warmer than the surrounding air.

25. Cumulus clouds usually mean an atmospheric state of
    a. stability.
    b. instability.
    c. cool, dry equilibrium.
    d. warm, humid equilibrium.

26. When water vapor condenses in a parcel of air rising in an unstable atmosphere, the parcel is
    a. forced to the ground.
    b. slowed.
    c. stopped.
    d. accelerated upward.

27. A parcel of air with a relative humidity of 50 percent is given an upward shove into the atmosphere. What is necessary before cloud droplets form in this air?
    a. cooling
    b. saturation
    c. condensation nuclei
    d. All of the above are correct.

28. A cloud is hundreds of tiny water droplets suspended in the air. The average density of liquid water in such a cloud is about
    a. 0.1 g/m^3.
    b. 1 g/m^3.
    c. 100 g/m^3.
    d. $1,000 \text{ g/m}^3$.

29. When water vapor in the atmosphere condenses to liquid water,
    a. dew falls to the ground.
    b. rain or snow falls to the ground.
    c. a cloud forms.
    d. All of the above are correct.

30. In order for liquid cloud droplets at the freezing point to freeze into ice crystals,
    a. condensation nuclei are needed.
    b. further cooling is required.
    c. ice-forming nuclei are needed.
    d. nothing more is required.

31. Which basic form of a cloud usually produces longer periods of drizzle, rain, or snow?
    a. stratus
    b. cumulus
    c. cirrus
    d. None of the above is correct.

32. Which basic form of a cloud usually produces brief periods of showers?
    a. stratus
    b. cumulus
    c. cirrus
    d. None of the above is correct.

33. The type of air mass weather that results after the arrival of polar continental air is
    a. frequent snowstorms with rapid changes.
    b. clear and cold with gradual changes.
    c. unpredictable but with frequent and rapid changes.
    d. much the same from day to day, with conditions depending on the air mass and the local conditions.

34. The appearance of high cirrus clouds, followed by thicker and lower stratus clouds, then continuous light rain over several days, probably means which of the following air masses has moved to your area?
    a. continental polar
    b. maritime tropical
    c. continental tropical
    d. maritime polar

35. A fully developed cyclonic storm is most likely to form
    a. on a stationary front.
    b. in a high-pressure center.
    c. from differential heating.
    d. over a cool ocean.

36. The basic difference between a tropical storm and a hurricane is
    a. size.
    b. location.
    c. wind speed.
    d. amount of precipitation.

37. Most of the great deserts of the world are located
    a. near the equator.
    b. 30° north or south latitude.
    c. 60° north or south latitude.
    d. anywhere, as there is no pattern to their location.

38. The average temperature of a location is made more even by the influence of
    a. a large body of water.
    b. elevation.
    c. nearby mountains.
    d. dry air.

39. The climate of a specific location is determined by
    a. its latitude.
    b. how much sunlight it receives.
    c. its altitude and nearby mountains and bodies of water.
    d. All of the above are correct.

40. The process that is not involved with how water cycles in and out of the atmosphere is
    a. evaporation.
    b. boiling.
    c. condensation.
    d. precipitation.

41. The process of water vapor cycling in and out of the atmosphere with evaporation of water from the surface and precipitation of water back to the surface is called the
    a. unusual weather cycle.
    b. atmospheric cycle.
    c. hydrologic cycle.
    d. reverse flow cycle.

42. Water that returns to Earth's surface in either solid or liquid form is called
    a. rain.
    b. snow.
    c. ice.
    d. precipitation.

43. You are standing on a warm beach in Tahiti. Rain begins to fall. The type of cloud must be
    a. cirrus.
    b. cumulus
    c. stratus.
    d. storm.

44. Precipitation that is formed by cycling repeatedly through a thunderstorm, falling, and then being returned to the upper of the thunderclouds by updrafts is
    a. rain.
    b. snow.
    c. hail.
    d. rain and snow.

45. The smallest most violent weather event is a (an)
  a. hurricane.
  b. tornado.
  c. hailstorm.
  d. ice storm.

Answers

1. c **2.** a **3.** b **4.** c **5.** c **6.** c **7.** b **8.** b **9.** c **10.** a **11.** b **12.** c **13.** a **14.** c **15.** c **16.** b **17.** b **18.** c **19.** a **20.** c **21.** a **22.** c **23.** a **24.** b **25.** b **26.** d **27.** d **28.** b **29.** c **30.** c **31.** a **32.** b **33.** d **34.** b **35.** a **36.** c **37.** b **38.** a **39.** d **40.** b **41.** c **42.** d **43.** b **44.** c **45.** b

QUESTIONS FOR THOUGHT

1. What is a cloud? Describe how a cloud forms.

2. What is atmospheric stability? What does this have to do with what kind of clouds may form on a given day?

3. Describe two ways that precipitation may form from the water droplets of a cloud.

4. What is an air mass?

5. What kinds of clouds and weather changes are usually associated with the passing of (a) a warm front? (b) A cold front?

6. Describe the wind direction, pressure, and weather conditions that are usually associated with (a) low-pressure centers and (b) high-pressure centers.

7. In which of the four basic types of air masses would you expect to find afternoon thunderstorms? Explain.

8. Describe the three main stages in the life of a thunderstorm cell, identifying the events that mark the beginning and end of each stage.

9. What is a tornado? When and where do tornadoes usually form?

10. What is a hurricane? Describe how the weather conditions change as a hurricane approaches, passes directly over, then moves away from a location.

11. How is climate different from the weather?

12. Describe the average conditions found in the three principal climate zones.

13. Identify the four major factors that influence the climate of a region and explain how each does its influencing.

14. Since heated air rises, why is snow found on top of a mountain and not at lower elevations?

FOR FURTHER ANALYSIS

1. Explain why dew is not considered to be a form of precipitation.

2. What are the significant similarities and differences between air mass weather and frontal weather?

3. Analyze and compare the potential damage caused by a hurricane to the potential damage caused by a tornado.

4. Describe several examples of regional climate factors completely overriding the expected weather in a given principal climatic zone. Explain how this happens.

INVITATION TO INQUIRY

Microclimate Experiments

The local climate of a small site or habitat is called a *microclimate*. Certain plants will grow within one microclimate but not another. For example, a north-facing slope of a hill is cooler and loses snow later than a south-facing slope. A different microclimate exists on each side of the hill, and some plants grow better on one side than on the other.

Investigate the temperature differences in the microclimate on the north side of a building and on the south side of the same building. Determine the heights, distances, and time of day that you will take temperature readings. Remember to shade the thermometer from direct sunlight for each reading. In your report, discuss the variables that may be influencing the temperature of a microclimate. Design experiments to test your hypotheses about each variable.

24

Earth's Waters

Vast oceans cover more than 70 percent of the surface of Earth. Freshwater is generally abundant on the land because the supply is replenished from ocean waters through the hydrologic cycle.

CORE **CONCEPT**

The hydrologic cycle is evaporation of water from the ocean, transport by moving air masses, precipitation on the land, and movement of water back to the ocean.

OUTLINE

Water on Earth
Most water on Earth is stored in its oceans.

Freshwater as a Resource
About two-thirds of municipal water supplies come from surface water and the other third from groundwater.

Movement of Seawater
The ocean is well-mixed by waves and currents.

Freshwater
Precipitation flows across the surface in streams and rivers or soaks in to become groundwater.

Seawater
More than 70 percent of Earth is covered by seawater.

Throughout history, humans have diverted rivers and reshaped the land to ensure a supply of freshwater. There is evidence, for example, that ancient civilizations along the Nile River diverted water for storage and irrigation some five thousand years ago. The ancient Greeks and Romans built systems of aqueducts to divert streams to their cities some two thousand years ago. Some of these aqueducts are still standing today. More recent water diversion activities were responsible for the name of Phoenix, Arizona. Phoenix was named after a mythical bird that arose from its ashes after being consumed by fire. The city was given this name because it is built on a system of canals that were first designed and constructed by ancient Native Americans, then abandoned hundreds of years before settlers reconstructed the ancient canal system (Figure 24.1). Water is and always has been an essential resource. Where water is in short supply, humans have historically turned to extensive diversion and supply projects to meet their needs.

Precipitation is the basic source of the water supply found today in streams, lakes, and beneath Earth's surface. Much of the precipitation that falls on the land, however, evaporates back into the atmosphere before it has a chance to become a part of this supply. The water that does not evaporate mostly moves directly to rivers and streams, flowing back to the ocean, but some soaks into the land. The evaporation of water, condensation of water vapor, and the precipitation-making processes were introduced in chapter 23 as important weather elements. They are also part of the generalized *hydrologic cycle* of evaporation from the ocean, transport through the atmosphere by moving air masses, precipitation on the land, and movement of water back to the ocean. Only part of this cycle was considered previously, however, and this was the part from evaporation through precipitation. This chapter is concerned with the other parts of the hydrologic cycle, that is, what happens to the water that falls on the land and makes it back to the ocean. It begins with a discussion of how water is distributed on Earth and a more detailed look at the hydrologic cycle. Then the travels of water across and into the land will be considered as streams, wells, springs, and other sources of usable water are discussed as limited resources. The tracing of the hydrologic cycle will be completed as the water finally makes it back to the ocean. This last part of the cycle will consider the nature of the ocean floor, the properties of seawater, and how waves and currents are generated. The water is now ready to evaporate, starting another one of Earth's never-ending cycles.

WATER ON EARTH

Some water is tied up in chemical bonds deep in Earth's interior, but free water is the most abundant chemical compound near the surface. Water is five or six times more abundant than the most abundant mineral in the outer 6 km (about 4 mi) of Earth, so it should be no surprise that water covers about 70 percent of the surface. On average, about 98 percent of this water exists in the liquid state in depressions on the surface and in sediments. Of the remainder, about 2 percent exists in the solid state as snow and ice on the surface in colder locations. Only a fraction of a percent exists as a variable amount of water vapor in the atmosphere at a given time. Water is continually moving back and forth between these "reservoirs," but the percentage found in each is essentially constant.

As shown in Figure 24.2, over 97 percent of Earth's water is stored in the oceans. This water contains a relatively high level of dissolved salts, which make ocean water unfit for human consumption and for most agricultural purposes. All other water, which is fit for human consumption and agriculture, is called **freshwater.** About two-thirds of Earth's freshwater supply is locked up in the ice caps of Greenland and the Antarctic and in glaciers. This leaves less than 1 percent of all the water found on Earth as available freshwater. There is a generally abundant supply, however, because the freshwater supply is continually replenished by the hydrologic cycle.

Evaporation of water from the ocean is an important process of the hydrologic cycle because (1) water vapor leaves the dissolved salts behind, forming precipitation that is freshwater, and (2) the gaseous water vapor is easily transported in the atmosphere from one part of Earth to another. Over a year, this natural desalination process produces and transports enough freshwater to cover the entire Earth with a layer about 85 cm (about 33 in) deep. Precipitation is not evenly distributed like this, of course, and some places receive much more, while other places receive almost none. Considering global averages, more water is evaporated from the ocean than returns directly to it by precipitation. On the other hand, more water is precipitated over the land than evaporates from the land back to the atmosphere. The net amount evaporated and precipitated over the land and over the ocean is balanced by

FIGURE 24.1 This is one of the water canals of the present-day system in Phoenix, Arizona. These canals were reconstructed from a system that was built by Native Americans, then abandoned. Phoenix is named after a mythical bird that was consumed by fire and then arose from its ashes.

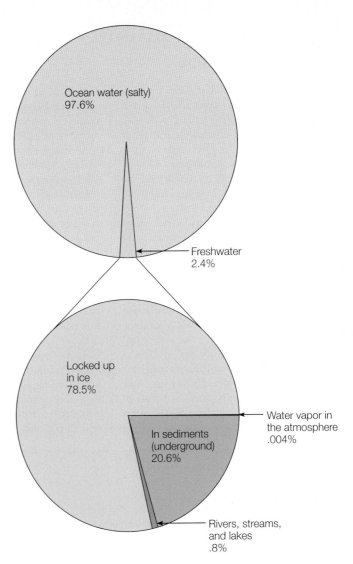

FIGURE 24.2 Estimates of the distribution of all the water found on Earth's surface.

the return of water to the ocean by streams and rivers. This cycle of evaporation, precipitation, and return of water to the ocean summarizes all aspects of the hydrologic cycle (Figure 24.3). This water returning on and under the land is the source of freshwater.

FRESHWATER

The basic source of freshwater is precipitation, but not all precipitation ends up as part of the freshwater supply. Liquid water is always evaporating, even as it falls. In arid climates, rain sometimes evaporates completely before reaching the surface, even from a fully developed thunderstorm. Evaporation continues from the water that does reach the surface. Puddles and standing water on the hard surface of city parking lots and streets, for example, gradually evaporate back to the atmosphere after a rain and the surface is soon dry. There are many factors that determine how much of a particular rainfall evaporates, but in general, more than two-thirds of the rain eventually returns to the atmosphere. The remaining amount either (1) flows downhill across the surface of the land toward a lower place or (2) soaks into the ground. Water moving across the surface is called *runoff*. Runoff begins as rain accumulates in thin sheets of water that move across the surface of the land. These sheets collect into a small body of running water called a *stream*. A stream is defined as any body of water that is moving across the land, from one so small that you could step across it to the widest river. Water that soaks into the ground moves down to a saturated zone, and it is now called *groundwater*. Groundwater moves through sediments and rocks beneath the surface, slowly moving in a downhill direction. Streams carry the runoff of a recent rainfall or

melting snow, but otherwise, most of the flow comes from groundwater that seeps into the stream channel. This explains how a permanent stream is able to continue flowing when it is not being fed by runoff or melting snow (Figure 24.4). Where or when the source of groundwater is in low supply, a stream may flow only part of the time, and it is designated as an *intermittent stream*.

The amount of a rainfall that becomes runoff or groundwater depends on a number of factors, including (1) the type of soil on the surface, (2) how dry the soil is, (3) the amount and type of vegetation, (4) the steepness of the slope, and (5) if the rainfall is a long, gentle one or a cloudburst. Different combinations of these factors can result in from 5 percent to almost 100 percent of a rainfall event running off, with the rest evaporating or soaking into the ground. On the average, however, about 70 percent of all precipitation evaporates back into the atmosphere, about 30 percent becomes runoff, and less than 1 percent soaks into the ground.

Science and Society

Water Quality

What do you think of when you see a stream? Do you wonder how deep it is? Do you think of something to do with the water, such as swimming, fishing, or having a good time? As you might imagine, not all people look at a stream and think about the same thing. A city engineer, for example, might wonder if the stream has enough water to serve as a source to supplement the city water supply. A rancher or farmer might wonder how the stream could be easily diverted to serve as a source of water for irrigation. An electric utility planner, on the other hand, might wonder if the stream could serve as a source of power.

Water in a stream is a resource that can be used many different ways, but using it requires knowing about the water quality as well as quantity. We need to know if the quality of the water is good enough for the intended use—and different uses have different requirements. Water fit for use in an electric power plant, for example, might not be suitable for use as a city water supply.

Indeed, water fit for use in a power plant might not be suitable for irrigation. Water quality is determined by the kinds and amounts of substances dissolved and suspended in the water and the consequences to users. Whether a source of water can be used for drinking water or not, for example, is regulated by stringent rules and guidelines about what cannot be in the water. These rules are designed to protect human health but do not call for absolutely pure water.

The water of even the healthiest stream is not absolutely pure. All water contains many naturally occurring substances such as ions of bicarbonates, calcium, and magnesium. A pollutant is not naturally occurring; it is usually a waste material that contaminates air, soil, or water. There are basically two types of water pollutants: degradable and persistent. Examples of degradable pollutants include sewage, fertilizers, and some industrial wastes. As the term implies, degradable pollutants can be broken down into simple, nonpolluting substances such

as carbon dioxide and nitrogen. Examples of persistent pollutants include some pesticides, petroleum and petroleum products, plastic materials, leached chemicals from landfill sites, oil-based paints, heavy metals and metal compounds such as lead, mercury, and cadmium, and certain radioactive materials. The damage they cause is either irreversible or reparable only over long periods of time.

QUESTIONS TO DISCUSS

Discuss with your group the following questions concerning water quality:

1. Which water use requires the purest water? Which use requires the least pure?

2. Create a hierarchy of possible water uses; for example, can water used in a power plant later be recycled for agriculture? Recycled for domestic use?

3. What can an individual do to improve water quality?

SURFACE WATER

If you could follow the smallest of streams downhill, you would find that it eventually merges with other streams until they merge to form a major river. The land area drained by a stream is known as the stream's drainage basin, or **watershed.** Each stream has its own watershed, but the watershed of a large river includes all the watersheds of the smaller streams that feed into the larger river. Figure 24.5 shows the watersheds of the Columbia River, the Colorado River, and the Mississippi River. Note that the water from the Columbia River and the Colorado River watersheds empties into the Pacific Ocean. The Mississippi River watershed drains into the Gulf of Mexico, which is part of the Atlantic Ocean.

Evaporation
from ocean 84%

Evaporation
from land 16%   Precipitation
on land 23%

Precipitation
on ocean 77%

Runoff from
land 7%

Ocean   Land

100% is based on a global average of 85 cm/yr precipitation.

FIGURE 24.3 On the average, more water is evaporated from the ocean than is returned by precipitation. More water is precipitated over the land than evaporates. The difference is returned to the ocean by rivers and streams.

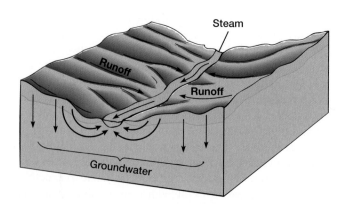

Steam

Runoff

Runoff

Groundwater

FIGURE 24.4 Some of the precipitation soaks into the ground to become groundwater. Groundwater slowly moves underground, and some of it emerges in streambeds, keeping the streams running during dry spells.

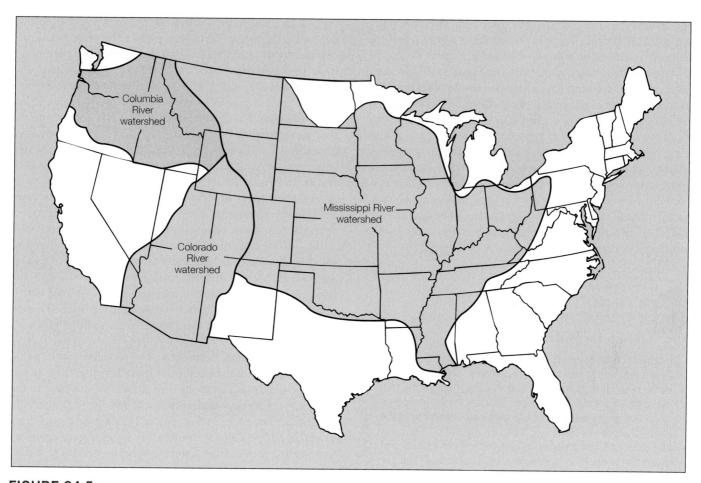

FIGURE 24.5 The approximate watersheds of the Columbia River, the Colorado River, and the Mississippi River.

Two adjacent watersheds are separated by a line called a **divide.** Rain that falls on one side of a divide flows into one watershed, and rain that falls on the other side flows into the other watershed. A *continental divide* separates river systems that drain into opposite sides of a continent. The North American continental divide trends northwestward through the Rocky Mountains. Imagine standing over this line with a glass of water in each hand, then pouring the water to the ground. The water from one glass will eventually end up in the Atlantic Ocean, and the water from the other glass will end up in the Pacific Ocean. Sometimes the Appalachian Mountains are considered to be an eastern continental divide, but water from both sides of this divide ends up on the same side of the continent, in the Atlantic Ocean.

Water moving downhill is sometimes stopped by a depression in a watershed, a depression where water temporarily collects as a standing body of freshwater. A smaller body of standing water is usually called a *pond,* and one of much larger size is called a *lake.*

A pond or lake can occur naturally in a depression, or it can be created by building a dam on a stream. A natural pond, a natural lake, or a pond or lake created by building a dam is called a *reservoir* if it is used for (1) water storage, (2) flood control, or (3) generating electricity. A reservoir can be used for one or two of these purposes but not generally for all three. A reservoir built for water storage, for example, is kept as full as possible to store water. This use is incompatible with use for flood control,

which would require a low water level in the reservoir in order to catch runoff, preventing waters from flooding the land. In addition, extensive use of reservoir water to generate electricity requires the release of water, which could be incompatible with water storage. The water of streams, ponds, lakes, and reservoirs is collectively called *surface water,* and all serve as sources of freshwater. The management of surface water, as you can see, can present some complicated problems.

GROUNDWATER

Precipitation soaks into the ground, or *percolates* slowly downward, until it reaches an area, or zone, where the open spaces between rock and soil particles are completely filled with water. Water from such a saturated zone is called **groundwater.** There is a tremendous amount of water stored as groundwater, which makes up a supply about twenty-five times larger than all the surface water on Earth. Groundwater is an important source of freshwater for human consumption and for agriculture. Groundwater is often found within 100 m (about 330 ft) of the surface, even in arid regions where little surface water is found. Groundwater is the source of water for wells in addition to being the source that keeps streams flowing during dry periods.

Water is able to percolate down to a zone of saturation because sediments contain open spaces between the particles called *pore spaces.* The more pore space a sediment has, the more water it will hold. The total amount of pore spaces in a given sample of sediment is a measure of its *porosity.* Sand and gravel sediments, for example, have large grains with large pore spaces between them, so these sediments have a high porosity. In order for water to move through a sediment, however, the pore spaces must be connected. The ability of a given sample of sediment to transmit water is a measure of its *permeability.* Sand and gravel have a high permeability because the grains do not fit tightly together, allowing water to move from one pore space to the next. Sand and gravel sediments thus have a high porosity as well as a high permeability. Clay sediments, on the other hand, have small, flattened particles that fit tightly together. Clay thus has a low permeability, and when saturated or compressed, clay becomes *impermeable,* meaning water cannot move through it at all (Figure 24.6).

CONCEPTS *Applied*

Room for Water

Compare the amount of porous space in sandstone, limestone, and other rocks. Dry samples of the rocks in a warm oven until they are thoroughly dry, then record the mass of each rock. Place each in a container of boiling water and watch until no more bubbles are coming from the rocks. Discard the water and blot dry. Again find the mass and calculate how much water entered the porous space of each rock. Note that 1g of water has a volume of 1mL, or 1cm^3. Report the volume of water that can occupy the pore space of a certain mass of each kind of rock tested.

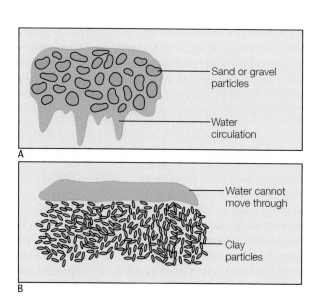

FIGURE 24.6 (*A*) Sand and gravel have large, irregular particles with large pore spaces, so they have a high porosity. Water can move from one pore space to the next, so they also have a high permeability. (*B*) Clay has small, flat particles, so it is practically impermeable because water cannot move from one pore to the next (not drawn to scale).

The amount of groundwater available in a given location depends on a number of factors, such as the present and past climate, the slope of the land, and the porosity and permeability of the sediments beneath the surface. Generally, sand and gravel sediments, along with solid sandstone, have the best porosity and permeability for transmitting groundwater. Other solid rocks, such as granite, can also transmit groundwater if they are sufficiently fractured by joints and cracks. In any case, groundwater will percolate downward until it reaches an area where pressure and other conditions have eliminated all pores, cracks, and joints. Above this impermeable layer, it collects in all available spaces to form a *zone of saturation.* Water from the zone of saturation is considered to be groundwater. Water from the zone above is not considered to be groundwater. The surface of the boundary between the zone of saturation and the zone above is called the **water table.** The surface of a water table is not necessarily horizontal, but it tends to follow the topography of the surface in a humid climate. A hole that is dug or drilled through Earth to the water table is called a well. The part of the well that is below the water table will fill with groundwater.

Precipitation falls on the land and percolates down to the zone of saturation, then begins to move laterally, or sideways, to lower and lower elevations until it finds its way back to the surface. This surface outflowing could take place at a stream, pond, lake, swamp, or spring (Figure 24.7). Groundwater flows gradually and very slowly through the tiny pore spaces, moving at a rate that ranges from kilometers (miles) per day to meters (feet) per year. Surface streams, on the other hand, move much faster at rates up to about 30 km per hour (about 20 mi/h).

An **aquifer** is a layer of sand, gravel, sandstone, or other highly permeable material beneath the surface that is capable of producing water in usable quantities. In some places, an aquifer carries water from a higher elevation, resulting in a pressure on water trapped by impermeable layers at lower elevations. Groundwater that is under such a confining pressure is in an *artesian* aquifer. "Artesian" refers to the pressure, and groundwater from an artesian well rises above the top of the aquifer but not necessarily to the surface. Some artesian wells

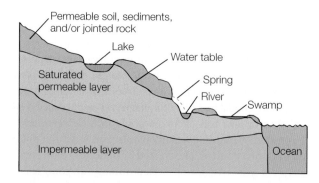

FIGURE 24.7 Groundwater from below the water table seeps into lakes, streams, and swamps and returns to the surface naturally at a spring. Groundwater eventually returns to the ocean, but the trip may take hundreds of years.

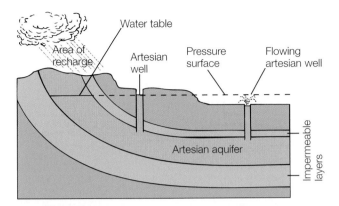

FIGURE 24.8 An artesian aquifer has groundwater that is under pressure because the groundwater is confined between two impermeable layers and has a recharge area at a higher elevation. The pressure will cause the water to rise in a well drilled into the aquifer, becoming a flowing well if the pressure is sufficiently high.

are under sufficient pressure to produce a fountainlike flow or spring (Figure 24.8). Some people call groundwater from any deep well "artesian water," which is technically incorrect.

CONCEPTS *Applied*

Deep Wells?

After doing some research, make a drawing to show the depth of the water table in your area and its relationship to local streams, lakes, swamps, and water wells.

FRESHWATER AS A RESOURCE

Water is an essential resource, not only because it is required for life processes but also because of its role in a modern industrialized society. Water is used in the home for drinking, cooking, and cleaning, as a carrier to remove wastes, and for maintaining lawns and gardens. These domestic uses lead to an equivalent consumption of about 570 L per person each day (about 150 gal/person/day), but this is only about 10 percent of the total consumed. Average daily use of water in the United States amounts to some 5,700 L per person each day (about 1,500 gal/person/day), or about enough water to fill a small swimming pool once a week. The bulk of the water is used by agriculture (about 40 percent), for the production of electricity (about 40 percent), and for industrial purposes (about 10 percent). These overall percentages of use vary from one region of the country to another, depending on (1) the relative proportions of industry, agriculture, and population, (2) the climate of the region, (3) the nature of the industrial or agricultural use, and (4) other variables. In an arid climate with a high proportion of farming and fruit growing, for example, up to two-thirds of the available water might be used for agriculture.

Most of the water supply is obtained from the surface water resources of streams, lakes, and reservoirs, and 37 percent of the

municipal water supply comes from groundwater. If you then add farms, villages, and many suburban areas, the percentage of groundwater used by humans is well above 40 percent. Surface water contains more sediments, bacteria, and possible pollutants than groundwater because it is more active and is directly exposed to the atmosphere. This means that surface water requires filtering to remove suspended particles, treatment to kill bacteria, and sometimes processing to remove pollution. In spite of the additional processing and treatment costs, surface water is less costly as a resource than groundwater. Groundwater is naturally filtered as it moves through the pore spaces of an aquifer, so it is usually relatively free of suspended particles and bacteria. Thus, the processing or treatment of groundwater is usually not necessary (Figure 24.9). Groundwater, on the other hand, will cost more to use as a resource because it must be pumped to the surface. The energy required for this pumping can be very expensive. In addition, groundwater generally contains more dissolved minerals (hard water), which may require additional processing or chemical treatment to remove the troublesome minerals.

The use of surface water as a source of freshwater means that the supply depends on precipitation. When a drought occurs, low river and lake resources may require curtailing

FIGURE 24.9 The filtering beds of a city water treatment facility. Surface water contains more sediments, bacteria, and other suspended materials because it is on the surface and is exposed to the atmosphere. This means that surface water must be filtered and treated when used as a domestic resource. Such processing is not required when groundwater is used as the resource.

A Closer Look

Water Quality and Wastewater Treatment

One of the most common forms of pollution control in the United States is wastewater treatment. The United States has a vast system of sewer pipes, pumping stations, and treatment plants, and about 74 percent of all Americans are served by such wastewater systems. Sewer pipelines collect the wastewater from homes, businesses, and many industries, and deliver it to treatment plants. Most of these plants were designed to make wastewater fit for discharge into streams or other receiving waters.

The basic function of a waste treatment plant is to speed up the natural process of purifying the water. There are two basic stages in the treatment, the *primary* stage and the *secondary* stage. The primary stage physically removes solids from the wastewater. The secondary stage uses biological processes to further purify wastewater. Sometimes, these stages are combined into one operation.

As raw sewage enters a treatment plant, it first flows through a screen to remove large floating objects such as rags and sticks that might cause clogs. After this initial screening, it passes into a grit chamber where cinders, sand, and small stones settle to the bottom (Box Figure 24.1). A grit chamber is particularly important in communities with combined sewer systems where sand or gravel may wash with storm water into the system along with rain, mud, and other stuff.

After screening and grit removal, the sewage is basically a mixture of organic and inorganic matter with other suspended solids. The solids are minute particles that

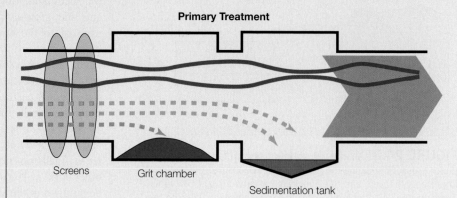

Primary Treatment

Screens   Grit chamber   Sedimentation tank

BOX FIGURE 24.1

can be removed in a sedimentation tank. The speed of the flow through the larger sedimentation tank is slower, and suspended solids gradually sink to the bottom of the tank. They form a mass of solids called *raw primary sludge,* which is usually removed from the tank by pumping. The sludge may be further treated for use as a fertilizer or disposed of through incineration, if necessary.

Once the effluent has passed through the primary treatment process, which is a physical filtering and settling, it enters a secondary stage that involves biological activities of microorganisms. The secondary stage of treatment removes about 85 percent of the organic matter in sewage by making use of the bacteria that are naturally a part of the sewage. Two principal techniques are used to provide secondary

treatment: (1) trickling filters or (2) activated sludge. A trickling filter is simply a bed of stones from 1 to 2 m (3 to 6 ft) deep through which the effluent from the sedimentation tank flows. Interlocking pieces of corrugated plastic or other synthetic media have also been used in trickling beds, but the important part is that it provides a place for bacteria to live and grow. Bacteria grow on the stones or synthetic media and consume most of the organic matter flowing by in the effluent. The now cleaner water trickles out through pipes to another sedimentation tank to remove excess bacteria. Disinfection of the effluent with chlorine generally completes this secondary stage of basic treatment.

The trend today is toward the use of an activated sludge process instead of trickling filters. The activated sludge process speeds up

water consumption. The curtailing of consumption occurs more often when a drought lasts for a longer period of time and when smaller lakes and reservoirs make the supply sensitive to rainfall amounts. In some parts of the western United States, such as the Colorado River watershed, *all* of the surface water is already being used, with certain percentages allotted for domestic, industrial, and irrigation uses. Groundwater is also used in this watershed, and in some locations, it is being pumped from the ground faster than it is being replenished by precipitation (Figure 24.10). As the population grows and new industries develop, more and more demands are placed on the surface water supply, which has already been committed to other uses, and on the diminishing supply of groundwater. This raises some very controversial issues about how

freshwater should be divided among agriculture, industries, and city domestic use. Agricultural interests claim they should have the water because they produce the food and fibers that people must have. Industrial interests claim they should have the water because they create the jobs and the products that people must have. Cities, on the other hand, claim that domestic consumption is the most important because people cannot survive without water. Yet others claim that no group has a right to use water when it is needed to maintain habitats. Who should have the first priority for water use?

Some have suggested that people should not try to live and grow food in areas that have a short water supply, that plenty of freshwater is available elsewhere. Others have suggested that humans have historically moved rivers and reshaped the

the work of the bacteria by bringing air and sludge heavily laden with bacteria into close contact with the effluent (Box Figure 24.2). After the effluent leaves the sedimentation tank in the primary stage, it is pumped into an aeration tank, where it is mixed with air and sludge loaded with bacteria and allowed to remain for several hours. During this time, the bacteria break down the organic matter into harmless by-products.

The sludge, now activated with additional millions of bacteria, can be used again by returning it to the aeration tank for mixing with new effluent and ample amounts of air. As with trickling, the final step is generally the addition of chlorine to the effluent, which kills more than 99 percent of the harmful bacteria. Some municipalities are now manufacturing chlorine solution on site to avoid having to transport and store large amounts of chlorine, sometimes in a gaseous form. Alternatives to chlorine disinfection, such as ultraviolet light or ozone, are being used in situations where chlorine in sewage effluents can be harmful to fish and other aquatic life.

New pollution problems have placed additional burdens on wastewater treatment systems. Today's pollutants may be more difficult to remove from water. Increased demands on the water supply only aggravate the problem. These challenges are being met through better and more complete methods of removing pollutants at treatment plants or through prevention of pollution at the source. Pretreatment of industrial waste, for example, removes many troublesome

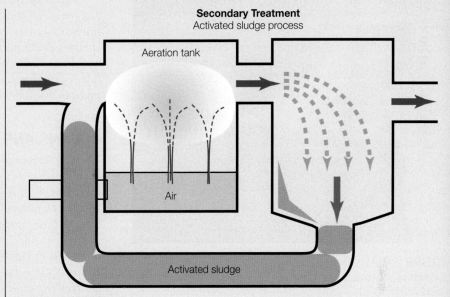

Secondary Treatment
Activated sludge process

Aeration tank

Air

Activated sludge

BOX FIGURE 24.2

pollutants at the beginning, rather than at the end, of the pipeline.

The increasing need to reuse water calls for better and better wastewater treatment. Every use of water—whether at home, in the factory, or on the farm—results in some change in its quality. New methods for removing pollutants are being developed to return water of more usable quality to receiving lakes and streams. Advanced waste treatment techniques in use or under development range from biological treatment capable of removing nitrogen and

phosphorus to physical-chemical separation techniques such as filtration, carbon adsorption, distillation, and reverse osmosis. These activities typically follow secondary treatment and are known as tertiary treatment.

These wastewater treatment processes, alone or in combination, can achieve almost any degree of pollution control desired. As waste effluents are purified to higher degrees by such treatment, the effluent water can be used for industrial, agricultural, or recreational purposes, or even as drinking water supplies.

Source: Drawings and some text, from "How Wastewater Treatment Works . . . The Basics," U.S. Environmental Protection Agency, Office of Water, www.epa.gov/owm/featinfo.html

land to obtain water, so perhaps one answer to the problem is to find new sources of freshwater. Possible sources include the recycling of wastewater and turning to the largest supply of water in the world, the ocean. About 90 percent of the water used by industries is presently dumped as a waste product. In some areas, treated city wastewater is already being recycled for use in power plants and for watering parks. A practically limitless supply of freshwater could be available by desalting ocean water, something which occurs naturally in the hydrologic cycle. The treatment of seawater to obtain a new supply of freshwater is presently too expensive because of the cost of energy to accomplish the task. New technologies, perhaps ones that use solar energy, may make this more practical in the future. In the meantime, the best sources of

extending the supply of freshwater appear to be the control of pollution, the recycling of wastewater, and conservation of the existing supply.

CONCEPTS *Applied*

Who Uses How Much?

Find out how much water is used in the industrial processes in your location. Compare this to the amount of water used in the home for drinking, cooking, cleaning, and so on.

FIGURE 24.10 This is groundwater pumped from the ground for irrigation. In some areas, groundwater is being removed from the ground faster than it is being replaced by precipitation, resulting in a water table that is falling. It is thus possible that the groundwater resource will soon become depleted in some areas.

SEAWATER

More than 70 percent of the surface of Earth is covered by seawater, with an average depth of 3,800 m (about 12,500 ft). The land areas cover 30 percent with an average elevation of only about 830 m (about 2,700 ft). With this comparison, you can see that humans live on and fulfill most of their needs by drawing from a small part of the total Earth. As populations continue to grow and as resources of the land continue to diminish, the ocean will be looked at more as a resource rather than a convenient place for dumping wastes. The ocean already provides some food and is a source of some minerals, but it can possibly provide freshwater, new sources of food, new sources of important minerals, and new energy sources in the future. There are vast deposits of phosphorite and manganese nodules on the ocean bottom, for example, that can provide valuable minerals. Phosphate is an important fertilizer needed in agriculture, and the land supplies are becoming depleted. Manganese nodules, which occur in great abundance on the ocean bottom, can be a source of manganese, iron, copper, cobalt, and nickel. Seawater contains enough deuterium to make it a feasible source of energy. One gallon of seawater contains about a spoonful of deuterium, with the energy equivalent of 300 gal of gasoline. It has been estimated there is sufficient deuterium in the oceans to supply power at one hundred times the present consumption for the next 10 billion years. The development of controlled nuclear fusion is needed, however, to utilize this potential energy source. The sea may provide new sources of food through *aquaculture,* the farming of the sea the way that the land is presently farmed. Some aquaculture projects have

already started with the farming of oysters, lobsters, shrimp, clams, and certain fishes, but these projects have barely begun to utilize the full resources that are possible (Figure 24.11).

Part of the problem of utilizing the ocean is that the ocean has remained mostly unexplored and a mystery until recent times. Only now are scientists beginning to understand the complex patterns of the circulation of ocean waters, the nature of the chemical processes at work in the ocean, and the interactions of the ocean and the atmosphere, and to chart the topography of the ocean floor.

OCEANS AND SEAS

The vast body of salt water that covers more than 70 percent of Earth's surface is usually called the *ocean* or the *sea.* Although there is really only one big ocean on Earth, specific regions have been given names for convenience in describing locations. For this purpose, three principal regions are recognized: (1) the Atlantic Ocean, (2) the Indian Ocean, and (3) the Pacific Ocean. As shown in Figure 24.12, these are not separate, independent bodies of salt water but are actually different parts of Earth's single, continuous ocean. In general, the **ocean** is a single, continuous body of salt water on the surface of Earth. Specific regions (Atlantic, Indian, and Pacific) are often subdivided further into North Atlantic Ocean, South Atlantic Ocean, and so on.

A **sea** is usually a smaller part of the ocean, a region with characteristics that distinguish it from the larger ocean of which it is a part. Often the term *sea* also is used in the name of certain inland bodies of salty water.

The Pacific Ocean is the largest of the three principal ocean regions. It has the largest surface area, covering 180 million km^2 (about 70 million mi^2) and the greatest average depth of 3.9 km (about 2.4 mi). The Pacific is circled by active converging plate boundaries, so it is sometimes described as being circled by a "rim of fire." It is called this because of the volcanoes associated with the converging plates. The "rim" also has the other associated features of converging plate boundaries such as oceanic trenches, island arcs, and earthquakes. The Atlantic Ocean is second in size, with a surface area of 107 million km^2 (about 41 million mi^2) and the shallowest average depth of only 3.3 km (about 2.1 mi). The Atlantic Ocean is bounded by nearly parallel continental margins with a diverging plate boundary between. It lacks the trench and island arc features of the Pacific, but it does have islands, such as Iceland, that are a part of the Mid-Atlantic Ridge at the plate boundary. The shallow seas of the Atlantic, such as the Mediterranean, Caribbean, and Gulf of Mexico, contribute to the shallow average depth of the Atlantic. The Indian Ocean has the smallest surface area, with 74 million km^2 (about 29 million mi^2) and an average depth of 3.8 km (about 2.4 mi).

As mentioned earlier, a sea is usually a part of an ocean that is identified because some characteristic sets it apart. For example, the Mediterranean, Gulf of Mexico, and Caribbean seas are bounded by land, and they are located in a warm, dry climate. Evaporation of seawater is greater than usual at these locations, which results in the seawater being saltier. Being bounded by land and having saltier seawater characterizes these locations as being different from the rest of the Atlantic. The Sargasso Sea,

FIGURE 24.11 A modern-day tuna fishing boat. Note the time-tested "crows nest" lookout on the mast and the more modern helicopter fastened on top. Both are used to look for schools of tuna, but the helicopter also does other jobs.

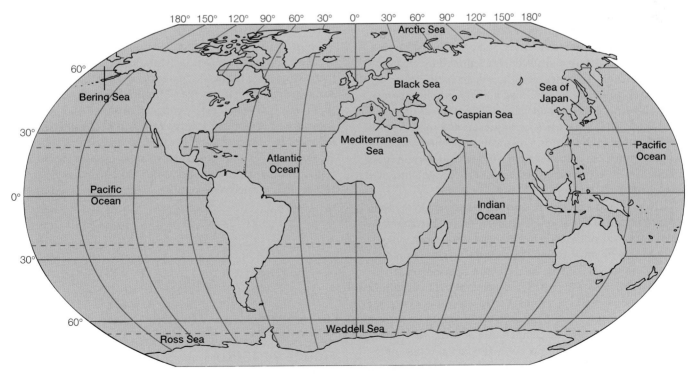

FIGURE 24.12 Distribution of the oceans and major seas on Earth's surface. There is really only one ocean; for example, where is the boundary between the Pacific, Atlantic, and Indian oceans in the Southern Hemisphere?

on the other hand, is a part of the Atlantic that is not bounded by land and has a normal concentration of sea salts. This sea is characterized by having an abundance of floating brown seaweeds that accumulate in this region because of the global wind and ocean current patterns. The Arctic Sea, which is also sometimes called the Arctic Ocean, is a part of the North Atlantic Ocean that is less salty. Thus, the terms *ocean* and *sea* are really arbitrary terms that are used to describe different parts of Earth's one continuous ocean.

THE NATURE OF SEAWATER

According to one theory, the ocean is an ancient feature of Earth's surface, formed at least 3 billion years ago as Earth cooled from its early molten state. The seawater and much of the dissolved materials are believed to have formed from the degassing of water vapor and other gases from molten rock materials. The degassed water vapor soon condensed, and over a period of time, it began collecting as a liquid in the depression of the early ocean basin. Ever since, seawater has continuously cycled through the hydrologic cycle, returning water to the ocean through the world's rivers. For millions of years, these rivers have carried large amounts of suspended and dissolved materials to the ocean. These dissolved materials, including salts, stay behind in the seawater as the water again evaporates, condenses, falls on the land, and then brings more dissolved materials much like a continuous conveyor belt.

You might wonder why the ocean basin has not become filled in by the continuous supply of sediments and dissolved materials that would accumulate over millions of years. The basin has not filled in because (1) accumulated sediments have been recycled to Earth's interior through plate tectonics and (2) dissolved materials are removed by natural processes just as fast as they are supplied by the rivers. Some of the dissolved materials, such as calcium and silicon, are removed by organisms to make solid shells, bones, and other hard parts. Other dissolved materials, such as iron, magnesium, and phosphorous, form solid deposits directly and also make sediments that settle to the ocean floor. Hard parts of organisms and solid deposits are cycled to Earth's interior along with suspended sediments that have settled out of the seawater. Studies of fossils and rocks indicate that the composition of seawater has changed little over the past 600 million years.

The dissolved materials of seawater are present in the form of ions because of the strong dissolving ability of water molecules. Almost all of the chemical elements are present, but only six ions make up more than 99 percent of any given sample of seawater. As shown in Table 24.1, chlorine and sodium are the most abundant ions. These are the elements of sodium chloride, or common table salt. As a sample of seawater evaporates, the positive metal ions join with the different negative ions to form a complex mixture of ionic compounds known as *sea salt*. Sea salt is mostly sodium chloride, but it also contains salts of the four metal ions (sodium, magnesium, calcium, and potassium) combined with the different negative ions of chlorine, sulfate, bicarbonate, and so on.

TABLE 24.1

Major dissolved materials in seawater

Ion	Percent (by weight)
Chloride (Cl^-)	55.05
Sodium (Na^+)	30.61
Sulfate (SO_4^{-2})	7.68
Magnesium (Mg^{+2})	3.69
Calcium (Ca^{+2})	1.16
Potassium (K^+)	1.10
Bicarbonate (HCO_3^-)	0.41
Bromine (Br^-)	0.19
Total	99.89

The amount of dissolved salts in seawater is measured as **salinity.** Salinity is defined as the mass of salts dissolved in 1.0 kg, or 1,000 g, of seawater. Since the salt content is reported in parts per thousand, the symbol ‰ is used (% means parts per hundred). Thus, 35‰ means that 1,000 g of seawater contains 35 g of dissolved salts (and 965 g of water). This is the same concentration as a 3.5 percent salt solution (Figure 24.13). Oceanographers use the salinity measure because the mass of a sample of seawater does not change with changes in the water temperature. Other measures of concentration are based on the volume of a sample, and the volume of a liquid does vary as it expands and contracts with changes in the temperature. Thus, by using the salinity measure, any corrections due to temperature differences are eliminated.

The average salinity of seawater is about 35‰, but the concentration varies from a low of about 32‰ in some locations up to a high of about 36‰ in other locations. The salinity of seawater in a given location is affected by factors that tend to increase or decrease the concentration. The concentration is increased by two factors: evaporation and the formation of sea ice. Evaporation increases the concentration because it is water vapor only that evaporates, leaving the dissolved salts behind in a greater concentration. Ice that forms from freezing seawater increases the concentration because when ice forms, the salts are excluded from the crystal structure. Thus, sea ice is freshwater, and the removal of this water leaves the dissolved salts behind in a greater concentration. The salinity of seawater is

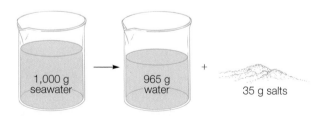

FIGURE 24.13 Salinity is defined as the mass of salts dissolved in 1.0 kg of seawater. Thus, if a sample of seawater has a salinity of 35‰, a 1,000 g sample would evaporate 965 g of water and leave 35 g of sea salts behind.

decreased by three factors: heavy precipitation, the melting of ice, and the addition of freshwater by a large river. All three of these factors tend to dilute seawater with freshwater, which lowers the concentration of salts.

Note that increases or decreases in the salinity of seawater are brought about by the addition or removal of freshwater. This changes only the amount of water present in the solution. The *kind* or *proportion* of the ions present (Table 24.1) in seawater does not change with increased or decreased amounts of freshwater. The same proportion, meaning the same chemical composition, is found in seawater of any salinity of any sample taken from any location anywhere in the world, from any depth of the ocean, or taken any time of the year. Seawater has a remarkably uniform composition that varies only in concentration. This means that the ocean is well mixed and thoroughly stirred around the entire Earth.

If you have ever allowed a glass of tap water to stand for a period of time, you may have noticed tiny bubbles collecting as the water warms. These bubbles are atmospheric gases, such as nitrogen and oxygen, that were dissolved in the water (Figure 24.14). Seawater also contains dissolved gases in addition to the dissolved salts. Near the surface, seawater contains mostly nitrogen and oxygen, in similar proportions to the mixture that is found in the atmosphere. There is more carbon dioxide than you would expect, however, as seawater contains a large amount of this gas. More carbon dioxide can dissolve in seawater because it reacts with water to form carbonic acid, H_2CO_3, the same acid that is found in a bubbly cola. In seawater, carbonic acid breaks down into bicarbonate and carbonate ions, which tend to remain in solution. Water temperature and the salinity have an influence on how much gas can be dissolved in seawater, and increasing either or both will reduce the amount of gases that can be dissolved. Cold, lower-salinity seawater in colder regions will dissolve more gases than the warm, higher-salinity seawater

FIGURE 24.14 Air will dissolve in water, and cooler water will dissolve more air than warmer water. The bubbles you see here are bubbles of carbon dioxide that came out of solution as the soda became warmer.

in tropical locations. Abundant algae and seaweeds in the upper, sunlit water tend to reduce the concentration of carbon dioxide and increase the concentration of dissolved oxygen through the process of photosynthesis. With increasing depth, less light penetrates the water, and below about 80 m (about 260 ft), there is insufficient light for photosynthesis. Thus, more algae and seaweeds and more dissolved oxygen are found above this depth. Below this depth, there are no algae or seaweeds, more dissolved carbon dioxide, and less dissolved oxygen. The oxygen-poor, deep ocean water does eventually circulate back to the surface, but the complete process may take several thousand years.

MOVEMENT OF SEAWATER

Consider the enormity of Earth's ocean, which has a surface area of some 361 million km^2 (about 139 million mi^2) and a volume of 1,370 million km^3 (about 328 million mi^3) of seawater. There must be a terrific amount of stirring in such an enormous amount of seawater to produce the well-mixed, uniform chemical composition that is found in seawater throughout the world. The amount of mixing required is more easily imagined if you consider the long history of the ocean, the very long period of time over which the mixing has occurred. Based on investigations of the movement of seawater, it has been estimated that there is a complete mixing of all Earth's seawater about every two thousand years or so. With an assumed age of 3 billion years, this means that Earth's seawater has been mixed 3,000,000,000 ÷ 2,000, or 1.5 million times. With this much mixing, you would be surprised if seawater were *not* identical all around Earth.

How does seawater move to accomplish such a complete mixing? Seawater is in a constant state of motion, both on the surface and below the surface. The surface has two types of motion: (1) *waves*, which have been produced by some disturbance, such as the wind, and (2) *currents*, which move water from one place to another. Waves travel across the surface as a series of wrinkles that range from a few centimeters high to more than 30 m (100 ft) high. Waves crash on the shore as booming breakers and make the surf. This produces local currents as water moves along the shore and back out to sea. There are also permanent, worldwide currents that move ten thousand times more water across the ocean than all the water moving in all the large rivers on the land. Beneath the surface, there are currents that move water up in some places and down in other places. Finally, there are enormous deep ocean currents that move tremendous volumes of seawater. The overall movement of many of the currents on the surface and their relationship to the deep ocean currents are not yet fully mapped or understood. The surface waves are better understood. The general trend and cause of permanent, worldwide currents in the ocean can also be explained.

Waves

Any slight disturbance will create ripples that move across a water surface. For example, if you gently blow on the surface of water in a glass, you will see a regular succession of small ripples moving across the surface. These ripples, which look like

A Closer Look

Estuary Pollution

Pollution is usually understood to mean something that is not naturally occurring and contaminates air, soil, or water to interfere with human health, well-being, or quality of the environment. An important factor in understanding pollution is the size of the human population and the amount of material that might become a pollutant. When the human population was small and produced few biological wastes, there was no pollution problem. The decomposers broke down the material into simpler nonpolluting substances such as water and carbon dioxide, and no harm was done. For example, suppose one person empties the tea leaves remaining from a cup of tea into a nearby river once a week. In this case, decomposer organisms in the water would break down the tea leaves almost as fast as they were added to the river. But imagine one hundred thousand people doing this every day. In this case, the tea leaves are released faster than they decompose, and the leaves become pollutants.

The part of the wide lower course of a river where the freshwater of the river mixes with the saline water from the oceans is called a coastal *estuary*. Estuary waters include bays and tidal rivers that serve as nursery areas for many fish and shellfish populations, including shrimp, oysters, crabs, and scallops. Unfortunately, the rivers carry pollution from their watersheds and adjacent wetlands to the estuary, where it affects the fish and shellfish industry, swimming, and recreation.

In 1996, the U.S. Environmental Protection Agency asked the coastal states to rate the general water quality in their estuaries. The states reported that pollutants affect aquatic life in 31 percent of the area surveyed, violate shellfish harvesting criteria in 27 percent of the area surveyed, and violate swimming-use criteria in 16 percent of the area surveyed.

The most common pollutants affecting the surveyed estuaries were excessive *nutrients,* which were found in 22 percent of all the estuaries surveyed. Excessive nutrients stimulate population explosions of algae. Fast-growing masses of algae block light from the habitat below, stressing the aquatic life. The algae die and eventually decompose, and this depletes the available oxygen supply, leading to further fish and shellfish kills.

The second most common pollutant was the presence of *bacteria,* which pollute 16 percent of all the estuary waters surveyed. Because *Escherichia coli* is a bacterium commonly found in the intestines of humans and other warm-blooded animals, the presence of *E. coli* is evidence that sewage is polluting the water. Bacteria interfere with recreational activities of people and can contaminate fish and shellfish.

The states also reported that *toxic organic chemicals* pollute 15 percent of the surveyed waters, *oxygen-depleting chemicals* pollute 12 percent, and *petroleum products* pollute another 8 percent of the surveyed waters. These pollutants impact the fish and shellfish industry, swimming, and recreational activities that require contact with the water.

The leading sources of the pollutants were identified as agriculture runoff, urban runoff, industrial discharges, municipal wastewater, and wastes from landfills.

---

small, moving wrinkles, are produced by the friction of the air moving across the water surface. The surface of the ocean is much larger, but a gentle wind produces patches of ripples in a similar way. These patches appear, then disappear as the wind begins to blow over calm water. If the wind continues to blow, larger and longer-lasting ripples are made, and the moving air can now push directly on the side of the ripples. A ripple may eventually grow into an **ocean wave,** a moving disturbance that travels across the surface of the ocean. In its simplest form, each wave has a ridge, or mound, of water called a *crest,* which is followed by a depression called a *trough.* Ocean waves are basically repeating series of these crests and troughs that move across the surface like wrinkles (Figure 24.15).

The simplest form of an ocean wave can be described by measurements of three distinct characteristics: (1) the *wave height,* which is the vertical distance between the top of a crest and the bottom of the next trough, (2) the *wavelength,* which is the horizontal distance between two successive crests (or other successive parts of the wave), and (3) the *wave period,* which is the time required for two successive crests (or other successive parts) of the wave to pass a given point (Figure 24.16).

The characteristics of an ocean wave formed by the wind depend on three factors: (1) the wind speed, (2) the length of time that the wind blows, and (3) the *fetch,* which is the distance the wind blows across the open ocean. As you can

FIGURE 24.15 The surface of the ocean is rarely, if ever, still. Any disturbance can produce a wave, but most waves on the open ocean are formed by a local wind.

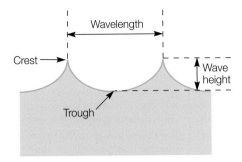

FIGURE 24.16 The simplest form of ocean waves, showing some basic characteristics. Most waves do not look like this representation because most are complicated mixtures of superimposed waves with a wide range of sizes and speeds.

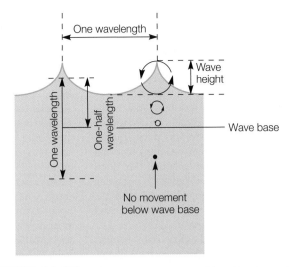

FIGURE 24.17 Water particles are moved in a circular motion by a wave passing in the open ocean. On the surface, a water particle traces out a circle with a diameter that is equal to the wave height. The diameters of the circles traced out by water particles decrease with depth to a depth that is equal to one-half the wavelength of the ocean wave. Bottom sediment cannot be moved by waves when the sediment is below the wave base.

imagine, larger waves are produced by strong winds that blow for a longer time over a long fetch. In general, longer-blowing, stronger winds produce waves with greater wave heights, longer wavelengths, and longer periods, but a given wind produces waves with a wide range of sizes and speeds. In addition, the wind does not blow in just one direction, and shifting winds produce a chaotic pattern of waves of many different heights and wavelengths. Thus, the surface of the ocean in the area of a storm or strong wind has a complicated mixture of many sizes and speeds of superimposed waves. The smaller waves soon die out from friction within the water, and the larger ones grow as the wind pushes against their crests. Ocean waves range in height from a few centimeters up to more than 30 m (about 100 ft), but giant waves more than 15 m (about 50 ft) are extremely rare.

The larger waves of the chaotic, superimposed mixture of waves in a storm area last longer than the winds that formed them, and they may travel for hundreds or thousands of kilometers from their place of origin. The longer wavelength waves travel faster and last longer than the shorter wavelength waves, so the longer wavelength waves tend to outrun the shorter wavelength waves as they die out from energy losses to water friction. Thus, the irregular, superimposed waves created in the area of a storm become transformed as they travel away from the area. They become regular groups of long-wavelength waves with a low wave height that are called **swell.** The regular waves of swell that you might observe near a shore may have been produced by a storm that occurred days before, thousands of kilometers across the ocean.

The regular crests and troughs of swell carry energy across the ocean, but they do not transport water across the open ocean. If you have ever been in a boat that is floating in swell, you know that you move in a regular pattern of up and forward on each crest, then backward and down on the following trough. The boat does not move along with the waves unless it is moved along by a wind or by some current. Likewise, a particle of water on the surface moves upward and forward with each wave crest, then backward and down on the following trough, tracing out a nearly circular path through this motion (Figure 24.17). The particle returns to its initial position, without any forward movement while tracing out the small circle. Note that the diameter

of the circular path is equal to the wave height. Water particles farther below the surface also trace out circular paths as a wave passes. The diameters of these circular paths below the surface are progressively smaller with increasing depth. Below a depth equal to about half the wavelength (wave base), there is no circular movement of the particles. Thus, you can tell how deeply the passage of a wave disturbs the water below if you measure the wavelength.

As swell moves from the deep ocean to the shore, the waves pass over shallower and shallower water depths. When a depth is reached that is equal to about half the wavelength, the circular motion of the water particles begins to reach the ocean bottom. The water particles now move across the ocean bottom, and the friction between the two results in the waves moving slower as the wave height increases. These important modifications result in a change in the direction of travel and in an increasingly unstable situation as the wave height increases.

Most waves move toward the shore at some angle. As the wave crest nearest the shore starts to slow, the part still over deep water continues on at the same velocity. The slowing at the shoreward side *refracts,* or bends, the wave so it is more parallel to the shore. Thus, waves always appear to approach the shore head-on, arriving at the same time on all parts of the shore.

After the waves reach water that is less than one-half the wavelength, friction between the bottom and the circular motion of the water particles progressively slow the bottom part of the wave. The wave front becomes steeper and steeper as the top overruns the bottom part of the wave. When the wave front becomes too steep, the top part breaks forward, and the wave is now called a **breaker** (Figure 24.18). In general, this occurs where the water depth is about one and one-third times the wave height. The zone where the breakers occur is called **surf** (Figure 24.19).

A Closer Look

Health of the Chesapeake Bay

Submerged aquatic vegetation—SAV for short—is vital to the Chesapeake Bay ecosystem and a key measuring stick for the Chesapeake's overall health. Eel and wigeon grasses grow in the shallows around the bay and across the bottom, providing food and shelter for baby blue crabs and fish, filtering pollutants, and providing life-sustaining oxygen for the water. So the health of the grasses provides an indication of the overall health of the Chesapeake.

The Chesapeake Bay once had an estimated 600,000 acres of grasses and provided an abundance of oysters, blue crab, shad, haddock, sturgeon, rockfish, and other prized sport fish and seafood. At this time, the bay water was clear, and watermen reported they could clearly see grasses on the bottom some 6 m (about 20 ft) below their boats. Then the water clouded, the grasses began to die, and the ecosystem of the bay began to decline. The low point was reached in 1984, when fewer than 40,000 acres of grasses could be found in the now murky waters of the Chesapeake. With the decline of the grasses came the decline of the aquatic species living in the bay. Shad, haddock, and sturgeon once supported large fisheries but are now scarce. The rockfish (striped bass) as well as the prized blue crab were once abundant but seem to decline and recover over the years. The decline of the blue crab population has shocked watermen because the crab is amazingly fertile. The crabs reach sexual maturity in about a year, and one female bears millions of eggs. Part of the problem is the loss of the underwater grasses, which shelter the baby crabs from predators. In addition to the loss of habitat, some believe the crab is being overfished.

What happened to the underwater grasses of the Chesapeake? Scientists believe it is a combination of natural erosion and pollutants that wash from farms in the watershed. The pollutants include nutrient-rich fertilizers, chemical residues, and overflow from sewage treatment plants. There are some 6,000 chicken houses around the Chesapeake, raising more than 600 million chickens a year and producing 750,000 tons of manure. The manure is used as fertilizer, and significant amounts of nitrogen and phosphates wash from these fields. These nutrients accelerate algae growth in the bay, which blocks sunlight. The grass dies and the loss results in muddying of the water, making it impossible for new grasses to begin growing.

There is some evidence to support this idea since the 600,000 acres of underwater grasses died back to a low of 40,000 acres in 1984, then began to rebound when sewage treatment plants were modernized and there were fewer pollutants from industry and farms. The grasses recovered to some 63,500 acres by 1999 but are evidently very sensitive to even slight changes in water quality. Thus, they may be expected to die back with unusual weather conditions that might bring more pollutants or muddy conditions but continue to rebound when the conditions are right. The abundance of blue crabs and other aquatic species can be expected to fluctuate with the health of the grasses. The trends of underwater grass growth do indeed provide a key measure for the health of the Chesapeake Bay.

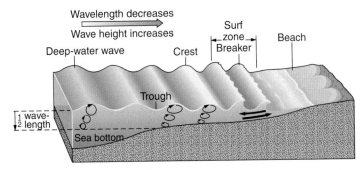

FIGURE 24.18 As a pattern of swell approaches a gently sloping beach, friction between the circular motion of the water particles and the bottom slows the wave, and the wave front becomes steeper and steeper. When the depth is about one and one-third times the wave height, the wave breaks forward, moving water toward the beach.

FIGURE 24.19 The white foam is in the surf zone, which is where the waves grow taller and taller, then break forward into a froth of turbulence. Do you see any evidence of rip currents in this picture?

Waves break in the foamy surf, sometimes forming smaller waves that then proceed to break in progressively shallower water. The surf may have several sets of breakers before the water is finally thrown on the shore as a surging sheet of seawater. The turbulence of the breakers in the surf zone and the final surge expend all the energy that the waves may have brought from thousands of kilometers away. Some of the energy does work in eroding the shoreline, breaking up rock masses into the sands that are carried by local currents back to the ocean. The rest of the energy goes into the kinetic energy of water molecules, which appears as a temperature increase.

A rogue wave is an unusually large wave that appears with smaller waves. The rogue wave has also been called a "freak" wave. Whatever the name, it is generally one or a group of two or three waves that are more than twice the size of the normal surrounding waves. The rogue wave is one or several very large "walls of water" and has unpredictable behavior, not following the wind direction, for example. Large rogue waves have been reported from 21 to 35 m (69 to 114 ft) tall and have been observed to almost capsize large ships.

It is believed that a rogue wave is an extreme storm wave. It probably forms during a storm from constructive interference (see p. 128) between smaller waves when the crests and troughs happen to match, coming together to form a mountainous wave that lasts several minutes before subsiding. Other processes, such as wave focusing by the shape of the coast or movement by currents, may play a part in forming rogue waves. The source of rogue waves continues to be a mystery and an active topic of research. This much is known—rogue waves do exist.

Swell does not transport water with the waves over a distance, but small volumes of water are moved as a growing wave is pushed to greater heights by the wind over the open ocean. A strong wind can topple such a wave on the open ocean, producing a foam-topped wave known as a *whitecap*. In general, whitecaps form when the wind is blowing at 30 km/h (about 20 mi/h) or more.

Waves do transport water where breakers occur in the surf zone. When a wave breaks, it tosses water toward the shore, where the water begins to accumulate. This buildup of water tends to move away in currents, or streams, as the water returns to a lower level. Some of the water might return directly to the sea by moving beneath the breakers. This direct return of water forms a weak current known as **undertow.** Other parts of the accumulated water might be pushed along by the waves, producing a **longshore current** that moves parallel to the shore in the surf zone. This current moves parallel to the shore until it finds a lower place or a channel that is deeper than the adjacent bottom. Where the current finds such a channel, it produces a **rip current,** a strong stream of water that bursts out against the waves and returns water through the surf to the sea (Figure 24.20). The rip current usually extends beyond the surf zone and then diminishes. A rip current, or where rip currents are occurring, can usually be located by looking for the combination of (1) a lack of surf, (2) darker looking water, which means a deeper channel, and (3) a turbid, or muddy, streak of water that extends seaward from the channel indicated by the darker water that lacks surf. See chapter 19 for information on how earthquakes can produce waves. See chapter 16 for information on tides.

Ocean Currents

Waves generated by the winds, earthquakes, and tidal forces keep the surface of the ocean in a state of constant motion. Local, temporary currents associated with this motion, such as rip currents or tidal currents, move seawater over a short distance. Seawater also moves in continuous **ocean currents,** streams of water that stay in about the same path as they move through other seawater over large distances. Ocean currents can be difficult to observe directly since they are surrounded by water that looks just like the water in the current. Wind is likewise difficult to observe directly since the moving air looks just like the rest of the atmosphere. Unlike the wind, an ocean current moves *continuously* in about the same path, often carrying water with different chemical and physical properties than the water it is moving through. Thus, an ocean current can be identified and tracked by measuring the physical and chemical characteristics of the current and the surrounding water. This shows where the current is coming from and where in the world it is going. In general, ocean currents are produced by (1) density differences in seawater and (2) winds that blow persistently in the same direction.

Density Currents. The density of seawater is influenced by three factors: (1) the water temperature, (2) salinity, and (3) suspended sediments. Cold water is generally more dense than warm water, thus sinking and displacing warmer water. Seawater of a high salinity has a higher relative density than less salty water, so it sinks and displaces water of less salinity. Likewise, seawater with a larger amount of suspended sediments has a higher relative density than clear water, so it sinks and displaces clear water. The following describes how these three ways of changing the density of seawater result in the ocean current

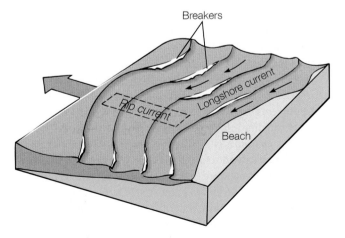

FIGURE 24.20 Breakers result in a buildup of water along the beach that moves as a longshore current. Where it finds a shore bottom that allows it to return to the sea, it surges out in a strong flow called a *rip current*.

known as a *density current,* which is an ocean current that flows because of density differences.

Earth receives more incoming solar radiation in the tropics than it does at the poles, which establishes a temperature difference between the tropical and polar oceans. The surface water in the polar ocean is often at or below the freezing point of freshwater, while the surface water in the tropical ocean averages about 26°C (about 79°F). Seawater freezes at a temperature below that of freshwater because the salt content lowers the freezing point. Seawater does not have a set freezing point, however, because as it freezes, the salinity is increased as salt is excluded from the ice structure. Increased salinity lowers the freezing point more, so the more ice that freezes from seawater, the lower the freezing point for the remaining seawater. Cold seawater near the poles is therefore the densest, sinking and creeping slowly as a current across the ocean floor toward the equator. Where and how such a cold, dense bottom current moves are influenced by the shape of the ocean floor, the rotation of Earth, and other factors. The size and the distance that cold bottom currents move can be a surprise. Cold, dense water from the Arctic, for example, moves in a 200 m (about 660 ft) diameter current on the ocean bottom between Greenland and Iceland. This current carries an estimated 5 million cubic meters of water per second (about 177 million cubic ft/s) of seawater to the 3.5 km (about 2.1 mi) deep water of the North Atlantic Ocean. This is a flow rate about 250 times larger than that of the Mississippi River. At about 30°N, the cold Arctic waters meet even denser water that has moved in currents all the way from the Antarctic to the deepest part of the North Atlantic Basin (Figure 24.21).

A second type of density current results because of differences in salinity. The water in the Mediterranean, for example, has a high salinity because it is mostly surrounded by land in a warm, dry climate. The Mediterranean seawater, with its higher salinity, is more dense than the seawater in the open Atlantic Ocean. This density difference results in two separate currents that flow in opposite directions between the Mediterranean and the Atlantic. The greater-density seawater flows from the bottom of the Mediterranean into the Atlantic, while the less dense Atlantic water flows into the Mediterranean near the surface. The dense Mediterranean seawater sinks to a depth of about 1,000 m (about 3,300 ft) in the Atlantic, where it spreads over a large part of the North Atlantic Ocean. This increases the salinity of this part of the ocean, making it one of the more saline areas in the world.

The third type of density current occurs when underwater sediments on a slope slide toward the ocean bottom, producing a current of muddy or turbid water called a *turbidity current.* Turbidity currents are believed to be a major mechanism that moves sediments from the continents to the ocean basin. They may also be responsible for some undersea features, such as submarine canyons. Turbidity currents are believed to occur only occasionally, however, and none has ever been directly observed or studied. There is thus no data or direct evidence of how they form or what effects they have on the ocean floor.

Surface Currents. There are broad and deep-running ocean currents that slowly move tremendous volumes of water relatively near the surface. As shown in Figure 24.22, each current is actually part of a worldwide system, or circuit, of currents. This system of ocean currents is very similar to the worldwide system of prevailing winds (see chapter 22). This similarity exists because it is the friction of the prevailing winds on the seawater surface that drives the ocean currents. The currents are modified by other factors, such as the rotation of Earth and the shape of the ocean basins, but they are basically maintained by the wind systems.

Each ocean has a great system of moving water called a **gyre** that is centered in the mid-latitudes. The gyres rotate to the right in the Northern Hemisphere and to the left in the Southern Hemisphere because of the Coriolis effect (see chapter 16). The movement of water around these systems, or gyres, plus some smaller systems form the surface circulation system of the world ocean. Each part of the system has a separate name, usually based on its direction of flow. All are called "currents," except one that is called a "stream" (the Gulf Stream) and those that are called "drifts." Both the Gulf Stream and the drifts are currents that are part of the connected system.

The major surface currents are like giant rivers of seawater that move through the ocean near the surface. You know that all the currents are connected, for a giant river of water cannot just start moving in one place, then stop in another. The Gulf Stream, for example, is a current about 100 km (about 60 mi) wide that may extend to a depth of 1 km (about 0.6 mi) below the surface, moving more than 75 million cubic meters of water per second (about 2.6 billion cubic ft/s). The Gulf Stream carries more than 370 times more water than the Mississippi River. The California Current is weaker and broader, carrying cool water southward at a relatively slow rate. The flow rate of all the currents must be equal, however, since all the ocean basins are connected and the sea level is changing very little, if at all, over long periods of time.

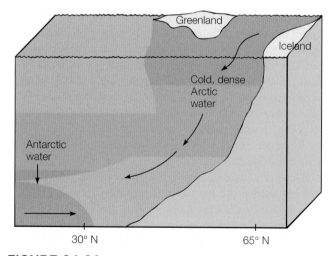

FIGURE 24.21 A cold-density current carries about 250 times more water than the Mississippi River from the Arctic and between Greenland and Iceland to the deep Atlantic Ocean. At about 30°N latitude, it meets water that has moved by cold-density currents all the way from the Antarctic.

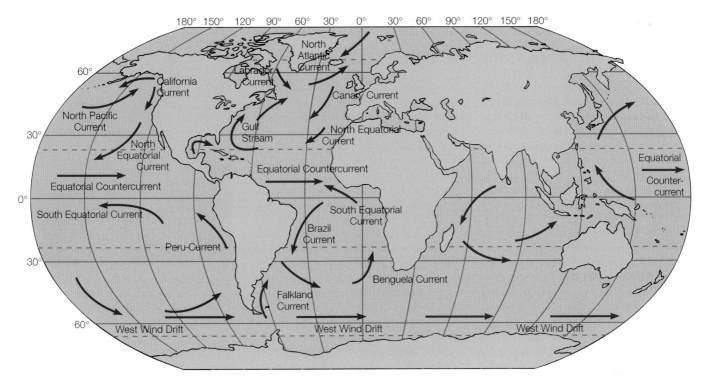

FIGURE 24.22 Earth's system of ocean currents.

THE OCEAN FLOOR

Some of the features of the ocean floor were discussed in chapter 18 because they were important in developing the theory of plate tectonics. Many features of the present ocean basins were created from the movement of large crustal plates, according to plate tectonics theory, and in fact, some ocean basins are thought to have originated with the movement of these plates. There is also evidence that some features of the ocean floor were modified during the ice ages of the past. During an ice age, much water becomes locked up in glacial ice, which lowers the sea level. The sea level dropped as much as 90 m (about 300 ft) during the most recent major ice age, exposing the margins of the continents to erosion. Today, these continental margins are flooded with seawater, forming a zone of relatively shallow water called the **continental shelf** (Figure 24.23). The continental shelf is considered

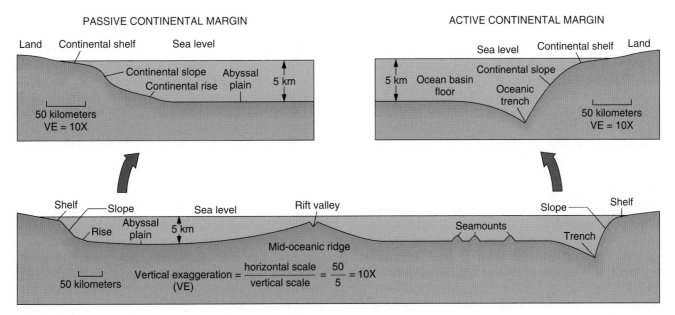

FIGURE 24.23 Profiles of seafloor topography. The vertical scales differ from the horizontal scales, causing vertical exaggeration, which makes slopes appear steeper than they really are. The bars for the horizontal scale are 50 km long, while the same distance vertically represents only 5 km, so the drawings have a vertical exaggeration of 10.

People Behind the Science

Rachel Louise Carson (1907–1964)

Rachel Louise Carson was a U.S. biologist, conservationist, and campaigner. Her writings on conservation and the dangers and hazards that many modern practices imposed on the environment inspired the creation of the modern environmental movement.

Carson was born in Springdale, Pennsylvania, on May 27, 1907, and educated at the Pennsylvania College for Women, studying English to achieve her ambition for a literary career. A stimulating biology teacher diverted her toward the study of science, and she went to Johns Hopkins University, graduating in zoology in 1929. She received her master's degree in zoology in 1932 and was then appointed to the department of zoology at the University of Maryland, spending her summers teaching and researching at the Woods Hole Marine Biological Laboratory in Massachusetts. Family commitments to her widowed mother and orphaned nieces forced her to abandon her academic career, and she worked for the U.S. Bureau of Fisheries, writing in her spare time articles on marine life and fish and producing her first book on the sea just before the Japanese attack on Pearl Harbor. During

World War II, she wrote fisheries information bulletins for the U.S. government and reorganized the publications department of what became known after the war as the U.S. Fish and Wildlife Service. In 1949, she was appointed chief biologist and editor of the service. She also became occupied with fieldwork and regularly wrote freelance articles on the natural world. During this period, she was also working on *The Sea Around Us.* Upon its publication in 1951, this book became an immediate best-seller, was translated into several languages, and won several literary awards. Given a measure of financial independence by this success, Carson resigned from her job in 1952 to become a professional writer. Her second book, *The Edge of the Sea* (1955), an ecological exploration of the seashore, further established her reputation as a writer on biological subjects. Her most famous book, *The Silent Spring* (1962), was a powerful indictment of the effects of the chemical poisons, especially DDT, with which humans were destroying Earth, sea, and sky. Despite denunciations from the influential agrochemical lobby, one immediate effect of Carson's book was the appointment of

a presidential advisory committee on the use of pesticides. By this time, Carson was seriously incapacitated by ill health, and she died in Silver Spring, Maryland, on April 14, 1964.

On a larger canvas, *The Silent Spring* alerted and inspired a new worldwide movement of environmental concern. While writing about broad scientific issues of pollution and ecological exploitation, Carson also raised important issues about the reckless squandering of natural resources by an industrial world.

Source: © Research Machines plc 2006. All Rights Reserved. Helicon Publishing is a division of Research Machines.

to be a part of the continent and not the ocean, even though it is covered with an average depth of about 130 m (about 425 ft) of seawater. The shelf slopes gently away from the shore for an average of 75 km (about 47 mi), but it is much wider on the edge of some parts of continents than other parts.

The continental shelf is a part of the continent that happens to be flooded by seawater at the present time. It still retains some of the general features of the adjacent land that is above water, such as hills, valleys, and mountains, but these features were smoothed off by the eroding action of waves when the sea level was lower. Today, a thin layer of sediments from the adjacent land covers these smoothed-off features.

Beyond the gently sloping continental shelf is a steeper feature called the **continental slope.** The continental slope is the transition between the continent and the deep ocean basin. The water depth at the top of the continental slope is about 120 m (about 390 ft), then plunges to a depth of about 3,000 m (about 10,000 ft) or more. The continental slope is generally 20 to 40 km (about 12 to 25 mi) wide, so the inclination is similar to that encountered driving down a steep mountain road on an interstate highway. At various places around the world, the continental slopes are cut by long, deep, and steep-sided **submarine canyons.** Some of these

canyons extend from the top of the slope and down the slope to the ocean basin. Such a submarine canyon can be similar in size and depth to the Grand Canyon on the Colorado River of Arizona. Submarine canyons are believed to have been eroded by turbidity currents, which were discussed in the section on ocean currents.

Beyond the continental slope is the bottom of the ocean floor, the **ocean basin.** Ocean basins are the deepest part of the ocean, covered by about 4 to 6 km (about 2 to 4 mi) of seawater. The basin is mostly a practically level plain called the **abyssal plain** and long, rugged mountain chains called **ridges** that rise thousands of meters above the abyssal plain. The Atlantic Ocean and Indian Ocean basins have ridges that trend north and south near the center of the basin. The Pacific Ocean basin has its ridge running north and south near the eastern edge. The Pacific Ocean basin also has more **trenches** than the Atlantic Ocean or the Indian Ocean basin (Figure 24.24). A trench is a long, relatively narrow, steep-sided trough that occurs along the edges of the ocean basins. Trenches range in depth from about 8 to 11 km (about 5 to 7 mi) deep below sea level. The origin of ridges and trenches was discussed in chapter 18.

The ocean basin and ridges of the ocean cover more than half of Earth's surface, accounting for more of the total surface than all the land of the continents. The plain of the ocean basin

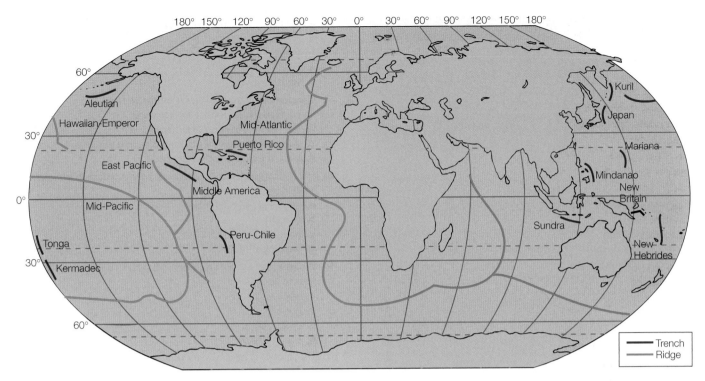

FIGURE 24.24 The location of some of the larger ridges and trenches of Earth's ocean floor.

alone, in fact, covers an area about equal to the area of the land. Scattered over the basin are more than ten thousand steep volcanic peaks called *seamounts*. By definition, seamounts rise more than 1 km (about 0.6 mi) above the ocean floor, sometimes higher than the sea-level surface of the ocean. A seamount that sticks above the water level makes an island. The Hawaiian Islands are examples of such giant volcanoes that have formed islands. Most seamount-formed islands are in the Pacific Ocean. Most islands in the Atlantic, on the other hand, are the tops of volcanoes of the Mid-Atlantic Ridge.

SUMMARY

Precipitation that falls on the land either evaporates, flows across the surface, or soaks into the ground. Water moving across the surface is called *runoff.* Water that moves across the land as a small body of running water is called a *stream.* A stream drains an area of land known as the stream drainage basin or *watershed.* The watershed of one stream is separated from the watershed of another by a line called a *divide.* Water that collects as a small body of standing water is called a *pond,* and a larger body is called a *lake.* A *reservoir* is a natural pond, a natural lake, or a lake or pond created by building a dam for water management or control. The water of streams, ponds, lakes, and reservoirs is collectively called *surface water.*

Precipitation that soaks into the ground *percolates* downward until it reaches a *zone of saturation.* Water from the saturated zone is called *groundwater.* The amount of water that a material will hold depends on its *porosity,* and how well the water can move through the material depends on its *permeability.* The surface of the zone of saturation is called the *water table.*

The *ocean* is the single, continuous body of salt water on the surface of Earth. A *sea* is a smaller part of the ocean with different characteristics. The dissolved materials in seawater are mostly the ions of six substances, but sodium ions and chlorine ions are the most abundant. *Salinity* is a measure of the mass of salts dissolved in 1,000 g of seawater.

An *ocean wave* is a moving disturbance that travels across the surface of the ocean. In its simplest form, a wave has a ridge called a *crest* and a depression called a *trough.* Waves have a characteristic *wave height, wavelength,* and *wave period.* The characteristics of waves made by the wind depend on the wind *speed,* the *time* the wind blows, and the *fetch.* Regular groups of low-profile, long-wavelength waves are called *swell.* When swell approaches a shore, the wave slows and increases in wave height. This slowing *refracts,* or bends, the waves so they approach the shore head-on. When the wave height becomes too steep, the top part breaks forward, forming *breakers* in the surf zone. Water accumulates at the shore from the breakers and returns to the sea as *undertow,* as *longshore currents,* or in *rip currents.*

Ocean currents are streams of water that move through other seawater over large distances. Some ocean currents are *density currents,* which are caused by differences in *water temperature, salinity,* or *suspended sediments.* Each ocean has a great system of moving water called a *gyre* that is centered in mid-latitudes. Different parts of a gyre are given different names such as the *Gulf Stream* or the *California Current.*

The ocean floor is made up of the *continental shelf,* the *continental slope,* and the *ocean basin.* The ocean basin has two main parts: the *abyssal plain* and mountain chains called *ridges.*

KEY TERMS

APPLYING THE CONCEPTS

1. What is the most abundant compound near or on the surface of Earth?
   a. O_2
   b. H_2O
   c. CO_2
   d. N_2

2. Evaporation, precipitation, and return of water to the oceans is called the
   a. rain cycle.
   b. aqueous cycle.
   c. hydrologic cycle.
   d. geologic cycle.

3. What is the major source of freshwater?
   a. streams
   b. underground water
   c. precipitation
   d. rivers

4. What is a small body of running water?
   a. runoff
   b. stream
   c. groundwater
   d. lake

5. The land that is drained by a stream is called the
   a. riverbank.
   b. watershed.
   c. river valley.
   d. mouth of the stream.

6. A small body of standing water is called a
   a. stream.
   b. pond.
   c. lake.
   d. watershed.

7. The amount of groundwater is estimated to be
   a. equal to the amount of surface water.
   b. less than the amount of surface water.
   c. twenty-five times the amount of surface water.
   d. twice the amount of surface water.

8. The total amount of pore spaces in a given sample of sediment is defined as
   a. grain size.
   b. porosity.
   c. permeability.
   d. concentration.

9. The ability of a given sample of sediment to transport water is defined as
   a. grain size.
   b. porosity.
   c. permeability.
   d. concentration.

10. The amount of groundwater in a specific area does not depend on
    a. porosity of the land.
    b. type of sediments.
    c. slope of the land.
    d. a nearby lake or reservoir.

11. The surface of the boundary between the zone of saturation and the zone above is called the
    a. water table.
    b. permeable zone.
    c. impermeable zone.
    d. groundwater table.

12. A layer of sand, gravel, sandstone, or any other highly permeable material beneath Earth's surface through which groundwater can move is a (an)
    a. well.
    b. spring.
    c. aquifer.
    d. artesian.

13. Water that usually does not have to be filtered is obtained from a
    a. lake.
    b. river.
    c. well.
    d. reservoir.

14. The treatment of seawater to replenish freshwater
    a. is readily available.
    b. kills ocean life.
    c. is energy-expensive.
    d. requires excessive filtration.

15. Deuterium from the oceans is considered a potential energy source for
    a. chemical power.
    b. thermal energy.
    c. controlled fusion.
    d. controlled fission.

16. Degradable pollution control is
    a. landfills.
    b. recycling.
    c. wastewater treatment.
    d. combustion.

17. The primary stage in wastewater treatment
    a. removes persistent pollutants.
    b. removes solids from wastewater.
    c. removes degradable pollutants.
    d. adds chlorine.

18. Chlorination of water is used to
    a. kill harmful bacteria.
    b. break down organic matter.
    c. remove persistent pollutants.
    d. provide needed nutrients.

19. The "rim of fire" surrounds the
    a. Atlantic Ocean.
    b. Indian Ocean.
    c. Pacific Ocean.
    d. Gulf of Mexico.

20. Dissolved materials and sediments are carried to oceans by
    a. rain.
    b. rivers.
    c. groundwater.
    d. pollutants.

21. What ion is found in seawater in the greatest amount?
    a. sodium ion, Na^+
    b. chloride ion, Cl^-
    c. calcium ion, $Ca1^-$
    d. potassium ion, K^+

22. The amount of dissolved salts in seawater is measured as
    a. salt concentration.
    b. solubility.
    c. salinity.
    d. percent.

23. What is the part of a river where the freshwater mixes with the salt water from the oceans?
    a. watershed
    b. estuary
    c. delta
    d. wetland

24. A regular group of long wavelength waves with a low wave height is called (a)
    a. fetch.
    b. swell.
    c. surf.
    d. breakers.

25. The distance that the wind blows across the open ocean is the
    a. fetch.
    b. swell.
    c. surf.
    d. breaker.

26. Steep volcanic peaks in the ocean basin are
    a. ridges.
    b. seamounts.
    c. steeds.
    d. quarter islands.

27. Of the total supply, the amount of water that is available for human consumption and agriculture is
    a. 97 percent.
    b. about two-thirds.
    c. about 3 percent.
    d. less than 1 percent.

28. Considering yearly global averages of precipitation that fall on and evaporate from the land,
    a. more is precipitated than evaporates.
    b. more evaporates than is precipitated.
    c. there is a balance between the amount precipitated and the amount evaporated.
    d. there is no pattern that can be generalized.

29. In general, how much of all the precipitation that falls on land ends up as runoff and groundwater?
    a. 97 percent
    b. about half
    c. about one-third
    d. less than 1 percent

30. Groundwater is
    a. any water beneath Earth's surface.
    b. water beneath Earth's surface from a saturated zone.
    c. water that soaks into the ground.
    d. Any of the above is correct.

31. How many different oceans are actually on Earth's surface?
    a. fourteen
    b. seven
    c. three
    d. one

32. The largest of the three principal ocean regions of Earth is the
    a. Atlantic Ocean.
    b. Pacific Ocean.
    c. Indian Ocean.
    d. South American Ocean.

33. The Gulf of Mexico is a shallow sea of the
    a. Atlantic Ocean.
    b. Pacific Ocean.
    c. Indian Ocean.
    d. South American Ocean.

34. Measurement of the salts dissolved in seawater taken from various locations throughout the world show that seawater has a
    a. uniform chemical composition and a variable concentration.
    b. variable chemical composition and a variable concentration.
    c. uniform chemical composition and a uniform concentration.
    d. variable chemical composition and a uniform concentration.

35. The percentage of dissolved salts in seawater averages about
    a. 35 percent.
    b. 3.5 percent.
    c. 0.35 percent.
    d. 0.035 percent.

36. The salinity of seawater is increased locally by
    a. the addition of water from a large river.
    b. heavy precipitation.
    c. the formation of sea ice.
    d. None of the above is correct.

37. Considering only the available light and the dissolving ability of gases in seawater, more abundant life should be found in a
    a. cool, relatively shallow ocean.
    b. warm, very deep ocean.
    c. warm, relatively shallow ocean.
    d. cool, very deep ocean.

38. The regular, low-profile waves called swell are produced from
    a. constant, prevailing winds.
    b. small, irregular waves becoming superimposed.
    c. longer wavelengths outrunning and outlasting shorter wavelengths.
    d. all wavelengths becoming transformed by gravity as they travel any great distance.

39. If the wavelength of swell is 10.0 m, then you know that the fish below the surface feel the waves to a depth of
    a. 5.0 m.
    b. 10.0 m.
    c. 20.0 m.
    d. however deep it is to the bottom.

40. In general, a breaker forms where the water depth is about one and one-third times the wave
    a. period.
    b. length.
    c. height.
    d. width.

41. Ocean currents are generally driven by
    a. the rotation of Earth.
    b. the prevailing winds.
    c. rivers from the land.
    d. All of the above are correct.

42. Of the following, the greatest volume of water is moved by the
    a. Mississippi River.
    b. California Current.
    c. Gulf Stream.
    d. Colorado River.

43. Water that is fit for human consumption and agriculture is called
    a. seawater.
    b. freshwater.
    c. distilled water.
    d. mountain water.

44. The continental divide that separates river systems in the United States is the
    a. Mississippi River.
    b. Rocky Mountains.
    c. Appalachian Mountains.
    d. Rio Grande.

45. The average domestic daily use of water per person is estimated to be
    a. 25 gal.
    b. 50 gal.
    c. 100 gal.
    d. 150 gal.

46. Groundwater costs more for consumer use because
    a. it must be pumped to the surface.
    b. it requires addition of bactericides.
    c. filtration is needed to remove salts.
    d. it is not chlorinated.

Answers

1. b 2. c 3. c 4. b 5. b 6. b 7. c 8. b 9. c 10. d 11. a 12. c 13. c 14. c 15. c 16. c 17. b 18. a 19. c 20. b 21. b 22. c 23. b 24. b 25. a 26. c 27. d 28. a 29. c 30. b 31. d 32. b 33. a 34. a 35. b 36. c 37. a 38. c 39. a 40. c 41. b 42. c 43. b 44. b 45. d 46. a

QUESTIONS FOR THOUGHT

1. How are the waters of Earth distributed as a solid, a liquid, and a gas at a given time? How much of the water is salt water and how much is freshwater?

2. Describe the hydrologic cycle. Why is the hydrologic cycle important in maintaining a supply of freshwater? Why is the hydrologic cycle called a cycle?

3. Describe in general all the things that happen to the water that falls on the land.

4. Explain how a stream can continue to flow even during a dry spell.

5. What is the water table? What is the relationship between the depth to the water table and the depth that a well must be drilled? Explain.

6. Compare the advantages and disadvantages of using (a) surface water and (b) groundwater as a source of freshwater.

7. Prepare arguments for (a) agriculture, (b) industries, and (c) cities each having first priority in the use of a limited water supply. Identify one of these arguments as being the "best case" for first priority, then justify your choice.

8. Discuss some possible ways of extending the supply of freshwater.

9. What is swell and how does it form?

10. Why do waves always seem to approach the shore head-on?

11. What factors determine the size of an ocean wave made by the wind?

12. Describe how a breaker forms from swell. What is surf?

13. Describe what you would look for to avoid rip currents at a beach.

FOR FURTHER ANALYSIS

1. Considering the distribution of all the water on Earth, which presently unavailable category would provide the most freshwater at the least cost for transportation, processing, and storage?

2. Describe a number of ways that you believe would increase the amount of precipitation going into groundwater rather than runoff.

3. Some people believe that constructing a reservoir for water storage is a bad idea because (1) it might change the downstream habitat below the dam and (2) the reservoir will eventually fill with silt and sediments. Write a "letter to the editor" that supports the idea that a reservoir is a bad idea for the reasons given. Now write a second letter that disagrees with the "bad idea letter" and supports the construction of the reservoir.

4. Explain how the average salinity of seawater has remained relatively constant over the past 600 million years in spite of the continuous supply of dissolved salts in the river waters of the world.

5. Can ocean waves or ocean currents be used as an energy source? Explain why or why not.

6. What are the significant similarities and differences between a river and an ocean current?

INVITATION TO INQUIRY

Water Use

Investigate the source of water and the amount used by industrial processes, agriculture, and homes in your area. Make pie graphs to compare your area to national averages, and develop explanations for any differences. What could be done to increase the supply of water in your area?

APPENDIX A
Mathematical Review

WORKING WITH EQUATIONS

Many of the problems of science involve an equation, a short-hand way of describing patterns and relationships that are observed in nature. Equations are also used to identify properties and to define certain concepts, but all uses have well-established meanings, symbols that are used by convention, and allowed mathematical operations. This appendix will assist you in better understanding equations and the reasoning that goes with the manipulation of equations in problem-solving activities.

BACKGROUND

In addition to a knowledge of rules for carrying out mathematical operations, an understanding of certain quantitative ideas and concepts can be very helpful when working with equations. Among these helpful concepts are (1) the meaning of inverse and reciprocal, (2) the concept of a ratio, and (3) fractions.

The term *inverse* means the opposite, or reverse, of something. For example, addition is the opposite, or inverse, of subtraction, and division is the inverse of multiplication. A *reciprocal* is defined as an inverse multiplication relationship between two numbers. For example, if the symbol n represents any number (except zero), then the reciprocal of n is $1/n$. The reciprocal of a number $(1/n)$ multiplied by that number (n) always gives a product of 1. Thus, the number multiplied by 5 to give 1 is 1/5 $(5 \times 1/5 = 5/5 = 1)$. So 1/5 is the reciprocal of 5, and 5 is the reciprocal of 1/5. Each number is the *inverse* of the other.

The fraction 1/5 means 1 divided by 5, and if you carry out the division, it gives the decimal 0.2. Calculators that have a $1/x$ key will do the operation automatically. If you enter 5, then press the $1/x$ key, the answer of 0.2 is given. If you press the $1/x$ key again, the answer of 5 is given. Each of these numbers is a reciprocal of the other.

A *ratio* is a comparison between two numbers. If the symbols m and n are used to represent any two numbers, then the ratio of the number m to the number n is the fraction m/n. This expression means to divide m by n. For example, if m is 10 and n is 5, the ratio of 10 to 5 is 10/5, or 2:1.

Working with *fractions* is sometimes necessary in problem-solving exercises, and an understanding of these operations is needed to carry out unit calculations. It is helpful in many of these operations to remember that a number (or a unit) divided by itself is equal to 1; for example,

$$\frac{5}{5} = 1 \qquad \frac{\text{inch}}{\text{inch}} = 1 \qquad \frac{5 \text{ inches}}{5 \text{inches}} = 1$$

When one fraction is divided by another fraction, the operation commonly applied is to "invert the denominator and multiply." For example, 2/5 divided by 1/2 is

$$\frac{\frac{2}{5}}{\frac{1}{2}} = \frac{2}{5} \times \frac{2}{1} = \frac{4}{5}$$

What you are really doing when you invert the denominator of the larger fraction and multiply is making the denominator (1/2) equal to 1. Both the numerator (2/5) and the denominator (1/2) are multiplied by 2/1, which does not change the value of the overall expression. The complete operation is

$$\frac{\frac{2}{5}}{\frac{1}{2}} \times \frac{\frac{2}{1}}{\frac{2}{1}} = \frac{\frac{2}{5} \times \frac{2}{1}}{\frac{1}{2} \times \frac{2}{1}} = \frac{\frac{4}{5}}{\frac{2}{2}} = \frac{\frac{4}{5}}{1} = \frac{4}{5}$$

SYMBOLS AND OPERATIONS

The use of symbols seems to cause confusion for some students because it seems different from their ordinary experiences with arithmetic. The rules are the same for symbols as they are for numbers, but you cannot do the operations with the symbols until you know what values they represent. The operation signs, such as $+$, $\div$, $\times$, and $-$ are used with symbols to indicate the operation that you *would* do if you knew the values. Some of the mathematical operations are indicated several ways. For example, $a \times b$, $a \cdot b$, and ab all indicate the same thing, that a is to be multiplied by b. Likewise, $a \div b$, a/b, and $a \times 1/b$ all indicate that a is to be divided by b. Since it is not possible to carry out the operations on symbols alone, they are called *indicated operations*.

OPERATIONS IN EQUATIONS

An equation is a shorthand way of expressing a simple sentence with symbols. The equation has three parts: (1) a left side, (2) an equal sign ($=$), which indicates the equivalence of the two sides, and (3) a right side. The left side has the same value and units as the right side, but the two sides may have a very different appearance. The two sides may also have the symbols that indicate mathematical operations ($+$, $-$, $\times$, and so forth) and may be in certain forms that indicate operations (a/b, ab, and so forth). In any case, the equation is a complete expression that states the left side has the same value and units as the right side.

Equations may contain different symbols, each representing some unknown quantity. In science, the phrase "solve the equation" means to perform certain operations with one symbol (which represents some variable) by itself on one side of the equation. This single symbol is usually, but not necessarily, on the left side and is not present on the other side. For example, the equation $F = ma$ has the symbol F on the left side. In science, you would say that this equation is solved for F. It could also be solved for m or for a, which will be considered shortly. The equation $F = ma$ is solved for F, and the *indicated operation* is to multiply m by a because they are in the form ma, which means the same thing as $m \times a$. This is the only indicated operation in this equation.

A solved equation is a set of instructions that has an order of indicated operations. For example, the equation for the relationship between a Fahrenheit and Celsius temperature, solved for °C, is $C = 5/9(F - 32)$. A list of indicated operations in this equation is as follows:

1. Subtract 32° from the given Fahrenheit temperature.
2. Multiply the result of (1) by 5.
3. Divide the result of (2) by 9.

Why are the operations indicated in this order? Because the bracket means 5/9 of the *quantity* $(F - 32°)$. In its expanded form, you can see that $5/9(F - 32°)$ actually means $5/9(F) - 5/9(32°)$. Thus, you cannot multiply by 5 or divide by 9 until you have found the quantity $(F - 32°)$. Once you have figured out the order of operations, finding the answer to a problem becomes almost routine as you complete the needed operations on both the numbers and the units.

SOLVING EQUATIONS

Sometimes it is necessary to rearrange an equation to move a different symbol to one side by itself. This is known as solving an equation for an unknown quantity. But you cannot simply move a symbol to one side of an equation. Since an equation is a statement of equivalence, the right side has the same value as the left side. If you move a symbol, you must perform the operation in a way that the two sides remain equivalent. This is accomplished by "canceling out" symbols until you have the unknown on one side by itself. One key to understanding the canceling operation is to remember that a fraction with the same number (or unit) over itself is equal to 1. For example, consider the equation $F = ma$, which is solved for F. Suppose you are considering a problem in which F and m are given, and the unknown is a. You need to solve the equation for a so it is on one side by itself. To eliminate the m, you do the *inverse* of the indicated operation on m, dividing both sides by m. Thus,

$$F = ma$$

$$\frac{F}{m} = \frac{ma}{m}$$

$$\frac{F}{m} = a$$

Since m/m is equal to 1, the a remains by itself on the right side. For convenience, the whole equation may be flipped to move the unknown to the left side,

$$a = \frac{F}{m}$$

Thus, a quantity that indicated a multiplication (ma) was removed from one side by an inverse operation of dividing by m.

Consider the following inverse operations to "cancel" a quantity from one side of an equation, moving it to the other side:

If the Indicated Operation of the Symbol You Wish to Remove Is:	Perform This Inverse Operation on Both Sides of the Equation
multiplication	division
division	multiplication
addition	subtraction
subtraction	addition
squared	square root
square root	square

EXAMPLE A.1

The equation for finding the kinetic energy of a moving body is $KE = 1/2mv^2$. You need to solve this equation for the velocity, v.

SOLUTION

The order of indicated operations in the equation is as follows:

1. Square v.
2. Multiply v^2 by m.
3. Divide the result of (2) by 2.

To solve for v, this order is *reversed* as the "canceling operations" are used:

Step 1: Multiply both sides by 2

$$KE = \frac{1}{2}mv^2$$

$$2KE = \frac{2}{2}mv^2$$

$$2KE = mv^2$$

Step 2: Divide both sides by m

$$\frac{2KE}{m} = \frac{mv^2}{m}$$

$$\frac{2KE}{m} = v^2$$

Step 3: Take the square root of both sides

$$\sqrt{\frac{2KE}{m}} = \sqrt{v^2}$$

$$\sqrt{\frac{2KE}{m}} = v$$

or

$$v = \sqrt{\frac{2KE}{m}}$$

The equation has been solved for v, and you are now ready to substitute quantities and perform the needed operations (see the example problem in chapter 1 on p. 13 for information on this topic).

SIGNIFICANT FIGURES

The numerical value of any measurement will always contain some uncertainty. Suppose, for example, that you are measuring one side of a square piece of paper as shown in Figure A.1. You could say that the paper is *about* 3.5 cm wide and you would be correct. This measurement, however, would be unsatisfactory for many purposes. It does not approach the true value of the length and contains too much uncertainty. It seems clear that the paper width is larger than 3.4 cm but narrower than 3.5 cm. But how much larger than 3.4 cm? You cannot be certain if the paper is 3.44, 3.45, or 3.46 cm wide. As your best estimate, you might say that the paper is 3.45 cm wide. Everyone would agree that you can be certain about the first two numbers (3.4) and they should be recorded. The last number (0.05) has been estimated and is not certain. The two certain numbers, together with one uncertain number, represent the greatest accuracy possible with the ruler being used. The paper is said to be 3.45 cm wide.

A *significant figure* is a number that is believed to be correct with some uncertainty only in the last digit. The value of the width of the paper, 3.45 cm, represents three significant figures. As you can see, the number of significant figures can be determined by the degree of accuracy of the measuring instrument being used. But suppose you need to calculate the area of the paper. You would multiply 3.45 cm × 3.45 cm and the product for the area would be 11.9025 cm². This is a greater precision than you were able to obtain with your measuring instrument. The result of a calculation can be no more accurate than the values being treated. Because the measurement had only three significant figures (two certain, one uncertain), then the answer can have only three significant figures. The area is correctly expressed as 11.9 cm².

There are a few simple rules that will help you determine how many significant figures are contained in a reported measurement:

1. All digits reported as a direct result of a measurement are significant.
2. Zero is significant when it occurs between nonzero digits. For example, 607 has three significant figures, and the zero is one of the significant figures.
3. In figures reported as *larger than the digit 1*, the digit zero is not significant when it follows a nonzero digit to indicate place. For example, in a report that "23,000 people attended the rock concert," the digits 2 and 3 are significant but the zeros are not significant. In this situation, the 23 is the measured part of the figure, and the three zeros tell you an estimate of how many attended the concert, that is, 23 thousand. If the figure is a measurement rather than an estimate, then it is written *with a decimal point after the last zero* to indicate that the zeros *are* significant. Thus 23,000 has *two* significant figures (2 and 3), but 23,000. has *five* significant figures. The figure 23,000 means "about 23 thousand," but 23,000. means 23,000. and not 22,999 or 23,001.
4. In figures reported as *smaller than the digit 1*, zeros after a decimal point that come before nonzero digits *are not* significant and serve only as place holders. For example, 0.0023 has two significant figures: 2 and 3. Zeros alone after a decimal point or zeros after a nonzero digit indicate a measurement, however, so these zeros *are* significant. The figure 0.00230, for example, has three significant figures since the 230 means 230 and not 229 or 231. Likewise, the figure 3.000 cm has four significant figures because the presence of the three zeros means that the measurement was actually 3.000 and not 2.999 or 3.001.

MULTIPLICATION AND DIVISION

When you multiply or divide measurement figures, the answer may have no more significant figures than the *least* number of significant figures in the figures being multiplied or divided. This simply means that an answer can be no more accurate than the least accurate measurement entering into the calculation, and that you cannot improve the accuracy of a measurement by doing a calculation. For example, in multiplying 54.2 mi/h × 4.0 h to find out the total distance traveled, the first figure (54.2) has three significant figures but the second (4.0) has only two significant figures. The answer can contain only two significant figures since this is the weakest number of those involved in the calculation. The correct answer is therefore 220 mi, not 216.8 mi. This may seem strange since multiplying the two numbers together gives the answer of 216.8 mi. This answer, however, means a greater accuracy than is possible, and the accuracy cannot be improved over the weakest number involved in the calculation. Since the weakest number (4.0) has only two significant figures the answer must also have only two significant figures, which is 220 mi.

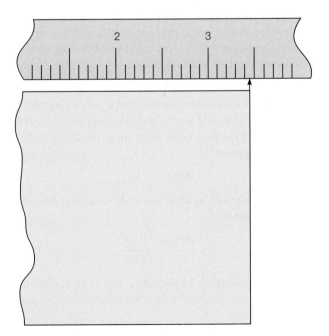

FIGURE A.1 How wide is this sheet of paper? Write your answer before reading the text _____.

The result of a calculation is *rounded* to have the same least number of significant figures as the least number of a measurement involved in the calculation. When rounding numbers, the last significant figure is increased by 1 if the number after it is 5 or larger. If the number after the last significant figure is 4 or less, the nonsignificant figures are simply dropped. Thus, if two significant figures are called for in the answer of the previous example, 216.8 is rounded up to 220 because the last number after the two significant figures is 6 (a number larger than 5). If the calculation result had been 214.8, the rounded number would be 210 miles.

Note that *measurement figures* are the only figures involved in the number of significant figures in the answer. Numbers that are counted or defined are not included in the determination of significant figures in an answer. For example, when dividing by 2 to find an average, the 2 is ignored when considering the number of significant figures. Defined numbers are defined exactly and are not used in significant figures. Since 1 kilogram is *defined* to be exactly 1,000 grams, such a conversion is not a measurement.

ADDITION AND SUBTRACTION

Addition and subtraction operations involving measurements, as with multiplication and division, cannot result in an answer that implies greater accuracy than the measurements had before the calculation. Recall that the last digit to the right in a measurement is uncertain, that is, it is the result of an estimate. The answer to an addition or subtraction calculation can have this uncertain number *no farther from the decimal place than it was in the weakest number involved in the calculation.* Thus, when 8.4 is added to 4.926, the weakest number is 8.4, and the uncertain number is .4, one place to the right of the decimal. The sum of 13.326 is therefore rounded to 13.3, reflecting the placement of this weakest doubtful figure.

The rules for counting zeros tell us that the numbers 203 and 0.200 both have three significant figures. Likewise, the numbers 230 and 0.23 only have two significant figures. Once you remember the rules, the counting of significant figures is straightforward. On the other hand, sometimes you find a number that seems to make it impossible to follow the rules. For example, how would you write 3,000 with two significant figures? There are several special systems in use for taking care of problems such as this, including the placing of a little bar over the last significant digit. One of the convenient ways of showing significant figures for difficult numbers is to use scientific notation, which is discussed in the section on scientific notation in this appendix. The convention for writing significant figures is to display one digit to the left of the decimal. The exponents are not considered when showing the number of significant figures in scientific notation. Thus, if you want to write three thousand showing one significant figure, you would write 3×10^3. To show two significant figures, it is 3.0×10^3, and for three significant figures, it becomes 3.00×10^3. As you can see, the correct use of scientific notation leaves little room for doubt about how many significant figures are intended.

EXAMPLE A.2

In a problem it is necessary to multiply 0.0039 km by 15.0 km. The result from a calculator is 0.0585 km^2. The least number of significant figures involved in this calculation is *two* (0.0039 is two significant figures; 15.0 is three—read the rules again to see why). The calculator result is therefore rounded off to have only two significant figures, and the answer is recorded as 0.059 km^2.

EXAMPLE A.3

The quantities of 10.3 calories, 10.15 calories, and 16.234 calories are added. The result from a calculator is 36.684 calories. The smallest number of decimal points is *one* digit to the right of the decimal, so the answer is rounded to 36.7 calories.

CONVERSION OF UNITS

The measurement of most properties results in both a numerical value and a unit. The statement that a glass contains 50 cm^3 of a liquid conveys two important concepts—the numerical value of 50 and the referent unit of cubic centimeters. Both the numerical value and the unit are necessary to communicate correctly the volume of the liquid.

When working with calculations involving measurement units, *both* the numerical value and the units are treated mathematically. As in other mathematical operations, there are general rules to follow.

1. Only properties with like units may be added or subtracted. It should be obvious that adding quantities such as 5 dollars and 10 dimes is meaningless. You must first convert to like units before adding or subtracting.
2. Like or unlike units may be multiplied or divided and treated in the same manner as numbers. You have used this rule when dealing with area (length $\times$ length = length^2, for example, cm $\times$ cm = cm^2) and when dealing with volume (length $\times$ length $\times$ length = length^3, for example, cm $\times$ cm $\times$ cm = cm^3).

You can use these two rules to create a *conversion ratio* that will help you change one unit to another. Suppose you need to convert 2.3 kg to grams. First, write the relationship between kilograms and grams:

$$1,000 \text{ g} = 1 \text{ kg}$$

Next, divide both sides by what you wish to convert *from* (kilograms in this example):

$$\frac{1,000 \text{ g}}{1} = \frac{1 \text{ kg}}{1 \text{ kg}}$$

One kilogram divided by 1 kg equals 1, just as 10 divided by 10 equals 1. Therefore, the right side of the relationship becomes 1 and the equation is:

$$\frac{1,000 \text{ g}}{1 \text{ kg}} = 1$$

The 1 is usually understood, that is, not stated, and the operation is called *canceling*. Canceling leaves you with the fraction 1,000 g/1 kg, which is a conversion ratio that can be used to convert from kilograms to grams. You simply multiply the conversion ratio by the numerical value and unit you wish to convert:

$$= 2.3 \text{ kg} \times \frac{1,000 \text{ g}}{1 \text{ kg}}$$

$$= \frac{2.3 \times 1,000}{1} \frac{\text{kg} \times \text{g}}{\text{kg}}$$

$$= \boxed{2,300 \text{ g}}$$

The kilogram units cancel. Showing the whole operation with units only, you can see how you end up with the correct unit of grams:

$$\text{kg} \times \frac{\text{g}}{\text{kg}} = \frac{\text{kg} \cdot \text{g}}{\text{kg}} = \text{g}$$

Since you did obtain the correct unit, you know that you used the correct conversion ratio. If you had blundered and used an inverted conversion ratio, you would obtain

$$2.3 \text{ kg} \times \frac{1 \text{ kg}}{1,000 \text{ g}} = .0023 \frac{\text{kg}^2}{\text{g}}$$

which yields the meaningless, incorrect units of kg^2/g. Carrying out the mathematical operations on the numbers and the units will always tell you whether or not you used the correct conversion ratio.

EXAMPLE A.4

A distance is reported as 100.0 km, and you want to know how far this is in miles.

SOLUTION

First, you need to obtain a *conversion factor* from a textbook or reference book, which usually lists the conversion factors by properties in a table. Such a table will show two conversion factors for kilometers and miles: (1) 1 km = 0.621 mi and (2) 1 mi = 1.609 km. You select the factor that is in the same form as your problem; for example, your problem is 100.0 km = ? mi. The conversion factor in this form is 1 km = 0.621 mi.

Second, you convert this conversion factor into a *conversion ratio* by dividing the factor by what you wish to convert *from*:

conversion factor:	1 km = 0.621
divide factor by what you want to convert from:	$\frac{1 \text{ km}}{1 \text{ km}} = \frac{0.621 \text{ mi}}{1 \text{ km}}$
resulting conversion rate:	$\frac{0.621 \text{ mi}}{\text{km}}$

Note that if you had used the 1 mi = 1.609 km factor, the resulting units would be meaningless. The conversion ratio is now multiplied by the numerical value *and unit* you wish to convert:

$$100.0 \text{ km} \times \frac{0.621 \text{ mi}}{\text{km}}$$

$$(100.0)(0.621)\frac{\text{km} \cdot \text{mi}}{\text{km}}$$

$$62.1 \text{ mi}$$

EXAMPLE A.5

A service station sells gasoline by the liter, and you fill your tank with 72 liters. How many gallons is this? (Answer: 19 gal)

SCIENTIFIC NOTATION

Most of the properties of things that you might measure in your everyday world can be expressed with a small range of numerical values together with some standard unit of measure. The range of numerical values for most everyday things can be dealt with by using units (1s), tens (10s), hundreds (100s), or perhaps thousands (1,000s). But the actual universe contains some objects of incredibly large size that require some very big numbers to describe. The Sun, for example, has a mass of about 1,970,000,000,000,000,000,000,000,000,000 kg. On the other hand, very small numbers are needed to measure the size and parts of an atom. The radius of a hydrogen atom, for example, is about 0.00000000005 m. Such extremely large and small numbers are cumbersome and awkward since there are so many zeros to keep track of, even if you are successful in carefully counting all the zeros. A method does exist to deal with extremely large or small numbers in a more condensed form. The method is called *scientific notation*, but it is also sometimes called *powers of ten* or *exponential notation*, since it is based on exponents of 10. Whatever it is called, the method is a compact way of dealing with numbers that not only helps you keep track of zeros but provides a simplified way to make calculations as well.

In algebra, you save a lot of time (as well as paper) by writing $(a \times a \times a \times a \times a)$ as a^5. The small number written to the right and above a letter or number is a superscript called an *exponent*. The exponent means that the letter or number is to be multiplied by itself that many times, for example, a^5 means a multiplied by itself five times, or $a \times a \times a \times a \times a$. As you can see, it is much easier to write the exponential form of this operation than it is to write it out in the long form. Scientific notation uses an exponent to indicate the power of the base 10. The exponent tells how many times the base, 10, is multiplied by itself. For example,

$$10,000 = 10^4$$
$$1,000 = 10^3$$
$$100 = 10^2$$
$$10 = 10^1$$
$$1 = 10^0$$
$$0.1 = 10^{-1}$$
$$0.01 = 10^{-2}$$
$$0.001 = 10^{-3}$$
$$0.0001 = 10^{-4}$$

This table could be extended indefinitely, but this somewhat shorter version will give you an idea of how the method works. The symbol 10^4 is read as "ten to the fourth power" and means

$10 \times 10 \times 10 \times 10$. Ten times itself four times is 10,000, so 10^4 is the scientific notation for 10,000. It is also equal to the number of zeros between the 1 and the decimal point, that is, to write the longer form of 10^4 you simply write 1, then move the decimal point four places to the *right;* 10 to the fourth power is 10,000.

The power of ten table also shows that numbers smaller than 1 have negative exponents. A negative exponent means a reciprocal:

$$10^{-1} = \frac{1}{10} = 0.1$$

$$10^{-2} = \frac{1}{100} = 0.01$$

$$10^{-3} = \frac{1}{1,000} = 0.001$$

To write the longer form of 10^{-4}, you simply write 1 then move the decimal point four places to the *left;* 10 to the negative fourth power is 0.0001.

Scientific notation usually, but not always, is expressed as the product of two numbers: (1) a number between 1 and 10 that is called the *coefficient* and (2) a power of ten that is called the *exponent.* For example, the mass of the sun that was given in long form earlier is expressed in scientific notation as

$$1.97 \times 10^{30} \, \text{kg}$$

and the radius of a hydrogen atom is

$$5.0 \times 10^{-11} \, \text{m}$$

In these expressions, the coefficients are 1.97 and 5.0, and the power of ten notations are the exponents. Note that in both of these examples, the exponent tells you where to place the decimal point if you wish to write the number all the way out in the long form. Sometimes scientific notation is written without a coefficient, showing only the exponent. In these cases, the coefficient of 1.0 is understood, that is, not stated. If you try to enter a scientific notation in your calculator, however, you will need to enter the understood 1.0, or the calculator will not be able to function correctly. Note also that 1.97×10^{30} kg and the expressions 0.197×10^{31} kg and 19.7×10^{29} kg are all correct expressions of the mass of the Sun. By convention, however, you will use the form that has one digit to the left of the decimal.

EXAMPLE A.6

What is 26,000,000 in scientific notation?

SOLUTION

Count how many times you must shift the decimal point until one digit remains to the left of the decimal point. For numbers larger than the digit 1, the number of shifts tells you how much the exponent is increased, so the answer is

$$2.6 \times 10^7$$

which means the coefficient 2.6 is multiplied by 10 seven times.

EXAMPLE A.7

What is 0.000732 in scientific notation? (Answer: 7.32×10^{-4})

It was stated earlier that scientific notation provides a compact way of dealing with very large or very small numbers, but it provides a simplified way to make calculations as well. There are a few mathematical rules that will describe how the use of scientific notation simplifies these calculations.

To *multiply* two scientific notation numbers, the coefficients are multiplied as usual, and the exponents are *added* algebraically. For example, to multiply (2×10^2) by (3×10^3), first separate the coefficients from the exponents,

$$(2 \times 3) \times (10^2 \times 10^3),$$

then multiply the coefficients and add the exponents,

$$6 \times 10^{(2 + 3)} = 6 \times 10^5$$

Adding the exponents is possible because $10^2 \times 10^3$ means the same thing as $(10 \times 10) \times (10 \times 10 \times 10)$, which equals $(100) \times (1,000)$, or 100,000, which is expressed as 10^5 in scientific notation. Note that two negative exponents add algebraically, for example $10^{-2} \times 10^{-3} = 10^{[(-2) + (-3)]} = 10^{-5}$. A negative and a positive exponent also add algebraically, as in $10^5 \times 10^{-3} = 10^{[(+5) + (-3)]} = 10^2$.

If the result of a calculation involving two scientific notation numbers does not have the conventional one digit to the left of the decimal, move the decimal point so it does, changing the exponent according to which way and how much the decimal point is moved. Note that the exponent increases by one number for each decimal point moved to the left. Likewise, the exponent decreases by one number for each decimal point moved to the right. For example, $938. \times 10^3$ becomes 9.38×10^5 when the decimal point is moved two places to the left.

To *divide* two scientific notation numbers, the coefficients are divided as usual and the exponents are *subtracted.* For example, to divide (6×10^6) by (3×10^2), first separate the coefficients from the exponents,

$$(6 \div 3) \times (10^6 \div 10^2)$$

then divide the coefficients and subtract the exponents,

$$2 \times 10^{(6 - 2)} = 2 \times 10^4$$

Note that when you subtract a negative exponent, for example, $10^{[(3) - (-2)]}$, you change the sign and add, $10^{(3 + 2)} = 10^5$.

EXAMPLE A.8

Solve the following problem concerning scientific notation:

$$\frac{(2 \times 10^4) \times (8 \times 10^{-6})}{8 \times 10^4}$$

SOLUTION

First, separate the coefficients from the exponents,

$$\frac{2 \times 8}{8} \times \frac{10^4 \times 10^{-6}}{10^4}$$

then multiply and divide the coefficients and add and subtract the exponents as the problem requires,

$$2 \times 10^{\{[(4) + (-6)] - [4]\}}$$

Solving the remaining additions and subtractions of the coefficients gives

$$2 \times 10^{-6}$$

THE SIMPLE LINE GRAPH

An equation describes a relationship between variables, and a graph helps you picture this relationship. A line graph pictures how changes in one variable correspond with changes in a second variable, that is, how the two variables change together. Usually one variable can be easily manipulated. The other variable is caused to change in value by the manipulation of the first variable. The **manipulated variable** is known by various names (*independent, input,* or *cause variable*), and the **responding variable** is known by various related names (*dependent, output,* or *effect variable*). The manipulated variable is usually placed on the horizontal axis, or *x*-axis, of the graph, so you could also identify it as the *x-variable*. The responding variable is placed on the vertical axis, or *y-axis*. This variable is identified as the *y-variable*.

Figure A.2 shows the mass of different volumes of water at room temperature. Volume is placed on the *x*-axis because the volume of water is easily manipulated, and the mass values change as a consequence of changing the values of volume. Note that both variables are named and that the measuring unit for each is identified on the graph.

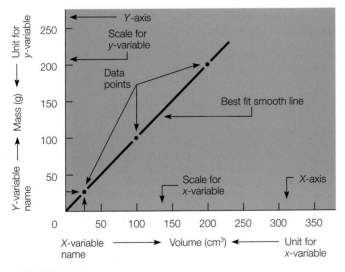

FIGURE A.2 The parts of a graph. On this graph, volume is placed on the *x*-axis and mass on the *y*-axis.

Figure A.2 also shows a number *scale* on each axis that represents changes in the values of each variable. The scales are usually, but not always, linear. A **linear scale** has equal intervals that represent equal increases in the value of the variable. Thus, a certain distance on the *x*-axis to the right represents a certain increase in the value of the *x*-variable. Likewise, certain distances up the *y*-axis represent certain increases in the value of the *y*-variable. The **origin** is the only point where both the *x*- and *y*-variables have a value of zero at the same time.

Figure A.2 shows three **data points.** A data point represents measurements of two related variables that were made at the same time. For example, a volume of 25 cm³ of water was found to have a mass of 25 g. Locate 25 cm³ on the *x*-axis and imagine a line moving straight up from this point on the scale. Locate 25 g on the *y*-axis and imagine a line moving straight out from this point on the scale. Where the lines meet is the data point for the 25 cm³ and 25 g measurements. A data point is usually indicated with a small dot or x (dots are used in the graph in Figure A.2).

A "best fit" smooth line is drawn through all the data points as close to them as possible. If it is not possible to draw the straight line *through* all the data points, then a straight line is drawn that has the same number of data points on both sides of the line. Such a line will represent a "best approximation" of the relationship between the two variables. The origin is also used as a data point in this example because a volume of zero will have a mass of zero.

The smooth line tells you how the two variables get larger together. With the same *x*- and *y*-axis scale, a 45° line means that they are increasing in an exact direct proportion. A more flat or more upright line means that one variable is increasing faster than the other. The more you work with graphs, the easier it will become for you to analyze what the "picture" means. There are more exact ways to extract information from a graph, and one of these techniques is discussed next.

One way to determine the relationship between two variables that are graphed with a straight line is to calculate the **slope.** The slope is a *ratio* between the changes in one variable and the changes in the other. The ratio is between the change in the value of the *x*-variable and the change in the value of the *y*-variable. Recall that the symbol Δ (Greek letter delta) means "change in," so the symbol Δx means the "change in *x*." The first step in calculating the slope is to find out how much the *x*-variable is changing (Δx) in relation to how much the *y*-variable is changing (Δy). You can find this relationship by first drawing a dashed line to the right of the straight *line* (not the data points), so that the *x*-variable has increased by some convenient unit (Figure A.3). Where you start or end this dashed line will not matter since the ratio between the variables will be the same everywhere on the graph line. The Δx is determined by subtracting the initial value of the *x*-variable on the dashed line (x_i) from the final value of the *x*-variable on the dashed line x_f, or $\Delta x = x_f - x_i$. In Figure A.3, the dashed line has an x_f of 200 cm³ and an x_i of 100 cm³, so Δx is 200 cm³ − 100 cm³, or 100 cm³. Note that Δx has both a number value and a unit.

Now you need to find Δy. The example in Figure A.3 shows a dashed line drawn back up to the graph line from the *x*-variable dashed line. The value of Δy is $y_f - y_i$. In the example, $\Delta y = 200$ g − 100 g, or 100 g.

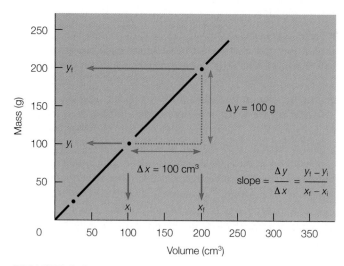

FIGURE A.3 The slope is a ratio between the changes in the y-variable and the changes in the x-variable, or $\Delta y/\Delta x$.

The slope of a straight graph line is the ratio of Δy to Δx, or

$$\text{slope} = \frac{\Delta y}{\Delta x}$$

In the example,

$$\text{slope} = \frac{100 \text{ g}}{100 \text{ cm}^3}$$

$$= 1 \frac{\text{g}}{\text{cm}^3} \text{ or } 1 \text{ g/cm}^3$$

Thus, the slope is 1 g/cm^3, and this tells you how the variables change together. Since g/cm^3 is also the unit of density, you know that you have just calculated the density of water from a graph.

Note that the slope can be calculated only for two variables that are increasing together, that is, for variables that are in direct proportion and have a line that moves upward and to the right. If variables change in any other way, mathematical operations must be performed to change the variables *into* this relationship. Examples of such necessary changes include taking the inverse of one variable, squaring one variable, taking the inverse square, and so forth.

Solubilities Chart

	Acetate	Bromide	Carbonate	Chloride	Fluoride	Hydroxide	Iodide	Nitrate	Oxide	Phosphate	Sulfate	Sulfide
Aluminum	S	S	—	S	s	i	S	S	i	i	S	d
Ammonium	S	S	S	S	S	S	S	S	—	S	S	S
Barium	S	S	i	S	s	S	S	S	S	i	s	d
Calcium	S	S	i	S	i	s	S	S	s	i	s	d
Copper (I)	—	s	i	s	i	—	i	—	i	—	d	i
Copper (II)	S	S	i	S	S	i	S	S	i	i	S	i
Iron (II)	S	S	i	S	s	i	S	S	i	i	S	i
Iron (III)	S	S	i	S	s	i	S	S	i	i	S	d
Lead	S	s	i	s	i	i	s	S	i	i	i	i
Magnesium	S	S	i	S	i	i	S	S	i	i	S	d
Mercury (I)	s	i	i	i	d	d	i	S	i	i	i	i
Mercury (II)	S	s	i	S	d	i	i	S	i	i	i	i
Potassium	S	S	S	S	S	S	S	S	S	S	S	i
Silver	s	i	i	i	S	—	i	S	i	i	i	i
Sodium	S	S	S	S	S	S	S	S	d	S	S	S
Strontium	S	S	s	S	i	s	S	S	—	i	i	i
Zinc	S	S	i	S	S	i	S	S	i	i	S	i

S–soluble
i–insoluble
s–slightly soluble
d–decomposes

Relative Humidity Table

Dry-Bulb Temperature (°C)	Difference Between Wet-Bulb and Dry-Bulb Temperatures (°C)																			
	1	2	3	4	5	6	7	8	9	10	11	12	13	14	15	16	17	18	19	20
0	81	64	46	29	13															
1	83	66	49	33	17															
2	84	68	52	37	22	7														
3	84	70	55	40	26	12														
4	86	71	57	43	29	16														
5	86	72	58	45	33	20	7													
6	86	73	60	48	35	24	11													
7	87	74	62	50	38	26	15													
8	87	75	63	51	40	29	19	8												
9	88	76	64	53	42	32	22	12												
10	88	77	66	55	44	34	24	15	6											
11	89	78	67	56	46	36	27	18	9											
12	89	78	68	58	48	39	29	21	12											
13	89	79	69	59	50	41	32	23	15	7										
14	90	79	70	60	51	42	34	26	18	10										
15	90	80	71	61	53	44	36	27	20	13	6									
16	90	81	71	63	54	46	38	30	23	15	8									
17	90	81	72	64	55	47	40	32	25	18	11									
18	91	82	73	65	57	49	41	34	27	20	14	7								
19	91	82	74	65	58	50	43	36	29	22	16	10								
20	91	83	74	66	59	51	44	37	31	24	18	12	6							
21	91	83	75	67	60	53	46	39	32	26	20	14	9							
22	92	83	76	68	61	54	47	40	34	28	22	17	11	6						
23	92	84	76	69	62	55	48	42	36	30	24	19	13	8						
24	92	84	77	69	62	56	49	43	37	31	26	20	15	10	5					
25	92	84	77	70	63	57	50	44	39	33	28	22	17	12	8					
26	92	85	78	71	64	58	51	46	40	34	29	24	19	14	10	5				
27	92	85	78	71	65	58	52	47	41	36	31	26	21	16	12	7				
28	93	85	78	72	65	59	53	48	42	37	32	27	22	18	13	9	5			
29	93	86	79	72	66	60	54	49	43	38	33	28	24	19	15	11	7			
30	93	86	79	73	67	61	55	50	44	39	35	30	25	21	17	13	9	5		
31	93	86	80	73	67	61	56	51	45	40	36	31	27	22	18	14	11	7		
32	93	86	80	74	68	62	57	51	46	41	37	32	28	24	20	16	12	9	5	
33	93	87	80	74	68	63	57	52	47	42	38	33	29	25	21	17	14	10	7	
34	93	87	81	75	69	63	58	53	48	43	39	35	30	28	23	19	15	12	8	5
35	94	87	81	75	69	64	59	54	49	44	40	36	32	28	24	20	17	13	10	7

Solutions for Follow-Up Example Exercises

Note: Solutions that involve calculations of measurements are rounded up or down to conform to the rules for signficant figures as described in Appendix A.

CHAPTER 1

Example 1.2, p. 9

$m = 15.0$ g
$V = 4.50$ cm^3
$\rho = ?$

$$\rho = \frac{m}{V}$$

$$= \frac{15.0 \text{ g}}{4.50 \text{ cm}^3}$$

$$= 3.33 \frac{\text{g}}{\text{cm}^3}$$

CHAPTER 2

Example 2.2, p. 28

$\bar{v} = 8.00$ km/h
$t = 10.0$ s
$d = ?$

The bicycle has a speed of 8.00 km/h and the time factor is 10.0 s, so km/h must be converted to m/s:

$$\bar{v} = \frac{0.2778 \frac{\text{m}}{\text{s}}}{\frac{\text{km}}{\text{h}}} \times 8.00 \frac{\text{km}}{\text{h}}$$

$$= (0.2778)(8.00) \frac{\text{m}}{\text{s}} \times \frac{\text{h}}{\text{km}} \times \frac{\text{km}}{\text{h}}$$

$$= 2.22 \frac{\text{m}}{\text{s}}$$

$$\bar{v} = \frac{d}{t}$$

$$\bar{v}t = \frac{dt}{t}$$

$$d = vt$$

$$= (2.22 \frac{\text{m}}{\text{s}})(10.0 \text{ s})$$

$$= (2.22)(10.0) \frac{\text{m}}{\text{s}} \times \frac{\text{s}}{1}$$

$$= 22.2 \text{ m}$$

Example 2.4, p. 30

$v_i = 0 \frac{\text{m}}{\text{s}}$

$v_f = ?$

$a = 5 \frac{\text{m}}{\text{s}^2}$

$t = 6$ s

$a = \frac{v_f - v_i}{t} \quad \therefore \quad v_f = at + v_i$

$$= (5 \frac{\text{m}}{\text{s}^2})(6 \text{ s})$$

$$= (5)(6) \ \frac{\text{m}}{\text{s}^2} \times \frac{\text{s}}{1}$$

$$= 30 \frac{\text{m}}{\text{s}}$$

Example 2.6, p. 32

$v_i = 25.0 \frac{\text{m}}{\text{s}}$

$v_f = 0 \frac{\text{m}}{\text{s}}$

$t = 10.0$ s

$a = ?$

$a = \frac{v_f - v_i}{t}$

$$= \frac{0 \frac{\text{m}}{\text{s}} - 25.0 \frac{\text{m}}{\text{s}}}{10.0 \text{ s}}$$

$$= \frac{-25.0}{10.0} \ \frac{\text{m}}{\text{s}} \times \frac{1}{\text{s}}$$

$$= -2.50 \frac{\text{m}}{\text{s}^2}$$

Example 2.9, p. 43

$m = 20$ kg
$F = 40$ N
$a = ?$

$$F = ma \quad \therefore \quad a = \frac{F}{m}$$

$$= \frac{40 \frac{\text{kg} \cdot \text{m}}{\text{s}^2}}{20 \text{ kg}}$$

$$= \frac{40}{20} \frac{\text{kg} \cdot \text{m}}{\text{s}^2} \times \frac{1}{\text{kg}}$$

$$= 2 \frac{\text{m}}{\text{s}^2}$$

Example 2.11, p. 44

$m = 60.0$ kg
$w = 100.0$ N
$g = ?$

$$w = mg \quad \therefore \quad g = \frac{w}{m}$$

$$= \frac{100.0 \frac{\text{kg} \cdot \text{m}}{\text{s}^2}}{60.0 \text{ kg}}$$

$$= \frac{100.0}{60.0} \quad \frac{\text{kg} \cdot \text{m}}{\text{s}^2} \times \frac{1}{\text{kg}}$$

$$= 1.67 \frac{\text{m}}{\text{s}^2}$$

Example 2.13, p. 46

Astronaut:

$a = 0.500 \text{ m/s}^2$
$t = 1.50 \text{ s}$
$v_i = 0 \text{ m/s}$
$v_f = ?$

$$a = \frac{v_f - v_i}{t} \quad \therefore \quad v_f = at + v_i$$

$$= \left(0.500 \frac{\text{m}}{\text{s}^2}\right)(1.50 \text{ s}) + 0 \frac{\text{m}}{\text{s}}$$

$$= (0.500)(1.50) \quad \frac{\text{m}}{\text{s}^2} \times \frac{\text{s}}{1}$$

$$= 0.750 \frac{\text{m}}{\text{s}}$$

Spacecraft:

$a = 0.250 \text{ m/s}^2$
$t = 1.50 \text{ s}$
$v_i = 0 \text{ m/s}$
$v_f = ?$

$$a = \frac{v_f - v_i}{t} \quad \therefore \quad v_f = at + v_i$$

$$= (0.250 \frac{\text{m}}{\text{s}^2})(1.50 \text{ s}) + 0 \frac{\text{m}}{\text{s}}$$

$$= (0.250)(1.50) \quad \frac{\text{m}}{\text{s}^2} \times \frac{\text{s}}{1}$$

$$= 0.375 \frac{\text{m}}{\text{s}}$$

Example 2.15, p. 47

Student and boat $m = 100.0 \text{ kg}$
Student and boat $v = ?$
Rock $m = 5.0 \text{ kg}$
Rock $v = 5 \text{ m/s}$
rock momentum − student and boat momentum = 0

$$(mv)_r - (mv)_{s\&b} = 0$$

$$(5.0 \text{ kg})(5.0 \frac{\text{m}}{\text{s}}) - (100.0 \text{ kg})(v_{s\&b}) = 0$$

$$(25 \text{ kg} \cdot \frac{\text{m}}{\text{s}}) - (100.0 \text{ kg})(v_{s\&b}) = 0$$

$$(25 \text{ kg} \cdot \frac{\text{m}}{\text{s}}) = (100.0 \text{ kg})(v_{s\&b})$$

$$v_{s\&b} = \frac{25 \text{ kg} \cdot \frac{\text{m}}{\text{s}}}{100.0 \text{ kg}}$$

$$= \frac{25}{100.0} \quad \frac{\text{kg}}{1} \times \frac{1}{\text{kg}} \times \frac{\text{m}}{\text{s}}$$

$$= 0.25 \frac{\text{m}}{\text{s}}$$

Example 2.17, p. 49

$m = 0.25 \text{ kg}$
$r = 0.25 \text{ m}$
$v = 2.0 \text{ m/s}$
$F = ?$

$$F = \frac{mv^2}{r}$$

$$= \frac{(0.25 \text{ kg})\left(2.0 \frac{\text{m}}{\text{s}}\right)^2}{0.25 \text{ m}}$$

$$= \frac{(0.25 \text{ kg})\left(\frac{4.0 \text{ m}^2}{\text{s}^2}\right)}{0.25 \text{ m}}$$

$$= \frac{(0.25)(4.0)}{0.25} \quad \frac{\text{kg} \cdot \text{m}^2}{\text{s}^2} \times \frac{1}{\text{m}}$$

$$= 4.0 \frac{\text{kg} \cdot \text{m}}{\text{s}^2}$$

$$= 4.0 \text{ N}$$

Example 2.20, p. 51

$G = 6.67 \times 10^{-11} \text{ N} \cdot \text{m}^2/\text{kg}^2$
$m_e = 6.0 \times 10^{24} \text{ kg}$
$d = 12.8 \times 10^6 \text{ m}$
$g = ?$

$$g = \frac{Gm_e}{d^2}$$

$$= \frac{\left(6.67 \times 10^{-11} \frac{\text{N} \cdot \text{m}^2}{\text{kg}^2}\right)(6.0 \times 10^{24} \text{ kg})}{(12.8 \times 10^6 \text{ m})^2}$$

$$= \frac{\left(6.67 \times 10^{-11} \frac{\text{N} \cdot \text{m}^2}{\text{kg}^2}\right)(6.0 \times 10^{24} \text{ kg})}{1.64 \times 10^{14} \text{ m}^2}$$

$$= \frac{(6.67 \times 10^{-11})(6.0 \times 10^{24})}{1.64 \times 10^{14}} \quad \frac{\text{N} \cdot \text{m}^2}{\text{kg}^2} \times \frac{\text{kg}}{1} \times \frac{1}{\text{m}^2}$$

$$= 2.44 \frac{\frac{\text{kg} \cdot \text{m}}{\text{s}^2}}{\text{kg}}$$

$$= 2.4 \frac{\text{m}}{\text{s}^2}$$

CHAPTER 3

Example 3.2, p. 63

$w = 50 \text{ lb}$
$d = 2 \text{ ft}$
$W = ?$

$$W = Fd$$
$$= (50 \text{ 1b}) (2 \text{ ft})$$
$$= (50) (2) \text{ ft} \cdot \text{lb}$$
$$= 100 \text{ ft} \cdot \text{lb}$$

Example 3.4, p. 66

$$w = 150 \text{ lb}$$
$$h = 15 \text{ ft}$$
$$t = 10.0 \text{ s}$$
$$P = ?$$

$$P = \frac{wh}{t}$$

$$= \frac{(150 \text{ lb}) (15 \text{ ft})}{10.0 \text{ s}}$$

$$= \frac{(150) (15)}{10.0} \quad \frac{\text{ft} \cdot \text{lb}}{\text{s}}$$

$$= 225 \frac{\text{ft} \cdot \text{lb}}{\text{s}}$$

$$= \frac{225 \dfrac{\text{ft} \cdot \text{lb}}{\text{s}}}{550 \dfrac{\dfrac{\text{ft} \cdot \text{lb}}{\text{s}}}{\text{hp}}}$$

$$= 0.41 \frac{\text{ft} \cdot \text{lb}}{\text{s}} \times \frac{\text{s}}{\text{ft} \cdot \text{lb}} \times \text{hp}$$

$$= 0.41 \text{ hp}$$

Example 3.6, p. 68

$$m = 5.00 \text{ kg}$$
$$g = 9.8 \text{ m/s}^2$$
$$h = 5.00 \text{ m}$$
$$W = ?$$

$$W = Fd$$

$$W = mgh$$

$$= (5.00 \text{ kg}) \left(9.8 \frac{\text{m}}{\text{s}^2}\right) (5.00 \text{ m})$$

$$= (5.00) (9.8) (5.00) \quad \frac{\text{kg} \cdot \text{m}}{\text{s}^2} \times \text{m}$$

$$= 245 \text{ N} \cdot \text{m}$$

$$= 250 \text{ J}$$

Example 3.8, p. 69

$$m = 100.0 \text{ kg}$$
$$v = 6.0 \text{ m/s}$$
$$W = ?$$

$$W = KE$$

$$KE = \frac{1}{2} mv^2$$

$$= \frac{1}{2} (100.0 \text{ kg}) \left(6.0 \frac{\text{m}}{\text{s}}\right)^2$$

$$= \frac{1}{2} (100.0 \text{ kg}) \left(36 \frac{\text{m}^2}{\text{s}^2}\right)$$

$$= \frac{1}{2} (100.0) (36) \quad \frac{\text{kg} \cdot \text{m}}{\text{s}^2} \times \text{m}$$

$$= 1,800 \text{ N} \cdot \text{m}$$

$$= 1,800 \text{ J}$$

Example 3.10, p. 73

$$m = 1.0 \text{ kg}$$
$$g = 9.8 \text{ m/s}^2$$
$$h = 1.0 \text{ m}$$
$$KE = ?$$

$$V_f = \sqrt{2gh}$$

$$= \sqrt{2 \left(9.8 \frac{\text{m}}{\text{s}^2}\right)(1.0 \text{ m})}$$

$$= \sqrt{2 (9.8) (1.0) \frac{\text{m}}{\text{s}^2} \times \text{m}}$$

$$= \sqrt{19.6 \frac{\text{m}^2}{\text{s}^2}}$$

$$= 4.4 \frac{\text{m}}{\text{s}}$$

$$KE = \frac{1}{2} mv^2$$

$$= \frac{1}{2} (1.0 \text{ kg}) \left(4.4 \frac{\text{m}}{\text{s}}\right)^2$$

$$= \frac{1}{2} (1.0 \text{ kg}) \left(19.4 \frac{\text{m}^2}{\text{s}^2}\right)$$

$$= \frac{1}{2} (1.0) (19.4) \frac{\text{kg} \cdot \text{m}}{\text{s}^2} \times \text{m}$$

$$= 9.7 \text{ N} \cdot \text{m}$$

$$= 9.7 \text{ J}$$

CHAPTER 4

Example 4.2, p. 91

$$T_C = 20°$$
$$T_F = ?$$

$$T_F = \frac{9}{5} T_C + 32°$$

$$= \frac{9}{5} 20° + 32°$$

$$= \frac{180°}{5} + 32°$$

$$= 36° + 32°$$

$$= 68°$$

Example 4.6, p. 95

$$m = 2 \text{ kg}$$
$$Q = 1.2 \text{ kcal}$$
$$\Delta T = 20°C$$
$$C = ?$$

$$Q = mc\Delta T \quad \therefore \quad C = \frac{Q}{m\Delta T}$$

$$= \frac{1.2 \text{ kcal}}{(2\text{kg})(20.0\text{C}°)}$$

$$= \frac{1.2}{(2)(20.0)} \frac{\text{kcal}}{\text{kgC}°}$$

$$= 0.03 \frac{\text{kcal}}{\text{kgC}°}$$

CHAPTER 5

Example 5.3, p. 121

$f = 2,500$ Hz

$v = 330$ m/s

$\lambda = ?$

$$v = \lambda f \quad \therefore \quad \lambda = \frac{v}{f}$$

$$= \frac{330 \frac{\text{m}}{\text{s}}}{2,500 \frac{1}{\text{s}}}$$

$$= \frac{330}{2,500} \frac{\text{m}}{\text{s}} \times \frac{\text{s}}{1}$$

$$= 0.13 \text{ m} \quad \text{or} \quad 13 \text{ cm}$$

Example 5.5, p. 124

$T_F = 86.0°$

$v = ?$

$$T_C = \frac{5}{9}(T_F - 32°)$$

$$= \frac{5}{9}(86.0° - 32°)$$

$$= \frac{5}{9}(54.0°)$$

$$= \frac{270.0°}{9}$$

$$= 30.0°$$

$$v_{T_p} = v_0 + \left(\frac{2.00 \text{ ft/s}}{°\text{C}}\right)(T_p)$$

$$= 1,087 \text{ ft/s} + \frac{2.00 \text{ ft/s}}{°\text{C}}(30.0°\text{C})$$

$$= 1,087 \text{ ft/s} + 60.0 \text{ ft/s}$$

$$= 1,147 \text{ ft/s}$$

Example 5.7, p. 125

$t = 1.00$ s

$v = 1,147$ ft/s

$d = ?$

$$v = \frac{d}{t} \quad \therefore \quad d = vt$$

$$= \left(1,147 \frac{\text{ft}}{\text{s}}\right)(1.00 \text{ s})$$

$$= (1,147)(1.00) \quad \frac{\text{ft}}{\text{s}} \times \frac{\text{s}}{1}$$

$$= 1,147 \text{ ft}$$

Sound traveled from the source to a reflecting surface, then back to the source, so the distance is

$$1,147 \text{ ft} \times \frac{1}{2} = 574 \text{ ft}$$

Example 5.9, p. 130

$L = 0.5$ m

$v = 400$ m/s

$F_2 = ?$

$$f_n = \frac{nv}{2L}$$

$$= \frac{(2)\left(400 \frac{\text{m}}{\text{s}}\right)}{(2)(0.5 \text{ m})}$$

$$= \frac{(2)(400)}{(2)(0.5)} \quad \frac{\text{m}}{\text{s}} \times \frac{1}{\text{m}}$$

$$= \frac{800}{1} \quad \frac{1}{\text{s}}$$

$$= 800 \text{ Hz}$$

CHAPTER 6

Example 6.4, p. 151

$V = 120$ V

$R = 30 \, \Omega$

$I = ?$

$$V = IR \quad \therefore \quad I = \frac{V}{R}$$

$$= \frac{120 \text{ V}}{30 \frac{\text{V}}{\text{A}}}$$

$$= \frac{120}{30} \quad \frac{\text{V}}{1} \times \frac{\text{A}}{\text{V}}$$

$$= 4\text{A}$$

Example 6.6, p. 152

$I = 0.5$ A

$V = 120$ V

$P = ?$

$$P = IV$$

$$= (0.5 \text{ A})(120 \text{ V})$$

$$= (0.5)(120) \quad \frac{\text{C}}{\text{s}} \times \frac{\text{J}}{\text{C}}$$

$$= 60 \frac{\text{J}}{\text{s}}$$

$$= 60 \text{ W}$$

Example 6.8, p. 154

$I = 0.5$ A

$V = 120$ V

$p = IV = 60$ W

Rate = $0.10/kWh

Cost = ?

$$\text{Cost} = \frac{(\text{watts})\,(\text{time})\,(\text{rate})}{1{,}000\,\frac{W}{kW}}$$

$$= \frac{(60\text{ W})\,(1.00\text{ h})\,\$0.10\,\frac{W}{kW}}{1{,}000\,\frac{W}{kW}}$$

$$= \frac{(60)\,(1.00)\,(0.10)}{1{,}000}\quad \frac{W}{1} \times \frac{h}{1} \times \frac{\$}{kWh} \times \frac{kW}{W}$$

$$= \$0.006 \text{ or } 0.6 \text{ of a cent per hour}$$

CHAPTER 7

Example 7.4, p. 194

$f = 7.00 \times 10^{14}$ Hz

$h = 6.63 \times 10^{-34}$ J $\cdot$ s

$E = ?$

$$E = hf$$

$$= (6.63 \times 10^{-34}\text{ J} \cdot \text{s})\left(7.00 \times 10^{14}\,\frac{1}{s}\right)$$

$$= (6.63 \times 10^{-34})\,(7.00 \times 10^{14})\quad \text{J} \cdot \text{s} \times \frac{1}{s}$$

$$= 4.64 \times 10^{-19}\text{ J}$$

CHAPTER 8

Example 8.2, p. 208

$f = 7.30 \times 10^{14}$ Hz

$h = 6.63 \times 10^{-34}$ J $\cdot$ s

$E = ?$

$$E = hf$$

$$= (6.63 \times 10^{-34}\text{ J} \cdot \text{s})\left(7.30 \times 10^{14}\,\frac{1}{s}\right)$$

$$= (6.63 \times 10^{-34})\,(7.30 \times 10^{14})\quad \text{J} \cdot \text{s} \times \frac{1}{s}$$

$$= 4.84 \times 10^{-19}\text{ J}$$

CHAPTER 10

Example 10.2, p. 253

Atoms	Atomic Weight	Totals
2 of C	2×12.0 u	24.0 u
6 of H	6×1.0 u	6.0 u
1 of O	1×16.0 u	16.0 u
		46.0 u

Example 10.4, p. 254

1. Formula: $C_{12}H_{22}O_{11}$
2. Formula weight

12 of C	12×12.0 u	= 144.0 u
22 of H	22×1.0 u	= 22.0 u
11 of O	11×16.0 u	= 176.0 u
		342.0 u

3. Percentage of carbon

$$\frac{(12.0\text{ u})\,(12)}{342.0\text{ u}} \times 100\% \text{ C}_{12}\text{H}_{22}\text{O}_{11}$$

$$\frac{144\text{ u}}{342.0\text{ u}} \times 100\% = 42.1\%$$

CHAPTER 11

Example 11.2, p. 282

% solute = 0.002%

V solution = 100,000

$$\frac{V_{\text{solute}}}{V_{\text{solution}}} \times 100\% \text{ solution} = \% \text{ solute} \; \therefore$$

$$V_{\text{solute}} = \frac{(V_{\text{solution}})\,(\% \text{ solute})}{100\% \text{ solution}}$$

$$= \frac{(100{,}000\text{ L})\,(0.002\%)}{100\% \text{ solution}}$$

$$= \frac{200\text{ L}}{100}$$

$$= 2\text{ L}$$

CHAPTER 13

Example 13.2, p. 330

$$^{226}_{88}\text{Ra} \rightarrow \, ^{4}_{2}\text{He} + \, ^{222}_{86}\text{Rn}$$

APPENDIX E

Solutions for Group A Parallel Exercises

Note: Solutions that involve calculations of measurements are rounded up or down to conform to the rules for significant figures described in Appendix A.

CHAPTER 1

1.1. Answers will vary but should have the relationship of 100 cm in 1 m, for example, 178 cm = 1.78 m.

1.2. Since density is given by the relationship $\rho = m/V$, then

$$\rho = \frac{m}{V} = \frac{272\text{g}}{20.0 \text{ cm}^3}$$

$$= \frac{272}{20.0} \frac{\text{g}}{\text{cm}^3}$$

$$= \boxed{13.6 \frac{\text{g}}{\text{cm}^3}}$$

1.3. The volume of a sample of lead is given and the problem asks for the mass. From the relationship of $\rho = m/V$, solving for the mass (m) tells you that the density (ρ) times the volume (V), or $m = \rho V$. The density of lead, 11.4 g/cm³, can be obtained from Table 1.3, so

$$\rho = \frac{m}{V}$$

$$V\rho = \frac{m\cancel{V}}{\cancel{V}}$$

$$m = \rho V$$

$$m = \left(11.4 \frac{\text{g}}{\text{cm}^3}\right)(10.0 \text{ cm}^3)$$

$$11.4 \times 10.0 \frac{\text{g}}{\text{cm}^3} \times \text{cm}^3$$

$$114 \frac{\text{g·cm}^{\cancel{3}}}{\cancel{\text{cm}}^3}$$

$$= \boxed{114 \text{ g}}$$

1.4. Solving the relationship $\rho = m/V$ for volume gives $V = m/\rho$, and

$$\rho = \frac{m}{V}$$

$$V\rho = \frac{m\cancel{V}}{\cancel{V}}$$

$$\frac{V\cancel{\rho}}{\cancel{\rho}} = \frac{m}{\rho}$$

$$V = \frac{m}{\rho}$$

$$V = \frac{600 \text{ g}}{3.00 \frac{\text{g}}{\text{cm}^3}}$$

$$= \frac{600}{3.00} \frac{\text{g}}{1} \times \frac{\text{cm}^3}{\text{g}}$$

$$= 200 \frac{\text{g·cm}^3}{\cancel{\text{g}}}$$

$$= \boxed{200 \text{ cm}^3}$$

1.5. A 50.0 cm³ sample with a mass of 34.0 grams has a density of

$$\rho = \frac{m}{V} = \frac{34.0 \text{ g}}{50.0 \text{ cm}^3}$$

$$= \frac{34.0}{50.0} \frac{\text{g}}{\text{cm}^3}$$

$$= \boxed{0.680 \frac{\text{g}}{\text{cm}^3}}$$

According to Table 1.3, 0.680 g/cm³ is the density of gasoline, so the substance must be gasoline.

1.6. The problem asks for a mass and gives a volume, so you need a relationship between mass and volume. Table 1.3 gives the density of water as 1.00 g/cm³, which is a density that is easily remembered. The volume is given in liters (L), which should first be converted to cm³ because this is the unit in which density is expressed. The relationship of $\rho = m/V$ solved for mass is ρV, so the solution is

$$\rho = \frac{m}{V} \quad \therefore \quad m = \rho V$$

$$m = \left(1.00 \frac{\text{g}}{\text{cm}^3}\right)(40,000 \text{ cm}^3)$$

$$= 1.00 \times 40,000 \frac{\text{g}}{\text{cm}^3} \times \text{cm}^3$$

$$= 40,000 \frac{\text{g·cm}^{\cancel{3}}}{\cancel{\text{cm}}^3}$$

$$= 40,000 \text{ g}$$

$$= \boxed{40 \text{ kg}}$$

1.7. From Table 1.3, the density of aluminum is given as 2.70 g/cm^3. Converting 2.1 kg to the same units as the density gives 2,100 g. Solving $\rho = m/V$ for the volume gives

$$V = \frac{m}{\rho} = \frac{2,100 \text{ g}}{2.70 \dfrac{\text{g}}{\text{cm}^3}}$$

$$= \frac{2,100}{2.70} \frac{\text{g}}{1} \times \frac{\text{cm}^3}{\text{g}}$$

$$= 777.78 \frac{\text{g·cm}^3}{\text{g}}$$

$$= \boxed{780 \text{ cm}^3}$$

1.8. The length of one side of the box is 0.1 m. Reasoning: Since the density of water is 1.00 g/cm^3, then the volume of 1,000 g of water is 1,000 cm^3. A cubic box with a volume of 1,000 cm^3 is 10 cm (since $10 \times 10 \times 10 = 1,000$). Converting 10 cm to m units, the cube is 0.1 m on each edge.

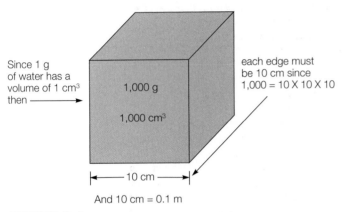

Since 1 g of water has a volume of 1 cm^3 then →

1,000 g

1,000 cm^3

each edge must be 10 cm since $1,000 = 10 \times 10 \times 10$

← 10 cm →

And 10 cm = 0.1 m

FIGURE A.4 Visualize the reasoning in 1.8.

1.9. The relationship between mass, volume, and density is $\rho = m/V$. The problem gives a volume but not a mass. The mass, however, can be assumed to remain constant during the compression of the bread so the mass can be obtained from the original volume and density, or

$$\rho = \frac{m}{V} \quad \therefore \quad m = \rho V$$

$$m = \left(0.2 \frac{\text{g}}{\text{cm}^3}\right)(3,000 \text{ cm}^3)$$

$$= 0.2 \times 3,000 \frac{\text{g}}{\text{cm}^3} \times \text{cm}^3$$

$$= 600 \frac{\text{g·cm}^3}{\text{cm}^3}$$

$$= 600 \text{ g}$$

A mass of 600 g and the new volume of 1,500 cm^3 means that the new density of the crushed bread is

$$\rho = \frac{m}{V}$$

$$= \frac{600 \text{ g}}{1,500 \text{ cm}^3}$$

$$= 600 \frac{\text{g}}{1,500 \text{ cm}^3}$$

$$= \boxed{0.4 \frac{\text{g}}{\text{cm}^3}}$$

1.10. According to Table 1.3, lead has a density of 11.4 g/cm^3. Therefore, a 1.00 cm^3 sample of lead would have a mass of

$$\rho = \frac{m}{V} \quad \therefore \quad m = \rho V$$

$$m = \left(11.4 \frac{\text{g}}{\text{cm}^3}\right)(1.00 \text{ cm}^3)$$

$$= 11.4 \times 1.00 \frac{\text{g}}{\text{cm}^3} \times \text{cm}^3$$

$$= 11.4 \frac{\text{g·cm}^3}{\text{cm}^3}$$

$$= 11.4 \text{ g}$$

Also according to Table 1.3, copper has a density of 8.96 g/cm^3. To balance a mass of 11.4 g of lead, a volume of this much copper would be required:

$$\rho = \frac{m}{V} \quad \therefore \quad V = \frac{m}{\rho}$$

$$V = \frac{11.4 \text{ g}}{8.96 \dfrac{\text{g}}{\text{cm}^3}}$$

$$= \frac{11.4}{8.96} \frac{\text{g}}{1} \times \frac{\text{cm}^3}{\text{g}}$$

$$= 1.27 \frac{\text{g·cm}^3}{\text{g}}$$

$$= \boxed{1.27 \text{ cm}^3}$$

CHAPTER 2

2.1 The distance and time are known and the problem asked for the average velocity. Listing these quantities with their symbols, we have

$$d = 160 \text{ km}$$
$$t = 2 \text{ h}$$
$$\bar{v} = ?$$

These are the quantities involved in the average speed equation, which is already solved for the unknown average speed:

$$\bar{v} = \frac{d}{t}$$

$$= \frac{160 \text{ km}}{2 \text{ h}}$$

$$= 80 \frac{\text{km}}{\text{h}}$$

2.2 Listing the quantities given in this problem, we have

$$d = 50.0 \text{ km}$$
$$t = 40.0 \text{ min}$$
$$\bar{v} = ?$$

The problem specifies that the answer should be in km/h. We see that 40 minutes is 40/60, or 2/3, or 0.667 of an hour, and the appropriate units are:

$$d = 50.0 \text{ km}$$

$$t = 0.667 \text{ h}$$

$$\bar{v} = ?$$

Substituting the known quantities, we have

$$\bar{v} = \frac{d}{t}$$

$$= \frac{50.0 \text{ km}}{0.667 \text{ h}}$$

$$= 75.0 \frac{\text{km}}{\text{h}}$$

2.3 Weight is the gravitational force on an object. Newton's second law of motion is $F = ma$, and since weight (w) is a force (F), then $F = w$ and the second law can be written as $w = ma$. The acceleration (a) is the acceleration due to gravity (g), so the equation for weight is $w = mg$. A kilogram is a unit of mass and g is known to be 9.8 m/s^2, so

$$w = mg$$

$$= (5.2 \text{ kg}) \left(9.8 \frac{\text{m}}{\text{s}^2} \right)$$

$$= (5.2)(9.8) \frac{\text{kg·m}}{\text{s}^2}$$

$$= 50.96 \text{ N}$$

$$= 51 \text{ N}$$

2.4 Listing the known and unknown quantities,

$$m = 40.0 \text{ kg}$$

$$a = 2.4 \frac{\text{m}}{\text{s}^2}$$

$$F = ?$$

These are the quantities found in Newton's second law of motion, $F = ma$, which is already solved for force (F). Thus,

$$F = ma$$

$$= (40.0 \text{ kg}) \left(2.4 \frac{\text{m}}{\text{s}^2} \right)$$

$$= (40.0)(2.4) \frac{\text{kg·m}}{\text{s}^2}$$

$$= 96 \text{ N}$$

2.5 Listing the known and unknown quantities,

$$F = 100.0 \text{ N}$$

$$m = 5.00 \text{ kg}$$

$$a = ?$$

$$F = ma \quad \therefore \quad a = \frac{F}{m}$$

$$= \frac{100.0 \frac{\text{kg·m}}{\text{s}^2}}{5.00 \text{ kg}}$$

$$= \frac{100.0}{5.00} \frac{\text{kg·m}}{\text{s}^2} \times \frac{1}{\text{kg}}$$

$$= 20.0 \frac{\text{m}}{\text{s}^2}$$

2.6. Listing these quantities given in this problem, with their symbols, we have

$$d = 22 \text{ km}$$

$$t = 15 \text{ min}$$

$$\bar{v} = ?$$

The usual units for a speed problem are km/h or m/s, and the problem specifies that the answer should be in km/h. We see that 15 minutes is 15/60, or 1/4, or 0.25 of an hour. We will now make a new list of the quantities with the appropriate units:

$$d = 22 \text{ km}$$

$$t = 0.25 \text{ h}$$

$$\bar{v} = ?$$

These quantities are related in the average speed equation, which is already solved for the unknown average velocity:

$$\bar{v} = \frac{d}{t}$$

Substituting the known quantities, we have

$$\bar{v} = \frac{22 \text{ km}}{0.25 \text{ h}}$$

$$= \boxed{88 \frac{\text{km}}{\text{h}}}$$

2.7. Listing the quantities with their symbols:

$$\bar{v} = 3.0 \times 10^8 \text{ m/s}$$

$$t = 20.0 \text{ min}$$

$$d = ?$$

We see that the velocity units are meters per second, but the time units are minutes. We need to convert minutes to seconds, and:

$$\bar{v} = 3.0 \times 10^8 \text{ m/s}$$

$$t = 1.20 \times 10^3 \text{ s}$$

$$d = ?$$

These relationships can be found in the average speed equation, which can be solved for the unknown:

$$\bar{v} = \frac{d}{t} \quad \therefore \quad d = \bar{v}t$$

$$d = \left(3.0 \times 10^8 \frac{\text{m}}{\text{s}} \right) (1.20 \times 10^3 \text{s})$$

$$= (3.0)(1.20) \times 10^{8+3} \frac{\text{m}}{\text{s}} \times \text{s}$$

$$= 3.6 \times 10^{11} \text{ m}$$

$$= \boxed{3.6 \times 10^8 \text{ km}}$$

2.8. Listing the quantities with their symbols, we can see the problem involves the quantities found in the definition of average speed:

$$\bar{v} = 350.0 \text{ m/s}$$
$$t = 5.00 \text{ s}$$
$$d = ?$$

$$\bar{v} = \frac{d}{t} \quad \therefore \quad d = \bar{v}t$$

$$d = \left(350.0 \frac{\text{m}}{\text{s}}\right)(5.00 \text{ s})$$

$$= (350.0)(5.00) \frac{\text{m}}{\text{s}} \times \text{s}$$

$$= \boxed{1{,}750 \text{ m}}$$

2.9. Note that the two speeds given (100.0 km/h and 50.0 km/h) are *average* speeds for two different legs of a trip. They are not initial and final speeds of an accelerating object, so you cannot add them together and divide by 2. The average speed for the total (entire) trip can be found from the definition of average speed, that is, average speed is the *total* distance covered divided by the *total* time elapsed. So, we start by finding the distance covered for each of the two legs of the trip:

$$\bar{v} = \frac{d}{t} \quad \therefore \quad d = \bar{v}t$$

$$\text{Leg 1 distance} = \left(100.0 \frac{\text{km}}{\text{h}}\right)(2.00 \text{ h})$$

$$= 200.0 \text{ km}$$

$$\text{Leg 2 distance} = \left(50.0 \frac{\text{km}}{\text{h}}\right)(1.00 \text{ h})$$

$$= 50.0 \text{ km}$$

Total distance (leg 1 plus leg 2) = 250.0 km
Total time = 3.00 h

$$\bar{v} = \frac{d}{t} = \frac{250.0 \text{ km}}{3.00 \text{ h}} = \boxed{83.3 \text{ km/h}}$$

2.10. The initial velocity, final velocity, and time are known and the problem asked for the acceleration. Listing these quantities with their symbols, we have

$$v_i = 0$$
$$v_f = 15.0 \text{ m/s}$$
$$t = 10.0 \text{ s}$$
$$a = ?$$

These are the quantities involved in the acceleration equation, which is already solved for the unknown:

$$a = \frac{v_f - v_i}{t}$$

$$a = \frac{15.0 \text{ m/s} - 0 \text{ m/s}}{10.0 \text{ s}}$$

$$= \frac{15.0}{10.0} \frac{\text{m}}{\text{s}} \times \frac{1}{\text{s}}$$

$$= \boxed{1.50 \frac{\text{m}}{\text{s}^2}}$$

2.11. The initial velocity, final velocity, and acceleration are known and the problem asked for the time. Listing these quantities with their symbols, we have

$$v_i = 20.0 \text{ m/s}$$
$$v_f = 25.0 \text{ m/s}$$
$$a = 3.0 \text{ m/s}^2$$
$$t = ?$$

These are the quantities involved in the acceleration equation, which must first be solved for the unknown time:

$$a = \frac{v_f - v_i}{t} \quad \therefore \quad t = \frac{v_f - v_i}{a}$$

$$t = \frac{25.0 \frac{\text{m}}{\text{s}} - 20.0 \frac{\text{m}}{\text{s}}}{3.0 \frac{\text{m}}{\text{s}^2}}$$

$$= \frac{5.00 \frac{\text{m}}{\text{s}}}{3.0 \frac{\text{m}}{\text{s}^2}}$$

$$= \frac{5.00}{3.0} \frac{\text{m}}{\text{s}} \times \frac{\text{s} \cdot \text{s}}{\text{m}}$$

$$= \boxed{1.7 \text{ s}}$$

2.12. The relationship between average velocity ($\bar{v}$), distance (d), and time (t) can be solved for time:

$$\bar{v} = \frac{d}{t}$$
$$\bar{v}t = d$$
$$t = \frac{d}{\bar{v}}$$

$$t = \frac{1{,}609 \text{ m}}{720 \text{ m/s}}$$

$$= \frac{1{,}609}{720} \frac{\text{m}}{1} \times \frac{\text{s}}{\text{m}}$$

$$= 2.23 \frac{\text{m} \cdot \text{s}}{\text{m}}$$

$$= \boxed{2.2 \text{ s}}$$

2.13. The relationship between average velocity ($\bar{v}$), distance (d), and time (t) can be solved for distance:

$$\bar{v} = \frac{d}{t} \quad \therefore \quad d = \bar{v}t$$

$$d = \left(40.0 \frac{\text{m}}{\text{s}}\right)(0.4625 \text{ s})$$

$$= 40.0 \times 0.4625 \frac{\text{m} \cdot \text{s}}{\text{s}}$$

$$= \boxed{18.5 \text{ m}}$$

2.14. "How many minutes . . ." is a question about time and the distance is given. Since the distance is given in km and the speed in m/s, a unit conversion is needed. The easiest thing to do is to convert km to m. There are 1,000 m in a km, and

$$(1.50 \times 10^8 \text{ km}) \times (1 \times 10^3 \text{ m/km}) = 1.50 \times 10^{11} \text{m}$$

The relationship between average velocity ($\bar{v}$), distance (d), and time (t) can be solved for time:

$$\bar{v} = \frac{d}{t} \qquad \therefore \qquad t = \frac{d}{\bar{v}}$$

$$t = \frac{1.50 \times 10^{11}\,\text{m}}{3.00 \times 10^{8}\,\frac{\text{m}}{\text{s}}}$$

$$= \frac{1.50}{3.00} \times 10^{11-8}\,\frac{\text{m}}{1} \times \frac{\text{s}}{\text{m}}$$

$$= 0.500 \times 10^{3}\,\frac{\text{m} \cdot \text{s}}{\text{m}}$$

$$= 5.00 \times 10^{2}\,\text{s}$$

$$\frac{500\,\text{s}}{60\,\frac{\text{s}}{\text{min}}} = \frac{500}{60}\,\frac{\text{s}}{1} \times \frac{\text{min}}{\text{s}}$$

$$= 8.33\,\frac{\text{s} \cdot \text{min}}{\text{s}}$$

$$= \boxed{8.33\,\text{min}}$$

(Information on how to use scientific notation [also called powers of ten or exponential notation] is located in appendix A, Mathematical Review.)
All significant figures are retained here because the units are defined exactly, without uncertainty.

2.15. The initial velocity (v_i) is given as 100.0 m/s, the final velocity (v_f) is given as 51.0 m/s, and the time is given as 5.00 s. Acceleration, including a deceleration or negative acceleration, is found from a change of velocity during a given time. Thus,

$$a = \frac{v_f - v_i}{t}$$

$$= \frac{\left(51.0\,\frac{\text{m}}{\text{s}}\right) - \left(100.0\,\frac{\text{m}}{\text{s}}\right)}{5.00\,\text{s}}$$

$$= \frac{-49.0\,\frac{\text{m}}{\text{s}}}{5.00\,\text{s}}$$

$$= -9.80\,\frac{\text{m}}{\text{s}} \times \frac{1}{\text{s}}$$

$$= \boxed{-9.80\,\frac{\text{m}}{\text{s}^2}}$$

(The negative sign means a negative acceleration, or deceleration.)

2.16. A ball thrown straight up decelerates to a velocity of zero, then accelerates back to the surface, just as a dropped ball would do from the height reached. Thus, the time required to decelerate upward is the same as the time required to accelerate downward. The ball returns to the surface with the same velocity with which it was thrown (neglecting friction). Therefore:

$$a = \frac{v_f - v_i}{t}$$

$$at = v_f - v_i$$

$$v_f = at + v_i$$

$$= \left(9.8\,\frac{\text{m}}{\text{s}^2}\right)(3.0\,\text{s})$$

$$= (9.8)\,(3.0)\,\frac{\text{m}}{\text{s}^2} \times \text{s}$$

$$= 29\,\frac{\text{m} \cdot \text{s}}{\text{s} \cdot \text{s}}$$

$$= \boxed{29\,\text{m/s}}$$

2.17. These three questions are easily answered by using the three sets of relationships, or equations, that were presented in this chapter:

(a) $v_f = at + v_i$, and when v_i is zero.

$$v_f = at$$

$$v_f = \left(9.8\,\frac{\text{m}}{\text{s}^2}\right)(4\,\text{s})$$

$$= 9.8 \times 4\,\frac{\text{m}}{\text{s}^2} \times \text{s}$$

$$= 39\,\frac{\text{m} \cdot \text{s}}{\text{s} \cdot \text{s}}$$

$$= \boxed{40\,\text{m/s}}$$

(b) $\bar{v} = \dfrac{v_f + v_i}{2} = \dfrac{40\,\text{m/s} + 0}{2} = 20c\,\text{m/s}$

(c) $\bar{v} = \dfrac{d}{t} \qquad \therefore \qquad d = \bar{v}t = \left(20\,\dfrac{\text{m}}{\text{s}}\right)(4\,\text{s})$

$$= 20 \times 4\,\frac{\text{m}}{\text{s}} \times \text{s}$$

$$= 80\,\frac{\text{m} \cdot \text{s}}{\text{s}}$$

$$= \boxed{80\,\text{m}}$$

2.18. Note that this problem can be solved with a series of three steps as in the previous problem. It can also be solved by the equation that combines all the relationships into one step. Either method is acceptable, but the following example of a one-step solution reduces the possibilities of error since fewer calculations are involved:

$$d = \frac{1}{2}gt^2 = \frac{1}{2}\left(9.8\,\frac{\text{m}}{\text{s}^2}\right)(5.00\,\text{s})^2$$

$$= \frac{1}{2}\left(9.8\,\frac{\text{m}}{\text{s}^2}\right)(25.0\,\text{s}^2)$$

$$= \left(\frac{1}{2}\right)(9.8)\,(25.0)\,\frac{\text{m}}{\text{s}^2} \times \text{s}^2$$

$$= 4.90 \times 25.0\,\frac{\text{m} \cdot \text{s}^2}{\text{s}^2}$$

$$= 122.5\,\text{m}$$

$$= \boxed{120\,\text{m}}$$

2.19. Listing the known and unknown quantities:

$$F = 100\,\text{N}$$

$$m = 5\,\text{kg}$$

$$a = ?$$

These are the quantities of Newton's second law of motion, $F = ma$, and

$$F = ma \quad \therefore \quad a = \frac{F}{m}$$

$$= \frac{100 \ \frac{kg \cdot m}{s^2}}{5 \ kg}$$

$$= \frac{100}{5} \ \frac{kg \cdot m}{s^2} \times \frac{1}{kg}$$

$$= \boxed{20 \ \frac{m}{s^2}}$$

2.20. Listing the known and unknown quantities:

$$m = 100 \ kg$$

$$v = 6 \ m/s$$

$$p = ?$$

These are the quantities found in the equation for momentum, $p = mv$, which is already solved for momentum (p). Thus,

$$p = mv$$

$$= (100 \ kg) \left(6 \frac{m}{s} \right)$$

$$= \boxed{600 \ \frac{kg \cdot m}{s}}$$

2.21. Listing the known and unknown quantities:

$$w = 13{,}720 \ N$$

$$v = 91 \ km/h$$

$$p = ?$$

The equation for momentum is $p = mv$, which is already solved for momentum (p). The weight unit must be first converted to a mass unit:

$$w = mg \quad \therefore \quad m = \frac{w}{g}$$

$$= \frac{13{,}720 \ \frac{kg \cdot m}{s^2}}{9.8 \ \frac{m}{s^2}}$$

$$= \frac{13{,}720}{9.8} \ \frac{kg \cdot m}{s^2} \times \frac{s^2}{m}$$

$$= 1{,}400 \ kg$$

The km/h unit should next be converted to m/s. Using the conversion factor from inside the front cover:

$$\frac{0.2778 \ \frac{m}{s}}{\frac{km}{h}} \times 91 \ \frac{km}{h}$$

$$0.2778 \times 91 \ \frac{m}{s} \times \frac{h}{km} \times \frac{km}{h}$$

$$25 \ \frac{m}{s}$$

Now, listing the converted known and unknown quantities:

$$m = 1{,}400 \ kg$$

$$v = 25 \ m/s$$

$$p = ?$$

and solving for momentum (p),

$$p = mv$$

$$= (1{,}400 \ kg) \left(25 \frac{m}{s} \right)$$

$$= \boxed{35{,}000 \ \frac{kg \cdot m}{s}}$$

2.22. Listing the known and unknown quantities:

Bullet $\rightarrow m = 0.015 \ kg$ Rifle $\rightarrow m = 6 \ kg$

Bullet $\rightarrow v = 200 \ m/s$ Rifle $\rightarrow v = ? \ m/s$

Note the mass of the bullet was converted to kg. This is a conservation of momentum question, where the bullet and rifle can be considered as a system of interacting objects:

Bullet momentum = rifle momentum

$$(mv)_b = (mv)_r$$

$$(mv)_b - (mv)_r = 0$$

$$(0.015 \ kg) \left(200 \frac{m}{s} \right) - (6 \ kg)v_r = 0$$

$$\left(3 \ kg \cdot \frac{m}{s} \right) - (6 \ kg \cdot v_r) = 0$$

$$\left(3 \ kg \cdot \frac{m}{s} \right) = (6 \ kg \cdot v_r)$$

$$v_r = \frac{3 \ kg \cdot \frac{m}{s}}{6 \ kg}$$

$$= \frac{3}{6} \ \frac{kg}{1} \times \frac{1}{kg} \times \frac{m}{s}$$

$$= \boxed{0.5 \ \frac{m}{s}}$$

The rifle recoils with a velocity of 0.5 m/s.

2.23. Listing the known and unknown quantities:

Astronaut $\rightarrow w = 2{,}156 \ N$ Wrench $\rightarrow m = 5.0 \ kg$

Astronaut $\rightarrow v = ? \ m/s$ Wrench $\rightarrow v = 5.0 \ m/s$

Note that the astronaut's weight is given, but we need mass for the conservation of momentum equation. Mass can be found because the weight on Earth was given, where we know $g = 9.8 \ m/s^2$. Thus, the mass is

$$w = mg \quad \therefore \quad m = \frac{w}{g}$$

$$= \frac{2{,}156 \ \frac{kg \cdot m}{s^2}}{9.8 \ \frac{m}{s^2}}$$

$$= \frac{2,156}{9.8} \frac{\text{kg·m}}{\text{s}^2} \times \frac{\text{s}^2}{\text{m}}$$

$$= 220 \text{ kg}$$

So the converted known and unknown quantities are:

Astronaut → $m = 220$ kg   Wrench → $m = 5.0$ kg
Astronaut → $v = ?$ m/s   Wrench → $v = 5.0$ m/s

This is a conservation of momentum question, where the astronaut and wrench can be considered as a system of interacting objects:

$$\text{Wrench momentum} = \text{astronaut momentum}$$

$$(mv)_\text{w} = (mv)_\text{a}$$

$$(mv)_\text{w} - (mv)_\text{a} = 0$$

$$(5.0 \text{ kg}) \left(5.0 \frac{\text{m}}{\text{s}}\right) - (220 \text{ kg}) v_\text{a} = 0$$

$$\left(25 \text{ kg} \cdot \frac{\text{m}}{\text{s}}\right) - (220 \text{ kg} \cdot v_\text{a}) = 0$$

$$v_\text{a} = \frac{25 \text{ kg} \cdot \frac{\text{m}}{\text{s}}}{220 \text{ kg}}$$

$$= \frac{25}{220} \frac{\text{kg}}{1} \times \frac{1}{\text{kg}} \times \frac{\text{m}}{\text{s}}$$

$$= \boxed{0.11 \frac{\text{m}}{\text{s}}}$$

The astronaut moves away with a velocity of 0.11 m/s.

2.24. **(a)** Weight (w) is a downward force from the acceleration of gravity (g) on the mass (m) of an object. This relationship is the same as Newton's second law of motion, $F = ma$, and

$$w = mg = (1.25 \text{ kg}) \left(9.8 \frac{\text{m}}{\text{s}^2}\right)$$

$$= (1.25)(9.8) \text{ kg} \times \frac{\text{m}}{\text{s}^2}$$

$$= 12.25 \frac{\text{kg·m}}{\text{s}^2}$$

$$= \boxed{12 \text{ N}}$$

(b) First, recall that a force (F) is measured in newtons (N) and a newton has units of $\text{N} = \frac{\text{kg·m}}{\text{s}^2}$. Second, the relationship between force (F), mass (m), and acceleration (a) is given by Newton's second law of motion, force = mass times acceleration, or $F = ma$. Thus,

$$F = ma \quad \therefore \quad a = \frac{F}{m} = \frac{10.0 \frac{\text{kg·m}}{\text{s}^2}}{1.25 \text{ kg}}$$

$$= \frac{10.0}{1.25} \frac{\text{kg·m}}{\text{s}^2} \times \frac{1}{\text{kg}}$$

$$= 8.00 \frac{\text{kg·m}}{\text{kg·s}^2}$$

$$= \boxed{8.00 \frac{\text{m}}{\text{s}^2}}$$

(Note how the units were treated mathematically in this solution and why it is necessary to show the units for a newton of force. The resulting unit in the answer *is* a unit of acceleration, which provides a check that the problem was solved correctly.

2.25.

$$F = ma = (1.25 \text{ kg}) \left(5.00 \frac{\text{m}}{\text{s}^2}\right)$$

$$= (1.25)(5.00) \text{ kg} \times \frac{\text{m}}{\text{s}^2}$$

$$= 6.25 \frac{\text{kg·m}}{\text{s}^2}$$

$$= \boxed{6.25 \text{ N}}$$

(Note that the solution is correctly reported in *newton* units of force rather than kg·m/s².)

2.26. The bicycle tire exerts a backward force on the road, and the equal and opposite reaction force of the road on the bicycle produces the forward motion. (The motion is always in the direction of the applied force.) Therefore,

$$F = ma = (70.0 \text{ kg}) \left(2.0 \frac{\text{m}}{\text{s}^2}\right)$$

$$= (70.0)(2.0) \text{ kg} \times \frac{\text{m}}{\text{s}^2}$$

$$= 140 \frac{\text{kg·m}}{\text{s}^2}$$

$$= \boxed{140 \text{ N}}$$

2.27. The question requires finding a force in the metric system, which is measured in newtons of force. Since newtons of force are defined in kg, m, and s, unit conversions are necessary, and these should be done first.

$$1 \frac{\text{km}}{\text{h}} = \frac{1,000 \text{ m}}{3,600 \text{ s}} = 0.2778 \frac{\text{m}}{\text{s}}$$

Dividing both sides of this conversion factor by what you are converting *from* gives the conversion ratio of

$$\frac{0.2778 \frac{\text{m}}{\text{s}}}{\frac{\text{km}}{\text{h}}}$$

Multiplying this conversion ratio times the two velocities in km/h will convert them to m/s as follows:

$$\left(\frac{0.2778 \frac{\text{m}}{\text{s}}}{\frac{\text{km}}{\text{h}}}\right) \left(80.0 \frac{\text{km}}{\text{h}}\right)$$

$$= (0.2778)(80.0) \frac{\text{m}}{\text{s}} \times \frac{\text{h}}{\text{km}} \times \frac{\text{km}}{\text{h}}$$

$$= 22.2 \frac{\text{m}}{\text{s}}$$

$$\left(\frac{0.2778 \frac{\text{m}}{\text{s}}}{\frac{\text{km}}{\text{h}}}\right) \left(44.0 \frac{\text{km}}{\text{h}}\right)$$

$$= (0.2778)(44.0)\frac{m}{s} \times \frac{h}{km} \times \frac{km}{h}$$

$$= 12.2\frac{m}{s}$$

Now you are ready to find the appropriate relationship between the quantities involved. This involves two separate equations: Newton's second law of motion and the relationship of quantities involved in acceleration. These may be combined as follows:

$$F = ma \text{ and } a = \frac{v_f - v_i}{t} \quad \therefore \quad F = m\left(\frac{v_f - v_i}{t}\right)$$

Now you are ready to substitute quantities for the symbols and perform the necessary mathematical operations:

$$= (1{,}500 \text{ kg})\left(\frac{22.2 \text{ m/s} - 12.2 \text{ m/s}}{10.0 \text{ s}}\right)$$

$$= (1{,}500 \text{ kg})\left(\frac{10.0 \text{ m/s}}{10.0 \text{ s}}\right)$$

$$= 1{,}500 \times 1.00 \frac{kg \cdot \frac{m}{s}}{s}$$

$$= 1{,}500 \frac{kg \cdot m}{s} \times \frac{1}{s}$$

$$= 1{,}500 \frac{kg \cdot m}{s \cdot s}$$

$$= 1{,}500 \frac{kg \cdot m}{s^2}$$

$$= 1{,}500 \text{ N} = \boxed{1.5 \times 10^3 \text{ N}}$$

2.28. A unit conversion is needed as in the previous problem:

$$\left(90.0\frac{km}{h}\right)\left(0.2778\frac{\frac{m}{s}}{\frac{km}{h}}\right) = 25.0 \text{ m/s}$$

(a) $F = ma \quad \therefore \quad m = \frac{F}{a}$ and $a = \frac{v_f - v_i}{t}$, so

$$m = \frac{F}{\frac{v_f - v_i}{t}} = \frac{5{,}000.0\frac{kg \cdot m}{s^2}}{\frac{25.0 \text{ m/s} - 0}{5.0 \text{ s}}}$$

$$= \frac{5{,}000.0\frac{kg \cdot m}{s^2}}{5.0\frac{m}{s^2}}$$

$$= \frac{5{,}000.0}{5.0}\frac{kg \cdot m}{s^2} \times \frac{s^2}{m}$$

$$= 1{,}000\frac{kg \cdot m \cdot s^2}{m \cdot s^2}$$

$$= \boxed{1.0 \times 10^3 \text{ kg}}$$

(b) $\qquad w = mg$

$$= (1.0 \times 10^3 \text{ kg})\left(9.8\frac{m}{s^2}\right)$$

$$= (1.0 \times 10^3)(9.8) \text{ kg} \times \frac{m}{s^2}$$

$$= 9.8 \times 10^3 \frac{kg \cdot m}{s^2}$$

$$= \boxed{9.8 \times 10^3 \text{ N}}$$

2.29. $\qquad w = mg$

$$= (70.0 \text{ kg})\left(9.8\frac{m}{s^2}\right)$$

$$= 70.0 \times 9.8 \text{ kg}\frac{m}{s^2}$$

$$= 686\frac{kg \cdot m}{s^2}$$

$$= \boxed{690 \text{ N}}$$

2.30. $\qquad F = \frac{mv^2}{r}$

$$= \frac{(0.20 \text{ kg})\left(3.0\frac{m}{s}\right)^2}{1.5 \text{ m}}$$

$$= \frac{(0.20 \text{ kg})\left(9.0\frac{m^2}{s^2}\right)}{1.5 \text{ m}}$$

$$= \frac{0.20 \times 9.0}{1.5}\frac{kg \cdot m^2}{s^2} \times \frac{1}{m}$$

$$= 1.2\frac{kg \cdot m \cdot m}{s^2 \cdot m}$$

$$= \boxed{1.2 \text{ N}}$$

2.31. **(a)** Newton's laws of motion consider the resistance to a change of motion, or mass, and not weight. The astronaut's mass is

$$w = mg \quad \therefore \quad m = \frac{w}{g} = \frac{1{,}960.0\frac{kg \cdot m}{s^2}}{9.8\frac{m}{s^2}}$$

$$= \frac{1{,}960.0}{9.8}\frac{kg \cdot m}{s^2} \times \frac{s^2}{m} = 200 \text{ kg}$$

(b) From Newton's second law of motion, you can see that the 100 N rocket gives the 200 kg astronaut an acceleration of:

$$F = ma \quad \therefore \quad a = \frac{F}{m} = \frac{100\frac{kg \cdot m}{s^2}}{200 \text{ kg}}$$

$$= \frac{100 \text{ kg} \cdot m}{200 s^2} \times \frac{1}{kg} = 0.5 \text{ m/s}^2$$

(c) An acceleration of 0.5 m/s² for 2.0 s will result in a final velocity of

$$a = \frac{v_f - v_i}{t} \quad \therefore \quad v_f = at + v_i$$

$$= (0.5 \text{ m/s}^2)(2.0 \text{ s}) + 0 \text{ m/s}$$

$$= \boxed{1 \text{ m/s}}$$

3.1. Listing the known and unknown quantities:

$$F = 200 \text{ N}$$
$$d = 3 \text{ m}$$
$$W = \text{?}$$

These are the quantities found in the equation for work, $W = Fd$, which is already solved for work (W). Thus,

$$W = Fd$$
$$= \left(200 \, \frac{\text{kg·m}}{\text{s}^2}\right)(3 \text{ m})$$
$$= (200)(3) \text{ N·m}$$
$$= \boxed{600 \text{ J}}$$

3.2. Listing the known and unknown quantities:

$$F = 440 \text{ N}$$
$$d = 5.0 \text{ m}$$
$$w = 880 \text{ N}$$
$$W = \text{?}$$

These are the quantities found in the equation for work, $W = Fd$, which is already solved for work (W). As you can see in the equation, the force exerted and the distance the box was moved are the quantities used in determining the work accomplished. The weight of the box is a different variable, and one that is not used in this equation. Thus,

$$W = Fd$$
$$= \left(440 \, \frac{\text{kg·m}}{\text{s}^2}\right)(5.0 \text{ m})$$
$$= 2,200 \text{ N·m}$$
$$= \boxed{2,200 \text{ J}}$$

3.3. Note that 10.0 kg is a mass quantity, and not a weight quantity. Weight is found from $w = mg$, a form of Newton's second law of motion. Thus, the force that must be exerted to lift the backpack is its weight, or $(10.0 \text{ kg}) \times (9.8 \text{ m/s}^2)$, which is 98 N. Therefore, a force of 98 N was exerted on the backpack through a distance of 1.5 m, and

$$W = Fd$$
$$= \left(98 \, \frac{\text{kg·m}}{\text{s}^2}\right)(1.5 \text{ m})$$
$$= 147 \text{ N·m}$$
$$= \boxed{150 \text{ J}}$$

3.4. Weight is defined as the force of gravity acting on an object, and the greater the force of gravity, the harder it is to lift the object. The force is proportional to the mass of the object, as the equation $w = mg$ tells you. Thus, the force you exert when lifting is $F = w = mg$, so the work you do on an object you lift must be $W = mgh$.

You know the mass of the box and you know the work accomplished. You also know the value of the acceleration due to gravity, g, so the list of known and unknown quantities is:

$$m = 102 \text{ kg}$$
$$g = 9.8 \text{ m/s}^2$$
$$W = 5,000 \text{ J}$$
$$h = \text{?}$$

The equation $W = mgh$ is solved for work, so the first thing to do is to solve it for h, the unknown height in this problem (note that height is also a distance):

$$W = mgh \qquad \therefore \qquad h = \frac{W}{mg}$$

$$= \frac{5,000 \, \frac{\text{kg·m}}{\text{s}^2} \times \text{m}}{(102 \text{ kg}) \left(9.8 \, \frac{\text{m}}{\text{s}^2}\right)}$$

$$= \frac{5,000.0}{102 \times 9.8} \, \frac{\text{kg·m}}{\text{s}^2} \times \frac{\text{m}}{1} \times \frac{1}{\text{kg}} \times \frac{\text{s}^2}{\text{m}}$$

$$= \frac{5,000}{999.6} \text{ m}$$

$$= \boxed{5 \text{ m}}$$

3.5. A student running up the stairs has to lift herself, so her weight is the required force needed. Thus, the force exerted is $F = w = mg$, and the work done is $W = mgh$. You know the mass of the student, the height, and the time. You also know the value of the acceleration due to gravity, g, so the list of known and unknown quantities is:

$$m = 60.0 \text{ kg}$$
$$g = 9.8 \text{ m/s}^2$$
$$h = 5.00 \text{ m}$$
$$t = 3.92 \text{ s}$$
$$P = \text{?}$$

The equation $p = \dfrac{mgh}{t}$ is already solved for power, so:

$$P = \frac{mgh}{t}$$

$$= \frac{(60.0 \text{ kg}) \left(9.8 \, \frac{\text{m}}{\text{s}^2}\right)(5.00 \text{ m})}{3.92 \text{ s}}$$

$$= \frac{(60.0)(9.8)(5.00)}{(3.92)} \, \frac{\left(\frac{\text{kg·m}}{\text{s}^2}\right) \times \text{m}}{\text{s}}$$

$$= \frac{2,940}{3.92} \, \frac{\text{N·m}}{\text{s}}$$

$$= 750 \, \frac{\text{J}}{\text{s}}$$

$$= \boxed{750 \text{ W}}$$

3.6. **(a)** $\dfrac{1.00\ \text{hp}}{746\ \text{W}} \times 1{,}400\ \text{W}$

$\dfrac{1{,}400}{746}\ \dfrac{\text{hp} \cdot \text{W}}{\text{W}}$

$\boxed{1.9\ \text{hp}}$

(b) $\dfrac{746\ \text{W}}{1.00\ \text{hp}} \times 3.5\ \text{hp}$

$746 \times 3.5\ \dfrac{\text{W} \cdot \text{hp}}{\text{hp}}$

$2{,}611\ \text{W}$

$\boxed{2{,}600\ \text{W}}$

3.7. Listing the known and unknown quantities:

$$m = 2{,}000\ \text{kg}$$
$$v = 72\ \text{km/h}$$
$$KE = ?$$

These are the quantities found in the equation for kinetic energy, $KE = 1/2mv^2$, which is already solved. However, note that the velocity is in units of km/h, which must be changed to m/s before doing anything else (it must be m/s because all energy and work units are in units of the joule [J]. A joule is a newton-meter, and a newton is a kg·m/s²). Using the conversion factor from inside the front cover of your text,

$$\dfrac{0.2778\ \frac{\text{m}}{\text{s}}}{\frac{\text{km}}{\text{h}}} \times 72\ \frac{\text{km}}{\text{h}}$$

$$(0.2778)\ (72)\ \frac{\text{m}}{\text{s}} \times \frac{\text{h}}{\text{km}} \times \frac{\text{km}}{\text{h}}$$

$$20\ \frac{\text{m}}{\text{s}}$$

and

$$KE = \frac{1}{2}\ mv^2$$

$$= \frac{1}{2}\ (2{,}000\ \text{kg}) \left(20\ \frac{\text{m}}{\text{s}}\right)^2$$

$$= \frac{1}{2}\ (2{,}000\ \text{kg}) \left(400\ \frac{\text{m}^2}{\text{s}^2}\right)$$

$$= \frac{1}{2} \times 2{,}000 \times 400\ \frac{\text{kg} \cdot \text{m}^2}{\text{s}^2}$$

$$= 400{,}000\ \frac{\text{kg} \cdot \text{m}}{\text{s}^2} \times \text{m}$$

$$= 400{,}000\ \text{N} \cdot \text{m}$$

$$= \boxed{4 \times 10^5\ \text{J}}$$

Scientific notation is used here to simplify a large number and to show one significant figure.

3.8. Recall the relationship between work and energy—that you do work on an object when you throw it, giving it kinetic enery, and the kinetic energy it has will do work on

something else when stopping. Because of the relationship between work and energy, you can calculate (1) the work you do, (2) the kinetic energy a moving object has as a result of your work, and (3) the work it will do when coming to a stop, and all three answers should be the same. Thus, you do not have a force or a distance to calculate the work needed to stop a moving car, but you can simply calculate the kinetic energy of the car. Both answers should be the same.

Before you start, note that the velocity is in units of km/h, which must be changed to m/s before doing anything else (it must be m/s because all energy and work units are in units of the joule [J]. A joule is a newton-meter, and a newton is a kg·m/s²). Using the conversion factor from inside the front cover,

$$\dfrac{0.2778\ \frac{\text{m}}{\text{s}}}{\frac{\text{km}}{\text{h}}} \times 54.0\ \frac{\text{km}}{\text{h}}$$

$$0.2778 \times 54.0\ \frac{\text{m}}{\text{s}} \times \frac{\text{h}}{\text{km}} \times \frac{\text{km}}{\text{h}}$$

$$15.0\ \frac{\text{m}}{\text{s}}$$

and

$$KE = \frac{1}{2}\ mv^2$$

$$= \frac{1}{2}\ (1{,}000.0\ \text{kg}) \left(15.0\ \frac{\text{m}}{\text{s}}\right)^2$$

$$= \frac{1}{2}\ (1{,}000.0\ \text{kg}) \left(225\ \frac{\text{m}^2}{\text{s}^2}\right)$$

$$= \frac{1}{2} \times 1{,}000.0 \times 225\ \frac{\text{kg} \cdot \text{m}^2}{\text{s}^2}$$

$$= 112{,}500\ \frac{\text{kg} \cdot \text{m}}{\text{s}^2} \times \text{m}$$

$$= 112{,}500\ \text{N} \cdot \text{m}$$

$$= \boxed{1.13 \times 10^5\ \text{J}}$$

Scientific notation is used here to simplify a large number and to easily show three significant figures. The answer could likewise be expressed as 113 kJ.

3.9. **(a)** $W = Fd$

$= (10\ \text{lb})(5\ \text{ft})$

$= (10)(5)\ \text{ft} \times \text{lb}$

$= 50\ \text{ft} \cdot \text{lb}$

(b) The distance of the bookcase from some horizontal reference level did not change, so the gravitational potential energy does not change.

3.10. The force (F) needed to lift the book is equal to the weight (w) of the book, or $F = w$. Since $w = mg$, then $F = mg$. Work is defined as the product of a force moved through a distance, or $W = Fd$. The work done in lifting the book is therefore $W = mgd$, and:

(a)
$$W = mgd$$
$$= (2.0 \text{ kg}) (9.8 \text{ m/s}^2) (2.00 \text{ m})$$
$$= (2.0) (9.8) (2.00) \frac{\text{kg·m}}{\text{s}^2} \times \text{m}$$
$$= 39.2 \frac{\text{kg·m}^2}{\text{s}^2}$$
$$= 39.2 \text{ J} = \boxed{39 \text{ J}}$$

(b) $PE = mgh = \boxed{39 \text{ J}}$

(c) $PE_{\text{lost}} = KE_{\text{gained}} = mgh = \boxed{39 \text{ J}}$

(or)
$$v = \sqrt{2gh} = \sqrt{(2) (9.8 \text{ m/s}^2) (2.00 \text{ m})}$$
$$= \sqrt{39.2 \text{ m}^2/\text{s}^2}$$
$$= 6.26 \text{ m/s}$$
$$KE = \frac{1}{2} mv^2 = \left(\frac{1}{2}\right) (2.0 \text{ kg}) (6.26 \text{ m/s})^2$$
$$= \left(\frac{1}{2}\right) (2.0 \text{ kg}) (39 \text{ m}^2/\text{s}^2)$$
$$= (1.0) (39) \frac{\text{kg·m}^2}{\text{s}^2}$$
$$= \boxed{39 \text{ J}}$$

3.11. Note that the gram unit must be converted to kg to be consistent with the definition of a newton-meter, or joule unit of energy:
$$KE = \frac{1}{2} mv^2 = \left(\frac{1}{2}\right) (0.15 \text{ kg}) (30.0 \text{ m/s})^2$$
$$= \left(\frac{1}{2}\right) (0.15 \text{ kg}) (900 \text{ m}^2/\text{s}^2)$$
$$= \left(\frac{1}{2}\right) (0.15) (900) \frac{\text{kg·m}^2}{\text{s}^2}$$
$$= 67.5 \text{ J} = \boxed{68 \text{ J}}$$

3.12. The km/h unit must first be converted to m/s before finding the kinetic energy. Note also that the work done to put an object in motion is equal to the energy of motion, or the kinetic energy that it has as a result of the work. The work needed to bring the object to a stop is also equal to the kinetic energy of the moving object:

Unit conversion:
$$1 \frac{\text{km}}{\text{h}} = 0.2778 \frac{\frac{\text{m}}{\text{s}}}{\frac{\text{km}}{\text{h}}} \quad \therefore \quad \left(90.0 \frac{\text{km}}{\text{h}}\right) \left(0.2778 \frac{\frac{\text{m}}{\text{s}}}{\frac{\text{km}}{\text{h}}}\right) = 25.0 \text{ m/s}$$

(a) $KE = \frac{1}{2} mv^2 = \frac{1}{2} (1,000.0 \text{ kg}) \left(25.0 \frac{\text{m}}{\text{s}}\right)^2$
$$= \frac{1}{2} (1,000.0 \text{ kg}) \left(625 \frac{\text{m}^2}{\text{s}^2}\right)$$
$$= \frac{1}{2} (1,000.0)(625) \frac{\text{kg·m}^2}{\text{s}^2}$$
$$= 312.5 \text{ kJ} = \boxed{313 \text{ kJ}}$$

(b) $W = Fd = KE = \boxed{313 \text{ kJ}}$

(c) $KE = W = Fd = \boxed{313 \text{ kJ}}$

3.13.
$$KE = \frac{1}{2} mv^2$$
$$= \frac{1}{2} (60.0 \text{ kg}) \left(2.0 \frac{\text{m}}{\text{s}}\right)^2$$
$$= \frac{1}{2} (60.0 \text{ kg}) \left(4.0 \frac{\text{m}^2}{\text{s}^2}\right)$$
$$= 30.0 \times 4.0 \text{ kg} \times \left(\frac{\text{m}^2}{\text{s}^2}\right)$$
$$= \boxed{120 \text{ J}}$$
$$KE = \frac{1}{2} mv^2$$
$$= \frac{1}{2} (60.0 \text{ kg}) \left(4.0 \frac{\text{m}}{\text{s}}\right)^2$$
$$= \frac{1}{2} (60.0 \text{ kg}) \left(16 \frac{\text{m}^2}{\text{s}^2}\right)$$
$$= 30.0 \times 16 \text{ kg} \times \left(\frac{\text{m}^2}{\text{s}^2}\right)$$
$$= \boxed{480 \text{ J}}$$

Thus, doubling the speed results in a fourfold increase in kinetic energy.

3.14.
$$KE = \frac{1}{2} mv^2$$
$$= \frac{1}{2} (70.0 \text{ kg}) (6.00 \text{ m/s})^2$$
$$= (35.0 \text{ kg}) (36.0 \text{ m}^2/\text{s}^2)$$
$$= 35.0 \times 36.0 \text{ kg} \times \frac{\text{m}^2}{\text{s}^2}$$
$$= \boxed{1,260 \text{ J}}$$
$$KE = \frac{1}{2} mv^2$$
$$= \frac{1}{2} (140.0 \text{ kg}) (6.00 \text{ m/s})^2$$
$$= (70.0 \text{ kg}) (36.0 \text{ m}^2/\text{s}^2)$$
$$= 70.0 \times 36.0 \text{ kg} \times \frac{\text{m}^2}{\text{s}^2}$$
$$= \boxed{2,520 \text{ J}}$$

Thus, doubling the mass results in a doubling of the kinetic energy.

3.15. **(a)** The force needed is equal to the weight of the student. The English unit of a pound is a force unit, so
$$W = Fd$$
$$= (170.0 \text{ lb}) (25.0 \text{ ft})$$
$$= \boxed{4,250 \text{ ft·lb}}$$

(b) Work (W) is defined as a force (F) moved through a distance (d), or $W = Fd$. Power (P) is defined as work (W) per unit of time (t), or $P = W/t$. Therefore,

$$P = \frac{Fd}{t}$$

$$= \frac{(170.0 \text{ lb})(25.0 \text{ ft})}{10.0 \text{ s}}$$

$$= \frac{(170.0)(25.0)}{10.0} \frac{\text{ft·lb}}{\text{s}}$$

$$= 425 \frac{\text{ft·lb}}{\text{s}}$$

One hp is defined as $550 \frac{\text{ft·lb}}{\text{s}}$ and

$$\frac{425 \text{ ft·lb/s}}{550 \frac{\text{ft·lb/s}}{\text{hp}}} = \boxed{0.77 \text{ hp}}$$

Note that the student's power rating (425 ft·lb/s) is less than the power rating defined as 1 horsepower (550 ft·lb/s). Thus, the student's horsepower must be *less* than 1 hp. A simple analysis such as this will let you know if you inverted the ratio or not.

3.16. **(a)** The force (F) needed to lift the elevator is equal to the weight of the elevator. Since the work (W) is equal to Fd and power (P) is equal to W/t, then

$$P = \frac{Fd}{t} \qquad \therefore \qquad t = \frac{Fd}{P}$$

$$= \frac{[2,000.0 \text{ lb}][20.0 \text{ ft}]}{\left(550 \frac{\text{ft·lb}}{\text{s}}\right)[20.0 \text{ hp}]}$$

$$= \frac{40,000}{11,000} \frac{\text{ft·lb}}{\frac{\text{ft·lb}}{\text{s}}} \times \frac{1}{\text{hp}} \times \text{hp}$$

$$= \frac{40,000}{11,000} \frac{\text{ft·lb}}{1} \times \frac{\text{s}}{\text{ft·lb}}$$

$$= 3.64 \frac{\text{ft·lb·s}}{\text{ft·lb}}$$

$$= \boxed{3.6 \text{ s}}$$

(b)

$$\bar{v} = \frac{d}{t}$$

$$= \frac{20.0 \text{ ft}}{3.6 \text{ s}}$$

$$= \boxed{5.6 \text{ ft/s}}$$

3.17. Since $PE_{\text{lost}} = KE_{\text{gained}}$, then $mgh = \frac{1}{2}mv^2$. Solving for v,

$$v = \sqrt{2gh} = \sqrt{(2)(32.0 \text{ ft/s}^2)(9.8 \text{ ft})}$$

$$= \sqrt{(2)(32.0)(9.8) \text{ ft}^2/\text{s}^2}$$

$$= \sqrt{627 \text{ ft}^2/\text{s}^2}$$

$$= 25 \text{ ft/s}$$

3.18. $KE = \frac{1}{2}mv^2 \qquad \therefore \qquad v = \sqrt{\frac{2KE}{m}}$

$$= \sqrt{\frac{(2)\left(200,000 \dfrac{\text{kg·m}^2}{\text{s}^2}\right)}{1,000.0 \text{ kg}}}$$

$$= \sqrt{\frac{400,000}{1,000.0} \frac{\text{kg·m}^2}{\text{s}^2} \times \frac{1}{\text{kg}}}$$

$$= \sqrt{\frac{400.000}{1,000.0} \frac{\text{kg·m}^2}{\text{kg·s}^2}}$$

$$= \sqrt{400 \text{ m}^2/\text{s}^2}$$

$$= \boxed{20 \text{ m/s}}$$

3.19. The maximum velocity occurs at the lowest point with a gain of kinetic energy equivalent to the loss of potential energy in falling 3.0 in (which is 0.25 ft), so

$$KE_{\text{gained}} = PE_{\text{lost}}$$

$$\frac{1}{2}mv^2 = mgh$$

$$v = \sqrt{2gh}$$

$$= \sqrt{(2)(32 \text{ ft/s}^2)(0.25 \text{ ft})}$$

$$= \sqrt{(2)(32)(0.25) \text{ ft/s}^2 \times \text{ft}}$$

$$= \sqrt{16 \text{ ft}^2/\text{s}^2}$$

$$= \boxed{4.0 \text{ ft/s}}$$

3.20. **(a)** $W = Fd$ and the force F that is needed to lift the load upward is mg, so $W = mgh$. Power is W/t, so

$$P = \frac{mgh}{t}$$

$$= \frac{(250.0 \text{ kg})(9.8 \text{ m/s}^2)(80.0 \text{ m})}{39.2 \text{ s}}$$

$$= \frac{(250.0)(9.8)(80.0)}{39.2} \frac{\text{kg}}{1} \times \frac{\text{m}}{\text{s}^2} \times \frac{\text{m}}{1} \times \frac{1}{\text{s}}$$

$$= \frac{196,000}{39.2} \frac{\text{kg·m}^2}{\text{s}^2} \times \frac{1}{\text{s}}$$

$$= 5,000 \frac{\text{J}}{\text{s}}$$

$$= \boxed{5.0 \text{ kW}}$$

(b) There are 746 watts per horsepower, so

$$\frac{5,000 \text{ W}}{746 \dfrac{\text{W}}{\text{hp}}} = \frac{5,000}{746} \frac{\text{W}}{1} \times \frac{\text{hp}}{\text{W}}$$

$$= 6.70 \frac{\text{W·hp}}{\text{W}}$$

$$= \boxed{6.7 \text{ hp}}$$

4.1. Listing the known and unknown quantities:

body temperature $\quad T_F = 98.6°$

$$T_C = ?$$

These are the quantities found in the equation for conversion of Fahrenheit to Celsius, $T_C = \dfrac{5}{9}(T_F - 32°)$, where T_F is the temperature in Fahrenheit and T_C is the temperature in Celsius. This equation describes a relationship between the two temperature scales and is used to convert a Fahrenheit temperature to Celsius. The equation is already solved for the Celsius temperature, T_C. Thus,

$$T_C = \frac{5}{9}(T_F - 32°)$$

$$= \frac{5}{9}(98.6° - 32°)$$

$$= \frac{333°}{9}$$

$$= \boxed{37°\ C}$$

4.2. $Q = mc\Delta T$

$$= (221\ \text{g})\left(0.093\ \frac{\text{cal}}{\text{gC°}}\right)(38.0°C - 20.0°C)$$

$$= (221)(0.093)(18.0)\ \text{g} \times \frac{\text{cal}}{\text{gC°}} \times °C$$

$$= 370\ \frac{\text{g·cal·°C}}{\text{gC°}}$$

$$= \boxed{370\ \text{cal}}$$

4.3. First, you need to know the energy of the moving bike and rider. Since the speed is given as 36.0 km/h, convert to m/s by multiplying times 0.2778 m/s per km/h:

$$\left(36.0\ \frac{\text{km}}{\text{h}}\right)\left(0.2778\ \frac{\text{m/s}}{\text{km/h}}\right)$$

$$= (36.0)(0.2778)\ \frac{\text{km}}{\text{h}} \times \frac{\text{h}}{\text{km}} \times \frac{\text{m}}{\text{s}}$$

$$= 10.0\ \text{m/s}$$

Then,

$$KE = \frac{1}{2}mv^2$$

$$= \frac{1}{2}(100.0\ \text{kg})(10.0\ \text{m/s})^2$$

$$= \frac{1}{2}(100.0\ \text{kg})(100\ \text{m}^2/\text{s}^2)$$

$$= \frac{1}{2}(100.0)(100)\frac{\text{kg·m}^2}{\text{s}^2}$$

$$= 5,000\ \text{J}$$

Second, this energy is converted to the calorie heat unit through the mechanical equivalent of heat relationship, that 1.0 kcal = 4,184 J, or that 1.0 cal = 4.184 J. Thus,

$$\frac{5,000\ \text{J}}{4,184\ \text{J/kcal}}$$

$$1.195\ \frac{\text{J}}{1} \times \frac{\text{kcal}}{\text{J}}$$

$$\boxed{1.20\ \text{kcal}}$$

4.4. First, you need to find the energy of the falling bag. Since the potential energy lost equals the kinetic energy gained, the energy of the bag just as it hits the ground can be found from

$$PE = mgh$$

$$= (15.53\ \text{kg})(9.8\ \text{m/s}^2)(5.50\ \text{m})$$

$$= (15.53)(9.8)(5.50)\ \frac{\text{kg·m}}{\text{s}^2} \times \text{m}$$

$$= 837\ \text{J}$$

In calories, this energy is equivalent to

$$\frac{837\ \text{J}}{4,184\ \text{J/kcal}} = 0.20\ \text{kcal}$$

Second, the temperature change can be calculated from the equation giving the relationship between a quantity of heat (Q), mass (m), specific heat of the substance (c), and the change of temperature:

$$Q = mc\Delta T \quad \therefore \quad \Delta T = \frac{Q}{mc}$$

$$= \frac{0.20\ \text{kcal}}{(15.53\ \text{kg})\left(0.200\ \frac{\text{kcal}}{\text{kgC°}}\right)}$$

$$= \frac{0.20}{(15.53)(0.200)}\ \frac{\text{kcal}}{1} \times \frac{1}{\text{kg}} \times \frac{\text{kgC°}}{\text{kcal}}$$

$$= 0.0644\ \frac{\text{kcal·kgC°}}{\text{kcal·kg}}$$

$$= \boxed{6.4 \times 10^{-2}\ °C}$$

4.5. The Calorie used by dietitians is a kilocalorie; thus, 250.0 Cal is 250.0 kcal. The mechanical energy equivalent is 1 kcal = 4,184 J, so (250.0 kcal)(4,184 J/kcal) = 1,046,250 J.

Since $W = Fd$ and the force needed is equal to the weight (mg) of the person, $W = mgh = (75.0\ \text{kg})(9.8\ \text{m/s}^2)(10.0\ \text{m}) = 7,350\ \text{J}$ for each stairway climb.

A total of 1,046,250 J of energy from the French fries would require (1,046,250 J) ÷ (7,350 J, per climb), or 142 trips up the stairs.

4.6. For unit consistency,

$$T_C = \frac{5}{9}(T_F - 32°) = \frac{5}{9}(68° - 32°) = \frac{5}{9}(36°) = 20°C$$

$$= \frac{5}{9}(32° - 32°) = \frac{5}{9}(0°) = 0°C$$

Glass bowl:

$$Q = mc\Delta T$$

$$= (0.5\text{ kg})\left(0.2\,\frac{\text{kcal}}{\text{kg}°\text{C}}\right)(20°\text{C})$$

$$= (0.5)\,(0.2)\,(20)\,\frac{\text{kg}}{1} \times \frac{\text{kcal}}{\text{kgC}°} \times \frac{°\text{C}}{1}$$

$$= \boxed{2\text{ kcal}}$$

Iron pan:

$$Q = mc\Delta T$$

$$= (0.5\text{ kg})\left(0.11\,\frac{\text{kcal}}{\text{kgC}°}\right)(20°\text{C})$$

$$= (0.5)\,(0.11)\,(20)\quad\text{kg} \times \frac{\text{kcal}}{\text{kgC}°} \times °\text{C}$$

$$= \boxed{1\text{ kcal}}$$

4.7. Note that a specific heat expressed in cal/gC° has the same numerical value as a specific heat expressed in kcal/kgC° because you can cancel the k units. You could convert 896 cal to 0.896 kcal, but one of the two conversion methods is needed for consistency with other units in the problem.

$$Q = mc\Delta T \quad \therefore \quad m = \frac{Q}{c\Delta T}$$

$$= \frac{896\text{ cal}}{\left(0.056\,\dfrac{\text{cal}}{\text{gC}°}\right)(80.0°\text{C})}$$

$$= \frac{896}{(0.056)\,(80.0)}\,\frac{\text{cal}}{1} \times \frac{\text{gC}°}{\text{cal}} \times \frac{1}{\text{C}°}$$

$$= 200\text{ g}$$

$$= \boxed{0.20\text{ kg}}$$

4.8. Since a watt is defined as a joule/s, finding the total energy in joules will tell the time:

$$Q = mc\Delta T$$

$$= (250.0\text{ g})\left(1.00\,\frac{\text{cal}}{\text{gC}°}\right)(60.0°\text{C})$$

$$= (250.0)\,(1.00)\,(60.0)\text{ g} \times \frac{\text{cal}}{\text{gC}°} \times °\text{C}$$

$$= 1.50 \times 10^4\text{ cal}$$

This energy in joules is

$$(1.50 \times 10^4\text{ cal})\left(4.184\,\frac{\text{J}}{\text{cal}}\right) = 62,800\text{ J}$$

A 300-watt heater uses energy at a rate of $300\,\frac{\text{J}}{\text{s}}$, so

$$\frac{62,800\text{ J}}{300\text{ J/s}} = 209\text{ s is required, which is } \frac{209\text{ s}}{60\,\dfrac{\text{s}}{\text{min}}} = 3.48\text{ min, or}$$

$$\boxed{\text{about } 3\frac{1}{2}\text{ min}}$$

4.9.
$$Q = mc\Delta T \quad \therefore \quad c = \frac{Q}{m\Delta T}$$

$$= \frac{60.0\text{ cal}}{(100.0\text{ g})\,(20.0°\text{C})}$$

$$= \frac{60.0}{(100.0)\,(20.0)}\,\frac{\text{cal}}{\text{gC}°}$$

$$= \boxed{0.0300\,\frac{\text{cal}}{\text{gC}°}}$$

4.10. Since the problem specified a solid changing to a liquid without a temperature change, you should recognize that this is a question about a phase change only. The phase change from solid to liquid (or liquid to solid) is concerned with the latent heat of fusion. For water, the latent heat of fusion is given as 80.0 cal/g, and

$$m = 250.0\text{ g} \qquad Q = mL_f$$

$$L_{f\,(water)} = 80.0\text{ cal/g} \qquad = (250.0\text{ g})\left(80.0\,\frac{\text{cal}}{\text{g}}\right)$$

$$Q = ?$$

$$= 250.0 \times 80.0\,\frac{\text{g·cal}}{\text{g}}$$

$$= 20{,}000\text{ cal} = \boxed{20.0\text{ kcal}}$$

4.11. To change water at 80.0°C to steam at 100.0°C requires two separate quantities of heat that can be called Q_1 and Q_2. The quantity Q_1 is the amount of heat needed to warm the water from 80.0°C to the boiling point, which is 100.0°C at sea level pressure ($\Delta T = 20.0°\text{C}$). The relationship between the variables involved is $Q_1 = mc\Delta T$. The quantity Q_2 is the amount of heat needed to take 100.0°C water through the phase change to steam (water vapor) at 100.0°C. The phase change from a liquid to a gas (or gas to liquid) is concerned with the latent heat of vaporization. For water, the latent heat of vaporization is given as 540.0 cal/g.

$$m = 250.0\text{ g} \qquad Q_1 = mc\Delta T$$

$$L_{v\,(water)} = 540.0\text{ cal/g}$$

$$Q = ? \qquad = (250.0\text{ g})\left(1.00\,\frac{\text{cal}}{\text{gC}°}\right)(20.0°\text{C})$$

$$= (250.0)(1.00)(20.0)\text{ g} \times \frac{\text{cal}}{\text{gC}°} \times °\text{C}$$

$$= 5{,}000\,\frac{\text{g·cal·}°\text{C}}{\text{gC}°}$$

$$= 5{,}000\text{ cal}$$

$$= 5.00\text{ kcal}$$

$$Q_2 = mL_v$$

$$= (250.0\text{ g})\left(540.0\,\frac{\text{cal}}{\text{g}}\right)$$

$$= 250.0 \times 540.0\,\frac{\text{g·cal}}{\text{g}}$$

$$= 135{,}000\text{ cal}$$

$$= 135.0\text{ kcal}$$

$$Q_{Total} = Q_1 + Q_2$$

$$= 5.00\text{ kcal} + 135.0\text{ kcal}$$

$$= \boxed{140.0\text{ kcal}}$$

4.12. To change 20.0°C water to steam at 125.0°C requires three separate quantities of heat. First, the quantity Q_1 is the amount of heat needed to warm the water from 20.0°C to 100.0°C ($\Delta T = 80.0°C$). The quantity Q_2 is the amount of heat needed to take 100.0°C water to steam at 100.0°C. Finally, the quantity Q_3 is the amount of heat needed to warm the steam from 100.0° to 125.0°C. According to Table 4.4, the c for steam is 0.480 cal/g°C.

$m = 100.0$ g

$\Delta T_{water} = 80.0°C$

$\Delta T_{steam} = 25.0°C$

$L_{v(water)} = 540.0$ cal/g

$c_{steam} = 0.480$ cal/gC°

$Q_1 = mc\Delta T$

$\quad = (100.0 \text{ g}) \left(1.00 \dfrac{\text{cal}}{\text{gC}°}\right)(80.0°C)$

$\quad = (100.0)(1.00)(80.0) \text{ g} \times \dfrac{\text{cal}}{\text{gC}°} \times °C$

$\quad = 8,000 \dfrac{\text{g·cal·°C}}{\text{gC°}}$

$\quad = 8,000$ cal

$\quad = 8.00$ kcal

$Q_2 = mL_v$

$\quad = (100.0 \text{ g}) \left(540.0 \dfrac{\text{cal}}{\text{g}}\right)$

$\quad = 100.0 \times 540.0 \dfrac{\text{g·cal}}{\text{g}}$

$\quad = 54,000$ cal

$\quad = 54.00$ kcal

$Q_3 = mc\Delta T$

$\quad = (100.0 \text{ g}) \left(0.480 \dfrac{\text{cal}}{\text{gC}°}\right)(25.0°C)$

$\quad = (100.0)(0.480)(25.0) \text{ g} \times \dfrac{\text{cal}}{\text{gC}°} \times °C$

$\quad = 1,200 \dfrac{\text{g·cal·°C}}{\text{gC°}}$

$\quad = 1,200$ cal

$\quad = 1.20$ kcal

$Q_{total} = Q_1 + Q_2 + Q_3$

$\quad = 8.00 \text{ kcal} + 54.00 \text{ kcal} + 1.20 \text{ kcal}$

$\quad = \boxed{63.20 \text{ kcal}}$

4.13. **(a) Step 1:** Cool the water from 18.0°C to 0°C.

$Q_1 = mc\Delta T$

$\quad = (400.0 \text{ g}) \left(1.00 \dfrac{\text{cal}}{\text{gC}°}\right)(18.0°C)$

$\quad = (400.0)(1.00)(18.0) \text{ g} \times \dfrac{\text{cal}}{\text{gC}°} \times °C$

$\quad = 7,200 \dfrac{\text{g·cal·°C}}{\text{gC°}}$

$\quad = 7,200$ cal

$\quad = 7.20$ kcal

Step 2: Find the energy needed for the phase change of water at 0°C to ice at 0°C.

$Q_2 = mL_f$

$\quad = (400.0 \text{ g}) \left(80.0 \dfrac{\text{cal}}{\text{g}}\right)$

$\quad = 400.0 \times 80.0 \times \dfrac{\text{g·cal}}{\text{g}}$

$\quad = 32,000$ cal

$\quad = 32.0$ kcal

Step 3: Cool the ice from 0°C to ice at −5.00°C.

$Q_3 = mc\Delta T$

$\quad = (400.0 \text{ g}) \left(0.500 \dfrac{\text{cal}}{\text{gC}°}\right)(5.00°C)$

$\quad = 400.0 \times 0.500 \times 5.00 \text{ g} \times \dfrac{\text{cal}}{\text{gC}°} \times °C$

$\quad = 1,000 \dfrac{\text{g·cal·°C}}{\text{g·C°}}$

$\quad = 1,000$ cal

$\quad = 1.00$ kcal

$Q_{total} = Q_1 + Q_2 + Q_3$

$\quad = 7.20 \text{ kcal} + 32.0 \text{ kcal} + 1.00 \text{ kcal}$

$\quad = \boxed{40.2 \text{ kcal}}$

(b) $Q = mL_v \quad \therefore \quad m = \dfrac{Q}{L_v}$

$\quad = \dfrac{40,200 \text{ cal}}{40.0 \dfrac{\text{cal}}{\text{g}}}$

$\quad = \dfrac{40,200}{40.0} \dfrac{\text{cal}}{1} \times \dfrac{\text{g}}{\text{cal}}$

$\quad = 1,005 \dfrac{\text{cal·g}}{\text{cal}}$

$\quad = \boxed{1.01 \times 10^3 \text{ g}}$

4.14. $W = J(Q_H - Q_L)$

$\quad = 4,184 \dfrac{\text{J}}{\text{kcal}} (0.3000 \text{ kcal} - 0.2000 \text{ kcal})$

$\quad = 4,184 \dfrac{\text{J}}{\text{kcal}} (0.1000 \text{ kcal})$

$\quad = 4,184 \times 0.1000 \dfrac{\text{J·kcal}}{\text{kcal}}$

$\quad = \boxed{418.4 \text{ J}}$

4.15. $W = J(Q_H - Q_L)$

$\quad = 4,184 \dfrac{\text{J}}{\text{kcal}} (55.0 \text{ kcal} - 40.0 \text{ kcal})$

$\quad = 4,184 \dfrac{\text{J}}{\text{kcal}} (15.0 \text{ kcal})$

$\quad = 4,184 \times 15.0 \dfrac{\text{J·kcal}}{\text{kcal}}$

$\quad = 62,760 \text{ J}$

$\quad = \boxed{62.8 \text{ kJ}}$

5.1. **(a)** $f = 3$ Hz
$\lambda = 2$ cm
$T = ?$

$T = \dfrac{1}{f}$

$= \dfrac{1}{3\frac{1}{s}}$

$= \dfrac{1}{3}\dfrac{s}{1}$

$= \boxed{0.3 \text{ s}}$

(b) $f = 3$ Hz
$\lambda = 2$ cm
$T = ?$

$v = \lambda f$

$= (2 \text{ cm})\left(3\frac{1}{s}\right)$

$= 2 \times 3 \text{ cm} \times \dfrac{1}{s}$

$= \boxed{6 \dfrac{\text{cm}}{\text{s}}}$

5.2. **Step 1:**

$t = 20.0°C$
$f = 20{,}000$ Hz
$v = ?$
$\lambda = ?$

$v_{T_{P(m/s)}} = v_0 + \left(\dfrac{0.600 \text{ m/s}}{°C}\right)(T_P)$

$= 331 \dfrac{\text{m}}{\text{s}} + \left(\dfrac{0.600 \text{ m/s}}{°C}\right)(20.0°C)$

$= 331 \dfrac{\text{m}}{\text{s}} + 12.0 \dfrac{\text{m}}{\text{s}}$

$= 343 \dfrac{\text{m}}{\text{s}}$

Step 2:

$v = \lambda f \quad \therefore \quad \lambda = \dfrac{v}{f}$

$= \dfrac{343 \frac{\text{m}}{\text{s}}}{20{,}000 \frac{1}{\text{s}}}$

$= \dfrac{343}{20{,}000} \dfrac{\text{m}}{\text{s}} \times \dfrac{\text{s}}{1}$

$= 0.01715 \text{ m}$

$= \boxed{0.02 \text{ m}}$

5.3. $f_1 = 440$ Hz
$f_2 = 446$ Hz
$f_b = ?$

$f_b = f_2 - f_1$

$= (446 \text{ Hz}) - (440 \text{ Hz})$

$= \boxed{6 \text{ Hz}}$

Note that the smaller frequency is subtracted from the larger one to avoid negative beats.

5.4. **Step 1:** Assume room temperature (20.0°C) to obtain the velocity:

$f = 2.00 \times 10^7$ Hz
$\lambda = ?$
$v = ?$

$v_{T_{P(m/s)}} = v_0 + \left(\dfrac{0.600 \frac{\text{m}}{\text{s}}}{°C}\right)(T_P)$

$= 331 \dfrac{\text{m}}{\text{s}} + \left(\dfrac{0.600 \frac{\text{m}}{\text{s}}}{°C}\right)(20.0°C)$

$= 331 \dfrac{\text{m}}{\text{s}} + 12.0 \dfrac{\text{m}}{\text{s}}$

$= 343 \dfrac{\text{m}}{\text{s}}$

Step 2:

$v = \lambda f \quad \therefore \quad \lambda = \dfrac{v}{f}$

$= \dfrac{343 \frac{\text{m}}{\text{s}}}{2.00 \times 10^7 \frac{1}{\text{s}}}$

$= \dfrac{343}{2.00 \times 10^7} \dfrac{\text{m}}{\text{s}} \times \dfrac{\text{s}}{1}$

$= 1.715 \times 10^{-5} \text{ m}$

$= \boxed{1.72 \times 10^{-5} \text{ m}}$

5.5. **Step 1:** Assume room temperature (20.0°C) to obtain the velocity (yes, you can have room temperature outside a room):

$d = 150$ m
$t = ?$
$v = ?$

$v_{T_{P(m/s)}} = v_0 + \left(\dfrac{0.600 \frac{\text{m}}{\text{s}}}{°C}\right)(T_P)$

$= 331 \dfrac{\text{m}}{\text{s}} + \left(\dfrac{0.600 \frac{\text{m}}{\text{s}}}{°C}\right)(20.0°C)$

$= 331 \dfrac{\text{m}}{\text{s}} + 12.0 \dfrac{\text{m}}{\text{s}}$

$= 343 \dfrac{\text{m}}{\text{s}}$

Step 2:

$v = \dfrac{d}{t} \quad \therefore \quad t = \dfrac{d}{v}$

$= \dfrac{150.0 \text{ m}}{343 \frac{\text{m}}{\text{s}}}$

$= \dfrac{150.0}{343} \dfrac{\text{m}}{1} \times \dfrac{\text{s}}{\text{m}}$

$= 0.4373177 \text{ s}$

$= \boxed{0.437 \text{ s}}$

5.6. **Step 1:** Find the velocity of sound in ft/s at a temperature of 20.0°C:

$t = 0.500$ s
$T = 20.0°C$

$d = ?$

$v = ?$

$v_{T_{P(ft/s)}} = v_0 + \left(\dfrac{2.00 \frac{\text{ft}}{\text{s}}}{°C}\right)(T_P)$

$= 1{,}087 \dfrac{\text{ft}}{\text{s}} + \left(\dfrac{2.00 \frac{\text{ft}}{\text{s}}}{°C}\right)(20.0°C)$

$= 1{,}087 \dfrac{\text{ft}}{\text{s}} + 40.0 \dfrac{\text{ft}}{\text{s}}$

$= 1{,}127 \dfrac{\text{ft}}{\text{s}}$

Step 2:

$$v = \frac{d}{t} \quad \therefore \quad d = vt$$

$$= \left(1,127 \frac{\text{ft}}{\text{s}}\right)(0.500 \text{ s})$$

$$= 1,127 \times 0.500 \frac{\text{ft}}{\text{s}} \times \text{s}$$

$$= 563.5 \text{ ft}$$

$$= 564 \text{ ft}$$

Step 3: The distance to the building is half the distance the sound traveled, so

$$\frac{564}{2} = \boxed{282 \text{ ft}}$$

5.7. **Step 1:**

$$t = 1.75 \text{ s} \qquad v = \frac{d}{t} \quad \therefore \quad d = vt$$

$$v = 1,530 \text{ m/s}$$

$$d = ? \qquad\qquad\qquad = \left(1,530 \frac{\text{m}}{\text{s}}\right)(1.75 \text{ s})$$

$$= 1,530 \times 1.75 \frac{\text{m}}{\text{s}} \times \text{s}$$

$$= 2,677.5 \text{ m}$$

Step 2: The sonar signal traveled from the ship to the bottom, then back to the ship, so the distance to the bottom is half of the distance traveled:

$$\frac{2,677.5 \text{ m}}{2} = 1,338.75 \text{ m}$$

$$= \boxed{1,340 \text{ m}}$$

5.8.

$$f = 660 \text{ Hz} \qquad v = \lambda f$$

$$\lambda = 9.0 \text{ m} \qquad = (9.0 \text{ m})\left(660 \frac{1}{\text{s}}\right)$$

$$v = ?$$

$$= 9.0 \times 660 \text{ m} \times \frac{1}{\text{s}}$$

$$= 5,940 \frac{\text{m}}{\text{s}}$$

$$= \boxed{5,900 \frac{\text{m}}{\text{s}}}$$

5.9. **Step 1:**

$$t = 2.5 \text{ s} \qquad v_{T\text{P(m/s)}} = v_0 + \left(\frac{0.600 \frac{\text{m}}{\text{s}}}{°\text{C}}\right)(T_\text{P})$$

$$T = 20.0°\text{C}$$

$$d = ? \qquad\qquad = 331 \frac{\text{m}}{\text{s}} + \left(\frac{0.600 \frac{\text{m}}{\text{s}}}{°\text{C}}\right)(20.0°\text{C})$$

$$v = ?$$

$$= 331 \frac{\text{m}}{\text{s}} + 12.0 \frac{\text{m}}{\text{s}}$$

$$= 343 \frac{\text{m}}{\text{s}}$$

Step 2:

$$v = \frac{d}{t} \quad \therefore \quad d = vt$$

$$= \left(343 \frac{\text{m}}{\text{s}}\right)(2.5 \text{ s})$$

$$= 343 \times 2.5 \frac{\text{m}}{\text{s}} \times \text{s}$$

$$= 857.5 \text{ m}$$

$$= \boxed{860 \text{ m}}$$

5.10. According to table 5.1, sound moves through air at 0°C with a velocity of 331 m/s and through steel with a velocity of 5,940 m/s. Therefore,

(a) $v = 331$ m/s $\qquad v = \frac{d}{t} \quad \therefore \quad d = vt$

$t = 8.00$ s

$d = ? \qquad\qquad\qquad = \left(331 \frac{\text{m}}{\text{s}}\right)(8.00 \text{ s})$

$$= 331 \times 8.00 \frac{\text{m}}{\text{s}} \times \text{s}$$

$$= 2,648 \text{ m}$$

$$= \boxed{2.65 \text{ km}}$$

(b) $v = 5,940$ m/s $\quad v = \frac{d}{t} \quad \therefore \quad d = vt$

$t = 8,00$ s

$d = ? \qquad\qquad\qquad = \left(5,940 \frac{\text{m}}{\text{s}}\right)(8.00 \text{ s})$

$$= 5,940 \times 8.00 \frac{\text{m}}{\text{s}} \times \text{s}$$

$$= 47,520 \text{ m}$$

$$= \boxed{47.5 \text{ km}}$$

5.11.

$$v = f\lambda$$

$$= \left(10 \frac{1}{\text{s}}\right)(0.50 \text{ m})$$

$$= 5 \frac{\text{m}}{\text{s}}$$

5.12. The distance between two *consecutive* condensations (or rarefactions) is one wavelength, so $\lambda = 3.00$ m and

$$v = f\lambda$$

$$= \left(112.0 \frac{1}{\text{s}}\right)(3.00 \text{ m})$$

$$= 336 \frac{\text{m}}{\text{s}}$$

5.13. **(a)** One complete wave every 4.0 s means that $T = 4.0$ s.

(b) $\qquad\qquad f = \frac{1}{T}$

$$= \frac{1}{4.0 \text{ s}}$$

$$= \frac{1}{4.0} \frac{1}{\text{s}}$$

$$= 0.25 \frac{1}{\text{s}}$$

$$= \boxed{0.25 \text{ Hz}}$$

5.14. The distance from one condensation to the next is one wavelength, so

$$v = f\lambda \qquad \therefore \qquad \lambda = \frac{v}{f}$$

$$= \frac{330 \frac{m}{s}}{260 \frac{1}{s}}$$

$$= \frac{330}{260} \frac{m}{s} \times \frac{s}{1}$$

$$= \boxed{1.3 \text{ m}}$$

5.15. **(a)** $v = f\lambda = \left(256 \frac{1}{s}\right)(1.34 \text{ m}) \qquad = \boxed{343 \text{ m/s}}$

(b) $= \left(440.0 \frac{1}{s}\right)(0.780 \text{ m}) \qquad = \boxed{343 \text{ m/s}}$

(c) $= \left(750.0 \frac{1}{s}\right)(0.457 \text{ m}) \qquad = \boxed{343 \text{ m/s}}$

(d) $= \left(2,500.0 \frac{1}{s}\right)(0.137 \text{ m}) \qquad = \boxed{343 \text{ m/s}}$

5.16. The speed of sound at 0.0°C is 1,087 ft/s, and

(a) $v_{T_F} = v_0 + \left[\frac{2.00 \text{ ft/s}}{°C}\right][T_P]$

$$= 1,087 \text{ ft/s} + \left[\frac{2.00 \text{ ft/s}}{°C}\right][0.0°C]$$

$$= 1,087 + (2.00)(0.0) \text{ ft/s} + \frac{\text{ft/s}}{°\cancel{C}} \times °\cancel{C}$$

$$= 1,0887 \text{ ft/s} + 0.0 \text{ ft/s}$$

$$= \boxed{1,087 \text{ ft/s}}$$

(b) $v_{20°} = 1,087 \text{ ft/s} + \left[\frac{2.00 \text{ ft/s}}{°C}\right][20.0°C]$

$$= 1,087 \text{ ft/s} + 40.0 \text{ ft/s}$$

$$= \boxed{1,127 \text{ ft/s}}$$

(c) $v_{40°} = 1,087 \text{ ft/s} + \left[\frac{2.00 \text{ ft/s}}{°C}\right][40.0°C]$

$$= 1,087 \text{ ft/s} + 80.0 \text{ ft/s}$$

$$= \boxed{1,167 \text{ ft/s}}$$

(d) $v_{80°} = 1,087 \text{ ft/s} + \left[\frac{2.00 \text{ ft/s}}{°C}\right][80.0°C]$

$$= 1,087 \text{ ft/s} + 160 \text{ ft/s}$$

$$= \boxed{1,247 \text{ ft/s}}$$

5.17. For consistency with the units of the equation given, 43.7°F is first converted to 6.50°C. The velocity of sound in this air is:

$$v_{T_F} = v_0 + \left[\frac{2.00 \text{ ft/s}}{°C}\right][T_P]$$

$$= 1,087 \text{ ft/s} + \left[\frac{2.00 \text{ ft/s}}{°C}\right][6.50°C]$$

$$= 1,087 \text{ ft/s} + 13.0 \text{ ft/s}$$

$$= 1,100 \text{ ft/s}$$

The distance that a sound with this velocity travels in the given time is

$$v = \frac{d}{t} \qquad \therefore \qquad d = vt$$

$$= (1,100 \text{ ft/s})(4.80 \text{ s})$$

$$= (1,100)(480) \frac{\text{ft} \cdot \cancel{s}}{\cancel{s}}$$

$$= 5,280 \text{ ft}$$

$$\frac{5,280 \text{ ft}}{2}$$

$$= 2,640 \text{ ft}$$

$$= \boxed{2,600 \text{ ft}}$$

Since the sound traveled from the rifle to the cliff and then back, the cliff must be about one-half mile away.

5.18. This problem requires three steps: (1) conversion of the °F temperature value to °C, (2) calculating the velocity of sound in air at this temperature, and (3) calculating the distance from the calculated velocity and the given time:

$$v_{T_F} = v_0 + \left[\frac{2.00 \text{ ft/s}}{°C}\right][T_P]$$

$$= 1,087 \text{ ft/s} + \left[\frac{2.00 \text{ ft/s}}{°C}\right][26.67°C]$$

$$= 1,087 \text{ ft/s} + 53.0 \text{ ft/s} = 1,140 \text{ ft/s}$$

$$v = \frac{d}{t} \qquad \therefore \qquad d = vt$$

$$= (1,140 \text{ ft/s})(4.63 \text{ s})$$

$$= \boxed{5,280 \text{ ft (one mile)}}$$

5.19. **(a)** $v = f\lambda \qquad \therefore \qquad \lambda = \frac{v}{f}$

$$= \frac{1,125 \frac{\text{ft}}{s}}{440 \frac{1}{s}}$$

$$= \frac{1,125}{440} \frac{\text{ft}}{s} \times \frac{s}{1}$$

$$= 2.56 \frac{\text{ft} \cdot \cancel{s}}{\cancel{s}}$$

$$= \boxed{2.6 \text{ ft}}$$

(b) $v = f\lambda \qquad \therefore \qquad \lambda = \frac{v}{f}$

$$= \frac{5,020}{440} \frac{\text{ft}}{s} \times \frac{s}{1}$$

$$= 11.4 \text{ ft} = \boxed{11 \text{ ft}}$$

6.1. First, recall that a negative charge means an excess of electrons. Second, the relationship between the total charge (q), the number of electrons (n), and the charge of a single electron (e) is $q = ne$. The fundamental charge of a single ($n = 1$) electron (e) is 1.60×10^{-19} C. Thus,

$$q = ne \quad \therefore \quad n = \frac{q}{e}$$

$$= \frac{1.00 \times 10^{-14}\,\text{C}}{1.60 \times 10^{-19}\,\dfrac{\text{C}}{\text{electron}}}$$

$$= \frac{1.00 \times 10^{-14}}{1.60 \times 10^{-19}}\,\frac{\text{C}}{1} \times \frac{\text{electron}}{\text{C}}$$

$$= 6.25 \times 10^{4}\,\frac{\text{C} \cdot \text{electron}}{\text{C}}$$

$$= \boxed{6.25 \times 10^{4}\,\text{electron}}$$

6.2. **(a)** Both balloons have negative charges so the force is repulsive, pushing the balloons away from each other.

(b) The magnitude of the force can be found from Coulomb's law:

$$F = \frac{kq_1 q_2}{d^2}$$

$$= \frac{(9.00 \times 10^9\,\text{N·m}^2/\text{C}^2)\,(3.00 \times 10^{-14}\,\text{C})\,(2.00 \times 10^{-12}\,\text{C})}{(2.00 \times 10^{-2}\,\text{m})^2}$$

$$= \frac{(9.00 \times 10^9)\,(3.00 \times 10^{-14})\,(2.00 \times 10^{-12})}{4.00 \times 10^{-4}}\,\frac{\dfrac{\text{N·m}^2}{\text{C}^2} \times \text{C} \times \text{C}}{\text{m}^2}$$

$$= \frac{5.40 \times 10^{-16}}{4.00 \times 10^{-4}}\,\frac{\text{N·m}^2}{\text{C}^2} \times \text{C}^2 \times \frac{1}{\text{m}^2}$$

$$= \boxed{1.35 \times 10^{-12}\,\text{N}}$$

6.3.

$$\frac{\text{potential}}{\text{difference}} = \frac{\text{work}}{\text{charge}}$$

or

$$V = \frac{W}{q}$$

$$= \frac{7.50\,\text{J}}{5.00\,\text{C}}$$

$$= 1.50\,\frac{\text{J}}{\text{C}}$$

$$= \boxed{1.50\,\text{V}}$$

6.4.

$$\frac{\text{electric}}{\text{current}} = \frac{\text{charge}}{\text{time}}$$

or

$$I = \frac{q}{t}$$

$$= \frac{6.00\,\text{C}}{2.00\,\text{s}}$$

$$= 3.00\,\frac{\text{C}}{\text{s}}$$

$$= \boxed{3.00\,\text{A}}$$

6.5. A current of 1.00 amp is defined as 1.00 coulomb/s. Since the fundamental charge of the electron is 1.60×10^{-19} C/electron,

$$\frac{1.00\,\dfrac{\text{C}}{\text{s}}}{1.60 \times 10^{-19}\,\dfrac{\text{C}}{\text{electron}}}$$

$$= 6.25 \times 10^{18}\,\frac{\text{C}}{\text{s}} \times \frac{\text{electron}}{\text{C}}$$

$$= \boxed{6.25 \times 10^{18}\,\frac{\text{electrons}}{\text{s}}}$$

6.6.

$$R = \frac{V}{I}$$

$$= \frac{120.0\,\text{V}}{4.00\,\text{A}}$$

$$= 30.0\,\frac{\text{V}}{\text{A}}$$

$$= \boxed{30.0\,\Omega}$$

6.7.

$$R = \frac{V}{I} \quad \therefore \quad I = \frac{V}{R}$$

$$= \frac{120.0\,\text{V}}{60.0\,\dfrac{\text{V}}{\text{A}}}$$

$$= \frac{120.0}{60.0}\,\text{V} \times \frac{\text{A}}{\text{V}}$$

$$= \boxed{2.00\,\text{A}}$$

6.8. **(a)** $R = \dfrac{V}{I} \quad \therefore \quad V = IR$

$$= (1.20\,\text{A})\left(10.0\,\frac{\text{V}}{\text{A}}\right)$$

$$= \boxed{12.0\,\text{V}}$$

(b) Power = (current) (potential difference)

or

$$P = IV$$

$$= \left(1.20\,\frac{\text{C}}{\text{s}}\right)\left(12.0\,\frac{\text{J}}{\text{C}}\right)$$

$$= (1.20)\,(12.0)\,\frac{\text{C}}{\text{s}} \times \frac{\text{J}}{\text{C}}$$

$$= 14.4\,\frac{\text{J}}{\text{s}}$$

$$= \boxed{14.4\,\text{W}}$$

6.9. Note that there are two separate electrical units that are rates: (1) the amp (coulomb/s) and (2) the watt (joule/s). The question asked for a rate of using energy. Energy is measured in joules, so you are looking for the power of the radio in watts. To find watts ($P = IV$), you will need to calculate the current (I) since it is not given. The current can be obtained from the relationship of Ohm's law:

$$I = \frac{V}{R}$$

$$= \frac{3.00 \text{ V}}{15.0 \frac{V}{A}}$$

$$= 0.200 \text{ A}$$

$$P = IV$$

$$= (0.200 \text{ C/s}) (3.00 \text{ J/C})$$

$$= \boxed{0.600 \text{ W}}$$

6.10. $\text{cost} = \dfrac{(\text{watts}) (\text{time}) (\text{rate})}{1,000 \frac{W}{kW}}$

$$= \frac{(1,200 \text{ W}) (0.25 \text{ h}) \left(\frac{\$0.10}{kWh}\right)}{1,000 \frac{W}{kW}}$$

$$= \frac{(1,200)(0.25) (0.10)}{1,000} \frac{W}{1} \times \frac{h}{1} \times \frac{\$}{kWh} \times \frac{kW}{W}$$

$$= \boxed{\$0.03} \text{ (3 cents)}$$

6.11.   The relationship between power (*P*), current (*I*), and voltage (*V*) will provide a solution. Since the relationship considers power in watts, the first step is to convert horsepower to watts. One horsepower is equivalent to 746 watts, so:

$$(746 \text{ W/hp})(2.00 \text{ hp}) = 1,492 \text{ W}$$

$$P = IV \quad \therefore \quad I = \frac{P}{V}$$

$$= \frac{1,492 \frac{J}{s}}{12.0 \frac{J}{C}}$$

$$= \frac{1,492}{12.0} \frac{J}{s} \times \frac{C}{J}$$

$$= 124.3 \frac{C}{s}$$

$$= \boxed{124 \text{ A}}$$

6.12.   **(a)** The rate of using energy is joule/s, or the watt. Since 1.00 hp = 746 W,

inside motor: (746 W/hp)(1/3 hp) = 249 W

outside motor: (746 W/hp)(1/3 hp) = 249 W

compressor motor: (746 W/hp)(3.70 hp) = 2,760 W

249 W + 249 W + 2,760 W = $\boxed{3,258 \text{ W}}$

(b) $\dfrac{(3,258 \text{ W}) (1.00 \text{ h}) \left(\frac{\$0.10}{kWh}\right)}{1,000 \frac{W}{kW}} = \$0.33 \text{ per hour}$

(c)   ($0.33/h) (12 h/day) (30 day/mo) = $\boxed{\$118.80}$

6.13.   The solution is to find how much current each device draws and then to see if the total current is less or greater than the breaker rating:

Toaster: $I = \dfrac{V}{R} = \dfrac{120 \text{ V}}{15 \text{ V/A}} = 8.0 \text{ A}$

Motor: (0.20 hp) (746 W/hp) = 150 W

$$I = \frac{P}{V} = \frac{150 \text{ J/s}}{120 \text{ J/C}} = 1.3 \text{ A}$$

Three 100 W bulbs: 3 × 100 W = 300 W

$$I = \frac{P}{V} = \frac{300 \text{ J/s}}{120 \text{ J/C}} = 2.5 \text{ A}$$

Iron $I = \dfrac{P}{V} = \dfrac{600 \text{ J/s}}{120 \text{ J/C}} = 5.0 \text{ A}$

The sum of the currents is 8.0 A + 1.3 A + 2.5 A + 5.0 A = 16.8 A, so the total current is greater than 15.0 amp and the circuit breaker will trip.

6.14.   **(a)** $V_P = 1,200 \text{ V}$

$N_P = 1 \text{ loop}$

$N_s = 200 \text{ loops}$

$V_s = ?$

$$\frac{V_P}{N_P} = \frac{V_s}{N_s} \quad \therefore \quad V_s = \frac{V_P N_s}{N_P}$$

$$V_s = \frac{(1,200 \text{ V}) (200 \text{ loop})}{1 \text{ loop}}$$

$$= \boxed{240,000 \text{ V}}$$

(b) $I_P = 40 \text{ A} \qquad V_P I_P = V_s I_s \quad \therefore \quad I_s = \dfrac{V_P I_P}{V_s}$

$I_s = ? \qquad I_s = \dfrac{1,200 \text{ V} \times 40 \text{ A}}{240,000 \text{ V}}$

$$= \frac{1,200 \times 40}{240,000} = \frac{V \cdot A}{V}$$

$$= \boxed{0.2 \text{ A}}$$

6.15.   **(a)** $V_s = 12 \text{ V}$

$I_s = 0.5 \text{ A}$

$V_P = 120 \text{ V}$

$$\frac{V_P}{N_P} = \frac{V_s}{N_s} \quad \therefore \quad \frac{N_P}{N_s} = \frac{V_P}{V_s}$$

$$\frac{N_P}{N_s} = \frac{120 \text{ V}}{12 \text{ V}} = \frac{10}{1}$$

$$\boxed{10 \text{ primary to 1 secondary}}$$

(b) $I_P = ? \qquad V_P I_P = V_s I_s \quad \therefore \quad I_P = \dfrac{V_s I_s}{V_P}$

$$I_P = \frac{(12 \text{ V}) (0.5 \text{ A})}{120 \text{ V}}$$

$$= \frac{12 \times 0.5}{120} \frac{V \cdot A}{V}$$

$$= \boxed{0.05 \text{ A}}$$

(c) $P_s = ? \qquad P_s = I_s V_s$

$$= (0.5 \text{ A}) (12 \text{ V})$$

$$= 0.5 \times 12 \frac{C}{s} \times \frac{J}{C}$$

$$= 6 \frac{J}{s}$$

$$= \boxed{6 \text{ W}}$$

6.16. **(a)** $V_p = 120$ V

$N_p = 50$ loops

$N_s = 150$ loops

$I_p = 5.0$ A

$V_s = ?$

$\dfrac{V_p}{N_p} = \dfrac{V_s}{N_s} \quad \therefore \quad V_s = \dfrac{V_p N_s}{N_p}$

$V_s = \dfrac{120 \text{ V} \times 150 \text{ loops}}{50 \text{ loops}}$

$= \dfrac{120 \times 150}{50} \dfrac{\text{V·loops}}{\text{loops}}$

$= \boxed{360 \text{ V}}$

(b) $I_s = ?$

$V_p I_p = V_s I_s \quad \therefore \quad I_s = \dfrac{V_p I_p}{V_s}$

$I_s = \dfrac{(120 \text{ V})(5.0 \text{ A})}{360 \text{ V}}$

$= \dfrac{120 \times 5.0}{360} \dfrac{\text{V·A}}{\text{V}}$

$= \boxed{1.7 \text{ A}}$

(c) $P_s = ?$

$P_s = I_s V_s$

$= \left(1.7 \dfrac{\text{C}}{\text{s}}\right)\left(360 \dfrac{\text{J}}{\text{C}}\right)$

$= 1.7 \times 360 \dfrac{\text{C}}{\text{s}} \times \dfrac{\text{J}}{\text{C}}$

$= 612 \dfrac{\text{J}}{\text{s}}$

$= \boxed{610 \text{ W}}$

6.17. Listing the known and unknown quantities:

Resistance (R)	8.0
Voltage (V)	12 V
Power (P)	? W
Current (I)	? A

For one bulb: The resistance (R) is given as 8.0 ohm (which is 8.0 volt/amp, by definition). The voltage source (V) is 12 V, and the power of the glowing bulb is to be determined. Power can be calculated from $P = IV$. This equation requires the current (I), which is not provided but can be calculated from Ohm's law, $V = IR$.

$V = IR \quad \therefore \quad I = \dfrac{V}{R}$

$= \dfrac{12 \text{ V}}{8.0 \dfrac{\text{V}}{\text{A}}}$

$= \dfrac{12}{8.0} \dfrac{\text{V}}{1} \times \dfrac{\text{A}}{\text{V}}$

$= 1.5 \text{ A}$

Now that we have the current, we can find the power:

$P = IV$

$= \left(1.5 \dfrac{\text{C}}{\text{s}}\right)\left(12 \dfrac{\text{J}}{\text{C}}\right)$

$= 1.5 \times 12 \dfrac{\text{C}}{\text{s}} \times \dfrac{\text{J}}{\text{C}}$

$= 18 \dfrac{\text{J}}{\text{s}}$

$= 18 \text{ W}$

For two 8.0 ohm bulbs in series: Since the bulbs are connected one after another, the current encounters resistance from each in turn, and the total resistance is equal to the sum of the individual resistances, or $R_{total} = R_1 + R_2$. Therefore,

$$R_{total} = 8.0 + 8.0 = 16.0$$

We can now find the current flowing through the entire circuit, which will be the same as the current flowing through each of the resistances.

$I = \dfrac{V}{R_{total}}$

$= \dfrac{12 \text{ V}}{16.0 \ \Omega}$

$= \dfrac{12}{16.0} \dfrac{\text{V}}{\dfrac{\text{V}}{\text{A}}}$

$= 0.75 \dfrac{\text{V}}{1} \times \dfrac{\text{A}}{\text{V}}$

$= 0.75 \text{ A}$

The current through each bulb is 0.75 amp. The voltage drop across each lamp can be calculated from Ohm's law, and $V = IR = (0.75 \text{ A})(8.0 \text{ V/A}) = 6.0$ V. As a check, you know this is correct because the voltage drop equals the voltage rise and 6.0 V + 6.0 V = 12.0 V. The power of each glowing bulb can now be calculated from $P = IV$ and $P = (0.75 \text{ A})(6.0 \text{ V}) = 4.5$ W. Two lamps connected in a series circuit, as compared to one lamp in a simple circuit, thus decreases the current by half and decreases the voltage by half, resulting in one-fourth the power for each lamp.

CHAPTER 7

7.1. The relationship between the speed of light in a transparent material (v), the speed of light in a vacuum ($c = 3.00 \times 10^8$ m/s) and the index of refraction (n) is $n = c/v$. According to Table 7.1, the index of refraction for water is $n = 1.33$ and for ice is $n = 1.31$.

(a) $c = 3.00 \times 10^8$ m/s

$n = 1.33$

$v = ?$

$n = \dfrac{c}{v} \quad \therefore \quad v = \dfrac{c}{n}$

$v = \dfrac{3.00 \times 10^8 \text{ m/s}}{1.33}$

$= \boxed{2.26 \times 10^8 \text{ m/s}}$

(b) $c = 3.00 \times 10^8$ m/s

$n = 1.31$

$v = ?$

$v = \dfrac{3.00 \times 10^8 \text{ m/s}}{1.31}$

$= \boxed{2.29 \times 10^8 \text{ m/s}}$

7.2.

$$d = 1.50 \times 10^8 \text{ km}$$
$$= 1.50 \times 10^{11}\text{m}$$
$$c = 3.00 \times 10^8 \text{ m/s}$$
$$t = ?$$

$$v = \frac{d}{t} \quad \therefore \quad t = \frac{d}{v}$$

$$t = \frac{1.50 \times 10^{11} \text{ m}}{3.00 \times 10^8 \frac{\text{m}}{\text{s}}}$$

$$= \frac{1.50 \times 10^{11}}{3.00 \times 10^8} \text{ m} \times \frac{\text{s}}{\text{m}}$$

$$= 5.00 \times 10^2 \frac{\text{m·s}}{\text{m}}$$

$$= \frac{5.00 \times 10^2 \text{s}}{60.0 \frac{\text{s}}{\text{min}}}$$

$$= \frac{5.00 \times 10^2}{60.0} \text{s} \times \frac{\text{min}}{\text{s}}$$

$$= \boxed{8.33 \text{ min}}$$

7.3.

$$d = 6.00 \times 10^9 \text{ km}$$
$$= 6.00 \times 10^{12}\text{m}$$
$$c = 3.00 \times 10^8 \text{ m/s}$$
$$t = ?$$

$$v = \frac{d}{t} \quad \therefore \quad t = \frac{d}{v}$$

$$t = \frac{6.00 \times 10^{12} \text{ m}}{3.00 \times 10^8 \frac{\text{m}}{\text{s}}}$$

$$= \frac{6.00 \times 10^{12}}{3.00 \times 10^8} \text{ m} \times \frac{\text{s}}{\text{m}}$$

$$= 2.00 \times 10^4 \text{ s}$$

$$= \frac{2.00 \times 10^4 \text{ s}}{3{,}600 \frac{\text{s}}{\text{h}}}$$

$$= \frac{2.00 \times 10^4}{3.600 \times 10^3} \text{s} \times \frac{\text{h}}{\text{s}}$$

$$= \boxed{5.56 \text{ h}}$$

7.4. From equation 7.1, note that both angles are measured from the normal and that the angle of incidence (θ_i) equals the angle of reflection (θ_r), or

$$\theta_i = \theta_r \quad \therefore \quad \boxed{\theta_i = 10°}$$

7.5.

$$v = 2.20 \times 10^8 \text{ m/s}$$
$$c = 3.00 \times 10^8 \text{ m/s}$$
$$n = ?$$

$$n = \frac{c}{v}$$

$$= \frac{3.00 \times 10^8 \frac{\text{m}}{\text{s}}}{2.20 \times 10^8 \frac{\text{m}}{\text{s}}}$$

$$= 1.36$$

According to Table 7.1, the substance with an index of refraction of 1.36 is $\boxed{\text{ethyl alcohol.}}$

7.6. **(a)** From equation 7.3:

$$\lambda = 6.00 \times 10^{-7} \text{ m} \qquad c = \lambda f \quad \therefore \quad f = \frac{c}{\lambda}$$
$$c = 3.00 \times 10^8 \text{ m/s}$$
$$f = ?$$

$$f = \frac{3.00 \times 10^8 \frac{\text{m}}{\text{s}}}{6.00 \times 10^{-7} \text{ m}}$$

$$= \frac{3.00 \times 10^8}{6.00 \times 10^{-7}} \frac{\text{m}}{\text{s}} \times \frac{1}{\text{m}}$$

$$= 5.00 \times 10^{14} \frac{1}{\text{s}}$$

$$= \boxed{5.00 \times 10^{14} \text{ Hz}}$$

(b) From equation 7.4:

$$f = 5.00 \times 10^{14} \text{ Hz}$$
$$h = 6.63 \times 10^{-34} \text{ J·s}$$
$$E = ?$$

$$E = hf$$
$$= (6.63 \times 10^{-34} \text{ J·s}) \left(5.00 \times 10^{14} \frac{1}{\text{s}}\right)$$
$$= (6.63 \times 10^{-34}) (5.00 \times 10^{14}) \text{ J·s} \times \frac{1}{\text{s}}$$
$$= \boxed{3.32 \times 10^{-19} \text{ J}}$$

7.7. First, you can find the energy of one photon of the peak intensity wavelength (5.60×10^{-7} m) by using equation 7.3 to find the frequency, then equation 7.4 to find the energy:

Step 1: $c = \lambda f \quad \therefore \quad f = \frac{c}{\lambda}$

$$= \frac{3.00 \times 10^8 \frac{\text{m}}{\text{s}}}{5.60 \times 10^{-7} \text{ m}}$$
$$= 5.36 \times 10^{14} \text{ Hz}$$

Step 2: $E = hf$

$$= (6.63 \times 10^{-34} \text{ J·s}) (5.36 \times 10^{14} \text{ Hz})$$
$$= 3.55 \times 10^{-19} \text{ J}$$

Step 3: Since one photon carries an energy of 3.55×10^{-19} J and the overall intensity is 1,000.0 W, each square meter must receive an average of

$$\frac{1{,}000.0 \frac{\text{J}}{\text{s}}}{3.55 \times 10^{-19} \frac{\text{J}}{\text{photon}}}$$

$$\frac{1.000 \times 10^3 \frac{\text{J}}{\text{s}}}{3.55 \times 10^{-19}} \times \frac{\text{photon}}{\text{J}}$$

$$\boxed{2.82 \times 10^{21} \frac{\text{photon}}{\text{s}}}$$

7.8. **(a)** $f = 4.90 \times 10^{14} \text{ Hz} \qquad c = \lambda f \quad \therefore \quad \lambda = \frac{c}{f}$
$$c = 3.00 \times 10^8 \text{ m/s}$$
$$\lambda = ?$$

$$\lambda = \frac{3.00 \times 10^8 \, \frac{m}{s}}{4.90 \times 10^{14} \, \frac{1}{s}}$$

$$= \frac{3.00 \times 10^8}{4.90 \times 10^{14}} \, \frac{m}{s} \times \frac{s}{1}$$

$$= \boxed{6.12 \times 10^{-7} \, m}$$

(b) According to Table 7.2, this is the frequency and wavelength of orange light.

7.9. $f = 5.00 \times 10^{20} \, Hz$

$h = 6.63 \times 10^{-34} \, J \cdot s$

$E = ?$

$$E = hf$$

$$= (6.63 \times 10^{-34} \, J \cdot s) \left(5.00 \times 10^{20} \, \frac{1}{s} \right)$$

$$= (6.63 \times 10^{-34}) \, (5.00 \times 10^{20}) \, J \cdot s \times \frac{1}{s}$$

$$= \boxed{3.32 \times 10^{-13} \, J}$$

7.10. $\lambda = 1.00 \, mm$

$= 0.001 \, m$

$f = ?$

$c = 3.00 \times 10^8 \, m/s$

$h = 6.63 \times 10^{-34} \, J \cdot s$

$E = ?$

Step 1: $c = \lambda f \quad \therefore \quad f = \frac{c}{\lambda}$

$$f = \frac{3.00 \times 10^8 \, \frac{m}{s}}{1.00 \times 10^{-3} \, m}$$

$$= \frac{3.00 \times 10^8}{1.00 \times 10^{-3}} \, \frac{m}{s} \times \frac{1}{m}$$

$$= 3.00 \times 10^{11} \, Hz$$

Step 2: $E = hf$

$$= (6.63 \times 10^{-34} \, J \cdot s) \left(3.00 \times 10^{11} \, \frac{1}{s} \right)$$

$$= (6.63 \times 10^{-34}) \, (3.00 \times 10^{11}) \, J \cdot s \times \frac{1}{s}$$

$$= \boxed{1.99 \times 10^{-22} \, J}$$

7.11. The index of refraction is found from $n = \frac{c}{v}$, where n is the index of refraction of a transparent material, c is the speed of light in a vaccum, and v is the speed of light in the material. The index of refraction of glass is found in Table 7.1 ($n = 1.50$).

$n = 1.50$

$c = 3.00 \times 10^8 \, m/s \qquad n = \frac{c}{v} \quad \therefore \quad v = \frac{c}{n}$

$v = ?$

$$= \frac{3.00 \times 10^8 \, \frac{m}{s}}{1.50}$$

$$= \boxed{2.00 \times 10^8 \, \frac{m}{s}}$$

7.12. Listing the known and unknown quantities:

Wavelength $\qquad\qquad \lambda = 5.60 \times 10^{-7} \, m$

Speed of light $\qquad\qquad c = 3.00 \times 10^{-8} \, m/s$

Frequency $\qquad\qquad f = ?$

The relationship between the wavelength (λ), frequency (f), and speed of light in a vacuum (c) is found in equation 7.3, $c = \lambda f$.

$$c = \lambda f \qquad \therefore \qquad f = \frac{c}{\lambda}$$

$$= \frac{3.00 \times 10^8 \, \frac{m}{s}}{5.60 \times 10^{-7} \, m}$$

$$= \frac{3.00 \times 10^8}{5.60 \times 10^{-7}} \, \frac{m}{s} \times \frac{1}{m}$$

$$= 5.40 \times 10^{14} \, \frac{1}{s}$$

$$= \boxed{5.40 \times 10^{14} \, Hz}$$

7.13. Listing the known and unknown quantities:

Frequency $\qquad\qquad f = 5.00 \times 10^{14} \, Hz$

Planck's constant $\qquad\qquad h = 6.63 \times 10^{-34} \, J \cdot s$

Energy $\qquad\qquad E = ?$

The relationship between the frequency (f) and energy (E) of a photon is found in equation 7.4, $E = hf$.

$$E = hf$$

$$= (6.63 \times 10^{-34} \, J \cdot s) \left(5.00 \times 10^{14} \, \frac{1}{s} \right)$$

$$= (6.63 \times 10^{-34}) \, (5.00 \times 10^{14}) \, J \cdot s \times \frac{1}{s}$$

$$= 3.32 \times 10^{-19} \, \frac{J \cdot s}{s}$$

$$= \boxed{3.32 \times 10^{-19} \, J}$$

7.14. Listing the known and unknown quantities:

Frequency $\qquad\qquad f = 6.50 \times 10^{14} \, Hz$

Planck's constant $\qquad\qquad h = 6.63 \times 10^{-34} \, J \cdot s$

Energy $\qquad\qquad E = ?$

The relationship between the frequency (f) and energy (E) of a photon is found in equation 7.4, $E = hf$.

$$E = hf$$

$$= (6.63 \times 10^{-34} \, J \cdot s) \left(6.50 \times 10^{14} \, \frac{1}{s} \right)$$

$$= (6.63 \times 10^{-34}) \, (6.50 \times 10^{14}) \, J \cdot s \times \frac{1}{s}$$

$$= 4.31 \times 10^{-19} \, \frac{J \cdot s}{s}$$

$$= \boxed{4.31 \times 10^{-19} \, J}$$

7.15 Listing the known and unknown quantities:

Wavelength $\qquad\qquad \lambda = 5.60 \times 10^{-7} \, m$

Planck's constant $\qquad\qquad h = 6.63 \times 10^{-34} \, J \cdot s$

Speed of light $\qquad\qquad c = 3.00 \times 10^8 \, m/s$

Frequency $\qquad\qquad f = ?$

First, you can find the energy of one photon of the peak intensity wavelength (5.60×10^{-7} m) by using the relationship between the wavelength (λ), frequency (f), and speed of light in a vacuum (c), $c = \lambda f$, then use the relationship between the frequency (f) and energy (E) of a photon, $E = hf$, to find the energy:

Step 1: $c = \lambda f \quad \therefore \quad f = \dfrac{c}{\lambda}$

$$= \frac{3.00 \times 10^8 \frac{m}{s}}{5.60 \times 10^{-7} m}$$

$$= \frac{3.00 \times 10^8}{5.60 \times 10^{-7}} \frac{m}{s} \times \frac{1}{m}$$

$$= 5.36 \times 10^{14} \frac{1}{s}$$

$$= 5.36 \times 10^{14} \text{ Hz}$$

Step 2: $E = hf$

$$= (6.63 \times 10^{-34} \text{ J·s}) \left(5.36 \times 10^{14} \frac{1}{s}\right)$$

$$= (6.63 \times 10^{-34})(5.36 \times 10^{14}) \text{ J·s} \times \frac{1}{s}$$

$$= 3.55 \times 10^{-19} \frac{\text{J·s}}{s}$$

$$= 3.55 \times 10^{-19} \text{ J}$$

Step 3: Since one photon carries an energy of 3.55×10^{-19} J and the overall intensity is 1,000.0 W, for each square meter there must be an average of

$$\frac{1,000.0 \frac{J}{s}}{3.55 \times 10^{-19} \frac{J}{\text{photon}}}$$

$$\frac{1.000 \times 10^{-3}}{3.55 \times 10^{-19}} \frac{J}{s} \times \frac{\text{photon}}{J}$$

$$\boxed{2.82 \times 10^{21} \frac{\text{photon}}{s}}$$

CHAPTER 8

8.1. $m = 1.68 \times 10^{-27}$ kg

$v = 3.22 \times 10^3$ m/s

$h = 6.63 \times 10^{-34}$ J·s

$\lambda = ?$

$$\lambda = \frac{h}{mv}$$

$$= \frac{6.63 \times 10^{-34} \text{ J·s}}{(1.68 \times 10^{-27} \text{ kg})\left(3.22 \times 10^3 \frac{m}{s}\right)}$$

$$= \frac{6.63 \times 10^{-34}}{(1.68 \times 10^{-27})(3.22 \times 10^3)} \frac{\text{J·s}}{\text{kg} \times \frac{m}{s}}$$

$$= \frac{6.63 \times 10^{-34}}{5.41 \times 10^{-24}} \frac{\frac{\text{kg·m}^2}{s \cdot s} \times s}{\text{kg} \times \frac{m}{s}}$$

$$= 1.23 \times 10^{-10} \frac{\text{kg·m·m}}{s} \times \frac{1}{\text{kg}} \times \frac{s}{m}$$

$$= \boxed{1.23 \times 10^{-10} \text{ m}}$$

8.2. (a) $n = 6$

$E_L = -13.6$ eV

$E_6 = ?$

$$E_n = \frac{E_L}{n^2}$$

$$E_6 = \frac{-13.6 \text{ eV}}{6^2}$$

$$= \frac{-13.6 \text{ eV}}{36}$$

$$= \boxed{-0.378 \text{ eV}}$$

(b)

$$= (-0.378 \text{ eV})\left(1.60 \times 10^{-19} \frac{J}{\text{eV}}\right)$$

$$= (-0.378)(1.60 \times 10^{-19}) \text{ eV} \times \frac{J}{\text{eV}}$$

$$= \boxed{-6.05 \times 10^{-20} \text{ J}}$$

8.3. (a) Energy is related to the frequency and Planck's constant in equation 8.1, $E = hf$. From equation 8.4,

$$hf = E_H - E_L \quad \therefore \quad E = E_H - E_L$$

For $n = 6$, $E_H = 6.05 \times 10^{-20}$ J

For $n = 2$, $E_L = 5.44 \times 10^{-19}$ J

$E = ?$ J

$E = E_H - E_L$

$$= (-6.05 \times 10^{-20} \text{ J}) - (-5.44 \times 10^{-19} \text{ J})$$

$$= \boxed{4.84 \times 10^{-19} \text{ J}}$$

(b) $E_H = -0.377$ eV*

$E_L = -3.40$ eV*

$E = ?$ eV

$E = E_H - E_L$

$$= (-0.377 \text{ eV}) - (-3.40 \text{ eV})$$

$$= \boxed{3.02 \text{ eV}}$$

*From figure 8.11

8.4. For $n = 6$, $E_H = -6.05 \times 10^{-20}$ J

For $n = 2$, $E_L = -5.44 \times 10^{-19}$ J

$h = 6.63 \times 10^{-34}$ J·s

$f = ?$

$$hf = E_H - E_L \quad \therefore \quad f = \frac{E_H - E_L}{h}$$

$$f = \frac{(-6.05 \times 10^{-20} \text{ J}) - (-5.44 \times 10^{-19} \text{ J})}{6.63 \times 10^{-34} \text{ J·s}}$$

$$= \frac{4.84 \times 10^{-19}}{6.63 \times 10^{-34}} \frac{J}{\text{J·s}}$$

$$= 7.29 \times 10^{14} \frac{1}{s}$$

$$= \boxed{7.29 \times 10^{14} \text{ Hz}}$$

8.5. $(n = 1) = -13.6 \text{ eV}$ $E_n = \dfrac{E_1}{n^2}$

$E = ?$

$$= \dfrac{-13.6 \text{ eV}}{1^2}$$

$$= -13.6 \text{ eV}$$

Since the energy of the electron is -13.6 eV, it will require 13.6 eV (or 2.17×10^{-18} J) to remove the electron.

8.6. $q/m = -1.76 \times 10^{11}$ C/kg

$q = -1.60 \times 10^{-19}$ C

$m = ?$

$$\text{mass} = \dfrac{\text{charge}}{\text{charge/mass}}$$

$$= \dfrac{-1.60 \times 10^{-19} \text{ C}}{-1.76 \times 10^{11} \dfrac{\text{C}}{\text{kg}}}$$

$$= \dfrac{-1.60 \times 10^{-19}}{-1.76 \times 10^{11}} \cancel{\text{C}} \times \dfrac{\text{kg}}{\cancel{\text{C}}}$$

$$= \boxed{9.09 \times 10^{-31} \text{ kg}}$$

8.7. $\lambda = 1.67 \times 10^{-10}$ m

$m = 9.11 \times 10^{-31}$ kg

$v = ?$

$$\lambda = \dfrac{h}{mv} \quad \therefore \quad v = \dfrac{h}{m\lambda}$$

$$v = \dfrac{6.63 \times 10^{-34} \text{ J·s}}{(9.11 \times 10^{-31} \text{ kg})(1.67 \times 10^{-10} \text{ m})}$$

$$= \dfrac{6.63 \times 10^{-34}}{(9.11 \times 10^{-31})(1.67 \times 10^{-10})} \dfrac{\text{J·s}}{\text{kg·m}}$$

$$= \dfrac{6.63 \times 10^{-34} \dfrac{\text{kg·m}^2}{\text{s·}\cancel{s}} \times \cancel{s}}{1.52 \times 10^{-40} \; \text{kg·m}}$$

$$= 4.36 \times 10^{6} \dfrac{\text{kg·m·m}}{\text{s}} \times \dfrac{1}{\text{kg}} \times \dfrac{1}{\text{m}}$$

$$= \boxed{4.36 \times 10^{6} \dfrac{\text{m}}{\text{s}}}$$

8.8. (a) Boron: $1s^2 2s^2 2p^1$
(b) Aluminum: $1s^2 2s^2 2p^6 3s^2 3p^1$
(c) Potassium: $1s^2 2s^2 2p^6 3s^2 3p^6 4s^1$

8.9. (a) Boron is atomic number 5 and there are 5 electrons.
(b) Aluminum is atomic number 13 and there are 13 electrons.
(c) Potassium is atomic number 19 and there are 19 electrons.

8.10. (a) Argon: $1s^2 2s^2 2p^6 3s^2 3p^6$
(b) Zinc: $1s^2 2s^2 2p^6 3s^2 3p^6 4s^2 3d^{10}$
(c) Bromine: $1s^2 2s^2 2p^6 3s^2 3p^6 4s^2 3d^{10} 4p^5$

8.11. Atomic weight is the weighted average of the isotopes as they occur in nature. Thus,

Lithium-6: 6.01512 u $\times$ 0.0742 = 0.446 u
Lithium-7: 7.016 u $\times$ 0.9258   = 6.4054 u

Lithium-6 contributes 0.446 u of the weighted average and lithium-7 contributes 6.4954 u. The atomic weight of lithium is therefore

0.446 u

+6.4954 u

6.941 u

8.12. Recall that the subscript is the atomic number, which identifies the number of protons. In a neutral atom, the number of protons equals the number of electrons, so the atomic number tells you the number of electrons, too. The superscript is the mass number, which identifies the number of neutrons and the number of protons in the nucleus. The number of neutrons is therefore the mass number minus the atomic number.

	Protons	Neutrons	Electrons
(a)	6	6	6
(b)	1	0	1
(c)	18	22	18
(d)	1	1	1
(e)	79	118	79
(f)	92	143	92

8.13.

		Period	Family
(a)	Radon (Rn)	6	VIIIA
(b)	Sodium (Na)	3	IA
(c)	Copper (Cu)	4	IB
(d)	Neon (Ne)	2	VIIIA
(e)	Iodine (I)	5	VIIA
(f)	Lead (Pb)	6	IVA

8.14. Recall that the number of outer-shell electrons is the same as the family number for the representative elements:
(a) Li: 1   (d) Cl: 7
(b) N: 5   (e) Ra: 2
(c) F: 7   (f) Be: 2

8.15. The same information that was used in question 8.14 can be used to draw the dot notation (see Figure 8.20):

(a) $\overset{\textstyle\cdot}{\underset{\textstyle\cdot}{\text{B}}}\cdot$ (c) $\overset{\textstyle\cdot}{\text{Ca}}\cdot$ (e) $\cdot\overset{\textstyle\cdot\cdot}{\underset{\textstyle\cdot\cdot}{\text{O}}}:$

(b) $:\overset{\textstyle\cdot\cdot}{\underset{\textstyle\cdot\cdot}{\text{Br}}}:$ (d) K$\cdot$ (f) $\cdot\overset{\textstyle\cdot\cdot}{\underset{\textstyle\cdot\cdot}{\text{S}}}:$

8.16. The charge is found by identifying how many electrons are lost or gained in achieving the noble gas structure:
(a) Boron 3+
(b) Bromine 1−
(c) Calcium 2+
(d) Potassium 1+
(e) Oxygen 2−
(f) Nitrogen 3−

8.17. Metals have one, two, or three outer electrons and are located in the left two-thirds of the periodic table. Semiconductors are adjacent to the line that separates the metals and nonmetals. Look at the periodic table on the inside back cover and you will see:

(a) Krypton—nonmetal
(b) Cesium—metal
(c) Silicon—semiconductor
(d) Sulfur—nonmetal
(e) Molybdenum—metal
(f) Plutonium—metal

8.18. (a) Bromine gained an electron to acquire a 1− charge, so it must be in family VIIA (the members of this family have seven electrons and need one more to acquire the noble gas structure).
(b) Potassium must have lost one electron, so it is in IA.
(c) Aluminum lost three electrons, so it is in IIIA.
(d) Sulfur gained two electrons, so it is in VIA.
(e) Barium lost two electrons, so it is in IIA.
(f) Oxygen gained two electrons, so it is in VIA.

8.19. (a) $^{16}_{8}O$ (c) $^{3}_{1}H$

     (b) $^{23}_{11}Na$ (d) $^{35}_{17}Cl$

CHAPTER 9

9.1.

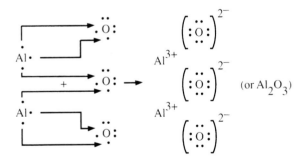

9.2. (a) Sulfur is in family VIA, so sulfur has six valence electrons and will need two more to achieve a stable outer structure like the noble gases. Two more outer shell electrons will give the sulfur atom a charge of 2−. Copper^{2+} will balance the 2− charge of sulfur, so the name is copper(II) sulfide. Note the -ide ending for compounds that have only two different elements.
(b) Oxygen is in family VIA, so oxygen has six valence electrons and will have a charge of 2−. Using the crossover technique in reverse, you can see that the charge on the oxygen is 2−, and the charge on the iron is 3−. Therefore, the name is iron(III) oxide.

(c) From information in (a) and (b), you know that oxygen has a charge of 2−. The chromium ion must have the same charge to make a neutral compound as it must be, so the name is chromium(II) oxide. Again, note the -ide ending for a compound with two different elements.
(d) Sulfur has a charge of 2−, so the lead ion must have the same positive charge to make a neutral compound. The name is lead(II) sulfide.

9.3. The name of some common polyatomic ions are in Table 9.3. Using this table as a reference, the names are
(a) hydroxide
(b) sulfite
(c) hypochlorite
(d) nitrate
(e) carbonate
(f) perchlorate

9.4. The Roman numeral tells you the charge on the variable-charge elements. The charges for the polyatomic ions are found in Table 9.3. The charges for metallic elements can be found in Tables 9.1 and 9.2. Using these resources and the crossover technique, the formulas are as follows:
(a) $Fe(OH)_3$ (d) NH_4NO_3
(b) $Pb_3(PO_4)_2$ (e) $KHCO_3$
(c) $ZnCO_3$ (f) K_2SO_3

9.5. Table 9.7 has information about the meaning of prefixes and stem names used in naming covalent compounds. (a), for example, asks for the formula of carbon tetrachloride. Carbon has no prefixes, so there is one carbon atom, and it comes first in the formula because it comes first in the name. The *tetra-* prefix means four, so there are four chlorine atoms. The name ends in *-ide*, so you know there are only two elements in the compound. The symbols can be obtained from the list of elements on the inside back cover of this text. Using all this information from the name, you can think out the formula for carbon tetrachloride. The same process is used for the other compounds and formulas:
(a) CCl_4 (d) SO_3
(b) H_2O (e) N_2O_5
(c) MnO_2 (f) As_2S_5

9.6. Again using information from Table 9.7, this question requires you to reverse the thinking procedure you learned in question 9.5.
(a) carbon monoxide
(b) carbon dioxide
(c) carbon disulfide
(d) dinitrogen monoxide
(e) tetraphosphorus trisulfide
(f) dinitrogen trioxide

9.7. The types of bonds formed are predicted by using the electronegativity scale in Table 9.5 and finding the absolute difference. On this basis:
(a) Difference = 1.7, which means ionic bond
(b) Difference = 0, which means covalent
(c) Difference = 0, which means covalent

(d) Difference = 0.4, which means covalent

(e) Difference = 3.0, which means ionic

(f) Difference = 1.6, which means polar covalent and almost ionic

CHAPTER 10

10.1. **(a)** $MgCl_2$ is an ionic compound, so the formula has to be empirical.

(b) C_2H_2 is a covalent compound, so the formula might be molecular. Since it is not the simplest whole number ratio (which would be CH), then the formula is molecular.

(c) BaF_2 is ionic; the formula is empirical.

(d) C_8H_{18} is not the simplest whole number ratio of a covalent compound, so the formula is molecular.

(e) CH_4 is covalent, but the formula might or might not be molecular (?).

(f) S_8 is a nonmetal bonded to a nonmetal (itself); this is a molecular formula.

10.2. **(a)** $CuSO_4$

$$1 \text{ of Cu} = 1 \times 63.5 \text{ u} = 63.5 \text{ u}$$
$$1 \text{ of S} = 1 \times 32.1 \text{ u} = 32.1 \text{ u}$$
$$4 \text{ of O} = 4 \times 16.0 \text{ u} = \underline{64.0 \text{ u}}$$
$$159.6 \text{ u}$$

(b) CS_2

$$1 \text{ of C} = 1 \times 12.0 \text{ u} = 12.0 \text{ u}$$
$$2 \text{ of S} = 2 \times 32.0 \text{ u} = \underline{64.0 \text{ u}}$$
$$76.0 \text{ u}$$

(c) $CaSO_4$

$$1 \text{ of Ca} = 1 \times 40.1 \text{ u} = 40.1 \text{ u}$$
$$1 \text{ of S} = 1 \times 32.0 \text{ u} = 32.0 \text{ u}$$
$$4 \text{ of O} = 4 \times 16.0 \text{ u} = \underline{64.0 \text{ u}}$$
$$136.1 \text{ u}$$

(d) Na_2CO_3

$$2 \text{ of Na} = 2 \times 23.0 \text{ u} = 46.0 \text{ u}$$
$$1 \text{ of C} = 1 \times 12.0 \text{ u} = 12.0 \text{ u}$$
$$3 \text{ of O} = 3 \times 16.0 \text{ u} = \underline{48.0 \text{ u}}$$
$$106.0 \text{ u}$$

10.3. **(a)** FeS_2

For Fe: $\dfrac{(55.9 \text{ u Fe}) (1)}{119.0 \text{ u } FeS_2} \times 100\% \, FeS_2 = 46.6\%$ Fe

For S: $\dfrac{(32.0 \text{ u S}) (2)}{119.9 \text{ u } FeS_2} \times 100\% \, FeS_2 = 53.4\%$ S

or $(100\% \, FeS_2) - (46.6\% \text{ Fe}) = 53.4\%$ S

(b) H_3BO_3

For H: $\dfrac{(1.0 \text{ u H}) (3)}{61.8 \text{ u } H_3BO_3} \times 100\% \, H_3BO_3 = 4.85\%$ H

For B: $\dfrac{(10.8 \text{ u B}) (1)}{61.8 \text{ u } H_3BO_3} \times 100\% \, H_3BO_3 = 17.5\%$ B

For O: $\dfrac{(16 \text{ u O}) (3)}{61.8 \text{ u } H_3BO_3} \times 100\% \, H_3BO_3 = 77.7\%$ O

(c) $NaHCO_3$

For Na: $\dfrac{(23.0 \text{ u Na}) (1)}{84.0 \text{ u } NaHCO_3} \times 100\% \, NaHCO_3 = 27.4\%$ Na

For H: $\dfrac{(1.0 \text{ u H}) (1)}{84.0 \text{ u } NaHCO_3} \times 100\% \, NaHCO_3 = 1.2\%$ H

For C: $\dfrac{(12.0 \text{ u C}) (1)}{84.0 \text{ u } NaHCO_3} \times 100\% \, NaHCO_3 = 14.3\%$ C

For O: $\dfrac{(16.0 \text{ u O}) (3)}{84.0 \text{ u } NaHCO_3} \times 100\% \, NaHCO_3 = 57.1\%$ O

(d) $C_9H_8O_4$

For C: $\dfrac{(12.0 \text{ u C}) (9)}{180.0 \text{ u } C_9H_8O_4} \times 100\% \, C_9H_8O_4 = 60.0\%$ C

For H: $\dfrac{(1.0 \text{ u H}) (8)}{180.0 \text{ u } C_9H_8O_4} \times 100\% \, C_9H_8O_4 = 4.4.\%$ H

For O: $\dfrac{(16.0 \text{ u O}) (4)}{180.0 \text{ u } C_9H_8O_4} \times 100\% \, C_9H_8O_4 = 35.6\%$ O

10.4. **(a)** $2 SO_2 + O_2 \rightarrow 2 SO_3$

(b) $4 P + 5 O_2 \rightarrow 2 P_2O_5$

(c) $2 Al + 6 HCl \rightarrow 2 AlCl_3 + 3 H_2$

(d) $2 NaOH + H_2SO_4 \rightarrow Na_2SO_4 + 2 H_2O$

(e) $Fe_2O_3 + 3 CO \rightarrow 2 Fe + 3 CO_2$

(f) $3 Mg(OH)_2 + 2 H_3PO_4 \rightarrow Mg_3(PO_4)_2 + 6 H_2O$

10.5. **(a)** General form of $XY + AZ \rightarrow XZ + AY$ with precipitate formed: Ion exchange reaction.

(b) General form of $X + Y \rightarrow XY$: Combination reaction.

(c) General form of $XY \rightarrow X + Y + \cdots$: Decomposition reaction.

(d) General form of $X + Y \rightarrow XY$: Combination reaction.

(e) General form of $XY + A \rightarrow AY + X$: Replacement reaction.

(f) General form of $X + Y \rightarrow XY$: Combination reaction.

10.6. **(a)** $C_5H_{12(g)} + 8 O_{2(g)} \rightarrow 5 CO_{2(g)} + 6 H_2O_{(g)}$

(b) $HCl_{(aq)} + NaOH_{(aq)} \rightarrow NaCl_{(aq)} + H_2O_{(l)}$

(c) $2 Al_{(s)} + Fe_2O_{3(s)} \rightarrow Al_2O_{3(s)} + 2 Fe_{(l)}$

(d) $Fe_{(s)} + CuSO_{4(aq)} \rightarrow FeSO_{4(aq)} + Cu_{(s)}$

(e) $MgCl_{(aq)} + Fe(NO_3)_{2(aq)} \rightarrow$ No reaction (all possible compounds are soluble and no gas or water was formed).

(f) $C_6H_{10}O_{5(s)} + 6 O_{2(g)} \rightarrow 6 CO_{2(g)} + 5 H_2O_{(g)}$

10.7. **(a)** $2 KClO_3 \overset{\Delta}{\rightarrow} 2 KCl_{(s)} + 3 O_2 \uparrow$

(b) $2 Al_2O_{3(l)} \overset{elec}{\rightarrow} 4 Al_{(s)} + 3 O_2 \uparrow$

(c) $CaCO_{3(s)} \overset{\Delta}{\rightarrow} CaO_{(s)} + CO_2 \uparrow$

10.8. **(a)** $2 Na_{(s)} + 2 H_2O_{(l)} \rightarrow 2 NaOH_{(aq)} + H_2 \uparrow$

(b) $Au_{(s)} + HCl_{(aq)} \rightarrow$ No reaction (gold is below hydrogen in the activity series).

(c) $Al_{(s)} + FeCl_{3(aq)} \rightarrow AlCl_{3(aq)} + Fe_{(s)}$

(d) $Zn_{(s)} + CuCl_{2(aq)} \rightarrow ZnCl_{2(aq)} + Cu_{(s)}$

10.9.
(a) $NaOH_{(aq)} + HNO_{3(aq)} \rightarrow NaNO_{3(aq)} + H_2O_{(l)}$

(b) $CaCl_{2(aq)} + KNO_{3(aq)} \rightarrow$ No reaction

(c) $3\,Ba(NO_3)_{2(aq)} + 2\,Na_3PO_{4(aq)} \rightarrow 6\,NaNO_{3(aq)} + Ba_3(PO_4)_2\downarrow$

(d) $2\,KOH_{(aq)} + ZnSO_{4(aq)} \rightarrow K_2SO_{4(aq)} + Zn(OH)_2\downarrow$

10.10. Five moles of oxygen combine with 2 moles of acetylene, so 2.5 moles of oxygen would be needed for 1 mole of acetylene. Therefore, 1 L of C_2H_2 requires 2.5 L of O_2.

CHAPTER 11

11.1. $m_{solute} = 1.75$ g

$m_{solution} = 50.0$ g

% weight = ?

$$\% \text{ solute} = \frac{m_{solute}}{m_{solution}} \times 100\% \text{ solution}$$

$$= \frac{1.75 \text{ g NaCl}}{50.0 \text{ g solution}} \times 100\% \text{ solution}$$

$$= \boxed{3.50\% \text{ NaCl}}$$

11.2. $m_{solution} = 103.5$ g

$m_{solute} = 3.50$ g

% weight = ?

$$\% \text{ solute} = \frac{m_{solute}}{m_{solution}} \times 100\% \text{ solution}$$

$$= \frac{3.50 \text{ g NaCl}}{103.5 \text{ g solution}} \times 100\% \text{ solution}$$

$$= \boxed{3.38\% \text{ NaCl}}$$

11.3. Since ppm is defined as the weight unit of solute in 1,000,000 weight units of solution, the percent by weight can be calculated just like any other percent. The weight of the dissolved sodium and chlorine ions is the part, and the weight of the solution is the whole, so

$$\% = \frac{\text{part}}{\text{whole}} \times 100\%$$

$$= \frac{30,113 \text{ g NaCl ions}}{1,000,000 \text{ g seawater}} \times 100\% \text{ seawater}$$

$$= \boxed{3.00\% \text{ NaCl ions}}$$

11.4. $m_{solution} = 250$ g

% solute = 3.0 g

$m_{solute} = ?$

$$\% \text{ solute} = \frac{m_{solute}}{m_{solution}} \times 100\% \text{ solution}$$

$$\therefore$$

$$m_{solute} = \frac{(m_{solution})\,(\% \text{ solute})}{100\% \text{ solution}}$$

$$= \frac{(250 \text{ g})\,(3.0\%)}{100\%}$$

$$= \boxed{7.5 \text{ g}}$$

11.5. % solution = 12% solution

$V_{solution} = 200$ mL

$V_{solute} = ?$

$$\% \text{ solution} = \frac{V_{solute}}{V_{solution}} \times 100\% \text{ solution}$$

$$\therefore$$

$$V_{solute} = \frac{(\% \text{ solution})\,(V_{solution})}{100\% \text{ solution}}$$

$$= \frac{(12\% \text{ solution})\,(200 \text{ mL})}{100\% \text{ solution}}$$

$$= \boxed{24 \text{ mL alcohol}}$$

11.6. % solution = 40%

$V_{solution} = 50$ mL

$V_{solute} = ?$

$$\% \text{ solution} = \frac{V_{solute}}{V_{solution}} \times 100\% \text{ solution}$$

$$\therefore$$

$$V_{solute} = \frac{(\% \text{ solution})\,(V_{solution})}{100\% \text{ solution}}$$

$$= \frac{(40\% \text{ solution})\,(50 \text{ mL})}{100\% \text{ solution}}$$

$$= \boxed{20 \text{ mL alcohol}}$$

11.7. (a) $\% \text{ concentration} = \dfrac{\text{ppm}}{1 \times 10^4}$

$$= \frac{5}{1 \times 10^4}$$

$$= \boxed{0.0005\% \text{ DDT}}$$

(b) $\% \text{ part} = \dfrac{\text{part}}{\text{whole}} \times 100\% \text{ whole}$

$$\therefore$$

$$\text{whole} = \frac{(100\%)\,\text{part}}{\% \text{ part}}$$

$$= \frac{(100\%)\,(17.0 \text{ g})}{0.00059\%} = \boxed{3,400,000 \text{ g or } 3,400 \text{ kg}}$$

11.8.

(a) $\underset{\text{acid}}{HC_2H_3O_{2(aq)}} + \underset{\text{base}}{H_2O_{(l)}} \rightarrow H_3O^+_{(aq)} + C_2H_3O_2^-_{(aq)}$

(b) $\underset{\text{base}}{C_6H_6NH_{2(l)}} + \underset{\text{acid}}{H_2O_{(l)}} \rightarrow C_6H_6NH_3^+_{(aq)} + OH^-_{(aq)}$

(c) $\underset{\text{acid}}{HClO_{4(aq)}} + \underset{\text{base}}{HC_2H_3O_{2(aq)}} \rightarrow H_2C_2H_3O_2^+_{(aq)} + ClO_4^-_{(aq)}$

(d) $\underset{\text{base}}{H_2O_{(l)}} + \underset{\text{acid}}{H_2O_{(l)}} \rightarrow H_3O^+_{(aq)} + OH^-_{(aq)}$

CHAPTER 12

12.1. **(a)**

```
    H   H   H   H   H
    |   |   |   |   |
H — C — C — C — C — C — H
    |   |   |   |   |
    H   H   H   H   H
```

(b)

```
          H
          |
      H — C — H
          |
      H   |   H
      |   |   |
  H — C — C — C — H
      |   |   |
      H   |   H
      H — C — H
          |
          H
```

(c) 2,2-dimethylpropane

12.2. *n*-hexane

```
    H   H   H   H   H   H
    |   |   |   |   |   |
H — C — C — C — C — C — C — H
    |   |   |   |   |   |
    H   H   H   H   H   H
```

3-methylpentane

```
    H   H   H   H   H
    |   |   |   |   |
H — C — C — C — C — C — H
    |   |   |   |   |
    H   H   |   H   H
        H — C — H
            |
            H
```

2-methylpentane

```
    H   H   H   H   H
    |   |   |   |   |
H — C — C — C — C — C — H
    |   |   |   |   |
    H   |   H   H   H
    H — C — H
        |
        H
```

2,2-dimethylbutane

```
    H   CH₃  H   H
    |   |    |   |
H — C — C —  C — C — H
    |   |    |   |
    H   CH₃  H   H
```

12.3. **(a)**

```
    H   H   CH₃ CH₃ H   H   H   H
    |   |   |   |   |   |   |   |
H — C — C — C — C — C — C — C — C — H
    |   |   |   |   |   |   |   |
    H   H   CH₃ H   H   H   H   H
```

(b)

```
    H   CH₃ H   H   H
    |   |   |   |   |
H — C = C — C — C — C — H
            |   |   |
            H   H   H
```

(c)

```
    H   H           CH₃ H   H
    |   |           |   |   |
H — C — C — C ≡ C — C — C — C — H
    |   |           |   |   |
    H   H           CH₃ H   H
```

12.4. **(a)** 2-chloro-4-methylpentane
(b) 2-methyl-l-pentene
(c) 3-ethyl-4-methyl-2-pentene

12.5. The 2,2,3-trimethylbutane is more highly branched, so it will have the higher octane rating.

2,2,3-trimethylbutane

```
    H   CH₃ CH₃ H
    |   |   |   |
H — C — C — C — C — H
    |   |   |   |
    H   CH₃ H   H
```

2,2,-dimethylpentane

```
    H   CH₃ H   H   H
    |   |   |   |   |
H — C — C — C — C — C — H
    |   |   |   |   |
    H   CH₃ H   H   H
```

12.6 **(a)** alcohol
(b) amide
(c) ether
(d) ester
(e) organic acid

CHAPTER 13

13.1. **(a)** cobalt-60: 27 protons, 33 neutrons
(b) potassium-40: 19 protons, 21 neutrons
(c) neon-24: 10 protons, 14 neutrons
(d) lead-208: 82 protons, 126 neutrons

13.2. **(a)** $^{60}_{27}Co$ **(c)** $^{24}_{10}Ne$
(b) $^{40}_{19}K$ **(d)** $^{204}_{82}Pb$

13.3.

(a) cobalt-60: Radioactive because odd numbers of protons (27) and odd numbers of neutrons (33) are usually unstable.

(b) potassium-40: Radioactive, again having an odd number of protons (19) and an odd number of neutrons (21).

(c) neon-24: Stable, because even numbers of protons and neutrons are usually stable.

(d) lead-208: Stable, because even numbers of protons and neutrons *and* because 82 is a particularly stable number of nucleons.

13.4.

(a) $^{56}_{26}\text{Fe} \rightarrow ^{0}_{-1}\text{e} + ^{56}_{27}\text{Co}$

(b) $^{7}_{4}\text{Be} \rightarrow ^{0}_{-1}\text{e} + ^{7}_{5}\text{B}$

(c) $^{64}_{29}\text{Cu} \rightarrow ^{0}_{-1}\text{e} + ^{64}_{30}\text{Zn}$

(d) $^{24}_{11}\text{Na} \rightarrow ^{0}_{-1}\text{e} + ^{24}_{12}\text{Mg}$

(e) $^{214}_{82}\text{Pb} \rightarrow ^{0}_{-1}\text{e} + ^{214}_{83}\text{Bi}$

(f) $^{32}_{15}\text{P} \rightarrow ^{0}_{-1}\text{e} + ^{32}_{16}\text{S}$

13.5.

(a) $^{235}_{92}\text{Fe} \rightarrow ^{4}_{2}\text{He} + ^{231}_{90}\text{Th}$

(b) $^{226}_{88}\text{Ra} \rightarrow ^{4}_{2}\text{He} + ^{222}_{86}\text{Rn}$

(c) $^{239}_{94}\text{Pu} \rightarrow ^{4}_{2}\text{He} + ^{235}_{92}\text{U}$

(d) $^{214}_{83}\text{Bi} \rightarrow ^{4}_{2}\text{He} + ^{210}_{81}\text{Tl}$

(e) $^{230}_{90}\text{Th} \rightarrow ^{4}_{2}\text{He} + ^{226}_{88}\text{Ra}$

(f) $^{210}_{84}\text{Po} \rightarrow ^{4}_{2}\text{He} + ^{206}_{82}\text{Pb}$

13.6. Thirty-two days is four half-lives. After the first half-life (8 days), 1/2 oz will remain. After the second half-life (8 + 8, or 16 days), 1/4 oz will remain. After the third half-life (8 + 8 + 8, or 24 days), 1/8 oz will remain. After the fourth half-life (8 + 8 + 8 + 8, or 32 days), 1/16 oz will remain, or 6.3×10^{-2} oz.

GLOSSARY

A

absolute humidity a measure of the actual amount of water vapor in the air at a given time—for example, in grams per cubic meter

absolute magnitude a classification scheme to compensate for the distance differences to stars; calculations of the brightness that stars would appear to have if they were all at a defined, standard distance of 10 parsecs

absolute zero the theoretical lowest temperature possible, which occurs when all random motion of molecules has ceased

abyssal plain the practically level plain of the ocean floor

acceleration a change in velocity per change in time; by definition, this change in velocity can result from a change in speed, a change in direction, or a combination of changes in speed and direction

accretion disk a fat bulging disk of gas and dust from the remains of the gas cloud that forms around a protostar

achondrites homogeneously textured stony meteorites

acid any substance that is a proton donor when dissolved in water; generally considered a solution of hydronium ions in water that can neutralize a base, forming a salt and water

acid-base indicator a vegetable dye used to distinguish acid and base solutions by a color change

air mass a large, more or less uniform body of air with nearly the same temperature and moisture conditions throughout

air mass weather the weather experienced within a given air mass; characterized by slow, gradual changes from day to day

alcohol an organic compound with a general formula of ROH, where R is one of the hydrocarbon groups—for example, methyl or ethyl

aldehyde an organic molecule with the general formula RCHO, where R is one of the hydrocarbon groups—for example, methyl or ethyl

alkali metals members of family IA of the periodic table, having common properties of shiny, low-density metals that can be cut

with a knife and that react violently with water to form an alkaline solution

alkaline earth metals members of family IIA of the periodic table, having common properties of soft, reactive metals that are less reactive than alkali metals

alkane a hydrocarbon with a single covalent bond between the carbon atoms

alkene a hydrocarbon with a double covalent carbon-carbon bond

alkyne a hydrocarbon with a carbon-carbon triple bond

allotropic forms elements that can have several different structures with different physical properties—for example, graphite and diamond are two allotropic forms of carbon

alpha particle the nucleus of a helium atom (two protons and two neutrons) emitted as radiation from a decaying heavy nucleus; also known as an alpha ray

alpine glaciers glaciers that form at high elevations in mountainous regions

alternating current an electric current that first moves one direction, then the opposite direction with a regular frequency

amino acids organic functional groups that form polypeptides and proteins

amp a unit of electric current; equivalent to C/s

ampere the full name of the unit amp

amplitude the extent of displacement from the equilibrium condition; the size of a wave from the rest (equilibrium) position

angle of incidence the angle of an incident (arriving) ray or particle to a surface; measured from a line perpendicular to the surface (the normal)

angle of reflection the angle of a reflected ray or particle from a surface; measured from a line perpendicular to the surface (the normal)

angular momentum quantum number in the quantum mechanics model of the atom, one of four descriptions of the energy state of an electron wave; this quantum number describes the energy sublevels of electrons within the main energy levels of an atom

angular unconformity a boundary in rock where the bedding planes above and below

the time interruption unconformity are not parallel, meaning probable tilting or folding followed by a significant period of erosion, which in turn was followed by a period of deposition

annular eclipse occurs when the penumbra reaches the surface of Earth; as seen from Earth, the Sun forms a bright ring around the disk of the new moon

Antarctic Circle the parallel identifying the limit toward the equator where the Sun appears above the horizon all day for at least one day during the summer; located at 66.5°S latitude

anticline an arch-shaped fold in layered bedrock

anticyclone a high-pressure center with winds flowing away from the center; associated with clear, fair weather

antinode the region of maximum amplitude between adjacent nodes in a standing wave

apogee the point at which the Moon's elliptical orbit takes the Moon farthest from Earth

apparent local noon the instant when the Sun crosses the celestial meridian at any particular longitude

apparent local solar time the time found from the position of the Sun in the sky; the shadow of the gnomon on a sundial

apparent magnitude a classification scheme for different levels of brightness of stars that you see; brightness values range from 1 to 6 with the number 1 (first magnitude) assigned to the brightest star and the number 6 (sixth magnitude) assigned to the faintest star that can be seen

apparent solar day the interval between two consecutive crossings of the celestial meridian by the Sun

aquifer a layer of sand, gravel, or other highly permeable material beneath the surface that is saturated with water and is capable of producing water in a well or spring

Archean a block of time of 4,000 to 2,500 million years before the present; one of three Precambrian eons before the time of visible life

Arctic Circle the parallel identifying the limit toward the equator where the Sun appears above the horizon all day for at least one day up to six months during the summer; located at 66.5°N latitude

area the extent of a surface

arid the dry climate classification; receives less than 25 cm (10 in) precipitation per year

aromatic hydrocarbon an organic compound with at least one benzene ring structure; cyclic hydrocarbons and their derivatives

artesian the term describing the condition where confining pressure forces groundwater from a well to rise above the aquifer

asteroids small rocky bodies left over from the formation of the solar system; most are accumulated in a zone between the orbits of Mars and Jupiter

asthenosphere a plastic, mobile layer of Earth's structure that extends around Earth below the lithosphere; ranges in thickness from a depth of 130 km to 160 km

astronomical unit the radius of Earth's orbit is defined as one astronomical unit (AU)

atmospheric stability the condition of the atmosphere related to the temperature of the air at increasing altitude compared to the temperature of a rising parcel of air at increasing altitude

atom the smallest unit of an element that can exist alone or in combination with other elements

atomic mass unit the relative mass unit (u) of an isotope based on the standard of the carbon-12 isotope, which is defined as a mass of exactly 12.00 u; one atomic mass unit (1 u) is 1/12 the mass of a carbon-12 atom

atomic number the number of protons in the nucleus of an atom

atomic weight the weighted average of the masses of stable isotopes of an element as they occur in nature, based on the abundance of each isotope of the element and the atomic mass of the isotope compared to C-12

autumnal equinox one of two times a year that daylight and night are of equal length; occurs on or about September 23 and identifies the beginning of the fall season

avalanche a mass movement of a wide variety of materials such as rocks, snow, trees, soils, and so forth in a single chaotic flow; also called debris avalanche

Avogadro's number the number of C-12 atoms in exactly 12.00 g of C; 6.02×10^{23} atoms or other chemical units; the number of chemical units in one mole of a substance

axis the imaginary line about which a planet or other object rotates

B

background radiation ionizing radiation (alpha, beta, gamma, etc.) from natural sources; between 100 and 500 millirems/yr of exposure to natural radioactivity from the environment

Balmer series a set of four line spectra, narrow lines of color emitted by hydrogen atom electrons as they drop from excited states to the ground state

band of stability a region of a graph of the number of neutrons versus the number of protons in nuclei; nuclei that have the neutron-to-proton ratios located in this band do not undergo radioactive decay

barometer an instrument that measures atmospheric pressure, used in weather forecasting and in determining elevation above sea level

base any substance that is a proton acceptor when dissolved in water; generally considered a solution that forms hydroxide ions in water that can neutralize an acid, forming a salt and water

basin a large, bowl-shaped fold in the land into which streams drain; also a small enclosed or partly enclosed body of water

batholith a large volume of magma that has cooled and solidified below the surface, forming a large mass of intrusive rock

beat rhythmic increases and decreases of volume from constructive and destructive interference between two sound waves of slightly different frequencies

beta particle a high-energy electron emitted as ionizing radiation from a decaying nucleus; also known as a beta ray

big bang theory the current model of galactic evolution in which the universe was created from an intense and brilliant explosion from a primeval fireball

binding energy the energy required to break a nucleus into its constituent protons and neutrons; also the energy equivalent released when a nucleus is formed

black hole the theoretical remaining core of a supernova that is so dense that even light cannot escape

blackbody radiation the electromagnetic radiation emitted by an ideal material (the blackbody) that perfectly absorbs and perfectly emits radiation

body wave a seismic wave that travels through the earth's interior, spreading outward from a disturbance in all directions

Bohr model a model of the structure of the atom that attempted to correct the deficiencies of the solar system model and account for the Balmer series

boiling point the temperature at which a phase change of liquid to gas takes place

through boiling; the same temperature as the condensation point

boundary the division between two regions of differing physical properties

Bowen's reaction series a crystallization series that occurs as a result of the different freezing point temperatures of various minerals present in magma

breaker a wave whose front has become so steep that the top part has broken forward of the wave, breaking into foam, especially against a shoreline

British thermal unit the amount of energy or heat needed to increase the temperature of 1 pound of water 1 degree Fahrenheit (abbreviated Btu)

C

calorie the amount of energy (or heat) needed to increase the temperature of 1 gram of water 1 degree Celsius

Calorie the dieter's "calorie"; equivalent to 1 kilocalorie

carbohydrates organic compounds that include sugars, starches, and cellulose; carbohydrates are used by plants and animals for structure, protection, and food

carbon film a type of fossil formed when the volatile and gaseous constituents of a buried organic structure are distilled away, leaving a carbon film as a record

carbonation in chemical weathering a reaction that occurs naturally between carbonic acid (H_2CO_3) and rock minerals

cast sediments deposited by groundwater in a mold, taking the shape and external features of the organism that was removed to form the mold, then gradually changing to sedimentary rock

cathode rays negatively charged particles (electrons) that are emitted from a negative terminal in an evacuated glass tube

celestial equator the line of the equator of Earth directly above Earth; the equator of Earth projected on the celestial sphere

celestial meridian an imaginary line in the sky directly above you that runs north through the north celestial pole, south through the south celestial pole, and back around the other side to make a big circle around Earth

celestial sphere a coordinate system of lines used to locate objects in the sky by imagining a huge turning sphere surrounding Earth with the stars and other objects attached to the sphere; the latitude and longitude lines of Earth's surface are projected to the celestial sphere

cellulose a polysaccharide abundant in plants that forms the fibers in cell walls that preserve the structure of plant materials

Celsius scale the referent scale that defines numerical values for measuring hotness or coldness, defined as degrees of temperature; based on the reference points of the freezing point of water and the boiling point of water at sea-level pressure, with 100 degrees between the two points

cementation a process by which spaces between buried sediment particles under compaction are filled with binding chemical deposits, binding the particles into a rigid, cohesive mass of a sedimentary rock

Cenozoic one of four geologic eras; the time of recent life, meaning the fossils of this era are identical to the life found on Earth today

centigrade an alternate name for the Celsius scale

centrifugal force an apparent outward force on an object following a circular path that is a consequence of the third law of motion

centripetal force the force required to pull an object out of its natural straight-line path and into a circular path; centripetal means "center seeking"

Cepheid variable a bright variable star that can be used to measure distance

chain reaction a self-sustaining reaction where some of the products are able to produce more reactions of the same kind; in a nuclear chain reaction, neutrons are the products that produce more nuclear reactions in a self-sustaining series

chemical bond an attractive force that holds atoms together in a compound

chemical change a change in which the identity of matter is altered and new substances are formed

chemical energy a form of energy involved in chemical reactions associated with changes in internal potential energy; a kind of potential energy that is stored and later released during a chemical reaction

chemical equation a concise way of describing what happens in a chemical reaction

chemical equilibrium occurs when two opposing reactions happen at the same time and at the same rate

chemical reaction a change in matter where different chemical substances are created by forming or breaking chemical bonds

chemical sediments ions from rock materials that have been removed from solution—for example, carbonate ions removed by crystallization or organisms to form calcium carbonate chemical sediments

chemical weathering the breakdown of minerals in rocks by chemical reactions with water, gases of the atmosphere, or solutions

chemistry the science concerned with the study of the composition, structure, and properties of substances and the transformations they undergo

Chinook a warm wind that has been warmed by compression; also called Santa Ana

chondrites a subdivision of stony meteorites containing small, spherical lumps of silicate minerals or glass

chondrules small, spherical lumps of silicate minerals or glass found in some meteorites

cinder cone volcano a volcanic cone that formed from cinders, sharp-edged rock fragments that cooled from frothy blobs of lava as they were thrown into the air

cirque a bowl-like depression in the side of a mountain, usually at the upper end of a mountain valley, formed by glacial erosion

clastic sediments weathered rock fragments that are in various states of being broken down from solid bedrock; boulders, gravel, sand, and silt

climate the general pattern of weather that occurs in a region over a number of years

climate change a departure from the expected average pattern of climate for a region over time

coalescence process (meteorology) the process by which large raindrops form from the merging and uniting of millions of tiny water droplets

cold front the front that is formed as a cold air mass moves into warmer air

combination chemical reaction a synthesis reaction in which two or more substances combine to form a single compound

comets celestial objects originating from the outer edges of the solar system that move about the Sun in highly elliptical orbits; solar heating and pressure from the solar wind form a tail on the comet that points away from the Sun

compaction the process of pressure from a depth of overlying sediments squeezing the deeper sediments together and squeezing water out

composite volcano a volcanic cone that formed from a buildup of alternating layers of cinders, ash, and lava flows

compound a pure chemical substance that can be decomposed by a chemical change into simpler substances with a fixed mass ratio

compressive stress a force that tends to compress the surface as Earth's plates move into each other

concentration an arbitrary description of the relative amounts of solute and solvent in a solution; a larger amount of solute makes a concentrated solution, and a small amount of solute makes a dilute concentration

condensation (sound) a compression of gas molecules; a pulse of increased density and pressure that moves through the air at the speed of sound

condensation (water vapor) where more vapor or gas molecules are returning to the liquid state than are evaporating

condensation nuclei tiny particles such as tiny dust, smoke, soot, and salt crystals that are suspended in the air on which water condenses

condensation point the temperature at which a gas or vapor changes back to a liquid

conduction the transfer of heat from a region of higher temperature to a region of lower temperature by increased kinetic energy moving from molecule to molecule

consistent law principle one of two basic principles of the special theory of relativity; the laws of physics are the same in all reference frames that move at a constant velocity with respect to each other

constancy of speed one of two basic principles of the special theory of relativity; the speed of light in empty space has the same value for all observers regardless of their velocity

constructive interference the condition in which two waves arriving at the same place, at the same time and in phase, add amplitudes to create a new wave

continental air mass a dry air mass that forms over a large land area

continental climate a climate influenced by air masses from large land areas; hot summers and cold winters

continental drift a concept that continents shift positions on Earth's surface, moving across the surface rather than being fixed, stationary landmasses

continental glaciers glaciers that cover a large area of a continent, for example, Greenland and the Antarctic

continental shelf a feature of the ocean floor; the flooded margins of the continents that form a zone of relatively shallow water adjacent to the continents

continental slope a feature of the ocean floor; a steep slope forming the transition between the continental shelf and the deep ocean basin

control rods rods inserted between fuel rods in a nuclear reactor to absorb neutrons and thus control the rate of the nuclear chain reaction

controlled experiment an experiment that allows for a comparison of two events that are identical in all but one respect

convection the transfer of heat from a region of higher temperature to a region of lower temperature by the displacement of high-energy molecules—for example, the displacement of warmer, less dense air (higher kinetic energy) by cooler, more dense air (lower kinetic energy)

convection cell a complete convective circulation pattern; also, slowly turning regions in the plastic asthenosphere that might drive the motion of plate tectonics

convection zone (of a star) a part of the interior of a star according to a model; the region directly above the radiation zone where gases are heated by the radiation zone below and move upward by convection to the surface, where they emit energy in the form of visible light, ultraviolet radiation, and infrared radiation

conventional current opposite to electron current—that is, considers an electric current to consist of a drift of positive charges that flow from the positive terminal to the negative terminal of a battery

convergent boundaries boundaries that occur between two plates moving toward each other

Copernican system heliocentric, or sun-centered solar system model developed by Nicholas Copernicus in 1543

core (of Earth) the center part of Earth, which consists of a solid inner part and liquid outer part, making up about 15-percent of Earth's total volume and about one-third of its mass

core (of a star) a dense, very hot region of a star where nuclear fusion reactions release gamma and X-ray radiation

Coriolis effect the apparent deflection due to the rotation of Earth; it is to the right in the Northern Hemisphere

correlation the determination of the equivalence in geologic age by comparing the rocks in two separate locations

coulomb the unit used to measure quantity of electric charge; equivalent to the charge resulting from the transfer of 6.24 billion billion particles such as the electron

Coulomb's law a relationship between charge, distance, and magnitude of the electrical force between two bodies

covalent bond a chemical bond formed by the sharing of a pair of electrons

covalent compound a chemical compound held together by a covalent bond or bonds

creep the slow downhill movement of soil down a steep slope

crest the high mound of water that is part of a wave; also refers to the condensation, or high-pressure part, of a sound wave

critical angle limit to the angle of incidence when all light rays are reflected internally

critical mass mass of fissionable material needed to sustain a chain reaction

crude oil petroleum pumped from the ground that has not yet been refined into usable products

crust the outermost part of Earth's interior structure; the thin, solid layer of rock that rests on top of the Mohorovicic discontinuity

curie the unit of nuclear activity defined as 3.70×10^{10} nuclear disintegrations per second

cycle a complete vibration

cyclone a low-pressure center where the winds move into the low-pressure center and are forced upward; a low-pressure center with clouds, precipitation, and stormy conditions

D

data measurement information used to describe something

data points points that may be plotted on a graph to represent simultaneous measurements of two related variables

daylight saving time setting clocks ahead one hour during the summer to more effectively utilize the longer days of summer, then setting the clocks back in the fall

decibel scale a nonlinear scale of loudness based on the ratio of the intensity level of a sound to the intensity at the threshold of hearing

decomposition chemical reaction a chemical reaction in which a compound is broken down into the elements that make up the compound, into simpler compounds, or into elements and simpler compounds

deep-focus earthquakes earthquakes that occur in the lower part of the upper mantle, between 350 and 700 km below the surface of Earth

deflation the widespread removal of base materials from the surface by the wind

degassing a process whereby gases and water vapor were released from rocks heated to melting during the early stages of the formation of a planet

delta a somewhat triangular deposit at the mouth of a river formed where a stream flowing into a body of water slowed and lost its sediment-carrying ability

density the compactness of matter described by a ratio of mass (or weight) per unit volume

density current an ocean current that flows because of density differences in seawater

destructive interference the condition in which two waves arriving at the same point at the same time out of phase add amplitudes to create zero total disturbance

dew the condensation of water vapor into droplets of liquid on surfaces

dew point the temperature at which condensation begins

diastrophism an all-inclusive term that means any and all possible movements of Earth's plates, including drift and any other process that deforms or changes Earth's surface by movement

diffuse reflection light rays reflected in many random directions, as opposed to the parallel rays reflected from a perfectly smooth surface such as a mirror

dike a tabular-shaped intrusive rock that formed when magma moved into joints or faults that cut across other rock bodies

direct current an electrical current that always moves in one direction

direct proportion when two variables increase or decrease together in the same ratio (at the same rate)

disaccharides two monosaccharides joined together with the loss of a water molecule; examples of disaccharides are sucrose (table sugar), lactose, and maltose

dispersion the effect of spreading colors of light into a spectrum with a material that has an index of refraction that varies with wavelength

divergent boundaries boundaries that occur between two plates moving away from each other

divide a line separating two adjacent watersheds

dome a large, upwardly bulging, symmetrical fold that resembles a dome

Doppler effect an apparent shift in the frequency of sound or light due to relative motion between the source of the sound or light and the observer

double bond a covalent bond formed when two pairs of electrons are shared by two atoms

dune a hill, low mound, or ridge of wind-blown sand or other sediments

dwarf planet an object that is orbiting the Sun, is nearly spherical, but has not cleared matter from its orbital zone and is not a satellite

E

earthflow a mass movement of a variety of materials such as soil, rocks, and water with a thick, fluidlike flow

earthquake a quaking, shaking, vibrating, or upheaval of Earth's surface

earthquake epicenter the point on Earth's surface directly above an earthquake focus

earthquake focus the place where seismic waves originate beneath the surface of Earth

echo a reflected sound that can be distinguished from the original sound, which usually arrives 0.1 s or more after the original sound

eclipse when the shadow of a celestial body falls on the surface of another celestial body

elastic rebound the sudden snap of stressed rock into new positions; the recovery from elastic strain that results in an earthquake

elastic strain an adjustment to stress in which materials recover their original shape after a stress is released

electric circuit consists of a voltage source that maintains an electrical potential, a continuous conducting path for a current to follow, and a device where work is done by the electrical potential; a switch in the circuit is used to complete or interrupt the conducting path

electric current the flow of electric charge

electric field a force field produced by an electrical charge

electric field lines a map of an electric field representing the direction of the force that a positive test charge would experience; the direction of an electric field shown by lines of force

electric generator a mechanical device that uses wire loops rotating in a magnetic field to produce electromagnetic induction in order to generate electricity

electric potential energy potential energy due to the position of a charge near other charges

electrical conductors materials that have electrons that are free to move throughout the material; for example, metals

electrical energy a form of energy from electromagnetic interactions; one of five forms of energy—mechanical, chemical, radiant, electrical, and nuclear

electrical force a fundamental force that results from the interaction of electrical charge and is billions and billions of times stronger than the gravitational force; sometimes called the "electromagnetic force" because of the strong association between electricity and magnetism

electrical insulators electrical nonconductors, or materials that obstruct the flow of electric current

electrical nonconductors materials that have electrons that are not moved easily within the material—for example, rubber; electrical nonconductors are also called electrical insulators

electrical resistance the property of opposing or reducing electric current

electrolyte a water solution of ionic substances that conducts an electric current

electromagnet a magnet formed by a solenoid that can be turned on and off by turning the current on and off

electromagnetic force one of four fundamental forces; the force of attraction or repulsion between two charged particles

electromagnetic induction a process in which current is induced by moving a loop of wire in a magnetic field or by changing the magnetic field

electron a subatomic particle that has the smallest negative charge possible, usually found in an orbital of an atom, but gained or lost when atoms become ions

electron configuration the arrangement of electrons in orbitals and suborbitals about the nucleus of an atom

electron current opposite to conventional current; that is, considers electric current to consist of a drift of negative charges that flows from the negative terminal to the positive terminal of a battery

electron dot notation a notation made by writing the chemical symbol of an element with dots around the symbol to indicate the number of outer orbital electrons

electronegativity the comparative ability of atoms of an element to attract bonding electrons

electron pair a pair of electrons with different spin quantum numbers that may occupy an orbital

electron volt the energy gained by an electron moving across a potential difference of one volt; equivalent to 1.60×10^{-19} J

electrostatic charge an accumulated electric charge on an object from a surplus or deficiency of electrons; also called "static electricity"

element a pure chemical substance that cannot be broken down into anything simpler by chemical or physical means; there are over one hundred known elements, the fundamental materials of which all matter is made

El Niño changes in atmospheric pressure systems, ocean currents, water temperatures, and wind patterns that seem to be linked to worldwide changes in the weather

empirical formula identifies the elements present in a compound and describes the simplest whole number ratio of atoms of these elements with subscripts

energy the ability to do work

English system a system of measurement that originally used sizes of parts of the human body as referents

entropy the measure of disorder in thermodynamics

eons major blocks of time in Earth's geologic history

epochs subdivisions of geologic periods

equation a statement that describes a relationship in which quantities on one side of the equal sign are identical to quantities on the other side

equation of time the cumulative variation between the apparent local solar time and the mean solar time

equinoxes Latin meaning "equal nights"; time when daylight and night are of equal length, which occurs during the spring equinox and the autumnal equinox

eras the major blocks of time in Earth's geologic history; the Cenozoic, Mesozoic, Paleozoic, and Precambrian

erosion the process of physically removing weathered materials; for example, rock fragments are physically picked up by an erosion agent such as a stream or a glacier

esters the class of organic compounds with the general structure of RCOORʲ, where R is one of the hydrocarbon groups—for example, methyl or ethyl; esters make up fats, oils, and waxes and some give fruit and flowers their taste and odor

ether the class of organic compounds with the general formula ROR′, where R is one of the hydrocarbon groups—for example, methyl or ethyl; mostly used as industrial and laboratory solvents

excited states as applied to an atom, describes the energy state of an atom that has electrons in a state above the minimum energy state for that atom; as applied to a nucleus, describes the energy state of a nucleus that has particles in a state above the minimum energy state for that nuclear configuration

exfoliation the fracturing and breaking away of curved, sheetlike plates from bare rock surfaces via physical or chemical weathering, resulting in dome-shaped hills and rounded boulders

exosphere the outermost layer of the atmosphere where gas molecules merge with the diffuse vacuum of space

experiment a re-creation of an event in a way that enables a scientist to gain valid and reliable empirical evidence

external energy the total potential and kinetic energy of an everyday-sized object

extrusive igneous rocks fine-grained igneous rocks formed as lava cools rapidly on the surface

F

Fahrenheit scale the referent scale that defines numerical values for measuring hotness or coldness, defined as degrees of temperature; based on the reference points of the freezing point of water and the boiling point of water at sea-level pressure, with 180 degrees between the two points

family vertical columns of the periodic table consisting of elements that have similar properties

fats organic compounds of esters formed from glycerol and three long-chain carboxylic acids that are also called triglycerides; called fats in animals and oils in plants

fault a break in the continuity of a rock formation along which relative movement has occurred between the rocks on either side

fault plane the surface along which relative movement has occurred between the rocks on either side; the surface of the break in continuity of a rock formation

ferromagnesian silicates silicates that contain iron and magnesium; examples include the dark-colored minerals olivine, augite, hornblende, and biotite

first law of motion every object remains at rest or in a state of uniform straight-line motion unless acted on by an unbalanced force

first law of thermodynamics a statement of the law of conservation of energy in the relationship between internal energy, work, and heat

first quarter the moon phase between the new phase and the full phase when the Moon is perpendicular to a line drawn through Earth and the Sun; one-half of the lighted Moon can be seen from Earth, so this phase is called the first quarter

floodplain the wide, level floor of a valley built by a stream; the river valley where a stream floods when it spills out of its channel

fluids matter that has the ability to flow or be poured; the individual molecules of a fluid are able to move, rolling over or by one another

focus the place beneath the surface where the waves of an earthquake originate

folds bends in layered bedrock as a result of stress or stresses that occurred when the rock layers were in a ductile condition, probably under considerable confining pressure from deep burial

foliation the alignment of flat crystal flakes of a rock into parallel sheets

force a push or pull capable of changing the state of motion of an object; a force has magnitude (strength) as well as direction

force field a model describing action at a distance by giving the magnitude and direction of force on a unit particle; considers a charge or a mass to alter the space surrounding it and a second charge or mass to interact with the altered space with a force

formula describes what elements are in a compound and in what proportions

formula weight the sum of the atomic weights of all the atoms in a chemical formula

fossil any evidence of former prehistoric life

fossil fuels organic fuels that contain the stored radiant energy of the Sun converted to chemical energy by plants or animals that lived millions of years ago; coal, petroleum, and natural gas are the common fossil fuels

Foucault pendulum a heavy mass swinging from a long wire that can be used to provide evidence about the rotation of Earth

fracture strain an adjustment to stress in which materials crack or break as a result of the stress

free fall when objects fall toward Earth with no forces acting upward; air resistance is neglected when considering an object to be in free fall

freezing point the temperature at which a phase change of liquid to solid takes place; the same temperature as the melting point for a given substance

frequency the number of cycles of a vibration or of a wave occurring in one second, measured in units of cycles per second (hertz)

freshwater water that is not saline and is fit for human consumption

front the boundary, or thin transition zone, between air masses of different temperatures

frost ice crystals formed by water vapor condensing directly from the vapor phase; frozen water vapor that forms on objects

frost wedging the process of freezing and thawing water in small rock pores and cracks that become larger and larger, eventually forcing pieces of rock to break off

fuel rod a long zirconium alloy tube containing fissionable material for use in a nuclear reactor

full moon the moon phase when Earth is between the Sun and the Moon and the entire side of the Moon facing Earth is illuminated by sunlight

functional group the atom or group of atoms in an organic molecule that is responsible for the chemical properties of a particular class or group of organic chemicals

fundamental charge the smallest common charge known; the magnitude of the charge of an electron and a proton, which is 1.60×10^{-19} coulomb

fundamental forces four forces that cannot be explained in terms of any other force; gravitational, electromagnetic, weak nuclear, and strong nuclear

fundamental frequency the lowest frequency (longest wavelength) that can set up standing waves in an air column or on a string

fundamental properties a property that cannot be defined in simpler terms other than to describe how it is measured; the fundamental properties are length, mass, time, and charge

G

g the symbol representing the acceleration of an object in free fall due to the force of gravity; its magnitude is 9.8 m/s² (32 ft/s²)

galactic clusters gravitationally bound subgroups of as many as one thousand stars that move together within the Milky Way galaxy

galaxy a group of billions and billions of stars that form the basic unit of the universe; for example, Earth is part of the solar system, which is located in the Milky Way galaxy

gamma ray very-short-wavelength electromagnetic radiation emitted by decaying nuclei

gases a phase of matter composed of molecules that are relatively far apart moving freely in a constant, random motion and having weak cohesive forces acting between them, resulting in the characteristic indefinite shape and indefinite volume of a gas

gasohol a solution of ethanol and gasoline

Geiger counter a device that indirectly measures ionizing radiation (beta and/or gamma) by detecting "avalanches" of electrons that are able to move because of the ions produced by the passage of ionizing radiation

general theory of relativity Einstein's geometric theory of gravity; gravity is an interaction between a mass and the space and time geometry of space

geologic time scale a "calendar" of geologic history based on the appearance and disappearance of particular fossils in the sedimentary rock record

geomagnetic time scale a time scale established from the number and duration of magnetic field reversals during the past 6 million years

geosynchronous satellite a satellite that turns with Earth and does not appear to move across the sky

geothermal energy heat from beneath Earth's surface, usually reaching the surface in the form of geysers, steam, or hot water

giant planets the large outer planets Jupiter, Saturn, Uranus, and Neptune that all have similar densities and compositions

glacier a large mass of ice on land that is formed from compacted snow and slowly moves under its own weight

globular clusters symmetrical and tightly packed clusters of as many as a million stars that move together as subgroups within the Milky Way galaxy

glycerol an alcohol with three hydroxyl groups per molecule; for example, glycerin (1,2,3-propanetriol)

glycogen a highly branched polysaccharide synthesized by the human body and stored in the muscles and liver; serves as a direct reserve source of energy

glycol an alcohol with two hydroxyl groups per molecule; for example, ethylene glycol that is used as an antifreeze

gram-atomic weight the mass in grams of one mole of an element that is numerically equal to its atomic weight

gram-formula weight the mass in grams of one mole of a compound that is numerically equal to its formula weight

gram-molecular weight the gram-formula weight of a molecular compound

granite a light-colored, coarse-grained igneous rock common on continents; igneous rocks formed by blends of quartz and feldspars, with small amounts of micas, hornblende, and other minerals

greenhouse effect the process of increasing the temperature of the lower parts of the atmosphere through redirecting energy back toward the surface; the absorption and reemission of infrared radiation by carbon dioxide, water vapor, and a few other gases in the atmosphere

ground state the energy state of an atom with electrons at the lowest energy state possible for that atom

groundwater water from a saturated zone beneath the surface; water from beneath the surface that supplies wells and springs

gyre the great circular systems of moving water in each ocean

H

hail a frozen form of precipitation, sometimes with alternating layers of clear and opaque, cloudy ice

hair hygrometer a device that measures relative humidity from changes in the length of hair

half-life the time required for one-half of the unstable nuclei in a radioactive substance to decay into a new element

halogen member of family VIIA of the periodic table, having common properties of very reactive nonmetallic elements common in salt compounds

hard water water that contains relatively high concentrations of dissolved salts of calcium and magnesium

heat the total internal energy of molecules, which is increased by gaining energy from a temperature difference (conduction, convection, radiation) or by gaining energy from a form conversion (mechanic, chemical, radiant, electrical, nuclear)

heat of formation energy released in a chemical reaction

Heisenberg uncertainty principle you cannot measure both the exact momentum and the exact position of a subatomic particle at the same time; the more exactly one of the two is known, the less certain you are of the value of the other

hertz unit of frequency; equivalent to one cycle per second

Hertzsprung-Russell diagram a diagram to classify stars with a temperature-luminosity graph

high short for high-pressure center (anticyclone), which is associated with clear, fair weather

high latitudes the latitudes close to the poles; those that sometimes receive no solar radiation at noon

high-pressure center another term for anticyclone

horsepower a measurement of power defined as a power rating of 550 ft·lb/s

hot spots sites on Earth's surface where plumes of hot rock materials rise from deep within the mantle

humid the moist climate classification; receives more than 50 cm (20 in) precipitation per year

humidity the amount of water vapor in the air; see *relative humidity*

hurricane a tropical cyclone with heavy rains and winds exceeding 120 km/h (75 mi/h)

hydration the attraction of water molecules for ions; a reaction that occurs between water and minerals that make up rocks

hydrocarbon an organic compound consisting of only the two elements hydrogen and carbon

hydrocarbon derivatives organic compounds that can be thought of as forming when one or more hydrogen atoms on a hydrocarbon have been replaced by an element or a group of elements other than hydrogen

hydrogen bond a weak to moderate bond between the hydrogen end ($+$) of a polar molecule and the negative end ($-$) of a second polar molecule

hydrologic cycle water vapor cycling into and out of the atmosphere through continuous evaporation of liquid water from the surface and precipitation of water back to the surface

hydronium ion a molecule of water with an attached hydrogen ion, H_3O^+

hypothesis a tentative explanation of a phenomenon that is compatible with the data and provides a framework for understanding and describing that phenomenon

I

ice-crystal process a precipitation-forming process that brings water droplets of a cloud together through the formation of ice crystals

ice-forming nuclei small, solid particles suspended in air; ice can form on the suspended particles

igneous rocks rocks that formed from magma, which is a hot, molten mass of melted rock materials

impulse a change of motion is brought about by an impulse; the product of the size of an applied force and the time the force is applied

incandescent matter emitting visible light as a result of high temperature; for example, a lightbulb, a flame from any burning source, and the Sun are all incandescent sources because of high temperature

incident ray a line representing the direction of motion of incoming light approaching a boundary

inclination of Earth's axis the tilt of Earth's axis measured from the plane of the ecliptic (23.5°); considered to be the same throughout the year

index fossils distinctive fossils of organisms that lived only a brief time; used to compare the age of rocks exposed in two different locations

index of refraction the ratio of the speed of light in a vacuum to the speed of light in a material

inertia a property of matter describing the tendency of an object to resist a change in its state of motion; an object will remain in unchanging motion or at rest in the absence of an unbalanced force

infrasonic sound waves having too low a frequency to be heard by the human ear; sound having a frequency of less than 20 Hz

inorganic chemistry the study of all compounds and elements in which carbon is not the principal element

insulators materials that are poor conductors of heat—for example, heat flows slowly through materials with air pockets because the molecules making up air are far apart; also, materials that are poor conductors of electricity—for example, glass or wood

intensity a measure of the energy carried by a wave

interference a phenomenon of light whereby the relative phase difference between two light waves produces light or dark spots, a result of light's wavelike nature

intermediate-focus earthquakes earthquakes that occur in the upper part of the mantle, between 70 and 350 km (43 and 217 mi) below the surface of Earth

intermolecular forces forces of interaction between molecules

internal energy the sum of all the potential energy and all the kinetic energy of all the molecules of an object

international date line the 180° meridian is arbitrarily called the international date line; used to compensate for cumulative time zone changes by adding or subtracting a day when the line is crossed

intertropical convergence zone a part of the lower troposphere in a belt from 10°N to 10°S of the equator where air is heated, expands, and becomes less dense and rises around the belt

intrusive igneous rocks coarse-grained igneous rocks formed as magma cools slowly deep below the surface

inverse proportion the relationship in which the value of one variable increases while the value of the second variable decreases at the same rate (in the same ratio)

inversion a condition of the troposphere when temperature increases with height rather than decreasing with height; a cap of cold air over warmer air that results in increased air pollution

ion an atom or a particle that has a net charge because of the gain or loss of electrons; polyatomic ions are groups of bonded atoms that have a net charge

ion exchange reaction a reaction that takes place when the ions of one compound interact with the ions of another, forming a solid that comes out of solution, a gas, or water

ionic bond the chemical bond of electrostatic attraction between negative and positive ions

ionic compounds chemical compounds that are held together by ionic bonds—that is, bonds of electrostatic attraction between negative and positive ions

ionization the process of forming ions from molecules

ionization counter a device that measures ionizing radiation (alpha, beta, gamma, etc.) by indirectly counting the ions produced by the radiation

ionized an atom or a particle that has a net charge because it has gained or lost electrons

ionosphere that part of the atmosphere—parts of the thermosphere and upper mesosphere—where free electrons and ions reflect radio waves around Earth and where the northern lights occur

iron meteorites the meteorite classification group whose members are composed mainly of iron

island arcs curving chains of volcanic islands that occur over belts of deep-seated earthquakes; for example, the Japanese and Indonesian islands

isomers chemical compounds with the same molecular formula but different molecular structure; compounds that are made from the same numbers of the same elements but have different molecular arrangements

isotope atoms of an element with identical chemical properties but with different masses; isotopes are atoms of the same element with different numbers of neutrons

J

jet stream a powerful, winding belt of wind near the top of the troposphere that tends to extend all the way around Earth, moving generally from the west in both hemispheres at speeds of 160 km/h or more

joint a break in the continuity of a rock formation without a relative movement of the rock on either side of the break

joule the metric unit used to measure work and energy; can also be used to measure heat; equivalent to newton-meter

K

Kelvin scale a temperature scale that does not have arbitrarily assigned referent points, and zero means nothing; the zero point on the Kelvin scale (also called absolute scale) is the lowest limit of temperature, where all random kinetic energy of molecules ceases

Kepler's first law describes how each planet of the solar system moves in an elliptical orbit, with the Sun located at one focus

Kepler's laws of planetary motion three laws describing the motion of the planets in the solar system

Kepler's second law describes how an imaginary line between the Sun and a planet moves over equal areas of the ellipse during equal time intervals

Kepler's third law a relationship in planetary motion that the square of the period of an orbit is directly proportional to the cube of the radius of the major axis of the orbit

ketone an organic compound with the general formula RCOR′, where R is one of the hydrocarbon groups; for example, methyl or ethyl

kilocalorie the amount of energy required to increase the temperature of 1 kilogram of water 1 degree Celsius: equivalent to 1,000 calories

kilogram the fundamental unit of mass in the metric system of measurement

kinetic energy the energy of motion; can be measured from the work done to put an object in motion, from the mass and velocity of the object while in motion, or from the amount of work the object can do because of its motion

kinetic molecular theory the collection of assumptions that all matter is made up of tiny atoms and molecules that interact physically, that explain the various states of matter, and that have an average kinetic energy that defines the temperature of a substance

Kuiper Belt a disk-shaped region of small icy bodies some 30 to 100 AU from the Sun; the source of short-period comets

L

L-wave seismic waves that move on the solid surface of Earth much as water waves move across the surface of a body of water

laccolith an intrusive rock feature that formed when magma flowed into the plane of contact between sedimentary rock layers, then raised the overlying rock into a blisterlike uplift

lake a large inland body of standing water

landforms the features of the surface of Earth such as mountains, valleys, and plains

landslide general term for rapid movement of any type or mass of materials

last quarter the moon phase between the full phase and the new phase when the Moon is perpendicular to a line drawn through Earth and the Sun; one-half of the lighted Moon can be seen from Earth, so this phase is called the last quarter

latent heat refers to the heat "hidden" in phase changes

latent heat of fusion the heat absorbed when 1 gram of a substance changes from the solid to the liquid phase, or the heat released by 1 gram of a substance when changing from the liquid phase to the solid phase

latent heat of vaporization the heat absorbed when 1 gram of a substance changes from the liquid phase to the gaseous phase, or the heat released when 1 gram of gas changes from the gaseous phase to the liquid phase

laterites highly leached soils of tropical climates; usually red with high iron and aluminum oxide content

latitude the angular distance from the equator to a point on a parallel that tells you how far north or south of the equator the point is located

lava magma, or molten rock, that is forced to the surface from a volcano or a crack in Earth's surface

law of conservation of energy energy is never created or destroyed; it can only be converted from one form to another as the total energy remains constant

law of conservation of mass same as the law of conservation of matter; mass, including single atoms, is neither created nor destroyed in a chemical reaction

law of conservation of matter matter is neither created nor destroyed in a chemical reaction

law of conservation of momentum the total momentum of a group of interacting objects remains constant in the absence of external forces

light ray model a model using lines to show the direction of motion of light to describe the travels of light

light-year the distance that light travels through empty space in one year, approximately 9.5×10^{12} km (5.86×10^{12} mi)

linear scale a scale, generally on a graph, where equal intervals represent equal changes in the value of a variable

lines of force lines drawn to make an electric field strength map, with each line originating on a positive charge and ending on a negative charge; each line represents a path on which a charge would experience a constant force, and lines closer together mean a stronger electric field

line spectrum   narrow lines of color in an otherwise dark spectrum; these lines can be used as "fingerprints" to identify gases

liquids   a phase of matter composed of molecules that have interactions stronger than those found in a gas but not strong enough to keep the molecules near the equilibrium positions of a solid, resulting in the characteristic definite volume but indefinite shape of a liquid

liter   a metric system unit of volume usually used for liquids

lithosphere   the solid layer of Earth's structure that is above the asthenosphere and includes the entire crust, the Moho, and the upper part of the mantle

loess   a very fine dust or silt that has been deposited by the wind over a large area

longitude   the angular distance of a point east or west from the prime meridian on a parallel

longitudinal wave   a mechanical disturbance that causes particles to move closer together and farther apart in the same direction that the wave is traveling

longshore current   a current that moves parallel to the shore, pushed along by waves that move accumulated water from breakers

loudness   a subjective interpretation of a sound that is related to the energy of the vibrating source, to the condition of the transmitting medium, and to the distance involved

low latitudes   latitudes close to the equator; those that sometimes receive vertical solar radiation at noon

luminosity   the total amount of energy radiated into space each second from the surface of a star

luminous   an object or objects that produce visible light; for example, the Sun, stars, lightbulbs, and burning materials are all luminous

lunar eclipse   occurs when the Moon is full and the Sun, Moon, and Earth are lined up so the shadow of Earth falls on the Moon

lunar highlands   light-colored mountainous regions of the Moon

M

macromolecule   a very large molecule, with a molecular weight of thousands or millions of atomic mass units, that is made up of a combination of many smaller, similar molecules

magma   a mass of molten rock material either below or on Earth's crust from which

igneous rock is formed by cooling and hardening

magnetic domain   tiny physical regions in permanent magnets, approximately 0.01 to 1 mm, that have magnetically aligned atoms, giving the domain an overall polarity

magnetic field   the model used to describe how magnetic forces on moving charges act at a distance

magnetic poles   the ends, or sides, of a magnet about which the force of magnetic attraction seems to be concentrated

magnetic quantum number   from the quantum mechanics model of the atom, one of four descriptions of the energy state of an electron wave; this quantum number describes the energy of an electron orbital as the orbital is oriented in space by an external magnetic field, a kind of energy sub-sublevel

magnetic reversal   the flipping of polarity of Earth's magnetic field as the north magnetic pole and the south magnetic pole exchange positions

main sequence stars   normal, mature stars that use their nuclear fuel at a steady rate; stars on the Hertzsprung-Russell diagram in a narrow band that runs from the top left to the lower right

manipulated variable   in an experiment, a quantity that can be controlled or manipulated; also known as the independent variable

mantle   the middle part of Earth's interior; a 2,870 km (about 1,780 mi) thick shell between the core and the crust

maria   smooth, dark areas on the Moon

marine climate   a climate influenced by air masses from over an ocean, with mild winters and cool summers compared to areas farther inland

maritime air mass   a moist air mass that forms over the ocean

mass   a measure of inertia, which means a resistance to a change of motion

mass defect   the difference between the sum of the masses of the individual nucleons forming a nucleus and the actual mass of that nucleus

mass movement   erosion caused by the direct action of gravity

mass number   the sum of the number of protons and neutrons in a nucleus defines the mass number of an atom; used to identify isotopes; for example, uranium-238

matter   anything that occupies space and has mass

matter waves   any moving object has wave properties, but at ordinary velocities, these properties are observed only for objects with a tiny mass; term for the wavelike properties of subatomic particles

meanders   winding, circuitous turns or bends of a stream

mean solar day   is twenty-four hours long and is averaged from the mean solar time

mean solar time   a uniform time averaged from the apparent solar time

measurement   the process of comparing a property of an object to a well-defined and agreed-upon referent

mechanical energy   the form of energy associated with machines, objects in motion, and objects having potential energy that results from gravity

mechanical weathering   the physical breaking up of rocks without any changes in their chemical composition

melting point   the temperature at which a phase change of solid to liquid takes place; the same temperature as the freezing point for a given substance

Mercalli scale   expresses the relative intensity of an earthquake in terms of effects on people and buildings using Roman numerals that range from I to XII

meridians   north-south running arcs that intersect at both poles and are perpendicular to the parallels

mesosphere   the term means "middle layer"— the solid, dense layer of Earth's structure below the asthenosphere but above the core; also the layer of the atmosphere below the thermosphere and above the stratosphere

Mesozoic   one of four geologic eras; the time of middle life, meaning some of the fossils for this time period are similar to the life found on Earth today, but many are different from anything living today

metal   matter having the physical properties of conductivity, malleability, ductility, and luster

metamorphic rocks   previously existing rocks that have been changed into a distinctly different rock by heat, pressure, or hot solutions

meteor   the streak of light and smoke that appears in the sky when a meteoroid is made incandescent by compression of Earth's atmosphere

meteorite   the solid iron or stony material of a meteoroid that survives passage through Earth's atmosphere and reaches the surface

meteoroids   remnants of comets and asteroids in space

meteorology   the science of understanding and predicting weather

meteor shower   an event in which many meteorites fall in a short period of time

meter   the fundamental metric unit of length

metric system   a system of referent units based on invariable referents of nature that have been defined as standards

microclimate a local, small-scale pattern of climate; for example, the north side of a house has a different microclimate than the south side

middle latitudes latitudes equally far from the poles and equator; between the high and low latitudes

mineral a naturally occurring, inorganic solid element or chemical compound with a crystalline structure

miscible fluids fluids that can mix in any proportion

mixture matter made of unlike parts that have a variable composition and can be separated into their component parts by physical means

model a mental or physical representation of something that cannot be observed directly that is usually used as an aid to understanding

moderator a substance in a nuclear reactor that slows fast neutrons so the neutrons can participate in nuclear reactions

Mohorovicic discontinuity the boundary between the crust and mantle that is marked by a sharp increase in the velocity of seismic waves as they pass from the crust to the mantle

molarity a measure of the concentration of a solution; the number of moles of a solute dissolved in one liter of solution

mold the preservation of the shape of an organism by the dissolution of the remains of a buried organism, leaving an empty space where the remains were

mole an amount of a substance that contains Avogadro's number of atoms, ions, molecules, or any other chemical unit; a mole is thus 6.02×10^{23} atoms, ions, or other chemical units

molecular formula a chemical formula that identifies the actual numbers of atoms in a molecule

molecular weight the formula weight of a molecular substance

molecule from the chemical point of view, a particle composed of two or more atoms held together by an attractive force called a chemical bond; from the kinetic theory point of view, the smallest particle of a compound or gaseous element that can exist and still retain the characteristic properties of a substance

momentum the product of the mass of an object times its velocity

monadnocks hills of resistant rock that are found on peneplains

monosaccharides simple sugars that are mostly 6-carbon molecules such as glucose and fructose

moraines deposits of bulldozed rocks and other mounded materials left behind by a melted glacier

mountain a natural elevation of Earth's crust that rises above the surrounding surface

mudflow a mass movement of a slurry of debris and water with the consistency of a thick milkshake

N

natural frequency the frequency of vibration of an elastic object that depends on the size, composition, and shape of the object

neap tide a period of less-pronounced high and low tides: occurs when the Sun and Moon are at right angles to one another

nebula a diffuse mass of interstellar clouds of hydrogen gas or dust

negative electric charge one of the two types of electric charge; repels other negative charges and attracts positive charges

negative ion an atom or particle that has a surplus, or imbalance, of electrons and, thus, a negative charge

net force the resulting force after all vector forces have been added; if a net force is zero, all the forces have canceled each other and there is not an unbalanced force

neutralized acid or base properties have been lost through a chemical reaction

neutron a neutral subatomic particle usually found in the nucleus of an atom

neutron star very small superdense remains of a supernova with a center core of pure neutrons

new crust zone the zone of a divergent boundary where new crust is formed by magma upwelling at the boundary

new moon the moon phase when the Moon is between Earth and the Sun and the entire side of the Moon facing Earth is dark

newton a unit of force defined as kg·m/s²; that is, a 1 newton force is needed to accelerate a 1 kg mass 1 m/s²

noble gas members of family VIII of the periodic table, having common properties of colorless, odorless, chemically inert gases; also known as rare gases or inert gases

node regions on a standing wave that do not oscillate

noise sounds made up of groups of waves of random frequency and intensity

nonelectrolytes water solutions that do not conduct an electric current; covalent compounds that form molecular solutions and cannot conduct an electric current

nonferromagnesian silicates silicates that do not contain iron or magnesium ions; examples include the minerals of muscovite (white mica), the feldspars, and quartz

nonmetal an element that is brittle (when a solid), does not have a metallic luster, is a poor conductor of heat and electricity, and is not malleable or ductile

nonsilicates minerals that do not have the silicon-oxygen tetrahedra in their crystal structure

noon the event of time when the Sun moves across the celestial meridian

normal a line perpendicular to the surface of a boundary

normal fault a fault where the hanging wall has moved downward with respect to the foot wall

north celestial pole a point directly above the North Pole of Earth; the point above the north pole on the celestial sphere

north pole the north pole of a magnet or lodestone is "north seeking," meaning that the pole of a magnet points northward when the magnet is free to turn

nova a star that explodes or suddenly erupts and increases in brightness

nuclear energy the form of energy from reactions involving the nucleus, the innermost part of an atom

nuclear fission the nuclear reaction of splitting a massive nucleus into more stable, less-massive nuclei with an accompanying release of energy

nuclear force one of four fundamental forces, a strong force of attraction that operates over very short distances between subatomic particles; this force overcomes the electric repulsion of protons in a nucleus and binds the nucleus together

nuclear fusion the nuclear reaction of low-mass nuclei fusing together to form more stable and more massive nuclei with an accompanying release of energy

nuclear reactor a steel vessel in which a controlled chain reaction of fissionable materials releases energy

nucleons the name used to refer to both the protons and neutrons in the nucleus of an atom

nucleus the tiny, relatively massive and positively charged center of an atom containing protons and neutrons; the small, dense center of an atom

numerical constant a constant without units; a number

O

oblate spheroid the shape of Earth—a somewhat squashed spherical shape

observed lapse rate the rate of change in temperature compared to change in altitude

occluded front a front that has been lifted completely off the ground into the atmosphere, forming a cyclonic storm

ocean the single, continuous body of salt water on the surface of Earth

ocean basin the deep bottom of the ocean floor, which starts beyond the continental slope

ocean currents streams of water within the ocean that stay in about the same path as they move over large distances; steady and continuous onward movement of a channel of water in the ocean

ocean wave a moving disturbance that travels across the surface of the ocean

oceanic ridges long, high, continuous, sub-oceanic mountain chains; for example, the Mid-Atlantic Ridge in the center of the Atlantic Ocean Basin

oceanic trenches long, narrow, deep troughs with steep sides that run parallel to the edges of continents

octet rule a generalization that helps keep track of the valence electrons in most representative elements; atoms of the representative elements (A families) attempt to acquire an outer orbital with eight electrons through chemical reactions

ohm the unit of resistance; equivalent to volts/amps

Ohm's law the electric potential difference is directly proportional to the product of the current times the resistance

oil field petroleum accumulated and trapped in extensive porous rock structure or structures

oils organic compounds of esters formed from glycerol and three long-chain carboxylic acids that are also called triglycerides; called fats in animals and oils in plants

Oort cloud a spherical "cloud" of small, icy bodies from 30,000 AU out to a light-year from the Sun; the source of long-period comets

opaque materials that do not allow the transmission of any light

orbital the region of space around the nucleus of an atom where an electron is likely to be found

organic acids acids derived from organisms; organic compounds with a general formula of RCOOH, where R is one of the hydrocarbon groups; for example, methyl or ethyl

organic chemistry the study of compounds in which carbon is the principal element

orientation of Earth's axis the direction that Earth's axis points; considered to be the same throughout the year

origin the only point on a graph where both the x and y variables have a value of zero at the same time

overtones higher resonant frequencies that occur at the same time as the fundamental frequency, giving a musical instrument its characteristic sound quality

oxbow lake a small body of water, or lake, that formed when two bends of a stream came together and cut off a meander

oxidation the process of a substance losing electrons during a chemical reaction; a reaction between oxygen and the minerals making up rocks

oxidation-reduction reaction a chemical reaction in which electrons are transferred from one atom to another; sometimes called "redox" for short

oxidizing agents substances that take electrons from other substances

P

Paleozoic one of four geologic eras; time of ancient life, meaning the fossils from this time period are very different from anything living on Earth today

parallel circuit for batteries, all positive terminals are connected and all negative terminals are connected; lightbulbs in a parallel circuit have alternate branches for the current to follow

parallels reference lines on Earth used to identify where in the world you are northward or southward from the equator; east and west running circles that are parallel to the equator on a globe with the distance from the equator called the latitude

parts per billion a concentration ratio of parts of solute in every 1 billion parts of solution (ppb); could be expressed as ppb by volume or as ppb by weight

parts per million a concentration ratio of parts of solute in every 1 million parts of solution (ppm); could be expressed as ppm by volume or as ppm by weight

Pauli exclusion principle no two electrons in an atom can have the same four quantum numbers; thus, a maximum of two electrons can occupy a given orbital

peneplain a nearly flat landform that is the end result of the weathering and erosion of the land surface

penumbra the zone of partial darkness in a shadow

percent by volume the volume of solute in 100 volumes of solution

percent by weight the weight of solute in 100 weight units of solution

perigee when the Moon's elliptical orbit brings the Moon closest to Earth

period (geologic time) subdivisions of geologic eras

period (periodic table) horizontal rows of elements with increasing atomic numbers; runs from left to right on the element table

period (wave) the time required for one complete cycle of a wave

periodic law similar physical and chemical properties recur periodically when the elements are listed in order of increasing atomic number

permeability the ability to transmit fluids through openings, small passageways, or gaps

permineralization the process that forms a fossil by alteration of an organism's buried remains by circulating groundwater depositing calcium carbonate, silica, or pyrite

petroleum oil that comes from oil-bearing rock, a mixture of hydrocarbons that is believed to have formed from ancient accumulations of buried organic materials such as remains of algae

Phanerozoic the eon of an abundant fossil record and living organisms

phase change the action of a substance changing from one state of matter to another; a phase change always absorbs or releases internal potential energy that is not associated with a temperature change

phases of matter the different physical forms that matter can take as a result of different molecular arrangements, resulting in characteristics of the common phases of a solid, liquid, or gas

photoelectric effect the movement of electrons in some materials as a result of energy acquired from absorbed light

photon a quanta of energy in a light wave; the particle associated with light

pH scale the scale that measures the acidity of a solution with numbers below 7 representing acids, 7 representing neutral, and numbers above 7 representing bases

physical change a change of the state of a substance but not the identity of the substance

pitch the frequency of a sound wave

Planck's constant the proportionality constant in the relationship between the energy of vibrating molecules and their frequency of vibration; a value of 6.63×10^{-34} Js

plane of the ecliptic the plane of Earth's orbit

planet an object that is orbiting the Sun, is nearly spherical, and is large enough to clear all matter from its orbital zone

plasma a phase of matter; a very hot gas consisting of electrons and atoms that have been stripped of their electrons because of high kinetic energies

plastic strain an adjustment to stress in which materials become molded or bent out of shape under stress and do not return to their original shape after the stress is released

plate tectonics the theory that Earth's crust is made of rigid plates that float on the upper mantle

plunging folds synclines and anticlines that are not parallel to the surface of Earth

polar air mass a cold air mass that forms in cold regions

polar climate zone the climate zone of the high latitudes; average monthly temperatures stay below 10°C (50°F), even during the warmest month of the year

polar covalent bond a covalent bond in which there is an unequal sharing of bonding electrons

polarized light whose constituent transverse waves are all vibrating in the same plane; also known as plane-polarized light

polar molecule a molecule with a dipole, with a negative side and a positive side

Polaroid a film that transmits only polarized light

polyatomic ion an ion made up of many atoms

polymers huge, chainlike molecules made of hundreds or thousands of smaller repeating molecular units called monomers

polysaccharides polymers consisting of monosaccharide units joined together in straight or branched chains; starches, glycogen, or cellulose

pond a small body of standing water, smaller than a lake

porosity the ratio of pore space to the total volume of a rock or soil sample, expressed as a percentage; freely admitting the passage of fluids through pores or small spaces between parts of the rock or soil

positive electric charge one of the two types of electric charge; repels other positive charges and attracts negative charges

positive ion an atom or particle that has a net positive charge due to an electron or electrons being torn away

potential energy energy due to position; energy associated with changes in position (e.g., gravitational potential energy) or changes in shape (e.g., compressed or stretched spring)

power the rate at which energy is transferred or the rate at which work is performed; defined as work per unit of time

Prearchean the earliest of the geologic eons before life

Precambrian one of four geologic eras; the time before the time of ancient life, meaning the rocks for this time period contain very few fossils

precession the slow wobble of the axis of Earth similar to the wobble of a spinning top

precipitation water that falls to the surface of Earth in the solid or liquid form

pressure defined as force per unit area; for example, pounds per square inch (lb/in^2)

primary coil part of a transformer; a coil of wire that is connected to a source of alternating current

primary loop part of the energy-converting system of a nuclear power plant; the closed pipe system that carries heated water from the nuclear reactor to a steam generator

prime meridian the referent meridian (0°) that passes through the Greenwich Observatory in England

principal quantum number from the quantum mechanics model of the atom, one of four descriptions of the energy state of an electron wave; this quantum number describes the main energy level of an electron in terms of its most probable distance from the nucleus

principle of crosscutting relationships a frame of reference based on the understanding that any geologic feature that cuts across or is intruded into a rock mass must be younger than the rock mass

principle of faunal succession a frame of reference based on the understanding that life forms have changed through time as old life forms disappear from the fossil record and new ones appear, but the same form is never exactly duplicated independently at two different times in history

principle of original horizontality a frame of reference based on the understanding that on a large scale, sediments are deposited in flat-lying layers, so any layers of sedimentary rocks that are not horizontal have been subjected to forces that have deformed Earth's surface

principle of superposition a frame of reference based on the understanding that an undisturbed sequence of horizontal rock layers is arranged in chronological order with the oldest layers at the bottom, and each consecutive layer will be younger than the one below it

principle of uniformity a frame of reference of slow, uniform changes in the earth's history; the processes changing rocks today are the processes that changed them in the past, or "the present is the key to the past"

proof a measure of ethanol concentration of an alcoholic beverage; proof is double the concentration by volume; for example, 50 percent by volume is 100 proof

properties qualities or attributes that, taken together, are usually unique to an object; for example, color, texture, and size

proportionality constant a constant applied to a proportionality statement that transforms the statement into an equation

proteins macromolecular polymers made of smaller molecules of amino acids, with molecular weight from about 6,000 to 50 million; proteins are amino acid polymers with roles in biological structures or functions; without such a function, they are known as polypeptides

Proterozoic the geologic eon before the Phanerozoic, meaning "beginning life"

protogalaxy a collection of gas, dust, and young stars in the process of forming a galaxy

proton the subatomic particle that has the smallest possible positive charge, usually found in the nucleus of an atom

protoplanet nebular model a model of the formation of the solar system that states that the planets formed from gas and dust left over from the formation of the Sun

protostar an accumulation of gases that will become a star

proxy data a natural source used to infer temperature change, rainfall, or some other climate condition of the past

psychrometer a two-thermometer device used to measure the relative humidity

Ptolemaic system geocentric model of the structure of the solar system that uses epicycles to explain retrograde motion

pulsars the source of regular, equally spaced pulsating radio signals believed to be the result of the magnetic field of a rotating neutron star

pure substance materials that are the same throughout and have a fixed definite composition

pure tone a sound made by very regular intensities and very regular frequencies from regular repeating vibrations

P-wave a pressure, or compressional, wave in which a disturbance vibrates materials back and forth in the same direction as the direction of wave movement

P-wave shadow zone a region on Earth between 103° and 142° of arc from an earthquake where no P-waves are received; believed to be explained by P-waves being refracted by the core

Q

quad 1 quadrillion Btu (10^{15} Btu); used to describe very large amounts of energy

quanta fixed amounts; usually referring to fixed amounts of energy absorbed or emitted by matter (*quanta* is plural, and *quantum* is singular)

quantities measured properties; includes the numerical value of the measurement and the unit used in the measurement

quantum mechanics a model of the atom based on the wave nature of subatomic particles, the mechanics of electron waves; also called wave mechanics

quantum numbers numbers that describe energy states of an electron; in the Bohr model of the atom, the orbit quantum numbers could be any whole number 1, 2, 3, and so on out from the nucleus; in the quantum mechanics model of the atom, four quantum numbers are used to describe the energy state of an electron wave

R

rad a measure of radiation received by a material (radiation absorbed dose)

radiant energy the form of energy that can travel through space; for example, visible light and other parts of the electromagnetic spectrum

radiation the transfer of heat from a region of higher temperature to a region of lower temperature by greater emission of radiant energy from the region of higher temperature

radiation zone part of the interior of a star according to a model; the region directly above the core where gamma and X rays from the core are absorbed and reemitted, with the radiation slowly working its way outward

radioactive decay the natural spontaneous disintegration or decomposition of a nucleus

radioactive decay constant a specific constant for a particular isotope that is the ratio of the rate of nuclear disintegration per unit of time to the total number of radioactive nuclei

radioactive decay series a series of decay reactions that begins with one radioactive nucleus that decays to a second nucleus that decays to a third nucleus and so on until a stable nucleus is reached

radioactivity the spontaneous emission of particles or energy from an atomic nucleus as it disintegrates

radiometric age the age of rocks determined by measuring the radioactive decay of unstable elements within the crystals of certain minerals in the rocks

rarefaction a thinning or pulse of decreased density and pressure of gas molecules

ratio a relationship between two numbers, one divided by the other; the ratio of distance per time is speed

real image an image generated by a lens or mirror that can be projected onto a screen

red giant stars one of two groups of stars on the Hertzsprung-Russell diagram that have a different set of properties than the main sequence stars; bright, low temperature giant stars that are enormously bright for their temperature

redox reaction the short name for oxidation-reduction reaction

reducing agent supplies electrons to the substance being reduced in a chemical reaction

referent referring to or thinking of a property in terms of another, more familiar object

reflected ray a line representing direction of motion of light reflected from a boundary

reflection the change when light, sound, or other waves bounce backward off a boundary

refraction a change in the direction of travel of light, sound, or other waves crossing a boundary

rejuvenation the process of uplifting land that renews the effectiveness of weathering and erosion processes

relative dating dating the age of a rock unit or geologic event relative to some other unit or event

relative humidity the ratio (times 100%) of how much water vapor is in the air to the maximum amount of water vapor that could be in the air at a given temperature

rem the measure of radiation that considers the biological effects of different kinds of ionizing radiation

replacement chemical reaction a reaction in which an atom or polyatomic ion is replaced in a compound by a different atom or polyatomic ion

replacement (fossil formation) a process in which an organism's buried remains are altered by circulating groundwaters carrying elements in solution; the removal of original materials by dissolutions and the replacement of new materials an atom or molecule at a time

representative elements the name given to the members of the A-group families of the periodic table; also called the main-group elements

reservoir a natural or artificial pond or lake used to store water, control floods, or generate electricity; a body of water stored for public use

resonance when the frequency of an external force matches the natural frequency and standing waves are set up

responding variable the variable that responds to changes in the manipulated variable; also known as the dependent variable because its value depends on the value of the manipulated variable

reverberation apparent increase in volume caused by reflections, usually arriving within 0.1 second after the original sound

reverse fault a fault where the hanging wall has moved upward with respect to the foot wall

revolution the motion of a planet as it orbits the Sun

Richter scale expresses the intensity of an earthquake in terms of a scale with each higher number indicating ten times more ground movement and about thirty times more energy released than the preceding number

ridges long, rugged mountain chains rising thousands of meters above the abyssal plains of the ocean basin

rift a split or fracture in a rock formation, in a land formation, or in the crust of Earth

rip current a strong, brief current that runs against the surf and out to sea

rock a solid aggregation of minerals or mineral materials that have been brought together into a cohesive solid

rock cycle the understanding of igneous, sedimentary, or metamorphic rock as a temporary state in an ongoing transformation of rocks into new types; the process of rocks continually changing from one type to another

rock flour rock pulverized by a glacier into powdery, silt-sized sediment

rockfall the rapid tumbling, bouncing, or free fall of rock fragments from a cliff or steep slope

rockslide a sudden, rapid movement of a coherent unit of rock along a clearly defined surface or plane

rotation the spinning of a planet on its axis

runoff water moving across the surface of Earth as opposed to soaking into the ground

S

salinity a measure of dissolved salts in seawater, defined as the mass of salts dissolved in 1,000 g of solution

salt any ionic compound except one with hydroxide or oxide ions

San Andreas fault in California, the boundary between the North American Plate and the Pacific Plate that runs north-south for some 1,300 km (800 mi) with the Pacific Plate moving northwest and the North American Plate moving southeast

saturated molecule an organic molecule that has the maximum number of hydrogen atoms possible

saturated solution the apparent limit to dissolving a given solid in a specified amount of water at a given temperature; a state of equilibrium that exists between dissolving solute and solute coming out of solution

saturation (of water vapor) an equilibrium condition that occurs when evaporation and condensation are in balance

scientific law a relationship between quantities, usually described by an equation in the physical sciences; is more important and describes a wider range of phenomena than a scientific principle

scientific principle a relationship between quantities concerned with a specific, or narrow range of observations and behavior

scintillation counter a device that indirectly measures ionizing radiation (alpha, beta, gamma, etc.) by measuring the flashes of light produced when the radiation strikes a phosphor

sea a smaller part of the ocean with characteristics that distinguish it from the larger ocean

sea breeze cool, dense air from over water moving over land as part of convective circulation

seafloor spreading the process by which hot, molten rock moves up from the interior of

Earth to emerge along mid-oceanic rifts, flowing out in both directions to create new rocks

seamounts steep, submerged volcanic peaks on the abyssal plain

second the standard unit of time in both the metric and English systems of measurement

secondary coil part of a transformer, a coil of wire in which the voltage of the original alternating current in the primary coil is stepped up or down by way of electromagnetic induction

secondary loop a part of nuclear power plant; the closed-pipe system that carries steam from a steam generator to the turbines, then back to the steam generator as feedwater

second law of motion the acceleration of an object is directly proportional to the net force acting on that object and inversely proportional to the mass of the object

second law of thermodynamics a statement that the natural process proceeds from a state of higher order to a state of greater disorder

sedimentary rocks rocks formed from particles or dissolved minerals from previously existing rocks

sediments accumulations of silt, sand, or gravel that settled out of the atmosphere or out of water

seismic waves vibrations that move as waves through any part of the Earth, usually associated with earthquakes, volcanoes, or large explosions

seismograph an instrument that measures and records seismic wave data

semiarid the climate classification between arid and humid; receives between 25 and 50 cm (10 and 20 in) precipitation per year

semiconductors elements that have properties between those of a metal and those of a nonmetal, sometimes conducting an electric current and sometimes acting like an electrical insulator depending on the conditions and their purity; also called metalloids

series circuit for batteries, the negative terminal of one cell is connected to the positive terminal of another cell; light bulbs in a series circuit are connected one after the other

shallow-focus earthquakes earthquakes that occur from the surface down to 70 km (43 mi) deep

shear stress produced when two plates slide past one another or one plate slides past another plate that is not moving

shell model of the nucleus a model of the nucleus that has protons and neutrons moving in energy levels or shells in the nucleus (similar to the shell structure of electrons in an atom)

shield volcano a broad, gently sloping volcanic cone constructed of solidified lava flows

shock wave a large, intense wave disturbance of very high pressure; the pressure wave created by an explosion, for example

short circuit a frayed or broken wire that provides a new path of lesser resistance that could allow a dangerous current

sidereal day the interval between two consecutive crossings of the celestial meridian by a particular star

sidereal month the time interval between two consecutive crossings of the Moon across any star

sidereal year the time interval required for Earth to move around its orbit so that the Sun is again in the same position against the stars

silicates minerals that contain silicon-oxygen tetrahedra either isolated or joined together in a crystal structure

sill a tabular-shaped intrusive rock that formed when magma moved into the plane of contact between sedimentary rock layers

simple harmonic motion the vibratory motion that occurs when there is a restoring force opposite to and proportional to a displacement

single bond covalent bond in which a single pair of electrons is shared by two atoms

slope the ratio of changes in the y variable to changes in the x variable or how fast the y-value increases as the x-value increases

small solar system bodies all objects other than planets or dwarf planets that are orbiting the Sun

soil a mixture of unconsolidated weathered earth materials and humus, which is altered, decay-resistant organic matter

solar constant the averaged solar power received by the outermost part of Earth's atmosphere when the sunlight is perpendicular to the outer edge and Earth is at an average distance from the Sun; about 1,370 watts per square meter

solenoid a cylindrical coil of wire that becomes electromagnetic when a current runs through it

solids a phase of matter with molecules that remain close to fixed equilibrium positions due to strong interactions between the molecules, resulting in the characteristic definite shape and definite volume of a solid

solstice the time when the Sun is at its maximum or minimum altitude in the sky, known as the summer solstice or the winter solstice, respectively

solubility the dissolving ability of a given solute in a specified amount of solvent, the concentration that is reached as a saturated solution is achieved at a particular temperature

solute the component of a solution that dissolves in the other component; the solvent

solution a homogeneous mixture of ions or molecules of two or more substances

solvent the component of a solution present in the larger amount; the solute dissolves in the solvent to make a solution

sonic boom sound waves that pile up into a shock wave when a source is traveling at or faster than the speed of sound

sound quality a characteristic of the sound produced by a musical instrument; determined by the presence and relative strengths of the overtones produced by the instrument

south celestial pole a point directly above the South Pole of Earth; the point above the south pole on the celestial sphere

south pole short for "south seeking"; the pole of a magnet that points southward when it is free to turn

specific heat each substance has its own specific heat, which is defined as the amount of energy (or heat) needed to increase the temperature of 1 gram of a substance 1 degree Celsius

speed a measure of how fast an object is moving—the rate of change of position per change in time; speed has magnitude only and does not include the direction of change

spin quantum number from the quantum mechanics model of the atom, one of four descriptions of the energy state of an electron wave; this quantum number describes the spin orientation of an electron relative to an external magnetic field

spring equinox one of two times a year that daylight and night are of equal length; occurs on or about March 21 and identifies the beginning of the spring season

spring tides unusually high and low tides that occur every two weeks because of the relative positions of Earth, Moon, and Sun

standard atmospheric pressure the average atmospheric pressure at sea level, which is also known as normal pressure; the standard pressure is 29.92 in or 760.0 mm of mercury (1,013.25 millibar)

standard time zones 15° wide zones defined to have the same time throughout the zone, defined as the mean solar time at the middle of each zone

standard unit a measurement unit established as the standard upon which the value of the other referent units of the same type are based

standing waves a condition where two waves of equal frequency traveling in opposite directions meet and form stationary regions of maximum displacement due to constructive interference and stationary regions of zero displacement due to destructive interference

starches complex carbohydrates (polysaccharides) that plants use as a stored food source and that serves as an important source of food for animals

stationary front occurs when the edge of a front is not advancing

steam generator a part of nuclear power plant; the heat exchanger that heats feedwater from the secondary loop to steam with the very hot water from the primary loop

step-down transformer a transformer that decreases the voltage of a current

step-up transformer a transformer that increases the voltage of a current

stony-iron meteorites meteorites composed of silicate minerals and metallic iron

stony meteorites meteorites composed mostly of silicate minerals that usually make up rocks on Earth

storm a rapid and violent weather change with strong winds, heavy rain, snow, or hail

strain adjustment to stress; a rock unit might respond to stress by changes in volume, changes in shape, or breaking

stratopause the upper boundary of the stratosphere

stratosphere the layer of the atmosphere above the troposphere where temperature increases with height

stream a large or small body of running water

stress a force that tends to compress, pull apart, or deform rock; stress on rocks in the earth's solid outer crust results as Earth's plates move into, away from, or alongside each other

strong acid an acid that ionizes completely in water, with all molecules dissociating into ions

strong base a base that is completely ionic in solution and has hydroxide ions

subduction zone the region of a convergent boundary where the crust of one plate is forced under the crust of another plate into the interior of Earth

sublimation the phase change of a solid directly into a vapor or gas

submarine canyons a feature of the ocean basin; deep, steep-sided canyons that cut through the continental slopes

summer solstice in the Northern Hemisphere, the time when the Sun reaches its maximum altitude in the sky, which occurs on or about June 22 and identifies the beginning of the summer season

superconductors some materials in which, under certain conditions, the electrical resistance approaches zero

supercooled water in the liquid phase when the temperature is below the freezing point

supernova a rare catastrophic explosion of a star into an extremely bright but short-lived phenomenon

supersaturated containing more than the normal saturation amount of a solute at a given temperature

surf the zone where breakers occur; the water zone between the shoreline and the outermost boundary of the breakers

surface wave a seismic wave that moves across Earth's surface, spreading across the surface as water waves spread on the surface of a pond from a disturbance

S-wave a sideways, or shear, wave in which a disturbance vibrates materials from side to side, perpendicular to the direction of wave movement

S-wave shadow zone a region of Earth more than 103° of arc away from the epicenter of an earthquake where S-waves are not recorded; believed to be the result of the core of Earth being a liquid, or at least acting like a liquid

swell regular groups of low-profile, long-wavelength waves that move continuously

syncline a trough-shaped fold in layered bedrock

synodic month the interval of time from new moon to new moon (or any two consecutive identical phases)

T

talus steep, conical or apronlike accumulations of rock fragments at the base of a slope

temperate climate zone climate zone of the middle latitudes; average monthly temperatures stay between 10°C and 18°C (50°F and 64°F) throughout the year

temperature how hot or how cold something is; a measure of the average kinetic energy of the molecules making up a substance

tensional stress the opposite of compressional stress; occurs when one part of a plate moves away from another part that does not move

terrestrial planets the planets Mercury, Venus, Earth, and Mars that have similar densities and compositions as compared to the outer giant planets

theory a broad, detailed explanation that guides the development of hypotheses and interpretations of experiments in a field of study

thermometer a device used to measure the hotness or coldness of a substance

thermosphere the thin, high, outer atmospheric layer of Earth where the molecules are far apart and have a high kinetic energy

third law of motion whenever two objects interact, the force exerted on one object is equal in size and opposite in direction to the force exerted on the other object; forces always occur in matched pairs that are equal and opposite

thrust fault a reverse fault with a low-angle fault plane

thunderstorm a brief, intense electrical storm with rain, lightning, thunder, strong winds, and sometimes hail

tidal bore a strong tidal current, sometimes resembling a wave, produced in very long, very narrow bays as the tide rises

tidal currents a steady and continuous onward movement of water produced in narrow bays by the tides

tides the periodic rise and fall of the level of the sea from the gravitational attraction of the Moon and Sun

tornado a long, narrow, funnel-shaped column of violently whirling air from a thundercloud that moves destructively over a narrow path when it touches the ground

total internal reflection the condition where all light is reflected back from a boundary between materials; occurs when light arrives at a boundary at the critical angle or beyond

total solar eclipse an eclipse that occurs when Earth, the Moon, and the Sun are lined up so the new moon completely covers the disk of the Sun; the umbra of the Moon's shadow falls on the surface of Earth

transform boundaries in plate tectonics, boundaries that occur between two plates sliding horizontally by each other along a long, vertical fault; sudden jerks along the boundary result in the vibrations of earthquakes

transformer a device consisting of a primary coil of wire connected to a source of alternating current and a secondary coil of wire in which electromagnetic induction increases or decreases the voltage of the source

transition elements members of the B-group families of the periodic table

transparent a term describing materials that allow the transmission of light; for example, glass and clear water are transparent materials

transportation the movement of eroded materials by agents such as rivers, glaciers, wind, or waves

transverse wave a mechanical disturbance that causes particles to move perpendicular to the direction that the wave is traveling

trenches a long, relatively narrow, steep-sided trough that occurs along the edges of the ocean basins

triglyceride an organic compound of esters formed from glycerol and three long-chain carboxylic acids; also called fats in animals and oil in plants

triple bond a covalent bond formed when three pairs of electrons are shared by two atoms

tropical air mass a warm air mass from warm regions

tropical climate zone the climate zone of the low latitudes; average monthly temperatures stay above 18°C (64°F), even during the coldest month of the year

tropical cyclone a large, violent circular storm that is born over the warm, tropical ocean near the equator; also called hurricane (Atlantic and eastern Pacific) and typhoon (in western Pacific)

tropical year the time interval between two consecutive spring equinoxes; used as standard for the common calendar year

tropic of Cancer parallel identifying the northern limit where the Sun appears directly overhead; located at 23.5°N latitude

tropic of Capricorn parallel identifying the southern limit where the Sun appears directly overhead; located at 23.5°S latitude

tropopause the upper boundary of the troposphere, identified by the altitude where the temperature stops decreasing and remains constant with increasing altitude

troposphere the layer of the atmosphere from the surface to where the temperature stops decreasing with height

trough the low mound of water that is part of a wave; also refers to the rarefaction, or low-pressure part, of a sound wave

tsunami a very large, fast, and destructive ocean wave created by an undersea earthquake, landslide, or volcanic explosion; a seismic sea wave

turbidity current a muddy current produced by underwater landslides

typhoon the name for hurricanes in the western Pacific

U

ultrasonic sound waves too high in frequency to be heard by the human ear; frequencies above 20,000 Hz

umbra the inner core of a complete shadow

unconformity a time break in the rock record

undertow a current beneath the surface of the water produced by the return of water from the shore to the sea

unit in measurement, a well-defined and agreed-upon referent

universal law of gravitation every object in the universe is attracted to every other object with a force directly proportional to the product of their masses and inversely proportional to the square of the distance between the centers of the two masses

unpolarized light light consisting of transverse waves vibrating in all conceivable random directions

unsaturated molecule an organic molecule that does not contain the maximum number of hydrogen atoms; a molecule that can add more hydrogen atoms because of the presence of double or triple bonds

V

valence the number of covalent bonds an atom can form

valence electrons electrons of the outermost orbital; the electrons that determine the chemical properties of an atom and the electrons that participate in chemical bonding

Van Allen belts belts of radiation caused by cosmic-ray particles becoming trapped and following Earth's magnetic field lines between the poles

vapor the gaseous state of a substance that is normally in the liquid state

variable a changing quantity usually represented by a letter or symbol

velocity describes both the speed and direction of a moving object; a change in velocity is a change in speed, in direction of travel, or both

ventifacts rocks sculpted by wind abrasion

vernal equinox another name for the spring equinox, which occurs on or about March 21 and marks the beginning of the spring season

vibration a back-and-forth motion that repeats itself

virtual image an image where light rays appear to originate from a mirror or lens; this image cannot be projected on a screen

volcanism volcanic activity; the movement of magma

volcano a hill or mountain formed by the extrusion of lava or rock fragments from a mass of magma below

volt a unit of potential difference equivalent to J/C

voltage drop the electric potential difference across a resistor or other part of a circuit that consumes power

voltage source a source of electric power in an electric circuit that maintains a constant voltage supply to the circuit

volume how much space something occupies

vulcanism volcanic activity; the movement of magma

W

warm front the front that forms when a warm air mass advances against a cool air mass

watershed the region or land area drained by a stream; a stream drainage basin

water table the boundary below which the ground is saturated with water

watt the metric unit for power; equivalent to J/s

wave a disturbance or oscillation that moves through a medium

wave equation the relationship of the velocity of a wave to the product of the wavelength and frequency of the wave

wave front a region of maximum displacement in a wave; a condensation in a sound wave

wave height the vertical distance of an ocean wave between the top of the wave crest and the bottom of the next trough

wavelength the horizontal distance between successive wave crests or other successive parts of the wave

wave mechanics an alternate name for quantum mechanics derived from the wavelike properties of subatomic particles

wave period the time required for two successive crests or other successive parts of the wave to pass a given point

weak acid an acid only partially ionized because of an equilibrium reaction with water

weak base a base only partially ionized because of an equilibrium reaction with water

weathering slow changes that result in the breaking up, crumbling, and destruction of any kind of solid rock

white dwarf stars one of two groups of stars on the Hertzsprung-Russell diagram that have a different set of properties than the main sequence stars; faint, white-hot stars that are very small and dense

wind a horizontal movement of air that moves along or parallel to the ground, sometimes in currents or streams

wind abrasion the natural sand-blasting process that occurs when wind particles break off small particles of rock and polish the rock they strike

windchill factor the cooling equivalent temperature that results from the wind making the air temperature seem much lower; the cooling power of wind

winter solstice in the Northern Hemisphere, the time when the Sun reaches its minimum altitude, which occurs on or about December 22 and identifies the beginning of the winter season

work the magnitude of applied force times the distance through which the force acts; can be thought of as the process by which one form of energy is transformed to another

Z

zone of saturation the zone of sediments beneath Earth's surface in which water has collected in all available spaces

CREDITS

Photographs

Chapter 1 Opener: © Werner H. Muller/Peter Arnold, Inc.; **1.1:** © Spencer Grant/Photo Researchers, Inc; **1.2:** © Rafael Macia/Photo Researchers, Inc; **1.3:** © Bill W. Tillery; **1.12:** © Chad Slattery; **1.15a:** © David Frazier/Photo Researchers, Inc; **People Behind the Science, p. 20:** © USGS Photo Library;

Chapter 2 Opener: © Douglas Faulkner/Photo Researchers; **2.1:** © Darrell Wong/Tony Stone/Getty; **2.7:** © Bill W. Tillery; **Box Fig 2.1:** © Eunice Harris/Photo Researchers; **2.14:** © Richard Megna/Fundamental Photographs; **2.18:** © Royalty-Free/Corbis; **2.23:** © Keith Jennings, Arizona State University Media Relations Office; **2.30:** © StockTrek/Getty Images; **2.32:** © NASA/Getty Images;

Chapter 3 Opener: © Hank Morgan/Science Source/Photo Researchers; **3.1, 3.10, 3.11a:** © Bill W. Tillery; **3.11b:** © Alan Oddie/PhotoEdit, Inc; **3.12, 3.14:** Arizona Public Service Company; **People Behind the Science, p. 76:** © Archive Photos/Getty; **3.20:** © Bill W. Tillery;

Chapter 4 Opener: © Lawrence Livermore/National Library/SPL/Photo Researchers; **4.1:** Photo courtesy of International Steel Group Inc; **4.2:** Courtesy of John C. Wheatly, Center for High Resolution Electron Microscopy; **4.6:** Honeywell, Inc; **4.14:** Manville Company; **4.22:** © Bill W. Tillery; **People Behind the Science, p. 107:** North Wind Picture Archives;

Chapter 5 Opener: © Clyde H. Smith/Peter Arnold, Inc.; **5.1:** © Amy Etra/PhotoEdit, Inc.; **5.5:** © Getty Images/Nature, Wildlife, Environment 2/Vol. 44.; **5.14:** © The McGraw-Hill Company, Inc. Bob Coyle, Photographer; **5.25, 5.26:** © NASA;

Chapter 6 Opener: © Keith Kent/Peter Arnold, Inc.; **6.1:** © Bill W. Tillery; **6.13:** Arizona Public Service Company; **6.17 (all):** © Bill W. Tillery; **People Behind the Science, p. 153:** © The Corcoran Gallery of Art/Corbis; **6.18:** © Steve Cole/PhotoDisc/Getty; **6.35a, 6.35b, Box Fig 6.3a, Box Fig 6.3b:** © Bill W. Tillery; **6.40:** © Brian Moeskau/Brian Moeskau Photography;

Chapter 7 Opener: © James L. Amos/Peter Arnold, Inc.; **7.1:** © John Kieffer/Peter Arnold, Inc.; **7.8:** © Bill W. Tillery; **7.9a:** 1987 Libbey-Owens-Ford Co.; **7.9b:** 1987 Libbey-Owens-Ford Co.; **7.13:** © Sergio Piumatti; **7.18:** © Bill W. Tillery; **7.19b:** © Jones & Childers; **7.25:** © Brock May/Photo Researchers, Inc; **People Behind the Science, p. 197:** © Baldwin H. Ward & Kathryn C. Ward/Corbis;

Chapter 8 Opener: © Astrid & Hanns-Frieder Michler/SPL/Photo Researchers, Inc.; **8.1:** © Ken Eward/Science Source/Photo Researchers, Inc.; **8.12:** © Weiss/Jerrican/Photo Researchers; **People Behind the Science, p. 220:** © Scala / Art Resource, NY;

Chapter 9 Opener: © David Weintraub/Photo Researchers, Inc.; **9.1:** © Bill W. Tillery; **9.3:** © Richard Megna/Fundamental Photographs; **9.7:** © Charles M. Falco/Photo Researchers, Inc.; **9.9:** © Arthur S. Aubry/Photodisc/Getty; **9.12:** © Bill W. Tillery; **People Behind the Science, p. 245:** © Roger Ressmeyer/Corbis;

Chapter 10 Opener: © Fundamental Photographs; **10.1:** © Visuals Unlimited; **10.3 (all):** © Bill W. Tillery; **10.4:** © Paul Sisul/Tony Stone/Getty; **10.7:** © Donovan Reese/Tony Stone/Getty; **10.8, 10.9:** © Bill W. Tillery; **10.10:** © Paul Silverman/Fundamental Photographs; **10.11, 10.13:** © Bill W. Tillery; **Box fig 10.1a:** © E. R. Degginger/Color Pic, Inc.; **People Behind the Science, p. 268:** © Bettmann/Corbis;

Chapter 11 Opener: © Blair Seitz/Photo Researchers; **11.1:** © Craig Tuttle/Corbis; **11.14a, 11.14b, 11.15, 11.16, Box fig 11.1:** © Bill W. Tillery;

Chapter 12 Opener: © Kenneth Eward/Bio Grafix-Science Source/Photo Researchers Inc.; **12.1:** © American Petroleum Institute Photograph and Film Collection, Archives Center, National Museum of American History, Behring Center, Smithsonian Institution; **12.7, 12.15:** © Bill W. Tillery; **12.18:** © R. J. Erwin 1988/Photo Researchers, Inc; **12.21:** © Bill W. Tillery;

Chapter 13 Opener: © Yann Arthus-Bertrand/Peter Arnold, Inc; **13.1:** © Bill W. Tillery; **13.2 (both):** © Fundamental Photographs; **13.10:** Arizona Public Service Company; **Box fig. 13.2a:** © SPL/Photo Researchers; **Box fig. 13.2b:** © SIU/Photo Researchers; **13.17a, 13.17b, 13.19, 13.20:** Arizona Public Service Company; **Box fig 13.3:** © Bill W. Tillery; **People Behind the Science, p. 350:** © Bettmann/Corbis;

Chapter 14 Opener: © NASA; **14.1a:** Lick Observatory, National Optical Astronomy Observatories; **14.2:** © J. Fuller/Visuals Unlimited; **14.11:** Lick Observatory; **14.14:** © National Optical Astronomy Observatory/Association of Universities for Research in Astronomy/National Science Foundation; **14.16:** Lick Observatory; **Box fig 14.1:** © Hencoup Enterprises/SPL/Photo Researchers;

Chapter 15 Opener, 15.1, 15.4, 15.5, 15.6, 15.7, 15.9a, 15.9b, 15.10: © NASA; **15.11 (both):** © Space Telescope Science Institute/NASA/SPL/Photo Researchers, Inc.; **15.12, 15.13:** © NASA; **15.15:** "Oort Cloud" by Don Davis reproduced with permission of Sky Publishing Corp.; **15.16:** Lick Observatory; **15.18 (both):** Center for Meteorite Studies, Arizona State University; **People Behind the Science, p. 400:** © Corbis;

Chapter 16 Opener: © Science VU/Visuals Unlimited; **16.1, 16.3:** © NASA; **16.19:** © Bill W. Tillery; **16.27:** Lick Observatory; **People Behind the Science, p. 425:** © Bettmann/Corbis;

Chapter 17 Opener: © Brian Parker/Tom Stack & Associates; **17.1:** © Bill W. Tillery; **17.4:** © Charles Falco/Photo Researchers; **17.5, 17.9, 17.10:** © Bill W. Tillery; **17.11 (both), 17.13, Box Fig 17.1, 17.15:** © C.C. Plummer; **17.16:** © Charles R. Belinky / Photo Researchers, Inc.; **17.17:** © Joyce Photographics / Photo Researchers, Inc.; **17.18:** © Andrew J. Martinez / Photo Researchers, Inc.; **17.21, 17.22:** © C.C. Plummer; **People Behind the Science, p. 449:** Courtesy of the Smithsonian Institution;

Chapter 18 Opener: © Jeff Foot/Tom Stack & Associates; **18.1:** © USGS Photo Library, Denver, CO; **People Behind the Science, 464:** © 2000 by the Trustees of Princeton University;

Chapter 19 Opener: © Douglas Cheeseman/Peter Arnold, Inc; **19.1:** © A. Post, USGS Photo Library, Denver, CO; **19.2:** © Bill W. Tillery; **19.4a:** © Robert W. Northrop, Photographer/Illustrator; **19.4b:** © Bill W. Tillery; **19.5:** © C.C. Plummer; **19.8a:** © National Park Service/Photo by Cecil W. Stoughton; **19.8b:** © D.E. Trimble, USGS Photo Library, Denver, CO; **19.9b:** © Frank M. Hanna; **19.13d:** © University of Colorado, Courtesy National Geophysical Data Center, Boulder CO; **19.17:** © NASA; **19.18b:**

© John S. Shelton; **19.20:** © Bill W. Tillery; **19.21b:** © D.W. Peterson, USGS; **19.22:** U.S. Geological Survey; **19.23b:** © B. Amundson; **People Behind the Science, p. 488:** Edinburgh University Library;

Chapter 20 Opener: © G Ziesler/Peter Arnold, Inc; **20.1:** © W.R. Hansen, USGS Photo Library, Denver CO.; **20.2:** National Park Service, Photo by Wm. Belnap, Jr.; **20.4a:** © A. J. Copley/Visuals Unlimited; **20.4b:** © L. Linkhart/Visuals Unlimited; **20.5:** © Ken Wagner/Visuals Unlimited; **20.6 (both):** © Bill W. Tillery; **20.8:** © William J. Weber/Visuals Unlimited; **20.9:** © Doug Sherman/Geofile; **20.11:** © B. Amundson; **20.12a:** © D.A. Rahm, photo courtesy of Rahm Memorial Collections, Western Washington University; **20.13:** © C.C. Plummer; **People Behind the Science, p. 506:** © The Granger Collection, NY; **20.16 (both):** © Bill W. Tillery;

Chapter 21 Opener: Courtesy Eldon Enger; **21.1:** © Andrew Gunners/Getty; **21.2:** © Robert W. Northrop, Photographer/Illustrator; **21.3:** © W.B. Saunders/Biological Photo Service; **21.4:** © PhotoLink/Getty Images; **21.6:** © B.A.E. Inc./ Alamy; **21.9:** © Bob Wallen; **21.12:** © Frank M. Hanna; **21.15:** U.S. Geological Survey; **21.16a:**

© Getty Images; **21.16b:** © James L. Amos/ Corbis; **People Behind the Science, p. 524:** Library of the Geological Survey of Austria;

Chapter 22 Opener: © Galen Rowell/ Mountain Light Photography; **22.1, 22.16, 22.18 (both), 22.19 (all):** © Bill W. Tillery;

Chapter 23 Opener: © Charles Mayer/Science Source/Photo Researchers; **23.1:** © Peter Arnold/Peter Arnold, Inc; **23.5, 23.7:** © NOAA; **23.8:** © Rachel Epstein/PhotoEdit, Inc; **23.17:** © Telegraph Herald/Photo by Patti Carr; **23.18, 23.19:** © NOAA; **23.21:** © David Parker/SPL/ Photo Researchers; **23.22:** © Bill W. Tillery; **23.25:** © Michael & Patricia Fogden/Minden Pictures; **23.26:** © Photodisc Collection/Getty; **23.27:** © Elizabeth Wallen; **23.31:** © Glen Allison/Getty Images; **23.34:** © Digital Vision/ PunchStock; **People Behind the Science, p. 576:** Courtesy of the Mathematical Institute at the University of St. Andrews;

Chapter 24 Opener: © James H. Karales/Peter Arnold, Inc; **24.1:** Salt River Project; **24.9:** City of Tempe, AZ; **24.10:** Salt River Project; **24.11, 24.14, 24.15:** © Bill W. Tillery; **24.19:** © John S. Shelton; **People Behind the Science, p. 600:** © Bettmann/Corbis; **Repeating Design Elements: Concepts Applied:** © Royalty-Free/

Corbis; **A Closer Look (man):** ©BananaStock/ PictureQuest; **A Closer Look (magnifying glass):** © Royalty-Free/Corbis; **People Behind the Science:** © Veer; **Myths, Mistakes, & Misunderstandings:** © PhotoLink/Getty Images; **Science and Society:** © Brand X Pictures/PunchStock; **Science and Society:** © Stockbyte/PunchStock;

Line Art/Text

Chapter 18: **18.11:** Pitman, W.C. III, Larson, R.L., and Herron, E.M., compilers, 1974. The age of the ocean basin: Boulder, Colorado, Geological Society of America Maps and Charts 6, 2 sheets. Reprinted by permission of The Geological Society of America, Boulder, CO. **18.13:** Source: After W. Hamilton, U.S. Geological Survey.

Chapter 19: **19.15:** Courtesy of Berkeley Seismological Laboratory, University of California Berkeley

Chapter 21: **21.17:** Modified from "Decade of North American Geology," 1983 Geologic Time Scale—Geological Society of America.

INDEX

Cementation, of sediments, 444
Centaurs (Kuiper Belt), 390
Centimeters, 6
Centrifugal force, 49
Centripetal force, 48–49, 51
Cenzoic era, 521, 523
Cepheid variable, 362
Ceres, 380
Cerium, 219, *342*
Cesium, *216*, *342*
Chadwick, James, 206
Chain reaction, 341
Chain silicates, 435
Charge, as fundamental property, 5
Charles, A. C., 17
Charley (hurricane, 2004), 564
Chemical bonds. *See also* Covalent bonds; Ionic bonds
    bond polarity, 238–40
    composition of compounds, 240–45
    formation, 231
    general classes of, 233
    microwave ovens, 243
Chemical energy, 70–71, 231–32
Chemical equations, 231–32, 255–60, 263–67
Chemical fertilizers, 255
Chemical formulas, 252–54
Chemical reactions
    chemical equations, 255–60, 263–67
    defined, 230
    energy, 231
    examples of common, 231
    types of, 260–63
    valence electrons and ions, 232–33
Chemical sediments, *443*, 444
Chemical weathering, 495, 496, 505, 506
Chemistry. *See* Organic chemistry
Chernobyl (Russia) nuclear accident, 346, 347
Chesapeake Bay, 596
Chile, and 1960 earthquake, 482
China, and history of science, 154, 186
Chinook, 537
Chitin, 515
Chlorate, *236*
Chloride, *613*, *592*
Chlorinated solvents, and pollution, *277*
Chlorine
    electron configuration, *215*
    freon and ozone layer, 534
    ions, *235*
    seawater, 592
    wastewater treatment, 589
Chlorofluorocarbons (CFCs), 545
Chloroform, *309*
Chondrites, 394
Chondrules, 394
Chromate, *236*
Chromium, *236*, *241*
Chrysotile, 442
Ciliary muscle, of eye, 186, *187*
Cinder cone volcano, 485
Circuit breakers, 168–69
Circuit connections (electrical), 165–70
Circular motion, 48–49
Cirque, and glaciers, 503
Cirrostratus clouds, *546*
Cirrus clouds, 545, *546*, 555
Citric acid, 287, 312

City (cities), and weather, 559, 571
Clastic sediments, 443
Clausius, Rudolf, 88
Clay, *443*, 586
Claystone, 443
Clean Air Act, 76
Cleavage, of minerals, 438
Clementine spacecraft, 418
Climate. *See also* Global warming; Greenhouse effect; Weather
    change, 573–76
    defined, 565–66
    description of, 569–70
    El Niño and *La Niña* events, 572
    major groups of, 566–67
    regional influences on, 568–69
    specific heat, 95
    volcanoes, 487
Climate zones, 570, *571*
Clouds, 544–45, *546*, 552–54
Coal, as energy source, 74, 75–76
Coalescence process, and precipitation, 554
Cobalt, 221, *236*, *342*
Coefficents, 256, 610
Coherent motion, 106
Cohesion, of molecules, 87
Cold air mass, *558*
Cold front, 556, *558*, 561
Color(s). *See also* Electromagnetic spectrum
    acids and bases, 288
    of light, 186, 188–89
    of minerals, 436, *437*
    of rainbow, 190
    referents, 3
    of sky, 192
    temperature of stars, 360, *361*
Colorado River, 584, *585*, 588
Columbia Plateau (Washington, Idaho, and Oregon), 484, *487*
Columbia River, 584, *585*
Columnar jointing, 476
Coma, of comet, 391
Combination reactions, 260–61
Combustion, as chemical reaction, 259–60
Comets, 388, 390–93
Common names, of compounds, 240
Commutator, of electric motor, 160–61
Compact disc (CD), 195
Compaction, of sediments, 444
Compass, 155, 156
Composite volcano, 485
Compound(s)
    chemical change, 230–32
    composition of, 240–45
    covalent, 237
    defined, 86
    ionic, 235
    percent composition, 253–54
Compound motion, 38–39
Compressive stress, 474, *477*
Computer models, of weather, 565
Concave lens, 186
Concave mirror, 183
Concentration, of solutions, 280–83
Concentration ratios, 281
Concepts, 2–3, 10
Condensation, 102–103, 120, 540–44
Condensation nuclei, 543–44, 554

Condensation point, 99
Conducting wires, 151
Conduction, of heat, 96–97
Conglomerate, 443
Conservation, of energy, 74, 77–78
Conservation of momentum, 46–48
Consistent law principle, 196
Constancy of speed, 196–97
Constant speed, 27
Constructive interference, 126, 127, 190
Contact time, 48
Containment building, of nuclear power plant, *344*
Continental air mass, 555
Continental climate, 570, *571*
Continental crust, 454
Continental divide, 585
Continental drift, *19*, 457
Continental glaciers, 502
Continental shelf, 599–600
Continental slope, 600
Continent-continent plate convergence, 462
Continuous spectrum, 208
Contrail, 542
Control group, 15
Controlled experiment, 15
Control rods, of nuclear reactor, 343–44
Convection, of heat, 97–98, 536, 552
Convection cells, 463
Convection current, 98
Convection zone, of star, 359
Conventional current, 148
Convergent boundaries, 460–62
Conversion
    of energy, 71–73
    of units, 608–609
Convex lens, 186
Convex mirror, 183
Copernican system, 399
Copernicus, Nicolas, 398–99
Copper
    density, *9*
    electrical conductors, *143*
    rate of conduction, *97*
    specific heat, *96*
    solubility, *613*
    variable-charge of ions, 221, *236*, *241*
Core
    of Earth, 452, 454, 455
    of star, 359
Corey, Robert, 245
Coriolis effect, 410, 560, 598
Correlation, and geologic events, 518–19
Corundum, *438*
Cosmic Background Explorer (COBE) spacecraft, 369
Cosmological constant, 371
Costs
    of mining mineral resources, 441
    of space exploration, 386
Coulomb (C), 143
Coulomb's law, 144
Covalent bonds, 233, 236–38, 239–40
Covalent compounds, 237, 242, 244, 252
Covalent molecule, 240
Crater(s)
    of moon, *419*
    of volcano, 484
Creep, and erosion, 499

Law of conservation of mass, 256
Law of conservation of momentum, 46
Law of inertia, 41
Law of reflection, 182–83
Lead
  density, *9*
  electrical conductors, *143*
  pollution, *277*
  products of nuclear fission, *342*
  rate of conduction, *97*
  solubility, *613*
  specific heat, *96*
  variable-charge ions, *236, 241*
Lead iodide, *251*
Lead nitrate, *251*
Leaning Tower of Pisa (Italy), 35
Length, 5
Lenses, and optics, 186
Leonid meteor shower, *394*
Lever, 64, 65
Life
  extraterrestrial, 368
  origins of, 425
Life cycle
  of galaxy, 368–71
  of star, 359, 362–65
Lignite, 76
Light. *See also* Light pollution
  particles, 189, 192–94
  present theory of, 194–96
  properties of, 180–89
  sources of, 178–80
  theory of relativity, 196–98
  waves, 189–92, 194
Lightning, *139*, 561–62
Light pollution, 365
Light ray model, 180
Light-year (ly), 356
Limestone
  carbon dioxide, 532
  caves, *498*
  composition, 444
  weathering, 496, 506
Linear model, of radiation exposure, 338
Linear scale, and simple line graph, 611
Linear wave front, 124
Lines of force, 145
Line spectrum, 208–209
Linoleic acid, *316*
Linolenic acid, *316*
Liquid(s), *87*, 88
Liquid-gas phase change, 99
Liquified natural gas (LPG), *258*, 307
Liter (L), 6
Lithification, 444
Lithium, *215, 216, 221, 235*
Lithosphere, 456, 460
Litmus paper, *287, 288*
Little ice age (1400–1860), 573
Little Sundance Mountain (Wyoming), *483*
Loam, 498
Loess, 504
Logan, Mount (Yukon Territory), 503
Lone pairs, 237
Longitude, 412
Longitudinal wave, 119, 122
Longshore current, 597

Los Angeles earthquake (1994), 479
Loudness, of sound, 127–28
Loudspeakers, 160
Lovelock, James Ephraim, 545
Love waves, 453
Low-density polyethylene, 318
Lowell, Percival, 385, 400
Low latitudes, 566, 567, *570, 571*
Low-level nuclear waste, 348
Low-quality energy, 106
Lowry, Thomas, 293
Lubricating oils, 308
Luminous objects, 178
Lunar eclipse, 423
Lunar highlands, 418, *419*
Lunar Prospector spacecraft, 418
Lure Observatory (Hawaii), 462
Luster, of minerals, 438
Lyell, Charles, 488, 513
Lysine, *313*

M

Mach, Ernst, 133
Mach number, 133
MACHOs (massive astrophysical compact halo
  objects), 372
Macromolecule, 313
Magma, 439–40, 484
Magnesium
  chemical reactions, *231*
  electron configuration, *215*
  ions, *235*
  seawater, *592*
  solubility, *613*
Magnesium bicarbonate, 291
Magnesium hydroxide, 242, *287*, 288, 289
Magnetic confinement, of nuclear waste, 349
Magnetic declination, 155
Magnetic dip, 156
Magnetic domain, 157
Magnetic fields
  compass, 156
  defined, 154–55
  of Earth, 157
  force fields, 144
  geomagnetic time scale, 520–21
  north and south poles, 155–56
  plate tectonics, 457–58
  source of, 156–57
Magnetic poles, 154
Magnetic quantum number, 214
Magnetic resonance imaging (MRI), 340, 341
Magnetic reversals, 157, 457–58
Magnetism, 154–61
Magnetite, 154, 458
Magnitude
  of earthquake, 481
  of refraction, 184
  of stars, 359, *360*
Main sequence stars, 362, 363
Maltose, 314
Mammoths, 514
Manganese, *236*, 590
Manipulated variable, 611
Mantle, of Earth, 452, 454, 455
Marble, 446
Maria (moon), 418, *419*

Marine climate, 570, *571*
Mariner 9 spacecraft, 385
Mariner 10 spacecraft, 382, *383*
Maritime air mass, 555
Mars, 380, *382*, 384–87
Mars Exploration Rovers, 386
Mars Odyssey spacecraft, 385
Mass. *See also* Mass number; Mass ratio
  metric system, 5
  Newton's second law of motion, 42, 43–44
  specific heat, 95
Mass defect, 339
Mass extinctions, 394, 523–24
Mass movement, 499
Mass number, 207, 329
Mass percentage, of compound, 253–54
Mass ratio, 266, *267*
Mathematical operations, 12. *See also* Equations
Matter
  light and interaction with, 180–81
  phases of, 87–88
Matter waves, 212–13
Matthews, D. H., 464
Mauna Loa (Hawaii), *485*
Maxwell, James, 144, 192–93, 197
McDonald Observatory (Texas), 462
Meanders, in rivers, 500
Mean solar day, 414
Mean solar time, 413
Measurement
  chemical equations, 265
  defined, 4
  earthquakes, 479–82
  heat, 94
  motion, 27–34
  plate tectonics, 462
  radiation, 335–38
  significant figures, 607–608
  systems, 4–6
  temperature, 89–91
  time, 6, 413–18
  understandings from, 7–13
Mechanical advantage, 64–65
Mechanical energy, 70
Mechanical waves, 119
Mechanical weathering, 495–96
Mechanical work, 62
Media, and pseudoscience, 16
Medicine, nuclear, 340–41. *See also* Health hazards
Mediterranean climate, *570, 571*
Mediterranean Sea, 590, 598
Medium, and transmission of sound waves, 122–23
Megawatt (MW), 66
Meitner, Lise, 340
Melting point, 99
Mendeleyev, Dmitri Ivanovich, 220
Mercalli scale, 481
Mercury (element)
  density, *9*
  solubility, *613*
  specific heat, *96*
  variable-charge ions, *236*
Mercury (planet), 157, 380, 381–82, *383*
Mercury oxide, 261
Meridians, 412
Mesosphere, 535
Mesozoic era, 521, 523

MESSENGER spacecraft, 382
Metabolism, 313
Metal(s)
    electrical conductors, 143
    physical properties, 218–21
    replacement reactions, 262
    temperature and expansion of, 89
Metal hydroxides, 241
Metallic hydrogen, 387
Metamorphic rocks, 444–46
Meteorology, 530. See also Climate; Weather
Meteors and meteorites, 394–95, 455, 524
Meteor shower, 394
Meter, 5
Methane
    alkanes, 301, 302, 303, 304
    composition of atmosphere, 532
    natural gas, 307, 308
Methanol, 310
Methyl alcohol, 286, 309, 310
Methyl group, 303
Metric system, of measurement, 4, 5–6, 44
Meyer, Lothar, 220
Mica, 437, 438
Microclimate, 570
Microscope, 187
Microwave ovens, 243
Mid-Atlantic Ridge, 458–59, 460, 590, 601
Middle ear infections, 123
Middle latitudes, 566, 567, 570, 571
Milankovitch, Milutin, 574
Milankovitch cycles, 574
Milky Way (galaxy), 366–67, 368, 369, 406
Miller, Stanley, 425
Millikan, Robert A., 205–206
Millirem (mrem), 337
Millisievert (MSv), 337
Mineral(s). See also Rocks
    costs of mining, 441
    crystal structures, 433–34
    defined, 432, 433
    formation processes, 439–40
    physical properties, 436–39
    recycling, 445
    silicates and nonsilicates, 434–35
Mineralization, of fossils, 515
Mineral oil, 308
Mining, of mineral resources, 441
Minor features, of landscape, 505
Mirages, 185
Miscible fluids, 279
Mississippi River, 501, 584, 585
Models, use of in science, 17–19
Moderator, of nuclear reactor, 344–45
Mohorovicic discontinuity, 454
Moh's hardness scale, 438
Molarity (M), 283
Mold, of fossil, 515
Mole, 265, 266, 282–83, 285
Molecular formula, 252, 253
Molecular orbital, 233
Molecular ratio, 266, 267
Molecular weight, 253
Molecules
    atmosphere, 530, 531, 532
    defined, 230
    diatomic, 86, 230, 231, 534

kinetic molecular theory, 86–88, 98–103
    monoatomic, 87, 230, 231, 534
    structure of water, 277–79
    triatomic, 230, 231
Mole ratio, 266, 267
Molina, Mario, 545
Momentum, 46–48
Monadnocks, 506
Monoatomic molecules, 87, 230, 231, 534
Monopole, 156
Monosaccharides, 314
Monthly time, 418
Montreal Protocol, 534
Moon
    composition and features, 420–21
    description of surface, 418, 419
    eclipses, 423
    formation of, 421
    gravity and acceleration toward Earth, 51
    movement away from Earth, 462
    phases of, 421–22
    tides, 424–25
    weather, 544
Moraines, 503
Motion
    compound, 38–39
    defined, 26
    of Earth, 406, 408–11
    energy, 62–69
    falling objects, 35–38
    horizontal on land, 34
    laws of, 40–46
    measurement, 27–34
    momentum, 46–48
Motor oil, 308
Mountains
    cloud-forming processes, 552
    influence on regional climate, 568
    origin of, 482–86
    stages of landform development, 506
    wind patterns, 537
Moving air masses, 552
Moving sources, of sound, 131–33
M type stars, 362
Mudflows, volcanic, 485
Multiple bonds, 238
Multiplication, and significant figures, 607–608
Muscovite, 435
Music, 129–31
Myopia, 186, 187

N

Names
    of alcohols, 309–10
    of alkane isomers, 302–304
    of compounds, 240–41, 242, 244
    of salts, 291
NASA (National Aeronautics and Space
        Administration), 389, 462, 571. See also Space
        exploration
National Institute of Standards and Technology, 4, 6
National Standard System (Canada), 4
National Weather Service, 538, 564
Native Americans
    irrigation canals, 582, 583
    measurement of time, 414
Natural frequency, of vibration, 128

Natural gas, 307, 308
Natural philosophy, 13–14
Natural radioactivity, 328
Nazca Plate, 461
Neap tides, 424
Near point, and vision, 186, 187
Nebulae, 358
Negative acceleration, 31
Negative electric charge, 140
Negative ion, 141
Neodymium, 219
Neon, 207, 215, 532
Neptune
    atmosphere, 392
    description, 380, 389–90
    Great Dark Spot, 379
    properties, 382
Neptunium, 342
Net force, 32, 33, 42
Neutralization, of acids and bases, 288
Neutrinos, 372
Neutron(s), 140, 141, 206–207
Neutron-to-proton ratio, 333
Neutron star, 364
New crust zone, 460
New Madrid earthquake (Missouri 1811), 479
New moon, 422
Newton, Isaac, 14, 46, 49, 50, 54, 188, 189, 401
Newton (N), 43, 44
Newton's law of gravitation, 49–54
Newton's laws of motion, 17, 41–46
Nichrome, 143
Nickel, 236, 342
Niobium, 342
Nitrate, 236, 613
Nitric acid, 287, 289, 292, 505
Nitrogen, 207, 215, 531
Nitrogen dioxide, 244
Nitrogen monoxide, 267
Nitrogen oxide, 292, 505
Nitroglycerine, 311
Nobel, Alfred, 311, 319
Nobel, Immanuel, 319
Noble gases, 218, 221
Nodes, 129
Noise, 129
Nonadecane, 308
Nonane, 303, 304, 308
Nonelectrolytes, 284
Nonferromagnesian silicates, 435, 437
Nonmetals, 218–21, 242
Nonsilicates, 434, 435, 436
Noon, 413
Normal fault, 477
North American Plate, 462, 463, 478
North celestial pole, 356
North Pole, 155, 356, 408
North Star, 357, 358, 411
N-type semiconductor, 167
Nuclear accidents, 346–47
Nuclear energy, 71, 74, 77, 338–50
Nuclear equations, 329–30
Nuclear fission, 340–43
Nuclear force, 331
Nuclear fusion, 340, 346–49
Nuclear medicine, 340–41

Nuclear power plants, 343–46
Nuclear reactor, 343
Nuclear Regulatory Commission (NRC), 349
Nuclear waste, 345–46, 348–49
Nuclear Waste Policy Act (1982), 349
Nucleation centers, 439
Nucleons, 329
Nucleus
    of atom, 140, *141*, 206–207, 331, 333
    of comet, 391
Numerical constant, 11
Nutrients, and water pollution, 594
Nutrition, and pseudoscience, 16. *See also* Diet;
    Food; Nutrients

O

Object(s), 2–3
Objective lens, of microscope, 187
Oblate spheroid (shape of Earth), 408
Observed lapse rate, 534
Observations, and scientific method, 14
Obsidian, 442
Occluded front, 559
Ocean(s). *See also* Currents; Hydrologic cycle;
    Ridges; Seafloor spreading hypothesis;
    Seawater; Trenches
    carbon dioxide, 532
    crust of Earth, 454
    definition of seas and, 590, 592
    deuterium and potential energy from, 347, 590
    floor, 599–601
    movement of seawater, 593–98, *599*
    plate tectonics, 458–59
    water on Earth's surface, 582
    waves, 593–97
Ocean basin, 600
Ocean-continent plate convergence, 461
Ocean-ocean plate convergence, 461–62
Ocean thermal energy conversion (OTEC), 79
Oct- (prefix), 302
Octadecane, *308*
Octane, 302, *303*, *304*, *308*
Octane number scale, 308
Octet rule, 232
Octyl acetate, *312*
Oersted, Jans Christian, 156–57
Ohm, G. S., 150
Ohm, 150
Ohmmeter, 159
Ohm's law, 150, 169
Oil(s), 316
Oil field, 307
-Ol (suffix), 310
Oleic acid, *316*
Olivine, 434–35, *437*, 455
Oort, Jan, 390–91
Oort cloud, 390, *393*, 396
Opaque materials, 180
Operational definition, 10, 43
Optics, 186–87
Orbital electrons, 214, 220
Orbits, of electrons, 210
Ordovician period, 522–23
Organic acid, *310*, 311–12
Organic chemistry
    defined, 300
    hydrocarbons, 300–306, 309–13

organic compounds, 313–19
    petroleum, 306–309
Organic compounds, 300, 313–19
Organic foods, and labeling, 16
Organic halides, 309, *310*
Orientation, of Earth's axis, 408
Original horizontality, principle of, 516
Orionid meteor shower, 394
Orthoclase, *437*, *438*
Oscillating theory, of universe, *371*
O type stars, 362
-*Ous* (suffix), 241
Overtones, 130
Oxbow lake, 500
Oxidation, 70, 496
Oxidation-reduction reaction, 260
Oxide, 435, *613*
Oxidizing agents, 260
Oxygen
    atmosphere, 531
    atomic weight, *207*
    chemical equation for water, 264
    diatomic and monoatomic, 534
    electron configuration, *215*
    ions, *235*
    triatomic molecules, 230
Oxygen-depleting chemicals, 594
Ozone, 230, 534
Ozone layer, 534

P

Paceline, 37
Pacific Ocean, 572, 590, 600, 601
Pacific Plate, 461–62, 463, 478
Painted Desert (Arizona), *513*
Paleomagnetics, 458, 459
Paleontology, 513
Paleozoic era, 521, 522, 523
Palmitic acid, 312, *316*
Palo Verde Nuclear Generating Station
    (Arizona), *328*
Parabola, 39
Paraffin, *308*, 309
Parallel(s), and Earth's axis, 411–12
Parallel circuit, 166, 168
Parallel forces, 32, *33*
Paraffin series, 301
Particle physics, 372
Particle theory of light, 189, 192–94
Parts per billion (ppb), 281
Parts per million (ppm), 281
Passel, Charles F., 538
Passive solar energy, 78, 101
Pathfinder spacecraft, 385–86
Pauli, Wolfgang, 215
Pauli exclusion principle, 215
Pauling, Linus Carl, 245
PCBs, *277*
Peat, 76
Pendulum, 71–73
Peneplain, 506
Pentadecane, *308*
Pentane, *303*, *304*, *308*
Per- (prefix), 241
Percent composition, of compounds, 253–54
Percent by volume, of solution, 281
Percent by weight, of solution, 281–82

Perchlorate, *236*
Perigee, 424
Perihelion, 399
Period
    of geologic time, 521–23
    of vibration, 117
    of wave, 121
Periodic disturbances, 119
Periodic motion, 116
Periodic table, 216–18
Permanent magnets, 157
Permanganate, *236*
Permeability, of sediments, 586
Permian period, 523
Perpetual motion machine, 106
Perseid meteor shower, *394*
Persistent pollutants, 584
Peru-Chile Trench, 461
PET scan (positron emission tomography), 341
Petrified Forest National Park (Arizona), 515
Petroleum. *See also* Fossil fuels; Hydrocarbons
    alkanes, 301
    as energy source, 74, 75
    organic chemistry, 306–309
    water pollution, 594
Petroleum jelly, 308
pH, 289–90, 505. *See also* Acid(s); Base(s)
Phanerozoic eon, 521
Phase(s), of moon, 421–22
Phase change, 99–100, 102
Phenyl, 306
Philosophae Naturalis Principia Mathematica
    (Newton 1686), 54
Phobos, 386–87
Phoenix, Arizona, and irrigation canals, 582, *583*
Phosphate, *236*, 445, *613*, 590
Phosphor(s), 336
Phosphoric acid, *287*
Phosphorus, *215*, 445
Photochemical smog, 267
Photoelectric effect, 193
Photons, 194, 208
Photosynthesis, 71, 231–32, 592
Photovoltaics, 78
Physical geology, 512
Physical science, defined, 1
Picocurie (pCi), 336
Pinatubo, Mount (Philippines), 487
Pitch, of sound, 122, 129
Place, identification of, 411–12
Planck, Max, 193, 208
Planck's constant, 194, 208
Plane of the ecliptic, 406
Planet(s). *See also* Earth; Jupiter; Mars; Mercury;
    Pluto; Saturn; Uranus; Venus
    extraterrestrial life, 368
    reflected light from, 356
    solar system, 380–90
Planetary nebula, *355*, 363, *364*
Plant(s), and carbohydrates, *315*. *See also*
    Agriculture; Grasses; Photosynthesis; Trees
Plasmas, 88, 349
Plastic(s), 316–17, *318*
Plastic deformation, 474
Plastic strain, 474
Plate boundary, 460
Plate tectonics, 457–65, 472

Plato, 204
Playfair, John, 488
Plucking, and glaciers, 503
Pluto, 380, 400
Plutonium, *334, 342*, 345
Polar air mass, 555, *557*
Polar climate zone, 567, *568, 570*
Polar covalent bond, 239, 277
Polaris, 358, 411
Polarization, of light, 190–92
Polarized plug, 169, *170*
Polar molecule, 277
Polaroid camera, 192
Poles, magnetic, 154, *155*, 156
Pollution. *See* Acid rain; Air pollution; Global warming; Light pollution; Water pollution
Polonium, 329
Polyatomic ions, 235, *236*, 241, 242
Polyethylene, 309, *317*, 318–19
Polymers, 309, 316–19
Polypeptide, 313
Polypropylene, *317, 318*
Polysaccharides, 314
Polystyrene, *317, 318*, 319
Polytetrafluoroethylene, *317*
Polyvinyl chloride (PVC), *317, 318*, 319
Polyvinylidene, *317*
Ponds, 585
Poor Richard's Almanac (Franklin 1732), 153
Porosity and pore spaces, in sediments, 586
Position, and potential energy, 67
Positive electric charge, 140
Positive ion, 141
Positive test charge, 144–45
Post meridiem (p.m.), 413
Potash, 240
Potassium
    electron structure, *215, 216*
    ions, *235*
    radioactive decay, *334*, 520
    seawater, *592*
    solubility, *613*
Potassium hydroxide, *287*, 288
Potassium iodide, *251*
Potential difference, of electrical charge, 146
Potential energy, 67, 68, 70, *100*
Pound, force unit of, 44
Powell, John Wesley, 506
Power, and work, 64–66, 68
Power tower, 78
Prearchean eon, 521
Precambrian period, 521–22
Precession, 411
Precipitation, 554–55, 582, *584*, 586
Prefixes
    metric system, 6
    names of alkanes, 302, 304
    names of compounds, 241, *244*
Pressure
    atmosphere, 532–33
    mechanical weathering of rocks, 495
    solubility of gases, 284
Primary coil, of transformer, 162
Primary loop, of nuclear reactor, 344
Prime meridian, 412
Principle(s), scientific. *See also* Law(s)
    consistent law principle, 196

of crosscutting relationships, 517
defined, 17
of faunal succession, 518–19
of original horizontality, 516
of superposition, 516–17
of uniformity, 472–73, 516, 517–18, 520
Prisms, 188–89
Problem solving, and equations, 11–13, 606–607
Promethium, *342*
Proof, of alcohol, 310–11
Propane
    hot air balloons, *258*
    molecular formula, *304*
    natural gas, 307, *308*
    structural formula, 302, *303*
Properties
    defined, 3
    equations and description of, 10
    quantifying, 3–4
    specific heat, 95
Proportionality constant, 11
Proportionality statements, 10–11
Proteins, 313–14
Proterozoic eon, 521
Proton(s), 140, *141*, 206–207
Proton acceptor, 288
Proton donor, 288
Protoplanet nebular model, 395–97
Protostar, 358, 362
Proxy data, and climate change, 573
Pseudoscience, 16
Psychrometer, 542
Ptolemaic system, 398
P-type semiconductor, 167
Pulley, 65
Pulsar, 364–65, 373
Pupil, of eye, 186, *187*
Pure tone, 129
P-wave, 453, 455, 479, *480*
P-wave shadow zone, 455
P/Wild 2 comet, 393

Q

Quad, 94
Quadrantid meteor shower, *394*
Quality of energy, 106
Quanta, 208
Quantification, of properties, 3–4
Quantitative uses, of chemical equations, 266
Quantities, symbols representing, 10
Quantization, of energy, 193–94
Quantum concept, 208
Quantum leaps, 210
Quantum mechanics, 212–15
Quantum numbers, 213, *214*
Quartz
    chemical weathering, 496
    colors, 436, *437*
    crystal structure, 433
    hardness, *438*
    nonferromagnesian silicates, 435, *437*
Quartzite, 446
Quaternary period, 524

R

Rad, 337
Radar, 133

Radiant energy, 71, 98
Radiation. *See also* Radioactive decay; Radioactive waste; Radioactivity; Radio radiation; Solar radiation; Ultraviolet radiation
    atmosphere and solar, 533–34
    exposure, *337*, 338, 534
    food preservation, 339
    heat transfer, 98
    measurement of, 335
    units of, 336–38
    zone of, 359
Radiationless orbits, 210
Radioactive decay, 329, 332–35, 336. *See also* Carbon-14
Radioactive isotopes, 520
Radioactive waste, 346
Radioactivity, 328–35. *See also* Radiation
Radiometric age, of rocks, 520
Radiopharmaceuticals, 340–41
Radio radiation, 194
Radium, 329, *342*
Radius, of Earth, 531
Radon, *342*
Rainbow, *18*, 190
Rainbow Bridge National Monument (Utah), *495*
Ranks, of coal, 76
Rare earths, 219
Rarefaction, 120
Rare gases, 218
Ratios, and measurement, 7–8, 605, 608
Rayleigh waves, 453
Rayon, 317
Reactor, nuclear, 77
Reagan, Ronald, 349
Real image, 183
Recycling
    of minerals, 445
    of plastics, *318*
    of wastewater, 589
Red giant stars, 362, 363
Redox reaction, 260
Red Sea, 460
Redshift, 133, 369
Reducing agent, 260
Reference frames, 196
Referent, 3
Reflected ray, 182
Reflecting telescopes, 187
Reflection
    of light, 180, 181, 182–83
    of waves, 124–25
Refracted ray, 183
Refracting telescopes, 187
Refraction, 124, 183–85, 188–89, 595
Rejuvenation, of erosion processes, 506
Relative dating, 519
Relative humidity, 541–42, *614*
Relative intensity, of earthquake, 481
Relativity, theory of, 196–98, 371, 389
Relay, electromagnetic, 159
Rem, 337
Renewable energy resources, 465
Replacement reactions, 261–62
Representative elements, 216
Reprocessing plants, for nuclear fuel, 345
Research, types of scientific, 15, 214
Resistance, in electrical circuits, 166, 168

Table of Atomic Weights (Based on Carbon-12)

Name	Symbol	Atomic Number	Atomic Weight	Name	Symbol	Atomic Number	Atomic Weight
Actinium	Ac	89	(227)	Mendelevium	Md	101	258.10
Aluminum	Al	13	26.9815	Mercury	Hg	80	200.59
Americium	Am	95	(243)	Molybdenum	Mo	42	95.94
Antimony	Sb	51	121.75	Neodymium	Nd	60	144.24
Argon	Ar	18	39.948	Neon	Ne	10	20.179
Arsenic	As	33	74.922	Neptunium	Np	93	(237)
Astatine	At	85	(210)	Nickel	Ni	28	58.71
Barium	Ba	56	137.34	Niobium	Nb	41	92.906
Berkelium	Bk	97	(247)	Nitrogen	N	7	14.0067
Beryllium	Be	4	9.0122	Nobelium	No	102	259.101
Bismuth	Bi	83	208.980	Osmium	Os	76	190.2
Bohrium	Bh	107	264	Oxygen	O	8	15.9994
Boron	B	5	10.811	Palladium	Pd	46	106.4
Bromine	Br	35	79.904	Phosphorus	P	15	30.9738
Cadmium	Cd	48	112.40	Platinum	Pt	78	195.09
Calcium	Ca	20	40.08	Plutonium	Pu	94	244.064
Californium	Cf	98	242.058	Polonium	Po	84	(209)
Carbon	C	6	12.0112	Potassium	K	19	39.098
Cerium	Ce	58	140.12	Praseodymium	Pr	59	140.907
Cesium	Cs	55	132.905	Promethium	Pm	61	144.913
Chlorine	Cl	17	35.453	Protactinium	Pa	91	(231)
Chromium	Cr	24	51.996	Radium	Ra	88	(226)
Cobalt	Co	27	58.933	Radon	Rn	86	(222)
Copper	Cu	29	63.546	Rhenium	Re	75	186.2
Curium	Cm	96	(247)	Rhodium	Rh	45	102.905
Dubnium	Db	105	(262)	Rubidium	Rb	37	85.468
Dysprosium	Dy	66	162.50	Ruthenium	Ru	44	101.07
Einsteinium	Es	99	(254)	Rutherfordium	Rf	104	(261)
Erbium	Er	68	167.26	Samarium	Sm	62	150.35
Europium	Eu	63	151.96	Scandium	Sc	21	44.956
Fermium	Fm	100	257.095	Seaborgium	Sg	106	(266)
Fluorine	F	9	18.9984	Selenium	Se	34	78.96
Francium	Fr	87	(223)	Silicon	Si	14	28.086
Gadolinium	Gd	64	157.25	Silver	Ag	47	107.868
Gallium	Ga	31	69.723	Sodium	Na	11	22.989
Germanium	Ge	32	72.59	Strontium	Sr	38	87.62
Gold	Au	79	196.967	Sulfur	S	16	32.064
Hafnium	Hf	72	178.49	Tantalum	Ta	73	180.948
Hassium	Hs	108	(269)	Technetium	Tc	43	(99)
Helium	He	2	4.0026	Tellurium	Te	52	127.60
Holmium	Ho	67	164.930	Terbium	Tb	65	158.925
Hydrogen	H	1	1.0079	Thallium	Tl	81	204.37
Indium	In	49	114.82	Thorium	Th	90	232.038
Iodine	I	53	126.904	Thulium	Tm	69	168.934
Iridium	Ir	77	192.2	Tin	Sn	50	118.69
Iron	Fe	26	55.847	Titanium	Ti	22	47.90
Krypton	Kr	36	83.80	Tungsten	W	74	183.85
Lanthanum	La	57	138.91	Uranium	U	92	238.03
Lawrencium	Lr	103	260.105	Vanadium	V	23	50.942
Lead	Pb	82	207.19	Xenon	Xe	54	131.30
Lithium	Li	3	6.941	Ytterbium	Yb	70	173.04
Lutetium	Lu	71	174.97	Yttrium	Y	39	88.905
Magnesium	Mg	12	24.305	Zinc	Zn	30	65.38
Manganese	Mn	25	54.938	Zirconium	Zr	40	91.22
Meitnerium	Mt	109	(268)				

Note: A value in parentheses denotes the number of the longest-lived or best-known isotope.